Mechanics of Flight

Mechanics of Flight

Second Edition

Warren F. Phillips

Professor
Mechanical and Aerospace Engineering
Utah State University

WILEY

John Wiley & Sons, Inc.

Library of Congress Cataloging-in-Publication Data:

Phillips, Warren F.
  Mechanics of flight / Warren F. Phillips. -- 2nd ed.
      p. cm.
  Includes bibliographical references and index.
  ISBN 978-0-470-53975-0 (cloth)
  1. Aerodynamics. 2. Flight. 3. Flight control. I. Title.
  TL570.P46 2010
  629.132'3--dc22
                          2009020809

Dedicated to my father
Frederick A. Phillips
1909−1971

CONTENTS

Preface

This book was written for aeronautical, aerospace, and/or mechanical engineers. The book is intended primarily as a textbook for an engineering class that is taught as an upper-division undergraduate or lower-division graduate class. Such an engineering class is typically required in aeronautical engineering programs and is a popular elective in many astronautical and mechanical engineering programs.

Collectively, the engineering topics covered in this book are usually referred to as *flight mechanics*. These topics build directly upon a related set of engineering topics that are typically grouped under the title of *aerodynamics*. Together the topics of aerodynamics and flight mechanics form the foundation of aeronautics, which is the science that deals with the design and operation of aircraft. Aerodynamics is the science of predicting and controlling the forces and moments that act on a craft moving through the atmosphere. Flight mechanics is the science of predicting and controlling the motion that results from the aerodynamic forces and moments acting on an aircraft.

This textbook was written with the assumption that the reader has completed the lower-division coursework that is required in any engineering program. Critical topics include calculus, differential equations, computer programming, numerical methods, statics, dynamics, and fluid mechanics. It is also assumed that the student has previously completed an introductory course in aerodynamics. The book does include an overview of aerodynamics. Thus, it would be possible for a discerning student who has completed the lower-division engineering coursework to comprehend the material in this book with no previous experience in aerodynamics. However, in the author's experience, most students require at least a full semester to comprehend the fundamentals of aerodynamics. Such comprehension is an essential prerequisite to the study of flight mechanics.

One important feature of this book is the inclusion of many worked examples, which are designed to help the student understand the process of applying the principles of flight mechanics to the solution of engineering problems. Another unique feature is inclusion of a detailed presentation of the quaternion formulation for six-degree-of-freedom flight simulation. Efficient numerical methods for integration of the quaternion formulation are also presented and discussed. This material is presented in sufficient detail to allow undergraduate students to write their own code. The quaternion formulation is not typically presented in other textbooks on atmospheric flight mechanics. In fact, many textbooks that deal with atmospheric flight mechanics do not even mention the quaternion formulation. Here it is shown that this formulation of the aircraft equations of motion, which has typically been implemented to eliminate a singularity associated with the Euler angle formulation, is far superior to the other commonly used formulations based on computational efficiency alone. Most practicing engineers working in the field of flight dynamics should find the chapter on flight simulation to be a valuable reference.

This book contains considerably more material than can be covered in a single one-semester class. This provides significant benefit for both the instructor and the student. The additional coverage gives the instructor some flexibility in choosing the material to be included in a particular one-semester course. The instructor may choose to cover all of the topics offered but omit some of the more advanced material on each topic. On the other hand, the instructor may wish to cover fewer topics in greater detail. Most of the material in this book can be covered in a two-semester undergraduate sequence. A convenient division of this material is to cover aircraft performance together with static stability and control during the first semester and then extend the treatment to include linearized flight dynamics and flight simulation in the second semester. From the student's point of view, the information contained in this book, beyond that which is covered in their formal coursework, provides an excellent source of reference material for future independent study.

There are two philosophical approaches that can be used to present the material associated with flight mechanics. One approach is to start with a development of the general six-degree-of-freedom aircraft equations of motion and then treat everything else as a special case of this general formulation. The other approach is to begin with simple problems that are formulated and solved starting from fundamental principles and then gradually work up to the more complex problems, leading eventually to the development of the general equations of motion. The latter approach is used in this textbook. This approach necessitates some repetition. As a result, the experienced reader may find some of the developments in the earlier chapters to be somewhat tedious. However, while the approach used here is not the most concise, the author believes that most students will find it more understandable. In general, the reader will find that throughout this book, conciseness is often sacrificed in favor of student convenience and understanding.

In Chapters 3 through 6, some of the basic principles of flight mechanics are explored by example, starting with simple problems such as steady level flight and building to more complex problems, including turns and spins. In Chapter 7, the more general rigid-body equations of motion are developed from fundamental principles. Since this chapter could serve as a starting point for a second-semester course in flight mechanics, it includes a review of coordinate systems and notation that were introduced gradually throughout Chapters 3 through 6. Chapter 7 was deliberately written to be independent of the presentation in Chapters 3 through 6. Thus, students or practicing engineers who are already familiar with the principles of static stability and control could start with Chapter 7 to begin a study of aircraft dynamics.

The second edition of this textbook contains considerable new material added to complement the material presented in the first edition. Throughout the book, many sections have been revised and/or expanded to provide students with better understanding of the related material. Moreover, 19 new sections or subsections and a new appendix have been added. These new sections include important material, which was not covered in the first edition and has not been covered in other textbooks on flight mechanics. For example, Chapter 4 includes a new section on the effects of nonlinear aerodynamics. Chapter 5 contains new sections on tail dihedral, lateral trim, and minimum-control airspeed. Chapter 6 has additions on dynamic stability constraints and center of gravity limits, and Chapter 11 includes new sections on flight simulation in geographic coordinates. A nomenclature section has also been added in this edition. Altogether, 169 pages of valuable new material have been added in the second edition of this textbook.

Warren F. Phillips

Acknowledgments

First, I would like to thank my good friend and counsel, Dr. Barry W. Santana, for the broad base of support and assistance that he provided during the writing of this book. In addition to reading drafts of the text and providing many insightful suggestions, he dedicated a great deal of time and travel to shooting and processing photographs especially for this book. I particularly appreciate the many hours of work that Barry invested in developing the index.

I also want to thank Captain Robert J. Niewoehner, who has contributed a great deal to the second edition of this textbook. After using the first edition in his classes at the United States Naval Academy, Professor Niewoehner was kind enough to provide me with many detailed suggestions that have led to significant improvements in the second edition. I worked closely with Dr. Niewoehner in developing some of the new material presented in the second edition, principally in Sections 1.11, 4.4, 4.8, 5.8–5.10, 6.1, and 10.3. His experience and insight have been very helpful.

While thanks are due a great number of my students for their feedback and many stimulating discussions, I am particularly grateful to Nicholas R. Alley for his input on the book and his assistance as coauthor of the solutions manual. I would also like to thank Drs. Glenn Gebert and Fred Lutze for reviewing a complete draft of the first edition and providing many helpful suggestions.

I wish to extend special thanks to Drs. John D. Anderson, John J. Bertin, Bernard Etkin, Barnes W. McCormick, Robert C. Nelson, Courtland D. Perkins, and Ludwig Prandtl, who have through their writings been my teachers and mentors. If I have contributed in some small way to this very exciting field, it is only because of what I learned from the writings of these great men.

Most of all, I want to thank my wife, Barbara, for her love, encouragement, and assistance, without which I could not have completed this work. I am especially grateful for the countless hours that she spent proofreading the many drafts that eventually led to the finished version of this book. I love you.

Warren F. Phillips

Chapter 1
Overview of Aerodynamics

1.1. Introduction and Notation

Flight mechanics is the science of predicting and controlling aircraft motion. From Newton's second law we know that the motion of any body depends on the forces and moments acting on the body. The forces and moments exerted on an aircraft in flight are the aerodynamic forces and moments acting on the aircraft's skin, the propulsive forces and moments created by the aircraft's engine or engines, and the gravitational force between the aircraft and the Earth. Because aerodynamic forces and moments are central to the study of aircraft motion, an understanding of the fundamentals of aerodynamics is a prerequisite to the study of flight mechanics. In this text it will be assumed that the reader has gained this prerequisite knowledge, either through the completion of at least one engineering course on aerodynamics or through independent study. In this chapter we review briefly some of the more important concepts that the reader should understand before proceeding with the material in the remainder of the book.

The aerodynamic forces and moments acting on any body moving through the atmosphere originate from only two sources,

1. The pressure distribution over the body surface.
2. The shear stress distribution over the body surface.

A resultant aerodynamic force, $\mathbf{F}_a$, and a resultant aerodynamic moment, $\mathbf{M}_a$, are the net effects of the pressure and shear stress distributions integrated over the entire surface of the body. To express these two vectors in terms of components, we must define a coordinate system. While several different coordinate systems will be used in our study of flight mechanics, the coordinate system commonly used in the study of aerodynamics is referred to here as *Cartesian aerodynamic coordinates*. When considering flow over a body such as an airfoil, wing, or airplane, the x-axis of this particular coordinate system is aligned with the body axis or *chord line*, pointing in the general direction of relative airflow. The origin is typically located at the front of the body or *leading edge*. The y-axis is chosen normal to the x-axis in an upward direction. Choosing a conventional right-handed coordinate system requires the z-axis to be pointing in the spanwise direction from right to left, as shown in Fig. 1.1.1. Here, the components of the resultant aerodynamic force and moment, described in this particular coordinate system, are denoted as

$$\mathbf{F}_a = A\mathbf{i}_x + N\mathbf{i}_y + B\mathbf{i}_z$$

$$\mathbf{M}_a = -\ell\mathbf{i}_x - n\mathbf{i}_y - m\mathbf{i}_z$$

where $\mathbf{i}_x$, $\mathbf{i}_y$, and $\mathbf{i}_z$ are the unit vectors in the x-, y-, and z-directions, respectively. The terminology that describes these components is

1

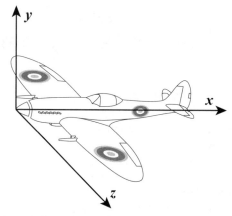

Figure 1.1.1. Cartesian aerodynamic coordinate system used in the study of aerodynamics.

A $\equiv$ aftward axial force $\equiv$ x-component of $\mathbf{F}_a$ (parallel to the chord)
N $\equiv$ upward normal force $\equiv$ y-component of $\mathbf{F}_a$ (normal to the chord and span)
B $\equiv$ leftward side force $\equiv$ z-component of $\mathbf{F}_a$ (parallel with the span)
ℓ $\equiv$ rolling moment (positive right wing down)
n $\equiv$ yawing moment (positive nose right)
m $\equiv$ pitching moment (positive nose up)

The traditional definitions for the moments in roll, pitch, and yaw do not follow the right-hand rule in this coordinate system. It is often convenient to split the resultant aerodynamic force into only two components,

D $\equiv$ drag $\equiv$ the component of $\mathbf{F}_a$ parallel to $\mathbf{V}_\infty$ $(D = \mathbf{F}_a \cdot \mathbf{i}_\infty)$
L $\equiv$ lift $\equiv$ the component of $\mathbf{F}_a$ perpendicular to $\mathbf{V}_\infty$ $(L = |\mathbf{F}_a - D\mathbf{i}_\infty|)$

where $\mathbf{V}_\infty$ is the freestream velocity or *relative wind* far from the body and $\mathbf{i}_\infty$ is the unit vector in the direction of the freestream.

For two-dimensional flow, it is often advantageous to define the *section force* and *section moment* to be the force and moment per unit span. For these definitions the notation used in this book will be

$\tilde{D}$ $\equiv$ section drag $\equiv$ drag force per unit span (parallel to $\mathbf{V}_\infty$)
$\tilde{L}$ $\equiv$ section lift $\equiv$ lift force per unit span (perpendicular to $\mathbf{V}_\infty$)
$\tilde{A}$ $\equiv$ section axial force $\equiv$ axial force per unit span (parallel to chord)
$\tilde{N}$ $\equiv$ section normal force $\equiv$ normal force per unit span (perpendicular to chord)
$\tilde{m}$ $\equiv$ section moment $\equiv$ pitching moment per unit span (positive nose up)

where the chord is a line extending from the leading edge to the trailing edge of the body. The chord length, c, is the length of this chord line.

The aerodynamic forces and moments are usually expressed in terms of dimensionless force and moment coefficients. For example,

$$C_D \equiv \text{drag coefficient} \equiv \frac{D}{\frac{1}{2}\rho_\infty V_\infty^2 S_{\text{ref}}}$$

$$C_L \equiv \text{lift coefficient} \equiv \frac{L}{\frac{1}{2}\rho_\infty V_\infty^2 S_{\text{ref}}}$$

$$C_A \equiv \text{axial force coefficient} \equiv \frac{A}{\frac{1}{2}\rho_\infty V_\infty^2 S_{\text{ref}}}$$

$$C_N \equiv \text{normal force coefficient} \equiv \frac{N}{\frac{1}{2}\rho_\infty V_\infty^2 S_{\text{ref}}}$$

$$C_m \equiv \text{pitching moment coefficient} \equiv \frac{m}{\frac{1}{2}\rho_\infty V_\infty^2 S_{\text{ref}} l_{\text{ref}}}$$

where ρ_∞ is the freestream density, S_{ref} is the reference area, and l_{ref} is the reference length. For a streamlined body such as a wing, S_{ref} is the planform area and l_{ref} is a reference chord length. For a bluff body, the frontal area is used as the reference. For two-dimensional flow, the section aerodynamic coefficients per unit span are defined:

$$\tilde{C}_D \equiv \text{section drag coefficient} \equiv \frac{\tilde{D}}{\frac{1}{2}\rho_\infty V_\infty^2 c}$$

$$\tilde{C}_L \equiv \text{section lift coefficient} \equiv \frac{\tilde{L}}{\frac{1}{2}\rho_\infty V_\infty^2 c}$$

$$\tilde{C}_A \equiv \text{section axial force coefficient} \equiv \frac{\tilde{A}}{\frac{1}{2}\rho_\infty V_\infty^2 c}$$

$$\tilde{C}_N \equiv \text{section normal force coefficient} \equiv \frac{\tilde{N}}{\frac{1}{2}\rho_\infty V_\infty^2 c}$$

$$\tilde{C}_m \equiv \text{section moment coefficient} \equiv \frac{\tilde{m}}{\frac{1}{2}\rho_\infty V_\infty^2 c^2}$$

where c is the section chord length defined in Fig. 1.1.2.

The resultant aerodynamic force acting on a two-dimensional airfoil section is completely specified in terms of either lift and drag or axial and normal force. These two equivalent descriptions of the resultant aerodynamic force are related to each other through the angle of attack, as shown in Fig. 1.1.2,

$$\alpha \equiv \text{angle of attack} \equiv \text{the angle from } \mathbf{V}_\infty \text{ to the chord line (positive nose up)}$$

If the normal and axial coefficients are known, the lift and drag coefficients can be found from the relations

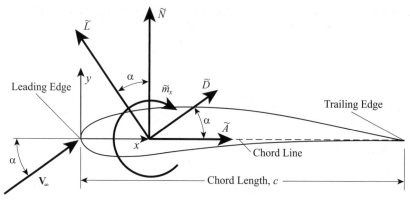

Figure 1.1.2. Section forces and moment.

$$\tilde{C}_L = \tilde{C}_N \cos\alpha - \tilde{C}_A \sin\alpha \qquad (1.1.1)$$

$$\tilde{C}_D = \tilde{C}_A \cos\alpha + \tilde{C}_N \sin\alpha \qquad (1.1.2)$$

and when the lift and drag coefficients are known, the normal and axial coefficients are found from

$$\tilde{C}_N = \tilde{C}_L \cos\alpha + \tilde{C}_D \sin\alpha \qquad (1.1.3)$$

$$\tilde{C}_A = \tilde{C}_D \cos\alpha - \tilde{C}_L \sin\alpha \qquad (1.1.4)$$

Because the lift is typically much larger than the drag, from Eq. (1.1.4) we see why the axial force is often negative even though the angle of attack is small.

The resultant aerodynamic force and moment acting on a body must have the same effect as the distributed loads. Thus, the resultant moment will depend on where the resultant force is placed on the body. For example, let x be the coordinate measured along the chord line of an airfoil, from the leading edge toward the trailing edge. If we place the resultant force and moment on the chord line, the value of the resultant moment will depend on the x-location of the resultant force. The resultant moment about some arbitrary point on the chord line a distance x from the leading edge, $\tilde{m}_x$, is related to the resultant moment about the leading edge, $\tilde{m}_{le}$, according to

$$\tilde{m}_{le} = \tilde{m}_x - x\tilde{N}$$

or in terms of dimensionless coefficients,

$$\tilde{C}_{m_{le}} = \tilde{C}_{m_x} - \frac{x}{c}\tilde{C}_N \qquad (1.1.5)$$

Two particular locations along the chord line are of special interest:

$x_{cp} \equiv$ center of pressure $\equiv$ the point about which the resultant moment is zero.

$x_{ac} \equiv$ aerodynamic center $\equiv$ the point about which the change in the resultant moment with respect to the angle of attack is zero.

Using the definition of center of pressure in Eq. (1.1.5), the section pitching moment coefficient about the leading edge can be written as

$$\tilde{C}_{m_{le}} = \tilde{C}_{m_x} - \frac{x}{c}\tilde{C}_N = -\frac{x_{cp}}{c}\tilde{C}_N$$

Solving for the **center of pressure** yields

$$\frac{x_{cp}}{c} = \frac{x}{c} - \frac{\tilde{C}_{m_x}}{\tilde{C}_N} \tag{1.1.6}$$

From this relation, the location of the center of pressure at any given angle of attack can be determined from the normal force coefficient and the moment coefficient about any point on the airfoil chord line. In general, the position of the center of pressure can vary significantly with angle of attack.

In a similar manner, using the aerodynamic center in Eq. (1.1.5) we can write

$$\tilde{C}_{m_{le}} = \tilde{C}_{m_x} - \frac{x}{c}\tilde{C}_N = \tilde{C}_{m_{ac}} - \frac{x_{ac}}{c}\tilde{C}_N$$

or after solving for the moment about the aerodynamic center, we have

$$\tilde{C}_{m_{ac}} = \tilde{C}_{m_x} + \left(\frac{x_{ac}}{c} - \frac{x}{c}\right)\tilde{C}_N \tag{1.1.7}$$

From the definition of the aerodynamic center, Eq. (1.1.7) requires

$$\frac{\partial \tilde{C}_{m_{ac}}}{\partial \alpha} = \frac{\partial \tilde{C}_{m_x}}{\partial \alpha} + \left(\frac{x_{ac}}{c} - \frac{x}{c}\right)\frac{\partial \tilde{C}_N}{\partial \alpha} = 0 \tag{1.1.8}$$

Solving Eq. (1.1.8) for the location of the **aerodynamic center** results in

$$\frac{x_{ac}}{c} = \frac{x}{c} - \frac{\partial \tilde{C}_{m_x}}{\partial \alpha} \bigg/ \frac{\partial \tilde{C}_N}{\partial \alpha} \tag{1.1.9}$$

The location of the aerodynamic center can be determined according to Eq. (1.1.9) from knowledge of how the normal force coefficient and the moment coefficient about any point on the chord line vary with angle of attack. For most airfoils, the position of the aerodynamic center is very nearly independent of angle of attack.

The concept of an aerodynamic center is extremely important in the study of flight mechanics. We see from Eq. (1.1.9) that the location of the aerodynamic center does not depend on the magnitude of the aerodynamic coefficients, as the center of pressure does; rather, it depends only on derivatives of the aerodynamic coefficients with respect to angle of attack. As we proceed with our study of flight mechanics, we shall discover many other such derivatives that are important. To avoid the somewhat cumbersome notation used in Eq. (1.1.9), here we shall use a shorthand notation for such derivatives. **In this text, a subscript preceded by a comma will be used to signify differentiation.** For example,

$$\tilde{C}_{L,\alpha} \equiv \frac{\partial \tilde{C}_L}{\partial \alpha}, \quad \tilde{C}_{D,\alpha} \equiv \frac{\partial \tilde{C}_D}{\partial \alpha}, \quad \tilde{C}_{N,\alpha} \equiv \frac{\partial \tilde{C}_N}{\partial \alpha}, \quad \tilde{C}_{m_x,\alpha} \equiv \frac{\partial \tilde{C}_{m_x}}{\partial \alpha}, \quad \text{etc.}$$

Using this shorthand notation, Eq. (1.1.9) is written as

$$\frac{x_{ac}}{c} = \frac{x}{c} - \frac{\tilde{C}_{m_x,\alpha}}{\tilde{C}_{N,\alpha}} \tag{1.1.9}$$

EXAMPLE 1.1.1. An airfoil has section lift, drag, and quarter-chord moment coefficients given by the following equations:

$$\tilde{C}_L = 6.0\alpha + 0.2, \quad \tilde{C}_D = 0.2\alpha^2 + 0.006, \quad \tilde{C}_{m_{c/4}} = -0.05 - 0.01\alpha$$

where α is the angle of attack in radians. Find the center of pressure and the aerodynamic center of the airfoil for angles of attack of −5, 0, 5, and 10 degrees.

Solution. Using Eq. (1.1.3) in Eq. (1.1.6), the center of pressure can be expressed in terms of the section lift, drag, and quarter-chord moment coefficients,

$$\frac{x_{cp}}{c} = 0.25 - \frac{\tilde{C}_{m_{c/4}}}{\tilde{C}_L \cos\alpha + \tilde{C}_D \sin\alpha}$$

$$= 0.25 - \frac{-0.05 - 0.01\alpha}{(6.0\alpha + 0.2)\cos\alpha + (0.2\alpha^2 + 0.006)\sin\alpha}$$

From Eq. (1.1.3), the change in the normal force coefficient with respect to angle of attack is

$$\tilde{C}_{N,\alpha} = \tilde{C}_{L,\alpha}\cos\alpha - \tilde{C}_L \sin\alpha + \tilde{C}_{D,\alpha}\sin\alpha + \tilde{C}_D \cos\alpha$$

Using this result in Eq. (1.1.9) and rearranging yields

$$\frac{x_{ac}}{c} = \frac{x}{c} - \frac{\tilde{C}_{m_x,\alpha}}{(\tilde{C}_{L,\alpha} + \tilde{C}_D)\cos\alpha - (\tilde{C}_L - \tilde{C}_{D,\alpha})\sin\alpha}$$

and for this particular airfoil we have

$$\frac{x_{ac}}{c} = 0.25 - \frac{\tilde{C}_{m_{c/4},\alpha}}{(\tilde{C}_{L,\alpha} + \tilde{C}_D)\cos\alpha - (\tilde{C}_L - \tilde{C}_{D,\alpha})\sin\alpha}$$

$$= 0.25 - \frac{-0.01}{(6.0 + 0.2\alpha^2 + 0.006)\cos\alpha - (6.0\alpha + 0.2 - 0.4\alpha)\sin\alpha}$$

At angles of attack of −5, 0, 5, and 10 degrees, these relations result in

α (deg)	α (rad)	x_{cp}/c	x_{ac}/c
−5.0	−0.08727	0.09791	0.25168
0.0	0.00000	0.50000	0.25167
5.0	0.08727	0.32051	0.25169
10.0	0.17453	0.29206	0.25175

Note that although the position of the center of pressure varies considerably with angle of attack, the position of the aerodynamic center is nearly independent of angle of attack. This result is typical for cambered airfoils, which are commonly used in the construction of aircraft wings.

Body-Fixed Stability and Control Coordinates
Although the Cartesian aerodynamic coordinate system shown in Fig. 1.1.1 is commonly used in the study of aerodynamics, there are several disadvantages to using this reference frame in the study of flight mechanics. Some of these disadvantages may not become apparent until later in our development of the aircraft equations of motion. However, we have already noted that the conventional definitions for the aerodynamic moments in roll, pitch, and yaw do not follow the usual right-hand rule in this traditional aerodynamic coordinate system.

Historically, the aerodynamic coordinate system originated with the development of airfoil theory. When formulating this two-dimensional problem in Cartesian coordinates, a conventional x-y coordinate system was chosen and it was convenient to have the x-axis aligned with the chord line pointing in the general direction of flow, with the origin at the leading edge. The y-axis was naturally chosen normal to the x-axis in the upward direction, as shown in Fig. 1.1.2. It was also quite natural to choose the angle of attack to be positive with a positive y-component of relative wind, because increasing the angle of attack in this direction increases the lift produced on the airfoil. Once the sign convention for angle of attack was established, a nose-up pitching moment was defined to be positive, because this is the direction of increasing angle of attack. When the study of aerodynamics was later extended to three-dimensional flow about finite wings, it was convenient to align the x- and y-axes with the airfoil sections as was done for the two-dimensional airfoil analysis. With this orientation of the x- and y-axes, choosing a conventional right-handed coordinate system requires the z-axis to be pointing in the spanwise direction to the pilot's left, as shown in Fig. 1.1.1. Retaining the original airfoil sign convention for pitch, the positive nose-up pitching moment is a left-hand moment about the z-axis. Thus, the left-handed moment convention in the Cartesian aerodynamic coordinate system was established.

There are significant advantages to using the traditional aerodynamic coordinate system for the study of lift and drag. Furthermore, the disadvantages of the left-handed moment convention for roll, pitch, and yaw were probably not fully recognized until early in the development of the theory of aircraft dynamics. By then, the convention was well established in the aerodynamics literature.

To avoid the left-handed moment convention while retaining the traditional sign convention for roll, pitch and yaw, the body-fixed coordinate system shown in Fig. 1.1.3 is commonly used in the study of flight mechanics. In this text, the notation (x_b, y_b, z_b) will always be used in reference to this particular body-fixed coordinate system. The origin of this Cartesian system is always located at the aircraft center of gravity. The x_b-axis points forward along some convenient fuselage reference line in the aircraft's plane of symmetry. The y_b-axis is normal to the plane of symmetry pointing in the direction of the right wing. The z_b-axis then points downward in the aircraft plane of symmetry, completing the right-handed Cartesian system. In this coordinate system, a positive angle of attack corresponds to a positive z_b-component of the airplane's velocity vector relative to the surrounding atmosphere. Similarly, a positive sideslip angle is defined to correspond to a positive y_b-component of the airplane's velocity relative to the air. With this coordinate system, the traditional sign conventions for roll, pitch, and yaw follow the right-hand rule relative to the x_b-, y_b-, and z_b-axes, respectively. The components of the resultant aerodynamic force and moment in this coordinate system are denoted throughout this textbook as

$$\mathbf{F}_a = X\mathbf{i}_{x_b} + Y\mathbf{i}_{y_b} + Z\mathbf{i}_{z_b}$$

$$\mathbf{M}_a = \ell\mathbf{i}_{x_b} + m\mathbf{i}_{y_b} + n\mathbf{i}_{z_b}$$

where $\mathbf{i}_{x_b}$, $\mathbf{i}_{y_b}$, and $\mathbf{i}_{z_b}$ are the unit vectors in the x_b-, y_b-, and z_b-directions, respectively.

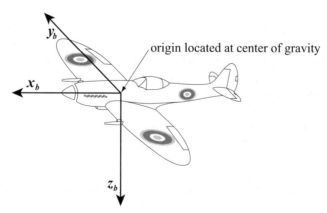

Figure 1.1.3. Body-fixed stability and control coordinates used in the study of flight mechanics.

The sign convention used in this textbook for control surface deflection also follows the right-hand rule relative to the (x_b, y_b, z_b) coordinate system shown in Fig. 1.1.3. As shown in Fig. 1.1.4, the control surfaces commonly used on a conventional airplane consist of ailerons, elevator, and rudder. The ailerons are used to control the rolling moment, the elevator is used to control the pitching moment, and the rudder is used to control the yawing moment. Downward deflection of the elevator is positive, deflection of the rudder to the left is positive, and aileron deflection is positive when the right aileron is deflected down and the left aileron is deflected up. Thus, positive elevator deflection is a right-hand rotation of the elevator relative to the y_b-axis. Similarly, positive rudder deflection corresponds to a right-hand rotation of the rudder relative to the z_b-axis. Positive aileron deflection can also be thought of as a right-hand rotation relative to the y_b-axis. For positive deflection of the ailerons, the aileron on the positive (right) wing has positive (right-hand) rotation and the aileron on the negative (left) wing has negative (left-hand) rotation.

For a positive deflection of the ailerons, the downward deflection of the right aileron increases lift on the right wing and the upward deflection of the left aileron decreases lift on the left wing. This produces a rolling moment to the left (negative ℓ). In a similar manner, a positive deflection of the elevator (downward) increases lift on the horizontal tail, producing a nose-down increment in the pitching moment (negative m). Likewise, a positive rudder deflection (to the left) produces a lift force to the right on the vertical tail and creates a nose-left yawing moment (negative n). In this text, angular deflections of the ailerons, elevator, and rudder are designated as δ_a, δ_e, and δ_r, respectively.

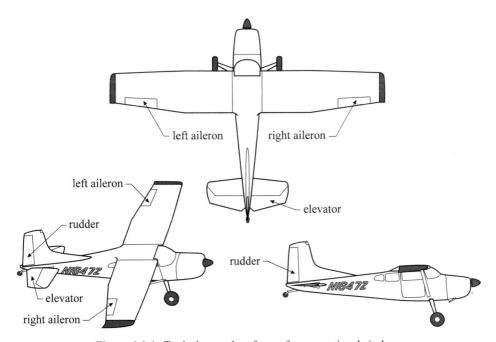

Figure 1.1.4. Typical control surfaces of a conventional airplane.

1.2. Fluid Statics and the Atmosphere

The lift and drag acting on a moving aircraft are strong functions of air density. Thus, to predict aircraft motion, we must be able to determine the density of the air at the aircraft's altitude. For this purpose, the atmosphere can be regarded as a static fluid.

In a static fluid, the change in pressure, p, with respect to geometric altitude, H, is

$$\frac{dp}{dH} = -\rho g \tag{1.2.1}$$

where ρ is the fluid density and g is the acceleration of gravity. For all practical purposes, the atmosphere can be assumed to behave as an ideal gas, which gives

$$\rho = \frac{p}{RT} \tag{1.2.2}$$

where R is the ideal gas constant for the gas and T is the absolute temperature. Thus, the pressure variation in the atmosphere is governed by

$$\frac{dp}{dH} = -\frac{g}{R}\frac{p}{T} \tag{1.2.3}$$

From Newton's law of gravitation, g varies with altitude according to the relation

$$g = g_o\left(\frac{R_E}{R_E + H}\right)^2$$

where g_o is the standard acceleration of gravity at sea level, $g_o = 9.806645$ m/s^2, and R_E is the radius of the Earth at sea level, $R_E = 6,356,766$ m. Using Newton's law of gravitation in Eq. (1.2.3) gives

$$\left(\frac{R_E + H}{R_E}\right)^2 \frac{dp}{dH} = -\frac{g_o}{R}\frac{p}{T} \tag{1.2.4}$$

The integration of Eq. (1.2.4) is simplified by introducing the change of variables

$$Z \equiv \frac{R_E H}{R_E + H} \tag{1.2.5}$$

With this change of variables, Eq. (1.2.4) can be written as

$$\frac{1}{p}\frac{dp}{dZ} = -\frac{g_o}{R}\frac{1}{T} \tag{1.2.6}$$

The new variable that is defined in Eq. (1.2.5), Z, is called the *geopotential altitude*. The difference between the geopotential and geometric altitudes is small in the first several thousand feet above sea level. However, at higher altitudes the difference is significant.

Since the temperature and pressure variations in the atmosphere are different from day to day, aircraft instruments make use of the concept of a *standard atmosphere*. Many different "standard" atmospheres have been defined. However, most are essentially indistinguishable below 100,000 feet, which encloses the domain of most aircraft.

The standard atmosphere commonly used in the calibration of aircraft instruments is divided into several regions, each of which is assumed to have a constant temperature gradient with respect to geopotential altitude. The temperature gradients defined for this standard atmosphere are given in Table 1.2.1. This information completely defines the temperature throughout the standard atmosphere. Other atmosphere definitions are quite commonly used for special area environments. These include the *polar atmosphere*, the *desert atmosphere*, and the *10 percent hot day*.

Since the temperature gradient is constant in each of the altitude ranges defined in Table 1.2.1, the temperature in each range is linear,

$$T(Z) = T(Z_i) + T_i'(Z - Z_i) \tag{1.2.7}$$

where Z_i is the minimum geopotential altitude in the range and T_i' is the temperature gradient for the range, defined as the change in temperature with respect to geopotential altitude, dT/dZ. The negative of this temperature gradient is commonly referred to as the *lapse rate*. Applying Eq. (1.2.7), the integration of Eq. (1.2.6) subject to the boundary condition, $p(Z_i) = p_i$, yields

$$p(Z) = \begin{cases} p_i \exp\left[-\dfrac{g_o(Z - Z_i)}{RT_i}\right], & T_i' = 0 \\[2em] p_i\left[\dfrac{T_i + T_i'(Z - Z_i)}{T_i}\right]^{\frac{-g_o}{RT_i'}}, & T_i' \neq 0 \end{cases} \tag{1.2.8}$$

From Eqs. (1.2.7) and (1.2.8), combined with the information given in Table 1.2.1, the temperature and pressure throughout the atmosphere can be determined from the ideal

Geopotential Altitude Range		Initial Temperature	Temperature Gradient
Z_i (m)	Z_{i+1} (m)	T_i (K)	T_i' (K/km)
0	11,000	288.150	−6.5
11,000	20,000	216.650	0.0
20,000	32,000	216.650	1.0
32,000	47,000	228.650	2.8
47,000	52,000	270.650	0.0
52,000	61,000	270.650	−2.0
61,000	79,000	252.650	−4.0
79,000	~90,000	180.650	0.0

Table 1.2.1. Temperature gradients with respect to geopotential altitude for the standard atmosphere commonly used in the calibration of aircraft instruments.

gas constant and the temperature and pressure at standard sea level. The gas constant for the standard atmosphere is defined to be 287.0528 N·m/kg·K. The standard atmosphere is defined to have a temperature at sea level of 288.150 K, and standard atmospheric pressure at sea level is defined as 101,325 N/m^2.

Once the temperature and pressure are known at any altitude the density can be determined from Eq. (1.2.2) and the speed of sound can be computed from

$$a = \sqrt{\gamma RT} \tag{1.2.9}$$

where γ is the ratio of constant pressure to constant volume specific heat, which is 1.4 for the standard atmosphere. Remember that since a newton is 1 kg·m/s^2, the ideal gas constant for the standard atmosphere can also be written as 287.0528 m^2/s^2·K.

EXAMPLE 1.2.1. Compute the absolute temperature, pressure, density, and speed of sound for the standard atmosphere defined in Table 1.2.1 at a geometric altitude of 30,000 meters.

Solution. We start by determining geopotential altitude from the known geometric altitude. From Eq. (1.2.5),

$$Z \equiv \frac{R_E H}{R_E + H} = \frac{6,356,766 \text{ m}(30,000 \text{ m})}{6,356,766 \text{ m} + 30,000 \text{ m}} = 29,859 \text{ m}$$

Since this is in the third of altitude ranges given in Table 1.2.1, we must first integrate across the first two ranges. At the boundary between the first and the second range ($Z_1 = 11$ km), from Eqs. (1.2.7) and (1.2.8) we obtain

$$T_1 = T_0 + T_0'(Z_1 - Z_0) = 288.150 \text{ K} + (-6.5 \text{ K/km})(11 - 0) \text{ km} = 216.650 \text{ K}$$

$$p_1 = p_0 \left[\frac{T_0 + T_0'(Z_1 - Z_0)}{T_0} \right]^{\frac{-g_o}{RT_0'}} = p_0 \left(\frac{T_1}{T_0} \right)^{\frac{-g_o}{RT_0'}}$$

$$= 101,325 \text{ N/m}^2 \left(\frac{216.650 \text{ K}}{288.150 \text{ K}} \right)^{\frac{-9.806645 \text{ m/s}^2}{287.0528 \text{ m}^2/\text{s}^2 \cdot \text{K}(-0.0065 \text{ K/m})}} = 22,632 \text{ N/m}^2$$

and at the boundary between the second and the third range ($Z_2 = 20$ km),

$$T_2 = T_1 + T_1'(Z_2 - Z_1) = 216.650 \text{ K} + (0.0 \text{ K/km})(20 - 11) \text{ km} = 216.650 \text{ K}$$

$$p_2 = p_1 \exp \left[-\frac{g_o(Z_2 - Z_1)}{RT_1} \right] = 22,632 \text{ N/m}^2 \exp \left[-\frac{9.806645(20 - 11) \times 10^3}{287.0528(216.650)} \right]$$

$$= 5,474.9 \text{ N/m}^2$$

Integrating from this point to the specified geopotential altitude of 29,859 meters, Eqs. (1.2.7) and (1.2.8) yield

$$T = T_2 + T_2'(Z - Z_2) = 216.650 \text{ K} + (1.0 \text{ K/km})(29.859 - 20) \text{ km} = \underline{226.509 \text{ K}}$$

$$p = p_2 \left[\frac{T_2 + T_2'(Z - Z_2)}{T_2} \right]^{\frac{-g_o}{RT_2'}} = p_2 \left(\frac{T}{T_2} \right)^{\frac{-g_o}{RT_2'}}$$

$$= 5,474.9 \text{ N/m}^2 \left(\frac{226.509 \text{ K}}{216.650 \text{ K}} \right)^{\frac{-9.806645 \text{ m/s}^2}{287.0528 \text{ m}^2/\text{s}^2 \cdot \text{K}(0.0010 \text{ K/m})}} = \underline{1,197.0 \text{ N/m}^2}$$

From Eq. (1.2.2), the density is

$$\rho = \frac{p}{RT} = \frac{1,197.0 \text{ N/m}^2}{287.0528 \text{ N} \cdot \text{m/kg} \cdot \text{K} (226.509 \text{ K})} = \underline{0.018410 \text{ kg/m}^3}$$

and from Eq. (1.2.9), the speed of sound is

$$a = \sqrt{\gamma RT} = \sqrt{1.4(287.0528 \text{ m}^2/\text{s}^2 \cdot \text{K}) 226.509 \text{ K}} = \underline{301.71 \text{ m/s}}$$

EXAMPLE 1.2.2. Compute the absolute temperature, pressure, density, and speed of sound, in English engineering units, for the standard atmosphere that is defined in Table 1.2.1 at a geometric altitude of 100,000 feet.

Solution. Since this standard atmosphere is defined in SI units, we start by converting the given geometric altitude to meters:

$$H = 100,000 \text{ ft} (0.304800 \text{ m/ft}) = 30,480 \text{ m}$$

Following the procedure used in Example 1.2.1, we obtain

$$Z = 30,335 \text{ m}/0.3048 \text{ m/ft} = 99,523 \text{ ft}$$

$$T = 226.985 \text{ K} (1.80000 \, ^\circ\text{R/K}) = \underline{408.572 \, ^\circ\text{R}}$$

$$p = 1,114.3 \text{ N/m}^2 (0.02088543 \text{ lbf/ft}^2/\text{N/m}^2) = 23.272 \text{ lbf/ft}^2 = \underline{0.16161 \text{ psi}}$$

$$\rho = 0.017102 \text{ kg/m}^3 (0.001940320 \text{ slug/ft}^3/\text{kg/m}^3) = \underline{0.000033182 \text{ slug/ft}^3}$$

$$a = 302.03 \text{ m/s}/0.3048 \text{ m/ft} = \underline{990.90 \text{ ft/sec}}$$

Atmospheric temperature and pressure vary substantially with time and location on the Earth. Nevertheless, for the correlation of test data taken at different times and locations and for the calibration of aircraft flight instruments, it is important to have an agreed-upon standard for atmospheric properties. From the assumption of an ideal gas, pressure, density, and the speed of sound can be determined directly from a defined temperature profile. The standard atmosphere, based on the temperature profile defined in Table 1.2.1 and tabulated in Appendices A and B, is commonly used for the purposes mentioned above. Slightly different temperature profiles have also been used to define other standard atmospheres, but the variation is small below 100,000 feet.

1.3. The Boundary Layer Concept

As pointed out in Sec. 1.1, there are only two types of aerodynamic forces acting on a body moving through the atmosphere: pressure forces and viscous shear forces. The Reynolds number provides a measure of the relative magnitude of the pressure forces in relation to the viscous shear forces. For the airspeeds typically encountered in flight, Reynolds numbers are quite high and viscous forces are usually small compared to pressure forces. This does not mean that viscous forces can be neglected. However, it does allow us to apply the simplifying concept of boundary layer theory.

For flow over a streamlined body at low angle of attack and high Reynolds number, the effects of viscosity are essentially confined to a thin layer adjacent to the surface of the body, as shown in Fig. 1.3.1. Outside the boundary layer, the shear forces can be neglected and since the boundary layer is thin, the change in pressure across the thickness of this layer is insignificant. With this flow model, the pressure forces can be determined from the inviscid flow outside the boundary layer, and the shear forces can be obtained from a solution to the boundary layer equations.

While boundary layer theory provides a tremendous simplification over the complete Navier-Stokes equations, solutions to the boundary layer equations are far from trivial, especially for the complex geometry that is often encountered in an aircraft. A thorough review of boundary layer theory is beyond the intended scope of this chapter and is not prerequisite to an understanding of the fundamental principles of flight mechanics. However, there are some important results of boundary layer theory that the reader should know and understand.

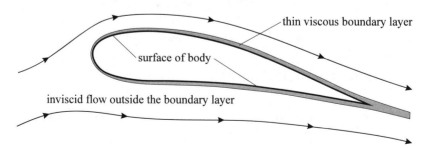

Figure 1.3.1. Boundary layer flow over a streamlined body at a low angle of attack.

1. For the high-Reynolds-number flows typically encountered in flight, the viscous shear forces are small compared to the pressure forces.

2. For flow at high Reynolds number, the pressure forces acting on a body can be closely approximated from an inviscid flow analysis, outside the boundary layer.

3. For two-dimensional flow about streamlined bodies at high Reynolds numbers and low angles of attack, pressure forces do not contribute significantly to drag.

4. For bluff bodies and streamlined bodies at high angles of attack, boundary layer separation occurs, as shown in Fig. 1.3.2, and pressure forces dominate the drag.

Boundary layer separation in the flow over airfoils and wings is commonly referred to as *stall*. An understanding of stall and its effect on lift and drag is critical in the study of flight mechanics. As the angle of attack for a wing is increased from zero, at first the boundary layer remains attached, as shown in Fig. 1.3.1, the lift coefficient increases as a nearly linear function of angle of attack, and the drag coefficient increases approximately with the angle of attack squared. As angle of attack continues to increase, the positive pressure gradient on the aft portion of the upper surface of the wing also increases. At some angle of attack this adverse pressure gradient may result in local boundary layer separation and the increase in lift with angle of attack will begin to diminish. At a slightly higher angle of attack, boundary layer separation becomes complete, as shown

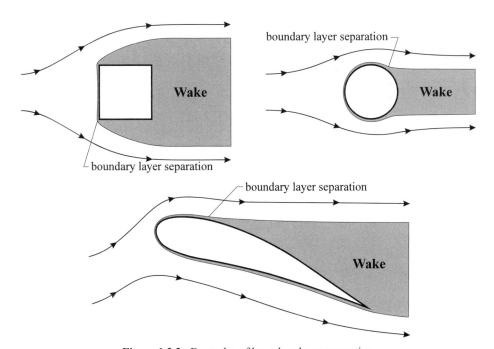

Figure 1.3.2. Examples of boundary layer separation.

in Fig. 1.3.2, and the lift rapidly decreases as the angle of attack is increased further. Boundary layer separation also greatly accelerates the increase in drag with angle of attack. The maximum lift coefficient and the exact shape of the lift curve for angles of attack near stall depend substantially on the airfoil section geometry. As the angle of attack increases beyond stall, the lift and drag coefficients become less sensitive to section geometry and for angles of attack beyond about 25 degrees, the lift and drag coefficients are nearly independent of the airfoil section.

The maximum lift coefficient that can be attained on a given wing before stall is quite important in flight mechanics. This parameter not only determines the maximum weight that can be carried by a given wing at a particular airspeed, it also affects takeoff distance, turning radius, and other measures of airplane performance.

1.4. Inviscid Aerodynamics

Since the pressure forces acting on an aircraft at high Reynolds number can be closely estimated from an inviscid analysis of the flow outside the boundary layer, inviscid aerodynamics is an important tool used in the estimation of the forces acting on an aircraft in flight. Written in vector notation, the general equations that govern inviscid fluid flow are the continuity equation

$$\frac{\partial \rho}{\partial t} + (\mathbf{V} \cdot \nabla)\rho + \rho \nabla \cdot \mathbf{V} = 0 \tag{1.4.1}$$

and the momentum equation

$$\frac{\partial \mathbf{V}}{\partial t} + (\mathbf{V} \cdot \nabla)\mathbf{V} = -\frac{\nabla p}{\rho} - g\nabla H \tag{1.4.2}$$

where $\mathbf{V}$ is the fluid velocity vector and t is time. Using the important mathematical identity

$$(\mathbf{V} \cdot \nabla)\mathbf{V} = \nabla\left(\tfrac{1}{2}V^2\right) - \mathbf{V} \times (\nabla \times \mathbf{V})$$

the momentum equation for inviscid flow can be written

$$\frac{\partial \mathbf{V}}{\partial t} + \nabla\left(\tfrac{1}{2}V^2\right) + \frac{\nabla p}{\rho} + g\nabla H = \mathbf{V} \times \mathbf{\Omega} \tag{1.4.3}$$

where $\mathbf{\Omega}$ is the curl of the velocity vector, traditionally called the *vorticity*:

$$\mathbf{\Omega} \equiv \nabla \times \mathbf{V} \tag{1.4.4}$$

It can be shown from the application of vector calculus to Eqs. (1.4.1) and (1.4.3) that vorticity cannot be generated in an inviscid flow. In the aerodynamics problem associated with flow over an aircraft in flight, the fluid velocity far upstream from the aircraft is uniform. From the definition of vorticity given in Eq. (1.4.4), this means that the vorticity far upstream from the aircraft is zero. Since vorticity cannot be generated in an inviscid flow, the vorticity must be zero everywhere in the inviscid flow outside the boundary layer. Vorticity is generated in the viscous boundary layer next to the skin of the aircraft and thus there is also vorticity in the boundary layer wake that trails behind an aircraft in flight. However, outside the boundary layer and trailing wake, the flow can be assumed to be inviscid and free of vorticity. Thus, for this flow, the right-hand side of Eq. (1.4.3) will be zero. This type of flow is called *inviscid irrotational flow*.

In summary, inviscid irrotational flow can be assumed to exist outside the boundary layer and trailing wake of an aircraft in flight. For this flow, the continuity and momentum equations are

$$\frac{\partial \rho}{\partial t} + (\mathbf{V} \cdot \nabla)\rho + \rho \nabla \cdot \mathbf{V} = 0 \tag{1.4.5}$$

$$\frac{\partial \mathbf{V}}{\partial t} + \nabla\left(\tfrac{1}{2}V^2\right) + \frac{\nabla p}{\rho} + g\nabla H = 0 \tag{1.4.6}$$

For the important special case of an aircraft in steady flight, the time derivatives are zero and Eq. (1.4.6) reduces to

$$\nabla\left(\tfrac{1}{2}V^2\right) + \frac{\nabla p}{\rho} + g\nabla H = 0 \tag{1.4.7}$$

For the freestream flow far from the aircraft, Eq. (1.4.7) gives

$$\nabla\left(\tfrac{1}{2}V_\infty^2\right) + \frac{\nabla p_\infty}{\rho_\infty} + g\nabla H = 0$$

Since the freestream velocity is uniform, the gradient of V_∞^2 is zero, which gives

$$g\nabla H = -\frac{\nabla p_\infty}{\rho_\infty} \tag{1.4.8}$$

This is simply a form of the hydrostatic equation, which is given by Eq. (1.2.1). Using Eq. (1.4.8) in Eq. (1.4.7) to eliminate the geometric altitude gradient, **the momentum equation for the inviscid flow about an aircraft in steady flight is**

$$\nabla\left(\tfrac{1}{2}V^2\right) + \frac{\nabla p}{\rho} - \frac{\nabla p_\infty}{\rho_\infty} = 0 \tag{1.4.9}$$

where p_∞ and ρ_∞ are evaluated at the same altitude as p and ρ.

A special case of Eq. (1.4.9) that is of particular interest is the case of flow with negligible variation in air density. This approximation can be applied to flight at low Mach numbers. Assuming that ρ is constant, Eq. (1.4.9) reduces to

$$\nabla\left(\tfrac{1}{2}V^2 + \frac{p - p_\infty}{\rho}\right) = 0 \tag{1.4.10}$$

Integrating this result from the freestream to some arbitrary point in the flow and solving for the local static pressure, **the momentum equation for incompressible inviscid flow about an aircraft in steady flight yields**

$$p = p_\infty + \tfrac{1}{2}\rho(V_\infty^2 - V^2) \tag{1.4.11}$$

Once the velocity field has been determined in some manner, Eq. (1.4.11) can be used to determine the pressure at any point in the flow. The net contribution that these pressure forces make to the resultant aerodynamic force is then computed from

$$\mathbf{F}_p = -\iint_{S} p\mathbf{n}\,dS = -\iint_{S} p_\infty\mathbf{n}\,dS + \tfrac{1}{2}\rho V_\infty^2 \iint_{S}[(V/V_\infty)^2 - 1]\mathbf{n}\,dS \tag{1.4.12}$$

where $\mathbf{n}$ is the unit outward normal and S is the surface area. The first integral on the far right-hand side of Eq. (1.4.12) is the buoyant force, and the second integral is the vector sum of the pressure contribution to the lift and drag. For most conventional airplanes the buoyant force is small and can be neglected.

From Eq. (1.4.5), the continuity equation for incompressible flow is

$$\nabla \cdot \mathbf{V} = 0 \tag{1.4.13}$$

Because the curl of the gradient of any scalar function is zero, a flow field is irrotational if the velocity field is written as the gradient of a scalar function

$$\mathbf{V} = \nabla\phi \tag{1.4.14}$$

where ϕ is normally called the *velocity potential* and could be any scalar function of space and time that satisfies the continuity equation and the required boundary conditions. Using Eq. (1.4.14) in Eq. (1.4.13), this requires that

$$\nabla \cdot \mathbf{V} = \nabla \cdot \nabla\phi = \nabla^2\phi = 0 \tag{1.4.15}$$

Thus the incompressible velocity potential must satisfy Laplace's equation and some appropriate boundary conditions. The far-field boundary condition is uniform flow. At the surface of the aircraft the normal component of velocity must go to zero. However, since the flow is inviscid, we do not require zero tangential velocity at the surface.

For flight at higher Mach numbers the result predicted by Eq. (1.4.9) is more complex. When the density of a fluid changes with pressure, the temperature changes as well. Thus, in a compressible fluid, the pressure gradients that accompany velocity gradients also produce temperature gradients. In general, temperature gradients result in heat transfer. However, the thermal conductivity of air is extremely low. Thus, it is commonly assumed that the flow outside the boundary layer is adiabatic as well as inviscid. Inviscid adiabatic flow is isentropic. From the fundamentals of thermodynamics recall that for isentropic flow of an ideal gas, density is related to pressure according to the relation

$$\rho = \rho_\infty \left(\frac{p}{p_\infty} \right)^{\frac{1}{\gamma}} \tag{1.4.16}$$

where again γ is the ratio of specific heats ($\gamma = 1.4$ for air). Substituting Eq. (1.4.16) into Eq. (1.4.9) and neglecting the gradients of p_∞ and ρ_∞ yields

$$\nabla \left(\frac{1}{2} V^2 + \frac{\gamma}{\gamma - 1} \frac{p}{\rho} \right) = 0 \tag{1.4.17}$$

Integrating Eq. (1.4.17) gives

$$\frac{1}{2} V^2 + \frac{\gamma}{\gamma - 1} \frac{p}{\rho} = \text{constant} \tag{1.4.18}$$

Equation (1.4.18) must apply at the stagnation state, where $V = 0$. Thus, the constant in Eq. (1.4.18) can be expressed in terms of the stagnation pressure and density, p_0 and ρ_0,

$$\frac{1}{2} V^2 + \frac{\gamma}{\gamma - 1} \frac{p}{\rho} = \frac{\gamma}{\gamma - 1} \frac{p_0}{\rho_0} \tag{1.4.19}$$

For an ideal gas $p/\rho = RT$ and $a^2 = \gamma RT$, where R is the gas constant and a is the speed of sound. Thus, Eqs. (1.4.16) and (1.4.19) written in terms of Mach number, M, require that

$$T = T_0 \left(1 + \frac{\gamma - 1}{2} M^2 \right)^{-1} \tag{1.4.20}$$

$$p = p_0 \left(1 + \frac{\gamma - 1}{2} M^2 \right)^{\frac{-\gamma}{\gamma - 1}} \tag{1.4.21}$$

$$\rho = \rho_0 \left(1 + \frac{\gamma - 1}{2} M^2 \right)^{\frac{-1}{\gamma - 1}} \tag{1.4.22}$$

Stagnation conditions are readily evaluated by applying Eqs. (1.4.20) through (1.4.22) to the freestream flow:

$$T_0 = T_\infty\left(1+\frac{\gamma-1}{2}M_\infty^2\right) \tag{1.4.23}$$

$$p_0 = p_\infty\left(1+\frac{\gamma-1}{2}M_\infty^2\right)^{\frac{\gamma}{\gamma-1}} \tag{1.4.24}$$

$$\rho_0 = \rho_\infty\left(1+\frac{\gamma-1}{2}M_\infty^2\right)^{\frac{1}{\gamma-1}} \tag{1.4.25}$$

Equation (1.4.14) applies to compressible flow as well as to incompressible flow. **For any irrotational flow, the velocity vector field can always be expressed as the gradient of a scalar potential field.** All such flows are called *potential flows*. The only requirement for potential flow is that the flow be irrotational. There are no further restrictions. It can be shown mathematically that the curl of the gradient of any scalar function is zero. Thus, it follows that **every potential flow is an irrotational flow.** It can also be shown mathematically that if the curl of any vector field is zero, that vector field can be expressed as the gradient of some scalar function. Therefore, **every irrotational flow is a potential flow.** For this reason the terms *potential flow* and *irrotational flow* are used synonymously.

1.5. Review of Elementary Potential Flows

Since the governing equations for incompressible potential flow are linear, solutions are additive. That is, if two or more solutions each satisfy the incompressible potential flow equations independently, their sum will satisfy these equations as well. As a result, the solution for incompressible potential flow over a body of complex geometry can be obtained by combining a number of elementary solutions.

In this section we review some of the more commonly used elementary solutions for incompressible potential flows. In your introductory classes on fluid mechanics and aerodynamics you probably developed at least some of these elementary solutions. Here, only the final results will be presented in summary form.

Uniform Flow
A uniform flow with constant but completely arbitrary velocity components will satisfy the potential flow equations and can be expressed as

$$\phi = V_x x + V_y y + V_z z + C \tag{1.5.1}$$

$$\mathbf{V} = \nabla\phi = V_x \mathbf{i}_x + V_y \mathbf{i}_y + V_z \mathbf{i}_z \tag{1.5.2}$$

Two-Dimensional Source Flow

For a two-dimensional (2-D) source located at the origin and having strength Λ, the scalar velocity potential and velocity vector fields in polar coordinates (r,θ) are given by

$$\phi \;=\; \frac{\Lambda}{2\pi}\ln r + C \tag{1.5.3}$$

$$\mathbf{V} \;=\; \nabla\phi \;=\; \frac{\Lambda}{2\pi r}\mathbf{i}_r \tag{1.5.4}$$

where C is an arbitrary constant and $\mathbf{i}_r$ is the unit vector in the r-direction.

Two-Dimensional Vortex Flow

A 2-D vortex with circulation Γ located at the origin is described by

$$\phi \;=\; -\frac{\Gamma}{2\pi}\theta + C \tag{1.5.5}$$

$$\mathbf{V} \;=\; \nabla\phi \;=\; -\frac{\Gamma}{2\pi r}\mathbf{i}_\theta \tag{1.5.6}$$

Two-Dimensional Doublet Flow

Combining a positive source with a negative source (called a *sink*) forms a doublet. This potential flow is described according to

$$\phi \;=\; \frac{\kappa}{2\pi}\frac{\cos\theta}{r} + C \tag{1.5.7}$$

$$\mathbf{V} \;=\; \nabla\phi \;=\; -\frac{\kappa\cos\theta}{2\pi r^2}\mathbf{i}_r - \frac{\kappa\sin\theta}{2\pi r^2}\mathbf{i}_\theta \tag{1.5.8}$$

Two-Dimensional Stagnation Flow

The flow near a 2-D stagnation point can be described as a potential flow. For the case where the freestream velocity is in the x-direction, the solid surface is coincident with the y-axis, and the stagnation point is located at the origin, the potential field and velocity field can be expressed as

$$\phi \;=\; \frac{B}{2}(x^2 - y^2) + C \tag{1.5.9}$$

$$\mathbf{V} \;=\; \nabla\phi \;=\; Bx\mathbf{i}_x - By\mathbf{i}_y \tag{1.5.10}$$

Two-Dimensional Source Panels

The source sheet is an important elementary potential flow that is not included in many introductory aerodynamics textbooks. While it is possible to synthesize flow over some bodies of very specific shape by combining a finite number of sources and sinks with a uniform flow, to synthesize flow over a body of arbitrary shape, greater flexibility is required. The 2-D source sheet helps provide such increased flexibility. Combining an infinite number of infinitesimally weak 2-D sources in side-by-side fashion as shown in Fig. 1.5.1 forms this potential flow.

While a source sheet with a completely arbitrary curved shape, such as that shown in Fig. 1.5.1, is required to synthesize flow exactly over a body of arbitrary shape, such a sheet can be closely approximated using a series of planar segments. A planar source sheet is usually referred to as a *source panel*. Approximating a source sheet with a series of source panels significantly simplifies the solution process.

Consider the differential segment of a 2-D source panel that lies on the x-axis at the location $x = x_o$, as shown in Fig. 1.5.2. The velocity induced by this differential source panel at an arbitrary point with coordinates (x,y) is given by

$$dV_x = dV_r \cos\theta = \frac{\lambda(x_o)}{2\pi r}\cos\theta\, dx_o = \frac{(x-x_o)\lambda(x_o)}{2\pi[(x-x_o)^2 + y^2]}dx_o$$

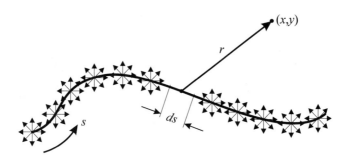

Figure 1.5.1. Edge view of a 2-D source sheet.

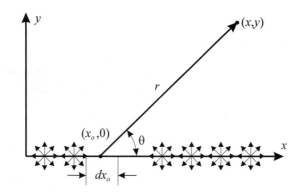

Figure 1.5.2. Edge view of a 2-D source panel on the x-axis.

and

$$dV_y = dV_r \sin\theta = \frac{\lambda(x_o)}{2\pi r}\sin\theta\, dx_o = \frac{y\,\lambda(x_o)}{2\pi[(x-x_o)^2 + y^2]}dx_o$$

where $\lambda(x_o)$ is the strength per unit length at x_o and the remaining notation is defined in Fig. 1.5.2. Integrating over a panel with a linear strength distribution that extends from $x_o = 0$ to $x_o = l$, the velocity induced at the point (x,y) by the entire source panel is

$$\begin{Bmatrix} V_x \\ V_y \end{Bmatrix} = \frac{1}{2\pi l}\begin{bmatrix} [-y\Phi + (l-x)\Psi + l] & (y\Phi + x\Psi - l) \\ [(l-x)\Phi + y\Psi] & (x\Phi - y\Psi) \end{bmatrix}\begin{Bmatrix} \lambda(0) \\ \lambda(l) \end{Bmatrix} \tag{1.5.11}$$

$$\Phi \equiv \mathrm{atan2}\left(yl,\; y^2 + x^2 - xl\right) \tag{1.5.12}$$

$$\Psi \equiv \frac{1}{2}\ln\left[\frac{x^2 + y^2}{(x-l)^2 + y^2}\right] \tag{1.5.13}$$

As can be seen in Fig. 1.5.3, the angle Φ can take any value between $-\pi$ and π. Note that when the point (x,y) is either left or right of the panel ($x < 0$ or $x > l$) the angle Φ is positive when y is above the axis, negative when y is below the axis, and approaches zero as y approaches zero from either side. However, when the point (x,y) is directly above the panel ($0 < x < l$ and $y > 0$) and y approaches zero from above, Φ approaches π. On the other hand, when the point (x,y) is directly below the panel ($0 < x < l$ and $y < 0$) and y approaches zero from below, Φ approaches $-\pi$. Thus we see that there is a step discontinuity of 2π in the angle Φ as we cross the panel itself,

$$\Phi(x,\pm 0) = \pm\pi, \quad 0 < x < l$$

where $y = +0$ indicates a point just above the surface and $y = -0$ indicates a point just below the surface.

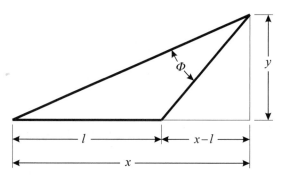

Figure 1.5.3. Geometry for a finite 2-D source panel.

Single-argument arctangent functions like the Fortran or C intrinsic function "atan" will always return values between $-\pi/2$ and $\pi/2$. Such functions cannot be used to compute the angle $\varPhi$ from Eq. (1.5.12) for all values of x and y. To obtain the correct result over the full range of x and y, the angle $\varPhi$ must always be computed using a two-argument arctangent function, such as the Fortran or C intrinsic function "atan2."

Two-Dimensional Vortex Panels

While 2-D source panels can be used to synthesize nonlifting flow over a 2-D body of arbitrary cross-section, a source panel produces no circulation. No potential flow that includes circulation can be synthesized from source panels alone. In addition, the Kutta-Joukowski law requires that for any closed 2-D body of arbitrary shape, the lift per unit span is always proportional to the circulation. Thus, no lifting potential flow can be synthesized from source panels alone.

For the synthesis of 2-D lifting flows, vortex panels are commonly used. A 2-D vortex panel is formed from 2-D vortices in the same way that a 2-D source panel is formed from 2-D sources. An infinite number of infinitesimally weak vortices are combined in side-by-side fashion as shown in Fig. 1.5.4. In this figure we are again looking at an edge view of the panel with the vortices all perpendicular to the page.

Now consider a differential segment of a vortex panel that lies on the x-axis at the location $x = x_o$ and has length dx_o. The velocity induced at any point (x,y) by this differential vortex panel is normal to a line drawn between the point (x,y) and the point $(x_o,0)$ and has a magnitude inversely proportional to the distance between the two points. The x- and y-components of the velocity induced at the point (x,y) by this infinitesimally small section of the vortex panel are then given by

$$dV_x = -dV_\theta \sin\theta = \frac{\gamma(x_o)}{2\pi r}\sin\theta\, dx_o = \frac{y\gamma(x_o)}{2\pi[(x-x_o)^2 + y^2]}dx_o$$

$$dV_y = dV_\theta \cos\theta = -\frac{\gamma(x_o)}{2\pi r}\cos\theta\, dx_o = -\frac{(x-x_o)\gamma(x_o)}{2\pi[(x-x_o)^2 + y^2]}dx_o$$

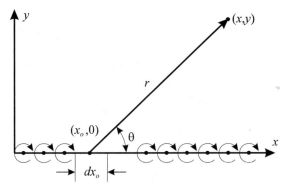

Figure 1.5.4. Edge view of a 2-D vortex panel on the x-axis.

where $\gamma(x_o)$ is the vortex strength per unit length at x_o and the remaining notation is defined in Fig. 1.5.4. For a linear vortex strength distribution on a panel extending from $x = 0$ to $x = l$, the total induced velocity at the point (x,y) is

$$\begin{Bmatrix} V_x \\ V_y \end{Bmatrix} = \frac{1}{2\pi l} \begin{bmatrix} [(l-x)\Phi + y\Psi] & (x\Phi - y\Psi) \\ [y\Phi - (l-x)\Psi - l] & (-y\Phi - x\Psi + l) \end{bmatrix} \begin{Bmatrix} \gamma(0) \\ \gamma(l) \end{Bmatrix} \qquad (1.5.14)$$

$$\Phi \equiv \text{atan2}\left(yl, \; y^2 + x^2 - xl\right) \qquad (1.5.15)$$

$$\Psi \equiv \frac{1}{2}\ln\left[\frac{x^2 + y^2}{(x-l)^2 + y^2}\right] \qquad (1.5.16)$$

At any point along a vortex panel, there is a local jump in tangential velocity across the panel that is equal to the local vortex strength. There is no normal velocity jump across a vortex panel.

Three-Dimensional Vortex Line Segments

Vortex filaments can be used to synthesize potential flow over finite wings and other lifting surfaces. Since the vortex filaments used for this purpose cannot be straight over their entire length, these vortex filaments are often broken up into finite segments. As a fundamental building block, consider an arbitrary straight segment of a vortex filament that extends from point 1 to point 2, as shown in Fig. 1.5.5. We wish to express the velocity vector induced at point 3 by this straight vortex segment. Let point 4 be an arbitrary point on the vortex segment between point 1 and point 2. Let $\mathbf{dV}$ be the differential velocity vector induced at point 3 by the directed differential segment of the filament, $\mathbf{dl}$, located at point 4. From the Biot-Savart law, we have

$$\mathbf{dV} = \frac{\Gamma \, \mathbf{dl} \times \mathbf{r}}{4\pi \, |\mathbf{r}|^3} \qquad (1.5.17)$$

where Γ is the vortex strength and $\mathbf{r}$ is the spatial vector from point 4 to point 3.

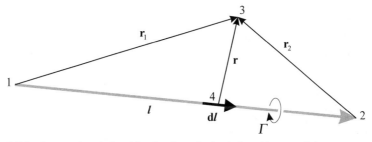

Figure 1.5.5. Geometric relationships for the velocity induced by a straight vortex segment.

After integrating Eq. (1.5.17) from point 1 to point 2, it can be shown that the velocity vector induced at an arbitrary point in space by **any straight vortex segment** is

$$\mathbf{V} = \frac{\Gamma}{4\pi} \frac{(r_1 + r_2)(\mathbf{r}_1 \times \mathbf{r}_2)}{r_1 r_2 (r_1 r_2 + \mathbf{r}_1 \cdot \mathbf{r}_2)} \qquad (1.5.18)$$

where Γ is the vortex strength, $\mathbf{r}_1$ is the spatial vector from the beginning of the segment to the arbitrary point in space, and $\mathbf{r}_2$ is the spatial vector from the end of the segment to the arbitrary point in space.

The velocity that is induced at an arbitrary point in space by a **straight semi-infinite vortex segment** can be determined as a special case of Eq. (1.5.18), by taking the limit as r_2 approaches infinity. This gives

$$\mathbf{V} = \frac{\Gamma}{4\pi} \frac{\mathbf{u}_\infty \times \mathbf{r}_1}{r_1 (r_1 - \mathbf{u}_\infty \cdot \mathbf{r}_1)} \qquad (1.5.19)$$

where Γ is the vortex strength, $\mathbf{u}_\infty$ is the unit vector in the direction from the beginning of the segment toward the end at infinity, and $\mathbf{r}_1$ is the spatial vector from the beginning of the segment to the arbitrary point in space.

1.6. Incompressible Flow over Airfoils

An airfoil is any 2-D cross-section of a wing or other lifting surface that lies in a plane perpendicular to the spanwise coordinate. An airfoil section is completely defined by the geometric shape of its boundary. However, the aerodynamic properties of an airfoil section are most profoundly affected by the shape of its centerline. This centerline is midway between the upper and lower surfaces of the airfoil and is called the *camber line*. If the airfoil is not symmetric, the camber line is not a straight line but rather a planar curve.

Because the shape of the camber line is such an important factor in airfoil design, it is critical that the student understand exactly how the camber line is defined. In addition, there are several other designations that will be used throughout this and following chapters when referring to the geometric attributes of airfoil sections. The student should be sure that he or she understands the following nomenclature as it applies to airfoil geometry such as that shown in Fig. 1.6.1.

The **camber line** is the locus of points midway between the upper and lower surfaces of an airfoil section as **measured perpendicular to the camber line itself**.

The **leading edge** is the most forward point on the camber line.

The **trailing edge** is the most rearward point on the camber line.

The **chord line** is a straight line connecting the leading edge and the trailing edge.

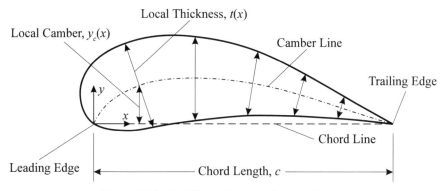

Figure 1.6.1. Airfoil coordinates and nomenclature.

The **chord length**, often referred to simply as the **chord**, is the distance between the leading edge and the trailing edge as measured along the chord line.

The **maximum camber**, often referred to simply as the **camber**, is the maximum distance between the chord line and the camber line as measured perpendicular to the chord line.

The **local thickness**, at any point along the chord line, is the distance between the upper and lower surfaces as **measured perpendicular to the camber line**.

The **maximum thickness**, often referred to simply as the **thickness**, is the maximum distance between the upper and lower surfaces as **measured perpendicular to the camber line**.

The **upper and lower surface coordinates** for an airfoil can be obtained explicitly from the camber line geometry, $y_c(x)$, and the thickness distribution, $t(x)$,

$$x_u(x) = x - \frac{t(x)}{2\sqrt{1+(dy_c/dx)^2}} \frac{dy_c}{dx} \qquad (1.6.1)$$

$$y_u(x) = y_c(x) + \frac{t(x)}{2\sqrt{1+(dy_c/dx)^2}} \qquad (1.6.2)$$

$$x_l(x) = x + \frac{t(x)}{2\sqrt{1+(dy_c/dx)^2}} \frac{dy_c}{dx} \qquad (1.6.3)$$

$$y_l(x) = y_c(x) - \frac{t(x)}{2\sqrt{1+(dy_c/dx)^2}} \qquad (1.6.4)$$

Thin Airfoil Theory
For airfoils with a maximum thickness of about 12 percent or less, the inviscid aerodynamic force and moment are only slightly affected by the thickness distribution. The resultant aerodynamic force and moment acting on such an airfoil depend almost exclusively on the angle of attack and the shape of the camber line. For this reason, the inviscid aerodynamics for these airfoils can be closely approximated by assuming that the airfoil thickness is zero everywhere along the camber line. Thus, airfoils with a thickness of about 12 percent or less can be approximated by combining a uniform flow with a vortex sheet placed along the camber line, as shown schematically in Fig. 1.6.2. The strength of this vortex sheet is allowed to vary with the distance, s, measured along the camber line. The variation in this strength, $\gamma(s)$, is determined so that the camber line becomes a streamline for the flow.

In the development of thin airfoil theory it is shown that the vortex strength distribution necessary to make the camber line a streamline is related to the camber line geometry according to

$$\frac{1}{2\pi} \int_{x_o=0}^{c} \frac{\gamma(x_o)}{x-x_o} dx_o = V_\infty\left(\alpha - \frac{dy_c}{dx}\right) \tag{1.6.5}$$

This is the fundamental equation of thin airfoil theory. Any vortex strength distribution, $\gamma(x_o)$, which satisfies Eq. (1.6.5) will make the camber line a streamline of the flow, at least within the accuracy of the approximations used in thin airfoil theory. For a given airfoil at a given angle of attack, the only unknown in Eq. (1.6.5) is the vortex strength distribution, $\gamma(x_o)$. This equation is subject to a boundary condition known as the *Kutta condition*, which requires that

$$\gamma(c) = 0 \tag{1.6.6}$$

Development of the general solution to Eq. (1.6.5), subject to (1.6.6), is presented in most undergraduate engineering textbooks on aerodynamics. This solution is found by using the change of variables, $x = c(1-\cos\theta)/2$. Only the final result is presented here.

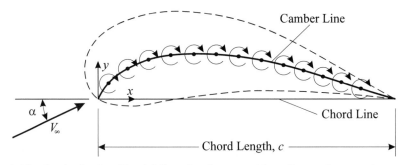

Figure 1.6.2. Synthesis of a thin airfoil section from a uniform flow and a curved vortex sheet distributed along the camber line.

The solution is in the form of an infinite series for the vortex strength distribution,

$$\gamma(\theta) = 2V_\infty \left(A_0 \frac{1+\cos\theta}{\sin\theta} + \sum_{n=1}^{\infty} A_n \sin(n\theta) \right) \tag{1.6.7}$$

$$A_0 = \alpha - \frac{1}{\pi} \int_{\theta=0}^{\pi} \frac{dy_c}{dx} d\theta \tag{1.6.8}$$

$$A_n = \frac{2}{\pi} \int_{\theta=0}^{\pi} \frac{dy_c}{dx} \cos(n\theta) d\theta \tag{1.6.9}$$

$$x(\theta) \equiv \frac{c}{2}(1-\cos\theta) \tag{1.6.10}$$

The aerodynamic force per unit span can be predicted from the **Kutta-Joukowski law**, which applies to all 2-D potential flows. This requires that the net aerodynamic force is always normal to the freestream and equal to

$$\tilde{L} = \rho V_\infty \Gamma \tag{1.6.11}$$

A direct consequence of the Kutta-Joukowski law is that the pressure drag for any 2-D flow without boundary layer separation is zero.

Applying the Kutta-Joukowski law to a differential segment of the vortex sheet that is used to synthesize a thin airfoil gives

$$d\tilde{L} = \rho V_\infty d\Gamma = \rho V_\infty \gamma(x_o) dx_o = \frac{\rho V_\infty c}{2} \gamma(\theta_o) \sin\theta_o d\theta_o$$

Applying Eq. (1.6.7) for the vortex strength distribution, this result can be used to evaluate the lift and moment coefficients as well as the center of pressure.

$$\tilde{C}_L = 2\pi(A_0 + \tfrac{1}{2}A_1) = 2\pi\left(\alpha - \frac{1}{\pi} \int_{\theta=0}^{\pi} \frac{dy_c}{dx}(1-\cos\theta) d\theta \right) = 2\pi(\alpha - \alpha_{L0}) \tag{1.6.12}$$

$$\tilde{C}_{m_{le}} = -\frac{\tilde{C}_L}{4} + \frac{\pi}{4}(A_2 - A_1) = -\frac{\tilde{C}_L}{4} + \frac{1}{2} \int_{\theta=0}^{\pi} \frac{dy_c}{dx}[\cos(2\theta) - \cos\theta] d\theta \tag{1.6.13}$$

$$\tilde{C}_{m_{c/4}} = \frac{\pi}{4}(A_2 - A_1) = \frac{1}{2} \int_{\theta=0}^{\pi} \frac{dy_c}{dx}[\cos(2\theta) - \cos\theta] d\theta \tag{1.6.14}$$

$$\frac{x_{cp}}{c} = \frac{1}{4} + \frac{\pi}{4\tilde{C}_L}(A_1 - A_2) = \frac{1}{4} + \frac{1}{2\tilde{C}_L} \int_{\theta=0}^{\pi} \frac{dy_c}{dx}[\cos\theta - \cos(2\theta)] d\theta \tag{1.6.15}$$

Note from Eq. (1.6.12) that thin airfoil theory predicts a section lift coefficient that is a linear function of angle of attack, and that the change in lift coefficient with respect to angle of attack is 2π per radian. Also note that the lift coefficient at zero angle of attack is a function only of the shape of the camber line. Thus, for thin airfoils,

$$\text{section lift slope} \equiv \frac{d\tilde{C}_L}{d\alpha} \equiv \tilde{C}_{L,\alpha} = 2\pi$$

$$\text{zero lift angle of attack} \equiv \alpha_{L0} = \frac{1}{\pi}\int_{\theta=0}^{\pi}\frac{dy_c}{dx}(1-\cos\theta)\,d\theta$$

Also notice that the leading-edge moment coefficient and center of pressure both depend on lift coefficient and hence on angle of attack. The quarter-chord moment coefficient, on the other hand, is independent of angle of attack and depends only on the shape of the camber line. Since the aerodynamic center is defined to be the point on the airfoil where the moment does not change with angle of attack, **for incompressible flow, the quarter chord is the aerodynamic center of a thin airfoil**. Thus, the quarter chord is usually referred to as the *theoretical* or *ideal aerodynamic center* of the cambered airfoil. Viscous effects and airfoil thickness can cause the quarter-chord moment coefficient to vary slightly with angle of attack, but this variation is small and the aerodynamic center is always close to the quarter chord for subsonic flow. For this reason, airfoil section moment data are usually reported in terms of the quarter-chord moment coefficient. For an airfoil with no camber, Eq. (1.6.15) shows that **the quarter chord is also the center of pressure for incompressible flow about a thin symmetric airfoil**.

As is the case with all 2-D potential flow, thin airfoil theory predicts a net aerodynamic force that is normal to the freestream. Thus, thin airfoil theory predicts a section drag coefficient that is exactly zero. This is not a function of the thin airfoil approximation. Numerical panel methods will also predict zero section drag, including the effects of thickness. Section drag in any 2-D subsonic flow results entirely from viscous effects, which are neglected in the potential flow equations. The viscous forces also have some effect on lift, but this effect is relatively small.

EXAMPLE 1.6.1. For the purpose of demonstrating the application of thin airfoil theory, consider a hypothetical airfoil section that has a parabolic camber line geometry defined by the equation

$$\frac{y_c}{c} = 4\frac{y_{mc}}{c}\left[\frac{x}{c}-\left(\frac{x}{c}\right)^2\right]$$

where c is the airfoil chord length and y_{mc} is the maximum camber. Using thin airfoil theory, obtain algebraic expressions for the zero-lift angle of attack and the quarter-chord moment coefficient as a function of maximum camber. For an airfoil of this geometry having 4 percent maximum camber, use this thin airfoil analysis to estimate the section lift coefficients at a 5-degree angle of attack and the section quarter-chord moment coefficient.

Solution. Differentiating the given camber line equation, the camber line slope is

$$\frac{dy_c}{dx} = 4y_{mc}\left(\frac{1}{c} - 2\frac{x}{c^2}\right) = 4\frac{y_{mc}}{c}\left(1 - 2\frac{x}{c}\right)$$

Using the change of variables

$$\frac{x}{c} \equiv \frac{1}{2}(1 - \cos\theta)$$

we have

$$\frac{dy_c}{dx} = 4\frac{y_{mc}}{c}\left[1 - 2\frac{1}{2}(1 - \cos\theta)\right] = 4\frac{y_{mc}}{c}\cos\theta$$

From Eqs. (1.6.8) and (1.6.9), the first three coefficients in the infinite series solution are

$$A_0 = \alpha - \frac{1}{\pi}\int_{\theta=0}^{\pi}\frac{dy_c}{dx}d\theta = \alpha - \frac{4}{\pi}\frac{y_{mc}}{c}\int_{\theta=0}^{\pi}\cos\theta\, d\theta = \alpha - \frac{4}{\pi}\frac{y_{mc}}{c}\left[\sin\theta\right]_{\theta=0}^{\pi} = \alpha$$

$$A_1 = \frac{2}{\pi}\int_{\theta=0}^{\pi}\frac{dy_c}{dx}\cos\theta\, d\theta = \frac{8}{\pi}\frac{y_{mc}}{c}\int_{\theta=0}^{\pi}\cos^2\theta\, d\theta = \frac{8}{\pi}\frac{y_{mc}}{c}\left[\frac{1}{2}\theta + \frac{1}{4}\sin(2\theta)\right]_{\theta=0}^{\pi}$$

$$= 4\frac{y_{mc}}{c}$$

$$A_2 = \frac{2}{\pi}\int_{\theta=0}^{\pi}\frac{dy_c}{dx}\cos(2\theta)\, d\theta = \frac{8}{\pi}\frac{y_{mc}}{c}\int_{\theta=0}^{\pi}\cos\theta\cos(2\theta)\, d\theta$$

$$= \frac{8}{\pi}\frac{y_{mc}}{c}\int_{\theta=0}^{\pi}\cos\theta(2\cos^2\theta - 1)\, d\theta = \frac{8}{\pi}\frac{y_{mc}}{c}\int_{\theta=0}^{\pi}(2\cos^3\theta - \cos\theta)\, d\theta$$

$$= \frac{8}{\pi}\frac{y_{mc}}{c}\left[\frac{2}{3}\sin\theta(\cos^2\theta + 2) - \sin\theta\right]_{\theta=0}^{\pi} = 0$$

From Eq. (1.6.12), the section lift coefficient is

$$\tilde{C}_L = 2\pi(A_0 + \frac{1}{2}A_1) = 2\pi\left(\alpha + 2\frac{y_{mc}}{c}\right) = 2\pi(\alpha - \alpha_{L0})$$

Thus, the zero-lift angle of attack can be expressed as

$$\alpha_{L0} = -2\frac{y_{mc}}{c}$$

Similarly, from Eq. (1.6.14), the quarter-chord moment coefficient is

$$\tilde{C}_{m_{c/4}} = \frac{\pi}{4}(A_2 - A_1) = \frac{\pi}{4}\left(0 - 4\frac{y_{mc}}{c}\right) = \underline{-\pi\frac{y_{mc}}{c}}$$

For this airfoil with 4 percent maximum camber and a 5-degree angle of attack,

$$\tilde{C}_L = 2\pi(\alpha - \alpha_{L0}) = 2\pi\left[\alpha - \left(-2\frac{y_{mc}}{c}\right)\right] = 2\pi\left[\frac{5\pi}{180} + 2(0.04)\right] = \underline{1.05}$$

$$\tilde{C}_{m_{c/4}} = -\pi\frac{y_{mc}}{c} = -\pi(0.04) = \underline{-0.126}$$

The Vortex Panel Method

Potential flow over an airfoil of arbitrary shape can be synthesized by combining uniform flow with a curved vortex sheet wrapped around the surface of the airfoil, as shown in Fig. 1.6.3. The vortex strength must vary along the surface such that the normal component of velocity induced by the entire sheet and the uniform flow is zero everywhere along the surface of the airfoil. In most cases, the strength distribution necessary to satisfy this condition is very difficult or impossible to determine analytically. However, for numerical computations, such a sheet can be approximated as a series of flat vortex panels wrapped around the surface of the airfoil. In the limit as the panel size becomes very small, the panel solution approaches that for the curved vortex sheet.

To define the vortex panels, a series of nodes is placed on the airfoil surface. For best results the nodes should be clustered more tightly near the leading and trailing edges. The most popular method for attaining this clustering is called *cosine clustering*. For this method we use the change of variables

$$\frac{x}{c} = \frac{1}{2}(1 - \cos\theta)$$

which is the same change of variables as that used in thin airfoil theory. Distributing the nodes uniformly in θ will provide the desired clustering in x. For best results near the

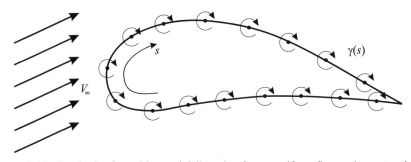

Figure 1.6.3. Synthesis of an arbitrary airfoil section from a uniform flow and a vortex sheet.

leading edge, an even number of nodes should always be used. For this particular distribution, the nodal coordinates are computed from the algorithm

$$\delta\theta = \frac{2\pi}{n-1} \tag{1.6.16}$$

$$\begin{Bmatrix} x_N(n/2+i) \\ y_N(n/2+i) \\ x_N(n/2+1-i) \\ y_N(n/2+1-i) \end{Bmatrix} = \begin{Bmatrix} x_u \\ y_u \\ x_l \\ y_l \end{Bmatrix}, \quad \frac{x}{c} = 0.5\{1.-\cos[(i-0.5)\delta\theta]\}, \quad i=1,n/2 \tag{1.6.17}$$

where n is the total number of nodes and the upper and lower surface coordinates are computed from Eqs. (1.6.1) through (1.6.4). A nodal distribution using cosine clustering with 12 nodes is shown in Fig. 1.6.4. Notice that both the first and last nodes are placed at the trailing edge. Between 50 and 200 nodes should be used for computation.

We can now synthesize an airfoil using $n-1$ vortex panels placed between these n nodes on the airfoil surface. The panels start at the trailing edge, are spaced forward along the lower surface, are wrapped up around the leading edge, and then run back along the upper surface to the trailing edge. The last panel ends at the trailing edge where the first panel began. The strength of each vortex panel is assumed to be linear along the panel and is required to be continuous from one panel to the next, i.e., the strength at the end of one panel must equal the strength at the beginning of the next panel. The strength is not required to be continuous across the trailing edge, i.e., $\gamma_1 \neq \gamma_n$. Each panel is assigned a local panel coordinate system (ξ,η), as shown in Fig. 1.6.5. The velocity induced by each of these panels is expressed in panel coordinates using Eq. (1.5.14), and the induced velocity components are then transformed to airfoil coordinates.

To solve for the n unknown nodal vortex strengths, control points are placed at the center of each of the $n-1$ panels. The coordinates of these control points are given by

$$\begin{Bmatrix} x_C(i) \\ y_C(i) \end{Bmatrix} = \begin{Bmatrix} \dfrac{x_N(i)+x_N(i+1)}{2} \\ \dfrac{y_N(i)+y_N(i+1)}{2} \end{Bmatrix}, \quad i=1,n-1 \tag{1.6.18}$$

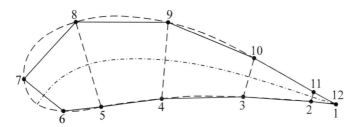

Figure 1.6.4. Vortex panel distribution with cosine clustering and an even number of nodes.

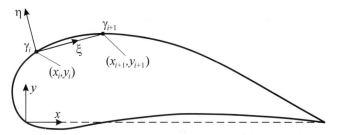

Figure 1.6.5. A vortex panel on the surface of an arbitrary airfoil section.

The normal velocity at each control point, induced by all $n-1$ panels and the uniform flow, must be zero. This gives $n-1$ equations for the n unknown nodal vortex strengths.

For the remaining equation, we know that the flow must leave the airfoil from the trailing edge. This means that the velocity just above the trailing edge must equal the velocity just below the trailing edge. If the angle between the upper and lower surfaces at the trailing edge is nonzero, the trailing edge is a stagnation point. If the angle between the upper and lower surfaces at the trailing edge is zero, the velocity at the trailing edge can be finite but the velocity must be continuous across the trailing edge. Since there is discontinuity in tangential velocity across a vortex sheet, we know that at the trailing edge, the discontinuity across the upper surface is equal and opposite to the discontinuity across the lower surface. Thus, the net discontinuity across both surfaces will be zero. Because the discontinuity in tangential velocity across any vortex sheet is equal to the local strength of the sheet, at the trailing edge, the strength of the upper surface must be exactly equal and opposite to the strength of the lower surface. That is,

$$\gamma_1 + \gamma_n = 0$$

This is called the *Kutta condition* and it provides the remaining equation necessary to solve for the n unknown nodal strengths.

The 2×2 panel coefficient matrix in airfoil coordinates, $[\mathbf{P}]_{i(x,y)}$, for the velocity induced at the arbitrary point (x,y) by panel i, extending from the node at (x_i, y_i) to the node at (x_{i+1}, y_{i+1}), is computed from the algorithm

$$l_i = \sqrt{(x_{i+1} - x_i)^2 + (y_{i+1} - y_i)^2} \tag{1.6.19}$$

$$\begin{Bmatrix} \xi \\ \eta \end{Bmatrix} = \frac{1}{l_i} \begin{bmatrix} (x_{i+1} - x_i) & (y_{i+1} - y_i) \\ -(y_{i+1} - y_i) & (x_{i+1} - x_i) \end{bmatrix} \begin{Bmatrix} (x - x_i) \\ (y - y_i) \end{Bmatrix} \tag{1.6.20}$$

$$\Phi = \text{atan2}\left(\eta l_i, \eta^2 + \xi^2 - \xi l_i\right) \tag{1.6.21}$$

$$\Psi = \frac{1}{2} \ln\left[\frac{\xi^2 + \eta^2}{(\xi - l_i)^2 + \eta^2}\right] \tag{1.6.22}$$

and

$$[P]_{i(x,y)} = \frac{1}{2\pi l_i^2}\begin{bmatrix}(x_{i+1}-x_i) & -(y_{i+1}-y_i)\\(y_{i+1}-y_i) & (x_{i+1}-x_i)\end{bmatrix}\begin{bmatrix}[(l_i-\xi)\varPhi+\eta\varPsi] & (\xi\varPhi-\eta\varPsi)\\[\eta\varPhi-(l_i-\xi)\varPsi-l_i] & (-\eta\varPhi-\xi\varPsi+l_i)\end{bmatrix}$$

(1.6.23)

The $n \times n$ airfoil coefficient matrix, $[A]$, is generated from the 2×2 panel coefficient matrix in airfoil coordinates, $[P]_{ji}$, for the velocity induced at control point i by panel j, extending from node j to node $j+1$, using the algorithm

$$A_{ij} = 0.0, \quad i = 1,n; \quad j = 1,n \tag{1.6.24}$$

$$\left.\begin{aligned}A_{ij} &= A_{ij} + \frac{x_{i+1}-x_i}{l_i}P_{21_{ji}} - \frac{y_{i+1}-y_i}{l_i}P_{11_{ji}}\\A_{ij+1} &= A_{ij+1} + \frac{x_{i+1}-x_i}{l_i}P_{22_{ji}} - \frac{y_{i+1}-y_i}{l_i}P_{12_{ji}}\end{aligned}\right\}, \quad i = 1,n-1; \quad j = 1,n-1 \tag{1.6.25}$$

$$A_{n1} = 1.0 \tag{1.6.26}$$

$$A_{nn} = 1.0 \tag{1.6.27}$$

The n **nodal vortex strengths**, γ_1 through γ_n, are then obtained by numerically solving the $n \times n$ linear system

$$[A]\begin{Bmatrix}\gamma_1\\\gamma_2\\\vdots\\\gamma_{n-1}\\\gamma_n\end{Bmatrix} = V_\infty\begin{Bmatrix}[(y_2-y_1)\cos\alpha-(x_2-x_1)\sin\alpha]/l_1\\[(y_3-y_2)\cos\alpha-(x_3-x_2)\sin\alpha]/l_2\\\vdots\\[(y_n-y_{n-1})\cos\alpha-(x_n-x_{n-1})\sin\alpha]/l_{n-1}\\0.0\end{Bmatrix} \tag{1.6.28}$$

Once the nodal strengths are known, the velocity and pressure at any point in space can be computed by adding the velocity induced by all $n-1$ vortex panels to the free-stream velocity,

$$\begin{Bmatrix}V_x\\V_y\end{Bmatrix} = V_\infty\begin{Bmatrix}\cos\alpha\\\sin\alpha\end{Bmatrix} + \sum_{i=1}^{n-1}[P]_{i(x,y)}\begin{Bmatrix}\gamma_i\\\gamma_{i+1}\end{Bmatrix} \tag{1.6.29}$$

$$V^2 = V_x^2 + V_y^2 \tag{1.6.30}$$

$$C_p \equiv \frac{p-p_\infty}{\frac{1}{2}\rho V_\infty^2} = 1 - \frac{V^2}{V_\infty^2} \tag{1.6.31}$$

The lift and moment coefficients for the entire airfoil are the sum of those induced by all of the $n-1$ vortex panels,

$$\tilde{C}_L = \sum_{i=1}^{n-1} \frac{l_i}{c} \frac{\gamma_i + \gamma_{i+1}}{V_\infty} \qquad (1.6.32)$$

$$\tilde{C}_{m_{le}} = -\frac{1}{3} \sum_{i=1}^{n-1} \frac{l_i}{c} \left[\frac{2x_i\gamma_i + x_i\gamma_{i+1} + x_{i+1}\gamma_i + 2x_{i+1}\gamma_{i+1}}{cV_\infty} \cos(\alpha) \right.$$
$$\left. + \frac{2y_i\gamma_i + y_i\gamma_{i+1} + y_{i+1}\gamma_i + 2y_{i+1}\gamma_{i+1}}{cV_\infty} \sin(\alpha) \right] \qquad (1.6.33)$$

Comparison with Experimental Data
Section lift and moment coefficients predicted by thin airfoil theory and panel codes are in good agreement with experimental data for low Mach numbers and small angles of attack. In Fig. 1.6.6, the inviscid lift coefficient for a NACA 2412 airfoil, as predicted by thin airfoil theory, is compared with the inviscid lift coefficient predicted by the vortex panel method and with experimental data for total lift coefficient as reported by Abbott and Von Doenhoff (1949). Thin airfoil theory predicts a section lift coefficient

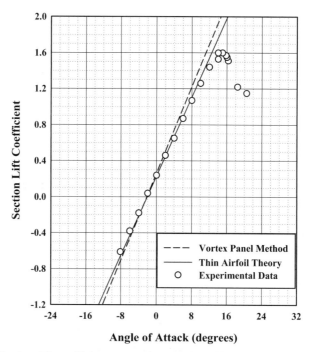

Figure 1.6.6. Section lift coefficient comparison among thin airfoil theory, the vortex panel method, and experimental data based on total lift, for the NACA 2412 airfoil.

that is independent of the thickness distribution and dependent only on angle of attack and camber line shape. At small angles of attack, the thin airfoil approximation agrees closely with experimental data based on total lift, for airfoils as thick as about 12 percent. The agreement seen in Fig. 1.6.6 is quite typical. For airfoils much thicker than about 12 percent, viscous effects become increasingly important, even at fairly low angles of attack, and the inviscid lift coefficient predicted by thin airfoil theory begins to deviate more from experimental observations based on total lift. This can be seen in Fig. 1.6.7.

The reason why thin airfoil theory predicts total lift so well over such a wide range of thickness is that thickness tends to increase the lift slope slightly, while viscous effects tend to decrease the lift slope. Since both thickness effects and viscous effects are neglected in thin airfoil theory, the resulting errors tend to cancel, giving the theory a broader range of applicability than would otherwise be expected. Coincidentally, because of these opposing errors, thin airfoil theory actually predicts a lift slope that agrees more closely with experimental data for total lift than does that predicted by the more elaborate vortex panel method. The vortex panel method accurately predicts the pressure distribution around the airfoil, including the effects of thickness. However, since the vortex panel method provides a potential flow solution, it does not account for viscous effects in any way.

After seeing that thin airfoil theory predicts total section lift better than the vortex panel method, one may wonder why we should ever be interested in the vortex panel

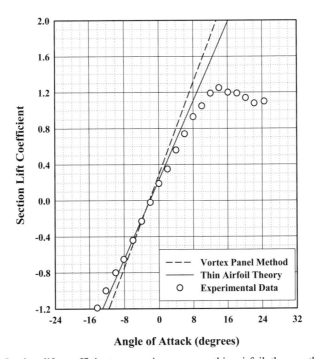

Figure 1.6.7. Section lift coefficient comparison among thin airfoil theory, the vortex panel method, and experimental data based on total lift, for the NACA 2421 airfoil.

method. For the answer, we must remember that both the vortex panel solution and the thin airfoil solution come from potential flow theory. Thus, neither of these two solutions can be expected to predict viscous forces. However, potential flow solutions are often used as boundary conditions for viscous flow analysis. Potential flow is used to predict the pressure distribution around the airfoil. The viscous forces are then computed from boundary layer theory. However, since boundary layer flow is greatly affected by the pressure distribution around the airfoil, our potential flow solution must accurately predict the section pressure distribution. Because thin airfoil theory does not account for airfoil thickness, it cannot be used to predict the surface pressure distribution on an airfoil section with finite thickness. The vortex panel method, on the other hand, accurately predicts the inviscid pressure distribution for airfoils of any thickness.

While the experimental data shown in Figs. 1.6.6 and 1.6.7 are based on total section lift, section lift data have been obtained that are based on pressure forces only. This is usually accomplished by spacing a large number of static pressure taps around the circumference of an airfoil. The net pressure force acting on the airfoil section is derived by numerically integrating the forces obtained from these pressure measurements. The lift coefficient predicted by thin airfoil theory and that predicted by the vortex panel method are compared with experimental data of this type in Fig. 1.6.8. Notice that the vortex panel method agrees very closely with these experimental data, while thin airfoil theory predicts a lift coefficient that is somewhat low.

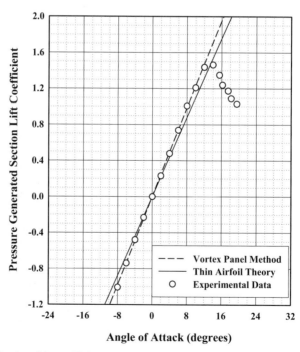

Figure 1.6.8. Section lift coefficient comparison among thin airfoil theory, the vortex panel method, and experimental data based on pressure lift, for the NACA 0015 airfoil.

In summary, thin airfoil theory can be used to obtain a first approximation for the total section lift coefficient, at small angles of attack, produced by airfoils of thickness less than about 12 percent. Thin airfoil theory gives no prediction for the section drag.

To improve on the results obtained from thin airfoil theory, we can combine a boundary layer solution with the velocity and pressure distribution obtained from the vortex panel method. Since this analytical procedure accounts for both thickness and viscous effects, it produces results that agree closely with experimental data for total lift and drag over a broad range of section thickness. Another alternative available with today's high-speed computers is the use of computational fluid dynamics (CFD). However, a CFD solution will increase the required computation time by several orders of magnitude.

In Figs. 1.6.6 through 1.6.8 it should be noticed that at angles of attack near the zero-lift angle of attack, the result predicted by thin airfoil theory agrees very well with that predicted by the vortex panel method and with that observed from all experimental data. Thus, we see that the thickness distribution has little effect on the lift produced by an airfoil at angles of attack near the zero-lift angle of attack. The thickness distribution does, however, have a significant effect on the maximum lift coefficient and on the stall characteristics of the airfoil section. This is seen by comparing Fig. 1.6.6 with Fig. 1.6.7. The two airfoils described in these two figures have exactly the same camber line. They differ only in thickness. Notice that the NACA 2412 section has a maximum lift coefficient of about 1.6, while the NACA 2421 airfoil produces a maximum lift coefficient of only about 1.2. Also notice that the thinner section has a sharper and more abrupt stall than that displayed by the thicker section.

1.7. Trailing-Edge Flaps and Section Flap Effectiveness

An attached trailing-edge flap is formed by hinging some aft portion of the airfoil section so that it can be deflected, either downward or upward, by rotating the flap about the hinge point as shown in Fig. 1.7.1. The deflection of a trailing-edge flap effectively changes the camber of the airfoil section, and in so doing changes the aerodynamic characteristics of the section. A downward deflection of the flap increases the effective camber and is thus usually considered to be a positive deflection.

For small angles of attack and small flap deflections, thin airfoil theory can be applied to an airfoil section with a deflected trailing-edge flap. Let y_d be the y-position

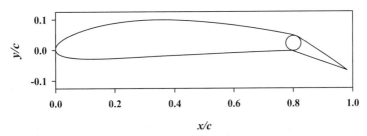

Figure 1.7.1. NACA 4412 airfoil section with a 20 percent trailing-edge flap.

of the section camber line with the trailing-edge flap deflected as shown in Fig. 1.7.2. All of the results obtained from thin airfoil theory must apply to this modified camber line geometry. Thus from Eq. (1.6.12), we can write

$$\tilde{C}_L = 2\pi(\alpha - \alpha_{L0}) = 2\pi\left(\alpha - \frac{1}{\pi}\int_{\theta=0}^{\pi}\frac{dy_d}{dx}(1-\cos\theta)\,d\theta\right) \tag{1.7.1}$$

However, within the small-angle approximation, the slope of the deflected camber line geometry can be related to the slope of the undeflected geometry according to

$$\frac{dy_d}{dx} = \begin{cases} \dfrac{dy_c}{dx}, & x \le c - c_f \\ \dfrac{dy_c}{dx} - \delta, & x \ge c - c_f \end{cases} \tag{1.7.2}$$

where y_c is the undeflected camber line ordinate, c_f is the flap chord length, and δ is the deflection of the flap in radians, with positive deflection being downward.

Using Eq. (1.7.2) in Eq. (1.7.1), we have

$$\tilde{C}_L = 2\pi(\alpha - \alpha_{L0}) = 2\pi\left(\alpha - \frac{1}{\pi}\int_{\theta=0}^{\pi}\frac{dy_c}{dx}(1-\cos\theta)\,d\theta + \frac{\delta}{\pi}\int_{\theta=\theta_f}^{\pi}(1-\cos\theta)\,d\theta\right) \tag{1.7.3}$$

or

$$\alpha_{L0} = \frac{1}{\pi}\int_{0-0}^{\pi}\frac{dy_c}{dx}(1-\cos\theta)\,d\theta - \frac{\delta}{\pi}\int_{\theta=\theta_f}^{\pi}(1-\cos\theta)\,d\theta \tag{1.7.4}$$

where α_{L0} is the zero-lift angle of attack for the airfoil section with the flap deflected and θ_f is given by

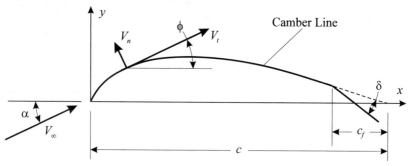

Figure 1.7.2. Camber line geometry for an airfoil section with attached trailing-edge flap and positive flap deflection.

$$\theta_f = \cos^{-1}\left(2\frac{c_f}{c} - 1\right) \tag{1.7.5}$$

From Eqs. (1.7.3) and (1.7.4), we see that the thin airfoil lift coefficient is affected by flap deflection only through a change in the zero-lift angle of attack. The lift coefficient is still a linear function of the angle of attack and the lift slope is not affected by the flap deflection. The first integral on the right-hand side of Eq. (1.7.4) is simply the zero-lift angle of attack with no flap deflection. The second integral in this equation is readily evaluated to yield what is commonly called the *ideal section flap effectiveness*,

$$\varepsilon_{fi} \equiv -\frac{\partial \alpha_{L0}}{\partial \delta} = \frac{1}{\pi}\int_{\theta=\theta_f}^{\pi}(1-\cos\theta)\,d\theta = 1 - \frac{\theta_f - \sin\theta_f}{\pi} \tag{1.7.6}$$

Notice that the ideal section flap effectiveness depends only on the ratio of flap chord to total chord and is independent of both camber line geometry and flap deflection. Using this definition in Eq. (1.7.4), the zero-lift angle of attack for a thin airfoil with an ideal trailing-edge flap is found to vary linearly with flap deflection,

$$\alpha_{L0}(\delta) = \alpha_{L0}(0) - \varepsilon_{fi}\,\delta \tag{1.7.7}$$

In a similar manner we can predict the quarter-chord moment coefficient for a thin airfoil with a deflected trailing-edge flap. From Eq. (1.6.14), we have

$$\tilde{C}_{m_{c/4}} = \frac{1}{2}\int_{\theta=0}^{\pi}\frac{dy_d}{dx}[\cos(2\theta) - \cos\theta]\,d\theta \tag{1.7.8}$$

Using Eq. (1.7.2) in Eq. (1.7.8) gives

$$\tilde{C}_{m_{c/4}} = \frac{1}{2}\int_{\theta=0}^{\pi}\frac{dy_c}{dx}[\cos(2\theta) - \cos\theta]\,d\theta - \frac{\delta}{2}\int_{\theta=\theta_f}^{\pi}[\cos(2\theta) - \cos\theta]\,d\theta \tag{1.7.9}$$

From Eq. (1.7.9) we see that the section quarter-chord moment coefficient for a thin airfoil is also a linear function of flap deflection. The first integral on the right-hand side of Eq. (1.7.9) is the quarter-chord moment coefficient for the airfoil section with no flap deflection, and the second integral can be evaluated to yield the section quarter-chord moment slope with respect to flap deflection. Thus, the quarter-chord moment coefficient for a thin airfoil section with an ideal trailing-edge flap can be written as

$$\tilde{C}_{m_{c/4}}(\delta) = \tilde{C}_{m_{c/4}}(0) + \tilde{C}_{m,\delta}\,\delta \tag{1.7.10}$$

where the change in the section quarter-chord moment coefficient with respect to flap deflection is given by

$$\widetilde{C}_{m,\delta} \equiv \frac{\partial \widetilde{C}_{m_{c/4}}}{\partial \delta} = -\frac{1}{2}\int_{\theta=\theta_f}^{\pi}[\cos(2\theta)-\cos\theta]d\theta = \frac{\sin(2\theta_f)-2\sin\theta_f}{4} \qquad (1.7.11)$$

Notice that the change in moment coefficient with respect to flap deflection depends only on the ratio of flap chord to total chord. Thus, as was the case with the ideal section flap effectiveness, the ideal section quarter-chord moment slope with respect to flap deflection is independent of both camber line geometry and flap deflection.

In summary, at angles of attack below stall, the lift coefficient for an airfoil section with a deflected trailing-edge flap is found to be very nearly a linear function of both the airfoil angle of attack, α, and the flap deflection, δ. This linear relation can be written

$$\widetilde{C}_L(\alpha,\delta) = \widetilde{C}_{L,\alpha}[\alpha-\alpha_{L0}(0)+\varepsilon_f\,\delta] \qquad (1.7.12)$$

where $\widetilde{C}_{L,\alpha}$ is the section lift slope, $\alpha_{L0}(0)$ is the zero-lift angle of attack with no flap deflection, and ε_f is called the *section flap effectiveness*. Previously we found that thin airfoil theory predicts a section lift slope of 2π. However, solutions obtained using the vortex panel method and experimental measurements have shown that the actual section lift slope can vary somewhat from this value. The zero-lift angle of attack with no flap deflection, as predicted by thin airfoil theory, was previously shown to be in excellent agreement with both the vortex panel method and experimental data. As we shall see, the section flap effectiveness predicted by thin airfoil theory agrees with results predicted using the vortex panel method, but deviates somewhat from experimental observation.

The actual section flap effectiveness is always less than the ideal section flap effectiveness given by Eq. (1.7.6). The hinge mechanism in a real trailing-edge flap always reduces the flap effectiveness. This reduction results from local boundary layer separation and from flow leakage through the hinge from the high-pressure side to the low-pressure side. In addition, at flap deflections greater than about ±10 degrees, the error associated with the small-angle approximation used to obtain Eq. (1.7.6) begins to become significant. This results in an additional decrease in the section flap effectiveness for larger flap deflections. A comparison among the section flap effectiveness predicted by thin airfoil theory, typical results predicted by the vortex panel method, and results observed experimentally is shown in Fig. 1.7.3. The data shown in this figure are from Abbott and Von Doenhoff (1949).

The discrepancy between the theoretical results and the experimental data shown in Fig. 1.7.3 is only about 7 percent with a flap chord fraction of 0.4, but at a flap chord fraction of 0.1 this discrepancy is nearly 25 percent. The deviation between actual section flap effectiveness and the ideal section flap effectiveness continues to increase as the flap chord fraction becomes smaller. The poor agreement at low flap chord fraction is attributed to the thickness of the boundary layer, which is much larger near the trailing edge. The trailing-edge flaps used to generate the data shown in Fig. 1.7.3 all had flap hinges that were sealed to prevent leakage from the high-pressure side to the low-pressure side. For unsealed trailing-edge flaps, an additional decrease in section flap effectiveness of about 20 percent is observed.

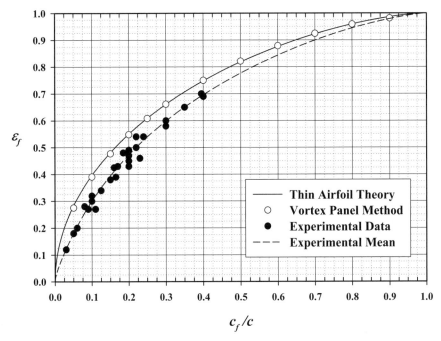

Figure 1.7.3. Section flap effectiveness comparison among thin airfoil theory, the vortex panel method, and experimental data.

The data shown in Fig. 1.7.3 were all taken at flap deflections of less than 10 degrees. For larger deflections, actual flap effectiveness deviates even more from the ideal, as a result of errors associated with the small-angle approximation. The actual section flap effectiveness for a trailing-edge flap can be expressed in terms of the ideal section flap effectiveness, ε_{fi}, given by Eq. (1.7.6), according to

$$\varepsilon_f = \eta_h \eta_d \varepsilon_{fi} \qquad (1.7.13)$$

where η_h and η_d are, respectively, the section flap hinge efficiency and the section flap deflection efficiency.

For well-designed sealed flaps, the section hinge efficiency can be approximated from Fig. 1.7.4. This figure is based on the mean line for the experimental data shown in Fig. 1.7.3. If the gap between the main wing and the trailing-edge flap is not sealed, it is recommended that the values found from Fig. 1.7.4 be reduced by 20 percent. Remember that the result given in this figure represents a mean experimental efficiency. The actual hinge efficiency for a specific trailing-edge flap installed in a particular airfoil section may deviate significantly from the value found in this figure. For flap deflections of more than ±10 degrees, Perkins and Hage (1949) recommend the flap deflection efficiency shown in Fig. 1.7.5. For flap deflections of less than ±10 degrees, a flap deflection efficiency of 1.0 should be used.

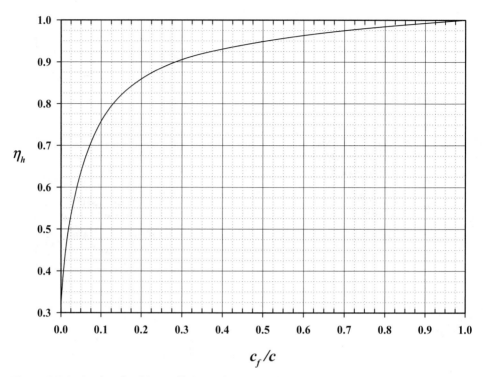

Figure 1.7.4. Section flap hinge efficiency for well-designed and sealed trailing-edge flaps. For unsealed flaps this hinge efficiency should be decreased by about 20 percent.

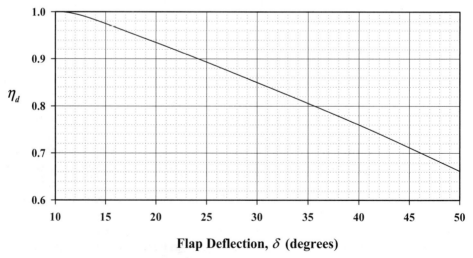

Figure 1.7.5. Section flap deflection efficiency for trailing-edge flaps with flap deflections of more than ±10 degrees. For flap deflections of less than ±10 degrees a deflection efficiency of 1.0 should be used.

At angles of attack below stall, the pitching moment coefficient for an airfoil section with a deflected trailing-edge flap is also found to be very nearly a linear function of both angle of attack, α, and flap deflection, δ. This linear relation can be written in the form

$$\tilde{C}_{m_{c/4}}(\alpha,\delta) \;=\; \tilde{C}_{m_{c/4}}(0,0) + \tilde{C}_{m,\alpha}\,\alpha + \tilde{C}_{m,\delta}\,\delta \qquad (1.7.14)$$

where $\tilde{C}_{m_{c/4}}(0,0)$, $\tilde{C}_{m,\alpha}$, and $\tilde{C}_{m,\delta}$ are, respectively, the section pitching moment coefficient at zero angle of attack and zero flap deflection, the moment slope with angle of attack, and the moment slope with flap deflection.

Thin airfoil theory predicts that that the quarter chord is the aerodynamic center of an airfoil section. Thus, thin airfoil theory predicts a zero quarter-chord moment slope with respect to angle of attack. In reality, solutions obtained from the vortex panel method and experimental observations have shown that the quarter chord is not exactly the aerodynamic center of all airfoil sections. Thus, in general, we should allow for a finite quarter-chord moment slope with angle of attack. However, for preliminary design, the quarter-chord moment slope with angle of attack is usually taken to be zero.

The section quarter-chord moment slope with respect to flap deflection is shown in Fig. 1.7.6. In this figure the ideal quarter-chord moment slope, as predicted by thin airfoil theory in Eq. (1.7.11), is compared with typical results predicted from the vortex panel method and with limited experimental data. Airfoil section thickness tends to increase the magnitude of the negative quarter-chord moment slope with respect to flap deflection. The hinge effects, on the other hand, tend to lessen the magnitude of this moment slope. The result from thin airfoil theory is often used for preliminary design.

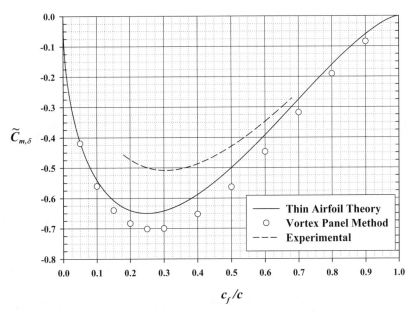

Figure 1.7.6. Quarter-chord moment slope with respect to flap deflection.

EXAMPLE 1.7.1. The airfoil section with 4 percent maximum camber, which is described in Example 1.6.1, has a sealed 20 percent trailing-edge flap. Estimate the section flap effectiveness and the section lift coefficient for a 5-degree flap deflection and a 0-degree angle of attack relative to the undeflected chord line.

Solution. From Eq. (1.7.5), a flap fraction of 0.20 gives

$$\theta_f = \cos^{-1}\left(2\frac{c_f}{c}-1\right) = \cos^{-1}[2(0.20)-1] = 2.214297$$

Using this in Eq. (1.7.6) yields an ideal section flap effectiveness of

$$\varepsilon_{fi} = 1 - \frac{\theta_f - \sin\theta_f}{\pi} = 1 - \frac{2.214297 - \sin(2.214297)}{\pi} = 0.55$$

Applying this and results from Figs. 1.7.4 and 1.7.5 to Eq. (1.7.13) yields

$$\varepsilon_f = \eta_h\,\eta_d\,\varepsilon_{fi} = (0.86)(1.00)(0.55) = \underline{0.47}$$

With this and the value of α_{L0} from Example 1.6.1, Eq. (1.7.12) results in

$$\tilde{C}_L = \tilde{C}_{L,\alpha}[\alpha - \alpha_{L0}(0) + \varepsilon_f\,\delta] = 2\pi[0.0 - (-0.08) + 0.47(5\pi/180)] = \underline{0.76}$$

1.8. Incompressible Flow over Finite Wings

In Sec. 1.6 we reviewed the aerodynamic properties of airfoils, which are the same as the aerodynamic properties of wings with infinite span. A wing of constant cross-section and infinite span would have no variation in aerodynamic forces in the spanwise direction. The aerodynamic forces per unit span acting on such a wing, at any point along the span, would be the same as those for an airfoil of the same cross-section. An airfoil or wing of infinite span is synthesized using vortex sheets that are made up of straight vortex filaments that extend to $\pm\infty$ in the direction of span. The vortex strength can vary over the sheet as we move in a chordwise direction from one vortex filament to another. However, there is no variation in vortex strength as we move along a vortex filament in the direction of span.

Any real wing, of course, must have finite span. At the tips of a finite wing, the air on the lower surface of the wing comes in direct contact with the air on the upper surface of the wing. Thus, at the tips of a finite wing, the pressure difference between the upper and lower surfaces must always go to zero. As a result, the lift on any finite wing must go to zero at the wingtips, as shown schematically in Fig. 1.8.1.

The pressure difference between the upper and lower surfaces of a finite wing is reduced near the wingtips, because some of the air from the high-pressure region below the wing spills outward, around the wingtip, and back inward toward the low-pressure region above the wing. Thus, while the flow around an infinite wing is entirely in the plane of the airfoil section, the flow around a finite wing is three-dimensional.

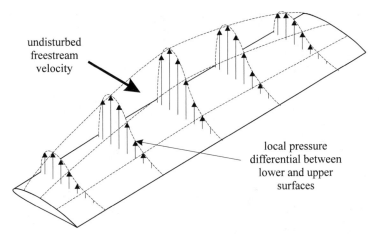

undisturbed
freestream
velocity

local pressure
differential between
lower and upper
surfaces

Figure 1.8.1. Lift distribution on a finite wing.

As air flows over a finite wing, the air below the wing moves outward toward the wingtip and the air above the wing moves inward toward the root, as shown in Fig. 1.8.2. Where the flows from the upper and lower surfaces combine at the trailing edge, the difference in spanwise velocity generates a trailing vortex sheet. Because this vortex sheet is free and not bound to the wing's surface, the flow field induced by the sheet tends to change the shape of the sheet as it moves downstream from the wing. As can be seen in Fig. 1.8.3, this vortex sheet rolls up around an axis trailing slightly inboard from each wingtip. At some distance behind the wing, the sheet becomes completely rolled up to form two large vortices, one trailing aft of each wingtip. For this reason,

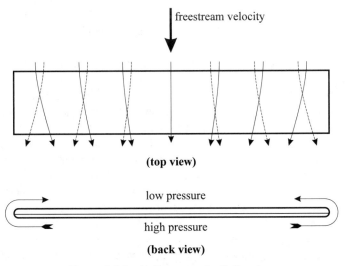

freestream velocity

(top view)

low pressure

high pressure

(back view)

Figure 1.8.2. Airflow around a finite wing.

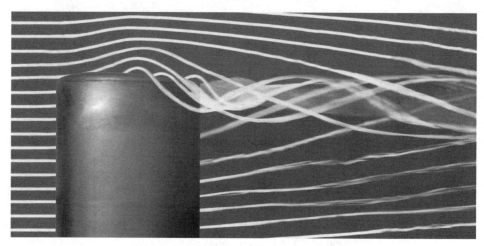

Figure 1.8.3. Vorticity trailing aft of a finite wing.

these vortices are referred to as *wingtip vortices*, even though they are generated over the full span of the wing. The downward velocity component that is induced between the wingtip vortices is called *downwash*. Potential flow theory predicts that the wingtip vortices must trail behind the wing for an infinite distance with no reduction in strength. In reality, viscous effects will eventually dissipate the energy in these vortices, but this is a slow process. These vortices will still have significant energy several miles behind a large aircraft and are of sufficient strength to cause control loss or structural damage to other aircraft following too closely.

From the discussion above, we see that a wing of finite span cannot be synthesized with a vortex sheet made up of vortex filaments that are always perpendicular to the airfoil sections of the wing. We can, however, still synthesize a finite wing with a vortex sheet, but the sheet must be made up of horseshoe-shaped vortex filaments. These filaments run out along the wing in the direction of span, curving back and eventually leaving the wing from the trailing edge at some point inboard of the wingtip. This is seen schematically in the plan view of the elliptic wing shown in Fig. 1.8.4. Because the strength of a vortex filament cannot vary along its length, the decrease in the circulation about the wing as we move out from the root toward the tip must result entirely from the vorticity that leaves the wing in the trailing vortex sheet.

The classical lifting-line theory developed by Prandtl (1918) was the first analytical method to satisfactorily predict the performance of a finite wing. While a general 3-D vortex lifting law was not available at the time of Prandtl's development, the 2-D vortex lifting law of Kutta (1902) and Joukowski (1906) was well known. Prandtl's lifting-line theory is based on the hypothesis that each spanwise section of a finite wing has a section lift that is equivalent to that acting on a similar section of an infinite wing having the same section circulation. However, to reconcile this theory with reality, the undisturbed freestream velocity in the Kutta-Joukowski law was intuitively replaced with the vector sum of the freestream velocity and the velocity induced by the trailing vortex sheet.

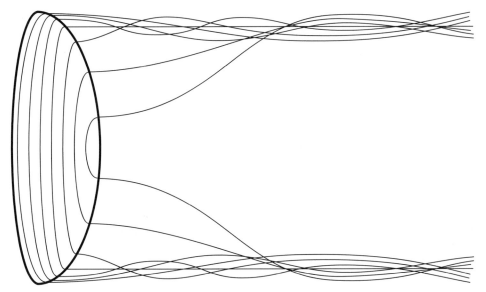

Figure 1.8.4. Schematic of the vorticity distribution on a finite wing with elliptic planform shape.

Today we know that Prandtl's hypothesis was correct. This can be shown as a direct consequence of the **vortex lifting law**, which is a three-dimensional counterpart to the two-dimensional Kutta-Joukowski law. This vortex lifting law states that for any potential flow containing vortex filaments, the force per unit length exerted on the surroundings at any point along a vortex filament is given by

$$\mathbf{dF} = \rho \Gamma \, \mathbf{V} \times \mathbf{d}l \qquad (1.8.1)$$

where $\mathbf{dF}$ is the differential aerodynamic force vector and ρ, Γ, $\mathbf{V}$, and $\mathbf{d}l$ are, respectively, the fluid density, vortex strength, local fluid velocity, and the directed differential vortex length vector (see Saffman 1992).

The force computed from Eq. (1.8.1) is called the *vortex force* or *Kutta lift*. The vortex lifting law is a very useful tool in the study of aerodynamics and flight mechanics. It provides the basis for much of finite wing theory. There are two important consequences of Eq. (1.8.1). First, *for a bound vortex filament, the vortex force is always perpendicular to both the local velocity vector and the vortex filament*. Second, since a free vortex filament can support no force, *free vortex filaments must be aligned everywhere with the streamlines of the flow*.

Prandtl's Classical Lifting-Line Theory
Prandtl's lifting-line theory gives good agreement with experimental data for straight wings of aspect ratio greater than about 4. Development of this theory is presented in any undergraduate engineering textbook on aerodynamics and will not be repeated here. Only summary results are presented in this section.

The model used by Prandtl to approximate the bound vorticity and trailing vortex sheet is shown in Fig. 1.8.5. All bound vortex filaments are assumed to follow the wing quarter-chord line, and all trailing vortex filaments are assumed to be straight and parallel with the freestream. Rollup of the trailing vortex sheet is ignored.

The foundation of lifting-line theory is the requirement that for each cross-section of the wing, the lift predicted from the vortex lifting law must be equal to that predicted from airfoil section theory, i.e.,

$$\tilde{C}_L(z) = \tilde{C}_{L,\alpha}[\alpha_{\text{eff}}(z) - \alpha_{L0}(z)]$$

where α_{eff} is the local section angle of attack, including the effects of velocity induced by the trailing vortex sheet. Because of the downwash induced on the wing by the trailing vortex sheet, the local relative wind is inclined at an angle, α_i, to the freestream, as shown in Fig. 1.8.6. This angle is called the *induced angle of attack*. Since the lift is always perpendicular to the local relative wind, the downwash tilts the lift vector back, creating a component of lift parallel to the freestream. This is called *induced drag*.

When the downwash is accounted for, Prandtl's hypothesis requires that

$$\frac{2\Gamma(z)}{V_\infty c(z)} + \frac{\tilde{C}_{L,\alpha}}{4\pi V_\infty} \int_{\zeta=-b/2}^{b/2} \frac{1}{z-\zeta}\left(\frac{d\Gamma}{dz}\right)_{z=\zeta} d\zeta = \tilde{C}_{L,\alpha}[\alpha(z) - \alpha_{L0}(z)] \qquad (1.8.2)$$

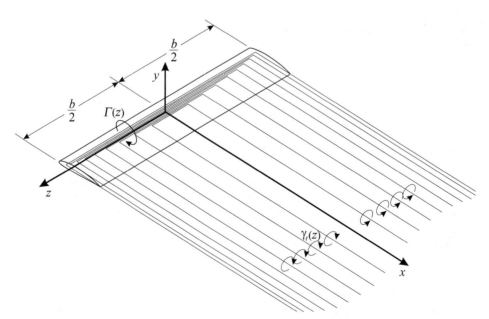

Figure 1.8.5. Prandtl's model for the bound vorticity and the trailing vortex sheet generated by a wing of finite span.

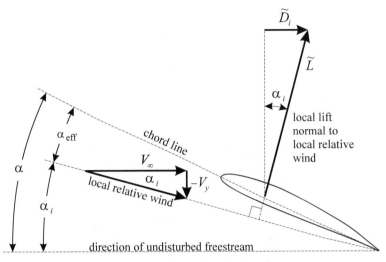

Figure 1.8.6. Induced angle of attack.

Equation (1.8.2) is the fundamental equation of Prandtl's lifting-line theory. It is a non-linear integrodifferential equation that involves only one unknown, the local section circulation as a function of the spanwise position, $\Gamma(z)$. All other parameters are known for a given wing design at a given geometric angle of attack and a given freestream velocity. Remember that the chord length, c, the geometric angle of attack, α, and the zero-lift angle of attack, α_{L0}, are all allowed to vary in the spanwise direction. The section lift slope could also vary with z but is usually assumed to be constant.

There is certain terminology associated with spanwise variation in wing geometry with which the reader should be familiar. If a wing has spanwise variation in geometric angle of attack as shown in Fig. 1.8.7, the wing is said to have *geometric twist*. The tip is commonly at a lower angle of attack than the root, in which case the geometric twist is referred to as *washout*. A spanwise variation in zero-lift angle of attack, like that shown in Fig. 1.8.8, is called *aerodynamic twist*. Deflecting a trailing-edge flap that extends over only part of the span is another form of aerodynamic twist (see Fig. 1.8.9).

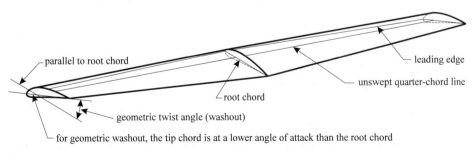

Figure 1.8.7. Geometric twist in an unswept rectangular wing.

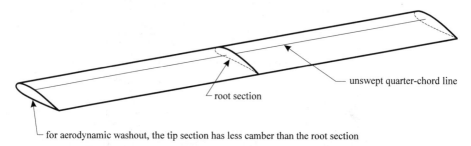

Figure 1.8.8. Aerodynamic twist in an unswept rectangular wing.

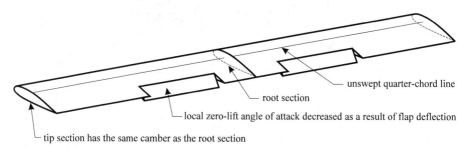

Figure 1.8.9. Aerodynamic twist resulting from deflection of a trailing-edge flap spanning only a portion of the wing.

For a single finite wing with no sweep or dihedral, an analytical solution to Prandtl's lifting-line equation can be obtained in terms of a Fourier sine series. For a given wing design, at a given angle of attack, the planform shape, $c(z)$, the geometric twist, $\alpha(z)$, and the aerodynamic twist, $\alpha_{L0}(z)$, are all known as functions of spanwise position. The circulation distribution is written as a Fourier series, the series is truncated to a finite number of terms (N), and the Fourier coefficients are determined by forcing the lifting-line equation to be satisfied at N specific sections along the span of the wing. From this solution the circulation distribution is given by

$$\Gamma(\theta) = 2bV_\infty \sum_{n=1}^{N} A_n \sin(n\theta); \qquad \theta = \cos^{-1}(-2z/b) \qquad (1.8.3)$$

where the Fourier coefficients, A_n, are obtained from

$$\sum_{n=1}^{N} A_n \left[\frac{4b}{\widetilde{C}_{L,\alpha}\, c(\theta)} + \frac{n}{\sin(\theta)} \right] \sin(n\theta) = \alpha(\theta) - \alpha_{L0}(\theta) \qquad (1.8.4)$$

Once the circulation distribution has been determined, the section lift distribution can be obtained from the vortex lifting law. The resulting lift and induced drag coefficients for the finite wing are

$$C_L = \pi R_A A_1, \qquad R_A \equiv \frac{b^2}{S} \qquad (1.8.5)$$

$$C_{D_i} = \pi R_A \sum_{n=1}^{N} n A_n^2 = \frac{C_L^2}{\pi R_A e_s}, \qquad e_s \equiv \frac{1}{1+\sigma}, \qquad \sigma \equiv \sum_{n=2}^{N} n \left(\frac{A_n}{A_1}\right)^2 \qquad (1.8.6)$$

The wingspan squared divided by the planform area, R_A, is called the *aspect ratio* and the parameter e_s is called the *span efficiency factor*.

Lifting-line theory predicts that an elliptic wing with no geometric or aerodynamic twist produces minimum possible induced drag for a given lift coefficient and aspect ratio. This planform has a chord that varies with the spanwise coordinate according to

$$c(z) = \frac{4b}{\pi R_A}\sqrt{1-(2z/b)^2} \quad \text{or} \quad c(\theta) = \frac{4b}{\pi R_A}\sin(\theta)$$

As shown in Fig. 1.8.10, an unswept elliptic wing has a straight quarter-chord line. This gives the leading edge less curvature and the trailing edge more curvature than a conventional ellipse, which has a straight half-chord line. Several aircraft have been designed and built with elliptic wings. One of the best known is the British Spitfire, shown in Fig. 1.8.11.

The lift and induced drag coefficients predicted from Eqs. (1.8.5) and (1.8.6) for an **elliptic wing with no geometric or aerodynamic twist** are

$$C_L = C_{L,\alpha}(\alpha - \alpha_{L0}), \qquad C_{L,\alpha} = \frac{\tilde{C}_{L,\alpha}}{1 + \tilde{C}_{L,\alpha}/(\pi R_A)} \qquad (1.8.7)$$

$$C_{D_i} = \frac{C_L^2}{\pi R_A} \qquad (1.8.8)$$

where $C_{L,\alpha}$ is the lift slope for the finite wing. Equation (1.8.7) shows that the lift slope for an untwisted elliptic wing is less than the section lift slope for the airfoil from which the wing was generated. However, as the aspect ratio for the wing becomes large, the

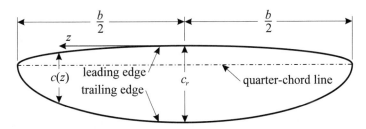

Figure 1.8.10. Planform shape of an unswept elliptic wing with an aspect ratio of 6.

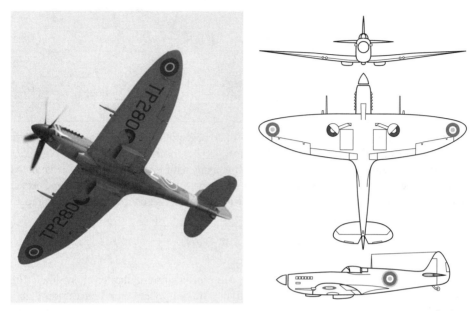

Figure 1.8.11. Elliptic wing used on the famous British Spitfire. (Photograph by Barry Santana)

lift slope for the finite wing approaches that of the airfoil section and the induced drag becomes small. At a given angle of attack, the untwisted elliptic wing produces more lift than any other untwisted wing of the same aspect ratio. Planform shape affects both the induced drag and the lift slope of a finite wing. However, the effect of planform shape is small compared to that of aspect ratio.

While untwisted elliptic wings produce minimum possible induced drag, they are more expensive to manufacture than simple rectangular wings. Untwisted rectangular wings are easy to manufacture, but they generate induced drag at a level that is less than optimum. The untwisted tapered wing, shown in Fig. 1.8.12, has commonly been used as a compromise. Tapered wings have a chord length that varies linearly from the root to the tip. They are nearly as easy to manufacture as rectangular wings, and they can be designed to produce induced drag close to the optimum value of an elliptic wing.

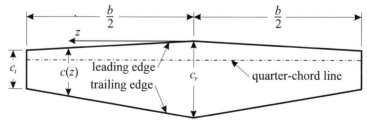

Figure 1.8.12. Planform shape of an unswept tapered wing with a taper ratio of 0.5 and an aspect ratio of 6.

The chord length for a tapered wing varies with the spanwise coordinate according to the relation

$$c(z) = \frac{2b}{R_A(1+R_T)}[1-(1-R_T)|2z/b|] \quad \text{or} \quad c(\theta) = \frac{2b}{R_A(1+R_T)}[1-(1-R_T)|\cos\theta|]$$

where R_T is the taper ratio, related to the root chord, c_r, and the tip chord, c_t, by

$$R_T \equiv c_t/c_r$$

For a wing of any planform having no sweep, dihedral, geometric twist, or aerodynamic twist, the circulation distribution predicted from Prandtl's lifting-line theory can be written in terms of a Fourier series with coefficients, a_n, that are independent of angle of attack. Under these conditions, Eq. (1.8.3) can be rearranged to give

$$\Gamma(\theta) = 2bV_\infty(\alpha - \alpha_{L0})\sum_{n=1}^{N} a_n \sin(n\theta); \qquad \theta = \cos^{-1}(-2z/b) \qquad (1.8.9)$$

where the Fourier coefficients are obtained from

$$\sum_{n=1}^{N} a_n \left[\frac{4b}{\widetilde{C}_{L,\alpha}\, c(\theta)} + \frac{n}{\sin(\theta)} \right] \sin(n\theta) = 1 \qquad (1.8.10)$$

The lift and induced drag coefficients predicted from Eqs. (1.8.5) and (1.8.6) for a **wing with no geometric or aerodynamic twist** can be written as

$$C_L = C_{L,\alpha}(\alpha - \alpha_{L0}), \quad C_{L,\alpha} = \frac{\widetilde{C}_{L,\alpha}}{[1+\widetilde{C}_{L,\alpha}/(\pi R_A)](1+\kappa_L)} \qquad (1.8.11)$$

$$C_{D_i} = \frac{C_L^2}{\pi R_A e_s}, \quad e_s = \frac{1}{1+\kappa_D} \qquad (1.8.12)$$

where the lift slope factor, κ_L, and the induced drag factor, κ_D, are given by

$$\kappa_L = \frac{1-(1+\pi R_A/\widetilde{C}_{L,\alpha})a_1}{(1+\pi R_A/\widetilde{C}_{L,\alpha})a_1} \qquad (1.8.13)$$

$$\kappa_D = \sum_{n=2}^{N} n\left(\frac{a_n}{a_1}\right)^2 \qquad (1.8.14)$$

For untwisted tapered wings, numerical results obtained for these two factors are shown in Figs. 1.8.13 and 1.8.14. These results were generated using a section lift slope of 2π.

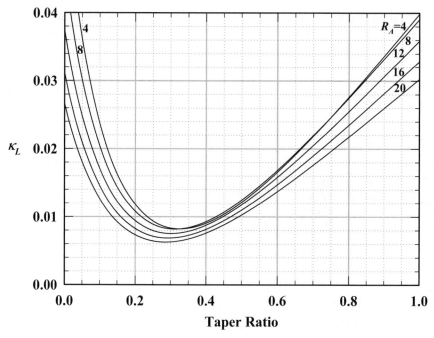

Figure 1.8.13. Lift slope factor for untwisted tapered wings from Prandtl's lifting-line theory.

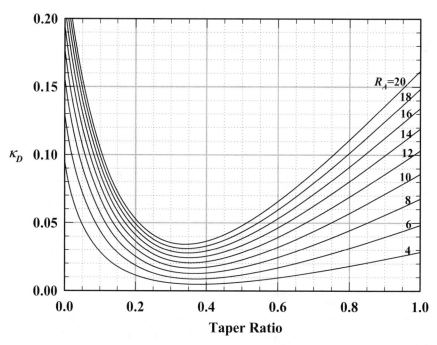

Figure 1.8.14. Induced drag factor for untwisted tapered wings from Prandtl's lifting-line theory.

Glauert (1926) first presented results similar to those shown in Fig. 1.8.14. Such results have sometimes led to the conclusion that a tapered wing with a taper ratio of about 0.4 always produces significantly less induced drag than a rectangular wing of the same aspect ratio developing the same lift. As a result, tapered wings are often used as a means of reducing induced drag. However, this reduction in drag usually comes at a price. Because a tapered wing has a lower Reynolds number at the wingtips than at the root, a tapered wing with no geometric or aerodynamic twist tends to stall first in the region of the wingtips. This wingtip stall commonly leads to poor handling qualities during stall recovery.

The results shown in Fig. 1.8.14 can be misleading if one loses sight of the fact that these results apply only for the special case of wings with no geometric or aerodynamic twist. This is only one of many possible twist distributions that could be used for a wing of any given planform. Furthermore, it is not the twist distribution that produces minimum induced drag with finite lift, except for the case of an elliptic wing. When the effects of twist are included, it can be shown that the conclusions sometimes reached from consideration of only those results shown in Fig. 1.8.14 are erroneous.

Effects of Wing Twist

For a wing with geometric and/or aerodynamic twist, the solution expressed in the form of Eqs. (1.8.3) through (1.8.6) is cumbersome for the evaluation of traditional wing properties, because the Fourier coefficients depend on angle of attack and must be reevaluated for each operating point studied. Furthermore, the definition of σ that is given in Eq. (1.8.6) is not practical for use at an arbitrary angle of attack, because the value becomes singular for a twisted wing when the lift coefficient approaches zero. A more useful form of the solution can be obtained by using the change of variables

$$\alpha(\theta) - \alpha_{L0}(\theta) \equiv (\alpha - \alpha_{L0})_{\text{root}} - \Omega\omega(\theta) \tag{1.8.15}$$

where Ω is defined to be the maximum total washout, geometric plus aerodynamic,

$$\Omega \equiv (\alpha - \alpha_{L0})_{\text{root}} - (\alpha - \alpha_{L0})_{\text{max}} \tag{1.8.16}$$

and $\omega(\theta)$ is the local washout distribution function, which is normalized with respect to maximum total washout

$$\omega(\theta) \equiv \frac{\alpha(\theta) - \alpha_{L0}(\theta) - (\alpha - \alpha_{L0})_{\text{root}}}{(\alpha - \alpha_{L0})_{\text{max}} - (\alpha - \alpha_{L0})_{\text{root}}} \tag{1.8.17}$$

The normalized washout distribution function, $\omega(\theta)$, is independent of angle of attack and always varies from 0.0 at the root to 1.0 at the point of maximum washout, which is commonly at the wingtips.

Using Eq. (1.8.15) in Eq. (1.8.4) gives

$$\sum_{n=1}^{N} A_n \left[\frac{4b}{\widetilde{C}_{L,\alpha} c(\theta)} + \frac{n}{\sin(\theta)} \right] \sin(n\theta) = (\alpha - \alpha_{L0})_{\text{root}} - \Omega\omega(\theta) \tag{1.8.18}$$

The Fourier coefficients in Eq. (1.8.18) for a **wing with geometric and/or aerodynamic twist** can be written conveniently as

$$A_n \equiv a_n(\alpha - \alpha_{L0})_{\text{root}} - b_n\Omega \qquad (1.8.19)$$

where

$$\sum_{n=1}^{N} a_n\left[\frac{4b}{\widetilde{C}_{L,\alpha}\, c(\theta)} + \frac{n}{\sin(\theta)}\right]\sin(n\theta) = 1 \qquad (1.8.20)$$

$$\sum_{n=1}^{N} b_n\left[\frac{4b}{\widetilde{C}_{L,\alpha}\, c(\theta)} + \frac{n}{\sin(\theta)}\right]\sin(n\theta) = \omega(\theta) \qquad (1.8.21)$$

Comparing Eq. (1.8.20) with Eq. (1.8.10), we see that the Fourier coefficients defined by Eq. (1.8.20) are those corresponding to the solution for a wing of the same planform shape but with no geometric or aerodynamic twist. The solution to Eq. (1.8.21) can be obtained in a similar manner and is also independent of angle of attack.

Using Eq. (1.8.19) in Eq. (1.8.5), the lift coefficient for a wing with washout can be expressed as

$$C_L = \pi R_A A_1 = \pi R_A\left[a_1(\alpha - \alpha_{L0})_{\text{root}} - b_1\Omega\right] \qquad (1.8.22)$$

Using Eq. (1.8.19) in Eq. (1.8.6), the induced drag coefficient is given by

$$C_{D_i} = \pi R_A \sum_{n-1}^{N} nA_n^2 = \pi R_A\left[a_1(\alpha - \alpha_{L0})_{\text{root}} - b_1\Omega\right]^2 + \pi R_A \sum_{n=2}^{N} n\left[a_n(\alpha - \alpha_{L0})_{\text{root}} - b_n\Omega\right]^2$$

or after using Eq. (1.8.22) to express the first term on the right-hand side in terms of the lift coefficient,

$$C_{D_i} = \frac{C_L^2}{\pi R_A} + \pi R_A \sum_{n=2}^{\infty} n\left[a_n^2(\alpha - \alpha_{L0})_{\text{root}}^2 - 2a_n b_n(\alpha - \alpha_{L0})_{\text{root}}\Omega + b_n^2\Omega^2\right] \qquad (1.8.23)$$

Equations (1.8.22) and (1.8.23) can be algebraically rearranged to yield a convenient expression for the lift and induced drag developed by a **finite wing with geometric and/or aerodynamic twist:**

$$C_L = C_{L,\alpha}\left[(\alpha - \alpha_{L0})_{\text{root}} - \varepsilon_\Omega\, \Omega\right] \qquad (1.8.24)$$

$$C_{D_i} = \frac{C_L^2(1 + \kappa_D) - \kappa_{DL}C_L C_{L,\alpha}\,\Omega + \kappa_{D\Omega}(C_{L,\alpha}\,\Omega)^2}{\pi R_A} \qquad (1.8.25)$$

where

$$C_{L,\alpha} = \pi R_A a_1 = \frac{\tilde{C}_{L,\alpha}}{[1 + \tilde{C}_{L,\alpha}/(\pi R_A)](1 + \kappa_L)} \tag{1.8.26}$$

$$\kappa_L \equiv \frac{1 - (1 + \pi R_A/\tilde{C}_{L,\alpha})a_1}{(1 + \pi R_A/\tilde{C}_{L,\alpha})a_1} \tag{1.8.27}$$

$$\varepsilon_\Omega \equiv \frac{b_1}{a_1} \tag{1.8.28}$$

$$\kappa_D \equiv \sum_{n=2}^{N} n \frac{a_n^2}{a_1^2} \tag{1.8.29}$$

$$\kappa_{DL} \equiv 2 \frac{b_1}{a_1} \sum_{n=2}^{N} n \frac{a_n}{a_1} \left(\frac{b_n}{b_1} - \frac{a_n}{a_1} \right) \tag{1.8.30}$$

$$\kappa_{D\Omega} \equiv \left(\frac{b_1}{a_1} \right)^2 \sum_{n=2}^{N} n \left(\frac{b_n}{b_1} - \frac{a_n}{a_1} \right)^2 \tag{1.8.31}$$

Comparing Eqs. (1.8.24) through (1.8.31) with Eqs. (1.8.11) and (1.8.14), we see that washout increases the zero-lift angle of attack for any wing but the lift slope for a wing of arbitrary planform shape is not affected by washout. Notice that the induced drag for a wing with washout is not zero at zero lift. In addition to the usual component of induced drag, which is proportional to the lift coefficient squared, a wing with washout produces a component of induced drag that is proportional to the washout squared, and this results in induced drag at zero lift. There is also a component of induced drag that varies with the product of the lift coefficient and the washout.

A commonly used washout distribution that is easy to implement is linear washout. For the special case of a linear variation in washout from the root to the tip, the normalized washout distribution function is simply

$$\omega(z) = |2z/b| \quad \text{or} \quad \omega(\theta) = |\cos(\theta)| \tag{1.8.32}$$

For tapered wings, the variation in chord length is also linear:

$$c(z) = \frac{2b}{R_A(1 + R_T)}[1 - (1 - R_T)|2z/b|] \quad \text{or} \quad c(\theta) = \frac{2b}{R_A(1 + R_T)}[1 - (1 - R_T)|\cos(\theta)|] \tag{1.8.33}$$

Using Eqs. (1.8.32) and (1.8.33) in Eqs. (1.8.20) and (1.8.21) yields the results for a tapered wing with linear washout,

$$\sum_{n=1}^{N} a_n \left[\frac{2R_A(1+R_T)}{\widetilde{C}_{L,\alpha}\left[1-(1-R_T)|\cos(\theta)|\right]} + \frac{n}{\sin(\theta)} \right] \sin(n\theta) = 1 \qquad (1.8.34)$$

$$\sum_{n=1}^{N} b_n \left[\frac{2R_A(1+R_T)}{\widetilde{C}_{L,\alpha}\left[1-(1-R_T)|\cos(\theta)|\right]} + \frac{n}{\sin(\theta)} \right] \sin(n\theta) = |\cos(\theta)| \qquad (1.8.35)$$

The solution obtained from Eq. (1.8.34) for the Fourier coefficients, a_n, is the familiar result that was used to produce Figs. 1.8.13 and 1.8.14. Induced drag for a wing with washout is readily predicted from Eq. (1.8.25) with the definitions given in Eqs. (1.8.29) through (1.8.31). For tapered wings with linear washout, the Fourier coefficients, b_n, can be obtained from Eq. (1.8.35) in exactly the same manner as the coefficients, a_n, are obtained from Eq. (1.8.34). Using the Fourier coefficients so obtained in Eqs. (1.8.28), (1.8.30), and (1.8.31) produces the results shown in Figs. 1.8.15 through 1.8.17. Notice from examining either Eq. (1.8.31) or Fig. 1.8.17 that $\kappa_{D\Omega}$ is always positive. Thus, the third term in the numerator on the right-hand side of Eq. (1.8.25) always contributes to an increase in induced drag. However, from the results shown in Fig. 1.8.16, we see that the second term in the numerator on the right-hand side of Eq. (1.8.25) can either increase or decrease the induced drag, depending on the signs of κ_{DL} and Ω. This raises an important question regarding wing efficiency. What level and distribution of washout will result in minimum induced drag for a given wing planform and lift coefficient?

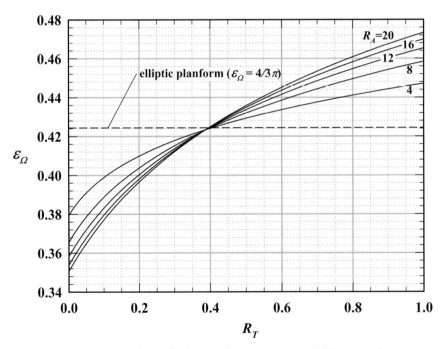

Figure 1.8.15. Washout effectiveness for tapered wings with linear washout.

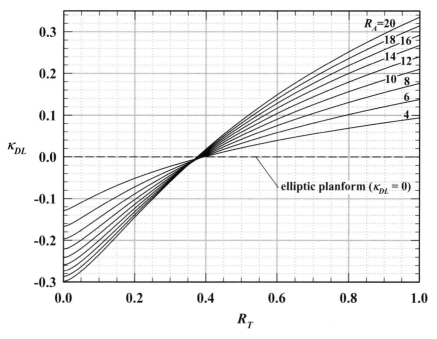

Figure 1.8.16. Lift-washout contribution to induced drag for tapered wings with linear washout.

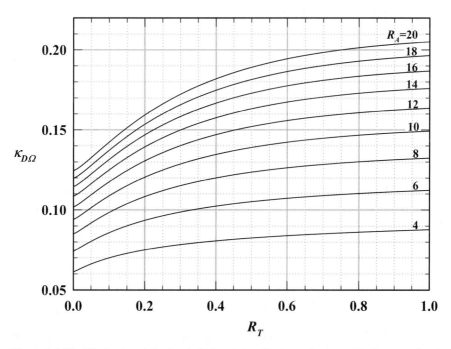

Figure 1.8.17. Washout contribution to induced drag for tapered wings with linear washout.

Minimizing Induced Drag with Washout

For a wing of any given planform shape having a fixed washout distribution, induced drag can be minimized with washout as a result of the trade-off between the second and third terms in the numerator on the right-hand side of Eq. (1.8.25). Thus, Eq. (1.8.25) can be used to determine the optimum value of total washout, which will result in minimum induced drag for any washout distribution function and any specified lift coefficient. Differentiating Eq. (1.8.25) with respect to Ω at constant lift coefficient gives

$$\frac{\partial C_{D_i}}{\partial \Omega} = \frac{-K_{DL}C_L C_{L,\alpha} + 2K_{D\Omega}C_{L,\alpha}^2 \,\Omega}{\pi R_A} \tag{1.8.36}$$

Setting the right-hand side of Eq. (1.8.36) to zero and solving for Ω, it can be seen that minimum induced drag is attained for any given wing planform, $c(z)$, any given washout distribution, $\omega(z)$, and any given design lift coefficient, C_{Ld}, by using an optimum total amount of washout, Ω_{opt}, which is given by the relation

$$\Omega_{opt} = \frac{K_{DL}C_{Ld}}{2K_{D\Omega}C_{L,\alpha}} \tag{1.8.37}$$

For the elliptic planform, all of the Fourier coefficients, a_n, are zero for n greater than 1. Thus, Eq. (1.8.30) shows that K_{DL} is zero for an elliptic wing. As a result, elliptic wings are always optimized with no washout. From consideration of Eq. (1.8.37) together with the results shown in Figs. 1.8.16 and 1.8.17, we see that tapered wings with linear washout and taper ratios greater than 0.4 are optimized with positive washout, whereas those with taper ratios less than about 0.4 are optimized with negative washout.

Using the value of optimum washout from Eq. (1.8.37) in the expression for induced drag coefficient given by Eq. (1.8.25), we find that the induced drag coefficient for a wing of arbitrary planform with a fixed washout distribution and optimum total washout is given by

$$\left(C_{D_i}\right)_{opt} = \frac{C_L^2}{\pi R_A}\left[1 + K_D - \frac{K_{DL}^2}{4K_{D\Omega}}\left(2 - \frac{C_{Ld}}{C_L}\right)\frac{C_{Ld}}{C_L}\right] \tag{1.8.38}$$

From Eq. (1.8.38) it can be seen that a wing with optimum washout will always produce less induced drag than a wing with no washout having the same planform and aspect ratio, provided that the actual lift coefficient is greater than one-half the design lift coefficient. When the actual lift coefficient is equal to the design lift coefficient, the induced drag coefficient for a wing with optimum washout is

$$\left(C_{D_i}\right)_{opt} = \frac{C_L^2}{\pi R_A}(1 + K_{Do}), \quad K_{Do} \equiv K_D - \frac{K_{DL}^2}{4K_{D\Omega}} \tag{1.8.39}$$

For tapered wings with linear washout, the variations in K_{Do} with aspect ratio and taper ratio are shown in Fig. 1.8.18. For comparison, the dashed lines in this figure show the same results for wings with no washout. Notice that when linear washout is used to further optimize tapered wings, taper ratios near 0.4 correspond closely to a maximum in induced drag, not to a minimum.

The choice of a linear washout distribution, which was used to generate the results shown in Fig. 1.8.18, is as arbitrary as the choice of no washout. While a linear variation in washout is commonly used and simple to implement, it is not the optimum washout distribution for wings with linear taper. Minimum possible induced drag for a finite lift coefficient always occurs when the local section lift varies with the spanwise coordinate in proportion to $\sin(\theta)$. This results in uniform downwash and requires that the product of the local chord length and local aerodynamic angle of attack, $\alpha - \alpha_{L0}$, varies elliptically with the spanwise coordinate, i.e.,

$$\frac{c(z)[\alpha(z)-\alpha_{L0}(z)]}{\sqrt{1-(2z/b)^2}} = \frac{c(\theta)[\alpha(\theta)-\alpha_{L0}(\theta)]}{\sin(\theta)} = \text{constant} \qquad (1.8.40)$$

There are many possibilities for wing geometry that will satisfy this condition. The elliptic planform with no geometric or aerodynamic twist is only one such geometry. Since the local aerodynamic angle of attack decreases along the span in direct proportion to the increase in washout, Eq. (1.8.40) can only be satisfied if the washout distribution satisfies the relation

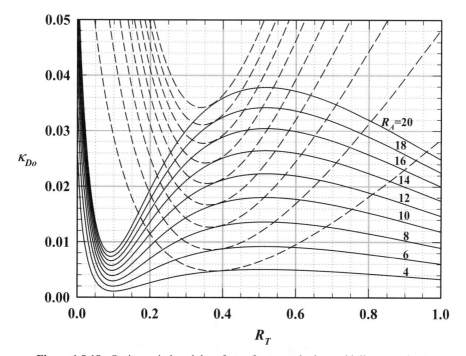

Figure 1.8.18. Optimum induced drag factor for tapered wings with linear washout.

$$\frac{c(z)[1-\omega(z)]}{\sqrt{1-(2z/b)^2}} = \frac{c(\theta)[1-\omega(\theta)]}{\sin(\theta)} = \text{constant} \tag{1.8.41}$$

Equation (1.8.41) is satisfied by the **optimum washout distribution**

$$\omega_{\text{opt}} = 1 - \frac{\sqrt{1-(2z/b)^2}}{c(z)/c_{\text{root}}} = 1 - \frac{\sin(\theta)}{c(\theta)/c_{\text{root}}}, \quad \Omega_{\text{opt}} = \frac{4b\,C_L}{\pi\,R_A \tilde{C}_{L,\alpha}\,c_{\text{root}}} \tag{1.8.42}$$

For wings with linear taper, this gives

$$\omega_{\text{opt}} = 1 - \frac{\sqrt{1-(2z/b)^2}}{1-(1-R_T)|2z/b|} = 1 - \frac{\sin(\theta)}{1-(1-R_T)|\cos(\theta)|}, \quad \Omega_{\text{opt}} = \frac{2(1+R_T)C_L}{\pi\tilde{C}_{L,\alpha}} \tag{1.8.43}$$

This optimum washout distribution is shown in Fig. 1.8.19 for several values of taper ratio. Results obtained for tapered wings having this washout distribution are presented in Figs. 1.8.20 through 1.8.22. It should be noted that computing Ω_{opt} from Eq. (1.8.42) or (1.8.43) is only valid for wings having the optimum washout distribution, ω_{opt}. On the other hand, Eq. (1.8.37) is valid for any washout distribution. For $\omega = \omega_{\text{opt}}$, Eq. (1.8.37) produces exactly the same result as Eq. (1.8.42) or (1.8.43).

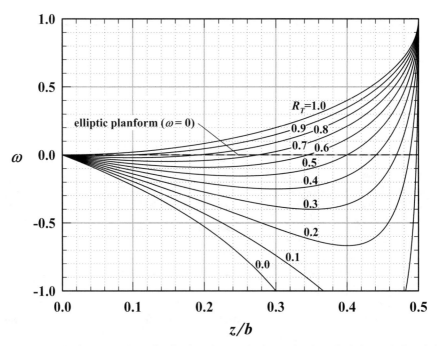

Figure 1.8.19. Optimum washout distribution that results in production of minimum induced drag for wings with linear taper, as defined in Eq. (1.8.43).

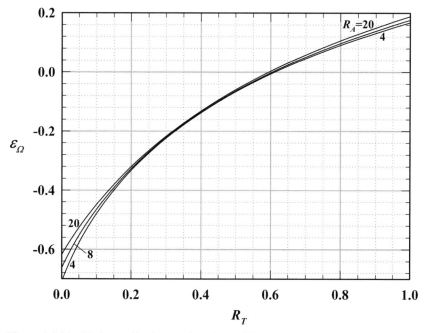

Figure 1.8.20. Washout effectiveness for wings with linear taper and optimum washout.

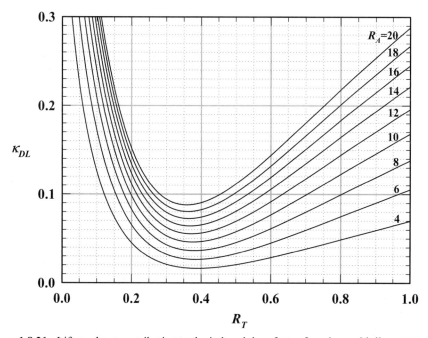

Figure 1.8.21. Lift-washout contribution to the induced drag factor for wings with linear taper and the optimum washout distribution specified by Eq. (1.8.43).

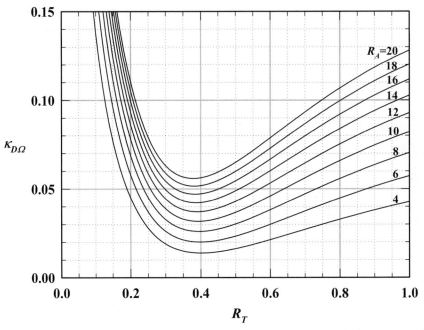

Figure 1.8.22. Washout contribution to the induced drag factor for wings with linear taper and the optimum washout distribution specified by Eq. (1.8.43).

When an unswept wing of arbitrary planform has the washout distribution specified by Eq. (1.8.42), the value of κ_{Do} as defined in Eq. (1.8.39) is always identically zero. With the washout distribution set to ω_{opt} and the total amount washout set to Ω_{opt}, an unswept wing of any planform shape can be designed to operate at a given lift coefficient with the same induced drag as that produced by an untwisted elliptic wing with the same aspect ratio and lift coefficient. This is demonstrated in the following examples.

EXAMPLE 1.8.1. When results similar to those shown in Fig. 1.8.14 were first published by Glauert (1926), all calculations were made without the aid of a computer. For this reason, only a few terms were retained in the Fourier series. For a rectangular planform with an aspect ratio of 8.0 and the optimum washout distribution specified by Eq. (1.8.42), solve Eqs. (1.8.20) and (1.8.21) by retaining only seven terms in the Fourier series and forcing Eqs. (1.8.20) and (1.8.21) to be satisfied at only seven sections along the span of the wing. Locate the first and last sections at the wingtips and space the intermediate sections equally in θ, i.e.,

$$\theta_i = \frac{(i-1)\pi}{N-1}, \quad i = 1, N$$

where N is 7. Assume an airfoil lift slope of 2π and use the solution to compute κ_D, κ_{DL}, $\kappa_{D\Omega}$, and κ_{Do}. Compare the results with those obtained for $N = 99$.

Solution. In order to obtain N independent equations for the N unknown Fourier coefficients, a_n, Eq. (1.8.20) can be written for each of N spanwise sections of the wing. With the first and last sections located at the wingtips and the intermediate sections spaced equally in θ, this gives the system of equations

$$
\begin{bmatrix}
C_{11} & C_{12} & C_{13} & \cdots & C_{1N} \\
C_{21} & C_{22} & C_{23} & \cdots & C_{2N} \\
C_{31} & C_{32} & C_{33} & \cdots & C_{3N} \\
\vdots & \vdots & \vdots & \ddots & \vdots \\
C_{N1} & C_{N2} & C_{N3} & \cdots & C_{NN}
\end{bmatrix}
\begin{Bmatrix}
a_1 \\ a_2 \\ a_3 \\ \vdots \\ a_N
\end{Bmatrix}
=
\begin{Bmatrix}
1 \\ 1 \\ 1 \\ \vdots \\ 1
\end{Bmatrix}
$$

where

$$
C_{ij} = \left[\frac{4b}{\widetilde{C}_{L,\alpha}\, c(\theta_i)} + \frac{j}{\sin(\theta_i)}\right]\sin(j\theta_i), \quad \theta_i = \frac{(i-1)\pi}{N-1}, \quad \begin{cases} i = 1,\, N \\ j = 1,\, N \end{cases}
$$

The components of the matrix, **[C]**, on the left-hand side of this linear system are indeterminate when evaluated at the wingtips, $\theta = 0$ and $\theta = \pi$. However, applying *l'Hospital's rule* gives

$$
\left(\frac{\sin(j\theta_i)}{\sin(\theta_i)}\right)_{\theta_i \to 0} = \left(\frac{j\cos(j\theta_i)}{\cos(\theta_i)}\right)_{\theta_i \to 0} = j
$$

$$
\left(\frac{\sin(j\theta_i)}{\sin(\theta_i)}\right)_{\theta_i \to \pi} = \left(\frac{j\cos(j\theta_i)}{\cos(\theta_i)}\right)_{\theta_i \to \pi} = (-1)^{j+1} j
$$

Thus, for the purpose of numerical evaluation, the matrix **[C]** is determined from the algorithm

$$
C_{1j} = j^2, \quad j = 1,\, N
$$

$$
C_{ij} = \left[\frac{4b}{\widetilde{C}_{L,\alpha}\, c(\theta_i)} + \frac{j}{\sin(\theta_i)}\right]\sin(j\theta_i), \quad \theta_i = \frac{(i-1)\pi}{N-1}, \quad \begin{cases} i = 2,\, N-1 \\ j = 1,\, N \end{cases}
$$

$$
C_{Nj} = (-1)^{j+1} j^2, \quad j = 1,\, N
$$

For this rectangular planform of aspect ratio of 8.0, the chord is independent of θ and $b/c(\theta) = 8$. Thus, for an airfoil section lift slope of 2π and $N = 7$, we have

$$
C_{ij} = \left[\frac{16}{\pi} + \frac{j}{\sin[(i-1)\pi/6]}\right]\sin[j(i-1)\pi/6], \quad \begin{cases} i = 2,\, 6 \\ j = 1,\, 7 \end{cases}
$$

Some example computations give

$$C_{21} = \left[\frac{16}{\pi} + \frac{1}{\sin[(2-1)\pi/6]}\right]\sin\left(\frac{1(2-1)\pi}{6}\right) = \left[\frac{16}{\pi} + \frac{1}{\sin(\pi/6)}\right]\sin\left(\frac{\pi}{6}\right) = 3.5465$$

$$C_{22} = \left[\frac{16}{\pi} + \frac{2}{\sin[(2-1)\pi/6]}\right]\sin\left(\frac{2(2-1)\pi}{6}\right) = \left[\frac{16}{\pi} + \frac{2}{\sin(\pi/6)}\right]\sin\left(\frac{2\pi}{6}\right) = 7.8747$$

$$C_{45} = \left[\frac{16}{\pi} + \frac{5}{\sin[(4-1)\pi/6]}\right]\sin\left(\frac{5(4-1)\pi}{6}\right) = \left[\frac{16}{\pi} + \frac{5}{\sin(3\pi/6)}\right]\sin\left(\frac{15\pi}{6}\right) = 10.093$$

After performing the remaining computations, we have

$$[C] = \begin{bmatrix} 1 & 4 & 9 & 16 & 25 & 36 & 49 \\ 3.5465 & 7.8747 & 11.093 & 11.339 & 7.5465 & 0 & -9.5465 \\ 5.4106 & 6.4106 & 0 & -8.4106 & -9.4106 & 0 & 11.411 \\ 6.0930 & 0 & -8.0930 & 0 & 10.093 & 0 & -12.093 \\ 5.4106 & -6.4106 & 0 & 8.4106 & -9.4106 & 0 & 11.411 \\ 3.5465 & -7.8747 & 11.093 & -11.339 & 7.5465 & 0 & -9.5465 \\ 1 & -4 & 9 & -16 & 25 & -36 & 49 \end{bmatrix}$$

The inverse of this matrix is

$$[C]^{-1} =$$

$$\begin{bmatrix} 0.00005 & 0.02061 & 0.04702 & 0.05661 & 0.04702 & 0.02061 & 0.00005 \\ 0.00000 & 0.03027 & 0.04081 & 0.00000 & -0.04081 & -0.03027 & 0.00000 \\ 0.00019 & 0.03016 & 0.00298 & -0.04047 & 0.00298 & 0.03016 & 0.00019 \\ 0.00000 & 0.02307 & -0.02834 & 0.00000 & 0.02834 & -0.02307 & 0.00000 \\ 0.00764 & 0.00286 & -0.01725 & 0.02481 & -0.01725 & 0.00286 & 0.00764 \\ 0.01389 & -0.01362 & 0.00806 & 0.00000 & -0.00806 & 0.01362 & -0.01389 \\ 0.00627 & -0.00742 & 0.00729 & -0.00638 & 0.00729 & -0.00742 & 0.00627 \end{bmatrix}$$

which yields

$$\begin{Bmatrix} a_1 \\ a_2 \\ a_3 \\ a_4 \\ a_5 \\ a_6 \\ a_7 \end{Bmatrix} = [C]^{-1}\begin{Bmatrix} 1 \\ 1 \\ 1 \\ 1 \\ 1 \\ 1 \\ 1 \end{Bmatrix} = \begin{Bmatrix} 0.191966 \\ 0 \\ 0.026191 \\ 0 \\ 0.011287 \\ 0 \\ 0.005921 \end{Bmatrix}$$

Comparing Eqs. (1.8.20) and (1.8.21), we see that the Fourier coefficients, b_n, can be obtained from a linear system of equations that is almost identical to that used to obtain the Fourier coefficients, a_n. The matrices on the left-hand sides of these two systems are identical. The only difference is that the right-hand side of the system used to obtain a_n is a vector having all components equal to 1, whereas the right-hand side of the system used to obtain b_n is a vector obtained from the washout distribution function,

$$
\begin{bmatrix}
C_{11} & C_{12} & C_{13} & \cdots & C_{1N} \\
C_{21} & C_{22} & C_{23} & \cdots & C_{2N} \\
C_{31} & C_{32} & C_{33} & \cdots & C_{3N} \\
\vdots & \vdots & \vdots & \ddots & \vdots \\
C_{N1} & C_{N2} & C_{N3} & \cdots & C_{NN}
\end{bmatrix}
\begin{Bmatrix}
b_1 \\ b_2 \\ b_3 \\ \vdots \\ b_N
\end{Bmatrix}
=
\begin{Bmatrix}
\omega(\theta_1) \\ \omega(\theta_2) \\ \omega(\theta_3) \\ \vdots \\ \omega(\theta_N)
\end{Bmatrix}
$$

Using the optimum washout distribution function from Eq. (1.8.42) yields

$$
\omega(\theta_i) = 1 - \frac{\sin(\theta_i)}{c(\theta_i)/c_{\text{root}}}, \quad \theta_i = \frac{(i-1)\pi}{N-1}, \quad i = 1, N
$$

For this rectangular planform, the chord is constant, so $c(\theta_i)/c_{\text{root}} = 1$. Thus, for $N = 7$, the Fourier coefficients, b_n, are found to be

$$
\begin{Bmatrix}
b_1 \\ b_2 \\ b_3 \\ b_4 \\ b_5 \\ b_6 \\ b_7
\end{Bmatrix}
= [\mathbf{C}]^{-1}
\begin{Bmatrix}
1 \\
1 - \sin(\pi/6) \\
1 - \sin(2\pi/6) \\
1 - \sin(3\pi/6) \\
1 - \sin(4\pi/6) \\
1 - \sin(5\pi/6) \\
1 - \sin(6\pi/6)
\end{Bmatrix}
= [\mathbf{C}]^{-1}
\begin{Bmatrix}
1 \\ 0.5 \\ 0.133975 \\ 0 \\ 0.133975 \\ 0.5 \\ 1
\end{Bmatrix}
=
\begin{Bmatrix}
0.033309 \\ 0 \\ 0.031334 \\ 0 \\ 0.013504 \\ 0 \\ 0.007084
\end{Bmatrix}
$$

Using these results in Eqs. (1.8.29), (1.8.30), (1.8.31), and (1.8.39) yields

$$
\kappa_D \equiv \sum_{n=2}^{N} n \frac{a_n^2}{a_1^2} = \underline{0.079791}
$$

$$
\kappa_{DL} \equiv 2 \frac{b_1}{a_1} \sum_{n=2}^{N} n \frac{a_n}{a_1} \left(\frac{b_n}{b_1} - \frac{a_n}{a_1} \right) = \underline{0.163225}
$$

$$
\kappa_{D\Omega} \equiv \left(\frac{b_1}{a_1} \right)^2 \sum_{n=2}^{N} n \left(\frac{b_n}{b_1} - \frac{a_n}{a_1} \right)^2 = \underline{0.083476}
$$

$$
\kappa_{Do} \equiv \kappa_D - \frac{\kappa_{DL}^2}{4\kappa_{D\Omega}} = \underline{0.000000}
$$

For comparison, repeating these computations with $N = 99$ gives

$$
\begin{Bmatrix} a_1 \\ a_2 \\ a_3 \\ a_4 \\ a_5 \\ a_6 \\ a_7 \\ \vdots \\ a_{98} \\ a_{99} \end{Bmatrix} = \begin{Bmatrix} 0.19248612 \\ 0 \\ 0.02740767 \\ 0 \\ 0.00656477 \\ 0 \\ 0.00202851 \\ \vdots \\ 0 \\ 0.00000144 \end{Bmatrix}, \qquad
\begin{Bmatrix} b_1 \\ b_2 \\ b_3 \\ b_4 \\ b_5 \\ b_6 \\ b_7 \\ \vdots \\ b_{98} \\ b_{99} \end{Bmatrix} = \begin{Bmatrix} 0.03393114 \\ 0 \\ 0.03278916 \\ 0 \\ 0.00785376 \\ 0 \\ 0.00242681 \\ \vdots \\ 0 \\ 0.00000172 \end{Bmatrix}
$$

$$K_D = \underline{0.067611}, \quad K_{DL} = \underline{0.137937}, \quad K_{D\Omega} = \underline{0.070353}, \quad K_{Do} = \underline{0.000000}$$

EXAMPLE 1.8.2.   We wish to repeat the computations in Example 1.8.1 for a wing of tapered planform with an aspect ratio of 8.0 and a taper ratio of 0.5.

Solution.   For a wing planform having linear taper, the section chord length varies with the spanwise coordinate as specified in Eq. (1.8.33). With an aspect ratio of 8.0 and a taper ratio of 0.5, this gives

$$\frac{c(\theta)}{b} = \frac{2}{R_A(1+R_T)}[1 - (1-R_T)|\cos(\theta)|] = \frac{1}{6}[1 - 0.5|\cos(\theta)|]$$

Thus, following Example 1.8.1, for this wing geometry and an airfoil section lift slope of 2π, the components of the matrix **[C]** are determined from

$$C_{1j} = j^2, \quad j = 1, N$$

$$C_{ij} = \left[\frac{12}{\pi[1 - 0.5|\cos(\theta_i)|]} + \frac{j}{\sin(\theta_i)} \right]\sin(j\theta_i), \quad \theta_i = \frac{(i-1)\pi}{N-1}, \quad \begin{cases} i = 2, N-1 \\ j = 1, N \end{cases}$$

$$C_{Nj} = (-1)^{j+1} j^2, \quad j = 1, N$$

For example, with $N = 7$,

$$C_{45} = \left[\frac{12}{\pi[1 - 0.5|\cos(3\pi/6)|]} + \frac{5}{\sin(3\pi/6)} \right]\sin(15\pi/6) = 8.8197$$

After performing the remaining computations, we have

$$[C] = \begin{bmatrix} 1 & 4 & 9 & 16 & 25 & 36 & 49 \\ 4.3684 & 9.2984 & 12.7369 & 12.7625 & 8.3684 & 0 & -10.3684 \\ 5.4106 & 6.4106 & 0 & -8.4106 & -9.4106 & 0 & 11.4106 \\ 4.8197 & 0 & -6.8197 & 0 & 8.8197 & 0 & -10.8197 \\ 5.4106 & -6.4106 & 0 & 8.4106 & -9.4106 & 0 & 11.4106 \\ 4.3684 & -9.2984 & 12.7369 & -12.7625 & 8.3684 & 0 & -10.3684 \\ 1 & -4 & 9 & -16 & 25 & -36 & 49 \end{bmatrix}$$

$$[C]^{-1} = \begin{bmatrix} 0.00010 & 0.01798 & 0.04785 & 0.06742 & 0.04785 & 0.01798 & 0.00010 \\ 0.00000 & 0.02628 & 0.03988 & 0.00000 & -0.03988 & -0.02628 & 0.00000 \\ 0.00003 & 0.02619 & 0.00055 & -0.04872 & 0.00055 & 0.02619 & 0.00003 \\ 0.00000 & 0.02003 & -0.02905 & 0.00000 & 0.02905 & -0.02003 & 0.00000 \\ 0.00767 & 0.00251 & -0.01663 & 0.02962 & -0.01663 & 0.00251 & 0.00767 \\ 0.01389 & -0.01182 & 0.00848 & 0.00000 & -0.00848 & 0.01182 & -0.01389 \\ 0.00628 & -0.00646 & 0.00741 & -0.00754 & 0.00741 & -0.00646 & 0.00628 \end{bmatrix}$$

$$\begin{Bmatrix} a_1 \\ a_2 \\ a_3 \\ a_4 \\ a_5 \\ a_6 \\ a_7 \end{Bmatrix} = [C]^{-1} \begin{Bmatrix} 1 \\ 1 \\ 1 \\ 1 \\ 1 \\ 1 \\ 1 \end{Bmatrix} = \begin{Bmatrix} 0.199278 \\ 0 \\ 0.004824 \\ 0 \\ 0.016713 \\ 0 \\ 0.006928 \end{Bmatrix}$$

For this planform with linear taper and a taper ratio of 0.5, the optimum washout distribution function from Eq. (1.8.43) is

$$\omega(\theta_i) = 1 - \frac{\sin(\theta_i)}{1 - 0.5|\cos(\theta_i)|}, \quad \theta_i = \frac{(i-1)\pi}{N-1}, \quad i = 1, N$$

$$\begin{Bmatrix} b_1 \\ b_2 \\ b_3 \\ b_4 \\ b_5 \\ b_6 \\ b_7 \end{Bmatrix} = [C]^{-1} \begin{Bmatrix} \omega(0) \\ \omega(\pi/6) \\ \omega(2\pi/6) \\ \omega(3\pi/6) \\ \omega(4\pi/6) \\ \omega(5\pi/6) \\ \omega(\pi) \end{Bmatrix} = [C]^{-1} \begin{Bmatrix} 1 \\ 0.118146 \\ -0.154701 \\ 0 \\ -0.154701 \\ 0.118146 \\ 1 \end{Bmatrix} = \begin{Bmatrix} -0.010351 \\ 0 \\ 0.006087 \\ 0 \\ 0.021088 \\ 0 \\ 0.008742 \end{Bmatrix}$$

Using these results in Eqs. (1.8.29), (1.8.30), (1.8.31), and (1.8.39) yields

$$\kappa_D = 0.045387, \quad \kappa_{DL} = 0.119253, \quad \kappa_{D\Omega} = 0.078334, \quad \kappa_{Do} = 0.000000$$

With $N = 99$ we obtain

$$
\begin{Bmatrix} a_1 \\ a_2 \\ a_3 \\ a_4 \\ a_5 \\ a_6 \\ a_7 \\ \vdots \\ a_{98} \\ a_{99} \end{Bmatrix} =
\begin{bmatrix} 0.19751337 \\ 0 \\ 0.00837113 \\ 0 \\ 0.00918923 \\ 0 \\ 0.00142163 \\ \vdots \\ 0 \\ 0.00000164 \end{bmatrix}, \quad
\begin{Bmatrix} b_1 \\ b_2 \\ b_3 \\ b_4 \\ b_5 \\ b_6 \\ b_7 \\ \vdots \\ b_{98} \\ b_{99} \end{Bmatrix} =
\begin{bmatrix} -0.01257714 \\ 0 \\ 0.01056269 \\ 0 \\ 0.01159497 \\ 0 \\ 0.00179381 \\ \vdots \\ 0 \\ 0.00000207 \end{bmatrix}
$$

$$\kappa_D = 0.017190, \quad \kappa_{DL} = 0.045569, \quad \kappa_{D\Omega} = 0.030200, \quad \kappa_{Do} = 0.000000$$

From Examples 1.8.1 and 1.8.2 it should be observed that all of the even Fourier coefficients, in both a_n and b_n, were identically zero for both the rectangular and tapered wings. This was realized regardless of how many terms were carried in the Fourier series and was a direct result of the spanwise symmetry of these two wings. If we were always willing to restrict our analysis to spanwise symmetric wing geometry, the computation time could be reduced by forcing all even coefficients to be zero and solving for the odd coefficients using sections distributed over only one side of the wing, i.e., $0 \le \theta \le \pi/2$. The requirement for a spanwise symmetric wing planform is not too restrictive. Most wings exhibit this symmetry. However, restricting our analysis to wings having a spanwise symmetric distribution in geometric and aerodynamic twist would eliminate the possibility of using lifting-line theory for the analysis of asymmetric control surface deflection. Such control surface deflection is almost always used in the wing of an airplane to provide roll control.

Solution with Control Surface Deflection and Rolling Rate
Trailing-edge flaps extending over only some portion of the wingspan are commonly used as control surfaces on an airplane wing. A spanwise symmetric control surface deflection can be used to provide pitch control, and spanwise asymmetric control surface deflection can be used to provide roll control. The control surfaces commonly used to provide roll control are called *ailerons*. These are small trailing-edge flaps located in the outboard sections of the wing. The ailerons are deflected asymmetrically to change the rolling moment. One aileron is deflected downward and the other is deflected upward, as

shown in Fig. 1.8.23. This increases lift on the semispan with the downward-deflected aileron and decreases lift on the semispan with the upward-deflected aileron. Aileron deflection is given the symbol δ_a, and the rolling moment is denoted as ℓ. The sign convention that is commonly used for aileron deflection is shown in Fig. 1.8.23. Aileron deflection is assumed positive when the right aileron is deflected downward and the left aileron is deflected upward. The traditional sign convention used for the rolling moment is positive to the right (i.e., a moment that would roll the right wing down). With these two sign conventions, a positive aileron deflection produces a rolling moment to the left (i.e., negative ℓ). The two ailerons are not necessarily deflected the same magnitude. The aileron angle, δ_a, is usually defined to be the average of the two angular deflections.

The rolling moment produced by aileron deflection results in a rolling acceleration, which in turn produces a rolling rate. This rolling rate changes the spanwise variation in geometric angle of attack as shown in Fig. 1.8.24. The symbol traditionally used to

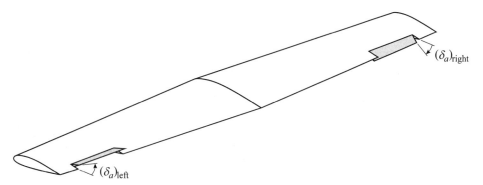

Figure 1.8.23. Asymmetric aileron deflection typically used for roll control.

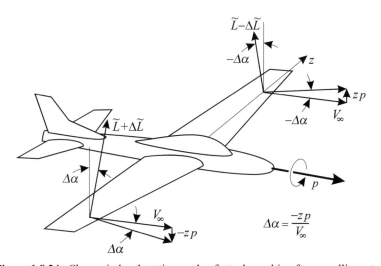

Figure 1.8.24. Change in local section angle of attack resulting from a rolling rate.

represent rolling rate is p. As shown in Fig. 1.8.24, p is taken to be positive when the right semispan is moving downward and the left semispan is moving upward. Thus, a positive rolling rate increases the local geometric angle of attack on the right semispan and decreases the local geometric angle of attack on the left semispan. This produces a negative rolling moment, which opposes the rolling rate.

When a small positive aileron deflection is first initiated, a negative rolling moment is produced by the asymmetric change in the wing's local aerodynamic angle of attack. This rolling moment imbalance results in negative rolling acceleration, which produces a negative rolling rate. As the magnitude of the rolling rate increases, an opposing rolling moment develops as a result of the asymmetric change in the local geometric angle of attack. This reduces the moment imbalance and slows the rolling acceleration. At some point, the positive rolling moment produced by the negative rolling rate will just balance the negative rolling moment produced by the positive aileron deflection, and a constant rolling rate will develop that is proportional to the aileron deflection.

As we will learn in Chapter 9, roll in a conventional airplane is what is referred to as *heavily damped motion*. This means that when a typical fixed-wing aircraft responds to aileron input, the period of rolling acceleration is very short and the airplane quickly approaches a steady rolling rate, which is proportional to aileron deflection. This gives the pilot the perception that aileron input commands the airplane's rolling rate. Thus, analysis of an airplane's roll response must include consideration of the rolling moment produced by the rolling rate as well as that produced by aileron deflection. Lifting-line theory provides the capability to do just that.

In general, a wing's spanwise variation in local aerodynamic angle of attack can be expressed as the value at the wing root plus the changes due to washout, control surface deflection, and rolling rate,

$$\alpha(z) - \alpha_{L0}(z) \equiv \left(\alpha - \alpha_{L0}\right)_{\text{root}} - \Omega\,\omega(z) + \delta_a\,\chi(z) - pz/V_\infty \qquad (1.8.44)$$

where Ω is defined to be the maximum total symmetric washout for the wing, geometric plus aerodynamic,

$$\Omega \equiv \left[\left(\alpha - \alpha_{L0}\right)_{\text{root}} - \left(\Delta\alpha - \Delta\alpha_{L0}\right)_{\text{max}}\right]_{\text{washout}} \qquad (1.8.45)$$

$\omega(z)$ is the symmetric washout distribution function,

$$\omega(z) \equiv \left[\frac{\Delta\alpha(z) - \Delta\alpha_{L0}(z) - \left(\alpha - \alpha_{L0}\right)_{\text{root}}}{\left(\Delta\alpha - \Delta\alpha_{L0}\right)_{\text{max}} - \left(\alpha - \alpha_{L0}\right)_{\text{root}}}\right]_{\text{washout}} \qquad (1.8.46)$$

and $\chi(z)$ is the control surface distribution function,

$$\chi(z) \equiv \left[\frac{\Delta\alpha(z) - \Delta\alpha_{L0}(z) - \left(\alpha - \alpha_{L0}\right)_{\text{root}}}{\delta_a}\right]_{\text{control}} \qquad (1.8.47)$$

For example, ailerons extending from the spanwise coordinate z_{ar} to z_{at} give

$$\chi(z) \equiv \begin{cases} 0, & z < -z_{at} \\ \mathcal{E}_f(z), & -z_{at} < z < -z_{ar} \\ 0, & -z_{ar} < z < z_{ar} \\ -\mathcal{E}_f(z), & z_{ar} < z < z_{at} \\ 0, & z > z_{at} \end{cases} \tag{1.8.48}$$

where $\mathcal{E}_f$ is the local section flap effectiveness.

Using the definition of θ from Eq. (1.8.3) together with Eq. (1.8.44) in the relation for the Fourier coefficients, A_n, specified by Eq. (1.8.4), gives

$$\sum_{n=1}^{N} A_n \left[\frac{4b}{\tilde{C}_{L,\alpha} c(\theta)} + \frac{n}{\sin(\theta)} \right] \sin(n\theta) = (\alpha - \alpha_{L0})_{\text{root}} - \Omega\,\omega(\theta) + \delta_a\,\chi(\theta) + \bar{p}\cos(\theta) \tag{1.8.49}$$

where $\bar{p}$ is a dimensionless rolling rate defined as $\bar{p} = pb/2V_\infty$. The Fourier coefficients in Eq. (1.8.49) can be conveniently written as

$$A_n = a_n(\alpha - \alpha_{L0})_{\text{root}} - b_n\Omega + c_n\delta_a + d_n\bar{p} \tag{1.8.50}$$

where

$$\sum_{n=1}^{N} a_n \left[\frac{4b}{\tilde{C}_{L,\alpha} c(\theta)} + \frac{n}{\sin(\theta)} \right] \sin(n\theta) = 1 \tag{1.8.51}$$

$$\sum_{n=1}^{N} b_n \left[\frac{4b}{\tilde{C}_{L,\alpha} c(\theta)} + \frac{n}{\sin(\theta)} \right] \sin(n\theta) = \omega(\theta) \tag{1.8.52}$$

$$\sum_{n=1}^{N} c_n \left[\frac{4b}{\tilde{C}_{L,\alpha} c(\theta)} + \frac{n}{\sin(\theta)} \right] \sin(n\theta) = \chi(\theta) \tag{1.8.53}$$

$$\sum_{n=1}^{N} d_n \left[\frac{4b}{\tilde{C}_{L,\alpha} c(\theta)} + \frac{n}{\sin(\theta)} \right] \sin(n\theta) = \cos(\theta) \tag{1.8.54}$$

Equations (1.8.51) and (1.8.52) are exactly Eqs. (1.8.20) and (1.8.21) and their solutions are obtained following Examples 1.8.1 and 1.8.2. The solutions to Eqs. (1.8.53) and (1.8.54) can be obtained in a similar manner and are both independent of angle of attack and rolling rate.

Once the Fourier coefficients are determined from Eqs. (1.8.50) through (1.8.54), the spanwise circulation distribution is known from Eq. (1.8.3) and the spanwise section lift distribution is given by

$$\tilde{L}(z) = \rho V_\infty \Gamma(z)$$

Thus, in view of Eq. (1.8.3), the rolling moment coefficient can be evaluated from

$$C_\ell = \frac{1}{\frac{1}{2}\rho V_\infty^2 Sb} \int\limits_{z=-b/2}^{b/2} \tilde{L}(z)z\,dz = \frac{2}{V_\infty Sb} \int\limits_{z=-b/2}^{b/2} \Gamma(z)z\,dz = -\frac{b^2}{S}\sum_{n=1}^{N} A_n \int\limits_{\theta=0}^{\pi} \sin(n\theta)\cos(\theta)\sin(\theta)d\theta$$

or after applying the trigonometric identity, $\sin(2\theta) = 2\sin(\theta)\cos(\theta)$, along with the definition of aspect ratio

$$C_\ell = -\frac{R_A}{2}\sum_{n=1}^{N} A_n \int\limits_{\theta=0}^{\pi} \sin(n\theta)\sin(2\theta)d\theta \tag{1.8.55}$$

The integral in Eq. (1.8.55) is evaluated from

$$\int\limits_{\theta=0}^{\pi} \sin(m\theta)\sin(n\theta)\,d\theta = \begin{cases} 0, & n \ne m \\ \pi/2, & n = m \end{cases} \tag{1.8.56}$$

After applying Eqs. (1.8.50) and (1.8.56), Eq. (1.8.55) becomes

$$C_\ell = -\frac{\pi R_A}{4}A_2 = -\frac{\pi R_A}{4}\left[a_2(\alpha - \alpha_{L0})_{\text{root}} - b_2\Omega + c_2\delta_a + d_2\bar{p}\right] \tag{1.8.57}$$

For a wing with a **spanwise symmetric planform and spanwise symmetric washout**, the solutions to Eqs. (1.8.51) and (1.8.52) give

$$a_n = b_n = 0, \quad n \text{ even} \tag{1.8.58}$$

and Eq. (1.8.57) reduces to

$$C_\ell = C_{\ell,\delta_a}\delta_a + C_{\ell,\bar{p}}\bar{p} \tag{1.8.59}$$

where

$$C_{\ell,\delta_a} = -\frac{\pi R_A}{4}c_2 \tag{1.8.60}$$

$$C_{\ell,\bar{p}} = -\frac{\pi R_A}{4}d_2 \tag{1.8.61}$$

The Fourier coefficients obtained from Eq. (1.8.53) depend on control surface geometry as well as the planform shape of the wing. Thus, the change in rolling moment coefficient with respect to aileron deflection depends on the size and shape of the ailerons and the wing planform. On the other hand, the Fourier coefficients that are evaluated from Eq. (1.8.54) are functions of only wing planform. For an elliptic planform, the spanwise variation in chord length is given by

$$c(y) = \frac{4b}{\pi R_A}\sqrt{1-(2y/b)^2} \quad \text{or} \quad c(\theta) = \frac{4b}{\pi R_A}\sin(\theta)$$

and Eq. (1.8.54) reduces to

$$\sum_{n=1}^{N} d_n \left(\frac{\pi R_A}{\tilde{C}_{L,\alpha}} + n \right) \sin(n\theta) = \sin(\theta)\cos(\theta) \tag{1.8.62}$$

The solution to Eq. (1.8.62) is given by the Fourier integral

$$d_n\left(\frac{\pi R_A}{\tilde{C}_{L,\alpha}} + n\right) = \frac{2}{\pi}\int_0^\pi \sin(\theta)\cos(\theta)\sin(n\theta)\,d\theta = \frac{1}{\pi}\int_0^\pi \sin(2\theta)\sin(n\theta)\,d\theta \tag{1.8.63}$$

which is readily evaluated from Eq. (1.8.56) to give

$$d_n = \begin{cases} \dfrac{\tilde{C}_{L,\alpha}}{2(\pi R_A + 2\tilde{C}_{L,\alpha})}, & n = 2 \\ 0, & n \neq 2 \end{cases} \tag{1.8.64}$$

Using Eq. (1.8.64) in Eq. (1.8.61) gives

$$C_{\ell,\bar{p}} = \frac{-\tilde{C}_{L,\alpha}}{8(1 + 2\tilde{C}_{L,\alpha}/\pi R_A)}$$

Thus, in view of Eq. (1.8.7), **the change in rolling moment coefficient with respect to dimensionless rolling rate for an elliptic wing** can be written as

$$C_{\ell,\bar{p}} = -\frac{K_{\ell\bar{p}}\, C_{L,\alpha}}{8}, \qquad K_{\ell\bar{p}} \equiv \frac{1 + \tilde{C}_{L,\alpha}/\pi R_A}{1 + 2\tilde{C}_{L,\alpha}/\pi R_A} \tag{1.8.65}$$

Similarly, in the general case of **a wing with arbitrary planform**,

$$C_{\ell,\bar{p}} = -\frac{K_{\ell\bar{p}}\, C_{L,\alpha}}{8}, \qquad K_{\ell\bar{p}} \equiv \frac{2d_2}{a_1} \tag{1.8.66}$$

Using an airfoil lift slope of 2π to determine a_1 and d_2 from Eqs. (1.8.51) and (1.8.54), the results shown in Fig. 1.8.25 are obtained for tapered wings.

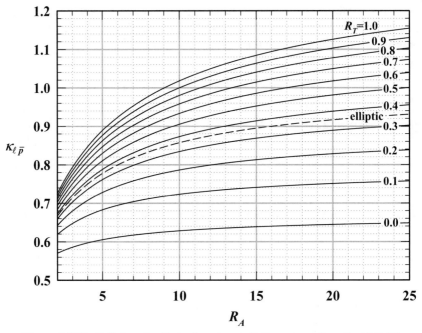

Figure 1.8.25. Roll damping factor for wings with linear taper from Eq. (1.8.66).

The asymmetric spanwise variations in aerodynamic angle of attack, which result from aileron deflection and roll, can also produce a yawing moment. The aerodynamic yawing moment is given the symbol n, and the traditional sign convention for yaw is positive to the right (i.e., a moment that would yaw the airplane's nose to the right). The yawing moment develops as a direct result of an asymmetric spanwise variation in drag. Thus, the yawing moment coefficient can be written as

$$C_n \equiv \frac{n}{\frac{1}{2}\rho V_\infty^2 Sb} = \frac{1}{\frac{1}{2}\rho V_\infty^2 Sb} \int_{z=-b/2}^{b/2} (-z)\widetilde{D}(z)\,dz \qquad (1.8.67)$$

As shown in Figs. 1.8.6 and 1.8.24, the asymmetric section drag results from tilting the section lift vector through the induced angle, α_i, and the roll angle, zp/V_∞. Thus, after expressing the section lift in terms of the section circulation, we have

$$\widetilde{D}(z) = \widetilde{L}\sin(\alpha_i + zp/V_\infty) \cong \widetilde{L}(\alpha_i + zp/V_\infty) = \rho V_\infty \Gamma(z)[\alpha_i(z) + zp/V_\infty] \quad (1.8.68)$$

The induced angle of attack as predicted by lifting-line theory is

$$\alpha_i(z) = \frac{1}{4\pi V_\infty} \int_{\zeta=-b/2}^{b/2} \frac{1}{z-\zeta}\left(\frac{d\Gamma}{dz}\right)_{z=\zeta} d\zeta \qquad (1.8.69)$$

Using Eqs. (1.8.68) and (1.8.69) in Eq. (1.8.67) gives

$$C_n = \int_{z=-b/2}^{b/2} \frac{-2z\Gamma(z)}{V_\infty^2 Sb} \left[\frac{1}{4\pi} \int_{\zeta=-b/2}^{b/2} \frac{1}{z-\zeta} \left(\frac{d\Gamma}{dz} \right)_{z=\zeta} d\zeta + zp \right] dz \qquad (1.8.70)$$

From Eq. (1.8.3) we have

$$\Gamma(\theta) = 2bV_\infty \sum_{n=1}^{N} A_n \sin(n\theta), \quad \cos(\theta) = -\frac{2z}{b}, \quad \frac{d\Gamma}{dz} = 4V_\infty \sum_{n=1}^{N} A_n \frac{n\cos(n\theta)}{\sin(\theta)}$$

and Eq. (1.8.70) can be written as

$$C_n = R_A \int_{\theta=0}^{\pi} \cos(\theta) \left[\sum_{n=1}^{N} A_n \sin(n\theta) \right] \left[\sum_{n=1}^{N} \frac{nA_n}{\pi} \int_{\phi=0}^{\pi} \frac{\cos(n\phi)d\phi}{\cos(\phi)-\cos(\theta)} \right] \sin(\theta)d\theta$$
$$- R_A \bar{p} \sum_{n=1}^{N} A_n \int_{\theta=0}^{\pi} \cos^2(\theta)\sin(\theta)\sin(n\theta)d\theta$$

Integrating in ϕ and using the trigonometric relation $\sin(2\theta) = 2\cos(\theta)\sin(\theta)$ yields

$$C_n = R_A \sum_{n=1}^{N} \sum_{m=1}^{N} nA_n A_m \int_{\theta=0}^{\pi} \cos(\theta)\sin(n\theta)\sin(m\theta)d\theta$$
$$- R_A \bar{p} \sum_{n=1}^{N} A_n \frac{1}{2} \int_{\theta=0}^{\pi} \cos(\theta)\sin(2\theta)\sin(n\theta)d\theta \qquad (1.8.71)$$

Since m and n are positive integers, the integrals with respect to θ are evaluated from

$$\int_{\theta=0}^{\pi} \cos(\theta)\sin(n\theta)\sin(m\theta)d\theta = \begin{cases} \pi/4, & m = n \pm 1 \\ 0, & m \neq n \pm 1 \end{cases}$$

and after some rearranging, we obtain

$$C_n = \frac{\pi R_A}{4} \sum_{n=2}^{N} (2n-1)A_{n-1}A_n - \frac{\pi R_A \bar{p}}{8}(A_1 + A_3) \qquad (1.8.72)$$

After applying Eq. (1.8.5), we find that the **yawing moment coefficient for a wing of arbitrary geometry** is given by

$$C_n = \frac{C_L}{8}(6A_2 - \bar{p}) + \frac{\pi R_A}{8}(10A_2 - \bar{p})A_3 + \frac{\pi R_A}{4} \sum_{n=4}^{N} (2n-1)A_{n-1}A_n \qquad (1.8.73)$$

where the Fourier coefficients, A_n, are related to Ω, δ_a, and $\bar{p}$ through Eq. (1.8.50).

If the wing planform and washout distribution are both spanwise symmetric, all even coefficients in both a_n and b_n are zero. The change in aerodynamic angle of attack that results from roll is always a spanwise odd function. Thus, all odd coefficients in d_n are zero for a spanwise symmetric wing. If the ailerons produce an equal and opposite change on each semispan, the control surface distribution function is a spanwise odd function as well and all odd coefficients in c_n are also zero. With this **wing symmetry** the Fourier coefficients must satisfy the relations $a_n = b_n = 0$ for n even and $c_n = d_n = 0$ for n odd. Thus, Eq. (1.8.50) reduces to

$$
A_n = \begin{cases} a_n(\alpha - \alpha_{L0})_{\text{root}} - b_n \Omega, & n \text{ odd} \\ c_n \delta_a + d_n \bar{p}, & n \text{ even} \end{cases} \tag{1.8.74}
$$

Using the Fourier coefficients from Eq. (1.8.74) in Eq. (1.8.73), the **yawing moment coefficient with the wing symmetry described above** is given by

$$
\begin{aligned}
C_n ={}& \frac{C_L}{8}[6c_2\delta_a - (1-6d_2)\bar{p}] + \frac{\pi R_A}{8}[10c_2\delta_a - (1-10d_2)\bar{p}][a_3(\alpha - \alpha_{L0})_{\text{root}} - b_3\Omega] \\
&+ \frac{\pi R_A}{4}\left[\sum_{n=4}^{N}(2n-1)[a_{n-1}(\alpha - \alpha_{L0})_{\text{root}} - b_{n-1}\Omega](c_n\delta_a + d_n\bar{p})\right]_{n \text{ even}} \\
&+ \frac{\pi R_A}{4}\left[\sum_{n=5}^{N}(2n-1)(c_{n-1}\delta_a + d_{n-1}\bar{p})[a_n(\alpha - \alpha_{L0})_{\text{root}} - b_n\Omega]\right]_{n \text{ odd}}
\end{aligned} \tag{1.8.75}
$$

For the special case of a **symmetric wing operating with optimum washout**, which is specified by Eq. (1.8.42), $a_n(\alpha - \alpha_{L0})_{\text{root}} - b_n\Omega$ is always zero for $n > 1$ and

$$
C_n = C_{n,\delta_a}\delta_a + C_{n,\bar{p}}\bar{p}, \quad C_{n,\delta_a} = \frac{3C_L}{4}c_2, \quad C_{n,\bar{p}} = -\frac{C_L}{8}(1-6d_2) \tag{1.8.76}
$$

For wings without optimum washout, the higher-order terms in Eq. (1.8.75) are not too large and the linear relation from Eq. (1.8.76) can be used as a rough approximation.

By comparing Eq. (1.8.59) with Eq. (1.8.76) we find that **within the accuracy of Eq. (1.8.76)**, the yawing moment can be expressed in terms of the rolling moment and the rolling rate,

$$
C_n = -\frac{3C_L}{\pi R_A}C_\ell - \frac{C_L}{8}\bar{p} \tag{1.8.77}
$$

When an airplane responds to aileron input, the rolling rate quickly reaches the steady-state value, which results in no rolling moment. Thus, we see from Eq. (1.8.77) that the magnitude of the steady yawing moment produced by aileron deflection is proportional to the lift coefficient developed by the wing and the steady rolling rate that develops as a result of the aileron deflection. From Eq. (1.8.77), we can also see that the sign of the yawing moment, which is induced on a wing by aileron deflection, is opposite to that of the rolling moment induced by the same aileron deflection. Positive aileron deflection induces a wing rolling moment to the left and a wing yawing moment to the right. For this reason, the yawing moment induced on the wing by aileron deflection is commonly called *adverse yaw*. We can also see from Eq. (1.8.77) that adverse yaw is more pronounced at low airspeeds, which require higher values of C_L.

EXAMPLE 1.8.3. The rectangular wing of aspect ratio 8.0, which was described in Example 1.8.1, has ailerons with sealed hinges that extend from the spanwise coordinate $z/b = \pm 0.25$ to $z/b = \pm 0.45$. The aileron chord length is constant and equal to 18 percent of the wing chord. Estimate the rolling and yawing moment coefficients that would be induced on this wing by a 5-degree deflection of these ailerons when the wing is operating at the design lift coefficient of 0.4 and there is no rolling rate. What steady dimensionless rolling rate could be sustained by this aileron deflection?

Solution. Since $c_f/c = 0.18$, from Fig. 1.7.4, the hinge efficiency for these sealed flaps is estimated to be 0.85. Because the flap deflection is less than 10 degrees, from Fig. 1.7.5, a deflection efficiency of 1.0 is used. Thus, by using Eqs. (1.7.5) and (1.7.6) in Eq. (1.7.13), the section flap effectiveness for these ailerons is constant and given by

$$\theta_f = \cos^{-1}\left(2\frac{c_f}{c} - 1\right) = 2.265, \quad \varepsilon_f = (0.85)(1.0)\left[1 - \frac{\theta_f - \sin(\theta_f)}{\pi}\right] = 0.445$$

With this result, the control surface distribution function from Eq. (1.8.48) is

$$\chi(z) \equiv \begin{cases} 0.000, & z/b < -0.45 \\ 0.445, & -0.45 < z/b < -0.25 \\ 0.000, & -0.25 < z/b < 0.25 \\ -0.445, & 0.25 < z/b < 0.45 \\ 0.000, & z/b > 0.45 \end{cases}$$

In view of Eq. (1.8.53), the Fourier coefficients, c_n, are obtained from

$$\begin{bmatrix} C_{11} & C_{12} & C_{13} & \cdots & C_{1N} \\ C_{21} & C_{22} & C_{23} & \cdots & C_{2N} \\ C_{31} & C_{32} & C_{33} & \cdots & C_{3N} \\ \vdots & \vdots & \vdots & \ddots & \vdots \\ C_{N1} & C_{N2} & C_{N3} & \cdots & C_{NN} \end{bmatrix} \begin{Bmatrix} c_1 \\ c_2 \\ c_3 \\ \vdots \\ c_N \end{Bmatrix} = \begin{Bmatrix} \chi(\theta_1) \\ \chi(\theta_2) \\ \chi(\theta_3) \\ \vdots \\ \chi(\theta_N) \end{Bmatrix}$$

where **[C]** is the same matrix as that used in Example 1.8.1. For $N = 99$, this gives

$$
\begin{Bmatrix} c_1 \\ c_2 \\ c_3 \\ c_4 \\ \vdots \\ c_{98} \\ c_{99} \end{Bmatrix} = [\mathbf{C}]^{-1} \begin{Bmatrix} \chi(\theta_1) \\ \chi(\theta_2) \\ \chi(\theta_3) \\ \chi(\theta_4) \\ \vdots \\ \chi(\theta_{98}) \\ \chi(\theta_{99}) \end{Bmatrix} = \begin{Bmatrix} 0 \\ 0.03853294 \\ 0 \\ 0.00335119 \\ \vdots \\ -0.00001777 \\ 0 \end{Bmatrix}
$$

From Eq. (1.8.60), the change in the rolling moment coefficient with respect to aileron deflection is

$$
C_{\ell,\delta_a} = -\frac{\pi R_A}{4} c_2 = -\frac{\pi(8.0)}{4} 0.0385 = -0.242
$$

and from Eq. (1.8.59) with no rolling rate,

$$
C_\ell = C_{\ell,\delta_a} \delta_a = -0.242 \frac{5.0\pi}{180} = \underline{-0.0211}
$$

Since this wing has optimum washout for this lift coefficient, from Eq. (1.8.77),

$$
C_n = -\frac{3C_L}{\pi R_A} C_\ell = -\frac{3(0.4)}{\pi(8.0)}(-0.0211) = \underline{0.00101}
$$

This result applies only for the case of optimum washout, which in this case requires 4.64 degrees of elliptic washout. For the same wing planform with no washout, carrying the higher-order terms using either Eq. (1.8.73) or (1.8.75) gives $C_n = 0.00123$, and with 4.5 degrees of linear washout we get $C_n = 0.00087$.

Similarly, the coefficients, d_n, are evaluated from Eq. (1.8.54) and the change in the rolling moment with respect to rolling rate is found from Eq. (1.8.61):

$$
\begin{Bmatrix} d_1 \\ d_2 \\ d_3 \\ d_4 \\ \vdots \\ d_{98} \\ d_{99} \end{Bmatrix} = [\mathbf{C}]^{-1} \begin{Bmatrix} \cos(\theta_1) \\ \cos(\theta_2) \\ \cos(\theta_3) \\ \cos(\theta_4) \\ \vdots \\ \cos(\theta_{98}) \\ \cos(\theta_{99}) \end{Bmatrix} = \begin{Bmatrix} 0 \\ 0.09411716 \\ 0 \\ 0.01326130 \\ \vdots \\ 0.00000241 \\ 0 \end{Bmatrix}, \quad C_{\ell,\bar{p}} = -\frac{\pi R_A}{4} d_2 = -0.591
$$

Setting the net rolling moment coefficient to zero, Eq. (1.8.59) is solved for the steady dimensionless rolling rate:

$$
\bar{p}_{steady} = -(C_{\ell,\delta_a}/C_{\ell,\bar{p}})\delta_a = -(-0.242/-0.591)(5.0\pi/180) = \underline{-0.0357}
$$

Wing Aspect Ratio and Geometric Mean Chord Length
We have seen that the aspect ratio of a finite wing has a profound effect on the wing's performance. As defined in Eq. (1.8.5), the aspect ratio for a wing of arbitrary planform can be computed as the square of the wingspan divided by the planform area,

$$R_A \equiv \frac{b^2}{S} \tag{1.8.78}$$

In a more general sense, aspect ratio is defined to be the ratio of the longer to the shorter dimension for any two-dimensional shape. The aspect ratio of a rectangle is simply the length of the long side divided by the length of the short side. Thus, the **aspect ratio of a rectangular wing** having a constant chord, c, and wingspan, b, can be defined as

$$R_A \equiv \frac{b}{c} \tag{1.8.79}$$

Because the planform area of a rectangular wing is simply the wingspan multiplied by the chord length, $S = bc$, the definitions given in Eqs. (1.8.78) and (1.8.79) are equivalent. For the more general case of a wing having a chord length that varies with the spanwise coordinate, z, the aspect ratio can be written as the ratio of the wingspan to the average or *mean chord length*, $\bar{c}$,

$$R_A \equiv \frac{b}{\bar{c}} \tag{1.8.80}$$

where, by equating the definitions of aspect ratio in Eqs. (1.8.78) and (1.8.80), we see that the **mean chord length** is defined as

$$\bar{c} \equiv \frac{S}{b} = \frac{1}{b} \int_{z=-b/2}^{b/2} c(z)\,dz \tag{1.8.81}$$

Because the wing aspect ratio defined by Eq. (1.8.78) has such a profound effect on wing performance, **both the wingspan and the mean chord length defined by Eq. (1.8.81) are important characteristic length scales for a finite wing**.

For a wing with linear taper, the planform area is one half the sum of the root chord and the tip chord multiplied by the wingspan. Thus, the mean chord length for a **wing with linear taper** is

$$\bar{c} = \frac{1}{b} \int_{z=-b/2}^{b/2} \left[c_{root} - (c_{root} - c_{tip}) \left| \frac{2z}{b} \right| \right] dz = \frac{c_{root} + c_{tip}}{2} = \frac{1+R_T}{2} c_{root} \tag{1.8.82}$$

For an **elliptic wing** the mean chord length is

$$\bar{c} = \frac{1}{b} \int_{z=-b/2}^{b/2} c_{root} \sqrt{1 - \left(\frac{2z}{b} \right)^2}\, dz = \frac{\pi}{4} c_{root} \tag{1.8.83}$$

Wing Camber and Mean Aerodynamic Chord

Because section lift does not contribute to the pitching moment about the lifting line of an unswept wing, the pitching moment coefficient about the wing's lifting line depends only on wing camber. From the definitions of wing and airfoil moment coefficients, the pitching moment coefficient about the origin of the coordinate system in Fig. 1.8.5 is

$$C_{m_0} = \frac{1}{S c_{ref}} \int_{z=-b/2}^{b/2} \widetilde{C}_{m_{ac}} c^2 dz \tag{1.8.84}$$

where $\widetilde{C}_{m_{ac}}$ is the section moment coefficient about the section aerodynamic center and c_{ref} is the reference chord length used to define the wing pitching moment coefficient. For **wings with no control surface deflection or any other form of aerodynamic twist**, the section moment coefficient is constant over the wingspan and Eq. (1.8.84) becomes

$$C_{m_0} = \frac{\widetilde{C}_{m_{ac}}}{S c_{ref}} \int_{z=-b/2}^{b/2} c^2 dz \tag{1.8.85}$$

Equation (1.8.85) leads to the definition of a characteristic length scale associated with wing camber, which is typically referred to as the **mean aerodynamic chord**

$$\overline{c}_{mac} \equiv \frac{1}{S} \int_{z=-b/2}^{b/2} c^2 dz \tag{1.8.86}$$

For **wings with linear taper** the mean aerodynamic chord and its location are given by

$$\overline{c}_{mac} = \frac{2}{3} \frac{1 + R_T + R_T^2}{1 + R_T} c_{root}, \text{ located at } \frac{z_{mac}}{b} = \frac{1}{6} \frac{1 + 2R_T}{1 + R_T} \tag{1.8.87}$$

For **wings of elliptic planform** Eq. (1.8.86) results in

$$\overline{c}_{mac} = \frac{8}{3\pi} c_{root}, \text{ located at } \frac{z_{mac}}{b} = \frac{1}{6\pi} \sqrt{9\pi^2 - 64} \tag{1.8.88}$$

Both $\overline{c}$ and $\overline{c}_{mac}$ are commonly used as the reference length for defining the wing pitching moment coefficient. In this textbook $\overline{c}$ is used for this purpose. This choice is completely arbitrary. To convert any moment coefficient based on $\overline{c}$ to that based on $\overline{c}_{mac}$, we simply multiply by the ratio $\overline{c}/\overline{c}_{mac}$, which is typically slightly less than unity. For example, the geometric relations for **wings with linear taper** produce the result

$$\overline{c}/\overline{c}_{mac} = 3(1 + R_T)^2 / [4(1 + R_T + R_T^2)] \tag{1.8.89}$$

which for taper ratios of 1.0 and 0.5 yields $\overline{c}/\overline{c}_{mac} = 1.00$ and $\overline{c}/\overline{c}_{mac} \cong 0.96$, respectively. For the **elliptic planform** the ratio $\overline{c}/\overline{c}_{mac}$ is

$$\overline{c}/\overline{c}_{mac} = 3\pi^2 / 32 \cong 0.93 \tag{1.8.90}$$

The Basic and Additional Section Lift Distributions
For many applications in aircraft design it is important to know how the lift is distributed over the span of a wing. From Eq. (1.8.3) combined with the vortex lifting law, lifting-line theory predicts that the spanwise variation in section lift for an unswept wing is

$$\tilde{L}(\theta) = \rho V_\infty \Gamma(\theta) = 2b\rho V_\infty^2 \sum_{n=1}^{\infty} A_n \sin(n\theta), \quad \theta = \cos^{-1}(-2z/b) \quad (1.8.91)$$

Thus, the spanwise variation in local section lift coefficient is given by

$$\tilde{C}_L(\theta) \equiv \frac{\tilde{L}(\theta)}{\frac{1}{2}\rho V_\infty^2 c(\theta)} = \frac{4b}{c(\theta)} \sum_{n=1}^{\infty} A_n \sin(n\theta) \quad (1.8.92)$$

The Fourier coefficients, A_n, in Eq. (1.8.92) can be written using the change of variables given by Eq. (1.8.19),

$$A_n \equiv a_n(\alpha - \alpha_{L0})_{\text{root}} - b_n\Omega$$

Furthermore, after applying Eqs. (1.8.26) and (1.8.28) to Eq. (1.8.24), we have

$$C_L = \pi R_A [a_1 (\alpha - \alpha_{L0})_{\text{root}} - b_1 \Omega]$$

or after solving for the root aerodynamic angle of attack

$$(\alpha - \alpha_{L0})_{\text{root}} = \frac{b_1}{a_1}\Omega + \frac{C_L}{\pi R_A a_1} \quad (1.8.93)$$

Using Eq. (1.8.93) in Eq. (1.8.19) yields

$$A_n = \left(\frac{b_1 a_n}{a_1} - b_n\right)\Omega + \frac{a_n}{\pi R_A a_1} C_L \quad (1.8.94)$$

Using this change of variables, Eq. (1.8.92) can be written as

$$\tilde{C}_L(\theta) = \Omega \sum_{n=1}^{\infty} 4\left(\frac{b_1 a_n}{a_1} - b_n\right)\frac{\sin(n\theta)}{c(\theta)/b} + C_L \sum_{n=1}^{\infty} \frac{4a_n}{\pi R_A a_1}\frac{\sin(n\theta)}{c(\theta)/b} \quad (1.8.95)$$

We see from Eq. (1.8.95) that the spanwise variation in local section lift coefficient can be divided conveniently into two components. The first term on the right-hand side of Eq. (1.8.95) is called the *basic section lift coefficient* and the second term is called the *additional section lift coefficient*. The basic section lift coefficient is independent of C_L and directly proportional to the total amount of wing twist, Ω. The additional section lift coefficient at any section of the wing is independent of wing twist and directly proportional to the net wing lift coefficient, C_L.

As can be seen from Eq. (1.8.95), **the basic section lift coefficient is the spanwise variation in local section lift coefficient that occurs when the total net lift developed by the wing is zero**. Examination of the first term on the right-hand side of Eq. (1.8.95) reveals that the basic section lift coefficient depends on all of the Fourier coefficients a_n and b_n. From Eq. (1.8.20) we see the Fourier coefficients a_n depend only on the wing planform. Equation (1.8.21) shows that the Fourier coefficients b_n depend on both the wing planform and the dimensionless twist distribution function, $\omega(\theta)$. Thus, **the spanwise variation in the basic section lift coefficient depends on wing planform and wing twist but is independent of the wing's angle of attack**.

Examination of the second term on the right-hand side of Eq. (1.8.95) discloses that the additional section lift coefficient depends only on the wing planform and the Fourier coefficients a_n. From Eq. (1.8.20) we have seen that the coefficients a_n do not depend on wing twist. Thus, for an unswept wing, Eq. (1.8.95) exposes the important fact that **the additional section lift coefficient is independent of wing twist**. Because the basic section lift coefficient is zero for an untwisted wing, we see that **the additional section lift coefficient is equivalent to the spanwise variation in local section lift coefficient that would be developed on an untwisted wing of the same planform operating at the same wing lift coefficient**.

Figure 1.8.26 shows how the net section lift coefficient and its two components obtained from Eq. (1.8.95) vary along the span of an unswept wing with linear taper of aspect ratio 8.0 and taper ratio 0.5. This figure shows the spanwise variation in wing section lift coefficient for several values of total linear twist with the net wing lift coefficient held constant at 1.0. Similar results are shown in Fig. 1.8.27 for three different values of wing lift coefficient with total linear twist held constant at 6 degrees. Notice that whereas the center of total lift on each semispan of this wing moves inboard

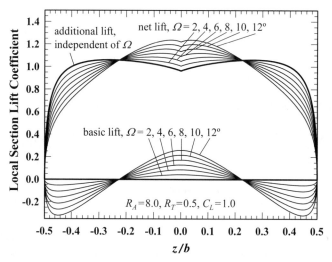

Figure 1.8.26. Spanwise variation in local section lift coefficient as a function of the total amount of linear twist with the net wing lift coefficient held constant at $C_L = 1.0$.

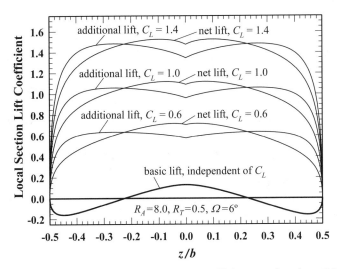

Figure 1.8.27. Spanwise variation in local section lift coefficient as a function of the net wing lift coefficient with the total amount of linear twist held constant at $\Omega = 6°$.

as washout is increased, the center of additional lift on each semispan does not change with either the amount of wing twist or the net wing lift coefficient.

Here $\widetilde{C}_{L_b}$ will be used to signify the **basic section lift coefficient**,

$$\widetilde{C}_{L_b}(\theta) \equiv \Omega \sum_{n=1}^{\infty} 4 \left(\frac{b_1 a_n}{a_1} - b_n \right) \frac{\sin(n\theta)}{c(\theta)/b} \tag{1.8.96}$$

and $\widetilde{C}_{L_a}$ will be used to denote the **additional section lift coefficient**,

$$\widetilde{C}_{L_a}(\theta) \equiv C_L \sum_{n=1}^{\infty} \frac{4 a_n}{\pi R_A a_1} \frac{\sin(n\theta)}{c(\theta)/b} \tag{1.8.97}$$

The total section lift coefficient at any spanwise section of the wing is simply the sum of the basic and additional section lift coefficients,

$$\widetilde{C}_L(\theta) = \widetilde{C}_{L_b}(\theta) + \widetilde{C}_{L_a}(\theta) \tag{1.8.98}$$

The important points to remember regarding these two components of lift are:

1. *The basic section lift coefficient is independent of angle of attack.*

2. *The additional section lift coefficient is independent of wing twist.*

3. *Both the basic and additional section lift coefficients depend on wing planform.*

Semispan Aerodynamic Center and Moment Components
The spanwise distribution of section aerodynamic loads acting on each semispan of a finite wing can be replaced with a resultant force acting at the aerodynamic center of the semispan and a resultant moment that does not change with angle of attack. Because drag is typically small compared with the lift, drag is commonly neglected in estimating the position of the aerodynamic center. This traditional approximation is presented here. The effect of drag on the position of the aerodynamic center is considered in Chapter 4.

As a first approximation, the aerodynamic center of each wing semispan is some-times assumed to be located at the section aerodynamic center of the airfoil section located at the spanwise coordinate of the semispan area centroid. Here the chord line that passes through the semispan area centroid is referred to as the *centroidal chord*. The spanwise coordinate of the wing semispan area centroid is given by

$$
\bar{z}_c \equiv \frac{2}{S} \int_{z=0}^{b/2} cz \, dz
\tag{1.8.99}
$$

For **wings with constant linear taper**, i.e., trapezoidal wings, Eq. (1.8.99) results in

$$
\frac{\bar{z}_c}{b} = \frac{1}{6} \frac{1 + 2R_T}{1 + R_T}
\tag{1.8.100}
$$

For **wings of elliptic planform**, the spanwise coordinate of the semispan centroid is

$$
\frac{\bar{z}_c}{b} = \frac{2}{3\pi}
\tag{1.8.101}
$$

The location specified by Eq. (1.8.100) is commonly referred to as the location of the mean aerodynamic chord. Referring to the centroidal chord of a trapezoidal wing as the mean aerodynamic chord can be misleading, because it could be taken to imply that the location of the mean aerodynamic chord is significant for other wing geometries as well. However, the mean aerodynamic chord passes through the semispan centroid only for the special case of a trapezoidal wing. For example, the mean aerodynamic chord of an elliptic wing is located according to Eq. (1.8.88) at $z_{mac}/b \cong 0.264$, whereas the location of the centroidal chord is given by Eq. (1.8.101) as $\bar{z}_c/b \cong 0.212$.

In general, the semispan aerodynamic center of a wing is not located along either the centroidal chord or the mean aerodynamic chord. For example, Fig. 1.8.28 shows the aerodynamic center, centroidal chord, and mean aerodynamic chord for several different semispan geometries. Equation (1.8.99) gives the true spanwise location of the semispan aerodynamic center only if the additional section lift coefficient is uniform across the wingspan. Because a uniform additional section lift coefficient is produced by an elliptic wing with no sweep or dihedral in the locus of airfoil section aerodynamic centers, the semispan aerodynamic center of such wings is located along the centroidal chord as specified by Eq. (1.8.101). However, wings with linear taper do not produce a uniform additional section lift coefficient. **Thus, Eq. (1.8.100) should be used only as a rough estimate for the semispan aerodynamic center of a trapezoidal wing.**

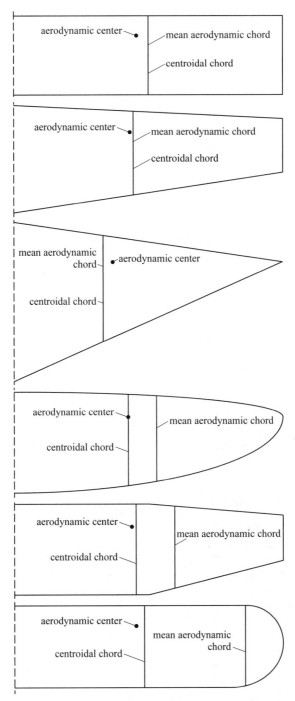

Figure 1.8.28. The aerodynamic center, centroidal chord, and mean aerodynamic chord for six different semispan geometries, all having the same aspect ratio and no quarter-chord sweep.

A more accurate estimate for the location of the semispan aerodynamic center of a wing with no sweep or dihedral in the locus of wing section aerodynamic centers can be obtained from Eqs. (1.8.96) through (1.8.98). Because we are neglecting drag, the resultant aerodynamic moment produced on each semispan of a wing about the origin of the coordinate system shown in Fig. 1.8.5 can be resolved into a pitching component about the z-axis and a rolling component about the freestream velocity vector.

Because section lift does not contribute to the pitching moment about the wing's lifting line, the contribution of the left semispan of an unswept wing to the pitching moment coefficient about the origin of the coordinate system shown in Fig. 1.8.5 is

$$(C_{m_0})_{\text{left}} = \frac{1}{S c_{\text{ref}}} \int_{z=0}^{b/2} \tilde{C}_{m_{ac}} c^2 dz \tag{1.8.102}$$

The moment coefficient specified by Eq. (1.8.102) is also the *root twisting moment coefficient* resulting from the aerodynamic load on this unswept wing. This moment results only from the effects of wing camber and is independent of geometric twist. For a wing with constant section pitching moment coefficient, Eq. (1.8.102) yields

$$(C_{m_0})_{\text{left}} = \frac{\bar{c}_{mac}}{c_{\text{ref}}} \frac{\tilde{C}_{m_{ac}}}{2} \tag{1.8.103}$$

In a more general sense, Eq. (1.8.102) could be thought of in terms of a mean section pitching moment coefficient,

$$(C_{m_0})_{\text{left}} = \frac{\bar{c}_{mac}}{c_{\text{ref}}} \frac{\bar{\bar{\tilde{C}}}_{m_{ac}}}{2}, \quad \bar{\bar{\tilde{C}}}_{m_{ac}} \equiv \frac{2}{S \bar{c}_{mac}} \int_{z=0}^{b/2} \tilde{C}_{m_{ac}} c^2 dz \tag{1.8.104}$$

The contribution of the left wing semispan to the rolling moment coefficient about the origin of the coordinate system shown in Fig. 1.8.5 is

$$(C_{\ell_0})_{\text{left}} = \frac{1}{Sb} \int_{z=0}^{b/2} \tilde{C}_L c z \, dz = \frac{1}{Sb} \int_{z=0}^{b/2} (\tilde{C}_{L_b} + \tilde{C}_{L_a}) c z \, dz \tag{1.8.105}$$

It is important to note that within the small-angle approximation, **the moment coefficient specified by Eq. (1.8.105) is the *root bending moment coefficient* resulting from the aerodynamic load on the wing semispan.** Equating the distributed wing-section loading to a resultant force and moment acting at the aerodynamic center of the wing semispan, we can also write

$$(C_{\ell_0})_{\text{left}} = (C_{\ell_{ac}})_{\text{left}} + \frac{\bar{z}_{ac}}{Sb} \int_{z=0}^{b/2} \tilde{C}_L c \, dz = (C_{\ell_{ac}})_{\text{left}} + \frac{\bar{z}_{ac}}{b} \frac{C_L}{2} \tag{1.8.106}$$

where $(C_{\ell_{ac}})_{\text{left}}$ is the left wing semispan rolling moment coefficient about the semispan aerodynamic center and $\bar{z}_{ac}$ is the z-coordinate of the semispan aerodynamic center. Combining Eqs. (1.8.105) and (1.8.106) to eliminate the moment about the origin yields

$$(C_{\ell_{ac}})_{\text{left}} + \frac{\overline{z}_{ac}}{b}\frac{C_L}{2} = \frac{1}{Sb}\int\limits_{z=0}^{b/2}(\widetilde{C}_{L_b} + \widetilde{C}_{L_a})cz\,dz \qquad (1.8.107)$$

Because the resultant moment about the aerodynamic center is invariant to small changes in angle of attack, differentiating Eq. (1.8.107) with respect to angle of attack, applying Eqs. (1.8.96) and (1.8.97), and solving for $\overline{z}_{ac}/b$ gives

$$\frac{\overline{z}_{ac}}{b} = \frac{2}{SbC_{L,\alpha}}\frac{\partial}{\partial\alpha}\int\limits_{z=0}^{b/2}(\widetilde{C}_{L_b}+\widetilde{C}_{L_a})cz\,dz = \frac{2}{SbC_{L,\alpha}}\frac{\partial}{\partial\alpha}\int\limits_{z=0}^{b/2}\widetilde{C}_{L_a}cz\,dz$$

$$= \frac{-2}{\pi}\sum_{n=1}^{\infty}\frac{a_n}{a_1}\int\limits_{\theta=\pi/2}^{\pi}\sin(n\theta)\cos\theta\sin\theta\,d\theta = \frac{1}{\pi}\sum_{n=1}^{\infty}\frac{a_n}{a_1}\int\limits_{\theta=\pi}^{\pi/2}\sin(n\theta)\sin(2\theta)\,d\theta \qquad (1.8.108)$$

Because the additional section lift coefficient is independent of wing twist, Eq. (1.8.108) discloses the important fact that **the spanwise position of the aerodynamic center of the wing semispan is not affected by wing twist.** Recognizing that the even terms in a_n are always zero for spanwise symmetric wings, the integration in Eq. (1.8.108) yields

$$\frac{\overline{z}_{ac}}{b} = \frac{2}{3\pi} + \frac{2}{\pi}\sum_{n=3}^{\infty}\frac{\sin[(n-2)\pi/2]}{n^2-4}\frac{a_n}{a_1} = \frac{2}{3\pi}\left(1+\sum_{n=1}^{\infty}\frac{(-1)^{n-1}3}{4n^2+4n-3}\frac{a_{2n+1}}{a_1}\right) \qquad (1.8.109)$$

Figure 1.8.29 shows how $\overline{z}_{ac}/b$ varies with taper ratio and aspect ratio for wings with linear taper. The reader should note from Fig. 1.8.29 that except for the special case of trapezoidal wings with a taper ratio near 0.35, **the spanwise location of the mean aerodynamic chord is not the semispan aerodynamic center**, as is commonly stated or implied in the literature. For example, with a linear taper ratio of 0.8 the mean aerodynamic chord is located at $z_{mac}/b \cong 0.24$, whereas the semispan aerodynamic center varies from about $\overline{z}_{ac}/b \cong 0.22$ to $\overline{z}_{ac}/b \cong 0.23$, depending on aspect ratio. Furthermore, the location of the semispan aerodynamic center also varies with wing sweep, whereas the spanwise location of the mean aerodynamic chord does not.

Using Eq. (1.8.97) in Eq. (1.8.108), it can be shown that the spanwise coordinate of the semispan aerodynamic center can also be expressed as

$$\overline{z}_{ac} = \frac{2}{SC_L}\int\limits_{z=0}^{b/2}\widetilde{C}_{L_a}cz\,dz \qquad (1.8.110)$$

Using Eq. (1.8.110) in Eq. (1.8.105) yields

$$(C_{\ell_0})_{\text{left}} = \frac{1}{Sb}\int\limits_{z=0}^{b/2}\widetilde{C}_{L_a}cz\,dz + \frac{1}{Sb}\int\limits_{z=0}^{b/2}\widetilde{C}_{L_b}cz\,dz$$

$$= \frac{\overline{z}_{ac}}{b}\frac{C_L}{2} + \frac{1}{Sb}\int\limits_{z=0}^{b/2}\widetilde{C}_{L_b}cz\,dz \qquad (1.8.111)$$

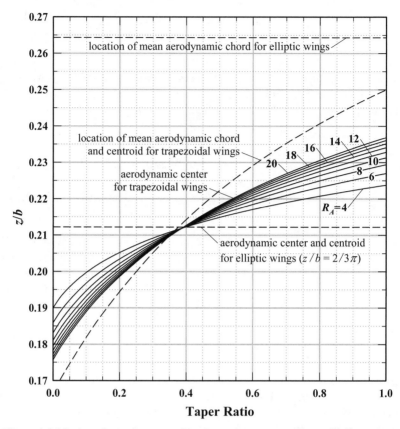

Figure 1.8.29. Aerodynamic center of semispan for unswept wings with linear taper.

For wings with no sweep or dihedral, the integral of the basic section lift coefficient on the right-hand side of Eq. (1.8.111) can be evaluated from Eq. (1.8.96). Following a development similar to that of Eq. (1.8.109) it is readily shown that for spanwise symmetric wings with spanwise symmetric twist, using Eq. (1.8.96) in Eq. (1.8.111) produces a useful relation for the **root wing bending moment coefficient**

$$(C_{\ell_0})_{\text{left}} = \frac{\bar{z}_{ac} C_L}{2b} - \kappa_{M\Omega} \frac{C_{L,\alpha} \Omega}{2} \qquad (1.8.112)$$

where

$$\kappa_{M\Omega} \equiv \frac{2}{\pi} \frac{b_1}{a_1} \sum_{n=3}^{\infty} \frac{\sin[(n-2)\pi/2]}{n^2 - 4} \left(\frac{b_n}{b_1} - \frac{a_n}{a_1} \right)$$
$$= \frac{2}{\pi} \frac{b_1}{a_1} \sum_{n=1}^{\infty} \frac{(-1)^{n-1}}{4n^2 + 4n - 3} \left(\frac{b_{2n+1}}{b_1} - \frac{a_{2n+1}}{a_1} \right) \qquad (1.8.113)$$

It is worth noting that the infinite series defined in Eq. (1.8.113) is dominated by the first term. In the case of a rectangular wing of aspect ratio 6.0, carrying only the first term on the right-hand side of Eq. (1.8.113) yields $\kappa_{M\Omega} = 0.024338$, whereas carrying 400 or more terms in this infinite series produces $\kappa_{M\Omega} = 0.024986$. For typical washout distributions $\kappa_{M\Omega}$ is positive. Figure 1.8.30 shows how $\kappa_{M\Omega}$ varies with taper ratio and aspect ratio for wings with constant linear taper and constant linear twist.

As expressed in Eq. (1.8.112), the root wing bending moment coefficient is composed of two components. The first proportional to the semispan lift acting through a moment arm of $\bar{z}_{ac}$ and the second is proportional to the product of the wing lift slope and the wing twist. For a given wing planform, the value of the proportionality constant $\kappa_{M\Omega}$ depends on the way in which the twist is distributed along the wingspan. This dependence enters into Eq. (1.8.113) through the Fourier coefficients b_n, which depend on the twist distribution through Eq. (1.8.21). For the typical case where washout is greatest near the wingtips, $\kappa_{M\Omega}$ is positive, as shown for the case of linear washout in Fig. 1.8.30. Thus, as might be expected, Eq. (1.8.112) shows that the root bending moment decreases linearly as washout is added at the wingtips. If a twist distribution were used that had the greatest washout near the wing root, then $\kappa_{M\Omega}$ would be negative and the root bending moment would increase in proportion to the amount of twist.

For a given planform and twist distribution, Eq. (1.8.112) shows that the change in bending moment with respect to Ω is directly proportional to the wing lift slope. This should be expected because $C_{L,\alpha}$ is a measure of the wing's lift response to a change in any aerodynamic angle, i.e., α, α_{L0}, or Ω. The lift slope for an unswept wing of arbitrary planform is given by Eq. (1.8.26) and is not affected by wing twist.

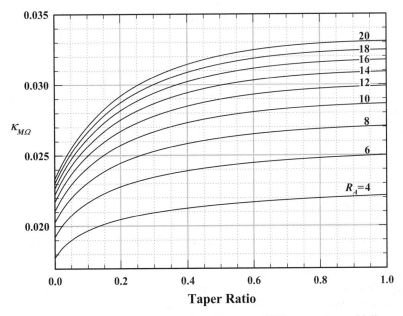

Figure 1.8.30. Twist factor for the root bending moment coefficient on wings with linear twist.

An Alternative View of Downwash

From Prandtl's lifting-line theory we have seen that a finite lifting wing results in the generation of downwash. This downwash has a significant effect on the performance of the wing and other surfaces of an airplane. Mathematically, we have described the downwash and its effects in terms of vorticity and circulation. However, the student may find it easier to understand the physics of downwash when it is described in terms of pressure and momentum.

When an airplane flies overhead, the downwash produced by the lifting wing could be measured on the ground if instruments of sufficient sensitivity were available. This downwash can be viewed as a result of the momentum imparted to the air by the force of the airplane's weight. The weight of the airplane is carried directly by the pressure difference between the upper and lower surfaces of the wing. However, the air also feels this pressure difference. The net effect of the pressure difference on the air is that a force is applied to the air equal to the airplane's weight. From Newton's second law, we know that such a force must result in a downward momentum imparted to the air. As the air directly beneath the airplane moves down, viscous forces from the adjacent air may slow its downward motion. However, these same viscous forces act on the adjacent air to impart downward momentum to that air as well. Newton's second law requires that the total downward momentum of the air cannot change unless acted upon by an external force. Ultimately, the Earth will stop this downward momentum, which began as the downwash from the airplane wing. The final manifestation of the downwash produced by a lifting wing is a very small pressure rise over a very large area of the Earth. The net effect of this pressure rise integrated over the total affected area is a force applied to the surface of the Earth equal to the airplane weight. Whether the airplane is resting on the ground or flying at 40,000 feet, the surface of the Earth must ultimately support the total weight of the aircraft. Because the airplane is moving rapidly over the ground, the downwash is spread over a large area of the Earth and usually goes undetected.

The vortex model that we have used to describe the effects of downwash on a finite wing is based on inviscid flow. This model predicts that the vorticity shed from the wing and the associated downwash will extend an infinite distance behind the wing. In reality, viscous effects in the air will eventually dissipate the wingtip vortices. However, the net downward momentum associated with these vortices does not change as a result of this viscous dissipation. No matter how large the viscosity, this downward momentum will eventually reach the ground, where it produces an aerodynamic force equal to the airplane's weight.

1.9. Flow over Multiple Lifting Surfaces

Prandtl's classical lifting-line equation, expressed in Eq. (1.8.2), applies only to a single lifting surface with no sweep or dihedral. In the development of this relation it was assumed that the only downwash induced on the wing was that induced by the trailing vortex sheet, which is shed from the wing itself. Airplanes are usually designed with more than one lifting surface. A typical configuration could be a wing combined with horizontal and vertical stabilizers. Each of these surfaces can generate lift and vorticity. The vorticity that is generated by one lifting surface will alter the flow about another.

For example, a lifting wing will induce downwash on an aft horizontal stabilizer, and a lifting aft stabilizer will induce upwash on the main wing. Such lifting surface interactions are not accounted for in Eq. (1.8.2).

Even a flying wing cannot be analyzed using Eq. (1.8.2) if the wing has sweep or dihedral. In a wing with sweep and/or dihedral, the quarter-chord line on one semispan makes an angle with the quarter-chord line of the other semispan, as show in Fig. 1.9.1. A wing is said to have *positive sweep* if the wing is swept back from the root to the tip, with the wingtips being aft of the root. A wing has *positive dihedral* if the wing slopes up from the root to the tip, with the wingtips being above the root. *Negative dihedral* is usually called *anhedral*. Because a wing can also have taper and/or geometric twist, defining the sweep and dihedral angles can be somewhat ambiguous. For a conventional tapered wing, the sweep angle for the leading edge is greater than that for the trailing edge. For a wing with geometric washout, the dihedral angle for the trailing edge is greater than that for the leading edge. Furthermore, neither the sweep nor the dihedral must remain constant over the semispan of a lifting surface. For this reason, local sweep and dihedral angles are defined in terms of the wing quarter-chord line and the orientation of the local streamwise airfoil section, as shown in Fig. 1.9.1.

When constant dihedral is added to a wing, each side of the wing is rotated about the root as a solid body. The rotation of the left-hand side is opposite to that of the right-hand side, so that both wingtips are raised above the root and brought closer together. Since dihedral is a solid-body rotation, the local airfoil sections rotate with the quarter-chord line. For a wing with no dihedral, a vector drawn normal to a local airfoil section is also normal to the aircraft plane of symmetry. As dihedral is added to the wing, both the quarter-chord line and the airfoil section normal are rotated upward, so that the normal to the local airfoil section always forms an angle equal to the dihedral angle with the normal to the aircraft plane of symmetry. For example, the vertical stabilizer of a conventional airplane usually has a dihedral angle of 90 degrees. For the rarely encountered case of an aircraft that does not have a plane of symmetry, the dihedral angle must be defined relative to some other defined reference plane.

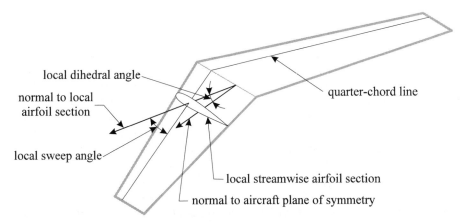

Figure 1.9.1. Local sweep and dihedral angles.

Typically, for an airplane with fixed wings, constant sweep is not a solid-body rotation of each semispan. For a wing with no sweep, the normal to each local airfoil section is aligned with the wing quarter-chord line, and the quarter-chord line is straight. As constant positive sweep is added to a wing, the quarter-chord line for each semispan is rotated back, moving the wingtips aft of the root. However, for constant planform area and aspect ratio, the local airfoil sections are not rotated with the quarter-chord line. Each airfoil section is translated straight back to the new quarter-chord location and the length of the quarter-chord line is increased, so that the distance from wingtip to wingtip does not change. A local sweep angle can be defined as the angle between the normal to the streamwise airfoil section and the quarter-chord line. Thus, the sweep angle for a wing with no dihedral is measured relative to a horizontal line, while the sweep angle for a vertical stabilizer with 90 degrees dihedral is measured relative to a vertical line.

Supersonic airplanes are sometimes designed with wings that can be rotated to vary the sweep angle in flight. This type of sweep variation is solid-body rotation. However, varying sweep in this manner changes not only the sweep but the shape of the streamwise airfoil section, the wingspan, and usually even the planform area of the wing.

Most airplanes are designed with some sweep and/or dihedral in the wing and/or the horizontal and vertical stabilizers. While the reasons for using sweep and dihedral will be left for a later discussion, it is clear from looking at the myriad of airplanes parked at any airport that a means for analyzing the effects of sweep and dihedral is needed. In addition, the downwash created by one lifting surface, such as the wing, has a dramatic effect on the performance of other lifting surfaces, such as the horizontal stabilizer. Thus, a means for predicting the effects of such interactions is also needed.

To predict the aerodynamic forces and moments acting on a complete airplane in flight, three-dimensional panel codes and CFD analysis are often used. However, these methods are very computationally intensive and may not be suitable for use in an undergraduate course on flight mechanics. An alternative method that provides an excellent educational tool at very low computational cost is the numerical lifting-line method, which is presented here.

The Numerical Lifting-Line Method

In developing Prandtl's lifting-line equation, a single lifting surface with no sweep or dihedral was assumed. The lifting line was confined to the z-axis and the trailing vortex sheet was assumed to be in the x-z plane (see Fig. 1.8.5). This significantly simplified the expressions for downwash and induced angle of attack. In order to use lifting-line theory to study the effects of sweep, dihedral, and/or the interactions between lifting surfaces, the theory must be generalized to allow for an arbitrary position and orientation of both the lifting line and the trailing vortex sheet. Here we shall examine a numerical lifting-line method that can be used to obtain a potential flow solution for the forces and moments acting on a system of lifting surfaces, each with arbitrary position and orientation. For a detailed development of this numerical method and a comparison with panel codes, CFD, and experimental data, see Phillips and Snyder (2000).

A first-order numerical lifting-line method can be obtained by synthesizing a system of lifting surfaces using a composite of horseshoe-shaped vortices. The continuous distribution of bound vorticity over the surface of each lifting surface, as well as the continuous distribution of free vorticity in the trailing vortex sheets, is approximated by a finite number of discrete horseshoe vortices, as shown in Fig. 1.9.2.

The bound portion of each horseshoe vortex is placed coincident with the *locus of wing section aerodynamic centers* called the *lifting line*, which is aligned with the local sweep and dihedral. As a first approximation, the lifting line is usually assumed to be the wing quarter-chord line. The trailing portion of each horseshoe vortex is aligned with the trailing vortex sheet. In Fig. 1.9.2, a small gap is shown between the left-hand trailing segment of one horseshoe vortex and the right-hand trailing segment of the next. This is for display purposes only. In reality, the left-hand corner of one horseshoe and the right-hand corner of the next are both placed at the same point. Thus, except at the wingtips, each trailing vortex segment is coincident with another trailing segment from the adjacent vortex. If two adjacent vortices have exactly the same strength, then the two coincident trailing segments exactly cancel, since one has clockwise rotation and the other has counterclockwise rotation. The net vorticity that is shed from the wing at any internal node is simply the difference in strength of the two adjacent vortices that share the node.

Each horseshoe vortex is composed of three straight segments, a finite bound segment and two semi-infinite trailing segments. From Eqs. (1.5.18) and (1.5.19), we can calculate the velocity induced at an arbitrary point in space (x, y, z) by a general horseshoe vortex. As shown in Fig. 1.9.3, a general horseshoe vortex is completely defined by two nodal points, (x_1, y_1, z_1) and (x_2, y_2, z_2), a trailing unit vector, $\mathbf{u}_\infty$, and a vortex strength, Γ. The horseshoe vortex starts at the fluid boundary, an infinite distance downstream. The inbound trailing vortex segment is directed along the vector $-\mathbf{u}_\infty$ to node 1 at (x_1, y_1, z_1). The bound vortex segment is directed along the wing lifting line from node 1 to node 2 at (x_2, y_2, z_2). The outbound trailing vortex segment is directed along the vector $\mathbf{u}_\infty$ from node 2 to the fluid boundary, an infinite distance back downstream. The velocity induced by the entire horseshoe vortex is simply the vector sum of the velocities induced by each of the three linear segments that make up the horseshoe.

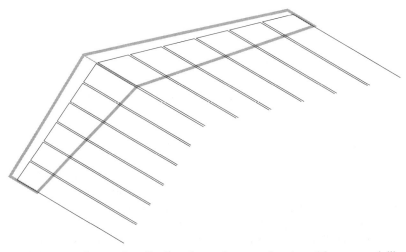

Figure 1.9.2. Horseshoe vortices distributed over the span of a wing with sweep and dihedral.

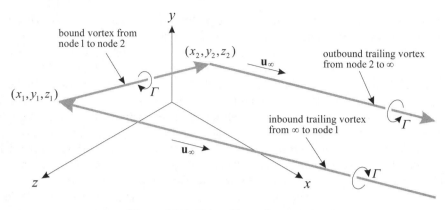

Figure 1.9.3. General horseshoe vortex.

In Eq. (1.5.19) the vorticity vector is assumed to point in the direction of $\mathbf{u}_\infty$ as determined by the right-hand rule. However, the inbound trailing vortex segment can be treated like an outbound segment with negative circulation. Thus, from Eqs. (1.5.18) and (1.5.19), the **velocity vector induced at an arbitrary point in space by a complete horseshoe vortex** is

$$\mathbf{V} = \frac{\Gamma}{4\pi}\left[\frac{\mathbf{u}_\infty \times \mathbf{r}_2}{r_2(r_2 - \mathbf{u}_\infty \cdot \mathbf{r}_2)} + \frac{(r_1 + r_2)(\mathbf{r}_1 \times \mathbf{r}_2)}{r_1 r_2(r_1 r_2 + \mathbf{r}_1 \cdot \mathbf{r}_2)} - \frac{\mathbf{u}_\infty \times \mathbf{r}_1}{r_1(r_1 - \mathbf{u}_\infty \cdot \mathbf{r}_1)}\right] \tag{1.9.1}$$

where $\mathbf{u}_\infty$ is the unit vector in the direction of the trailing vortex sheet, $\mathbf{r}_1$ is the spatial vector from (x_1, y_1, z_1) to (x, y, z), and $\mathbf{r}_2$ is the spatial vector from (x_2, y_2, z_2) to (x, y, z),

$$\mathbf{r}_1 = (x - x_1)\mathbf{i}_x + (y - y_1)\mathbf{i}_y + (z - z_1)\mathbf{i}_z \tag{1.9.2}$$

$$\mathbf{r}_2 = (x - x_2)\mathbf{i}_x + (y - y_2)\mathbf{i}_y + (z - z_2)\mathbf{i}_z \tag{1.9.3}$$

In obtaining the classical lifting-line solution for a single wing without sweep or dihedral, the trailing vortex sheet was assumed to be aligned with the wing chord. This was done to facilitate obtaining an analytical solution. In obtaining a numerical solution, there is little advantage in aligning the trailing vortex sheet with a vehicle axis such as the chord line. More correctly, the trailing vortex sheet should be aligned with the freestream. This is easily done in the numerical solution by setting $\mathbf{u}_\infty$ equal to the unit vector in the direction of the freestream. When using this method to predict the forces and moments on a single lifting surface, there is very little difference between the results obtained from slightly different orientations of the trailing vortex sheet.

When a system of lifting surfaces is synthesized using N of these horseshoe vortices, in a manner similar to that shown in Fig. 1.9.2, Eq. (1.9.1) can be used to determine the resultant velocity induced at any point in space if the vortex strength of each horseshoe vortex is known. However, these strengths are not known a priori. To compute the strengths of the N vortices, we must have a system of N equations relating these N strengths to some known properties of the wing. For these relations we turn to the three-dimensional vortex lifting law given in Eq. (1.8.1).

If flow over a finite wing is synthesized from a uniform flow combined with horseshoe vortices placed along the locus of wing section aerodynamic centers, from Eq. (1.9.1), the local velocity induced at a control point placed anywhere along the bound segment of horseshoe vortex i is

$$\mathbf{V}_i = \mathbf{V}_\infty + \sum_{j=1}^{N} \frac{\Gamma_j \mathbf{v}_{ji}}{\overline{c}_j} \tag{1.9.4}$$

where $\mathbf{V}_\infty$ is the velocity of the uniform flow, Γ_j is the strength of horseshoe vortex j, $\mathbf{v}_{ji}$ is a dimensionless velocity that would be induced at control point i by horseshoe vortex j, having a unit strength

$$\mathbf{v}_{ji} = \begin{cases} \dfrac{\overline{c}_j}{4\pi} \left[\dfrac{\mathbf{u}_\infty \times \mathbf{r}_{j2i}}{r_{j2i}(r_{j2i} - \mathbf{u}_\infty \cdot \mathbf{r}_{j2i})} + \dfrac{(r_{j1i} + r_{j2i})(\mathbf{r}_{j1i} \times \mathbf{r}_{j2i})}{r_{j1i} r_{j2i}(r_{j1i} r_{j2i} + \mathbf{r}_{j1i} \cdot \mathbf{r}_{j2i})} - \dfrac{\mathbf{u}_\infty \times \mathbf{r}_{j1i}}{r_{j1i}(r_{j1i} - \mathbf{u}_\infty \cdot \mathbf{r}_{j1i})} \right], & j \neq i \\[4mm] \dfrac{\overline{c}_j}{4\pi} \left[\dfrac{\mathbf{u}_\infty \times \mathbf{r}_{j2i}}{r_{j2i}(r_{j2i} - \mathbf{u}_\infty \cdot \mathbf{r}_{j2i})} - \dfrac{\mathbf{u}_\infty \times \mathbf{r}_{j1i}}{r_{j1i}(r_{j1i} - \mathbf{u}_\infty \cdot \mathbf{r}_{j1i})} \right], & j = i \end{cases} \tag{1.9.5}$$

$\mathbf{r}_{j1i}$ is the spatial vector from node 1 of horseshoe vortex j to the control point of horseshoe vortex i, $\mathbf{r}_{j2i}$ is the spatial vector from node 2 of horseshoe vortex j to the control point of horseshoe vortex i, and $\mathbf{u}_\infty$ is the unit vector in the direction of the freestream. At this point, $\overline{c}_j$ could be any characteristic length associated with the wing section aligned with horseshoe vortex j. This characteristic length is simply used to nondimensionalize Eq. (1.9.5) and has no effect on the induced velocity. An appropriate choice for $\overline{c}_j$ will be addressed at a later point. The bound vortex segment is excluded from Eq. (1.9.5) when $j = i$, because a straight vortex segment induces no downwash along its own length. However, the second term in Eq. (1.9.5), for $j \neq i$, is indeterminate when used with $j = i$, because $r_{j1i} r_{j2i} + \mathbf{r}_{j1i} \cdot \mathbf{r}_{j2i} = 0$.

From Eqs. (1.8.1) and (1.9.4), the aerodynamic force acting on a spanwise differential section of the wing located at control point i is given by

$$d\mathbf{F}_i = \rho \Gamma_i \left(\mathbf{V}_\infty + \sum_{j=1}^{N} \frac{\Gamma_j}{\overline{c}_j} \mathbf{v}_{ji} \right) \times d\boldsymbol{l}_i \tag{1.9.6}$$

Allowing for the possibility of flap deflection, the local section lift coefficient for the airfoil section located at control point i is a function of local angle of attack and local flap deflection,

$$\widetilde{C}_{Li} = \widetilde{C}_{Li}(\alpha_i, \delta_i) \tag{1.9.7}$$

where $\widetilde{C}_{Li}$, α_i, and δ_i are, respectively, the local airfoil section lift coefficient, the local angle of attack, and the local flap deflection, all evaluated for the airfoil section aligned with control point i. Defining $\mathbf{u}_{ni}$ and $\mathbf{u}_{ai}$ to be the local unit normal and axial vectors for the airfoil section located at control point i, as shown in Fig. 1.9.4, the local angle of attack at control point i can be written as

$$\alpha_i = \tan^{-1}\left(\frac{\mathbf{V}_i \cdot \mathbf{u}_{ni}}{\mathbf{V}_i \cdot \mathbf{u}_{ai}}\right) \tag{1.9.8}$$

If the relation implied by Eq. (1.9.7) is known at each section of the wing, the magnitude of the aerodynamic force acting on a spanwise differential section of the wing located at control point i can be written as

$$\left|\mathbf{dF}_i\right| = \tfrac{1}{2}\rho V_\infty^2\, \widetilde{C}_{Li}(\alpha_i,\delta_i)\, dS_i \tag{1.9.9}$$

where dS_i is a spanwise differential planform area element located at control point i. Setting the magnitude of the force obtained from Eq. (1.9.6) equal to that obtained from Eq. (1.9.9), applying Eq. (1.9.4) to Eq. (1.9.8), and rearranging, we can write

$$2\left|\left(\mathbf{u}_\infty + \sum_{j=1}^{N}\mathbf{v}_{ji}\, G_j\right)\times\boldsymbol{\zeta}_i\right| G_i - \widetilde{C}_{Li}(\alpha_i,\delta_i) = 0 \tag{1.9.10}$$

$$\mathbf{u}_\infty \equiv \frac{\mathbf{V}_\infty}{V_\infty}, \qquad \boldsymbol{\zeta}_i \equiv \overline{c}_i\frac{\mathbf{dl}_i}{dS_i}, \qquad G_i \equiv \frac{\Gamma_i}{\overline{c}_i V_\infty}$$

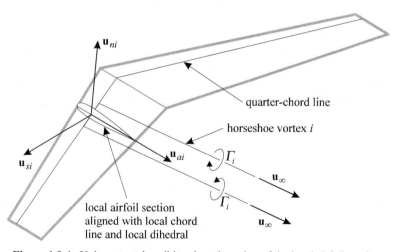

Figure 1.9.4. Unit vectors describing the orientation of the local airfoil section.

where the local section angle of attack is determined from

$$
\alpha_i = \tan^{-1}\left[\frac{\left(\mathbf{u}_\infty + \sum\limits_{j=1}^{N}\mathbf{v}_{ji}\,G_j\right)\cdot\mathbf{u}_{ni}}{\left(\mathbf{u}_\infty + \sum\limits_{j=1}^{N}\mathbf{v}_{ji}\,G_j\right)\cdot\mathbf{u}_{ai}}\right]
\tag{1.9.11}
$$

Equation (1.9.10) can be written for N different control points, one associated with each of the N horseshoe vortices used to synthesize the lifting surface or system of lifting surfaces. This provides a system of N nonlinear equations relating the N unknown dimensionless vortex strengths, G_i, to known properties of the wing. At angles of attack below stall, this system of nonlinear equations surrenders quickly to Newton's method.

To apply Newton's method, the system of equations is written in the vector form

$$
\mathbf{f}(\mathbf{G}) = \mathbf{R}
\tag{1.9.12}
$$

where

$$
f_i(\mathbf{G}) = 2\left|\left(\mathbf{u}_\infty + \sum\limits_{j=1}^{N}\mathbf{v}_{ji}\,G_j\right)\times\zeta_i\right|G_i - \tilde{C}_{Li}(\alpha_i,\delta_i)
\tag{1.9.13}
$$

and $\mathbf{R}$ is a vector of residuals. We wish to find the vector of dimensionless vortex strengths, $\mathbf{G}$, that will make all components of the residual vector, $\mathbf{R}$, go to zero. Thus, we want the change in the residual vector to be $-\mathbf{R}$. We start with some initial estimate for the $\mathbf{G}$ vector and iteratively refine this estimate by applying the Newton corrector equation

$$
[\mathbf{J}]\Delta\mathbf{G} = -\mathbf{R}
\tag{1.9.14}
$$

where $[\mathbf{J}]$ is an $N\times N$ Jacobian matrix of partial derivatives, which is obtained by differentiating Eq. (1.9.13),

$$
J_{ij} = \frac{\partial f_i}{\partial G_j} = \begin{cases} \dfrac{2\mathbf{w}_i\cdot(\mathbf{v}_{ji}\times\zeta_i)}{|\mathbf{w}_i|}G_i - \dfrac{\partial\tilde{C}_{Li}}{\partial\alpha_i}\dfrac{v_{ai}(\mathbf{v}_{ji}\cdot\mathbf{u}_{ni})-v_{ni}(\mathbf{v}_{ji}\cdot\mathbf{u}_{ai})}{v_{ai}^2+v_{ni}^2}, & j\neq i \\[3ex] 2|\mathbf{w}_i|+\dfrac{2\mathbf{w}_i\cdot(\mathbf{v}_{ji}\times\zeta_i)}{|\mathbf{w}_i|}G_i - \dfrac{\partial\tilde{C}_{Li}}{\partial\alpha_i}\dfrac{v_{ai}(\mathbf{v}_{ji}\cdot\mathbf{u}_{ni})-v_{ni}(\mathbf{v}_{ji}\cdot\mathbf{u}_{ai})}{v_{ai}^2+v_{ni}^2}, & j=i \end{cases}
\tag{1.9.15}
$$

$$
\mathbf{w}_i \equiv \mathbf{v}_i\times\zeta_i, \quad v_{ni} \equiv \mathbf{v}_i\cdot\mathbf{u}_{ni}, \quad v_{ai} \equiv \mathbf{v}_i\cdot\mathbf{u}_{ai}, \quad \mathbf{v}_i \equiv \mathbf{u}_\infty + \sum\limits_{j=1}^{N}\mathbf{v}_{ji}\,G_j
$$

Using Eq. (1.9.15) in Eq. (1.9.14), we compute the correction vector, $\Delta\mathbf{G}$. This correction vector is used to obtain an improved estimate for the dimensionless vortex strength vector, $\mathbf{G}$, according to the relation

$$\mathbf{G} = \mathbf{G} + \Psi \Delta\mathbf{G} \qquad (1.9.16)$$

where Ψ is a relaxation factor. The process is repeated until the magnitude of the largest residual is less than some convergence criteria. For angles of attack below stall, this method converges very rapidly using almost any initial estimate for the $\mathbf{G}$ vector with a relaxation factor of unity. At angles of attack beyond stall, the method must be highly underrelaxed and is very sensitive to the initial estimate for the $\mathbf{G}$ vector.

For the fastest convergence of Newton's method, we require an accurate initial estimate for the dimensionless vortex strength vector. For this purpose, a linearized version of Eq. (1.9.10) is useful. Furthermore, for wings of high aspect ratio at small angles of attack, the nonlinear terms in Eq. (1.9.10) are quite small and the linearized system can be used directly to give an accurate prediction of aerodynamic performance.

Applying the small-angle approximation to Eq. (1.9.10) and neglecting all other second-order terms, we obtain

$$2|\mathbf{u}_\infty \times \boldsymbol{\zeta}_i| G_i - \tilde{C}_{Li,\alpha} \sum_{j=1}^N \mathbf{v}_{ji} \cdot \mathbf{u}_{ni} \, G_j = \tilde{C}_{Li,\alpha}\left(\mathbf{u}_\infty \cdot \mathbf{u}_{ni} - \alpha_{L0i} + \varepsilon_i \, \delta_i\right) \qquad (1.9.17)$$

The linearized system of equations given by Eq. (1.9.17) gives good results, at small angles of attack, for wings of reasonably high aspect ratio and little sweep. For larger angles of attack or highly swept wings, the nonlinear system given by Eq. (1.9.10) should be used. However, at angles of attack below stall, Eq. (1.9.17) still provides a reasonable initial estimate for the dimensionless vortex strength vector, to be used with Newton's method for obtaining a solution to the nonlinear system.

Once the vortex strengths are determined from either Eq. (1.9.10) or Eq. (1.9.17), the total aerodynamic force vector can be determined from Eq. (1.9.6). If the lifting surface or surfaces are synthesized from a large number of horseshoe vortices, each covering a small spanwise increment of one lifting surface, we can approximate the aerodynamic force as being constant over each spanwise increment. Then, from Eq. (1.9.6), the total aerodynamic force is given by

$$\mathbf{F}_a = \rho \sum_{i=1}^N \left(\Gamma_i \mathbf{V}_\infty + \sum_{j=1}^N \frac{\Gamma_i \Gamma_j}{\overline{c}_j} \mathbf{v}_{ji} \right) \times \delta\boldsymbol{l}_i \qquad (1.9.18)$$

where $\delta\boldsymbol{l}_i$ is the spatial vector along the bound segment of horseshoe vortex i from node 1 to node 2. Nondimensionalizing this result, the **total nondimensional aerodynamic force vector** is

$$\frac{\mathbf{F}_a}{\frac{1}{2}\rho V_\infty^2 S_r} = 2 \sum_{i=1}^N \left(G_i \mathbf{u}_\infty + \sum_{j=1}^N G_i G_j \mathbf{v}_{ji} \right) \times \boldsymbol{\zeta}_i \frac{\delta S_i}{S_r} \qquad (1.9.19)$$

where S_r is the global reference area and δS_i is the planform area of the spanwise increment of the lifting surface covered by horseshoe vortex i. If we assume a linear variation in chord length over each spanwise increment, we have

$$\delta S_i \equiv \int_{s=s_1}^{s_2} c\, ds = \frac{c_{i_1} + c_{i_2}}{2} (s_{i_2} - s_{i_1}) \tag{1.9.20}$$

where c is the local chord length and s is the spanwise coordinate.

The aerodynamic moment generated about the center of gravity is

$$\mathbf{M}_a = \rho \sum_{i=1}^{N} \left\{ \mathbf{r}_i \times \left[\left(\Gamma_i \mathbf{V}_\infty + \sum_{j=1}^{N} \frac{\Gamma_i \Gamma_j}{\bar{c}_j} \mathbf{v}_{ji} \right) \times \delta l_i \right] + \delta \mathbf{M}_i \right\} \tag{1.9.21}$$

where $\mathbf{r}_i$ is the spatial vector from the center of gravity to control point i and $\delta \mathbf{M}_i$ is the moment generated about the lifting line by the spanwise increment of the wing covered by horseshoe vortex i. If we assume a constant section moment coefficient over each spanwise increment, then

$$\delta \mathbf{M}_i \cong -\frac{1}{2} \rho V_\infty^2 \tilde{C}_{mi} \int_{s=s_1}^{s_2} c^2 ds\, \mathbf{u}_{si} \tag{1.9.22}$$

where $\tilde{C}_{mi}$ is the local section moment coefficient and $\mathbf{u}_{si}$ is the local unit vector in the spanwise direction as shown in Fig. 1.9.4 and defined by

$$\mathbf{u}_{si} = \mathbf{u}_{ai} \times \mathbf{u}_{ni} \tag{1.9.23}$$

Using Eq. (1.9.22) in Eq. (1.9.21) and nondimensionalizing gives

$$\frac{\mathbf{M}_a}{\frac{1}{2}\rho V_\infty^2 S_r l_r} = \sum_{i=1}^{N} \left\{ 2\mathbf{r}_i \times \left[\left(G_i \mathbf{u}_\infty + \sum_{j=1}^{N} G_i G_j \mathbf{v}_{ji} \right) \times \zeta_i \right] - \frac{\tilde{C}_{mi}}{\delta S_i} \int_{s=s_1}^{s_2} c^2 ds\, \mathbf{u}_{si} \right\} \frac{\delta S_i}{S_r l_r} \tag{1.9.24}$$

where l_r is the global reference length.

To this point, the local characteristic length, $\bar{c}_i$, has not been defined. It could be any characteristic length associated with the spanwise increment of the wing covered by horseshoe vortex i. From Eq. (1.9.24), the most natural choice for this local characteristic length is the integral of the chord length squared, with respect to the spanwise coordinate, divided by the incremental area. For a linear variation in chord length over each spanwise increment, this gives

$$\bar{c}_i = \frac{1}{\delta S_i} \int_{s=s_1}^{s_2} c^2 ds = \frac{2}{3} \frac{c_{i_1}^2 + c_{i_1} c_{i_2} + c_{i_2}^2}{c_{i_1} + c_{i_2}} \tag{1.9.25}$$

With this definition, the **dimensionless aerodynamic moment vector** is

$$\frac{\mathbf{M}_a}{\frac{1}{2}\rho V_\infty^2 S_r l_r} = \sum_{i=1}^{N} \left\{ 2\mathbf{r}_i \times \left[\left(G_i \mathbf{u}_\infty + \sum_{j=1}^{N} G_i G_j \mathbf{v}_{ji} \right) \times \zeta_i \right] - \tilde{C}_{mi}\, \bar{c}_i\, \mathbf{u}_{si} \right\} \frac{\delta S_i}{S_r l_r} \qquad (1.9.26)$$

Once the dimensionless vortex strengths, G_i, are known, Eqs. (1.9.19) and (1.9.26) are used to evaluate the components of the aerodynamic force and moment.

Each lifting surface must, of course, be divided into spanwise elements, in a manner similar to that shown symbolically in Fig. 1.9.2. In this figure, the wing is divided into elements of equal spanwise increment. However, this is not the most efficient way in which to grid a wing. Since the spanwise derivative of shed vorticity is greater in the region near the wingtips, for best computational efficiency, the nodal points should be clustered more tightly in this region. Conventional cosine clustering has been found to be quite efficient for this purpose. For straight wings, clustering is only needed near the wingtips and the cosine distribution can be applied across the entire span of the wing. However, for wings with sweep or dihedral, there is a step change in the slope of the quarter-chord line at the root of the wing. This step change produces an increase in the spanwise variation of downwash in the region near the root. Thus, in general, it is recommended that cosine clustering be applied independently over each semispan of each lifting surface. This clusters the nodes more tightly at both the tip and root. This clustering is based on the change of variables,

$$\frac{s}{b} = \frac{1-\cos(\theta)}{4} \qquad (1.9.27)$$

where s is the spanwise coordinate and b is twice the semispan. Over each semispan, θ varies from zero to π as s varies from zero to $b/2$. Distributing the nodes uniformly in θ will provide the desired clustering in s. If the total number of horseshoe elements desired on each semispan is n, the spanwise nodal coordinates are computed from

$$\frac{s_i}{b} = \frac{1}{4}\left[1 - \cos\left(\frac{i\pi}{n}\right)\right], \quad 0 \le i \le n \qquad (1.9.28)$$

where the bound segment of horseshoe vortex i extends from node i to node $i-1$ on any left semispan and from node $i-1$ to node i on any right semispan. Using this nodal distribution with about 40 horseshoe elements per semispan gives the best compromise between speed and accuracy. Figure 1.9.5 shows a system of lifting surfaces overlaid with a grid of this type using 20 elements per semispan.

For maximum accuracy and computational efficiency, some attention must also be paid to the location of control points. At first thought, it would seem most reasonable to place control points on the bound segment of each vortex, midway between the two trailing legs. However, it has been found that this does not give the best accuracy. A significant improvement in accuracy, for a given number of elements, can be achieved by placing the control points midway in θ rather than midway in s. Thus, the spanwise control point coordinates should be computed from

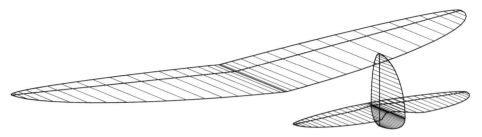

Figure 1.9.5. Lifting-line grid with cosine clustering and 20 elements per semispan.

$$\frac{s_i}{b} = \frac{1}{4}\left[1-\cos\left(\frac{i\pi}{n}-\frac{\pi}{2n}\right)\right], \quad 1\le i\le n \tag{1.9.29}$$

This distribution places control points very near the spatial midpoint of each bound vortex segment over most of the wing. However, near the root and the tip, these control points are significantly offset from the spatial midpoint.

This numerical lifting-line method can be used to predict the aerodynamic forces and moments acting on a system of lifting surfaces with arbitrary position and orientation. Each lifting surface is synthesized by distributing horseshoe vortices along the lifting line in the manner shown in Fig. 1.9.5. Because all of the horseshoe vortices used to synthesize the complete system of lifting surfaces are combined and forced to satisfy either Eq. (1.9.10) or Eq. (1.9.17) as a single system of coupled equations, all of the interactions between lifting surfaces are accounted for directly.

Unlike the closed-form solution to Prandtl's classical lifting-line theory, the numerical lifting-line method can be applied to wings with sweep and/or dihedral. To examine how well the numerical lifting-line method predicts the effects of sweep, Phillips and Snyder (2000) compared results obtained from this method, a numerical panel method, and an inviscid CFD solution. These results were also compared with experimental data for two different wings. Some of the results from this comparison are shown in Fig. 1.9.6. The solid line and filled symbols correspond to a straight wing of aspect ratio 6.57, with experimental data obtained from McAlister and Takahashi (1991). The dashed line and open symbols are for a 45-degree swept wing of aspect ratio 5.0, having experimental data reported by Weber and Brebner (1958). Both wings have symmetric airfoil sections and constant chord throughout the span, with no geometric twist. The straight wing has a thickness of 15 percent and the swept wing has a thickness of 12 percent. From the results shown in Fig. 1.9.6, we see that the lift coefficient predicted by all four methods is in good agreement with experimental observations for both wings. However, the computational time required to obtain a solution using the numerical panel method was about 2.5×10^4 times that required for the lifting-line method, and the inviscid CFD solutions required approximately 2.7×10^6 times as long as the lift-line solutions. The accuracy of the numerical lifting-line method for predicting dihedral effects was also investigated by Phillips and Snyder (2000), and the results obtained were similar to those shown in Fig. 1.9.6.

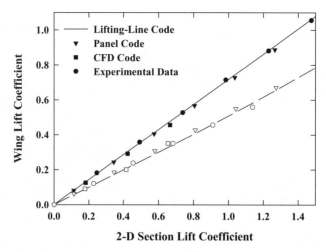

Figure 1.9.6. Comparison between the lift coefficient predicted by the numerical lifting-line method, a numerical panel method, and an inviscid CFD solutions with data obtained from wind tunnel tests for an unswept wing and a wing with 45 degrees of sweep.

The insight of Ludwig Prandtl (1875–1953) was nothing short of astonishing. This was never more dramatically demonstrated than in the development of his classical lifting-line theory, during the period 1911 through 1918. The utility of this simple and elegant theory is so great that it is still widely used today. Furthermore, with a few minor alterations and the use of a modern computer, the model proposed by Prandtl can be used to predict the inviscid forces and moments acting on lifting surfaces of aspect ratio greater than about 4 with an accuracy as good as that obtained from modern panel codes or CFD, but at a small fraction of the computational cost.

Like panel methods, lifting-line theory provides only a potential flow solution. Thus, the forces and moments computed from this method do not include viscous effects. In addition to this restriction, which also applies to panel methods, lifting-line theory imposes an additional restriction, which does not apply to panel methods. For lifting surfaces with low aspect ratio, Prandtl's hypothesis breaks down and the usual relationship between local section lift and local section angle of attack no longer applies. It has long been established that lifting-line theory gives good agreement with experimental data for lifting surfaces of aspect ratio greater than about 4. For lifting surfaces of lower aspect ratio, panel methods or CFD solutions should be used.

The numerical lifting-line method contains no inherent requirement for a linear relationship between section lift and section angle of attack. Thus, the method can be applied, with caution, to account approximately for the effects of stall. The lifting-line method requires a known relationship for the section lift coefficient as a function of section angle of attack. Since such relationships are often obtained experimentally beyond stall, the numerical lifting-line method predicts stall by using a semiempirical correction to an otherwise potential flow solution. For this reason, the method should be used with extreme caution for angles of attack beyond stall.

The effects of viscous parasitic drag can be approximately accounted for in the present numerical lifting-line method through the vector addition of a parasitic drag component to the contribution from each segment in Eqs. (1.9.19) and (1.9.26). The parasitic segment contribution to the net aerodynamic force vector is

$$\left(\delta \mathbf{F}_i\right)_{\text{parasite}} \cong \tfrac{1}{2} \rho V_\infty^2 \, \widetilde{C}_{Di}(\alpha_i) \delta S_i \mathbf{u}_i$$

where $\widetilde{C}_{Di}(\alpha_i)$ represents a relation for the local section drag coefficient as a function of angle of attack and $\mathbf{u}_i$ is the unit vector in the direction of the local velocity vector,

$$\mathbf{u}_i \equiv \frac{\mathbf{V}_i}{V_i}$$

A simple polynomial fit to experimental airfoil section data could be used to describe the relation for section drag coefficient as a function of angle of attack. Similarly, the parasitic segment contribution to the net aerodynamic moment vector is

$$\left(\delta \mathbf{M}_i\right)_{\text{parasite}} \cong \tfrac{1}{2} \rho V_\infty^2 \, \widetilde{C}_{Di}(\alpha_i) \delta S_i (\mathbf{r}_i \times \mathbf{u}_i)$$

Thus, the **dimensionless parasitic contributions to the force and moment vectors** are

$$\left(\frac{\mathbf{F}_a}{\tfrac{1}{2}\rho V_\infty^2 S_r}\right)_{\text{parasite}} = \sum_{i=1}^{N} \widetilde{C}_{Di}(\alpha_i) \frac{\delta S_i}{S_r} \mathbf{u}_i \qquad (1.9.30)$$

$$\left(\frac{\mathbf{M}_a}{\tfrac{1}{2}\rho V_\infty^2 S_r l_r}\right)_{\text{parasite}} = \sum_{i=1}^{N} \widetilde{C}_{Di}(\alpha_i) \frac{\delta S_i}{S_r l_r} (\mathbf{r}_i \times \mathbf{u}_i) \qquad (1.9.31)$$

In our study of flight mechanics, we will find it necessary to know how the aerodynamic force and moment components for a complete aircraft vary over a broad range of operating conditions. Ultimately, this information is commonly gathered from wind tunnel testing. However, such testing is very time consuming and expensive. Another alternative that is often used to obtain the required information is CFD computations. Although modern computers are very fast and the CFD algorithms available today are quite accurate, the volume of data needed to define the required aerodynamic parameters is large. Even with the fastest available computers, gathering the desired information from CFD computations requires considerable time. Thus, for preliminary design, it is important to have a more computationally efficient means for the estimation of aerodynamic force and moment components acting on a complete aircraft. When parasitic drag is included, the numerical lifting-line method described in this section provides such an analytical tool. This method reduces computation time by more than four orders of magnitude over that required for inviscid panel codes and by more than six orders of magnitude compared with CFD computations.

1.10. Wing Stall and Maximum Lift Coefficient

Many aspects of aircraft design and performance analysis depend on the maximum lift coefficient that can be attained on a finite wing prior to stall. Because in general the local section lift coefficient is not constant along the span of a finite wing, it is of interest to know the value of the maximum section lift coefficient and the position along the span at which this maximum occurs. Such knowledge allows us to predict the onset of wing stall from known airfoil section properties, including the maximum airfoil section lift coefficient.

From Eqs. (1.8.96) through (1.8.98), lifting-line theory predicts that the spanwise variation in local section lift coefficient for an unswept wing is given by

$$\tilde{C}_L(\theta) = \Omega \sum_{n=1}^{\infty} 4\left(\frac{b_1 a_n}{a_1} - b_n\right)\frac{\sin(n\theta)}{c(\theta)/b} + C_L \sum_{n=1}^{\infty} \frac{4a_n}{\pi R_A a_1}\frac{\sin(n\theta)}{c(\theta)/b} \quad (1.10.1)$$

where $\theta \equiv \cos^{-1}(-2z/b)$ and Ω is the total amount of wing twist,

$$\Omega \equiv (\alpha - \alpha_{L0})_{\text{root}} - (\alpha - \alpha_{L0})_{\text{max}} \quad (1.10.2)$$

As defined by Eqs. (1.8.20) and (1.8.21) the Fourier coefficients a_n and b_n are obtained from the relations

$$\sum_{n=1}^{N} a_n \left[\frac{4b}{\tilde{C}_{L,\alpha} c(\theta)} + \frac{n}{\sin(\theta)}\right]\sin(n\theta) = 1 \quad (1.10.3)$$

$$\sum_{n=1}^{N} b_n \left[\frac{4b}{\tilde{C}_{L,\alpha} c(\theta)} + \frac{n}{\sin(\theta)}\right]\sin(n\theta) = \omega(\theta) \quad (1.10.4)$$

where $\omega(\theta)$ is the dimensionless twist distribution function,

$$\omega(\theta) \equiv \frac{\alpha(\theta) - \alpha_{L0}(\theta) - (\alpha - \alpha_{L0})_{\text{root}}}{(\alpha - \alpha_{L0})_{\text{max}} - (\alpha - \alpha_{L0})_{\text{root}}} \quad (1.10.5)$$

The first term on the right-hand side of Eq. (1.10.1) is the *basic section lift coefficient* and the second term is the *additional section lift coefficient*. The basic section lift coefficient is independent of angle of attack and directly proportional to the total amount of wing twist, Ω. The additional section lift coefficient is independent of wing twist and directly proportional to the net wing lift coefficient, C_L.

In Fig. 1.8.26 we saw how the net section lift coefficient and its two components obtained from Eq. (1.10.1) vary along the span of a linearly tapered wing of aspect ratio 8.0 and taper ratio 0.5. This figure shows the spanwise variation in section lift coefficient for several values of total linear twist with the net wing lift coefficient held constant at 1.0. Similar results were shown in Fig. 1.8.27 for three different values of the net wing lift coefficient with the total linear twist held constant at 6 degrees. Notice that the spanwise coordinates of the maximums in both the basic and additional section lift coefficients do not change with either the amount of wing twist or the net wing lift coefficient. However, the maximum in the net section lift coefficient moves inboard as the total amount of twist is increased, and for wings with positive twist (i.e., washout), this maximum moves outboard as the wing lift coefficient increases. Thus, **the spanwise position of the maximum section lift coefficient on each semispan of a twisted wing varies with both wing twist and angle of attack**.

Combining terms from the basic and additional section lift coefficients on the right-hand side of Eq. (1.10.1) and rearranging, it can be shown that the maximum section lift coefficient occurs at a value of θ that satisfies the relation

$$\frac{d\tilde{C}_L}{d\theta} = \frac{4C_L}{\pi R_A} \sum_{n=1}^{\infty} \left(\frac{a_n}{a_1} + \frac{b_1 a_n - a_1 b_n}{a_1^2} \frac{\pi R_A a_1 \Omega}{C_L} \right)$$
$$\times \left[b \frac{n\cos(n\theta)c(\theta) - \sin(n\theta)\,dc/d\theta}{c^2(\theta)} \right] = 0 \qquad (1.10.6)$$

Following the development of Phillips and Alley (2007) and applying Eq. (1.8.26), the spanwise location of the airfoil section that supports the largest section lift coefficient is found to be a root of the equation

$$\sum_{n=1}^{\infty} \left(a_n + \frac{b_1 a_n - a_1 b_n}{a_1} \frac{C_{L,\alpha} \Omega}{C_L} \right) \left[n\cos(n\theta)\frac{c(\theta)}{b} - \sin(n\theta)\frac{d(c/b)}{d\theta} \right] = 0 \quad (1.10.7)$$

In the most general case, this root must be found numerically.

After finding the root of Eq. (1.10.7) to obtain the value of θ, which corresponds to the airfoil section that supports the maximum section lift coefficient, this value of θ can be used in Eq. (1.10.1) to determine the maximum section lift coefficient for the wing at the specified operating condition. Dividing Eq. (1.10.1) by C_L and applying Eq. (1.8.26), the ratio of the local section lift coefficient to the total wing lift coefficient is

$$\frac{\tilde{C}_L(\theta)}{C_L} = \frac{4b}{\pi R_A c(\theta)} \left[\sum_{n=1}^{\infty} \frac{a_n}{a_1}\sin(n\theta) + \frac{C_{L,\alpha}\Omega}{C_L} \sum_{n=2}^{\infty} \frac{b_1 a_n - a_1 b_n}{a_1^2} \sin(n\theta) \right] \quad (1.10.8)$$

Examination of Eqs. (1.10.7) and (1.10.8) reveals that for $\Omega=0$, these equations are independent of the net wing lift coefficient. This means that **for an untwisted wing of any planform, the ratio of the maximum section lift coefficient to the total wing lift coefficient and the position along the span at which this maximum occurs are independent of operating conditions and functions of the wing planform only**. Figure 1.10.1 shows how the ratio of total wing lift coefficient to maximum section lift coefficient varies with aspect ratio and taper ratio for untwisted wings with linear taper. The spanwise coordinate of the maximum section lift coefficient for such untwisted wings is shown in Fig. 1.10.2 as a function of aspect ratio and taper ratio. Notice that the spanwise location of the airfoil section that supports the maximum section lift coefficient is quite insensitive to the aspect ratio and nearly a linear function of taper ratio.

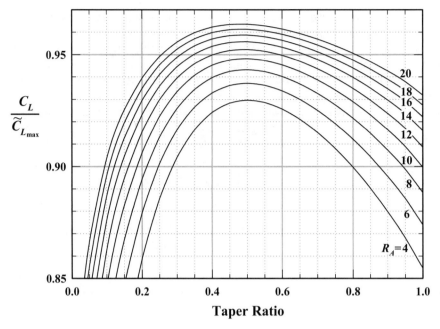

Figure 1.10.1. Maximum lift coefficient for tapered wings with no sweep or twist.

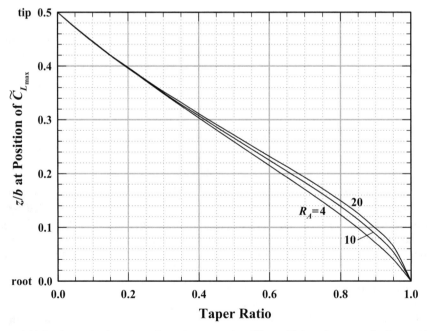

Figure 1.10.2. Spanwise location of maximum section lift coefficient for tapered wings with no sweep or twist.

For unswept wings of arbitrary planform and arbitrary twist, the maximum section lift coefficient is obtained by evaluating Eq. (1.10.8) at the value of θ corresponding to the root of Eq. (1.10.7). This result can be algebraically rearranged to obtain a relation for the ratio of the total wing lift coefficient to the maximum section lift coefficient, which yields

$$\left(\frac{C_L}{\tilde{C}_{L_{\max}}}\right)_{\Lambda=0} = \left(\frac{C_L}{\tilde{C}_{L_{\max}}}\right)_{\substack{\Omega=0 \\ \Lambda=0}}\left(1 - \kappa_{L\Omega}\frac{C_{L,\alpha}\Omega}{\tilde{C}_{L_{\max}}}\right) \tag{1.10.9}$$

where

$$\left(\frac{C_L}{\tilde{C}_{L_{\max}}}\right)_{\substack{\Omega=0 \\ \Lambda=0}} = \frac{\pi R_A c(\theta_{\max})}{4b}\bigg/\sum_{n=1}^{\infty}\frac{a_n}{a_1}\sin(n\theta_{\max}) \tag{1.10.10}$$

$$\kappa_{L\Omega} \equiv \frac{4b}{\pi R_A c(\theta_{\max})}\sum_{n=2}^{\infty}\frac{b_1 a_n - a_1 b_n}{a_1^2}\sin(n\theta_{\max}) \tag{1.10.11}$$

and $\theta_{\max}$ is the value of θ at the wing section that supports the maximum airfoil section lift coefficient, which is obtained from the root of Eq. (1.10.7).

For unswept rectangular wings with positive washout ($\Omega \geq 0$), the maximum section lift coefficient occurs at the wing root ($\theta_{\max} = \pi/2$) and the constant chord length is given by $c(\theta) = b/R_A$. Using these results in Eqs. (1.10.10) and (1.10.11) and simplifying, for **unswept rectangular wings with positive washout** we obtain

$$\left(\frac{C_L}{\tilde{C}_{L_{\max}}}\right)_{\substack{\Omega=0 \\ \Lambda=0}} = \frac{\pi}{4}\bigg/\sum_{i=0}^{\infty}(-1)^i\frac{a_{2i+1}}{a_1}, \qquad \kappa_{L\Omega} = \frac{4}{\pi}\sum_{i=1}^{\infty}(-1)^i\frac{b_1 a_{2i+1} - a_1 b_{2i+1}}{a_1^2} \tag{1.10.12}$$

For an unswept elliptic wing with $\Omega \geq 0$, the maximum section lift coefficient occurs at the wing root ($\theta_{\max} = \pi/2$), the section chord length is $c(\theta) = [4b/(\pi R_A)]\sin(\theta)$, and the planform Fourier coefficients are $a_1 = \tilde{C}_{L,\alpha}/(\pi R_A + \tilde{C}_{L,\alpha})$ and $a_n = 0$ for $n > 1$. Using these results in Eqs. (1.10.10) and (1.10.11), Phillips and Alley (2007) have shown that for **unswept elliptic wings with positive linear washout**

$$\left(\frac{C_L}{\tilde{C}_{L_{\max}}}\right)_{\substack{\Omega=0 \\ \Lambda=0}} = 1, \qquad \kappa_{L\Omega} = \frac{1}{\pi}\sum_{i=1}^{\infty}\frac{4}{(2i+1)^2 - 4}\left(\frac{\pi R_A + \tilde{C}_{L,\alpha}}{\pi R_A + (2i+1)\tilde{C}_{L,\alpha}}\right) \tag{1.10.13}$$

Notice that for unswept rectangular and elliptic wings with positive washout, $\kappa_{L\Omega}$ is independent of both Ω and C_L and the ratio of wing lift coefficient to maximum section lift coefficient is a linear function of $C_{L,\alpha}\Omega/\tilde{C}_{L_{\max}}$. Results predicted from Eqs. (1.10.12) and (1.10.13) are shown in Fig. 1.10.3 as a function of aspect ratio.

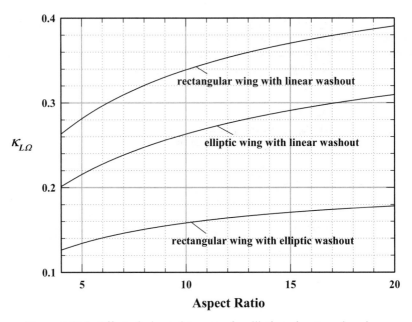

Figure 1.10.3. Effect of wing twist on $\kappa_{L\Omega}$ for elliptic and rectangular wings.

As was seen in Figs. 1.8.26 and 1.8.27, for wings of arbitrary planform the maximum section lift coefficient does not necessarily occur at the wing root. Thus, $\kappa_{L\Omega}$ may vary with $C_{L,\alpha}\Omega/\tilde{C}_{L_{\max}}$ as well as the wing planform. The results plotted in Figs. 1.10.4 and 1.10.5 show how $\kappa_{L\Omega}$ varies with $C_{L,\alpha}\Omega/\tilde{C}_{L_{\max}}$ for wings with linear taper and linear washout.

For **unswept wings with twist optimized to produce minimum induced drag**, the optimum twist distribution and optimum total amount of twist are given in Eq. (1.8.42). For wings with linear taper that are twisted in this manner, Phillips and Alley (2007) obtained a closed-form solution to Eqs. (1.10.7) and (1.10.8). From this solution, we find that the z-coordinate of the wing section that supports the maximum airfoil section lift coefficient is a linear function of the taper ratio and independent of operating conditions,

$$z_{\max} \equiv z(\theta_{\max}) = \pm(1-R_T)b/2 \tag{1.10.14}$$

The ratio of the total wing lift coefficient to the maximum section lift coefficient for such wings is found to be independent of operating conditions and given by

$$\left(\frac{C_L}{\tilde{C}_{L_{\max}}}\right)_{\Lambda=0} = \frac{\pi(2R_T - R_T^2)^{1/2}}{2(1+R_T)} \tag{1.10.15}$$

The reader is cautioned that results predicted from Eqs. (1.10.14) and (1.10.15) are only valid if the wing twist is maintained in proportion to the wing lift coefficient according to the relations provided by Eq. (1.8.42).

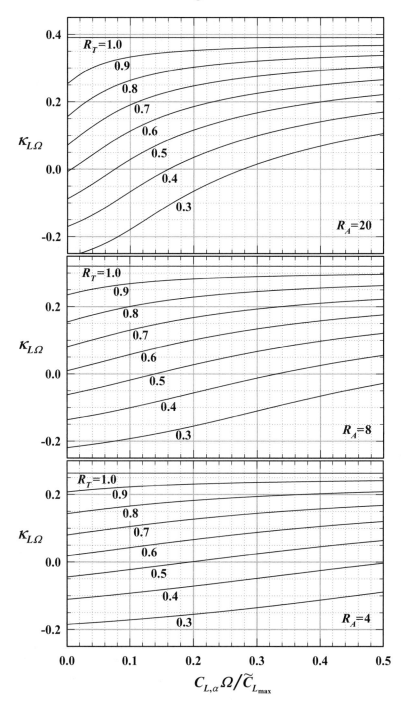

Figure 1.10.4. Effect of wing twist and taper ratio on $\kappa_{L\Omega}$ for wings with linear taper and linear washout.

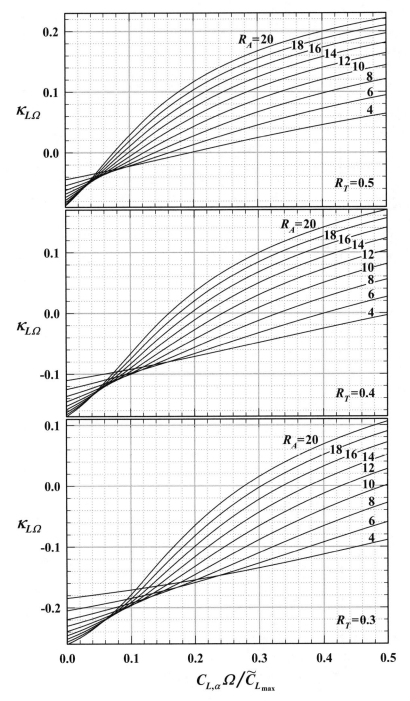

Figure 1.10.5. Effect of wing twist and aspect ratio on $\kappa_{L\Omega}$ for wings with linear taper and linear washout.

When Eq. (1.10.9) and Figs. 1.10.1 through 1.10.5 are used to estimate the total wing lift coefficient that corresponds to a given maximum airfoil section lift coefficient, the results apply only to wings without sweep. As a lifting wing of any planform is swept back, the lift near the root of each semispan is reduced as a result of the downwash induced by the bound vorticity generated on the opposite semispan. This tends to move the point of maximum section lift outboard. For wings with significant taper, this outboard shift causes the point of maximum section lift to occur at an airfoil section having a smaller section chord length, which increases the maximum airfoil section lift coefficient that is produced for a given wing lift coefficient. Because the series solution to Prandtl's lifting-line equation applies only to unswept wings, a numerical solution is required to predict the effects of wing sweep. The numerical lifting-line method presented in Sec. 1.9 can be used for this purpose.

The reader should also note that using the maximum airfoil section lift coefficient in Eq. (1.10.9) will give an estimate for the wing lift coefficient at the onset of airfoil section stall. At higher angles of attack, separated flow will exist over some sections of the wing and drag will be substantially increased. However, the wing lift coefficient predicted from Eq. (1.10.9) is not exactly the maximum wing lift coefficient. Viscous interactions between adjacent sections of the wing can initiate flow separation at slightly lower angles of attack than predicted by Eq. (1.10.9). Furthermore, as the angle of attack is increased somewhat beyond that which produces the onset of airfoil section stall, the section lift coefficient on the stalled section of the wing will decrease. However, the section lift coefficient on the unstalled sections of the wing will continue to increase with angle of attack until the maximum section lift coefficient is reached on these sections as well. Thus, the maximum wing lift coefficient could differ slightly from that which is predicted by Eq. (1.10.9). Because boundary layer separation is a viscous phenomenon, the maximum wing lift coefficient for a given wing geometry must be determined from experimental methods or computational fluid dynamics (CFD).

To account for the effects of wing sweep and stall, experimental data and results obtained from CFD computations suggest that predictions from Eq. (1.10.9) can be modified by including sweep and stall correction factors,

$$C_{L_{\max}} = \left(\frac{C_L}{\widetilde{C}_{L_{\max}}}\right)_{\substack{\Omega=0 \\ \Lambda=0}} K_{LS} K_{L\Lambda}\left(\widetilde{C}_{L_{\max}} - K_{L\Omega} C_{L,\alpha}\Omega\right) \tag{1.10.16}$$

The sweep factor $K_{L\Lambda}$ depends on the wing sweep angle and wing planform. The stall factor K_{LS} depends on wing aspect ratio and the wing twist parameter $C_{L,\alpha}\Omega/\widetilde{C}_{L_{\max}}$.

For wings with linear taper, Phillips and Alley (2007) presented numerical lifting-line results for the sweep factor, which correlate well with the approximation

$$K_{L\Lambda} \cong 1 + K_{\Lambda 1}\Lambda - K_{\Lambda 2}\Lambda^{1.2} \tag{1.10.17}$$

where Λ is the wing quarter-chord sweep angle in radians. The empirical coefficients $K_{\Lambda 1}$ and $K_{\Lambda 2}$ depend on aspect ratio and taper ratio as shown in Figs. 1.10.6 and 1.10.7, respectively.

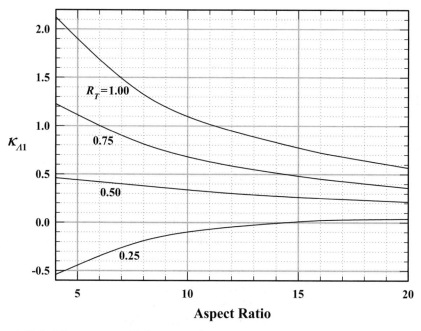

Figure 1.10.6. Wing sweep coefficient $K_{\Lambda 1}$ to be used in Eq. (1.10.17) for wings with linear taper.

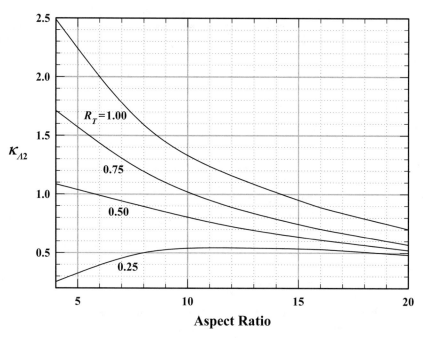

Figure 1.10.7. Wing sweep coefficient $K_{\Lambda 2}$ to be used in Eq. (1.10.17) for wings with linear taper.

For wings with linear taper and linear twist, CFD results obtained by Alley, Phillips, and Spall (2007) for the stall factor in Eq. (1.10.16) were found to correlate well with the approximate algebraic relation

$$\kappa_{Ls} \cong 1 + \left(0.0042R_A - 0.068\right)\left(1 + 2.3\frac{C_{L,\alpha}\Omega}{\tilde{C}_{L_{max}}}\right) \tag{1.10.18}$$

Equation (1.10.18) was obtained from computations of the maximum wing lift coefficient for 25 different wing geometries. These wings had aspect ratios ranging from 4 to 20, taper ratios from 0.5 to 1.0, quarter-chord sweep angles from 0 to 30 degrees, and linear geometric washout ranging from 0 to 8 degrees.

When estimating a maximum wing lift coefficient from Eq. (1.10.16), it is essential that the maximum 2-D airfoil section lift coefficient be evaluated at the same Reynolds number and Mach number as those for the 3-D wing. For the case of a rectangular wing this is rather straightforward. However, for an elliptic or tapered wing the section Reynolds number is not constant across the wingspan. This gives rise to an important question regarding what characteristic length should be used to define the Reynolds number associated with predicting the maximum lift coefficient for a wing of arbitrary planform. The simplest choice would be the mean chord length defined by Eq. (1.8.81) or the mean aerodynamic chord defined by Eq. (1.8.86). However, neither of these chord lengths provides a particularly suitable characteristic length for defining the Reynolds number associated with wing stall. A more appropriate characteristic length for this purpose is the chord length of the wing section supporting the maximum section lift coefficient.

The spanwise coordinate of the maximum airfoil section lift coefficient at the onset of stall depends on wing taper, twist, and sweep. For an untwisted rectangular wing with no sweep, the onset of airfoil section stall always occurs at the wing root. As wing taper ratio is decreased from 1.0, the point of maximum airfoil section lift coefficient moves outboard from the root. Adding sweep to the wing also moves the point of maximum section lift coefficient outboard. On the other hand, adding washout to a wing with taper and/or sweep moves the point of maximum airfoil section lift coefficient inboard. Lifting-line theory can be used to predict the spanwise coordinate of the wing section that supports the maximum airfoil section lift coefficient. For tapered wings with no sweep or twist, the coordinate of the maximum section lift coefficient may be obtained from Fig. 1.10.2.

Results obtained from Fig. 1.10.2 can be modified to account for the effects of twist and sweep. The spanwise coordinate of the wing section that supports the maximum section lift coefficient can be estimated from

$$\frac{z_{max}}{b} = 0.5 - \kappa_{Z\Lambda}\left[0.5 - \kappa_{Z\Omega}\left(\frac{z_{max}}{b}\right)_{\substack{\Omega=0 \\ \Lambda=0}}\right] \tag{1.10.19}$$

For wings with linear taper and linear twist, $\kappa_{Z\Omega}$ and $\kappa_{Z\Lambda}$ are obtained from Figs. 1.10.8 and 1.10.9, respectively.

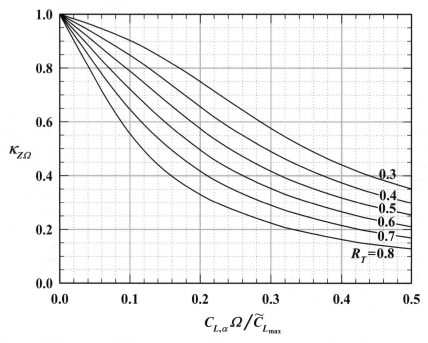

Figure 1.10.8. Twist coefficient $K_{Z\Omega}$ to be used in Eq. (1.10.19) for wings with linear twist.

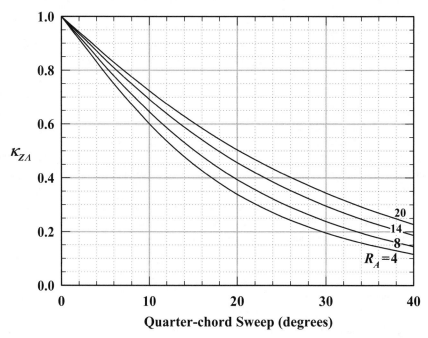

Figure 1.10.9. Sweep coefficient $K_{Z\Lambda}$ to be used in Eq. (1.10.19) for wings with linear taper.

EXAMPLE 1.10.1. Consider a wing with linear taper of aspect ratio 8.0 and taper ratio 0.5. The wing has a span of 196 feet and a thin airfoil section having a constant maximum section lift coefficient of 1.6. We wish to estimate the maximum wing lift coefficient and the section chord length at the onset of airfoil section stall for this wing planform with the following amounts of quarter-chord sweep and linear washout:

$$\Lambda = 0, \ \Omega = 0; \qquad \Lambda = 0, \ \Omega = 4°; \qquad \Lambda = 30°, \ \Omega = 0; \qquad \Lambda = 30°, \ \Omega = 4°$$

Solution. Using an airfoil section lift slope of 2π in Eq. (1.8.26) with κ_L obtained from Fig. 1.8.13, for the unswept wing we have

$$\kappa_L = 0.0125, \qquad C_{L,\alpha} = \frac{\tilde{C}_{L,\alpha}}{[1 + \tilde{C}_{L,\alpha}/(\pi R_A)](1 + \kappa_L)} = 4.964$$

From Figs. 1.10.1, 1.10.2, 1.10.6, and 1.10.7 we obtain

$$\left(\frac{C_L}{\tilde{C}_{L_{max}}}\right)_{\substack{\Omega=0 \\ \Lambda=0}} = 0.943, \qquad \left(\frac{z_{max}}{b}\right)_{\substack{\Omega=0 \\ \Lambda=0}} = 0.264, \qquad \kappa_{\Lambda 1} = 0.38, \qquad \kappa_{\Lambda 2} = 0.89$$

With 4 degrees of linear washout, Figs. 1.10.5 and 1.10.8 and Eq. (1.10.18) yield

$$\frac{C_{L,\alpha}\Omega}{\tilde{C}_{L_{max}}} = 0.2166, \qquad \kappa_{L\Omega} = 0.034, \qquad \kappa_{Z\Omega} = 0.54, \qquad \kappa_{Ls} = 0.948$$

For 30 degrees sweep and no washout, Eqs. (1.10.17) and (1.10.18) and Fig. 1.10.9 give

$$\Lambda = 0.5236, \qquad \kappa_{L\Lambda} = 0.79, \qquad \kappa_{Ls} = 0.966, \qquad \kappa_{Z\Lambda} = 0.24$$

The maximum wing lift coefficient is obtained from Eq. (1.10.16), the z-coordinate of the wing section supporting the maximum section lift coefficient is estimated from Eq. (1.10.19), and the chord length at the position of the maximum section lift coefficient is found from

$$c_{max} = \frac{2b}{R_A(1 + R_T)} \Big[1 - (1 - R_T)|2z_{max}/b| \Big]$$

Thus we obtain

Λ	Ω	κ_{Ls}	$\kappa_{L\Lambda}$	$\kappa_{L\Omega}$	$C_{L_{max}}$	$\kappa_{Z\Lambda}$	$\kappa_{Z\Omega}$	z_{max}/b	c_{max}
0°	0°	0.966	1.00		1.46	1.00	1.00	0.264	24.0
0°	4°	0.948	1.00	0.034	1.42	1.00	0.54	0.143	28.0
30°	0°	0.966	0.79		1.15	0.24	1.00	0.443	18.2
30°	4°	0.948	0.79	0.034	1.12	0.24	0.54	0.414	19.1

1.11. Wing Aerodynamic Center and Pitching Moment

The distribution of section aerodynamic loads acting on a spanwise symmetric wing with spanwise symmetric loading can be replaced with a resultant force vector acting at the aerodynamic center of the wing and a pitching moment that does not vary with small changes in angle of attack. For a spanwise symmetric wing with no sweep in the locus of airfoil section aerodynamic centers, the section aerodynamic center of each airfoil section falls at the same axial coordinate. Thus, for unswept wings, the aerodynamic center of the complete wing lies in the plane of symmetry at the same axial coordinate as the section aerodynamic centers of the root airfoil section, independent of the spanwise section lift distribution. However, for swept wings, the axial coordinate of the section aerodynamic center is a function of the spanwise coordinate, and the axial position of the wing's aerodynamic center is not obvious from simple inspection.

Wing sweep affects the position of the aerodynamic center of a wing in two ways. First and most obvious, when the wing is swept back, the locus of airfoil section aerodynamic centers on the outboard sections of the wing are moved aft of the aerodynamic center of the root airfoil section. Thus, lift developed on a swept wing contributes significantly to the pitching moment about the aerodynamic center of the root airfoil section. In addition, sweep alters the vorticity-induced downwash distribution over the wing planform. Moving the wingtip vortex aft of the wing root tends to reduce the downwash induced on the inboard sections of the wing. On the other hand, the bound vorticity on one semispan of a swept wing induces downwash on the opposite semispan. This tends to increase the wing downwash, more so on the inboard sections of the wing. Thus, not only does sweep alter the geometry of the locus of airfoil section aerodynamic centers, it changes the spanwise section lift distribution as well.

To examine how the aerodynamic center of a swept wing can be located, we first consider the pitching moment developed by an arbitrary wing. From Fig. 1.11.1, the pitching moment about the origin, $x = 0$, $y = 0$, for a wing of arbitrary planform and dihedral can be written as

$$m_0 = \int_{z=-b/2}^{b/2} \tilde{m}_{ac}\, dz - \int_{z=-b/2}^{b/2} (\tilde{L}\cos\alpha + \tilde{D}\sin\alpha)\tilde{x}_{ac}\, dz - \int_{z=-b/2}^{b/2} (\tilde{L}\sin\alpha - \tilde{D}\cos\alpha)\tilde{y}_{ac}\, dz \quad (1.11.1)$$

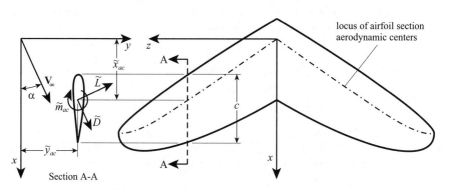

Figure 1.11.1. Section lift, drag, and pitching moment acting on a section of an arbitrary wing.

where $\tilde{x}_{ac}$ and $\tilde{y}_{ac}$ denote the x- and y-coordinates of the locus of airfoil section aerodynamic centers measured relative to the root section aerodynamic center. Note that in Eq. (1.11.1), $\tilde{m}_{ac}$, $\tilde{L}$, and $\tilde{D}$ are section moment and force components per unit span, not per unit distance measured parallel with the local dihedral. Dividing Eq. (1.11.1) by the dynamic pressure, planform area, and reference chord length yields

$$
\begin{aligned}
C_{m_0} \equiv \frac{m_0}{\frac{1}{2}\rho_\infty V_\infty^2 S c_{\mathrm{ref}}} = & \frac{1}{S c_{\mathrm{ref}}} \int\limits_{z=-b/2}^{b/2} \tilde{C}_{m_{ac}} c^2 dz \\
& - \frac{1}{S c_{\mathrm{ref}}} \int\limits_{z=-b/2}^{b/2} (\tilde{C}_L \cos\alpha + \tilde{C}_D \sin\alpha) c \tilde{x}_{ac}\, dz \\
& - \frac{1}{S c_{\mathrm{ref}}} \int\limits_{z=-b/2}^{b/2} (\tilde{C}_L \sin\alpha - \tilde{C}_D \cos\alpha) c \tilde{y}_{ac}\, dz
\end{aligned}
\tag{1.11.2}
$$

Because drag is typically small compared with the lift, drag is commonly neglected when estimating the position of the aerodynamic center at small angles of attack. Furthermore, the trigonometric functions that appear in Eq. (1.11.2) are traditionally linearized using the small-angle approximations $\cos(\alpha) \cong 1$ and $\sin(\alpha) \cong \alpha$. Thus, for small angles of attack, Eq. (1.11.2) can be approximated as

$$
C_{m_0} \cong \frac{1}{S c_{\mathrm{ref}}} \left(\int\limits_{z=-b/2}^{b/2} \tilde{C}_{m_{ac}} c^2 dz - \int\limits_{z=-b/2}^{b/2} \tilde{C}_L c \tilde{x}_{ac}\, dz - \int\limits_{z=-b/2}^{b/2} \tilde{C}_L \alpha c \tilde{y}_{ac}\, dz \right)
\tag{1.11.3}
$$

The dihedral angle for most wings is small enough so that we can neglect the product of the angle of attack and the y-offset between the outboard airfoil sections and the root airfoil section, i.e., $\alpha \tilde{y}_{ac} \cong 0$. Additionally, for a spanwise symmetric wing with spanwise symmetric lift, each semispan contributes equally to the pitching moment. Thus, for symmetric wings with small dihedral angles operating at small angles of attack, the wing pitching moment coefficient about the root airfoil section aerodynamic center can be approximated as

$$
C_{m_0} \cong \frac{2}{S c_{\mathrm{ref}}} \int\limits_{z=0}^{b/2} \tilde{C}_{m_{ac}} c^2 dz - \frac{2}{S c_{\mathrm{ref}}} \int\limits_{z=0}^{b/2} \tilde{C}_L c \tilde{x}_{ac}\, dz
\tag{1.11.4}
$$

The distributed aerodynamic loads acting on the wing can be replaced with a resultant force and moment acting at the aerodynamic center of the wing ($\bar{x}_{ac}$, $\bar{y}_{ac}$). Thus, the pitching moment about the origin can also be written as

$$
m_0 = m_{ac} - \bar{x}_{ac} (L \cos\alpha + D \sin\alpha) - \bar{y}_{ac} (L \sin\alpha - D \cos\alpha)
\tag{1.11.5}
$$

where m_{ac} is the pitching moment about the aerodynamic center of the wing. Dividing by the dynamic pressure, planform area, and reference chord length, Eq. (1.11.5) can be written in dimensionless form as

$$C_{m_0} = C_{m_{ac}} - \frac{\bar{x}_{ac}}{c_{ref}}(C_L \cos\alpha + C_D \sin\alpha) - \frac{\bar{y}_{ac}}{c_{ref}}(C_L \sin\alpha - C_D \cos\alpha) \quad (1.11.6)$$

We now apply the same small-angle approximations to Eq. (1.11.6) that were used to obtain Eq. (1.11.4) from Eq. (1.11.2). That is, we neglect drag, $C_D \cong 0$, assume small angles of attack, $\cos(\alpha) \cong 1$ and $\sin(\alpha) \cong \alpha$, and assume small dihedral angles, $\alpha\bar{y}_{ac} \cong 0$. Thus, Eq. (1.11.6) is traditionally approximated as

$$C_{m_0} \cong C_{m_{ac}} - \frac{\bar{x}_{ac}}{c_{ref}}C_L \quad (1.11.7)$$

Combining Eqs. (1.11.4) and (1.11.7) to eliminate the pitching moment coefficient about the origin gives

$$C_{m_{ac}} - \frac{\bar{x}_{ac}}{c_{ref}}C_L \cong \frac{2}{Sc_{ref}}\int_{z=0}^{b/2}\tilde{C}_{m_{ac}}c^2 dz - \frac{2}{Sc_{ref}}\int_{z=0}^{b/2}\tilde{C}_L c\tilde{x}_{ac} dz \quad (1.11.8)$$

By definition, $C_{m_{ac}}$, $\bar{x}_{ac}$, $\tilde{C}_{m_{ac}}$, and $\tilde{x}_{ac}$ do not vary with small changes in angle of attack. Accordingly, differentiating Eq. (1.11.8) with respect to angle of attack and rearranging yields a relation for the axial position of the wing aerodynamic center in terms of the spanwise section lift distribution. Thus, **for the common case of symmetric wings with small dihedral angles operating at small angles of attack, the axial position of the wing aerodynamic center is traditionally approximated as**

$$\bar{x}_{ac} \cong \frac{1}{C_{L,\alpha}}\frac{\partial}{\partial\alpha}\left(\frac{2}{S}\int_{z=0}^{b/2}\tilde{C}_L c\tilde{x}_{ac} dz\right) \quad (1.11.9)$$

From Eq. (1.11.9) we see that determination of the axial position of the aerodynamic center of a swept wing requires knowledge of the spanwise section lift distribution. Because an analytical solution for the section lift distribution acting on a swept wing does not exist, predictions for the position of the aerodynamic center of swept wings require numerical solutions. Inviscid panel codes and computational fluid dynamics (CFD) are commonly used for this purpose.

As a first approximation, the aerodynamic center of a swept wing is sometimes assumed to be located at the same axial position as the section aerodynamic center of the airfoil section located at the spanwise coordinate of the semispan area centroid. For wings with constant quarter-chord sweep and linear taper, i.e., trapezoidal wings, we have $\bar{x}_c = \bar{z}_c \tan\Lambda$ and from Eq. (1.8.100) this approximation results in

$$\frac{\bar{x}_{ac}}{\bar{c}} \cong \frac{\bar{x}_c}{\bar{c}} = \frac{R_A}{6}\frac{1+2R_T}{1+R_T}\tan\Lambda \quad (1.11.10)$$

The approximation given by Eq. (1.11.10) is equivalent to assuming that the section lift coefficient is constant over the wingspan.

Another commonly used approximation was first suggested by Anderson (1937). **This approximation neglects any changes in the section lift distribution that result from wing sweep.** For wings with constant quarter-chord sweep, $\bar{x}_{ac} = \bar{z}_{ac} \tan \Lambda$, where $\bar{z}_{ac}$ is the z-coordinate of the semispan aerodynamic center. Thus, from Eq. (1.8.109) Anderson's approximation yields

$$\frac{\bar{x}_{ac}}{\bar{c}} \cong \frac{2R_A}{3\pi}\left(1 + \sum_{n=1}^{\infty} \frac{(-1)^{n-1}3}{4n^2 + 4n - 3}\frac{a_{2n+1}}{a_1}\right)\tan \Lambda \qquad (1.11.11)$$

Predictions from Eqs. (1.11.10) and (1.11.11) are compared with CFD solutions in Fig. 1.11.2. Results for 236 wings with constant linear taper and constant quarter-chord sweep are shown in Fig. 1.11.2. Wing aspect ratio was varied from 4.0 to 20 and taper ratios from 0.25 to 1.0 were investigated. For a given taper and aspect ratio, the quarter-chord sweep angle was varied from 0 to 50 degrees. All wings had airfoil sections from the NACA 4-digit airfoil series with camber varied from 0 to 4 percent and thickness ranging from 6 to 18 percent. To investigate the effects of wing twist, linear geometric washout was varied from −4.0 to +8.0 degrees. For further details regarding the CFD solutions presented in Fig. 1.11.2, see Phillips, Hunsaker, and Niewoehner (2008).

In Fig. 1.11.2, the location of each aerodynamic center is presented as a deviation from the result predicted by Eq. (1.11.10). This deviation is plotted as a function of the

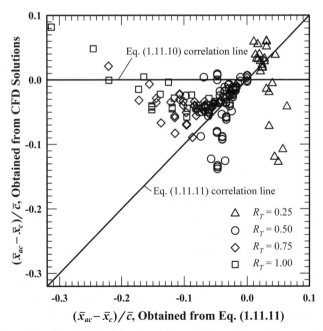

Figure 1.11.2. Deviation of the wing aerodynamic center from the section aerodynamic center of the airfoil located at the semispan centroidal chord as predicted from computational fluid dynamics results vs. the same deviation predicted from Eq. (1.11.11).

same deviation as predicted from Eq. (1.11.11). To see how the data that are plotted in Fig. 1.11.2 are used to assess the accuracy of Eqs. (1.11.10) and (1.11.11), we first recognize that if Eq. (1.11.10) were precise, each aerodynamic center would have the same axial coordinate as the airfoil section aerodynamic center of the semispan centroidal chord. Thus, exact correlation of Eq. (1.11.10) with the CFD results would cause all points in Fig. 1.11.2 to fall along a horizontal line with the vertical ordinate of zero. This is the line denoted as the Eq. (1.11.10) correlation line in Fig. 1.11.2. On the other hand, if Eq. (1.11.11) were to match the CFD predictions exactly, all points in Fig. 1.11.2 would fall along the 45-degree line, which is labeled as the Eq. (1.11.11) correlation line. From the results plotted in Fig. 1.11.2, we see that neither Eq. (1.11.10) nor Eq. (1.11.11) is accurate over a wide range of wing geometry.

Notice from Fig. 1.11.2 that Eq. (1.11.10) seems to be more accurate for the majority of the rectangular wings, whereas the results for many of the wings having a taper ratio of 0.5 agree more closely with Eq. (1.11.11). The reader should particularly notice the heavy concentration of circular symbols just below the intersection of the Eq. (1.11.10) and Eq. (1.11.11) correlation lines. Most of these data are for wings having a taper ratio of 0.5 with quarter-chord sweep angles in the range between 25 and 35 degrees. These results agree closely with Eq. (1.11.11) and show that for such commonly used wing geometries, the lifting-line result presented in Eq. (1.11.11) gives a reasonable first approximation for the position of the aerodynamic center of the wing. However, **for the case of more general wing geometry, neither Eq. (1.11.10) nor Eq. (1.11.11) should be used to predict the aerodynamic center of a swept wing.** These approximations are included here only for historical reasons and because they are likely to be encountered in the literature.

For swept trapezoidal wings, an improved approximation for the axial position of the aerodynamic center is obtained by multiplying the right-hand side of Eq. (1.11.11) by an empirical sweep correction factor

$$\frac{\overline{x}_{ac}}{\overline{c}} \cong \kappa_{ac} \frac{2R_A}{3\pi} \left(1 + \sum_{n=1}^{\infty} \frac{(-1)^{n-1} 3}{4n^2 + 4n - 3} \frac{a_{2n+1}}{a_1} \right) \tan\Lambda = \kappa_{ac} R_A \left(\frac{\overline{z}_{ac}}{b} \right)_{\Lambda=0} \tan\Lambda \quad (1.11.12)$$

where $(\overline{z}_{ac})_{\Lambda=0}$ is the z-coordinate of the semispan aerodynamic center for an unswept wing of the same planform, which can be obtained from Eq. (1.8.109) or Fig. 1.8.29. Results for κ_{ac}, which were presented by Phillips, Hunsaker, and Niewoehner (2008), are shown in Fig. 1.11.3 as a function of wing taper ratio, aspect ratio, and quarter-chord sweep angle. Other results obtained in the same study show that **wing camber, thickness, and twist have no significant effect on the position of the aerodynamic center of a swept wing.** In Fig. 1.11.3 notice that for wings of taper ratio near 0.5, aspect ratios in the range of 6 to 8, and quarter-chord sweep angles near 30 degrees, all values of κ_{ac} are close to unity. This means that the lifting-line solution presented in Eq. (1.11.11) provides a good approximation for this commonly used wing geometry, without using the empirical correction factor. On the other hand, both Fig. 1.11.2 and Fig. 1.11.3 show that **some wing geometries result in very large discrepancies between Eq. (1.11.11) and results obtained from CFD solutions.**

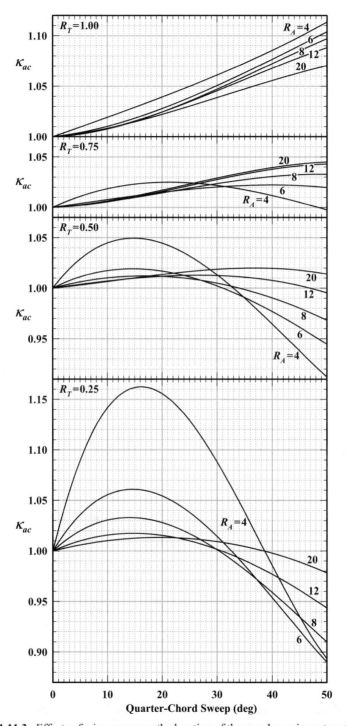

Figure 1.11.3. Effects of wing sweep on the location of the aerodynamic center of a wing.

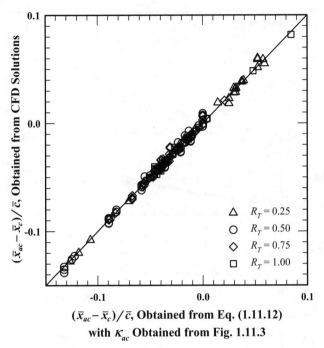

Figure 1.11.4. Deviation of the wing aerodynamic center from the section aerodynamic center of the airfoil located at the semispan centroidal chord as predicted from computational fluid dynamics results vs. the same deviation predicted from Eq. (1.11.12).

Figure 1.11.4 shows the same CFD solutions presented in Fig. 1.11.2 compared with results predicted from Eq. (1.11.12) using the values for κ_{ac} that are plotted in Fig. 1.11.3. Comparing Fig. 1.11.4 with Fig. 1.11.2, we see that using the empirical sweep correction factor plotted in Fig. 1.11.3 provides a very significant improvement over Eq. (1.11.11).

Once the aerodynamic center of a wing is located, the pitching moment coefficient about that aerodynamic center can be determined from the lift coefficient and pitching moment coefficient about the origin. Rearranging Eq. (1.11.7) yields

$$C_{m_{ac}} \cong C_{m_0} + \frac{\bar{x}_{ac}}{c_{\text{ref}}} C_L \qquad (1.11.13)$$

The pitching moment coefficient about the origin is approximated from Eq. (1.11.4). Using this result in Eq. (1.11.13) we obtain

$$C_{m_{ac}} \cong \frac{2}{S c_{\text{ref}}} \int_{z=0}^{b/2} \widetilde{C}_{m_{ac}} c^2 dz - \frac{1}{c_{\text{ref}}} \left(\frac{2}{S} \int_{z=0}^{b/2} \widetilde{C}_L \, c \, \widetilde{x}_{ac} \, dz - \bar{x}_{ac} C_L \right) \qquad (1.11.14)$$

The total section lift coefficient is the sum of the basic and additional section lift coefficients, $\widetilde{C}_L = \widetilde{C}_{L_b} + \widetilde{C}_{L_a}$, and the axial coordinate of the wing aerodynamic center can be expressed in terms of the additional section lift coefficient in a manner similar to that used to obtain Eq. (1.8.110),

$$\bar{x}_{ac} \cong \frac{2}{SC_L} \int_{z=0}^{b/2} \widetilde{C}_{L_a} c \widetilde{x}_{ac} \, dz \qquad (1.11.15)$$

Thus, the pitching moment coefficient about the aerodynamic center of the wing can be expressed in terms of only the airfoil section pitching moment coefficient and the spanwise variation in local section lift coefficient that occurs when the net lift developed by the wing is zero. Using Eq. (1.11.15) in Eq. (1.11.14), **for symmetric wings with small dihedral angles operating at small angles of attack, the pitching moment coefficient about the wing aerodynamic center can be approximated as**

$$C_{m_{ac}} \cong \frac{2}{Sc_{\mathrm{ref}}} \int_{z=0}^{b/2} \widetilde{C}_{m_{ac}} c^2 \, dz - \frac{2}{Sc_{\mathrm{ref}}} \int_{z=0}^{b/2} \widetilde{C}_{L_b} c \widetilde{x}_{ac} \, dz \qquad (1.11.16)$$

The first term on the right-hand side of Eq. (1.11.16) results from the effects of camber and is simply twice the semispan contribution for an unswept wing of the same planform, which is given by Eq. (1.8.102) or Eq. (1.8.104). The second term on the right-hand side of Eq. (1.11.16) results only from wing twist.

For wings with constant quarter-chord sweep, the x-coordinate of the locus of airfoil section aerodynamic centers is proportional to the z-coordinate, $\widetilde{x}_{ac} = z \tan \Lambda$, and after using Eq. (1.8.104) in Eq. (1.11.16) we have

$$C_{m_{ac}} \cong \frac{\bar{c}_{mac}}{c_{\mathrm{ref}}} \overline{\widetilde{C}}_{m_{ac}} - \frac{2 \tan \Lambda}{Sc_{\mathrm{ref}}} \int_{z=0}^{b/2} \widetilde{C}_{L_b} c z \, dz \qquad (1.11.17)$$

For unswept wings the integral on the right-hand side of Eq. (1.11.17) can be evaluated from Eq. (1.8.96). Results presented by Anderson (1937) suggest that this same result can be used in Eq. (1.11.17) as a first approximation for swept wings. Following a procedure similar to that used to develop Eq. (1.11.12), Phillips, Hunsaker, and Niewoehner (2008) have shown that improved results can be obtained from the relation

$$C_{m_{ac}} \cong \frac{\bar{c}_{mac}}{c_{\mathrm{ref}}} \overline{\widetilde{C}}_{m_{ac}} + \kappa_{M\Omega} \frac{\tan(\kappa_{MA} \Lambda)}{c_{\mathrm{ref}}/b} C_{L,\alpha} \Omega \qquad (1.11.18)$$

where $\kappa_{M\Omega}$ is the twist factor for an unswept wing of the same planform, which can be evaluated from Eq. (1.8.113) or Fig. 1.8.30. The wing lift slope in Eq. (1.11.18) is that for the swept wing, which can be estimated from

$$C_{L,\alpha} \cong \pi R_A a_1 \kappa_{L\alpha} = (C_{L,\alpha})_{\Lambda=0} \kappa_{L\alpha} \qquad (1.11.19)$$

where $\kappa_{L\alpha}$ can be obtained from Fig. 1.11.5. The sweep factor κ_{MA} in Eq. (1.11.18) is obtained from Fig. 1.11.6.

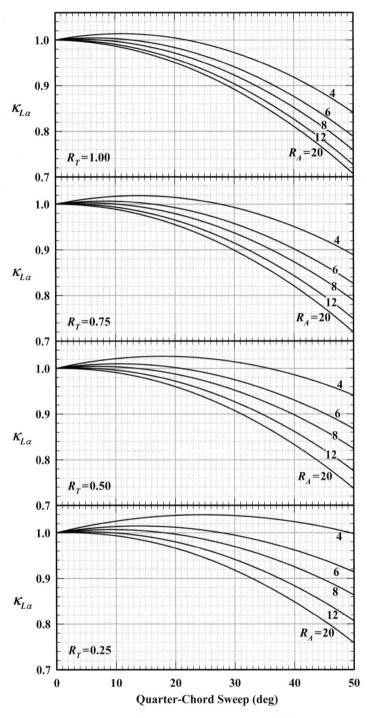

Figure 1.11.5. Effects of wing sweep on the lift slope.

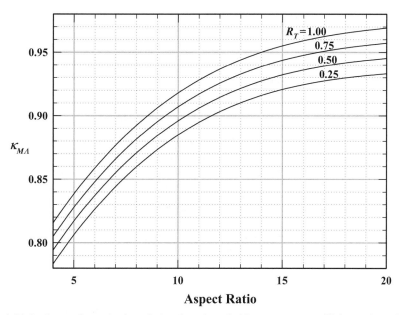

Figure 1.11.6. Sweep factor in the relation for wing pitching moment coefficients about the wing aerodynamic center.

EXAMPLE 1.11.1. Consider a wing with linear taper of aspect ratio 8.0 and taper ratio 0.5. The wing has a span of 196 feet and a thin symmetric airfoil section. We wish to estimate the axial position of the wing aerodynamic center and the pitching moment coefficient about that aerodynamic center for this wing planform with the following amounts of quarter-chord sweep and linear washout:

$$\Lambda = 15°, \; \Omega = 0; \quad \Lambda = 15°, \; \Omega = 4°; \quad \Lambda = 30°, \; \Omega = 0; \quad \Lambda = 30°, \; \Omega = 4°$$

Solution. Using an airfoil section lift slope of 2π in Eq. (1.8.26) with κ_L obtained from Fig. 1.8.13, for an unswept wing of the same planform we have

$$\kappa_L = 0.0125 \quad \text{and} \quad (C_{L,\alpha})_{\Lambda=0} = \frac{\tilde{C}_{L,\alpha}}{[1 + \tilde{C}_{L,\alpha}/(\pi R_A)](1 + \kappa_L)} = 4.964$$

From Figs. 1.8.29 and 1.8.30 we obtain

$$\left(\frac{\bar{z}_{ac}}{b}\right)_{\Lambda=0} = 0.216 \quad \text{and} \quad \kappa_{M\Omega} = 0.0262$$

With 15 degrees of sweep, Figs. 1.11.3 and 1.11.5 and Eq. (1.11.19) yield

$$\kappa_{ac} = 1.012, \quad \kappa_{L\alpha} = 0.998, \quad \text{and} \quad C_{L,\alpha} \cong (C_{L,\alpha})_{\Lambda=0} \, \kappa_{L\alpha} = 4.954$$

Similarly, with 30 degrees of sweep

$$\kappa_{ac} = 1.005, \quad \kappa_{L\alpha} = 0.953, \quad \text{and} \quad C_{L,\alpha} \cong (C_{L,\alpha})_{\Lambda=0} \, \kappa_{L\alpha} = 4.731$$

From Figs. 1.11.6 we obtain

$$\kappa_{M\Lambda} = 0.871$$

The axial position of the wing aerodynamic center measured aft of the root quarter chord is estimated from Eq. (1.11.12). This result is independent of wing twist and for 15 degrees of sweep we obtain

$$\bar{x}_{ac} \cong \frac{S}{b} \kappa_{ac} R_A \left(\frac{\bar{z}_{ac}}{b} \right)_{\Lambda=0} \tan \Lambda = \kappa_{ac} b \left(\frac{\bar{z}_{ac}}{b} \right)_{\Lambda=0} \tan \Lambda$$
$$= 1.012 \times 196 \times 0.216 \times \tan(15°) = 11.48 \text{ ft}$$

Similarly, with 30 degrees of sweep

$$\bar{x}_{ac} \cong \kappa_{ac} b \left(\frac{\bar{z}_{ac}}{b} \right)_{\Lambda=0} \tan \Lambda = 1.005 \times 196 \times 0.216 \times \tan(30°) = 24.56 \text{ ft}$$

The pitching moment coefficient about the wing aerodynamic center can be estimated from Eq. (1.11.18). Because this wing has a symmetric airfoil section, the airfoil section pitching moment and the first term in Eq. (1.11.18) are zero. Furthermore, because the second term in Eq. (1.11.18) is proportional to the wing twist, the pitching moment coefficient about the aerodynamic center of the untwisted wings is zero. Using Eq. (1.11.18) and choosing the mean chord as the reference length, for the wing with 15 degrees of sweep and 4 degrees of washout we obtain

$$C_{m_{ac}} \cong \frac{\bar{c}_{mac}}{c_{ref}} \tilde{C}_{m_{ac}} + \kappa_{M\Omega} \frac{\tan(\kappa_{M\Lambda} \Lambda)}{\bar{c}/b} C_{L,\alpha} \Omega$$
$$= 0.0 + 0.0262 \frac{\tan(0.871 \times 15°)}{1/8} 4.954 (4 \times \pi/180) = 0.0168$$

Similarly, with 30 degrees of sweep and 4 degrees of washout

$$C_{m_{ac}} \cong 0.0 + 0.0262 \frac{\tan(0.871 \times 30°)}{1/8} 4.731 (4 \times \pi/180) = 0.0340$$

Λ	Ω	κ_{ac}	$(\bar{z}_{ac}/b)_0$	$\bar{x}_{ac}$ (ft)	$\tilde{C}_{m_{ac}}$	$C_{L,\alpha}$	$\kappa_{M\Omega}$	$\kappa_{M\Lambda}$	$C_{m_{ac}}$
15°	0°	1.012	0.216	11.48	0.0000	4.954	0.0262	0.871	0.0000
15°	4°	1.012	0.216	11.48	0.0000	4.954	0.0262	0.871	0.0168
30°	0°	1.005	0.216	24.56	0.0000	4.731	0.0262	0.871	0.0000
30°	4°	1.005	0.216	24.56	0.0000	4.731	0.0262	0.871	0.0340

1.12. Inviscid Compressible Aerodynamics

In the analytical methods that were reviewed in Secs. 1.6 through 1.10, the variation in air density was assumed to be negligible. This is a reasonable approximation for flight Mach numbers less than about 0.3. For higher flight Mach numbers, compressibility effects become increasingly important. For flight speeds near or exceeding the speed of sound, the effects of compressibility vastly alter the airflow about an aircraft in flight. In this and the following sections, we review some of the important concepts associated with compressible aerodynamics.

As discussed in Sec. 1.4, inviscid flow over any body immersed in a uniform flow is irrotational. Any irrotational flow is a potential flow. There is no requirement that the flow be incompressible. Thus, the velocity field for inviscid flow about a body immersed in uniform flow can always be expressed according to Eq. (1.4.14):

$$\mathbf{V} = \nabla\phi \tag{1.12.1}$$

where ϕ is the scalar velocity potential. From Eq. (1.4.5), the continuity equation for steady compressible flow is

$$(\mathbf{V}\cdot\nabla)\rho + \rho\nabla\cdot\mathbf{V} = 0 \tag{1.12.2}$$

Inviscid adiabatic flow is isentropic. For an ideal gas this requires that

$$\frac{p}{p_0} = \left(\frac{\rho}{\rho_0}\right)^{\gamma} \tag{1.12.3}$$

where p_0 and ρ_0 are the stagnation pressure and density. For steady isentropic flow of an ideal gas, the momentum equation is expressed in Eq. (1.4.19):

$$\tfrac{1}{2}V^2 + \frac{\gamma}{\gamma-1}\frac{p}{\rho} = \frac{\gamma}{\gamma-1}\frac{p_0}{\rho_0} \tag{1.12.4}$$

For an ideal gas

$$p/\rho = RT \tag{1.12.5}$$

$$a^2 = \gamma RT \tag{1.12.6}$$

Substituting Eqs. (1.12.5) and (1.12.6) into Eq. (1.12.3) gives

$$\frac{p}{p_0} = \frac{RT\rho}{RT_0\rho_0} = \frac{\gamma RT\rho}{\gamma RT_0\rho_0} = \frac{a^2\rho}{a_0^2\rho_0} = \left(\frac{\rho}{\rho_0}\right)^{\gamma} \tag{1.12.7}$$

Similarly, applying Eqs. (1.12.5) and (1.12.6) to Eq. (1.12.4) yields

$$\frac{1}{2}V^2 + \frac{a^2}{\gamma-1} = \frac{a_0^2}{\gamma-1} \tag{1.12.8}$$

Solving Eq. (1.12.7) for the air density and solving Eq. (1.12.8) for the speed of sound squared results in

$$\rho = \rho_0 \left(\frac{a^2}{a_0^2}\right)^{\frac{1}{\gamma-1}} \tag{1.12.9}$$

$$a^2 = a_0^2 - \frac{\gamma-1}{2}V^2 \tag{1.12.10}$$

Substituting Eq. (1.12.10) into Eq. (1.12.9), we have

$$\rho = \rho_0 \left(1 - \frac{\gamma-1}{2}\frac{V^2}{a_0^2}\right)^{\frac{1}{\gamma-1}} \tag{1.12.11}$$

This specifies the local air density as a function of the local velocity and the stagnation conditions, which are known from the freestream conditions according to Eqs. (1.4.23) through (1.4.25).

Applying Eq. (1.12.1) to express the velocity in Eq. (1.12.11) in terms of the velocity potential, the local air density is found to be a function only of ϕ and known stagnation properties of the flow:

$$\rho = \rho_0 \left(1 - \frac{\gamma-1}{2a_0^2}\mathbf{V}\cdot\mathbf{V}\right)^{\frac{1}{\gamma-1}} = \rho_0 \left(1 - \frac{\gamma-1}{2a_0^2}\nabla\phi\cdot\nabla\phi\right)^{\frac{1}{\gamma-1}} \tag{1.12.12}$$

After using Eqs. (1.12.1) and (1.12.12) in Eq. (1.12.2), the **continuity equation for steady compressible potential flow** can be written as

$$\left(a_0^2 - \frac{\gamma-1}{2}\nabla\phi\cdot\nabla\phi\right)\nabla^2\phi - \frac{1}{2}(\nabla\phi\cdot\nabla)(\nabla\phi\cdot\nabla\phi) = 0 \tag{1.12.13}$$

This equation contains only one unknown, the velocity potential ϕ. Both γ and a_0 are known constants of the flow. Since Eq. (1.12.13) is a single scalar equation in only one scalar unknown, it provides a tremendous simplification over the more general Navier-Stokes equations. However, like the Navier-Stokes equations but unlike the Laplace equation that governs the velocity potential for incompressible flow, Eq. (1.12.13) is nonlinear. Once the velocity potential has been determined from Eq. (1.12.13), the velocity field can be evaluated from Eq. (1.12.1). With the velocity known, the speed of sound at any point in the flow can be determined from Eq. (1.12.10). Knowing the local velocity and speed of sound at every point in the flow allows us to compute the

Mach number and then, using Eqs. (1.4.20) through (1.4.22), to evaluate the pressure, temperature, and air density.

It is sometimes useful to apply a change of variables in Eq. (1.12.13). Here we shall define a new velocity vector to be the difference between the local velocity and the freestream velocity,

$$\mathbf{V}_p \equiv \mathbf{V} - \mathbf{V}_\infty \tag{1.12.14}$$

This is commonly called the *perturbation velocity*. From this definition, we also define the *perturbation velocity potential*,

$$\phi_p \equiv \phi - \phi_\infty \tag{1.12.15}$$

where ϕ_∞ is the velocity potential for the uniform flow, which was presented in Sec. 1.5. For the uniform flow potential we have

$$\nabla \phi_\infty = \mathbf{V}_\infty \tag{1.12.16}$$

$$\nabla \phi_\infty \cdot \nabla \phi_\infty = V_\infty^2 \tag{1.12.17}$$

$$\nabla^2 \phi_\infty = 0 \tag{1.12.18}$$

Using Eqs. (1.12.15) through (1.12.18) in Eq. (1.12.13) results in

$$\left[a_0^2 - \frac{\gamma - 1}{2} V_\infty^2 \left(1 + 2\mathbf{u}_\infty \cdot \nabla \hat{\phi}_p + \nabla \hat{\phi}_p \cdot \nabla \hat{\phi}_p \right) \right] \nabla^2 \hat{\phi}_p$$
$$- \frac{1}{2} V_\infty^2 \left[\left(\nabla \hat{\phi}_p + \mathbf{u}_\infty \right) \cdot \nabla \right] \left(2\mathbf{u}_\infty \cdot \nabla \hat{\phi}_p + \nabla \hat{\phi}_p \cdot \nabla \hat{\phi}_p \right) = 0 \tag{1.12.19}$$

where $\mathbf{u}_\infty$ is the unit vector in the direction of the freestream

$$\mathbf{u}_\infty \equiv \frac{\mathbf{V}_\infty}{V_\infty} \tag{1.12.20}$$

and

$$\hat{\phi}_p \equiv \frac{\phi_p}{V_\infty} \tag{1.12.21}$$

Equation (1.12.19) is the general equation for the perturbation velocity potential. To this point in the development, no approximation has been made in going from Eq. (1.12.13) to Eq. (1.12.19). This result applies to any irrotational flow.

Equation (1.12.19) can be linearized under the assumption that the perturbation velocity is small compared to the freestream velocity. This approximation results in

$$\left(a_0^2 - \frac{\gamma - 1}{2} V_\infty^2 \right) \nabla^2 \hat{\phi}_p - V_\infty^2 \left(\mathbf{u}_\infty \cdot \nabla \right) \left(\mathbf{u}_\infty \cdot \nabla \hat{\phi}_p \right) = 0 \tag{1.12.22}$$

Applying Eq. (1.12.10) to the freestream gives

$$a_\infty^2 = a_0^2 - \frac{\gamma - 1}{2} V_\infty^2 \tag{1.12.23}$$

Substituting Eq. (1.12.23) in Eq. (1.12.22) and dividing through by a_∞^2, the **linearized equation for the perturbation velocity potential** is

$$\nabla^2 \phi_p - M_\infty^2 \left(\mathbf{u}_\infty \cdot \nabla \right) \left(\mathbf{u}_\infty \cdot \nabla \phi_p \right) = 0 \tag{1.12.24}$$

Notice that as the flight Mach number approaches zero this result reduces to Laplace's equation, which applies to incompressible flow. Equation (1.12.24) provides reasonable predictions for slender bodies at low angles of attack with subsonic Mach numbers ($M_\infty < 0.8$) and supersonic Mach numbers ($1.2 < M_\infty < 5$). It is not valid for transonic Mach numbers ($0.8 < M_\infty < 1.2$) or hypersonic Mach numbers ($M_\infty > 5$).

1.13. Compressible Subsonic Flow

The aerodynamic theory of incompressible flow over airfoil sections was reviewed in Secs. 1.6 and 1.7. The analytical methods that have been developed for incompressible flow over airfoils can be applied to subsonic compressible flow through a simple change of variables. However, the reader should recall that subsonic flow does not exist all the way to a flight Mach number of 1.0. For an airfoil section producing positive lift, the flow velocities just outside the boundary layer on the upper surface are greater than the freestream velocity. Thus, at some flight speeds below Mach 1.0, supersonic flow will be encountered in some region above the upper surface of the airfoil. The flight Mach number at which sonic flow is first encountered at some point on the upper surface of the airfoil is called the *critical Mach number*. Here we review the theory of compressible flow over airfoils at flight speeds below the critical Mach number.

The Prandtl-Glauert Compressibility Correction

For thin airfoils at small angles of attack and flight Mach numbers below critical, the linearized potential flow approximation given by Eq. (1.12.24) can be applied. For the special case of two-dimensional flow and a coordinate system having the x-axis aligned with the freestream velocity vector, $\mathbf{u}_\infty$ is simply the unit vector in the x-direction, and the continuity equation as expressed in Eq. (1.12.24) becomes

$$\left(\frac{\partial^2 \phi_p}{\partial x^2} + \frac{\partial^2 \phi_p}{\partial y^2} \right) - M_\infty^2 \left(\frac{\partial}{\partial x} \right) \left(\frac{\partial \phi_p}{\partial x} \right) = 0$$

or after rearranging,

$$\frac{\partial^2 \phi_p}{\partial x^2} + \frac{1}{1-M_\infty^2} \frac{\partial^2 \phi_p}{\partial y^2} = 0 \qquad (1.13.1)$$

The nature of this partial differential equation depends on the sign of $1-M_\infty^2$. Recall from your introductory course on differential equations that if the flight Mach number is greater than 1.0, Eq. (1.13.1) is hyperbolic. However, for subsonic Mach numbers this partial differential equation is elliptic.

For subsonic flight, Eq. (1.13.1) is simplified by using the change of independent variables

$$\hat{y} \equiv y\sqrt{1-M_\infty^2} \qquad (1.13.2)$$

With this change of variables, Eq. (1.13.1) becomes

$$\frac{\partial^2 \phi_p}{\partial x^2} + \frac{\partial^2 \phi_p}{\partial \hat{y}^2} = 0 \qquad (1.13.3)$$

which is exactly Laplace's equation that governs incompressible potential flow. Thus, with this coordinate transformation, the incompressible flow solution can be used for compressible flow.

The local pressure coefficient obtained from any solution to Eq. (1.13.3) is given by

$$C_p = \frac{C_{pM0}}{\sqrt{1-M_\infty^2}} \qquad (1.13.4)$$

where C_{pM0} is the pressure coefficient for zero Mach number, which is the solution obtained for incompressible flow. Equation (1.13.4) relates the pressure coefficient for subsonic compressible flow to that for incompressible flow and is commonly referred to as the *Prandtl-Glauert compressibility correction*.

For inviscid flow, the section lift and moment coefficients can be found from simple integrals of the pressure distribution over the surface of the airfoil. As a result, the final result for section lift and moment coefficients look much like Eq. (1.13.4):

$$\tilde{C}_L = \frac{\tilde{C}_{LM0}}{\sqrt{1-M_\infty^2}} \qquad (1.13.5)$$

$$\tilde{C}_m = \frac{\tilde{C}_{mM0}}{\sqrt{1-M_\infty^2}} \qquad (1.13.6)$$

where $\tilde{C}_{LM0}$ and $\tilde{C}_{mM0}$ are the section lift and moment coefficients obtained for incompressible flow. The development of Eqs. (1.13.4) through (1.13.6) can be found in any undergraduate engineering textbook on aerodynamics and will not be repeated here.

Critical Mach Number

For potential flow over an airfoil producing positive lift, the flow velocities along some portion of the upper surface are greater than the freestream velocity. Also recall that for incompressible potential flow, the ratio of local velocity to the freestream velocity at any point in the flow does not vary with the magnitude of the freestream velocity. In other words, if we double the freestream velocity in an incompressible potential flow, the velocity at every point in the flow field will double as well. This is a direct result of the linear nature of the Laplace equation, which governs incompressible potential flow. Thus, within the approximation of Eq. (1.13.1), the position of the point of maximum velocity on the upper surface of the airfoil does not change with freestream velocity.

From the definition of pressure coefficient, we can write

$$
C_p \equiv \frac{p - p_\infty}{\frac{1}{2}\rho_\infty V_\infty^2} = \frac{2}{(\rho_\infty/p_\infty)V_\infty^2}\left(\frac{p}{p_\infty} - 1\right)
$$

$$
= \frac{2}{(\gamma/\gamma RT_\infty)V_\infty^2}\left(\frac{p}{p_\infty} - 1\right) = \frac{2}{\gamma M_\infty^2}\left(\frac{p}{p_\infty} - 1\right)
$$

(1.13.7)

From Eq. (1.4.21),

$$
\frac{p}{p_\infty} = \left(\frac{1+[(\gamma-1)/2]M_\infty^2}{1+[(\gamma-1)/2]M^2}\right)^{\frac{\gamma}{\gamma-1}}
$$

(1.13.8)

Substituting Eq. (1.13.8) into Eq. (1.13.7), the pressure coefficient can be expressed as a function of Mach number,

$$
C_p = \frac{2}{\gamma M_\infty^2}\left[\left(\frac{1+[(\gamma-1)/2]M_\infty^2}{1+[(\gamma-1)/2]M^2}\right)^{\frac{\gamma}{\gamma-1}} - 1\right]
$$

(1.13.9)

This relation allows us to compute the local pressure coefficient at any point in the flow from the local Mach number and the freestream Mach number.

The critical Mach number, M_{cr}, is the freestream Mach number that results in a local Mach number of 1.0 at that point on the upper surface where the pressure coefficient is lowest. Using this fact together with Eqs. (1.13.4) and (1.13.9) gives the relation

$$
\frac{(C_{pM0})_{min}}{\sqrt{1-M_{cr}^2}} = \frac{2}{\gamma M_{cr}^2}\left[\left(\frac{1+[(\gamma-1)/2]M_{cr}^2}{1+[(\gamma-1)/2]}\right)^{\frac{\gamma}{\gamma-1}} - 1\right]
$$

(1.13.10)

The minimum pressure coefficient on the airfoil surface for incompressible flow can be evaluated using a panel code. With this pressure coefficient known, Eq. (1.13.10) contains only one unknown, the critical Mach number. This is easily solved for the value of the critical Mach number by using the secant method. Since the secant method is used frequently in this text for the solution of flight mechanics problems, its use is demonstrated in the following example.

EXAMPLE 1.13.1. For a particular airfoil at a particular angle of attack, a vortex panel code solution predicts that the minimum pressure coefficient on the surface of the airfoil is –0.43. Estimate the critical Mach number.

Solution. We start by writing Eq. (1.13.10) in residual form, i.e.,

$$R \equiv \frac{2\sqrt{1-M_{cr}^2}}{\gamma M_{cr}^2} \left[\left(\frac{1+[(\gamma-1)/2]M_{cr}^2}{1+[(\gamma-1)/2]} \right)^{\frac{\gamma}{\gamma-1}} -1 \right] - \left(C_{pM0} \right)_{min}$$

where R is the residual. We wish to find the value of M_{cr} that will make R go to zero. We start with some initial estimate for the Mach number, say $M_1 = 0.80$. With $\gamma = 1.4$, this gives a residual value of

$$R_1 = \frac{\sqrt{1-M_1^2}}{0.7M_1^2} \left[\left(\frac{1+0.2M_1^2}{1.2} \right)^{3.5} -1 \right] - (-0.43) = 0.169216$$

For the secant iteration, we need a second estimate, say $M_2 = 0.70$. This gives

$$R_2 = \frac{\sqrt{1-M_2^2}}{0.7M_2^2} \left[\left(\frac{1+0.2M_2^2}{1.2} \right)^{3.5} -1 \right] - (-0.43) = -0.126364$$

With these two results we compute our next estimate from the secant algorithm,

$$M_3 = M_2 - R_2 \frac{M_2-M_1}{R_2-R_1} = 0.742751, \quad R_3 = 0.017349$$

The process is repeated until the residual is less than some convergence criterion. On each iteration, the oldest estimate is discarded. This gives

$$M_4 = 0.737590, \quad R_4 = 0.001507$$
$$M_5 = 0.737099, \quad R_5 = 0.000020$$

Since this residual is essentially zero, our solution is

$$M_{cr} = \underline{0.74}$$

Drag Divergence

While the lift and pitching moment produced on an airfoil section are generated primarily from pressure forces, at subsonic speeds the drag is attributed almost entirely to viscous forces generated in the boundary layer. Below the critical Mach number, airfoil section drag coefficient does not vary substantially with Mach number. However, beyond the critical Mach number the drag coefficient begins to increase very rapidly with increasing Mach number, reaching a maximum at Mach 1 as shown in Fig. 1.13.1. This figure is intended only to show the general shape of the drag curve that might be expected for some airfoil. The exact shape of this drag curve and the total drag rise for a particular airfoil is quite sensitive to the shape of the airfoil section. There is no simple theory capable of predicting the variation in section drag coefficient with Mach number in the transonic region. At Mach numbers near 1.0, the flow over an airfoil is extremely complex and very sensitive to section shape. In the transonic region we commonly rely on experimental data.

The large increase in drag that occurs at Mach numbers just below 1.0, called *drag divergence*, is due primarily to the shock waves that form in transonic flow and the premature boundary layer separation caused by these shocks. The dramatic increase in drag that is experienced by an airplane as it approaches Mach 1 is what gave rise to the concept and terminology of the *sound barrier*. Some early researchers believed that the drag would become infinite as the speed of sound was approached. Thus, it was thought by some that it would be impossible to exceed the speed of sound in an aircraft. As can be imagined from inspection of Fig. 1.13.1, the speed of sound does present a substantial barrier to an aircraft with limited thrust. Nevertheless, the drag at Mach 1 is finite, and like most things in this universe, the sound barrier will yield to sufficient power. It will yield more easily, however, if finesse is applied with the power.

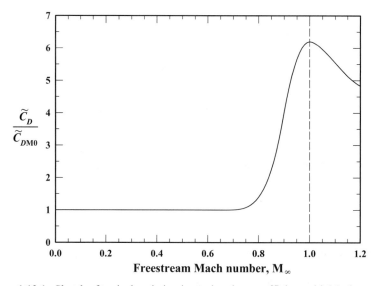

Figure 1.13.1. Sketch of typical variation in section drag coefficient with Mach number.

In the last half-century a great deal of "engineering finesse" has been applied to help "lower the sound barrier" by increasing the critical Mach number and decreasing the peak drag experienced near Mach 1. Contributions that have led to the design of improved supersonic aircraft include the use of thinner airfoils and swept wings, the introduction of the *area rule* in the supersonic design philosophy, and development of the supercritical airfoil. All of these topics are covered in most aerodynamics textbooks and the reader is encouraged to review this material, if necessary.

1.14. Supersonic Flow

In an irrotational supersonic flow where the streamlines make only small angles with the freestream, the linearized potential flow approximation given by Eq. (1.12.24) can be applied. Again consider two-dimensional flow with the *x*-axis aligned with the freestream velocity vector. These are exactly the conditions that were used in Sec. 1.13 to study subsonic compressible flow over an airfoil. Thus, Eq. (1.13.1) applies to supersonic flow at small angles as well as to subsonic flow. However, since our interest is now in Mach numbers greater than 1.0, Eq. (1.13.1) is more conveniently written as

$$\frac{\partial^2 \phi_p}{\partial x^2} - \frac{1}{M_\infty^2 - 1} \frac{\partial^2 \phi_p}{\partial y^2} = 0 \qquad (1.14.1)$$

Solutions to this hyperbolic partial differential equation are characteristically very different from solution to the elliptic equation encountered for subsonic flow. Thus, we should expect supersonic flow to be very different from subsonic flow.

In obtaining a solution to Eq. (1.14.1), a change of variables is again useful. We wish to find a variable change that will allow us to separate the independent variables, *x* and *y*, in Eq. (1.14.1). This is a common technique used in obtaining solutions to partial differential equations. We want to find a change of variables, say $\xi(x,y)$, that will make the perturbation potential a function of ξ only, i.e., $\phi_p(\xi)$. With this in mind, Eq. (1.14.1) can be written as

$$\frac{d^2 \phi_p}{d\xi^2}\left(\frac{\partial \xi}{\partial x}\right)^2 + \frac{d\phi_p}{d\xi}\frac{\partial^2 \xi}{\partial x^2} - \frac{1}{M_\infty^2 - 1}\left[\frac{d^2 \phi_p}{d\xi^2}\left(\frac{\partial \xi}{\partial y}\right)^2 + \frac{d\phi_p}{d\xi}\frac{\partial^2 \xi}{\partial y^2}\right] = 0 \qquad (1.14.2)$$

Equation (1.14.2) is satisfied for ϕ_p equal to any function of ξ if

$$\frac{\partial^2 \xi}{\partial x^2} = \frac{1}{M_\infty^2 - 1}\frac{\partial^2 \xi}{\partial y^2} \quad \text{and} \quad \frac{\partial \xi}{\partial x} = \pm\frac{1}{\sqrt{M_\infty^2 - 1}}\frac{\partial \xi}{\partial y}$$

Both of these relations are satisfied if we set both sides of the second relation equal to an arbitrary constant, say C_1. This gives

$$\frac{\partial \xi}{\partial x} = C_1 \quad \text{and} \quad \frac{\partial \xi}{\partial y} = \pm C_1\sqrt{M_\infty^2 - 1}$$

These equations are easily integrated to yield

$$\xi(x, y) = C_1 x + f_1(y) \quad \text{and} \quad \xi(x, y) = \pm C_1 \sqrt{M_\infty^2 - 1}\, y + f_2(x)$$

which require that

$$\xi(x, y) = C_1\left(x \pm \sqrt{M_\infty^2 - 1}\, y\right) + C_2$$

Because C_1 and C_2 are arbitrary, we can choose $C_1 = 1$ and $C_2 = 0$, which gives

$$\xi(x, y) = x \pm \sqrt{M_\infty^2 - 1}\, y \qquad (1.14.3)$$

Either of the two solutions for ξ that are given in Eq. (1.14.3) will provide the desired variable change. With either definition, any function $\phi_p(\xi)$ will satisfy Eq. (1.14.1). Thus, any problem associated with small-angle potential flow at supersonic Mach numbers reduces to that of finding the function $\phi_p(\xi)$ that will also satisfy the required boundary conditions.

From Eq. (1.14.3), a great deal can be deduced about supersonic potential flow at small angles without applying any boundary conditions. Since the perturbation potential is a function only of ξ, lines of constant perturbation potential are lines of ξ. Thus, the **constant perturbation potential lines are described by the equation**

$$x \pm \sqrt{M_\infty^2 - 1}\, y = \text{constant} \qquad (1.14.4)$$

Since this is the equation of a straight line, the constant perturbation potential lines are all straight lines. Furthermore, solving Eq. (1.14.4) for y and differentiating with respect to x, we find that **the slope of any constant perturbation potential line is**

$$\left(\frac{\partial y}{\partial x}\right)_{\phi_p = \text{const}} = \mp 1 / \sqrt{M_\infty^2 - 1} \qquad (1.14.5)$$

The angle that the constant perturbation potential lines makes with the freestream (i.e., the x-axis) is called the **Mach angle** and is given by

$$\mu = \mp \tan^{-1}\left(1 / \sqrt{M_\infty^2 - 1}\right) \qquad (1.14.6)$$

Notice that for supersonic potential flow at small angles, the Mach angle is independent of surface geometry. It depends only on the freestream Mach number. Also notice that there are two solutions for the Mach angle and that the signs in Eqs. (1.14.5) and (1.14.6) are opposite to those in Eqs. (1.14.3) and (1.14.4). Thus, choosing the negative sign in Eq. (1.14.3) results in a positive Mach angle, and vice versa.

From the definition of perturbation potential, given in Eq. (1.12.15), the local fluid velocity vector is given by

$$\mathbf{V} = \nabla\phi = \nabla\phi_\infty + \nabla\phi_p = \mathbf{V}_\infty + \nabla\phi_p = V_\infty \mathbf{i}_x + \nabla\phi_p$$

Using the relation

$$\phi_p = \phi_p(\xi) = \phi_p(x \pm \sqrt{M_\infty^2 - 1}\, y)$$

we have

$$\mathbf{V} = \left(V_\infty + \frac{d\phi_p}{d\xi}\frac{\partial\xi}{\partial x}\right)\mathbf{i}_x + \frac{d\phi_p}{d\xi}\frac{\partial\xi}{\partial y}\mathbf{i}_y = \left(V_\infty + \frac{d\phi_p}{d\xi}\right)\mathbf{i}_x \pm \sqrt{M_\infty^2 - 1}\,\frac{d\phi_p}{d\xi}\mathbf{i}_y \quad (1.14.7)$$

The slope of a streamline at any point in the flow can be evaluated from Eq. (1.14.7),

$$\frac{V_y}{V_x} = \pm\sqrt{M_\infty^2 - 1}\,\frac{d\phi_p}{d\xi}\bigg/\left(V_\infty + \frac{d\phi_p}{d\xi}\right)$$

Because we are considering the case of small perturbation velocity, only the first-order term in this result should be retained. Since the streamlines are tangent to the velocity vector, **the slope of the local streamline relative to the x-axis is**

$$\left(\frac{\partial y}{\partial x}\right)_{streamline} = \pm\frac{\sqrt{M_\infty^2 - 1}}{V_\infty}\frac{d\phi_p}{d\xi} \quad (1.14.8)$$

From Eq. (1.14.7), the local velocity squared is

$$V^2 = \left(V_\infty + \frac{d\phi_p}{d\xi}\right)^2 + (M_\infty^2 - 1)\left(\frac{d\phi_p}{d\xi}\right)^2 = V_\infty^2 + 2V_\infty\frac{d\phi_p}{d\xi} + M_\infty^2\left(\frac{d\phi_p}{d\xi}\right)^2$$

Again, since we are considering the case of small perturbation velocity, the second-order term should be ignored and the **velocity at any point in the flow** is given by

$$V^2 = V_\infty^2 + 2V_\infty\frac{d\phi_p}{d\xi} \quad (1.14.9)$$

From the momentum equation for a compressible potential flow, which is given by Eq. (1.4.19), we can express the local temperature in terms of the local velocity. Using the ideal gas law in Eq. (1.4.19) at an arbitrary point and the freestream gives

$$\frac{1}{2}V^2 + \frac{\gamma RT}{\gamma - 1} = \frac{1}{2}V_\infty^2 + \frac{\gamma RT_\infty}{\gamma - 1}$$

Solving this for temperature, using Eq. (1.14.9) to eliminate the velocity, and applying the relation $a^2 = \gamma RT$ gives the **temperature at any point in the flow**:

$$T = T_\infty \left[1 - (\gamma - 1) \frac{M_\infty}{a_\infty} \frac{d\phi_p}{d\xi} \right] \qquad (1.14.10)$$

From the isentropic relations given in Eqs. (1.4.20) and (1.4.21), the pressure ratio is easily related to the temperature ratio and after applying Eq. (1.14.10), we obtain

$$\frac{p}{p_\infty} = \left(\frac{T}{T_\infty} \right)^{\frac{\gamma}{\gamma - 1}} = \left[1 - (\gamma - 1) \frac{M_\infty}{a_\infty} \frac{d\phi_p}{d\xi} \right]^{\frac{\gamma}{\gamma - 1}}$$

Expanding this result in a Taylor series and retaining only the first-order term results in an expression for the **pressure at any point in the flow**:

$$p = p_\infty \left(1 - \gamma \frac{M_\infty}{a_\infty} \frac{d\phi_p}{d\xi} \right) \qquad (1.14.11)$$

Using Eq. (1.14.11) in the definition of pressure coefficient as expressed in Eq. (1.13.7) results in

$$C_p = \frac{2}{\gamma M_\infty^2} \left(\frac{p}{p_\infty} - 1 \right) = -\frac{2}{V_\infty} \frac{d\phi_p}{d\xi} \qquad (1.14.12)$$

Solving Eq. (1.14.8) for the derivative of the perturbation potential and using Eq. (1.14.6) to eliminate the freestream Mach number in favor of the Mach angle gives

$$\frac{d\phi_p}{d\xi} = \pm \frac{V_\infty}{\sqrt{M_\infty^2 - 1}} \left(\frac{\partial y}{\partial x} \right)_{\text{streamline}} = -V_\infty \tan(\mu) \left(\frac{\partial y}{\partial x} \right)_{\text{streamline}} \qquad (1.14.13)$$

Substituting Eq. (1.14.13) into Eq. (1.14.12), **the pressure coefficient at any point in a small-angle potential flow at supersonic Mach numbers can be expressed as a function of only the Mach angle and the slope of the local streamline**:

$$C_p = 2 \tan(\mu) \left(\frac{\partial y}{\partial x} \right)_{\text{streamline}} \qquad (1.14.14)$$

Remember that the Mach angle can be chosen as either positive or negative as needed to satisfy the required boundary conditions. However, the positive sign in Eq. (1.14.14) holds regardless of the sign chosen for the Mach angle. If the streamline slope has the same sign as the Mach angle, the pressure coefficient is positive. If these signs are opposite, the pressure coefficient is negative.

For those of us interested in computing aerodynamic forces, this is a very simple result in comparison to most solutions that have been obtained for fluid flow. In this respect, supersonic potential flow at small angles is less complex than subsonic flow at small angles. It rivals Bernoulli's equation for raw simplicity. The Mach angle depends only on the freestream Mach number. Furthermore, at the surface of a solid body, the streamlines must be tangent to the surface. Thus, the slope of the surface streamlines depends only on the surface geometry and the angle of attack.

Supersonic Thin Airfoils

The supersonic small-angle potential flow equations can be used to predict the aero-dynamic force and moment components acting on a thin airfoil at small angles of attack, provided that the airfoil has a sharp leading edge as shown in Fig. 1.14.1. This type of leading edge is commonly used on supersonic airfoils because a blunt leading edge produces a strong bow shock in supersonic flight.

Consider supersonic flow over the cambered airfoil shown in Fig. 1.14.1. Because the flow must be symmetric for a symmetric airfoil at zero angle of attack, the positive slope is chosen for the Mach lines above the airfoil and the negative slope is chosen for those Mach lines below the airfoil. The slope of the upper surface relative to the freestream can be expressed as the slope of the chord line relative to the freestream, plus the slope of the camber line relative to the chord line, plus the slope of the thickness line relative to the camber line. Thus, for small angles of attack, **the upper surface slope is**

$$\frac{dy_u}{dx} = -\alpha + \frac{dy_c}{dx} + \frac{dy_t}{dx} \qquad (1.14.15)$$

where y_c is the local camber and y_t is one-half the local thickness. Similarly, **the lower surface slope is**

$$\frac{dy_l}{dx} = -\alpha + \frac{dy_c}{dx} - \frac{dy_t}{dx} \qquad (1.14.16)$$

Since the streamlines must be tangent to the airfoil surface, the results above can be used in Eq. (1.14.14) to evaluate the pressure coefficient on the airfoil surface.

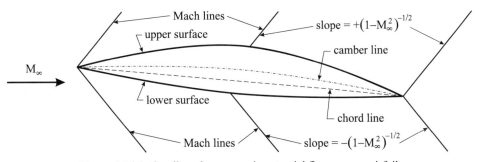

Figure 1.14.1. Small-angle supersonic potential flow over an airfoil.

The contribution of pressure on the lower surface to the section lift is the surface pressure multiplied by a differential area per unit span projected on the x-axis. Similarly, the upper surface contributes to the lift in just the opposite direction. Thus, the section lift coefficient is

$$\tilde{C}_L = \frac{\tilde{L}}{\frac{1}{2}\rho_\infty V_\infty^2 c} = \int_0^c \frac{p_l - p_u}{\frac{1}{2}\rho_\infty V_\infty^2} \frac{dx}{c} = \int_0^c (C_{pl} - C_{pu}) \frac{dx}{c} \qquad (1.14.17)$$

The slope of the Mach lines is positive on the upper surface and negative on the lower surface. Thus, from Eq. (1.14.6),

$$\tan(\mu_l) = -1\big/\sqrt{M_\infty^2 - 1} \quad \text{and} \quad \tan(\mu_u) = +1\big/\sqrt{M_\infty^2 - 1}$$

Using these results with Eqs. (1.14.14) through (1.14.16) applied to Eq. (1.14.17) gives

$$\tilde{C}_L = \int_0^c \left[2\frac{-1}{\sqrt{M_\infty^2 - 1}}\left(-\alpha + \frac{dy_c}{dx} - \frac{dy_t}{dx}\right) - 2\frac{+1}{\sqrt{M_\infty^2 - 1}}\left(-\alpha + \frac{dy_c}{dx} + \frac{dy_t}{dx}\right) \right] \frac{dx}{c}$$

$$= \frac{4}{\sqrt{M_\infty^2 - 1}} \int_0^c \left(\alpha - \frac{dy_c}{dx} \right)\frac{dx}{c} = \frac{4}{\sqrt{M_\infty^2 - 1}}\left[\alpha - \frac{y_c(c) - y_c(0)}{c} \right]$$

By definition, the camber of any airfoil is zero at both the leading and trailing edges. Thus, **the section lift coefficient is**

$$\tilde{C}_L = \frac{4\alpha}{\sqrt{M_\infty^2 - 1}} = \tilde{C}_{L,\alpha}\,\alpha \quad \text{where} \quad \tilde{C}_{L,\alpha} = \frac{4}{\sqrt{M_\infty^2 - 1}} \qquad (1.14.18)$$

From this result, we see that the small-angle potential flow equations predict that neither camber nor thickness makes a contribution to the lift at supersonic airspeeds.

Similarly, the pitching moment coefficient about the leading edge is evaluated from

$$\tilde{C}_{m_{le}} = \frac{\tilde{m}_{le}}{\frac{1}{2}\rho_\infty V_\infty^2 c^2} = \int_0^c (C_{pu} - C_{pl})\frac{x\,dx}{c^2}$$

which gives

$$\tilde{C}_{m_{le}} = \frac{-4}{\sqrt{M_\infty^2 - 1}} \int_0^c \left(\alpha - \frac{dy_c}{dx} \right)\frac{x\,dx}{c^2} = \frac{-2\alpha}{\sqrt{M_\infty^2 - 1}} + \frac{4}{\sqrt{M_\infty^2 - 1}}\left(\int_0^c \frac{dy_c}{dx}\frac{x\,dx}{c^2} \right)$$

or

$$\tilde{C}_{m_{le}} = -\frac{1}{2}\tilde{C}_L + \tilde{C}_{L,\alpha} \int_0^c \frac{dy_c}{dx}\frac{x\,dx}{c^2}$$

From this result and the material that was reviewed in Sec. 1.1, it can be seen that **the aerodynamic center is located at the half-chord and the pitching moment coefficient about the aerodynamic center is**

$$\tilde{C}_{m_{ac}} = \tilde{C}_{L,\alpha} \int_0^c \frac{dy_c}{dx} \frac{x\,dx}{c^2} \quad \text{where} \quad \frac{x_{ac}}{c} = \frac{1}{2} \tag{1.14.19}$$

The moment coefficient about the aerodynamic center depends only on the camber line shape. Since camber does not contribute to lift in small-angle supersonic potential flow, symmetric airfoils are often used for supersonic flight. Notice from Eq. (1.14.19) that a thin symmetric airfoil produces no moment about its half-chord. Thus, **the half-chord is also the center of pressure for a thin symmetric airfoil in supersonic flight.**

The section drag for supersonic potential flow over a thin airfoil is not zero as it is for subsonic flow. This section drag is called *wave drag* and is computed from

$$\tilde{C}_D = \frac{\tilde{D}}{\frac{1}{2}\rho_\infty V_\infty^2 c} = \int_0^c \left(\frac{p_u - p_\infty}{\frac{1}{2}\rho_\infty V_\infty^2} \frac{dy_u}{dx} - \frac{p_l - p_\infty}{\frac{1}{2}\rho_\infty V_\infty^2} \frac{dy_l}{dx} \right) \frac{dx}{c} = \int_0^c \left(C_{pu} \frac{dy_u}{dx} - C_{pl} \frac{dy_l}{dx} \right) \frac{dx}{c}$$

Applying Eqs. (1.14.6) and (1.14.14) through (1.14.16) gives

$$\tilde{C}_D = \frac{2}{\sqrt{M_\infty^2 - 1}} \int_0^c \left[\left(-\alpha + \frac{dy_c}{dx} + \frac{dy_t}{dx} \right)^2 + \left(-\alpha + \frac{dy_c}{dx} - \frac{dy_t}{dx} \right)^2 \right] \frac{dx}{c}$$

$$= \frac{4}{\sqrt{M_\infty^2 - 1}} \int_0^c \left[\alpha^2 - 2\alpha \frac{dy_c}{dx} + \left(\frac{dy_c}{dx} \right)^2 + \left(\frac{dy_t}{dx} \right)^2 \right] \frac{dx}{c}$$

$$= \frac{4}{\sqrt{M_\infty^2 - 1}} \left\{ \alpha^2 + \int_0^c \left[\left(\frac{dy_c}{dx} \right)^2 + \left(\frac{dy_t}{dx} \right)^2 \right] \frac{dx}{c} \right\}$$

This result can be written in terms of the lift slope that was evaluated previously and presented in Eq. (1.14.18). Thus, **the section wave drag coefficient is**

$$\tilde{C}_D = \tilde{C}_{L,\alpha} \alpha^2 + \tilde{C}_{L,\alpha} \int_0^c \left[\left(\frac{dy_c}{dx} \right)^2 + \left(\frac{dy_t}{dx} \right)^2 \right] \frac{dx}{c} \tag{1.14.20}$$

From the results presented in Eqs. (1.14.18), (1.14.19), and (1.14.20), we see that small-angle potential flow theory predicts that there is no advantage to using camber in a supersonic airfoil. It only contributes to the pitching moment and the drag, making no predicted contribution to the lift. Furthermore, we see that airfoil thickness makes a predicted contribution to wave drag that is proportional to thickness squared. This is one reason why thin airfoil sections are commonly used on supersonic aircraft. Most modern airplanes that are designed to fly at Mach 1.5 and above use airfoils having a thickness on the order of 4 percent. This approximate linearized supersonic airfoil theory agrees reasonably well with experimental observations for airfoils as thick as about 10 percent and angles of attack as large as 20 degrees. As should be expected, this inviscid theory underpredicts the drag and moment coefficients and overpredicts the lift coefficient. However, the discrepancy is only a few percent. Most of the discrepancy can be attributed to viscous effects in the boundary layer. The drag that is predicted by this linear theory includes only wave drag. If a contribution for viscous drag is included from boundary layer computations, the predictions are improved. However, the viscous drag is quite small compared to the wave drag and for many purposes can be ignored.

1.15. Problems

1.1. The following is a tabulation of the section lift, drag, and quarter-chord moment coefficients taken from test data for a particular airfoil section.

α (deg)	$\tilde{C}_L$	$\tilde{C}_D$	$\tilde{C}_{m_{c/4}}$
−6.0	−0.38	0.0077	−0.0446
−5.0	−0.28	0.0071	−0.0440
−4.0	−0.18	0.0066	−0.0434
−3.0	−0.07	0.0063	−0.0428
−2.0	0.04	0.0060	−0.0422
−1.0	0.14	0.0059	−0.0416
0.0	0.24	0.0058	−0.0410
1.0	0.35	0.0058	−0.0404
2.0	0.46	0.0060	−0.0398
3.0	0.56	0.0062	−0.0392
4.0	0.65	0.0066	−0.0386
5.0	0.76	0.0070	−0.0380
6.0	0.87	0.0075	−0.0374
7.0	0.97	0.0081	−0.0368
8.0	1.07	0.0088	−0.0362
9.0	1.16	0.0096	−0.0356
10.0	1.26	0.0105	−0.0350
11.0	1.35	0.0115	−0.0344
12.0	1.44	0.0126	−0.0338

On one graph, plot both the section lift coefficient and the section normal force coefficient as a function of angle of attack. On another graph, plot both the section drag coefficient and the section axial force coefficient as a function of angle of attack.

1.2. From the data presented in problem 1.1, on one graph, plot the location of the center of pressure, x_{cp}/c, and the aerodynamic center, x_{ac}/c, as a function of angle of attack.

1.3.   As predicted from thin airfoil theory, the section lift and leading-edge moment coefficients for a NACA 2412 airfoil section are given by

$$\widetilde{C}_L \;=\; 2\pi(\alpha+0.03625) \quad\text{and}\quad \widetilde{C}_{m_{le}} \;=\; -\frac{\pi}{2}(\alpha+0.07007)$$

where α is in radians. On one graph, plot both the section lift and normal force coefficients as a function of angle of attack, from -6 to $+12$ degrees. On another graph, plot both the section drag and axial force coefficients as a function of angle of attack, from -6 to $+12$ degrees.

1.4.   From the thin airfoil relations for the airfoil section presented in problem 1.3, on one graph plot the location of the center of pressure, x_{cp}/c, and the aerodynamic center, x_{ac}/c, as a function of angle of attack, from -6 to $+12$ degrees.

1.5.   Write a computer subroutine to compute the geopotential altitude, temperature, pressure, and air density as a function of geometric altitude for the standard atmosphere that is defined in Table 1.2.1. The only input variable should be the geometric altitude in meters. All output variables should be in standard SI units. Test this subroutine by comparing its output with the results that were obtained in Example 1.2.1 and those presented in Appendix A.

1.6.   Write a computer subroutine to compute the geopotential altitude, temperature, pressure, and air density as a function of geometric altitude for the standard atmosphere that is defined in Table 1.2.1. The only input variable should be the geometric altitude in feet. All output variables should be in English engineering units. Test the subroutine by comparison with Example 1.2.2 and the results presented in Appendix B. Since this standard atmosphere is defined in SI units, this subroutine could be written using simple unit conversion and a single call to the subroutine written for problem 1.5.

1.7.   Show that the velocity distribution for the two-dimensional source flow, which is expressed in Eq. (1.5.4), is irrotational.

1.8.   Show that the velocity distribution for the two-dimensional vortex flow, which is expressed in Eq. (1.5.6), is irrotational.

1.9.   Show that the velocity distribution for the two-dimensional doublet flow, which is expressed in Eq. (1.5.8), is irrotational.

1.10.   Show that the velocity distribution for the two-dimensional stagnation flow, which is expressed in Eq. (1.5.10), is irrotational.

1.11.   Show that the velocity distribution for the two-dimensional vortex flow, which is expressed in Eq. (1.5.6), satisfies the incompressible continuity equation.

1.12. The section geometry for a NACA 4-digit series airfoil is completely fixed by the maximum camber, the location of maximum camber, the maximum thickness, and the chord length. The first digit indicates the maximum camber in percent of chord. The second digit indicates the distance from the leading edge to the point of maximum camber in tenths of the chord. The last two digits indicate the maximum thickness in percent of chord. The camber line is given by

$$
y_c(x) = \begin{cases}
y_{mc}\left[2\left(\dfrac{x}{x_{mc}}\right)-\left(\dfrac{x}{x_{mc}}\right)^2\right], & 0 \le x \le x_{mc} \\[3mm]
y_{mc}\left[2\left(\dfrac{c-x}{c-x_{mc}}\right)-\left(\dfrac{c-x}{c-x_{mc}}\right)^2\right], & x_{mc} \le x \le c
\end{cases}
$$

where y_{mc} is the maximum camber and x_{mc} is the position of maximum camber. The thickness about the camber line, measured perpendicular to the camber line itself, is given by

$$
t(x) = t_m\left[2.969\sqrt{\frac{x}{c}} - 1.260\left(\frac{x}{c}\right) - 3.516\left(\frac{x}{c}\right)^2 + 2.843\left(\frac{x}{c}\right)^3 - 1.015\left(\frac{x}{c}\right)^4\right]
$$

where t_m is the maximum thickness. For the NACA 2412 airfoil section, determine x/c and y/c for the camber line, upper surface, and lower surface at chordwise positions of $x/c = 0.10$.

1.13. Using the formulas given in problem 1.12, for the NACA 0012 airfoil section, determine x/c and y/c of the upper and lower surfaces at chordwise positions of $x/c = 0.00, 0.001, 0.10, 0.40, 0.50, 0.90$, and 1.00.

1.14. Using the formulas given in problem 1.12, for the NACA 4421 airfoil section, determine x/c and y/c of the upper and lower surfaces at chordwise positions of $x/c = 0.00, 0.001, 0.10, 0.40, 0.50, 0.90$, and 1.00.

1.15. Using the formulas given in problem 1.12, for the NACA 0021 airfoil section, determine the angle of the upper surface relative to the chord line in degrees at a chordwise position of $x/c = 0.10$.

1.16. Using the formulas given in problem 1.12, for the NACA 4421 airfoil section, determine the angle of the upper surface relative to the chord line in degrees at a chordwise position of $x/c = 0.10$.

1.17. The formulas for the geometry of a NACA 4-digit series airfoil are given in problem 1.12. For the NACA 2412 airfoil section, use thin airfoil theory to determine the zero-lift angle of attack, the lift coefficients at 0- and 5-degree angles of attack, and the quarter-chord moment coefficient. Also plot the center of pressure as a function of angle of attack from −5 to 10 degrees.

1.18. The formulas for the geometry of a NACA 4-digit series airfoil are given in problem 1.12. Using thin airfoil theory, for the NACA 4412 airfoil section, determine the zero-lift angle of attack, the lift coefficients at 0- and 5-degree angles of attack, the quarter-chord moment coefficient, and the center of pressure at the 0- and 5-degree angles of attack.

1.19. The NACA 23012 airfoil section has a camber line defined by the equation

$$\frac{y_c}{c} = \begin{cases} 2.6595\left(\frac{x}{c}\right)^3 - 1.6156\left(\frac{x}{c}\right)^2 + 0.3051\left(\frac{x}{c}\right), & 0.0 \leq \frac{x}{c} \leq 0.2025 \\ 0.02208\left(1 - \frac{x}{c}\right), & 0.2025 \leq \frac{x}{c} \leq 1.0 \end{cases}$$

Using thin airfoil theory, determine the zero-lift angle of attack, the lift coefficients at 0- and 5-degree angles of attack, the quarter-chord moment coefficient, and the center of pressure at the 0- and 5-degree angles of attack.

1.20. The NACA 24012 airfoil section has a camber line defined by the equation

$$\frac{y_c}{c} = \begin{cases} 1.1072\left(\frac{x}{c}\right)^3 - 0.9632\left(\frac{x}{c}\right)^2 + 0.2523\left(\frac{x}{c}\right), & 0.0 \leq \frac{x}{c} \leq 0.2900 \\ 0.02700\left(1 - \frac{x}{c}\right), & 0.2900 \leq \frac{x}{c} \leq 1.0 \end{cases}$$

Using thin airfoil theory, determine the zero-lift angle of attack, the lift coefficients at 0- and 5-degree angles of attack, the quarter-chord moment coefficient, and the centers of pressure at the 0- and 5-degree angles of attack.

1.21. Write a computer program that uses the vortex panel method that was described in Sec. 1.6 to compute the pressure distribution on a NACA 4-digit series airfoil. Use the program to compute the lift coefficient of the NACA 2412 airfoil section. Compare the results of your program with those plotted in Fig. 1.6.6.

1.22. Use the computer program written for problem 1.21 to compute the lift coefficient of the NACA 2421 airfoil section. Compare the results of your program with those plotted in Fig. 1.6.7.

1.23. Use the computer program written for problem 1.21 to compute the lift coefficient of the NACA 0015 airfoil section. Compare the results of your program with those plotted in Fig. 1.6.8.

1.24. Using thin airfoil theory, plot the ideal section flap effectiveness as a function of flap chord fraction. Compare your results with those plotted in Fig. 1.7.3.

1.25. Using thin airfoil theory, plot the quarter-chord moment slope as a function of flap chord fraction. Compare your results with those plotted in Fig. 1.7.6.

1.26. Consider an elliptic wing of aspect ratio 8 having no sweep or dihedral and no geometric or aerodynamic twist. The wing is constructed with a thin symmetric airfoil section. For an angle of attack of 5 degrees, estimate the lift and induced drag coefficients for the wing.

1.27. Consider a rectangular wing of aspect ratio 8 having no sweep or dihedral and no geometric or aerodynamic twist. The wing is constructed with a thin symmetric airfoil section. For an angle of attack of 5 degrees, estimate the lift and induced drag coefficients for the wing.

1.28. Consider a tapered wing of aspect ratio 8 and taper ratio 0.5 having no sweep or dihedral and no geometric or aerodynamic twist. The wing is constructed with a thin symmetric airfoil section. For an angle of attack of 5 degrees, estimate the lift and induced drag coefficients for the wing.

1.29. Write a computer program that uses the series solution to Prandtl's lifting-line equation to predict the lift and induced drag coefficients for an elliptic wing having linear geometric and aerodynamic twist. Use your program to rework problem 1.26 for a wing with 5 degrees of linear washout.

1.30. Use the computer program that was written for problem 1.29 to plot the lift and induced drag coefficients as a function of angle of attack for elliptic wings of aspect ratio 8 having 0 and 5 degrees of linear washout.

1.31. Write a computer program that uses the series solution to Prandtl's lifting-line equation to predict the lift and induced drag coefficients for a tapered wing having linear geometric and aerodynamic twist. Use your program to rework problems 1.27 and 1.28 for wings with 5 degrees of linear washout.

1.32. Use the program from problem 1.31 to plot lift coefficient as a function of angle of attack and induced drag coefficient as a function of lift coefficient for a rectangular wing of aspect ratio 8 and having 5 degrees of linear washout.

1.33. Use the program from problem 1.31 to plot lift coefficient as a function of angle of attack and induced drag coefficient as a function of lift coefficient for a tapered wing of aspect ratio 8 and taper ratio 0.5 having 5 degrees of linear washout.

1.34. Write a computer program that uses the series solution to Prandtl's lifting-line equation to predict rolling and yawing moment coefficients for a tapered wing with linear washout, aileron deflection, and roll. Use your program to rework Example 1.8.3 for wings with 0 and 4.5 degrees of linear washout.

1.35. Use the program written for problem 1.34 to plot the rolling and yawing moment coefficients as a function of the linear washout angle for the wing and aileron planform in Example 1.8.3. Use a 5-degree aileron deflection with no roll.

1.36. Write a computer program that uses the numerical lifting-line method presented in Sec. 1.9 to predict the lift and induced drag for a tapered wing with linear geometric and aerodynamic twist. Test your program by comparing results with those obtained in problems 1.27, 1.28, 1.32, and 1.33.

1.37. Use the computer program from problem 1.36 to compute the lift at 10 degrees angle of attack for a rectangular wing with a thin symmetric airfoil section and an aspect ratio of 6. The root chord is 5.5 ft. Use $V = 176$ ft/sec at standard sea level.

1.38. Use the computer program that was written for problem 1.36 to compute the lift at 10 degrees angle of attack for a tapered wing having aspect ratio 4.0, taper ratio 0.5, and a thin symmetric airfoil section. The wing quarter chord is swept such that there is no sweep in the trailing edge and the planform area of the wing is 36 ft^2. Use $V = 176$ ft/sec at standard sea level.

1.39. The wing described in problem 1.38 is to be used as an aft horizontal stabilizer for the main wing described in problem 1.37. The root quarter chord of the horizontal stabilizer is 15 feet aft of the quarter chord of the main wing. Use the computer program from problem 1.36 to compute the lift at 10 degrees angle of attack for each lifting surface in this wing-tail combination. Use $V = 176$ ft/sec at standard sea level. Compare the results obtained with those obtained in problems 1.37 and 1.38 for each lifting surface in isolation. Discuss your findings.

1.40. For the NACA 2412 airfoil section, which was described in problem 1.17, plot the lift slope and quarter-chord moment coefficient as a function of Mach number from 0.0 to 0.8.

1.41. For the NACA 2412 airfoil section described in problem 1.17, plot the derivative of lift per unit area with respect to airspeed as a function of flight Mach number from 0.0 to 0.8. Neglect the Reynolds number dependence and use a 5-degree angle of attack with standard conditions at sea level.

1.42. Using the panel code written for problem 1.21 and the Prandtl-Glauert compressibility correction, estimate the critical Mach number of the NACA 2412 airfoil section for angles of attack of 0 and 5 degrees.

1.43. Using the panel code written for problem 1.21 and the Prandtl-Glauert compressibility correction, estimate the critical Mach number of the NACA 2421 airfoil section for angles of attack of 0 and 5 degrees.

1.44. A symmetric supersonic airfoil section has a parabolic thickness distribution with the point of maximum thickness located at the half-chord. For a maximum thickness of 4 percent, plot the section lift, drag, and half-chord moment coefficients as a function of freestream Mach number from 1.2 to 3.0. Use a fixed angle of attack of 5 degrees.

1.45. A symmetric supersonic airfoil section has a parabolic thickness distribution with the point of maximum thickness located at the half-chord. For a maximum thickness of 12 percent, plot the section lift, drag, and half-chord moment coefficients as a function of freestream Mach number from 1.2 to 3.0. Use a fixed angle of attack of 5 degrees.

1.46. A symmetric supersonic airfoil section has a parabolic thickness distribution with the point of maximum thickness located at the half-chord. For a maximum thickness of 4 percent, plot the section drag per unit area as a function of free-stream Mach number from 1.2 to 3.0 for a constant wing loading of 100 lbf/ft^2 and standard conditions at 30,000 feet.

1.47. A symmetric supersonic airfoil section has a parabolic thickness distribution with the point of maximum thickness located at the half-chord. For a maximum thickness of 12 percent, plot the section drag per unit area as a function of free-stream Mach number from 1.2 to 3.0 for a constant wing loading of 100 lbf/ft^2 and standard conditions at 30,000 feet.

1.48. A symmetric diamond-wedge airfoil section has a thickness distribution that increases linearly from the leading edge to the point of maximum thickness located at the half-chord. The thickness then decreases linearly from the half-chord to the sharp trailing edge. For a maximum thickness of 4 percent, plot the section lift and drag coefficients as a function of freestream Mach number from 1.2 to 3.0. Use a fixed angle of attack of 5 degrees.

1.49. A symmetric diamond-wedge airfoil section has a thickness distribution that increases linearly from the leading edge to the point of maximum thickness located at the half-chord. The thickness then decreases linearly from the half-chord to the sharp trailing edge. For a maximum thickness of 4 percent, plot the section drag per unit area as a function of Mach number from 1.2 to 3.0 for a constant wing loading of 100 lbf/ft^2 and standard conditions at 30,000 feet.

1.50. Consider a rectangular wing of aspect ratio 8.0 with a constant thin airfoil section having $\widetilde{C}_{m_{ac}} = 0.053$ and $\widetilde{C}_{L_{max}} = 1.6$. Estimate $C_{L_{max}}$, $\bar{x}_{ac}/\bar{c}$, and $C_{m_{ac}}$ for this wing planform with the following amounts of quarter-chord sweep and linear washout:

$$\Lambda = 0,\ \Omega = 0; \qquad \Lambda = 0,\ \Omega = 4°; \qquad \Lambda = 30°,\ \Omega = 0; \qquad \Lambda = 30°,\ \Omega = 4°$$

1.51. Consider a wing with linear taper of aspect ratio 4.0 and taper ratio 0.5. The wing has a constant thin airfoil section with $\widetilde{C}_{m_{ac}} = 0.053$ and $\widetilde{C}_{L_{max}} = 1.6$. Estimate $C_{L_{max}}$, $\bar{x}_{ac}/\bar{c}$, and $C_{m_{ac}}$ for this wing with the following amounts of quarter-chord sweep and linear washout:

$$\Lambda = 0,\ \Omega = 0; \qquad \Lambda = 0,\ \Omega = 5°; \qquad \Lambda = 30°,\ \Omega = 0; \qquad \Lambda = 30°,\ \Omega = 5°$$

Chapter 2
Overview of Propulsion

2.1. Introduction

Since the performance, stability, control, and dynamics of an aircraft are affected by the propulsion system, the study of flight mechanics requires some knowledge of propulsion. However, from the beginning it should be understood that this chapter is not intended to provide a detailed treatment of aircraft propulsion systems. Furthermore, this very cursory treatment of propulsion is not intended to make light of the importance of propulsion in aeronautics and airplane design. More than any other single factor, the evolution of the modern airplane has depended on the development and improvement of power plant technology. In fact, propulsion has led the way for most major advancements in aeronautics. Except for the very special case of a glider, the propulsion system is a very critical and important part of any aircraft. The performance and design of aircraft propulsion systems is extremely complex and complete textbooks are devoted to this topic. Furthermore, propulsion is not just an important topic in aeronautics, it is one of the major disciplines within the field of aerospace engineering. No single text-book, let alone one chapter, could do justice to this topic. The material in this chapter is intended only as a very brief introduction to the topic of aircraft propulsion.

Most aircraft propulsion systems produce thrust in a somewhat similar manner. Air enters the device through an inlet surface, and as a result of power that is applied to the device in some form, the kinetic energy of that air is increased. The associated increase in the momentum of the air that passes through the propulsion system results in a net reaction force, which is what we call *thrust*. To examine how this works, consider the reversible, incompressible flow through an ideal propulsion system, such as that shown schematically in Fig. 2.1.1.

For our analysis we choose a control volume that extends everywhere sufficiently far from the propulsion system so that ambient pressure exists on the entire enclosing surface. Also, the velocity of the air at every point on the surface of the control volume is equal to the freestream velocity, V_∞, except over that section where the slipstream exits the control volume. The velocity in the slipstream, at the cross-section where it exits the control volume, is assumed uniform and will be designated as V_s. Applying Newton's second law to this control volume, we have

$$T = \dot{m}(V_s - V_\infty) \tag{2.1.1}$$

where T is the thrust developed and $\dot{m}$ is the mass flow rate of the air that passes through the propulsion system. The first law of thermodynamics applied to this same control volume gives

$$P_{in} = \dot{m}\left(h_s - h_\infty + \frac{V_s^2}{2} - \frac{V_\infty^2}{2}\right)$$

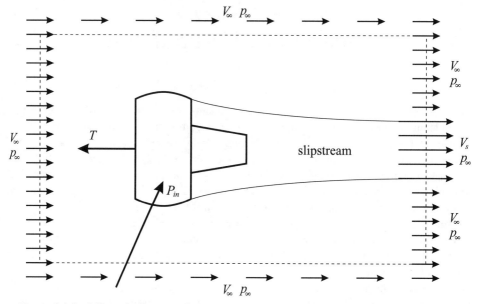

Figure 2.1.1. Schematic diagram of a control volume surrounding an ideal propulsion system.

where P_{in} is the power input, h_s and h_∞ are the enthalpy (internal energy plus p/ρ) of the air in the ultimate slipstream and freestream, respectively. Since in this simple analysis we are assuming reversible, incompressible flow, there will be no change in internal energy. Furthermore, since the freestream pressure is equal to the ultimate slipstream pressure, the enthalpy in the ultimate slipstream is equal to the enthalpy in the freestream. Thus, the first law of thermodynamics reduces to

$$P_{in} = \dot{m}\left(\frac{V_s^2}{2} - \frac{V_\infty^2}{2}\right) = \frac{\dot{m}}{2}(V_s - V_\infty)(V_s + V_\infty) \qquad (2.1.2)$$

The propulsive efficiency for an aircraft engine is defined as the product of the thrust and the freestream velocity divided by the power input,

$$\eta_p \equiv \frac{TV_\infty}{P_{in}} \qquad (2.1.3)$$

Using Eqs. (2.1.1) and (2.1.2) in Eq. (2.1.3) results in

$$\eta_p = \frac{\dot{m}(V_s - V_\infty)V_\infty}{(\dot{m}/2)(V_s - V_\infty)(V_s + V_\infty)} = \frac{2V_\infty}{V_s + V_\infty}$$

$$= \frac{2V_\infty}{2V_\infty + (V_s - V_\infty)} = \frac{1}{1 + (V_s - V_\infty)/2V_\infty} \qquad (2.1.4)$$

From Eq. (2.1.1) we see that the thrust developed by an ideal propulsion system can be increased either by increasing the mass flow rate or by increasing the velocity increment, which is imparted to the air that passes through the propulsion system. However, from Eq. (2.1.4), we see that the efficiency of an ideal propulsion system decreases as the velocity increment is increased. Thus, **the most efficient production of thrust is attained by using a large mass flow rate with a small velocity increment.** Unfortunately, the size and weight of turbo-machinery is generally proportional to the flow rate that must be handled. For this reason, **using a larger velocity increment with a smaller mass flow rate will generally result in a higher thrust-to-weight ratio.** This is also an important consideration for aircraft propulsion.

Because of their large diameter, engine-propeller systems generally produce relatively high flow rates with correspondingly small velocity increments. For this reason, engine-propeller systems are usually the most efficient of the aircraft propulsion systems commonly used for low-speed subsonic flight. On the other hand, the jet engine produces a fairly large velocity increment with a relatively low mass flow rate. Thus, at subsonic airspeeds, the jet engine is the least efficient of the commonly used aircraft engines, but it usually has the highest thrust-to-weight ratio. A *turbofan engine* is a compromise between the propeller and the jet. It has a diameter and flow rate that are typically less than those for an equivalent engine-propeller combination but greater than those for a jet engine of similar power. Most of the engines that are used on commercial transports, and commonly referred to as jet engines, are actually turbofan engines. Turbojet engines are more commonly used on military fighters, where thrust-to-weight ratio is typically more important than efficiency. An understanding of these simple concepts should help the student better comprehend the following historical introduction to the development and application of modern aircraft propulsion systems.

At the beginning of the twentieth century, both the steam engine and the internal combustion engine were available. However, at that time, the existing engines were too heavy to be practical for aircraft propulsion. The major contributions made by the Wright brothers, which made possible their first powered flight, were the development of a lighter internal combustion engine and the design of a more efficient propeller. The 1903 Wright Flyer, which is shown in Fig. 2.1.2, was powered by a four-cylinder, liquid-cooled engine that weighed 200 pounds and developed 12 horsepower. This aircraft first carried a human pilot aloft using an internal combustion engine having a power-to-weight ratio of only 0.06 hp/lb. The flight marked the beginning of a 40-year period of rapid development for the piston engine-propeller combination as an aircraft power plant. During this period the power-to-weight ratio was dramatically increased and propeller efficiency was significantly improved. The liquid-cooled piston engine continued to be used as a means of aircraft propulsion through the end of World War II.

The first air-cooled piston engine to be used for aircraft propulsion was the French Gnome, originally built in 1908. This engine was what is called a *rotary-radial configuration*. With this design, the cylinders displaced radially while rotating about the engine centerline with the propeller. Such engines had a power-to-weight ratio of about 0.3 hp/lb, which was a significant improvement over the liquid-cooled piston engines of that time. However, the gyroscopic effects that were induced by the rotation of the massive cylinders caused severe problems during maneuvers.

Figure 2.1.2. The 1903 Wright Flyer was powered by a four-cylinder, liquid-cooled internal combustion engine that weighed 200 pounds and developed 12 horsepower, providing a power-to-weight ratio of 0.06 hp/lb.

These rotary-radial engines were soon replaced with what is called the *static-radial configuration*. With this air-cooled design, the cylinders are still arranged radially about the axis of rotation but remain fixed relative to the airframe. The pistons move radially within the cylinders and only the crankshaft rotates with the propeller. The first radial engine of this type to be produced commercially was the Whirlwind, developed by the Wright Company. It was a nine-cylinder Wright Whirlwind that powered the *Spirit of St. Louis* when Charles Lindbergh made the first nonstop Atlantic crossing in 1927. The Wright Whirlwind was also used on the famous Ford Tri-motor, shown in Fig. 2.1.3.

Figure 2.1.3. This 1929 Ford Tri-motor was powered by three J-5C Wright Whirlwind radial piston engines that each weighed 500 pounds and developed 220 horsepower, producing a power-to-weight ratio of 0.44 hp/lb. (Photograph by Barry Santana)

The model J-5C Wright Whirlwind developed 220 horsepower with a dry weight of 500 pounds, which gave the engine a power-to-weight ratio of 0.44 hp/lb. Once again, this was a dramatic increase in power-to-weight ratio over any engine that was previously used for aircraft propulsion. Further improvements soon led to air-cooled radial piston engines having power-to-weight ratios in excess of 0.6 hp/lb.

In addition to providing a significant increase in power-to-weight ratio, the early static-radial engines were very reliable by the standards of that day, and many are still flying at the present time. In 1912, when Robert G. Fowler completed his crossing of the continent by air, it took 121 calendar days in a Model B Wright Flyer from start to finish. In that time, the spark plugs were changed 96 times and Bob made 65 forced landings (an average of one out of every three flights). Only 15 years later, an air-cooled radial engine pulled Charles Lindbergh across the Atlantic in a nonstop flight. Just two years after that, three of these same engines took Admiral Richard Byrd to Antarctica in a Ford Tri-motor.

The large static-radial engines of the late 1920s and early 1930s truly earned their place in history. These big radials had a melodious deep-throated sound that has never been equaled by any other engine. In less than 30 years from the time of Wright brothers' first powered flight at Kitty Hawk, the power-to-weight ratio of the aircraft engine had been increased by more than an order of magnitude, and reliability was increased to the point where commercial aviation was now a reality.

Starting in 1935, air-cooled radial engines were also used to power large long-range bombers. The B-17 Flying Fortress, shown in Fig. 2.1.4, was powered by four Wright R-1820-97 radial engines that each delivered 1,200 horsepower. Later, the B-29 Superfortress was powered by larger air-cooled radial engines, the 2,200-horsepower

Figure 2.1.4. The B-17 Flying Fortress was powered by four Wright R-1820-97 nine-cylinder, air-cooled radial engines, which each delivered 1,200 horsepower. (Photograph by Barry Santana)

Figure 2.1.5. The early Spitfire Mark XVIII was powered by a liquid-cooled, 12-cylinder, Rolls-Royce Griffon engine that weighed 2,050 pounds, developed 2,035 horsepower, and provided a power-to-weight ratio of almost 1.0 hp/lb. (Photograph by Barry Santana)

Wright R-3350-23 Cyclone. Fitted with four of these 18-cylinder engines, each with a pair of General Electric B-11 superchargers, the B-29 had a range of 3,250 miles with a liftoff payload of 5,000 pounds and 8,198 gallons of fuel.

An extreme in power-to-weight ratio for reciprocating piston engines was attained in the fighter aircraft of World War II. For example, the early version of the Spitfire Mark XVIII shown in Fig. 2.1.5 was powered by a liquid-cooled, 12-cylinder, Rolls-Royce Griffon engine developing 2,035 horsepower at 2,750 rpm, while turning a 132-inch, five-blade, Rotol propeller. The engine had a dry weight of 2,050 pounds and a power-to-weight ratio of nearly 1.0 hp/lb. At 24,500 feet, the airplane had a top speed in level flight of 439 mph, and the rate of climb at sea level was 4,580 ft/min.

In the first four decades of the twentieth century, not only was the power-to-weight ratio of the internal combustion engine increased by a factor of nearly 17, propeller efficiency was dramatically improved as well. At first, propeller efficiency was improved only as a result of improved aerodynamic shapes for propeller blades. However, for reasons that shall later be made apparent, propellers with low pitch-to-diameter ratio perform best at low airspeed, while propellers with high pitch-to-diameter ratio perform best at high airspeed. Since an airplane must operate over a wide range of airspeeds, overall propeller efficiency was greatly improved with the invention of the first variable-pitch propeller in 1924. Further improvement came with the introduction of the constant-speed propeller in 1935. A constant-speed propeller is simply a variable-pitch propeller that is controlled by a governor mechanism, which automatically varies the propeller pitch to maintain the proper torque on the engine so that the rotational speed of the engine is held constant. This is advantageous because the efficiency of a piston engine is maximized at only one particular rotational speed. By controlling the engine speed with a constant-speed propeller, the engine can be made to operate at maximum efficiency over a wide range of aerodynamic operating conditions. The introduction of the variable-pitch and constant-speed propellers was one of the most important developments in aviation history. With such mechanisms, the propulsive efficiency for modern propellers can usually be maintained at about 80 percent or above.

By the end of World War II, the aircraft propulsion system consisting of a propeller driven by a reciprocating piston engine had nearly reached its performance limit. Power-to-weight ratios had increased to nearly 1 horsepower per pound, overall efficiency of the engine-propeller combination had increased to almost 30 percent, airspeeds had increased to about 450 mph, and reciprocating piston engines had been produced with brake power outputs as large as 5,000 horsepower. The great advancements that were achieved in this power plant technology during the first four decades of the twentieth century came about as a result of improved materials and engine structure design, advanced fuel injection systems, development of engine superchargers, design of improved propeller blade shapes, and the development of variable-pitch and constant-speed propellers.

To partly understand why performance limits were encountered with the piston engine-propeller combination, we return to the simplified analysis of the ideal propulsion system that was presented in Eqs. (2.1.1) through (2.1.4). In Eq. (2.1.1) we have seen that the thrust developed is proportional to the mass flow rate multiplied by the velocity increment. From Eq. (2.1.4), we found that the efficiency decreases as the velocity increment is increased. Thus, to maintain efficiency as engine power was increased, it was necessary to increase the mass flow rate rather than the velocity increment. Since mass flow rate is proportional to the product of area and velocity, to increase the flow rate without increasing the velocity increment, propeller diameters were increased as engine power was increased. For example, the Spitfire Mark XVIII turned a 132-inch, five-blade propeller, and four 139-inch, three-blade propellers pulled the B-17. As propeller diameters were increased, the propeller blade-tip speeds were increased as well. A limit to this increase in engine power and propeller diameter is encountered when propeller blade tips approach or exceed the speed of sound. At this point, compressibility effects drastically reduce propeller efficiency. To avoid increasing propeller diameter as engine power is increased, the number of propeller blades can be increased. However, this has a limit as well, since increasing the number of blades increases the velocity increment and blade interference, which also decrease propeller efficiency. These limits essentially stopped the development of the piston engine-propeller combination as a means of high-speed aircraft propulsion.

Today, reciprocating engine-propeller combinations are produced only for the smaller low-speed aircraft that are used in general aviation. The best piston engine-propeller combinations in production today usually consist of a horizontally opposed, air-cooled piston engine turning a constant-speed propeller. These power plants typically have a maximum rated brake power-to-weight ratio of about 0.6 hp/lb, a brake power-specific fuel consumption of about 0.5 lb-fuel/hp·hr, and a propeller efficiency of about 80 percent. For such engines, the operating brake power at cruise is typically about 70 percent of maximum rated power.

The limitations of the reciprocating engine-propeller combination were removed with the development of jet propulsion. The jet engine also has its origins in the early part of the twentieth century. A reciprocating jet engine was patented in 1908, the ramjet was patented in 1913, and the first turbojet was patented in 1921. However, significant development of the jet engine did not take place until the late 1930s. The first patent for a turbojet engine that was actually developed and produced was granted to Frank Whittle in 1930. Unfortunately, it was some time before Whittle was able to get support for the development of his invention. Whittle first sought support for the development of his engine from the British Air Ministry, Department of Engine Development. The department declined to support the project, taking the view that no form of the gas turbine would ever be practical because of a lack of materials capable of enduring the high temperatures. From the beginning, Frank Whittle correctly believed that the turbojet engine had great advantages over other forms of aircraft propulsion for high-speed flight. However, those in a position to support development of his turbojet would not share this view for another several years.

In 1935, Hans von Ohain was granted a German patent for a jet engine. In contrast to the British effort, German development of this new propulsion system began immediately. The first demonstration engine was completed by the end of February 1937, and the first test run took place in early March of that same year. An experimental flight engine was completed and tested in the spring of 1939. On August 27, 1939, the world's first experimental aircraft to be powered by a turbojet engine made its inaugural flight with Erich Warsitz as the pilot. This aircraft, the Heinkel He-178 shown in Fig. 2.1.6, demonstrated the improved power-to-weight ratio that could be achieved with the jet engine. As first flown, the single turbojet engine that powered the He-178 developed about 1,000 pounds of thrust. Later improvements increased this thrust to about 1,100 pounds. The engine attained a net power-to-weight ratio that was about 2.5 times that of the best engine-propeller combinations of that time.

German development of the turbojet engine proceeded very rapidly. A prototype jet fighter, the He-280, was flight tested in late March of 1941 by test pilot Fritz Schaefer.

Figure 2.1.6. The first jet-powered aircraft, the Heinkel He-178, was flown on August 27, 1939. (Warren M. Bodie, via National Air and Space Museum, Smithsonian Institution, SI 84-10658)

Figure 2.1.7. The first mass-produced jet fighter, the Messerschmitt Me-262, was powered by two turbojet engines, each weighing 1,650 pounds and developing 2,000 pounds of thrust. (Photo by Dale Hrabak, National Air and Space Museum, Smithsonian Institution, SI 79-4620)

By 1944, mass-produced jet fighters like the Me-262 shown in Fig. 2.1.7 were flying at speeds in excess of 550 mph. The Me-262 was the first operational jet fighter. It was powered by two Jumo 004B turbojet engines. Each engine weighed 1,650 pounds and developed approximately 2,000 pounds of thrust, giving the engine a thrust-to-weight ratio of more than 1.2. While this thrust-to-weight ratio far exceeded that for the best engine-propeller combinations, the thrust-specific fuel consumption of 1.4 lb-fuel/hr/lb-thrust was much higher than that for reciprocating engines. For example, the Spitfire Mark XVIII powered by the Rolls-Royce Griffon 65 engine had a thrust-specific fuel consumption of about 0.65 lb-fuel/hr/lb-thrust with an airspeed of 439 mph. In addition to its high fuel consumption rate, the early turbojet had a very short engine life. The Jumo 004B turbojet had an engine life of about 25 hours. However, since the average combat life of a German fighter in World War II was well below 25 hours, this was not considered to be a problem at the time. The major disadvantage of the Me-262 was its high fuel consumption rate. This severely limited its range and kept it from becoming a major factor in the outcome of the war.

Spurred by German success, British development of Frank Whittle's turbojet engine began about six years after Whittle was granted the original patent. It was not until May 15, 1941 that the first experimental British aircraft was flown with a turbojet engine of Whittle's design. This was the Gloster/Whittle E28/39 shown in Fig. 2.1.8. Even after demonstration of the Gloster/Whittle, British Air Ministry officials showed little interest in jet propulsion. It was 1945 before mass production of a British jet fighter began with the Gloster Meteor, shown in Fig. 2.1.9.

The development of jet-powered aircraft in the United States followed closely on the heels of the British effort and was also based on Frank Whittle's turbojet designs. On October 1, 1942, the first U.S. jet-propelled aircraft was flown. This was the Bell Airacomet, shown in Fig. 2.1.10. This experimental aircraft was powered by two GE 1-A engines, which were U.S. copies of Whittle's W2B turbojet. Each engine weighed approximately 1,000 pounds and developed 1,250 pounds of thrust.

Figure 2.1.8. The first experimental British jet-powered aircraft was the Gloster/Whittle E28/39. (Royal Aerospace Establishment, Crown copyright)

Figure 2.1.9. The Gloster Meteor was the first British jet fighter to be mass-produced.

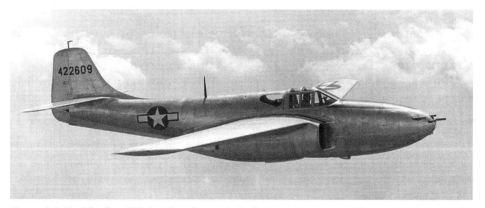

Figure 2.1.10. The first U.S. jet aircraft was the Bell Airacomet, powered by two GE 1-A engines each weighing 1,000 pounds and developing 1,250 pounds of thrust. (U.S. Air Force)

Figure 2.1.11. The first production U.S. jet fighter was the Lockheed P-80, powered by a single 2,000-pound engine developing approximately 4,000 pounds of thrust.

The earliest production jet fighter developed in the United States, the Lockheed P-80, first flew in January of 1944. This aircraft, shown in Fig. 2.1.11, was also powered by an advanced version of one of Whittle's turbojet designs. The single 2,000-pound engine developed 4,000 pounds of thrust and propelled the P-80 (subsequently renamed F-80) to a new speed record of more than 620 mph.

The United States entered the age of jet-powered commercial aviation when the Boeing 707 prototype, shown in Fig. 2.1.12, made its inaugural flight on July 15, 1954. The first commercial version of the 707 was the 120-series, powered by four Pratt and Whitney JT3 turbojet engines that each developed 13,500 pounds of static thrust. Boeing quickly developed the larger 320-series. Four Pratt and Whitney JT4 turbojet engines each developing 16,800 pounds of thrust powered this version of the 707. With fuel capacity increased from 15,000 gallons to more than 23,000 gallons, the Boeing 707-320 had an intercontinental range of more than 4,000 miles.

Figure 2.1.12. The Boeing 707 prototype, nicknamed the Dash-80, is shown at Moffett Field in Mountain View, California on December 12, 1965. (NASA/Ames photograph by Jim Remington)

Figure 2.1.13. The North American XB-70 reached Mach 3 powered by six General Electric YJ-93 turbojet engines that each delivered 30,000 pounds of thrust with afterburners employed. (U.S. Air Force)

In the early 1950s, jet-powered aircraft began to exceed the speed of sound. Improvements in the turbojet engine powered a rapid increase in aircraft size and flight speed. On October 14, 1965 the North American XB-70 exceeded Mach 3 at 73,000 feet (2,056 mph). The XB-70, shown in Fig. 2.1.13, weighed more than 250 tons and was powered by six General Electric YJ-93 turbojet engines, each delivering 30,000 pounds of thrust with afterburners employed.

While the efficiency of an engine-propeller combination decreases dramatically at high airspeeds, the turbojet engine actually has improved efficiency at higher airspeeds. To understand the reason for this, we return again to the simplified analysis of the ideal propulsion system that was presented in Eqs. (2.1.1) through (2.1.4). Recall from this analysis that thrust is the product of the mass flow rate and the velocity increment that is imparted to the air passing through the propulsion system. Since propulsive efficiency decreases as the velocity increment is increased, the most efficient production of thrust is achieved with a high mass flow rate and a small velocity increment. The mass flow rate of air passing through a jet engine is proportional to the inlet velocity, which increases with increasing flight speed. As the flight speed is increased, the mass flow rate through a turbojet engine is increased, while the increment between the flight speed and the exhaust gas velocity is decreased. Thus, at supersonic speeds, the propulsive efficiency of the turbojet engine is outstanding. However, at subsonic speeds, the exhaust gas velocity for a turbojet engine is too high to provide good efficiency.

For flight at high subsonic speeds, the turbofan engine provides an excellent alternative for improving propulsive efficiency. The turbofan engine also has its origins with the British inventor Frank Whittle. In 1936, Whittle filed a patent application for an engine that he called a *turbofan* or *bypass engine*. The turbofan engine has a larger turbine than a turbojet engine of similar power. As the exhaust gas passes through this enlarged turbine, more energy is extracted and the velocity of the slipstream is reduced. The extra power produced by the turbine is used to turn a ducted fan, which adds mass to the slipstream. As we have seen from Eq. (2.1.4), decreasing the slipstream velocity while increasing the total mass flow rate will increase the propulsive efficiency at subsonic speeds. Whittle's work on turbofan engines was far ahead of his time but was of great importance for the future of turbo-propulsion systems.

Significant development of the turbofan engine did not occur until the 1960s. The first turbofan engines had a relatively low bypass ratio of about 2 to 1. The bypass ratio is defined as the ratio of the mass flow rate bypassing the turbine to the mass flow rate passing through the turbine. In the early 1960s, the U.S. Air Force established requirements for military transports having extremely long range at high subsonic airspeeds. To meet these requirements, the high-bypass-ratio turbofan engine was developed. The General Electric TF39-1 was the first turbofan engine of this type. The TF39-1 had a bypass ratio of 8 to 1, with a pressure ratio across the compressor of about 25 to 1. The Lockheed C5 Galaxy, which is shown in Fig. 2.1.14, was first deployed in December of 1969 powered by four of the TF39-1 turbofan engines that each developed 41,000 pounds of thrust. General Electric TF39-1C turbofan engines, each providing 43,000 pounds of thrust, powered a later version of the C5 Galaxy. These engines gave the C5 a maximum range of more than 6,000 miles and a maximum takeoff weight of 837,000 pounds. At high subsonic airspeeds, these newly developed turbofan engines had a much higher overall efficiency than the earlier turbojet engines.

The 1960s military experience with high-bypass-ratio turbofan engines paved the way for the second generation of commercial jet aircraft. These were the *widebody* aircraft, typified by the Boeing 747 shown in Fig. 2.1.15. The Boeing 747-100 entered service on January 22, 1970 powered by four Pratt and Whitney JT9D-7A turbofan

Figure 2.1.14. When powered by four General Electric TF39-1C turbofan engines, each providing 43,000 pounds of thrust, the Lockheed C5 Galaxy had a range of about 6,000 miles and a maximum takeoff weight of 837,000 pounds. (U.S. Air Force)

Figure 2.1.15. The Boeing 747-100 entered service on January 22, 1970 powered by four Pratt and Whitney JT9D-7A turbofan engines that each provided a thrust of 46,500 pounds. (NASA photograph)

engines that provided a total thrust of 186,000 pounds. The later 400-series, which went into service in 1988, develops more than 253,000 pounds of thrust when powered by four Pratt and Whitney PW4062 turbofan engines. The PW4062 has a maximum thrust-to-weight ratio of more than 6.0 and a minimum thrust-specific fuel consumption of less than 0.45 lb-fuel/hr/lb-thrust. Even at 40,000 feet with a Mach number of 0.8, the PW4062 has a thrust-specific fuel consumption of less than 0.6 lb-fuel/hr/lb-thrust. With four of these engines and a 57,285-gallon fuel capacity, the 747-400 has a maximum takeoff weight of 875,000 pounds and a range of more than 8,000 miles.

During the last three decades of the twentieth century, the continuous improvement of turbomachinery components led to the evolution of turbojet and turbofan engines with greater thrust-to-weight ratio and higher overall efficiency. This evolution produced higher compressor pressure ratios, higher turbine inlet temperatures, and higher bypass ratios. During this period, the small to medium-sized aircraft market also benefited tremendously from improvements in turbomachinery technology. Today, the turboshaft or turboprop engine is commonly used to power short-range commuter aircraft like the de Havilland Dash 8 shown in Fig. 2.1.16.

A turboprop engine can be designed to turn at constant rotational speed during most phases of aircraft operation. The design is typified by the Allison T56-A-15 engines that power the C-130 Hercules shown in Fig. 2.1.17. The T56-A-15 turns continuously at 13,820 rpm, which allows the turbine to maintain its most efficient operating speed. Since this speed is too fast for the efficient operation of any propeller, in the C-130 installation, each engine turns a 162-inch propeller through a gear reduction assembly having a total reduction ratio of 13.54 to 1. To maintain the constant rotational speed, changes in power requirements are met by changing the fuel flow and propeller pitch.

Figure 2.1.16. Two Pratt and Whitney PW121, 2000-horsepower turboprop engines power the de Havilland Dash 8-100 regional airliner. (Photograph by John Martin)

Figure 2.1.17. Four Allison T56-A-15 constant-speed turboprop engines turning at 13,820 rpm power the C-130H Hercules. With a power-to-weight ratio of 2.65 hp/lb, the T56-A-15 is capable of developing 4,910 horsepower and weighs only 1,850 pounds. (U.S. Air Force)

From the pilot's point of view, controlling a constant-speed turboprop engine is remarkably simple. A single throttle lever controls the engine's propeller, electrical, and fuel systems. This is accomplished through an electromechanical control system. The throttle is linked to the fuel control system, which meters the fuel flow to maintain constant engine power. A power setting is established by the position of the throttle, and a fuel-trimming system maintains the power at a constant level by monitoring the turbine inlet temperature and changing the fuel flow accordingly. As the fuel flow is changed, the automatic control system very precisely adjusts the propeller pitch to maintain constant engine rpm.

The constant-speed turboprop engine offers excellent fuel efficiency at a relatively high power-to-weight ratio. At low subsonic Mach numbers, a thrust-specific fuel consumption on the order of 0.2 lb-fuel/hr/lb-thrust is attainable and the power-to-weight ratio for these engines can be in excess of 2.5 hp/lb. In addition, the constant-speed turboprop engine provides another important operational feature. Since the engine is always turning at full speed, response to a change in power demand is almost instantaneous. This provides excellent performance characteristics during takeoff, landing, and emergency situations.

In effect, a turboprop engine is an extension of the turbofan engine to higher bypass ratios. At low subsonic airspeeds, the modern turboprop provides excellent overall efficiency. However, conventional propellers operating at high subsonic Mach numbers suffer a substantial loss in efficiency, due to compressibility effects. Turboprops have been designed to provide high propulsive efficiency at Mach numbers as high as 0.8, but no such design has yet been placed in commercial service. Improvements in the power-to-weight ratio and thermodynamic efficiency of turboshaft engines also played a major role in the development of modern helicopters.

In the first 100 years since the maiden flight of the 1903 Wright Flyer, great strides have been made in the development of aircraft propulsion systems. A 12-horsepower piston engine with a power-to-weight ratio of approximately 0.06 hp/lb powered the 1903 Wright Flyer. Today, modern gas turbine engines have a power-to-weight ratio in excess of 2.5 hp/lb, and the largest gas turbines develop close to 100,000 horsepower. The 1903 Wright Flyer first lifted from the Earth with a gross weight of 750 pounds and covered a total distance of 120 feet. Today, transport aircraft powered by modern turbofan engines have a maximum takeoff weight approaching 1,000,000 pounds and a maximum range of more than 8,000 miles. Based on these astounding advancements, it is indeed difficult to imagine what the next century holds in store.

2.2. The Propeller

An airplane propeller has much in common with a finite wing. In fact, each blade of the propeller can be viewed as a rotating wing. Like a wing, the cross-section of a propeller blade is an airfoil section, as shown in Fig. 2.2.1. Both the wing and the propeller blade are designed to produce lift. The lift developed by a wing is usually directed to support the weight of the airplane and keep it aloft. The lift developed by a propeller is typically aligned more or less with the direction of flight and is intended to support the airplane drag and accelerate the airplane in the direction of motion. To a person not familiar with the terminology of aeronautics, it might seem strange to say that we can direct lift to produce a horizontal force in the direction of motion. However, recall that the aerodynamic definition of lift is not a force opposing gravity but a force normal to the relative airflow. It is primarily the motion of the airplane through the air that provides the relative airflow over an airplane wing. Thus, the lift on a wing is normal to the direction of flight. The relative airflow over a propeller blade is provided primarily by the rotation of the propeller. Thus, the lift on a propeller blade is directed approximately along the axis of rotation. The component of the aerodynamic force that is parallel to the axis of rotation is the *thrust*.

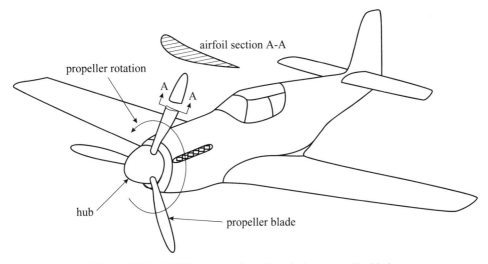

Figure 2.2.1. Airfoil cross-section of an airplane propeller blade.

Because the velocity of each section of a rotating propeller blade depends on the distance of the section from the axis of rotation, a propeller blade usually has much more twist or geometric washout than a typical wing. As shown in Fig. 2.2.2, the sections close to the axis are moving more slowly and form a larger angle with the plane of rotation, while the sections close to the tip are moving faster and form a smaller angle with the plane of rotation. The angle that the section zero-lift line makes with the plane of rotation will be called the *aerodynamic pitch angle*. This pitch angle varies with the radial distance, r, and will be denoted as $\beta(r)$.

The reader should be cautioned that pitch angles are often tabulated relative to the section chord line or to a flat lower surface of the airfoil section. However, the aerodynamic pitch angle, which is used in this text, is more convenient for aerodynamic analysis. If the airfoil section geometry is known and a pitch angle is given in terms of some other definition, the aerodynamic pitch angle can be determined. For example,

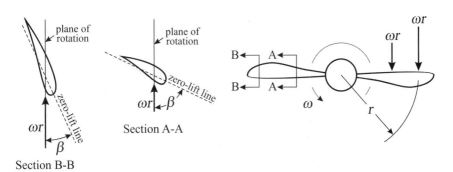

Figure 2.2.2. Radial variation in aerodynamic pitch angle along a propeller blade.

if β_c is the pitch angle relative to the chord line, the aerodynamic pitch angle, β, can be written simply as

$$\beta(r) = \beta_c(r) - \alpha_{L0}(r) \tag{2.2.1}$$

where α_{L0} is the zero-lift angle of attack for the airfoil section of the blade that is located at radius, r.

The pitch of a propeller is not commonly specified in terms of a pitch angle, but rather in terms of a pitch length, which is usually referred to simply as the *pitch*. The pitch length for a propeller is similar to that for a common screw. In fact, propellers were originally called *airscrews*. If the propeller were turned through the air without slipping, the distance that the propeller would move forward in each revolution is the pitch length. In this text, the aerodynamic pitch length is given the symbol λ and is defined relative to the section zero-lift line as

$$\lambda(r) = 2\pi r \tan\beta \tag{2.2.2}$$

This geometric relation is shown schematically in Fig. 2.2.3. Here again the reader is cautioned that propeller pitch is normally tabulated relative to the section chord line or to a flat lower surface of the airfoil section. The aerodynamic pitch, λ, is determined from the chord-line pitch, λ_c, according to the relation

$$\lambda(r) = 2\pi r \frac{\lambda_c - 2\pi r \tan\alpha_{L0}}{2\pi r + \lambda_c \tan\alpha_{L0}} \tag{2.2.3}$$

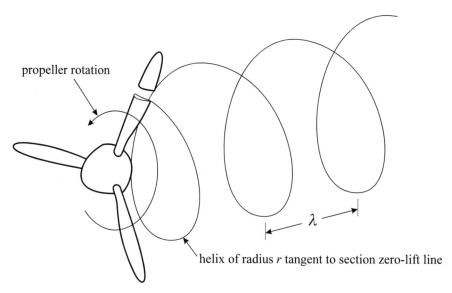

propeller rotation

helix of radius r tangent to section zero-lift line

Figure 2.2.3. The helix that defines aerodynamic pitch length.

For the special case of a *constant-pitch propeller*, the pitch length does not vary with the radial coordinate, *r*. Thus, in a constant-pitch propeller, the tangent of the pitch angle decreases with increasing radius in proportion to 1/*r*. The term *constant-pitch* should not be confused with the term *fixed-pitch*, which is related to the term *variable-pitch*. In a *variable-pitch propeller* the propeller pitch can be changed by rotating each blade about an axis that runs along the length of the blade. A *fixed-pitch propeller* is one in which the pitch cannot be varied. As previously stated, a *constant-pitch propeller* is one in which the pitch length does not vary with *r*.

Like a wing, each blade of the propeller will be subject to a drag force as well as a lift force. Since the propeller blade length is finite, there will be both parasitic drag and induced drag. Because drag is defined as the component of the aerodynamic force that is parallel to the relative airflow, the drag on a rotating propeller blade produces a moment about the propeller axis that opposes the propeller rotation. The torque necessary to counter this moment and the power required to sustain the rotation must be provided by the airplane's engine. Since the axis of propeller rotation is typically aligned closely with the direction of motion, this propeller-induced moment is felt by the airplane as a rolling moment, which must be countered with an aerodynamic moment produced by the airframe or by another propeller rotating in the opposite direction.

Maximizing thrust while minimizing the torque necessary to turn a propeller is obviously one important aspect of good propeller design. The ratio of the thrust developed to the torque required for a propeller is analogous with the lift-to-drag ratio for a wing. The torque required to turn the propeller multiplied by the angular velocity is called the propeller *brake power*. It is this power that must be supplied by the engine. The thrust developed by the propeller multiplied by the airspeed of the airplane is called the *propulsive power*. This is the useful power that is provided to propel the airplane forward against the airframe drag. The ratio of the propulsive power to the brake power for a propeller is called the *propulsive efficiency*. This is one important measure of propeller performance. However, it is not the only important measure. The thrust that is developed by a propeller when the airplane is not moving is called the *static thrust*. It is important for a propeller to produce high static thrust when accelerating an airplane from brake release on takeoff. Since the airspeed is zero for the case of static thrust, the propulsive power and the propulsive efficiency are both zero. Thus, propulsive efficiency is not a particularly good measure of a propeller's ability to accelerate an airplane from a standing start.

The thrust force and the rolling moment are not the only aerodynamic force and moment produced by a rotating propeller. Additional aerodynamic reactions are produced if the propeller rotation axis is not aligned perfectly with the direction of flight. The angle that the propeller axis makes with the freestream airflow is called the *thrust angle* or the *propeller angle of attack*. Since the engine and propeller are typically mounted directly to the airframe, the propeller angle of attack will change with the airplane angle of attack. Because the airplane angle of attack can change during flight, the propeller angle of attack can also change.

When a rotating propeller is at some positive angle of attack relative to the freestream flow, there is a component of the freestream in the plane of propeller rotation, as seen in Fig. 2.2.4. This component of the freestream changes the relative airflow over each blade of the propeller. It increases the relative airspeed for the downward-moving blades and decreases the relative airspeed for the upward-moving blades. Thus, both the lift and drag are increased on the downward-moving side of the propeller and decreased on the upward-moving side. The difference in thrust generated by the two sides of the propeller produces a yawing moment and the difference in circumferential force produces a net normal force in the plane of rotation, as seen in Fig. 2.2.4.

The axial component of the airplane's airspeed will also affect the aerodynamic forces and moments acting on a rotating propeller. Since this component of the airspeed is normal to the plane of propeller rotation, it changes the angle of attack for the individual blades. This acts very much like the downwash on a finite wing, and like downwash on a wing, this normal component of airflow changes the aerodynamic forces generated on each blade of the propeller.

Even in the static case, when the airplane is not moving, there is airflow normal to the plane of propeller rotation. As is the case with any finite wing, the vorticity shed from each blade of the propeller produces downwash along its own length. However, in addition to being subjected to its own downwash, each blade of a propeller is "flying directly behind" the blade that precedes it in the rotation sequence. The blades of a rotating propeller act very much like an infinite series of finite wings, flying in a row one behind the other. The downwash on any one such wing would be increased by the vorticity shed from every other wing that proceeds it in the flight line. Similarly, the downwash on each blade of a rotating propeller is amplified by its proximity to the other blades. In the case of a rotating propeller, the total downwash generated by all of the propeller blades combined is usually called the propeller's *induced velocity*.

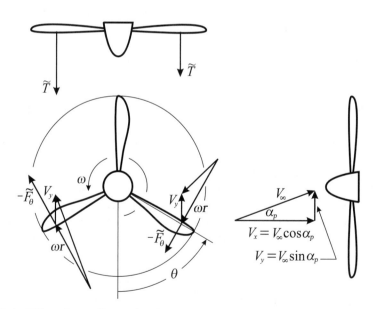

Figure 2.2.4. Effect of propeller angle of attack on the aerodynamic forces acting on a rotating propeller.

A propeller, or even a common house fan, provides a good means for the student to get a feel for the downwash produced by a lifting surface. The downwash produced by a rotating propeller or a common house fan is concentrated in a much smaller area than is that produced by a moving wing. Because the blades are passing repeatedly through the same small section of the fluid, the downwash is amplified with each successive pass. This concentrated downwash is readily detected by the human senses. Although the downwash induced by a rotating propeller is much stronger than the downwash induced by the lifting wing of an airplane, it is conceptually no different in physical origin.

Similar to the lift on a wing, the thrust developed by a propeller is imparted to the individual blades directly through a pressure difference between the upstream and downstream sides of the blade. However, also like a wing, the same pressure difference is imposed on the fluid as well. The pressure difference acting on the fluid generates fluid momentum in a direction opposite to the aerodynamic force exerted on the propeller. The velocity increase induced at the plane of the propeller disk is the downwash that is usually referred to as the propeller's induced velocity.

Optimizing the performance of an airplane powered by an engine-propeller combination depends on properly matching the propeller with the engine, as well as matching the engine-propeller combination with the airframe. Understanding propeller performance is critical to this optimization process. For this purpose, it is useful to be able to predict the aerodynamic forces and moments acting on a rotating propeller as a function of the operating conditions.

2.3. Propeller Blade Theory

In order to quantitatively understand and predict the performance of a rotating propeller, it is necessary to analyze the aerodynamics of the blade in detail. To this end, consider the cross-section of the propeller blade that is shown in Fig. 2.3.1. The blade is rotating with an angular velocity of ω and is advancing through the air with a relative airspeed of V_∞. Initially we shall assume that the forward velocity vector is aligned with the axis of propeller rotation so that there is no component of the forward airspeed in the plane of rotation. The angle that the zero-lift line for the blade section makes with the plane of rotation is what we have called the *aerodynamic pitch angle*, β. In general, this pitch angle varies with the radial distance, r.

The total downwash angle, ε_b, for the blade cross-section located at radius r is the sum of two parts, the downwash angle that results from the propeller's forward motion, ε_∞, and the induced downwash angle, ε_i,

$$\varepsilon_b(r) = \varepsilon_\infty(r) + \varepsilon_i(r) \tag{2.3.1}$$

In this text, ε_∞ is named the *advance angle*, ε_i is called the *induced angle*, and ε_b is simply referred to as the *downwash angle*. This downwash angle reduces the angle of attack for the airfoil section and tilts the lift vector back, as is shown in Fig. 2.3.1. This tilting of the lift vector through the angle ε_b increases the torque required to turn the propeller in much the same way that downwash adds induced drag to a lifting wing. From the geometry shown in Fig. 2.3.1, the section thrust and circumferential force are related to the section lift and drag forces through this local downwash angle according to the relations

$$\widetilde{T} = \widetilde{L}\cos\varepsilon_b - \widetilde{D}\sin\varepsilon_b \tag{2.3.2}$$

$$\widetilde{F}_\theta = -\widetilde{D}\cos\varepsilon_b - \widetilde{L}\sin\varepsilon_b \tag{2.3.3}$$

From the axial and circumferential components of induced velocity, V_{xi} and $V_{\theta i}$, the downwash angle, ε_b, is determined from the geometry shown in Fig. 2.3.1,

$$\varepsilon_b(r) = \tan^{-1}\left(\frac{V_\infty + V_{xi}}{\omega r - V_{\theta i}}\right) \tag{2.3.4}$$

From the known rotational speed and forward airspeed, the advance angle, ε_∞, is determined directly from this same geometry as

$$\varepsilon_\infty(r) = \tan^{-1}\left(\frac{V_\infty}{\omega r}\right) \tag{2.3.5}$$

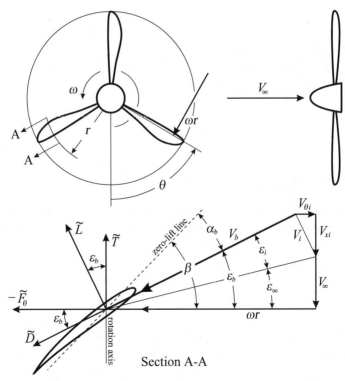

Figure 2.3.1. Section forces and velocities acting on a rotating propeller blade.

Thus, using Eq. (2.3.1), the induced angle, ε_i, is

$$\varepsilon_i(r) = \tan^{-1}\left(\frac{V_\infty + V_{xi}}{\omega r - V_{\theta i}}\right) - \tan^{-1}\left(\frac{V_\infty}{\omega r}\right) \tag{2.3.6}$$

The aerodynamic angle of attack for the blade section, α_b, measured relative to the section zero-lift line is simply the aerodynamic pitch angle, β, less the downwash angle, ε_b. From Eq. (2.3.1), this can be written as

$$\alpha_b(r) = \beta(r) - \varepsilon_b(r) = \beta(r) - \varepsilon_\infty(r) - \varepsilon_i(r) \tag{2.3.7}$$

The fluid velocity relative to the blade cross-section located at radius r is designated as V_b. This is the vector sum of the velocity created by the rotation of the blade, the forward velocity of the propeller, and the propeller induced velocity. The largest component of V_b is usually the circumferential component, ωr, which results from the blade rotation. In addition, there is an axial component of the airspeed, relative to the blade section, which results from the propeller's forward airspeed, V_∞. There is also a component of velocity relative to the blade section that results from the propeller's induced velocity. The induced velocity results from the same aerodynamic pressure difference that produces the lift on the propeller blades. In general, the propeller's induced velocity will have both an axial component, V_{xi}, and a circumferential component, $V_{\theta i}$. The induced velocity will vary with the radial coordinate, r. From the geometry shown in Fig. 2.3.1, the total relative airspeed, V_b, at the plane of the blade section is given by

$$V_b^2 = (\omega r - V_{\theta i})^2 + (V_\infty + V_{xi})^2 = \omega^2 r^2 \left[\left(1 - \frac{V_{\theta i}}{\omega r}\right)^2 + \left(\frac{V_\infty}{\omega r} + \frac{V_{xi}}{\omega r}\right)^2\right] \tag{2.3.8}$$

Using Eqs. (2.3.8), the section lift for the blade section located at radius r is expressed as

$$\tilde{L} = \frac{1}{2}\rho V_b^2 c_b \tilde{C}_L = \frac{1}{2}\rho \omega^2 r^2 c_b \tilde{C}_L \left[\left(1 - \frac{V_{\theta i}}{\omega r}\right)^2 + \left(\frac{V_\infty}{\omega r} + \frac{V_{xi}}{\omega r}\right)^2\right] \tag{2.3.9}$$

where c_b is the local section chord length and $\tilde{C}_L$ is the local section lift coefficient, which is a function of the local section aerodynamic angle of attack given by Eq. (2.3.7). Also using Eqs. (2.3.8), the local section drag can be written

$$\tilde{D} = \frac{1}{2}\rho V_b^2 c_b \tilde{C}_D = \frac{1}{2}\rho \omega^2 r^2 c_b \tilde{C}_D \left[\left(1 - \frac{V_{\theta i}}{\omega r}\right)^2 + \left(\frac{V_\infty}{\omega r} + \frac{V_{xi}}{\omega r}\right)^2\right] \tag{2.3.10}$$

where $\tilde{C}_D$ is the local two-dimensional section drag coefficient, which includes only parasitic drag. The induced drag is included as part of the lift vector. In general, the section drag coefficient is also a function of the local section angle of attack.

Applying Eqs. (2.3.9) and (2.3.10) to Eq. (2.3.2), the axial component of section force is written

$$\tilde{T} = \frac{1}{2}\rho\omega^2 r^2 c_b \left[\left(1-\frac{V_{\theta i}}{\omega r}\right)^2 + \left(\frac{V_\infty}{\omega r}+\frac{V_{xi}}{\omega r}\right)^2\right]\left(\tilde{C}_L \cos \varepsilon_b - \tilde{C}_D \sin \varepsilon_b\right) \quad (2.3.11)$$

Similarly, from Eq. (2.3.3), the circumferential component of section force is

$$\tilde{F}_\theta = -\frac{1}{2}\rho\omega^2 r^2 c_b \left[\left(1-\frac{V_{\theta i}}{\omega r}\right)^2 + \left(\frac{V_\infty}{\omega r}+\frac{V_{xi}}{\omega r}\right)^2\right]\left(\tilde{C}_D \cos \varepsilon_b + \tilde{C}_L \sin \varepsilon_b\right) \quad (2.3.12)$$

The thrust per unit radial distance for the full prop-circle is the axial section force per blade multiplied by the number of blades, k. Thus, from Eq. (2.3.11),

$$\frac{dT}{dr} = k\tilde{T}$$

$$= \frac{k}{2}\rho\omega^2 r^2 c_b \left[\left(1-\frac{V_{\theta i}}{\omega r}\right)^2 + \left(\frac{V_\infty}{\omega r}+\frac{V_{xi}}{\omega r}\right)^2\right]\left(\tilde{C}_L \cos \varepsilon_b - \tilde{C}_D \sin \varepsilon_b\right) \quad (2.3.13)$$

The torque required to turn the propeller is simply the aerodynamic rolling moment, ℓ, that is produced about the axis of propeller rotation. Thus, the torque per unit radial distance, for the full prop-circle, is the circumferential section force per blade multiplied by the radius and the number of blades. From Eq. (2.3.12), this results in

$$\frac{d\ell}{dr} = -kr\tilde{F}_\theta$$

$$= \frac{k}{2}\rho\omega^2 r^3 c_b \left[\left(1-\frac{V_{\theta i}}{\omega r}\right)^2 + \left(\frac{V_\infty}{\omega r}+\frac{V_{xi}}{\omega r}\right)^2\right]\left(\tilde{C}_D \cos \varepsilon_b + \tilde{C}_L \sin \varepsilon_b\right) \quad (2.3.14)$$

If the propeller geometry were completely defined, the section chord length and section pitch angle would be known functions of the radial coordinate, r. The section lift and drag coefficients would be known functions of the radial coordinate, r, and the local section angle of attack, α_b. If the rotational speed and the forward airspeed are known, the advance angle, ε_∞, is known from Eq. (2.3.5). Even so, the two differential equations expressed in Eqs. (2.3.13) and (2.3.14) contain four unknowns: the thrust, T, the torque, ℓ, and the two components of the propeller induced velocity, V_{xi}, and $V_{\theta i}$. If these two components were known, the downwash angle, ε_b, could be determined from Eq. (2.3.4). With only two equations and four unknowns, obviously, additional information is needed before Eqs. (2.3.13) and (2.3.14) can be solved. For this additional information we will now consider the vortex lifting law and the vorticity that is shed from the rotating propeller blades.

Just as is the case for a lifting wing, lift cannot be generated on a rotating propeller blade without the simultaneous generation of vorticity. Furthermore, as was previously done for a finite wing, the lift on a finite propeller blade can be related to the bound vorticity through the vortex lifting law. This requires that for any cross-section of a propeller blade, the section lift, $\tilde{L}$, is related to the fluid density, ρ, the relative airspeed, V_b, and the total section circulation, Γ, according to

$$\tilde{L} = \rho V_b \Gamma \tag{2.3.15}$$

Thus, using Eqs. (2.3.8) and (2.3.9) with Eq. (2.3.15), the local section circulation for any cross-section of a propeller blade can be related to previously defined variables,

$$\Gamma = \tfrac{1}{2} V_b c_b \tilde{C}_L = \tfrac{1}{2} \omega r c_b \tilde{C}_L \sqrt{\left(1 - \frac{V_{\theta i}}{\omega r}\right)^2 + \left(\frac{V_\infty}{\omega r} + \frac{V_{xi}}{\omega r}\right)^2} \tag{2.3.16}$$

Since the lift on a propeller blade is generated directly from a pressure difference between the two sides of the blade, the lift must go to zero at the blade tip, where such a pressure difference cannot be supported. This fact, combined with Eq. (2.3.16), requires that vorticity must be shed from the blade tips of a rotating propeller, just as it is from a lifting wing. It is this shed vorticity that produces the induced downwash on the propeller blades. The bound vorticity does not influence the induced downwash. Because of the symmetry of the blades in the prop-circle, each blade receives as much upwash from the bound vorticity on other blades as it does downwash.

The primary difference between the vorticity shed from a propeller blade and that shed from a lifting wing is the path that the vorticity takes as it moves downstream. Whereas the wingtip vortices shed from a lifting wing move downstream in a fairly linear fashion, the blade-tip vortices shed from a rotating propeller follow a helical path, as is seen in Fig. 2.3.2. In this photograph, the path of each blade-tip vortex can be visualized as a result of contrails, which are caused by moisture condensing in the low-pressure region near the core of each vortex. Since the blade-tip vortices all rotate in the same direction, the region inside the propeller's helical trailing vortex system is a region of very strong downwash. This downwash is the air movement that can be felt directly behind a rotating propeller or in front of a common house fan. This cylindrical region of strong downwash is called the *slipstream*. In the region just outside the slipstream, there is upwash. It should be remembered that in aerodynamics terms, "downwash" and "upwash" have nothing to do with "up" and "down" in the conventional sense. *Downwash* is airflow in a direction opposite to the lift vector, and *upwash* is airflow in the same direction as the lift vector. Thus, the induced upwash in the region outside the slipstream is an upstream flow that replaces the air that has been removed from the region forward of the propelled disk by the induced downwash. This flow pattern is shown schematically in Fig. 2.3.3.

Figure 2.3.2. Contrails showing the helical path of the blade-tip vortices shed from a rotating propeller. (U.S. Air Force)

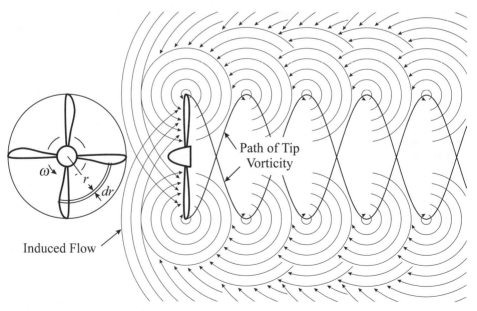

Figure 2.3.3. Schematic cross-section of the vortex-induced flow in the vicinity of a rotating propeller.

Because the vortex lines follow a helical path rather than a circular path, the velocity induced by each vortex, which is normal to the vorticity vector, has a component in the circumferential direction. Thus, the air in the slipstream rotates as it moves downstream. This rotation is not obvious from the feel of the air in the slipstream of a propeller or a common house fan. However, by holding a piece of light yarn in the slipstream of a fan, this rotation is easily seen.

In Fig. 2.3.3, an attempt has been made to show separately the velocity induced by each section of each blade-tip vortex. At any point in space, the resultant induced velocity is the vector sum of the velocity induced by the entire length of all vortex filaments in the slipstream. Since the velocity induced by a vortex filament at any point in space decreases with the distance from the filament, the closest filaments have the greatest influence on the velocity induced at a particular point. However, the induced velocity at any point is influenced by the entire helical vortex system. In the slipstream downstream from the propeller, the upstream portion of the vortex system as well as the downstream portion produces downwash. As a result, downstream from the propeller, the induced velocity is greater than it is in the plane of the propeller, where the only vorticity is downstream. Far downstream from the propeller, the helical vortex system extends essentially from negative infinity to positive infinity, while at the plane of the propeller itself, the vortex system extends only from zero to infinity. Thus, far downstream from the propeller, the induced velocity is about twice what it is in the plane of the propeller disk.

As one might imagine, computing the velocity induced by the helical vortex system trailing downstream from a rotating propeller is considerably more complex than computing the velocity induced by the vortex sheet trailing from a finite lifting wing. One method for predicting this induced velocity is known as *Goldstein's vortex theory*. To predict the velocity induced in the plane of the propeller disk, Goldstein (1929) made two simplifying hypotheses. First, **the vortex sheet trailing from a rotating propeller blade was assumed to lie along a helical surface of constant pitch.** Second, **the induced velocity was assumed to be normal to the resultant velocity, which is the vector sum of the rotation velocity, the forward velocity, and the induced velocity itself.** These assumptions were not made arbitrarily. It can be shown that these conditions are both satisfied in the ultimate slipstream of an optimum propeller. This is the so-called *Betz condition* (see Betz 1919). However, Goldstein's assumptions are difficult to justify in the plane of a propeller having blades of arbitrary pitch and planform shape. Nevertheless, McCormick (1995) states that "studies have been performed that support normality at the plane of the propeller" and he has shown that this theory gives results that are in reasonable agreement with experimental data.

From the normality hypothesis, shown in Fig. 2.3.4, the total relative airspeed, V_b, is given by

$$V_b = \sqrt{\omega^2 r^2 + V_\infty^2} \, \cos \varepsilon_i = \frac{\omega r}{\cos \varepsilon_\infty} \cos \varepsilon_i \qquad (2.3.17)$$

Similarly, from Fig. 2.3.4, the induced velocity and its axial and circumferential components can be written as

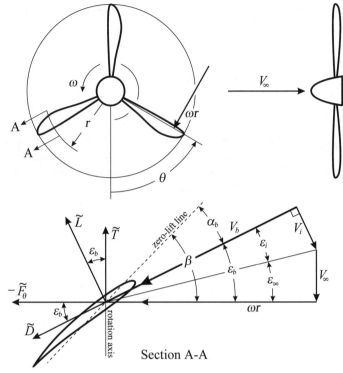

Figure 2.3.4. Section forces and velocities acting on a rotating propeller blade under Goldstein's hypothesis.

$$V_i = \sqrt{\omega^2 r^2 + V_\infty^2} \sin \varepsilon_i = \frac{\omega r}{\cos \varepsilon_\infty} \sin \varepsilon_i \qquad (2.3.18)$$

$$V_{xi} = V_i \cos \varepsilon_b = \frac{\omega r}{\cos \varepsilon_\infty} \sin \varepsilon_i \cos(\varepsilon_i + \varepsilon_\infty) \qquad (2.3.19)$$

$$V_{\theta i} = V_i \sin \varepsilon_b = \frac{\omega r}{\cos \varepsilon_\infty} \sin \varepsilon_i \sin(\varepsilon_i + \varepsilon_\infty) \qquad (2.3.20)$$

From the hypothesis of constant pitch in the trailing helical vortex sheet, Goldstein's vortex theory predicts that the local circumferential component of induced velocity, $V_{\theta i}$, in the plane of the propeller disk is related to the local section circulation, Γ, by

$$k\Gamma = 4\pi \kappa r V_{\theta i} \qquad (2.3.21)$$

The proportionality constant, κ, known as *Goldstein's kappa factor*, is available in graphical form but has never been presented in closed form.

A close approximation to Goldstein's result can be obtained from the relation

$$k\Gamma \cong 4\pi f r V_{\theta i} \qquad (2.3.22)$$

The parameter f is known as *Prandtl's tip loss factor* (see Prandtl and Betz 1927), which can be expressed as

$$f = \frac{2}{\pi}\cos^{-1}\left\{\exp\left[-\frac{k(1-2r/d_p)}{2\sin\beta_t}\right]\right\} \qquad (2.3.23)$$

where β_t is the aerodynamic pitch angle at the propeller blade tip. Using Eq. (2.3.17) in Eq. (2.3.16), the section circulation can also be written as

$$\Gamma = \tfrac{1}{2}V_b c_b \tilde{C}_L = \tfrac{1}{2}\omega r c_b \tilde{C}_L \frac{\cos\varepsilon_i}{\cos\varepsilon_\infty} \qquad (2.3.24)$$

After applying Eqs. (2.3.20) and (2.3.24) to Eq. (2.3.22), we obtain the relation

$$\frac{kc_b}{16r}\tilde{C}_L - \cos^{-1}\left\{\exp\left[-\frac{k(1-2r/d_p)}{2\sin\beta_t}\right]\right\}\tan\varepsilon_i\sin(\varepsilon_\infty+\varepsilon_i) = 0 \qquad (2.3.25)$$

With knowledge of the propeller geometry, the rotational speed, and the forward speed, Eq. (2.3.25) contains only a single unknown, the induced angle, ε_i. This equation is easily solved numerically to determine the induced angle as a function of the radial coordinate, r.

Once the induced angle is known as a function of radius, the total thrust developed by the propeller is determined by using Eq. (2.3.13) with Eq. (2.3.17) and integrating from the hub radius, r_h, to the tip radius, r_o,

$$
\begin{aligned}
T &= \int_{r=r_h}^{r_o}\frac{dT}{dr}dr \\
&= \frac{k\rho\omega^2}{2}\int_{r=r_h}^{r_o}r^2 c_b \frac{\cos^2\varepsilon_i}{\cos^2\varepsilon_\infty}\left[\tilde{C}_L\cos(\varepsilon_\infty+\varepsilon_i)-\tilde{C}_D\sin(\varepsilon_\infty+\varepsilon_i)\right]dr
\end{aligned} \qquad (2.3.26)
$$

In a similar manner, the total torque required to turn the propeller is found by integrating Eq. (2.3.14) with Eq. (2.3.17),

$$
\begin{aligned}
\ell &= \int_{r=r_h}^{r_o}\frac{d\ell}{dr}dr \\
&= \frac{k\rho\omega^2}{2}\int_{r=r_h}^{r_o}r^3 c_b \frac{\cos^2\varepsilon_i}{\cos^2\varepsilon_\infty}\left[\tilde{C}_D\cos(\varepsilon_\infty+\varepsilon_i)+\tilde{C}_L\sin(\varepsilon_\infty+\varepsilon_i)\right]dr
\end{aligned} \qquad (2.3.27)
$$

Equations (2.3.26) and (2.3.27) can be integrated numerically to determine the propeller thrust and torque. The brake power required to turn the propeller is just the torque multiplied by the angular velocity,

$$P_b \equiv \ell\omega \tag{2.3.28}$$

As is the case with lift and drag, the aerodynamic forces and moments for a propeller are normally expressed in terms of dimensionless coefficients. Whereas the freestream velocity was used as the characteristic velocity in defining the lift and drag coefficients for a wing, the freestream velocity is not the most important velocity associated with the production of thrust by a rotating propeller. The velocity that has the greatest influence on propeller thrust is the rotational velocity of the blades. While there are several characteristic dimensions associated with a propeller, the one most commonly used in the definition of dimensionless propeller coefficients is the propeller diameter, d_p. Although other definitions have occasionally been used, by far the most common definitions for the propeller coefficients are:

Thrust coefficient $\qquad\qquad C_T \equiv \dfrac{T}{\rho(\omega/2\pi)^2 d_p^4}$ $\qquad\qquad$ (2.3.29)

Torque coefficient $\qquad\qquad C_\ell \equiv \dfrac{\ell}{\rho(\omega/2\pi)^2 d_p^5}$ $\qquad\qquad$ (2.3.30)

Power coefficient $\qquad\qquad C_P \equiv \dfrac{\ell\omega}{\rho(\omega/2\pi)^3 d_p^5}$ $\qquad\qquad$ (2.3.31)

With these definitions, Eqs. (2.3.25) through (2.3.28) are readily nondimensionalized to yield

$$\frac{\hat{c}_b}{8\zeta}\tilde{C}_L(\alpha_b,\zeta) - \cos^{-1}\left\{\exp\left[-\frac{k(1-\zeta)}{2\sin\beta_t}\right]\right\}\tan\varepsilon_i\,\sin(\varepsilon_\infty + \varepsilon_i) = 0 \tag{2.3.32}$$

$$C_T = \frac{\pi^2}{4}\int_{\zeta=\zeta_h}^{1}\zeta^2\hat{c}_b\frac{\cos^2\varepsilon_i}{\cos^2\varepsilon_\infty}\left[\tilde{C}_L\cos(\varepsilon_\infty + \varepsilon_i) - \tilde{C}_D\sin(\varepsilon_\infty + \varepsilon_i)\right]d\zeta \tag{2.3.33}$$

$$C_\ell = \frac{\pi^2}{8}\int_{\zeta=\zeta_h}^{1}\zeta^3\hat{c}_b\frac{\cos^2\varepsilon_i}{\cos^2\varepsilon_\infty}\left[\tilde{C}_D\cos(\varepsilon_\infty + \varepsilon_i) + \tilde{C}_L\sin(\varepsilon_\infty + \varepsilon_i)\right]d\zeta \tag{2.3.34}$$

$$C_P = 2\pi C_\ell \tag{2.3.35}$$

where β_t is the aerodynamic pitch angle at the blade tip. The local section lift and drag coefficients are evaluated at the local section angle of attack,

$$\alpha_b(\zeta) = \beta - \varepsilon_\infty - \varepsilon_i \tag{2.3.36}$$

and

$$\zeta \equiv \frac{r}{d_p/2} \tag{2.3.37}$$

$$\hat{c}_b \equiv \frac{kc_b}{d_p} \tag{2.3.38}$$

$$\beta \equiv \tan^{-1}\left(\frac{K}{\pi\zeta}\right) \tag{2.3.39}$$

$$\varepsilon_\infty \equiv \tan^{-1}\left(\frac{J}{\pi\zeta}\right) \tag{2.3.40}$$

$$K \equiv \frac{\lambda}{d_p} \tag{2.3.41}$$

$$J \equiv \frac{2\pi V_\infty}{\omega d_p} \tag{2.3.42}$$

The local *chord length ratio*, $\hat{c}_b$, is the total chord length for all blades combined divided by the propeller diameter. The dimensionless quantity, K, is the local *pitch-to-diameter ratio*. The dimensionless quantity, J, is called the *advance ratio* and is the distance that the propeller is advanced in one revolution divided by the propeller diameter. In general, both $\hat{c}_b$ and K will vary with ζ. However, for a propeller with constant aerodynamic pitch, the aerodynamic pitch length does not vary with radius and thus K is constant.

Knowing the propeller diameter and the planform shape of the blade, we can determine the dimensionless chord length ratio, $\hat{c}_b$, as a function of ζ from Eq. (2.3.38). From the known variation in pitch length and propeller diameter, the pitch-to-diameter ratio, K, is also determined as a function of ζ using Eq. (2.3.41). With knowledge of the forward speed and rotational speed, the advance ratio, J, is found from Eq. (2.3.42). Knowing the variation in pitch-to-diameter ratio and advance ratio we can now compute both $\beta(\zeta)$ and $\varepsilon_\infty(\zeta)$ from Eqs. (2.3.39) and (2.3.40). From the known section properties of the propeller blade, the section lift and drag coefficients are known functions of α_p and ζ. With this knowledge, the induced angle, $\varepsilon_i(\zeta)$, is determined by solving Eq. (2.3.32) numerically. With the induced angle known, the aerodynamic coefficients are determined numerically from Eqs. (2.3.33) through (2.3.35).

Having determined the advance ratio, the thrust coefficient, and the power coefficient, the propulsive efficiency for the propeller can be determined. The propulsive efficiency is defined as the propulsive power divided by the brake power,

$$\eta_p \equiv \frac{TV_\infty}{\ell\omega} = \left[\frac{T}{\rho(\omega/2\pi)^2 d_p^4}\right]\left[\frac{V_\infty}{(\omega/2\pi)d_p}\right]\left[\frac{\rho(\omega/2\pi)^3 d_p^5}{\ell\omega}\right] \tag{2.3.43}$$

Thus, the propulsive efficiency is

$$\eta_p = \frac{C_T J}{C_P} \tag{2.3.44}$$

Notice that the propeller thrust coefficient, C_T, the torque coefficient, C_ℓ, the power coefficient, C_P, and the propulsive efficiency, η_p, are all functions of the chord length ratio, $\hat{c}_b(\zeta)$, the pitch-to-diameter ratio, $K(\zeta)$, and advance ratio, J. These coefficients also depend on the blade cross-section through the section lift and drag coefficients. Since results must be determined numerically, to examine the nature of propeller performance we must turn to a specific example using some particular geometry.

EXAMPLE 2.3.1. To examine how propeller performance depends on pitch-to-diameter ratio and advance ratio, we shall consider a family of fixed-pitch propellers having two blades with an elliptic planform shape. The blade chord length varies with radial position according to the relation

$$\frac{c_b}{d_p} = 0.075\sqrt{1-\zeta^2}$$

The airfoil section as well as the chord-line pitch length, λ_c, for this family of propellers is constant along the blade. The hub diameter is 10 percent of the propeller diameter. The zero-lift angle of attack for the airfoil section is -2.1 degrees and the section lift and drag coefficients can be approximated as

$$\tilde{C}_L = \begin{cases} 2\pi\alpha_b, & \alpha_b \leq 0.25 \\[2mm] \dfrac{\pi}{2}\dfrac{\cos\alpha_b}{\cos(0.25)}, & \alpha_b > 0.25 \end{cases}$$

$$\tilde{C}_D = \begin{cases} 0.224\alpha_b^2 + 0.006, & \alpha_b \leq 0.25 \\[2mm] 16.6944\alpha_b^2 - 1.0234, & 0.25 \leq \alpha_b \leq 0.30 \\[2mm] \dfrac{\pi}{2}\dfrac{\sin\alpha_b}{\cos(0.25)}, & \alpha_b \geq 0.30 \end{cases}$$

where α_b is the aerodynamic angle of attack, measured relative to the zero-lift line. We wish to study the performance of this family of propellers, over a range of operating conditions, based on the propeller blade theory presented in this section.

Solution. For this propeller, the chord-line pitch length, λ_c, is said to be constant. However, this does not mean that the aerodynamic pitch length is constant. The aerodynamic pitch length is related to the chord-line pitch length according to Eq. (2.2.3),

$$\lambda = 2\pi r \frac{\lambda_c - 2\pi r \tan \alpha_{L0}}{2\pi r + \lambda_c \tan \alpha_{L0}}$$

or

$$K = \pi\zeta \frac{K_c - \pi\zeta \tan \alpha_{L0}}{\pi\zeta + K_c \tan \alpha_{L0}}$$

For this airfoil the zero-lift angle of attack is -2.1 degrees ($\tan\alpha_{L0} = -0.0367$).

The geometry, the aerodynamic angles, and thrust and power distribution, as predicted by Goldstein's theory, are presented in Figs. 2.3.5 and 2.3.6 for one of these propellers at two different operating conditions. There are two things that should be observed from these figures. First, notice that even though this propeller would typically be called a constant-pitch propeller, the aerodynamic pitch varies considerably over the length of the blade. Also notice that at an advance ratio of 0.25, despite an aerodynamic pitch angle as high as 60 degrees, the forward motion and induced velocity keep the angle of attack low and the propeller blades do not stall. However, at an advance ratio of zero, the blades are stalled in the region near the hub, with angles of attack approaching 40 degrees.

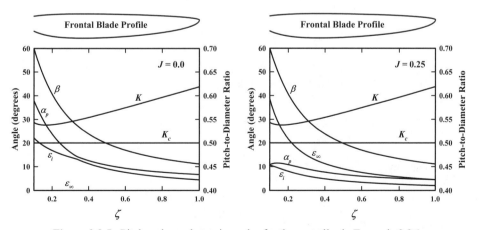

Figure 2.3.5. Pitch and aerodynamic angles for the propeller in Example 2.3.1.

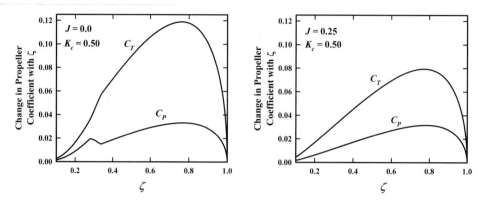

Figure 2.3.6. Thrust and power distribution for the propeller in Example 2.3.1.

As an example calculation, consider the 75 percent radius section of one of these propellers with a chord-line pitch-to-diameter ratio of 0.5 operating at an advance ratio of 0.25. The aerodynamic pitch-to-diameter ratio for this cross-section is

$$K = \pi\zeta\frac{K_c - \pi\zeta\tan\alpha_{L0}}{\pi\zeta + K_c\tan\alpha_{L0}} = \pi(0.75)\frac{0.5 - \pi(0.75)(-0.0367)}{\pi(0.75) + 0.5(-0.0367)} = 0.591$$

From Eq. (2.3.39), the aerodynamic pitch angle for this cross-section is

$$\beta \equiv \tan^{-1}\left(\frac{K}{\pi\zeta}\right) = \tan^{-1}\left(\frac{0.591}{\pi 0.75}\right) = 0.2458 = 14.08°$$

From Eq. (2.3.40), the advance angle for this cross-section is

$$\varepsilon_\infty \equiv \tan^{-1}\left(\frac{J}{\pi\zeta}\right) = \tan^{-1}\left(\frac{0.25}{\pi 0.75}\right) = 0.1057 = 6.06°$$

From Eq. (2.3.38), the chord length ratio for this cross-section is

$$\hat{c}_b \equiv \frac{kc_b}{d_p} = k\,0.075\sqrt{1-\zeta^2} = (2)0.075\sqrt{1-0.75^2} = 0.09922$$

Using these values, Eq. (2.3.32) is solved numerically to give

$$\varepsilon_i = 0.04815 = 2.76°$$

With these results, the integrands in Eqs. (2.3.33) and (2.3.34) are

$$\frac{dC_T}{d\zeta} = 0.0791 \text{ and } \frac{dC_P}{d\zeta} = 0.0315$$

Integrating similar results from $\zeta = 0.1$ to 1.0, we obtain

$$C_T = \underline{0.0465}, \quad C_P = \underline{0.0182}, \text{ and } \eta_p = \frac{C_T J}{C_P} = \underline{0.637}$$

Based on Goldstein's theory and this particular propeller geometry, the results of a parametric study showing how the thrust coefficient, power coefficient, and propulsive efficiency depend on the advance ratio and pitch-to-diameter ratio are presented in Figs. 2.3.7 through 2.3.9. These figures are for a family of propellers with similar geometry. Because these are fixed-pitch propellers, each curve shown in each of these figures is for a different propeller. The pitch-to-diameter ratio for one of these propellers is fixed by the design and manufacture of the blades. It does not vary with operating conditions. The advance ratio, on the other hand, depends primarily on the operating conditions and is directly proportional to the forward airspeed. Notice that for any given propeller, the thrust and power both decrease with increasing advance ratio, going to zero at some particular advance ratio. However, the propulsive efficiency increases with advance ratio to some point of maximum efficiency and then rapidly decreases to zero as the advance ratio is further increased.

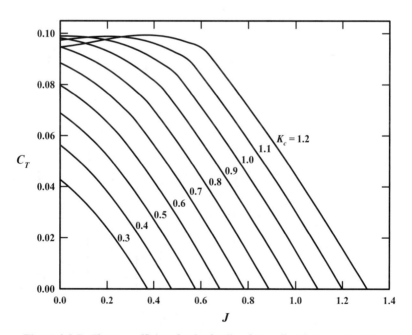

Figure 2.3.7. Thrust coefficient for the family of propellers in Example 2.3.1.

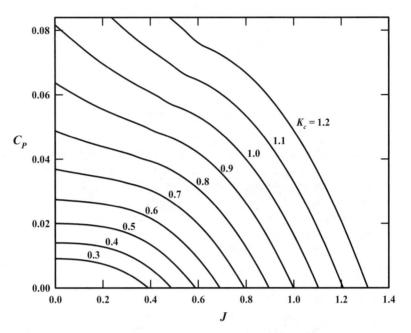

Figure 2.3.8. Power coefficient for the family of propellers in Example 2.3.1.

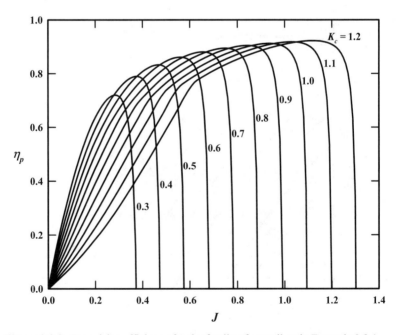

Figure 2.3.9. Propulsive efficiency for the family of propellers in Example 2.3.1.

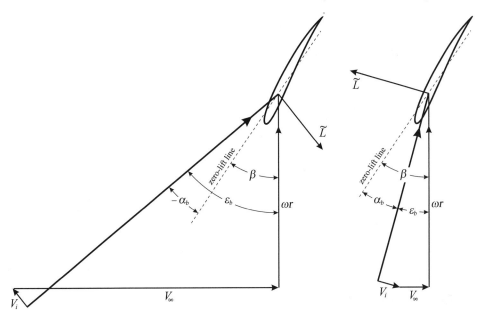

Figure 2.3.10. Variation of blade angle of attack and section lift with advance ratio.

Several important points can be observed from Example 2.3.1. First of all, notice from Fig. 2.3.7 that in the range of advance ratio where the blades are not stalled, the thrust coefficient decreases in a nearly linear fashion with increasing advance ratio. This is because for small angles, the angle of attack for each blade section is nearly a linear function of advance ratio. This can be seen in Fig. 2.3.10. When the advance ratio becomes large enough, the aerodynamic angle of attack becomes negative and both the thrust and the brake power required to turn the propeller become negative as well. Under such conditions the propeller is operating as a wind turbine. A wind turbine produces power that can be used to do work but it also produces a negative thrust (i.e., drag). When the engine of an airplane is not running, a fixed-pitch propeller will operate as a wind turbine. When an airplane propeller is turning in this fashion it is said to be *windmilling*. The negative thrust produced by a windmilling propeller can greatly increase the drag on an airplane when one engine or more is inoperative. When a variable-pitch propeller is used, the pitch can be adjusted to minimize this added drag. This is called *feathering the prop*.

The power produced by the windmilling propeller of an airplane has sometimes been put to good use. In the early days of aviation, airplane engines were hand started by turning the engine over with the prop. The pilot or an assistant would stand behind the propeller and slowly turn the engine to the beginning of a compression stroke. One blade of the propeller would then be pulled down rapidly to turn over and hopefully start the engine. Although it is not common, the engines of some small single-engine airplanes are sometimes still started in this manner today. Years ago, when hand starting was the only means of starting an airplane engine, these engines were much less reliable than aircraft engines are today. If an engine stalled in flight, putting the airplane into a dive would sometimes restart the engine. With the added airspeed attained in the dive, the power generated by the windmilling propeller would turn the engine over more rapidly and the engine would sometimes restart. This was a risky tactic, however, because should the engine fail to restart, much precious altitude was lost.

Understanding the relationship between advance ratio and blade angle of attack that is shown in Fig. 2.3.10 makes it easy to understand the relationship between advance ratio and propeller efficiency, which is presented in Fig. 2.3.9. Once we realize that very large advance ratios will produce negative angles of attack on the propeller blades, we realize that the thrust developed by a propeller must go to zero for some particular value of the advance ratio. Since the propulsive power is just the thrust multiplied by the airspeed, when the thrust goes to zero the propulsive power must go to zero as well. At the advance ratio that results in exactly zero thrust, some power is still required to turn the propeller against the blade drag. Thus, the point of zero thrust is a point of zero propulsive efficiency. The point of zero advance ratio corresponds to the condition of zero forward airspeed. Thus, this is also a point of zero propulsive power and zero propulsive efficiency. Regardless of how much static thrust is developed by the propeller, the thrust multiplied by the airspeed is always zero at the static condition. This explains the two points of zero propulsive efficiency for the curves shown in Fig. 2.3.9. The point on the right corresponds to zero thrust, and the point on the left corresponds to zero airspeed.

A very important observation should be made from Fig. 2.3.9. Since each curve in this figure is for a different fixed-pitch propeller, we see that maximum efficiency is attained with a fixed-pitch propeller only at one specific value of advance ratio. Furthermore, the advance ratio that results in maximum efficiency is a function of the pitch-to-diameter ratio. Propellers with low pitch-to-diameter ratio perform best at low advance ratio while propellers with high pitch-to-diameter ratio perform best at high advance ratio. Considering conditions during takeoff as well as cruise, an airplane must operate over a wide range of airspeeds. Thus, in the early days of aviation when only fixed-pitch propellers were available, the performance of airplanes was severely limited by this propeller characteristic. This problem was remedied in 1924 when H. S. Hele-Shaw and T. E. Beacham patented the first variable-pitch propeller in England. The blades of a variable-pitch propeller are attached to a mechanism in the hub that rotates the entire blade about an axis that runs along the length of the blade, as shown in Fig. 2.3.11. This allows the blade pitch to be varied continuously by the pilot to maintain optimum efficiency at any airspeed. By the early 1930s such propellers were in production in the United States.

A further improvement along these same lines was the introduction of the *constant-speed propeller* in 1935. A constant-speed propeller is simply a variable-pitch propeller that is controlled by a governor mechanism, which continuously and automatically varies the propeller pitch to maintain the torque on the engine so that the rotational speed of the engine is held constant. This is advantageous because the brake power output from a piston engine is optimized at some particular rotational speed. By controlling the engine speed with a constant-speed propeller, the engine can be made to operate at either maximum efficiency or maximum power, depending on how the governor is adjusted by the pilot. On takeoff the pilot will typically adjust the governor for maximum power. In cruise, the governor can be adjusted for maximum efficiency or for some compromise between airspeed and efficiency.

The introduction of the variable-pitch and constant-speed propellers was one of the most important developments in aviation history. With such mechanisms, the propulsive efficiency for modern propellers can be maintained in the range from approximately 80 to 90 percent over a wide range of airspeed.

Example 2.3.1 shows how propeller efficiency is related to propeller pitch and forward speed. However, efficiency is not always the primary concern in aircraft propulsion. Sometimes it is more important to maximize thrust. From the results shown

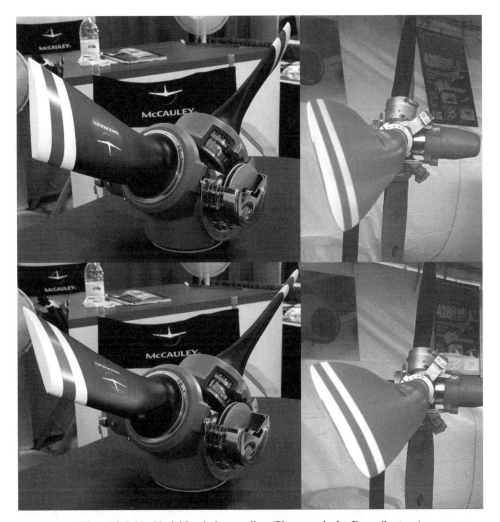

Figure 2.3.11. Variable-pitch propeller. (Photographs by Barry Santana)

in Fig. 2.3.7 we have seen that for a given advance ratio, the thrust coefficient for a propeller is always increased by increasing the pitch-to-diameter ratio, provided that the propeller blades do not stall. From this figure, a student might reach the erroneous conclusion that a pitch-to-diameter ratio of about 1.0 would result in the greatest static thrust. Figure 2.3.7 can be somewhat misleading in this regard, because the thrust coefficient is nondimensionalized with respect to the rotational speed of the propeller. Both the thrust and rotational speed depend on the brake power that is applied to the propeller shaft. For a given brake power input, as pitch-to-diameter ratio is increased, the rotational speed of the propeller will decrease. Thus, it is not clear from Fig. 2.3.7 how the static thrust will vary with pitch-to-diameter ratio for a given brake power input. This is examined in the following example.

EXAMPLE 2.3.2. Consider a fixed-pitch propeller having three blades, with the same blade geometry as that described in Example 2.3.1. We wish to examine how the static thrust developed by such a propeller depends on propeller pitch and diameter. For this purpose, we wish to plot the static thrust at standard sea level as a function of pitch-to-diameter ratio for a constant brake power input of 180 horsepower. This is to be done for several different propeller diameters.

Solution. Since we are concerned with static thrust, the advance ratio is zero for any value of rotational speed. As an example calculation, consider a 72-inch propeller having a chord-line pitch-to-diameter ratio of 0.6. Following the procedure used in Example 2.3.1, for $J = 0.0$ and $K_c = 0.6$, we obtain

$$C_T = 0.11004$$

and

$$C_P = 0.03952$$

Since we know that the brake power is 180 horsepower, the rotational speed of the propeller can be determined from the known power coefficient and the definition of power coefficient, which is given by Eq. (2.3.31). This gives

$$\frac{\omega}{2\pi} = \left(\frac{\ell\omega}{\rho d_p^5 C_P} \right)^{\frac{1}{3}} = \left[\frac{(180 \times 550)\ \text{ft}\cdot\text{lbf/sec}}{0.0023769\ \text{slug/ft}^3 (72/12)^5\ \text{ft}^5 (0.03952)} \right]^{\frac{1}{3}}$$

$$= 51.37\ \text{rps} = 3{,}082\ \text{rpm}$$

We can now determine the static thrust from the known thrust coefficient and the definition of thrust coefficient, which is given by Eq. (2.3.29). Thus we have

$$T = \rho(\omega/2\pi)^2 d_p^4 C_T$$

$$= 0.0023769\ \text{slug/ft}^3 (51.37\ \text{sec}^{-1})^2 (72/12)^4\ \text{ft}^4 (0.11004) = \underline{894\ \text{lbf}}$$

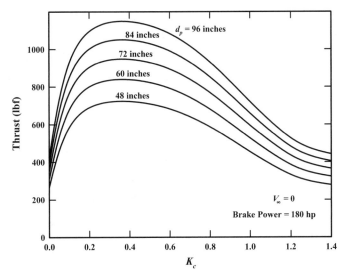

Figure 2.3.12. Variation in static thrust with pitch-to-diameter ratio for the propellers in Example 2.3.2, with a brake power input of 180 horsepower.

Repeating these calculations for different values of pitch-to-diameter ratio and different propeller diameters, we obtain the results that are shown in Fig. 2.3.12. Notice that for this blade geometry, maximum static thrust is realized for a chord-line pitch-to-diameter ratio of about 0.36. The existence of this maximum is not obvious from direct examination of the dimensionless results that are plotted in Fig. 2.3.7.

Similar results are shown in Fig. 2.3.13, for the same propellers with a 300-horsepower input. Notice that while the increased power increases the magnitude of the static thrust, it does not change the shape of the curves. For incompressible flow and this particular blade geometry, a chord-line pitch-to-diameter ratio of about 0.36 will produce the greatest static thrust for any fixed power input and any propeller diameter.

A very important point is demonstrated in Example 2.3.2. Notice from Figs. 2.3.12 and 2.3.13 that for incompressible flow with fixed power input and fixed pitch-to-diameter ratio, the static thrust is always greatest for the propeller with the largest diameter. With the brake power and pitch-to-diameter ratio held constant, increasing propeller diameter requires decreasing the rotational speed. When the diameter is increased and the rotational speed is decreased to maintain constant power input, the mass flow rate of the air drawn through the propeller disk is increased, but the slip-stream velocity that is imparted to that air is decreased. As was pointed out in Sec. 2.1, thrust is produced with the greatest efficiency when the mass flow rate is large and the velocity increment is small. Thus, large-diameter propellers with correspondingly lower rotational speeds result in the most efficient production of thrust. Similar results are shown in Figs. 2.3.14 and 2.3.15 for forward speeds of 50 and 100 mph. In all cases, **for incompressible flow with fixed power input, maximum thrust is attained with the largest propeller diameter.**

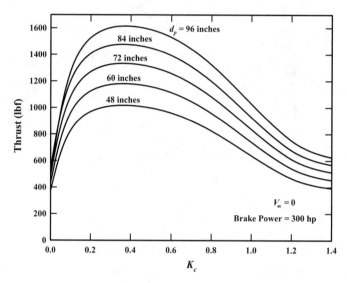

Figure 2.3.13. Variation in static thrust with pitch-to-diameter ratio for the propellers in Example 2.3.2 with a brake power input of 300 horsepower.

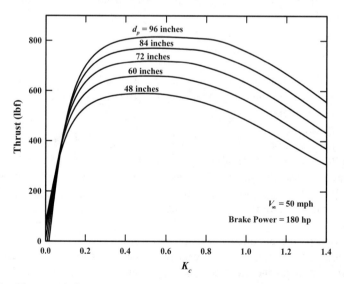

Figure 2.3.14. Thrust variation for the propellers in Example 2.3.2 with a forward airspeed of 50 mph.

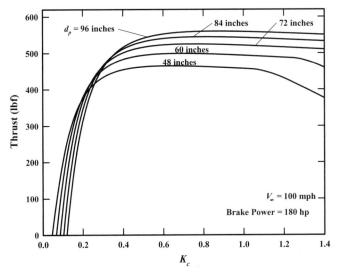

Figure 2.3.15. Thrust variation for the propellers in Example 2.3.2 with a forward airspeed of 100 mph.

In Example 2.3.2, propeller diameter was held constant and the rotational speed necessary to give a specified brake power was computed. However, a piston engine produces maximum power at only one particular rotational speed. When selecting a propeller for a particular piston engine, both the diameter and the pitch can be varied to match the propeller to the engine and the application. This is demonstrated in the following two examples.

EXAMPLE 2.3.3. At sea level, a particular piston engine develops maximum brake power of 180 horsepower at 2,900 rpm. For this example, we wish to select a fixed-pitch propeller for this engine that will maximize the static thrust at sea level. We assume a three-blade configuration with the same blade geometry as that described in Example 2.3.1.

Solution. As an example calculation, we again consider a propeller having a chord-line pitch-to-diameter ratio of 0.6. From Example 2.3.2, for $J = 0.0$ and $K_c = 0.6$, we obtain

$$C_T = 0.11004$$

and

$$C_P = 0.03952$$

Since we require a brake power of 180 horsepower and a rotational speed of 2,900 rpm, Eq. (2.3.31) can be used to fix the propeller diameter. This gives

$$d_p = \left(\frac{\ell\omega}{\rho(\omega/2\pi)^3 C_P}\right)^{\frac{1}{5}} = \left[\frac{(180\times550)\,\text{ft}\cdot\text{lbf/sec}}{0.0023769\,\text{slug/ft}^3(2,900/60)^3\,\text{sec}^{-3}(0.03952)}\right]^{\frac{1}{5}}$$

$$= 6.223\,\text{ft} = 74.68\,\text{inches}$$

From Eq. (2.3.29) and the known thrust coefficient, the sea-level static thrust developed by this engine-propeller combination would be

$$T = \rho(\omega/2\pi)^2 d_p^4 C_T$$

$$= 0.0023769\,\text{slug/ft}^3(2,900/60)^2\,\text{sec}^{-2}(6.223\,\text{ft})^4(0.11004) = \underline{916\,\text{lbf}}$$

Repeating these calculations for different values of pitch-to-diameter ratio, we obtain the results that are plotted in Fig. 2.3.16. Thus, Goldstein's vortex theory for incompressible flow predicts that maximum static thrust for this engine and propeller geometry is realized at a chord-line pitch-to-diameter ratio of about 0.19 with a 105-inch propeller. However, this propeller and operating condition would produce a tip speed of 1,329 ft/sec, which corresponds to a tip Mach number of nearly 1.2. **This incompressible flow theory is not valid for a tip Mach number this large, so the solution is not realistic.** When the tip Mach number is greater than unity, the brake power required to turn the propeller at a certain speed goes up dramatically. Thus, for fixed power input, the speed or the diameter would need to be decreased and the thrust would decrease accordingly.

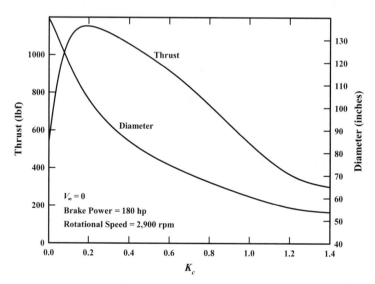

Figure 2.3.16. Variation in thrust with pitch-to-diameter ratio for the engine-propeller combination in Example 2.3.3.

If compressibility effects were accounted for, maximum static thrust would be realized with a smaller propeller having a larger pitch-to-diameter ratio. Partly to avoid supersonic tip speeds, aircraft engines are typically designed to develop maximum power at a rotational speed that is considerably less than that for comparable automobile engines.   Considerations other than aerodynamic performance, such as noise level, ground clearance, and structural considerations, can also limit the maximum propeller diameter.

Although high static thrust is important for good takeoff performance, an airplane propeller is not typically optimized for maximum static thrust. A similar procedure can be used to select a propeller that will optimize some other performance criterion. However, when the forward speed associated with the selected performance criterion is nonzero, an additional iterative procedure is usually required. This is demonstrated in the following example.

EXAMPLE 2.3.4. For best efficiency and engine life, the manufacturer of the piston engine in Example 2.3.3 recommends that for continuous operation, the engine should be operated at 65 percent of maximum power and 2,350 rpm. In this example we wish to select a fixed-pitch propeller for this engine and operating condition that will maximize the propulsive efficiency at standard sea level with an airspeed of 120 mph. Again, we will assume a three-blade propeller configuration with the same blade geometry as that described in Example 2.3.1.

Solution. As an example calculation, we again consider a propeller having a chord-line pitch-to-diameter ratio of 0.6. For this case we cannot compute the advance ratio directly, because the propeller diameter is unknown. We must start with an assumed propeller diameter: for example, 100 inches. For this assumed diameter the advance ratio would be

$$J = \frac{V_\infty}{(\omega/2\pi)d_p} = \frac{(120\times 5{,}280/3{,}600)\,\text{ft/sec}}{(2{,}350/60)\,\text{sec}^{-1}(100/12)\,\text{ft}} = 0.539234$$

Following the procedure used in Example 2.3.1, with this value of J and $K_c = 0.6$ the power coefficient for this propeller and operating condition is found to be

$$C_P = 0.019498$$

and

$$\ell\omega = \rho(\omega/2\pi)^3 d_p^5 C_P = (0.0023769)(2{,}350/60)^3(100/12)^5(0.019498)$$
$$= 111{,}852 \ \text{ft·lbf/sec} = 203.4 \ \text{hp}$$

Since we require a brake power of 117 horsepower, the assumed diameter is too large. As a second estimate we will use

$$d_p = 90 \text{ inches} = 7.5 \text{ ft}$$

$$J = \frac{V_\infty}{(\omega/2\pi)d_p} = \frac{(120 \times 5,280/3,600) \text{ ft/sec}}{(2,350/60)\sec^{-1}(7.5)\text{ ft}} = 0.599149$$

$$C_P = 0.012830$$

$$\ell\omega = \rho(\omega/2\pi)^3 d_p^5 C_P = (0.0023769)(2,350/60)^3(7.5)^5(0.012830)$$
$$= 43,480 \text{ ft·lbf/sec} = 79.1 \text{ hp}$$

Continuing with this iteration, using the secant method, we rapidly converge on

$$d_p = 93.6416 \text{ inches} = 7.80347 \text{ ft}$$

$$J = \frac{V_\infty}{(\omega/2\pi)d_p} = \frac{(120 \times 5,280/3,600) \text{ ft/sec}}{(2,350/60)\sec^{-1}(7.80347)\text{ ft}} = 0.575849$$

$$C_P = 0.015572$$

$$\ell\omega = \rho(\omega/2\pi)^3 d_p^5 C_P = (0.0023769)(2,350/60)^3(7.80347)^5(0.015572)$$
$$= 64,350 \text{ ft·lbf/sec} = 117.0 \text{ hp}$$

$$C_T = 0.022818$$

From Eq. (2.3.29) and the known thrust coefficient, the sea-level thrust developed by this engine-propeller combination at 120 mph would be

$$T = \rho(\omega/2\pi)^2 d_p^4 C_T = (0.0023769)(2,350/60)^2(7.80347 \text{ ft})^4(0.022818)$$
$$= 308.5 \text{ lbf}$$

The propulsive efficiency for this propeller and operating condition is then computed as the propulsive power divided by the brake power,

$$\eta_p = \frac{TV_\infty}{\ell\omega} = \frac{308.5 \text{ lbf} (120 \times 5,280/3,600) \text{ ft/sec}}{64,350 \text{ ft·lbf/sec}} = \underline{0.844}$$

Repeating these calculations for other pitch-to-diameter ratios, we obtain the results plotted in Fig. 2.3.17. These results predict that a maximum propulsive efficiency of 0.867 is achieved with an 80-inch propeller having a chord-line pitch-to-diameter ratio of about 0.79.

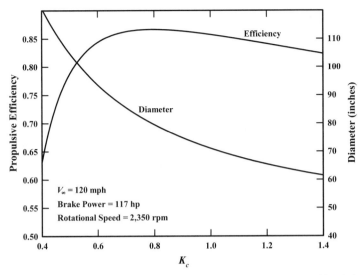

Figure 2.3.17. Variation in propeller diameter and propulsive efficiency with pitch-to-diameter ratio for the propeller in Example 2.3.4.

It should be noted that the penalty for slightly decreasing the diameter and increasing the pitch is not too great. A 72-inch propeller having a chord-line pitch-to-diameter ratio of about 0.97 would provide the same load at 2,350 rpm and would decrease the efficiency by less than 1 percent. The propeller actually selected for this engine and some particular airframe would probably be a compromise between cruise and takeoff performance, plus other considerations, such as noise level, ground clearance, and structural considerations.

A few comments are in order concerning the accuracy of the propeller theory presented in this section. First of all, we should not expect the theory to be accurate in the region of advance ratio where the propeller blades are stalled. The basis for this propeller analysis is potential flow theory, which is an inviscid theory. When the blades are not stalled, the effects of viscosity are confined to the boundary layers at the blade surfaces. These effects were added to the basic potential flow model by including parasitic drag as an additional aerodynamic force. This technique works quite well when the blades are not stalled, because the viscous forces are small. However, when the blades are stalled, the viscous effects in the blade wakes dominate the flow. Under these conditions, the flow pattern is significantly altered by the presence of the viscous wakes, and the method used to predict the induced velocity is not expected to be accurate.

As already stated, another source of error in this propeller model is Goldstein's original hypothesis. The model used for predicting the velocity induced on the propeller blades by the trailing vortex sheet is based on the Betz condition, which is only valid in the ultimate wake of an optimum propeller. In the plane of the prop-circle of a propeller having blades of arbitrary planform shape, this will clearly introduce some error.

Another source of error in this model is the assumption of incompressible flow. The highest relative airspeeds are encountered at the blade tips. For many high-performance aircraft engines, propeller tip speeds approach or exceed the speed of sound. Obviously, as local Mach numbers approach 1.0 at the blade tips, the assumption of incompressible flow is not reasonable. As the blade tip speed approaches the speed of sound, shock waves develop on the blades and the blade drag goes up considerably. This increases the power required to turn the propeller at a given speed and reduces the propeller efficiency. When the propeller blade-tip speed exceeds the speed of sound, the propeller efficiency begins to drop dramatically. This is the primary reason why propellers were never successfully used for supersonic flight. The effects of compressibility can be reasonably accounted for in Goldstein's model by including Mach number dependence in the relations for the section lift and drag coefficients. This is left as a programming exercise for the student.

In spite of some sources of error, if the blade geometry and aerodynamic section lift and drag coefficients are accurately described as a function of radial position, angle of attack, and Mach number, Goldstein's theory gives results that are in reasonable agreement with experimental data. Furthermore, the trends and insight into propeller performance, which are shown by this model, are valid even when the actual performance deviates somewhat from that predicted by the model.

Perhaps the greatest source of error encountered when using Goldstein's model to predict the performance of a commercial propeller is a lack of knowledge of the blade geometry. Propeller manufacturers guard such information closely and are very reluctant to share information on blade geometry with customers. The propeller and hub diameters as well as the section chord length as a function of radial position are easily measured. However, the variations in chord-line pitch and airfoil section geometry with radial position are not so readily determined.

Manufacturers typically provide propeller pitch as a single number specified in inches. This could lead the user to believe that the propeller has a pitch length that is constant along the axis of the blade. This is seldom actually the case. Figure 2.3.18 shows how chord-line and aerodynamic pitch vary with radial position for one particular commercial propeller. The geometry shown in this figure was obtained from a commercially purchased propeller using a laser coordinate measuring machine. For comparison, the pitch specified by the manufacturer is also shown in this figure as a flat line. Notice that the measured chord-line pitch is in close agreement with the pitch specified by the manufacturer only at one location along the axis of the blade. Such variation in pitch with radial position is not unusual in commercial propellers. The pitch length specified by the manufacturer for any particular propeller should be viewed only as a manufacturer's naming convention. This specified pitch length is, at best, only a relative measure of propeller pitch and is typically quite useless as input to Goldstein's analytical model. Determining the actual variation in chord-line pitch and airfoil section geometry with radial position may be the most difficult part of predicting the performance of a commercial propeller.

Once the propeller blade geometry has been accurately determined, the section lift and drag coefficients for any airfoil cross-section of the blade can be determined numerically as a function of angle of attack and Mach number. With such input,

Goldstein's model provides a reasonable means for predicting propeller performance. This is demonstrated in Figs. 2.3.19 and 2.3.20. The data shown in this figure are from Phillips, Anderson, and Kelly (2003) and were taken using the same propeller for which the geometry is shown in Fig. 2.3.18. The data shown in these figures were taken at propeller angles of attack of –6, 0, and +6 degrees. Notice that within the accuracy of the experimental measurements, the propeller thrust and power coefficients are independent of the propeller angle of attack.

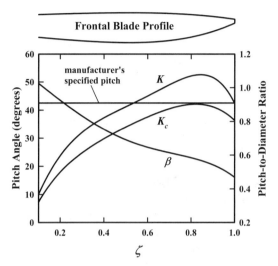

Figure 2.3.18. Variation in propeller pitch with radial position for a commercial propeller.

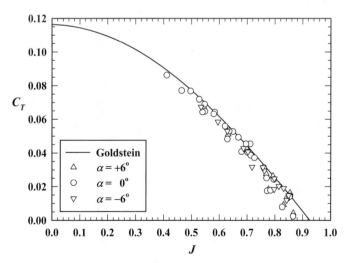

Figure 2.3.19. Comparison between the thrust coefficient predictions from Goldstein's theory and experimental data for the propeller geometry shown in Fig. 2.3.18.

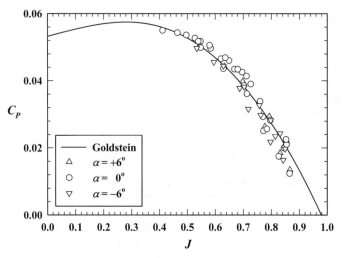

Figure 2.3.20. Comparison between the power coefficient predictions from Goldstein's theory and experimental data for the propeller geometry shown in Fig. 2.3.18.

2.4. Propeller Momentum Theory

Another method that has been used for predicting propeller induced velocity is known as *propeller momentum theory*. Like Goldstein's vortex theory, momentum theory is based on inviscid, incompressible flow. However, this flow model is very different from that hypothesized by Goldstein. Momentum theory is based on the hypothesis of a streamtube, which exactly encloses the propeller disk as shown in Fig. 2.4.1. This streamtube is assumed to extend from a plane infinitely far upstream from the propeller disk to a plane infinitely far downstream. Both far planes are assumed parallel with the plane of the propeller disk, and the static pressure in both far planes is assumed constant and equal to the freestream static pressure. All of the fluid that enters this streamtube on the far upstream side must pass through the propeller disk and exit on the far downstream side.

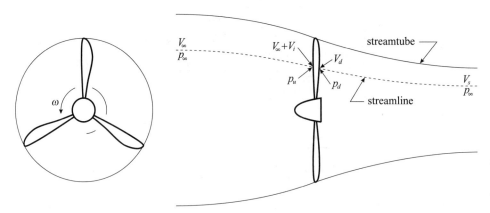

Figure 2.4.1. Classical momentum theory model for the fluid flows about a rotating propeller.

In addition to the foundation hypothesis of streamtube flow as shown in Fig. 2.4.1, classical propeller momentum theory imposes four simplifying approximations, i.e.,

1. The flow is assumed to be inviscid and incompressible.
2. All rotation of the fluid within the streamtube is neglected.
3. The flow velocity is assumed uniform over each cross-section of the streamtube.
4. The pressure is assumed uniform over each cross-section of the streamtube.

These assumptions result in an idealized one-dimensional solution that is normally considered to represent an upper limit for propeller performance.

To evaluate the induced velocity from classical propeller momentum theory, we consider the flow along the streamline shown in Fig. 2.4.1. The fluid far upstream from the propeller disk has pressure p_∞ and velocity V_∞. If the propeller is developing positive thrust, the pressure just upstream from the propeller disk is below ambient, while the pressure just downstream from the propeller disk is above ambient. The near upstream pressure is denoted as p_u and the near downstream pressure is represented as p_d. By definition, the velocity on the near upstream side of the propeller disk is the freestream velocity plus the induced velocity, $V_\infty + V_i$, and the velocity on the near downstream side is designated as V_d.

Since we are neglecting all rotation of the fluid within the streamtube, the fluid velocity on the downstream side of the propeller disk is readily determined as a direct result of continuity combined with the assumption of incompressible flow. If we apply conservation of mass to the propeller disk area, A_p, we have

$$\rho V_d A_p = \rho(V_\infty + V_i)A_p$$

For the assumed incompressible flow, this reduces to the rather obvious result that the fluid velocity cannot change in passing through the propeller disk,

$$V_d = V_\infty + V_i \tag{2.4.1}$$

The velocity does change, however, as the fluid approaches and leaves the propeller disk.

Applying Bernoulli's equation along the streamline shown in Fig. 2.4.1, upstream from the propeller disk we obtain

$$\frac{p_\infty}{\rho} + \frac{V_\infty^2}{2} = \frac{p_u}{\rho} + \frac{(V_\infty + V_i)^2}{2} \tag{2.4.2}$$

Likewise, for the streamline on the aft side of the propeller disk, from a point just behind the propeller to a point in the far slipstream, using Eq. (2.4.1), we can write

$$\frac{p_\infty}{\rho} + \frac{V_s^2}{2} = \frac{p_d}{\rho} + \frac{(V_\infty + V_i)^2}{2} \tag{2.4.3}$$

where V_s is the ultimate slipstream velocity at a downstream point where the pressure has returned to ambient. Subtracting Eq. (2.4.2) from Eq. (2.4.3) results in

$$\frac{V_s^2}{2} - \frac{V_\infty^2}{2} = \frac{p_d}{\rho} - \frac{p_u}{\rho} \tag{2.4.4}$$

The axial force on the propeller disk is expressed in terms of the pressure difference across the disk and in terms of the momentum imparted to the fluid. The thrust acting on the prop-circle is expressed as the downstream pressure, p_d, minus the upstream pressure, p_u, multiplied by the area, A_p,

$$T = A_p(p_d - p_u) \tag{2.4.5}$$

Since this force is applied directly to the fluid, it will result in a fluid momentum increase, which must satisfy

$$T = \dot{m}(V_s - V_\infty) \tag{2.4.6}$$

where $\dot{m}$ is the mass flow rate through the area, A_p, which is given by

$$\dot{m} = A_p \rho(V_\infty + V_i) \tag{2.4.7}$$

Using Eq. (2.4.7) in Eq. (2.4.6), the thrust can be written as

$$T = A_p \rho(V_\infty + V_i)(V_s - V_\infty) \tag{2.4.8}$$

Combining Eqs. (2.4.5) and (2.4.8) to eliminate the thrust, we obtain

$$A_p(p_d - p_u) = A_p \rho(V_\infty + V_i)(V_s - V_\infty)$$

or

$$\frac{p_d}{\rho} - \frac{p_u}{\rho} = (V_\infty + V_i)(V_s - V_\infty) \tag{2.4.9}$$

Using Eq. (2.4.4) to eliminate the pressure from Eq. (2.4.9), the ultimate slipstream velocity can be related to the freestream velocity and the propeller induced velocity,

$$\frac{V_s^2}{2} - \frac{V_\infty^2}{2} = \frac{(V_s - V_\infty)(V_s + V_\infty)}{2} = (V_\infty + V_i)(V_s - V_\infty)$$

Solving this for V_s gives

$$V_s = V_\infty + 2V_i \tag{2.4.10}$$

Thus, we see that classical momentum theory predicts that the difference between the ultimate slipstream velocity and the freestream velocity is twice the induced velocity.

Similarly, from the energy equation, the brake power, P, that is required to turn the propeller must satisfy the relation

$$P = \dot{m}\left(h_s - h_\infty + \frac{V_s^2}{2} - \frac{V_\infty^2}{2}\right)$$

where h_s and h_∞ are, respectively, the ultimate slipstream and freestream enthalpy. Since the fluid is assumed incompressible and the ultimate slipstream pressure is assumed equal to the freestream pressure, the enthalpy in the ultimate slipstream is equal to that in the freestream. Thus, after applying Eq. (2.4.7), the energy equation becomes

$$P = A_p \rho (V_\infty + V_i)\left(\frac{V_s^2}{2} - \frac{V_\infty^2}{2}\right) \tag{2.4.11}$$

Using Eq. (2.4.10) in Eqs. (2.4.8) and (2.4.11), the thrust can be written as

$$T = 2A_p \rho (V_\infty + V_i)V_i \tag{2.4.12}$$

and brake power required to turn the propeller is

$$P = 2A_p \rho (V_\infty + V_i)^2 V_i \tag{2.4.13}$$

Equation (2.4.12) is quadratic in V_i and can be solved for the induced velocity to give

$$V_i = \sqrt{\frac{V_\infty^2}{4} + \frac{T}{2A_p \rho}} - \frac{V_\infty}{2} \tag{2.4.14}$$

Using Eq. (2.4.14) in Eq. (2.4.13), the brake power required to turn the propeller can be expressed as a function of the thrust and the freestream airspeed,

$$P = T\left(\frac{V_\infty}{2} + \sqrt{\frac{V_\infty^2}{4} + \frac{T}{2A_p \rho}}\right) \tag{2.4.15}$$

The propulsive efficiency for the propeller is the propulsive power output, TV_∞, divided by the brake power input, P. Since this is an inviscid, incompressible flow model, the propulsive efficiency obtained from classical propeller momentum theory is usually called the *ideal efficiency*, η_i. From Eqs. (2.4.12) and (2.4.13) we obtain

$$\eta_i \equiv \frac{TV_\infty}{P} = \frac{2A_p \rho (V_\infty + V_i)V_i V_\infty}{2A_p \rho (V_\infty + V_i)^2 V_i} = \frac{V_\infty}{V_\infty + V_i} = \frac{1}{1 + V_i/V_\infty}$$

or with Eq. (2.4.14),

$$\eta_i = \left(\frac{1}{2} + \sqrt{\frac{1}{4} + \frac{T}{2A_p \rho V_\infty^2}} \right)^{-1} \tag{2.4.16}$$

For convenience, we can express Eq. (2.4.14) in terms of the usual dimensionless propeller variables. This gives

$$\frac{V_i}{(\omega/2\pi)d_p} = \sqrt{\frac{J^2}{4} + \frac{2C_T}{\pi}} - \frac{J}{2} \tag{2.4.17}$$

where $C_T \equiv T/[\rho(\omega/2\pi)^2 d_p^4]$ and $J \equiv V_\infty/[(\omega/2\pi)d_p]$. Similarly, Eq. (2.4.15) can be written in dimensionless form as

$$C_P = C_T \left(\frac{J}{2} + \sqrt{\frac{J^2}{4} + \frac{2C_T}{\pi}} \right) \tag{2.4.18}$$

where $C_P \equiv P/[\rho(\omega/2\pi)^3 d_p^5]$. From Eq. (2.4.16), the ideal propulsive efficiency predicted by classical propeller momentum theory can be written in terms of these same dimensionless variables,

$$\eta_i = \frac{C_T J}{C_P} = \left(\frac{1}{2} + \sqrt{\frac{1}{4} + \frac{2C_T}{\pi J^2}} \right)^{-1} \tag{2.4.19}$$

Classical propeller momentum theory is relatively simple and quite appealing. As a result, this theory has been commonly applied in the literature. The major objection to the theory has been its failure to account for rotation of the fluid within the slipstream. There is no physical basis for neglecting slipstream rotation. Clearly, torque must be applied to turn the propeller and that torque must result in rotation of the fluid within the slipstream. Since some of the power supplied to the propeller must go to support this rotation, the thrust and propulsive efficiency are reduced as a result of slipstream rotation. In the following analysis, we examine the magnitude of this effect.

The exclusion of slipstream rotation is not the only objection that can be raised against classical propeller momentum theory. The assumptions of uniform flow and uniform pressure result in a one-dimensional solution that is not consistent with results predicted from propeller vortex theory. However, here we wish to concentrate on examining the effects of slipstream rotation on the results predicted by propeller momentum theory. Thus, we shall now consider the incorporation of the angular momentum equation into this propeller model. To isolate the effects of slipstream rotation, we continue with the assumptions of inviscid, incompressible, uniform flow but we now allow for rotation of the fluid within the streamtube.

Accordingly, we continue to assume the existence of a streamtube, which encloses the complete propeller disk as shown in Fig. 2.4.2. This streamtube is still assumed to extend infinitely far upstream from the propeller disk, to a plane where the static pressure is constant and equal to the freestream static pressure, p_∞. In this plane, the axial velocity is the freestream velocity, V_∞, and there is no circumferential velocity. Likewise, the streamtube is assumed to extend infinitely far downstream from the propeller disk, to a plane where the velocity in the slipstream is no longer changing in the axial direction. Consistent with the uniform flow assumption used in classical propeller momentum theory, we continue to assume uniform axial velocity but now allow for a uniform angular velocity as well. Since we wish to examine the effect of slipstream rotation on the results predicted by propeller momentum theory, at first thought one might be tempted to continue with the assumption of uniform pressure, which is also imposed in classical propeller momentum theory. However, this assumption is not consistent with rotation in the ultimate slipstream.

The solution from classical propeller momentum theory does not satisfy the angular momentum equation. If we continue with the uniform pressure assumption from classical momentum theory while allowing fluid rotation, we find that there is only one possible case for which the angular momentum equation together with the other laws of Newtonian mechanics can all be satisfied with a finite propeller rotational speed. That is the case of zero thrust. Since we are now allowing for fluid rotation in the streamtube and considering a propeller developing finite thrust, we can no longer assume uniform pressure over each cross-section of the streamtube. We must instead allow the pressure in the streamtube to be a function of the radial position, r, as well as the axial position, x.

On the near upstream side of the propeller disk, the axial fluid velocity is equal to the sum of the forward airspeed and the axial component of induced velocity, $V_\infty + V_{xi}$. On the near downstream side of the propeller disk the axial fluid velocity is designated V_{xd}. The circumferential velocity on the near upstream side of the propeller disk is just the circumferential component of the propeller induced velocity, $V_{\theta i}$, and the circumferential velocity on the near downstream side of the propeller disk is denoted as $V_{\theta d}$.

The circumferential component of fluid velocity just upstream from the propeller disk can be deduced by applying the angular momentum equation to the section of

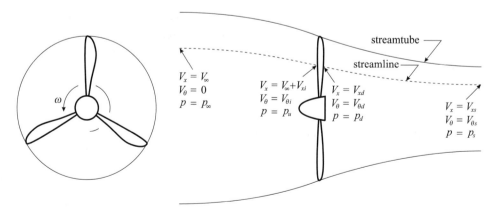

Figure 2.4.2. Momentum theory model for the pressures and velocities of the fluid that flows through the disk of a rotating propeller, including the effects of fluid rotation.

streamtube that is upstream from the propeller. Because there is no torque acting on this section of the streamtube, there can be no change in the angular momentum of the air as it passes through this section. Since the air has no angular momentum far upstream from the propeller disk, it cannot have angular momentum when it reaches the upstream side of the propeller disk. This requires that

$$V_{\theta i} = 0 \qquad (2.4.20)$$

and thus

$$V_{xi} = V_i \qquad (2.4.21)$$

There is angular momentum imparted to the air as it flows through the propeller disk. This results from the torque applied to the propeller. In the section of streamtube that is downstream from the propeller disk the angular momentum flux must again remain constant, since there is no torque applied to this section of the streamtube. At first thought, one might be led to conclude that the angular velocity remains constant in the downstream section of the streamtube, just as it did in the upstream section. However, this is not the case. The angular velocity remains constant in the upstream section of the streamtube only because the angular velocity is zero at the inlet. As air flows from the region far upstream toward the low-pressure side of the propeller disk, the axial velocity is increased due to the pressure decrease. Likewise, as air flows away from the high-pressure side of the propeller disk, the axial velocity continues to increase as a result of the pressure difference between the downstream side of the propeller disk and the far slipstream. This acceleration causes the streamtube enclosing the slipstream to contract in diameter with increasing distance downstream from the propeller disk. This is shown in Fig. 2.4.2. Since the diameter of the streamtube decreases and the angular momentum flux for the air remains constant, the angular velocity of the air in the slipstream must increase along with the axial velocity as it moves away from the downstream side of the propeller disk.

The axial component of fluid velocity just downstream from the propeller disk is related to the axial component on the upstream side through the continuity equation. If we use the notation shown in Fig. 2.4.2 and apply conservation of mass across any segment of the propeller disk, we have

$$\rho V_{xd} = \rho(V_\infty + V_{xi})$$

Since the fluid density is assumed constant, this fixes the axial component of fluid velocity just downstream from the propeller disk. In view of Eq. (2.4.21), this gives

$$V_{xd} = V_\infty + V_i \qquad (2.4.22)$$

The pressure and velocity just upstream from the propeller disk are related to the freestream pressure and velocity through Bernoulli's equation. Applying Bernoulli's equation along the streamline shown in Fig. 2.4.2 and using Eqs. (2.4.20) and (2.4.21), upstream from the propeller disk we obtain

$$\frac{p_\infty}{\rho} + \frac{V_\infty^2}{2} = \frac{p_u}{\rho} + \frac{(V_\infty + V_i)^2}{2} \tag{2.4.23}$$

Here we see that since there is no rotation of the fluid in the upstream section of the streamtube, the uniform flow approximation requires uniform pressure upstream from the propeller disk. For the streamline on the aft side of the propeller disk, from a point just behind the propeller to a point in the far slipstream, using Eq. (2.4.22) we can write

$$\frac{p_s}{\rho} + \frac{V_{xs}^2 + V_{\theta s}^2}{2} = \frac{p_d}{\rho} + \frac{(V_\infty + V_i)^2 + V_{\theta d}^2}{2} \tag{2.4.24}$$

where V_{xs} and $V_{\theta s}$ are the axial and circumferential components of the ultimate slipstream velocity far downstream from the propeller disk. Subtracting Eq. (2.4.23) from Eq. (2.4.24) results in

$$\frac{p_s}{\rho} - \frac{p_\infty}{\rho} + \frac{V_{xs}^2 + V_{\theta s}^2}{2} - \frac{V_\infty^2}{2} = \frac{p_d}{\rho} - \frac{p_u}{\rho} + \frac{V_{\theta d}^2}{2} \tag{2.4.25}$$

From the uniform flow approximation, the circumferential velocity in the ultimate slipstream can be written as the product of the local angular velocity and the local radius. Thus, Eq. (2.4.25) can be written as

$$\frac{p_s}{\rho} - \frac{p_\infty}{\rho} + \frac{V_{xs}^2 + \omega_s^2 r_s^2}{2} - \frac{V_\infty^2}{2} = \frac{p_d}{\rho} - \frac{p_u}{\rho} + \frac{V_{\theta d}^2}{2} \tag{2.4.26}$$

where ω_s and r_s are the angular velocity and radial coordinate at the location of the streamline far downstream from the propeller.

In the ultimate slipstream this axisymmetric flow becomes independent of the axial position, x. Thus, the continuity equation in the ultimate slipstream requires that

$$\frac{d(r_s V_{rs})}{dr_s} = 0 \tag{2.4.27}$$

where r_s is the radial position and V_{rs} is the radial velocity at a point in the ultimate slipstream. Similarly, the three components of the momentum equation in the ultimate slipstream reduce to

$$\rho \left(V_{rs} \frac{dV_{rs}}{dr_s} - \frac{V_{\theta s}^2}{r_s} \right) = -\frac{\partial p_s}{\partial r_s} \tag{2.4.28}$$

$$\rho\left(V_{rs}\frac{\partial V_{\theta s}}{\partial r_s} + \frac{V_{rs}V_{\theta s}}{r_s}\right) = 0 \tag{2.4.29}$$

$$\rho V_{rs}\frac{\partial V_{xs}}{\partial r_s} = 0 \tag{2.4.30}$$

Equation (2.4.27) combined with the boundary condition, which specifies no radial velocity at the outer edge of the ultimate slipstream, gives

$$V_{rs} = 0 \tag{2.4.31}$$

As a result of Eq. (2.4.31), Eqs. (2.4.29) and (2.4.30) are identically satisfied, whereas Eq. (2.4.28) reduces to

$$\frac{\partial p_s}{\partial r_s} = \rho\frac{V_{\theta s}^2}{r_s} = \rho\omega_s^2 r_s \tag{2.4.32}$$

The solution to Eq. (2.4.32), subject to the boundary condition of freestream static pressure at the outer edge of the slipstream, $r_s = R_s$, is

$$p_s(r_s) = p_\infty - \frac{\rho\omega_s^2}{2}(R_s^2 - r_s^2) \tag{2.4.33}$$

Using Eq. (2.4.33) in Eq. (2.4.26) gives

$$\omega_s^2\left(r_s^2 - \frac{R_s^2}{2}\right) + \frac{V_{xs}^2}{2} - \frac{V_\infty^2}{2} = \frac{p_d}{\rho} - \frac{p_u}{\rho} + \frac{V_{\theta d}^2}{2} \tag{2.4.34}$$

Applying the continuity equation to the flow inside the stream surface that contains the streamline under consideration, we have

$$\pi r_p^2\rho(V_\infty + V_i) = \pi r_s^2\rho V_{xs}$$

where r_p is the radial position of the streamline at the propeller. After rearranging, this results in

$$r_s^2 = \frac{V_\infty + V_i}{V_{xs}}r_p^2 \tag{2.4.35}$$

Using Eq. (2.4.35) in Eq. (2.4.34) results in

$$\omega_s^2\left(r_p^2 - \frac{R_p^2}{2}\right)\frac{V_\infty + V_i}{V_{xs}} + \frac{V_{xs}^2}{2} - \frac{V_\infty^2}{2} = \frac{p_d}{\rho} - \frac{p_u}{\rho} + \frac{V_{\theta d}^2}{2} \tag{2.4.36}$$

where R_p is the propeller's outside radius. Since angular momentum remains constant downstream from the propeller, we have

$$r_p V_{\theta d} = r_s V_{\theta s} = r_s^2 \omega_s$$

or in view of Eq. (2.4.35),

$$V_{\theta d} = \frac{V_\infty + V_i}{V_{xs}} r_p \omega_s \tag{2.4.37}$$

After applying Eq. (2.4.37), Eq. (2.4.36) becomes

$$\omega_s^2 \left(r_p^2 - \frac{R_p^2}{2} \right) \frac{V_\infty + V_i}{V_{xs}} + \frac{V_{xs}^2}{2} - \frac{V_\infty^2}{2} = \frac{p_d}{\rho} - \frac{p_u}{\rho} + \left(\frac{V_\infty + V_i}{V_{xs}} \right)^2 \frac{r_p^2 \omega_s^2}{2}$$

or

$$\frac{p_d}{\rho} - \frac{p_u}{\rho} = \omega_s^2 \left[\left(1 - \frac{V_\infty + V_i}{2V_{xs}} \right) r_p^2 - \frac{R_p^2}{2} \right] \frac{V_\infty + V_i}{V_{xs}} + \frac{V_{xs}^2}{2} - \frac{V_\infty^2}{2} \tag{2.4.38}$$

The axial velocity in the ultimate slipstream is related to the thrust through the axial momentum equation,

$$T + \int_{A_s} (p_\infty - p_s) dA_s = \int_{A_s} (V_{xs} - V_\infty) d\dot{m}$$

In view of Eq. (2.4.33), this gives

$$T + \int_{r_s=0}^{R_s} \frac{\rho \omega_s^2}{2} (R_s^2 - r_s^2)(2\pi r_s dr_s) = \int_{r_s=0}^{R_s} (V_{xs} - V_\infty)(\rho V_{xs} 2\pi r_s dr_s)$$

which is easily integrated to yield

$$T = \pi R_s^2 \rho V_{xs} (V_{xs} - V_\infty) - \frac{\pi R_s^4 \rho \omega_s^2}{4}$$

or with Eq. (2.4.35),

$$T = \pi R_p^2 \rho \left[(V_\infty + V_i)(V_{xs} - V_\infty) - \left(\frac{V_\infty + V_i}{2V_{xs}} \right)^2 \omega_s^2 R_p^2 \right] \tag{2.4.39}$$

Since there is no change in axial velocity across the propeller disk, the thrust acting on the prop-circle can also be expressed as

$$T = \int_{A_p}(p_d - p_u)dA_p = \int_{r_p=0}^{R_p}(p_d - p_u)2\pi r_p dr_p$$

With Eq. (2.4.38), this is readily integrated to yield

$$T = \pi R_p^2 \rho\left[\frac{V_{xs}^2}{2} - \frac{V_\infty^2}{2} - \left(\frac{V_\infty + V_i}{2V_{xs}}\right)^2 \omega_s^2 R_p^2\right] \tag{2.4.40}$$

Combining Eqs. (2.4.39) and (2.4.40) to eliminate the thrust, we have

$$\frac{V_{xs}^2}{2} - \frac{V_\infty^2}{2} = (V_\infty + V_i)(V_{xs} - V_\infty)$$

This is solved for the axial component of the ultimate slipstream velocity to give

$$V_{xs} = V_\infty + 2V_i \tag{2.4.41}$$

Thus we see that the relation given by Eq. (2.4.10), which was developed from classical momentum theory neglecting rotation of the slipstream, is valid with slipstream rotation, provided that only the axial component of the ultimate slipstream velocity is used.

The rotation in the ultimate slipstream is related to the propeller torque through the angular momentum equation

$$\ell = \int_{A_s} r_s V_{\theta s} d\dot{m} = \int_{r_s=0}^{R_s} r_s^2 \omega_s (\rho V_{xs} 2\pi r_s dr_s) = \frac{\pi}{2}\rho V_{xs}\omega_s R_s^4$$

Again, applying Eq. (2.4.35), this becomes

$$\ell = \pi\rho\frac{(V_\infty + V_i)^2}{2V_{xs}}\omega_s R_p^4 \tag{2.4.42}$$

The brake power required to turn the propeller must satisfy the energy equation,

$$\omega\ell = \int_{A_s}\left(h_s - h_\infty + \frac{V_{xs}^2 + V_{\theta s}^2}{2} - \frac{V_\infty^2}{2}\right)d\dot{m}$$

For the assumed uniform, incompressible flow this reduces to

$$\omega\ell = \int_{r_s=0}^{R_s}\left(\frac{p_s}{\rho} - \frac{p_\infty}{\rho} + \frac{V_{xs}^2 + \omega_s^2 r_s^2}{2} - \frac{V_\infty^2}{2}\right)\rho V_{xs}(2\pi r_s dr_s)$$

Applying Eq. (2.4.33) and integrating, this gives

$$\omega \ell = \pi R_s^2 \rho V_{xs} \left(\frac{V_{xs}^2}{2} - \frac{V_\infty^2}{2} \right)$$

or after using Eq. (2.4.35),

$$\omega \ell = \pi R_p^2 \rho (V_\infty + V_i) \left(\frac{V_{xs}^2}{2} - \frac{V_\infty^2}{2} \right) \tag{2.4.43}$$

Combining Eqs. (2.4.42) and (2.4.43) to eliminate the torque, we can solve for the angular velocity in the ultimate slipstream. This gives

$$\omega_s = \frac{V_{xs} (V_{xs}^2 - V_\infty^2)}{\omega R_p^2 (V_\infty + V_i)}$$

Using Eq. (2.4.41) to eliminate the axial component of the ultimate slipstream velocity, we have

$$\omega_s = \frac{4(V_\infty + 2V_i)V_i}{\omega R_p^2} \tag{2.4.44}$$

Applying Eqs. (2.4.41) and (2.4.44) to either Eq. (2.4.39) or (2.4.40), the thrust developed by the propeller can be expressed as

$$T = 2\pi R_p^2 \rho (V_\infty + V_i)V_i \left[1 - \frac{2(V_\infty + V_i)V_i}{\omega^2 R_p^2} \right] \tag{2.4.45}$$

Likewise, applying Eqs. (2.4.41) and (2.4.44) to either Eq. (2.4.42) or (2.4.43), the brake power required to turn the propeller can be expressed as

$$P = \omega \ell = 2\pi R_p^2 \rho (V_\infty + V_i)^2 V_i \tag{2.4.46}$$

From Eqs. (2.4.45) and (2.4.46), the ideal propulsive efficiency predicted by propeller momentum theory, including the effects of slipstream rotation, is given by

$$\eta_i \equiv \frac{TV_\infty}{P} = \frac{V_\infty}{V_\infty + V_i} - \frac{2V_\infty V_i}{\omega^2 R_p^2} \tag{2.4.47}$$

Equation (2.4.45) is quadratic in $(V_\infty + V_i)V_i$. Thus, we have

$$(V_\infty + V_i)V_i = \frac{\omega^2 R_p^2}{4} \left(1 \pm \sqrt{1 - \frac{4T}{\pi \rho \omega^2 R_p^4}} \right)$$

This result is quadratic in V_i and can be solved for the induced velocity,

$$V_i = -\frac{V_\infty}{2} \pm \sqrt{\frac{V_\infty^2}{4} + \frac{\omega^2 R_p^2}{4}\left(1 \pm \sqrt{1 - \frac{4T}{\pi\rho\omega^2 R_p^4}}\right)}$$

In the limit as the rotational speed of the propeller becomes large, the induced velocity must be positive and finite. This specifies the signs in the previous relation to give

$$V_i = \sqrt{\frac{V_\infty^2}{4} + \frac{\omega^2 R_p^2}{4}\left(1 - \sqrt{1 - \frac{4T}{A_p\rho\omega^2 R_p^2}}\right)} - \frac{V_\infty}{2} \tag{2.4.48}$$

Notice that in the limit as the angular velocity of the propeller approaches infinity, the result predicted by Eq. (2.4.48) reduces to that predicted by Eq. (2.4.14).

In terms of the usual dimensionless propeller variables, Eq. (2.4.48) becomes

$$\frac{V_i}{(\omega/2\pi)d_p} = \sqrt{\frac{J^2}{4} + \frac{\pi^2}{4}\left(1 - \sqrt{1 - \frac{16C_T}{\pi^3}}\right)} - \frac{J}{2} \tag{2.4.49}$$

and from Eq. (2.4.47), the ideal propulsive efficiency can be written as

$$\eta_i = \left[\frac{1}{2} + \sqrt{\frac{1}{4} + \frac{\pi^2}{4J^2}\left(1 - \sqrt{1 - \frac{16C_T}{\pi^3}}\right)}\right]^{-1} \frac{J^2}{\pi^2}\left[\sqrt{1 + \frac{\pi^2}{J^2}\left(1 - \sqrt{1 - \frac{16C_T}{\pi^3}}\right)} - 1\right] \tag{2.4.50}$$

By comparing Eqs. (2.4.49) and (2.4.50) with Eqs. (2.4.17) and (2.4.19) it can be shown that slipstream rotation increases induced velocity and decreases propulsive efficiency. However, it is not clear from examination of these equations what the magnitudes of these effects are. To display the effect of slipstream rotation on propeller induced velocity, Fig. 2.4.3a shows the percent error in Eq. (2.4.17) as compared with results predicted from Eq. (2.4.49). The effect of slipstream rotation on propulsive efficiency is shown in Fig. 2.4.3b through a similar comparison of results predicted from Eqs. (2.4.19) and (2.4.50). From these two figures we see that the effect of slipstream rotation on the induced velocity and propulsive efficiency, as predicted by propeller momentum theory, is on the order of 5 percent or less for the range of thrust coefficient and advance ratio that is normally encountered in airplane propellers.

The induced velocity predicted by either Eq. (2.4.17) or Eq. (2.4.49) can be used with propeller blade theory to replace the relation for induced velocity obtained from Goldstein's vortex theory. However, this method is not being recommended and will not be discussed in detail. The major difference between the induced velocity predicted from momentum theory and that predicted from vortex theory is that momentum theory predicts a zero circumferential component of induced velocity on the upstream side of the propeller. This is a direct result of the hypothesized streamtube flow.

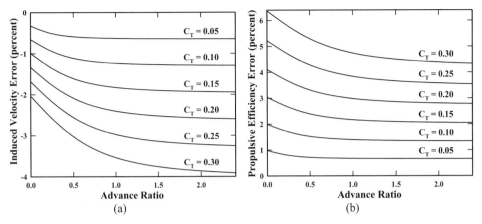

Figure 2.4.3. Errors in Eqs. (2.4.17) and (2.4.19) that result from neglecting slipstream rotation, as predicted by Eqs. (2.4.49) and (2.4.50).

The airflow that is induced by a rotating propeller results directly from the lift developed on the blades. The work done by the rotating blades moves air from the forward side of the propeller disk to the aft side. This creates a region of low pressure in front of the propeller disk. Clearly, air must flow into this low-pressure region, to replace the air that is evacuated by the propeller. The foundation of propeller momentum theory is the assumption that this induced airflow comes from a region far upstream from the propeller disk. This assumption is simply not valid. In reality, the induced airflow entering the low-pressure region forward of the propeller disk comes from the downstream side of the propeller, not from the upstream side.

A more realistic description of the flow induced by a rotating propeller was shown schematically in Fig. 2.3.3. The induced airflow on the aft side of the propeller disk is divided into two regions by the helical blade-tip vortices. In the slipstream, which is enclosed by this helix, the induced flow is moving in the downstream direction, away from the propeller disk. In the region outside the slipstream, the induced flow is moving back upstream. This upstream flow continues into the region slightly ahead of the propeller disk. In this region the induced flow becomes radial, moving inward to replace the air that is being evacuated from this region by the propeller. Near the blade tips the streamlines for this induced flow become almost circular, going from the aft side of the propeller out around the blade tips and back into the upstream side of the propeller disk. In the region upstream from the propeller disk, the induced flow rapidly goes to zero with increasing distance from the propeller. Because the downstream velocity of the induced flow within the slipstream is much greater than the upstream velocity outside the slipstream, there is a net momentum flux, which is balanced by the thrust.

This flow pattern is easily verified by interrogating the flow around a common house fan with a piece of light yarn. The fan should be placed in the center of a large room. If the yarn is held in the slipstream, both the axial velocity and the rotation of the air within the slipstream can be observed. If the yarn is held slightly downstream and slightly outboard from the blade tips, it will be drawn back upstream, around the blade tips, and into the fan. If the yarn is held upstream at various distances from the fan, it will be observed that the air a short distance upstream from the fan is essentially undisturbed.

Even though propeller momentum theory does not provide a particularly accurate description of flow in the vicinity of a rotating propeller, the effects of circumferential induced velocity are fairly small for lightly loaded propellers, and momentum theory provides some valuable insight into propeller performance with comparatively little effort. Furthermore, because propeller momentum theory is widely presented in the literature, the student should be aware of its existence and its shortcomings.

2.5. Off-Axis Forces and Moments Developed by a Propeller

In the propeller analysis that we have considered to this point, we have assumed that the direction of forward motion is perfectly aligned with the axis of propeller rotation. For this case, a thrust force and a rolling moment are produced, which are both aligned with the axis of rotation. No other aerodynamic forces or moments are produced when the propeller axis is aligned in this manner. Additional aerodynamic reactions are produced if the propeller rotation axis is not aligned perfectly with the direction of flight. The angle that the propeller axis makes with the freestream airflow is the propeller angle of attack. Since the engine and propeller are typically mounted directly to the airframe, the propeller angle of attack will change with the airplane angle of attack, which can change during flight.

When a rotating propeller is at a positive angle of attack relative to the freestream flow, there is a component of the freestream in the plane of propeller rotation, as seen in Fig. 2.5.1. This component of the freestream changes the relative airflow over each blade of the propeller. It increases the relative airspeed for the downward-moving blades and decreases the relative airspeed for the upward-moving blades. Thus, both the lift and drag are increased on the downward-moving side of the propeller and decreased on the upward-moving side. The difference in thrust generated by the two sides of the propeller produces a yawing moment and the difference in circumferential force produces a net normal force in the plane of propeller rotation, as shown in Fig. 2.2.4. Similarly, sideslip will cause a pitching moment and a side force. These off-axis forces and moments can have a significant effect on airplane trim and stability.

To see how propeller angle of attack affects the propeller forces and moments, consider the cross-section of the propeller blade that is shown in Fig. 2.5.1. The blade is rotating counterclockwise with an angular velocity of ω and is advancing through the air with a relative airspeed of V_∞. The angle of attack that the forward velocity vector makes with the axis of propeller rotation is denoted as α_p. When this angle is nonzero, the velocity of the air relative to the blade cross-section depends on the angular coordinate, θ, as well as the radial coordinate, r.

The airspeed relative to the blade cross-section, located at radius, r, and angular position, θ, is designated as V_b. This is the vector sum of the velocity created by the rotation of the blade, the forward velocity of the propeller, and the propeller induced velocity. The circumferential component of this relative airspeed varies slightly with the angle, θ, because of the component of forward airspeed that is in the plane of propeller rotation. For convenience, we shall define V_e to be the magnitude of that portion of this vector sum that includes only the velocity created by the rotation of the blade and that resulting from the forward motion. This definition is shown graphically in Fig. 2.5.1. From the geometry shown in this figure, we can write

$$V_e^2(r,\theta) = (\omega r - V_\infty \sin\alpha_p \sin\theta)^2 + (V_\infty \cos\alpha_p)^2 \tag{2.5.1}$$

Because the propeller angle of attack, α_p, is typically small, we apply the small-angle approximation to this angle only and Eq. (2.5.1) is closely approximated as

$$V_e^2(r,\theta) \cong \omega^2 r^2 + V_\infty^2 - 2\omega r V_\infty \alpha_p \sin\theta \tag{2.5.2}$$

If we now apply the Betz condition according to Goldstein's vortex theory, the total airspeed relative to the blade cross-section is

$$V_b(r,\theta) = V_e \cos\varepsilon_i = \sqrt{\omega^2 r^2 + V_\infty^2 - 2\omega r V_\infty \alpha_p \sin\theta} \cos\varepsilon_i \tag{2.5.3}$$

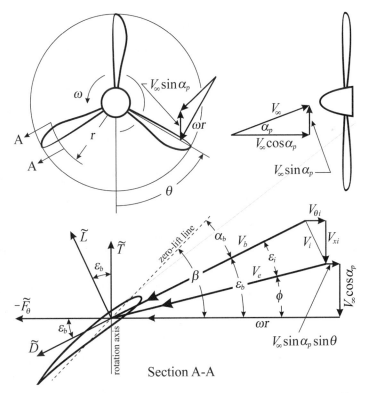

Figure 2.5.1. Effect of propeller angle of attack on the section forces and velocities acting on a rotating propeller blade.

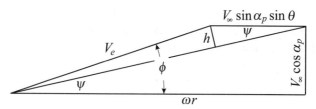

Figure 2.5.2. Trigonometric details for the angle, ϕ, that results from the propeller's total forward motion and is shown in Fig. 2.5.1.

The blade section angle of attack, α_b, measured relative to the section zero-lift line, is equal to the aerodynamic pitch angle, β, less the downwash angle, ε_b. In Fig. 2.5.1 the downwash angle for the blade section is shown to be the sum of two parts, the induced angle, ε_i, and the angle that results from the propeller's total forward motion, ϕ. From the geometry shown in Fig. 2.5.2, we can write

$$\phi = \psi + \sin^{-1}\left(\frac{h}{V_e}\right) = \psi + \sin^{-1}\left[\frac{(V_\infty \sin\alpha_p \sin\theta)\sin\psi}{V_e}\right]$$

$$= \psi + \sin^{-1}\left[\left(\frac{V_\infty \sin\alpha_p \sin\theta}{V_e}\right)\frac{V_\infty \cos\alpha_p}{\sqrt{(\omega r)^2 + (V_\infty \cos\alpha_p)^2}}\right] \tag{2.5.4}$$

After applying Eq. (2.5.1), the result in Eq. (2.5.4) can be written as

$$\phi = \tan^{-1}\left(\frac{V_\infty \cos\alpha_p}{\omega r}\right)$$

$$+ \sin^{-1}\left[\frac{(V_\infty \sin\alpha_p \sin\theta)(V_\infty \cos\alpha_p)}{\sqrt{(\omega r - V_\infty \sin\alpha_p \sin\theta)^2 + (V_\infty \cos\alpha_p)^2}\sqrt{(\omega r)^2 + (V_\infty \cos\alpha_p)^2}}\right] \tag{2.5.5}$$

Again assuming that the propeller angle of attack is small, Eq. (2.5.5) is very closely approximated as

$$\phi \cong \tan^{-1}\left(\frac{V_\infty}{\omega r}\right) + \frac{V_\infty^2}{(\omega r)^2 + V_\infty^2}\alpha_p \sin\theta \tag{2.5.6}$$

The total downwash angle, ε_b, for the blade cross-section, located at radius, r, and angular position, θ, is the sum of the angle, ϕ, and the induced angle, ε_i. Thus, we can write

$$\varepsilon_b(r,\theta) = \varepsilon_\infty + \varepsilon_i + \alpha_p \sin^2\varepsilon_\infty \sin\theta \tag{2.5.7}$$

where ε_∞ is the previously defined advance angle,

$$\varepsilon_\infty \equiv \tan^{-1}\left(\frac{V_\infty}{\omega r}\right) \tag{2.5.8}$$

The aerodynamic angle of attack is simply the aerodynamic pitch angle less the total downwash angle. Thus, from Eq. (2.5.7), this can be written as

$$\alpha_b(r,\theta) = \beta - \varepsilon_b = \beta - \varepsilon_\infty - \varepsilon_i - \alpha_p \sin^2\varepsilon_\infty \sin\theta \tag{2.5.9}$$

From Eqs. (2.5.3) and (2.5.9) we see that a positive propeller angle of attack will decrease both the relative airspeed and the section angle of attack for $0 < \theta < \pi$ and will increase both of these values in the range $\pi < \theta < 2\pi$. This will clearly result in a difference in both the lift and drag between the propeller blades on the right and left sides of the propeller disk.

From this point we follow a development similar to that presented in Sec. 2.3. The section thrust and circumferential forces are related to the section lift and drag forces through the local downwash angle according to the relations

$$\tilde{T} = \tilde{L}\cos\varepsilon_b - \tilde{D}\sin\varepsilon_b \tag{2.5.10}$$

and

$$\tilde{F}_\theta = -\tilde{D}\cos\varepsilon_b - \tilde{L}\sin\varepsilon_b \tag{2.5.11}$$

Using the result given by Eq. (2.5.7) along with the small-angle approximation for α_p, the cosine of the total downwash angle, ε_b, can be written as

$$\cos\varepsilon_b = \cos(\varepsilon_\infty + \varepsilon_i)\cos(\alpha_p \sin^2\varepsilon_\infty \sin\theta)$$
$$- \sin(\varepsilon_\infty + \varepsilon_i)\sin(\alpha_p \sin^2\varepsilon_\infty \sin\theta)$$

$$\cos\varepsilon_b \cong \cos(\varepsilon_\infty + \varepsilon_i) - \alpha_p \sin(\varepsilon_\infty + \varepsilon_i)\sin^2\varepsilon_\infty \sin\theta \tag{2.5.12}$$

Likewise, the sine of ε_b is written

$$\sin\varepsilon_b = \sin(\varepsilon_\infty + \varepsilon_i)\cos(\alpha_p \sin^2\varepsilon_\infty \sin\theta)$$
$$+ \cos(\varepsilon_\infty + \varepsilon_i)\sin(\alpha_p \sin^2\varepsilon_\infty \sin\theta)$$

$$\sin\varepsilon_b \cong \sin(\varepsilon_\infty + \varepsilon_i) + \alpha_p \cos(\varepsilon_\infty + \varepsilon_i)\sin^2\varepsilon_\infty \sin\theta \tag{2.5.13}$$

Using Eqs. (2.5.3) and (2.5.9) while continuing to apply the small-angle approxima-tion for α_p, the section lift is

$$\tilde{L} = \frac{1}{2}\rho V_b^2 c_b \tilde{C}_L = \frac{1}{2}\rho V_b^2 c_b \tilde{C}_{L,\alpha} \alpha_b$$

$$= \frac{1}{2}\rho c_b \tilde{C}_{L,\alpha}(\omega^2 r^2 + V_\infty^2 - 2\omega r V_\infty \alpha_p \sin\theta)\cos^2\varepsilon_i(\beta - \varepsilon_\infty - \varepsilon_i - \alpha_p \sin^2\varepsilon_\infty \sin\theta)$$

$$\cong \frac{1}{2}\rho\omega^2 r^2 c_b \tilde{C}_{L,\alpha}\left[\frac{\omega^2 r^2 + V_\infty^2}{\omega^2 r^2}(\beta - \varepsilon_\infty - \varepsilon_i - \alpha_p \sin^2\varepsilon_\infty \sin\theta)\right.$$

$$\left. -2\frac{V_\infty}{\omega r}(\beta - \varepsilon_\infty - \varepsilon_i)\alpha_p \sin\theta\right]\cos^2\varepsilon_i$$

$$\tilde{L} \cong \frac{1}{2}\rho\omega^2 r^2 c_b \tilde{C}_{L,\alpha}\left[\frac{\beta - \varepsilon_\infty - \varepsilon_i - \alpha_p \sin^2\varepsilon_\infty \sin\theta}{\cos^2\varepsilon_\infty}\right.$$

$$\left. -2(\beta - \varepsilon_\infty - \varepsilon_i)\alpha_p \tan\varepsilon_\infty \sin\theta\right]\cos^2\varepsilon_i \tag{2.5.14}$$

where c_b is the local section chord length and $\tilde{C}_{L,\alpha}$ is the local section lift slope. Using Eq. (2.5.3), the section drag can be written

$$\tilde{D} = \frac{1}{2}\rho V_b^2 c_b \tilde{C}_D = \frac{1}{2}\rho c_b \tilde{C}_D(\omega^2 r^2 + V_\infty^2 - 2\omega r V_\infty \alpha_p \sin\theta)\cos^2\varepsilon_i$$

$$= \frac{1}{2}\rho\omega^2 r^2 c_b \tilde{C}_D\left(\frac{\omega^2 r^2 + V_\infty^2}{\omega^2 r^2} - 2\frac{V_\infty}{\omega r}\alpha_p \sin\theta\right)\cos^2\varepsilon_i$$

$$\tilde{D} = \frac{1}{2}\rho\omega^2 r^2 c_b \tilde{C}_D\left(\frac{1}{\cos^2\varepsilon_\infty} - 2\alpha_p \tan\varepsilon_\infty \sin\theta\right)\cos^2\varepsilon_i \tag{2.5.15}$$

where $\tilde{C}_D$ is the local 2-D section parasitic drag coefficient. Remember that induced drag is included in the tilting of the lift vector.

Applying Eqs. (2.5.12), (2.5.13), (2.5.14), and (2.5.15) to Eq. (2.5.10), the axial component of section force is written

$$\tilde{T} = \tilde{L}\cos\varepsilon_b - \tilde{D}\sin\varepsilon_b = \frac{1}{2}\rho\omega^2 r^2 c_b \tilde{C}_{L,\alpha}$$

$$\times\left[\frac{\beta - \varepsilon_\infty - \varepsilon_i - \alpha_p \sin^2\varepsilon_\infty \sin\theta}{\cos^2\varepsilon_\infty} - 2(\beta - \varepsilon_\infty - \varepsilon_i)\alpha_p \tan\varepsilon_\infty \sin\theta\right]\cos^2\varepsilon_i$$

$$\times\left[\cos(\varepsilon_\infty + \varepsilon_i) - \alpha_p \sin(\varepsilon_\infty + \varepsilon_i)\sin^2\varepsilon_\infty \sin\theta\right]$$

$$-\frac{1}{2}\rho\omega^2 r^2 c_b \tilde{C}_D\left(\frac{1}{\cos^2\varepsilon_\infty} - 2\alpha_p \tan\varepsilon_\infty \sin\theta\right)\cos^2\varepsilon_i$$

$$\times\left[\sin(\varepsilon_\infty + \varepsilon_i) + \alpha_p \cos(\varepsilon_\infty + \varepsilon_i)\sin^2\varepsilon_\infty \sin\theta\right]$$

or after rearranging and applying the small-angle approximation for α_p,

$$\tilde{T} \cong \tfrac{1}{2}\rho\omega^2 r^2 c_b \cos^2\varepsilon_i \left\{ \left[\frac{\tilde{C}_{L,\alpha}(\beta - \varepsilon_\infty - \varepsilon_i)\cos(\varepsilon_\infty + \varepsilon_i) - \tilde{C}_D \sin(\varepsilon_\infty + \varepsilon_i)}{\cos^2\varepsilon_\infty} \right] \right.$$

$$- \left\{ \tan^2\varepsilon_\infty \left[(\tilde{C}_{L,\alpha} + \tilde{C}_D)\cos(\varepsilon_\infty + \varepsilon_i) + \tilde{C}_{L,\alpha}(\beta - \varepsilon_\infty - \varepsilon_i)\sin(\varepsilon_\infty + \varepsilon_i) \right] \right. \tag{2.5.16}$$

$$\left. \left. + 2\tan\varepsilon_\infty \left[\tilde{C}_{L,\alpha}(\beta - \varepsilon_\infty - \varepsilon_i)\cos(\varepsilon_\infty + \varepsilon_i) - \tilde{C}_D \sin(\varepsilon_\infty + \varepsilon_i) \right] \right\} \alpha_p \sin\theta \right\}$$

Similarly, the circumferential component of section force is

$$\tilde{F}_\theta = -\tilde{L}\sin\varepsilon_b - \tilde{D}\cos\varepsilon_b = -\tfrac{1}{2}\rho\omega^2 r^2 c_b \tilde{C}_{L,\alpha}$$

$$\times \left[\frac{\beta - \varepsilon_\infty - \varepsilon_i - \alpha_p \sin^2\varepsilon_\infty \sin\theta}{\cos^2\varepsilon_\infty} - 2(\beta - \varepsilon_\infty - \varepsilon_i)\alpha_p \tan\varepsilon_\infty \sin\theta \right] \cos^2\varepsilon_i$$

$$\times \left[\sin(\varepsilon_\infty + \varepsilon_i) + \alpha_p \cos(\varepsilon_\infty + \varepsilon_i)\sin^2\varepsilon_\infty \sin\theta \right]$$

$$- \tfrac{1}{2}\rho\omega^2 r^2 c_b \tilde{C}_D \left(\frac{1}{\cos^2\varepsilon_\infty} - 2\alpha_p \tan\varepsilon_\infty \sin\theta \right) \cos^2\varepsilon_i$$

$$\times \left[\cos(\varepsilon_\infty + \varepsilon_i) - \alpha_p \sin(\varepsilon_\infty + \varepsilon_i)\sin^2\varepsilon_\infty \sin\theta \right]$$

$$\tilde{F}_\theta \cong -\tfrac{1}{2}\rho\omega^2 r^2 c_b \cos^2\varepsilon_i \left\{ \left[\frac{\tilde{C}_D \cos(\varepsilon_\infty + \varepsilon_i) + \tilde{C}_{L,\alpha}(\beta - \varepsilon_\infty - \varepsilon_i)\sin(\varepsilon_\infty + \varepsilon_i)}{\cos^2\varepsilon_\infty} \right] \right.$$

$$+ \left\{ \tan^2\varepsilon_\infty \left[\tilde{C}_{L,\alpha}(\beta - \varepsilon_\infty - \varepsilon_i)\cos(\varepsilon_\infty + \varepsilon_i) - (\tilde{C}_{L,\alpha} + \tilde{C}_D)\sin(\varepsilon_\infty + \varepsilon_i) \right] \right. \tag{2.5.17}$$

$$\left. \left. - 2\tan\varepsilon_\infty \left[\tilde{C}_D \cos(\varepsilon_\infty + \varepsilon_i) + \tilde{C}_{L,\alpha}(\beta - \varepsilon_\infty - \varepsilon_i)\sin(\varepsilon_\infty + \varepsilon_i) \right] \right\} \alpha_p \sin\theta \right\}$$

The thrust per unit radial distance, for the full prop-circle, is the average axial force integrated over a full revolution, multiplied by the number of blades, k. Thus, Eq. (2.5.16) is readily integrated with respect to θ to give

$$\frac{dT}{dr} = \frac{k}{2\pi} \int_{\theta=0}^{2\pi} \tilde{T}\, d\theta$$

$$= \frac{k}{4\pi}\rho\omega^2 r^2 c_b \left[\frac{\tilde{C}_{L,\alpha}(\beta - \varepsilon_\infty - \varepsilon_i)\cos(\varepsilon_\infty + \varepsilon_i) - \tilde{C}_D \sin(\varepsilon_\infty + \varepsilon_i)}{\cos^2\varepsilon_\infty} \right] \cos^2\varepsilon_i \int_{\theta=0}^{2\pi} d\theta$$

$$- \frac{k}{4\pi}\rho\omega^2 r^2 c_b \left\{ \tan^2\varepsilon_\infty \left[(\tilde{C}_{L,\alpha} + \tilde{C}_D)\cos(\varepsilon_\infty + \varepsilon_i) + \tilde{C}_{L,\alpha}(\beta - \varepsilon_\infty - \varepsilon_i)\sin(\varepsilon_\infty + \varepsilon_i) \right] \right.$$

$$\left. + 2\tan\varepsilon_\infty \left[\tilde{C}_{L,\alpha}(\beta - \varepsilon_\infty - \varepsilon_i)\cos(\varepsilon_\infty + \varepsilon_i) - \tilde{C}_D \sin(\varepsilon_\infty + \varepsilon_i) \right] \right\} \alpha_p \cos^2\varepsilon_i \int_{\theta=0}^{2\pi} \sin\theta\, d\theta$$

$$\frac{dT}{dr} = \frac{k}{2}\rho\omega^2 r^2 c_b \frac{\cos^2\varepsilon_i}{\cos^2\varepsilon_\infty}\left[\tilde{C}_{L,\alpha}(\beta-\varepsilon_\infty-\varepsilon_i)\cos(\varepsilon_\infty+\varepsilon_i)-\tilde{C}_D\sin(\varepsilon_\infty+\varepsilon_i)\right] \quad (2.5.18)$$

and

$$T = \int_{r=r_h}^{r_o}\frac{dT}{dr}dr$$

$$= \frac{k\rho\omega^2}{2}\int_{r=r_h}^{r_o} r^2 c_b \frac{\cos^2\varepsilon_i}{\cos^2\varepsilon_\infty}\left[\tilde{C}_{L,\alpha}(\beta-\varepsilon_\infty-\varepsilon_i)\cos(\varepsilon_\infty+\varepsilon_i)-\tilde{C}_D\sin(\varepsilon_\infty+\varepsilon_i)\right]dr$$

$$(2.5.19)$$

Similarly, the torque required to turn the propeller is

$$\ell = \int_{r=r_h}^{r_o}\frac{k}{2\pi}\int_{\theta=0}^{2\pi}-\bar{F}_\theta\, r\, d\theta\, dr$$

Applying Eq. (2.5.17), the integration with respect to θ is readily carried out to yield

$$\ell = \frac{k\rho\omega^2}{2}\int_{r=r_h}^{r_o} r^3 c_b \frac{\cos^2\varepsilon_i}{\cos^2\varepsilon_\infty}\left[\tilde{C}_D\cos(\varepsilon_\infty+\varepsilon_i)+\tilde{C}_{L,\alpha}(\beta-\varepsilon_\infty-\varepsilon_i)\sin(\varepsilon_\infty+\varepsilon_i)\right]dr$$

$$(2.5.20)$$

Comparing Eqs. (2.5.19) and (2.5.20) with Eqs. (2.3.26) and (2.3.27), we find that within the accuracy of the small-angle approximation used for α_p, the propeller thrust and torque are not affected by small changes in the propeller angle of attack.

In a similar manner we can determine the normal component of the aerodynamic force acting on a rotating propeller. This is the component of force that acts in the plane of propeller rotation and in a direction parallel to the in-plane component of the freestream velocity. From Fig. 2.5.1 we can see that the normal component of section force is equal to the circumferential component, in the direction of propeller rotation, multiplied by the sine of the rotation angle, θ,

$$\tilde{N} = \bar{F}_\theta \sin\theta$$

Applying Eq. (2.5.17) to this result, we have

$$\tilde{N} = -\frac{1}{2}\rho\omega^2 r^2 c_b\left[\frac{\tilde{C}_D\cos(\varepsilon_\infty+\varepsilon_i)+\tilde{C}_{L,\alpha}(\beta-\varepsilon_\infty-\varepsilon_i)\sin(\varepsilon_\infty+\varepsilon_i)}{\cos^2\varepsilon_\infty}\right]\cos^2\varepsilon_i\sin\theta$$

$$-\frac{1}{2}\rho\omega^2 r^2 c_b\left\{\tan^2\varepsilon_\infty\left[\tilde{C}_{L,\alpha}(\beta-\varepsilon_\infty-\varepsilon_i)\cos(\varepsilon_\infty+\varepsilon_i)-(\tilde{C}_{L,\alpha}+\tilde{C}_D)\sin(\varepsilon_\infty+\varepsilon_i)\right]\right.$$

$$\left.-2\tan\varepsilon_\infty\left[\tilde{C}_D\cos(\varepsilon_\infty+\varepsilon_i)+\tilde{C}_{L,\alpha}(\beta-\varepsilon_\infty-\varepsilon_i)\sin(\varepsilon_\infty+\varepsilon_i)\right]\right\}\alpha_p\cos^2\varepsilon_i\sin^2\theta$$

The normal force per unit radial distance, for the full prop-circle, is the average section normal force per blade integrated over a full revolution, multiplied by the number of blades. Integrating from θ equals zero to 2π and multiplying by k gives

$$\frac{dN}{dr} = \frac{k}{2\pi}\int_{\theta=0}^{2\pi}\tilde{N}\,d\theta$$

$$= -\frac{k}{4\pi}\rho\omega^2 r^2 c_b \left[\frac{\tilde{C}_D\cos(\varepsilon_\infty+\varepsilon_i)+\tilde{C}_{L,\alpha}(\beta-\varepsilon_\infty-\varepsilon_i)\sin(\varepsilon_\infty+\varepsilon_i)}{\cos^2\varepsilon_\infty}\right]\cos^2\varepsilon_i\int_{\theta=0}^{2\pi}\sin\theta\,d\theta$$

$$-\frac{k}{4\pi}\rho\omega^2 r^2 c_b\left\{\tan^2\varepsilon_\infty\left[\tilde{C}_{L,\alpha}(\beta-\varepsilon_\infty-\varepsilon_i)\cos(\varepsilon_\infty+\varepsilon_i)-(\tilde{C}_{L,\alpha}+\tilde{C}_D)\sin(\varepsilon_\infty+\varepsilon_i)\right]\right.$$

$$\left.-2\tan\varepsilon_\infty\left[\tilde{C}_D\cos(\varepsilon_\infty+\varepsilon_i)+\tilde{C}_{L,\alpha}(\beta-\varepsilon_\infty-\varepsilon_i)\sin(\varepsilon_\infty+\varepsilon_i)\right]\right\}\alpha_p\cos^2\varepsilon_i\int_{\theta=0}^{2\pi}\sin^2\theta\,d\theta$$

$$\frac{dN}{dr} = \frac{k}{4}\rho\omega^2 r^2 c_b\left\{2\tan\varepsilon_\infty\left[\tilde{C}_D\cos(\varepsilon_\infty+\varepsilon_i)+\tilde{C}_{L,\alpha}(\beta-\varepsilon_\infty-\varepsilon_i)\sin(\varepsilon_\infty+\varepsilon_i)\right]\right.$$

$$\left.-\tan^2\varepsilon_\infty\left[\tilde{C}_{L,\alpha}(\beta-\varepsilon_\infty-\varepsilon_i)\cos(\varepsilon_\infty+\varepsilon_i)-(\tilde{C}_{L,\alpha}+\tilde{C}_D)\sin(\varepsilon_\infty+\varepsilon_i)\right]\right\}\alpha_p\cos^2\varepsilon_i$$

Integrating this result with respect to r from the hub radius to the tip radius, the total normal force acting on the propeller is found to be

$$N = \int_{r=r_h}^{r_o}\frac{dN}{dr}\,dr = \frac{k\rho\omega^2}{4}\alpha_p$$

$$\times\int_{r=r_h}^{r_o}r^2 c_b\cos^2\varepsilon_i\left\{2\tan\varepsilon_\infty\left[\tilde{C}_D\cos(\varepsilon_\infty+\varepsilon_i)+\tilde{C}_{L,\alpha}(\beta-\varepsilon_\infty-\varepsilon_i)\sin(\varepsilon_\infty+\varepsilon_i)\right]\right. \tag{2.5.21}$$

$$\left.-\tan^2\varepsilon_\infty\left[\tilde{C}_{L,\alpha}(\beta-\varepsilon_\infty-\varepsilon_i)\cos(\varepsilon_\infty+\varepsilon_i)-(\tilde{C}_{L,\alpha}+\tilde{C}_D)\sin(\varepsilon_\infty+\varepsilon_i)\right]\right\}dr$$

While this normal force is typically quite small and is insignificant compared to the lift of the wing, it can have a very significant effect on aircraft stability.

Using Eq. (2.5.17) and following a similar procedure, it is readily shown that without sideslip, the aerodynamic side force in the plane of propeller rotation is zero,

$$Y = \int_{r=r_h}^{r_o}\frac{dY}{dr}\,dr = \int_{r=r_h}^{r_o}\frac{k}{2\pi}\int_{\theta=0}^{2\pi}\tilde{Y}\,d\theta\,dr = \frac{k}{2\pi}\int_{r=r_h}^{r_o}\int_{\theta=0}^{2\pi}-\tilde{F}_\theta\cos\theta\,d\theta\,dr = 0 \tag{2.5.22}$$

From Eq. (2.5.16), without sideslip, the aerodynamic pitching moment about the center of the propeller disk is also zero,

$$m = \int_{r=r_h}^{r_o}\frac{k}{2\pi}\int_{\theta=0}^{2\pi}\tilde{T}\,r\cos\theta\,d\theta\,dr = 0 \tag{2.5.23}$$

and the aerodynamic yawing moment about the center of the propeller disk is

$$n = \int_{r=r_h}^{r_o} \frac{k}{2\pi} \int_{\theta=0}^{2\pi} \tilde{T} \, r \sin\theta \, d\theta \, dr = -\frac{k\rho\omega^2}{4} \alpha_p$$

$$\times \int_{r=r_h}^{r_o} r^3 c_b \cos^2\varepsilon_i \Big\{ 2\tan\varepsilon_\infty \Big[\tilde{C}_{L,\alpha} (\beta - \varepsilon_\infty - \varepsilon_i) \cos(\varepsilon_\infty + \varepsilon_i) - \tilde{C}_D \sin(\varepsilon_\infty + \varepsilon_i) \Big] \qquad (2.5.24)$$

$$+ \tan^2\varepsilon_\infty \Big[(\tilde{C}_{L,\alpha} + \tilde{C}_D) \cos(\varepsilon_\infty + \varepsilon_i) + \tilde{C}_{L,\alpha} (\beta - \varepsilon_\infty - \varepsilon_i) \sin(\varepsilon_\infty + \varepsilon_i) \Big] \Big\} \, dr$$

The dimensionless off-axis aerodynamic coefficients for a propeller are defined in exactly the same way as the thrust and torque coefficients. With these traditional definitions, Eqs. (2.5.21) through (2.5.24) are readily nondimensionalized to yield the following:

Side force coefficient

$$C_Y \equiv \frac{Y}{\rho(\omega/2\pi)^2 d_p^4} = 0 \qquad (2.5.25)$$

Normal force coefficient

$$C_N \equiv \frac{N}{\rho(\omega/2\pi)^2 d_p^4}$$

$$= \frac{\pi^2}{8} \alpha_p \int_{\zeta=\zeta_h}^{1} \zeta^2 \hat{c}_b \cos^2\varepsilon_i \Big\{ 2\tan\varepsilon_\infty \Big[\tilde{C}_D \cos(\varepsilon_\infty + \varepsilon_i) + \tilde{C}_L \sin(\varepsilon_\infty + \varepsilon_i) \Big] \qquad (2.5.26)$$

$$- \tan^2\varepsilon_\infty \Big[\tilde{C}_L \cos(\varepsilon_\infty + \varepsilon_i) - (\tilde{C}_{L,\alpha} + \tilde{C}_D) \sin(\varepsilon_\infty + \varepsilon_i) \Big] \Big\} d\zeta$$

Pitching moment coefficient

$$C_m \equiv \frac{m}{\rho(\omega/2\pi)^2 d_p^5} = 0 \qquad (2.5.27)$$

Yawing moment coefficient

$$C_n \equiv \frac{n}{\rho(\omega/2\pi)^2 d_p^5}$$

$$= -\frac{\pi^2}{16} \alpha_p \int_{\zeta=\zeta_h}^{1} \zeta^3 \hat{c}_b \cos^2\varepsilon_i \Big\{ 2\tan\varepsilon_\infty \Big[\tilde{C}_L \cos(\varepsilon_\infty + \varepsilon_i) - \tilde{C}_D \sin(\varepsilon_\infty + \varepsilon_i) \Big] \qquad (2.5.28)$$

$$+ \tan^2\varepsilon_\infty \Big[(\tilde{C}_{L,\alpha} + \tilde{C}_D) \cos(\varepsilon_\infty + \varepsilon_i) + \tilde{C}_L \sin(\varepsilon_\infty + \varepsilon_i) \Big] \Big\} d\zeta$$

where again

$$\zeta \equiv \frac{r}{d_p/2} \qquad (2.3.37)$$

$$\hat{c}_b \equiv \frac{kc_b}{d_p} \qquad (2.3.38)$$

$$\beta \equiv \tan^{-1}\left(\frac{K}{\pi\zeta}\right) \qquad (2.3.39)$$

$$\varepsilon_\infty \equiv \tan^{-1}\left(\frac{J}{\pi\zeta}\right) \qquad (2.3.40)$$

$$K \equiv \frac{\lambda}{d_p} \qquad (2.3.41)$$

$$J \equiv \frac{2\pi V_\infty}{\omega d_p} \qquad (2.3.42)$$

and ε_i is determined from

$$\frac{\hat{c}_b}{8\zeta}\tilde{C}_L - \cos^{-1}\left\{\exp\left[-\frac{k(1-\zeta)}{2\sin\beta_t}\right]\right\}\tan\varepsilon_i\sin(\varepsilon_\infty + \varepsilon_i) = 0 \qquad (2.3.32)$$

Notice that both the normal force coefficient and yawing moment coefficient are directly proportional to the propeller angle of attack. By symmetry, it is easily shown that the side force and pitching moment are related to the sideslip angle in exactly the same way that the normal force and yawing moment are related to the angle of attack. The variations in the normal force and side force with angle of attack and sideslip angle become particularly important in the study of aircraft stability and control. Because of these forces, a rotating propeller that is forward of the airplane's center of gravity has a significant destabilizing effect while an aft propeller is stabilizing. This will be covered in detail in Chapter 4.

EXAMPLE 2.5.1. For the family of two-blade propellers described in Example 2.3.1, we wish to examine how the off-axis forces and moments depend on advance ratio and pitch-to-diameter ratio. For this purpose, we want to plot the change in the normal force coefficient with propeller angle of attack as a function of advance ratio. These results are to be compared for several different values of pitch-to-diameter ratio. Similar plots are to be obtained for the change in the yawing moment coefficient with respect to propeller angle of attack.

Solution. As an example calculation, we again consider the 75 percent radius section of one of these propellers with a chord-line pitch-to-diameter ratio of 0.5, operating at an advance ratio of 0.25. From Example 2.3.1, at this cross-section of the propeller we have

$$\zeta = 0.75$$

$$K = 0.591$$

$$\beta = 0.2458 = 14.08°$$

$$\varepsilon_\infty = 0.1057 = 6.06°$$

$$\hat{c}_b = 0.09922$$

$$\varepsilon_i = 0.04815 = 2.76°$$

With these results, from Eq. (2.5.26),

$$\frac{dC_{N,\alpha}}{d\zeta} = \frac{\pi^2}{8}\zeta^2\hat{c}_b\cos^2\varepsilon_i\left\{2\tan\varepsilon_\infty\left[\tilde{C}_D\cos(\varepsilon_\infty+\varepsilon_i)+\tilde{C}_L\sin(\varepsilon_\infty+\varepsilon_i)\right]\right.$$
$$\left.-\tan^2\varepsilon_\infty\left[\tilde{C}_L\cos(\varepsilon_\infty+\varepsilon_i)-(\tilde{C}_{L,\alpha}+\tilde{C}_D)\sin(\varepsilon_\infty+\varepsilon_i)\right]\right\}$$
$$= 0.0017079$$

Integrating similar results from $\zeta = 0.1$ to 1.0, we obtain

$$C_{N,\alpha} \equiv \frac{dC_N}{d\alpha_p} = \underline{0.00273}$$

Similarly, from Eq. (2.5.28),

$$\frac{dC_{n,\alpha}}{d\zeta} = -\frac{\pi^2}{16}\zeta^3\hat{c}_b\cos^2\varepsilon_i\left\{2\tan\varepsilon_\infty\left[\tilde{C}_L\cos(\varepsilon_\infty+\varepsilon_i)-\tilde{C}_D\sin(\varepsilon_\infty+\varepsilon_i)\right]\right.$$
$$\left.+\tan^2\varepsilon_\infty\left[(\tilde{C}_{L,\alpha}+\tilde{C}_D)\cos(\varepsilon_\infty+\varepsilon_i)+\tilde{C}_L\sin(\varepsilon_\infty+\varepsilon_i)\right]\right\}$$
$$= -0.0049410$$

$$C_{n,\alpha} \equiv \frac{dC_n}{d\alpha_p} = \underline{-0.00300}$$

After repeating these calculations for various values of the advance ratio and several different pitch-to-diameter ratios, we obtain the results that are plotted in Figs. 2.5.3 and 2.5.4.

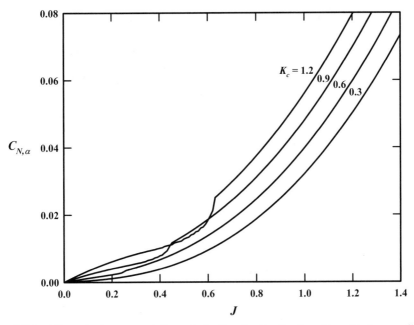

Figure 2.5.3. Dimensionless normal force derivative for the family of two-blade propellers in Example 2.5.1.

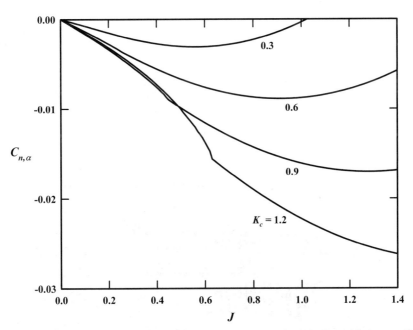

Figure 2.5.4. Dimensionless yawing-moment derivative for the family of two-blade propellers in Example 2.5.1.

2.6. Turbojet Engines: The Thrust Equation

A complete analysis for the flow of air and products of combustion through the various components of a turbojet engine is beyond the intended scope of this chapter. Here we confine our study of the turbojet engine to what is commonly called *thermodynamic cycle analysis*, which examines the thermodynamic changes in the working fluid as it flows through the components of the engine. While this type of analysis does not provide the information needed for the detailed design of individual components such as turbine and compressor blades, it does provide some valuable insight that can be useful in the study of flight mechanics. For a more detailed study of gas turbine propulsion systems, the reader is referred to the excellent text by Jack Mattingly (1996).

As shown schematically in Fig. 2.6.1, the major internal components of an installed turbojet engine are the inlet duct, compressor, combustion chamber, gas turbine, and exit nozzle. Before considering a thermodynamic cycle analysis for this propulsion system, we shall apply the momentum equation to examine the axial forces acting on such an engine mounted to the wing of a moving airplane with a strut.

Consider the control volume shown in Fig. 2.6.2. This control volume encloses the complete engine and nacelle and cuts through the mounting strut between the engine and wing. The axial forces acting on this control volume include the reaction shear force in the strut, F_r, the drag force acting on the nacelle, D_{nac}, the differential pressure force, $(p_i-p_\infty)dA_i$, integrated over the inlet area, and the differential pressure force, $(p_e-p_\infty)dA_e$, integrated over the exit area. Here dA_i and dA_e are, respectively, the differential cross-sectional areas for the inlet and exit. The pressure forces on the inlet and exit areas are defined relative to the freestream pressure, p_∞, because this is the standard convention used to define the drag force acting on the nacelle. The freestream pressure integrated over the surface of the control volume is the buoyant force, which in this case is negligible. The reaction shear force in the strut is equal to the net thrust developed by the installed engine, including the losses due to drag. This is conventionally called the *installed thrust* or *available thrust* and is traditionally denoted as T.

From Newton's second law, the summation of forces in the axial direction is equal to the net axial momentum flux. The axial momentum flux crossing the boundary of the control volume is the mass flow rate multiplied by the axial momentum per unit mass,

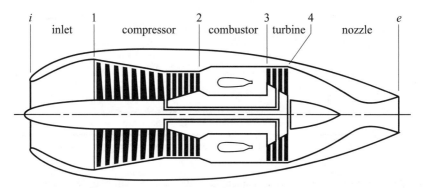

Figure 2.6.1. Components and station labeling for the turbojet engine.

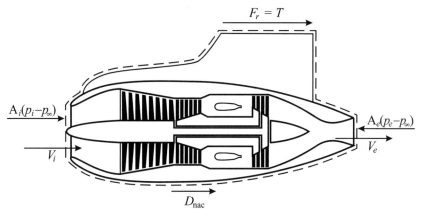

Figure 2.6.2. Control volume forces and momentum fluxes for a turbojet engine.

which is simply the axial velocity component. In the exit plane the axial momentum flux, which is considered positive leaving the control volume, is $(\rho_e V_e dA_e)V_e$ and in the inlet plane it is $(-\rho_i V_i dA_i)V_i$. Assuming uniform flow over the inlet and exit planes and neglecting the momentum of incoming fuel, Newton's second law applied to the control volume shown in Fig. 2.6.2 can be written as

$$F_r + D_{\text{nac}} + A_i(p_i - p_\infty) - A_e(p_e - p_\infty) = \rho_e A_e V_e^2 - \rho_i A_i V_i^2$$

Using the notation for mass flow rate, $\dot{m} \equiv \rho A V$, and solving for available thrust gives

$$T \equiv F_r = \dot{m}_e V_e + A_e(p_e - p_\infty) - \dot{m}_i V_i - A_i(p_i - p_\infty) - D_{\text{nac}} \qquad (2.6.1)$$

In order to separate the effects of the engine and nacelle, Eq. (2.6.1) is traditionally rearranged to the form

$$T = [\dot{m}_e V_e - \dot{m}_i V_\infty + A_e(p_e - p_\infty)] - [\dot{m}_i(V_i - V_\infty) + A_i(p_i - p_\infty)] - D_{\text{nac}} \qquad (2.6.2)$$

The grouping of terms in the first set of square brackets in Eq. (2.6.2) is usually called the *uninstalled thrust* and will simply be given the symbol F,

$$F \equiv \dot{m}_e V_e - \dot{m}_i V_\infty + A_e(p_e - p_\infty) \qquad (2.6.3)$$

Here F is used to denote uninstalled thrust to avoid later confusion with temperature. The terms in the second set of square brackets in Eq. (2.6.2) have sometimes been called the *pre-entry drag* but are more commonly referred to as the *additive drag*, D_{add},

$$D_{\text{add}} \equiv \dot{m}_i(V_i - V_\infty) + A_i(p_i - p_\infty) \qquad (2.6.4)$$

With this terminology, the installed or available thrust is equal to the uninstalled thrust less the sum of the nacelle drag and additive drag,

$$T = F - (D_{nac} + D_{add}) \tag{2.6.5}$$

Some insight into the nature of additive drag can be provided by rearranging the definition in Eq. (2.6.4). Expressing the inlet mass flow rate in terms of density, velocity, and area results in

$$D_{add} = A_i[\rho_i(V_i^2 - V_iV_\infty) + p_i - p_\infty] \tag{2.6.6}$$

Assuming that air is an ideal gas, the air density can be expressed in terms of pressure, temperature, and the ideal gas constant, R. Thus, Eq. (2.6.6) can be written

$$D_{add} = A_i\left[\frac{p_i}{RT_i}(V_i^2 - V_iV_\infty) + p_i - p_\infty\right] \tag{2.6.7}$$

The speed of sound in an ideal gas is

$$a = \sqrt{\gamma RT} \tag{2.6.8}$$

where γ is the ratio of constant-pressure specific heat to constant-volume specific heat. In view of Eq. (2.6.8), it is convenient to rearrange Eq. (2.6.7) as

$$D_{add} = A_ip_\infty\left[\gamma\frac{p_i}{p_\infty}\left(\frac{V_i^2}{\gamma RT_i} - \frac{V_iV_\infty}{\gamma R\sqrt{T_iT_\infty}}\sqrt{\frac{T_\infty}{T_i}} + \frac{1}{\gamma}\right) - 1\right]$$

or

$$D_{add} = A_ip_\infty\left[\gamma\frac{p_i}{p_\infty}\left(M_i^2 - M_iM_\infty\sqrt{\frac{T_\infty}{T_i}} + \frac{1}{\gamma}\right) - 1\right] \tag{2.6.9}$$

where T_i and T_∞ are the inlet and freestream temperatures, respectively. **(Be careful not to confuse temperature and thrust.)** The pressure and temperature ratios in Eq. (2.6.9) depend on the inlet and flight Mach numbers. Assuming isentropic flow from the freestream to the inlet, the stagnation pressure, p_0, and the stagnation temperature, T_0, remain constant. Thus, Eq. (2.6.9) can be written

$$D_{add} = A_ip_\infty\left[\gamma\frac{p_i/p_{0i}}{p_\infty/p_{0\infty}}\left(M_i^2 - M_iM_\infty\sqrt{\frac{T_\infty/T_{0\infty}}{T_i/T_{0i}}} + \frac{1}{\gamma}\right) - 1\right] \tag{2.6.10}$$

where, from compressible flow theory,

$$\frac{p}{p_0} = \left(1 + \frac{\gamma-1}{2}M^2\right)^{\frac{-\gamma}{\gamma-1}} \quad \text{and} \quad \frac{T}{T_0} = \left(1 + \frac{\gamma-1}{2}M^2\right)^{-1}$$

From Eq. (2.6.10) it can be seen that the additive drag will be zero if the inlet is designed such that the inlet Mach number, M_i, is equal to the flight Mach number, M_∞. Turbojet inlets are commonly designed to provide this match at the design cruise Mach number.

EXAMPLE 2.6.1. The inlet of a particular turbojet engine has a cross-sectional area of 46 ft^2 and is operating at standard sea level with an inlet Mach number of 0.6. Assuming that the inlet Mach number remains constant, plot the additive drag predicted from Eq. (2.6.10) as a function of flight Mach number from 0 to 1.

Solution. For this engine and the specified operating conditions, we have

$$A_i = 46 \text{ ft}^2, \quad p_\infty = 14.696 \text{ psi}, \quad M_i = 0.6, \quad \gamma = 1.4$$

and

$$\frac{p_i}{p_{0i}} = \left(1 + \frac{\gamma-1}{2}M_i^2\right)^{\frac{-\gamma}{\gamma-1}} = [1+0.2(0.6)^2]^{-3.5} = 0.784004$$

$$\frac{T_i}{T_{0i}} = \left(1 + \frac{\gamma-1}{2}M_i^2\right)^{-1} = [1+0.2(0.6)^2]^{-1} = 0.932836$$

As an example calculation, for a flight Mach number of 0.1, we obtain

$$\frac{p_\infty}{p_{0\infty}} = \left(1 + \frac{\gamma-1}{2}M_\infty^2\right)^{\frac{-\gamma}{\gamma-1}} = [1+0.2(0.1)^2]^{-3.5} = 0.993031$$

$$\frac{T_\infty}{T_{0\infty}} = \left(1 + \frac{\gamma-1}{2}M_\infty^2\right)^{-1} = [1+0.2(0.1)^2]^{-1} = 0.998004$$

$$D_{\text{add}} = A_i p_\infty \left[\gamma \frac{p_i/p_{0i}}{p_\infty/p_{0\infty}} \left(M_i^2 - M_i M_\infty \sqrt{\frac{T_\infty/T_{0\infty}}{T_i/T_{0i}}} + \frac{1}{\gamma}\right) - 1 \right]$$

$$= 46(14.696 \times 144)\left[1.4\frac{0.784004}{0.993031}\left(0.6^2 - 0.6 \times 0.1\sqrt{\frac{0.998004}{0.932836}} + \frac{1}{1.4}\right) - 1\right]$$

$$= 11{,}567 \text{ lbf}$$

Repeating these calculations for different values of the flight Mach number with the inlet Mach number held constant, the results shown in Fig. 2.6.3 are obtained. The inlet static pressure is also shown in this figure as a function of the flight Mach number. Notice that the inlet pressure is below the freestream pressure for airspeeds below the design flight Mach number and above the freestream for higher Mach numbers.

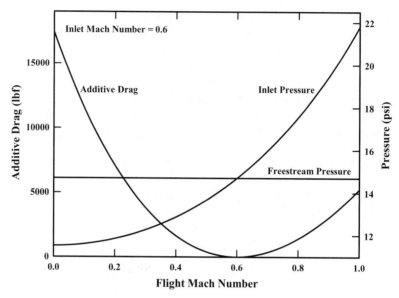

Figure 2.6.3. Inlet pressure and additive drag for Example 2.6.1.

Because the inlet area of a turbojet engine is typically larger than the exit area, the pressure component of nacelle drag will be negative, provided that the flow does not separate. In fact, it can be shown that in a completely inviscid fluid the nacelle drag would exactly offset the additive drag, and the installed thrust would be equal to the uninstalled thrust. In viscous flow, when the inlet is operating near the design Mach number, the pressure component of nacelle drag will very nearly offset the additive drag, and the installed thrust will be less than the uninstalled thrust only by an amount approximately equal to the viscous component of nacelle drag. However, when the flight Mach number is significantly different from the inlet Mach number, the airflow near the inlet must be turned through large angles. This results in flow separation with a resulting increase in nacelle drag. When the flight Mach number differs considerably from the inlet Mach number, the additive drag will dominate and the installed thrust will be substantially reduced. This is particularly significant during takeoff, when the airspeed is low and the additive drag is large. To reduce additive drag at low flight Mach numbers and increase static thrust, some turbojet installations are designed so that the inlet area can be increased for full-throttle operation on takeoff.

For efficient performance, the inlet of a turbojet engine must be designed specifically for the intended operating condition. The inlet duct must ingest air at or near the flight Mach number and reduce the velocity to a level suitable for efficient operation of the compressor. For subsonic flight, the simple divergent geometry shown in Fig. 2.6.1 is satisfactory. However, for supersonic flight, shock waves will occur as the velocity of the ingested flow is reduced. The compression process is quite inefficient if the airflow is reduced to subsonic speed through a single normal shock wave. Thus, supersonic inlets are quite different from that shown in Fig. 2.6.1 (see Mattingly 1996).

2.7. Turbojet Engines: Cycle Analysis

To gain some insight into the performance characteristics of a turbojet engine, we shall now consider the steady operation of the engine shown schematically in Fig. 2.6.1. For this analysis we consider an ideal engine having all thermodynamically reversible components. For this engine, the flow through the inlet and compressor is modeled as reversible adiabatic compression. The process in the combustion chamber is modeled as heat addition at constant stagnation pressure. The flow through the turbine and exhaust nozzle is modeled as a reversible adiabatic expansion back to ambient pressure. In terms of the station labeling shown in Fig. 2.6.1, the following assumptions are made for this ideal cycle analysis.

1. The working fluid is air, which is assumed to behave as an ideal gas with constant specific heat.

2. The flow from the freestream to the compressor inlet is assumed to be isentropic. This assumption requires that

$$\frac{T_{0\infty}}{T_\infty} = \left(\frac{p_{0\infty}}{p_\infty}\right)^{\frac{\gamma-1}{\gamma}} = 1 + \frac{\gamma-1}{2}M_\infty^2 \qquad (2.7.1)$$

$$\frac{T_{01}}{T_{0\infty}} = \left(\frac{p_{01}}{p_{0\infty}}\right)^{\frac{\gamma-1}{\gamma}} \qquad (2.7.2)$$

3. The flow through the compressor is also assumed to be isentropic, which requires that

$$\frac{T_{02}}{T_{01}} = \left(\frac{p_{02}}{p_{01}}\right)^{\frac{\gamma-1}{\gamma}} \qquad (2.7.3)$$

4. The process in the combustion chamber is assumed to take place at constant stagnation pressure, and the fuel flow rate is assumed to be negligible compared to that of the air. Thus, from the combustion model we have

$$\dot{m}_e = \dot{m}_i \equiv \dot{m} \qquad (2.7.4)$$

$$p_{03} = p_{02} \qquad (2.7.5)$$

5. The turbine flow is assumed to be isentropic and the turbine shaft work rate, $\dot{W}_{3-4}$, is input directly to the compressor, with no loss. This requires that

$$\frac{T_{04}}{T_{03}} = \left(\frac{p_{04}}{p_{03}}\right)^{\frac{\gamma-1}{\gamma}} \qquad (2.7.6)$$

$$\dot{W}_{1-2} + \dot{W}_{3-4} = 0 \qquad (2.7.7)$$

6. The exhaust nozzle is modeled as an isentropic expansion from the turbine outlet pressure to ambient pressure, which requires that

$$\frac{T_{0e}}{T_{04}} = \left(\frac{p_{0e}}{p_{04}}\right)^{\frac{\gamma-1}{\gamma}} \tag{2.7.8}$$

$$\frac{T_{0e}}{T_e} = \left(\frac{p_{0e}}{p_e}\right)^{\frac{\gamma-1}{\gamma}} = 1 + \frac{\gamma-1}{2}M_e^2 \tag{2.7.9}$$

$$p_e = p_\infty \tag{2.7.10}$$

Since there is no shaft work or heat transfer associated with the airflow from the freestream to the compressor inlet, the first law of thermodynamics applied to this portion of the cycle requires that

$$\dot{m}\left[C_p(T_1 - T_\infty) + \tfrac{1}{2}(V_1^2 - V_\infty^2)\right] = \dot{m}C_p(T_{01} - T_{0\infty}) = 0$$

Combining this result with Eqs. (2.7.1) and (2.7.2), from known values of the operating conditions, T_∞, p_∞, and M_∞, the stagnation temperature and pressure at the compressor inlet are readily determined:

$$T_{01} = T_{0\infty} = T_\infty\left(1 + \frac{\gamma-1}{2}M_\infty^2\right) \tag{2.7.11}$$

$$p_{01} = p_{0\infty} = p_\infty\left(1 + \frac{\gamma-1}{2}M_\infty^2\right)^{\frac{\gamma}{\gamma-1}} \tag{2.7.12}$$

A very important design parameter for the turbojet engine is the pressure ratio across the compressor. Because the stagnation conditions at the compressor inlet are now known, the stagnation temperature and pressure at the compressor outlet can be determined directly from knowledge of this design parameter and Eq. (2.7.3):

$$T_{02} = T_{01}\left(\frac{p_{02}}{p_{01}}\right)^{\frac{\gamma-1}{\gamma}} \tag{2.7.13}$$

$$p_{02} = p_{01}\frac{p_{02}}{p_{01}} \tag{2.7.14}$$

Since there is no shaft work during combustion, using Eq. (2.7.4), the first law of thermodynamics applied to the combustion chamber yields

$$\dot{Q}_{2-3} = \dot{m}[C_p(T_3 - T_2) + \tfrac{1}{2}(V_3^2 - V_2^2)] = \dot{m}C_p(T_{03} - T_{02})$$

where $\dot{Q}_{2-3}$ is the rate of heat transfer in the combustion chamber, which results directly from the fuel that is burned. Assuming complete combustion, this heat transfer rate is known from the fuel flow rate imposed by the throttle setting and the enthalpy of combustion for the fuel. With this knowledge, stagnation temperature and pressure at the turbine inlet are obtained from this first law energy balance combined with Eq. (2.7.5) and the known combustion chamber inlet conditions:

$$T_{03} = T_{02} + \frac{\dot{Q}_{2-3}}{\dot{m}C_p} \qquad (2.7.15)$$

$$p_{03} = p_{02} \qquad (2.7.16)$$

Applying the first law of thermodynamics from the compressor inlet to the turbine outlet and separately for the exhaust nozzle alone yields

$$\dot{m}C_p(T_{04} - T_{01}) + \dot{W}_{1-2} + \dot{W}_{3-4} = \dot{Q}_{2-3}$$

$$\dot{m}C_p(T_{0e} - T_{04}) = 0$$

Combining these relations with Eqs. (2.7.6) through (2.7.8), the stagnation temperature and pressure in the exhaust nozzle are easily determined,

$$T_{0e} = T_{04} = T_{01} + \frac{\dot{Q}_{2-3}}{\dot{m}C_p} \qquad (2.7.17)$$

$$p_{0e} = p_{04} = p_{03}\left(\frac{T_{04}}{T_{03}}\right)^{\frac{\gamma}{\gamma-1}} \qquad (2.7.18)$$

Now that the stagnation pressure is known at the nozzle exit, the Mach number and static temperature at the exit can be obtained directly from Eqs. (2.7.9) and (2.7.10),

$$M_e^2 = \frac{2}{\gamma-1}\left[\left(\frac{p_{0e}}{p_\infty}\right)^{\frac{\gamma-1}{\gamma}} - 1\right] \qquad (2.7.19)$$

$$T_e = T_{0e}\left(\frac{p_\infty}{p_{0e}}\right)^{\frac{\gamma-1}{\gamma}} \qquad (2.7.20)$$

Three separate efficiencies are commonly defined for a turbojet engine. Because a turbojet has no net shaft work output, the thermal efficiency for a turbojet engine is traditionally defined as the net rate of increase in the kinetic energy of the working fluid divided by the rate of energy input. In view of Eq. (2.7.4), the **thermal efficiency** for the ideal turbojet engine is defined as

$$\eta_T \equiv \frac{\dot{m}(V_e^2 - V_\infty^2)/2}{\dot{Q}_{2-3}} \tag{2.7.21}$$

For an aircraft engine, propulsive power is a more significant measure of useful power output. Work is force multiplied by distance, and power is force multiplied by the rate of change of distance (i.e., velocity). Thus, the uninstalled propulsive power for a turbojet engine is defined as the uninstalled thrust multiplied by the freestream velocity. The uninstalled **propulsive efficiency** for a turbojet engine is therefore traditionally defined as the uninstalled propulsive power divided by the net rate of change in the kinetic energy of the working fluid,

$$\eta_P \equiv \frac{FV_\infty}{\dot{m}(V_e^2 - V_\infty^2)/2} \tag{2.7.22}$$

The uninstalled **overall efficiency** for a turbojet engine is defined as the uninstalled propulsive power divided by the rate of energy input,

$$\eta_O \equiv \frac{FV_\infty}{\dot{Q}_{2-3}} = \eta_T \eta_P \tag{2.7.23}$$

The net rate of increase in kinetic energy imparted to the working fluid that passes through an ideal turbojet engine is

$$\frac{\dot{m}}{2}(V_e^2 - V_\infty^2) = \frac{\dot{m}}{2}\gamma RT_\infty \left(M_e^2 \frac{T_e}{T_\infty} - M_\infty^2 \right) \tag{2.7.24}$$

Using Eqs. (2.7.11) through (2.7.20) and some algebraic manipulation, this result can be rearranged to give

$$\frac{\dot{m}}{2}(V_e^2 - V_\infty^2) = \dot{Q}_{2-3}\left(1 - \frac{T_\infty}{T_{02}} \right) = \dot{Q}_{2-3}\left[1 - \left(1 + \frac{\gamma-1}{2}M_\infty^2 \right)^{-1}\left(\frac{p_{01}}{p_{02}} \right)^{\frac{\gamma-1}{\gamma}} \right]$$

and from the definition in Eq. (2.7.21), the thermal efficiency is

$$\eta_T = 1 - \frac{T_\infty}{T_{02}} = 1 - \left(1 + \frac{\gamma-1}{2}M_\infty^2 \right)^{-1}\left(\frac{p_{01}}{p_{02}} \right)^{\frac{\gamma-1}{\gamma}} \tag{2.7.25}$$

From Eq. (2.7.25), it can be seen that the thermal efficiency of an ideal turbojet engine depends only on the freestream Mach number and the pressure ratio across the compressor. At any given flight Mach number, the thermal efficiency increases as the pressure ratio across the compressor is increased. Furthermore, for any specified compressor pressure ratio, the thermal efficiency increases with flight Mach number. One important observation that should be made from Eq. (2.7.25) is that a high pressure ratio is required to produce a reasonable thermal efficiency at low flight Mach numbers. Even with a pressure ratio of 30, the static thermal efficiency of an ideal turbojet engine is only about 62 percent. For this reason turbojet engines do not provide a particularly efficient means of propulsion for low-speed flight. However, at high Mach numbers the thermal efficiency is excellent.

Using Eqs. (2.7.4) and (2.7.10) with Eq. (2.6.3), the uninstalled thrust for the ideal turbojet engine is

$$F = \dot{m}(V_e - V_\infty) = \dot{m}\sqrt{\gamma R T_\infty}\left(M_e\sqrt{\frac{T_e}{T_\infty}} - M_\infty\right) \quad (2.7.26)$$

The size and weight of a turbojet engine is approximately proportional to the flow rate that must be handled. Since thrust-to-weight ratio is an important consideration for aircraft engines, an important performance parameter for turbojet engines is thrust per unit mass flow rate, which is commonly called *specific thrust*. After applying Eqs. (2.7.11) through (2.7.20) to the result expressed in Eq. (2.7.26), the uninstalled specific thrust for the ideal turbojet engine can be written as

$$\frac{F}{\dot{m}} = \sqrt{\gamma R T_\infty}\left\{\sqrt{\frac{2}{\gamma-1}\left[\left(\frac{p_{04}}{p_{03}}\right)^{\frac{\gamma-1}{\gamma}} - \left(\frac{p_{01}}{p_{02}}\right)^{\frac{\gamma-1}{\gamma}}\left(1+\frac{\gamma-1}{2}M_\infty^2\right)^{-1}\right]\frac{T_{03}}{T_\infty}} - M_\infty\right\} \quad (2.7.27)$$

The propulsive efficiency for an ideal turbojet engine is found by applying Eqs. (2.7.24) and (2.7.26) to the definition in Eq. (2.7.22). This gives

$$\eta_P = \frac{2\,M_\infty}{M_e\sqrt{T_e/T_\infty} + M_\infty} = \frac{2V_\infty}{V_e + V_\infty} \quad (2.7.28)$$

We see from Eq. (2.7.25) that for a given ambient temperature, only by increasing the compressor outlet temperature can the thermal efficiency of an ideal turbojet engine be increased. From Eq. (2.7.27), we see that increasing turbine inlet temperature will increase specific thrust for a given pressure ratio and operating condition. Thus, high thermal efficiency and specific thrust require high operating temperatures. The maximum temperature in a turbojet engine without afterburning occurs at the turbine inlet. As a result, the performance of turbojet engines is limited by the maximum temperature that can be endured by the materials used in the construction of the turbine blades. This limits the amount of fuel that can be burned in the combustion chamber.

EXAMPLE 2.7.1. From Eqs. (2.7.25) and (2.7.27) it is seen that the pressure ratio across the compressor is an important parameter in determining the performance characteristics of a turbojet engine. However, with everything else held constant, increasing this pressure ratio will increase the turbine inlet temperature. Since maximum turbine temperature is limited by material considerations, we wish to examine the variation in ideal turbojet performance with compressor pressure ratio, under the restriction of fixed turbine inlet temperature. Consider an ideal turbojet operating under standard temperature and pressure at 30,000 feet. Plot the thermal efficiency and specific thrust as a function of compressor pressure ratio for flight Mach numbers ranging from 0.0 to 3.0. Assume that the stagnation temperature at the turbine inlet is limited to 3,000 °R.

Solution. For the specified operating conditions we have

$$T_\infty = 411.9\,°R, \quad p_\infty = 4.37\text{ psi}, \quad T_{03} = 3{,}000\,°R,$$
$$C_p = 0.24\text{ Btu/lbm·°R}, \quad \gamma = 1.4$$

As an example calculation, consider a flight Mach number of 0.5 and a compressor pressure ratio of 10. From Eqs. (2.7.11) through (2.7.20) we obtain

$$T_{01} = T_{0\infty} = T_\infty\left(1+\frac{\gamma-1}{2}M_\infty^2\right) = 432.5\,°R$$

$$p_{01} = p_{0\infty} = p_\infty\left(1+\frac{\gamma-1}{2}M_\infty^2\right)^{\frac{\gamma}{\gamma-1}} = 5.18\text{ psi}$$

$$T_{02} = T_{01}\left(\frac{p_{02}}{p_{01}}\right)^{\frac{\gamma-1}{\gamma}} = 835.0\,°R$$

$$p_{02} = p_{01}\frac{p_{02}}{p_{01}} = 51.84\text{ psi}$$

$$\frac{\dot{Q}_{2-3}}{\dot{m}C_p} = T_{03} - T_{02} = 2{,}165.0\,°R$$

$$p_{03} = p_{02} = 51.84\text{ psi}$$

$$T_{0e} = T_{04} = T_{01} + \frac{\dot{Q}_{2-3}}{\dot{m}C_p} = 2{,}597.5\,°R$$

$$p_{0e} = p_{04} = p_{03}\left(\frac{T_{04}}{T_{03}}\right)^{\frac{\gamma}{\gamma-1}} = 31.31\text{ psi}$$

$$M_e = \sqrt{\frac{2}{\gamma-1}\left[\left(\frac{p_{0e}}{p_\infty}\right)^{\frac{\gamma-1}{\gamma}}-1\right]} = 1.94$$

$$T_e = T_{0e}\left(\frac{p_\infty}{p_{0e}}\right)^{\frac{\gamma-1}{\gamma}} = 1,479.9\,°R$$

From Eq. (2.7.25), the thermal efficiency is

$$\eta_T = 1-\frac{T_\infty}{T_{02}} = \underline{0.507}$$

The ideal gas constant is

$$R = \frac{\gamma-1}{\gamma}C_p = 0.06857\frac{\text{Btu}}{\text{lbm·°R}} = 1,716\frac{\text{ft·lbf}}{\text{slug·°R}} = 1,716\frac{\text{ft}^2}{\text{sec}^2\text{·°R}}$$

From Eq. (2.7.26), the specific thrust is

$$\frac{F}{\dot{m}} = \sqrt{\gamma RT_\infty}\left(M_e\sqrt{\frac{T_e}{T_\infty}}-M_\infty\right) = 3,166\frac{\text{ft}}{\text{sec}} = 3,166\frac{\text{lbf}}{\text{slug/sec}} = \underline{98.4\frac{\text{lbf}}{\text{lbm/sec}}}$$

Similarly, we obtain the results shown in Figs. 2.7.1 and 2.7.2.

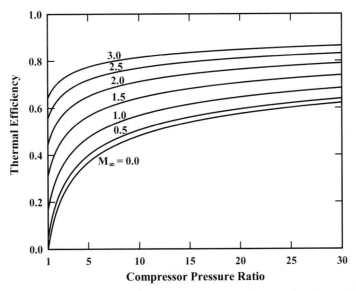

Figure 2.7.1. Thermal efficiency versus compressor pressure ratio for Example 2.7.1.

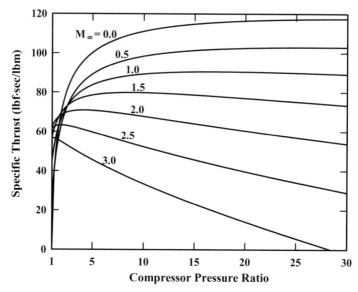

Figure 2.7.2. Specific thrust versus compressor pressure ratio for Example 2.7.1.

From the results of Example 2.7.1 it is seen that when the turbine inlet temperature is held constant, specific thrust exhibits a maximum in its variation with compressor pressure ratio. This is because as the compressor pressure ratio is increased, the burner inlet temperature is also increased. This means that the fuel-air ratio must be decreased to avoid overheating the turbine. If the compressor pressure ratio were large enough, the maximum allowable temperature would be attained at the compressor outlet and any addition of fuel would overheat the turbine. Thus, a turbojet engine with high compressor pressure ratio cannot produce thrust at high Mach numbers without exceeding the maximum allowable temperature in the turbine. Even though thermal efficiency is increased with high compressor pressure ratio, the accompanying decrease in specific thrust at high Mach number makes turbojet engines with high compressor pressure ratio impractical for supersonic flight.

From Eq. (2.7.27) it can be shown with simple calculus that the compressor pressure ratio that produces a maximum in specific thrust for a particular turbine inlet temperature and flight Mach number is given by

$$
\left(\frac{p_{02}}{p_{01}}\right)_{\max(F/\dot{m})} = \left[\left(1+\frac{\gamma-1}{2}M_\infty^2\right)^{-1}\sqrt{\frac{T_{03}}{T_\infty}}\right]^{\frac{\gamma}{\gamma-1}} \tag{2.7.29}
$$

This optimum compressor pressure ratio decreases rapidly with increasing Mach number in supersonic flight. For subsonic flight, high compressor pressure ratio is desirable to attain better thermal efficiency and high specific thrust. However, for supersonic flight lower compressor pressure ratios are typically used to obtain higher specific thrust.

2.8. The Turbojet Engine with Afterburner

Burning additional fuel aft of the turbine in a turbojet engine will increase the specific thrust that can be developed without exceeding maximum allowable turbine temperature. The aft combustion chamber shown schematically in Fig. 2.8.1 is called an *afterburner*. The maximum temperature in the afterburner can be higher than that leaving the primary combustion chamber, because the products of combustion from the afterburner do not pass through a turbine.

Thermodynamic cycle analysis for the ideal turbojet engine with an afterburner is identical to that presented in Sec. 2.7, up through the turbine outlet at state 4. Therefore, Eqs. (2.7.11) through (2.7.16) still apply and the stagnation temperature and pressure at the turbine outlet are

$$T_{04} = T_{01} + \frac{\dot{Q}_{2-3}}{\dot{m}C_p} \tag{2.8.1}$$

$$p_{04} = p_{03}\left(\frac{T_{04}}{T_{03}}\right)^{\frac{\gamma}{\gamma-1}} \tag{2.8.2}$$

If we continue to assume an ideal or thermodynamically reversible engine, the stagnation pressure remains constant aft of the turbine and the stagnation temperature is determined from the first law. Thus, the stagnation conditions aft of the turbine are

$$T_{0e} = T_{05} = T_{04} + \frac{\dot{Q}_{4-5}}{\dot{m}C_p} \tag{2.8.3}$$

$$p_{0e} = p_{05} = p_{04} \tag{2.8.4}$$

Since the afterburner does not affect either the static or stagnation pressure at the exit of an ideal turbojet engine, the exit Mach number is not changed as a result of afterburning.

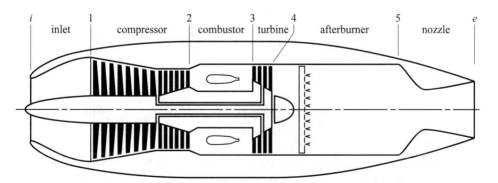

Figure 2.8.1. Components and station labeling for the turbojet engine with afterburner.

The exit velocity and thrust are increased by the afterburner only as a result of the increase in exit temperature and the accompanying increase in the speed of sound. Thus, Eqs. (2.7.19) and (2.7.20) still apply to the ideal turbojet engine with afterburner, provided that the increased stagnation temperature at the exit, which is computed from Eq. (2.8.3), is used in Eq. (2.7.20).

When the increased exit temperature is used, the kinetic energy increase can still be computed from Eq. (2.7.24), and the thermal efficiency for an ideal turbojet engine with afterburner is

$$\eta_T \equiv \frac{\dot{m}(V_e^2 - V_\infty^2)/2}{\dot{Q}_{2-3} + \dot{Q}_{4-5}} = \frac{\dot{m}\gamma RT_\infty [M_e^2(T_e/T_\infty) - M_\infty^2]/2}{\dot{Q}_{2-3} + \dot{Q}_{4-5}}$$

From Eqs. (2.8.1) and (2.8.3) the total energy input can be expressed in terms of the stagnation temperature difference between states 1 and 5,

$$\dot{Q}_{2-3} + \dot{Q}_{4-5} = \dot{m}C_p(T_{05} - T_{01})$$

and the thermal efficiency is written

$$\eta_T = \frac{\gamma RT_\infty [M_e^2(T_e/T_\infty) - M_\infty^2]/2}{C_p(T_{05} - T_{01})} = \frac{\gamma - 1}{2}\left[M_e^2 \frac{T_e}{T_\infty} - M_\infty^2\right]\frac{T_\infty}{T_{05} - T_{01}} \qquad (2.8.5)$$

Similarly, the uninstalled thrust for an ideal turbojet engine with afterburner can still be computed from Eq. (2.7.26). Thus, specific thrust is readily computed from the known exit temperature and Mach number,

$$\frac{F}{\dot{m}} = \sqrt{\gamma RT_\infty}\left(M_e\sqrt{\frac{T_e}{T_\infty}} - M_\infty\right) \qquad (2.8.6)$$

EXAMPLE 2.8.1. To examine the effects of an afterburner on the performance of an ideal turbojet engine, repeat Example 2.7.1 for the same operating conditions but with afterburner included. Assume that the maximum stagnation temperature aft of the turbine is limited to 4,000 °R.

Solution. For the specified operating conditions we have

$$T_\infty = 411.9\,°R, \quad p_\infty = 4.37\text{ psi}, \quad T_{03} = 3,000\,°R, \quad T_{05} = 4,000\,°R,$$
$$C_p = 0.24\text{ Btu/lbm·°R}, \quad \gamma = 1.4$$

As an example calculation, again we consider a flight Mach number of 0.5 and a compressor pressure ratio of 10. Following Example 2.7.1, from Eqs. (2.7.11) through (2.7.16) we obtain

$$T_{01} = 432.5\,°\text{R} \quad \text{and} \quad p_{01} = 5.18\,\text{psi}$$

$$T_{02} = 835.0\,°\text{R} \quad \text{and} \quad p_{03} = p_{02} = 51.84\,\text{psi}$$

$$\frac{\dot{Q}_{2-3}}{\dot{m}C_p} = T_{03} - T_{02} = 2{,}165.0\,°\text{R}$$

From Eqs. (2.8.1) through (2.8.4) we obtain

$$T_{04} = T_{01} + \frac{\dot{Q}_{2-3}}{\dot{m}C_p} = 2{,}597.5\,°\text{R}$$

$$p_{0e} = p_{05} = p_{04} = p_{03}\left(\frac{T_{04}}{T_{03}}\right)^{\frac{\gamma}{\gamma-1}} = 31.31\,\text{psi}$$

$$\frac{\dot{Q}_{4-5}}{\dot{m}C_p} = T_{05} - T_{04} = 1{,}402.5\,°\text{R}$$

$$T_{0e} = T_{05} = 4{,}000\,°\text{R}$$

From Eqs. (2.7.19) and (2.7.20) we obtain

$$M_e = \sqrt{\frac{2}{\gamma-1}\left[\left(\frac{p_{0e}}{p_\infty}\right)^{\frac{\gamma-1}{\gamma}} - 1\right]} = 1.94$$

$$T_e = T_{0e}\left(\frac{p_\infty}{p_{0e}}\right)^{\frac{\gamma-1}{\gamma}} = 2{,}278.9\,°\text{R}$$

From Eq. (2.8.5), the thermal efficiency is

$$\eta_T = \frac{\gamma-1}{2}\left[M_e^2\frac{T_e}{T_\infty} - M_\infty^2\right]\frac{T_\infty}{T_{05} - T_{01}} = \underline{0.477}$$

and from Eq. (2.8.6), the specific thrust is

$$\frac{F}{\dot{m}} = \sqrt{\gamma R T_\infty}\left(M_e\sqrt{\frac{T_e}{T_\infty}} - M_\infty\right) = 4{,}049\,\frac{\text{lbf}}{\text{slug/sec}} = \underline{125.8\,\frac{\text{lbf}}{\text{lbm/sec}}}$$

Repeating similar computations for different values of compressor pressure ratio and different values of flight Mach number, we obtain the results that are shown in Figs. 2.8.2 and 2.8.3.

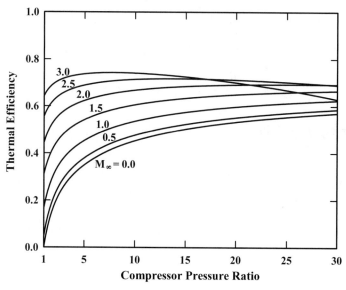

Figure 2.8.2. Thermal efficiency versus compressor pressure ratio for Example 2.8.1.

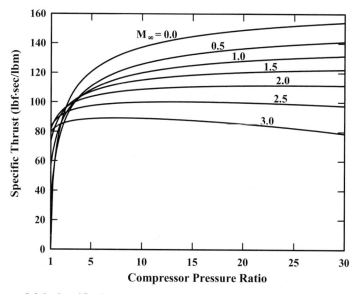

Figure 2.8.3. Specific thrust versus compressor pressure ratio for Example 2.8.1.

By comparing the results presented in Example 2.8.1 with those that were presented in Example 2.7.1, we see that afterburning increases specific thrust but reduces thermal efficiency. It can also be seen that afterburning increases the value of compressor pressure ratio that results in maximum specific thrust.

It can be shown that the compressor pressure ratio that produces maximum specific thrust for an ideal turbojet engine with afterburning depends only on the flight Mach number, ambient temperature, and stagnation temperature at the turbine inlet. The result is given by

$$
\left(\frac{p_{02}}{p_{01}}\right)_{\max(F/\dot{m})} = \left\{\frac{1}{2}\left[\left(1+\frac{\gamma-1}{2}M_\infty^2\right)^{-1}\frac{T_{03}}{T_\infty}+1\right]\right\}^{\frac{\gamma}{\gamma-1}}
\tag{2.8.7}
$$

This optimum compressor pressure ratio is significantly higher than the optimum value without afterburning.

2.9. Turbofan Engines

The thrust developed by any propulsion system can be increased either by increasing the mass flow rate passing through the propulsion system or by increasing the velocity increment imparted to this flow. A large mass flow rate with a small velocity increment provides the best propulsive efficiency, while a larger velocity increment with a smaller mass flow rate will generally result in a higher thrust-to-weight ratio. Because of their large diameter, engine-propeller systems produce relatively high flow rates with small velocity increments and have correspondingly high propulsive efficiency. The turbojet engine, on the other hand, produces a fairly large velocity increment with a relatively low mass flow rate, providing high specific thrust. A turbofan engine is a compromise between the propeller and the turbojet. It has a diameter and flow rate that are typically less than those for an equivalent engine-propeller combination but greater than those for a turbojet of similar power. Turbofan engines are commonly used on subsonic transport aircraft, which require high efficiency for longer range. Turbojet engines are more commonly used on fighters, where thrust-to-weight ratio is typically more important than propulsive efficiency and higher speed is desired.

As shown schematically in Fig. 2.9.1, the turbofan engine has a larger turbine than a turbojet engine of similar power. As the products of combustion pass through this larger turbine, more energy is extracted and the exhaust gas velocity is reduced. The extra power produced by the turbine is used to turn a ducted fan, which adds mass to the slipstream. Increasing the total mass flow rate while decreasing the velocity increment decreases specific thrust and increases propulsive efficiency. However, a larger turbine, fan, and second nozzle add weight to the engine. It can be shown that the ideal turbofan engine has the same thermal efficiency as the ideal turbojet engine. Thus, the turbofan engine has improved overall efficiency at the cost of reduced thrust-to-weight ratio.

The airflow through a turbofan engine is typically characterized in terms of a design parameter called the *bypass ratio*. This is the ratio of the mass flow rate bypassing the core, $\dot{m}_b$, to the mass flow rate passing through the core, $\dot{m}_c$,

$$
\text{bypass ratio} \equiv R_b \equiv \frac{\dot{m}_b}{\dot{m}_c}
\tag{2.9.1}
$$

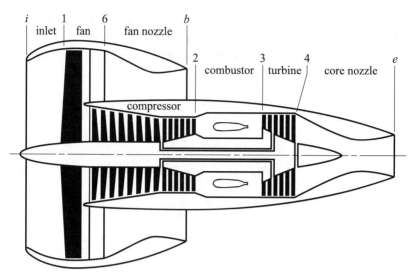

Figure 2.9.1. Components and station labeling for the turbofan engine.

Bypass ratio is sometimes given the symbol α. However, in this text, the symbol R_b is used to avoid confusion with angle of attack.

For thermodynamic cycle analysis of the ideal turbofan engine, we assume isentropic flow through the fan and fan nozzle as well as through the compressor, turbine, and core nozzle. As was the case with the ideal turbojet analysis, we assume complete expansion back to ambient pressure for both the core nozzle and the fan nozzle.

For the flow through the core, Eqs. (2.7.11) through (2.7.16) apply to the turbofan engine as well as to the turbojet. Like the compressor, the fan design is characterized by the stagnation pressure ratio. The stagnation conditions in the fan nozzle can easily be determined from this design parameter and the assumption of isentropic flow,

$$T_{0b} = T_{06} = T_{01}\left(\frac{p_{06}}{p_{01}}\right)^{\frac{\gamma-1}{\gamma}} \qquad (2.9.2)$$

$$p_{0b} = p_{06} = p_{01}\frac{p_{06}}{p_{01}} \qquad (2.9.3)$$

The stagnation temperature at the turbine outlet is obtained for the first law of thermodynamics applied to both the core flow and the fan flow,

$$\dot{m}_c C_p(T_{04} - T_{01}) + \dot{m}_b C_p(T_{06} - T_{01}) = \dot{Q}_{2-3}$$

Combining this relation with the assumption of isentropic flow in the core nozzle, the stagnation conditions in this nozzle are

$$T_{0e} = T_{04} = T_{01} - R_b(T_{06} - T_{01}) + \frac{\dot{Q}_{2-3}}{\dot{m}_c C_p} \qquad (2.9.4)$$

$$p_{0e} = p_{04} = p_{03} \left(\frac{T_{04}}{T_{03}} \right)^{\frac{\gamma}{\gamma-1}} \qquad (2.9.5)$$

The Mach number and temperature at the core and fan nozzle exits are obtained from the assumption of isentropic flow with complete expansion. From the known stagnation conditions, this gives

$$M_e^2 = \frac{2}{\gamma-1} \left[\left(\frac{p_{0e}}{p_\infty} \right)^{\frac{\gamma-1}{\gamma}} - 1 \right] \qquad (2.9.6)$$

$$T_e = T_{0e} \left(\frac{p_\infty}{p_{0e}} \right)^{\frac{\gamma-1}{\gamma}} \qquad (2.9.7)$$

$$M_b^2 = \frac{2}{\gamma-1} \left[\left(\frac{p_{0b}}{p_\infty} \right)^{\frac{\gamma-1}{\gamma}} - 1 \right] \qquad (2.9.8)$$

$$T_b = T_{0b} \left(\frac{p_\infty}{p_{0b}} \right)^{\frac{\gamma-1}{\gamma}} \qquad (2.9.9)$$

The total rate of increase in kinetic energy is the sum of that for the core stream and that for the bypass stream,

$$\frac{\dot{m}_c(V_e^2 - V_\infty^2) + \dot{m}_b(V_b^2 - V_\infty^2)}{2} = \frac{\dot{m}_c}{2} \gamma R T_\infty \left[M_e^2 \frac{T_e}{T_\infty} - M_\infty^2 + R_b \left(M_b^2 \frac{T_b}{T_\infty} - M_\infty^2 \right) \right] \qquad (2.9.10)$$

The only energy input for this engine is in the combustion chamber of the core. Thus, after applying Eq. (2.9.4) and recalling that $R = C_p(\gamma-1)/\gamma$, the thermal efficiency for the ideal turbofan engine is

$$\eta_T \equiv \frac{\dot{m}_c(V_e^2 - V_\infty^2) + \dot{m}_b(V_b^2 - V_\infty^2)}{2\dot{Q}_{2-3}} = \frac{(\gamma-1)[M_e^2 T_e + R_b M_b^2 T_b - (1+R_b) M_\infty^2 T_\infty]}{2[T_{04} + R_b T_{06} - (1+R_b) T_{01}]}$$

Applying Eqs. (2.7.11) through (2.7.16) and Eqs. (2.9.2) through (2.9.9), this result can be expressed as

$$\eta_T = 1 - \left(1 + \frac{\gamma-1}{2}M_\infty^2\right)^{-1}\left(\frac{p_{01}}{p_{02}}\right)^{\frac{\gamma-1}{\gamma}} \qquad (2.9.11)$$

which is exactly the same result as that obtained in Sec. 2.7 for the turbojet engine. Notice from Eq. (2.9.11) that the thermal efficiency of an ideal turbofan engine does not depend on either the bypass ratio or the pressure ratio across the fan. At first the reader may find this surprising, since the net rate of increase in kinetic energy for the turbofan engine depends on the output of the fan. This result is easily justified from simple thermodynamics. The turbine of a turbofan engine extracts power from the core stream and transfers it to the bypass stream by driving the fan. Since the engine is assumed to be thermodynamically reversible, this power transfer takes place without loss and the net power output defined in terms of kinetic energy is unchanged.

Since the ideal turbofan engine has complete expansion in both the core and fan nozzles, the uninstalled thrust for this engine is

$$F = \dot{m}_c(V_e - V_\infty) + \dot{m}_b(V_b - V_\infty)$$
$$= \dot{m}_c\sqrt{\gamma RT_\infty}\left[M_e\sqrt{\frac{T_e}{T_\infty}} - M_\infty + R_b\left(M_b\sqrt{\frac{T_b}{T_\infty}} - M_\infty\right)\right] \qquad (2.9.12)$$

and the specific thrust is

$$\frac{F}{\dot{m}_c + \dot{m}_b} = \frac{\sqrt{\gamma RT_\infty}}{1 + R_b}\left[M_e\sqrt{\frac{T_e}{T_\infty}} - M_\infty + R_b\left(M_b\sqrt{\frac{T_b}{T_\infty}} - M_\infty\right)\right] \qquad (2.9.13)$$

From Eqs. (2.9.10) and (2.9.12), the propulsive efficiency for an ideal turbofan engine is

$$\eta_P \equiv \frac{2FV_\infty}{\dot{m}_c(V_e^2 - V_\infty^2) + \dot{m}_b(V_b^2 - V_\infty^2)} = \frac{2V_\infty[V_e - V_\infty + R_b(V_b - V_\infty)]}{V_e^2 - V_\infty^2 + R_b(V_b^2 - V_\infty^2)} \qquad (2.9.14)$$

Since turbine inlet temperature is limited by material considerations, the available energy per unit mass in the core flow is limited as well. This places an upper limit on the fan pressure ratio that can be used with a particular bypass ratio. This upper limit is reached when expansion in the turbine is sufficient to reduce the Mach number at the core nozzle exit to zero. Once this condition has been reached, no additional power could possibly be extracted from the core flow by using a larger turbine to drive a larger fan. From Eq. (2.9.6), it can be seen that this limiting condition requires that

$$p_{0e} = p_\infty$$

Using this limit with Eqs. (2.7.11) through (2.7.16) and Eqs. (2.9.2) through (2.9.5), it is readily shown that the upper limit for the fan pressure ratio is given by

$$\left(\frac{p_{06}}{p_{01}}\right)_{\max} = \left[\frac{1+R_b-\left(p_{02}/p_{01}\right)^{\frac{\gamma-1}{\gamma}}}{R_b} + \frac{\frac{T_{03}}{T_\infty}\left[1+\frac{\gamma-1}{2}M_\infty^2-\left(p_{01}/p_{02}\right)^{\frac{\gamma-1}{\gamma}}\right]}{R_b\left(1+\frac{\gamma-1}{2}M_\infty^2\right)^2}\right]^{\frac{\gamma}{\gamma-1}} \qquad (2.9.15)$$

The turbofan engine has three important design parameters that can be varied to help meet the requirements of any particular application. These are the compressor pressure ratio, the fan pressure ratio, and the bypass ratio. The turbojet engine can be considered to be a special case of the turbofan engine, which has a bypass ratio of zero. While the thermal efficiency of the ideal turbofan engine depends only on compressor pressure ratio and flight Mach number, the bypass ratio and fan pressure ratio will significantly affect the propulsive efficiency and thus the overall efficiency of the engine. The manner in which the propulsive efficiency of an ideal turbofan engine varies with bypass ratio and fan pressure ratio is not at all obvious from inspection of Eq. (2.9.14). This is examined in the following example.

EXAMPLE 2.9.1. To explore how turbofan performance varies with bypass ratio and fan pressure ratio, consider an ideal turbofan engine operating under standard temperature and pressure at 30,000 feet. To isolate the effects of the chosen design parameters, assume a fixed flight Mach number of 0.8, a fixed compressor pressure ratio of 20, and a stagnation temperature at the turbine inlet of 3,000 °R. For these conditions, plot overall efficiency and specific thrust as a function of fan pressure ratio for several bypass ratios ranging from 0 to 12.

Solution. From the specified operating conditions and fixed design parameters, the following values are to be held constant:

$$T_\infty = 411.8\,°R, \quad p_\infty = 4.37\ \text{psi}, \quad M_\infty = 0.8, \quad p_{02}/p_{01} = 20,$$
$$T_{03} = 3,000\,°R, \quad C_p = 0.24\ \text{Btu/lbm}\cdot°R, \quad \gamma = 1.4$$

As an example calculation, consider a bypass ratio of 4.0 and a fan pressure ratio of 3.0. From Eqs. (2.7.11) through (2.7.16) we obtain

$$T_{01} = T_{0\infty} = T_\infty\left(1+\frac{\gamma-1}{2}M_\infty^2\right) = 464.5\,°R$$

$$p_{01} = p_{0\infty} = p_\infty\left(1+\frac{\gamma-1}{2}M_\infty^2\right)^{\frac{\gamma}{\gamma-1}} = 6.66\ \text{psi}$$

$$T_{02} = T_{01}\left(\frac{p_{02}}{p_{01}}\right)^{\frac{\gamma-1}{\gamma}} = 1,093.2\,°R$$

$$p_{02} = p_{01} \frac{p_{02}}{p_{01}} = 133.23 \text{ psi}$$

$$\frac{\dot{Q}_{2-3}}{\dot{m}_c C_p} = T_{03} - T_{02} = 1{,}906.8 \text{ }^\circ\text{R}$$

$$p_{03} = p_{02} = 133.23 \text{ psi}$$

and Eqs. (2.9.2) through (2.9.9) result in

$$T_{0b} = T_{06} = T_{01}\left(\frac{p_{06}}{p_{01}}\right)^{\frac{\gamma-1}{\gamma}} = 635.8 \text{ }^\circ\text{R}$$

$$p_{0b} = p_{06} = p_{01}\frac{p_{06}}{p_{01}} = 19.98 \text{ psi}$$

$$T_{0e} = T_{04} = T_{01} - R_b(T_{06} - T_{01}) + \frac{\dot{Q}_{2-3}}{\dot{m}_c C_p} = 1{,}686.1 \text{ }^\circ\text{R}$$

$$p_{0e} = p_{04} = p_{03}\left(\frac{T_{04}}{T_{03}}\right)^{\frac{\gamma}{\gamma-1}} = 17.73 \text{ psi}$$

$$M_e = \sqrt{\frac{2}{\gamma-1}\left[\left(\frac{p_{0e}}{p_\infty}\right)^{\frac{\gamma-1}{\gamma}} - 1\right]} = 1.57$$

$$T_e = T_{0e}\left(\frac{p_\infty}{p_{0e}}\right)^{\frac{\gamma-1}{\gamma}} = 1{,}130.0 \text{ }^\circ\text{R}$$

$$M_b = \sqrt{\frac{2}{\gamma-1}\left[\left(\frac{p_{0b}}{p_\infty}\right)^{\frac{\gamma-1}{\gamma}} - 1\right]} = 1.65$$

$$T_b = T_{0b}\left(\frac{p_\infty}{p_{0b}}\right)^{\frac{\gamma-1}{\gamma}} = 411.8 \text{ }^\circ\text{R}$$

From Eq. (2.9.11), the thermal efficiency is

$$\eta_T = 1 - \left(1 + \frac{\gamma-1}{2}M_\infty^2\right)^{-1}\left(\frac{p_{01}}{p_{02}}\right)^{\frac{\gamma-1}{\gamma}} = 0.623$$

From Eq. (2.9.13), the specific thrust is

$$\frac{F}{\dot{m}_c + \dot{m}_b} = \frac{\sqrt{\gamma R T_\infty}}{1 + R_b}\left[M_e \sqrt{\frac{T_e}{T_\infty}} - M_\infty + R_b \left(M_b \sqrt{\frac{T_b}{T_\infty}} - M_\infty \right) \right] = 32.1\frac{\text{lbf}}{\text{lbm}/\text{sec}}$$

From Eq. (2.9.14), the propulsive efficiency is

$$\eta_P = \frac{2V_\infty[V_e - V_\infty + R_b(V_b - V_\infty)]}{V_e^2 - V_\infty^2 + R_b(V_b^2 - V_\infty^2)} = 0.576$$

and the overall efficiency is

$$\eta_O = \eta_T \eta_P = \underline{0.359}$$

Similarly, we obtain the results shown in Figs. 2.9.2 and 2.9.3. The termination of each curve with a bypass ratio greater than 3 corresponds to the maximum possible fan pressure ratio computed from Eq. (2.9.15).

Notice from Fig. 2.9.2 that for each value of bypass ratio, there is an optimum fan pressure ratio that results in the highest overall efficiency. Also notice that the overall efficiency of an engine having the optimum fan pressure ratio is improved as the bypass ratio is increased. For a bypass ratio of 12 the optimum fan pressure ratio is about 1.88, which provides an overall efficiency that is more than 220 percent of that for an ideal

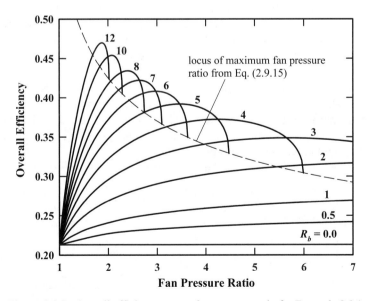

Figure 2.9.2. Overall efficiency versus fan pressure ratio for Example 2.9.1.

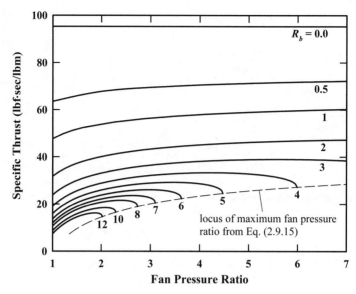

Figure 2.9.3. Specific thrust versus fan pressure ratio for Example 2.9.1.

turbojet engine with the same compressor pressure ratio. The results shown in Fig. 2.9.2 apply only for the ambient conditions, flight Mach number, compressor pressure ratio, and turbine inlet temperature specified in Example 2.9.1. However, the trends shown in this figure apply in general.

From Fig. 2.9.3 it can also be seen that for each value of bypass ratio, there is an optimum fan pressure ratio that will result in the greatest specific thrust. However, the maximum specific thrust that can be attained decreases as the bypass ratio is increased. The maximum specific thrust for the ideal turbofan engine with a bypass ratio of 12 is only about 17 percent of that produced by the ideal turbojet engine with the same compressor pressure ratio.

As it turns out, the fan pressure ratio that produces maximum specific thrust for a fixed bypass ratio is exactly the same as that which gives maximum overall efficiency. This optimum fan pressure ratio can be found by taking the partial derivative of either propulsive efficiency or specific thrust with respect to fan pressure ratio. It can be shown that this optimum fan pressure ratio is given by

$$\left(\frac{p_{06}}{p_{01}}\right)_{opt} = \left[1 - \frac{\left(p_{02}/p_{01}\right)^{\frac{\gamma-1}{\gamma}}}{1+R_b} + \frac{1+\frac{\gamma-1}{2}M_\infty^2 + \frac{T_{03}}{T_\infty}\left[1+\frac{\gamma-1}{2}M_\infty^2 - \left(p_{01}/p_{02}\right)^{\frac{\gamma-1}{\gamma}}\right]}{(1+R_b)\left(1+\frac{\gamma-1}{2}M_\infty^2\right)^2}\right]^{\frac{\gamma}{\gamma-1}}$$

$$(2.9.16)$$

where R_b, M_∞, p_{02}/p_{01}, and T_{03}/T_∞ are all assumed to be fixed.

Choosing the fan pressure ratio according to Eq. (2.9.16) will provide the greatest specific thrust and overall efficiency for a fixed set of the remaining design and operating parameters. However, the best choice for bypass ratio and compressor pressure ratio depends substantially on the specific application. The design requirements that usually have the greatest effect on these two choices are the flight Mach number and desired range. If an airplane is to be designed for very long range, the engine weight will be small compared to the weight of fuel that must be carried. For such designs, overall efficiency is more important than engine weight. Thus, turbofan engines with high bypass ratios and high compressor pressure ratios are typically the optimal solution for long-range subsonic transports. On the other hand, airplanes designed for high supersonic airspeeds and short range are better fitted with turbojets or turbofans having a low bypass ratio. High supersonic Mach numbers also usually dictate lower compressor pressure ratios, because of the reduction in specific thrust with increasing compressor pressure ratios that is shown in Fig. 2.7.2.

2.10. Concluding Remarks

The study of flight mechanics requires some knowledge of propulsion, and this textbook was not written under the assumption that the reader has previously completed an engineering course on propulsion. The very brief introduction to the topic of aircraft propulsion that was presented in this chapter should provide the reader with sufficient background to understand the material in the remainder of this text. However, in this chapter we have only touched on some of the concepts that are important in the design and selection of aircraft propulsion systems. For those students who intend to specialize in flight mechanics, the author would strongly suggest at least one or two engineering courses in the area of propulsion.

In this chapter, more emphasis was placed on propeller-based aircraft propulsion systems than on turbojet and turbofan systems. This may seem inappropriate in view of the fact that turbojet and turbofan propulsion systems are more commonly used on commercial and military aircraft. However, some flight testing is usually very helpful in gaining an understanding of the fundamental principles of flight mechanics. With funding for education being what it is today, it is not likely that many students will have access to turbo-powered aircraft. For this reason, the author has chosen to emphasize propellers and propeller-driven aircraft in the hope that students will experimentally investigate some of the flight mechanics concepts, which are presented in the following chapters. Knowledge gained from flight tests conducted in small general aviation airplanes is easily extended to the larger turbo-powered aircraft. This same philosophy has been applied in the remainder of this textbook.

Flight testing with small propeller-driven airplanes is more affordable than one may think. Many universities have facilities available for such flight testing. Even if your university does not, you will probably be able to locate a general aviation flight school that will be willing to provide this experience for a fee. A small general aviation airplane, with instructor included, can usually be rented at a reasonable hourly rate. Depending on the size of the airplane, two or three students could share the cost and a great deal can be learned from a well-planned flight of 20 to 30 minutes.

2.11. Problems

2.1. Write a computer program to predict the performance of a propeller using Goldstein's vortex theory as presented in Sec. 2.3. Use the secant method to solve Eq. (2.3.32) and apply an appropriate numerical integration method to integrate Eqs. (2.3.33) and (2.3.34). To define the propeller geometry and section aerodynamic coefficients, the program should use a functional relationship that can be easily changed.

2.2. Run the program written for problem 2.1 for the propeller geometry used in Example 2.3.1. Regenerate all of the figures that are presented for this example and compare your results with those presented in the text.

2.3. Run the program written for problem 2.1 for a propeller geometry that has a rectangular planform shape. The geometry is identical to that in Example 2.3.1, except that the blade has a constant chord with the same planform area as the elliptic blade used in the example ($c_b/d_p = 0.0573$). Generate all of the figures that are presented in Example 2.3.1 and compare your results with those obtained for the elliptic blade.

2.4. Modify the computer program written for problem 2.1 so that it also computes the change in normal force coefficient with propeller angle of attack and the change in yawing moment coefficient with respect to propeller angle of attack.

2.5. Run the program written for problem 2.4 for the propeller geometry used in Example 2.3.1. Regenerate the figures that are presented for Example 2.5.1 and compare your results with those presented in the text.

2.6. Run the program written for problem 2.4 for a propeller geometry that has a rectangular planform shape. The geometry is identical to that in Example 2.3.1 except that the blade has a constant chord with the same planform area as the elliptic blade used in the example ($c_b/d_p = 0.0573$). Generate the figures that are presented in Example 2.5.1 and compare your results with those obtained for the elliptic blade.

2.7. Modify the computer program written for problem 2.1 so that it will compute the rotational speed and thrust developed by the propeller as a function of propeller diameter, pitch, input brake power, and forward airspeed. To test your program, duplicate the results for the 72-inch propeller that are plotted in Fig. 2.3.14.

2.8. Use the computer program written for problem 2.7 to predict the performance for one of the three-blade propellers described in Example 2.3.2. Consider a 72-inch-diameter propeller with a 48-inch chord-line pitch, operating at standard sea level. Assuming constant brake power input of 180 horsepower, plot the rotational speed in rpm and the thrust developed by the propeller in lbf as a function of airspeed from zero to 150 mph.

2.9. Consider the propeller used in problem 2.8, operating at standard sea level. Assuming a constant rotational speed of 2,900 rpm, plot the brake power input in horsepower and the thrust developed by the propeller in lbf as a function of airspeed from zero to 200 mph.

2.10. An experimental radio-controlled airplane is to be powered by a DC electric motor turning a fixed-pitch two-blade propeller, which has the same blade geometry as that described in Example 2.3.1. The propeller is to be attached directly to the output shaft of the motor. The current that is drawn by the DC motor, I_m, is given by

$$I_m = \frac{E_m - \omega/K_v}{R_m}$$

where E_m is the voltage applied to the motor, ω is the angular velocity, K_v is a motor voltage constant, and R_m is the motor armature resistance. The brake power developed by the motor can be expressed as

$$P_m = (I_m - I_o)(E_m - I_m R_m)$$

where I_o is a motor constant normally referred to as the *no-load current*. Modify the computer program written for problem 2.1 so that it will compute the rotational speed, motor current, and thrust developed by the propeller as a function of motor voltage, propeller diameter, pitch, and forward airspeed. Assume that the motor constants, K_v, R_m, and I_o, are all known.

2.11. Use the computer program written for problem 2.10 to predict the performance for one of the motor-propeller combinations that were described in that problem. Consider a 22-inch-diameter propeller with a 19-inch chord-line pitch, operating at standard sea level with constant motor voltage of 54 volts. Use the motor constants, $K_v = 9.7$ rad/sec/volt, $R_m = 0.11$ ohm, and $I_o = 3.0$ amps, and plot the motor current in amps, rotational speed in rpm, brake horsepower, and propeller thrust in lbf, as a function of airspeed from zero to 75 mph.

2.12. The motor-propeller combination in problem 2.11 is fitted with a control system that adjusts the motor voltage to maintain a constant motor current of 30 amps. For this configuration, plot the motor voltage, rotational speed, brake horsepower, and propeller thrust, as a function of airspeed from zero to 75 mph.

2.13. Since the mechanism of a variable-pitch propeller rotates the entire blade as shown in Fig. 2.3.11, rotation of this mechanism applies an incremental pitch angle variation, $\Delta\beta$, that does not vary along the length of the blade. Thus, the incremental change in pitch length, $\Delta\lambda$, does vary with the radial coordinate according to Eq. (2.2.2). Modify the computer program written for problem 2.1 so that it can be used to predict the performance of a variable-pitch propeller.

As a test case, use a two-blade propeller with the same blade geometry and section properties that were used for the fixed-pitch propeller in Example 2.3.1 with a chord-line pitch-to-diameter ratio of 1.0. The local section chord-line pitch length for this variable-pitch propeller will be independent of ζ for only one particular setting of the variable-pitch mechanism. Thus, as an input to the program, the pitch setting is to be identified in terms of a nominal pitch length, which is defined to be equal to the local section chord-line pitch length evaluated at the 75 percent radius. With this program, generate figures similar to those that were presented in Example 2.3.1. In all figures, chord-line pitch-to-diameter ratio, K_c, should be replaced with nominal pitch-to-diameter ratio, K_n. Compare your results with those presented in Example 2.3.1. The results should be identical only when the nominal pitch-to-diameter ratio for the variable-pitch propeller is 1.0.

2.14. Modify the computer program that was written for problem 2.13 so that it will predict the performance of a 72-inch, three-blade, constant-speed propeller with the same blade geometry as that described in problem 2.13. The propeller is operating at standard sea level and the control system maintains the fuel flow and propeller pitch such that the rotational speed and brake power are held constant at 2,900 rpm and 180 hp, respectively. Plot the nominal chord-line pitch in inches, the thrust developed by the propeller in lbf, and the propulsive efficiency, as a function of airspeed from zero to 150 mph.

2.15. Use propeller momentum theory to predict the performance of the 72-inch constant-speed propeller in problem 2.14. Including the effect of slipstream rotation, plot the thrust developed by the propeller and the propulsive efficiency as a function of airspeed from zero to 150 mph. Compare the results with those predicted from propeller blade theory in problem 2.14.

2.16. Modify the computer program that was written for problem 2.1 to include the effects of section Mach number on the section lift and drag coefficients. For the section lift use Eq. (1.13.5) for $M_\infty \leq 0.8$ and Eq. (1.14.18) for $M_\infty \geq 1.2$. As a rough approximation for the transonic region, use a simple linear interpolation between $M_\infty = 0.8$ and $M_\infty = 1.2$. To account for subsonic drag divergence, use the approximation

$$\frac{\widetilde{C}_D}{\widetilde{C}_{DM0}} = \begin{cases} 1.0, & M_\infty \leq 0.80 \\ 1.0 + 160{,}000(M_\infty - 0.8)^4/27, & 0.80 \leq M_\infty \leq 0.95 \\ 6.0 - 800(1.0 - M_\infty)^2, & 0.95 \leq M_\infty \leq 1.00 \end{cases}$$

For $M_\infty \geq 1.2$, use Eq. (1.14.20) with the section geometry that was described in problem 1.45 and use simple linear interpolation between $M_\infty = 1.0$ and $M_\infty = 1.2$. Use the modified program to rework Examples 2.3.3 and 2.3.4, including this approximation for the effects of compressibility.

2.17. A large turbofan engine has an inlet area of 65 ft^2 and is operating with an inlet Mach number of 0.50. Determine the additive drag for static operation at sea level and for a flight Mach number of 0.80 at 40,000 feet. Assume standard atmospheric conditions and the same inlet Mach number for both of these operating points.

2.18. For the turbofan engine described in problem 2.17, determine the inlet mass flow rate for both of the stated operating conditions.

2.19. Determine the additive drag for an inlet area of 5.2 m^2, an inlet Mach number of 0.8, a flight Mach number of 0.2, and a standard altitude of 10,000 meters.

2.20. Consider an ideal turbojet engine with a compressor pressure ratio of 12.5 and a maximum turbine inlet temperature of 1,640 °F. With the engine running at full throttle, determine the thermal efficiency and uninstalled specific thrust for static operation at standard sea level and for a flight Mach number of 2.0 at 40,000 feet with standard atmospheric conditions.

2.21. The turbojet engine described in problem 2.20 is operating with a fuel-air ratio that is 60 percent of that at full throttle. For the same two operating points, determine the thermal efficiency and uninstalled specific thrust.

2.22. The turbojet engine described in problem 2.20 has an inlet area of 10.0 ft^2. If the inlet Mach number is 0.30 at standard sea level, determine the installed static thrust. Assume zero nacelle drag.

2.23. Assuming complete combustion, determine the fuel consumption rate in pounds per hour and gallons per minute for the turbojet engine and operating condition that are described in problem 2.22. The enthalpy of combustion for the fuel is 18,400 Btu/lbm and the fuel weighs 6.7 pounds per gallon.

2.24. For the same conditions that were used in Example 2.7.1 to generate Figs. 2.7.1 and 2.7.2, plot the thrust-specific fuel consumption in pounds of fuel per hour per pound of uninstalled thrust, as a function of compressor pressure ratio for different flight Mach numbers. Assume complete combustion with an enthalpy of combustion for the fuel of 18,400 Btu/lbm.

2.25. For the conditions used in Example 2.7.1, plot the compressor pressure ratio that results in maximum specific thrust as a function of Mach number.

2.26. The turbojet engine described in problem 2.20 has an afterburner installed and is operating at full throttle with the afterburner engaged. The stagnation temperature at the nozzle inlet is 3,000 °F. For the same two operating points that were specified in problem 2.20, determine the thermal efficiency and uninstalled specific thrust.

2.27. For the same conditions that were used in Example 2.8.1 to generate Figs. 2.8.2 and 2.8.3, plot the thrust-specific fuel consumption in pounds of fuel per hour per pound of uninstalled thrust as a function of compressor pressure ratio for different flight mach numbers. Assume complete combustion with an enthalpy of combustion for the fuel of 18,400 Btu/lbm.

2.28. For the conditions used in Example 2.8.1, plot the compressor pressure ratio that results in maximum specific thrust as a function of Mach number.

2.29. Consider an ideal turbofan engine with a compressor pressure ratio of 26.8, fan pressure ratio of 2.31, bypass ratio 1.91, turbine inlet temperature of 2,550 °F, and airflow of 356 lbm/sec. If the flight Mach number is 0.80 with standard conditions at 40,000 feet, determine the uninstalled thrust, overall efficiency, and fuel consumption rate in lbm/hr. Assume complete combustion with an enthalpy of combustion for the fuel of 18,400 Btu/lbm. What is the total fuel consumption for two of these engines in gallons per minute? The fuel weight is 6.7 lbf/gal.

2.30. Consider an ideal turbofan engine with compressor pressure ratio of 39.3, bypass ratio of 8.40, and maximum turbine inlet temperature of 2,350 °F. The fuel-air ratio at cruise is 30 percent of full throttle and the airflow is 3,037 lbm/sec. If the cruise Mach number is 0.80 at 35,000 feet and the fan pressure ratio is optimized for this operating condition, determine the uninstalled thrust, overall efficiency, and fuel consumption rate. Assume complete combustion with an enthalpy of combustion for the fuel of 18,400 Btu/lbm. If an airplane uses four of these engines, what is the total fuel consumption at cruise in gallons per minute? What is the specific range in miles per gallon? The fuel weight is 6.7 lbf/gal.

2.31. To examine how optimal turbofan performance varies with bypass ratio and Mach number, consider the ideal turbofan engine operating under standard temperature and pressure at 30,000 feet. For this comparison, assume a fixed compressor pressure ratio of 20 and a fixed stagnation temperature at the turbine inlet of 3,000 °R. Assuming that each engine has a fan pressure ratio equal to optimum value for the given bypass ratio and operating conditions, plot overall efficiency and specific thrust as a function of Mach number for the same bypass ratios used in Fig. 2.9.2.

2.32. A significant performance parameter for the turbofan engine is the ratio of the uninstalled specific thrust for the core flow to that for the bypass flow. Here, this thrust ratio will be given the symbol R_F,

$$R_F \equiv \frac{F_c/\dot{m}_c}{F_b/\dot{m}_b}$$

For the design parameters and operating conditions used in problem 2.31, plot thrust ratio as a function of Mach number for the bypass ratios used in Fig. 2.9.2.

Chapter 3
Aircraft Performance

3.1. Introduction

Now that we have some understanding of the forces and moments acting on a moving airplane, we can turn our attention to a topic that students often find more interesting, aircraft performance. In this chapter we begin to answer some important questions about how airplanes fly. How fast and how high can an airplane fly? How slow can it fly? How rapidly can it climb? How far can a given airplane fly, and how long can it remain in the air? What happens if the engines fail? How quickly and in what distance can an airplane turn? What distance is required for takeoff and landing? The engineering analysis used to answer these questions and more is usually referred to as *aircraft performance analysis* and is the topic of this chapter. The answer to each of the questions above depends on the airplane design. How large are the wings? How much does the airplane weigh? What power and thrust are developed by the engines? Airplane performance analysis is very closely related to airplane design. In performance analysis, the airplane design parameters, such as wing area and gross weight, are assumed to be known, and engineering analysis is applied to determine some performance parameter, such as rate of climb or takeoff distance. In the design process, certain performance parameters, for example minimum airspeed and maximum range, are specified as engineering design requirements, and the design parameters and operating conditions necessary to meet these requirements must be determined. Essentially the same engineering analysis is used in either case.

In the preceding two chapters we have reviewed the physical phenomena associated with the production of lift, drag, and thrust as well as the moments produced along with these aerodynamic forces. We now combine what we have learned about the aerodynamic forces and moments with what the student has learned from his or her study of calculus, differential equations, statics, and dynamics to begin our study of the movement of an aircraft as it responds to these forces and moments. The important discipline within aerospace engineering that deals with aircraft motion is commonly referred to as *flight mechanics*. This broad field of study is normally broken up into three subareas: *aircraft performance*, *aircraft stability and control*, and *flight simulation*. While these three subdisciplines are very closely related, in this text we divide our study of flight mechanics along these traditional lines. As the title indicates, this chapter deals with the study of aircraft performance. The topics of aircraft stability and control as well as flight simulation are covered in Chapters 4 through 11.

The material presented in this chapter should be thought of as only a preliminary study of airplane performance. Here, emphasis is placed on obtaining closed-form analytic solutions suitable for preliminary design. Such solutions are invaluable in helping the student to understand how a particular performance parameter is affected by different aircraft design and operating parameters. The analysis presented in this chapter makes use of some simplifying assumptions that, though reasonable, are nevertheless only approximate. While the accuracy of these approximations is quite adequate for most preliminary design, for the final design and development of any aircraft, detailed computer simulations and wind tunnel tests are needed.

This chapter does not address all possible measures of airplane performance. The chapter is simply a collection of important examples that demonstrate how engineering analysis and mathematics can be used to evaluate and optimize certain airplane performance parameters. The methods and tools used to analyze these particular measures of performance could be used as well to study other measures of aircraft performance, and in studying this chapter students should be preparing to do exactly that. This chapter is not intended as an all-inclusive "cookbook" of aircraft performance equations. It is, rather, intended to help the student gain experience in formulating and solving engineering problems in general and flight mechanics problems in particular. The student should strive to recognize what is common to all of these examples. In each case the airplane is considered to be a rigid body on which is exerted four fundamental forces: lift, drag, thrust, and weight. The movement of the airplane as it responds to these forces is governed by Newton's second law, *force equals mass times acceleration*. Since Newton's second law is a vector equation, it provides three scalar equations that can be used to relate aircraft performance parameters to design and operating parameters. Or in the design scenario, design parameters can be related to performance and operating parameters.

The student should also become intimately familiar with another analysis technique used in many of the examples in this chapter. Once a closed-form solution for some dependent performance parameter has been obtained in terms of the associated independent design and operating parameters, the performance parameter can be optimized with respect to a particular design or operating parameter. This is done by differentiating with respect to the independent parameter of interest and setting the result to zero. For example, the rate of climb for any airplane is a function of airspeed as well as other design and operating parameters. Furthermore, the rate of climb exhibits a maximum at some particular airspeed. The airspeed that results in the maximum rate of climb can be found by differentiating the expression for rate of climb with respect to airspeed and setting the result to zero. When the resulting equation is solved for airspeed we obtain an expression for the airspeed that will produce the maximum rate of climb. A similar approach can be used to optimize other airplane performance parameters. This analysis technique is used throughout the chapter and the student is expected to be able to use the technique to solve many of the homework problems at the end of the chapter.

3.2. Thrust Required

One important performance parameter for any aircraft is the thrust required to maintain steady level flight. By *steady flight*, we mean flight at constant speed. Here, the term *level flight* will require that both the climb angle and the bank angle are zero. For steady level flight, there is no acceleration and no change in altitude. Thus, Newton's second law requires that the vector summation of forces acting on an airplane in steady level flight is zero. During level unaccelerated flight, if the thrust vector is aligned with the direction of flight, the wing must provide enough lift to balance the weight of the aircraft, and the engine or engines must provide sufficient thrust to balance the aerodynamic drag on the aircraft. Because the aerodynamic drag varies greatly with airspeed, the thrust required to maintain steady level flight is also a strong function of airspeed. More completely, the thrust required is a function of airspeed, altitude, size, shape, and weight of the airplane.

In general, the thrust vector will be inclined upward at some angle, α_T, with respect to the direction of flight, as shown in Fig. 3.2.1. For level unaccelerated flight, the drag force, D, must equal the component of thrust in the direction of flight,

$$D = T_R \cos \alpha_T \qquad (3.2.1)$$

and the lift force, L, must equal the weight of the aircraft less the component of thrust normal to the direction of flight,

$$L = W - T_R \sin \alpha_T \qquad (3.2.2)$$

where T_R is the thrust required for steady level flight and W is the gross weight of the aircraft. From Eqs. (3.2.1) and (3.2.2) we can write

$$\frac{L}{D} = \frac{W - T_R \sin \alpha_T}{T_R \cos \alpha_T} \qquad (3.2.3)$$

or solving for the thrust required,

$$T_R = \frac{W}{(L/D)\cos \alpha_T + \sin \alpha_T} \qquad (3.2.4)$$

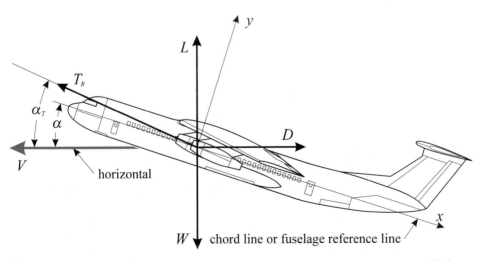

Figure 3.2.1. Forces acting on an aircraft in steady level flight, with angles exaggerated for better visibility.

Thus, we see that the thrust required for steady level flight is directly proportional to the gross weight of the aircraft. Furthermore, we see that the thrust required also depends on the ratio of lift to drag and on the thrust angle. The lift-to-drag ratio, commonly referred to as L over D, is a very important measure of an airplane's aerodynamic efficiency, and this ratio is a strong function of airspeed. Since the weight is clearly independent of airspeed, for any given value of α_T, the airspeed that results in the greatest L/D will also result in the lowest thrust required. To see how the thrust required varies with airspeed, we must examine how the ratio L/D varies with airspeed.

The total drag force can be written as the parasitic drag plus the induced drag,

$$D = \tfrac{1}{2}\rho V^2 S_w C_D = \tfrac{1}{2}\rho V^2 S_w \left(C_{Dp} + \frac{C_L^2}{\pi e_s R_A} \right) \tag{3.2.5}$$

where V is the airspeed, S_w is the area of the main wing, e_s is the span efficiency factor, and R_A is the aspect ratio. The parasitic drag coefficient, C_{Dp}, is very nearly a parabolic function of angle of attack. Since lift coefficient is a linear function of angle of attack, the parasitic drag coefficient can be written as a parabolic function of C_L,

$$C_{Dp} = C_{D0} + C_{D0,L} C_L + C_{D0,L^2} C_L^2 \tag{3.2.6}$$

Using Eq. (3.2.6) in Eq. (3.2.5) and combining the quadratic term from the parasitic drag with the induced drag, the total drag can be written as

$$D = \tfrac{1}{2}\rho V^2 S_w \left(C_{D0} + C_{D0,L} C_L + \frac{C_L^2}{\pi e R_A} \right) \tag{3.2.7}$$

where e is called the *Oswald efficiency factor*. The Oswald efficiency factor for the entire aircraft is considerably less than the span efficiency factor for the wing, because it accounts for part of the parasitic drag in addition to the induced drag. From Eq. (3.2.7) we can write

$$\frac{L}{D} = \frac{\tfrac{1}{2}\rho V^2 S_w C_L}{\tfrac{1}{2}\rho V^2 S_w C_D} = \frac{C_L}{C_D} = \frac{C_L}{C_{D0} + C_{D0,L} C_L + \dfrac{C_L^2}{\pi e R_A}} \tag{3.2.8}$$

The aerodynamic coefficients in Eq. (3.2.8) can all depend on Mach number. However, at low Mach numbers this dependence is slight.

Even at low Mach numbers, the lift coefficient for steady level flight is a function of airspeed. For level unaccelerated flight at any given airspeed, the aircraft must be trimmed to fly at an angle of attack that will produce the correct amount of lift to just satisfy Eq. (3.2.2). If the angle of attack is too large, too much lift is generated at the given airspeed and the airplane will climb. If the angle of attack is too small, not enough lift is generated and the airplane will begin to sink.

Using the definition of lift coefficient with Eq. (3.2.2), we have

$$C_L = \frac{L}{\frac{1}{2}\rho V^2 S_w} = \frac{W - T_R \sin \alpha_T}{\frac{1}{2}\rho V^2 S_w} \tag{3.2.9}$$

Using Eq. (3.2.4) to eliminate the thrust required from Eq. (3.2.9) gives

$$C_L = \frac{W}{\frac{1}{2}\rho V^2 S_w}\left[1 - \frac{\sin \alpha_T}{(L/D)\cos \alpha_T + \sin \alpha_T}\right] = \frac{W}{\frac{1}{2}\rho V^2 S_w}\left[\frac{1}{1 + (D/L)\tan \alpha_T}\right] \tag{3.2.10}$$

For most conventional airplanes, the bracketed term on the right side of Eq. (3.2.10) is very close to unity, and the lift coefficient is very nearly inversely proportional to the square of the airspeed. So we see that for steady level flight, increasing the airspeed will require the angle of attack and the corresponding lift coefficient to decrease.

Using Eqs. (3.2.8) and (3.2.10), we can numerically determine both C_L and L/D as a function of airspeed. A plot of the lift-to-drag ratio versus airspeed for a typical general aviation aircraft at sea level is shown in Fig. 3.2.2. Notice that L/D has a well-defined maximum at some particular airspeed. This characteristic is common to all airplanes. While the maximum value of L/D and the airspeed at which this maximum occurs vary with aircraft design, all airplanes exhibit a maximum in the lift-to-drag ratio at some particular airspeed.

The maximum lift-to-drag ratio and the airspeed that results in this maximum are both important aircraft design parameters. We can find the lift coefficient that will result in maximum L/D by differentiating Eq. (3.2.8) with respect to C_L and setting the result to zero. This gives

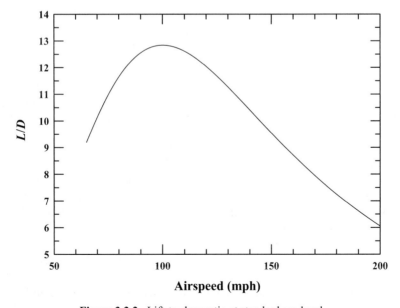

Figure 3.2.2. Lift-to-drag ratio at standard sea level.

$$\frac{\partial(L/D)}{\partial C_L} = \frac{\left(C_{D_0} + C_{D_0,L}\, C_L + \dfrac{C_L^2}{\pi e R_A}\right) - C_L\left(C_{D_0,L} + \dfrac{2C_L}{\pi e R_A}\right)}{\left(C_{D_0} + C_{D_0,L}\, C_L + \dfrac{C_L^2}{\pi e R_A}\right)^2} = 0 \qquad (3.2.11)$$

or

$$C_{D_0} = \frac{C_L^2}{\pi e R_A} \qquad (3.2.12)$$

The term on the right-hand side of Eq. (3.2.12) is often referred to as the *induced drag* even though, strictly speaking, it is the induced drag plus the quadratic portion of the parasitic drag. Using this terminology, it is said that maximum lift-to-drag ratio occurs at the airspeed that makes the induced drag equal to the parasitic drag at zero lift. Using Eq. (3.2.12) to eliminate the lift coefficient from Eq. (3.2.8), we obtain an expression for the maximum lift-to-drag ratio,

$$(L/D)_{\max} = \frac{\sqrt{\pi e R_A}}{2\sqrt{C_{D_0}} + C_{D_0,L}\sqrt{\pi e R_A}} \qquad (3.2.13)$$

The airspeed at which an aircraft must be flown in order to attain maximum lift-to-drag ratio is found by solving Eq. (3.2.10) for the airspeed and using Eqs. (3.2.12) and (3.2.13) to eliminate C_L and L/D. This gives

$$V_{MD} - \frac{\sqrt{2}}{\sqrt[4]{\pi e R_A C_{D_0}}}\sqrt{\frac{W/S_w}{\rho}}\sqrt{1\bigg/\left[1 + \left(2\sqrt{\frac{C_{D_0}}{\pi e R_A}} + C_{D_0,L}\right)\tan\alpha_T\right]} \qquad (3.2.14)$$

where V_{MD} is usually referred to as the *minimum drag airspeed* or the *best L over D airspeed*. The parameter W/S_w, called the *wing loading*, is another very important parameter in aircraft design. Note that the best L over D airspeed for any airplane is proportional to the square root of the wing loading divided by the air density. This means that the higher an airplane flies and the greater its wing loading, the faster it must be flown to achieve best L over D. We can see from Eq. (3.2.13), however, that the maximum lift-to-drag ratio for any airplane is independent of both wing loading and altitude. The maximum lift-to-drag ratio is a function only of the aircraft design. No matter how heavily the plane is loaded or how high it is flown, the maximum lift-to-drag ratio is always the same. For this reason maximum lift-to-drag ratio is a commonly used measure of aerodynamic efficiency for an airplane.

This maximum value for L/D is often referred to simply as the *lift-to-drag ratio* for the aircraft. For example, you may hear it said that a particular airplane has a lift-to-drag ratio of 12. This manner of speaking could erroneously lead the student to believe that this airplane will always produce 1 pound of drag for each 12 pounds of aircraft weight. Nothing could be further from the truth. As we have seen in Fig. 3.2.2, the ratio of lift to drag for an airplane varies greatly with airspeed. It must be remembered that when an aerodynamicist specifies an unqualified lift-to-drag ratio for an aircraft, he or she is referring to the maximum lift-to-drag ratio for that aircraft.

For most conventional airplanes, the change in drag coefficient with respect to lift coefficient at zero lift is either zero or a very small negative value. Thus, from Eq. (3.2.13) we see that the maximum lift-to-drag ratio can be increased only by decreasing the parasitic drag coefficient or by increasing the aspect ratio and Oswald efficiency factor. For this reason, the problem of designing an aircraft for high L over D is as much a structures problem as it is an aerodynamics problem. From an aerodynamics viewpoint, we want to build a very streamlined airplane with a very high aspect ratio. From a structures viewpoint, the longer and thinner the wing must be, the more difficult it is to design the wing so that it will support the weight of the aircraft and payload.

From Eq. (3.2.4) we see that because maximum lift-to-drag ratio is independent of both wing area and altitude, the minimum thrust required is also independent of wing area and altitude. Figure 3.2.3 shows how the thrust required varies with airspeed and altitude for a typical general aviation aircraft. Wing area affects the thrust required in the same way as air density. Thus, a plot of thrust required as a function of airspeed for constant weight and different values of wing area would look very much like Fig. 3.2.3, with higher wing loading moving the curve to the right.

We have now seen that there is an optimum airspeed that results in maximum L/D and minimum thrust required. Similarly, we can show from Eq. (3.2.4) that there is an optimum thrust angle that can be used to minimize the thrust required. To find the value

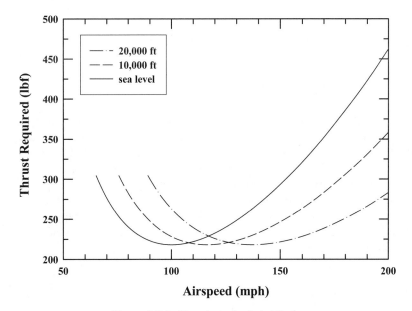

Figure 3.2.3. Thrust required at altitude.

of α_T that will result in minimum thrust required, we differentiate Eq. (3.2.4) with respect to α_T and set the result to zero. This gives

$$\frac{\partial T_R}{\partial \alpha_T} = \frac{W\left[(L/D)\sin\alpha_T - \cos\alpha_T\right]}{\left[(L/D)\cos\alpha_T + \sin\alpha_T\right]^2} = 0 \qquad (3.2.15)$$

or

$$\alpha_T = \tan^{-1}(D/L) \qquad (3.2.16)$$

So we see that to completely minimize the thrust required, the thrust vector should not be aligned exactly with the direction of flight. Rather, the thrust vector should be tilted upward slightly, at an angle equal to the arctangent of the inverse of the lift-to-drag ratio for the aircraft. However, for a well-designed airplane, the lift-to-drag ratio is fairly high. Most modern airplanes will have maximum lift-to-drag ratios between 8 and 24, with even higher values for some specialty aircraft. Thus, the optimum thrust angle for most conventional airplanes is very small (between 2 and 7 degrees, with even smaller values for airplanes with very high aerodynamic efficiency).

Since α_T and D/L are both small for conventional airplanes, for the purpose of making preliminary calculations for airplane performance, we often use the small-angle approximations

$$\cos\alpha_T = 1 - \frac{\alpha_T^2}{2!} + \frac{\alpha_T^4}{4!} - \frac{\alpha_T^6}{6!} + \cdots \cong 1 \qquad (3.2.17)$$

$$\sin\alpha_T = \alpha_T - \frac{\alpha_T^3}{3!} + \frac{\alpha_T^5}{5!} - \frac{\alpha_T^7}{7!} + \cdots \cong \alpha_T \qquad (3.2.18)$$

$$(D/L)\sin\alpha_T \cong \alpha_T^2 \cong 0 \qquad (3.2.19)$$

This will be called the *small-thrust-angle approximation* and is based on neglecting all terms of order α_T squared and smaller when compared to unity. Notice that we are not making the approximation $\sin\alpha_T \cong 0$, but rather, $\sin\alpha_T \cong \alpha_T$ and $(D/L)\sin\alpha_T \cong 0$. Since both $\sin\alpha_T$ and D/L are of the same order of magnitude as the angle of attack (10^{-1} for most conventional airplanes), we clearly should not neglect either $\sin\alpha_T$ or D/L when compared to unity. The product of D/L with $\sin\alpha_T$, on the other hand, is of the same order as the angle of attack squared, and neglecting this product compared to unity is consistent with the other small-angle approximations that we used previously. With this approximation, Eqs. (3.2.1) and (3.2.10) reduce to

$$T_R = D \qquad (3.2.20)$$

$$L = W \qquad (3.2.21)$$

Thus, for small thrust angles, the equations of motion for level unaccelerated flight simply require that the thrust must equal the drag and the lift must equal the weight. Therefore, the thrust required is simply the gross weight divided by the lift-to-drag ratio for the airplane:

$$T_R = \frac{W}{L/D} \tag{3.2.22}$$

Using Eq. (3.2.8) in Eq. (3.2.22), for small thrust angles we can write the thrust required for steady level flight as

$$T_R = \frac{C_D}{C_L}W = \left(\frac{C_{D_0}}{C_L} + C_{D_0,L} + \frac{C_L}{\pi e R_A}\right)W \tag{3.2.23}$$

Using the small-thrust-angle approximation with Eq. (3.2.10) gives

$$C_L = \frac{W}{\frac{1}{2}\rho V^2 S_w} \tag{3.2.24}$$

Using Eq. (3.2.24) in Eq. (3.2.23), we can write the thrust required for small thrust angles as a function of airspeed:

$$T_R = \left(\frac{\frac{1}{2}\rho V^2 C_{D_0}}{W/S_w} + C_{D_0,L} + \frac{W/S_w}{\frac{1}{2}\pi e R_A \rho V^2}\right)W \tag{3.2.25}$$

Using Eq. (3.2.13) in Eq. (3.2.22), the minimum thrust required for small thrust angles can be closely approximated as

$$T_{R\,min} = \left(2\sqrt{\frac{C_{D_0}}{\pi e R_A}} + C_{D_0,L}\right)W \tag{3.2.26}$$

Using Eq. (3.2.19) in Eq. (3.2.14), the minimum drag airspeed for small thrust angles becomes

$$V_{MD} = \left(\frac{4}{\pi e R_A C_{D_0}}\right)^{1/4}\sqrt{\frac{W/S_w}{\rho}} \tag{3.2.27}$$

As mentioned previously, the change in drag coefficient with respect to lift coefficient at zero lift is either zero or a very small negative value. Thus, for the purpose of making preliminary performance calculations, the value of $C_{D_0,L}$ in Eqs. (3.2.25) and (3.2.26) is often assumed to be zero.

EXAMPLE 3.2.1. A turbojet engine produces a thrust that is nearly independent of airspeed. As a first approximation, we can model a turbojet engine as a device that produces an available thrust, T_A, that is constant with airspeed, V, and directly proportional to the air density, ρ,

$$T_A = \tau \frac{\rho}{\rho_0} T_{A0}$$

where τ is a throttle setting expressed as a fraction of the full-throttle thrust, ρ_0 is the air density at standard sea level, and T_{A0} is the full-throttle thrust developed at standard sea level.

Consider an executive business jet having the following properties:

$$S_w = 320 \text{ ft}^2, \quad b_w = 54 \text{ ft}, \quad W = 20{,}000 \text{ lbf}, \quad T_{A0} = 6{,}500 \text{ lbf},$$
$$C_{D_0} = 0.023, \quad C_{D_0,L} = 0.0, \quad e = 0.82$$

Using the small-thrust-angle approximation with the foregoing approximation for the thrust available, determine the airspeed for this aircraft in steady level flight with a 60 percent throttle setting, for both sea level and 20,000 feet.

Solution. For steady level flight, the thrust required must equal the thrust available. For small thrust angles, the thrust required is given by Eq. (3.2.25), and for this example, the thrust available is given by the relation above. Thus we can write

$$\left(\frac{\frac{1}{2} \rho V^2 C_{D_0}}{W/S_w} + C_{D_0,L} + \frac{W/S_w}{\frac{1}{2} \pi e R_A \rho V^2} \right) W = \tau \frac{\rho}{\rho_0} T_{A0}$$

This equation can be written in terms of a dimensionless velocity ratio

$$R_V \equiv V \bigg/ \sqrt{\frac{W/S_w}{\rho}}$$

and a dimensionless thrust-to-weight ratio

$$R_{T/W} \equiv \frac{T_A}{W} = \tau \frac{\rho}{\rho_0} \frac{T_{A0}}{W}$$

With these definitions, the balance between thrust required and thrust available can be written as

$$C_{D_0} R_V^4 - 2(R_{T/W} - C_{D_0,L}) R_V^2 + \frac{4}{\pi e R_A} = 0$$

This gives a quadratic equation for the dimensionless velocity ratio squared and we have

$$R_V^2 = \frac{R_{T/W} - C_{D_0,L}}{C_{D_0}} \pm \sqrt{\left(\frac{R_{T/W} - C_{D_0,L}}{C_{D_0}}\right)^2 - \frac{4}{\pi e R_A C_{D_0}}}$$

The airspeed for steady level flight is then

$$V = \sqrt{\frac{R_{T/W} - C_{D_0,L}}{C_{D_0}} \pm \sqrt{\left(\frac{R_{T/W} - C_{D_0,L}}{C_{D_0}}\right)^2 - \frac{4}{\pi e R_A C_{D_0}}} \sqrt{\frac{W/S_w}{\rho}}}$$

From this equation we see that in the absence of stall, there may be two steady level airspeeds that can be sustained by the aircraft with a given thrust. The reason for this is made clear by reviewing Fig. 3.2.3. The airspeed of interest is the larger of these two roots. For the aircraft in this example, the aspect ratio is

$$R_A = \frac{b_w^2}{S_w} = \frac{(54 \text{ ft})^2}{320 \text{ ft}^2} = 9.1125$$

the wing loading is

$$\frac{W}{S_w} = \frac{20,000 \text{ lbf}}{320 \text{ ft}^2} = 62.5 \text{ lbf/ft}^2$$

and the thrust-to-weight ratio for 60 percent of maximum thrust at standard sea level is

$$R_{T/W} = \frac{T_A}{W} = \tau \frac{\rho}{\rho_0} \frac{T_{A0}}{W} = 0.6 \frac{0.0023769 \text{ slug/ft}^3}{0.0023769 \text{ slug/ft}^3} \frac{6,500 \text{ lbf}}{20,000 \text{ lbf}} = 0.195$$

Thus, at sea level, the steady airspeed that can be sustained in level flight with 60 percent of maximum thrust is

$$V = $$

$$\sqrt{\frac{0.195 - 0.0}{0.023} + \sqrt{\left(\frac{0.195 - 0.0}{0.023}\right)^2 - \frac{4}{\pi(0.82)9.1125(0.023)}} \sqrt{\frac{62.5 \text{ lbf/ft}^2}{0.0023769 \text{ slug/ft}^3}}}$$

$$= 659 \text{ ft/sec} = 449 \text{ mph} = 390 \text{ knots} = 723 \text{ km/hr}$$

At 20,000 feet, the thrust-to-weight ratio for 60 percent of maximum thrust is

$$\frac{T_A}{W} = \tau \frac{\rho}{\rho_0} \frac{T_{A0}}{W} = 0.6 \frac{0.0012673 \text{ slug/ft}^3}{0.0023769 \text{ slug/ft}^3} \frac{6,500 \text{ lbf}}{20,000 \text{ lbf}} = 0.104$$

and

$$V =$$

$$\sqrt{\frac{0.104-0.0}{0.023} + \sqrt{\left(\frac{0.104-0.0}{0.023}\right)^2 - \frac{4}{\pi(0.82)9.1125(0.023)}} \sqrt{\frac{62.5 \text{ lbf/ft}^2}{0.0012673 \text{ slug/ft}^3}}}$$

$$= 633 \text{ ft/sec} = 432 \text{ mph} = 375 \text{ knots} = 695 \text{ km/hr}$$

Thus we see that while the available thrust produced by this turbojet engine is reduced by nearly 50 percent in going from sea level to 20,000 feet, the corresponding airspeed is only reduced by about 4 percent.

3.3. Power Required

Another important performance parameter for any aircraft is the power required to maintain level, unaccelerated flight. This parameter is also a strong function of airspeed. The power delivered by a jet engine or any engine-propeller combination is just the rate of work that results from the thrust force acting on the aircraft moving through the air. The work is the force multiplied by the distance through which it acts as the aircraft moves through the air. The work rate, or power, is then the dot product of the force vector and velocity vector. The power required is then the thrust required multiplied by the airspeed and the cosine of the thrust angle,

$$P_R = T_R V \cos \alpha_T = DV \tag{3.3.1}$$

where P_R is defined as the power required for steady level flight. Using the small-thrust-angle approximation, for steady level flight we have

$$C_L = \frac{W}{\frac{1}{2}\rho V^2 S_w} \tag{3.3.2}$$

Solving Eq. (3.3.2) for the airspeed, we obtain

$$V = \sqrt{\frac{2(W/S_w)}{\rho C_L}} \tag{3.3.3}$$

Using Eq. (3.3.3) in Eq. (3.3.1) along with the small-thrust-angle approximation gives

$$P_R = T_R \sqrt{\frac{2(W/S_w)}{\rho C_L}} \tag{3.3.4}$$

From Eq. (3.2.23), the thrust required for small thrust angles is given by

$$T_R = \frac{C_D}{C_L} W = \left(\frac{C_{D_0}}{C_L} + C_{D_0,L} + \frac{C_L}{\pi e R_A} \right) W$$ (3.3.5)

Using Eq. (3.3.5) in Eq. (3.3.4), we have

$$P_R = \sqrt{2} \left(\frac{C_{D_0}}{C_L^{3/2}} + \frac{C_{D_0,L}}{C_L^{1/2}} + \frac{C_L^{1/2}}{\pi e R_A} \right) W \sqrt{\frac{W/S_w}{\rho}}$$ (3.3.6)

Using Eq. (3.3.2) in Eq. (3.3.6), we can write the power required for steady level flight as a function of airspeed,

$$P_R = \left(\frac{C_{D_0} \rho V^3}{2(W/S_w)} + C_{D_0,L} V + \frac{2(W/S_w)}{\pi e R_A \rho V} \right) W$$ (3.3.7)

Since the first two terms on the right of Eq. (3.3.7) increase with airspeed while the third term decreases with airspeed, the power required will also have a minimum at some particular airspeed. Figure 3.3.1 shows the power required as a function of airspeed for a typical general aviation aircraft at sea level.

We can find the aircraft lift coefficient that results in minimum power required by differentiating Eq. (3.3.6) with respect to C_L and setting the result to zero. This gives

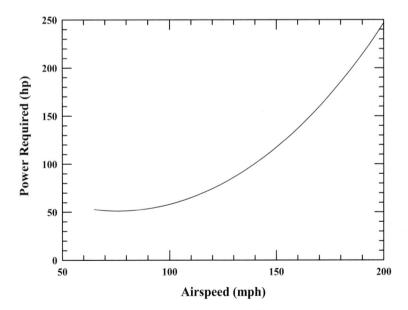

Figure 3.3.1. Power required at standard sea level.

$$\frac{\partial P_R}{\partial C_L} = \sqrt{2}\left(-\frac{3}{2}\frac{C_{D_0}}{C_L^{5/2}} - \frac{1}{2}\frac{C_{D_0,L}}{C_L^{3/2}} + \frac{1}{2}\frac{1}{\pi e R_A C_L^{1/2}}\right)W\sqrt{\frac{W/S_w}{\rho}} = 0 \qquad (3.3.8)$$

or

$$3C_{D_0} + C_{D_0,L}\,C_L = \frac{C_L^2}{\pi e R_A} \qquad (3.3.9)$$

Because $C_{D_0,L}$ is very nearly zero, from Eq. (3.3.9) we see that at the airspeed which requires minimum power, the induced drag is approximately three times the zero-lift drag. After solving Eq. (3.3.9) for the lift coefficient that is required when the airplane is flying at minimum power, we have

$$C_L = \frac{\pi e R_A}{2}\left(C_{D_0,L} + \sqrt{C_{D_0,L}^2 + \frac{12C_{D_0}}{\pi e R_A}}\right) \qquad (3.3.10)$$

Using the lift coefficient from Eq. (3.3.10) in Eq. (3.3.6), we can obtain the minimum power required. The basic features of this relationship can be more easily examined by considering the usual approximation $C_{D_0,L} = 0$. Using this approximation, the minimum power required becomes

$$P_{R\,\text{min}} = (DV)_{\text{min}} \cong \frac{4\sqrt{2}\,C_{D_0}^{1/4}}{(3\pi e R_A)^{3/4}}W\sqrt{\frac{W/S_w}{\rho}} \qquad (3.3.11)$$

From this equation we see that as was the case for minimum thrust required, the minimum power required for a given weight can be made smaller by decreasing the parasitic drag or increasing the aspect ratio and Oswald efficiency factor. However, the minimum thrust required was independent of both wing loading and air density. The minimum power required, on the other hand, increases with both wing loading and altitude. Thus, when designing an aircraft for very high payload-to-power ratio, we not only want the highest possible aspect ratio and the lowest possible parasitic drag but the lowest possible wing loading as well. Without a doubt, the ultimate in low-powered aircraft designs have been the human-powered aircraft (Gossamer Condor and Gossamer Albatross) designed and built by Paul MacCready and his colleagues. The wing loading for these aircraft was on the order of ¼ pound per square foot. This is in contrast to a wing loading on the order of 100 pounds per square foot for modern high-powered fighter aircraft.

The airspeed at which an aircraft must fly in order to attain minimum power required is found by combining Eqs. (3.3.3) and (3.3.10). This gives

$$V_{MDV} = \frac{2}{\sqrt{\pi e R_A C_{D_0,L} + \sqrt{(\pi e R_A C_{D_0,L})^2 + 12\pi e R_A C_{D_0}}}}\sqrt{\frac{W/S_w}{\rho}} \qquad (3.3.12)$$

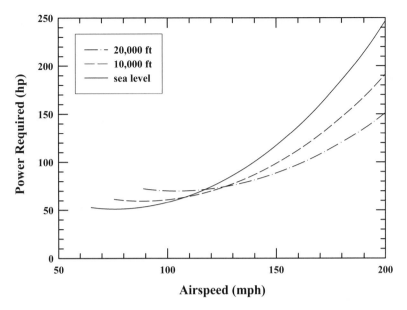

Figure 3.3.2. Power required at altitude.

where V_{MDV} is the airspeed for minimum power required, which means minimum DV product. Note that as was the case for the minimum thrust airspeed, the minimum power airspeed is proportional to the square root of the wing loading divided by the air density. The minimum power airspeed is typically about 75 to 80 percent of the minimum thrust airspeed. Note that both the minimum power airspeed and the minimum power required increase with altitude. Figure 3.3.2 shows how the power required varies with altitude for a typical general aviation aircraft.

EXAMPLE 3.3.1. For small thrust angles, the propulsive power available from an aircraft power plant is equal to the thrust available multiplied by the airspeed. An engine combined with a variable-pitch propeller can be adjusted, either manually by the pilot or automatically with a governor mechanism, to produce an available power that is nearly independent of airspeed over the range of airspeeds encountered in normal flight. As a first approximation, we can model this type of power plant as a device that produces an available power, P_A, that is constant with airspeed, V, and directly proportional to the air density, ρ,

$$P_A = \tau \frac{\rho}{\rho_0} P_{A0}$$

where τ is a throttle setting expressed as a fraction of the full-throttle power available, ρ_0 is the air density at standard sea level, and P_{A0} is the full-throttle power available at standard sea level.

Consider a general aviation aircraft having such a power plant and the following properties:

$$S_w = 180 \text{ ft}^2, \quad b_w = 33 \text{ ft}, \quad W = 2{,}700 \text{ lbf}, \quad P_{A0} = 180 \text{ hp},$$
$$C_{D_0} = 0.023, \quad C_{D_0,L} = 0.0, \quad e = 0.82$$

Using the small-thrust-angle approximation, determine the maximum airspeed that can be sustained by this aircraft in steady level flight for both sea level and 10,000 feet.

Solution. For steady level flight, the power required must equal the power available. For small thrust angles, the power required is given by Eq. (3.3.7), and for this example, the power available is given by the relation above. Thus we can write

$$\left(\frac{C_{D_0} \rho V^3}{2(W/S_w)} + C_{D_0,L} V + \frac{2(W/S_w)}{\pi e R_A \rho V} \right) W = \tau \frac{\rho}{\rho_0} P_{A0}$$

This equation can be written in terms of the dimensionless velocity ratio

$$R_V \equiv V \Big/ \sqrt{\frac{W/S_w}{\rho}}$$

and a dimensionless power-to-weight ratio

$$R_{P/W} \equiv P_A \Big/ \left(W \sqrt{\frac{W/S_w}{\rho}} \right) = \tau \frac{\rho}{\rho_0} P_{A0} \Big/ \left(W \sqrt{\frac{W/S_w}{\rho}} \right)$$

Thus, the balance between power required and power available can be written

$$\frac{C_{D_0}}{2} R_V^3 + C_{D_0,L} R_V + \frac{2}{\pi e R_A} R_V^{-1} = R_{P/W} \tag{3.3.13}$$

This does not reduce to a simple quadratic equation, as was the case for the turbojet engine in Example 3.2.1. However, the equation can quickly be solved for the velocity ratio by using Newton's method.

Let the left-hand side of Eq. (3.3.13) be $f(R_V)$. Then we have

$$f'(R_V) \equiv \frac{\partial f}{\partial R_V} = \frac{3 C_{D_0}}{2} R_V^2 + C_{D_0,L} - \frac{2}{\pi e R_A} R_V^{-2}$$

Newton's method allows us to obtain an improved estimate for the dimensionless velocity ratio, from some initial estimate, by using Newton's corrector equation,

$$R_{Vi+1} = R_{Vi} + \frac{R_{P/W} - f(R_{Vi})}{f'(R_{Vi})} \qquad (3.3.14)$$

As was the case for the turbojet engine in Example 3.2.1, there may be two positive roots to Eq. (3.3.13). The airspeed of interest is associated with the larger of these two roots. To ensure that Newton's method converges on the larger root, we obtain our initial estimate for the dimensionless velocity ratio by neglecting all but the highest-order term on the left-hand side of Eq. (3.3.13). Retaining only the cubic term, this gives

$$R_{V0} = \left(\frac{2R_{P/W}}{C_{D_0}} \right)^{1/3} \qquad (3.3.15)$$

For the aircraft in this example, the aspect ratio is

$$R_A = \frac{b_w^2}{S_w} = \frac{(33 \text{ ft})^2}{180 \text{ ft}^2} = 6.05$$

the wing loading is

$$\frac{W}{S_w} = \frac{2,700 \text{ lbf}}{180 \text{ ft}^2} = 15.0 \text{ lbf/ft}^2$$

and the full-throttle power available at standard sea level is

$$P_A = P_{A0} = 180 \text{ hp} = 99,000 \text{ ft·lbf/sec}$$

This gives a dimensionless power-to-weight ratio of

$$R_{P/W} \equiv P_A \Big/ \left(W \sqrt{\frac{W/S_w}{\rho}} \right)$$

$$= 99,000 \text{ ft·lbf/sec} \Big/ \left(2,700 \text{ lbf} \sqrt{\frac{15.0 \text{ lbf/ft}^2}{0.0023769 \text{ slug/ft}^3}} \right)$$

$$= \frac{99,000 \text{ ft·lbf/sec}}{214,488 \text{ ft·lbf/sec}} = 0.462$$

and from Eq. (3.3.15), the initial estimate for the dimensionless velocity ratio at sea level is

$$R_{V0} = \left(\frac{2R_{P/W}}{C_{D_0}} \right)^{1/3} = \left(\frac{2(0.462)}{0.023} \right)^{1/3} = 3.4249$$

Using this initial estimate and iteratively applying Eq. (3.3.14) gives

i	R_{Vi}	$f(R_{Vi})$	$f'(R_{Vi})$	R_{Vi+1}
0	3.4249	0.49947	0.39374	3.3297
1	3.3297	0.46309	0.37093	3.3268
2	3.3268	0.46200	0.37024	3.3268

Thus, at standard sea level, the maximum airspeed that can be sustained by the aircraft in steady level flight is

$$V = R_V \sqrt{\frac{W/S_w}{\rho}} = 3.3268 \sqrt{\frac{15.0 \text{ lbf/ft}^2}{0.0023769 \text{ slug/ft}^3}}$$

$$= 264 \text{ ft/sec} = 180 \text{ mph} = 156 \text{ knots} = 290 \text{ km/hr}$$

Repeating the calculations for 10,000 feet, we have

$$P_A = \tau \frac{\rho}{\rho_0} P_{A0}$$

$$= 1.0 \frac{0.0017556 \text{ slug/ft}^3}{0.0023769 \text{ slug/ft}^3} 180 \text{ hp} = 132.95 \text{ hp} = 73,122 \text{ ft lbf/sec}$$

$$R_{P/W} \equiv P_A \bigg/ \left(W \sqrt{\frac{W/S_w}{\rho}} \right)$$

$$= 73,122 \cdot \text{ft lbf/sec} \bigg/ \left(2,700 \text{ lbf} \sqrt{\frac{15.0 \text{ lbf/ft}^2}{0.0017556 \text{ slug/ft}^3}} \right)$$

$$= \frac{73,122 \text{ ft} \cdot \text{lbf/sec}}{249,572 \text{ ft} \cdot \text{lbf/sec}} = 0.293$$

$$R_{V0} = \left(\frac{2R_{P/W}}{C_{D0}} \right)^{1/3} = \left(\frac{2(0.293)}{0.023} \right)^{1/3} = 2.9425$$

i	R_{Vi}	$f(R_{Vi})$	$f'(R_{Vi})$	R_{Vi+1}
0	2.9425	0.33661	0.28390	2.7889
1	2.7889	0.29548	0.25185	2.7791
2	2.7791	0.29300	0.24984	2.7791

$$V = R_V \sqrt{\frac{W/S_w}{\rho}} = 2.7791 \sqrt{\frac{15.0 \text{ lbf/ft}^2}{0.0017556 \text{ slug/ft}^3}}$$

$$= 257 \text{ ft/sec} = 175 \text{ mph} = 152 \text{ knots} = 282 \text{ km/hr}$$

For this propeller-driven airplane, the available power is reduced by more than 25 percent in going from sea level to 10,000 feet, while the corresponding maximum airspeed is reduced by only about 3 percent. This is because, as is seen in Fig. 3.3.2, the power required at high airspeeds is also significantly reduced with altitude.

3.4. Rate of Climb and Power Available

The rate at which an airplane can climb is another important performance parameter. As we should expect, the rate of climb is a function not only of the aerodynamic design and weight of the aircraft but is also a function of the power available from the airplane's engines. This power depends on engine type and design as well as on airspeed, air density, and throttle setting.

To determine the steady rate of climb, we consider an airplane in unaccelerated climbing flight, as is shown in Fig. 3.4.1. The flight path relative to the surrounding air is inclined at an angle of γ to the horizontal, and the airplane's airspeed along this flight path is V. The drag force must equal the component of thrust parallel to the flight path less the component of the airplane's weight that is aligned with the flight path,

$$D = T_A \cos\alpha_T - W \sin\gamma \qquad (3.4.1)$$

where T_A is the thrust available. The lift must support the airplane's weight component that is normal to the flight path less the component of thrust normal to the flight path,

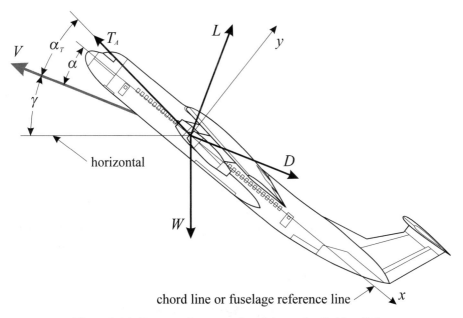

Figure 3.4.1. Forces acting on an aircraft in steady climbing flight.

$$L = W \cos\gamma - T_A \sin\alpha_T \tag{3.4.2}$$

Combining Eqs. (3.4.1) and (3.4.2) with Eq. (3.2.7) and the definition of lift coefficient, we have the **general formulation for steady climbing flight,**

$$\tfrac{1}{2}\rho V^2 S_w\left(C_{D_0} + C_{D_0,L}\, C_L + \frac{C_L^2}{\pi e R_A}\right) = T_A \cos\alpha_T - W \sin\gamma \tag{3.4.3}$$

$$\tfrac{1}{2}\rho V^2 S_w C_L = W \cos\gamma - T_A \sin\alpha_T \tag{3.4.4}$$

Equations (3.4.3) and (3.4.4) can be used to solve for the unknown lift coefficient and one additional unknown. If the airspeed and thrust available are known, we could use this formulation to solve for the climb angle. If the thrust available is known, we could also use the formulation to obtain the equilibrium airspeed for a given climb angle. On the other hand, we could also employ Eqs. (3.4.3) and (3.4.4) to determine the thrust necessary to support a given climb angle at a given airspeed.

The drag experienced during climbing flight is somewhat less than that for level flight. This is because, as can be seen from Eq. (3.4.4), the lift for climbing flight is less than the lift for level flight. Since the lift is reduced during climbing flight, the induced drag is also reduced and, as a result, the total drag is reduced. For example, consider an aircraft in a vertical climb having the thrust vector aligned with the direction of flight. For this special case, Eq. (3.4.4) shows that the lift, and hence the induced drag, are exactly zero. Of course, as Eq. (3.4.3) predicts, a vertical climb requires a thrust-to-weight ratio greater than 1. However, most airplanes have a thrust-to-weight ratio much less than 1. For such aircraft, the climb angle is always small and we can obtain an approximation for the rate of climb by using the level flight drag in Eq. (3.4.1).

Solving Eq. (3.4.1) for $\sin\gamma$, we have

$$\sin\gamma = \frac{T_A \cos\alpha_T - D}{W} \tag{3.4.5}$$

The rate of climb, V_c, is the vertical component of the airplane's velocity, i.e., the airspeed multiplied by the sine of the flight path angle. Thus, from Eq. (3.4.5) we have

$$V_c \equiv V \sin\gamma = \frac{VT_A \cos\alpha_T - VD}{W} \tag{3.4.6}$$

For an aircraft with low thrust-to-weight ratio, the climb angle is always small and we can obtain a conservative estimate for the rate of climb by using the level flight drag in Eq. (3.4.6). The level flight drag is related to the thrust required for level flight, T_R, according to Eq. (3.2.1). Thus, the rate of climb is approximated as

$$V_c = \frac{VT_A \cos\alpha_T - VT_R \cos\alpha_T}{W} \tag{3.4.7}$$

The airspeed multiplied by the thrust available and the cosine of the thrust angle is the dot product of the thrust vector with the velocity vector, which is the power available, P_A. The airspeed multiplied by the thrust required for level flight and the cosine of the thrust angle is the power required for level flight, as defined by Eq. (3.3.1). Thus we can write Eq. (3.4.7) as

$$V_c = \frac{P_A - P_R}{W} \tag{3.4.8}$$

The difference between the power available and the power required for level flight is called the *excess power*. For small thrust-to-weight ratio, the rate of climb can be approximated as the excess power divided by the total aircraft weight. This equation is not as simple as it looks. In general, both the power available and the power required are functions of airspeed and air density. The power required for steady level flight has been discussed and can be computed from either Eq. (3.3.6) or Eq. (3.3.7). In general, the power available and its dependence on airspeed, density, and throttle setting needs to be determined from engine tests.

In any case, at very high airspeeds the power required always exceeds the power available and the rate of climb is negative. This means that such airspeeds can only be achieved in a dive. The airspeed at which the full throttle power available matches the power required is the maximum level flight airspeed at the given altitude. The maximum rate of climb can be found graphically by plotting both the power available and the power required as a function of airspeed and finding the point at which the difference is maximum. If the power available is known in the form of an equation, we can find the maximum climb rate and the maximum climb rate airspeed analytically or numerically. Figures 3.4.2 and 3.4.3 show how the thrust and power available might vary with airspeed and throttle setting for a typical general aviation aircraft at sea level.

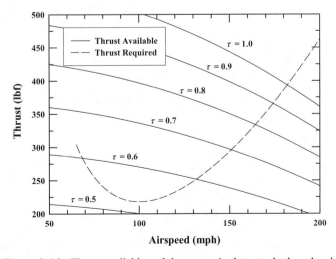

Figure 3.4.2. Thrust available and thrust required at standard sea level.

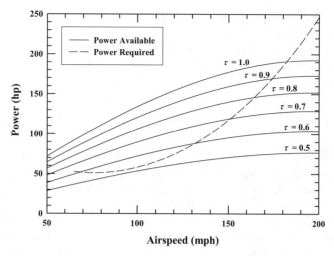

Figure 3.4.3. Power available and power required at standard sea level.

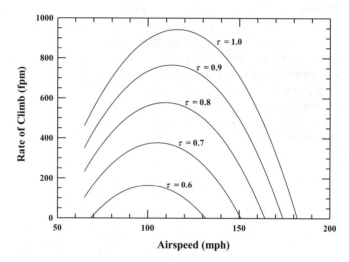

Figure 3.4.4. Rate of climb at standard sea level.

At any throttle setting, in steady level flight, the airplane will fly at the airspeed where the power available is just equal to the power required. If the power available exceeds the power required, the aircraft will accelerate or climb. If the power required exceeds the power available, the aircraft must decelerate or descend. The rate of climb, from Eq. (3.4.8), is the excess power divided by the weight and is shown in Fig. 3.4.4. Notice that the airspeed that results in the maximum rate of climb is greater than that for minimum power required. This is because for this particular power plant and operating point, the power available is increasing with airspeed.

As altitude increases, the power required increases and the power available decreases. Thus, the rate of climb will decrease with altitude. Figure 3.4.5 shows the power available and the power required, for the present aircraft, at 10,000 feet and Fig. 3.4.6 shows the rate of climb at the same altitude.

Notice that for this aircraft and power plant, the maximum rate of climb is reduced from more than 900 ft/min at sea level to about 500 ft/min at 10,000 feet. Since the rate of climb decreases with altitude, at some altitude the power available at the best climb airspeed will just match the power required and the airplane will not be able to climb

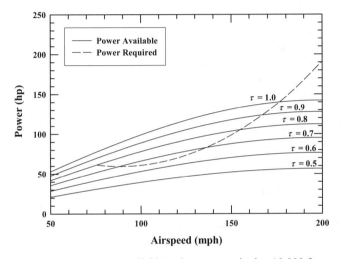

Figure 3.4.5. Power available and power required at 10,000 feet.

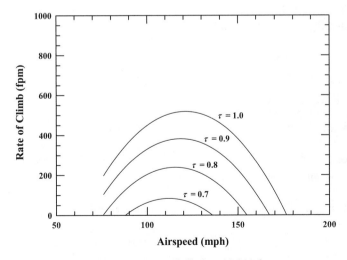

Figure 3.4.6. Rate of climb at 10,000 feet.

higher. This altitude is called the *absolute ceiling* for the aircraft. The *service ceiling* for an aircraft is defined to be the altitude at which the maximum rate of climb is reduced to 100 ft/min.

Figures 3.4.4 and 3.4.6 show that at a given altitude and throttle setting, there are two airspeeds which can result in zero rate of climb. In most cases, only the faster of these two airspeeds is possible because the lower airspeed is usually less than stall.

When an airplane is flying in steady level flight at any given throttle setting, the power available is just equal to the power required, making the rate of climb zero. If the pilot slowly pulls back on the stick, the aircraft will slow down and climb. As the pilot continues to pull back on the stick, the airspeed will continue to decrease and the climb rate will increase until the maximum rate of climb for that throttle setting is reached. If the pilot continues to pull back on the stick after reaching the maximum rate of climb, the airspeed will continue to decrease but now the climb rate will begin to decrease as well. If the process continues, the wing will eventually stall.

Of course, this analysis was based on the assumption of a small climb angle, which means low thrust-to-weight ratio. Some fighter aircraft have a thrust-to-weight ratio greater than 1 and are capable of accelerating at a 90-degree climb angle. When the thrust-to-weight ratio is large, a climb analysis can still be carried out but Eq. (3.4.8) is no longer valid. For an aircraft with large thrust-to-weight ratio, Eqs. (3.4.3) and (3.4.4) must be used to compute the lift and induced drag. The thrust required for steady level flight cannot be substituted for the drag force in Eq. (3.4.6). The resulting analysis is somewhat more complex but still manageable.

For both jet engines and reciprocating engine-propeller combinations, power available is a function of throttle setting, airspeed, and air density. The actual maximum rate of climb for any aircraft must be determined either numerically or graphically from experimental data for the particular engine or engines used.

> **EXAMPLE 3.4.1.** Consider the turbojet aircraft described in Example 3.2.1. For this aircraft the thrust available was approximated as being constant with airspeed and directly proportional to air density,
>
> $$T_A = \tau \frac{\rho}{\rho_0} T_{A0}$$
>
> where τ is a throttle setting expressed as a fraction of the full-throttle thrust and T_{A0} is the full-throttle thrust available at standard sea level. Assuming a thrust vector aligned with the flight path, determine the full-throttle rate of climb for an airspeed of 360 mph at sea level. Compute the rate of climb both with and without using the small-climb-angle approximation.
>
> **Solution.** For this executive business jet we have
>
> $$S_w = 320 \text{ ft}^2, \quad b_w = 54 \text{ ft}, \quad W = 20,000 \text{ lbf}, \quad T_{A0} = 6,500 \text{ lbf},$$
> $$C_{D_0} = 0.023, \quad C_{D_0,L} = 0.0, \quad e = 0.82, \quad R_A = 9.1125,$$

$$V = 360 \text{ mph} = 528 \text{ ft/sec},$$

$$\sqrt{\frac{W/S_w}{\rho}} = \sqrt{\frac{20,000 \text{ lbf}/320 \text{ ft}^2}{0.0023769 \text{ slug/ft}^3}} = 162.16 \text{ ft/sec}, \quad R_V \equiv V \Big/ \sqrt{\frac{W/S_w}{\rho}} = 3.2561$$

Using the small-climb-angle approximation, the rate of climb computed from Eq. (3.4.8) is

$$V_c = \frac{P_A - P_R}{W}$$

For small thrust angles, the power available is the thrust available multiplied by the airspeed and the power required for steady level flight is given by Eq. (3.3.7). Thus we have

$$
\begin{aligned}
V_c &= \frac{T_A}{W} V - \frac{C_{D_0}\rho V^3}{2(W/S_w)} - C_{D_0,L}\, V - \frac{2(W/S_w)}{\pi e R_A \rho V} \\
&= \left(\frac{T_A}{W} R_V - \frac{C_{D_0}}{2} R_V^3 - C_{D_0,L}\, R_V - \frac{2}{\pi e R_A R_V} \right) \sqrt{\frac{W/S_w}{\rho}} \\
&= \left[\frac{6,500}{20,000} 3.2561 - \frac{0.023}{2} 3.2561^3 - 0.0 - \frac{2}{\pi (0.82) 9.1125 (3.2561)} \right] 162.16 \\
&= 103.0 \text{ ft/sec} = 6,179 \text{ ft/min}
\end{aligned}
$$

To obtain the rate of climb for this airplane and operating condition without using the small-climb-angle approximation, we start with Eqs. (3.4.3) and (3.4.4). Equation (3.4.4) can be solved for the cosine of the climb angle. For a thrust vector aligned with the flight path, this gives

$$\cos\gamma = \frac{\frac{1}{2}\rho V^2 S_w}{W} C_L \tag{3.4.9}$$

Using this result in Eq. (3.4.3) and rearranging, we have

$$\frac{\frac{1}{2}\rho V^2 S_w}{W}\left(C_{D_0} + C_{D_0,L} C_L + \frac{C_L^2}{\pi e R_A} \right) + \sqrt{1 - \left(\frac{\frac{1}{2}\rho V^2 S_w}{W} C_L \right)^2} = \frac{T_A}{W}$$

Using the known values for the aircraft, this equation reduces to

$$\sqrt{1.0 - 28.102\, C_L^2} = 0.2031 - 0.2258\, C_L^2$$

This is easily solved to yield the lift coefficient

$$C_L = 0.1850$$

Using this lift coefficient in Eq. (3.4.9), we have

$$\cos \gamma = \frac{\frac{1}{2}\rho V^2 S_w}{W} C_L = 0.9807$$

and

$$\gamma = 11.27°$$

The rate of climb is the airspeed multiplied by the sine of the climb angle. Thus,

$$V_c = V \sin \gamma = \underline{103.2 \text{ ft/sec} = 6,194 \text{ ft/min}}$$

Notice that for this example, with a thrust-to-weight ratio of 0.325, the small-climb-angle approximation used in Eq. (3.4.8) is accurate to within 1 percent.

EXAMPLE 3.4.2. Consider the general aviation airplane described in Example 3.3.1. For this propeller-driven aircraft the power available was approximated as being constant with airspeed and directly proportional to air density,

$$P_A = \tau \frac{\rho}{\rho_0} P_{A0}$$

where τ is a throttle setting expressed as a fraction of the full-throttle power available and P_{A0} is the full throttle power available at standard sea level. Using the small-climb-angle approximation, determine the maximum rate of climb for this airplane at both sea level and 10,000 feet.

Solution. For this airplane

$$S_w = 180 \text{ ft}^2, \quad b_w = 33 \text{ ft}, \quad W = 2,700 \text{ lbf}, \quad P_{A0} = 180 \text{ hp} = 99,000 \text{ ft·lbf/sec},$$
$$C_{D_0} = 0.023, \quad C_{D_0,L} = 0.0, \quad e = 0.82, \quad R_A = 6.05, \quad W/S_w = 15.0 \text{ lbf/ft}^2$$

With the small-climb-angle approximation, the rate of climb is computed by using the power required from Eq. (3.3.7) in Eq. (3.4.8). For this airplane, we have

$$V_c = \frac{P_A - P_R}{W} = \tau \frac{\rho}{\rho_0} \frac{P_{A0}}{W} - \frac{C_{D_0}\rho V^3}{2(W/S_w)} - \frac{2(W/S_w)}{\pi e R_A \rho V}$$

Here again it is convenient to use the dimensionless velocity ratio

$$R_V \equiv V \Big/ \sqrt{\frac{W/S_w}{\rho}}$$

and the dimensionless power-to-weight ratio

$$R_{P/W} \equiv P_A \Big/ \left(W \sqrt{\frac{W/S_w}{\rho}} \right) = \tau \frac{\rho}{\rho_0} P_{A0} \Big/ \left(W \sqrt{\frac{W/S_w}{\rho}} \right)$$

Thus, for this airplane, the rate of climb can be written as

$$V_c = \left(R_{P/W} - \frac{C_{D_0}}{2} R_V^3 - \frac{2}{\pi e R_A R_V} \right) \sqrt{\frac{W/S_w}{\rho}} \tag{3.4.10}$$

The velocity ratio that results in maximum rate of climb is found by differentiating the rate of climb with respect to velocity ratio and setting the result equal to zero. This gives

$$\frac{\partial V_c}{\partial R_V} = \left(-\frac{3 C_{D_0}}{2} R_V^2 + \frac{2}{\pi e R_A R_V^2} \right) \sqrt{\frac{W/S_w}{\rho}} = 0$$

or

$$R_V = \left(\frac{4}{3 \pi e R_A C_{D_0}} \right)^{1/4} = \left(\frac{4}{3\pi(0.82)(6.05)(0.023)} \right)^{1/4} = 1.3887$$

The maximum rate of climb is found by using this result in Eq. (3.4.10). For this airplane at sea level

$$R_{P/W} \equiv P_A \Big/ \left(W \sqrt{\frac{W/S_w}{\rho}} \right) = 0.462$$

$$\left(V_c \right)_{max} =$$
$$\left(0.462 - \frac{0.023}{2}(1.3887)^3 - \frac{2}{\pi(0.82)(6.05)(1.3887)} \right) \sqrt{\frac{15.0 \text{ lbf/ft}^2}{0.0023769 \text{ slug/ft}^3}}$$
$$= 26.91 \text{ ft/sec} = 1,615 \text{ ft/min}$$

and at 10,000 feet

$$R_{P/W} \equiv P_A \Big/ \left(W \sqrt{\frac{W/S_w}{\rho}} \right) = 0.293$$

$$\left(V_c \right)_{max} =$$

$$\left(0.293 - \frac{0.023}{2}(1.3887)^3 - \frac{2}{\pi(0.82)(6.05)(1.3887)} \right) \sqrt{\frac{15.0 \text{ lbf/ft}^2}{0.0017556 \text{ slug/ft}^3}}$$

$$= 15.69 \text{ ft/sec} = 942 \text{ ft/min}$$

EXAMPLE 3.4.3. For the turbojet aircraft described in Examples 3.2.1 and 3.4.1, use the small-climb-angle approximation and determine the maximum rate of climb at both sea level and 20,000 feet.

Solution. For this executive business jet at full throttle

$$R_{T/W} \equiv \frac{T_A}{W} = \frac{\rho}{\rho_0} \frac{T_{A0}}{W},$$

$$S_w = 320 \text{ ft}^2, \quad b_w = 54 \text{ ft}, \quad W = 20{,}000 \text{ lbf}, \quad T_{A0} = 6{,}500 \text{ lbf},$$

$$C_{D_0} = 0.023, \quad C_{D_0,L} = 0.0, \quad e = 0.82, \quad R_A = 9.1125,$$

$$W/S_w = 62.5 \text{ lbf/ft}^2$$

Using the small-climb-angle approximation, the rate of climb for this aircraft is

$$V_c = \frac{T_A}{W} V - \frac{C_{D_0} \rho V^3}{2(W/S_w)} - C_{D_0,L} V - \frac{2(W/S_w)}{\pi e R_A \rho V}$$

$$= \left(R_{T/W} R_V - \frac{C_{D_0}}{2} R_V^3 - C_{D_0,L} R_V - \frac{2}{\pi e R_A R_V} \right) \sqrt{\frac{W/S_w}{\rho}}$$

The velocity ratio at maximum climb rate is found from

$$\frac{\partial V_c}{\partial R_V} = \left(R_{T/W} - \frac{3 C_{D_0}}{2} R_V^2 + \frac{2}{\pi e R_A R_V^2} \right) \sqrt{\frac{W/S_w}{\rho}} = 0$$

or

$$\frac{3 C_{D_0}}{2} R_V^4 - R_{T/W} R_V^2 - \frac{2}{\pi e R_A} = 0$$

This equation is quadratic in the velocity ratio squared. Thus, at maximum rate of climb,

$$R_V^2 = \frac{R_{T/W}}{3 C_{D_0}} \pm \sqrt{\left(\frac{R_{T/W}}{3 C_{D_0}} \right)^2 + \frac{4}{3 \pi e R_A C_{D_0}}}$$

where only the positive root is of interest. For this airplane at standard sea level,

$$R_{T/W} \equiv \frac{T_A}{W} = \frac{\rho}{\rho_0} \frac{T_{A0}}{W} = 0.325$$

$$R_V = \sqrt{\frac{R_{T/W}}{3C_{D_0}} + \sqrt{\left(\frac{R_{T/W}}{3C_{D_0}}\right)^2 + \frac{4}{3\pi e R_A C_{D_0}}}} = 3.11055$$

$$\left(V_c\right)_{max} = \left(R_{T/W} R_V - \frac{C_{D_0}}{2} R_V^3 - C_{D_0,L} R_V - \frac{2}{\pi e R_A R_V}\right)\sqrt{\frac{W/S_w}{\rho}} = \underline{8,487 \text{ ft/min}}$$

and at 20,000 feet,

$$R_{T/W} \equiv \frac{T_A}{W} = \frac{\rho}{\rho_0} \frac{T_{A0}}{W} = 0.173$$

$$R_V = \sqrt{\frac{R_{T/W}}{3C_{D_0}} + \sqrt{\left(\frac{R_{T/W}}{3C_{D_0}}\right)^2 + \frac{4}{3\pi e R_A C_{D_0}}}} = 2.33800$$

$$\left(V_c\right)_{max} = \left(R_{T/W} R_V - \frac{C_{D_0}}{2} R_V^3 - C_{D_0,L} R_V - \frac{2}{\pi e R_A R_V}\right)\sqrt{\frac{W/S_w}{\rho}} = \underline{4,066 \text{ ft/min}}$$

3.5. Fuel Consumption and Endurance

Another important performance parameter for an aircraft is fuel consumption during steady level flight. Fuel consumption is sometimes specified in terms of endurance. Endurance is defined as the total time that an aircraft can stay in the air on a tank of fuel. We can maximize endurance, for a given aircraft and fuel capacity, by minimizing the fuel consumption per unit time.

For fuel consumption analysis it is convenient to define the power-specific fuel consumption for a power plant, q_P, to be the weight of fuel consumption per unit time divided by the propulsive power available. In general, the power-specific fuel consumption for a given power plant is a function of air density, throttle setting, and airspeed,

$$q_P \equiv \frac{\dot{Q}}{P_A} = q_P(\rho, \tau, V) \tag{3.5.1}$$

where $\dot{Q}$ is the total weight of fuel consumed per unit time and P_A is the propulsive power available. Power-specific fuel consumption and its dependence on air density, throttle setting, and airspeed for any particular power plant must be determined from power plant tests. Figure 3.5.1 shows how this power-specific fuel consumption for an internal combustion engine with a fixed-pitch propeller might vary with airspeed and throttle setting at sea level.

The empirical relation implied by Eq. (3.5.1) and shown hypothetically in Fig. 3.5.1 provides one equation relating the fuel consumption rate to the air density, throttle setting, and airspeed. Steady level flight imposes a second relationship between air density, throttle setting, and airspeed. The power available from the power plant can also be expressed as an empirical function of air density, throttle setting, and airspeed,

$$P_A = P_A(\rho, \tau, V) \tag{3.5.2}$$

In steady level flight, the power available must equal the power required. Using the small-thrust-angle approximation, from Eq. (3.3.7) the power required is given by

$$P_R = \left(\frac{C_{D_0} \rho V^3}{2(W/S_w)} + C_{D_0,L} V + \frac{2(W/S_w)}{\pi e R_A \rho V} \right) W \tag{3.5.3}$$

Thus, combining Eqs. (3.5.2) and (3.5.3), steady level flight requires that

$$P_A(\rho, \tau, V) = \left(\frac{C_{D_0} \rho V^3}{2(W/S_w)} + C_{D_0,L} V + \frac{2(W/S_w)}{\pi e R_A \rho V} \right) W \tag{3.5.4}$$

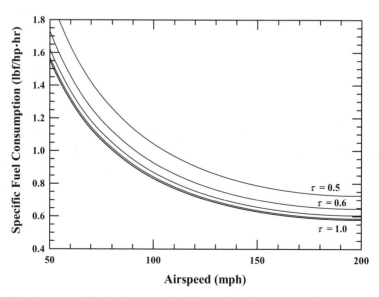

Figure 3.5.1. Power-specific fuel consumption for an aircraft power plant at standard sea level.

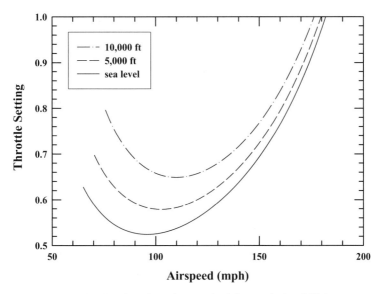

Figure 3.5.2. Throttle setting required for steady level flight.

Thus, steady level flight imposes a second relationship among the air density, throttle setting, and airspeed, as given by Eq. (3.5.4). At any given altitude and airspeed, this empirical relation determines the throttle setting necessary to maintain steady level flight. Figure 3.5.2 shows how the throttle setting required for steady level flight might vary with airspeed and altitude for a typical general aviation aircraft.

Combining the power-specific fuel consumption relation implied by Eq. (3.5.1) and the steady level flight requirement implied by Eq. (3.5.4), for a given aircraft and power plant we can eliminate the throttle setting and express power-specific fuel consumption for steady level flight as a function of only the airspeed and air density. Combining the power-specific fuel consumption relation shown in Fig. 3.5.1 with the steady level flight requirement shown in Fig. 3.5.2, we obtain the power-specific fuel consumption relation for steady level flight shown in Fig. 3.5.3.

The total fuel consumption rate is then found by multiplying the power-specific fuel consumption for steady level flight by the power required for steady level flight at the same airspeed. Figure 3.5.4 shows the total fuel consumption rate as a function of airspeed and altitude for the present example.

We notice that for this example, the fuel consumption rate has a minimum at some value of airspeed that is somewhat faster than the airspeed for minimum power required. This is because, for this example, the power-specific fuel consumption is decreasing with airspeed. If the power-specific fuel consumption were independent of airspeed, the airspeed for minimum fuel consumption rate would be equal to the airspeed for minimum power required. We also notice that the minimum fuel consumption rate and the airspeed required to attain minimum fuel consumption rate both increase with altitude for this example.

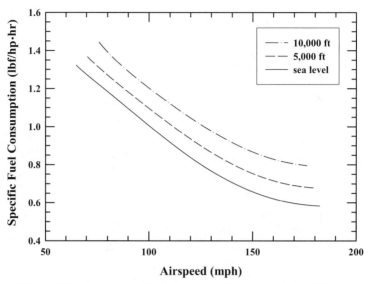

Figure 3.5.3. Power-specific fuel consumption for steady level flight.

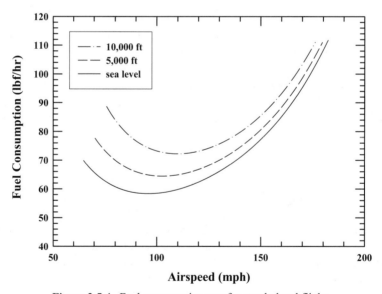

Figure 3.5.4. Fuel consumption rate for steady level flight.

It is also important to recognize that because fuel consumption depends on power required and power required depends on weight, the fuel consumption rate will vary with the gross weight of the aircraft. For the example considered here, Figure 3.5.5 shows the fuel consumption at sea level as a function of airspeed for three different values of gross vehicle weight. Notice that the fuel consumption rate, at any airspeed, increases with gross vehicle weight (GVW).   Also notice that the airspeed required to attain minimum fuel consumption rate increases with gross vehicle weight. This means that to maximize endurance, the pilot should gradually decrease airspeed as fuel is burned off and the gross vehicle weight is decreased. Figure 3.5.6 shows how the airspeed at sea level should be varied with gross weight for the aircraft under consideration.

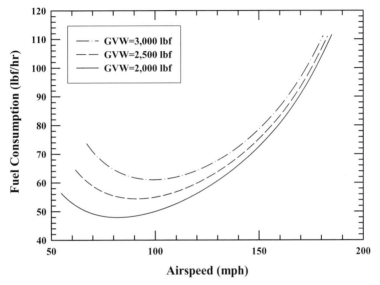

Figure 3.5.5. Effect of gross weight on the fuel consumption rate at sea level.

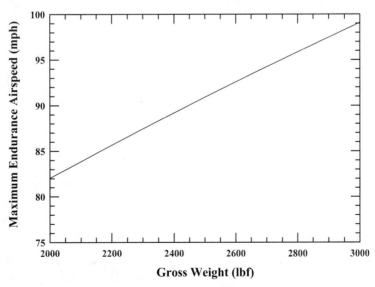

Figure 3.5.6. Maximum endurance airspeed as a function of gross weight at sea level.

The weight of the aircraft decreases with time as a result of the fuel that is burned. Thus, the time rate of change in the gross aircraft weight is the negative of the fuel consumption rate,

$$\frac{dW}{dt} = -\dot{Q} \tag{3.5.5}$$

The maximum endurance, E_{max}, is the time required to decrease the aircraft weight from the gross weight with fuel tanks full, W_f, to the gross weight with fuel tanks empty, W_e. This is found by rearranging Eq. (3.5.5) and integrating,

$$E_{max} = \int_0^{E_{max}} dt = -\int_{W_f}^{W_e} \frac{1}{\dot{Q}} dW = \int_{W_e}^{W_f} \frac{1}{\dot{Q}} dW \tag{3.5.6}$$

The inverse of the total fuel consumption rate, which appears in Eq. (3.5.6), is the instantaneous specific endurance, i.e., the flight time per unit weight of fuel burned. Figure 3.5.7 shows how the specific endurance varies with gross weight at sea level for this same hypothetical aircraft. The result shown in this figure is based on maintaining the maximum endurance airspeed at all times. Therefore, it represents the maximum possible specific endurance. As specified by Eq. (3.5.6), the total maximum endurance is simply the area under this curve between W_e and W_f.

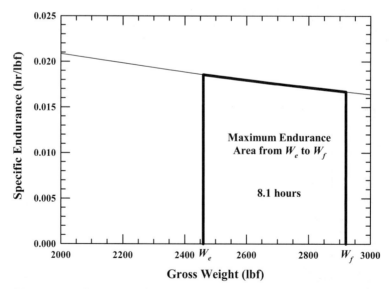

Figure 3.5.7. Maximum specific endurance as a function of gross weight at sea level. The maximum total endurance is the area under this curve between the aircraft gross weight with fuel tanks empty and the aircraft gross weight with fuel tanks full.

EXAMPLE 3.5.1. As discussed in Example 3.3.1, an internal combustion engine combined with a variable-pitch propeller can be adjusted, either manually by the pilot or automatically with a governor mechanism, to produce an available power that is nearly independent of airspeed over the range of airspeeds encountered in normal flight. For such a power plant, the power-specific fuel consumption based on propulsive power available can be written as

$$q_P \equiv \dot{Q}/P_A = \dot{Q}/\eta_P P_b = q_b/\eta_P$$

where η_p is the propeller efficiency, P_b is the brake power developed by the engine, and q_b is the power-specific fuel consumption for the engine based on brake power. With the fuel mixture and propeller pitch properly adjusted, both the power-specific fuel consumption for the engine and the propeller efficiency are very nearly independent of airspeed and altitude. Thus, as a first approximation, this type of power plant is commonly modeled as having a power-specific fuel consumption based on propulsive power available that is constant with both airspeed and air density.

Consider the general aviation aircraft described in Example 3.3.1. The power-specific fuel consumption based on brake power is 0.5 lbf/hp·hr and the propeller efficiency is 80 percent. The dry weight for the airplane is 2,300 lbf and the fuel capacity is 400 lbf. Assuming that the pilot maintains the maximum endurance airspeed at all times, determine the total endurance at sea level.

Solution. For this airplane

$$S_w = 180 \text{ ft}^2, \quad b_w = 33 \text{ ft}, \quad W_f = 2,700 \text{ lbf}, \quad W_e = 2,300 \text{ lbf},$$
$$C_{D_0} = 0.023, \quad C_{D_0,L} = 0.0, \quad e = 0.82, \quad R_A = 6.05,$$
$$q_P = q_b/\eta_P = 0.625 \text{ lbf/hp·hr} = 3.157 \times 10^{-7} \text{ ft}^{-1}$$

The specific endurance is the inverse of the total fuel consumption rate and, for steady level flight, the power available must equal the power required,

$$1/\dot{Q} = 1/q_P P_A = 1/q_P P_R$$

For this airplane the power required is

$$P_R = DV$$

$$= \tfrac{1}{2}\rho V^2 S_w \left(C_{D_0} + \frac{C_L^2}{\pi e R_A} \right) V = \tfrac{1}{2}\rho V^2 S_w \left[C_{D_0} + \frac{1}{\pi e R_A} \left(\frac{W}{\tfrac{1}{2}\rho V^2 S_w} \right)^2 \right] V$$

$$= \left(\frac{C_{D_0}\rho V^3}{2(W/S_w)} + \frac{2(W/S_w)}{\pi e R_A \rho V} \right) W = \left(\frac{C_{D_0}}{2} R_V^3 + \frac{2}{\pi e R_A R_V} \right) W \sqrt{\frac{W/S_w}{\rho}}$$

Thus, for this airplane, the specific endurance is

$$
\frac{1}{\dot{Q}} = \frac{1}{q_P P_R} = \frac{1}{q_P \left(\dfrac{C_{D_0}}{2} R_V^3 + \dfrac{2}{\pi e R_A R_V} \right) W \sqrt{\dfrac{W/S_w}{\rho}}}
\tag{3.5.7}
$$

The velocity ratio that results in maximum specific endurance is found by differentiating the specific endurance with respect to velocity ratio and setting the result equal to zero. This gives

$$
R_V = \left(\frac{4}{3 \pi e R_A C_{D_0}} \right)^{1/4}
$$

Using this result in Eq. (3.5.7), the maximum instantaneous specific endurance for this airplane is

$$
\frac{1}{\dot{Q}} = \frac{(3 \pi e R_A)^{3/4} \sqrt{\rho}}{\sqrt{32} \, q_P \, C_{D_0}^{1/4} \, W \sqrt{W/S_w}}
$$

From Eq. (3.5.6) the maximum constant-altitude endurance at sea level is

$$
\begin{aligned}
E_{max} &= \int_{W_e}^{W_f} \frac{1}{\dot{Q}} \, dW = \int_{W_e}^{W_f} \frac{(3 \pi e R_A)^{3/4} \sqrt{\rho}}{\sqrt{32} \, q_P \, C_{D_0}^{1/4} \, W \sqrt{W/S_w}} \, dW \\[2mm]
&= \frac{(3 \pi e R_A)^{3/4} \sqrt{\rho S_w}}{\sqrt{32} \, q_P \, C_{D_0}^{1/4}} \int_{W_e}^{W_f} \frac{dW}{W^{3/2}} \\[2mm]
&= \frac{2(3 \pi e R_A)^{3/4}}{\sqrt{32} \, C_{D_0}^{1/4}} \left(\frac{1}{\sqrt{W_e}} - \frac{1}{\sqrt{W_f}} \right) \frac{\sqrt{\rho S_w}}{q_P} \\[2mm]
&= \frac{2[3 \pi (0.82)(6.05)]^{3/4}}{\sqrt{32} \, (0.023)^{1/4}} \left(\frac{1}{\sqrt{2,300 \text{ lbf}}} - \frac{1}{\sqrt{2,700 \text{ lbf}}} \right) \\[2mm]
&\quad \times \frac{\sqrt{0.0023769 \text{ slug/ft}^3 \, (180 \text{ ft}^2)}}{3.157 \times 10^{-7} \text{ ft}^{-1}} \\[2mm]
&= \underline{54{,}030 \text{ sec}} = \underline{15.0 \text{ hr}}
\end{aligned}
$$

It should be noted that for this propeller-driven aircraft, the maximum endurance is proportional to the square root of the air density. Thus, endurance is shortened as altitude is increased.

3.6. Fuel Consumption and Range

For most aircraft applications, it is not appropriate to optimize fuel consumption in terms of endurance. Fuel consumption is usually specified in terms of range. *Range* is defined as the total distance that an aircraft can travel on a tank of fuel. We can maximize range for a given aircraft and fuel capacity by minimizing fuel consumption per unit distance. While fuel consumption with time is normally expressed as a fuel consumption rate, for example in pounds per hour, the consumption of fuel with distance is not usually expressed in terms of fuel consumption per unit distance but rather, in terms of distance per unit fuel consumption. For an automobile this is commonly expressed in miles per gallon. For an aircraft, weight is more important than volume, so it is convenient to express the fuel consumption with distance in terms of distance traveled per unit weight of fuel consumed. This is the specific range for the aircraft, r_s, or the inverse of the fuel consumption per unit distance. Distance traveled over the ground per unit weight of fuel consumed is the distance traveled per unit time or the ground speed, V_g, divided by the weight of fuel consumed per unit time,

$$r_s \equiv \frac{V_g}{\dot{Q}} \tag{3.6.1}$$

In the performance measures that we have considered to this point, the only velocity that has been important has been the airspeed of the aircraft. Here we see that the range of an aircraft depends not only on airspeed but on ground speed as well. For level flight, the relationship among ground speed, airspeed, and wind speed is shown in Fig. 3.6.1.

We see from Fig. 3.6.1 that ground speed for an aircraft depends on the speed and direction of the wind as well as airspeed and heading. If an aircraft is flying directly into a headwind, the ground speed will be less than the airspeed by an amount equal to the wind speed. If an aircraft is flying directly with a tailwind, the ground speed will be equal to the sum of the airspeed and wind speed. A crosswind will also affect the ground speed of an aircraft because, to maintain a specified track over the ground, the pilot must *crab* into the wind at an angle to the desired line of flight. When flying in a direct crosswind, only one component of the airspeed contributes to the ground speed. The other component must balance the crosswind to keep the aircraft from drifting off track.

In Fig. 3.6.1 the angle ψ_g is called the *ground track*, ψ is called the *heading*, and ψ_w is the wind direction. The angle φ_c is called the *crab angle*, and is equal to the difference between the heading and the ground track. The angle φ_w, the *wind-track angle*, is equal to the difference between the wind direction and the ground track. From this figure we see that the ground speed, V_g, is related to the airspeed, V, and the wind speed, V_w, according to

$$V_g = \sqrt{V^2 - V_w^2 \sin^2 \varphi_w} - V_w \cos \varphi_w \tag{3.6.2}$$

The fuel consumption rate is the power-specific fuel consumption multiplied by the power available. For an aircraft in steady level flight, the power available must equal the power required, which is the product of the drag and the airspeed. Thus, the fuel consumption per unit time for steady level flight can be written as

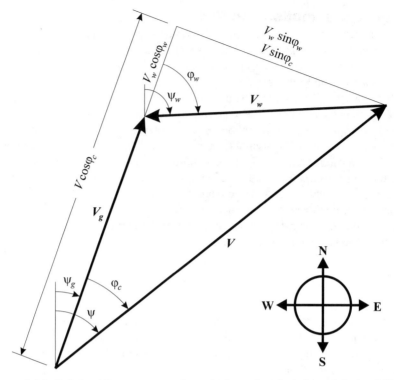

Figure 3.6.1. Relationship among ground speed, airspeed, and wind speed for level flight.

$$\dot{Q} = q_P D V \tag{3.6.3}$$

Using the small-thrust-angle approximation, the drag is equal to the thrust required as given by Eq. (3.2.25),

$$D = \left(\frac{\frac{1}{2}\rho V^2 C_{D_0}}{W/S_w} + C_{D_0,L} + \frac{W/S_w}{\frac{1}{2}\pi e R_A \rho V^2} \right) W \tag{3.6.4}$$

Combining Eqs. (3.6.1) and (3.6.3), the specific range can be written

$$r_s = \frac{V_g}{q_P D V} \tag{3.6.5}$$

From Eqs. (3.6.4) and (3.6.5) we see that because the drag has a minimum at some particular value of airspeed, the specific range will exhibit a maximum with airspeed. However, the maximum range airspeed is not necessarily the minimum drag airspeed.

To find the airspeed that will maximize the specific range for any particular wind speed and direction, we differentiate Eq. (3.6.5) with respect to airspeed and set the result to zero,

$$\frac{\partial r_s}{\partial V} = \frac{1}{q_P DV}\frac{\partial V_g}{\partial V} - \frac{V_g}{q_P^2 DV}\frac{\partial q_P}{\partial V} - \frac{V_g}{q_P D^2 V}\frac{\partial D}{\partial V} - \frac{V_g}{q_P DV^2} = 0 \qquad (3.6.6)$$

or

$$\frac{V}{D}\frac{\partial D}{\partial V} = \frac{V}{V_g}\frac{\partial V_g}{\partial V} - \frac{V}{q_P}\frac{\partial q_P}{\partial V} - 1 \qquad (3.6.7)$$

From Eq. (3.6.2) we can write

$$\frac{V}{V_g}\frac{\partial V_g}{\partial V} = \left(\frac{V}{\sqrt{V^2 - V_w^2\sin^2\varphi_w} - V_w\cos\varphi_w}\right)\left(\frac{V}{\sqrt{V^2 - V_w^2\sin^2\varphi_w}}\right) = \frac{V}{V - V_{eh}} \qquad (3.6.8)$$

where V_{eh} can be thought of as an *effective headwind*, that is, the velocity of a pure headwind that would produce the same ground speed as the actual wind. This effective headwind speed is given by

$$V_{eh} = V_w\left[\cos\varphi_w\sqrt{1 - (V_w/V)^2\sin^2\varphi_w} + (V_w/V)\sin^2\varphi_w\right] \qquad (3.6.9)$$

Using Eqs. (3.6.8) and (3.6.4) in Eq. (3.6.7), at the airspeed that gives maximum specific range, we require that

$$\frac{V}{D}\frac{\partial D}{\partial V} = \left(\frac{\rho V^2 C_{D_0}}{W/S_w} - \frac{4W/S_w}{\pi e R_A \rho V^2}\right)\bigg/\left(\frac{\frac{1}{2}\rho V^2 C_{D_0}}{W/S_w} + C_{D_0,L} + \frac{W/S_w}{\frac{1}{2}\pi e R_A \rho V^2}\right)$$

$$= \frac{V}{V - V_{eh}} - \frac{V}{q_P}\frac{\partial q_P}{\partial V} - 1 = \frac{V_{eh}}{V - V_{eh}} - \frac{V}{q_P}\frac{\partial q_P}{\partial V} \qquad (3.6.10)$$

Figure 3.6.2 shows how the term on the left-hand side of Eq. (3.6.10) varies with airspeed for a typical general aviation aircraft. From this figure we see that the left-hand side of Eq. (3.6.10) is a monotonically increasing function of airspeed. We also see that increasing altitude moves the curve to the right, in the direction of higher airspeed. The exact shape and position of the curves will vary with aircraft design and operating conditions, but these general characteristics hold for any aircraft. Since the left-hand side of Eq. (3.6.10) is always increasing with airspeed, anything that causes the right-hand side of this equation to increase will cause the maximum range airspeed to increase, and anything that will cause the right-hand side to decrease will cause the maximum range airspeed to decrease.

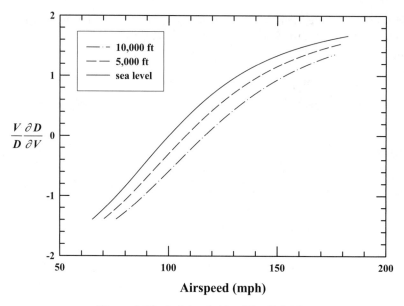

Figure 3.6.2. Left-hand side of Eq. (3.6.10).

Thus we observe from Eq. (3.6.10) that if both the change in power-specific fuel consumption with respect to airspeed and the wind speed are zero, the maximum range airspeed will be equal to the airspeed that results in minimum drag. We see further that the maximum range airspeed will increase as the change in power-specific fuel consumption with respect to airspeed decreases. Furthermore, a headwind will increase the maximum range airspeed, a tailwind will decrease the maximum range airspeed, and any crosswind will always increase the airspeed required for maximum range.

Before we can determine the maximum range airspeed and the maximum specific range for any particular aircraft, we must know how the thrust available and the power-specific fuel consumption depend on air density, throttle setting, and airspeed for the specific power plant that is being used. As mentioned previously, these relationships need to be determined from engine tests. Once these experimental relationships are available, we can find the throttle setting that will result in steady level flight at any given airspeed and altitude, by setting the power available from the power plant equal to the power required by the airframe for steady level flight. For the example that we have been considering, this relationship among throttle setting, airspeed, and altitude is shown in Fig. 3.5.2.

Using this relationship for throttle setting as a function of airspeed and air density, which is imposed by the condition of steady level flight, we can eliminate throttle setting from the relationship for power-specific fuel consumption as a function of air density, throttle setting, and airspeed. This gives the power-specific fuel consumption for steady level flight as a function of only airspeed and air density. Figure 3.5.3 shows this relationship for the present example.

We can now use this relationship with Eqs. (3.6.2) and (3.6.4) in Eq. (3.6.5) to obtain the specific range as a function of airspeed, air density, wind speed, and wind direction. For the aircraft that we have been using, Fig. 3.6.3 shows how the specific range depends on airspeed and altitude for zero wind. Figure 3.6.4 shows the specific range at sea level for four different wind conditions. Notice that the wind affects both the maximum specific range and the maximum range airspeed. As the wind speed approaches the minimum drag airspeed for the airframe, this effect becomes very large.

The airspeed that results in maximum specific range can be found by solving Eq. (3.6.10) for the airspeed. As was the case for the maximum endurance airspeed, the maximum range airspeed is a function of aircraft gross weight. In general, the maximum range airspeed will be a function of weight, wind, and air density. For the current example, Fig. 3.6.5 shows the maximum range airspeed at sea level as a function of gross weight under four different wind conditions. For the purpose of comparison, the maximum endurance airspeed is also included here.

Comparing Figs. 3.2.2 and 3.6.4 we see that the airspeed required for maximum range, even with a 50-mph tailwind, is considerably higher than the minimum drag airspeed. This is a function of the performance of the power plant. For this example, with a gross weight of 3,000 pounds, the airframe has minimum drag at about 100 mph. If the power-specific fuel consumption were independent of airspeed, this would be the maximum range airspeed for no wind. However, as can be seen from Fig. 3.5.1, the power-specific fuel consumption for this power plant is minimum at an airspeed of almost 200 mph. The net result is a trade-off between minimizing the drag and minimizing the power-specific fuel consumption, which results in a no-wind maximum range airspeed of about 135 mph.

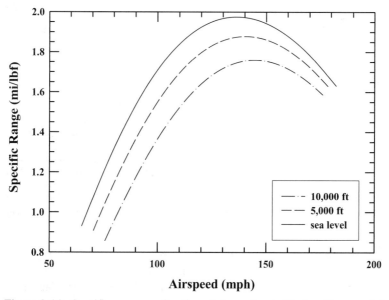

Figure 3.6.3. Specific range as a function of airspeed and altitude with zero wind.

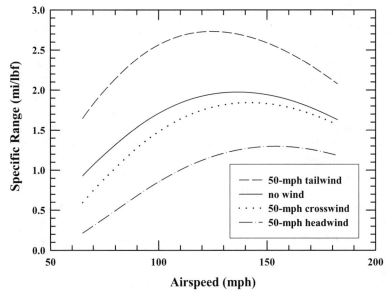

Figure 3.6.4. Specific range at sea level as a function of airspeed and wind.

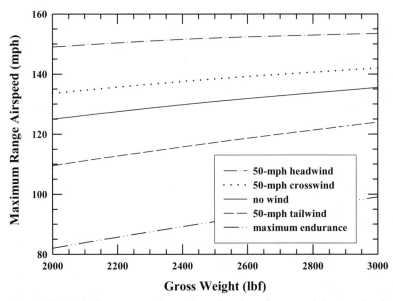

Figure 3.6.5. Maximum range airspeed as a function of gross weight at sea level.

It is also important to notice that the airspeed required for maximum range is considerably higher than that required for maximum endurance, with the same engine and airframe. This is true in general, independent of the power plant performance characteristics. The zero-wind maximum range airspeed is related to the minimum drag airspeed in the same way that the maximum endurance airspeed is related to the airspeed for minimum power required. Since the airspeed for minimum power required is always less then the airspeed for minimum drag, the maximum endurance airspeed will always be less than the zero wind maximum range airspeed.

Using the maximum range airspeed in Eq. (3.6.5), we obtain the maximum possible specific range for a given weight, wind, and air density. For this example, Fig. 3.6.6 shows how the maximum specific range at sea level varies with gross weight and wind.

As the aircraft travels, fuel is burned and the gross weight of the aircraft decreases. The specific range is the negative of the change in distance with respect to gross weight,

$$\frac{dx_f}{dW} = -r_s \tag{3.6.11}$$

where x_f is the distance traveled over the ground. The maximum range, R_{max}, is found by integrating Eq. (3.6.11) from the gross weight with fuel tanks full, W_f, to the gross weight with fuel tanks empty, W_e,

$$R_{max} = \int_0^{R_{max}} dx_f = -\int_{W_f}^{W_e} r_s \, dW = \int_{W_e}^{W_f} r_s \, dW \tag{3.6.12}$$

Thus, the maximum range for any of the wind conditions shown in Fig. 3.6.6 would be the area under the curve for that wind condition between W_e and W_f. This is shown for no wind at standard sea level in Fig. 3.6.7. Figure 3.6.8 shows how the maximum range at sea level is affected by wind.

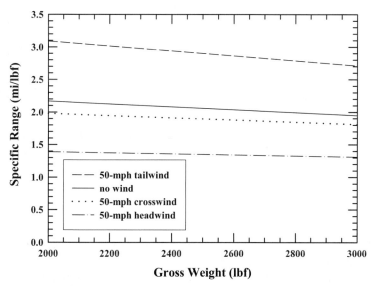

Figure 3.6.6. Maximum specific range as a function of gross weight at sea level.

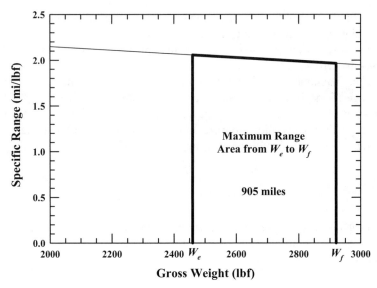

Figure 3.6.7.   Maximum range for no wind at sea level.

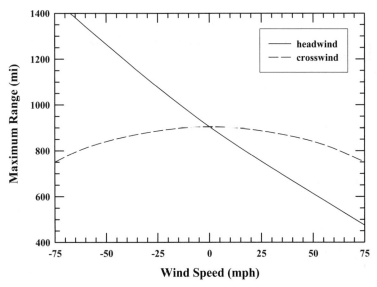

Figure 3.6.8.   Effect of wind on maximum range.

EXAMPLE 3.6.1.   For the propeller-driven general aviation aircraft and power plant described in Example 3.5.1, assuming that the pilot maintains the maximum range airspeed at all times, determine the maximum range for this aircraft with no wind at sea level.

Solution. For this airplane

$$S_w = 180 \text{ ft}^2, \quad b_w = 33 \text{ ft}, \quad W_f = 2{,}700 \text{ lbf}, \quad W_e = 2{,}300 \text{ lbf},$$
$$C_{D_0} = 0.023, \quad C_{D_0,L} = 0.0, \quad e = 0.82, \quad R_A = 6.05,$$
$$q_P = q_b/\eta_p = 0.625 \text{ lbf/hp·hr} = 3.157 \times 10^{-7} \text{ ft}^{-1}$$

The specific range is the ground speed divided by the total fuel consumption rate. The total fuel consumption rate is the power-specific fuel consumption multiplied by the power available. However, for steady level flight with no wind, the power available is equal to the power required and the ground speed is equal to the airspeed. Thus, we have

$$r_s \equiv \frac{V_g}{\dot{Q}} = \frac{V_g}{q_P P_A} = \frac{V}{q_P P_R}$$

For this airplane the power required is

$$
\begin{aligned}
P_R &= DV = \tfrac{1}{2}\rho V^2 S_w \left(C_{D_0} + \frac{C_L^2}{\pi e R_A} \right) V \\
&= \tfrac{1}{2}\rho V^2 S_w \left[C_{D_0} + \frac{1}{\pi e R_A} \left(\frac{W}{\frac{1}{2}\rho V^2 S_w} \right)^2 \right] V \\
&= \left(\frac{C_{D_0}\rho V^3}{2(W/S_w)} + \frac{2(W/S_w)}{\pi e R_A \rho V} \right) W = \left(\frac{C_{D_0}}{2} R_V^3 + \frac{2}{\pi e R_A R_V} \right) W \sqrt{\frac{W/S_w}{\rho}}
\end{aligned}
$$

Thus, for this airplane, the no-wind specific range is

$$
\begin{aligned}
r_s &= \frac{V}{q_P P_R} = \frac{V}{q_P \left(\dfrac{C_{D_0}}{2} R_V^3 + \dfrac{2}{\pi e R_A R_V} \right) W \sqrt{\dfrac{W/S_w}{\rho}}} \\
&= \frac{R_V}{q_P \left(\dfrac{C_{D_0}}{2} R_V^3 + \dfrac{2}{\pi e R_A R_V} \right) W}
\end{aligned} \tag{3.6.13}
$$

The velocity ratio that results in maximum specific range is found by differentiating the specific range with respect to velocity ratio and setting the result equal to zero. This gives

$$R_V = \left(\frac{4}{\pi e R_A C_{D_0}} \right)^{1/4}$$

Using this result in Eq. (3.6.13), the maximum instantaneous specific range for this airplane is

$$r_s = \sqrt{\frac{\pi e R_A}{4 C_{D_0}}} \frac{1}{q_P W}$$

From Eq. (3.6.12) the maximum range is

$$R_{\max} = \int_{W_e}^{W_f} r_s \, dW = \int_{W_e}^{W_f} \sqrt{\frac{\pi e R_A}{4 C_{D_0}}} \frac{1}{q_P W} dW = \sqrt{\frac{\pi e R_A}{4 C_{D_0}}} \frac{1}{q_P} \int_{W_e}^{W_f} \frac{dW}{W}$$

$$= \sqrt{\frac{\pi e R_A}{4 C_{D_0}}} \frac{1}{q_P} \ln\left(\frac{W_f}{W_e}\right)$$

$$= \sqrt{\frac{\pi (0.82)(6.05)}{4(0.023)}} \frac{1}{3.157 \times 10^{-7} \text{ ft}^{-1}} \ln\left(\frac{2,700 \text{ lbf}}{2,300 \text{ lbf}}\right)$$

$$= \underline{6.611 \times 10^6 \text{ ft}} = 1,252 \text{ mi}$$

It should be noted that while the maximum endurance for this particular propeller-driven aircraft is proportional to the square root of air density, the maximum range is completely independent of air density. This is a result of the performance characteristics of the particular power plant being used. A very different result is obtained for a turbojet engine, as seen in the next example.

EXAMPLE 3.6.2. As discussed in Example 3.2.1, a turbojet engine produces a thrust that is nearly independent of airspeed and directly proportional to air density. The fuel consumption rate for a turbojet engine is very nearly proportional to the thrust produced by the engine, independent of airspeed or altitude. For this reason, fuel consumption for a turbojet engine is often expressed in terms of a thrust-specific fuel consumption, q_T,

$$q_T \equiv \frac{\dot{Q}}{T_A}$$

As a first approximation, we can assume that the thrust-specific fuel consumption for a turbojet engine is constant. Thus, for this engine, the power-specific fuel consumption based on propulsive power available can be conveniently written as

$$q_P \equiv \frac{\dot{Q}}{P_A} = \frac{\dot{Q}}{T_A V} = \frac{q_T}{V}$$

Consider the executive business jet described in Example 3.2.1. The thrust-specific fuel consumption for the engine is 0.6 lbf-fuel/hr/lbf-thrust, the dry weight of the airplane is 12,000 lbf, and the fuel capacity is 8,000 lbf. Assuming that the pilot maintains the maximum range airspeed at all times, determine the maximum range for this aircraft with no wind at both sea level and 20,000 feet.

Solution. For this airplane

$$S_w = 320 \text{ ft}^2, \quad b_w = 54 \text{ ft}, \quad W_f = 20,000 \text{ lbf}, \quad W_e = 12,000 \text{ lbf},$$
$$C_{D_0} = 0.023, \quad C_{D_0,L} = 0.0, \quad e = 0.82, \quad R_A = 9.1125,$$
$$q_T = 0.6 \text{ lbf/hr·lbf} = 1.667 \times 10^{-4} \text{ sec}^{-1}$$

With no wind, the specific range for this aircraft is

$$r_s \equiv \frac{V_g}{\dot{Q}} = \frac{V_g}{q_P P_A} = \frac{V}{q_P P_R} = \frac{V^2}{q_T P_R}$$

Following Example 3.6.1, the power required is

$$P_R = \left(\frac{C_{D_0} \rho V^3}{2(W/S_w)} + \frac{2(W/S_w)}{\pi e R_A \rho V} \right) W = \left(\frac{C_{D_0}}{2} R_V^3 + \frac{2}{\pi e R_A R_V} \right) W \sqrt{\frac{W/S_w}{\rho}}$$

Thus, for this jet, the specific range is

$$r_s = \frac{V^2}{q_T P_R} = \frac{V^2}{q_T \left(\dfrac{C_{D_0}}{2} R_V^3 + \dfrac{2}{\pi e R_A R_V} \right) W \sqrt{\dfrac{W/S_w}{\rho}}}$$

$$= \frac{R_V^2}{q_T \left(\dfrac{C_{D_0}}{2} R_V^3 + \dfrac{2}{\pi e R_A R_V} \right) W} \sqrt{\frac{W/S_w}{\rho}}$$

(3.6.14)

and the velocity ratio that results in maximum specific range is

$$R_V = \left(\frac{12}{\pi e R_A C_{D_0}} \right)^{1/4}$$

Using this result in Eq. (3.6.14), the maximum instantaneous specific range for the turbojet is

$$r_s = \frac{3}{2} \left(\frac{\pi e R_A}{12 C_{D_0}^3} \right)^{1/4} \frac{1}{q_T W} \sqrt{\frac{W/S_w}{\rho}}$$

From Eq. (3.6.12), for constant altitude, the maximum range at sea level is

$$R_{\max} = \int_{W_e}^{W_f} r_s\, dW = \int_{W_e}^{W_f} \frac{3}{2}\left(\frac{\pi e R_A}{12 C_{D_0}^3}\right)^{1/4} \frac{1}{q_T W}\sqrt{\frac{W/S_w}{\rho}}\, dW$$

$$= \frac{3}{2}\left(\frac{\pi e R_A}{12 C_{D_0}^3}\right)^{1/4}\frac{1}{q_T\sqrt{\rho S_w}}\int_{W_e}^{W_f}\frac{dW}{\sqrt{W}} = 3\left(\frac{\pi e R_A}{12 C_{D_0}^3}\right)^{1/4}\frac{\sqrt{W_f}-\sqrt{W_e}}{q_T\sqrt{\rho S_w}}$$

$$= 3\left[\frac{\pi(0.82)(9.1125)}{12(0.023)^3}\right]^{1/4}\frac{\sqrt{20,000\ \text{lbf}}-\sqrt{12,000\ \text{lbf}}}{1.667\times10^{-4}\ \text{sec}^{-1}\sqrt{0.0023769\ \text{slug/ft}^3}\,(320\,\text{ft}^2)}$$

$$= 1.317\times10^7\ \text{ft} = 2,495\ \text{mi} = 4,015\ \text{km}$$

Repeating these calculation for 20,000 feet ($\rho = 0.0012673$ slug/ft^3) and again assuming constant altitude, the maximum range at this altitude is found to be

$$R_{\max} = 3\left(\frac{\pi e R_A}{12 C_{D_0}^3}\right)^{1/4}\frac{\sqrt{W_f}-\sqrt{W_e}}{q_T\sqrt{\rho S_w}} = 1.804\times10^7\ \text{ft} = 3,416\ \text{mi} = 5,498\ \text{km}$$

Note that for the turbojet in Example 3.6.2, the predicted maximum range was inversely proportional to the square root of air density. At 20,000 feet the maximum range for this turbojet was about 37 percent greater than that at sea level. Clearly, there is a significant advantage to flying such airplanes at higher altitudes. Typical cruising altitudes for subsonic commercial jets are in the range of 30,000 to 40,000 feet. In the example it was assumed that the density altitude remained constant for the entire flight. This neglects the fuel required for takeoff, climb, descent, and landing. When all phases of the flight are accounted for, the optimal flight profile (speed and altitude) depends on the length of the flight as well as the variations in thrust-specific fuel consumption with airspeed and altitude, which are encountered for the more common turbofan engine.

3.7. Power Failure and Gliding Flight

We have now seen that the steady rate of climb that an airplane can sustain depends on the power produced by the airplane's engines. We have also seen that the velocity that an airplane can sustain in steady level flight is dependent on engine power. An obvious question that one might ask is: What happens if the engines fail?

In the previous analysis for rate of climb, there was no restriction that the power available be positive and nonzero. Equation (3.4.8) can be used to predict the rate of climb for any aircraft with low thrust-to-weight ratio. Thus, this equation applies equally well to the case of an airplane with zero thrust-to-weight ratio. From Eqs. (3.3.1) and (3.4.8), for zero power available we have

$$V_c = -\frac{P_R}{W} = -\frac{DV}{W} \tag{3.7.1}$$

Since the drag, the airspeed, and the weight are always positive, the steady rate of climb in powerless flight is always negative. The negative of the rate of climb is called the sink rate, V_s,

$$V_s = \frac{P_R}{W} = \frac{DV}{W} \tag{3.7.2}$$

So we see that to minimize sink rate, we minimize power required or the DV product. Using Eq. (3.3.7) for the power required in Eq. (3.7.2), we have

$$V_s = \frac{P_R}{W} = \frac{C_{D_0} \rho V^3}{2(W/S_w)} + C_{D_0,L} V + \frac{2(W/S_w)}{\pi e R_A \rho V} \tag{3.7.3}$$

The sink rate as a function of airspeed and altitude for a typical general aviation aircraft is shown in Fig. 3.7.1.

Since the weight is independent of airspeed, the minimum sink airspeed, V_{MS}, is the same as the airspeed for minimum power required, and from Eq. (3.3.12) we have

$$V_{MS} = V_{MDV} = \frac{2}{\sqrt{\pi e R_A C_{D_0,L} + \sqrt{(\pi e R_A C_{D_0,L})^2 + 12\pi e R_A C_{D_0}}}} \sqrt{\frac{W/S_w}{\rho}} \tag{3.7.4}$$

The minimum sink rate is found by using this airspeed in Eq. (3.7.3). Using the common approximation that $C_{D_0,L}$ is zero, this gives

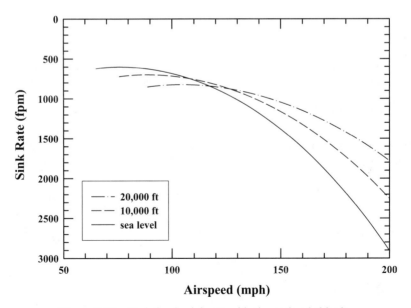

Figure 3.7.1. Variation in sink rate with airspeed and altitude.

$$(V_s)_{min} = \frac{(DV)_{min}}{W} \cong \frac{4\sqrt{2}\, C_{D_0}^{1/4}}{(3\pi e R_A)^{3/4}} \sqrt{\frac{W/S_w}{\rho}} \tag{3.7.5}$$

If the pilot flies at the minimum sink airspeed after the engine fails, he or she can maximize the time that the airplane will remain in the air. If the plane is very near a good landing field, this may be a good strategy. It would give the pilot the maximum possible time to restart the engine and it would give people on the ground the maximum possible time to prepare for an emergency landing. However, the pilot may not be near a good landing field when the engine fails. In this case the pilot may want to maximize distance traveled over the ground, not time in the air.

To consider distance traveled over the ground, we return to Eqs. (3.4.1) and (3.4.2). Setting the thrust available to zero, we have

$$D = -W \sin\gamma \tag{3.7.6}$$

and

$$L = W \cos\gamma \tag{3.7.7}$$

The horizontal component of the airplane's airspeed, V_h, is

$$V_h = V \cos\gamma \tag{3.7.8}$$

The vertical component of the airplane's airspeed, V_s, is the sink rate

$$V_s = -V \sin\gamma \tag{3.7.9}$$

Here we should recall that the sink rate is positive in the downward direction and γ, the angle of climb relative to the surrounding air, is negative for powerless flight. Thus, Eq. (3.7.9) gives positive values of V_s for a gliding plane.

If there is no wind, the airplane's velocity over the ground is just the horizontal component of the airspeed. However, when a wind is present, the ground speed, V_g, is related to the horizontal component of the airplane's airspeed, V_h, the wind speed, V_w, and wind-track angle, φ_w, as shown in Fig. 3.7.2. From this figure, the ground speed is

$$V_g = \sqrt{V_h^2 - V_w^2 \sin^2\varphi_w} - V_w \cos\varphi_w \tag{3.7.10}$$

The glide ratio, R_G, is defined as the airplane's horizontal velocity over the ground divided by the sink rate,

$$R_G \equiv \frac{V_g}{V_s} \tag{3.7.11}$$

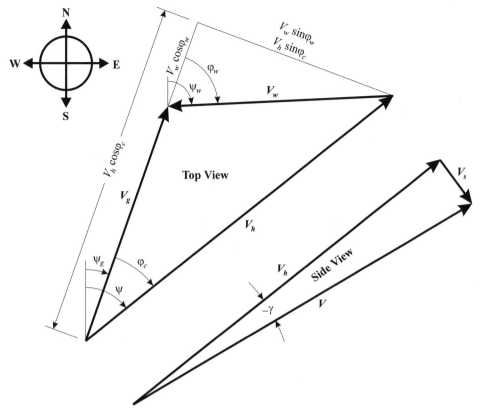

Figure 3.7.2. Relationship among ground speed, airspeed, and wind speed for gliding flight.

Using Eqs. (3.7.8), (3.7.9), and (3.7.10) in Eq. (3.7.11), the glide ratio is related to the airspeed, climb angle, and the wind:

$$R_G = \frac{\sqrt{V_h^2 - V_w^2 \sin^2 \varphi_w} - V_w \cos \varphi_w}{V_s} = \frac{\sqrt{V^2 \cos^2 \gamma - V_w^2 \sin^2 \varphi_w} - V_w \cos \varphi_w}{-V \sin \gamma} \quad (3.7.12)$$

From Eq. (3.7.6) we can write

$$\sin \gamma = -\frac{D}{W} \quad (3.7.13)$$

and from Eq. (3.7.7) we can write

$$\cos \gamma = \frac{L}{W} \quad (3.7.14)$$

Using Eqs. (3.7.13) and (3.7.14) in Eq. (3.7.12) gives

$$
\begin{aligned}
R_G &= \frac{\sqrt{V^2(L/W)^2 - V_w^2 \sin^2 \varphi_w} - V_w \cos \varphi_w}{V(D/W)} \\
&= \frac{L}{D}\left(\sqrt{1 - \frac{V_w^2 \sin^2 \varphi_w}{V^2(L/W)^2}} - \frac{V_w \cos \varphi_w}{V(L/W)} \right)
\end{aligned}
\tag{3.7.15}
$$

For most aircraft, the lift-to-drag ratio is large enough so that the glide angle is small and the lift is very nearly equal to the weight. Thus, we can closely approximate the glide ratio as

$$
\begin{aligned}
R_G &\cong \frac{L}{D}\left(\sqrt{1 - \frac{V_w^2 \sin^2 \varphi_w}{V^2}} - \frac{V_w \cos \varphi_w}{V} \right) \\
&= \frac{C_L}{C_D}\left(\sqrt{1 - \frac{V_w^2 \sin^2 \varphi_w}{V^2}} - \frac{V_w \cos \varphi_w}{V} \right)
\end{aligned}
\tag{3.7.16}
$$

From Eq. (3.7.15) or Eq. (3.7.16) we see that the zero-wind glide ratio, R_{G0}, is equal to the lift-to-drag ratio. We further see that a headwind or crosswind will decrease the glide ratio while a tailwind will increase it. From Eq. (3.2.8),

$$
\frac{L}{D} = \frac{C_L}{C_D} = \frac{C_L}{C_{D0} + C_{D0,L}\,C_L + \dfrac{C_L^2}{\pi e R_A}} = \left(\frac{C_{D0}}{C_L} + C_{D0,L} + \frac{C_L}{\pi e R_A} \right)^{-1}
\tag{3.7.17}
$$

For small glide angles, the lift is very nearly equal to the weight and

$$
\frac{C_L}{C_D} \cong \left(\frac{\frac{1}{2}\rho V^2 C_{D0}}{W/S_w} + C_{D0,L} + \frac{W/S_w}{\frac{1}{2}\pi e R_A \rho V^2} \right)^{-1}
\tag{3.7.18}
$$

The maximum zero-wind glide ratio is simply the maximum lift-to-drag ratio for the aircraft. From Eq. (3.2.13) this gives

$$
(R_{G0})_{\max} = (C_L/C_D)_{\max} = \frac{\sqrt{\pi e R_A}}{2\sqrt{C_{D0}} + C_{D0,L}\sqrt{\pi e R_A}}
\tag{3.7.19}
$$

The zero-wind best glide airspeed, V_{BG0}, is the airspeed that gives maximum lift-to-drag ratio. This is the airspeed resulting in the lift coefficient that will satisfy Eq. (3.2.12). For small glide angles, the lift is very nearly equal to the weight, and we can write

$$V_{BG0} = V_{MD} \cong \frac{\sqrt{2}}{\sqrt[4]{\pi e R_A C_{D_0}}} \sqrt{\frac{W/S_w}{\rho}} \qquad (3.7.20)$$

We see from Eqs. (3.7.19) and (3.7.20) that the zero-wind maximum glide ratio is independent of wing loading and air density, while the zero-wind best glide airspeed is proportional to the square root of the wing loading divided by the air density. The glide ratio as a function of airspeed and altitude for a typical general aviation aircraft is shown in Fig. 3.7.3, for the case of zero wind.

Including the effects of wind, we compute the glide ratio for the airplane by using Eq. (3.7.18) in Eq. (3.7.16). This gives

$$R_G \cong \frac{\sqrt{1 - \dfrac{V_w^2 \sin^2 \varphi_w}{V^2}} - \dfrac{V_w \cos \varphi_w}{V}}{\dfrac{\frac{1}{2}\rho V^2 C_{D_0}}{W/S_w} + C_{D_0,L} + \dfrac{W/S_w}{\frac{1}{2}\pi e R_A \rho V^2}} \qquad (3.7.21)$$

Figure 3.7.4 shows the glide ratio at sea level, for a typical general aviation aircraft, as a function of airspeed for several different wind conditions.

To find the best glide airspeed, we differentiate Eq. (3.7.21) with respect to airspeed and set the result to zero. This gives an equation that can be solved numerically for the best glide airspeed,

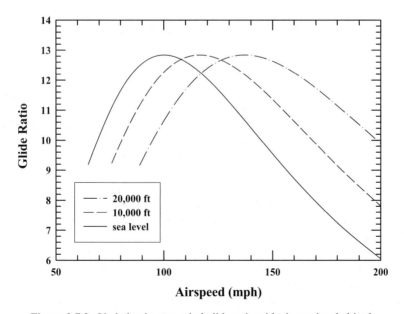

Figure 3.7.3. Variation in zero-wind glide ratio with airspeed and altitude.

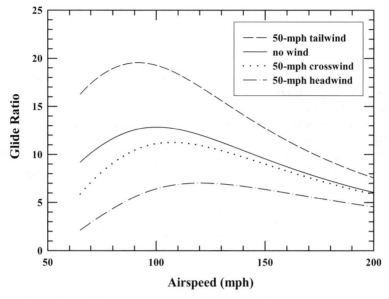

Figure 3.7.4. Variation in glide ratio at sea level with airspeed and wind.

$$2\left(V^4 - V_{MD}^4\right)V - \left(3V^4 - V_{MD}^4\right)V_w\left(\cos\varphi_w\sqrt{1 - \frac{V_w^2\sin^2\varphi_w}{V^2}} + \frac{V_w\sin^2\varphi_w}{V}\right) = 0 \quad (3.7.22)$$

To obtain the maximum glide ratio, we use the best glide airspeed in Eq. (3.7.21). Figure 3.7.5 shows the best glide airspeed at sea level as a function of wind speed. Figure 3.7.6 shows the maximum glide ratio as a function of wind speed for the same general aviation aircraft.

We can see from Eq. (3.7.21) that, in general, the glide ratio is a function of wing loading. Figure 3.7.7 shows how the maximum glide ratio varies with headwind and gross weight for the same general aviation aircraft that was used for the other figures in this chapter. Notice that the heaviest wing loading results in the largest glide ratio when there is a headwind. It may seem counterintuitive that increasing the weight of an aircraft could increase its glide ratio under any conditions. However, this is in fact the case. Added weight affects the maximum glide ratio of an aircraft in two ways. Added weight, of course, increases the induced drag, but it also increases the potential energy of the aircraft at any given altitude. In gliding flight, it is the decrease in potential energy of the aircraft that provides the energy required to move the aircraft through the air. A heavier aircraft gives up more potential energy for each foot of altitude lost. Thus, a heavier aircraft can do more work against the drag for each foot of altitude lost. Even though the heavier weight produces greater drag, it also provides additional potential energy needed to move the aircraft against that drag. Since increasing drag at constant weight would decrease the glide ratio and increasing the weight at constant drag would increase the glide ratio, it is not immediately obvious what net effect increased wing loading will have on the glide ratio. As it turns out, increasing wing loading improves maximum glide ratio in a headwind and degrades maximum glide ratio in a tailwind. With no wind, the maximum glide ratio for any aircraft is independent of wing loading.

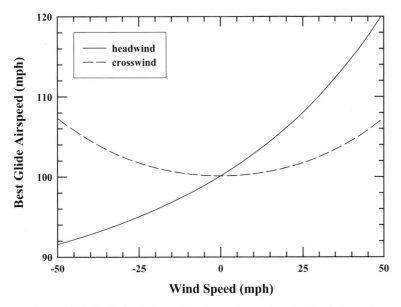

Figure 3.7.5. Variation in best glide airspeed at sea level with wind speed.

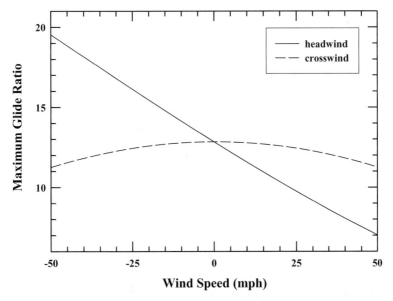

Figure 3.7.6. Variation in maximum glide ratio at sea level with wind speed.

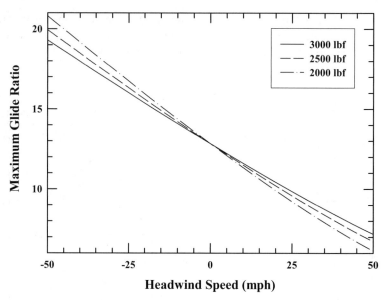

Figure 3.7.7. Variation in the maximum glide ratio at sea level with headwind speed and aircraft gross weight.

EXAMPLE 3.7.1. For the general aviation aircraft described in Example 3.5.1, determine the best-glide airspeed at sea level and the maximum glide ratio with fuel tanks full, for the case of no wind, with a 60-mph headwind, and with a 60-mph tailwind.

Solution. For this airplane

$$S_w = 180 \text{ ft}^2, \quad b_w = 33 \text{ ft}, \quad W_f = 2,700 \text{ lbf}, \quad W_f/S_w = 15.0 \text{ lbf/ft}^2,$$
$$C_{D_0} = 0.023, \quad C_{D_0,L} = 0.0, \quad e = 0.82, \quad R_A = 6.05$$

With no wind, the glide ratio is equal to the lift-to-drag ratio for the airplane. Using the small-glide-angle approximation, from Eq. (3.7.18) we have

$$
\begin{aligned}
R_G &= \frac{C_L}{C_D} = \left(\frac{\frac{1}{2}\rho V^2 C_{D_0}}{W/S_w} + C_{D_0,L} + \frac{W/S_w}{\frac{1}{2}\pi e R_A \rho V^2} \right)^{-1} \\
&= \left(\frac{C_{D_0}}{2} R_V^2 + \frac{2}{\pi e R_A R_V^2} \right)^{-1}
\end{aligned}
\tag{3.7.23}
$$

Differentiating with respect to velocity ratio and setting the result to zero, we obtain the velocity ratio for maximum glide ratio,

$$R_V = \left(\frac{4}{\pi e R_A C_{D_0}} \right)^{1/4} \tag{3.7.24}$$

Thus, the best glide airspeed with no wind is

$$
\begin{aligned}
V_{BG} &= R_V \sqrt{\frac{W/S_w}{\rho}} = \left(\frac{4}{\pi e R_A C_{D_0}} \right)^{1/4} \sqrt{\frac{W/S_w}{\rho}} \\
&= \left(\frac{4}{\pi (0.82)(6.05)(0.023)} \right)^{1/4} \sqrt{\frac{15.0 \text{ lbf/ft}^2}{0.0023769 \text{ slug/ft}^3}} \\
&= \underline{145 \text{ ft/sec} = 99 \text{ mph} = 86 \text{ knots} = 159 \text{ km/hr}}
\end{aligned}
$$

Using Eq. (3.7.24) in Eq. (3.7.23), the maximum glide ratio with no wind is

$$R_G = \sqrt{\frac{\pi e R_A}{4 C_{D_0}}} = \sqrt{\frac{\pi (0.82)(6.05)}{4(0.023)}} = \underline{13.0}$$

With a headwind, V_{hw}, from Eq. (3.7.16),

$$
\begin{aligned}
R_G &= \frac{C_L}{C_D} \left(1 - \frac{V_{hw}}{V} \right) = \left(\frac{C_{D_0}}{2} R_V^2 + \frac{2}{\pi e R_A R_V^2} \right)^{-1} \left(1 - \frac{V_{hw}}{V} \right) \\
&= \left(\frac{C_{D_0}}{2} R_V^2 + \frac{2}{\pi e R_A R_V^2} \right)^{-1} \left(1 - \frac{R_{V_{hw}}}{R_V} \right)
\end{aligned}
\tag{3.7.25}
$$

where

$$R_{V_{hw}} \equiv V_{hw} \bigg/ \sqrt{\frac{W/S_w}{\rho}}$$

Differentiating the glide ratio with respect to velocity ratio and setting the result to zero, we obtain an equation for the best glide velocity ratio with a headwind,

$$C_{D_0} R_V^5 - \frac{3 C_{D_0} R_{V_{hw}}}{2} R_V^4 - \frac{4}{\pi e R_A} R_V + \frac{2 R_{V_{hw}}}{\pi e R_A} = 0 \tag{3.7.26}$$

For a 60-mph headwind,

$$R_{V_{hw}} \equiv V_{hw} \bigg/ \sqrt{\frac{W/S_w}{\rho}} = 88 \text{ ft/sec} \bigg/ \sqrt{\frac{15.0 \text{ lbf/ft}^2}{0.0023769 \text{ slug/ft}^3}} = 1.108$$

and Eq. (3.7.26) becomes

$$0.023R_V^5 - 0.038226R_V^4 - 0.25665R_V + 0.142184 = 0$$

This is easily solved using Newton's method to give the best glide velocity ratio,

$$R_V = 2.3325$$

Thus the best glide airspeed with a 60-mph headwind is

$$V_{BG} = R_V \sqrt{\frac{W/S_w}{\rho}} = 2.3325 \sqrt{\frac{15.0 \text{ lbf/ft}^2}{0.0023769 \text{ slug/ft}^3}}$$

$$= 185 \text{ ft/sec} = 126 \text{ mph} = 110 \text{ knots} = 203 \text{ km/hr}$$

Using this result in Eq. (3.7.25), the maximum glide ratio with a 60-mph headwind is found to be

$$(R_G)_{max} = \left(\frac{C_{D_0}}{2} R_V^2 + \frac{2}{\pi e R_A R_V^2} \right)^{-1} \left(1 - \frac{R_{V_{hw}}}{R_V} \right)$$

$$= \left(\frac{0.023}{2}(2.3325)^2 + \frac{2}{\pi(0.82)(6.05)(2.3325)^2} \right)^{-1} \left(1 - \frac{1.108}{2.3325} \right) = \underline{6.1}$$

For a 60-mph tailwind,

$$R_{V_{hw}} \equiv V_{hw} \sqrt{\frac{W/S_w}{\rho}} = -88 \text{ ft/sec} \bigg/ \sqrt{\frac{15.0 \text{ lbf/ft}^2}{0.0023769 \text{ slug/ft}^3}} = -1.108$$

Eq. (3.7.26) becomes

$$0.023R_V^5 + 0.038226R_V^4 - 0.25665R_V - 0.142184 = 0$$

and

$$R_V = 1.6508$$

Thus the best glide airspeed with a 60-mph tailwind is

$$V_{BG} = R_V \sqrt{\frac{W/S_w}{\rho}} = 1.6508 \sqrt{\frac{15.0 \text{ lbf/ft}^2}{0.0023769 \text{ slug/ft}^3}}$$

$$= 131 \text{ ft/sec} = 89 \text{ mph} = 78 \text{ knots} = 144 \text{ km/hr}$$

and the maximum glide ratio with a 60-mph tailwind is

$$(R_G)_{max} = \left(\frac{C_{D_0}}{2} R_V^2 + \frac{2}{\pi e R_A R_V^2} \right)^{-1} \left(1 - \frac{R_{V_{hw}}}{R_V} \right)$$

$$= \left(\frac{0.023}{2}(1.6508)^2 + \frac{2}{\pi(0.82)(6.05)(1.6508)^2} \right)^{-1} \left(1 + \frac{1.108}{1.6508} \right) = \underline{21.3}$$

In summary,

V_{hw} (mph)	$R_{V_{hw}}$	R_V	V_{BG} (mph)	$(R_g)_{max}$
−60	−1.108	1.6508	89	21.3
0	0.0	1.8277	99	13.0
60	1.108	2.3325	126	6.1

3.8. Airspeed, Wing Loading, and Stall

In previous sections we have discussed several important airspeeds. Two of these airspeeds are independent of engine thrust. One is the airspeed that results in the minimum value of the drag-airspeed product for steady level flight when the thrust is aligned with the direction of flight,

$$V_{MDV} = \frac{2}{\sqrt{\pi e R_A C_{D0,L} + \sqrt{(\pi e R_A C_{D0,L})^2 + 12 \pi e R_A C_{D0}}}} \sqrt{\frac{W/S_w}{\rho}} \qquad (3.8.1)$$

Another important airspeed is that which results in minimum level flight drag when the thrust is aligned with the direction of flight,

$$V_{MD} = \frac{\sqrt{2}}{\sqrt[4]{\pi e R_A C_{D0}}} \sqrt{\frac{W/S_w}{\rho}} \qquad (3.8.2)$$

The airspeed, V_{MDV}, will result in the minimum power required for steady level flight and the minimum sink rate for gliding flight. This airspeed would also result in maximum endurance for an aircraft if the power-specific fuel consumption were independent of airspeed. The airspeed, V_{MD}, will result in minimum thrust required for steady level flight and maximum zero-wind glide ratio for gliding flight. The V_{MD} airspeed will also result in maximum range for no wind and an aircraft with power-specific fuel consumption independent of airspeed.

There is another important airspeed that is independent of engine thrust. This is the level flight stall speed for thrust aligned with the direction of flight. The stall speed is the minimum airspeed at which the airplane can fly. It can be evaluated from Eq. (3.3.3) by setting the lift coefficient equal to the maximum possible lift coefficient,

$$V_{\min} = \sqrt{\frac{2}{C_{L\max}}} \sqrt{\frac{W/S_w}{\rho}} \tag{3.8.3}$$

Notice that all three of the airspeeds above are proportional to the square root of the wing loading divided by the air density. These airspeeds also depend on the parasitic drag coefficient, the Oswald efficiency, the aspect ratio, and the maximum lift coefficient. However, these other parameters are limited by factors, such as structural strength and wing section design, and cannot be varied over a wide range to adjust the design airspeed of an airplane. The primary factor that the designer uses to control airspeed is wing loading.

The aspect ratio for modern subsonic airplanes and sailplanes varies from about 5 to 24, but is typically between 6 and 8 for most general aviation and transport aircraft. The Oswald efficiency for modern aircraft can vary from about 0.7 to as high as 0.95 but is typically in the range of 0.8 for most subsonic airplanes. The zero-lift parasitic drag coefficient for today's subsonic airplanes and sailplanes ranges from about 0.01 to 0.10, with typical values usually between 0.02 and 0.05 for general aviation and transport aircraft. The maximum lift coefficient for modern airfoils can be as low as 0.8 for thin symmetric airfoils without flaps, and as high as 5 for airfoils with both leading edge and trailing edge high-lift mechanisms in landing configuration. However, most modern airfoils in cruise configuration have a maximum lift coefficient in the range between 1.2 and 1.6. The wing loading, on the other hand, can vary from less than 1 pound per square foot for slow-flying ultralights to over 100 pounds per square foot for high-speed jet transports.

We now see that airplane wing loading is the only design parameter that can be varied widely to control the design airspeed of a particular airplane. We also see from Eqs. (3.8.1) through (3.8.3) that all three critical airspeeds are proportional to the square root of wing loading. Thus, if an airplane is to be designed for very low takeoff and landing speeds, suitable for rough landing fields, it will need to have a low wing loading. If, on the other hand, a high cruise speed is desired, the plane will need to be designed with a high wing loading.

If any one of the three airspeeds given by Eqs. (3.8.1) through (3.8.3) is known as a design constraint, then we can use the appropriate equation for that airspeed to help fix the wing area. For example, if we were designing for minimum drag at a particular cruise speed, Eq. (3.8.2) could be used to determine the required wing loading.

EXAMPLE 3.8.1. We wish to design an airplane to have a minimum drag cruise speed of 100 miles per hour at sea level with a gross weight of 3,000 pounds. Assume that for structural reasons, the aspect ratio has been fixed at 7. For these initial calculations, use typical values of 0.8 for the Oswald efficiency and 0.03 for the zero-lift parasitic drag coefficient. Estimate the wing area required for this airplane.

Solution. From Eq. (3.8.2) we can compute an initial estimate for the required wing area,

$$S_w = \frac{2}{\sqrt{\pi e R_A C_{D_0}}} \frac{W}{\rho V_{MD}^2}$$

$$= \frac{2}{\sqrt{\pi (0.8)(7.0)(0.03)}} \frac{3,000 \text{ lbf}}{0.0023769 \text{ slug/ft}^3 \left(100 \frac{5,280}{3,600} \text{ ft/sec}\right)^2} = \underline{162 \text{ ft}^2}$$

As the design is refined and the parasitic drag coefficient and Oswald efficiency become known more accurately, the wing area may change somewhat. However, if the airplane is to have minimum drag at 100 miles per hour, with a gross weight of 3,000 pounds, the wing area cannot be far from this value.

3.9. The Steady Coordinated Turn
In a steady coordinated turn, the wings of an airplane are banked at a constant angle, ϕ, as shown in Fig. 3.9.1. In a perfectly coordinated turn there is no side force. Since the wings are banked at the angle ϕ, the lift vector is also inclined at the angle ϕ toward the center of the turn. The horizontal component of lift produces a constant acceleration

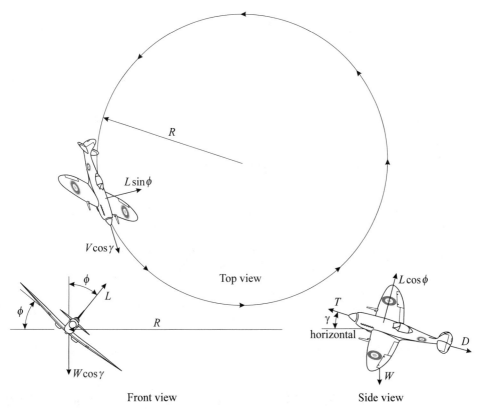

Figure 3.9.1. The steady coordinated turn.

normal to the flight path, causing the airplane to turn at constant airspeed along a circular or spiral path of radius R. For this steady turn, the radial acceleration is equal to the circumferential velocity squared divided by the turning radius. Thus, summing forces in the radial direction, we can write

$$L \sin \phi = \frac{W}{g} \frac{(V \cos \gamma)^2}{R}$$ (3.9.1)

Using the small-thrust-angle approximation and summing forces in the direction of flight gives

$$T = D + W \sin \gamma = \frac{1}{2} \rho V^2 S_w \left(C_{D_0} + C_{D_0,L} C_L + \frac{C_L^2}{\pi e R_A} \right) + W \sin \gamma$$ (3.9.2)

and summing forces in a direction normal to both the direction of flight and the radial direction,

$$L \cos \phi = W \cos \gamma$$ (3.9.3)

Solving Eq. (3.9.3) for the lift, we have

$$L = \frac{W \cos \gamma}{\cos \phi}$$ (3.9.4)

and the lift coefficient is

$$C_L \equiv \frac{L}{\frac{1}{2} \rho V^2 S_w} = \frac{W \cos \gamma}{\frac{1}{2} \rho V^2 S_w \cos \phi}$$ (3.9.5)

Using Eq. (3.9.4) to eliminate the lift from Eq. (3.9.1) and solving for the turning radius, we have

$$R = \frac{V^2 \cos \gamma}{g \tan \phi}$$ (3.9.6)

Using Eq. (3.9.5) in Eq. (3.9.2), the thrust required to maintain constant airspeed is

$$T = \frac{1}{2} \rho V^2 S_w C_{D_0} + C_{D_0,L} \frac{W \cos \gamma}{\cos \phi} + \frac{1}{\frac{1}{2} \pi e R_A \rho V^2 S_w} \left(\frac{W \cos \gamma}{\cos \phi} \right)^2 + W \sin \gamma$$ (3.9.7)

Solving Eq. (3.9.7) for $\sin\gamma$ gives

$$\sin\gamma = \frac{T - \left[\frac{1}{2}\rho V^2 S_w C_{D_0} + C_{D_0,L}\frac{W\cos\gamma}{\cos\phi} + \frac{1}{\frac{1}{2}\pi e R_A \rho V^2 S_w}\left(\frac{W\cos\gamma}{\cos\phi}\right)^2\right]}{W} \tag{3.9.8}$$

For any **steady coordinated turn**, Eqs. (3.9.6) and (3.9.8) can be used to obtain the turning radius and the climb angle as a function of bank angle, airspeed, and thrust.

The sine of γ multiplied by the airspeed is the rate of climb, V_c, and the thrust multiplied by the airspeed is the power available, P_A. Thus, multiplying Eq. (3.9.8) by the airspeed, we have

$$V_c = \frac{P_A - \left[\frac{1}{2}\rho V^3 S_w C_{D_0} + C_{D_0,L} V\frac{W\cos\gamma}{\cos\phi} + \frac{1}{\frac{1}{2}\pi e R_A \rho V S_w}\left(\frac{W\cos\gamma}{\cos\phi}\right)^2\right]}{W} \tag{3.9.9}$$

Most turns executed in normal flight are close to being level, i.e., γ is small. Thus, applying the usual small-angle approximation, $\cos\gamma \cong 1$, to Eqs. (3.9.6) and (3.9.9), for a **nearly level** steady coordinated turn, the turning radius and the rate of climb are given by

$$R = \frac{V^2}{g\tan\phi} \tag{3.9.10}$$

and

$$V_c = \frac{P_A - \left[\frac{1}{2}\rho V^3 S_w C_{D_0} + C_{D_0,L} V\frac{W}{\cos\phi} + \frac{1}{\frac{1}{2}\pi e R_A \rho V S_w}\left(\frac{W}{\cos\phi}\right)^2\right]}{W} \tag{3.9.11}$$

Notice from Eq. (3.9.10) that turning radius for a nearly level coordinated turn is independent of the thrust or power available. The turning radius depends only on bank angle and airspeed. However, from Eq. (3.9.11), we see that the rate of climb in such a turn is a function of power available as well as bank angle and airspeed. The bracketed term in Eq. (3.9.11) is the power that would be required for a steady level coordinated turn with the same bank angle and airspeed. The difference between the power available and the power required for a level turn is called the *excess power* for the turn. Thus, as was the case for straight flight, the rate of climb for turning flight is equal to the excess power divided by the gross weight. However, since the cosine of ϕ is always less than unity in a turn, the power required for level flight is always greater in a turn than it is for straight flight. Because the lift required in a turn is always greater than the lift required for straight flight at the same rate of climb, the drag and the power required to maintain constant airspeed are also greater in a turn. The excess power in a turn can be either positive or negative. If the power required for steady level turning flight is greater than the power available, then the rate of climb for the steady turn will simply be negative.

In executing a turn, the pilot can manipulate the throttle and elevator as well as the ailerons and rudder. The throttle and elevator can be used together to control the rate of climb and the airspeed. The rudder is normally used only to eliminate the side force and coordinate the turn. The ailerons are used primarily to control the bank angle, which controls the turning radius at any given airspeed. From Eq. (3.9.10) we see that for a given airspeed, the turning radius is made smaller by increasing the bank angle. There are, however, limits to how large a bank angle may be used in a steady coordinated turn.

From Eq. (3.9.5) we see that as the bank angle goes to 90 degrees, the lift coefficient would need to go to infinity. Since the lift coefficient has a maximum value, the bank angle will have a maximum value as well. Using the maximum lift coefficient in Eq. (3.9.5) and applying the small-climb-angle approximation, for a nearly level coordinated turn we can write

$$\cos \phi_{max} = \frac{W}{\frac{1}{2} \rho V^2 S_w C_{L_{max}}} \tag{3.9.12}$$

This is one obvious limit to the bank angle. If this bank angle is exceeded, the airplane will stall. Hence, the maximum bank angle computed from Eq. (3.9.12) is called the *stall-limited bank angle*.

There is also another important limit to the bank angle in a coordinated turn. This is related to the structural strength of the aircraft. During any maneuver in which the airplane travels along a curved path, there is a component of acceleration normal to the flight path. If the vector sum of the airplane's acceleration and the acceleration of gravity falls in a plane normal to the wing span, there is no spanwise component of total acceleration and the maneuver is said to be *coordinated*. The component of the airplane's acceleration that is normal to the flight path can either add to or subtract from the acceleration of gravity. Such maneuvers load the aircraft in a manner similar to what would be encountered if the airplane were flying straight and level in a gravitational field with a strength equal to the vector sum of the airplane's acceleration and the acceleration of gravity. Therefore, such maneuvers are usually rated in terms of what is called the *load factor*, n, which is defined as the lift generated during the maneuver divided by the aircraft weight,

$$n \equiv \frac{L}{W} = \frac{\frac{1}{2} \rho V^2 S_w C_L}{W} = \frac{\frac{1}{2} \rho V^2}{W/S_w} C_L \tag{3.9.13}$$

Load factor is usually stated in terms of "g's." For example, if a maneuver produces a lift force of five times the aircraft weight, the maneuver is said to be a "5-g maneuver."

From Eq. (3.9.3) we see that the load factor for a steady coordinated turn is a function only of bank angle and climb angle. For a nearly level turn, we use the small-climb-angle approximation and rearrange Eq. (3.9.3) to give

$$n \equiv \frac{L}{W} = \frac{1}{\cos\phi} \tag{3.9.14}$$

Obviously, there is a limit to the lift force that can be applied to the wings without causing structural damage. This limit is usually specified in terms of a load factor at maximum gross weight, W_{max}. Because the wings are not always symmetric with respect to positive and negative loading, we usually specify both a positive load limit,

$$n_{pll} \equiv \frac{L_{pll}}{W_{max}} \tag{3.9.15}$$

and a negative load limit,

$$n_{nll} \equiv \frac{L_{nll}}{W_{max}} \tag{3.9.16}$$

where L_{pll} is the maximum allowable positive lift and L_{nll} is the minimum allowable negative lift. In a conventional coordinated turn, the lift and load factor are both positive. Thus, rearranging Eq. (3.9.14), we find the maximum bank angle that can be used without exceeding the structural limit is given by

$$\cos\phi_{max} = \frac{1}{n_{max}} = \frac{W}{L_{pll}} = \frac{W}{n_{pll} W_{max}} \tag{3.9.17}$$

The bank angle limit imposed by Eq. (3.9.17) is the load-limited bank angle. The maximum allowable bank angle in a nearly level coordinated turn is the smaller of the stall-limited bank angle, given by Eq. (3.9.12), and the load-limited bank angle, given by Eq. (3.9.17). Figure 3.9.2 shows the stall-limited bank angle and the load-limited bank angle, at maximum gross weight, as a function of airspeed for a typical general aviation aircraft with a 5-g positive load limit.

The turning radius for a steady coordinated turn can be expressed in terms of load factor as well as bank angle. Squaring Eqs. (3.9.1) and (3.9.3) and adding the two equations together, we have

$$L^2(\sin^2\phi + \cos^2\phi) = L^2 = \left(\frac{WV^2\cos^2\gamma}{gR}\right)^2 + W^2\cos^2\gamma \tag{3.9.18}$$

Solving Eq. (3.9.18) for the turning radius gives

$$R = \frac{WV^2\cos^2\gamma}{g\sqrt{L^2 - W^2\cos^2\gamma}} = \frac{V^2\cos^2\gamma}{g\sqrt{n^2 - \cos^2\gamma}} \tag{3.9.19}$$

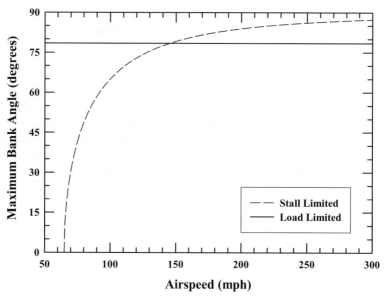

Figure 3.9.2. Stall-limited and load-limited maximum bank angles for a coordinated turn at maximum gross weight.

Using the small-climb-angle approximation in Eq. (3.9.19), a nearly level coordinated turn has a turning radius that can be expressed as a function of airspeed and load factor,

$$R = \frac{V^2}{g\sqrt{n^2-1}} \tag{3.9.20}$$

Load factor is an important parameter for all maneuvers involving curved flight. At low airspeed, the maximum possible load factor is limited by stall,

$$n_{max} = \frac{L_{max}}{W} = \frac{\frac{1}{2}\rho V^2}{W/S_w}C_{L\,max} \tag{3.9.21}$$

Since the wing can also stall at negative angles of attack, there is a low-speed negative limit for the load factor that will also result in stall,

$$n_{min} = \frac{L_{min}}{W} = \frac{\frac{1}{2}\rho V^2}{W/S_w}C_{L\,min} \tag{3.9.22}$$

At higher airspeeds the load factor is limited by structural constraints,

$$n_{max} = \frac{L_{pll}}{W} = \frac{L_{pll}}{W_{max}}\frac{W_{max}}{W} = n_{pll}\frac{W_{max}}{W} \tag{3.9.23}$$

and

$$n_{\min} = \frac{L_{nll}}{W} = \frac{L_{nll}}{W_{\max}}\frac{W_{\max}}{W} = n_{nll}\frac{W_{\max}}{W} \tag{3.9.24}$$

The stall-limited and load-limited load factors are shown as a function of airspeed in Fig. 3.9.3, for the same aircraft and weight as that used for Fig. 3.9.2. This plot, called a *V-n diagram*, graphically displays the allowable flight envelope for the aircraft. Any flight maneuver, corresponding to an operating point that is below both the positive stall limit and the positive load limit while being above both negative limits, is an allowable maneuver, provided that some maximum allowable airspeed is not exceeded. Maneuvers that result in operating points above the positive stall limit or below the negative stall limit are not possible, because the wing will stall. Maneuvers that result in operating points above the positive load limit and below the positive stall limit are aerodynamically possible but may result in structural damage. Likewise, structural damage may result from any maneuver with an operating point that is below the negative load limit and above the negative stall limit. The velocity corresponding to the vertical line on the right-hand side of the *V-n* diagram is called the *VNE airspeed* or the *velocity to never exceed*. At flight velocities higher than this limit, the dynamic pressure is above the design limit for the aircraft. The VNE airspeed is a "red-line" airspeed that, by design, should always be higher than the maximum level flight airspeed by at least a factor of 1.2. For some aircraft, such as fighters and stunt planes, the VNE airspeed may be as high as the terminal airspeed in a vertical dive.

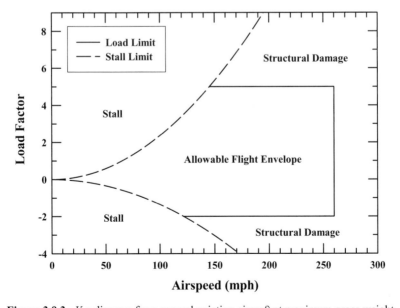

Figure 3.9.3. *V-n* diagram for a general aviation aircraft at maximum gross weight.

The airspeed at which the maximum allowable load factor changes from the positive stall limit to the positive load limit is often called the *corner velocity* for positive-*g* maneuvers. More commonly this airspeed is called the *maneuvering speed*. This terminology comes about because this is the airspeed at which the airplane can perform the tightest and fastest maneuvers without stalling the wing or risking structural damage to the aircraft. Furthermore, if the maneuvering speed is not exceeded during a positive-*g* maneuver, structural damage to the aircraft cannot occur as a result of aerodynamic lift. Here this airspeed will be designated as V_M. Setting the maximum load factor obtained from Eq. (3.9.21) equal to that obtained from Eq. (3.9.23) and solving for the airspeed, we have

$$V_M = \sqrt{\frac{2 n_{pll}}{C_{L_{max}}}} \sqrt{\frac{W_{max}/S_w}{\rho}} = \sqrt{n_{pll}} \, V_{smgw} \qquad (3.9.25)$$

where V_{smgw} is the level-flight stall speed at maximum gross weight.

The allowable flight envelope for any aircraft changes with actual gross weight. As the gross weight of the aircraft is decreased below the maximum gross weight, the allowable flight envelope is broadened. Decreasing gross weight at any airspeed increases the allowable load factor based on both stall and structural strength. This is shown in Fig. 3.9.4. In this figure, the *V-n* diagram is shown for both maximum gross weight and 75 percent of maximum gross weight. Notice that the maneuvering speed and the VNE airspeed do not change with the actual gross weight of the aircraft. When the gross weight of an airplane is reduced below the maximum gross weight, the stall

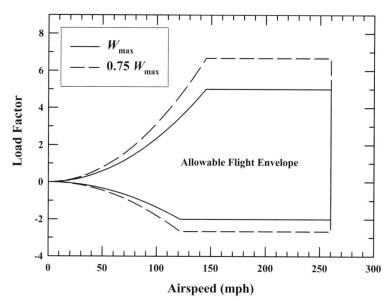

Figure 3.9.4. *V-n* diagram at two different gross weights for a typical general aviation aircraft, showing that the maneuvering speed is independent of gross weight.

speed at any given load factor is reduced. In addition, since the strength of the airplane remains constant with weight, the maximum allowable load factor based on structural strength will always increase when the gross weight is decreased. The net effect of these two changes in the *V-n* diagram is that the maneuvering speed does not change with the aircraft gross weight.

As seen from Eq. (3.9.25), the true airspeed at the maneuvering speed does change with air density. However, airspeed indicators are always calibrated for standard sea level and a density error is experienced unless the actual air density matches the air density at standard sea level. This density error is inversely proportional to the square root of air density. Because the maneuvering speed is also inversely proportional to the square root of air density, the indicated airspeed at the maneuvering speed does not change with either gross weight or altitude. Thus, indicated airspeed at the maneuvering speed is a design feature, independent of operating conditions.

The maneuvering speed given by Eq. (3.9.25) represents a critical dividing line. At airspeeds below V_M, the airplane cannot be structurally damaged by the production of positive lift. The wing will always stall before the structural limit is reached. At airspeeds above V_M, lift can be generated that may structurally damage the aircraft, and the pilot must be careful to avoid maneuvers that will generate such lift. The maneuvering speed is the highest airspeed at which the pilot can manipulate the controls arbitrarily without any risk of damaging the aircraft through excessive aerodynamic loading. The maneuvering speed is also the highest airspeed at which structural damage cannot occur from excessive aerodynamic loading due to turbulent gusts. Whenever severe turbulence is encountered, the pilot is always well advised to adjust the airspeed to match the maneuvering speed. This gives the best possible control without risk of structural damage.

EXAMPLE 3.9.1. For both the general aviation aircraft described in Example 3.5.1 and the executive business jet described in Example 3.6.2, determine the minimum turning radius for a nearly level turn at sea level with fuel tanks full and with fuel tanks empty. Assume that both airplanes are designed for a positive load limit of 4 *g*'s with fuel tanks full and have a maximum lift coefficient in cruise configuration of 1.2.

Solution. From Eq. (3.9.20), the turning radius is

$$R = \frac{V^2}{g\sqrt{n^2 - 1}}$$

The stall-limited maximum load factor is given by Eq. (3.9.21) and the load-limited maximum load factor is given by Eq. (3.9.23). Thus, the stall-limited turning radius is

$$\left(R_{\min}\right)_{\text{stall}} = 1 \bigg/ g\sqrt{\left(\frac{\rho C_{L_{\max}}}{2W/S_w}\right)^2 - \frac{1}{V^4}} \qquad (3.9.26)$$

and the load-limited turning radius is

$$\left(R_{min}\right)_{load} = V^2 \Big/ g\sqrt{\left(n_{pll}\frac{W_{max}}{W}\right)^2 - 1} \qquad (3.9.27)$$

The stall-limited turning radius decreases with increasing airspeed, while the load-limited turning radius increases with increasing airspeed. This can be seen in Fig. 3.9.5 for the present general aviation airplane with fuel tanks full.

From Fig. 3.9.5 we see that the minimum possible turning radius is achieved at the airspeed where the stall-limited turning radius is equal to the load-limited turning radius. This is the maneuvering speed. Notice that increasing the strength of this airplane would not significantly decrease the minimum turning radius.

Setting the turning radius from Eq. (3.9.26) equal to that from Eq. (3.9.27) and solving for airspeed,

$$V_{R_{min}} = V_M = \sqrt{\frac{2n_{pll}}{C_{L_{max}}}}\sqrt{\frac{W_{max}/S_w}{\rho}}$$

Using this result for the airspeed in either Eq. (3.9.26) or Eq. (3.9.27), we obtain the minimum turning radius that can be achieved at any airspeed, without stalling the wing or structurally damaging the airplane,

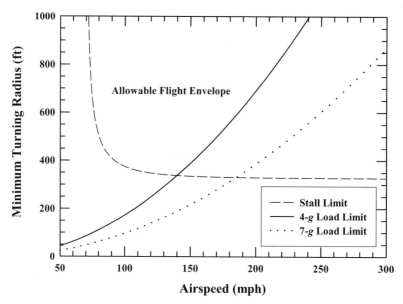

Figure 3.9.5. Minimum turning radius in a coordinated turn at sea level for the general aviation airplane in Example 3.9.1 with fuel tanks full.

$$R_{\min} = \frac{2n_{pll}}{C_{L_{\max}}\sqrt{n_{pll}^2 - \left(W/W_{\max}\right)^2}} \frac{W/S_w}{\rho g} \qquad (3.9.28)$$

For the general aviation airplane,

$$C_{L_{\max}} = 1.2, \quad n_{pll} = 4.0, \quad W_f/S_w = 15.0 \text{ lbf/ft}^2, \quad W_e/S_w = 12.8 \text{ lbf/ft}^2$$

$$\left(R_{\min}\right)_{\text{full}} = \frac{2(4.0)}{1.2\sqrt{4.0^2 - (15.0/15.0)^2}} \frac{15.0 \text{ lbf/ft}^2}{0.0023769 \text{ slug/ft}^3(32.2 \text{ ft/sec}^2)} = \underline{337 \text{ ft}}$$

$$\left(R_{\min}\right)_{\text{empty}} = \frac{2(4.0)}{1.2\sqrt{4.0^2 - (12.8/15.0)^2}} \frac{12.8 \text{ lbf/ft}^2}{0.0023769 \text{ slug/ft}^3(32.2 \text{ ft/sec}^2)} = \underline{285 \text{ ft}}$$

and for the business jet,

$$C_{L_{\max}} = 1.2, \quad n_{pll} = 4.0, \quad W_f/S_w = 62.5 \text{ lbf/ft}^2, \quad W_e/S_w = 37.5 \text{ lbf/ft}^2$$

$$\left(R_{\min}\right)_{\text{full}} = \frac{2(4.0)}{1.2\sqrt{4.0^2 - (62.5/62.5)^2}} \frac{62.5 \text{ lbf/ft}^2}{0.0023769 \text{ slug/ft}^3(32.2 \text{ ft/sec}^2)} = \underline{1{,}406 \text{ ft}}$$

$$\left(R_{\min}\right)_{\text{empty}} = \frac{2(4.0)}{1.2\sqrt{4.0^2 - (37.5/62.5)^2}} \frac{37.5 \text{ lbf/ft}^2}{0.0023769 \text{ slug/ft}^3(32.2 \text{ ft/sec}^2)} = \underline{826 \text{ ft}}$$

While turning radius is an important consideration in analyzing the turning performance of an aircraft, it is not the only important consideration. We are often also concerned with turning rate. *Turning rate* is simply the angular velocity of the airplane as it executes a turn. While turning radius is a measure of the distance required to turn an aircraft, turning rate is a measure of the time required to turn the aircraft.

The angular velocity of the airplane as it moves along a curved path is simply the translational velocity divided by the radius of curvature. Thus, using the result given in Eq. (3.9.20), the turning rate for a nearly level steady coordinated turn is

$$\Omega = \frac{V \cos\gamma}{R} \cong \frac{V}{R} = \frac{g\sqrt{n^2 - 1}}{V} \qquad (3.9.29)$$

Using the stall-limited load factor from Eq. (3.9.21) in Eq. (3.9.29), we obtain the stall-limited maximum turning rate,

$$\left(\Omega_{\max}\right)_{\text{stall}} = g\sqrt{\left(\frac{\rho C_{L_{\max}}}{2W/S_w}\right)^2 V^2 - \frac{1}{V^2}} \qquad (3.9.30)$$

and using the load-limited load factor from Eq. (3.9.23) in Eq. (3.9.29), we obtain the load-limited maximum turning rate,

$$\left(\Omega_{\max}\right)_{\text{load}} = \frac{g}{V}\sqrt{\left(n_{pll}\frac{W_{\max}}{W}\right)^2 - 1}$$ (3.9.31)

Figure 3.9.6 shows both the stall-limited and load-limited turning rates plotted as a function of airspeed for the general aviation airplane used in Example 3.9.1. Note that as was the case for the minimum turning radius, the maximum turning rate is achieved at the airspeed where the stall limit is equal to the load limit, i.e., the maneuvering speed. Setting the maximum turning rate from Eq. (3.9.30) equal to that from Eq. (3.9.31), we have

$$V_{\Omega_{\max}} = V_M = \sqrt{\frac{2n_{pll}}{C_{L_{\max}}}}\sqrt{\frac{W_{\max}/S_w}{\rho}}$$ (3.9.32)

Using this result for the airspeed in either Eq. (3.9.30) or Eq. (3.9.31), we obtain the maximum attainable turning rate,

$$\Omega_{\max} = \sqrt{\frac{C_{L_{\max}}}{2}\left(\frac{n_{pll}W_{\max}}{W} - \frac{W}{n_{pll}W_{\max}}\right)g\sqrt{\frac{\rho}{W/S_w}}}$$ (3.9.33)

Notice that both the minimum turning radius and the maximum turning rate are functions of the maximum lift coefficient, the positive load limit, the air density, and the

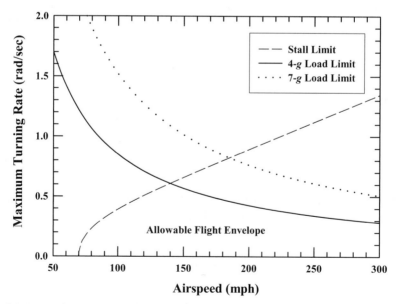

Figure 3.9.6. Maximum turning rate in a coordinated turn at sea level for the general aviation airplane in Example 3.9.1 with fuel tanks full.

wing loading. The designer has no control over air density and little control over the maximum lift coefficient. Most airplanes in cruise configuration will have a maximum lift coefficient between about 1.2 and 1.6. Thus, the primary features that the designer must use to control the turning performance of a conventional airplane are the wing loading and the load limit. Furthermore, as can be seen from Fig. 3.9.5, the minimum turning radius is not significantly affected by increasing the load limit beyond about 4. The only design parameter that can be used to significantly change the minimum turning radius of a conventional airplane is the wing loading. To provide a short turning radius in a conventional coordinated turn, the wing loading must be kept low.

Turning rate is not usually an important consideration in civilian aircraft. If an airplane needs to maneuver, take off, and land in small confined spaces, such as a small mountain valley, a short turning radius can be important. The maximum turning rate, however, has little effect on an airplane's ability to maneuver in tight places. Since increasing the load limit beyond about 4 cannot significantly reduce the minimum turning radius, load limit has little effect on the turning performance of most civilian airplanes. The positive load limit for civilian transports and general aviation airplanes is normally in the range from about 2.5 to 4.5.

On the other hand, turning rate can be a very important consideration for combat airplanes. As can be seen in Fig. 3.9.6, the maximum turning rate for an airplane is significantly improved by increasing the load limit well beyond 4. At large load factors, the maximum turning rate increases in proportion to the square root of the positive load limit. Thus, good turning performance for a combat airplane requires a high positive load limit as well as a low wing loading. The absolute upper limit for turning rate in a human-piloted aircraft is affected more by the pilot than by our ability to build an airplane capable of withstanding high load factor. A 9-g maneuver is about the maximum that a human pilot can endure without passing out. The positive load limit for modern fighter airplanes is normally in the range from about 6.5 to 9.0. Wing loading is always a trade-off in fighter airplanes because high wing loading facilitates high speed, but low wing loading facilitates agility.

EXAMPLE 3.9.2. For both the general aviation airplane and the business jet that were considered in Example 3.9.1, determine the indicated airspeed that should be used to achieve the highest turning rate. Also compute the maximum turning rate at 10,000 feet, for both aircraft, with fuel tanks full and with fuel tanks empty.

Solution. Assuming a nearly level coordinated turn, from Eq. (3.9.32) the airspeed that allows the highest turning rate is the maneuvering speed,

$$V_{\Omega_{max}} = V_M = \sqrt{\frac{2n_{pll}}{C_{L_{max}}}} \sqrt{\frac{W_{max}/S_w}{\rho}}$$

From Eq. (3.9.33) the maximum turning rate that can be achieved without stalling the wing or structurally damaging the airplane is

$$\Omega_{max} = \sqrt{\frac{C_{L_{max}}}{2}\left(\frac{n_{pll}W_{max}}{W} - \frac{W}{n_{pll}W_{max}}\right)g\sqrt{\frac{\rho}{W/S_w}}}$$

Airspeed indicators are always calibrated for the air density at standard sea level. Thus, for the general aviation airplane,

$$C_{L_{max}} = 1.2, \quad n_{pll} = 4.0, \quad W_f/S_w = 15.0 \text{ lbf/ft}^2, \quad W_e/S_w = 12.8 \text{ lbf/ft}^2$$

$$(V_M)_{indicated} = \sqrt{\frac{2(4.0)}{1.2}\sqrt{\frac{15.0 \text{ lbf/ft}^2}{0.0023769 \text{ slug/ft}^3}}} = 205 \text{ ft/sec} = 140 \text{ mph} = 122 \text{ knots}$$

$$(\Omega_{max})_{full} = \sqrt{\frac{1.2}{2}\left(\frac{(4.0)15.0}{15.0} - \frac{15.0}{(4.0)15.0}\right)(32.2 \text{ ft/sec}^2)\sqrt{\frac{0.0017556 \text{ slug/ft}^3}{15.0 \text{ lbf/ft}^2}}}$$

$$= 0.5225 \text{ rad/sec}$$

$$(\Omega_{max})_{empty} = \sqrt{\frac{1.2}{2}\left(\frac{(4.0)15.0}{12.8} - \frac{12.8}{(4.0)15.0}\right)(32.2 \text{ ft/sec}^2)\sqrt{\frac{0.0017556 \text{ slug/ft}^3}{12.8 \text{ lbf/ft}^2}}}$$

$$= 0.6179 \text{ rad/sec}$$

and for the business jet,

$$C_{L_{max}} = 1.2, \quad n_{pll} = 4.0, \quad W_f/S_w = 62.5 \text{ lbf/ft}^2, \quad W_e/S_w = 37.5 \text{ lbf/ft}^2$$

$$(V_M)_{indicated} = \sqrt{\frac{2(4.0)}{1.2}\sqrt{\frac{62.5 \text{ lbf/ft}^2}{0.0023769 \text{ slug/ft}^3}}} = 419 \text{ ft/sec} = 285 \text{ mph} = 248 \text{ knots}$$

$$(\Omega_{max})_{full} = \sqrt{\frac{1.2}{2}\left(\frac{(4.0)62.5}{62.5} - \frac{62.5}{(4.0)62.5}\right)(32.2 \text{ ft/sec}^2)\sqrt{\frac{0.0017556 \text{ slug/ft}^3}{62.5 \text{ lbf/ft}^2}}}$$

$$= 0.2560 \text{ rad/sec}$$

$$(\Omega_{max})_{empty} = \sqrt{\frac{1.2}{2}\left(\frac{(4.0)62.5}{37.5} - \frac{37.5}{(4.0)62.5}\right)(32.2 \text{ ft/sec}^2)\sqrt{\frac{0.0017556 \text{ slug/ft}^3}{37.5 \text{ lbf/ft}^2}}}$$

$$= 0.4357 \text{ rad/sec}$$

For both of the examples that were presented in this section, the turning analysis was based on the assumption of a nearly level turn. That is, the small-climb-angle approximation, $\cos\gamma = 1$, was used. This approximation simplified the turning analysis by making both the turning radius and the turning rate independent of the power or thrust available from the engines. For steeply climbing or diving turns we must return to the more general result that was given by Eq. (3.9.19). Using Eqs. (3.9.21) and (3.9.23) in Eq. (3.9.19), we obtain the stall-limited and load-limited turning radii for steeply climbing or diving turns,

$$\left(R_{min}\right)_{stall} = \cos^2\gamma \left/ g\sqrt{\left(\frac{\rho C_{L_{max}}}{2W/S_w}\right)^2 - \frac{\cos^2\gamma}{V^4}}\right. \tag{3.9.34}$$

$$\left(R_{min}\right)_{load} = V^2\cos^2\gamma \left/ g\sqrt{\left(n_{pll}\frac{W_{max}}{W}\right)^2 - \cos^2\gamma}\right. \tag{3.9.35}$$

Using these results in the general form of Eq. (3.9.29), we obtain the stall-limited and load-limited turning rates for steeply climbing or diving turns,

$$\left(\Omega_{max}\right)_{stall} = \frac{g}{\cos\gamma}\sqrt{\left(\frac{\rho C_{L_{max}}}{2W/S_w}\right)^2 V^2 - \frac{\cos^2\gamma}{V^2}} \tag{3.9.36}$$

$$\left(\Omega_{max}\right)_{load} = \frac{g}{V\cos\gamma}\sqrt{\left(n_{pll}\frac{W_{max}}{W}\right)^2 - \cos^2\gamma} \tag{3.9.37}$$

Thus, we see that for steeply climbing or diving turns, the minimum turning radius is smaller and the maximum turning rate is larger than for nearly level turns. However, for steady turns of this type, the minimum turning radius and the maximum turning rate depend on the climb angle, which in turn depends on the thrust available. The general relation between the turning climb angle and the thrust available for a steady turn is given by Eq. (3.9.8). The general relation for load factor in a steady climbing or diving turn, as obtained from Eq. (3.9.3), is

$$n \equiv \frac{L}{W} = \frac{\cos\gamma}{\cos\phi} \tag{3.9.38}$$

Using Eq. (3.9.38) in Eq. (3.9.8), the turning climb angle is related to the thrust available and the load factor according to

$$\sin\gamma = \frac{T - \left(\frac{1}{2}\rho V^2 S_w C_{D_0} + C_{D_0,L}\,nW + \frac{n^2 W^2}{\frac{1}{2}\pi e R_A \rho V^2 S_w}\right)}{W} \tag{3.9.39}$$

Using Eqs. (3.9.21) and (3.9.23) in Eq. (3.9.39), we obtain the stall-limited and load-limited climb angles for steady climbing or diving turns,

$$\left(\sin\gamma\right)_{stall-limit} = \frac{T - \frac{1}{2}\rho V^2 S_w\left(C_{D_0} + C_{D_0,L}\,C_{L_{max}} + \frac{C_{L_{max}}^2}{\pi e R_A}\right)}{W} \tag{3.9.40}$$

$$\left(\sin\gamma\right)_{\text{load-limit}} = \frac{T - \left(\frac{1}{2}\rho V^2 S_w C_{D_0} + C_{D_0,L}\, n_{pll}\, W_{\max} + \dfrac{n_{pll}^2\, W_{\max}^2}{\frac{1}{2}\pi e R_A \rho V^2 S_w}\right)}{W} \quad (3.9.41)$$

Thus, for a steeply climbing or diving turn with constant airspeed, Eqs. (3.9.40) and (3.9.41) can be used to eliminate the climb angle from Eqs. (3.9.34) through (3.9.37) and express the limiting values for turning radius and turning rate as a function of thrust and airspeed.

While a steeply climbing or diving turn can result in reduced turning radius and increased turning rate, such turns are not typically used with civilian aircraft. However, such turns are routinely employed with combat airplanes. Furthermore, pilots of combat airplanes often execute what are called *dynamic turns* in which the airspeed and angle of attack change very rapidly. In such maneuvers the aircraft may reach velocities near zero and angles of attack as high as 90 degrees. Maneuvers that involve angles of attack greater than stall are called *post-stall maneuvers*. Such maneuvers, of course, violate the constant-airspeed assumption that has been used throughout this section.

An example of a post-stall dynamic turn called the *cobra maneuver* is shown in Fig. 3.9.7. In this maneuver the pilot pitches the aircraft up rapidly and decelerates, slowing to extremely low speeds while executing the turn. As the angle of attack approaches 90 degrees the aircraft is rolled approximately 90 degrees about the velocity

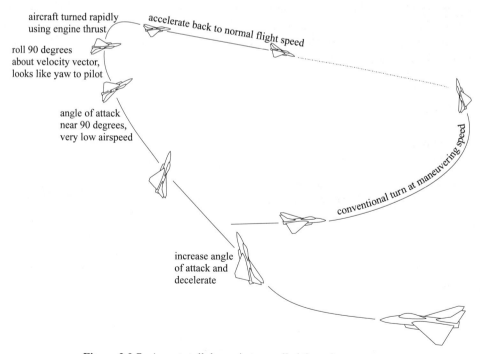

Figure 3.9.7. A post-stall dynamic turn called the cobra maneuver.

vector. This roll looks like a 90-degree yaw to the pilot because of the nearly 90-degree angle of attack. At the top of the maneuver, the aircraft is turned very rapidly using engine thrust until the nose is pointing at the target aircraft. The pilot then takes the shot and accelerates out of the high angle of attack and back to normal flight speed. Such maneuvers, of course, cannot be treated using a constant-airspeed turning analysis.

It should also be noted that there is another somewhat more subtle assumption implied in the turning analysis that was presented in this section. This assumption was implicitly introduced with Eq. (3.9.5). In using this equation for the lift coefficient we have implied that the airspeed is the same at every spanwise section of the wing.

In reality, when an airplane is turning, the outer wingtip is moving faster than the center of gravity, and the inner wingtip is moving slower than the center of gravity. If the turning radius is large compared to the semispan, then this effect is very small and can be neglected in the turning analysis. For most airplanes, with typical values of wing loading and span, this assumption is well justified. However, the minimum turning radius is proportional to wing loading, and some specialty aircraft with low wing loading and large span can violate this approximation.

The spanwise distribution of airspeed across the wing of a turning airplane varies linearly with distance from the axis of the turn,

$$V(r) = V_{cg}\, r/R \tag{3.9.42}$$

where V_{cg} is the airspeed for the center of gravity, r is the local distance from the turn axis, and R is the distance from the turn axis to the center of gravity, i.e., the turning radius. Thus, the airspeed at the inner wingtip is

$$V_{\text{inner}} = V_{cg}\left(1 - \frac{b}{2R}\cos\phi\right) \tag{3.9.43}$$

and the airspeed at the outer wingtip is

$$V_{\text{outer}} = V_{cg}\left(1 + \frac{b}{2R}\cos\phi\right) \tag{3.9.44}$$

For airplanes with typical wing loading and span, this airspeed difference is slight and is easily countered with a little aileron deflection to increase the lift coefficient on the inside wing. However, when flying an airplane with low wing loading and large span, the decrease in airspeed at the inside wingtip can cause this wingtip to stall at a load factor considerably less than that predicted by Eq. (3.9.21). This type of wingtip stall often leads to a spin.

A modification to Eq. (3.9.21) that can approximately account for the spanwise variation in airspeed is

$$n_{\max} = \frac{\frac{1}{2}\rho V_{\text{inner}}^2}{W/S_w}C_{L_{\max}} = \frac{\frac{1}{2}\rho V^2}{W/S_w}\left(1 - \frac{b}{2R}\cos\phi_{\max}\right)^2 C_{L_{\max}} \tag{3.9.45}$$

Using Eqs. (3.9.14) and (3.9.20) to eliminate the bank angle and the turning radius from this result, Eq. (3.9.45) can be rearranged to yield

$$n_{max} - \frac{bg}{2V^2}\sqrt{n_{max}^2 - 1} - \sqrt{\frac{2W/S_w}{\rho V^2 C_{L\,max}}}\sqrt{n_{max}^3} = 0 \qquad (3.9.46)$$

Equation (3.9.46) can be solved numerically to yield the stall-limited load factor as a function of airspeed. This result can be used in Eq. (3.9.20) to obtain the stall-limited turning radius as a function of airspeed. Combining the stall-limited load factor from Eq. (3.9.46) with the load-limited load factor from Eq. (3.9.23), we obtain an equation for the airspeed at which the minimum turning radius is attained,

$$V^2 - \sqrt{\frac{2n_{pll}W_{max}/S_w}{\rho C_{L\,max}}}\,V - \frac{bg}{2}\sqrt{1 - \left(\frac{W}{n_{pll}W_{max}}\right)^2} = 0 \qquad (3.9.47)$$

The airspeed that allows for the minimum turning radius without stalling the inside wingtip or causing structural damage to the aircraft is approximated as the positive root of Eq. (3.9.47),

$$V_{R_{min}} = \sqrt{\frac{n_{pll}W_{max}/S_w}{2\rho C_{L\,max}}} + \sqrt{\frac{n_{pll}W_{max}/S_w}{2\rho C_{L\,max}} + \frac{bg}{2}\sqrt{1 - \left(\frac{W}{n_{pll}W_{max}}\right)^2}} \qquad (3.9.48)$$

The minimum turning radius is approximated by using this airspeed in Eq. (3.9.27) and the maximum turning rate is approximated by using this airspeed in Eq. (3.9.31).

Accounting for wingtip stall always increases the stall-limited turning radius. For most airplanes this increase is small and can be ignored. Figure 3.9.8 shows the effect of wingtip stall on the minimum turning radius for the general aviation airplane described in Example 3.9.1. The solid line shows the stall-limited turning radius predicted by Eq. (3.9.26), and the dashed line shows the stall-limited turning radius obtained from the solution to Eq. (3.9.46). For this airplane, with a relatively low wing loading of 15 lbf/ft^2, the effect of the spanwise variation in airspeed is perceptible but small. Most commercial and military airplanes have significantly higher wing loading, and the effect of wingtip stall on the turning performance of these airplanes is even smaller. The effect of wingtip stall becomes more important in the design of airplanes with very low wing loading and large span.

The turning characteristics of an airplane with large span and low wing loading are much different from those of more typical airplanes. For such aircraft, the turning analysis must account for the variation in airspeed with position along the span. For example, a turn is usually initiated with a deflection of the ailerons. To initiate a right turn the left aileron is deflected down and the right aileron is deflected up. This increases the lift and drag coefficients on the outboard portion of the left wing and decreases the lift and drag coefficients on the outboard portion of the right wing. For airplanes with typical wing loading and span, this increases the lift and drag on the left wing and decreases the lift and drag on the right wing, banking the airplane to the right and yawing it to the left. The tendency of an airplane to yaw left when being banked right is called *adverse yaw*. This leftward yawing motion decreases the airspeed over the outboard portion of the left wing and increases the airspeed over the outboard portion of the right wing. While the differential in lift coefficient between the outboard portions of the right and left wings will tend to bank the plane right, the differential in airspeed between the outboard portions of the right and left wings will tend to bank the plane left. For an airplane with typical wing loading and span, the lift coefficient differential dominates and the plane banks right. However, for an airplane with very large span and very low wing loading, the airspeed differential may actually dominate the lift coefficient differential, causing the airplane to bank left and yaw left as the result of what would normally be thought of as a right-turning aileron deflection. This is only one of many possible effects that may result from spanwise airspeed variation during a turn.

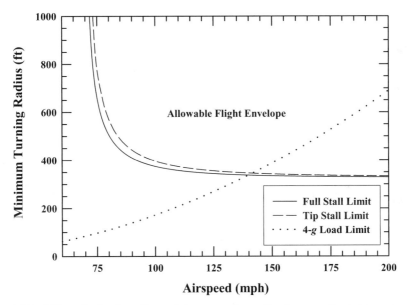

Figure 3.9.8. Effect of wingtip stall on minimum turning radius in a coordinated turn at sea level for the general aviation airplane in Example 3.9.1 with fuel tanks full.

3.10. Takeoff and Landing Performance

One important aspect of airplane design and performance analysis is takeoff and landing distances. The liftoff distance is defined as the distance along the ground required for an airplane, starting with zero ground speed, to accelerate to flight speed, rotate to liftoff attitude, and lift from the ground. This total distance, referred to as the *ground roll*, is designated here as s_g. This total ground roll distance can be divided into two segments, the acceleration distance, s_a, and the rotation distance, s_r.

For aircraft with the commonly used tricycle landing gear, during the acceleration phase of the ground roll, the angle of attack and lift coefficient are fixed by the landing gear. The case of tail-dragger landing gear is addressed later. Once the airplane has accelerated to flight speed, the pilot initiates an elevator deflection to rotate the airplane and increase the angle of attack. During the rotation phase of the ground roll, the angle of attack and lift coefficient are constantly increasing as a result of the moment applied to the airplane through the elevator deflection. As the airplane rotates, the angle of attack and lift coefficient continue to increase until the lift is equal to the weight and the airplane lifts from the ground.

Maximum thrust is nearly always used for takeoff, so once the extent of engagement for the high-lift devices is fixed, the lift and drag coefficients during the acceleration run are functions only of airplane design. Thus, the acceleration distance is controlled almost entirely by the design of the airplane and can easily be determined. The rotation distance, on the other hand, depends mostly on the pilot. Maximum elevator deflection is seldom used. However, the rotation distance is a small fraction of the total ground roll, and by making some appropriate assumption concerning pilot input, a reasonable estimate for the rotation distance can be made.

To determine the acceleration distance, we apply Newton's second law in the direction of motion. The mass of the airplane multiplied by its acceleration must equal the summation of forces in the direction of motion. For a zero thrust angle and constant headwind speed, V_{hw}, we have

$$\frac{W}{g}\frac{dV_g}{dt} = \frac{W}{g}\frac{d}{dt}(V - V_{hw}) = \frac{W}{g}\frac{dV}{dt} = T - D - F_r \qquad (3.10.1)$$

where V_g is ground speed, V is airspeed, and the force terms on the far right-hand side are, respectively, the engine thrust, the aerodynamic drag, and the rolling friction between the tires and the ground. Each of these forces is shown schematically in Fig. 3.10.1.

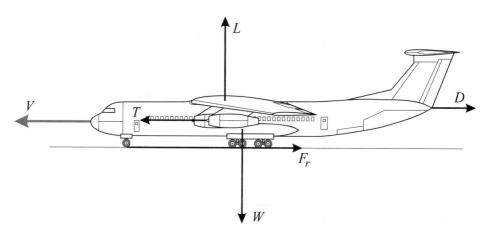

Figure 3.10.1. Forces acting on an airplane with tricycle landing gear during the acceleration run or landing ground roll.

The change in airspeed with respect to time can be written as

$$\frac{dV}{dt} = \frac{ds}{dt}\frac{dV}{ds} = (V - V_{hw})\frac{dV}{ds} \tag{3.10.2}$$

where s is distance traveled along the ground. Using Eq. (3.10.2) in Eq. (3.10.1) and rearranging, we obtain

$$\frac{ds}{dV} = \frac{W(V - V_{hw})}{g(T - D - F_r)} \tag{3.10.3}$$

Separating variables and integrating between any two points along the runway gives

$$s_{i+1} - s_i = \frac{W}{g}\int_{V_i}^{V_{i+1}} \frac{(V - V_{hw})dV}{T - D - F_r} \tag{3.10.4}$$

Here, we have neglected the fuel that is burned off during takeoff and are assuming constant weight. The integral in Eq. (3.10.4) can be evaluated numerically, if we know how the thrust, the drag, and the rolling friction depend on airspeed. Figure 3.10.2 shows how these forces might vary with airspeed for a typical general aviation airplane.

The total drag on the airplane is the parasitic drag plus the induced drag. However, during the takeoff run, the induced drag is diminished by a phenomenon known as *ground effect*. When a moving wing is close to the ground, the strength of the trailing vortices is decreased as a result of interaction with the ground. Since these vortices

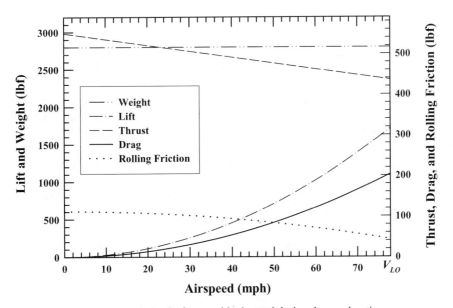

Figure 3.10.2. Variation in forces with airspeed during the acceleration run.

induce the downwash that generates induced drag, the downwash and associated induced drag are reduced when an airplane is close to the ground. An approximate equation for the drag on an airplane in ground effect is

$$D = \frac{1}{2}\rho V^2 S_w C_D \cong \frac{1}{2}\rho V^2 S_w \left(C_{D_0} + C_{D_0,L}\, C_L + \frac{(16 h_w/b_w)^2}{1+(16 h_w/b_w)^2} \frac{C_L^2}{\pi e R_A} \right) \qquad (3.10.5)$$

where h_w is the height of the wing above the ground and b_w is the wingspan. During ground roll, the zero-lift drag coefficient is typically higher than in cruise configuration, because the landing gear is extended and high-lift devices are engaged.

During the ground roll, the rolling friction force is given by

$$F_r = \mu_r(W - L) = \mu_r(W - \tfrac{1}{2}\rho V^2 S_w C_L) \qquad (3.10.6)$$

where μ_r is the coefficient of rolling friction and $W - L$ is the net normal force between the tires and the ground. Experiments have shown that the coefficient of rolling friction, on takeoff, varies from about 0.04 for paved surfaces to 0.10 for grass landing fields. For rough landing fields with soft or loose soil, even higher values of μ_r should be used.

The lift coefficient remains fairly constant during the acceleration phase of the takeoff run, so both the lift and the drag vary approximately as the square of the airspeed. The actual value of the lift coefficient during the acceleration run is controlled by the angle of attack that is maintained by the landing gear and by the extent to which high-lift devices are engaged during takeoff. Hence, the lift coefficient for the acceleration run is essentially a design feature of the airplane. Increasing the lift coefficient for the acceleration run will reduce the rolling friction but will increase the drag. Depending on the airplane design, the lift coefficient for the acceleration run could vary from less than 0.1 to as much as 70 percent of $C_{L_{max}}$. In any case, as the airplane accelerates down the runway, the lift coefficient remains constant at this design value until the airplane reaches the rotation speed. Thus, during the acceleration run, both the drag and the rolling friction are linear functions of the airspeed squared.

The manner in which thrust depends on airspeed is a function of the power plant being used. For turbojet engines, the thrust remains relatively constant with airspeed. For a piston engine combined with a constant-pitch propeller, the thrust decreases nearly linearly with airspeed. A turbofan engine produces thrust that varies almost as a parabolic function of airspeed. Normally, the relation between engine thrust and airspeed is determined experimentally and provided in tabular form. For most power plants, the variation in thrust with airspeed can be approximated as parabolic over a fairly broad range of airspeed and, in general, we could describe thrust as a piecewise parabolic function of airspeed. If the acceleration run is divided into small segments and the thrust is assumed parabolic with airspeed over each segment, then for segment i,

$$T = (T_0)_i + (T')_i V + (T'')_i V^2 \qquad (3.10.7)$$

where T_0, T', and T'' are experimentally determined coefficients.

Applying Eqs. (3.10.5) through (3.10.7) to Eq. (3.10.4), we have

$$s_{i+1} - s_i = \frac{1}{g} \int_{V_i}^{V_{i+1}} \frac{(V - V_{hw}) \, dV}{(K_0)_i + (K_1)_i V + (K_2)_i V^2} \tag{3.10.8}$$

where

$$(K_0)_i = \frac{(T_0)_i}{W} - \mu_r \tag{3.10.9}$$

$$(K_1)_i = \frac{(T')_i}{W} \tag{3.10.10}$$

$$(K_2)_i = \frac{(T'')_i}{W} + \frac{\rho}{2W/S_w} (C_L \mu_r - C_D) \tag{3.10.11}$$

The integration on the right-hand side of Eq. (3.10.8) is easily carried out to yield

$$s_{i+1} - s_i = \frac{(K_T)_i - V_{hw}(K_W)_i}{g} \tag{3.10.12}$$

where

$$(K_T)_i = \begin{cases} \dfrac{(V^2)_{i+1} - (V^2)_i}{2(K_0)_i}, & (K_1)_i = 0 \\[2ex] \dfrac{(K_0)_i}{(K_1)_i^2} \ln\left(\dfrac{f_i}{f_{i+1}}\right) + \dfrac{(V)_{i+1} - (V)_i}{(K_1)_i}, & (K_1)_i \neq 0 \end{cases} \quad (K_2)_i = 0 \\[4ex] \dfrac{1}{2(K_2)_i} \ln\left(\dfrac{f_{i+1}}{f_i}\right) - \dfrac{(K_1)_i (K_W)_i}{2(K_2)_i}, \quad (K_2)_i \neq 0 \tag{3.10.13}$$

$$(K_W)_i = \begin{cases} \dfrac{(V)_{i+1} - (V)_i}{(K_0)_i}, & (K_1)_i = 0 \\[2ex] \dfrac{1}{(K_1)_i} \ln\left(\dfrac{f_{i+1}}{f_i}\right), & (K_1)_i \neq 0 \end{cases} \quad (K_2)_i = 0 \\[4ex] \dfrac{1}{\sqrt{-(K_R)_i}} \ln\left[\dfrac{\left(f'_{i+1} - \sqrt{-(K_R)_i}\right)\left(f'_i + \sqrt{-(K_R)_i}\right)}{\left(f'_{i+1} + \sqrt{-(K_R)_i}\right)\left(f'_i - \sqrt{-(K_R)_i}\right)}\right], \quad (K_R)_i < 0 \\[4ex] \dfrac{2}{f'_i} - \dfrac{2}{f'_{i+1}}, \quad (K_R)_i = 0 \\[4ex] \dfrac{2}{\sqrt{(K_R)_i}}\left[\tan^{-1}\left(\dfrac{f'_{i+1}}{\sqrt{(K_R)_i}}\right) - \tan^{-1}\left(\dfrac{f'_i}{\sqrt{(K_R)_i}}\right)\right], \quad (K_R)_i > 0 \tag{3.10.14}$$

$$f_i = (K_0)_i + (K_1)_i (V)_i + (K_2)_i (V^2)_i \tag{3.10.15}$$

$$f_{i+1} = (K_0)_i + (K_1)_i (V)_{i+1} + (K_2)_i (V^2)_{i+1} \tag{3.10.16}$$

$$f_i' = (K_1)_i + 2(K_2)_i (V)_i \tag{3.10.17}$$

$$f_{i+1}' = (K_1)_i + 2(K_2)_i (V)_{i+1} \tag{3.10.18}$$

$$(K_R)_i = 4(K_0)_i (K_2)_i - (K_1)_i^2 \tag{3.10.19}$$

To obtain the total acceleration distance, Eq. (3.10.4) must be integrated from zero ground speed to the rotation speed. The change in airspeed during rotation is usually quite small, so the rotation speed is very close to the liftoff speed. Thus, for computing the ground roll, the rotation speed is usually assumed to be equal to the liftoff speed. The airspeed at liftoff must be at least 1.1 times the stall speed for the airplane in takeoff configuration,

$$V_{LO} = 1.1 \sqrt{\frac{2}{C_{L_{max}}}} \sqrt{\frac{W/S_w}{\rho}} \tag{3.10.20}$$

Here $C_{L_{max}}$ is the maximum lift coefficient for the airplane in takeoff configuration. Remember that the landing gear and tail configuration may limit the maximum angle of attack, and hence the maximum lift coefficient for takeoff.

For best accuracy, the acceleration distance should be broken up into many small segments with the length of each segment computed from Eq. (3.10.12). However, for most power plants, the variation in thrust with airspeed is nearly parabolic over the entire acceleration run. Thus, a good approximation for the acceleration distance is obtained by applying Eq. (3.10.12) over the full acceleration run as a single step. To obtain a parabolic approximation for the thrust over the full acceleration run, the coefficients in Eq. (3.10.7) can be computed as

$$T_0 = T_S \tag{3.10.21}$$

$$T' = \frac{6\overline{T} - 4T_S - 2T_{LO}}{V_{LO}} \tag{3.10.22}$$

$$T'' = \frac{3T_S + 3T_{LO} - 6\overline{T}}{V_{LO}^2} \tag{3.10.23}$$

where T_S is the static thrust, T_{LO} is the thrust at liftoff, and $\overline{T}$ is the average thrust,

$$\overline{T} \equiv \frac{1}{V_{LO}} \int_0^{V_{LO}} T \, dV \tag{3.10.24}$$

For the critical case of **no wind**, this gives

$$s_a = \frac{K_T}{g} \tag{3.10.25}$$

where

$$K_T = \begin{cases} \dfrac{V_{LO}^2}{2K_0}, & K_1 = 0 \\[2mm] \dfrac{K_0}{K_1^2} \ln\left(\dfrac{f_S}{f_{LO}}\right) + \dfrac{V_{LO}}{K_1}, & K_1 \neq 0 \\[2mm] \dfrac{1}{2K_2} \ln\left(\dfrac{f_{LO}}{f_S}\right) - \dfrac{K_1 K_W}{2K_2}, & K_2 \neq 0 \end{cases} \quad \begin{matrix} K_2 = 0 \\ \\ \end{matrix} \tag{3.10.26}$$

$$K_W = \begin{cases} \dfrac{1}{\sqrt{-K_R}} \ln\left[\dfrac{\left(f'_{LO} - \sqrt{-K_R}\right)\left(f'_S + \sqrt{-K_R}\right)}{\left(f'_{LO} + \sqrt{-K_R}\right)\left(f'_S - \sqrt{-K_R}\right)}\right], & K_R < 0 \\[4mm] \dfrac{2}{f'_S} - \dfrac{2}{f'_{LO}}, & K_R = 0 \\[4mm] \dfrac{2}{\sqrt{K_R}}\left[\tan^{-1}\left(\dfrac{f'_{LO}}{\sqrt{K_R}}\right) - \tan^{-1}\left(\dfrac{f'_S}{\sqrt{K_R}}\right)\right], & K_R > 0 \end{cases} \tag{3.10.27}$$

$$f_S = K_0 = \frac{T_S}{W} - \mu_r \tag{3.10.28}$$

$$f_{LO} = K_0 + K_1 V_{LO} + K_2 V_{LO}^2 = \frac{T_{LO}}{W} - \mu_r + \frac{\rho V_{LO}^2}{2W/S_w}(\mu_r C_L - C_D) \tag{3.10.29}$$

$$f'_S = K_1 = \frac{6\overline{T} - 4T_S - 2T_{LO}}{WV_{LO}} \tag{3.10.30}$$

$$f'_{LO} = K_1 + 2K_2 V_{LO} = \frac{2T_S + 4T_{LO} - 6\overline{T}}{WV_{LO}} + \frac{\rho V_{LO}}{W/S_w}(C_L \mu_r - C_D) \tag{3.10.31}$$

$$K_0 = \frac{T_S}{W} - \mu_r \tag{3.10.32}$$

$$K_1 = \frac{T'}{W} = \frac{6\overline{T} - 4T_S - 2T_{LO}}{WV_{LO}} \tag{3.10.33}$$

$$K_2 = \frac{T''}{W} + \frac{\rho}{2W/S_w}(C_L \mu_r - C_D)$$

$$= \frac{3T_S + 3T_{LO} - 6\overline{T}}{WV_{LO}^2} + \frac{\rho}{2W/S_w}(C_L \mu_r - C_D) \qquad (3.10.34)$$

$$K_R = 4K_0 K_2 - K_1^2 \qquad (3.10.35)$$

When the airplane has accelerated to flight speed, the pilot initiates an elevator deflection to rotate the airplane, increasing the angle of attack and lift coefficient to the point where the lift is equal to the weight and the airplane lifts from the ground. The rotation distance is a small fraction of the total ground roll and, as previously mentioned, the change in airspeed during rotation is usually quite small. Thus, for computing the ground roll, the rotation is usually assumed to take place at a constant airspeed equal to the liftoff speed given by Eq. (3.10.20). The rotation rate depends almost entirely on the pilot. Maximum elevator deflection is almost never used during rotation. In computing ground roll, the designer should allow the pilot a rotation time in proportion to the size of the airplane. For small general aviation aircraft and fighter planes, a rotation time on the order of 1 second is adequate, but for large transports and bombers, the rotation time should be increased to about 3 seconds. For the very largest airplanes, as much as 4 seconds should probably be allowed for rotation. Thus, the rotation distance is computed simply as the time allowed for rotation multiplied by the ground speed at the liftoff airspeed,

$$s_r = (V_{LO} - V_{hw})t_r \qquad (3.10.36)$$

where t_r is the assumed rotation time. The total ground roll is the sum of the acceleration distance and the rotation distance,

$$s_g = s_a + s_r \qquad (3.10.37)$$

To gain some additional insight into the parameters that affect takeoff performance, consider a further simplification to Eq. (3.10.25). As we have seen from Fig. 3.10.2, the drag increases with airspeed while the rolling friction decreases with airspeed. Furthermore, for many power plants, the change in thrust during takeoff is small. Thus, as a first approximation for the acceleration distance, we can neglect the variation in net force with airspeed (i.e., $K_1 \cong K_2 \cong 0$). With this approximation, the no-wind acceleration distance from Eq. (3.10.25) reduces to

$$s_a = \frac{K_T}{g} \cong \frac{V_{LO}^2}{2gK_0} \cong \frac{WV_{LO}^2}{2g(T - D - F_r)} \qquad (3.10.38)$$

Equation (3.10.38) gives a reasonable first estimate for the acceleration distance even when there is some variation in the net force with airspeed, provided that the net force is evaluated at about 70 percent of the liftoff airspeed. Using the liftoff speed from Eq. (3.10.20) and adding the rotation distance from Eq. (3.10.36), an approximate first estimate for the no-wind ground roll on takeoff is

$$s_g \cong \frac{1.21W^2}{\rho g S_w C_{L_{max}} (T - D - F_r)_{0.7V_{LO}}} + 1.1 t_r \sqrt{\frac{2}{C_{L_{max}}}} \sqrt{\frac{W/S_w}{\rho}} \qquad (3.10.39)$$

From Eq. (3.10.39) we note that the acceleration distance increases as the square of the gross weight. If we double the weight of an airplane, the acceleration distance will quadruple. We also see that acceleration distance varies inversely with air density. Since the thrust also decreases with decreasing density, this effect is much greater than would at first appear from looking at Eq. (3.10.39). In fact, the acceleration distance varies almost as the inverse square of the air density. As a result, hot summer days and high-altitude airports require significantly longer acceleration runs. We can also see from Eq. (3.10.39) that the acceleration distance is inversely proportional to the wing area. Thus, if we require a short takeoff distance, low wing loading is important.

The no-wind acceleration distance computed from Eqs. (3.10.25) through (3.10.35) is for an airplane with tricycle landing gear, as shown in Fig. 3.10.3a and b. In obtaining this relation, it was assumed that the angle of attack remains constant for the entire acceleration run. This is a good assumption for tricycle landing gear, because this type of gear holds the attitude of the airplane constant prior to rotation. For a tail dragger, however, this assumption is not valid. A tail dragger starts the acceleration run at a high angle of attack with the tail wheel on the ground, as shown in Fig. 3.10.3c. During this initial phase of the acceleration run, the lift and drag coefficients are quite large,

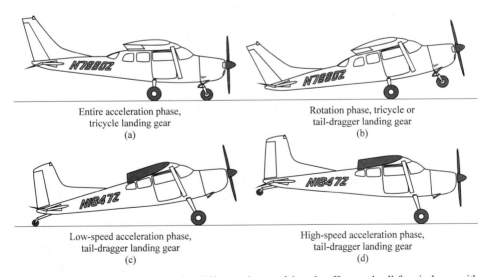

Entire acceleration phase,
tricycle landing gear
(a)

Rotation phase, tricycle or
tail-dragger landing gear
(b)

Low-speed acceleration phase,
tail-dragger landing gear
(c)

High-speed acceleration phase,
tail-dragger landing gear
(d)

Figure 3.10.3. Aircraft attitude during different phases of the takeoff ground roll for airplanes with tricycle and tail-dragger landing gear.

due to this high angle of attack. The tail must remain on the ground to provide steering until sufficient airspeed is developed to afford adequate steering with the rudder. This airspeed is called the *minimum ground control speed*. At this airspeed the pilot initiates an elevator deflection to lift the tail, as shown in Fig. 3.10.3d, decreasing the angle of attack and the associated lift and drag coefficients. The remainder of the acceleration run takes place at this reduced angle of attack to keep the drag low. Once the airplane has accelerated to flight speed, the pilot initiates a second elevator deflection, this time to increase the angle of attack and rotate the plane to liftoff, as shown in Fig. 3.10.3b. This final rotation phase of the ground roll is similar to that for an airplane with tricycle landing gear.

While Eq. (3.10.25) does not apply directly to a tail-dragger, Eq. (3.10.12) can still be used independently over each phase of the acceleration run. The low-speed phase starts at a ground speed of zero (i.e., an airspeed equal to the headwind speed) and continues until the airspeed is equal to the minimum ground control speed. During this phase, the angle of attack, the lift coefficient, and the drag coefficient are held constant at the values associated with the tail-down attitude. The lift coefficient for the low-speed phase of the acceleration run should be very close to $C_{L_{max}}$. The high-speed phase of the acceleration run starts at an airspeed equal to the minimum ground control speed and continues to the liftoff speed. During this phase, we assume constant lift and drag coefficients associated with the lower angle of attack in the tail-up attitude. The total acceleration distance is the sum of that for both the low-speed phase and the high-speed phase. The details of this calculation are left as an exercise for the student.

EXAMPLE 3.10.1. Consider the takeoff performance of the general aviation airplane described in Example 3.3.1. For this airplane in takeoff configuration, the zero-lift drag coefficient is 0.036, the maximum lift coefficient is 1.40, and the wing is 6.0 feet above the ground. During takeoff, the propeller pitch is fixed at the value that gives best takeoff performance. The thrust is proportional to air density, and with this pitch setting, the power plant produces a static thrust at sea level of 1,200 lbf. The sea-level thrust decreases linearly with airspeed at the rate of 4.0 lbf for each 1.0 ft/sec of airspeed.

As previously discussed, the angle of attack and lift coefficient for the acceleration run are fixed by the design of the tricycle landing gear. Depending on the relative height of the nose wheel and main gear, the angle of attack for the acceleration run can be fixed at any desired value. For this airplane, determine the lift coefficient that could be used to minimize the acceleration distance. Using this lift coefficient, determine the total no-wind ground roll required for takeoff at both sea level and at 5,000 feet. Also compute the total no-wind ground roll at sea level for both a lift coefficient of zero and a lift coefficient of two times the optimum value. Assume that μ_r is 0.04.

Solution. The lift coefficient that will result in the shortest acceleration distance is the lift coefficient that results in the lowest value of drag and rolling friction combined. Adding the values given by Eqs. (3.10.5) and (3.10.6) and rearranging, we have

$$D + F_r = \mu_r W + \frac{1}{2}\rho V^2 S_w \left[C_{D_0} + (C_{D_0,L} - \mu_r)C_L + \frac{(16h_w/b_w)^2}{1+(16h_w/b_w)^2} \frac{C_L^2}{\pi e R_A} \right]$$

Differentiating with respect to lift coefficient and setting the result to zero gives the optimum lift coefficient,

$$(C_L)_{opt} = \pi e R_A \frac{1+(16h_w/b_w)^2}{2(16h_w/b_w)^2}(\mu_r - C_{D_0,L})$$

For this airplane in takeoff configuration,

$$S_w = 180\ \text{ft}^2, \quad b_w = 33\ \text{ft}, \quad h_w = 6.0\ \text{ft}, \quad W = 2{,}700\ \text{lbf},$$
$$W/S_w = 15\ \text{lbf/ft}^2, \quad R_A = 6.05, \quad C_{D_0} = 0.036, \quad C_{D_0,L} = 0.0,$$
$$e = 0.82, \quad C_{L_{max}} = 1.4, \quad \mu_r = 0.04$$

The optimum lift coefficient for the acceleration run is

$$(C_L)_{opt} = \pi(0.82)(6.05)\frac{1+[16(6/33)]^2}{2[16(6/33)]^2}(0.04 - 0.0) = 0.3485$$

and the corresponding total drag coefficient for the acceleration run is

$$C_D = C_{D_0} + C_{D_0,L}\,C_L + \frac{(16h_w/b_w)^2}{1+(16h_w/b_w)^2}\frac{C_L^2}{\pi e R_A}$$

$$= 0.036 + 0.0 + \frac{[16(6/33)]^2}{1+[16(6/33)]^2}\frac{0.3485^2}{\pi(0.82)(6.05)} = 0.042969$$

From Eq. (3.10.20), the liftoff velocity at sea level is

$$V_{LO} = 1.1\sqrt{\frac{2}{C_{L_{max}}}}\sqrt{\frac{W/S_w}{\rho}}$$

$$= 1.1\sqrt{\frac{2}{1.4}}\sqrt{\frac{15\ \text{lbf/ft}^2}{0.0023769\ \text{slug/ft}^3}} = 104.444\ \text{ft/sec}$$

For this airplane the thrust is a linear function of the airspeed, and at standard sea level, the thrust coefficients are

$$T_S = 1{,}200\ \text{lbf}, \qquad T' = -4.0\ \text{lbf}\cdot\text{sec/ft}, \qquad T'' = 0.0$$

Using Eq. (3.10.25) to compute the acceleration distance gives

$$K_0 = \frac{T_S}{W} - \mu_r = \frac{1{,}200 \text{ lbf}}{2{,}700 \text{ lbf}} - 0.04 = 0.40444$$

$$K_1 = \frac{T'}{W} = \frac{-4.0 \text{ lbf·sec/ft}}{2{,}700 \text{ lbf}} = -0.0014815 \text{ sec/ft}$$

$$K_2 = \frac{T''}{W} + \frac{\rho}{2W/S_w}(C_L \mu_r - C_D) = \frac{\rho}{2W/S_w}(\mu_r C_L - C_D)$$

$$= \frac{0.0023769 \text{ slug/ft}^3}{2(15 \text{ lbf/ft}^2)}[0.04(0.3485) - 0.042969]$$

$$= -2.3000 \times 10^{-6} \text{ sec}^2/\text{ft}^2$$

$$K_R = 4K_0 K_2 - K_1^2$$

$$= 4(0.40444)(-2.3000 \times 10^{-6} \text{ sec}^2/\text{ft}^2) - (-0.0014815 \text{ sec/ft})^2$$

$$= -5.9157 \times 10^{-6} \text{ sec}^2/\text{ft}^2$$

$$\sqrt{-K_R} = 0.0024322 \text{ sec/ft}$$

$$f_S = K_0 = 0.40444$$

$$f_{LO} = K_0 + K_1 V_{LO} + K_2 V_{LO}^2$$

$$= 0.40444 - 0.0014815(104.444) - 2.3000 \times 10^{-6}(104.444)^2$$

$$= 0.22462$$

$$f_S' = K_1 = -0.0014815 \text{ sec/ft}$$

$$f_{LO}' = K_1 + 2K_2 V_{LO}$$

$$= -0.0014815 + 2(-2.3000 \times 10^{-6} \text{ sec}^2/\text{ft}^2)(104.444 \text{ ft/sec})$$

$$= -0.0019619 \text{ sec/ft}$$

$$K_W = \frac{1}{\sqrt{-K_R}} \ln\left[\frac{\left(f_{LO}' - \sqrt{-K_R}\right)\left(f_S' + \sqrt{-K_R}\right)}{\left(f_{LO}' + \sqrt{-K_R}\right)\left(f_S' - \sqrt{-K_R}\right)}\right]$$

$$= \frac{1}{0.0024322 \text{ sec/ft}} \ln\left[\frac{\left(-0.001962 - 0.002432\right)\left(-0.001482 + 0.002432\right)}{\left(-0.001962 + 0.002432\right)\left(-0.001482 - 0.002432\right)}\right]$$

$$= 336.98 \text{ ft/sec}$$

$$K_T = \frac{1}{2K_2} \ln\left(\frac{f_{LO}}{f_S}\right) - \frac{K_1 K_W}{2K_2}$$

$$= \frac{\ln(0.22462/0.40444) + 0.0014815 \text{ sec/ft}(336.98 \text{ ft/sec})}{2(-2.3000 \times 10^{-6} \text{ sec}^2/\text{ft}^2)}$$

$$= 19{,}317 \text{ ft}^2/\text{sec}^2$$

$$s_a = \frac{K_T}{g} = \frac{19{,}317 \text{ ft}^2/\text{sec}^2}{32.2 \text{ ft/sec}^2} = 600 \text{ ft}$$

The total ground roll is the acceleration distance plus the rotation distance,

$$s_g = s_a + V_{LO} t_r = 600 \text{ ft} + (104 \text{ ft/sec})(1 \text{ sec}) = \underline{704 \text{ ft}}$$

Repeating the calculations for 5,000 feet, we have

$$V_{LO} = 112.513 \text{ ft/sec}$$
$$T_S = 1{,}034.05 \text{ lbf}$$
$$T' = -3.4468 \text{ lbf·sec/ft}$$
$$K_0 = 0.34298$$
$$K_1 = -0.0012766 \text{ sec/ft}$$
$$K_2 = -1.9819 \times 10^{-6} \text{ sec}^2/\text{ft}^2$$
$$K_R = -4.3487 \times 10^{-6} \text{ sec}^2/\text{ft}^2$$
$$\sqrt{-K_R} = 0.0020854 \text{ sec/ft}$$
$$f_S = 0.34298$$
$$f_{LO} = 0.17426$$
$$f_S' = -0.0012766 \text{ sec/ft}$$
$$f_{LO}' = -0.0017226 \text{ sec/ft}$$
$$K_W = 444.17 \text{ ft/sec}$$
$$K_T = 27{,}777 \text{ ft}^2/\text{sec}^2$$
$$s_a = 863 \text{ ft}$$
$$s_g = \underline{975 \text{ ft}}$$

Repeating the calculations for sea level and $C_L = 0.0$, we obtain

$$s_g = \underline{712 \text{ ft}}$$

and repeating the calculations for sea level and $C_L = 2(0.3485)$ gives

$$s_g = \underline{712 \text{ ft}}$$

Two things should be noted from Example 3.10.1. First, notice that there is a very significant increase in the ground roll required for takeoff at 5,000 feet over that required at sea level. Both the designer and the pilot should always remember this fact. Second, notice that while there is an optimum angle of attack for the acceleration run, this optimum is quite flat. The lift coefficient can be widely varied on either side of this optimum without increasing the ground roll a great deal. In Example 3.10.1, doubling this lift coefficient or reducing it to zero increases the ground roll by only about 1 percent. This is because increasing the angle of attack for the acceleration run will decrease the rolling friction and increase the drag. Thus, the net effect of changing the angle of attack is quite small.

EXAMPLE 3.10.2. For the general aviation airplane considered in Example 3.10.1, determine the total ground roll required for takeoff at sea level with a 20-mph headwind. Assume the optimum lift coefficient and a μ_r value of 0.04.

Solution. With a headwind the initial airspeed for the acceleration run is not zero but equal to the headwind speed. Thus, we cannot use Eq. (3.10.25) but must return to the more general result given by Eq. (3.10.12). From Example 3.10.1 and the given headwind,

$$S_w = 180 \text{ ft}^2, \quad b_w = 33 \text{ ft}, \quad h_w = 6.0 \text{ ft}, \quad W = 2{,}700 \text{ lbf}, \quad W/S_w = 15 \text{ lbf/ft}^2,$$
$$R_A = 6.05, \quad C_{D_0} = 0.036, \quad C_{D_0,L} = 0.0, \quad e = 0.82, \quad C_{L_{\max}} = 1.4,$$
$$C_L = 0.3485, \quad \mu_r = 0.04, \quad C_D = 0.042969, \quad V_{LO} = 104.44 \text{ ft/sec},$$
$$V_{hw} = 29.33 \text{ ft/sec}, \quad V_1 = 29.33 \text{ ft/sec}, \quad V_2 = 104.44 \text{ ft/sec},$$
$$T_0 = 1{,}200 \text{ lbf}, \quad T' = -4.0 \text{ lbf} \cdot \text{sec/ft}, \quad T'' = 0.0$$

Using Eq. (3.10.12) to compute the acceleration distance, we have

$$K_0 = \frac{T_0}{W} - \mu_r = 0.40444$$

$$K_1 = \frac{T'}{W} = -0.0014815 \text{ sec/ft}$$

$$K_2 = \frac{T''}{W} + \frac{\rho}{2W/S_w}(C_L\mu_r - C_D) = -2.3000 \times 10^{-6} \text{ sec}^2/\text{ft}^2$$

$$K_R = 4K_0 K_2 - K_1^2 = -5.9157 \times 10^{-6} \text{ sec}^2/\text{ft}^2$$

$$\sqrt{-K_R} = 0.0024322 \text{ sec/ft}$$

$$f_1 = K_0 + K_1 V_1 + K_2 V_1^2 = 0.35901$$

$$f_2 = K_0 + K_1 V_2 + K_2 V_2^2 = 0.22462$$

$$f_1' = K_1 + 2K_2 V_1 = -0.0016164 \text{ sec/ft}$$

$$f_2' = K_1 + 2K_2 V_2 = -0.0019619 \text{ sec/ft}$$

$$K_W = \frac{1}{\sqrt{-K_R}} \ln\left[\frac{\left(f_2' - \sqrt{-K_R}\right)\left(f_1' + \sqrt{-K_R}\right)}{\left(f_2' + \sqrt{-K_R}\right)\left(f_1' - \sqrt{-K_R}\right)}\right] = 260.14 \text{ ft/sec}$$

$$K_T = \frac{1}{2K_2} \ln\left(\frac{f_2}{f_1}\right) - \frac{K_1 K_W}{2K_2} = 18{,}160 \text{ ft}^2/\text{sec}^2$$

$$s_2 - s_1 = \frac{K_T - V_{hw} K_W}{g} = 327 \text{ ft}$$

The total ground roll is this 327-ft acceleration distance plus the rotation distance. The rotation distance is the time allowed for rotation multiplied by the ground speed at liftoff,

$$s_g = s_2 - s_1 + (V_{LO} - V_{hw})t_r = \underline{402 \text{ ft}}$$

This compares to a ground roll of 704 ft with no headwind.

The landing ground roll can be analyzed in a manner similar to that used for takeoff. The main differences are that during deceleration, the thrust is set to the idle value or in the case of large airplanes, some reverse thrust, and the coefficient of rolling friction is increased to about 0.4 to account for braking. Sometimes spoilers are also used to increase the drag and reduce the lift, imposing greater weight on the tires and increasing the friction force. With these differences, Eqs. (3.10.8) through (3.10.19) still apply and can be used to obtain the braking distance, s_b.

For best accuracy, the braking distance can be broken up into small segments with the length of each segment computed from Eq. (3.10.12). Normally, thrust reversal can be applied only above some cutoff speed, and the braking distance must be divided into at least two segments. The total braking distance is the distance required to reduce the ground speed from the touchdown value to zero. To allow for a margin of safety, the touchdown airspeed, V_{TD}, should be about 15 percent above the stall speed for the aircraft in landing configuration (a margin of only 10 percent is normally used for military aircraft). The maximum lift coefficient for the landing configuration is typically larger than that for takeoff configuration, because the high-lift devices are fully engaged for landing. Thus the stall speed for landing is typically less than that for takeoff.

For airplanes without thrust reversal, the idle thrust is quite small and can usually be ignored. For this type of aircraft, the braking distance can be determined by applying Eq. (3.10.12) over the full braking distance as a single step. Thus, for the critical case of **no wind and no thrust reversal**, this gives

$$s_b = \begin{cases} \dfrac{V_{TD}^2}{2g\mu_r}, & C_L = \dfrac{C_D}{\mu_r} \\[2ex] \dfrac{W/S_w}{\rho g(C_D - \mu_r C_L)} \ln\left[1 + \dfrac{\rho V_{TD}^2}{2W/S_w}\left(\dfrac{C_D}{\mu_r} - C_L\right)\right], & C_L \neq \dfrac{C_D}{\mu_r} \end{cases} \qquad (3.10.40)$$

In computing the total ground roll for landing, the designer should allow for a short "free roll" to account for the reaction time required for the pilot to apply the brakes and engage the spoilers and the thrust reversers. This reaction time, t_f, is typically on the order of 1 to 3 seconds, depending on the complexity of the braking system. The free-roll distance is usually approximated as this assumed reaction time multiplied by the ground speed at the touchdown airspeed,

$$s_f = (V_{TD} - V_{hw})t_f \qquad (3.10.41)$$

The total ground roll for landing is the sum of the free-roll and the braking distances,

$$s_g = s_f + s_b \tag{3.10.42}$$

EXAMPLE 3.10.3. Consider the landing performance of the executive business jet described in Example 3.2.1. For this airplane, during the landing ground roll, the total drag coefficient is 0.040, the lift coefficient is 0.10, and the maximum lift coefficient is 2.10. The airplane is equipped with a thrust reversal system that delivers a constant thrust of $-2,000$ lbf at sea level. However, to avoid reingestion of the exhaust gases, this can only be used at airspeeds above 100 ft/sec. Estimate the total no-wind ground roll for this aircraft on landing at sea level. Because of the complexity of this braking system, allow a full 3 seconds for pilot reaction time. Assume that μ_r is 0.4.

Solution. For this airplane,

$$S_w = 320 \text{ ft}^2, \quad b_w = 54 \text{ ft}, \quad W = 20{,}000 \text{ lbf}, \quad W/S_w = 62.5 \text{ lbf/ft}^2,$$
$$C_D = 0.040, \quad C_L = 0.10, \quad C_{L_{max}} = 2.10, \quad \mu_r = 0.4$$

$$V_{TD} = 1.15 \sqrt{\frac{2}{C_{L_{max}}}} \sqrt{\frac{W/S_w}{\rho}} = 1.15 \sqrt{\frac{2}{2.1}} \sqrt{\frac{62.5 \text{ lbf/ft}^2}{0.0023769 \text{ slug/ft}^3}} = 181.99 \text{ ft/sec}$$

$$s_f = V_{TD}\, t_f = 181.99 \text{ ft/sec}(3\,\text{sec}) = 546.0 \text{ ft}$$

After this initial free roll, the thrust force remains constant at $-2,000$ lbf from the touchdown velocity to 100 ft/sec. At that point the engines are throttled back to idle and the thrust reversers are disengaged. For the remainder of the ground roll, the thrust is assumed to be zero. Thus, the braking distance is determined by using Eq. (3.10.12) over two segments.

$$V_1 = 181.99 \text{ ft/sec}, \quad V_2 = 100.00 \text{ ft/sec}, \quad V_3 = 0.00 \text{ ft/sec},$$
$$T_{1-2} = -2{,}000.0 \text{ lbf}, \quad T_{2-3} = 0.0 \text{ lbf}$$
$$(K_0)_1 = -0.5, \quad (K_0)_2 = -0.4$$
$$(K_1)_1 = 0.0, \quad (K_1)_2 = 0.0$$
$$(K_2)_1 = 0.0, \quad (K_2)_2 = 0.0$$

Because K_2 is zero for both segments and since there is no wind, there is no need to compute K_W. In addition, since K_1 is also zero for both segments, K_T is given by

$$(K_T)_1 = \frac{V_2^2 - V_1^2}{2(K_0)_1} = 23{,}120 \text{ ft}^2/\text{sec}^2, \quad (K_T)_2 = \frac{V_3^2 - V_2^2}{2(K_0)_2} = 12{,}500 \text{ ft}^2/\text{sec}^2$$

and the braking distance for each segment is

$$s_2 - s_1 = \frac{(K_T)_1}{g} = 718.0 \text{ ft}, \quad s_3 - s_2 = \frac{(K_T)_2}{g} = 388.2 \text{ ft}$$

The total ground roll distance for the landing is the sum of the free roll and the two segments of the braking roll,

$$s_g = s_f + (s_2 - s_1) + (s_3 - s_2) = 546.0 \text{ ft} + 718.0 \text{ ft} + 388.2 \text{ ft} = \underline{1,652 \text{ ft}}$$

3.11. Accelerating Climb and Balanced Field Length

The takeoff ground-roll distance discussed in Sec. 3.10 is the actual distance traveled along the runway, from brake release until the wheels of the aircraft leave the ground. Obviously, for safety reasons, the wheels of the aircraft should not leave the ground right at the end of the runway. This quite naturally raises another question. What is the minimum runway length required for the safe operation of any particular aircraft? This depends not only on the ability of the aircraft to accelerate along the ground, but also on the ability of the aircraft to climb. Since there could be an obstacle at the end of the runway, the aircraft should climb to some minimum *obstacle clearance altitude* before passing over the end of the runway, as shown in Fig. 3.11.1. This minimum obstacle clearance altitude is usually 35 feet for commercial airplanes and 50 feet for small general aviation airplanes and military aircraft. Transport aircraft, which are certified under Federal Air Regulations (FAR) Part 25, are required to accelerate during this initial climb, from an airspeed at liftoff that is at least 10 percent above stall to an airspeed that is at least 20 percent above stall at the obstacle clearance altitude. The total running distance measured along the ground, s_{OC}, from brake release until the airplane has reached the obstacle clearance altitude is usually called the *FAR takeoff distance*.

The minimum runway length required for a transport aircraft operating under FAR Part 25 is also influenced by the possibility of engine failure. If one engine were to fail just prior to liftoff, the airplane should be capable of continuing the takeoff with one engine inoperative, because the distance required to stop the aircraft from a speed close to the liftoff speed may be too great. However, if one engine were to fail just after the acceleration run was started, the pilot could more easily abort the takeoff and stop the

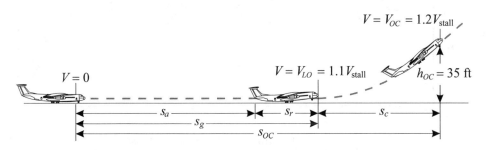

Figure 3.11.1. FAR Part 25 takeoff.

aircraft. This is because in the early part of the acceleration run, the ground speed is still low and the distance required to stop the aircraft is much less than the distance required to continue the takeoff with one engine inoperative. As the point of engine failure moves further into the acceleration run, the speed of the aircraft at engine failure is increased. As the engine failure speed increases, the distance required to continue the takeoff with one engine inoperative is decreased and the distance required to stop the airplane from the point of engine failure is increased. At some point along the acceleration run, the airplane reaches a speed where the distance required to continue the takeoff and climb to the obstacle clearance altitude with one engine inoperative is exactly equal to the distance required to stop the airplane from that same speed. This speed is called the *critical engine failure speed*. The distance required to accelerate an airplane from rest to the critical engine failure speed, using all engines, and then to continue the takeoff and climb to the obstacle clearance altitude, with one engine inoperative, is called the *balanced field length*. The two equilateral options that determine the balanced field length are shown in Fig. 3.11.2. This balanced field length is considered to be the minimum runway length required for safe takeoff of the aircraft and is significantly longer than the FAR takeoff distance with all engines operative.

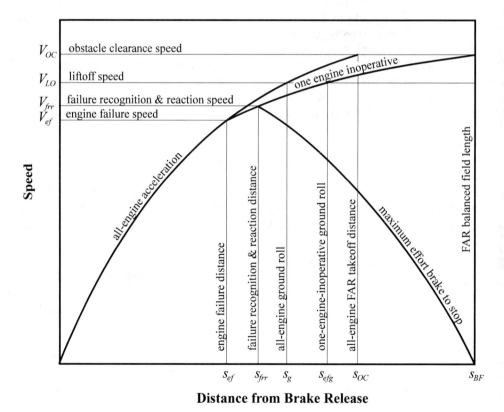

Figure 3.11.2. The two equilateral options defining balanced field length.

If one engine fails at any speed below the critical engine failure speed, the takeoff should always be aborted. If one engine fails at any speed above the critical engine failure speed, the takeoff should be continued with one engine inoperative. In any case, if these requirements are adhered to, the distance required to complete the takeoff and climb to the obstacle clearance altitude or brake the aircraft to a complete stop, in the event of engine failure, will always be less than or equal to the balanced field length.

In determining the balanced field length for transport aircraft operating under FAR Part 25, the use of reverse thrust is not permitted. In addition, we must account for the time required for the pilot to recognize the engine failure and react. For example, the time required for the pilot to realize that an engine has failed might be on the order of 1 second, with an additional second required for the pilot to react and apply the maximum braking force. Notice that the balanced field length for a twin-engine aircraft will be significantly longer than that for, say, a four-engine aircraft having the same total thrust. This is because the twin loses half its thrust on engine failure, while the quad loses only 25 percent of its thrust on engine failure.

To estimate the FAR takeoff distance and the balanced field length for an aircraft, we must be able to estimate the distance traveled while climbing to the obstacle clearance altitude. This phase of the takeoff is usually referred to as the *transition phase*, because it encompasses the transition from horizontal motion along the runway to the steady climbing flight that takes place following the takeoff. Since FAR Part 25 requires the airspeed to increase from at least 1.1 V_{stall} at liftoff to 1.2 V_{stall} at the obstacle clearance altitude, the airplane must accelerate in both the axial and normal directions during transition. The forces acting on an airplane during such an accelerating climb are shown in Fig. 3.11.3.

For the critical case of **no wind**, Newton's second law applied to the direction of flight requires that

$$\frac{W}{g}\frac{dV}{dt} = T - D - W\sin\gamma \tag{3.11.1}$$

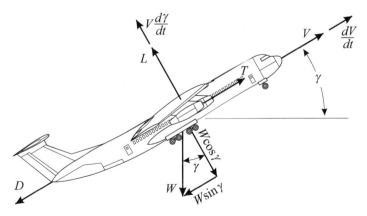

Figure 3.11.3. Forces acting on an airplane during accelerating climb.

In the direction normal to the curved flight path, the centripetal acceleration is supported by the difference between the lift and the normal component of weight,

$$\frac{W}{g}\frac{V^2}{R} = \frac{W}{g}\frac{V^2}{V/\Omega} = \frac{W}{g}V\frac{d\gamma}{dt} = L - W\cos\gamma \tag{3.11.2}$$

where R is the radius of curvature of the flight path and $\Omega = d\gamma/dt$ is the angular velocity. Solving Eq. (3.11.1) for $\sin\gamma$, we have

$$\sin\gamma = \frac{T-D}{W} - \frac{1}{g}\frac{dV}{dt} \tag{3.11.3}$$

The vertical component of velocity is

$$\frac{dh}{dt} = V\sin\gamma \tag{3.11.4}$$

and the horizontal component of velocity is

$$\frac{ds}{dt} = V\cos\gamma \tag{3.11.5}$$

Using Eq. (3.11.3) in Eq. (3.11.4) gives

$$\frac{dh}{dt} = \frac{T-D}{W}V - \frac{V}{g}\frac{dV}{dt} \tag{3.11.6}$$

Applying Eq. (3.11.5) to Eq. (3.11.6) and rearranging gives

$$\frac{T-D}{\cos\gamma}\frac{ds}{dt} = W\frac{dh}{dt} + \frac{W}{g}V\frac{dV}{dt} \tag{3.11.7}$$

After integrating both sides of Eq. (3.11.7) from liftoff to the obstacle clearance altitude, we obtain

$$\int_{S_{LO}}^{S_{OC}} \frac{T-D}{\cos\gamma}ds = \int_0^{h_{OC}} W\,dh + \frac{1}{g}\int_{V_{LO}}^{V_{OC}} WV\,dV \tag{3.11.8}$$

where the subscripts LO and OC refer to the liftoff point and the obstacle clearance point, respectively. Assuming constant aircraft weight, the integration on the right-hand side of Eq. (3.11.8) is readily carried out to yield

$$\int_{S_{LO}}^{S_{OC}} \frac{T-D}{\cos\gamma}ds = Wh_{OC} + \frac{W}{2g}(V_{OC}^2 - V_{LO}^2) \tag{3.11.9}$$

From Eq. (3.11.9) we see that the work done by the aircraft engines, in excess of that dissipated by the drag force, can be used to increase potential energy, kinetic energy, or both. The exact flight path that is followed during the climb to obstacle clearance altitude depends on pilot technique. Both the climb angle, γ, and the velocity, V, are changing with time during this portion of the takeoff. The thrust force depends on velocity, and the drag force depends on both velocity and climb angle. Thus, if we knew exactly how the velocity and the climb angle varied with position during the climb from liftoff to obstacle clearance altitude, we could evaluate the integral on the left-hand side of Eq. (3.11.9) numerically. Unfortunately, this variation in velocity and climb angle depends on pilot technique and is not known. However, by examining the integrand on the left-hand side of Eq. (3.11.9), we can make a reasonable assumption that allows us to perform the required integration.

Because of induced drag, the drag coefficient increases with lift coefficient, which increases with load factor. Solving Eq. (3.11.2) for the load factor gives

$$\frac{L}{W} = \cos\gamma + \frac{V}{g}\frac{d\gamma}{dt} \tag{3.11.10}$$

From Eq. (3.11.10), we see that the load factor is a function of both the climb angle and the time rate of change of the climb angle. The first term on the right in Eq. (3.11.10) tends to reduce the load factor and induced drag during transition, while the second term tends to increase the load factor and induced drag. In addition, the decrease in ground effect, which occurs as the airplane climbs, tends to increase the drag somewhat. The small increase in velocity that takes place during transition reduces the thrust and increases the drag. The net effect is that the increase in climb angle and airspeed together with the decrease in ground effect, which all take place during transition, combine to cause a slight decrease in the numerator of the integrand in Eq. (3.11.9). However, since the climb angle is increasing, the denominator in this integrand is also decreasing during transition. Thus, because of these counteracting effects and because climb angle and velocity changes are typically small, the integrand in Eq. (3.11.9) changes only slightly during the transition phase. As a result, it is reasonable to assume that the integrand on the left-hand side of Eq. (3.11.9) is a linear function of the distanced traveled, s. With this assumption, the integration can be performed to give

$$\overline{F}_c s_c = W h_{OC} + \frac{W}{2g}(V_{OC}^2 - V_{LO}^2) \tag{3.11.11}$$

where $\overline{F}_c$ is an effective net force during transition,

$$\overline{F}_c = \frac{T_{LO} - D_{LO} + (T_{OC} - D_{OC})/\cos(\gamma_{OC})}{2} \tag{3.11.12}$$

and s_c is the distance traveled over the ground during the climb from liftoff to obstacle clearance altitude, $s_{OC} - s_{LO}$. From Eq. (3.11.11) the distance required to climb to obstacle clearance altitude is

$$s_c = \frac{W}{\overline{F}_c}\left[h_{OC} + \frac{V_{OC}^2 - V_{LO}^2}{2g}\right] \tag{3.11.13}$$

The climb angle at obstacle clearance should be the steady climb angle for the aircraft in takeoff configuration, corresponding to the airspeed at the obstacle clearance altitude. From Eq. (3.11.3) this gives

$$\gamma_{OC} = \sin^{-1}\left(\frac{T_{OC} - D_{OC}}{W}\right) \tag{3.11.14}$$

From the requirements of FAR Part 25, the altitude and the airspeed are known at both liftoff and obstacle clearance. Thus, the thrust force and the drag can be determined at these two points, including the effects of climb angle and ground effect. Using these results with Eqs. (3.11.12) and (3.11.14), the effective net force during transition is readily determined. This effective net force is then used in Eq. (3.11.13) to determine the distance traveled over the ground during the climb from liftoff to the obstacle clearance altitude.

Since the effective net force during transition is decreased by the angular acceleration and increased by the climb angle and ground effect, a reasonable first approximation for this effective net force can be obtained by using the values for steady level flight in the absence of ground effect,

$$\overline{F}_c \cong \frac{(T_A - T_R)_{LO} + (T_A - T_R)_{OC}}{2} \tag{3.11.15}$$

where T_A is the thrust available and T_R is the thrust required for steady level flight as determined in Sec. 3.2.

EXAMPLE 3.11.1. For the executive business jet described in Example 3.2.1, determine the FAR takeoff distance at standard sea level with no wind. This airplane is fitted with turbojet engines that produce a thrust that is independent of airspeed. For the takeoff acceleration run at standard sea level, use the following properties:

$$S_w = 320 \text{ ft}^2, \quad b_w = 54 \text{ ft}, \quad h_w = 6.0 \text{ ft}, \quad W = 20{,}000 \text{ lbf}, \quad T_A = 6{,}500 \text{ lbf},$$
$$C_{D_0} = 0.033, \quad C_{D_0,L} = 0.0, \quad e = 0.74, \quad C_{L_{max}} = 1.6, \quad C_L = 0.4,$$
$$\mu_r = 0.04, \quad t_r = 1 \text{ sec}$$

Solution. The ground roll is found following the solution to Example 3.10.1,

$$C_D = C_{D_0} + C_{D_0,L} C_L + \frac{(16 h_w/b_w)^2}{1 + (16 h_w/b_w)^2} \frac{C_L^2}{\pi e R_A} = 0.038737$$

$$V_{LO} = 1.1\sqrt{\frac{2}{C_{L\,max}}}\sqrt{\frac{W/S_w}{\rho}} = 199.43 \text{ ft/sec}$$

$$T_S = 6{,}500 \text{ lbf,} \qquad T' = 0.0, \qquad T'' = 0.0$$

$$K_0 = \frac{T_S}{W} - \mu_r = 0.28500$$

$$K_1 = \frac{T'}{W} = 0.0$$

$$K_2 = \frac{T''}{W} + \frac{\rho}{2W/S_w}(C_L\mu_r - C_D) = -4.323552\times10^{-7} \text{ sec}^2/\text{ft}^2$$

$$f_S = K_0 = 0.28500$$

$$f_{LO} = K_0 + K_1 V_{LO} + K_2 V_{LO}^2 = 0.2678049$$

$$K_T = \frac{1}{2K_2}\ln\left(\frac{f_{LO}}{f_S}\right) - \frac{K_1 K_W}{2K_2} = 71{,}967 \text{ ft}^2/\text{sec}^2$$

$$s_a = \frac{K_T}{g} = 2{,}235.0 \text{ ft}$$

$$s_g = s_a + V_{LO}\,t_r = 2{,}434.4 \text{ ft}$$

The velocity at the obstacle clearance altitude is

$$V_{OC} = 1.2\sqrt{\frac{2}{C_{L\,max}}}\sqrt{\frac{W/S_w}{\rho}} = 217.56 \text{ ft/sec}$$

As a first approximation, the lift and drag coefficients at obstacle clearance are taken as those for steady level flight in the absence of ground effect,

$$C_{L_{OC}} \cong \frac{W}{\frac{1}{2}\rho V_{OC}^2 S_w} = 1.1111$$

$$C_{D_{OC}} \cong C_{D_0} + C_{D_0,L}\,C_{L_{OC}} + \frac{C_{L_{OC}}^2}{\pi e R_A} = 0.091272$$

Our first approximation for the drag force at obstacle clearance is then

$$D_{OC} \cong \tfrac{1}{2}\rho V_{OC}^2 S_w C_{D_{OC}} = 1{,}643 \text{ lbf}$$

and, from Eq. (3.11.14), our first estimate for the climb angle at obstacle clearance altitude is

$$\gamma_{OC} \cong \sin^{-1}\left(\frac{T_{OC} - D_{OC}}{W}\right) = 14.1°$$

Using this value for the climb angle, we can obtain an improved estimate for the lift and drag coefficients at the obstacle clearance altitude of 35 feet,

$$C_{L_{OC}} = \frac{W \cos(\gamma_{OC})}{\frac{1}{2}\rho V_{OC}^2 S_w} = 1.0778$$

$$C_{D_{OC}} = C_{D_0} + C_{D_0,L} \, C_{L_{OC}} + \frac{[16(h_w + h_{OC})/b_w]^2}{1 + [16(h_w + h_{OC})/b_w]^2} \frac{C_{L_{OC}}^2}{\pi e R_A} = 0.087467$$

The improved estimate for the drag force at obstacle clearance is

$$D_{OC} = \frac{1}{2}\rho V_{OC}^2 S_w C_{D_{OC}} = 1{,}574 \text{ lbf}$$

and the improved estimate for the climb angle at obstacle clearance is

$$\gamma_{OC} = \sin^{-1}\left(\frac{T_{OC} - D_{OC}}{W}\right) = 14.3°$$

The lift and drag coefficients at liftoff are

$$C_{L_{LO}} = \frac{W}{\frac{1}{2}\rho V_{LO}^2 S_w} = 1.3223$$

$$C_{D_{LO}} = C_{D_0} + C_{D_0,L} \, C_{L_{LO}} + \frac{(16 h_w/b_w)^2}{1 + (16 h_w/b_w)^2} \frac{C_{L_{LO}}^2}{\pi e R_A} = 0.095694$$

The drag force at liftoff is then

$$D_{LO} = \frac{1}{2}\rho V_{LO}^2 S_w C_{D_{LO}} = 1{,}447 \text{ lbf}$$

The integrand in Eq. (3.11.9), evaluated at both liftoff and obstacle clearance, is

$$\left(\frac{T - D}{\cos\gamma}\right)_{LO} = \frac{6{,}500 \text{ lbf} - 1{,}447 \text{ lbf}}{1.0} = 5{,}053 \text{ lbf}$$

$$\left(\frac{T - D}{\cos\gamma}\right)_{OC} = \frac{6{,}500 \text{ lbf} - 1{,}574 \text{ lbf}}{0.96954} = 5{,}082 \text{ lbf}$$

From Eq. (3.11.12), the average effective net force during transition is then

$$\overline{F}_c = \frac{T_{LO} - D_{LO} + (T_{OC} - D_{OC})/\cos(\gamma_{OC})}{2} = 5{,}067 \text{ lbf}$$

and from Eq. (3.11.13), the distance required to climb to obstacle clearance altitude is

$$S_c = \frac{W}{F_c}\left[h_{OC} + \frac{V_{OC}^2 - V_{LO}^2}{2g}\right] = 601.5 \text{ ft}$$

This gives a FAR takeoff distance of

$$S_{OC} = S_g + S_c = \underline{3,036 \text{ ft}}$$

EXAMPLE 3.11.2. The executive business jet in Example 3.11.1 is powered by two turbojet engines. Assuming that one engine fails during a no-wind takeoff at standard sea level, plot the total field length required to stop the aircraft after engine failure and the total field length required to continue the FAR takeoff on the remaining engine, as a function of the engine failure recognition speed. Use an obstacle clearance altitude of 35 feet and assume 1 second for engine failure recognition and another second for the pilot to react and apply the brakes. What are the *critical engine failure recognition speed* and the balanced field length for this aircraft?

Solution. The acceleration distance prior to engine failure and the acceleration distance after engine failure are found using Eq. (3.10.12). From Example 3.11.1, we have

$$S_w = 320 \text{ ft}^2, \quad b_w = 54 \text{ ft}, \quad h_w = 6.0 \text{ ft}, \quad W = 20,000 \text{ lbf}, \quad C_{D_0} = 0.033,$$
$$C_{D_0,L} = 0.0, \quad e = 0.74, \quad C_{L_{max}} = 1.6, \quad C_L = 0.4, \quad (\mu_r)_a = 0.04,$$
$$(\mu_r)_b = 0.4, \quad t_r = 1 \text{ sec}, \quad C_D = 0.038737, \quad V_{LO} = 199.43 \text{ ft/sec},$$
$$(T_A)_{2-engines} = 6,500 \text{ lbf}, \quad (T_A)_{1-engine} = 3,250 \text{ lbf}$$

Equation (3.10.12) is applied separately over the two segments of the acceleration run, from zero velocity to engine failure and from engine failure to liftoff speed. For example, an engine failure speed of 50 ft/sec would give

$$V_1 = 0.00 \text{ ft/sec}, \quad V_2 = 50.00 \text{ ft/sec}, \quad V_3 = 199.43 \text{ ft/sec},$$
$$T_{1-2} = 6,500 \text{ lbf}, \quad T_{2-3} = 3,250 \text{ lbf}$$
$$(K_0)_1 = 0.28500, \quad (K_0)_2 = 0.12250$$
$$(K_1)_1 = 0.0, \quad (K_1)_2 = 0.0$$
$$(K_2)_1 = -4.323552\times10^{-7} \text{ sec}^2/\text{ft}^2, \quad (K_2)_2 = -4.323552\times10^{-7} \text{ sec}^2/\text{ft}^2$$
$$(K_T)_1 = 4,394 \text{ ft}^2/\text{sec}^2, \quad (K_T)_2 = 164,666 \text{ ft}^2/\text{sec}^2$$
$$(s_a)_1 = 136.5 \text{ ft}, \quad (s_a)_2 = 5,113.9 \text{ ft}$$
$$S_g = (s_a)_1 + (s_a)_2 + V_{LO} t_r = 5,449.8 \text{ ft}$$

Following Example 3.11.1 and computing the distance required to climb to 35 feet with only one engine, we obtain

$$V_{OC} = 217.56 \text{ ft/sec}$$

$$\gamma_{OC} = 4.6°$$

$$s_c = 1,777.8 \text{ ft}$$

Thus, the FAR takeoff distance for engine failure at 50 ft/sec is

$$s_{OC} = s_g + s_c = \underline{7,228 \text{ ft}}$$

To determine the engine failure recognition speed we return to Newton's second law stated in Eq. (3.10.1),

$$\frac{W}{g}\frac{dV}{dt} = T - D - F_r$$

Rearranging, we have

$$\frac{dV}{dt} = g\frac{T - D - F_r}{W}$$

Since the engine failure recognition time is very short, we can assume a constant net force and this can be integrated to yield the engine failure recognition speed, V_{fr},

$$V_{fr} = V_{ef} + g\frac{T - D - F_r}{W}(t_{fr} - t_{ef}) = V_{ef} + g(K_0 + K_1 V + K_2 V^2)(t_{fr} - t_{ef})$$

For the present example with engine failure at 50 ft/sec, the engine failure recognition speed, V_{fr}, is

$$V_{fr} = V_{ef} + g(K_0 + K_1 V_{ef} + K_2 V_{ef}^2)(t_{fr} - t_{ef})$$

$$= 50 \text{ ft/sec} + 32.2 \text{ ft/sec}^2 [0.12250 + 0.0 - 4.323552 \times 10^{-7}(50^2)]1.0 \text{ sec}$$

$$= 53.9 \text{ ft/sec}$$

In a similar manner the engine failure recognition and reaction speed, V_{frr}, is

$$V_{frr} = V_{fr} + g(K_0 + K_1 V_{fr} + K_2 V_{fr}^2)(t_{frr} - t_{fr})$$

$$= 53.9 \text{ ft/sec} + 32.2 \text{ ft/sec}^2 [0.12250 + 0.0 - 4.323552 \times 10^{-7}(53.9^2)]1.0 \text{ sec}$$

$$= 57.8 \text{ ft/sec}$$

The distance traveled during the failure recognition and reaction time is also found by using Eq. (3.10.12). This gives

$$s_{frr} - s_{ef} = 107.8 \text{ ft}$$

Similarly, using Eq. (3.10.12), the distance required to brake to a full stop from the failure recognition and reaction speed of 57.8 ft/sec is

$$s_b = 131.0 \text{ ft}$$

The total field length required to abort a takeoff after engine failure at 50 ft/sec is then, the distance required to accelerate to 50 ft/sec with both engines, plus the distance traveled during the failure recognition and reaction time with one engine inoperative, plus the distance required to brake to a full stop from the failure recognition and reaction speed,

$$s_{abort} = (s_a)_1 + (s_{frr} - s_{ef}) + s_b = \underline{375 \text{ ft}}$$

These calculations are repeated for engine failure speeds ranging from zero to the liftoff speed, in increments of 1 ft/sec. Sample results are shown in Table 3.11.1, and all results are plotted in Fig. 3.11.4. The speed at which the field length required to abort the takeoff is equal to the field length required to complete the FAR takeoff is the critical engine failure recognition speed, and the corresponding FAR takeoff distance is the balanced field length. For this airplane at sea level,

V_{ef}	V_{fr}	s_{OC}	s_{abort}
0.0000	3.9445	7,409.3660	10.3049
1.0000	4.9445	7,409.2950	13.0142
2.0000	5.9444	7,409.0770	15.9040
3.0000	6.9444	7,408.7170	18.9874
4.0000	7.9443	7,408.2090	22.2509
⋮	⋮	⋮	⋮
50.0000	53.9097	7,227.5340	375.3876
⋮	⋮	⋮	⋮
193.0000	196.4259	4,436.2270	4,244.8180
194.0000	197.4205	4,402.0580	4,289.1900
195.0000	198.4151	4,367.6350	4,333.8480
196.0000	199.4097	4,332.9600	4,378.7940
197.0000	200.4042	4,298.0250	4,424.0300
198.0000	201.3987	4,262.8370	4,469.5600
199.0000	202.3932	4,227.3920	4,515.3830

Table 3.11.1. Sample data used to construct the plot in Fig. 3.11.4.

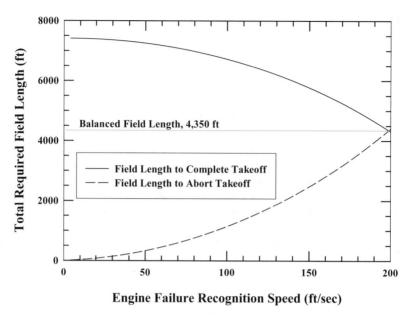

Figure 3.11.4. Graphical determination of balanced field length for the twin-engine airplane in Example 3.11.2.

the critical engine failure recognition speed is about 199 ft/sec and the balanced field length is about 4,350 ft, which is 43 percent greater than the FAR takeoff distance for the same aircraft with both engines operational, as determined in Example 3.11.1.

For the twin-engine executive business jet in Example 3.11.2, the critical engine failure recognition speed was found to be almost exactly equal to the liftoff speed. This means that if the pilot recognizes an engine failure while the aircraft is still on the ground, the takeoff should always be aborted. Even if the airplane has reached liftoff speed when the failure is detected, the pilot could still stop the aircraft in the same distance that would be required to complete the takeoff to obstacle clearance altitude. The decision to stop the aircraft rather than continue the takeoff with one engine inoperative is seen to be even more appropriate when one considers the relative consequences associated with coming up short. If the pilot elects to continue the takeoff with one engine inoperative and does not quite clear an obstacle, a very serious airborne collision will result. On the other hand, if the pilot decides to abort the takeoff and is not able to completely stop the aircraft before the end of the runway, the consequence of going off the runway at this much lower speed will be substantially less severe.

The high critical engine failure recognition speed that was found in Example 3.11.2 is typical of twin-engine aircraft. This is a direct consequence of the simple fact that when one engine fails on a twin, the airplane loses half of its power. When one engine fails on an airplane with three or more engines, the fraction of total power lost is much less. This significantly reduces the critical engine failure recognition speed and the balanced field length. For example, Fig. 3.11.5 shows how the aircraft in Example 3.11.2 would perform under conditions of engine failure if the same total thrust were provided by four engines instead of two. In this case the critical engine failure recognition speed would be reduced to about 180 ft/sec, and the balanced field length would be reduced to about 3,570 feet.

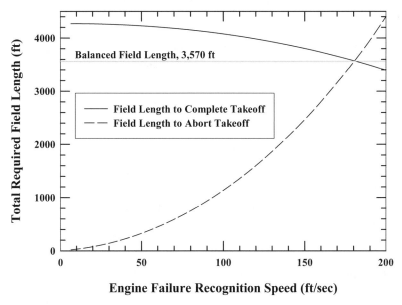

Figure 3.11.5. Balanced field length for the airplane in Example 3.11.2 assuming four engines instead of two.

3.12. Problems

3.1. Consider the executive business jet described in Example 3.2.1. Using the small-thrust-angle approximation and assuming that the wing does not stall, compute the lift coefficient, drag coefficient, thrust required, and power required for steady level flight at sea level for airspeeds of 200, 300, 400, and 500 mph.

3.2. For the executive business jet described in Example 3.2.1, using the small-thrust-angle approximation, compute the minimum thrust required and the airspeed at which it occurs for both sea level and 20,000 feet.

3.3. For the executive business jet described in Example 3.2.1, using the small-thrust-angle approximation, compute the minimum power required and the airspeed at which it occurs for both sea level and 20,000 feet.

3.4. Consider the general aviation airplane described in Example 3.3.1. Using the small-thrust-angle approximation and assuming that the wing does not stall, compute the lift coefficient, drag coefficient, thrust required, and power required for steady level flight at sea level for airspeeds of 50, 100, 150, and 200 mph.

3.5. For the general aviation airplane described in Example 3.3.1, using the small-thrust-angle approximation, compute the minimum thrust required and the airspeed at which it occurs for both sea level and 10,000 feet.

3.6. For the general aviation airplane described in Example 3.3.1, using the small-thrust-angle approximation, compute the minimum power required and the airspeed at which it occurs for both sea level and 10,000 feet.

3.7. A small jet transport has the following properties in cruise configuration:

$$S_w = 950 \text{ ft}^2, \quad b_w = 75 \text{ ft}, \quad W = 73{,}000 \text{ lbf},$$
$$C_{D_0} = 0.020, \quad C_{D_0,L} = 0.0, \quad e = 0.67$$

Using the small-thrust-angle approximation and assuming that the wing does not stall, compute the lift coefficient, the drag coefficient, the thrust required, and the power required for steady level flight at 30,000 feet for airspeeds of 400, 600, and 800 ft/sec.

3.8. For the jet transport described in problem 3.7, using the small-thrust-angle approximation, compute the minimum thrust required and the airspeed at which it occurs for both sea level and 30,000 feet.

3.9. For the jet transport described in problem 3.7, using the small-thrust-angle approximation, compute the minimum power required and the airspeed at which it occurs for both sea level and 30,000 feet.

3.10. A large jet transport, similar to the Boeing 747, has the following properties in cruise configuration:

$$S_w = 5{,}500 \text{ ft}^2, \quad b_w = 196 \text{ ft}, \quad W = 636{,}600 \text{ lbf},$$
$$C_{D_0} = 0.020, \quad C_{D_0,L} = 0.0, \quad e = 0.72$$

Using the small-thrust-angle approximation and assuming that the wing does not stall, compute the lift coefficient, the drag coefficient, the thrust required, and the power required for steady level flight at 30,000 feet for airspeeds of 400, 600, and 800 ft/sec.

3.11. For the jet transport described in problem 3.10, using the small-thrust-angle approximation, compute the minimum thrust required and the airspeed at which it occurs for both sea level and 30,000 feet.

3.12. For the jet transport described in problem 3.10, using the small-thrust-angle approximation, compute the minimum power required and the airspeed at which it occurs for both sea level and 30,000 feet.

3.13. The jet transport in problem 3.7 is fitted with turbojet engines that produce a thrust that is independent of airspeed and directly proportional to air density. The total full-throttle thrust available from these engines is given by the equation

$$T_A = \frac{\rho}{\rho_0} 27,700 \text{ lbf}$$

where ρ is the air density and ρ_0 is the air density at standard sea level. Using the small-climb-angle approximation, find the rate of climb at sea level and 30,000 feet for airspeeds of 400, 600, and 800 ft/sec.

3.14. For the jet transport described in problem 3.13, using the small-climb-angle approximation, compute the maximum rate of climb and the airspeed at which it occurs for both sea level and 30,000 feet.

3.15. Assume that the jet transport described in problem 3.10 is fitted with turbojet engines that produce a thrust that is independent of airspeed. Suppose that the total full-throttle thrust available from these engines is given by the equation

$$T_A = \left(\frac{\rho}{\rho_0}\right)^{0.8} 250,000 \text{ lbf}$$

where ρ is the air density and ρ_0 is the air density at standard sea level. Using the small-climb-angle approximation, find the rate of climb at sea level and 30,000 feet for airspeeds of 400, 600, and 800 ft/sec.

3.16. For the jet transport described in problem 3.15, using the small-climb-angle approximation, compute the maximum rate of climb and the airspeed at which it occurs for both sea level and 30,000 feet.

3.17. A propeller-driven transport has the following properties in cruise configuration:

$$S_w = 945 \text{ ft}^2, \quad b_w = 96 \text{ ft}, \quad W = 40,000 \text{ lbf}, \quad C_{D_0} = 0.025, \quad C_{D_0,L} = 0.0,$$
$$e = 0.80, \quad T_A = (\rho/\rho_0)(10,000 \text{ lbf} - 12.0 \text{ lbf·sec/ft } V)$$

Using the small-climb-angle approximation, compute the rate of climb at both sea level and 10,000 feet for airspeeds of 200, 250, 300, and 350 ft/sec.

3.18. For the propeller-driven transport described in problem 3.17, using the small-climb-angle approximation, compute the maximum rate of climb and the airspeed at which it occurs for both sea level and 10,000 feet.

3.19. A human athlete can produce more than 1 horsepower for a few seconds. However, in sustained aerobic exercise, a human athlete is capable of delivering only about 0.33 horsepower. In the development of the human-powered airplane (Gossamer Condor) designed by Paul MacCready and his colleagues, the power delivered by the pilot was converted to propulsive power available with an efficiency of about 70 percent. Assuming that a human-powered airplane could be built with the following reasonable properties,

$$R_A = 12, \quad W = 200 \text{ lbf}, \quad P_A = 0.70(0.33 \text{ hp})$$
$$C_{D_0} = 0.010, \quad C_{D_0,L} = 0.0, \quad e = 0.80$$

estimate the wing area required for the airplane and the airspeed at which it should be flown at standard sea level. What is the minimum power that the pilot would need to provide to fly this airplane in Denver, Colorado, at an elevation of 5,000 feet?

3.20. For the human-powered airplane in problem 3.19, assuming that the pilot can produce 0.6 horsepower for a short time during takeoff and climb, estimate the rate of climb for this aircraft at sea level.

3.21. For the small jet transport described in problem 3.7, the maximum usable fuel weight is 29,500 lbf and the thrust-specific fuel consumption is 0.69 lbf of fuel per hour per lbf of thrust. This airplane has the following properties in cruise configuration:

$$S_w = 950 \text{ ft}^2, \quad b_w = 75 \text{ ft}, \quad W_f = 73,000 \text{ lbf}, \quad W_e = 43,500 \text{ lbf},$$
$$q_T = 0.69 \text{ lbf/lbf-hr}, \quad C_{D_0} = 0.020, \quad C_{D_0,L} = 0.0, \quad e = 0.67$$

Assuming that the thrust-specific fuel consumption is independent of both airspeed and altitude, compute the maximum endurance airspeed with maximum fuel weight and with minimum fuel weight for both sea level and 30,000 feet.

3.22. For the jet transport described in problem 3.21, assuming that the pilot maintains the maximum endurance airspeed at all times, determine the maximum endurance for this airplane at 30,000 feet.

3.23. For the large jet transport described in problem 3.10, the total fuel capacity is 330,700 pounds and thrust-specific fuel consumption is 0.66 pound of fuel per hour per pound of thrust. Because some fuel must be held in reserve, the maximum usable fuel weight is 304,200 lbf. This airplane has the following properties in cruise configuration:

$$S_w = 5,500 \text{ ft}^2, \quad b_w = 196 \text{ ft}, \quad W_f = 636,600 \text{ lbf}, \quad W_e = 332,400 \text{ lbf},$$
$$q_T = 0.66 \text{ lbf/lbf} \cdot \text{hr}, \quad C_{D_0} = 0.020, \quad C_{D_0,L} = 0.0, \quad e = 0.72$$

Assuming that the thrust-specific fuel consumption is independent of both airspeed and altitude, compute the minimum fuel consumption rate, in gallons per minute, with maximum fuel weight and with minimum fuel weight for both sea level and 30,000 feet. Jet fuel weighs 6.67 pounds per gallon.

3.24. For the jet transport described in problem 3.23, assuming that the pilot maintains the no-wind maximum range airspeed at all times, compute the fuel consumption rate, in gallons per minute, with maximum fuel weight and with minimum fuel weight for both sea level and 30,000 feet.

3.25. For the propeller-driven airplane and power plant described in Example 3.5.1, if the pilot maintains the maximum endurance airspeed at all times, determine the total distance that the airplane will travel on a tank of fuel at sea level.

3.26. For the propeller-driven airplane and power plant described in Example 3.5.1, if the pilot maintains the maximum range airspeed at all times, determine the total time required for the airplane to fly a distance equal to the maximum range with no wind at sea level.

3.27. To attain maximum range, the airspeed of an airplane must be decreased as fuel is burned off and gross weight is decreased. For the executive business jet that is described in Example 3.6.2, determine the maximum range airspeed with fuel tanks full and with fuel tanks empty, for no wind at 20,000 feet.

3.28. For the executive business jet described in Example 3.6.2, determine the maximum range at 20,000 feet, with a 100-mph headwind and with a 100-mph tailwind.

3.29. For the jet transport described in problem 3.21, assuming that the pilot maintains the maximum range airspeed at all times, determine the no-wind maximum range for this airplane at 30,000 feet.

3.30. For the jet transport described in problem 3.23, assuming that the pilot maintains the maximum range airspeed at all times, determine the no-wind maximum range for this airplane at 30,000 feet. Also determine the average fuel consumption rate in gallons per minute and the average specific range in miles per gallon. If this airplane is carrying 500 passengers, what is the average specific range in passenger miles per gallon? Compare this with an automobile carrying one person that gets 20 miles per gallon.

3.31. For the jet transport in problem 3.30, assuming that the pilot maintains a constant airspeed equal to the maximum range airspeed based on average weight, determine the no-wind range for this airplane flying at 30,000 feet. Also determine the no-wind range for the same altitude, if the pilot maintains a constant airspeed of 100 ft/sec more than the maximum range airspeed based on average weight. Compare these values with the maximum range found in problem 3.30.

3.32. A commercial airplane such as the Boeing 747 is not usually flown at the maximum range airspeed. The reason for this is purely economic. The jet transport in problem 3.31 costs $150,000,000 and is flown 14 hr/day. Using a 30-year life and a 7 percent interest rate, what is the investment cost per hour of flight time? The total personnel cost for the crew, including fringe benefits, is $1,200 per hour and the cost of jet fuel is $1.00 per gallon. Determine the sum of the investment cost, the personnel cost, and the fuel cost in dollars per hour and in dollars per mile, assuming no wind and a constant airspeed equal to the maximum range airspeed based on average weight. Also determine the same costs for a constant airspeed equal to 100 ft/sec more than the maximum range airspeed based on average weight. Repeat the problem assuming that the cost of jet fuel goes to $2.00 per gallon.

3.33. For the propeller-driven transport described in problem 3.17, the maximum usable fuel weight is 8,000 lbf and the power-specific fuel consumption based on propulsive power is 0.52 lbf/hp·hr. This airplane has the following properties in cruise configuration:

$$S_w = 945 \text{ ft}^2, \quad b_w = 96 \text{ ft}, \quad W_f - 40,000 \text{ lbf}, \quad W_e = 32,000 \text{ lbf},$$
$$q_P = 0.52 \text{ lbf/hp·hr}, \quad C_{D_0} = 0.025, \quad C_{D_0,L} = 0.0, \quad e = 0.80$$

Assuming that the power-specific fuel consumption is independent of both airspeed and altitude, and that the pilot maintains the maximum range airspeed at all times, determine the no-wind maximum range for this airplane at 10,000 feet. Also determine the average specific range in miles per gallon. Use a fuel weight of 5.64 pounds per gallon.

3.34. For the general aviation aircraft described in Example 3.5.1, determine the best glide airspeed at sea level and the maximum glide ratio with *fuel tanks empty*, for no wind, for a 60-mph headwind, and for a 60-mph tailwind.

3.35. For the executive business jet described in Example 3.6.2, determine the best glide airspeed and maximum glide ratio at 20,000 feet with *fuel tanks full*, for no wind, for a 100-mph headwind, and for a 100-mph tailwind.

3.36. For the executive business jet described in Example 3.6.2, determine the best glide airspeed and maximum glide ratio at 20,000 feet with *fuel tanks empty*, for no wind, for a 100 mph-headwind, and for a 100-mph tailwind.

3.37. For the small jet transport described in problem 3.7, determine the following performance parameters for no-power gliding flight:
(a) Calculate the minimum sink airspeed at sea level and 30,000 feet.
(b) Calculate the minimum sink rate at sea level and 30,000 feet.
(c) Calculate the no-wind best glide airspeed at sea level and 30,000 feet.
(d) Calculate the no-wind maximum glide ratio at sea level and 30,000 feet.

3.38. For the large jet transport described in problem 3.10, determine the following performance parameters for no-power gliding flight:
(a) Calculate the minimum sink airspeed at sea level and 30,000 feet.
(b) Calculate the minimum sink rate at sea level and 30,000 feet.
(c) Calculate the no-wind best glide airspeed at sea level and 30,000 feet.
(d) Calculate the no-wind maximum glide ratio at sea level and 30,000 feet.

3.39. For the propeller-driven transport described in problem 3.17, determine the following performance parameters for no-power gliding flight:
(a) Calculate the minimum sink airspeed at sea level and 10,000 feet.
(b) Calculate the minimum sink rate at sea level and 10,000 feet.
(c) Calculate the no-wind best glide airspeed at sea level and 10,000 feet.
(d) Calculate the no-wind maximum glide ratio at sea level and 10,000 feet.

3.40. For the human-powered airplane described in problem 3.19, determine the following performance parameters for no-power gliding flight:
(a) Calculate the minimum sink airspeed at sea level.
(b) Calculate the minimum sink rate at sea level.
(c) Calculate the no-wind best glide airspeed at sea level.
(d) Calculate the no-wind maximum glide ratio at sea level.

3.41. For the small jet transport described in problem 3.7, determine the best glide airspeed and the maximum glide ratio at sea level with a 50-mph headwind, with no wind, and with a 50-mph tailwind.

3.42. For the large jet transport described in problem 3.10, determine the best glide airspeed and the maximum glide ratio at sea level with a 50-mph headwind, with no wind, and with a 50-mph tailwind.

3.43. For the propeller-driven transport described in problem 3.17, determine the best glide airspeed and the maximum glide ratio at sea level with a 50-mph headwind, with no wind, and with a 50-mph tailwind.

3.44. For the human-powered airplane described in problem 3.19, determine the best glide airspeed and the maximum glide ratio at sea level with a 10-mph headwind, with no wind, and with a 10-mph tailwind.

3.45. Determine the stall speed at standard sea level for each of the airplanes described in problems 3.7, 3.10, 3.17, and 3.19. For the sake of this comparison, assume that all four airplanes have a maximum lift coefficient of 1.2.

3.46. For the jet transport described in problem 3.7, compute the maneuvering speed, the minimum turning radius, and the maximum turning rate for maximum gross weight at both sea level and 30,000 feet. Assume a maximum lift coefficient of 1.2 and a positive load factor limit of 4.0.

3.47. For the jet transport described in problem 3.10, compute the maneuvering speed, the minimum turning radius, and the maximum turning rate for maximum gross weight at both sea level and 30,000 feet. Assume a maximum lift coefficient of 1.2 and a positive load factor limit of 4.0.

3.48. For the propeller-driven transport described in problem 3.17, compute the maneuvering speed, the minimum turning radius, and the maximum turning rate for maximum gross weight at both sea level and 30,000 feet. For this airplane, assume a maximum lift coefficient of 1.6 and a positive load factor limit of 4.0.

3.49. The propeller-driven transport in problem 3.48 is flying at an airspeed equal to the minimum power airspeed for straight and level flight. For this airspeed, compute the minimum turning radius and the maximum bank angle at both sea level and 30,000 feet.

3.50. For the human-powered airplane described in problem 3.19, compute the maneuvering speed, the minimum turning radius, and the maximum turning rate for maximum gross weight at sea level. Assume a maximum lift coefficient of 1.2 and a positive load factor limit of 1.1.

3.51. The human-powered airplane in problem 3.50 is flying at an airspeed equal to the minimum power airspeed for straight and level flight. For this airspeed, compute the minimum turning radius and the maximum bank angle at sea level.

3.52. The jet transport described in problem 3.7 is fitted with turbofan engines. Determine the total no-wind ground roll required for takeoff at sea level and at 5,000 feet. For this airplane in takeoff configuration,

$$S_w = 950 \text{ ft}^2, \quad b_w = 75 \text{ ft}, \quad h_w = 6.0 \text{ ft}, \quad W = 73{,}000 \text{ lbf}, \quad \mu_r = 0.04,$$
$$C_{D_0} = 0.033, \quad C_{D_0,L} = 0.0, \quad e = 0.64, \quad C_L = 0.10, \quad C_{L_{max}} = 1.86,$$
$$T_A = (\rho/\rho_0)[27{,}700 \text{ lbf} - (21.28 \text{ lbf·sec/ft}) V + (0.01117 \text{ lbf·sec}^2/\text{ft}^2) V^2]$$

3.53. The large jet transport described in problem 3.10 is fitted with turbofan engines. Determine the total no-wind ground roll required for takeoff at sea level and at 5,000 feet. For this airplane in takeoff configuration,

$$S_w = 5{,}500 \text{ ft}^2, \quad b_w = 196 \text{ ft}, \quad h_w = 16 \text{ ft}, \quad W = 636{,}600 \text{ lbf}, \quad \mu_r = 0.04,$$
$$C_{D_0} = 0.033, \quad C_{D_0,L} = 0.0, \quad e = 0.64, \quad C_L = 0.30, \quad C_{L_{max}} = 1.94,$$
$$T_A = (\rho/\rho_0)^{0.741}[199{,}000 \text{ lbf} - (177.6 \text{ lbf·sec/ft}) V + (0.1193 \text{ lbf·sec}^2/\text{ft}^2) V^2]$$

3.54. Consider the takeoff performance of the propeller-driven transport that was described in problem 3.17. Determine the total ground roll required for takeoff at sea level and at 5,000 feet, with no wind. For this airplane in takeoff configuration,

$$S_w = 945 \text{ ft}^2, \quad b_w = 96 \text{ ft}, \quad h_w = 12 \text{ ft}, \quad W = 40,000 \text{ lbf}, \quad \mu_r = 0.04,$$
$$C_{D_0} = 0.038, \quad C_{D_0,L} = 0.0, \quad e = 0.74, \quad C_L = 0.40, \quad C_{L\max} = 1.60,$$
$$T_A = (\rho/\rho_0)[10,000 \text{ lbf} - (12.0 \text{ lbf} \cdot \text{sec/ft}) V]$$

3.55. Consider the takeoff performance of the human-powered airplane described in problem 3.19. Determine the total ground roll required for takeoff at sea level with no wind and with a headwind of 5 mph. For this airplane in takeoff configuration,

$$S_w = 775 \text{ ft}^2, \quad b_w = 96.4 \text{ ft}, \quad h_w = 11 \text{ ft}, \quad W = 200 \text{ lbf}, \quad \mu_r = 0.05,$$
$$C_{D_0} = 0.010, \quad C_{D_0,L} = 0.0, \quad e = 0.80, \quad C_L = 0.40, \quad C_{L\max} = 1.40,$$
$$T_A = (\rho/\rho_0)[31.0 \text{ lbf} - (1.0 \text{ lbf} \cdot \text{sec/ft}) V]$$

3.56. Consider the takeoff performance of a general aviation airplane with tail-dragger landing gear. Determine the total ground roll required for takeoff at sea level with no wind. Assume a lift coefficient equal to the maximum lift coefficient during the tail-down portion of the ground roll and a lift coefficient of 0.40 for the tail-up portion of the ground roll. The minimum ground control speed for this airplane is 30 mph. For this airplane in takeoff configuration,

$$S_w = 180 \text{ ft}^2, \quad b_w = 33 \text{ ft}, \quad h_w = 6.0 \text{ ft}, \quad W = 2,700 \text{ lbf}, \quad \mu_r = 0.04,$$
$$C_{D_0} = 0.036, \quad C_{D_0,L} = 0.0, \quad e = 0.82, \quad C_{L\max} = 1.40,$$
$$T_A = (\rho/\rho_0)[1,200 \text{ lbf} - (4.0 \text{ lbf} \cdot \text{sec/ft}) V]$$

3.57. One advantage of the tail-dragger landing gear is that it allows the pilot to control the angle of attack during the high-speed phase of the takeoff ground roll. This gives better takeoff performance on a wider variety of ground surfaces. On rough, soft, or loose ground the coefficient of rolling friction is much higher than that for a paved surface and adjusting the angle of attack to maximize acceleration on any given surface can reduce takeoff distance. For the tail-dragger described in problem 3.56, compute the optimum lift coefficient for the high-speed phase of the takeoff ground roll for a coefficient of rolling friction of 0.04 and for a coefficient of rolling friction of 0.15.

3.58. Consider the landing performance of the jet transport described in problem 3.10. Determine the total no-wind ground roll required for landing at sea level and at 5,000 feet. For this airplane in landing configuration,

$$S_w = 5,500 \text{ ft}^2, \quad b_w = 196 \text{ ft}, \quad h_w = 16 \text{ ft}, \quad W = 636,600 \text{ lbf}, \quad \mu_r = 0.4,$$
$$C_{D_0} = 0.033, \quad C_{D_0,L} = 0.0, \quad e = 0.64, \quad C_L = 0.10, \quad C_{L\max} = 2.30,$$
$$T_A = -(\rho/\rho_0)^{0.8} \, 70,000 \text{ lbf} \, (V > 100 \text{ ft/sec}), \quad T_A = 0 \, (V < 100 \text{ ft/sec})$$

3.59. Consider the landing performance of the propeller-driven transport in problem 3.17. Determine the total no-wind ground roll required for landing at sea level and at 5,000 feet. For this airplane in landing configuration,

$$S_w = 945 \text{ ft}^2, \ b_w = 96 \text{ ft}, \ h_w = 12 \text{ ft}, \ W = 40,000 \text{ lbf}, \ \mu_r = 0.4, \ C_{D_0} = 0.038,$$
$$C_{D_0,L} = 0.0, \ e = 0.74, \ C_L = 0.10, \ C_{L_{max}} = 2.10, \ T_A = 0.0$$

3.60. Consider the no-brake landing performance of the human-powered airplane described in problem 3.19. Determine the total ground roll required for landing at sea level with no wind and with a headwind of 5 mph. For this airplane in landing configuration,

$$S_w = 775 \text{ ft}^2, \ b_w = 96.4 \text{ ft}, \ h_w = 11 \text{ ft}, \ W = 200 \text{ lbf}, \ \mu_r = 0.05, \ C_{D_0} = 0.010,$$
$$C_{D_0,L} = 0.0, \ e = 0.80, \ C_L = 0.40, \ C_{L_{max}} = 1.40, \ T_A = 0$$

3.61. Assume that the small jet transport described in problem 3.7 has the following properties in cruise configuration:

$$S_w = 950 \text{ ft}^2, \quad b_w = 75 \text{ ft}, \quad W = 73,000 \text{ lbf},$$
$$C_{D_0} = 0.020, \quad C_{D_0,L} = 0.0, \quad e = 0.67,$$
$$T_A = \frac{\rho}{\rho_0} 27,700 \text{ lbf}$$

Using the small-thrust-angle approximation, determine the service ceiling.

3.62. Assume that the large jet transport described in problem 3.10 has the following properties in cruise configuration:

$$S_w = 5,500 \text{ ft}^2, \quad b_w = 196 \text{ ft}, \quad W = 636,600 \text{ lbf},$$
$$C_{D_0} = 0.020, \quad C_{D_0,L} = 0.0, \quad e = 0.72,$$
$$T_A = \left(\frac{\rho}{\rho_0}\right)^{0.8} 250,000 \text{ lbf}$$

Using the small-thrust-angle approximation, determine the service ceiling.

3.63. Assume that the propeller-driven transport described in problem 3.17 has the following properties in cruise configuration:

$$S_w = 945 \text{ ft}^2, \quad b_w = 96 \text{ ft}, \quad W = 40,000 \text{ lbf},$$
$$C_{D_0} = 0.025, \quad C_{D_0,L} = 0.0, \quad e = 0.80,$$
$$T_A = (\rho/\rho_0)[10,000 \text{ lbf} - (12.0 \text{ lbf} \cdot \text{sec/ft}) V]$$

Using the small-thrust-angle approximation, determine the service ceiling.

3.64. Assume that the human-powered airplane described in problem 3.19 has the following properties:

$$S_w = 775 \text{ ft}^2, \quad b_w = 96.4 \text{ ft}, \quad W = 200 \text{ lbf},$$
$$C_{D_0} = 0.010, \quad C_{D_0,L} = 0.0, \quad e = 0.80,$$
$$P_A = (\rho/\rho_0)0.42 \text{ hp}$$

Using the small-thrust-angle approximation, determine the absolute ceiling for this aircraft.

3.65. For the large jet transport described in problem 3.53, determine the FAR Part 25 takeoff distance at standard sea level with no wind. This airplane is fitted with four turbofan engines that produce a thrust that varies significantly with airspeed. Assume a 3-second rotation time and use the following properties for takeoff configuration:

$$S_w = 5,500 \text{ ft}^2, \quad b_w = 196 \text{ ft}, \quad h_w = 16 \text{ ft}, \quad W = 636,600 \text{ lbf}, \quad \mu_r = 0.04,$$
$$C_{D_0} = 0.033, \quad C_{D_0,L} = 0.0, \quad e = 0.64, \quad C_L = 0.30, \quad C_{L_{max}} = 1.94,$$
$$T_A = (\rho/\rho_0)^{0.741}[199,000 \text{ lbf} - (177.6 \text{ lbf}\cdot\text{sec/ft}) V + (0.1193 \text{ lbf}\cdot\text{sec}^2/\text{ft}^2) V^2]$$

3.66. Consider the jet transport described in problem 3.65. Assuming that one of the four engines fails during a no-wind takeoff at standard sea level, plot the total field length required to stop the aircraft after engine failure and the total field length required to continue the FAR Part 25 takeoff using the remaining three engines as a function of the engine failure recognition speed. Use an obstacle clearance altitude of 35 feet and assume 1 second for engine failure recognition and another second for the pilot to react and apply the brakes. What are the critical engine failure recognition speed and the balanced field length?

3.67. While the distance traveled from brake release to obstacle clearance altitude is the most important factor associated with airplane takeoff, it is sometimes necessary to estimate the takeoff time. For example, to estimate the fuel consumed during takeoff, it is necessary to know how long the engines must be operated at full throttle during the acceleration, rotation, and climb to obstacle clearance altitude. The acceleration time can be computed in much the same way that the acceleration distance was computed in Sec. 3.10. Starting with Newton's second law, as expressed in Eq. (3.10.1), and assuming a parabolic variation in thrust with airspeed, show that the time required to accelerate from airspeed V_i to airspeed V_{i+1} is

$$t_{i+1} - t_i = (K_W)_i/g$$

where $(K_W)_i$ is given by Eq. (3.10.14).

3.68. For the jet transport in problem 3.52, determine the total time required for the takeoff ground roll with no wind at sea level and at 5,000 feet.

3.69. For the large jet transport in problem 3.53, determine the total time required for the takeoff ground roll with no wind at sea level and at 5,000 feet.

3.70. For the propeller-driven transport in problem 3.54, determine the total time required for the takeoff ground roll with no wind at sea level and at 5,000 feet.

3.71. For the human-powered airplane in problem 3.55, determine the total time required for the takeoff ground roll at sea level with no wind and with a headwind of 5 mph.

3.72. For the tail-dragger in problem 3.56, determine the total time required for the takeoff ground roll with no wind at sea level.

3.73. Write a computer program to determine the total ground roll distance and time required for takeoff. Assume a thrust variation with airspeed and density of the form

$$T = (\rho/\rho_0)^a (T_0 - T' V + T'' V^2)$$

where the air density ρ and the constants a, T_0, T', and T'' are specified by the user. The user should also be allowed to specify the following parameters,

$$S_w, \quad b_w, \quad h_w, \quad W, \quad \mu_r, \quad C_{D_0}, \quad C_{D_0,L}, \quad e, \quad C_L, \quad C_{L_{max}}, \quad V_{hw}, \quad t_r$$

Test your program by resolving problems 3.53, 3.55, 3.69, and 3.71.

3.74. Write a computer program to determine the critical engine failure recognition speed and the balanced field length for an aircraft as a function of the parameters specified in problems 3.65 and 3.66. Use a secant method to numerically find the engine failure speed that gives the same total field length required to either continue or abort the takeoff. Test your program by resolving Example 3.11.2 and problem 3.66.

Chapter 4
Longitudinal Static Stability and Trim

4.1. Fundamentals of Static Equilibrium and Stability

Much of the performance analysis in Chapter 3 was based on the concept of static equilibrium, which for an aircraft is often called *trimmed flight*. Static equilibrium occurs whenever there is no acceleration of the aircraft. Unaccelerated flight requires that the summations of forces and moments acting on the aircraft are zero. For example, in steady level flight with the thrust vector aligned with the direction of flight, the lift must equal the weight, $L = W$, and the thrust must equal the drag, $T = D$. These two relations were used to determine the thrust required for steady level flight as a function of aircraft weight and airspeed. Static equilibrium also requires that the side force acting on the aircraft is zero. This condition is often assumed to be satisfied as a result of aircraft symmetry. Additionally, the summation of moments about the center of gravity (CG) in roll, pitch, and yaw must all be zero for trimmed flight. Aircraft symmetry will often result in zero rolling and yawing moments, but the pitching moment is usually zeroed with control input. **When the controls are set so that the resultant forces and the moments about the center of gravity are all zero, the aircraft is said to be in *trim*, which simply means static equilibrium.**

As shown in Fig. 4.1.1, the control surfaces on a conventional airplane usually consist of ailerons, elevator, and rudder. The ailerons are used to control the rolling moment, the elevator is used to control the pitching moment, and the rudder is used to control the yawing moment. These aircraft control surfaces provide two functions. First, the control surfaces must be able to maintain static trim over the entire range of airspeed and altitude for which the aircraft is able to fly. This includes being able to trim the aircraft against any asymmetric thrust force, which may occur when one or more of the engines has failed on a multiengine aircraft. Additionally, the control surfaces must provide the moments necessary to adequately maneuver the aircraft over this same range of airspeed and altitude. Since trim is closely coupled with static stability, trim is covered in this and the following chapter. Maneuverability is covered in Chapter 6.

The ease with which a pilot is able to maintain trim is one important aspect of the aircraft characteristics known as *handling qualities*. If the pilot cannot maintain trim with relative ease, the aircraft will be difficult or even dangerous to fly. Any airplane that is difficult to fly will not be popular with pilots, no matter how well it performs in other regards. This point is very important for the designer to understand and remember. The student should commit it to memory now and never forget it.

The ease of maintaining static trim is related to a property of the equilibrium state, which is known as *static stability*. The static stability of any equilibrium state is related to the response of the system to a small disturbance from that equilibrium state. **If a system in an equilibrium state returns to equilibrium following a small disturbance, the state is said to be a *stable equilibrium*.** On the other hand, **if the system diverges from equilibrium when slightly disturbed, the state is said to be an *unstable equilibrium*.**

377

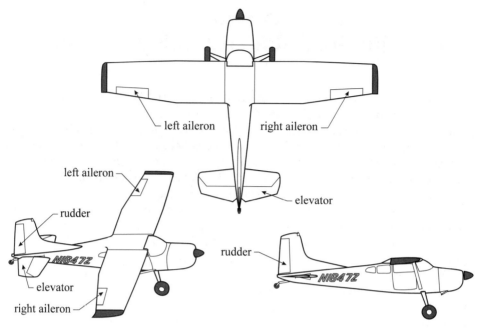

Figure 4.1.1. Typical control surfaces of a conventional airplane.

The most commonly used example of a stable equilibrium state is a ball in a bowl, as shown in Fig. 4.1.2a. When the ball is at rest in the bottom of the bowl, it is in an equilibrium state, since the force between the ball and the bowl exactly balances the force of gravity and there is no moment about the ball's center of gravity. If the ball were to be displaced from the bottom of the bowl, it would roll back to the bottom as a consequence of the resulting force and moment imbalance. For this reason, when the ball is at rest in the bottom of the bowl, it is said to be in a stable equilibrium state. Conversely, the situation shown in Fig. 4.1.2c is not a stable equilibrium. When the ball is at rest at the exact apex of the convex surface, it is in static equilibrium, because there is no force or moment imbalance at that exact position. However, any disturbance of the ball from the exact apex, no matter how small, would produce a force and moment imbalance, which would in turn cause the ball to roll away from the apex. Thus, the situation where a ball is resting exactly at the apex of a convex surface is classified as an unstable equilibrium state.

An equilibrium situation can also exist that is neither stable nor unstable. For example, consider a ball resting on a perfectly flat and horizontal surface, as shown in Fig. 4.1.2b. This situation is clearly an equilibrium state, since the summation of forces and moments acting on the ball are zero. When this ball is displaced from the equilibrium position, there is no force or moment imbalance produced. Thus, there is no tendency for the ball to either return to its original equilibrium point or to diverge from it. This condition is classified as a neutrally stable equilibrium state. Neutral stability is the dividing line between stable equilibrium and unstable equilibrium.

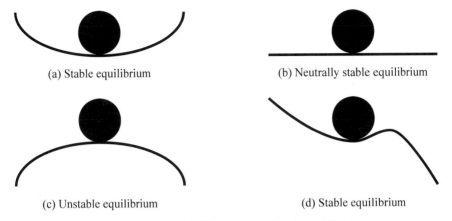

(a) Stable equilibrium   (b) Neutrally stable equilibrium

(c) Unstable equilibrium   (d) Stable equilibrium

Figure 4.1.2. Several different types of static equilibrium.

The situation shown in Fig. 4.1.2d is also worth some discussion at this point. Strictly speaking, this is a stable equilibrium, because a very small disturbance from the position shown would result in a force and moment imbalance that would return the ball to its original equilibrium state. However, a larger disturbance to the right could cause the ball to move past the apex, which would then produce a force and moment imbalance that would cause the ball to move away from its original equilibrium state. This type of stable equilibrium can sometimes occur with an aircraft in trimmed flight. This is a particularly dangerous situation and will be discussed in greater detail at a later point.

As shown in Fig. 4.1.3, a rigid airplane in free flight has six degrees of freedom, three translational degrees of freedom and three rotational degrees of freedom. For the airplane to be in fully stable trim, there can be no instability in any of the six degrees of freedom. Small translational disturbances in axial, normal, or sideslip velocity must all result in a return to the original trimmed equilibrium condition. Similarly, rotational disturbances in roll, pitch, and yaw must all result in a return to the original equilibrium attitude. With few exceptions, if the rotational degrees of freedom for an airplane are stable, translational stability will not be a problem.

Any object moving through the air will experience drag that opposes the motion. If the angle of attack remains fixed, this drag increases with increasing airspeed. Except for the case of a ramjet, which is not normally used for aircraft propulsion, the thrust developed by an aircraft engine is either constant with airspeed or decreases with increasing airspeed. When an airplane is in static equilibrium with regard to translation in the direction of motion, the forward component of thrust must balance the drag. At constant angle of attack, a small increase in airspeed will result in an increase in drag and either a decrease in thrust or no change in thrust. In either case, the increase in airspeed results in a drag that is greater than the thrust. This force imbalance in the axial direction will result in a deceleration, which will restore the airspeed to the original static equilibrium value. Conversely, if the airspeed is decreased by a small disturbance with no change in angle of attack, the drag will become less than the thrust and the aircraft will accelerate back to the equilibrium airspeed.

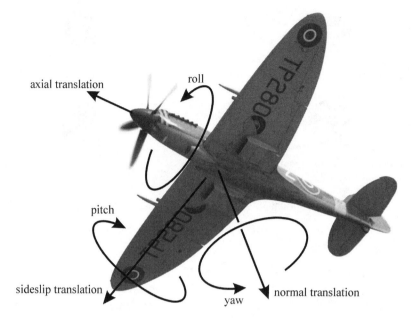

axial translation

roll

pitch

sideslip translation

yaw

normal translation

Figure 4.1.3. The six degrees of freedom for a rigid airplane in free flight.

In a similar manner, any disturbance in velocity in a direction normal to the equilibrium flight path will result in an aerodynamic force that opposes the disturbance, provided that the flow remains attached. Thus, when the attitude of an airplane is held constant, the aerodynamic forces quite naturally maintain translational static stability, in the absence of stall. It should be reemphasized that this translational stability is contingent on being able to maintain the equilibrium attitude of the aircraft. Since induced drag depends on angle of attack, if the angle of attack were to change along with the airspeed, it is possible that the total drag could decrease as the airspeed increases. In addition, this natural translational stability depends on maintaining attached flow. At some angles of attack beyond stall, the airplane may be unstable for disturbances in normal velocity. However, since rotational stability is most critical to maintaining trimmed flight in the absence of stall, static aircraft stability in roll, pitch, and yaw will be the primary emphasis of this and the following chapter.

A thorough understanding of the fundamental principles of static stability is very important to the airplane designer. However, it is also important to understand the influence of the pilot on the human-machine system that constitutes a modern airplane. Most of the technical problems associated with manned powered flight were solved well before the Wright brothers' first successful flight on December 17, 1903. What was lacking was a sufficient understanding of the influence of the pilot on aircraft stability and control. From an analytical point of view, it is easier to design an unmanned aircraft with an autopilot than it is to design a good manned airplane. The human pilot is a very complex component of the modern airplane, and whenever human ability and behavior become part of an engineered system, the designer's job takes on a new dimension.

While static stability is usually desired for most systems under human control, it is not an absolute requirement in all cases. Most humans can easily learn to ride a bicycle, which is inherently unstable. With a greater level of training and concentration, a human can even ride a unicycle, which is very unstable. Furthermore, after some initial practice, riding a bicycle does not even require conscious control by the human rider. The rider is still providing unconscious control. This is evidenced by the fact that a riderless bicycle will not travel a long distance without tipping over. Once a riderless bicycle is moving, however, it will usually travel for some short distance before tipping over. This is because even though the upright moving bicycle is in an unstable equilibrium, the divergence from that equilibrium is very slow at first. It is this slow initial divergence, associated with the instability, which makes it possible for a human rider to maintain the bicycle's equilibrium with no conscious effort. Flying an airplane is similar to riding a bicycle. Many of the control inputs furnished by the pilot are provided without conscious thought. Whereas the ability of a human to provide such control inputs with little or no conscious effort makes the pilot's job easier, such human control is very difficult or impossible to model analytically. The effects of having a human in the control loop have always been investigated experimentally. In some ways this makes the designer's job more difficult.

In this and the following chapter we discuss how to design an airplane for static stability, how to determine airplane trim requirements, and how to assure some aspects of good handling qualities when the airplane is interfaced with its human pilot.

4.2. Pitch Stability of a Cambered Wing

As discussed previously, static stability in the rotational degrees of freedom is of primary importance for maintaining aircraft trim. For an airplane to be statically stable in rotation, any disturbances in roll, pitch, or yaw must all result in the production of a restoring moment that will return the aircraft to the original equilibrium attitude. To begin our study of static aircraft stability, we shall consider the pitch stability of a simple cambered wing, the cross-section of which is shown in Fig. 4.2.1. We will first explore the requirements for trim and then examine the pitch stability of the equilibrium state.

Assuming that the wing is symmetric in the spanwise direction and that the motion of the wing through the air is in a direction normal to the span, there will be no side force and no rolling or yawing moment. For this symmetric flight condition, the aerodynamic forces acting on the wing can be resolved into a lift force, L, a drag force, D, and a pitching moment about the aerodynamic center of the wing, m_{ac}. To simplify this initial stability analysis we shall assume that the aerodynamic center and the center of gravity are aligned with the thrust vector, which is aligned with the direction of flight. For the wing to be in trim (i.e., static equilibrium), the summation of forces in both the horizontal and vertical directions must be zero. This requires that

$$T = D \tag{4.2.1}$$

and

$$L = W \tag{4.2.2}$$

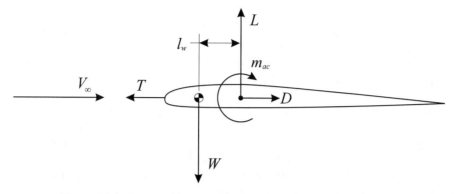

Figure 4.2.1. Forces and moments acting on a cambered wing in flight.

Additionally, trim requires that the summation of pitching moments about the center of gravity, m, must be zero. From Fig. 4.2.1, this yields

$$m = m_{ac} - l_w L = 0 \tag{4.2.3}$$

where l_w is the distance that the aerodynamic center of the wing is aft of the center of gravity. After applying the definitions of the lift and moment coefficients, Eq. (4.2.3) can be written as

$$\tfrac{1}{2}\rho V_\infty^2 S_w \bar{c} C_m = \tfrac{1}{2}\rho V_\infty^2 S_w(\bar{c} C_{m_{ac}} - l_w C_L) = 0 \tag{4.2.4}$$

or

$$C_m = C_{m_{ac}} - \frac{l_w}{\bar{c}} C_L = 0 \tag{4.2.5}$$

For a given weight and airspeed, the lift coefficient is fixed by the trim requirement in Eq. (4.2.2). The moment coefficient about the aerodynamic center is fixed by the wing geometry. Thus, for a given geometry, weight, and airspeed, Eq. (4.2.5) requires that

$$l_w = \frac{C_{m_{ac}}}{C_L} \bar{c} \tag{4.2.6}$$

Since the lift coefficient is positive and the moment coefficient about the aerodynamic center of a simple cambered wing is negative, we see that trim requires $l_w < 0$. Thus, for static equilibrium the aerodynamic center must be forward of the center of gravity.

If this static equilibrium state is stable in pitch, then a small increase in angle of attack must produce a negative pitching moment about the center of gravity, to decrease the angle of attack back toward trim. Conversely, a small decrease in angle of attack must produce a positive pitching moment, to increase the angle of attack and restore trim. Thus, the pitching moment about the center of gravity must vary with angle of attack such that any change in angle of attack produces a change of opposite sign in the pitching moment about the center of gravity. The mathematical criterion for pitch stability is then

$$\frac{\partial m}{\partial \alpha} = \frac{1}{2} \rho V_\infty^2 S_w \bar{c} \frac{\partial C_m}{\partial \alpha} < 0 \qquad (4.2.7)$$

Since the dynamic pressure, the wing area, and the mean chord length are always positive, pitch stability requires that

$$\frac{\partial C_m}{\partial \alpha} \equiv C_{m,\alpha} < 0 \qquad (4.2.8)$$

which is **the general pitch stability criterion for any aircraft**. For this reason, the derivative in Eq. (4.2.8) is usually called the *pitch stability derivative* or *pitch stiffness*. The word "stiffness" comes from an analogy with a torsional spring, which produces a restoring moment that is proportional to the angular displacement from equilibrium.

Using Eq. (4.2.5) in Eq. (4.2.8) gives

$$\frac{\partial C_{m_{ac}}}{\partial \alpha} - \frac{l_w}{\bar{c}} \frac{\partial C_L}{\partial \alpha} < 0 \qquad (4.2.9)$$

By definition, the aerodynamic center is a point on the wing where the change in pitching moment coefficient with respect to angle of attack is zero,

$$\frac{\partial C_{m_{ac}}}{\partial \alpha} = 0 \qquad (4.2.10)$$

and thus pitch stability requires

$$-\frac{l_w}{\bar{c}} \frac{\partial C_L}{\partial \alpha} < 0 \qquad (4.2.11)$$

Since the change in lift coefficient with angle of attack is positive for angles of attack less than stall, we see that stability requires $l_w > 0$. This means that for static stability, the aerodynamic center must be aft of the center of gravity. This condition is opposite to that required for static equilibrium. Thus, **a simple cambered wing is not statically stable in free flight.**

Using the trim requirement in Eq. (4.2.6) to eliminate the distance l_w from the stability requirement in Eq. (4.2.11) gives

$$-\frac{C_{m_{ac}}}{C_L} \frac{\partial C_L}{\partial \alpha} < 0 \qquad (4.2.12)$$

This is the requirement for stable trim in this tailless flying wing. The lift coefficient must be positive to support the weight of the aircraft and the lift slope is positive for any wing at angles of attack below stall. Thus, **if stable trim is to be maintained, a single wing with no tail must always produce a positive pitching moment coefficient about its aerodynamic center.**

In our study of airfoil sections, we found that a symmetric airfoil produces no moment about the aerodynamic center, and an airfoil with simple positive camber produces a negative pitching moment about the aerodynamic center. Neither of these common airfoil sections meets the criterion required for stable trim in a flying wing.

To produce a positive pitching moment about the aerodynamic center, an airfoil section must have negative camber over at least some fraction of the chord. An airfoil with negative camber over the entire chord is inefficient when producing positive lift and has a low maximum lift coefficient. A better choice is an airfoil that has negative camber over only some portion of the chord near the trailing edge. Such an airfoil section is shown in Fig. 4.2.2. An airfoil of this type is said to have a *reflexed trailing edge*. It can be designed to produce a positive pitching moment about the aerodynamic center and can be used to satisfy the criterion for stable trim specified by Eq. (4.2.12).

It is possible to design an aircraft, consisting of only a single flying wing with no tail, so that stable trimmed flight can be achieved. However, such designs are difficult at best and usually do not provide good handling qualities without a significant reduction in performance or the implementation of some type of computer control. A better option usually is to combine a wing with a conventional aft tail. This is the design that is most often seen in nature and it is difficult to improve upon.

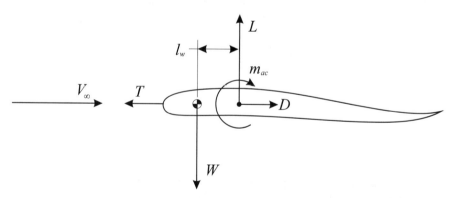

Figure 4.2.2. Forces and moments acting on a cambered wing with a reflexed trailing edge.

4.3. Simplified Pitch Stability Analysis for a Wing-Tail Combination

Trim and static stability in pitch are readily achieved in an airplane by combining a wing with a conventional aft tail, as shown in Fig. 4.3.1. The complete tail assembly, consisting of both the horizontal and vertical tail, is called the *empennage*. The purpose of the horizontal tail is to provide both stability and control in pitch. As can be seen in Fig. 4.3.2, the horizontal tail usually consists of a fixed forward section, called the *horizontal stabilizer* or *tailplane*, and a rotating aft section, called the elevator. Both the horizontal stabilizer and the elevator contribute to pitch stability, but only the elevator provides pitch control. In some cases the entire horizontal tail rotates to provide pitch control. This is called an *all-flying tail*. On some airplanes both the front stabilizer section and the aft elevator section of the horizontal tail rotate separately to provide pitch control. This configuration is called a *stabilator*. The stabilator looks and responds very much like the conventional horizontal tail, but it differs somewhat in its mechanical operation.

Figure 4.3.1. British Spitfire, showing the conventional aft tail. (Photograph by Barry Santana)

Figure 4.3.2. Empennage of the British Spitfire. (Photographs by Barry Santana)

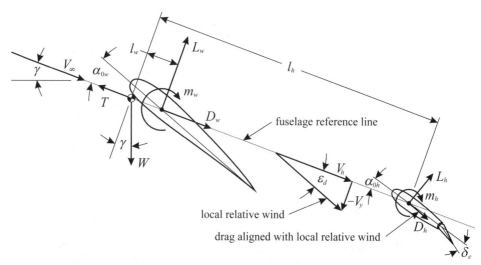

Figure 4.3.3. Forces and moments acting on a wing combined with a horizontal tail.

For the approximate analysis to be presented in this section, consider the geometry shown in Fig. 4.3.3. Because the angle of attack for the wing and tail need not be the same, the angle of attack for the airplane is specified relative to a fuselage reference line. For a spanwise symmetric aircraft, the fuselage reference line is chosen to be in the plane of symmetry, but otherwise, its orientation is quite arbitrary. Sometimes the fuselage reference line is taken to be the zero-lift line, but this is not always the case. For this analysis we shall take the fuselage reference line to pass through the center of gravity parallel to the freestream velocity at trim. To simplify this initial analysis, we shall also assume that the aerodynamic center of both the wing and the tail, as well as the center of thrust, all fall on this fuselage reference line. The more general case will be examined later. We will also assume that the thrust vector is aligned with the direction of motion, which is inclined relative to the horizontal by the climb angle γ. Trim and static stability are often presented in terms of level flight $(\gamma = 0)$. However, it is important that we also understand trim and stability as they apply to climbing or descending flight.

Because static stability is always defined relative to an equilibrium state, we will begin our analysis by examining the requirements for static equilibrium. From Fig. 4.3.3, the force balance required for trim (remember, *trim* is read *static equilibrium*) gives

$$T = D_w + D_h \cos \varepsilon_d + L_h \sin \varepsilon_d + W \sin \gamma \tag{4.3.1}$$

$$L = L_w + L_h \cos \varepsilon_d - D_h \sin \varepsilon_d = W \cos \gamma \tag{4.3.2}$$

where L is the total lift and ε_d is the angle between the local relative wind at the tail and the freestream. The pitching moment about the center of gravity, m, must be zero at trim,

$$m = m_w + m_h - l_w L_w - l_h L_h \cos \varepsilon_d + l_h D_h \sin \varepsilon_d = 0 \tag{4.3.3}$$

where m_w and m_h are, respectively, the pitching moments for the wing and horizontal tail about their aerodynamic centers. The lengths l_w and l_h are the distances from the center of gravity to the aerodynamic center of the wing and horizontal tail, respectively, both measured positive aft of the CG.

Equations (4.3.1) through (4.3.3) provide three of the six requirements for static equilibrium. The remaining three requirements are simply the trivial relations that the side force, rolling moment, and yawing moment must all be zero. From Eq. (4.3.1) we can determine the thrust required to maintain constant airspeed. From the requirements specified by Eqs. (4.3.2) and (4.3.3), we can relate the aerodynamic forces developed by the wing and horizontal tail to the weight of the aircraft and the position of the center of gravity. Equation (4.3.2) simply requires that at trim, the aerodynamic forces developed by the wing and horizontal tail combined must support the normal component of the aircraft weight. Equation (4.3.3) specifies how the aerodynamic force must be divided between the wing and horizontal tail, to maintain the attitude required for trim. Thus, if we are to be able to attain trim for a variety of flight conditions, we must have some means to control the lift developed by both the wing and the horizontal tail.

The downwash angle ε_d is typically small, so we can apply the usual small-angle approximations, $\cos\varepsilon_d \cong 1$ and $D_h \sin\varepsilon_d << L_h$. Using these approximations and dividing Eq. (4.3.2) by the freestream dynamic pressure and wing area gives

$$\frac{L}{\frac{1}{2}\rho V_\infty^2 S_w} = \frac{L_w}{\frac{1}{2}\rho V_\infty^2 S_w} + \frac{\frac{1}{2}\rho V_h^2 S_h}{\frac{1}{2}\rho V_\infty^2 S_w}\frac{L_h}{\frac{1}{2}\rho V_h^2 S_h} = \frac{W\cos\gamma}{\frac{1}{2}\rho V_\infty^2 S_w} \quad (4.3.4)$$

or

$$C_L = C_{L_w} + \frac{S_h}{S_w}\frac{\frac{1}{2}\rho V_h^2}{\frac{1}{2}\rho V_\infty^2}C_{L_h} = \frac{W\cos\gamma}{\frac{1}{2}\rho V_\infty^2 S_w} \quad (4.3.5)$$

where C_{L_w} and C_{L_h} are, respectively, the lift coefficients for the wing and isolated tail, each based on their own planform area, and C_L is the total lift coefficient based on the planform area of the main wing. The dynamic pressure for the tail in Eq. (4.3.5) is distinguished from the freestream dynamic pressure because on average over the span of the tail, the dynamic pressure may be less than or greater than the freestream value. The average dynamic pressure for the tail depends on the location of the tail. If the tail is located in the wake of the wing or the fuselage, the dynamic pressure on the tail will be less than the freestream value. On the other hand, if the tail is located in the slipstream of a propeller or in the exhaust of a jet engine, the dynamic pressure on the tail will be greater than that for the freestream. Equation (4.3.5) is usually written as

$$C_L = C_{L_w} + \frac{S_h}{S_w}\eta_h C_{L_h} = \frac{W\cos\gamma}{\frac{1}{2}\rho V_\infty^2 S_w} \quad (4.3.6)$$

$$\eta_h \equiv \frac{\frac{1}{2}\rho V_h^2}{\frac{1}{2}\rho V_\infty^2} \quad (4.3.7)$$

This ratio of dynamic pressures is called the *tail efficiency*, and it can take values from about 0.8 to 1.2. However, the distribution of dynamic pressure over the span of the tail is very difficult to determine and is typically quite close to the freestream value. Thus, at least for the purpose of preliminary design, the tail efficiency is usually assumed to be unity.

In a manner similar to that used to obtain Eq. (4.3.6), we apply the small-angle approximations for ε_d to Eq. (4.3.3) and divide by the freestream dynamic pressure, the area of the main wing, and the mean chord length of the main wing:

$$
\frac{m}{\frac{1}{2}\rho V_\infty^2 S_w \bar{c}_w} = \frac{m_w}{\frac{1}{2}\rho V_\infty^2 S_w \bar{c}_w} + \frac{S_h \bar{c}_h}{S_w \bar{c}_w}\eta_h \frac{m_h}{\frac{1}{2}\rho V_h^2 S_h \bar{c}_h}
$$
$$
- \frac{l_w}{\bar{c}_w}\frac{L_w}{\frac{1}{2}\rho V_\infty^2 S_w} - \frac{S_h l_h}{S_w \bar{c}_w}\eta_h \frac{L_h}{\frac{1}{2}\rho V_h^2 S_h} = 0
\tag{4.3.8}
$$

In terms of the definitions of the lift and moment coefficients, this gives

$$
C_m = C_{m_w} + \frac{S_h \bar{c}_h}{S_w \bar{c}_w}\eta_h C_{m_h} - \frac{l_w}{\bar{c}_w}C_{L_w} - \frac{S_h l_h}{S_w \bar{c}_w}\eta_h C_{L_h} = 0
\tag{4.3.9}
$$

Solving Eq. (4.3.6) for the wing lift coefficient at trim gives

$$
C_{L_w} = \frac{W\cos\gamma}{\frac{1}{2}\rho V_\infty^2 S_w} - \frac{S_h}{S_w}\eta_h C_{L_h}
\tag{4.3.10}
$$

Using this result in Eq. (4.3.9), we can solve for the tail lift coefficient that is required to balance the airplane at trim. Thus, we obtain

$$
C_{L_h} = \frac{S_w \bar{c}_w}{S_h(l_h - l_w)\eta_h}C_{m_w} + \frac{\bar{c}_h}{(l_h - l_w)}C_{m_h} - \frac{S_w l_w}{S_h(l_h - l_w)\eta_h}\frac{W\cos\gamma}{\frac{1}{2}\rho V_\infty^2 S_w}
\tag{4.3.11}
$$

Using Eq. (4.3.11) in Eq. (4.3.10), the lift coefficient required for the wing at trim can be written as

$$
C_{L_w} = \frac{l_h}{l_h - l_w}\frac{W\cos\gamma}{\frac{1}{2}\rho V_\infty^2 S_w} - \frac{\bar{c}_w}{l_h - l_w}C_{m_w} - \frac{S_h \bar{c}_h}{S_w(l_h - l_w)}\eta_h C_{m_h}
\tag{4.3.12}
$$

To gain some additional insight into the trim requirements provided by Eqs. (4.3.11) and (4.3.12), consider the special case where both the wing and the tail have symmetric airfoil sections at every cross section along the span (remember, a symmetric airfoil section has no flap deflection). Since a symmetric airfoil produces no moment about its aerodynamic center, the moment coefficients in Eqs. (4.3.11) and (4.3.12) now go to zero. This gives

$$C_{L_h} = -\frac{S_w l_w}{S_h (l_h - l_w) \eta_h} \frac{W \cos \gamma}{\frac{1}{2} \rho V_\infty^2 S_w} \qquad (4.3.13)$$

and

$$C_{L_w} = \frac{l_h}{(l_h - l_w)} \frac{W \cos \gamma}{\frac{1}{2} \rho V_\infty^2 S_w} \qquad (4.3.14)$$

Since we are examining the particular case of an aft tail, the horizontal tail is aft of both the CG and the main wing. Thus, both l_h and $(l_h - l_w)$ are positive. The only parameter on the right-hand side of either Eq. (4.3.13) or Eq. (4.3.14) that could be negative is l_w. The center of gravity could be either forward or aft of the aerodynamic center of the main wing. Notice from Eq. (4.3.13) that the lift on the horizontal tail must be negative if the center of gravity is forward of the aerodynamic center of the main wing ($l_w > 0$). If l_w is positive, Eq. (4.3.14) requires that the lift produced by the main wing must be greater than the normal component of aircraft weight. This does not give the best performance in terms of aerodynamic efficiency. When the horizontal tail is carrying lift, positive or negative, induced drag is produced on the tail. If the lift on the horizontal tail is negative, the main wing must support this added load along with the normal component of aircraft weight. This increases the induced drag on the main wing as well. Thus, a double induced drag penalty is paid for negative lift carried on the horizontal tail. To avoid negative lift on the horizontal tail, Eq. (4.3.13) requires that the center of gravity be located at or behind the aerodynamic center of the main wing ($l_w \le 0$). **It should be emphasized that the avoidance of negative lift on the horizontal tail is only for improved aerodynamic efficiency. It is not required for either trim or stability.** Most airplanes will carry some negative lift on the horizontal tail over some portion of the allowable range of operating conditions.

Even carrying positive lift on the tail will usually result in some reduction in aerodynamic efficiency. This is because the horizontal tail will usually have a lower aspect ratio and a lower aerodynamic efficiency than the main wing. If this is the case, any lift carried on the tail will result in more induced drag than if that same lift were carried on the more efficient wing. For this reason, a good preliminary design is often achieved by placing the center of gravity such that the lift on the tail is zero when the airplane is trimmed with no elevator deflection at the design cruise airspeed. From Eq. (4.3.13), this requires that $l_w = 0$, which means that the center of gravity is placed at the aerodynamic center of the main wing if both the wing and the tail have symmetric airfoil sections.

For the more general trim condition specified by Eq. (4.3.11), best aerodynamic efficiency is achieved when the position of the CG relative to the aerodynamic center of the main wing is specified by setting the lift on the tail to zero and solving for l_w. **For the case of level flight, zero lift on the aft tail requires**

$$\frac{l_w}{\bar{c}_w} = \frac{\frac{1}{2} \rho V_\infty^2 S_w}{W} \left(C_{m_w} + \frac{S_h \bar{c}_h}{S_w \bar{c}_w} \eta_h C_{m_h} \right) \qquad (4.3.15)$$

Because the pitching moment coefficient about the aerodynamic center is zero for a symmetric airfoil and negative for an airfoil with positive camber, this places the center of gravity at or slightly aft of the aerodynamic center of the main wing. For this reason, many conventional airplanes with cambered wings have the center of gravity located slightly aft of the wing quarter chord.

Now that we have a better understanding of the requirements for static equilibrium in pitch, let us examine how the elevator of a conventional airplane is used to attain trim for a given flight condition. For small angles of attack, the lift coefficient for the main wing is a linear function of angle of attack,

$$C_{L_w} = C_{L_w,\alpha} (\alpha + \alpha_{0w} - \alpha_{L0w}) \qquad (4.3.16)$$

where $C_{L_w,\alpha}$ is the lift slope for the main wing, α is the angle of attack for the airplane as measured relative to the fuselage reference line, α_{0w} is the angle that the wing chord makes with the fuselage reference line, and α_{L0w} is the zero-lift angle of attack for the wing (Sec. 1.8). For small angles of attack and elevator deflections, the lift coefficient for the tail varies linearly with both angle of attack and elevator deflection. However, the presence of the wing and other components of the aircraft also significantly affect the lift coefficient for the tail through the downwash angle, ε_d.

In our study of finite wings, we found that the vorticity trailing behind a lifting wing produces downwash in the region aft of the wing, as shown in Fig. 4.3.4. **This downwash has a significant effect on the lift developed by an aft tail and should never be ignored when designing the tail of an airplane.** The downwash induced on

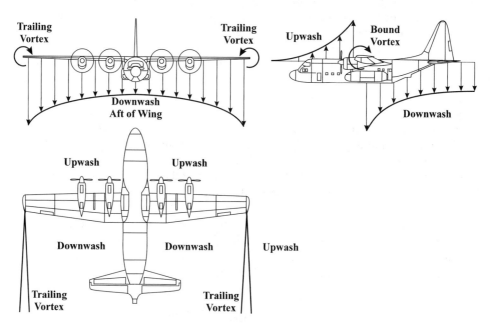

Figure 4.3.4. Downwash distribution created by the wing of a conventional airplane.

the tail by the main wing reduces the effective angle of attack for the horizontal tail and thus affects the trim. Additionally, since the downwash varies with wing angle of attack, it reduces the tail's effectiveness in stabilizing the airplane. This downwash varies along the span of the horizontal tail and is affected by the planform shape of the wing as well as the presence of the fuselage and nacelles. In reality, the performance of the wing is also affected by the presence of the tail but this effect is less significant. The only two ways to estimate accurately the interactions between the different surfaces of the airplane are by making use of computer simulations and/or wind tunnel tests. Such methods should always be employed in the later phases of the airplane design process. The effects of the interactions between the main wing and the horizontal tail can be adequately estimated using either the numerical lifting-line method or the vortex panel method.

For the purpose of preliminary design and to gain some useful insight into the basic principles involved, it is useful to have an approximate closed-form solution for estimating the downwash induced on the tail by the wing. Such an approximate method can be developed by assuming that the flow field in the vicinity of the airplane is affected only by the main wing. From Prandtl's lifting-line theory, it can be shown that the downwash at any point in the proximity of a finite lifting wing is proportional to the freestream velocity multiplied by the lift coefficient for the wing divided by the aspect ratio for the wing,

$$V_y = -K_d \frac{V_\infty C_{L_w}}{R_{A_w}} \qquad (4.3.17)$$

where the proportionality constant, K_d, depends on the planform shape of the wing and the position of the tail relative to the wing. Because the downwash is small compared to the freestream velocity, the downwash angle, ε_d, can be closely approximated as the downwash velocity divided by the freestream velocity,

$$\varepsilon_d \cong \frac{-V_y}{V_\infty} = K_d \frac{C_{L_w}}{R_{A_w}} \qquad (4.3.18)$$

Since the downwash angle at the tail is proportional to the lift coefficient for the wing, which is a linear function of angle of attack, the downwash angle can also be written as a linear function of the angle of attack,

$$\varepsilon_d = \varepsilon_{d0} + \varepsilon_{d,\alpha}\,\alpha \qquad (4.3.19)$$

where ε_{d0} is the downwash angle when the fuselage reference line is at zero angle of attack and $\varepsilon_{d,\alpha}$ is the change in the downwash angle with respect to angle of attack.

For small angles of attack and elevator deflection and assuming that the tail has a symmetric airfoil section when the elevator is not deflected, the lift coefficient for the tail in the presence of the downwash that is specified by Eq. (4.3.19) can be written as

$$C_{L_h} = C_{L_h,\alpha}(\alpha + \alpha_{0h} - \varepsilon_d + \varepsilon_e \delta_e) = C_{L_h,\alpha}[(1 - \varepsilon_{d,\alpha})\alpha + \alpha_{0h} - \varepsilon_{d0} + \varepsilon_e \delta_e] \qquad (4.3.20)$$

Here $C_{L_h,\alpha}$ is the lift slope for the isolated tail, α_{0h} is the angle that the tail chord makes with the fuselage reference line, ε_e is the elevator effectiveness, and δ_e is the elevator deflection as shown in Fig. 4.3.3.

For small elevator deflection, the moment coefficient about the aerodynamic center of the horizontal tail is also a linear function of the elevator deflection. Since we are assuming the tail to have a symmetric airfoil section when the elevator is not deflected, we can write

$$C_{m_h} = C_{m_h,\delta_e}\,\delta_e \tag{4.3.21}$$

where C_{m_h,δ_e} is the change in the pitching moment coefficient for the isolated tail about its aerodynamic center with respect to elevator deflection. When the elevator is deflected downward, a negative pitching moment is produced about the aerodynamic center. Thus, the moment slope, C_{m_h,δ_e}, is always negative when δ_e is defined to be positive downward, as shown in Fig. 4.3.3.

Using Eqs. (4.3.16), (4.3.20), and (4.3.21) in Eqs. (4.3.6) and (4.3.9) results in

$$\left[C_{L_w,\alpha} + \frac{S_h}{S_w}\eta_h C_{L_h,\alpha}(1-\varepsilon_{d,\alpha})\right]\alpha + \left[\frac{S_h}{S_w}\eta_h C_{L_h,\alpha}\varepsilon_e\right]\delta_e$$
$$= \frac{W\cos\gamma}{\frac{1}{2}\rho V_\infty^2 S_w} - C_{L_w,\alpha}(\alpha_{0w}-\alpha_{L0w}) - \frac{S_h}{S_w}\eta_h C_{L_h,\alpha}(\alpha_{0h}-\varepsilon_{d0}) \tag{4.3.22}$$

$$\left[-\frac{l_w}{\bar{c}_w}C_{L_w,\alpha} - \frac{S_h l_h}{S_w \bar{c}_w}\eta_h C_{L_h,\alpha}(1-\varepsilon_{d,\alpha})\right]\alpha$$
$$+ \left[\frac{S_h \bar{c}_h}{S_w \bar{c}_w}\eta_h C_{m_h,\delta_e} - \frac{S_h l_h}{S_w \bar{c}_w}\eta_h C_{L_h,\alpha}\varepsilon_e\right]\delta_e$$
$$= -C_{m_w} + \frac{l_w}{\bar{c}_w}C_{L_w,\alpha}(\alpha_{0w}-\alpha_{L0w}) + \frac{S_h l_h}{S_w \bar{c}_w}\eta_h C_{L_h,\alpha}(\alpha_{0h}-\varepsilon_{d0}) \tag{4.3.23}$$

These two equations provide the requirements necessary to solve for the angle of attack and the elevator deflection at trim. At any given airspeed, the elevator deflection controls the equilibrium angle of attack for the aircraft and thus controls the lift developed by both the wing and the tail. When the elevator is deflected downward, additional lift is generated on the tail. Since the tail is aft of the CG, this produces a negative pitching moment about the CG, which rotates the nose of the aircraft downward and reduces the angle of attack. As the angle of attack is reduced, the lift on both the wing and the tail are reduced, as is the magnitude of the negative pitching moment. The angle of attack will continue to decrease until the pitching moment about the CG is returned to zero. **A stable airplane in free flight will always seek the angle of attack that makes the pitching moment about the CG equal to zero.** At any given airspeed there is only one elevator deflection that will satisfy both Eqs. (4.3.22) and (4.3.23). This is the elevator deflection that results in trimmed flight at that particular airspeed and climb angle. For any other deflection of the elevator, the aircraft climb angle will either increase or decrease if the airspeed remains fixed. Thus, at any airspeed above stall, the deflection of the elevator can be used to trim the airplane, provided that the elevator has sufficient size and effectiveness.

The angle of attack at trim, as computed from Eqs. (4.3.22) and (4.3.23), depends on the mounting angles for both the wing and the tail as well as the definition of the fuselage reference line. One convenient way to define the fuselage reference line is to define it as the minimum drag axis for the fuselage. With the fuselage reference line so defined, minimum drag on the fuselage and the tail are achieved when both the fuselage angle of attack and the elevator deflection are set to zero, with no lift on the horizontal tail. Using these three conditions at some design airspeed, say V_d, to fix the location of the center of gravity and the mounting angles for the wing and the tail will often result in a good preliminary design.

Since we are assuming the tail to have a symmetric airfoil section when there is no elevator deflection, the condition for **zero lift on the tail in level flight**, which is specified by Eq. (4.3.15), reduces to

$$l_w = \frac{\frac{1}{2}\rho V_d^2 S_w C_{m_w}}{W} \bar{c}_w \tag{4.3.24}$$

when the elevator is not deflected at the design airspeed. **With both the fuselage angle of attack and the elevator deflection set to zero for level flight at the design airspeed,** Eqs. (4.3.22) and (4.3.23) reduce to

$$S_w C_{L_w,\alpha}(\alpha_{0w} - \alpha_{L0w}) + S_h \eta_h C_{L_h,\alpha}(\alpha_{0h} - \varepsilon_{d0}) = \frac{W}{\frac{1}{2}\rho V_d^2} \tag{4.3.25}$$

and

$$S_w l_w C_{L_w,\alpha}(\alpha_{0w} - \alpha_{L0w}) + S_h l_h \eta_h C_{L_h,\alpha}(\alpha_{0h} - \varepsilon_{d0}) = S_w \bar{c}_w C_{m_w} \tag{4.3.26}$$

After applying Eq. (4.3.24), Eqs. (4.3.25) and (4.3.26) are readily solved for the mounting angles of the wing and the tail. This gives

$$\alpha_{0w} = \frac{W}{\frac{1}{2}\rho V_d^2 S_w C_{L_w,\alpha}} + \alpha_{L0w} \tag{4.3.27}$$

and

$$\alpha_{0h} = \varepsilon_{d0} \tag{4.3.28}$$

With the center of gravity located to satisfy Eq. (4.3.24) and the wing and the tail mounted to satisfy Eqs. (4.3.27) and (4.3.28), the drag will be minimized for level flight at the design airspeed, V_d. At this design operating condition, the wing supports the entire weight of the airplane and the tail provides only stability.

To see how the tail provides this stability, we return to the general pitch stability criterion that was specified in Eq. (4.2.8). For pitch stability, we require that the derivative of the pitching moment coefficient for the complete airplane with respect to angle of attack must be negative. For a symmetric tail with small angles of attack and elevator deflection, the total pitching moment coefficient can be expressed by applying Eqs. (4.3.16), (4.3.20), and (4.3.21) to Eq. (4.3.9). This gives

$$
\begin{aligned}
C_m = C_{m_w} &+ \frac{S_h \bar{c}_h}{S_w \bar{c}_w} \eta_h C_{m_h, \delta_e} \, \delta_e - \frac{l_w}{\bar{c}_w} C_{L_w, \alpha} (\alpha + \alpha_{0w} - \alpha_{L0w}) \\
&- \frac{S_h l_h}{S_w \bar{c}_w} \eta_h C_{L_h, \alpha} [(1 - \varepsilon_{d,\alpha}) \alpha + \alpha_{0h} - \varepsilon_{d0} + \varepsilon_e \delta_e]
\end{aligned}
\tag{4.3.29}
$$

For small changes in angle of attack, the aerodynamic centers of the wing and tail are fixed points that do not vary with angle of attack. Thus, differentiating Eq. (4.3.29) with angle of attack, we see that pitch stability for this wing-tail combination requires

$$
C_{m,\alpha} \equiv \frac{\partial C_m}{\partial \alpha} = -\frac{l_w}{\bar{c}_w} C_{L_w, \alpha} - \frac{S_h l_h}{S_w \bar{c}_w} \eta_h C_{L_h, \alpha} (1 - \varepsilon_{d,\alpha}) < 0
\tag{4.3.30}
$$

The change in the downwash angle with respect to angle of attack is typically less than 1.0. Furthermore, the lift slope, the planform area, and the mean chord length are always positive for both the wing and the tail. Because we are by definition considering an aft tail, the aerodynamic center of the tail is aft of the center of gravity ($l_h > 0$). Thus, from Eq. (4.3.30), we see that this wing-tail combination will be unconditionally stable if the center of gravity is forward of the aerodynamic center of the main wing ($l_w > 0$). This is a good thing to remember when loading an airplane for the first test flight. Even if the airplane is designed to have the CG aft of the wing quarter chord, it should probably be loaded to move the CG forward of the wing quarter chord for the first test flight. For the geometry of the present case, this will assure stable flight but will require negative lift on the aft tail for all operating conditions. For the more general case, even this will not absolutely assure stability, because the fuselage, nacelles, and thrust deck can all have a significant destabilizing effect.

If the center of gravity is aft of the aerodynamic center of the main wing ($l_w < 0$), then the main wing is destabilizing and the product of the tail area and the tail length, $S_h l_h$, must be sufficiently large so that the total change in pitching moment with respect to angle of attack is still negative. Thus, we see that by varying the size and/or length of an aft tail we can control the pitch stability of a conventional subsonic airplane.

The second term in Eq. (4.3.30) is the contribution of the horizontal tail to the pitch stability of the airplane. It is this term that typically makes the major contribution to the overall pitch stability of the aircraft. Notice that this term is proportional to the product of the tail area and the tail length, $S_h l_h$, divided by the product of the wing area and the wing chord, $S_w \bar{c}_w$. Each of these products represents a characteristic *volume* associated with the airplane. The product, $S_h l_h$, is sometimes called the *tail volume*, and increasing this volume for an aft tail will always increase the stability of the airplane. The ratio, $S_h l_h / S_w \bar{c}_w$, is called the *horizontal tail volume ratio*,

$$\mathcal{V}_h \equiv \frac{S_h l_h}{S_w \bar{c}_w} \tag{4.3.31}$$

The designer can easily change this tail volume ratio by either changing the area of the horizontal tail or by changing its distance aft of the center of gravity. It is the magnitude of this tail volume ratio that can be used to most directly control the pitch stability of any airplane with an aft tail. Notice that the horizontal tail volume ratio, as well as the first term in Eq. (4.3.30), are affected by the location of the CG and are thus affected by the way in which the airplane is loaded. **Aircraft loading has a very dramatic effect on pitch stability.**

At this point it is worth reiterating some of the simplifying assumptions that were made in the development of Eq. (4.3.30). To simplify the analysis, we assumed that the aerodynamic center of both the wing and the tail, as well as the center of thrust, are all aligned with the direction of flight on the fuselage reference line. Additionally, we assumed that the thrust vector is always aligned with the center of gravity. With these assumptions, neither the thrust nor the drag contributes to the pitching moment. Such conditions are almost never exactly satisfied in a real airplane. However, in many cases these conditions are nearly satisfied and the two terms on the right-hand side of the equal sign in Eq. (4.3.30) are the dominant terms in determining pitch stability. The influence of thrust on trim and static stability can be important and very complex. The drag can also have a significant effect on trim and static stability, particularly if the wing is mounted either far above or below the center of gravity. Such effects will be discussed later in this chapter.

EXAMPLE 4.3.1. Consider a general aviation airplane with an aft tail. The airplane has a gross weight of 2,700 lbf. The wing and horizontal tail have the following properties:

$$S_w = 180 \text{ ft}^2, \quad b_w = 33 \text{ ft}, \quad C_{L_w,\alpha} = 4.44, \quad \alpha_{L0w} = -2.20°, \quad C_{m_w} = -0.053,$$

$$S_h = 36 \text{ ft}^2, \quad b_h = 12 \text{ ft}, \quad C_{L_h,\alpha} = 3.97, \quad \varepsilon_e = 0.60, \quad C_{m_h,\delta_e} = -0.55,$$

$$l_h - l_w = 15 \text{ ft}$$

The horizontal tail has a symmetric airfoil section when the elevator is not deflected. The aerodynamic centers of the wing and the tail both fall on the fuselage reference line, which is the minimum drag axis of the fuselage. The wing and the tail are mounted so that when the airplane is in level flight with an airspeed of 120 mph at standard sea level, trim is attained with no elevator deflection, no lift on the horizontal tail, and a zero angle of attack for the fuselage reference line. Neglecting any moments associated with the thrust deck and the fuselage, determine the location of the center of gravity, the mounting angles for the wing and the tail, and the pitch stability derivative for this wing-tail combination. Using the simplified analysis presented in this section, plot the elevator deflection and the angle of attack for trimmed level flight as a function of airspeed from 80 to 160 mph. Also plot the lift on the horizontal tail at level trim for the same range of airspeed. Assume a tail efficiency of 1.0 and a downwash angle at the location of the tail given by the relation

$$\varepsilon_d = 0.6 \frac{C_{L_w}}{R_{A_w}}$$

Solution. Since there is no lift on the tail at 120 mph, the lift on the wing must exactly support the weight of the airplane at this airspeed,

$$\frac{1}{2}\rho V_d^2 S_w C_{L_w 0} = W$$

where the notation $C_{L_w 0}$ is used to indicate that this is a lift coefficient evaluated at zero fuselage angle of attack and zero elevator deflection. Solving for this wing lift coefficient gives

$$C_{L_w 0} = \frac{W}{\frac{1}{2}\rho V_d^2 S_w} = \frac{2,700}{\frac{1}{2}0.0023769(120 \times 5,280/3,600)^2 180} = 0.4075$$

The lift coefficient is related to angle of attack,

$$C_{L_w} = C_{L_w,\alpha}(\alpha + \alpha_{0w} - \alpha_{L0w})$$

Setting the fuselage reference line at zero angle of attack and solving for the mounting angle of the wing, we have

$$\alpha_{0w} = \frac{C_{L_w 0}}{C_{L_w,\alpha}} + \alpha_{L0w} = \frac{0.4075}{4.44} + \frac{(-2.20)\pi}{180} = 0.05338 \text{ rad} = \underline{3.06°}$$

For this operating condition, the downwash angle at the tail is

$$\varepsilon_{d0} = 0.6\frac{C_{L_w 0}}{R_{A_w}} = 0.6\frac{0.4075}{33^2/180} = 0.04041 \text{ rad} = 2.315°$$

There is to be no lift on the horizontal tail at this operating condition and

$$C_{L_h} = C_{L_h,\alpha}[(1-\varepsilon_{d,\alpha})\alpha + \alpha_{0h} - \varepsilon_{d0} + \varepsilon_e \delta_e]$$

Solving for the mounting angle of the horizontal tail that will give zero lift on the tail with the fuselage angle of attack and the elevator deflection set to zero, we obtain

$$\alpha_{0h} = \varepsilon_{d0} = \underline{2.3°}$$

Because the pitching moment about the CG must be zero at trim,

$$C_m = C_{m_w} + \frac{S_h \bar{c}_h}{S_w \bar{c}_w} \eta_h C_{m_h} - \frac{l_w}{\bar{c}_w} C_{L_w} - \frac{S_h l_h}{S_w \bar{c}_w} \eta_h C_{L_h} = 0$$

At this operating condition, the lift and moment for the tail are both zero, so we have

$$C_{m0} = C_{m_w} - \frac{l_w}{\bar{c}_w} C_{L_w 0} = 0$$

or

$$l_w = \frac{C_{m_w}}{C_{L_w 0}} \bar{c}_w = \frac{-0.053}{0.4075} \frac{180 \text{ ft}^2}{33 \text{ ft}} = \underline{-0.71 \text{ ft}}$$

Thus, the CG is about 8.5 inches aft of the wing quarter chord.
The downwash gradient at the tail is

$$\varepsilon_{d,\alpha} = \frac{\partial \varepsilon_d}{\partial \alpha} = 0.6 \frac{1}{R_{A_w}} \frac{\partial C_{L_w}}{\partial \alpha} = 0.6 \frac{C_{L_w,\alpha}}{R_{A_w}} = 0.6 \frac{4.44}{33^2/180} = 0.44$$

The pitch stability derivative for this simple wing-tail combination is given by Eq. (4.3.30),

$$\begin{aligned} C_{m,\alpha} \equiv \frac{\partial C_m}{\partial \alpha} &= -\frac{l_w}{\bar{c}_w} C_{L_w,\alpha} - \frac{S_h l_h}{S_w \bar{c}_w} \eta_h C_{L_h,\alpha} (1 - \varepsilon_{d,\alpha}) \\ &= -\frac{(-0.7094)}{180/33} 4.44 - \frac{36(15 - 0.7094)}{180(180/33)} (1.0)(3.97)(1 - 0.44) \\ &= \underline{-0.59 \text{ rad}^{-1}} \end{aligned}$$

Since the pitch stability derivative is negative, this wing-tail combination is stable.
For level trim at any other airspeed, the angle of attack and the elevator deflection must be such that the total lift is equal to the weight and the total pitching moment about the CG is zero,

$$C_{L_w,\alpha} (\alpha + \alpha_{0w} - \alpha_{L0w}) + \frac{S_h}{S_w} \eta_h C_{L_h,\alpha} [(1 - \varepsilon_{d,\alpha})\alpha + \alpha_{0h} - \varepsilon_{d0} + \varepsilon_e \delta_e] = \frac{W}{\frac{1}{2}\rho V_\infty^2 S_w}$$

$$\begin{aligned} C_{m_w} &+ \frac{S_h \bar{c}_h}{S_w \bar{c}_w} \eta_h C_{m_h,\delta_e} \delta_e - \frac{l_w}{\bar{c}_w} C_{L_w,\alpha} (\alpha + \alpha_{0w} - \alpha_{L0w}) \\ &- \frac{S_h l_h}{S_w \bar{c}_w} \eta_h C_{L_h,\alpha} [(1 - \varepsilon_{d,\alpha})\alpha + \alpha_{0h} - \varepsilon_{d0} + \varepsilon_e \delta_e] = 0 \end{aligned}$$

These two equations can be written as

$$\begin{bmatrix} C_{L,\alpha} & C_{L,\delta_e} \\ C_{m,\alpha} & C_{m,\delta_e} \end{bmatrix} \begin{Bmatrix} \alpha \\ \delta_e \end{Bmatrix} = \begin{Bmatrix} C_L - C_{L0} \\ -C_{m0} \end{Bmatrix}$$

where

$$C_{L,\alpha} = C_{L_w,\alpha} + \frac{S_h}{S_w} \eta_h C_{L_h,\alpha}(1 - \varepsilon_{d,\alpha}) = 4.8844$$

$$C_{L,\delta_e} = \frac{S_h}{S_w} \eta_h C_{L_h,\alpha} \varepsilon_e = 0.4764$$

$$C_L = \frac{W}{\frac{1}{2}\rho V_\infty^2 S_w}$$

$$C_{L0} = C_{L_w,\alpha}(\alpha_{0w} - \alpha_{L0w}) + \frac{S_h}{S_w} \eta_h C_{L_h,\alpha}(\alpha_{0h} - \varepsilon_{d0}) = 0.4075$$

$$C_{m,\alpha} = -\frac{l_w}{\bar{c}_w} C_{L_w,\alpha} - \frac{S_h l_h}{S_w \bar{c}_w} \eta_h C_{L_h,\alpha}(1 - \varepsilon_{d,\alpha}) = -0.5867$$

$$C_{m,\delta_e} = \frac{S_h \bar{c}_h}{S_w \bar{c}_w} \eta_h C_{m_h,\delta_e} - \frac{S_h l_h}{S_w \bar{c}_w} \eta_h C_{L_h,\alpha} \varepsilon_e = -1.3086$$

$$C_{m0} = C_{m_w} - \frac{l_w}{\bar{c}_w} C_{L_w,\alpha}(\alpha_{0w} - \alpha_{L0w}) - \frac{S_h l_h}{S_w \bar{c}_w} \eta_h C_{L_h,\alpha}(\alpha_{0h} - \varepsilon_{d0}) = 0.0$$

This 2×2 linear system of equations is easily solved for the angle of attack and elevator deflection:

$$\begin{Bmatrix} \alpha \\ \delta_e \end{Bmatrix} = \frac{1}{C_{L,\alpha} C_{m,\delta_e} - C_{L,\delta_e} C_{m,\alpha}} \begin{Bmatrix} (C_L - C_{L0})C_{m,\delta_e} + C_{L,\delta_e} C_{m0} \\ -(C_L - C_{L0})C_{m,\alpha} - C_{L,\alpha} C_{m0} \end{Bmatrix}$$

After substituting the values for this particular airplane, we have

$$\begin{Bmatrix} \alpha \\ \delta_e \end{Bmatrix} = \begin{Bmatrix} 0.214096 C_L - 0.087236 \\ -0.095986 C_L + 0.039111 \end{Bmatrix}$$

At 80 mph, the total lift coefficient, angle of attack, and elevator deflection are

$$C_L = \frac{W}{\frac{1}{2}\rho V_\infty^2 S_w} = \frac{2{,}700}{\frac{1}{2}0.0023769(80 \times 5{,}280/3{,}600)^2 180} = 0.916785$$

$$\begin{Bmatrix} \alpha \\ \delta_e \end{Bmatrix} = \begin{Bmatrix} 0.214096(0.916785) - 0.087236 \\ -0.095986(0.916785) + 0.039111 \end{Bmatrix} = \begin{Bmatrix} 0.10904 \\ -0.04889 \end{Bmatrix} = \begin{Bmatrix} 6.2° \\ -2.8° \end{Bmatrix}$$

With these values of angle of attack and elevator deflection, the lift developed on the wing and the tail of this airplane are readily computed from the known properties of each lifting surface. This gives

$$L_w = \tfrac{1}{2}\rho V_\infty^2 S_w C_{L_w,\alpha}(\alpha + \alpha_{0w} - \alpha_{L0w}) = \underline{2{,}626 \text{ lbf}}$$

$$L_h = \tfrac{1}{2}\rho V_\infty^2 S_h \eta_h C_{L_h,\alpha}[(1 - \varepsilon_{d,\alpha})\alpha + \alpha_{0h} - \varepsilon_{d0} + \varepsilon_e \delta_e] = \underline{74 \text{ lbf}}$$

Repeating these calculations at other airspeeds yields the data plotted in Fig. 4.3.5.

Notice from Example 4.3.1 that as airspeed is decreased, more negative elevator deflection is required for trim. With the center of gravity located about 8.5 inches aft of the wing quarter chord, at airspeeds above 120 mph the lift on the tail is negative, whereas for airspeeds below 120 mph the lift on the tail is positive. Figure 4.3.6 shows how these trim conditions change when the center of gravity is moved forward to the wing quarter chord. Notice that moving the center of gravity forward requires more negative elevator deflection and results in more negative lift on the tail. As the center of gravity is moved even farther forward, still more negative elevator deflection is required for trim. Since the elevator deflection is limited, there is some forward limit to the CG location, beyond which the elevator is no longer able to trim the aircraft. The airplane would not be able to take off with the center of gravity located forward of this point.

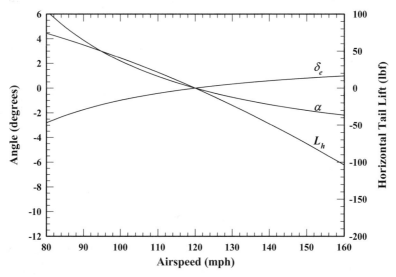

Figure 4.3.5. Conditions at level trim for the airplane in Example 4.3.1, with the center of gravity located 8.5 inches aft of the wing quarter chord.

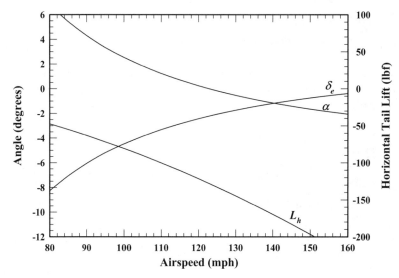

Figure 4.3.6. Conditions at level trim for the airplane in Example 4.3.1, with the center of gravity located exactly at the wing quarter chord.

This section should not be viewed as a "cookbook" of equations for analyzing pitch stability and trim. The symmetric aft tail with a trailing-edge elevator, which was considered here, is only one of several options available for providing longitudinal trim and static stability. Moreover, the simplified approximations used in this section ignore contributions to trim and static stability that can be important. **What students should understand and remember from this section are the fundamental requirements for longitudinal trim and static stability. Trim requires a zero summation of forces and moments in the longitudinal plane (i.e., the aircraft's plane of symmetry). Static stability requires a negative derivative of the pitching moment with respect to angle of attack.** The explicit details of the mathematical formulation of these requirements will vary somewhat with the particular aircraft configuration being considered. Thus, **students should concentrate on learning to formulate such problems from basic principles.** To reduce the possibility of making errors, **a detailed free-body diagram should be drawn and a consistent sign convention should always be used.**

4.4. Stick-Fixed Neutral Point and Static Margin

Some point on an airplane exists about which the total pitching moment does not vary with small changes in angle of attack. This is similar to the aerodynamic center of an airfoil or wing. With the center of gravity located at this point, the airplane is neutrally stable in pitch. Thus, this point is commonly referred to as the *stick-fixed neutral point* rather than the aerodynamic center. Nevertheless, **the stick-fixed neutral point is the aerodynamic center of the complete airplane.**

To locate the stick-fixed neutral point for the simplified wing-tail combination discussed in Sec. 4.3, we replace the inequality in Eq. (4.3.30) with the equality

$$C_{m_{np},\alpha} \equiv \frac{\partial C_{m_{np}}}{\partial \alpha} = -\frac{l_{wn}}{\bar{c}_w} C_{L_w,\alpha} - \frac{S_h l_{hn}}{S_w \bar{c}_w} \eta_h C_{L_h,\alpha} (1 - \varepsilon_{d,\alpha}) = 0 \qquad (4.4.1)$$

Here the variables l_w and l_h in Eq. (4.3.30) have been replaced with l_{wn} and l_{hn}, to indicate a distance measured aft of the neutral point rather than a distance measured aft of the actual center of gravity. However, to be consistent with our previous notation, l should always indicate a distance measured aft of the actual center of gravity. Thus, a distance measured aft of the neutral point can be written as a distance measured aft of the CG less the distance that the neutral point is aft of the CG (e.g., from Fig. 4.4.1, $l_{hn} = l_h - l_{np}$). Using this more consistent notation, Eq. (4.4.1) can be written in the form

$$C_{m_{np},\alpha} = -\frac{l_w - l_{np}}{\bar{c}_w} C_{L_w,\alpha} - \frac{S_h (l_h - l_{np})}{S_w \bar{c}_w} \eta_h C_{L_h,\alpha} (1 - \varepsilon_{d,\alpha}) = 0 \qquad (4.4.2)$$

where l_{np} is the distance that the stick-fixed neutral point is aft of the CG, as shown in Fig. 4.4.1. Equation (4.4.2) is readily rearranged to yield

$$l_{np} - l_w = \frac{\dfrac{S_h}{S_w} \eta_h C_{L_h,\alpha} (1 - \varepsilon_{d,\alpha})}{C_{L_w,\alpha} + \dfrac{S_h}{S_w} \eta_h C_{L_h,\alpha} (1 - \varepsilon_{d,\alpha})} (l_h - l_w) \qquad (4.4.3)$$

which expresses the distance the neutral point lies aft of the wing's aerodynamic center as a fraction of the distance that the tail's aerodynamic center lies aft of the aerodynamic center of the wing. **The position of the neutral point does not depend on the location of the CG, the definition of the reference chord length, or the choice of a datum.**

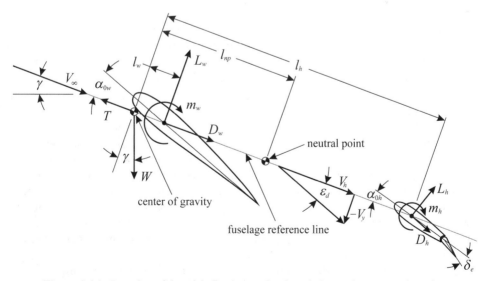

Figure 4.4.1. Location of the stick-fixed neutral point relative to the center of gravity.

Using Eqs. (4.3.16) and (4.3.20) in Eq. (4.3.6), yields

$$C_L = C_{L_w,\alpha}(\alpha + \alpha_{0w} - \alpha_{L0w}) + \frac{S_h}{S_w}\eta_h C_{L_h,\alpha}[(1-\varepsilon_{d,\alpha})\alpha + \alpha_{0h} - \varepsilon_{d0} + \varepsilon_e \delta_e] \quad (4.4.4)$$

Thus, the lift slope for the wing-tail combination is

$$C_{L,\alpha} \equiv \frac{\partial C_L}{\partial \alpha} = C_{L_w,\alpha} + \frac{S_h}{S_w}\eta_h C_{L_h,\alpha}(1-\varepsilon_{d,\alpha}) \quad (4.4.5)$$

From Eq. (4.4.5), note that the denominator on the right-hand side of Eq. (4.4.3) is the lift slope for the wing-tail combination. Thus, after rearranging Eq. (4.4.3), we have

$$\frac{l_{np}}{\bar{c}_w} = \left[\frac{l_w}{\bar{c}_w}C_{L_w,\alpha} + \frac{S_h l_h}{S_w \bar{c}_w}\eta_h C_{L_h,\alpha}(1-\varepsilon_{d,\alpha})\right] \Big/ C_{L,\alpha} \quad (4.4.6)$$

From Eq. (4.3.30), the pitch stability derivative for this wing-tail combination is

$$C_{m,\alpha} \equiv \frac{\partial C_m}{\partial \alpha} = -\frac{l_w}{\bar{c}_w}C_{L_w,\alpha} - \frac{S_h l_h}{S_w \bar{c}_w}\eta_h C_{L_h,\alpha}(1-\varepsilon_{d,\alpha}) \quad (4.4.7)$$

Thus, the numerator on the right-hand side of Eq. (4.4.6) is equal to the negative of the pitch stability derivative for the wing-tail combination,

$$l_{np}/\bar{c}_w = -C_{m,\alpha}/C_{L,\alpha} \quad (4.4.8)$$

In view of Eq. (4.4.8), the pitch stability derivative can be rewritten as

$$C_{m,\alpha} = -(l_{np}/\bar{c}_w)C_{L,\alpha} \quad (4.4.9)$$

Because static stability requires a negative pitch stability derivative and the lift slope is positive for angles of attack below stall, it follows from Eq. (4.4.9) that, **for an airplane to be statically stable in pitch, the center of gravity must be forward of the stick-fixed neutral point**. This is an important result that should be remembered.

The ratio $l_{np}/\bar{c}_w$ is called the *stick-fixed static margin*. It is simply the distance that the CG is forward of the stick-fixed neutral point, expressed as a fraction of the arbitrary reference chord length. A rule of thumb sometimes used for airplanes is that to provide good handling qualities for a human pilot, a static margin of at least 5 percent should be maintained. **This 5-percent-static-margin rule should be used with caution only as a rough initial estimate for airplanes of traditional geometric proportions, because the effects of pitch stability do not scale with wing chord length.** Wing chord appears in the dimensionless ratio on the left-hand side of Eq. (4.4.8) only because it was arbitrarily chosen to nondimensionalize the pitch stability derivative on the right-hand side. Clearly, the relation described by Eq. (4.4.8) is independent of the reference length that is used to nondimensionalize both the static margin and pitch stability derivative. Thus, **minimum pitch stability constraints should always be set based on the dynamic considerations presented in Chapters 6, 8, and 10, not based on aircraft static margin.**

While Eq. (4.4.9) was obtained for the wing-tail combination shown in Fig. 4.4.1, the result is more general. The aerodynamic pressure and shear forces acting on any object moving through the atmosphere can always be resolved into a lift force, a drag force, and an aerodynamic moment, all acting at the center of gravity, as is shown in Fig. 4.4.2. According to common notation, we have used m to represent the aerodynamic pitching moment about the center of gravity. If we now let m_{np} denote the aerodynamic pitching moment about the neutral point, then from the free-body diagram shown in Fig. 4.4.2, a summation of moments about the neutral point gives

$$m_{np} = m + l_{np}(L\cos\alpha + D\sin\alpha) + h_{np}(L\sin\alpha - D\cos\alpha) \qquad (4.4.10)$$

For a typical airplane, the vertical offset between the neutral point and the center of gravity is small and the drag is much less than the lift. Furthermore, in normal flight operations, the angle of attack is small. Thus, we can apply the traditional small-angle approximations $h_{np}\sin\alpha \cong 0$, $D \cong 0$, and $\cos\alpha \cong 1$. After using these approximations in Eq. (4.4.10) and nondimensionalizing the result in the usual manner, we have

$$C_{m_{np}} \cong C_m + \frac{l_{np}}{\bar{c}_w}C_L \qquad (4.4.11)$$

Differentiating Eq. (4.4.11) with respect to angle of attack results in

$$\frac{\partial C_{m_{np}}}{\partial\alpha} \cong \frac{\partial C_m}{\partial\alpha} + \frac{l_{np}}{\bar{c}_w}\frac{\partial C_L}{\partial\alpha} \qquad (4.4.12)$$

By definition, the stick-fixed neutral point is the aerodynamic center of the airplane, so the moment about the stick-fixed neutral point does not change with α. Thus, **within the approximations used in this simplified model**, the result in Eq. (4.4.12) reduces exactly to the more specific relation given in Eq. (4.4.9),

$$C_{m,\alpha} \cong -\frac{l_{np}}{\bar{c}_w}C_{L,\alpha} \qquad (4.4.13)$$

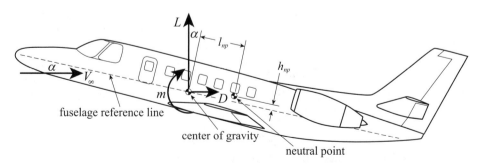

Figure 4.4.2. Stick-fixed neutral point for a complete airplane.

EXAMPLE 4.4.1. Determine the static margin in the range of linear lift for the wing-tail combination that is described in Example 4.3.1.

Solution. For this airplane we have

$$S_w = 180 \text{ ft}^2, \quad b_w = 33 \text{ ft}, \quad C_{L_w,\alpha} = 4.44, \quad l_w = -0.71 \text{ ft},$$
$$S_h = 36 \text{ ft}^2, \quad b_h = 12 \text{ ft}, \quad C_{L_h,\alpha} = 3.97, \quad l_h = 14.29 \text{ ft},$$
$$\eta_h = 1.0, \quad \varepsilon_{d,\alpha} = 0.44$$

By definition, the moment slope relative to the neutral point is zero. Thus,

$$C_{m_{np},\alpha} \equiv \frac{\partial C_{m_{np}}}{\partial \alpha} = -\frac{l_{wn}}{\bar{c}_w} C_{L_w,\alpha} - \frac{S_h l_{hn}}{S_w \bar{c}_w} \eta_h C_{L_h,\alpha} (1 - \varepsilon_{d,\alpha}) = 0$$

or in terms of the distances measured aft of the CG,

$$-\frac{l_w - l_{np}}{\bar{c}_w} C_{L_w,\alpha} - \frac{S_h(l_h - l_{np})}{S_w \bar{c}_w} \eta_h C_{L_h,\alpha} (1 - \varepsilon_{d,\alpha}) = 0$$

Solving for the static margin gives

$$\frac{l_{np}}{\bar{c}_w} = \frac{l_w C_{L_w,\alpha} + \dfrac{S_h l_h}{S_w} \eta_h C_{L_h,\alpha} (1 - \varepsilon_{d,\alpha})}{\bar{c}_w \left[C_{L_w,\alpha} + \dfrac{S_h}{S_w} \eta_h C_{L_h,\alpha} (1 - \varepsilon_{d,\alpha}) \right]}$$

$$= \frac{(-0.71)4.44 + \dfrac{36(14.29)}{180} 1.0(3.97)(1 - 0.44)}{\dfrac{180}{33} \left[4.44 + \dfrac{36}{180} 1.0(3.97)(1 - 0.44) \right]} = 0.1202 = \underline{12\%}$$

It is important to recognize that static margin and pitch stability are not entirely under the control of the designer. The pitch stability of an airplane can be dramatically affected by the way in which the airplane is loaded. The difference between good handling qualities and poor handling qualities (or even static instability) can be a matter of a shift in the center of gravity of only a few inches. Such a shift in the CG can easily result from insufficient care in loading the airplane. The designer must correctly specify and communicate the allowable CG range for an airplane to the end user. However, from that point it is the responsibility of the pilot and crew to see that the airplane is properly loaded so as to keep the CG within design specifications.

Equation (4.4.13) specifies how the pitch stability of an airplane depends on the location of the center of gravity. The farther that the CG is moved forward, the more stable the airplane becomes. However, we have also seen from the results plotted for Example 4.3.1 that as the center of gravity is moved forward, more negative elevator deflection is required for trim. Thus, there is a limit to how far the CG can be moved forward and still maintain trim. In Example 4.3.1 we found that the elevator deflection required for trim is given by

$$(\delta_e)_{\text{trim}} = -\frac{(C_L - C_{L0})C_{m,\alpha} + C_{L,\alpha} C_{m0}}{C_{L,\alpha} C_{m,\delta_e} - C_{L,\delta_e} C_{m,\alpha}} \tag{4.4.14}$$

Using Eq. (4.4.13) to express the pitch stability derivative in terms of the lift slope and static margin, Eq. (4.4.14) can be written as

$$(\delta_e)_{\text{trim}} = -\frac{C_{m0} - (C_L - C_{L0})(l_{np}/\overline{c}_w)}{C_{m,\delta_e} + C_{L,\delta_e}(l_{np}/\overline{c}_w)} \tag{4.4.15}$$

In addition to the static margin, both the pitching moment coefficient at zero angle of attack and the change in pitching moment coefficient with respect to elevator deflection depend on the position of the center of gravity. Thus, it is not obvious from Eq. (4.4.15) exactly how the elevator deflection required for trim will vary with the position of the center of gravity. This is examined in the following example.

EXAMPLE 4.4.2. For the wing-tail combination described in Examples 4.3.1 and 4.4.1, plot the elevator deflection required to trim the airplane at 80 mph as a function of the location of the center of gravity measured relative to the quarter chord of the main wing.

Solution. Let x denote the distance measured aft of the quarter chord of the main wing. For this airplane we have

$$S_w = 180 \text{ ft}^2, \quad b_w = 33 \text{ ft}, \quad C_{L_w,\alpha} = 4.44, \quad C_{m_w} = -0.053, \quad W = 2{,}700 \text{ lbf},$$
$$S_h = 36 \text{ ft}^2, \quad b_h = 12 \text{ ft}, \quad C_{L_h,\alpha} = 3.97, \quad C_{m_h,\delta_e} = -0.55,$$
$$\varepsilon_e = 0.60, \quad x_h = 15 \text{ ft}, \quad \eta_h = 1.0,$$
$$\alpha_{L0w} = -2.20°, \quad \alpha_{0w} = 3.06°, \quad \alpha_{L0h} = 0.0°, \quad \alpha_{0h} = \varepsilon_{d0} = 2.3°$$

From Example 4.3.1,

$$C_{L0} = C_{L_w,\alpha}(\alpha_{0w} - \alpha_{L0w}) + \frac{S_h}{S_w}\eta_h C_{L_h,\alpha}(\alpha_{0h} - \varepsilon_{d0}) = 0.4075$$

$$C_{L,\delta_e} = \frac{S_h}{S_w}\eta_h C_{L_h,\alpha}\varepsilon_e = 0.4764$$

and at 80 mph, the total lift coefficient was found to be

$$C_L = \frac{W}{\frac{1}{2}\rho V_\infty^2 S_w} = 0.9168$$

The pitching moment coefficient at zero angle of attack is

$$C_{m0} = C_{m_w} - \frac{l_w}{\bar{c}_w} C_{L_w,\alpha} (\alpha_{0w} - \alpha_{L0w}) - \frac{S_h l_h}{S_w \bar{c}_w} \eta_h C_{L_h,\alpha} (\alpha_{0h} - \varepsilon_{d0})$$

and the change in pitching moment coefficient with respect to elevator deflection for this wing-tail combination is

$$C_{m,\delta_e} = \frac{S_h \bar{c}_h}{S_w \bar{c}_w} \eta_h C_{m_h,\delta_e} - \frac{S_h l_h}{S_w \bar{c}_w} \eta_h C_{L_h,\alpha} \varepsilon_e$$

Any distance, l, measured aft of the CG, is the distance measured aft of the wing quarter chord, x, less the distance that the CG is aft of the wing quarter chord, x_{CG}. So we can write

$$l_w = -x_{CG}, \quad l_h = x_h - x_{CG}, \quad l_{np} = x_{np} - x_{CG}$$

Thus we have

$$C_{m0} = C_{m_w} + \frac{x_{CG}}{\bar{c}_w} C_{L_w,\alpha} (\alpha_{0w} - \alpha_{L0w}) - \frac{S_h(x_h - x_{CG})}{S_w \bar{c}_w} \eta_h C_{L_h,\alpha} (\alpha_{0h} - \varepsilon_{d0})$$

$$C_{m,\delta_e} = \frac{S_h \bar{c}_h}{S_w \bar{c}_w} \eta_h C_{m_h,\delta_e} - \frac{S_h(x_h - x_{CG})}{S_w \bar{c}_w} \eta_h C_{L_h,\alpha} \varepsilon_e$$

or after substituting the values for this particular wing-tail combination,

$$C_{m0} = 0.4075 \frac{x_{CG}}{\bar{c}_w} - 0.053$$

$$C_{m,\delta_e} = 0.4764 \frac{x_{CG}}{\bar{c}_w} - 1.3706$$

From Example 4.4.1,

$$\frac{x_{np}}{\bar{c}_w} = \frac{l_{np}}{\bar{c}_w} - \frac{l_w}{\bar{c}_w} = 0.1202 - \frac{-0.71}{180/33} = 0.2504$$

Thus, for this wing-tail combination at 80 mph, Eq. (4.4.15) reduces to

$$
\begin{aligned}
(\delta_e)_{trim} &= -\frac{C_{m0} - (C_L - C_{L0})[(x_{np} - x_{CG})/\bar{c}_w]}{C_{m,\delta_e} + C_{L,\delta_e}[(x_{np} - x_{CG})/\bar{c}_w]} \\
&= -\frac{0.4075(x_{CG}/\bar{c}_w) - 0.053 - (0.9168 - 0.4075)[0.2504 - (x_{CG}/\bar{c}_w)]}{0.4764(x_{CG}/\bar{c}_w) - 1.3706 + 0.4764[0.2504 - (x_{CG}/\bar{c}_w)]} \\
&= -\frac{0.9168(x_{CG}/\bar{c}_w) - 0.1805}{-1.2513} = \underline{0.7327(x_{CG}/\bar{c}_w) - 0.1442}
\end{aligned}
$$

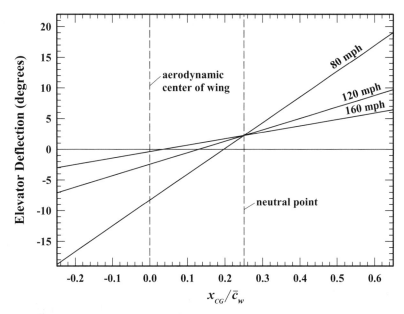

Figure 4.4.3. Elevator angle required for level trim as a function of CG location measured aft of the wing quarter chord for the wing-tail combination in Example 4.4.2.

Within the approximations used for this example, the elevator angle required to trim the airplane is a linear function of CG location. This is not obvious from Eq. (4.4.15). Results for the particular wing-tail combination used in the present example are shown in Fig. 4.4.3, for airspeeds of 80, 120, and 160 mph.

From Example 4.4.2 we see that at low airspeed, large negative elevator deflections are required to trim the airplane if the CG is moved very far forward of the neutral point. When the elevator deflection becomes larger than about 10 degrees, the elevator effectiveness begins to decrease. Since there is a limit to how much elevator deflection can be used without stalling the tail, there is a limit to how far the CG can be moved forward of the neutral point. Thus, we have now seen that there are both forward and aft limits for the allowable position of the center of gravity. If the CG is moved too far aft, the airplane becomes unstable. If the CG is moved too far forward, the airplane cannot be trimmed at low airspeed. An absolute forward CG limit is that requiring maximum *up elevator* (i.e., negative δ_e) to trim the airplane at the airspeed, which requires the maximum allowable lift coefficient. As we shall see later, there can be more restrictive constraints on the forward CG limit, based on control forces and maneuverability.

Another interesting point can be observed from Example 4.4.2. It can be seen from Fig. 4.4.3 that at any given airspeed, there is a CG location for which the airplane is in trim with no elevator deflection at all. This illustrates another possible method of trim control. An airplane with no elevator can be trimmed at any airspeed simply by moving the center of gravity. This is called *weight-shift control*, and it was actually used for the very first controlled human flight in a craft heavier than air. This took place in 1891 with an unpowered aircraft, designed, built, and flown by Otto Lilienthal, one of the true giants in aviation history. A few of Lilienthal's many flights are shown in Fig. 4.4.4.

Otto Lilienthal was truly the first aviator and the father of modern aviation. His nineteenth-century predecessors felt that the key to flight was to *build an engine with sufficient power and attach it to an airframe of sufficient strength and man could propel himself into aviation history with brute force alone.* By contrast, Lilienthal was the first to recognize the need to get in the air and fly with gliders before an engine could be used successfully for powered flight. The Wright brothers carried on his philosophy, which ultimately led to the control solutions that made powered flight possible. During the period from 1891 to 1896, Lilienthal made more than 2,000 successful glider flights. He

--Otto Lilienthal 1848-1896.

Figure 4.4.4. Otto Lilienthal, the true father of modern aviation, was the first human to pilot an aircraft in stable controlled flight. (Archives of the Otto-Lilienthal-Museum)

published papers on flight that were read throughout the world, and these papers inspired other aviation pioneers, including the Wright brothers. At the time of his death, Lilienthal was working on powered flight and it is more than likely that had he lived a few more years, he would have achieved powered flight before the Wright brothers.

On Sunday, August 9, 1896, Otto Lilienthal was injured in a glider crash near Stollen, Germany. He died the next day in Berlin. According to a witness, Lilienthal did not know he was fatally injured. The witness gave Lilienthal's last words as: "Ich muss etwas ausruhen, dann machen wir weiter" (I must rest a little, then we continue). During his life, Lilienthal used the phrase "Opfer müssen gebracht werden" (Sacrifices must be made). This is the epitaph now carved on his tombstone in Lichterfelde cemetery.

Trim Drag and Weight-Shift Control

With the exception of hang gliders, Lilienthal's method of weight-shift control is not used as the sole means for providing trim control on any modern aircraft. However, weight-shift control is sometimes used in combination with conventional elevator control to trim an airplane in pitch. To see why this is sometimes done, we return to the results obtained in Example 4.4.2.

The total drag on this wing-tail combination is the sum of three components: the drag produced directly by the wing, the drag produced directly by the tail, and the drag induced on the tail by the wing, which results from the tail's lift vector being rotated through the downwash angle as shown in Fig. 4.4.1. The drag coefficients for the wing and tail can be written as parabolic functions of their respective lift coefficients. Thus, using the small-angle approximation for the downwash, the total drag coefficient for this wing-tail combination is written as

$$
\begin{aligned}
C_D \;=\; & C_{D_{0w}} + C_{D_{0,L_w}} C_{L_w} + \frac{C_{L_w}^2}{\pi e_w R_{A_w}} \\
& + \frac{S_h}{S_W} \eta_h \left(C_{D_{0h}} + C_{D_{0,L_h}} C_{L_h} + \frac{C_{L_h}^2}{\pi e_h R_{A_h}} \right) + \frac{S_h}{S_W} \eta_h C_{L_h} \varepsilon_d
\end{aligned}
\tag{4.4.16}
$$

Notice that the last term in Eq. (4.4.16) is negative for any lifting configuration carrying a download on the aft tail. We have seen in Examples 4.3.1 and 4.4.2 that the required elevator deflection and the lift on the wing and tail are functions of both airspeed and CG location. It is rather straightforward to use these results together with Eq. (4.4.16) to obtain the results plotted in Figs. 4.4.5 and 4.4.6.

Locating the aircraft CG forward of the wing quarter chord produces a very stable configuration. However, as seen in Fig. 4.4.5, this requires negative lift on the aft tail at all airspeeds. This download on the tail produces induced drag on the tail and it must be countered by additional lift on the main wing, which in turn increases the induced drag produced by the wing. **The total additional induced drag resulting from the tail load required to trim an aircraft is called *trim drag*.** Notice from Figs. 4.4.5 and 4.4.6 that as the CG is moved aft, the tail download and associated trim drag are decreased until a minimum in the trim drag is reached at a point where the tail load is slightly positive. Carrying excess positive lift on the tail will also cause an increase in trim drag, resulting from the last term on the right-hand side of Eq. (4.4.16).

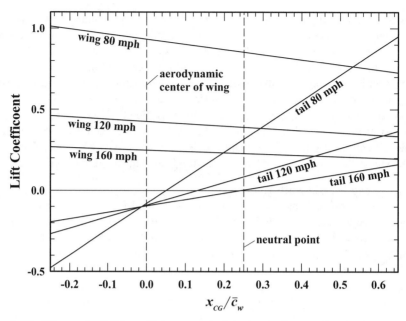

Figure 4.4.5. Wing and tail lift coefficients required for level trim as a function of CG location measured aft of the wing quarter chord for the wing-tail combination in Example 4.4.2.

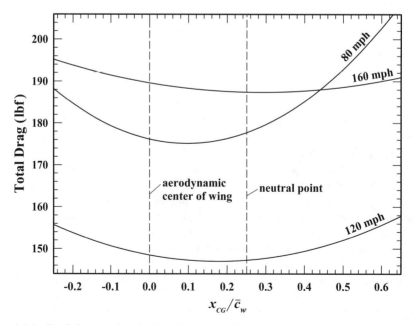

Figure 4.4.6. Total drag produced at level trim as a function of CG location measured aft of the wing quarter chord for the wing-tail combination in Example 4.4.2.

As a means of reducing fuel consumption in some modern aircraft, weight-shift control is used to help trim the aircraft once cruise altitude is reached. This is typically implemented by pumping fuel from the wing tanks into an aft holding tank, usually in the tail. The aft shift in the CG reduces stability, but this is easily handled by the stability augmentation function of the autopilot. The minimum-drag location for the CG moves farther aft as airspeed is increased. At high airspeeds the minimum-drag CG location is often aft of the neutral point. If increased stability is required for landing, fuel in the aft holding tank can be burned or pumped back into the main fuel tanks. Reducing trim drag is a significant motivation behind the shift to fly-by-wire in modern transport aircraft.

Figure 4.4.5 points out another advantage of moving the CG aft, i.e., reduced lift on the wing at low airspeeds. With a properly designed stability augmentation system the CG can even be moved aft of the neural point to reduce the liftoff, landing, and approach speeds for short takeoff and landing applications. A good example of this was related to the author by Captain Robert Niewoehner, who was the Navy's lead test pilot on the Super Hornet program. Early A–D models of the F/A-18 have a maximum landing weight of 34,000 lbf with a 400-ft^2 wing (85 lbf/ft^2). The larger Super Hornet carries its maximum landing weight of 44,000 lbf with a 500-ft^2 wing (88 lbf/ft^2). The increased wing loading would suggest a higher landing speed. However, the larger model has a landing speed 10 knots slower, a profound improvement when landing on aircraft carriers at sea. The difference results from a 5 percent aft shift in CG. Trimmed for approach, the early F/A-18 models require a strong download on the tail, reducing the net lift. The latter models land with a tail upload, working with the wing rather than against.

4.5. Estimating the Downwash Angle on an Aft Tail

We have seen that the downwash induced on an aft tail by the main wing has a significant effect on the trim and static stability of an airplane. This downwash reduces the effective angle of attack for the horizontal tail and also reduces its effectiveness in stabilizing the airplane. The downwash varies along the span of the horizontal tail and is affected by the planform shape of the wing as well as the presence of the fuselage and nacelles. To accurately estimate the interactions between the main wing and the tail, we must make use of computer simulations and/or wind tunnel tests. Such methods are always employed in the later phases of the airplane design process. However, for the purpose of preliminary design, it is useful to have an approximate closed-form solution for estimating the downwash induced on the tail by the main wing. Such an approximate solution can be developed by assuming that the flow field in the vicinity of the airplane is affected only by the main wing.

The downwash induced on an aft tail is a direct result of the vorticity generated by the main wing. At any given point in space, the downwash velocity, V_y, is directly proportional to the wingtip vortex strength, Γ_{wt}. From the vortex lifting law, for any given wing geometry, the strength of the wingtip vortex is proportional to the lift generated by the wing divided by the air density and the freestream velocity. Thus, we see that the downwash velocity induced by a lifting wing of some given geometry, at any given point in space, is directly proportional to the freestream velocity multiplied by the lift coefficient for the wing,

$$V_y \sim \Gamma_{wt} \sim \frac{L_w}{\rho V_\infty} \sim V_\infty C_{L_w} \tag{4.5.1}$$

The downwash velocity is independent of air density and the proportionality constant between the far right- and left-hand sides of Eq. (4.5.1) varies with position and the geometry of the wing.

More specifically, from Prandtl's lifting-line theory and the Biot-Savart law, the downwash induced a few chord lengths or more downstream from an unswept wing can be approximated as

$$
\begin{aligned}
V_y(\bar{x},\bar{y},\bar{z}) = &-\frac{V_\infty C_{L_w} \kappa_v}{\pi^2 R_{A_w}} \left\{ \frac{\kappa_b - \bar{z}}{\bar{y}^2 + (\kappa_b - \bar{z})^2} \left[1 + \frac{\bar{x}}{\sqrt{\bar{x}^2 + \bar{y}^2 + (\kappa_b - \bar{z})^2}} \right] \right. \\
&+ \frac{\bar{x}}{\bar{x}^2 + \bar{y}^2} \left[\frac{\kappa_b - \bar{z}}{\sqrt{\bar{x}^2 + \bar{y}^2 + (\kappa_b - \bar{z})^2}} + \frac{\kappa_b + \bar{z}}{\sqrt{\bar{x}^2 + \bar{y}^2 + (\kappa_b + \bar{z})^2}} \right] \\
&+ \left. \frac{\kappa_b + \bar{z}}{\bar{y}^2 + (\kappa_b + \bar{z})^2} \left[1 + \frac{\bar{x}}{\sqrt{\bar{x}^2 + \bar{y}^2 + (\kappa_b + \bar{z})^2}} \right] \right\}
\end{aligned}
\tag{4.5.2}
$$

where R_{A_w} is the aspect ratio of the wing,

$$\bar{x} = \frac{x}{b_w/2}, \quad \bar{y} = \frac{y}{b_w/2}, \quad \bar{z} = \frac{z}{b_w/2}$$

x is the axial coordinate in the direction of the freestream, y is the upward normal coordinate, and z is the spanwise coordinate, all measured relative to the wing quarter chord at the aircraft plane of symmetry as shown in Fig. 4.5.1.

The parameters κ_v and κ_b depend on wing geometry. The vortex strength factor, κ_v, is a ratio of the wingtip vortex strength to that generated by an elliptic wing having the same lift coefficient and aspect ratio. The vortex span factor, κ_b, is defined as the spacing between the wingtip vortices divided by the wingspan. Both κ_v and κ_b can be determined analytically from the series solution to Prandtl's lifting-line equation. For the details of this solution, see Phillips, Anderson, Jenkins, and Sunouchi (2002). The results are

$$\kappa_v = 1 + \sum_{n=2}^{\infty} \frac{A_n}{A_1} \sin(n\pi/2) \tag{4.5.3}$$

$$\kappa_b = \frac{\dfrac{\pi}{4} + \displaystyle\sum_{n=2}^{\infty} \frac{nA_n}{(n^2-1)A_1} \cos(n\pi/2)}{1 + \displaystyle\sum_{n=2}^{\infty} \frac{A_n}{A_1} \sin(n\pi/2)} \tag{4.5.4}$$

where the coefficients, A_n, are the Fourier coefficients in the series solution to Prandtl's lifting-line equation, which was presented in Sec. 1.8.

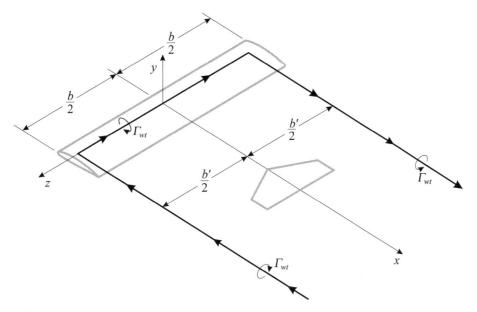

Figure 4.5.1. Vortex model for estimating the downwash on an aft tail behind an unswept wing.

For an elliptic wing with no sweep, dihedral, or twist, κ_v is 1.0 and κ_b is $\pi/4$. For an unswept tapered wing with no dihedral or twist, κ_v and κ_b are related to aspect ratio and taper ratio as shown in Figs. 4.5.2 and 4.5.3. The effects of washout are handled using Eqs. (4.5.3) and (4.5.4) with the Fourier coefficients evaluated as described in Sec. 1.8.

As a first approximation, the variation in downwash along the span of the horizontal tail is usually neglected. The downwash for the entire tail is typically taken to be that evaluated at the aerodynamic center. For a symmetric airplane, the aerodynamic center of the tail is in the plane of symmetry. The change in the downwash with respect to the spanwise coordinate is zero at the aircraft plane of symmetry. Furthermore, the span of the horizontal tail is usually small compared to that of the wing. Thus, the downwash is often fairly uniform over this span, and a reasonable first approximation for the downwash on an aft tail is found by setting the dimensionless spanwise coordinate, $\bar{z}$, equal to zero in Eq. (4.5.2).

Because the downwash is small compared to the freestream velocity, the downwash angle, ε_d, can be closely approximated as the downwash velocity divided by the freestream velocity. Thus, the downwash angle in the aircraft plane of symmetry a few chord lengths or more downstream from an unswept wing is approximated from Eq. (4.5.2) as

$$\varepsilon_d(\bar{x},\bar{y},0) = \frac{-V_y(\bar{x},\bar{y},0)}{V_\infty} = \frac{\kappa_v\,\kappa_p}{\kappa_b}\frac{C_{L_w}}{R_{A_w}} \qquad (4.5.5)$$

where

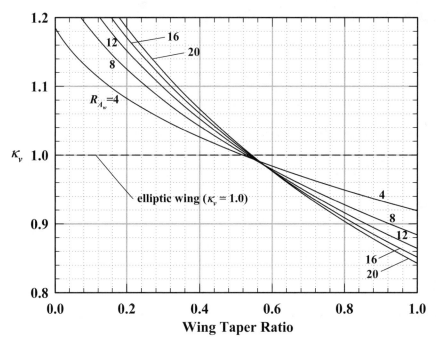

Figure 4.5.2. Wingtip vortex strength factor from Prandtl's lifting-line theory.

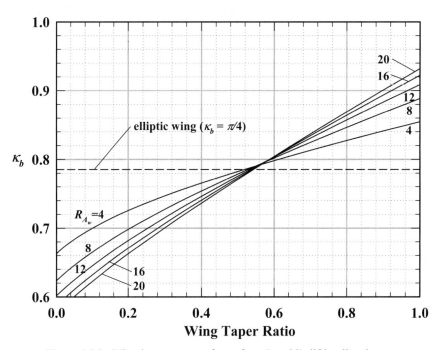

Figure 4.5.3. Wingtip vortex span factor from Prandtl's lifting-line theory.

$$\kappa_p = \frac{2\kappa_b^2}{\pi^2(\bar{y}^2 + \kappa_b^2)}\left[1 + \frac{\bar{x}(\bar{x}^2 + 2\bar{y}^2 + \kappa_b^2)}{(\bar{x}^2 + \bar{y}^2)\sqrt{\bar{x}^2 + \bar{y}^2 + \kappa_b^2}}\right] \tag{4.5.6}$$

The tail position factor, κ_p, depends on the planform shape of the wing and the position of the tail relative to the wing. However, for a wing with no twist, κ_v, κ_b, and κ_p are all independent of angle of attack. Thus, as a first approximation, the downwash angle at the tail is estimated as being linearly proportional to the lift coefficient for the wing. The variation in κ_p with tail position is shown in Fig. 4.5.4. The planform shape of the wing affects the value of κ_p only through its effect on κ_b. Thus, for a main wing with no sweep or dihedral, κ_p is a unique function of $\bar{x}/\kappa_b$ and $\bar{y}/\kappa_b$, as shown in Fig. 4.5.4.

Sweep in the main wing also has a significant effect on the downwash induced on an aft tail. Sweep affects this downwash in three ways. First, because sweep changes the spanwise distribution of vorticity on the wing, it changes the strength of the wingtip vortices for a given lift coefficient and aspect ratio. This same change in the vorticity distribution will also change the location of the center of vorticity in the vortex sheet trailing behind each semispan. Since each wingtip vortex rolls up around the center of vorticity from one side of the wing, wing sweep affects the spacing of the wingtip vortices. Because wing sweep affects both the strength and spacing of the wingtip

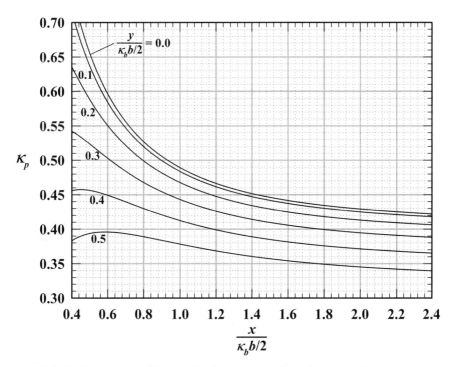

Figure 4.5.4. Effect of tail position on the downwash angle in the plane of symmetry aft of an unswept wing.

vortices, sweep in the main wing will affect both the vortex strength factor, κ_v, and the vortex span factor, κ_b. More significantly, sweep in the main wing affects the downwash on an aft tail through a simple change in proximity of the wing surface to the tail. As the wing is swept back, the outboard portions of the wing are brought closer to the tail as is shown in Fig. 4.5.5. This places the bound portion of the vortex system closer to the aft tail, and thus changes the downwash induced on the tail.

Unfortunately, the series solution to Prandtl's classical lifting-line equation does not apply to a swept wing. No closed-form solution for the spanwise vorticity distribution on a swept wing has ever been obtained. In the absence of such a solution for this vorticity distribution, it is not possible to obtain a closed-form solution for the variation of κ_v and κ_b with sweep. Neglecting the effects of sweep on κ_v and κ_b is quite restrictive and such results should be used with extreme caution for highly swept wings. Nevertheless, if we are willing to neglect the effect of sweep on κ_v and κ_b, it is possible to obtain a closed-form approximation for the proximity effect that results from moving the bound vortex closer to the aft tail when the wing is swept back.

This approximation is based on the vortex model suggested by McCormick (1995), which has been shown here in Fig. 4.5.5. With this model, the bound vorticity is approximated as two straight vortex line segments, one aligned with the quarter chord of each semispan. Each wingtip vortex is modeled as a semi-infinite line vortex trailing from the wing at the center of shed vorticity. In reality, the direction of the trailing wingtip vortices is determined by the geometry of the airplane, the angle of attack, and the downward deflection of the vortex system, which is caused by the downwash that is induced on one vortex by the other. However, this level of sophistication is hardly

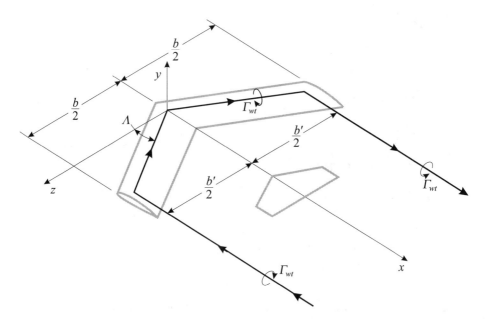

Figure 4.5.5. Vortex model for estimating the downwash on an aft tail behind a swept wing.

justified when one considers the approximate nature of the other aspects of this vortex model. Instead, we shall simply assume that wingtip vortices trail downstream from the wing quarter chord at the center of shed vorticity in a direction parallel to the x-axis. This same approximation was made in the development of Prandtl's classical theory.

With these approximations, the vorticity generated by a lifting swept wing is modeled as four straight vortex line segments, all of which fall in the x-z plane as shown in Fig. 4.5.5. Applying the Biot-Savart law to this vortex system, the downwash in the aircraft plane of symmetry is found to be

$$
V_y(x,y,0) = \frac{\Gamma_{wt}}{2\pi} \left\{ \frac{-(b'/2)}{(x-s')^2 + y^2 + (b'/2)^2 - (x-s')\sqrt{(x-s')^2 + y^2 + (b'/2)^2}} \right.
$$
$$
\left. + \frac{-(b'/2)x\left(\sqrt{x^2+y^2} + \sqrt{(x-s')^2 + y^2 + (b'/2)^2}\right)}{(x^2+y^2)[(x-s')^2 + y^2 + (b'/2)^2] + (x^2 - xs' + y^2)\sqrt{x^2+y^2}\sqrt{(x-s')^2 + y^2 + (b'/2)^2}} \right\}
$$

$$(4.5.7)$$

where

$$
s' \equiv (b'/2)\tan\Lambda \tag{4.5.8}
$$

and Λ is the quarter-chord sweep angle. The series solution to Prandtl's classical lifting-line equation predicts that the strength of each wingtip vortex is related to the lift coefficient developed by the wing according to the relation

$$
\Gamma_{wt} = \kappa_v \frac{2bV_\infty C_{L_w}}{\pi R_{A_w}} \tag{4.5.9}
$$

where κ_v is 1.0 for an elliptic wing and is expressed in terms of an infinite series for any other planform shape. For a tapered wing, the results shown Fig. 4.5.2 can be used. Also from Prandtl's lifting-line theory, the spacing between the wingtip vortices is related to the wingspan according to

$$
b' = \kappa_b b \tag{4.5.10}
$$

where κ_b takes a value of $\pi/4$ for an elliptic planform shape and is expressed as an infinite series for other planforms. For the tapered wing, Fig. 4.5.3 can be used.

Applying Eqs. (4.5.9) and (4.5.10) to Eq. (4.5.7), the downwash angle in the aircraft plane of symmetry a few chord lengths or more downstream from a swept wing, can be approximated as

$$
\varepsilon_d(\bar{x},\bar{y},0) = \frac{-V_y(\bar{x},\bar{y},0)}{V_\infty} = \frac{\kappa_v \kappa_p \kappa_s}{\kappa_b} \frac{C_{L_w}}{R_{A_w}} \tag{4.5.11}
$$

where

$$\kappa_s = \frac{1 + \dfrac{\bar{x} - \bar{s}}{\bar{t}} + \dfrac{\bar{x}(\bar{r} + \bar{t})(\bar{t}_0^2 - \bar{x}^2)}{\bar{r}\bar{t}(\bar{r}\bar{t} + \bar{r}^2 - \overline{xs})}}{1 + \dfrac{\bar{x}(\bar{r}^2 + \bar{t}_0^2 - \bar{x}^2)}{\bar{r}^2\bar{t}_0}}$$ (4.5.12)

$$\bar{r} \equiv \sqrt{\bar{x}^2 + \bar{y}^2}$$ (4.5.13)

$$\bar{s} \equiv \kappa_b \tan\Lambda$$ (4.5.14)

$$\bar{t} \equiv \sqrt{(\bar{x} - \bar{s})^2 + \bar{y}^2 + \kappa_b^2}$$ (4.5.15)

$$\bar{t}_0 \equiv \sqrt{\bar{x}^2 + \bar{y}^2 + \kappa_b^2}$$ (4.5.16)

The wing sweep factor, κ_s, depends on the planform shape of the wing and the position of the tail in addition to the quarter-chord sweep angle, Λ. However, as was the case for κ_p, the planform shape of the wing affects the value of κ_s as predicted by Eq. (4.5.12), only through its effect on κ_b. Thus, κ_s is found to be a unique function of Λ, $\bar{x}/\kappa_b$, and $\bar{y}/\kappa_b$. The variation in κ_s with axial tail position is shown in Fig. 4.5.6 for several values of quarter-chord sweep and tail height. The results shown in Fig. 4.5.6 for the case $y = 0$ agree exactly with the results presented by McCormick (1995) for the special case of an elliptic planform shape (i.e., $\kappa_b = \pi/4$). However, McCormick states that the sweep correction "does not depend significantly on the tail height," and he suggests that the zero-height solution be used in general. Figure 4.5.6 does not support that statement.

The downwash computed using the analytical method presented in this section agrees very well with experimental data and the commonly used empirical correlation recommended in the U.S. Air Force Stability and Control DATCOM (Hoak, 1960). For a detailed comparison, see Phillips, Anderson, Jenkins, and Sunouchi (2002).

If the position of the tail and the planform shape of the main wing are known, Eq. (4.5.11) can be combined with Eq. (4.5.6), Eq. (4.5.12), and the results presented in Figs. 4.5.2 and 4.5.3 to estimate the downwash angle. However, at the beginning of the preliminary design process the position of the tail is not typically known. When the tail is mounted sufficiently aft of the main wing, the downwash angle is independent of axial position. As the dimensionless axial position of the tail approaches infinity, from Eq. (4.5.6), the tail position factor along the x-axis reduces to

$$\kappa_p(\bar{x},0,0) \underset{\bar{x}\to\infty}{=} \frac{4}{\pi^2}$$ (4.5.17)

and from Eq. (4.5.12), the wing sweep factor becomes

$$\kappa_s(\bar{x},0,0) \underset{\bar{x}\to\infty}{=} 1.0$$ (4.5.18)

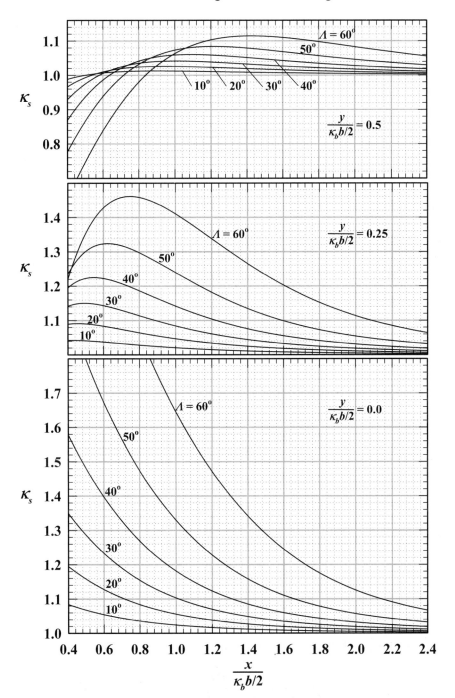

Figure 4.5.6. Effect of wing sweep on the downwash angle in the plane of symmetry aft of the main wing.

Using Eqs. (4.5.17) and (4.5.18) in Eq. (4.5.11), the downwash angle along the x-axis far behind the main wing can be approximated as

$$\varepsilon_d(\bar{x},0,0) \underset{\bar{x}\to\infty}{=} \frac{4\kappa_v}{\pi^2\kappa_b}\frac{C_{L_w}}{R_{A_w}} \tag{4.5.19}$$

For an elliptic wing ($\kappa_v = 1.0$ and $\kappa_b = \pi/4$), this gives

$$\varepsilon_d(\bar{x},0,0) \underset{\bar{x}\to\infty}{=} \frac{16}{\pi^3}\frac{C_{L_w}}{R_{A_w}} \tag{4.5.20}$$

Equation (4.5.20) can be used as a rough estimate for the downwash angle on an aft tail.

EXAMPLE 4.5.1. Consider the general aviation aircraft, which was described in Example 4.3.1. The wing has a taper ratio of 0.4 and a quarter-chord sweep of 10 degrees. If the horizontal tail were mounted 3.4 feet above the aerodynamic center of the wing, estimate the downwash gradient using the approximation presented in this section.

Solution. For this wing-tail combination, we have

$$S_w = 180 \text{ ft}^2, \quad b_w = 33 \text{ ft}, \quad R_{A_w} = 33^2/180 = 6.05, \quad C_{L_w,\alpha} = 4.44, \quad R_{T_w} = 0.40,$$
$$\Lambda = 10°, \quad l_h - l_w = 15 \text{ ft}, \quad \bar{x} = 15/(33/2) = 0.909, \quad \bar{y} = 3.4/(33/2) = 0.206$$

From Figs. 4.5.2 and 4.5.3 (or using a computer program based on the series solution to Prandtl's lifting-line equation), for a wing aspect ratio of 6.05 and a wing taper ratio of 0.40, the following results are obtained:

$$\kappa_v = 1.035$$

$$\kappa_b = 0.759$$

From Eqs. (4.5.6) and (4.5.12), the position factor and the sweep factor are found to be

$$\kappa_p = 0.433$$

$$\kappa_s = 1.015$$

From Eq. (4.5.11), the downwash at the position of the horizontal tail is

$$\varepsilon_d = \frac{\kappa_v\,\kappa_p\,\kappa_s}{\kappa_b}\frac{C_{L_w}}{R_{A_w}} = \frac{1.035(0.433)1.015}{0.759}\frac{C_{L_w}}{R_{A_w}} = 0.599\frac{C_{L_w}}{R_{A_w}}$$

The downwash gradient is then

$$\varepsilon_{d,\alpha} = \frac{\partial \varepsilon_d}{\partial \alpha} = 0.599 \frac{1}{R_{A_w}} \frac{\partial C_{L_w}}{\partial \alpha} = 0.599 \frac{C_{L_w,\alpha}}{R_{A_w}} = 0.599 \frac{4.44}{6.05} = \underline{0.44}$$

This means that for each degree of increase in the angle of attack, the downwash on the tail is increased by 0.44 degree. Thus, only 56 percent of any change in angle of attack is actually felt by the tail. This reduces the effectiveness of the horizontal stabilizer by 44 percent. For comparison, as a rough estimate for this example, Eq. (4.5.20) gives $\varepsilon_{d,\alpha} = 0.38$.

To reiterate, **the downwash induced on an aft tail by the main wing has a critical effect and should never be ignored when designing the tail of an airplane.**

4.6. Simplified Pitch Stability Analysis for a Wing-Canard Combination

Trim and static pitch stability can also be achieved by combining a wing with a forward canard, as shown in Fig. 4.6.1. A canard is a horizontal lifting surface mounted ahead of the main wing rather than behind it. While the canard configuration is not nearly as common as the aft tail, the first successful powered airplane, the 1903 Wright Flyer shown in Fig. 4.6.2, was a canard design. Shortly after the Wright Brothers, most airplane designers abandoned the canard in favor of the aft tail. However, this probably had more to do with avoiding patent difficulties with the Wrights than with any intrinsic inferiority of the canard design. In fact, as we shall see, the canard design has some inherent advantages, and the student should understand the workings of this design.

Figure 4.6.1. Beech Starship with a forward canard design. (Photograph by Barry Santana)

Figure 4.6.2. The 1903 Wright Flyer in the Smithsonian National Air and Space Museum.

Here we shall proceed with our analysis of the wing-canard configuration just as we did with the wing-tail configuration. First we consider the requirements for trim and then proceed to an examination of the pitch stability requirement. For the analysis presented in this section, we use the simplified configuration shown in Fig. 4.6.3. We again take the fuselage reference line to pass through the center of gravity parallel to the freestream velocity at trim. Following the simplified analysis for the aft horizontal tail, we assume that the aerodynamic centers of the wing and canard, as well as the center of thrust, all fall on this fuselage reference line. Again, we also assume the thrust vector to be aligned with the direction of motion.

The position of the CG shown in Fig. 4.6.3 may at first be confusing to the student who is already familiar with the canard configuration. Here, for consistency, the CG is shown as being forward of both the canard and the wing. In this textbook, the length, l, for any lifting surface is always defined to be the distance that the aerodynamic center of the surface is aft of the center of gravity. If this notation is consistently used, we need not be concerned beforehand as to the location of any surface relative to the CG. If our analysis shows that a negative value of l is required for any surface, this simply means that this particular surface must be forward of the CG.

As was the case with the simplified analysis for the aft horizontal tail, the force balance in the direction of flight simply allows us to determine the thrust required to maintain constant airspeed. The trim requirements for pitch are the same as they were for the simplified aft tail configuration. Simply restated, the total lift must balance the normal component of weight, and the total pitching moment about the CG must be zero. Using the small-angle approximations for the upwash angle ε_u, this gives

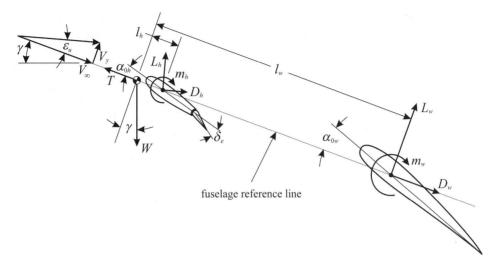

Figure 4.6.3. Forces and moments acting on a wing combined with a horizontal canard.

$$C_{L_w} + \frac{S_h}{S_w} C_{L_h} = \frac{W \cos \gamma}{\frac{1}{2} \rho V_\infty^2 S_w} \tag{4.6.1}$$

$$C_m = C_{m_w} + \frac{S_h \bar{c}_h}{S_w \bar{c}_w} C_{m_h} - \frac{l_w}{\bar{c}_w} C_{L_w} - \frac{S_h l_h}{S_w \bar{c}_w} C_{L_h} = 0 \tag{4.6.2}$$

where a dynamic pressure ratio of 1.0 has been assumed for both the horizontal canard and the main wing. Here the subscript h is used to denote the horizontal canard, just as was done for the aft horizontal tail. A canard is distinguished from an aft horizontal tail simply by its position relative to the main wing (i.e., the sign of $l_h - l_w$).

As was the case with the wing-tail analysis, the requirements that are specified by Eqs. (4.6.1) and (4.6.2) relate the lift forces developed by the wing and the canard to the weight of the aircraft and the position of the center of gravity. Equation (4.6.1) requires that at trim, the combined lift developed by the wing and the canard must support the normal component of the weight of the aircraft. Equation (4.6.2) specifies how the total lift must be divided between the wing and the canard in order to maintain the attitude required for trim. These two equations can be solved for the lift on the wing and the horizontal canard to give

$$C_{L_w} = \frac{\bar{c}_w}{l_w - l_h} C_{m_w} + \frac{S_h \bar{c}_h}{S_w (l_w - l_h)} C_{m_h} - \frac{l_h}{l_w - l_h} \frac{W \cos \gamma}{\frac{1}{2} \rho V_\infty^2 S_w} \tag{4.6.3}$$

$$C_{L_h} = \frac{S_w l_w}{S_h (l_w - l_h)} \frac{W \cos \gamma}{\frac{1}{2} \rho V_\infty^2 S_w} - \frac{S_w \bar{c}_w}{S_h (l_w - l_h)} C_{m_w} - \frac{\bar{c}_h}{l_w - l_h} C_{m_h} \tag{4.6.4}$$

By definition, the horizontal canard is forward of the main wing and $(l_w - l_h)$ is positive. Furthermore, the area and chord length for both the wing and the canard are always positive. Thus, if neither the wing nor the canard have negative camber, both moment coefficients and the first two terms on the right-hand side of Eq. (4.6.3) must be less than or equal to zero. Since the lift on the main wing must be positive to support a major portion of the weight, we see that the last term on right-hand side of Eq. (4.6.3) must make a positive contribution to the lift. This requires $l_h < 0$. In other words, **the canard must always be mounted forward of the CG to maintain trim**.

For pitch stability, we require that the change in the total pitching moment coefficient with respect to angle of attack is negative. Because C_{m_w} and C_{m_h} are the moment coefficients relative to the aerodynamic centers of the wing and the canard, the moment terms in Eq. (4.6.2) do not change with angle of attack, and differentiating Eq. (4.6.2), the stability condition for this wing-canard combination requires

$$\frac{\partial C_m}{\partial \alpha} = -\frac{l_w}{\bar{c}_w}\frac{\partial C_{L_w}}{\partial \alpha} - \frac{S_h l_h}{S_w \bar{c}_w}\frac{\partial C_{L_h}}{\partial \alpha} < 0 \qquad (4.6.5)$$

The aerodynamic interactions between the wing and the canard are very complex. Because the canard is forward of the lifting wing, the wing induces upwash on the canard. Any lift that is generated on the canard will also produce wingtip vorticity that interacts with the main wing. Because the span of the canard is typically less than that of the wing, these wingtip vortices can even pass directly over the wing. As one might imagine, this produces a very complex flow about the wing. This interaction can only be evaluated from detailed computer simulations or wind tunnel tests. For preliminary calculations, we have no choice but to neglect the effects of the canard's vorticity on the performance of the main wing. The effect of the upwash induced on the canard by the main wing is typically small but is easily accounted for by using the method presented in Sec. 4.5. With these approximations, the lift coefficients for the wing and the canard can be expressed in terms of angle of attack. For small angles of attack, the lift coefficient for the main wing is a linear function of angle of attack,

$$C_{L_w} = C_{L_w,\alpha}(\alpha + \alpha_{0w} - \alpha_{L0w}) \qquad (4.6.6)$$

where $C_{L_w,\alpha}$ is the lift slope for the wing, α is the angle of attack for the airplane as measured relative to the fuselage reference line, α_{0w} is the angle that the wing chord makes with the fuselage reference line, and α_{L0w} is the zero-lift angle of attack for the wing. Assuming small angles of attack and small control surface deflections, the lift coefficient for the canard varies linearly with both the angle of attack and control surface deflection,

$$C_{L_h} = C_{L_h,\alpha}(\alpha + \alpha_{0h} - \alpha_{L0h} + \varepsilon_u + \varepsilon_e \delta_e) \qquad (4.6.7)$$

where $C_{L_h,\alpha}$ is the lift slope for the isolated canard, α_{0h} is the angle that the canard chord makes with the fuselage reference line, α_{L0h} is the zero-lift angle of attack for the canard with no control surface deflection, ε_u is the upwash angle, ε_e is the control surface effectiveness, and δ_e is the control surface deflection. For small angles of attack, the upwash angle is also a linear function of angle of attack,

$$\varepsilon_u = \varepsilon_{u0} + \varepsilon_{u,\alpha}\,\alpha \tag{4.6.8}$$

Using Eqs. (4.6.6) through (4.6.8) in Eq. (4.6.5), the stability criterion for this wing-canard combination is

$$C_{m,\alpha} = -\frac{l_w}{\bar{c}_w}C_{L_w,\alpha} - \frac{S_h l_h}{S_w \bar{c}_w}C_{L_h,\alpha}(1+\varepsilon_{u,\alpha}) < 0 \tag{4.6.9}$$

The first term to the right of the equal sign in Eq. (4.6.9) is the wing's contribution to the airplane pitch stability derivative and the second term is the canard's contribution. We have already seen that trim requires $l_h < 0$ for a forward canard. Thus, since the lift slope is always positive for angles of attack below stall, Eq. (4.6.9) shows that **a canard always has a destabilizing effect on the airplane**. At this point the student might ask: How is it possible to replace an aft stabilizer with a destabilizing canard? The answer is also found in Eq. (4.6.9).

Since the second term to the right of the equal sign in Eq. (4.6.9) always makes a positive contribution to the pitch stability derivative, the first term must make a negative contribution if the airplane is to be stable in pitch. For angles of attack less than stall, the lift slope is positive and this requires $l_w > 0$, which means that **the aerodynamic center of the main wing must always be aft of the airplane's center of gravity for a stable wing-canard configuration**. Furthermore, it must be far enough aft of the CG to counter the destabilizing effects of the fuselage and canard. Thus, **it is the main wing and not the canard that provides stability for the wing-canard configuration**. Even though the canard is destabilizing, it provides the essential function of generating a positive pitching moment about the center of gravity, which is necessary to trim the airplane against the negative pitching moment caused by having the CG ahead of the main wing. From Eq. (4.6.4) we see that because the moment coefficients for both the wing and canard are typically less than or equal to zero and l_w must be positive, **the lift on the canard must be positive** as well.

In summary, **to maintain both trim and static stability, the center of gravity for any wing-canard airplane must be located between the wing and the canard**. Additionally, in trimmed flight, **both the wing and the canard will always carry positive lift**. This is one of the intrinsic advantages of the canard design. With a conventional aft tail, the lift on the horizontal tail is negative over at least some portion of the airplane's operating range. This requires additional lift on the main wing and produces additional induced drag on both the wing and the tail. With the canard configuration, the positive lift that is always generated by the canard reduces the lift required on the main wing and thus results in less induced drag than would be generated by a similar conventional airplane carrying negative lift on the aft tail.

Notice that in both Eqs. (4.6.2) and (4.6.9), a dimensionless ratio appears that looks like the volume ratio, which was defined for the horizontal aft tail. Since the length, l_h, is always negative for a canard, we shall define the *canard volume ratio* as

$$\mathcal{V}_c \equiv \frac{S_h(-l_h)}{S_w \bar{c}_w} \tag{4.6.10}$$

Using this definition in Eq. (4.6.2) and rearranging gives

$$\mathcal{V}_c = \frac{l_w}{\bar{c}_w} \frac{C_{L_w}}{C_{L_h}} - \frac{C_{m_w}}{C_{L_h}} - \frac{S_h \bar{c}_h}{S_w \bar{c}_w} \frac{C_{m_h}}{C_{L_h}} \tag{4.6.11}$$

Using this same volume ratio definition, for angles of attack less than stall, Eq. (4.6.9) can be rearranged to yield

$$\mathcal{V}_c < \frac{l_w}{\bar{c}_w} \frac{C_{L_w,\alpha}}{C_{L_h,\alpha}(1+\varepsilon_{u,\alpha})} \tag{4.6.12}$$

From Eq. (4.6.11), we see that trim requires a canard volume ratio large enough to counter the negative pitching moments generated by the lift of the main wing and any camber in the wing and/or canard. On the other hand, from Eq. (4.6.12), stability requires a canard volume ratio small enough so as not to overpower the stability provided by the main wing. Thus, the trim and stability analysis for the design of a canard configuration is more critical than it is for the design of an aft tail. With the aft tail, pitch control and pitch stability are both improved by increasing the volume ratio of the horizontal tail. However, sizing the volume ratio for a canard requires a trade-off between pitch control and pitch stability, which makes sizing a canard more critical than sizing an aft tail.

Using the volume ratio requirement for trim, as given by Eq. (4.6.11), we can eliminate the volume ratio from Eq. (4.6.12), and after some rearranging, we obtain

$$\frac{C_{L_w}}{C_{L_h}} < \frac{C_{L_w,\alpha}}{C_{L_h,\alpha}(1+\varepsilon_{u,\alpha})} + \frac{\bar{c}_w}{l_w} \frac{C_{m_w}}{C_{L_h}} + \frac{S_h \bar{c}_h}{S_w l_w} \frac{C_{m_h}}{C_{L_h}} \tag{4.6.13}$$

Because the canard must always support a portion of the aircraft weight, for best aerodynamic efficiency, the canard should be designed with an aspect ratio as high as possible, just like the wing. As a result, the lift slope for the canard should be about the same as that for the wing. Since the upwash gradient, $\varepsilon_{u,\alpha}$, is positive ahead of the main wing, the first term on the right of Eq. (4.6.13) is typically slightly less than 1.0. Furthermore, if both the wing and the canard have positive camber, the last two terms in Eq. (4.6.13) have small negative values. Thus, **to maintain both trim and static stability with a typical wing-canard combination, the lift coefficient for the canard must be greater than that for the wing**. This underscores the importance of designing the canard with a fairly high aspect ratio.

The fact that the canard must support a higher lift coefficient than the wing points out what is sometimes considered to be another advantage of the canard configuration. As the trim speed is reduced, the lift coefficients for both the wing and the canard must be increased. However, since the lift coefficient for the canard must be greater than that for the wing, the maximum lift coefficient is typically reached for the canard before the main wing stalls. Once the canard has stalled, it is not possible to generate the additional positive pitching moment necessary to further increase the angle of attack and stall the main wing. Thus, with a wing-canard combination so designed, it is almost impossible to stall the main wing in a gradual maneuver. The wing can still be stalled in a rapidly accelerating pitch-up maneuver known as a *whip stall*.

The higher lift coefficient carried by the canard also results in what can sometimes be a disadvantage of the wing-canard design. If a very high maximum lift coefficient is required for short takeoff and landing, it does no good to fit the main wing with complex high-lift devices, unless the canard is similarly fitted. The maximum lift coefficient for the complete airplane is controlled primarily by the maximum lift coefficient for the canard. Furthermore, when high-lift devices are deployed, the magnitude of the negative pitching moment coefficient for both the main wing and the canard is increased significantly. From Eq. (4.6.13), we see that this decreases even further the ratio of the allowable lift coefficient for the wing to that for the canard. Thus, even though the main wing must carry the largest portion of the airplane's weight, the canard must produce a significantly larger lift coefficient to provide the positive pitching moment necessary to rotate the airplane to liftoff. For this reason, the canard design is not particularly well suited for short takeoff and landing applications.

EXAMPLE 4.6.1. Consider a general aviation airplane of a size and weight similar to the one described in Example 4.3.1. This airplane is a wing-canard configuration and weighs 2,700 lbf. The wing and the canard are geometrically similar, having the same aspect ratio and airfoil section with no sweep. The following properties are known:

$$S_w = 180 \text{ ft}^2, \quad b_w = 33 \text{ ft}, \quad R_{T_w} = 0.40, \quad C_{L_w,\alpha} = 4.44,$$
$$\alpha_{L0w} = -2.20°, \quad C_{m_w} = -0.053,$$
$$S_h = 36 \text{ ft}^2, \quad \varepsilon_e = 0.60, \quad C_{m_h,\delta_e} = -0.55, \quad l_w - l_h = 15 \text{ ft}$$

The aerodynamic centers of the wing and the canard both fall on the fuselage reference line, which is the minimum drag axis of the fuselage. The wing and the canard are mounted so that when the airplane is in level flight with an airspeed of 120 mph at standard sea level, trim is attained with no control surface deflection and a zero angle of attack for the fuselage reference line. In the range of linear lift, the static margin for this wing-canard combination is 5 percent. Neglecting any moments associated with the thrust deck and the fuselage, determine the location of the center of gravity and the mounting angles for the wing and the canard. Plot the control surface deflection and the angle of attack at trim as a function of airspeed from 80 to 160 mph. Also plot the lift coefficient on the wing and canard at trim for the same range of airspeed.

Solution. From Figs. 4.5.2 and 4.5.3 (or using a computer program based on the series solution to Prandtl's lifting-line equation), for a wing aspect ratio of 6.05 and a wing taper ratio of 0.40, the following results are obtained:

$$\kappa_v = 1.035$$

$$\kappa_b = 0.759$$

From Eqs. (4.5.6) and (4.5.12), the position factor and the sweep factor for the position $\bar{x} = -0.909$ and $\bar{y} = 0.0$ are found to be

$$\kappa_p = -0.061$$

$$\kappa_s = 1.0$$

From Eq. (4.5.11), the upwash at the position of the canard is

$$\varepsilon_u = -\varepsilon_d = -\frac{\kappa_v \kappa_p \kappa_s}{\kappa_b} \frac{C_{L_w}}{R_{A_w}} = -\frac{1.035(-0.061)1.0}{0.759} \frac{C_{L_w}}{R_{A_w}} = 0.083 \frac{C_{L_w}}{R_{A_w}}$$

The upwash gradient is then

$$\varepsilon_{u,\alpha} = \frac{\partial \varepsilon_u}{\partial \alpha} = 0.083 \frac{1}{R_{A_w}} \frac{\partial C_{L_w}}{\partial \alpha} = 0.083 \frac{C_{L_w,\alpha}}{R_{A_w}} = 0.083 \frac{4.44}{6.05} = 0.0609$$

Since the canard and the wing have the same aspect ratio,

$$\frac{b_h^2}{S_h} = \frac{b_w^2}{S_w}$$

$$b_h = b_w \sqrt{\frac{S_h}{S_w}} = 33 \text{ ft} \sqrt{\frac{36}{180}} = 14.758 \text{ ft}$$

and

$$\bar{c}_w = \frac{S_w}{b_w} = \frac{180 \text{ ft}^2}{33 \text{ ft}} = 5.455 \text{ ft}$$

$$\bar{c}_h = \frac{S_h}{b_h} = \frac{36 \text{ ft}^2}{14.758 \text{ ft}} = 2.439 \text{ ft}$$

By definition, the moment slope relative to the neutral point is zero. Thus,

$$C_{m_{np},\alpha} \equiv \frac{\partial C_{m_{np}}}{\partial \alpha} = -\frac{l_w - l_{np}}{\bar{c}_w} C_{L_w,\alpha} - \frac{S_h(l_h - l_{np})}{S_w \bar{c}_w} C_{L_h,\alpha}(1 + \varepsilon_{u,\alpha}) = 0$$

Because the wing and the canard are geometrically similar, the lift slope is the same for both surfaces. Thus, solving for the wing position,

$$\frac{l_w - l_{np}}{\bar{c}_w} C_{L_w, \alpha} + \frac{S_h(l_h - l_w + l_w - l_{np})}{S_w \bar{c}_w} C_{L_h, \alpha}(1 + \varepsilon_{u, \alpha}) = 0$$

$$\frac{l_w - l_{np}}{\bar{c}_w} + \frac{S_h(l_w - l_{np})}{S_w \bar{c}_w}(1 + \varepsilon_{u, \alpha}) = \frac{S_h(l_w - l_h)}{S_w \bar{c}_w}(1 + \varepsilon_{u, \alpha})$$

$$\frac{l_w}{\bar{c}_w} = \frac{\dfrac{S_h(l_w - l_h)}{S_w \bar{c}_w}(1 + \varepsilon_{u, \alpha})}{1 + \dfrac{S_h}{S_w}(1 + \varepsilon_{u, \alpha})} + \frac{l_{np}}{\bar{c}_w}$$

$$= \frac{\dfrac{36(15)}{180(5.455)}(1 + 0.0609)}{1 + \dfrac{36}{180}(1 + 0.0609)} + 0.05 = \underline{0.5314}$$

Thus, the CG is nearly 35 inches forward of the wing quarter chord. The canard volume ratio is then

$$\frac{S_h(-l_h)}{S_w \bar{c}_w} = \frac{S_h}{S_w}\left(\frac{l_w - l_h}{\bar{c}_w} - \frac{l_w}{\bar{c}_w}\right) = \frac{36}{180}\left(\frac{15}{180/33} - 0.5314\right) = 0.4437$$

For this configuration, trim requires that the total lift must equal the weight and the total pitching moment about the CG must be zero,

$$C_{L_w} + \frac{S_h}{S_w} C_{L_h} = \frac{W}{\frac{1}{2}\rho V_\infty^2 S_w}$$

$$C_{m_w} + \frac{S_h \bar{c}_h}{S_w \bar{c}_w} C_{m_h} - \frac{l_w}{\bar{c}_w} C_{L_w} - \frac{S_h l_h}{S_w \bar{c}_w} C_{L_h} = 0$$

At the design airspeed of 120 mph, the airplane is trimmed with no control surface deflection at zero angle of attack. Since the wing and the canard are geometrically similar, they both have the same moment coefficient and lift slope, with no control surface deflection. Thus, with zero fuselage angle of attack and zero control surface deflection,

$$C_{L_w 0} + \frac{36}{180} C_{L_h 0} = \frac{2,700}{\frac{1}{2}0.0023769(120 \times 5,280/3,600)^2 180}$$

$$-0.053 - \frac{36(2.439)}{180(5.455)}0.053 - 0.5314 C_{L_w 0} + 0.4437 C_{L_h 0} = 0$$

or

$$
\begin{bmatrix} 1 & 0.2 \\ -0.53136 & 0.44373 \end{bmatrix} \begin{Bmatrix} C_{L_w0} \\ C_{L_h0} \end{Bmatrix} = \begin{Bmatrix} 0.40746 \\ 0.05774 \end{Bmatrix}
$$

The lift coefficients for the wing and the canard, with the fuselage reference line at zero angle of attack and no control surface deflection, are then

$$
\begin{Bmatrix} C_{L_w0} \\ C_{L_h0} \end{Bmatrix} = \begin{Bmatrix} 0.30773 \\ 0.49864 \end{Bmatrix}
$$

With this lift coefficient on the wing, the upwash angle on the canard is

$$
\varepsilon_{u0} = 0.083 \frac{C_{L_w0}}{R_{A_w}} = 0.083 \frac{0.30772}{6.05} = 0.00422 \text{ rad}
$$

For the range of linear lift, the lift coefficients are

$$
C_{L_w} = C_{L_w,\alpha} (\alpha + \alpha_{0w} - \alpha_{L0w})
$$

$$
C_{L_h} = C_{L_h,\alpha} [(1 + \varepsilon_{u,\alpha})\alpha + \alpha_{0h} - \alpha_{L0h} + \varepsilon_{u0} + \varepsilon_e \delta_e]
$$

For zero angle of attack and control surface deflection, this gives

$$
C_{L_w0} = C_{L_w,\alpha} (\alpha_{0w} - \alpha_{L0w}) = 0.30772
$$

$$
C_{L_h0} = C_{L_h,\alpha} (\alpha_{0h} - \alpha_{L0h} + \varepsilon_{u0}) = 0.49868
$$

The mounting angles for the wing and the canard are then found to be

$$
\alpha_{0w} = \frac{C_{L_w0}}{C_{L_w,\alpha}} + \alpha_{L0w} = \frac{0.30772}{4.44} + (-2.20)\frac{\pi}{180} = 0.0309 \text{ rad} = \underline{1.8°}
$$

$$
\alpha_{0h} = \frac{C_{L_h0}}{C_{L_h,\alpha}} + \alpha_{L0h} - \varepsilon_{u0} = \frac{0.49868}{4.44} + (-2.20)\frac{\pi}{180} - 0.00422
$$

$$
= 0.0697 \text{ rad} = \underline{3.99°}
$$

For the range of linear lift, the canard moment coefficient about the aerodynamic center is

$$
C_{m_h} = C_{m_h0} + C_{m_h,\delta_e} \delta_e
$$

where $C_{m_h 0}$ is the moment coefficient with no control surface deflection. The trim requirements for pitch are then written in terms of angle of attack and control surface deflection,

$$C_{L_w,\alpha}(\alpha + \alpha_{0w} - \alpha_{L0w}) + \frac{S_h}{S_w} C_{L_h,\alpha}[(1 + \varepsilon_{u,\alpha})\alpha + \alpha_{0h} - \alpha_{L0h} + \varepsilon_{u0} + \varepsilon_e \delta_e]$$

$$= \frac{W}{\frac{1}{2}\rho V_\infty^2 S_w}$$

$$C_{m_w} + \frac{S_h \bar{c}_h}{S_w \bar{c}_w}(C_{m_h 0} + C_{m_h,\delta_e}\delta_e) - \frac{l_w}{\bar{c}_w} C_{L_w,\alpha}(\alpha + \alpha_{0w} - \alpha_{L0w})$$

$$- \frac{S_h l_h}{S_w \bar{c}_w} C_{L_h,\alpha}[(1 + \varepsilon_{u,\alpha})\alpha + \alpha_{0h} - \alpha_{L0h} + \varepsilon_{u0} + \varepsilon_e \delta_e] = 0$$

These two equations can be written as

$$\begin{bmatrix} C_{L,\alpha} & C_{L,\delta_e} \\ C_{m,\alpha} & C_{m,\delta_e} \end{bmatrix} \begin{Bmatrix} \alpha \\ \delta_e \end{Bmatrix} = \begin{Bmatrix} C_L - C_{L0} \\ -C_{m0} \end{Bmatrix}$$

where

$$C_{L,\alpha} = C_{L_w,\alpha} + \frac{S_h}{S_w} C_{L_h,\alpha}(1 + \varepsilon_{u,\alpha}) = 5.38209$$

$$C_{L,\delta_e} = \frac{S_h}{S_w} C_{L_h,\alpha}\varepsilon_e = 0.53280$$

$$C_L = \frac{W}{\frac{1}{2}\rho V_\infty^2 S_w}$$

$$C_{L0} = C_{L_w,\alpha}(\alpha_{0w} - \alpha_{L0w}) + \frac{S_h}{S_w} C_{L_h,\alpha}(\alpha_{0h} - \alpha_{L0h} + \varepsilon_{u0}) = 0.40746$$

$$C_{m,\alpha} = -\frac{l_w}{\bar{c}_w} C_{L_w,\alpha} - \frac{S_h l_h}{S_w \bar{c}_w} C_{L_h,\alpha}(1 + \varepsilon_{u,\alpha}) = -0.26910$$

$$C_{m,\delta_e} = \frac{S_h \bar{c}_h}{S_w \bar{c}_w} C_{m_h,\delta_e} - \frac{S_h l_h}{S_w \bar{c}_w} C_{L_h,\alpha}\varepsilon_e = 1.13289$$

and

$$C_{m0} = C_{m_w} + \frac{S_h \bar{c}_h}{S_w \bar{c}_w} C_{m_h 0} - \frac{l_w}{\bar{c}_w} C_{L_w,\alpha}(\alpha_{0w} - \alpha_{L0w})$$

$$- \frac{S_h l_h}{S_w \bar{c}_w} C_{L_h,\alpha}(\alpha_{0h} - \alpha_{L0h} + \varepsilon_{u0}) = 0.0$$

After substituting the values for this particular airplane, we obtain a linear system of equations that can be solved for the angle of attack and elevator deflection, which are needed to support a given total lift coefficient:

$$\begin{bmatrix} 5.38209 & 0.53280 \\ -0.26910 & 1.13289 \end{bmatrix} \begin{Bmatrix} \alpha \\ \delta_e \end{Bmatrix} = \begin{Bmatrix} C_L - 0.40746 \\ 0 \end{Bmatrix}$$

At 80 mph, the total lift coefficient is

$$C_L = \frac{W}{\frac{1}{2}\rho V_\infty^2 S_w} = \frac{2,700}{\frac{1}{2}0.0023769(80\times5,280/3,600)^2 180}$$

$$= 0.916785$$

and

$$\begin{bmatrix} 5.38209 & 0.53280 \\ -0.26910 & 1.13289 \end{bmatrix} \begin{Bmatrix} \alpha \\ \delta_e \end{Bmatrix} = \begin{Bmatrix} 0.916785 - 0.40746 \\ 0 \end{Bmatrix}$$

$$\begin{Bmatrix} \alpha \\ \delta_e \end{Bmatrix} = \begin{Bmatrix} 0.09246 \\ 0.02196 \end{Bmatrix} = \begin{Bmatrix} 5.3° \\ 1.3° \end{Bmatrix}$$

With these values of angle of attack and control surface deflection, the lift coefficients for the wing and the canard are

$$C_{L_w} = C_{L_w,\alpha}(\alpha + \alpha_{0w} - \alpha_{L0w}) = \underline{0.72}$$

and

$$C_{L_h} = C_{L_h,\alpha}[(1 + \varepsilon_{u,\alpha})\alpha + \alpha_{0h} - \alpha_{L0h} + \varepsilon_{u0} + \varepsilon_e\delta_e] = \underline{0.99}$$

Repeating these calculations at different airspeeds, the data plotted in Fig. 4.6.4 are obtained. Notice that the lift coefficient for both the wing and the canard are always positive and the lift coefficient for the canard is always greater than that for the wing. These two characteristics are common to all canard configurations and it is these characteristics that provide both the advantage and the disadvantage of the canard design.

Figure 4.6.5 shows how the trim conditions change when the CG is moved forward to give a static margin of 10 percent. Notice that while this CG location provides greater stability, more control surface deflection is needed to trim the airplane. This increases the lift on the canard, and the ratio of the lift coefficient for the canard to that of the wing becomes even greater. Because moving the CG forward increases the lift coefficient for the canard at any given airspeed, it increases the stall speed for the airplane.

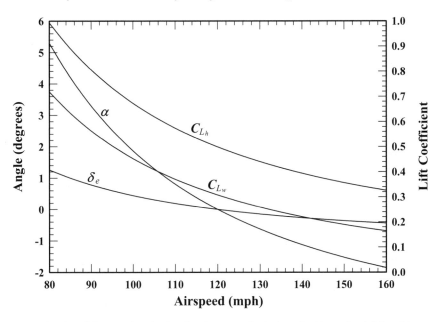

Figure 4.6.4.   Conditions at level trim for the canard airplane in Example 4.6.1, with a static margin of 5 percent.

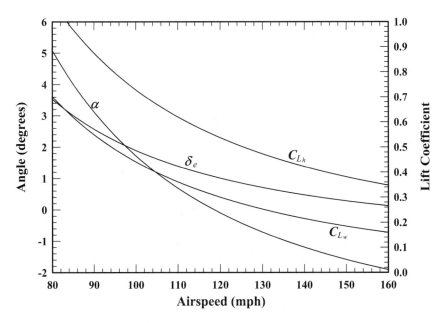

Figure 4.6.5.   Conditions at level trim for the canard airplane in Example 4.6.1, with a static margin of 10 percent.

EXAMPLE 4.6.2. Estimate the maximum total lift coefficient and the stall speed for the wing-canard combination described in Example 4.6.1. Assume that the CG is located to give a static margin of 5 percent. The maximum lift coefficient for the main wing is 1.40. Because of control surface deflection, the maximum lift coefficient for the canard is increased slightly to 1.50. However, this control surface deflection also increases the magnitude of the negative pitching moment coefficient for the canard to –0.077.

Solution. The maximum lift coefficient for the wing-canard combination is reached when either the wing or the canard stalls. Since the lift coefficient for the canard is higher than that for the wing, we will start by assuming that the canard stalls first. Thus, for this airplane and operating condition, we have

$$S_w = 180 \text{ ft}^2, \quad \bar{c}_w = 5.455 \text{ ft}, \quad l_w = 2.899 \text{ ft}, \quad C_{L_w \max} = 1.40, \quad C_{m_w} = -0.053,$$

$$S_h = 36 \text{ ft}^2, \quad \bar{c}_h = 2.439 \text{ ft}, \quad l_h = -12.101 \text{ ft}, \quad C_{L_h \max} = 1.50, \quad C_{m_h} = -0.077$$

At trim, the total moment about the CG must be zero,

$$C_{m_w} + \frac{S_h \bar{c}_h}{S_w \bar{c}_w} C_{m_h} - \frac{l_w}{\bar{c}_w} C_{L_w} - \frac{S_h l_h}{S_w \bar{c}_w} C_{L_h} = 0$$

When the canard is at maximum lift coefficient, this gives

$$
\begin{aligned}
C_{L_w} &= \frac{\bar{c}_w}{l_w} C_{m_w} + \frac{S_h \bar{c}_h}{S_w l_w} C_{m_h} - \frac{S_h l_h}{S_w l_w} C_{L_h \max} \\
&= \frac{5.455}{2.899}(-0.053) + \frac{36(2.439)}{180(2.899)}(-0.077) - \frac{36(-12.101)}{180(2.899)} 1.50 = 1.14
\end{aligned}
$$

This is less than the maximum lift coefficient for the wing, so our assumption that the canard stalls first was correct. At trim the total lift must also equal the weight,

$$C_L = C_{L_w} + \frac{S_h}{S_w} C_{L_h} = \frac{W}{\frac{1}{2}\rho V_\infty^2 S_w}$$

When the canard is at the maximum lift coefficient, this gives

$$C_{L \max} = C_{L_w} + \frac{S_h}{S_w} C_{L_h \max} = 1.14 + \frac{36}{180}1.50 = \underline{1.44}$$

The stall speed is then

$$V_{\text{stall}} = \sqrt{\frac{W}{\frac{1}{2}\rho S_w C_{L \max}}} = \sqrt{\frac{2,700}{\frac{1}{2}0.0023769(180)1.44}} = 93.6 \text{ ft/sec} = \underline{63.8 \text{ mph}}$$

EXAMPLE 4.6.3. The main wing of the wing-canard combination described in Example 4.6.2 is fitted with high lift flaps. When these flaps are deflected by 10 degrees, the maximum lift coefficient for the main wing is increased to 1.7 and the negative pitching moment coefficient for the wing is increased to –0.149. Determine the maximum total lift coefficient and the stall speed for the wing-canard combination with these flaps so engaged.

Solution. Again, we will start by assuming that the canard stalls first. Thus, for this airplane and operating condition, we have

$$S_w = 180 \text{ ft}^2, \quad \bar{c}_w = 5.455 \text{ ft}, \quad l_w = 2.899 \text{ ft}, \quad C_{L_w \max} = 1.70, \quad C_{m_w} = -0.149,$$

$$S_h = 36 \text{ ft}^2, \quad \bar{c}_h = 2.439 \text{ ft}, \quad l_h = -12.101 \text{ ft}, \quad C_{L_h \max} = 1.50, \quad C_{m_h} = -0.077$$

When the canard is at maximum lift coefficient, trim requires that

$$
\begin{aligned}
C_{L_w} &= \frac{\bar{c}_w}{l_w}C_{m_w} + \frac{S_h\bar{c}_h}{S_w l_w}C_{m_h} - \frac{S_h l_h}{S_w l_w}C_{L_h \max} \\
&= \frac{5.455}{2.899}(-0.149) + \frac{36(2.439)}{180(2.899)}(-0.077) - \frac{36(-12.101)}{180(2.899)}1.50 = 0.96
\end{aligned}
$$

This is again less than the maximum lift coefficient for the wing, which means that our original assumption was correct. With the canard at maximum lift coefficient, the second trim requirement gives

$$C_{L \max} = C_{L_w} + \frac{S_h}{S_w}C_{L_h \max} = 0.96 + \frac{36}{180}1.50 = \underline{1.26}$$

and the stall speed is

$$V_{stall} = \sqrt{\frac{W}{\frac{1}{2}\rho S_w C_{L \max}}} = \sqrt{\frac{2,700}{\frac{1}{2}0.0023769(180)1.26}} = 100.1 \text{ ft/sec} = \underline{68.2 \text{ mph}}$$

Thus, we see that deploying flaps on the main wing will actually reduce the maximum lift coefficient and increase the stall speed for this wing-canard combination.

In summary, the canard configuration provides an alternative to the aft tail for maintaining trim and static pitch stability in an airplane. To maintain both trim and static stability with the canard configuration, the center of gravity must be located between the wing and the canard, and the lift coefficient for the canard must be greater than that for the main wing. A canard always has a destabilizing effect on an airplane and it is only the main wing that provides pitch stability for the wing-canard design. One advantage of the canard over the aft tail is that both the wing and the canard will always carry positive lift. In addition, the stall characteristics for the wing-canard configuration are usually quite forgiving. This is because the canard typically stalls before the wing, making it almost impossible to stall the main wing in a gradual maneuver. The major disadvantage of the canard design is the difficulty in achieving the high overall lift coefficient needed for short takeoff and landing applications.

4.7. Effects of Drag and Vertical Offset

In the pitch stability analyses that we have considered to this point, we have assumed that the aerodynamic center of both the wing and the tail or canard, as well as the center of thrust and the center of gravity, all fall on the fuselage reference line. Additionally, the fuselage reference line was assumed to be parallel with the freestream velocity at trim. These simplifying assumptions allowed us to examine the major contributions to aircraft pitch stability, without unduly complicating the formulation. However, in reality, these conditions are almost never exactly satisfied. In this section we examine some of the secondary effects that result from relaxing these assumptions.

In general, the wing and the horizontal control surface, as well as the center of thrust, can each be offset from the center of gravity in both the horizontal and vertical directions. A schematic of the forces and moments acting on such a system is shown in Fig. 4.7.1. This figure allows for downwash or upwash (negative downwash) on the main wing as well as the horizontal control surface. In the more approximate analyses presented in Secs. 4.3 and 4.6, downwash or upwash was included only for the horizontal control

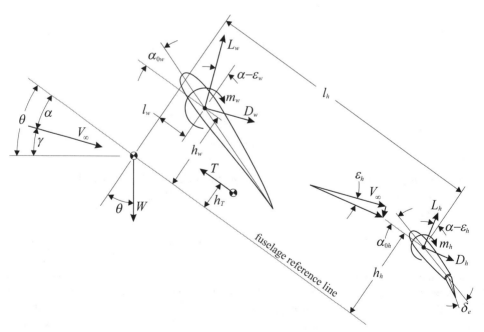

Figure 4.7.1. Forces and moments acting on a wing combined with a horizontal control surface.

surface. In Fig. 4.3.3 the downwash on the aft tail was labeled ε_d, whereas in Fig. 4.6.3 the upwash on the forward canard was labeled ε_u. To make the results presented in this section more general, the downwash or upwash on any horizontal control surface is simply labeled ε_h and downwash is assumed to be positive. Similarly, Fig. 4.7.1 allows for the possibility of downwash or upwash on the main wing, which is denoted ε_w. If we choose to neglect downwash or upwash on the main wing, ε_w can be set to zero.

In Fig. 4.7.1, the horizontal control surface was drawn in the configuration of an aft tail. However, the results of the following analysis can be applied to either an aft tail or a forward canard. The canard is distinguished from the aft tail only by its positions relative to the wing and the signs of ε_w and ε_h. For an aft tail, l_h is greater than l_w, whereas for the canard, l_h is less than l_w. A lifting wing induces downwash on an aft tail, and when the tail is producing positive lift, it induces upwash on the wing. Thus, for the aft tail configuration, ε_h has the same sign as the lift on the wing and ε_w has the opposite sign from that of the lift on the tail. Conversely, for the wing-canard configuration, ε_w has the same sign as the lift on the canard and ε_h takes the sign opposite to that for the lift on the wing. As has been our usual practice, the variable l is used to denote a distance aft of the CG in a direction parallel with the fuselage reference line. Similarly, the variable h is used to indicate a distance measured above the CG in a direction normal to the fuselage reference line. If this sign convention is adhered to rigorously, we need not make any distinction between the wing-tail and wing-canard configurations. The numerical values used in the formulation will take care of the differences.

In the previous stability analyses, we assumed steady flight, with the thrust vector and the fuselage reference line parallel with the direction of motion. For the present analysis, we consider the more general case where the airplane is in a steady climb at some climb angle, γ, and has an angle of attack, α, relative to the fuselage reference line. For this case, the fuselage reference line makes an angle, θ, with the horizontal, where $\theta = \gamma + \alpha$. The angle, θ, is commonly referred to as the *elevation angle*. **Since the orientation of the fuselage reference line is arbitrary, for the analysis presented in this section it is chosen to be parallel with the thrust vector.**

From Fig. 4.7.1, an axial force balance along the fuselage reference line requires

$$
\begin{aligned}
A = \; & D_w \cos(\alpha - \varepsilon_w) - L_w \sin(\alpha - \varepsilon_w) \\
& + D_h \cos(\alpha - \varepsilon_h) - L_h \sin(\alpha - \varepsilon_h) = T - W \sin\theta
\end{aligned}
\tag{4.7.1}
$$

Similarly, a balance of forces in the direction normal to the fuselage reference line gives

$$
\begin{aligned}
N = \; & L_w \cos(\alpha - \varepsilon_w) + D_w \sin(\alpha - \varepsilon_w) \\
& + L_h \cos(\alpha - \varepsilon_h) + D_h \sin(\alpha - \varepsilon_h) = W \cos\theta
\end{aligned}
\tag{4.7.2}
$$

A net moment balance in pitch about the center of gravity requires that the sum of the aerodynamic moment, m, and the thrust moment, $-h_T T$, is zero. Thus, we have

$$
\begin{aligned}
m = \; & m_w - (l_w L_w - h_w D_w)\cos(\alpha - \varepsilon_w) - (h_w L_w + l_w D_w)\sin(\alpha - \varepsilon_w) \\
& + m_h - (l_h L_h - h_h D_h)\cos(\alpha - \varepsilon_h) - (h_h L_h + l_h D_h)\sin(\alpha - \varepsilon_h) = h_T T
\end{aligned}
\tag{4.7.3}
$$

Notice that when the thrust vector does not pass through the center of gravity, the aerodynamic moment, m, is not zero at trim but must balance the moment produced by the thrust of the engine. Because the thrust at trim must balance the drag, the thrust required to trim such an airplane is a nonlinear function of angle of attack, and significant nonlinearity is introduced into the trim equations. Equation (4.7.1) specifies the thrust required for steady flight, while Eqs. (4.7.2) and (4.7.3) provide the trim requirements.

Nondimensionalizing Eqs. (4.7.1) through (4.7.3) in the usual manner, we have

$$
\begin{aligned}
C_A &= C_{D_w}\cos(\alpha-\varepsilon_w)-C_{L_w}\sin(\alpha-\varepsilon_w) \\
&+\frac{S_h}{S_w}\eta_h\left[C_{D_h}\cos(\alpha-\varepsilon_h)-C_{L_h}\sin(\alpha-\varepsilon_h)\right] = \frac{T}{\frac{1}{2}\rho V_\infty^2 S_w}-\frac{W}{\frac{1}{2}\rho V_\infty^2 S_w}\sin\theta \quad (4.7.4)
\end{aligned}
$$

$$
\begin{aligned}
C_N &= C_{L_w}\cos(\alpha-\varepsilon_w)+C_{D_w}\sin(\alpha-\varepsilon_w) \\
&+\frac{S_h}{S_w}\eta_h\left[C_{L_h}\cos(\alpha-\varepsilon_h)+C_{D_h}\sin(\alpha-\varepsilon_h)\right] = \frac{W}{\frac{1}{2}\rho V_\infty^2 S_w}\cos\theta \quad (4.7.5)
\end{aligned}
$$

$$
\begin{aligned}
C_m &= C_{m_w}-\frac{l_w C_{L_w}-h_w C_{D_w}}{\overline{c}_w}\cos(\alpha-\varepsilon_w)-\frac{h_w C_{L_w}+l_w C_{D_w}}{\overline{c}_w}\sin(\alpha-\varepsilon_w) \\
&+\frac{S_h}{S_w}\eta_h\left[\frac{\overline{c}_h}{\overline{c}_w}C_{m_h}-\frac{l_h C_{L_h}-h_h C_{D_h}}{\overline{c}_w}\cos(\alpha-\varepsilon_h)-\frac{h_h C_{L_h}+l_h C_{D_h}}{\overline{c}_w}\sin(\alpha-\varepsilon_h)\right] \quad (4.7.6) \\
&= \frac{h_T}{\overline{c}_w}\frac{T}{\frac{1}{2}\rho V_\infty^2 S_w}
\end{aligned}
$$

where C_A, C_N, and C_m are, respectively, the total axial force, normal force, and pitching moment coefficients based on the planform area and mean chord length of the main wing. In the absence of stall, the lift coefficient for the main wing can be expressed as a linear function of angle of attack and the lift coefficients for the horizontal control surface can be written as a linear function of both angle of attack and elevator deflection,

$$
C_{L_w} = C_{L_w,\alpha}(\alpha+\alpha_{0w}-\alpha_{L0w}-\varepsilon_w) \quad (4.7.7)
$$

$$
C_{L_h} = C_{L_h,\alpha}(\alpha+\alpha_{0h}-\alpha_{L0h}-\varepsilon_h+\varepsilon_e\delta_e) \quad (4.7.8)
$$

The drag coefficients for the main wing and horizontal control surface can be written as parabolic functions of their respective lift coefficients,

$$
C_{D_w} = C_{D0w}+C_{D0,L_w}C_{L_w}+\frac{C_{L_w}^2}{\pi e_w R_{A_w}} \quad (4.7.9)
$$

$$
C_{D_h} = C_{D0h}+C_{D0,L_h}C_{L_h}+\frac{C_{L_h}^2}{\pi e_h R_{A_h}} \quad (4.7.10)
$$

The moment coefficient for the main wing, about its aerodynamic center, does not vary with small changes in angle of attack and that for the horizontal control surface can be approximated as a simple linear function of elevator deflection. In the range of linear lift, we can write

$$C_{m_h} = C_{m_h 0} + C_{m_h,\delta_e} \delta_e \tag{4.7.11}$$

Equation (4.7.4) and Eqs. (4.7.7) through (4.7.11) can be used in the trim relations given by Eqs. (4.7.5) and (4.7.6) to evaluate the angle of attack and elevator deflection required for trim. If the geometry of the airplane is completely known, the vortex model described in Sec. 4.5 could be used to approximate ε_w and ε_h as functions of the lift coefficients for the wing and horizontal control surface. Thus, Eqs. (4.7.5) and (4.7.6) provide two equations containing only two unknowns, the angle of attack, α, and the elevator deflection, δ_e. However, these equations are nonlinear, so in general, they must be solved numerically. From such a solution, we can determine the angle of attack and elevator deflection required for trim.

Notice from Eqs. (4.7.5) and (4.7.6) that even for the special case when h_w, h_h, and h_T are all zero, the equations that specify the trim requirements are still nonlinear. The simplified linear analyses presented in Secs. 4.3 and 4.6 were based on the assumption that the fuselage reference line was parallel with the direction of flight. Because the aircraft angle of attack depends on airspeed, this condition can be exactly satisfied only at one particular airspeed. However, since the angle of attack is small during normal flight operations, this condition is nearly satisfied at all airspeeds and the approximate linear analysis gives good results when the wing, the horizontal surface, and the thrust vector are all in close vertical alignment with the center of gravity. This is demonstrated in the following example.

EXAMPLE 4.7.1. Consider the general aviation airplane with an aft tail that was described in Example 4.3.1. Neglecting the upwash on the main wing and any moments associated with the thrust deck and the fuselage, but accounting for the nonlinear trim requirements, plot the elevator deflection and the angle of attack for trimmed level flight as a function of airspeed from 80 to 160 mph. Also plot the lift on the horizontal tail at level trim for the same range of airspeed. Compare the results with those obtained in Example 4.3.1. For the wing and horizontal tail of this airplane use the following properties:

$$S_w = 180 \text{ ft}^2, \quad \overline{c}_w = 5.4545 \text{ ft}, \quad R_{A_w} = 6.05, \quad e_w = 0.95, \quad l_w = -0.7094 \text{ ft},$$
$$h_w = 0.0 \text{ ft}, \quad W = 2{,}700 \text{ lbf}, \quad C_{L_w,\alpha} = 4.44, \quad \alpha_{0w} = 0.05338, \quad \varepsilon_w \cong 0.00,$$
$$\alpha_{L0w} = -0.03840, \quad C_{m_w} = -0.053, \quad C_{D_{0w}} = 0.01, \quad C_{D_0,L_w} = 0.00,$$
$$S_h = 36 \text{ ft}^2, \quad \overline{c}_h = 3.0 \text{ ft}, \quad R_{A_h} = 4.0, \quad e_h = 0.97, \quad l_h = 14.2906 \text{ ft},$$
$$h_h = 0.0 \text{ ft}, \quad \eta_h = 1.0, \quad C_{L_h,\alpha} = 3.97, \quad \alpha_{0h} = 0.04041,$$
$$\alpha_{L0h} = 0.0, \quad \varepsilon_e = 0.60, \quad C_{m_h 0} = 0.0, \quad C_{m_h,\delta_e} = -0.55, \quad C_{D_{0h}} = 0.015,$$
$$C_{D_0,L_h} = 0.00, \quad \varepsilon_{h0} = 0.04041, \quad \varepsilon_{h,\alpha} = 0.44033, \quad h_T = 0.0 \text{ ft}$$

Solution. For level flight, $\gamma = 0.0$ and thus $\theta = \alpha$. A free-body diagram of this wing-tail combination is shown in Fig. 4.7.2. A balance of forces normal to the fuselage reference line requires that

$$
C_{L_w} \cos\alpha + C_{D_w} \sin\alpha
$$
$$
+ \frac{S_h}{S_w}\left[C_{L_h} \cos(\alpha - \varepsilon_h) + C_{D_h} \sin(\alpha - \varepsilon_h)\right] = \frac{W}{\frac{1}{2}\rho V_\infty^2 S_w} \cos\alpha
$$

and a moment balance in pitch about the center of gravity gives

$$
C_{m_w} - \frac{l_w}{\bar{c}_w}\left(C_{L_w} \cos\alpha + C_{D_w} \sin\alpha\right)
$$
$$
+ \frac{S_h \bar{c}_h}{S_w \bar{c}_w} C_{m_h} - \frac{S_h l_h}{S_w \bar{c}_w}\left[C_{L_h} \cos(\alpha - \varepsilon_h) + C_{D_h} \sin(\alpha - \varepsilon_h)\right] = 0
$$

where

$$
\varepsilon_h = \varepsilon_{h0} + \varepsilon_{h,\alpha}\, \alpha
$$

$$
C_{L_w} = C_{L_w,\alpha}\,(\alpha + \alpha_{0w} - \alpha_{L0w})
$$

$$
C_{L_h} = C_{L_h,\alpha}\,(\alpha + \alpha_{0h} - \alpha_{L0h} - \varepsilon_h + \varepsilon_e \delta_e)
$$

$$
C_{D_w} = C_{D0w} + C_{D0,L_w} C_{L_w} + C_{L_w}^2\big/(\pi e_w R_{A_w})
$$

$$
C_{D_h} = C_{D0h} + C_{D0,L_h} C_{L_h} + C_{L_h}^2\big/(\pi e_h R_{A_h})
$$

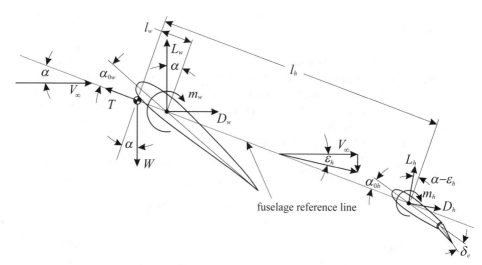

Figure 4.7.2. Free-body diagram of the wing-tail combination in Example 4.7.1.

$$C_{m_h} = C_{m_h 0} + C_{m_h, \delta_e} \delta_e$$

Because the trim conditions are nonlinear in both α and δ_e, we use an iterative procedure to find the angle of attack and elevator deflection that will satisfy both conditions simultaneously. This nonlinear system is easily solved with Newton's method. The solution can be obtained quickly with a digital computer or a programmable calculator. Here we shall outline the procedure and give some example computations. The reader is cautioned that since Newton's method will involve numerical differentiation, round-off error will be amplified by several orders of magnitude and could result in numerical instability. Thus, high precision is required for the intermediate computations.

To facilitate finding the solution, it is convenient to write the trim conditions in the form

$$C_{L_w} \cos\alpha + C_{D_w} \sin\alpha + \frac{S_h}{S_w} \left[C_{L_h} \cos(\alpha - \varepsilon_h) + C_{D_h} \sin(\alpha - \varepsilon_h) \right]$$

$$- \frac{W}{\frac{1}{2}\rho V_\infty^2 S_w} \cos\alpha \equiv R_L(\alpha, \delta_e)$$

$$C_{m_w} - \frac{l_w}{\bar{c}_w} \left(C_{L_w} \cos\alpha + C_{D_w} \sin\alpha \right) + \frac{S_h \bar{c}_h}{S_w \bar{c}_w} C_{m_h}$$

$$- \frac{S_h l_h}{S_w \bar{c}_w} \left[C_{L_h} \cos(\alpha - \varepsilon_h) + C_{D_h} \sin(\alpha - \varepsilon_h) \right] \equiv R_m(\alpha, \delta_e)$$

where R_L and R_m are called *residuals*. We wish to find the values of α and δ_e that will make R_L and R_m simultaneously go to zero. For any other values of α and/or δ_e one or both of the residuals will be nonzero.

To demonstrate how the trim computations are carried out, consider the case of level flight at 80 mph. For this airspeed, the weight coefficient is

$$\frac{W}{\frac{1}{2}\rho V_\infty^2 S_w} = \frac{2,700}{\frac{1}{2}0.0023769(80 \times 5,280/3,600)^2 180} = 0.916785$$

We start with initial estimates for α and δ_e, say, $\alpha = 0$ and $\delta_e = 0$. This gives

$$\varepsilon_h = 0.04041 + 0.44033(0.0) = 0.040410$$

$$C_{L_w} = 4.44(0.0 + 0.05338 + 0.03840) = 0.407503$$

$$C_{L_h} = 3.97[0.0 + 0.04041 - 0.0 - 0.04041 + 0.60(0.0)] = 0.000000$$

$$C_{D_w} = 0.01 + \frac{(0.4075032)^2}{\pi(0.95)(6.05)} = 0.019197$$

$$C_{D_h} = 0.015 + \frac{(0.0)^2}{\pi(0.97)(4.0)} = 0.015000$$

$$C_{m_h} = 0.0 - 0.55(0.0) = 0.000000$$

Using these values, the residuals are found to be

$$R_L(0.0,0.0) = -0.509403$$

$$R_m(0.0,0.0) = 0.000316$$

To make an intelligent refinement to our initial guess, we must estimate the change in both residuals, R_L and R_m, with respect to both of the independent variables, α and δ_e. These four derivatives can be estimated numerically as

$$R_{L,\alpha}(\alpha,\delta_e) \equiv \frac{\partial R_L}{\partial \alpha}(\alpha,\delta_e) \cong \frac{R_L(\alpha+0.001,\delta_e)-R_L(\alpha-0.001,\delta_e)}{0.002}$$

$$R_{L,\delta_e}(\alpha,\delta_e) \equiv \frac{\partial R_L}{\partial \delta_e}(\alpha,\delta_e) \cong \frac{R_L(\alpha,\delta_e+0.001)-R_L(\alpha,\delta_e-0.001)}{0.002}$$

$$R_{m,\alpha}(\alpha,\delta_e) \equiv \frac{\partial R_m}{\partial \alpha}(\alpha,\delta_e) \cong \frac{R_m(\alpha+0.001,\delta_e)-R_m(\alpha-0.001,\delta_e)}{0.002}$$

$$R_{m,\delta_e}(\alpha,\delta_e) \equiv \frac{\partial R_m}{\partial \delta_e}(\alpha,\delta_e) \cong \frac{R_m(\alpha,\delta_e+0.001)-Rm(\alpha,\delta_e-0.001)}{0.002}$$

Recalculating the residuals for $\alpha = 0.001$ and $\delta_e = 0$ gives

$$R_L(0.001,0.0) = -0.504498$$

$$R_m(0.001,0.0) = -0.000271$$

and for $\alpha = -0.001$ and $\delta_e = 0$, we have

$$R_L(-0.001,0.0) = -0.514307$$

$$R_m(-0.001,0.0) = 0.000904$$

Thus

$$R_{L,\alpha}(0.0,0.0) \cong [-0.504498-(-0.514307)]/0.002 = 4.904888$$

$$R_{m,\alpha}(0.0,0.0) \cong [-0.000271-0.000904]/0.002 = -0.587747$$

In a similar manner, we obtain

$$R_{L,\delta_e}(0.0,0.0) \cong 0.476011$$

$$R_{m,\delta_e}(0.0,0.0) \cong -1.307633 \,^*$$

If both R_L and R_m were linear functions of α and δ_e, then the change in angle of attack, $\Delta\alpha$, and the change in elevator deflection, $\Delta\delta_e$, required to make both residuals zero would be found from

$$\begin{bmatrix} R_{L,\alpha} & R_{L,\delta_e} \\ R_{m,\alpha} & R_{m,\delta_e} \end{bmatrix} \begin{Bmatrix} \Delta\alpha \\ \Delta\delta_e \end{Bmatrix} = \begin{Bmatrix} -R_L \\ -R_m \end{Bmatrix}$$

Even though the residuals are not linear functions of α and δ_e, an improved estimate for the solution can be found by computing these changes and updating our estimates according to the relation

$$\begin{Bmatrix} \alpha \\ \delta_e \end{Bmatrix}_{new} = \begin{Bmatrix} \alpha \\ \delta_e \end{Bmatrix}_{old} + \begin{Bmatrix} \Delta\alpha \\ \Delta\delta_e \end{Bmatrix}$$

For the present case this gives

$$\begin{bmatrix} 4.904888 & 0.476011 \\ -0.587747 & -1.307633 \end{bmatrix} \begin{Bmatrix} \Delta\alpha \\ \Delta\delta_e \end{Bmatrix} = \begin{Bmatrix} 0.509403 \\ -0.000316 \end{Bmatrix}$$

$$\begin{Bmatrix} \Delta\alpha \\ \Delta\delta_e \end{Bmatrix} = \begin{Bmatrix} 0.108569 \\ -0.048557 \end{Bmatrix}$$

$$\begin{Bmatrix} \alpha \\ \delta_e \end{Bmatrix} = \begin{Bmatrix} 0.000000 \\ 0.000000 \end{Bmatrix} + \begin{Bmatrix} 0.108569 \\ -0.048557 \end{Bmatrix} = \begin{Bmatrix} 0.108569 \\ -0.048557 \end{Bmatrix}$$

With these improved estimates for the angle of attack and elevator deflection, the residuals can be recomputed. This gives

$$\begin{Bmatrix} R_L \\ R_m \end{Bmatrix} = \begin{Bmatrix} 0.003929 \\ -0.000248 \end{Bmatrix}$$

---

[*] The computations used to obtain the numbers in this example were carried out with much greater precision than the reported six-place accuracy. By comparing these reported results with those obtained using six-place computation, it can be seen that three significant digits of accuracy are lost in this computation. This is one example of why intermediate computations should always be carried out with the full precision of the tools at hand. Only in reporting the final result should the output be rounded back to the precision of the input. If computations in this example were carried out to only two-place accuracy, these computations would produce only random numbers.

With the improved estimates for α and δ_e, the residuals are now smaller but not zero. Repeating the process to refine the estimates a second time, we obtain

$$\begin{Bmatrix} \alpha \\ \delta_e \end{Bmatrix} = \begin{Bmatrix} 0.107760 \\ -0.048379 \end{Bmatrix} = \begin{Bmatrix} 6.2° \\ -2.8° \end{Bmatrix}, \quad \begin{Bmatrix} R_L \\ R_m \end{Bmatrix} = \begin{Bmatrix} 0.000000 \\ 0.000000 \end{Bmatrix}$$

The solution found in Example 4.3.1 using the simplified linear analysis was

$$\begin{Bmatrix} \alpha \\ \delta_e \end{Bmatrix} = \begin{Bmatrix} 0.10904 \\ -0.04889 \end{Bmatrix} = \begin{Bmatrix} 6.2° \\ -2.8° \end{Bmatrix}$$

Using the computed values for the angle of attack and elevator deflection at trim, the lift on the horizontal tail of this airplane in trimmed flight at 80 mph is found to be

$$L_h = \tfrac{1}{2}\rho V_\infty^2 S_h C_{L_h,\alpha}\left[(1-\varepsilon_{h,\alpha})\alpha + \alpha_{0h} - \alpha_{L0h} - \varepsilon_{h0} + \varepsilon_e\delta_e\right] = \underline{73\text{ lbf}}$$

This compares to a value of 74 lbf, which was computed in Example 4.3.1 using the simplified linear analysis.

Repeating these calculations at different airspeeds, the data plotted in Fig. 4.7.3 are obtained. In this figure, the dashed lines are the results computed in Example 4.3.1 using the simplified linear analysis. The solid lines are the results computed in the present example, using the full nonlinear trim conditions. The difference is insignificant.

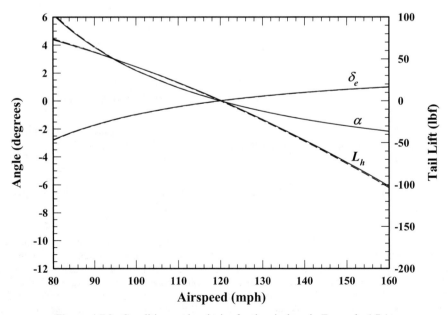

Figure 4.7.3. Conditions at level trim for the airplane in Example 4.7.1.

For the airplane in Example 4.7.1 with h_w, h_h, and h_T all zero, the simplified linear analysis agrees with the more exact solution to within about 1 percent. This is well within the accuracy to which we could expect to estimate the aerodynamic parameters from which this solution was obtained. In general, the approximate linear analysis gives good agreement with the more exact analysis when the wing, horizontal control surface, and thrust vector are all in close vertical alignment with the center of gravity. When these conditions are not met, the nonlinear effects become more significant. This is demonstrated in the following example.

EXAMPLE 4.7.2. Repeat Example 4.7.1 for the case where the thrust vector is 5 feet above the center of gravity (h_T=5.0 feet). Assume that all other properties of the airplane remain unchanged.

Solution. A free-body diagram of this wing-tail combination is shown in Fig. 4.7.4. The force balance for this example is exactly the same as that for Example 4.7.1. The only difference is that the aerodynamic pitching moment is no longer zero at trim but must balance the moment produced by the thrust of the engine acting through the vertical moment arm h_T. In dimensionless form, the trim requirement for pitch now becomes

$$C_{m_w} - \frac{l_w}{\bar{c}_w}\left(C_{L_w}\cos\alpha + C_{D_w}\sin\alpha\right) + \frac{S_h\bar{c}_h}{S_w\bar{c}_w}C_{m_h}$$
$$-\frac{S_h l_h}{S_w\bar{c}_w}\left[C_{L_h}\cos(\alpha-\varepsilon_h)+C_{D_h}\sin(\alpha-\varepsilon_h)\right] = \frac{h_T}{\bar{c}_w}\frac{T}{\frac{1}{2}\rho V_\infty^2 S_w}$$

For this geometry and level flight, the trim residual equations are

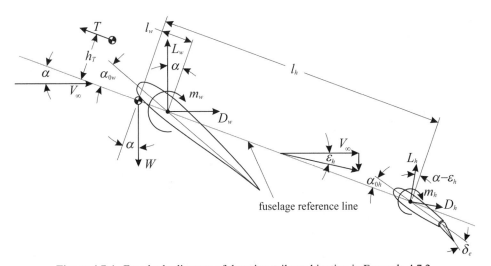

Figure 4.7.4. Free-body diagram of the wing-tail combination in Example 4.7.2.

$$C_{L_w} \cos\alpha + C_{D_w} \sin\alpha + \frac{S_h}{S_w}\left[C_{L_h}\cos(\alpha-\varepsilon_h)+C_{D_h}\sin(\alpha-\varepsilon_h)\right]$$

$$-\frac{W}{\frac{1}{2}\rho V_\infty^2 S_w}\cos\alpha \equiv R_L(\alpha,\delta_e)$$

$$C_{m_w} - \frac{l_w}{\bar{c}_w}\left(C_{L_w}\cos\alpha + C_{D_w}\sin\alpha\right)+\frac{S_h\bar{c}_h}{S_w\bar{c}_w}C_{m_h}$$

$$-\frac{S_h l_h}{S_w\bar{c}_w}\left[C_{L_h}\cos(\alpha-\varepsilon_h)+C_{D_h}\sin(\alpha-\varepsilon_h)\right]-\frac{h_T}{\bar{c}_w}\frac{T}{\frac{1}{2}\rho V_\infty^2 S_w} \equiv R_m(\alpha,\delta_e)$$

where

$$\varepsilon_h = \varepsilon_{h0} + \varepsilon_{h,\alpha}\alpha$$

$$C_{L_w} = C_{L_w,\alpha}(\alpha+\alpha_{0w}-\alpha_{L0w})$$

$$C_{L_h} = C_{L_h,\alpha}(\alpha+\alpha_{0h}-\alpha_{L0h}-\varepsilon_h+\varepsilon_e\delta_e)$$

$$C_{D_w} = C_{D_{0w}}+C_{D_0,L_w}C_{L_w}+\frac{C_{L_w}^2}{\pi e_w R_{A_w}}$$

$$C_{D_h} = C_{D_{0h}}+C_{D_0,L_h}C_{L_h}+\frac{C_{L_h}^2}{\pi e_h R_{A_h}}$$

$$C_{m_h} = C_{m_h 0}+C_{m_h,\delta_e}\delta_e$$

and

$$\frac{T}{\frac{1}{2}\rho V_\infty^2 S_w} = \frac{W}{\frac{1}{2}\rho V_\infty^2 S_w}\sin\alpha + C_{D_w}\cos\alpha - C_{L_w}\sin\alpha$$

$$+\frac{S_h}{S_w}\left[C_{D_h}\cos(\alpha-\varepsilon_h)-C_{L_h}\sin(\alpha-\varepsilon_h)\right]$$

Following the procedure used in Example 4.7.1 with an initial estimate of $\alpha = 0$ and $\delta_e = 0$, for an airspeed of 80 mph we have

$$\begin{Bmatrix}\alpha\\\delta_e\end{Bmatrix} = \begin{Bmatrix}0.000000\\0.000000\end{Bmatrix}, \quad \begin{Bmatrix}R_L\\R_m\end{Bmatrix} = \begin{Bmatrix}-0.509403\\-0.020028\end{Bmatrix}$$

After one refinement we obtain

$$\begin{Bmatrix}\alpha\\\delta_e\end{Bmatrix} = \begin{Bmatrix}0.115980\\-0.124927\end{Bmatrix}, \quad \begin{Bmatrix}R_L\\R_m\end{Bmatrix} = \begin{Bmatrix}0.004343\\0.040916\end{Bmatrix}$$

Two more iterations produce

$$\begin{Bmatrix} \alpha \\ \delta_e \end{Bmatrix} = \begin{Bmatrix} 0.111922 \\ -0.091746 \end{Bmatrix} = \begin{Bmatrix} 6.4° \\ -5.3° \end{Bmatrix}, \quad \begin{Bmatrix} R_L \\ R_m \end{Bmatrix} = \begin{Bmatrix} 0.000000 \\ 0.000000 \end{Bmatrix}$$

Thus, the lift on the horizontal tail at trim is

$$L_h = \tfrac{1}{2}\rho V_\infty^2 S_h C_{L_h,\alpha} [(1-\varepsilon_{h,\alpha})\alpha + \alpha_{0h} - \alpha_{L0h} - \varepsilon_{h0} + \varepsilon_e \delta_e] = \underline{18\ \text{lbf}}$$

which differs greatly from the 74 lbf predicted from the simplified linear analysis. Repeating the calculations at different airspeeds, the data plotted in Fig. 4.7.5 are obtained. Again, the dashed lines are from the simplified linear analysis and the solid lines come from the full nonlinear trim conditions.

In general, drag and other nonlinear effects can have a significant effect on aircraft trim whenever the wing, horizontal control surface, and thrust vector are not vertically aligned with the center of gravity. These nonlinearities can have a significant effect on aircraft stability as well. For the more general airplane configuration being considered in this section, the pitch stability derivative is evaluated by differentiating Eq. (4.7.6) with respect to α. Because C_{m_w} and C_{m_h} are moment coefficients about the aerodynamic centers of the wing and horizontal control surface, these terms in Eq. (4.7.6) do not change with angle of attack. From Eqs. (4.7.9) and (4.7.10), the changes in the drag coefficients with respect to angle of attack are

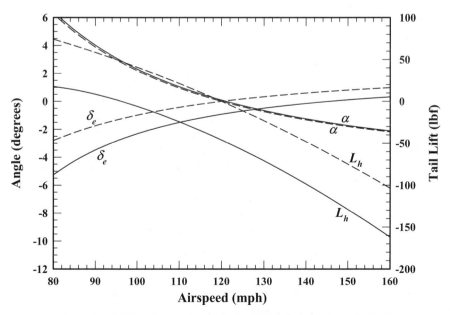

Figure 4.7.5. Conditions at level trim for the airplane in Example 4.7.2.

$$\frac{\partial C_{D_w}}{\partial \alpha} = \left(C_{D_0,L_w} + \frac{2C_{L_w}}{\pi e_w R_{A_w}}\right)\frac{\partial C_{L_w}}{\partial \alpha} = \left(C_{D_0,L_w} + \frac{2C_{L_w}}{\pi e_w R_{A_w}}\right)C_{L_w,\alpha}(1-\varepsilon_{w,\alpha}) \quad (4.7.12)$$

$$\frac{\partial C_{D_h}}{\partial \alpha} = \left(C_{D_0,L_h} + \frac{2C_{L_h}}{\pi e_h R_{A_h}}\right)\frac{\partial C_{L_h}}{\partial \alpha} = \left(C_{D_0,L_h} + \frac{2C_{L_h}}{\pi e_h R_{A_h}}\right)C_{L_h,\alpha}(1-\varepsilon_{h,\alpha}) \quad (4.7.13)$$

Thus, assuming that the thrust does not change with α, the pitch stability derivative is

$$\begin{aligned}
\frac{\partial C_m}{\partial \alpha} &= \left[\frac{l_w C_{L_w} - h_w C_{D_w}}{\bar{c}_w}\sin(\alpha-\varepsilon_w) - \frac{h_w C_{L_w} + l_w C_{D_w}}{\bar{c}_w}\cos(\alpha-\varepsilon_w)\right](1-\varepsilon_{w,\alpha}) \\
&- \left[\frac{l_w}{\bar{c}_w}\cos(\alpha-\varepsilon_w) + \frac{h_w}{\bar{c}_w}\sin(\alpha-\varepsilon_w)\right]C_{L_w,\alpha}(1-\varepsilon_{w,\alpha}) \\
&+ \left[\frac{h_w}{\bar{c}_w}\cos(\alpha-\varepsilon_w) - \frac{l_w}{\bar{c}_w}\sin(\alpha-\varepsilon_w)\right]\left(C_{D_0,L_w} + \frac{2C_{L_w}}{\pi e_w R_{A_w}}\right)C_{L_w,\alpha}(1-\varepsilon_{w,\alpha}) \\
&+ \frac{S_h}{S_w}\eta_h\left[\frac{l_h C_{L_h} - h_h C_{D_h}}{\bar{c}_w}\sin(\alpha-\varepsilon_h) - \frac{h_h C_{L_h} + l_h C_{D_h}}{\bar{c}_w}\cos(\alpha-\varepsilon_h)\right](1-\varepsilon_{h,\alpha}) \\
&- \frac{S_h}{S_w}\eta_h\left[\frac{l_h}{\bar{c}_w}\cos(\alpha-\varepsilon_h) + \frac{h_h}{\bar{c}_w}\sin(\alpha-\varepsilon_h)\right]C_{L_h,\alpha}(1-\varepsilon_{h,\alpha}) \\
&+ \frac{S_h}{S_w}\eta_h\left[\frac{h_h}{\bar{c}_w}\cos(\alpha-\varepsilon_h) - \frac{l_h}{\bar{c}_w}\sin(\alpha-\varepsilon_h)\right]\left(C_{D_0,L_h} + \frac{2C_{L_h}}{\pi e_h R_{A_h}}\right)C_{L_h,\alpha}(1-\varepsilon_{h,\alpha})
\end{aligned} \quad (4.7.14)$$

For this more general case, notice that even in the range of linear lift, the stability derivative is not constant but varies with angle of attack. Thus, it is possible that an airplane could be stable for some angles of attack and be unstable for others.

EXAMPLE 4.7.3. Consider a wing-canard airplane that is similar to the one described in Example 4.6.1, except that this airplane has a high wing. The wing and the canard are geometrically similar, having the same aspect ratio and airfoil section with no sweep. The wing, canard, and engine are mounted so that, when the airplane is in level flight at the design airspeed and altitude, trim is attained with no elevator deflection and a zero angle of attack for the fuselage reference line, which passes through the center of gravity parallel to the thrust vector. For this geometry and operating condition, the following properties are known:

$$S_w = 180 \text{ ft}^2, \quad S_h = 36 \text{ ft}^2, \quad b_w = 33.0 \text{ ft}, \quad b_h = 14.758 \text{ ft}, \quad \bar{c}_w = 5.455 \text{ ft},$$

$$\bar{c}_h = 2.439 \text{ ft}, \quad R_{A_w} = R_{A_h} = 6.05, \quad R_{T_w} = R_{T_h} = 0.40, \quad C_{L_w,\alpha} = C_{L_h,\alpha} = 4.44,$$

$$C_{D_{0w}} = C_{D_{0h}} = 0.008, \quad C_{D_0,L_w} = C_{D_0,L_h} = 0.0, \quad e_w = e_h = 0.99,$$

$$C_{m_w} = C_{m_h} = -0.053, \quad \alpha_{L0w} = \alpha_{L0h} = -2.2°, \quad \alpha_{0w} = 1.8°, \quad \alpha_{0h} = 4.0°,$$

$$l_w = 3.0 \text{ ft}, \quad l_h = -12.0 \text{ ft}, \quad h_w = 4.0 \text{ ft}, \quad h_h = 0.0,$$

$$\varepsilon_w = 0.017 C_{L_h}, \quad \varepsilon_h = -0.012 C_{L_w}, \quad \eta_h = 1.0$$

With the controls fixed as required for this trim condition, plot the aerodynamic pitching moment coefficient as a function of angle of attack from –20 degrees to +10 degrees. Assume that the force produced by the propeller is independent of angle of attack and that neither the wing nor the canard stalls.

Solution. The geometry for this wing-canard configuration is shown in Fig. 4.7.6. At the specified trim point there is no elevator deflection. With the controls fixed in this position the lift coefficients for the wing and the canard at an arbitrary angle of attack are given by

$$C_{L_w} = C_{L_w,\alpha}(\alpha + \alpha_{0w} - \alpha_{L0w} - \varepsilon_w) = C_{L_w,\alpha}(\alpha + \alpha_{0w} - \alpha_{L0w} - 0.017C_{L_h})$$

$$C_{L_h} = C_{L_h,\alpha}(\alpha + \alpha_{0h} - \alpha_{L0h} - \varepsilon_h) = C_{L_h,\alpha}(\alpha + \alpha_{0h} - \alpha_{L0h} + 0.012C_{L_w})$$

These mathematical relations provides two coupled equations involving the two unknown lift coefficients, C_{L_w} and C_{L_h}. These equations can be rearranged as

$$\begin{bmatrix} 1 & 0.017C_{L_w,\alpha} \\ -0.012C_{L_h,\alpha} & 1 \end{bmatrix}\begin{Bmatrix} C_{L_w} \\ C_{L_h} \end{Bmatrix} = \begin{Bmatrix} C_{L_w,\alpha}(\alpha + \alpha_{0w} - \alpha_{L0w}) \\ C_{L_h,\alpha}(\alpha + \alpha_{0h} - \alpha_{L0h}) \end{Bmatrix} \equiv \begin{Bmatrix} C_{0w} \\ C_{0h} \end{Bmatrix}$$

This coupled system of two linear equations in two unknowns is readily solved for the wing and canard lift coefficients to yield

$$C_{L_w} = \frac{C_{0w} - (0.017C_{L_w,\alpha})C_{0h}}{1 - (0.017C_{L_w,\alpha})(-0.012C_{L_h,\alpha})} = \frac{C_{0w} - 0.07548C_{0h}}{1.004022}$$

$$C_{L_h} = \frac{C_{0h} - (-0.012C_{L_h,\alpha})C_{0w}}{1 - (0.017C_{L_w,\alpha})(-0.012C_{L_h,\alpha})} = \frac{C_{0h} + 0.05328C_{0w}}{1.004022}$$

At the trim point, the fuselage angle of attack is zero and we obtain

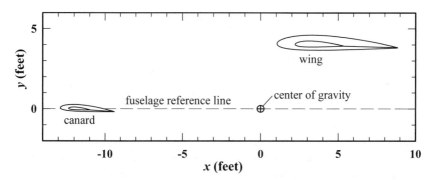

Figure 4.7.6. Geometry for the wing-canard configuration in Example 4.7.3.

$$\begin{Bmatrix} C_{0_w} \\ C_{0_h} \end{Bmatrix} = \begin{Bmatrix} 4.44\left(0 + \dfrac{1.8\pi}{180} - \dfrac{-2.2\pi}{180}\right) \\ 4.44\left(0 + \dfrac{4.0\pi}{180} - \dfrac{-2.2\pi}{180}\right) \end{Bmatrix} = \begin{Bmatrix} 0.30997 \\ 0.48045 \end{Bmatrix}, \quad \begin{Bmatrix} C_{L_w} \\ C_{L_h} \end{Bmatrix} = \begin{Bmatrix} 0.27261 \\ 0.49498 \end{Bmatrix}$$

Using these lift coefficients, the drag coefficients at trim are found to be

$$C_{D_w} = C_{D_{0w}} + \frac{C_{L_w}^2}{\pi e_w R_{A_w}} = 0.008 + \frac{(0.27261)^2}{\pi(0.99)(6.05)} = 0.01195$$

$$C_{D_h} = C_{D_{0h}} + \frac{C_{L_h}^2}{\pi e_h R_{A_h}} = 0.008 + \frac{(0.49498)^2}{\pi(0.99)(6.05)} = 0.02102$$

and the downwash angles are

$$\begin{Bmatrix} \varepsilon_w \\ \varepsilon_h \end{Bmatrix} = \begin{Bmatrix} 0.017 C_{L_h} \\ -0.012 C_{L_w} \end{Bmatrix} = \begin{Bmatrix} 0.017(0.49498) \\ -0.012(0.27261) \end{Bmatrix} = \begin{Bmatrix} 0.008415 \\ -0.003271 \end{Bmatrix}$$

Using these aerodynamic coefficients and downwash angles in Eq. (4.7.6), the total aerodynamic pitching moment coefficient at trim is

$$\begin{aligned}
C_m &= C_{m_w} - \frac{l_w C_{L_w} - h_w C_{D_w}}{\bar{c}_w}\cos(\alpha - \varepsilon_w) \\
&\quad - \frac{h_w C_{L_w} + l_w C_{D_w}}{\bar{c}_w}\sin(\alpha - \varepsilon_w) \\
&\quad + \frac{S_h \bar{c}_h}{S_w \bar{c}_w} C_{m_h} - \frac{S_h(l_h C_{L_h} - h_h C_{D_h})}{S_w \bar{c}_w}\cos(\alpha - \varepsilon_h) \\
&\quad - \frac{S_h(h_h C_{L_h} + l_h C_{D_h})}{S_w \bar{c}_w}\sin(\alpha - \varepsilon_h) \\
&= -0.053 - \frac{3(0.27261) - 4(0.01195)}{5.455}\cos(0 - 0.008415) \\
&\quad - \frac{4(0.27261) + 3(0.01195)}{5.455}\sin(0 - 0.008415) \\
&\quad - \frac{36(2.439)}{180(5.455)}0.053 - \frac{36[-12(0.49498) - 0]}{180(5.455)}\cos(0 + 0.003271) \\
&\quad - \frac{36[0 - 12(0.02102)]}{180(5.455)}\sin(0 + 0.003271) = \underline{0.02065}
\end{aligned}$$

Because this corresponds to a trim point, this moment must exactly balance the pitching moment that is produced by the thrust vector. From this moment and the computations for the drag, we could compute the vertical offset for the thrust vector if that were desired.

Following the same procedure that was used for zero angle of attack, the aerodynamic coefficients and downwash angles for the wing and canard with the fuselage reference line at a 1.0-degree angle of attack are

$$
\begin{Bmatrix} C_{L_w} \\ C_{L_h} \end{Bmatrix} = \begin{Bmatrix} 0.34397 \\ 0.57627 \end{Bmatrix}, \quad \begin{Bmatrix} C_{D_w} \\ C_{D_h} \end{Bmatrix} = \begin{Bmatrix} 0.01429 \\ 0.02565 \end{Bmatrix}, \quad \begin{Bmatrix} \varepsilon_w \\ \varepsilon_h \end{Bmatrix} = \begin{Bmatrix} 0.009797 \\ -0.004128 \end{Bmatrix}
$$

Thus, with the elevator and throttle fixed at the trim settings but with a fuselage angle of attack of 1.0 degree, the pitching moment coefficient would be

$$
\begin{aligned}
C_m &= C_{m_w} - \frac{l_w C_{L_w} - h_w C_{D_w}}{\bar{c}_w} \cos(\alpha - \varepsilon_w) - \frac{h_w C_{L_w} + l_w C_{D_w}}{\bar{c}_w} \sin(\alpha - \varepsilon_w) \\
&\quad + \frac{S_h \bar{c}_h}{S_w \bar{c}_w} C_{m_h} - \frac{S_h (l_h C_{L_h} - h_h C_{D_h})}{S_w \bar{c}_w} \cos(\alpha - \varepsilon_h) - \frac{S_h (h_h C_{L_h} + l_h C_{D_h})}{S_w \bar{c}_w} \sin(\alpha - \varepsilon_h) \\
&= \underline{0.01531}
\end{aligned}
$$

It is important to recognize that this operating condition does not correspond to a trim point. At this angle of attack the airplane is not in trim, because the lift is not equal to the weight and the aerodynamic pitching moment does not balance the pitching moment that is produced as a result of the thrust developed by the engine. This does not mean that the airplane could not be at this angle of attack with the controls in this position. It simply means that since this is not a trim point, the airplane must be accelerating. Atmospheric turbulence can cause the angle of attack to deviate significantly from trim with no control input.

Repeating these calculations at different angles of attack, the data plotted in Fig. 4.7.7 are obtained. In addition to the moment coefficient, this figure also shows the total lift and drag coefficients as well as the individual lift coefficients for the wing and the canard over the same range of angle of attack. Notice that the maximum lift coefficient for the canard is about 1.3, while the minimum lift coefficient for the wing is approximately -1.15. Thus, the angle of attack range that is shown in Fig. 4.7.7 corresponds roughly to the range between negative and positive stall.

The slope of the moment coefficient curve in Fig. 4.7.7 is the pitch stability derivative. Note that the change in the moment coefficient with respect to angle of attack is negative for angles of attack greater than about -3.5 degrees. However, the slope becomes positive when the angle of attack is less than -3.5 degrees. This does not mean that the airplane is unstable according to our mathematical definition of stability. Stability is always defined relative to a trim point. The only trim point on this curve is that for zero fuselage angle of attack. The moment slope is negative at this trim point, so this equilibrium flight condition is mathematically stable in pitch. However, this simply means that the airplane is stable with respect to an infinitesimal pitch disturbance. Unfortunately, the atmosphere cares little for mathematics, and as anyone who has spent much time in an airplane knows, disturbances are not always infinitesimal.

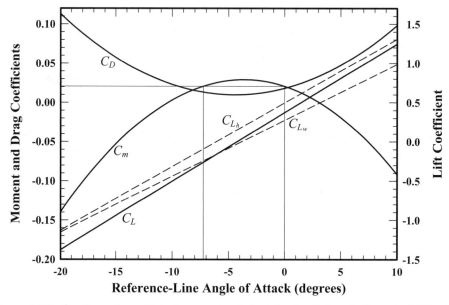

Figure 4.7.7. Aerodynamic coefficients vs. angle of attack for the airplane in Example 4.7.3, with a vertical wing offset of 4.0 feet.

The stable trimmed flight condition described in Example 4.7.3 is like the stable equilibrium that is shown in Fig. 4.1.2d. From Fig. 4.7.7, we see that a large negative disturbance in angle of attack does not produce a restoring moment. When the airplane is flying at trim ($\alpha = 0$), a pitch disturbance of about -3.5 degrees produces a maximum in the restoring moment. For larger negative disturbances, the restoring moment decreases. A pitch disturbance of about -7.25 degrees produces no restoring moment, and larger negative disturbances will produce divergence. This trait could not be exposed using the simplified linear stability analysis presented in Sec. 4.6.

The condition shown in Fig. 4.7.7 can be dangerous. The airplane is stable for the angles of attack that are usually encountered in normal flight, but it is divergent for large negative disturbances in angles of attack. Under smooth flying conditions, the airplane could be flown successfully without encountering the negative angles of attack that would expose this "instability." However, in turbulence, an airplane's angle of attack can change vary rapidly, with no control input from the pilot. Under such conditions, large negative angles of attack may be unavoidable. An airplane that displays the characteristic shown in Fig. 4.7.7 could suddenly and unexpectedly become inverted as a result of turbulence, even though the airplane does not exhibit any unstable tendencies in smooth air.

A similar situation sometimes occurs in ultralight hang gliders like the one shown in Fig. 4.7.8. For the angle of attack range encountered in normal flight, the glider is completely stable. However, when exiting a strong thermal the glider can suddenly pass from a region of very high lift into heavy sink, which is caused by the displaced cold air spilling down over the outside of the warm air rising in the thermal. For obvious reasons, this is commonly called "going over the falls." When this happens, the glider can suddenly undergo a very large decrease in angle of attack. If the glider is stable at all angles of attack, the experience will simply result in a sudden decrease in the pitch attitude and a rapid downward acceleration that increases the angle of attack and eventually returns the glider to trimmed flight. However, if the glider is not stable at large negative angles of attack, which are not normally encountered in flight, "going over the falls" can cause the glider to flip and incur significant structural damage. Hang gliders are naturally susceptible to this trait because of the large vertical separation between the aerodynamic center of the wing and the center of gravity. This design flaw can be avoided with proper use of sweep and washout.

The characteristic that is shown in Fig. 4.7.7 is most likely to occur in a canard or tailless aircraft having a high-wing design and is sometimes called *canard tuck*. This condition can be avoided with proper design, even with a moderately high wing. However, the easiest way to avoid the characteristic is by keeping both the center of gravity and the aerodynamic center of the wing closely aligned with an axis that is approximately parallel with the direction of flight. To demonstrate this, Example 4.7.3 can be reworked with the wing moved down to the fuselage reference line (i.e., $h_w = 0$). The results of such computations are shown Fig. 4.7.9. Notice that for this configuration, the pitching moment coefficient remains quite linear with angle of attack, from negative to positive stall.

Figure 4.7.8. Ultralight hang gliders sometimes exhibit pitch instability at large negative angles of attack. (Photograph by Barry Santana)

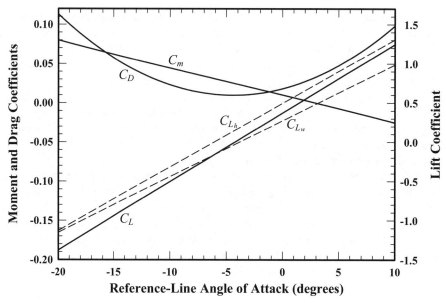

Figure 4.7.9. Aerodynamic coefficients vs. angle of attack for the airplane in Example 4.7.3, with no vertical wing offset.

The most serious approximation associated with the analysis presented in this section is that we have approximated the vortex interactions between the two lifting surfaces by simply using an average downwash angle for each surface, which is assumed to be proportional to the lift developed on the other surface. In reality, the downwash can vary substantially over the span of each lifting surface and it is not exactly a linear function of lift. For example, with the airplane described in Example 4.7.3, the downwash induced on the main wing by the canard varies from positive near the wing root to negative in the regions outboard from the tip vortices shed by the canard. Not only does the downwash vary over the span of this wing, but the downwash gradient changes with angle of attack, due to the fact that the canard vortices pass closer to the wing at the higher angles of attack. The numerical lifting-line method presented in Sec. 1.9 is capable of capturing these additional nonlinearities. Results obtained from such computations are presented in Fig. 4.7.10.

We see from Fig. 4.7.10 that at the higher angles of attack, the numerical lifting-line method predicts a decrease in pitch stability relative to the results obtained from the closed-form model used in Example 4.7.3. This is a consequence of the fact that the downwash gradient on the wing increases with increasing angle of attack, because the canard vortices pass closer to the wing. This increased downwash gradient reduces the stabilizing contribution from the aft wing. However, the differences between the numerical lifting-line results and those obtained in Example 4.7.3 are small. Reasonably accurate results can be obtained from the approximation used here and it yields important insight into how and why the nonlinearities in the trigonometric functions and the aerodynamic force and moment components affect aircraft stability.

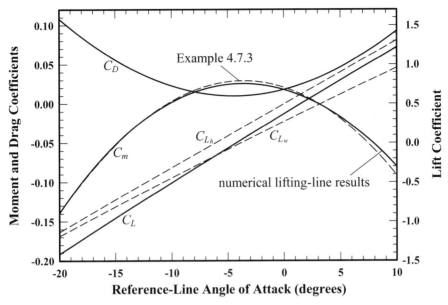

Figure 4.7.10. Aerodynamic coefficients vs. angle of attack for the airplane in Example 4.7.3, as predicted from numerical lifting-line computations.

A significant nonlinear variation in the pitching moment coefficient with angle of attack can occur whenever the center of gravity is not in close vertical alignment with the aerodynamic center of the airplane. The aerodynamic pressure and shear forces acting on an airplane can always be resolved into a normal force, an axial force, and an aerodynamic moment, all acting at the aerodynamic center, as is shown in Fig. 4.7.11. Here we let m_{ac} denote the aerodynamic pitching moment about the aerodynamic center. According to our usual notation, we use m to represent the aerodynamic pitching moment about the center of gravity. Then from the free-body diagram in Fig. 4.7.11, a summation of moments about the center of gravity gives

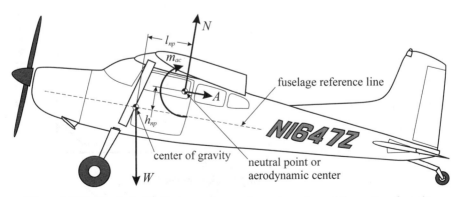

Figure 4.7.11. Vertical offset between the aerodynamic center and the center of gravity.

$$m = m_{ac} - l_{np}N + h_{np}A \qquad (4.7.15)$$

Nondimensionalizing Eq. (4.7.15), we have

$$C_m = C_{m_{ac}} - \frac{l_{np}}{\bar{c}_w}C_N + \frac{h_{np}}{\bar{c}_w}C_A \qquad (4.7.16)$$

By definition, the aerodynamic center is a point on the airplane about which the change in the aerodynamic moment with angle of attack is zero. Thus, differentiating Eq. (4.7.16) with respect to angle of attack and expressing the normal and axial force components in terms of lift and drag results in

$$
\begin{aligned}
C_{m,\alpha} &= -\frac{l_{np}}{\bar{c}_w}C_{N,\alpha} + \frac{h_{np}}{\bar{c}_w}C_{A,\alpha} \\
&= -\frac{l_{np}}{\bar{c}_w}[(C_{L,\alpha}+C_D)\cos\alpha - (C_L - C_{D,\alpha})\sin\alpha] \qquad (4.7.17) \\
&\quad -\frac{h_{np}}{\bar{c}_w}[(C_{L,\alpha}+C_D)\sin\alpha + (C_L - C_{D,\alpha})\cos\alpha]
\end{aligned}
$$

Equation (4.7.17) provides a more general relation for the pitch stability derivative than does the more commonly used approximation given by Eq. (4.4.13). Figure 4.7.12 shows how the force coefficients on the left-hand side of Eq. (4.7.17) vary with angle of attack for a typical wing. Notice that even in the range of linear lift, it is not reasonable to approximate the axial force coefficient as a linear function of the angle of attack.

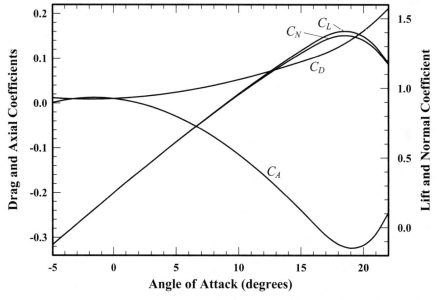

Figure 4.7.12. Aerodynamic force coefficients for a finite wing as predicted from CFD.

Thus, we see from Eq. (4.7.17) that a vertical offset between the aerodynamic center and the center of gravity can produce a significant nonlinear variation in the pitching moment coefficient with angle of attack.

Potential problems can also arise when the thrust vector has a significant vertical offset from the center of gravity, as it does on the amphibian shown in Fig. 4.7.13. This condition frequently occurs with amphibians, because of the need to keep the propeller out of the water. When the thrust vector has a large vertical offset from the CG, a large aerodynamic pitching moment is required to trim the airplane against the pitching moment produced by the engine. In the case of the airplane that is shown in Fig. 4.7.13, the thrust developed by the engine produces a negative pitching moment about the center of gravity. To offset this negative pitching moment, the airplane must be trimmed so that the airframe produces a balancing positive moment. At takeoff, when the engine is producing maximum thrust, the positive pitching moment needed to trim the airplane is greatest. If the engine should suddenly fail just after liftoff, the thrust developed by the engine would be replaced with drag. Thus, the negative pitching moment produced by the engine would be replaced with a positive pitching moment. This moment, combined with the additional positive pitching moment that is being applied to trim the airplane at full throttle, will cause a rapid nose-up rotation of the airplane. If the pilot does not respond with a rapid nose-down control input, the wing could easily stall. Engine failure just after liftoff is always dangerous. However, the danger is greatly amplified if the wing should stall.

Such a high thrust vector can also cause some potential difficulties on landing, should a go-around become necessary. On a typical landing the thrust is considerably less than maximum thrust. When the pilot decides that a go-around is prudent, full throttle is applied as rapidly as possible. With a high thrust vector, this rapid increase in thrust causes a negative pitching moment, which puts the airplane out of trim and pitches the nose down. Rapid nose-up control input is needed just to keep the airplane in trim. This tendency for the airplane to pitch down when the throttle is increased makes it more

Figure 4.7.13. Example of an airplane with a high thrust vector. (Photographs by Barry Santana)

difficult for the pilot to climb out quickly after the go-around is initiated. Greater pitch control authority is needed for such designs.

As we have seen from the material presented in this section, the moments associated with the vertical offset of the wing, tail, and thrust vector can be important. While accounting for these moments will increase the complexity of the trim and stability analysis, it is not conceptually difficult. However, the added bookkeeping does increase the possibility of making an error. To minimize this possibility, **the student should always draw a good free-body diagram**. It also helps to use a consistent notation for labeling the axial and vertical offsets. In this text it is suggested that **the variable *l* should always be used to denote a distance measured aft of the center of gravity and the variable *h* should always be used to denote a distance measured above the center of gravity**.

4.8. Effects of Nonlinearities on the Aerodynamic Center

The concept of an aerodynamic center is extremely important in the study of flight mechanics. The stick-fixed neutral point is the aerodynamic center of a complete airplane. Furthermore, the methods that have been presented in this chapter for estimating the pitch stability of an airplane depend on knowing the location of the aerodynamic center of each lifting surface used in the construction of the aircraft. At small angles of attack, the location of the aerodynamic center of a lifting surface can be approximated using the method presented in Sec. 1.11. However, that method neglects the nonlinear effects discussed in the preceding section. The traditional approximations used in Sec. 1.11 neglect the contribution of drag to the pitching moment. Furthermore, in this traditional approach, the trigonometric functions that appear in the relation between lift and pitching moment are linearized using the small-angle approximations $\cos(\alpha) \approx 1$ and $\sin(\alpha) \approx \alpha$. In this section, we shall now relax these approximations to examine how drag and the trigonometric nonlinearities influence the position of the aerodynamic center of a wing or complete airplane.

The aerodynamic center of an airfoil section was defined in Chapter 1 to be the point on the chord line about which the change in the pitching moment with respect to angle of attack is zero. The position of this aerodynamic center is nearly independent of angle of attack. Thin airfoil theory predicts that the subsonic aerodynamic center of an airfoil is the quarter chord, and experimental data show that the actual aerodynamic center of most airfoils is very near the quarter chord at low Mach numbers and small angles of attack.

For a finite wing or complete airplane, *the aerodynamic center is traditionally defined to be the point about which the pitching moment is independent of angle of attack*. Furthermore, *traditional formulations for the pitch stability derivative tacitly neglect the change in position of the aerodynamic center with respect to angle of attack*. However, outside the traditional approximation of linear aerodynamics, there may be no fixed point on a wing or complete aircraft about which the pitching moment is totally independent of the angle of attack. Thus, a less restrictive definition for the aerodynamic center of a wing or complete airplane is needed if we are to account for the trigonometric and aerodynamic nonlinearities.

In general, the aerodynamic center $(\bar{x}_{ac}, \bar{y}_{ac})$ of a spanwise symmetric wing or complete airplane lies in the plane of symmetry and is located so as to satisfy two constraints.

1. *The pitching moment about the aerodynamic center must be invariant to small changes in angle of attack.*

$$\frac{\partial C_{m_{ac}}}{\partial \alpha} \equiv 0 \qquad (4.8.1)$$

2. *The location of the aerodynamic center must be invariant to small changes in angle of attack.*

$$\frac{\partial \bar{x}_{ac}}{\partial \alpha} \equiv 0, \quad \frac{\partial \bar{y}_{ac}}{\partial \alpha} \equiv 0 \qquad (4.8.2)$$

Note that Eq. (4.8.1) does not necessarily require the pitching moment about the aerodynamic center to be independent of angle of attack, as is commonly stated. In order to require $C_{m_{ac}}$ to be independent of α, we would need to force all derivatives of $C_{m_{ac}}$ with respect to α to be zero, not just the first derivative. Because there are only two degrees of freedom associated with the position of the aerodynamic center, we cannot force $C_{m_{ac}}$ to be independent of α for an arbitrary wing or complete airplane. By similar reasoning, Eq. (4.8.2) does not require the position of the aerodynamic center to be independent of angle of attack. Although we are requiring the first derivative of $\bar{x}_{ac}$ and $\bar{y}_{ac}$ with respect to α to be zero, the higher-order derivatives may not be zero, allowing the position of the aerodynamic center to change with angle of attack.

To examine how the aerodynamic center is located from Eqs. (4.8.1) and (4.8.2), we first consider the pitching moment developed by an arbitrary wing. From Fig. 4.8.1, the pitching moment about the origin, $x = 0$, $y = 0$, for a wing of arbitrary planform and dihedral can be written as

$$m_0 = \int_{z=-b/2}^{b/2} \tilde{m}_{ac}\, dz - \int_{z=-b/2}^{b/2} (\tilde{L}\cos\alpha + \tilde{D}\sin\alpha)\tilde{x}_{ac}\, dz - \int_{z=-b/2}^{b/2} (\tilde{L}\sin\alpha - \tilde{D}\cos\alpha)\tilde{y}_{ac}\, dz \qquad (4.8.3)$$

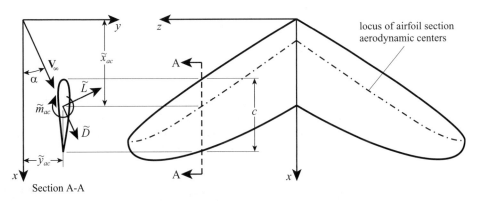

Figure 4.8.1. Section lift, drag, and pitching moment acting on a section of an arbitrary wing.

where $\tilde{x}_{ac}$ and $\tilde{y}_{ac}$ denote the x- and y-coordinates of the locus of airfoil section aero-dynamic centers. Note that, in Eq. (4.8.3), $\tilde{m}_{ac}$, $\tilde{L}$, and $\tilde{D}$ are section moment and force components per unit span, not per unit distance measured parallel with the local dihedral. Dividing Eq. (4.8.3) by dynamic pressure yields

$$\frac{m_0}{\frac{1}{2}\rho V_\infty^2} = \int_{z=-b/2}^{b/2}\tilde{C}_{m_{ac}}c^2 dz - \int_{z=-b/2}^{b/2}(\tilde{C}_L \cos\alpha + \tilde{C}_D \sin\alpha)c\tilde{x}_{ac} dz$$
$$- \int_{z=-b/2}^{b/2}(\tilde{C}_L \sin\alpha - \tilde{C}_D \cos\alpha)c\tilde{y}_{ac} dz$$

(4.8.4)

In general, $\tilde{C}_{m_{ac}}$, $\tilde{C}_L$, $\tilde{C}_D$, c, $\tilde{x}_{ac}$, and $\tilde{y}_{ac}$ all vary with the spanwise coordinate z. Note that, as used in Eq. (4.8.4), $\tilde{C}_{m_{ac}}$, $\tilde{C}_L$, and $\tilde{C}_D$ are local in situ wing section aerodynamic coefficients, which include the effects of wing downwash. Thus, $\tilde{C}_L$ is reduced by the induced angle of attack and $\tilde{C}_D$ includes both parasitic and induced drag.

The wing reference area is traditionally taken to be the horizontal projected planform area, which for a spanwise symmetric wing can be written as

$$S \equiv \int_{z=-b/2}^{b/2} c\, dz = 2\int_{z=0}^{b/2} c\, dz$$

(4.8.5)

With this definition and the traditional definitions for the wing lift and drag coefficients, we now define what we shall call the *semispan aerodynamic center of lift* $(\bar{x}_L, \bar{y}_L, \bar{z}_L)$,

$$\bar{x}_L \equiv \frac{2}{C_L S}\int_{z=0}^{b/2}\tilde{C}_L c\tilde{x}_{ac} dz, \quad \bar{y}_L \equiv \frac{2}{C_L S}\int_{z=0}^{b/2}\tilde{C}_L c\tilde{y}_{ac} dz,$$
$$\bar{z}_L \equiv \frac{2}{C_L S}\int_{z=0}^{b/2}\tilde{C}_L cz\, dz, \quad C_L \equiv \frac{2}{S}\int_{z=0}^{b/2}\tilde{C}_L c\, dz$$

(4.8.6)

and the *semispan aerodynamic center of drag* $(\bar{x}_D, \bar{y}_D, \bar{z}_D)$,

$$\bar{x}_D \equiv \frac{2}{C_D S}\int_{z=0}^{b/2}\tilde{C}_D c\tilde{x}_{ac} dz, \quad \bar{y}_D \equiv \frac{2}{C_D S}\int_{z=0}^{b/2}\tilde{C}_D c\tilde{y}_{ac} dz,$$
$$\bar{z}_D \equiv \frac{2}{C_D S}\int_{z=0}^{b/2}\tilde{C}_D cz\, dz, \quad C_D \equiv \frac{2}{S}\int_{z=0}^{b/2}\tilde{C}_D c\, dz$$

(4.8.7)

Note that these definitions differ from the traditional definitions for the centers of lift and drag, because the section aerodynamic center is used in Eqs. (4.8.6) and (4.8.7) rather than the section center of pressure. In a similar manner, we define a *mean section moment coefficient* and the *mean aerodynamic chord length*

$$\bar{C}_{m_{ac}} \equiv \frac{2}{S\bar{c}_{mac}}\int_{z=0}^{b/2}\tilde{C}_{m_{ac}}c^2 dz, \quad \bar{c}_{mac} \equiv \frac{2}{S}\int_{z=0}^{b/2}c^2 dz$$

(4.8.8)

Using these definitions and dividing by the planform area of the wing, Eq. (4.8.4) can be written as

$$C_{m_0} c_{ref} \equiv \frac{m_0}{\frac{1}{2}\rho V_\infty^2 S} = \widetilde{\overline{C}}_{m_{ac}} \overline{c}_{mac} - C_L \overline{x}_L \cos\alpha - C_D \overline{x}_D \sin\alpha$$
$$- C_L \overline{y}_L \sin\alpha + C_D \overline{y}_D \cos\alpha$$
(4.8.9)

where c_{ref} is the arbitrary reference length chosen to nondimensionalize the pitching moment coefficient.

The distributed aerodynamic loads acting on the wing surface can be replaced with a resultant force and moment acting at the aerodynamic center of the wing. Thus, the pitching moment about the origin can also be written as

$$m_0 = m_{ac} - \overline{x}_{ac}(L\cos\alpha + D\sin\alpha) - \overline{y}_{ac}(L\sin\alpha - D\cos\alpha) \quad (4.8.10)$$

where m_{ac} is the pitching moment about the aerodynamic center of the wing. Dividing by the dynamic pressure and planform area, Eq. (4.8.10) becomes

$$C_{m_0} c_{ref} = C_{m_{ac}} c_{ref} - \overline{x}_{ac}(C_L \cos\alpha + C_D \sin\alpha) - \overline{y}_{ac}(C_L \sin\alpha - C_D \cos\alpha) \quad (4.8.11)$$

Combining Eqs. (4.8.9) and (4.8.11) to eliminate the pitching moment coefficient about the origin and rearranging gives

$$\overline{x}_{ac}(C_L \cos\alpha + C_D \sin\alpha) + \overline{y}_{ac}(C_L \sin\alpha - C_D \cos\alpha) - C_{m_{ac}} c_{ref}$$
$$= C_L(\overline{x}_L \cos\alpha + \overline{y}_L \sin\alpha) + C_D(\overline{x}_D \sin\alpha - \overline{y}_D \cos\alpha) - \widetilde{\overline{C}}_{m_{ac}} \overline{c}_{mac}$$
(4.8.12)

From Eq. (4.8.8) and the definition of section aerodynamic center we can write

$$\frac{\partial \widetilde{\overline{C}}_{m_{ac}}}{\partial \alpha} = 0 \quad (4.8.13)$$

With application of Eqs. (4.8.1), (4.8.2), and (4.8.13), the derivatives of Eqs. (4.8.9) and (4.8.12) with respect to angle of attack yield

$$\overline{x}_{ac} \frac{\partial}{\partial\alpha}(C_L \cos\alpha + C_D \sin\alpha) + \overline{y}_{ac} \frac{\partial}{\partial\alpha}(C_L \sin\alpha - C_D \cos\alpha)$$
$$= \frac{\partial}{\partial\alpha}[C_L(\overline{x}_L \cos\alpha + \overline{y}_L \sin\alpha) + C_D(\overline{x}_D \sin\alpha - \overline{y}_D \cos\alpha)] \quad (4.8.14)$$
$$= -c_{ref} \frac{\partial C_{m_0}}{\partial\alpha}$$

The relation between $\overline{x}_{ac}$ and $\overline{y}_{ac}$, which is prescribed by Eq. (4.8.14), is necessary and sufficient to require the pitching moment about the point $(\overline{x}_{ac}, \overline{y}_{ac})$ to have a zero first derivative with respect to angle of attack. However, **Eq. (4.8.14) does not define a point in the plane of symmetry; rather, it defines a line.** This line in the wing's plane of symmetry will be referred to here as the *neutral axis* of the wing, which is formally defined to be *the locus of points about which the change in pitching moment with respect to angle of attack is zero*, thus satisfying the first criterion in our definition for the aerodynamic center.

In general, Eq. (4.8.14) shows that both the slope and x-axis intercept for the neutral axis of a wing or complete airplane are nonlinear functions of angle of attack. However, at any given angle of attack the neutral axis is always a straight line. For a small change in angle of attack relative to any given operating condition, the neutral axis will be rotated through a small angle about some point along its length. That point is the aerodynamic center as defined by Eqs. (4.8.1) and (4.8.2). Every point along the neutral axis satisfies Eq. (4.8.1) for the specified angle of attack. However, only one point on the neutral axis also satisfies Eq. (4.8.2) for the same angle of attack. Because only the first derivative of $\bar{x}_{ac}$ and $\bar{y}_{ac}$ with respect to α is forced to zero, we cannot say that the aerodynamic center is a fixed point, independent of angle of attack. Because there are only two degrees of freedom associated with the position of the aerodynamic center, the variation in its position with angle of attack is a matter to be evaluated, not defined. When all nonlinearities are included, the aerodynamic center can be uniquely defined only for small perturbations about any given angle of attack.

The Traditional Approximation
Even considering a wing with no dihedral, neglecting drag, and applying the traditional approximation for small angles of attack, Eq. (4.8.14) still defines a line in the x-y plane, not a point. Using the approximations $C_D \approx 0$, $\cos(\alpha) \approx 1$, and $\sin(\alpha) \approx \alpha$, Eq. (4.8.14) reduces to

$$\bar{x}_{ac}\frac{\partial C_L}{\partial \alpha} + \bar{y}_{ac}\frac{\partial}{\partial \alpha}(C_L \alpha) = -c_{\text{ref}}\frac{\partial C_{m0}}{\partial \alpha} \tag{4.8.15}$$

For wings with little or no dihedral at small angles of attack, both the lift and the pitching moment are traditionally approximated as linear functions of angle of attack,

$$C_L = C_{L,\alpha}(\alpha - \alpha_{L0}), \quad C_{L,\alpha,\alpha} = 0, \quad C_{m0,\alpha,\alpha} = 0 \tag{4.8.16}$$

where the subscript notation "$,\alpha,\alpha$" denotes a second derivative with respect to α. Thus, using Eq. (4.8.16) in Eq. (4.8.15) and rearranging, we obtain

$$\bar{x}_{ac} + \bar{y}_{ac}(2\alpha - \alpha_{L0}) = -c_{\text{ref}}\, C_{m0,\alpha}/C_{L,\alpha} \tag{4.8.17}$$

This further demonstrates the importance of our refined definition for the aerodynamic center, which is specified by Eqs. (4.8.1) and (4.8.2). **Even after neglecting drag and applying the small-angle approximation, defining the aerodynamic center according to Eq. (4.8.1) yields a line rather than a point.** The second criterion set by Eq. (4.8.2) is necessary in order to isolate the aerodynamic center as a point along this line.

For the small-angle approximation in the absence of drag and dihedral effects, the aerodynamic center can be located along the neutral axis by differentiating Eq. (4.8.17) with respect to α, subject to Eqs. (4.8.2) and (4.8.16). This gives

$$\bar{y}_{ac} = 0 \tag{4.8.18}$$

Using this result in Eq. (4.8.17) yields the traditional approximation

$$\bar{x}_{ac} = -c_{\text{ref}}\, C_{m_0,\alpha}\big/C_{L,\alpha} \tag{4.8.19}$$

In the usual development of Eq. (4.8.19), the result given by Eq. (4.8.18) is not typically determined mathematically but is simply assumed a priori. Although it may seem obvious that the aerodynamic center of a planar wing should always fall in the plane of the wing, as we shall see, this is not necessarily the case. In extending our analysis for the location of the aerodynamic center to the realm of nonlinear aerodynamics, including the effects of drag and relaxing of the small-angle approximation, we can no longer simply assume that Eq. (4.8.18) is valid. The y-coordinate of the aerodynamic center must be isolated from the neutral axis defined in Eq. (4.8.14) by applying the second criterion of our refined definition for the aerodynamic center, i.e., Eq. (4.8.2).

Wings with Constant Sweep and Dihedral
Additional insight into the nature of Eq. (4.8.14) can be gleaned by considering the special case of a spanwise symmetric wing with constant sweep and constant dihedral in the locus of wing section aerodynamic centers. For wings with this geometry, the axial and normal coordinates of the locus of wing section aerodynamic centers are linear functions of the spanwise coordinate z. Thus, we have

$$\tilde{x}_{ac} = \tilde{x}_{\text{root}} + z\tan\Lambda, \quad \tilde{y}_{ac} = \tilde{y}_{\text{root}} + z\tan\Gamma \tag{4.8.20}$$

where $\tilde{x}_{\text{root}}$ and $\tilde{y}_{\text{root}}$ are the x- and y-coordinates of the root section aerodynamic center. Using Eq. (4.8.20) in Eqs. (4.8.6) and (4.8.7) gives

$$\bar{x}_L = \tilde{x}_{\text{root}} + \bar{z}_L\tan\Lambda, \quad \bar{y}_L = \tilde{y}_{\text{root}} + \bar{z}_L\tan\Gamma, \quad \bar{z}_L \equiv \frac{2}{C_L S}\int\limits_{z=0}^{b/2}\tilde{C}_L cz\, dz \tag{4.8.21}$$

$$\bar{x}_D \equiv \tilde{x}_{\text{root}} + \bar{z}_D\tan\Lambda, \quad \bar{y}_D \equiv \tilde{y}_{\text{root}} + \bar{z}_D\tan\Gamma, \quad \bar{z}_D \equiv \frac{2}{C_D S}\int\limits_{z=0}^{b/2}\tilde{C}_D cz\, dz \tag{4.8.22}$$

With the simplifications introduced by Eqs. (4.8.21) and (4.8.22), Eq. (4.8.14) can be written as

$$(\bar{x}_{ac} - \tilde{x}_{\text{root}})\frac{\partial}{\partial\alpha}(C_L\cos\alpha + C_D\sin\alpha) + (\bar{y}_{ac} - \tilde{y}_{\text{root}})\frac{\partial}{\partial\alpha}(C_L\sin\alpha - C_D\cos\alpha)$$
$$= \tan\Lambda\frac{\partial}{\partial\alpha}(\bar{z}_L C_L\cos\alpha + \bar{z}_D C_D\sin\alpha) \tag{4.8.23}$$
$$+ \tan\Gamma\frac{\partial}{\partial\alpha}(\bar{z}_L C_L\sin\alpha - \bar{z}_D C_D\cos\alpha)$$

In general, C_L, C_D, $\bar{z}_L$, and $\bar{z}_D$ all vary with angle of attack. Furthermore, $\bar{z}_L$ and $\bar{z}_D$ are not necessarily coincident.

The General Relations for Aerodynamic Center
Equation (4.8.14) defines the neutral axis of an arbitrary wing in terms of the lift and drag coefficients for the wing and the locations of the aerodynamic centers of lift and drag. Because no analytical solution exists for the spanwise distributions of lift and drag on a wing of arbitrary geometry, the position of the aerodynamic center must be evaluated numerically or experimentally for wings of arbitrary sweep and/or dihedral. Although Eq. (4.8.14) provides valuable insight into how the position of the aerodynamic center is affected by lift and drag, for experimental or numerical evaluation of the aerodynamic center location it is more convenient to write Eq. (4.8.14) in terms of the axial and normal force coefficients,

$$C_A = C_D \cos\alpha - C_L \sin\alpha \tag{4.8.24}$$

$$C_N = C_L \cos\alpha + C_D \sin\alpha \tag{4.8.25}$$

Applying Eqs. (4.8.24) and (4.8.25) to Eqs. (4.8.11) and (4.8.14) yields

$$C_{m_0} c_{ref} = C_{m_{ac}} c_{ref} - \bar{x}_{ac} C_N + \bar{y}_{ac} C_A \tag{4.8.26}$$

$$\bar{x}_{ac} C_{N,\alpha} - \bar{y}_{ac} C_{A,\alpha} = -C_{m_0,\alpha} c_{ref} \tag{4.8.27}$$

Equation (4.8.27) defines the wing's neutral axis at some arbitrary angle of attack α. Every point along this neutral axis satisfies the first criterion in our refined definition for the aerodynamic center, which is given by Eq. (4.8.1). The aerodynamic center of the wing can be isolated from the neutral axis by differentiating Eq. (4.8.27) with respect to angle of attack and applying the second criterion in our definition for the aerodynamic center, which is given by Eq. (4.8.2). This gives

$$\bar{x}_{ac} C_{N,\alpha,\alpha} - \bar{y}_{ac} C_{A,\alpha,\alpha} = -C_{m_0,\alpha,\alpha} c_{ref} \tag{4.8.28}$$

The aerodynamic center of the wing is the only point common to the two lines defined by Eqs. (4.8.27) and (4.8.28). Solving Eqs. (4.8.27) and (4.8.28) for the position of the aerodynamic center and using the result in Eq. (4.8.26) yields

$$\bar{x}_{ac} = \frac{C_{A,\alpha} C_{m_0,\alpha,\alpha} - C_{m_0,\alpha} C_{A,\alpha,\alpha}}{C_{N,\alpha} C_{A,\alpha,\alpha} - C_{A,\alpha} C_{N,\alpha,\alpha}} c_{ref} \tag{4.8.29}$$

$$\bar{y}_{ac} = \frac{C_{N,\alpha} C_{m_0,\alpha,\alpha} - C_{m_0,\alpha} C_{N,\alpha,\alpha}}{C_{N,\alpha} C_{A,\alpha,\alpha} - C_{A,\alpha} C_{N,\alpha,\alpha}} c_{ref} \tag{4.8.30}$$

$$C_{m_{ac}} c_{ref} = C_{m_0} c_{ref} + \bar{x}_{ac} C_N - \bar{y}_{ac} C_A \tag{4.8.31}$$

Because the choice of coordinate system origin is arbitrary, Eqs. (4.8.29) through (4.8.31) can be used to determine the position of the aerodynamic center and the pitching moment about the aerodynamic center from analytical results, numerical computations, or experimental measurements for the resultant aerodynamic force components, pitching moment, and their first and second derivatives, obtained relative to any point in the x-y plane.

Application of Eq. (4.8.29) **through** (4.8.31) is necessary only for wings having significant sweep and/or dihedral in the locus of airfoil section aerodynamic centers. In the case where $\tilde{x}_{ac}$ and $\tilde{y}_{ac}$ do not vary with the spanwise coordinate z, Eqs. (4.8.6) and (4.8.7) are easily integrated to yield

$$\bar{x}_L = \bar{x}_D = \tilde{x}_{ac}, \quad \bar{y}_L = \bar{y}_D = \tilde{y}_{ac} \tag{4.8.32}$$

If $\tilde{x}_{ac}$ and $\tilde{y}_{ac}$ are also independent of α, Eq. (4.8.14) reduces to

$$\begin{aligned}
\bar{x}_{ac} \frac{\partial}{\partial \alpha} (C_L \cos\alpha + C_D \sin\alpha) + \bar{y}_{ac} \frac{\partial}{\partial \alpha} (C_L \sin\alpha - C_D \cos\alpha) \\
= \tilde{x}_{ac} \frac{\partial}{\partial \alpha} (C_L \cos\alpha + C_D \sin\alpha) + \tilde{y}_{ac} \frac{\partial}{\partial \alpha} (C_L \sin\alpha - C_D \cos\alpha)
\end{aligned} \tag{4.8.33}$$

or after rearranging

$$\begin{aligned}
(\bar{x}_{ac} - \tilde{x}_{ac}) \frac{\partial}{\partial \alpha} (C_L \cos\alpha + C_D \sin\alpha) \\
+ (\bar{y}_{ac} - \tilde{y}_{ac}) \frac{\partial}{\partial \alpha} (C_L \sin\alpha - C_D \cos\alpha) = 0
\end{aligned} \tag{4.8.34}$$

Equation (4.8.34) describes a straight line that passes through the point $\bar{x}_{ac} = \tilde{x}_{ac}$ and $\bar{y}_{ac} = \tilde{y}_{ac}$ for any value of α, C_L, and C_D. Thus, from Eqs. (4.8.8), (4.8.9), (4.8.11), and (4.8.33), **for a spanwise symmetric wing having no sweep or dihedral in the locus of airfoil section aerodynamic centers**, we obtain the rather trivial result

$$\bar{x}_{ac} = \tilde{x}_{ac}, \quad \bar{y}_{ac} = \tilde{y}_{ac}, \quad \text{and} \quad C_{m_{ac}} C_{\text{ref}} = \frac{2}{S} \int_{z=0}^{b/2} \tilde{C}_{m_{ac}} c^2 \, dz \tag{4.8.35}$$

For the more general case of wings with sweep and/or dihedral, Eqs. (4.8.29) through (4.8.31) can be used as demonstrated in the following example.

EXAMPLE 4.8.1. The horizontal stabilizer of a jet transport is constructed with a thin symmetric airfoil section having its aerodynamic center at the quarter chord. This lifting surface has constant linear taper and a constant quarter-chord sweep angle of 30 degrees with no dihedral or twist. The stabilizer has a span of 72 feet, a root chord of 16 feet, and a tip chord of 8 feet. The following data were obtained for this geometry from CFD results:

α (deg)	C_L	C_D	C_{m_0}
−1.00	−0.074343	0.0092649	0.054459
0.00	0.000000	0.0089496	0.000000
1.00	0.074343	0.0092649	−0.054459
14.00	0.988355	0.0672777	−0.709675
15.00	1.050595	0.0753134	−0.751961
16.00	1.111105	0.0837338	−0.792508

The origin and reference length used for computing the moment coefficients were the wing root quarter chord and geometric mean chord length, respectively. Using second-order central-difference approximations for the first and second derivatives, estimate the position of the aerodynamic center and associated pitching moment coefficient for this lifting surface at angles of attack of 0 and 15 degrees.

Solution. Using Eqs. (4.8.24) and (4.8.25), the aerodynamic force components are easily transformed from lift and drag to axial and normal coordinates:

α (deg)	C_N	C_A	C_{m_0}
14.00	0.975273	-0.1738255	-0.709675
15.00	1.034289	-0.1991668	-0.751961
16.00	1.091143	-0.2257719	-0.792508

Using the second-order central-difference approximation, the first derivative of the normal force coefficient with respect to angle of attack at 15 degrees is estimated as

$$C_{N,\alpha} \cong \frac{C_N(\alpha+\Delta\alpha)-C_N(\alpha-\Delta\alpha)}{2\Delta\alpha} = \frac{1.091143-0.975273}{2(\pi/180)} = 3.319$$

where $\Delta\alpha$ is the angle of attack increment (i.e., 1 degree). Also at 15 degrees, the second-order central-difference approximation for the second derivative of C_N with respect to α produces the result

$$C_{N,\alpha,\alpha} \cong \frac{C_N(\alpha+\Delta\alpha)-2C_N(\alpha)+C_N(\alpha-\Delta\alpha)}{\Delta\alpha^2}$$

$$= \frac{1.091143-2(1.034289)+0.975273}{(\pi/180)^2} = -7.102$$

The derivatives of other aerodynamic coefficients can be approximated in a similar manner,

$$C_{A,\alpha} \cong -1.488, \quad C_{A,\alpha,\alpha} \cong -4.149, \quad C_{m_0,\alpha} \cong -2.373, \quad C_{m_0,\alpha,\alpha} \cong 5.705$$

Once the necessary force and moment coefficients and their first and second derivatives are evaluated, the position of the aerodynamic center can be obtained from Eqs. (4.8.29) and (4.8.30). This gives

$$\frac{\bar{x}_{ac}}{c_{ref}} = \frac{C_{A,\alpha} C_{m_0,\alpha,\alpha} - C_{m_0,\alpha} C_{A,\alpha,\alpha}}{C_{N,\alpha} C_{A,\alpha,\alpha} - C_{A,\alpha} C_{N,\alpha,\alpha}} = 0.754$$

$$\frac{\bar{y}_{ac}}{c_{ref}} = \frac{C_{N,\alpha} C_{m_0,\alpha,\alpha} - C_{m_0,\alpha} C_{N,\alpha,\alpha}}{C_{N,\alpha} C_{A,\alpha,\alpha} - C_{A,\alpha} C_{N,\alpha,\alpha}} = -0.086$$

Now that the position of the aerodynamic center is known, the moment coefficient about this location can be obtained from Eq. (4.8.31), which yields

$$C_{m_{ac}} = C_{m_0} + \frac{\overline{x}_{ac}}{c_{ref}}C_N - \frac{\overline{y}_{ac}}{c_{ref}}C_A = 0.0102$$

Repeating the computations for $\alpha = 0$ degrees yields

$$\frac{\overline{x}_{ac}}{c_{ref}} = 0.731, \quad \frac{\overline{y}_{ac}}{c_{ref}} = 0.000, \quad C_{m_{ac}} = 0.0000$$

The CFD results used in this example were obtained by Phillips, Alley, and Niewoehner (2008), who presented results for the aerodynamic center of this lifting surface for angles of attack from -5 to $+18$ degrees. These results are shown in Fig. 4.8.2. For comparison, the wing's neutral axis is also shown in this figure, for angles of attack of $-5, 0, 5, 10,$ and 15 degrees. The neutral axis at any given angle of attack passes through the aerodynamic center for the same angle of attack and is tangent to the curve defined by the locus of aerodynamic centers.

An interesting characteristic of swept wings is revealed by the results obtained in the previous example. At zero angle of attack, the neutral axis for this uncambered and untwisted wing is vertical and the aerodynamic center is located at the coordinates $\overline{x}_{ac}/c_{ref} = 0.731$ and $\overline{y}_{ac}/c_{ref} = 0.000$, which agrees very closely with the traditional approximation given by Eqs. (4.8.18) and (4.8.19). As the angle of attack is increased from 0 to 15 degrees the aerodynamic center of this particular surface moves aft by slightly more than 2 percent of the mean chord. The aerodynamic center also moves down by more then 8 percent of the mean chord. This movement is significant in view of the fact that the design static margin is often on the order of 5 percent of the mean chord. At first thought, it may seem impossible for the aerodynamic center of a wing with no dihedral to fall outside the plane of the wing. It is easier to understand the physics of this phenomenon by considering the description of the neutral axis in terms of lift and drag.

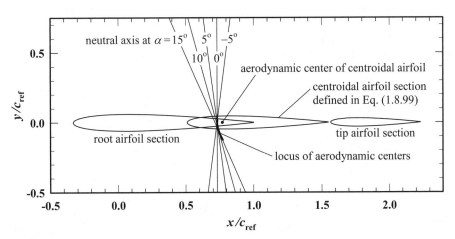

Figure 4.8.2. The locus of aerodynamic centers as predicted from computational fluid dynamics for the swept lifting surface in Example 4.8.1.

To gain some insight into the physics behind this phenomenon, we recognize that at zero angle of attack the position of the aerodynamic center is dominated by lift. Thus, the aerodynamic center at zero angle of attack is nearly coincident with the center of additional lift. Because this lifting surface has no twist, the center of additional lift is also the center of total lift. For this geometry the aerodynamic center of lift is located at $\bar{x}_L/c_{ref} = 0.733$ and $\bar{y}_L/c_{ref} = 0.000$.

Because this lifting surface has no camber or twist, the drag at zero angle of attack is all parasitic drag and the section drag coefficient is nearly uniform over the span. Thus, the aerodynamic center of drag at zero angle of attack should be located very near the aerodynamic center of the centroidal chord, which for this particular geometry is located at $\bar{x}_c/c_{ref} = 0.770$ and $\bar{y}_c/c_{ref} = 0.000$, as shown in Fig. 4.8.2.

With this line of reasoning, we see that the aerodynamic center of drag is slightly aft of the aerodynamic center of lift at zero angle of attack. Because this lifting surface has no twist, the center of total lift does not move significantly with small changes in angle of attack. However, as the angle of attack is increased, induced drag becomes an increasing fraction of the total drag. Because induced drag is greatest near the wingtips, increasing the angle of attack moves the aerodynamic center of drag outboard and aft. Thus, we see that for this geometry at angles of attack below stall, the aerodynamic center of drag is always aft of the aerodynamic center of lift. This requires that the center of aerodynamic force, which is defined in Fig. 4.8.3, is always below the centerline plane at positive angles of attack. Although the aerodynamic center of a wing does not necessarily fall at the center of aerodynamic force, we should not be too surprised to learn from the CFD results that the aerodynamic center for this geometry also falls below the centerline plane at positive angles of attack below stall.

It is significant to note that the aft movement of the aerodynamic center, which is shown in Fig. 4.8.2, is stabilizing at angles of attack below stall, because the normal force increases with increasing angle of attack. However, the vertical movement of the aerodynamic center shown in Fig. 4.8.2 also affects pitch stability, and the nature of this effect may not be quite so obvious. To examine the net effect of this aerodynamic center movement on pitch stability, consider the variations in the aerodynamic force coefficients with angle of attack for the lifting surface in Example 4.8.1, which are shown in Fig. 4.8.4. From Eq. (4.7.17), the pitch stability derivative can be written as

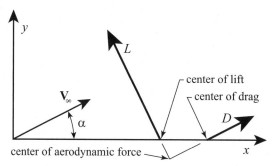

Figure 4.8.3. Center of aerodynamic force, defined here to be the intersection of the wing lift and drag vectors.

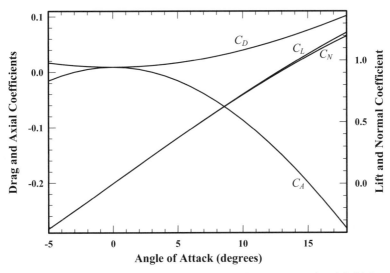

Figure 4.8.4. Aerodynamic force coefficients as predicted from computational fluid dynamics for the lifting surface in Example 4.8.1.

$$C_{m,\alpha} = -\frac{\overline{x}_{ac} - x_{CG}}{c_{ref}} C_{N,\alpha} + \frac{\overline{y}_{ac} - y_{CG}}{c_{ref}} C_{A,\alpha} \qquad (4.8.36)$$

where x_{CG} and y_{CG} are the x- and y-coordinates of the center of gravity. From Fig. 4.8.4, we see that at angles of attack below stall, $C_{N,\alpha}$ is always positive. Thus, the first term on the right-hand side of Eq. (4.8.36) confirms previous results, which have shown that moving the aerodynamic center of a wing aft (i.e., increasing $\overline{x}_{ac}$) is stabilizing. On the other hand, Fig. 4.8.4 also shows that at positive angles of attack below stall, $C_{A,\alpha}$ is negative. Thus, the second term on the right-hand side of Eq. (4.8.36) shows that moving the aerodynamic center down (i.e., decreasing $\overline{y}_{ac}$) is destabilizing at positive angles of attack below stall. When the angle of attack is small, the effect of the aft movement of the aerodynamic center with increasing angle of attack tends to cancel the effect of the downward movement, and the net effect is slight. At higher angles of attack the net effect is destabilizing. This is demonstrated in Fig. 4.8.5, which shows the pitching moment about the point $x/c_{ref} = 0.681$ and $y/c_{ref} = 0.000$. This point is forward of the aerodynamic center at $\alpha = 0$ by 5 percent of the mean chord. At zero angle of attack, the pitch stability derivative taken from Fig. 4.8.5 is -0.128. However, as the angle of attack approaches maximum lift, the movement of the aerodynamic center changes the pitch stability derivative to about -0.047. This represents a reduction in pitch stability of more than 63 percent, which is clearly significant.

The method of using Eqs. (4.8.29)–(4.8.31) to locate the aerodynamic center and determine its pitching moment is not limited to use with computational fluid dynamics. The required aerodynamic coefficients and derivatives could be evaluated from other numerical methods, analytical solutions, or experimental data. Likewise, the method is not limited to a single lifting surface. It can also be used to locate the aerodynamic center

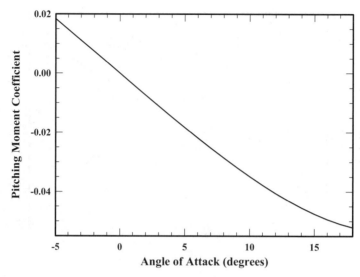

Figure 4.8.5. Pitching moment coefficient for Example 4.8.1 about a point forward of the zero-angle-of-attack aerodynamic center by 5 percent of the mean chord.

and determine its pitching moment for a system of lifting surfaces or a complete airplane. This is demonstrated in the following example.

> **EXAMPLE 4.8.2.** Consider the same wing-canard configuration and operating conditions described in Example 4.7.3. Using the numerical lifting-line method that is presented in Sec. 1.9, plot the neutral axis and the position of the aerodynamic center as a function of angle of attack from −20 degrees to +10 degrees.
>
> ***Solution.*** Based on the aerodynamic coefficients given for the wing and canard in Example 4.7.3, we shall use the constant airfoil section coefficients $\tilde{C}_{L,\alpha}=5.87$, $\tilde{C}_D=0.008$, and $\tilde{C}_m=-0.053$ for both the wing and canard. Since the choice of coordinate system is arbitrary, we choose a system with the origin at the center of gravity, the x-axis pointing aft along the fuselage reference line, and the y-axis pointing upward in the plane of symmetry. At any angle of attack the neutral axis can be plotted from Eq. (4.8.27), which can be rearranged as
>
> $$ y = \frac{C_{N,\alpha}}{C_{A,\alpha}} x + \frac{C_{m_0,\alpha}\, c_{\text{ref}}}{C_{A,\alpha}} $$
>
> The x- and y-coordinates of the aerodynamic center at any angle of attack are evaluated from Eqs. (4.8.29) and (4.8.30). Using the geometry that is specified in Example 4.7.3, we follow the procedure outlined in Example 4.8.1 with double-precision computations, cosine clustering, and 100 elements per semispan. The required first and second derivatives are approximated using second-order central difference with $\Delta\alpha=0.5$ degrees. Thus, with $\alpha=-20$ degrees we obtain

$$C_A = -0.382570 \qquad C_N = -1.367244 \qquad C_{m_0}\, c_{\text{ref}} = -0.760411 \text{ ft}$$

$$C_{A,\alpha} = 2.522930 \qquad C_{N,\alpha} = 4.841755 \qquad C_{m_0,\alpha}\, c_{\text{ref}} = 6.550200 \text{ ft}$$

$$C_{A,\alpha,\alpha} = -8.369767 \qquad C_{N,\alpha,\alpha} = 1.655486 \qquad C_{m_0,\alpha,\alpha}\, c_{\text{ref}} = -25.692999 \text{ ft}$$

$$\overline{x}_{ac} = 0.223663 \text{ ft} \qquad \overline{y}_{ac} = 3.025500 \text{ ft} \qquad C_{m_{ac}}\, c_{\text{ref}} = 0.091253 \text{ ft}$$

Similar computations for $\alpha = +10$ degrees yield

$$\overline{x}_{ac} = 0.166615 \text{ ft} \qquad \overline{y}_{ac} = 2.845059 \text{ ft} \qquad C_{m_{ac}}\, c_{\text{ref}} = 0.105938 \text{ ft}$$

Repeating the computations at other angles of attack we obtain the results plotted in Fig. 4.8.6. For this example with no sweep, we see only slight variations in the position of the aerodynamic center with angle of attack. As seen in Example 4.8.1, larger vertical variations in the position of the aerodynamic center with angle of attack should be expected for geometries with substantial sweep.

A Two-Surface Small-Angle Approximation
Using Eqs. (4.7.4)–(4.7.11) combined with the downwash model presented in Sec. 4.5 provides an analytical formulation that can be used with Eqs. (4.8.29) and (4.8.30) to evaluate the position of the aerodynamic center of the lifting-surface combination shown in Fig. 4.7.1, including the effects of the nonlinearities in the trigonometric functions and the drag. Phillips, Alley, and Niewoehner (2008) have shown that for aircraft without highly swept wings, the x-coordinate of the aerodynamic center as predicted from this nonlinear formulation does not differ greatly from that predicted from the traditional approximation given by Eq. (4.8.19). The problem with the traditional small-angle approximation is not that it provides an unreasonable estimate for the axial position of the aerodynamic center. The problem is that it provides no estimate for the vertical position.

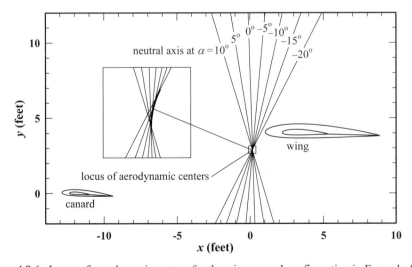

Figure 4.8.6. Locus of aerodynamic centers for the wing-canard configuration in Example 4.8.2.

A small-angle approximation that does provide an estimate for the y-coordinate of the aerodynamic center can be obtained from Eqs. (4.8.29) and (4.8.30). This is bases on approximating the average downwash angle for each lifting surface as being linearly proportional to the lift developed on the other surface,

$$\varepsilon_w \cong C_{\varepsilon w} C_{L_h}, \qquad \varepsilon_h \cong C_{\varepsilon h} C_{L_w} \tag{4.8.37}$$

Following the analysis presented in Example 4.7.3, using Eq. (4.8.37) in Eqs. (4.7.4) through (4.7.11), and applying the results together with the traditional approximations $C_D \approx 0$, $\cos(\alpha) \approx 1$, and $\sin(\alpha) \approx \alpha$ to Eqs. (4.8.29) and (4.8.30) yields

$$\bar{x}_{ac} = \frac{x_w C_{w,\alpha} + x_h C_{h,\alpha}}{C_{w,\alpha} + C_{h,\alpha}} + \frac{(y_h - y_w)[C_{w,\alpha} C_{h0} - C_{h,\alpha} C_{w0}]}{(C_{w,\alpha} + C_{h,\alpha})^2} \tag{4.8.38}$$

$$\bar{y}_{ac} = \frac{y_w C_{w,\alpha} + y_h C_{h,\alpha}}{C_{w,\alpha} + C_{h,\alpha}} \tag{4.8.39}$$

$$C_{w,\alpha} \equiv \frac{C_{L_w,\alpha}(1 - C_{\varepsilon w} C_{L_h,\alpha})}{1 - (C_{\varepsilon w} C_{L_w,\alpha})(C_{\varepsilon h} C_{L_h,\alpha})} \tag{4.8.40}$$

$$C_{h,\alpha} \equiv \frac{S_h}{S_w} \eta_h \frac{C_{L_h,\alpha}(1 - C_{\varepsilon h} C_{L_w,\alpha})}{1 - (C_{\varepsilon w} C_{L_w,\alpha})(C_{\varepsilon h} C_{L_h,\alpha})} \tag{4.8.41}$$

$$C_{w0} \equiv \frac{C_{L_w,\alpha}[(\alpha_{0w} - \alpha_{L0w}) - C_{\varepsilon w} C_{L_h,\alpha}(\alpha_{0h} - \alpha_{L0h} + \varepsilon_e \delta_e)]}{1 - (C_{\varepsilon w} C_{L_w,\alpha})(C_{\varepsilon h} C_{L_h,\alpha})} \tag{4.8.42}$$

$$C_{h0} \equiv \frac{S_h}{S_w} \eta_h \frac{C_{L_h,\alpha}[(\alpha_{0h} - \alpha_{L0h} + \varepsilon_e \delta_e) - C_{\varepsilon h} C_{L_w,\alpha}(\alpha_{0w} - \alpha_{L0w})]}{1 - (C_{\varepsilon w} C_{L_w,\alpha})(C_{\varepsilon h} C_{L_h,\alpha})} \tag{4.8.43}$$

where (x_w, y_w) and (x_h, y_h) are the coordinates of the aerodynamic center of the main wing and the horizontal control surface, respectively. For a detailed development of this result, see Phillips, Alley, and Niewoehner (2008).

4.9. Effects of the Fuselage, Nacelles, and External Stores

The contribution of the fuselage, nacelles, and external stores to the static stability of an airplane is typically destabilizing. Furthermore, in some cases these effects can be quite significant. Thus, the fuselage, nacelles, and external stores should be included in the trim and stability analysis during all phases of the airplane design process.

The aerodynamic forces and moments generated on the fuselage, nacelles, and external stores are extremely complex, and are significantly affected by aerodynamic interactions with the wing and tail. To estimate these forces and moments accurately, we must make use of computer simulations and/or wind tunnel tests. Such methods should always be employed in the later phases of the airplane design process. However, for the purpose of preliminary design, it is useful to have some method that can be used to easily estimate these effects.

As a first approximation, the effect of the fuselage, nacelles, and external stores on lift is usually neglected. However, even for preliminary design, some estimation of the effect of the fuselage, nacelles, and external stores on airplane moments should be made. The method to be presented here is taken from Hoak (1960), and is presented in terms of the fuselage, but the same approach can be used for the nacelles and external stores.

Consider the fuselage geometry shown in Fig. 4.9.1. As has been our usual convention, l will indicate a distance measured aft of the CG and c will indicate the total length of an object in the chordwise direction. While the reference area for a lifting surface such as the wing or tail is always taken as the planform area, the reference area for a fuselage or nacelle is typically defined to be the projected frontal area. This is the maximum cross-sectional area of the body in a plane normal to the axial chord.

A rough estimate of the fuselage pitching moment coefficient about the airplane center of gravity can be made from the experimental correlation,

$$C_{m_f} \equiv \frac{m_{cg_f}}{\frac{1}{2}\rho V_\infty^2 S_f c_f} = -2\frac{l_f}{c_f}\left[1 - 1.76\left(\frac{d_f}{c_f}\right)^{3/2}\right]\alpha_f \tag{4.9.1}$$

where S_f is the maximum cross-sectional area of the fuselage, c_f is the chord length of the fuselage, l_f is the distance that the fuselage center of pressure is aft of the airplane center of gravity, d_f is the diameter of a circle having the same area as the maximum cross-sectional area of the fuselage,

$$d_f \equiv 2\sqrt{S_f/\pi} \tag{4.9.2}$$

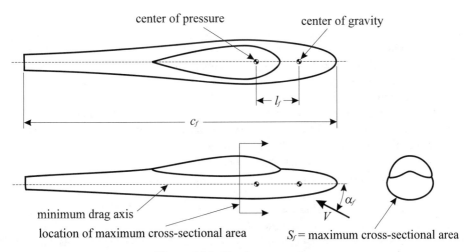

Figure 4.9.1. Fuselage geometry.

and α_f is the angle of attack for the minimum drag axis of the fuselage. As a first estimate, the fuselage center of pressure can be assumed to be halfway between the nose and the point of maximum cross-sectional area. Since this point is normally forward of the airplane's center of gravity, l_f is typically negative.

The pitching moment computed from Eq. (4.9.1) can be added to the contributions from the wing and the tail to obtain an estimate for the complete airframe. However, the reference area for the complete airplane was defined to be the planform area of the main wing, and the reference length was defined to be the mean chord length of the main wing. Thus, the contribution of the fuselage to the total airplane pitching moment coefficient can be approximated by multiplying the result from Eq. (4.9.1) by the product $S_f c_f$ and dividing by the product $S_w \bar{c}_w$:

$$(\Delta C_m)_f \equiv \frac{m_{cg_f}}{\frac{1}{2}\rho V_\infty^2 S_w \bar{c}_w} = \frac{S_f c_f}{S_w \bar{c}_w} C_{m_f} = -2\frac{S_f l_f}{S_w \bar{c}_w}\left[1 - 1.76\left(\frac{d_f}{c_f}\right)^{3/2}\right]\alpha_f \quad (4.9.3)$$

This can be written as

$$(\Delta C_m)_f = (\Delta C_{m0})_f + (\Delta C_{m,\alpha})_f \alpha \quad (4.9.4)$$

where

$$(\Delta C_{m0})_f = -2\frac{S_f l_f}{S_w \bar{c}_w}\left[1 - 1.76\left(\frac{d_f}{c_f}\right)^{3/2}\right]\alpha_{0f} \quad (4.9.5)$$

$$(\Delta C_{m,\alpha})_f = -2\frac{S_f l_f}{S_w \bar{c}_w}\left[1 - 1.76\left(\frac{d_f}{c_f}\right)^{3/2}\right] \quad (4.9.6)$$

α is the airplane angle of attack relative to the fuselage reference line, and α_{0f} is the angle that the minimum drag axis of the fuselage makes with the fuselage reference line. Because the fuselage center of pressure is usually forward of the airplane center of gravity, making l_f negative, the fuselage contribution to the pitch stability derivative is typically destabilizing.

It is important to remember that this experimental correlation gives only a rough estimate of the aerodynamic pitching moment generated on the fuselage, nacelles, and external stores. It should only be used as a first approximation for preliminary design. Wind tunnel tests and/or computer simulations should always be used in the later phases of design. Because the aerodynamic moments generated on the fuselage, nacelles, and external stores are primarily the result of pressure forces, panel methods will give a reasonable estimate for these moments. However, for best results, viscous computational fluid dynamics (CFD) or wind tunnel tests should be used. Although nacelles and external stores are typically much smaller than the fuselage, their effect on stability can be very significant and should not be neglected.

EXAMPLE 4.9.1. The airplane described in Example 4.4.1 has a fuselage that is 23 feet long. The point of maximum cross-section is located 9 feet aft of the nose and the airplane center of gravity is 8 feet aft of the nose. The maximum cross-sectional area of the fuselage is 21 ft². Estimate the static margin in the linear lift range for the complete airframe, including the effect of the fuselage.

Solution. For this airplane we have

$$S_w = 180 \text{ ft}^2, \quad b_w = 33 \text{ ft}, \quad C_{L_w,\alpha} = 4.44, \quad l_w = -0.71 \text{ ft},$$

$$S_h = 36 \text{ ft}^2, \quad b_h = 12 \text{ ft}, \quad C_{L_h,\alpha} = 3.97, \quad l_h = 14.29 \text{ ft},$$

$$\eta_h = 1.0, \quad \varepsilon_{d,\alpha} = 0.44, \quad S_f = 21 \text{ ft}^2, \quad d_f = 2\sqrt{21 \text{ ft}^2/\pi} = 5.17 \text{ ft},$$

$$c_f = 23 \text{ ft}, \quad l_f = (9 \text{ ft}/2) - 8 \text{ ft} = -3.5 \text{ ft}$$

By definition, the moment slope relative to the stick-fixed neutral point is zero. Thus, using Eq. (4.9.6) to estimate the fuselage contribution yields

$$C_{m_{np},\alpha} \equiv \frac{\partial C_{m_{np}}}{\partial \alpha} = -\frac{l_w - l_{np}}{\overline{c}_w} C_{L_w,\alpha} - \frac{S_h(l_h - l_{np})}{S_w \overline{c}_w} \eta_h (1 - \varepsilon_{d,\alpha}) C_{L_h,\alpha}$$

$$- 2\frac{S_f(l_f - l_{np})}{S_w \overline{c}_w}\left[1 - 1.76\left(\frac{d_f}{c_f}\right)^{3/2}\right] = 0$$

Solving for the static margin gives

$$\frac{l_{np}}{\overline{c}_w} = \frac{l_w C_{L_w,\alpha} + \dfrac{S_h l_h}{S_w}\eta_h(1 - \varepsilon_{d,\alpha})C_{L_h,\alpha} + 2\dfrac{S_f l_f}{S_w}\left[1 - 1.76\left(\dfrac{d_f}{c_f}\right)^{3/2}\right]}{\overline{c}_w\left\{C_{L_w,\alpha} + \dfrac{S_h}{S_w}\eta_h(1 - \varepsilon_{d,\alpha})C_{L_h,\alpha} + 2\dfrac{S_f}{S_w}\left[1 - 1.76\left(\dfrac{d_f}{c_f}\right)^{3/2}\right]\right\}}$$

$$= \frac{(-0.71)4.44 + \dfrac{36(14.29)}{180}1.0(1 - 0.44)3.97 + 2\dfrac{21(-3.5)}{180}\left[1 - 1.76\left(\dfrac{5.17}{23}\right)^{3/2}\right]}{\dfrac{180}{33}\left\{4.44 + \dfrac{36}{180}1.0(1 - 0.44)3.97 + 2\dfrac{21}{180}\left[1 - 1.76\left(\dfrac{5.17}{23}\right)^{3/2}\right]\right\}}$$

$$= 9\%$$

Comparing this result with that from Example 4.4.1, we see that the destabilizing effect of the fuselage was to move the neutral point forward by about 3 percent of the wing chord. This is clearly significant in view of the fact that the design static margin is often on the order of 5 percent of the wing chord.

4.10. Contribution of Running Propellers

Running propellers can have a profound effect on the trim and static stability of an airplane. The contributions of a running propeller to trim and static stability can be divided into two main categories. These are the direct effects, arising from the aerodynamic forces and moments on the propeller itself, and the indirect effects, which result from the interaction of the propeller slipstream with other surfaces of the airplane.

The effects of the propeller slipstream on other airplane surfaces such as the wing, tail, and fuselage are very complex and do not lend themselves to accurate analytical treatment. These effects can be accurately evaluated only from powered wind tunnel tests. Such tests are commonly performed in the final phase of the airplane design process. Simply because of complexity, the slipstream effects are usually neglected in the preliminary design phase. It should be remembered, however, that the slipstream effects can be significant.

The direct effects of the propeller forces and moments, on the other hand, can be estimated analytically using the method presented in Chapter 2. When a rotating propeller is advancing through the air so that the axis of rotation make some angle of attack with the freestream, in addition to the thrust, a normal force is produced that is proportional to the angle of attack. This is shown in Figs. 4.10.1 and 4.10.2. The pitching moment that the propeller produces about the airplane's center of gravity is a result of the thrust and normal force, each multiplied by the appropriate moment arm,

$$m_{cg_p} = -h_p T - l_p N_p \qquad (4.10.1)$$

where T and N_p are, respectively, the propeller thrust and normal force. For consistency, we are defining l_p as the axial distance that the propeller is mounted aft of the CG and h_p is the vertical distance that the propeller axis is above the CG. For a conventional tractor prop, l_p is typically negative, and for a pusher prop, l_p is usually positive.

Figure 4.10.1. Forces generated by an aft propeller. (Photograph by Barry Santana)

Figure 4.10.2. Forces generated by a forward propeller. (Photograph by Barry Santana)

The pitching moment computed from Eq. (4.10.1) can be added to the pitching moment obtained for all of the other airplane components to determine the pitching moment for the complete airplane. Thus, the contribution of the propeller to the total airplane pitching moment coefficient is found by nondimensionalizing Eq. (4.10.1),

$$\left(\Delta C_m\right)_p = \frac{m_{cg_p}}{\frac{1}{2}\rho V_\infty^2 S_w \overline{c}_w} = -\frac{h_p}{\overline{c}_w}\frac{T}{\frac{1}{2}\rho V_\infty^2 S_w} - \frac{l_p}{\overline{c}_w}\frac{N_p}{\frac{1}{2}\rho V_\infty^2 S_w} \tag{4.10.2}$$

For small thrust angles the equilibrium thrust is nearly equal to the drag and Eq. (4.10.2) can be closely approximated as

$$\left(\Delta C_m\right)_p = \frac{m_{cg_p}}{\frac{1}{2}\rho V_\infty^2 S_w \overline{c}_w} = -\frac{h_p}{\overline{c}_w} C_D - \frac{l_p}{\overline{c}_w}\frac{N_p}{\frac{1}{2}\rho V_\infty^2 S_w} \tag{4.10.3}$$

While the equilibrium thrust developed by an airplane's propeller can be related to the airplane drag, this cannot be said for the propeller normal force. To evaluate the propeller normal force, we return to the material presented in Chapter 2.

Propeller forces are conventionally expressed in terms of the thrust coefficient and the normal force coefficient,

$$C_T \equiv \frac{T}{\rho(\omega/2\pi)^2 d_p^4} \tag{4.10.4}$$

$$C_{N_p} \equiv \frac{N_p}{\rho(\omega/2\pi)^2 d_p^4} \tag{4.10.5}$$

where d_p is the propeller diameter and ω is the angular velocity. The normal force coefficient is linearly proportional to the angle of attack, α_p, that the propeller axis makes with the direction of relative airflow,

$$C_{N_p} = C_{N_p,\alpha} \, \alpha_p \qquad (4.10.6)$$

The proportionality constant is the dimensionless normal force gradient, defined as

$$C_{N_p,\alpha} \equiv \frac{1}{\rho(\omega/2\pi)^2 d_p^4} \frac{\partial N_p}{\partial \alpha_p} \qquad (4.10.7)$$

For a given propeller geometry both the thrust coefficient and the dimensionless normal force gradient are strong functions of the propeller advance ratio, J, which was defined in Chapter 2 as

$$J \equiv \frac{2\pi V_\infty}{\omega d_p} \qquad (4.10.8)$$

With a fixed-pitch propeller of given geometry, to attain some particular level flight airspeed, the pilot must adjust the propeller's rotational speed so that the propeller thrust will balance the drag at the desired airspeed. The advance ratio that is required to balance the drag with the thrust determines the propeller normal force gradient for a particular operating condition. Since the thrust coefficient decreases with increasing advance ratio while the dimensionless normal force gradient increases with increasing advance ratio, the propeller's effect on static stability and trim will vary significantly with propeller advance ratio.

Using Eqs. (4.10.5) through (4.10.8) in Eq. (4.10.3), the contribution of the propeller to the total airplane pitching moment coefficient is

$$\left(\Delta C_m\right)_p = -\frac{h_p}{\bar{c}_w} C_D - \frac{2 d_p^2 l_p}{S_w \bar{c}_w} \frac{C_{N_p,\alpha}}{J^2} \alpha_p \qquad (4.10.9)$$

The dimensionless normal force gradient, which appears in Eq. (4.10.9), depends on the number and geometry of the propeller blades, as well as the advance ratio. For known propeller geometry, the normal force gradient can be determined as a function of advance ratio using the method presented in Chapter 2. For example, Fig. 4.10.3 shows the dimensionless normal force gradient divided by the advance ratio squared, plotted as a function of advance ratio and pitch-to-diameter ratio, K_c, for one particular propeller blade geometry.

The propeller angle of attack in Eq. (4.10.9) is affected by the airflow around the wing. If the propeller is aft of the wing, it is located in a region of downwash and the propeller angle of attack is decreased. If the propeller is forward of the wing, it is in a region of upwash and the propeller angle of attack is increased. The angle of attack that the propeller axis makes with the local relative airflow can be written as

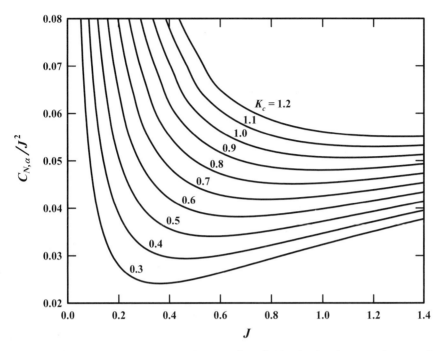

Figure 4.10.3. Dimensionless normal force gradient divided by the advance ratio squared.

$$\alpha_p = \alpha - \varepsilon_{dp} + \alpha_{0p} \tag{4.10.10}$$

where α is the airplane angle of attack relative to the fuselage reference line, ε_{dp} is the local downwash angle at the position of the propeller, and α_{0p} is the angle that the propeller axis makes with the fuselage reference line. It should be noted that in some textbooks, a propeller upwash angle is used in Eq. (4.10.10), because propellers are most commonly mounted ahead of the wing. For consistency with the material presented in Sec. 4.5, in this textbook we are defining ε_{dp} as the propeller downwash angle. With the present definition, if the propeller is forward of the wing, ε_{dp} is negative. If the propeller is aft of the wing, ε_{dp} is positive.

Using Eq. (4.10.10) with Eq. (4.10.9), the contribution of the propeller to the total airplane pitching moment coefficient can be written as

$$\left(\Delta C_m\right)_p = \left(\Delta C_{m0}\right)_p + \left(\Delta C_{m,\alpha}\right)_p \alpha \tag{4.10.11}$$

where

$$\left(\Delta C_{m0}\right)_p = -\frac{h_p}{\overline{c}_w} C_D - \frac{2 d_p^2 l_p}{S_w \overline{c}_w} \frac{C_{N_p,\alpha}}{J^2} (\alpha_{0p} - \varepsilon_{d0p}) \tag{4.10.12}$$

$$(\Delta C_{m,\alpha})_p = -\frac{2 d_p^2 l_p}{S_w \bar{c}_w}(1 - \varepsilon_{d,\alpha})_p \frac{C_{N_p,\alpha}}{J^2} \qquad (4.10.13)$$

ε_{d0p} is the propeller downwash angle with the fuselage reference line at zero angle of attack and $(\varepsilon_{d,\alpha})_p$ is the downwash gradient at the position of the propeller. Since the change in the normal force coefficient with angle of attack is positive, if the propeller is forward of the center of gravity, l_p is negative and the propeller contribution to the pitch stability derivative is destabilizing. If the propeller is aft of the center of gravity, l_p is positive and the propeller contribution to the pitch stability derivative is stabilizing.

EXAMPLE 4.10.1. The airplane that is described in Examples 4.4.1 and 4.9.1 is powered by a piston engine turning a 74-inch propeller at 2,350 rpm and is flying at 80 mph. The change in normal force coefficient with propeller angle of attack for this propeller and operating condition is 0.04. The propeller is mounted in front of the fuselage and is 9 feet forward of the center of gravity. Estimate the static margin in the linear lift range for the complete airplane, including the effect of the fuselage and propeller.

Solution. For this airplane we have

$$S_w = 180 \text{ ft}^2, \quad b_w = 33 \text{ ft}, \quad C_{L_w,\alpha} = 4.44, \quad l_w = -0.71 \text{ ft},$$
$$S_h = 36 \text{ ft}^2, \quad b_h = 12 \text{ ft}, \quad C_{L_h,\alpha} = 3.97, \quad l_h = 14.29 \text{ ft},$$
$$R_{T_w} = 0.40, \quad \Lambda_w = 0°, \quad \eta_h = 1.0, \quad (\varepsilon_{d,\alpha})_h = 0.44,$$
$$S_f = 21 \text{ ft}^2, \quad d_f = 2\sqrt{21 \text{ ft}^2/\pi} = 5.17 \text{ ft}, \quad c_f = 23 \text{ ft}, \quad l_f = -3.5 \text{ ft},$$
$$d_p = 74/12 \text{ ft}, \quad \omega/2\pi = 2,350 \text{ rpm}, \quad l_p = -9 \text{ ft}, \quad V_\infty = 80 \text{ mph}, \quad C_{N_p,\alpha} = 0.04$$

For this propeller and airspeed the advance ratio is

$$J = \frac{V_\infty}{(\omega/2\pi)d_p} = \frac{80 \times 5,280/3,600}{(2,350/60)(74/12)} = 0.4858$$

The downwash angle for the propeller can be estimated using the method presented in Sec. 4.5. For this wing and propeller position, we have

$$R_{A_w} = 6.05, \quad R_{T_w} = 0.40, \quad \bar{x} = \frac{l_p - l_w}{b_w/2} = -0.502, \quad \bar{y} = 0.0, \quad \Lambda_w = 0°$$

and the propeller downwash angle is

$$\varepsilon_{dp} = \frac{\kappa_v \kappa_p \kappa_s}{\kappa_b} \frac{C_{L_w}}{R_{A_w}} = \frac{1.035(-0.165)1.0}{0.759} \frac{C_{L_w}}{R_{A_w}} = -0.225 \frac{C_{L_w}}{R_{A_w}}$$

The propeller downwash gradient is then

$$(\varepsilon_{d,\alpha})_p = -0.225\frac{C_{L_w,\alpha}}{R_{A_w}} = -0.225\frac{4.44}{6.05} = -0.165$$

By definition, the moment slope relative to the neutral point is zero. Thus, using Eqs. (4.9.6) and (4.10.13) to estimate the fuselage and propeller contributions to the moment slope,

$$\begin{aligned}
C_{m_{np},\alpha} &= -\frac{l_w - l_{np}}{\bar{c}_w}C_{L_w,\alpha} - \frac{S_h(l_h - l_{np})}{S_w\bar{c}_w}\eta_h(1-\varepsilon_{d,\alpha})_hC_{L_h,\alpha} \\
&\quad + \frac{l_f - l_{np}}{l_f}(\Delta C_{m,\alpha})_f - \frac{2d_p^2(l_p - l_{np})}{J^2 S_w\bar{c}_w}(1-\varepsilon_{d,\alpha})_p\, C_{N_p,\alpha} \\
&= 0
\end{aligned}$$

where

$$\begin{aligned}
(\Delta C_{m,\alpha})_f &= -2\frac{S_f l_f}{S_w\bar{c}_w}\left[1-1.76\left(\frac{d_f}{c_f}\right)^{3/2}\right] \\
&= -2\frac{21(-3.5)}{180(180/33)}\left[1-1.76\left(\frac{5.17}{23}\right)^{3/2}\right] \\
&= 0.1216
\end{aligned}$$

Solving for the static margin gives

$$\frac{l_{np}}{\bar{c}_w} = \frac{l_w C_{L_w,\alpha} + \dfrac{S_h l_h}{S_w}\eta_h(1-\varepsilon_{d,\alpha})_h C_{L_h,\alpha} - \bar{c}_w(\Delta C_{m,\alpha})_f + \dfrac{2d_p^2 l_p}{J^2 S_w}(1-\varepsilon_{d,\alpha})_p C_{N_p,\alpha}}{\bar{c}_w\left\{C_{L_w,\alpha} + \dfrac{S_h}{S_w}\eta_h(1-\varepsilon_{d,\alpha})_h C_{L_h,\alpha} - \dfrac{\bar{c}_w}{l_f}(\Delta C_{m,\alpha})_f + \dfrac{2d_p^2}{J^2 S_w}(1-\varepsilon_{d,\alpha})_p C_{N_p,\alpha}\right\}}$$

$$= \frac{(-.71)4.44 + \dfrac{36(14.29)}{180}1.(1-.44)3.97 - \dfrac{180}{33}.1216 + \dfrac{2(74/12)^2(-9)}{(.4858)^2 180}(1.165).04}{\dfrac{180}{33}\left\{4.44 + \dfrac{36}{180}1.(1-.44)3.97 - \dfrac{180/33}{-3.5}.1216 + \dfrac{2(74/12)^2}{(.4858)^2 180}(1.165).04\right\}}$$

$$= \underline{6\%}$$

Comparing this result with that from Example 4.9.1, we see that the destabilizing effect of the propeller was to move the neutral point forward by about 3 percent of the wing chord. This is quite typical of what might be expected for an airplane with low thrust-to-weight ratio. However, an even greater destabilizing effect can be encountered for airplanes with higher thrust-to-weight ratio, such as the fighter planes of World War II.

4.11. Contribution of Jet Engines

Turbojet or turbofan engines can also make a significant contribution to the trim and static stability of an airplane. These effects are also divided into two categories, those arising directly from the forces and moments produced by the engine and those which result from the interaction of the exhaust jet with other surfaces of the airplane.

Even if the exhaust jet does not impinge directly upon any surface of the airplane, the jet can still have significant indirect effects resulting from the entrained flow. Because the velocity of the exhaust gas is much greater than that of the surrounding air, a turbulent shear layer develops between the exhaust jet and the freestream. The viscous shear and turbulent mixing that occur between the exhaust jet and the surrounding air decrease the velocity of the exhaust gas and increase the velocity of the adjacent air. This entrainment of freestream air into the exhaust jet causes the jet to broaden as it moves downstream and induces a flow of ambient air toward the axis of the jet, as is shown in Fig. 4.11.1. If the tail or any other surface of the airplane is placed in this entrainment flow field, the local angle of attack will be modified by the entrained flow. For example, if the horizontal tail is above the exhaust jet, as shown in Fig. 4.11.1, the angle of attack will be decreased as a result of the entrainment.

The indirect effects of the exhaust jet on other airplane surfaces, such as the wing, the tail, and the fuselage, are very complex and can only be evaluated analytically with the use of computational methods on a digital computer. Because entrainment is a viscous phenomenon, inviscid methods such as panel codes cannot be used for this analysis. A viscous CFD code must be used. Such methods can be used in the final phase of the airplane design process. However, because of complexity, the indirect effects of jet exhaust are usually neglected in the preliminary design phase. It should be remembered, however, that such effects could be significant.

The direct effects of the forces and moments produced by a jet or turbofan engine are essentially identical to those produced by an engine-propeller combination. Like a propeller, a turbojet or turbofan engine produces both an axial thrust and an off-axis normal force. The normal force is proportional to the angle that the axis of the exhaust jet makes with the local freestream at the engine inlet. As was the case for a propeller, the pitching moment that a jet engine produces about the airplane center of gravity is

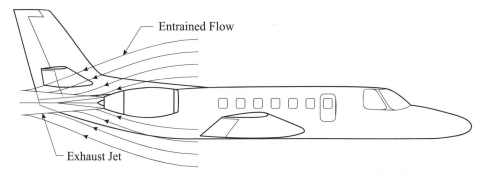

Figure 4.11.1. Flow entrained by the exhaust of a turbojet or turbofan engine.

simply a result of the thrust and the normal force, each multiplied by the appropriate moment arm. The only difference between computing the pitching moment contribution for a jet engine and that for propeller is the manner in which the normal force is computed.

The velocity of the air passing through a turbojet or turbofan engine can undergo a change in direction as well as a change in magnitude. The thrust and normal forces exerted on the airplane by the engine result directly from these changes in the velocity vector. Consider a control volume that encloses the engine of the airplane that is shown in Fig. 4.11.2. This control volume extends everywhere sufficiently far from the engine so that ambient pressure exists on the entire enclosing surface. Writing the vector form of Newton's second law for this nonaccelerating control volume, we have

$$\mathbf{F}_j = \dot{m}_j (\mathbf{V}_i - \mathbf{V}_j) \tag{4.11.1}$$

where $\mathbf{F}_j$ is the resultant force exerted on the airplane by the engine, $\mathbf{V}_i$ and $\mathbf{V}_j$ are, respectively, the inlet and exit velocities, and $\dot{m}_j$ is the mass flow rate passing through the engine. The axial and normal components of this vector equation result in

$$T = \dot{m}_j (V_j - V_i \cos \alpha_j) \tag{4.11.2}$$

$$N_j = \dot{m}_j V_i \sin \alpha_j \tag{4.11.3}$$

where V_j is the magnitude of the exhaust jet velocity, V_i is the magnitude of the local freestream velocity at the inlet, and α_j is the angle of attack that the axis of the exhaust jet makes with the direction of the local freestream at the inlet. For small angles, Eqs. (4.11.2) and (4.11.3) can be approximated as

$$T = \dot{m}_j (V_j - V_\infty) \tag{4.11.4}$$

$$N_j = \dot{m}_j V_\infty \alpha_j \tag{4.11.5}$$

where V_∞ is the far-field airspeed. Solving Eq. (4.11.4) for the mass flow rate and using the result in Eq. (4.11.5) gives

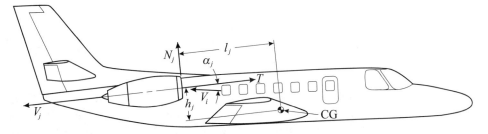

Figure 4.11.2. Forces generated by a turbojet or turbofan engine.

$$N_j = \frac{V_\infty T}{V_j - V_\infty} \alpha_j \tag{4.11.6}$$

From the geometry in Fig. 4.11.2, the pitching moment that the engine produces about the airplane center of gravity is

$$m_{cg_j} = -Th_j - N_j l_j \tag{4.11.7}$$

Nondimensionalizing Eq. (4.11.7) and applying Eq. (4.11.6), the contribution of the jet engine to the total airplane pitching moment coefficient is

$$(\Delta C_m)_j = \frac{m_{cg_j}}{\frac{1}{2}\rho V_\infty^2 S_w \bar{c}_w} = -\frac{T}{\frac{1}{2}\rho V_\infty^2 S_w}\left(\frac{h_j}{\bar{c}_w} + \frac{V_\infty l_j}{(V_j - V_\infty)\bar{c}_w}\alpha_j\right) \tag{4.11.8}$$

The local angle of attack in Eq. (4.11.8) is affected by the airflow around the wing. If the engine inlet is aft of the wing, it is located in a region of downwash and the local angle of attack is decreased. If the inlet is forward of the wing, it is in a region of upwash and the local angle of attack is increased. The angle of attack that the exhaust jet axis makes with the local relative airflow at the engine inlet can be written as

$$\alpha_j = \alpha - \varepsilon_{dj} + \alpha_{0j} \tag{4.11.9}$$

where α is the airplane angle of attack relative to the fuselage reference line, ε_{dj} is the local downwash angle at the position of the engine inlet, and α_{0j} is the angle that the exhaust jet axis makes with the fuselage reference line.

Using Eq. (4.11.9) in Eq. (4.11.8), the contribution of a jet engine to the total airplane pitching moment coefficient can be written as

$$(\Delta C_m)_j = (\Delta C_{m0})_j + (\Delta C_{m,\alpha})_j \alpha \tag{4.11.10}$$

where

$$(\Delta C_{m0})_j = -\frac{T}{\frac{1}{2}\rho V_\infty^2 S_w}\left[\frac{h_j}{\bar{c}_w} + \frac{V_\infty l_j}{(V_j - V_\infty)\bar{c}_w}(\alpha_{0j} - \varepsilon_{d0j})\right] \tag{4.11.11}$$

$$(\Delta C_{m,\alpha})_j = -\frac{T}{\frac{1}{2}\rho V_\infty^2 S_w}\frac{V_\infty l_j}{(V_j - V_\infty)\bar{c}_w}(1 - \varepsilon_{d,\alpha})_j \tag{4.11.12}$$

ε_{d0j} is the inlet downwash angle with the fuselage reference line at zero angle of attack, and $(\varepsilon_{d,\alpha})_j$ is the downwash gradient at the position of the engine inlet. Since the thrust is positive in normal flight, if the inlet is forward of the center of gravity, l_j is negative and the engine's contribution to the pitch stability derivative is destabilizing. If the inlet is aft of the center of gravity, l_j is positive and the engine's contribution to the pitch stability derivative is stabilizing.

As discussed in Sec. 2.1, the ideal propulsive efficiency for a jet engine is a function only of the exhaust jet velocity and the freestream velocity,

$$\eta_{p_i} = \frac{2V_\infty}{V_\infty + V_j} = \frac{2}{1 + V_j/V_\infty} \tag{4.11.13}$$

Thus, we can express the ratio of the exhaust jet velocity to the freestream velocity as a function of ideal propulsive efficiency. Rearranging Eq. (4.11.13), the velocity ratio for a jet engine can be expressed as

$$\frac{V_j}{V_\infty} = \frac{2}{\eta_{p_i}} - 1 \tag{4.11.14}$$

or

$$\frac{V_\infty}{V_j - V_\infty} = \frac{\eta_{p_i}}{2(1 - \eta_{p_i})} \tag{4.11.15}$$

The velocity ratio in Eq. (4.11.15), which also appears in Eqs. (4.11.11) and (4.11.12), is plotted in Fig. 4.11.3. This figure combined with Eqs. (4.11.11) and (4.11.12) shows that increasing the propulsive efficiency of a jet engine will also increase the engine's effect on static stability and trim.

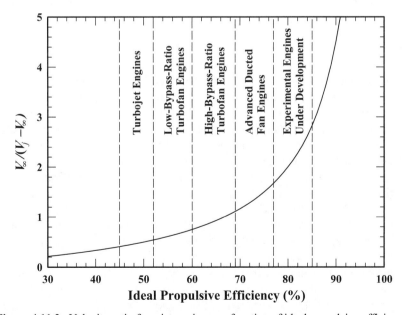

Figure 4.11.3. Velocity ratio for a jet engine as a function of ideal propulsive efficiency.

Using Eq. (4.11.15) in Eqs. (4.11.11) and (4.11.12), we have

$$(\Delta C_{m0})_j = -\frac{T}{\frac{1}{2}\rho V_\infty^2 S_w}\left[\frac{h_j}{\bar{c}_w} + \frac{\eta_{p_i} l_j}{2(1-\eta_{p_i})\bar{c}_w}(\alpha_{0j} - \varepsilon_{d0j})\right] \qquad (4.11.16)$$

$$(\Delta C_{m,\alpha})_j = -\frac{T}{\frac{1}{2}\rho V_\infty^2 S_w}\frac{\eta_{p_i} l_j}{2(1-\eta_{p_i})\bar{c}_w}(1-\varepsilon_{d,\alpha})_j \qquad (4.11.17)$$

For small thrust angles, the thrust is very nearly equal to the drag and Eqs. (4.11.16) and (4.11.17) can be closely approximated as

$$(\Delta C_{m0})_j = -C_D\left[\frac{h_j}{\bar{c}_w} + \frac{\eta_{p_i} l_j}{2(1-\eta_{p_i})\bar{c}_w}(\alpha_{0j} - \varepsilon_{d0j})\right] \qquad (4.11.18)$$

$$(\Delta C_{m,\alpha})_j = -C_D\frac{\eta_{p_i} l_j}{2(1-\eta_{p_i})\bar{c}_w}(1-\varepsilon_{d,\alpha})_j \qquad (4.11.19)$$

From Eqs. (4.11.18) and (4.11.19) we can estimate the effects of a jet engine on airplane trim and stability if we know the ideal propulsive efficiency for the engine. From knowledge of the engine type, a rough estimate for this efficiency can be made from the information in Fig. 4.11.3. After replacing Eqs. (4.10.12) and (4.10.13) with Eqs. (4.11.18) and (4.11.19), the direct contribution of a turbojet or turbofan engine to the trim and static stability of an airplane is identical to that of an engine-propeller combination, as described in Sec. 4.10.

4.12. Problems

4.1.   An airplane is to weigh 2,000 pounds and its fuselage is to be 15 feet long. The center of gravity is to be 7.5 feet aft of its nose. The airplane is to be a wing-tail configuration. It is to be designed to fly at 15,000 feet at 150 mph. For this first analysis assume that all lifting surfaces are symmetric thin airfoils with infinite aspect ratio. Neglecting changes in weight with the size of the wing and tail, determine the following (there are many possible solutions):
(a)   The size of the wing and tail required to maintain trimmed cruise.
(b)   The location of the wing and tail in order to place the airplane's neutral point 0.5 feet aft of the CG.
(c)   The angle that the wing and the tail make with the upstream flow, taking the fuselage to make an angle of 0 degrees with the upstream flow.

4.2.   For the airplane in problem 4.1, assume that the wing has an aspect ratio of 6.0, a taper ratio of 0.4, and no sweep. Also assume that the horizontal tail has an aspect ratio of 4.0, a taper ratio of 0.4, and no sweep. Refine your design to account for downwash.

4.3.   The British Spitfire Mark XVIII that is shown in Figs. 4.12.1 and 4.12.2 weighs 8,375 lbf and both the wing and horizontal tail have elliptic planform shapes. The horizontal tail has a symmetric airfoil section when the elevator is not deflected. The wing and the horizontal tail have the following properties:

$$S_w = 244 \text{ ft}^2, \quad b_w = 36.83 \text{ ft}, \quad C_{L_w,\alpha} = 4.62, \quad \alpha_{L0w} = -2.20°, \quad C_{m_w} = -0.053,$$

$$S_h = 31 \text{ ft}^2, \quad b_h = 10.64 \text{ ft}, \quad C_{L_h,\alpha} = 4.06, \quad \varepsilon_e = 0.60, \quad C_{m_h,\delta_e} = -0.55,$$

$$l_h - l_w = 18.16 \text{ ft}$$

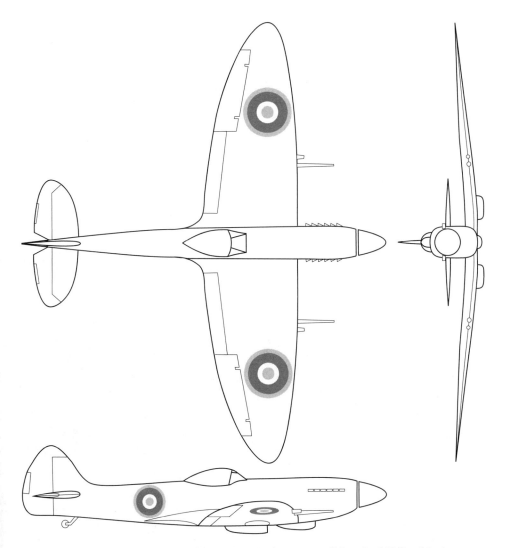

Figure 4.12.1. The British Spitfire Mark XVIII has an overall length of 32 feet 8 inches.

Figure 4.12.2. The British Spitfire Mark XVIII has an elliptic wing with a planform area of 244 ft². The wing spans 36 feet 10 inches and supports a normal gross weight of 8,375 pounds. Later versions were powered by a liquid-cooled, 12-cylinder, Rolls-Royce Griffon engine developing 2,375 horsepower at 2,750 rpm. The 132-inch, five-blade, Rotol propeller is turned with 1.961:1 gear reduction. The airplane has a top speed in level flight of 450 mph and has a rate of climb at sea level of over 5,000 ft/min. (Photographs by Barry Santana)

For the CG located at the wing quarter chord, the wing and the tail are mounted so that in level flight with an airspeed of 200 mph at standard sea level, trim is attained with no elevator deflection and a zero angle of attack for the fuselage reference line. Neglecting any effects associated with the propeller and fuselage, determine the mounting angles for the wing and the tail and the pitch stability derivative for the wing-tail combination. Use the simplified analysis presented in Sec. 4.3 and neglect the effects of wing downwash on the tail. In this and the following problems that deal with the spitfire, any dimensions that are not otherwise given may be scaled from Figs. 4.12.1 and 4.12.2.

4.4.   Excluding the effects of wing downwash, determine the static margin in the range of linear lift for the wing-tail combination of the Spitfire Mark XVIII that is described in problem 4.3.

4.5.   Using the method that was presented in Sec. 4.5, determine the relation between the downwash on the horizontal tail and the lift coefficient for the wing of the Spitfire Mark XVIII that is described in problem 4.3.

4.6.   Using the results that were obtained from problem 4.5 for the downwash, recompute the mounting angles for both the wing and tail of the Spitfire Mark XVIII that is described in problem 4.3. Compare the result to that obtained in problem 4.3.

4.7.   Including the effects of wing downwash but neglecting any effects associated with the propeller and fuselage, determine the static margin in the range of linear lift for the wing-tail combination of the Spitfire Mark XVIII that is described in problem 4.3. Compare the result to that obtained in problem 4.4.

4.8.   Assume that the wing and the horizontal tail of the Spitfire Mark XVIII have the following properties:

$$S_w = 244 \text{ ft}^2, \quad b_w = 36.83 \text{ ft}, \quad C_{L_w,\alpha} = 4.62, \quad C_{m_w} = -0.053,$$
$$\alpha_{L0w} = -2.20°, \quad \alpha_{0w} = 2.0°, \quad l_w = 0.0 \text{ ft},$$
$$S_h = 31 \text{ ft}^2, \quad b_h = 10.64 \text{ ft}, \quad C_{L_h,\alpha} = 4.06, \quad C_{m_h0} = 0.000,$$
$$\alpha_{L0h} = 0.0°, \quad \alpha_{0h} = 0.0°, \quad l_h = 18.16 \text{ ft},$$
$$W = 8,375 \text{ lbf}, \quad \varepsilon_e = 0.60, \quad C_{m_h,\delta_e} = -0.55$$

Including the effects of wing downwash but neglecting the effects of the propeller and fuselage, find the elevator deflection and the angle of attack required for trimmed level flight at sea level for airspeeds of 100, 200, and 300 mph. Also determine the lift on the horizontal tail at level trim for the same airspeeds. Assume a tail efficiency of 1.0 and use the simplified analysis that was presented in Sec. 4.3.

4.9.   For the Spitfire Mark XVIII described in problem 4.8, including the effects of downwash, plot the elevator deflection required to trim the airplane at 100 mph as a function of CG location measured relative to the quarter chord of the main wing. Neglect the effects of the propeller and fuselage.

4.10.   The Spitfire described in problem 4.8 has a fuselage 32 feet 8 inches long. The point of maximum cross-section is 16 feet aft of the nose and the CG is 11 feet aft of the nose. The maximum cross-sectional area of the fuselage is 15 ft². Estimate the static margin in the linear lift range for the complete airframe.

4.11.   A 2,375-horsepower piston engine turning a 132-inch, five-blade propeller powers the Spitfire Mark XVIII described in problem 4.8. When the airplane is flying at 200 mph the propeller is turning at 1,300 rpm. The change in normal force coefficient with propeller angle of attack for this propeller and operating condition is 0.270. The propeller is mounted 9 feet 4 inches forward of the center of gravity. Estimate the static margin in the linear lift range for the complete airplane, including the effect of the fuselage and propeller.

4.12.   Assume the following properties for the Spitfire Mark XVIII:

$$S_w = 244 \text{ ft}^2, \quad b_w = 36.83 \text{ ft}, \quad C_{L_w,\alpha} = 4.62, \quad C_{m_w} = -0.053,$$
$$\alpha_{L0w} = -2.20°, \quad \alpha_{0w} = 2.0°, \quad l_w = 0.0 \text{ ft}, \quad h_w = -2.6 \text{ ft},$$
$$C_{D0w} = 0.01, \quad C_{D_0,L_w} = 0.00, \quad e_w = 1.00,$$
$$S_h = 31 \text{ ft}^2, \quad b_h = 10.64 \text{ ft}, \quad C_{L_h,\alpha} = 4.06, \quad C_{m_h0} = 0.000,$$
$$\alpha_{L0h} = 0.0°, \quad \alpha_{0h} = 0.0°, \quad l_h = 18.16 \text{ ft}, \quad h_h = 0.0 \text{ ft},$$
$$C_{D0h} = 0.01, \quad C_{D_0,L_h} = 0.00, \quad e_h = 1.00,$$
$$W = 8,375 \text{ lbf}, \quad \varepsilon_e = 0.60, \quad C_{m_h,\delta_e} = -0.55, \quad l_p = -9.33 \text{ ft}, \quad h_p = 0.0 \text{ ft}$$

Including the effects of downwash but neglecting the propeller and fuselage, find the elevator deflection and angle of attack required for trimmed level flight at sea level for airspeeds of 100, 200, and 300 mph. Also determine the lift on the horizontal tail at level trim for the same airspeeds. Assume a tail efficiency of 1.0 and account for the effects of drag and vertical offset as discussed in Sec. 4.7. Compare the results with those obtained in problem 4.8.

4.13.   Assume that the Spitfire Mark XVIII has the properties listed in problem 4.12. With the controls fixed at the 200-mph trim condition, plot the pitching moment coefficient as a function of angle of attack from –20 to +12 degrees. Include the effects of wing downwash but neglect the effects of the propeller and fuselage. Assume that neither the wing nor the horizontal tail stalls. Assume a tail efficiency of 1.0 and account for the effects of drag and vertical offset as discussed in Sec. 4.7. On the same graph plot the pitching moment coefficient as obtained from the simplified analysis presented in Sec. 4.3.

4.14. A small jet transport with an aft tail weighs 73,000 pounds. The wing and the horizontal tail have the following properties:

$$S_w = 950 \text{ ft}^2, \quad b_w = 75 \text{ ft}, \quad C_{L_w,\alpha} = 4.30, \quad \alpha_{L0w} = -2.0°, \quad C_{m_w} = -0.050,$$

$$S_h = 207 \text{ ft}^2, \quad b_h = 31 \text{ ft}, \quad C_{L_h,\alpha} = 4.00, \quad \varepsilon_e = 0.51, \quad C_{m_h,\delta_e} = -0.49,$$

$$l_h - l_w = 42.5 \text{ ft}$$

The airplane's main wing has a taper ratio of 0.35 and a quarter-chord sweep angle of 15 degrees. The horizontal tail is mounted 17.5 feet above the wing and has a symmetric airfoil section when the elevator is not deflected. The wing and the horizontal tail are mounted so that when the airplane is in level flight with an airspeed of 400 mph at an altitude of 30,000 feet, trim is attained with no elevator deflection, no lift on the horizontal tail, and a zero angle of attack for the fuselage reference line. Neglecting any moments associated with the thrust deck and the fuselage, determine the location of the center of gravity, the mounting angles for the wing and the tail, and the pitch stability derivative for the wing-tail combination. Use the simplified analysis presented in Sec. 4.3 and assume a tail efficiency of 1.0.

4.15. Including the effects of wing downwash, determine the static margin in the range of linear lift for the wing-tail combination of the small jet transport that is described in problem 4.14. Neglect any effects associated with the thrust deck and the fuselage.

4.16. Two high-bypass-ratio turbofan engines power a small jet transport. Assume that the airplane has the following properties:

$$S_w = 950 \text{ ft}^2, \quad b_w = 75 \text{ ft}, \quad R_{T_w} = 0.35, \quad C_{L_w,\alpha} = 4.30, \quad C_{m_w} = -0.050,$$

$$\Lambda_w = 15°, \quad \alpha_{L0w} = -2.0°, \quad \alpha_{0w} = 1.5°, \quad l_w = 0.7 \text{ ft}, \quad h_w = -3.8 \text{ ft},$$

$$C_{D_{0w}} = 0.01, \quad C_{D_0,L_w} = 0.00, \quad e_w = 0.95,$$

$$S_h = 207 \text{ ft}^2, \quad b_h = 31 \text{ ft}, \quad R_{T_h} = 0.35, \quad C_{L_h,\alpha} = 4.00, \quad C_{m_h 0} = 0.000,$$

$$\Lambda_h = 15°, \quad \alpha_{L0h} = 0.0°, \quad \alpha_{0h} = 0.0°, \quad l_h = 43.2 \text{ ft}, \quad h_h = 13.7 \text{ ft},$$

$$C_{D_{0h}} = 0.01, \quad C_{D_0,L_h} = 0.00, \quad e_h = 0.96,$$

$$W = 73,000 \text{ lbf}, \quad \varepsilon_e = 0.51, \quad C_{m_h,\delta_e} = -0.49, \quad l_j = 9.4 \text{ ft}, \quad h_j = 2.4 \text{ ft},$$

$$C_{D_{L0}} = 0.020, \quad C_{D_0,L} = 0.0, \quad e = 0.67$$

Including the effects of wing downwash but neglecting effects of the thrust deck and fuselage, find the elevator deflection and the angle of attack required for trimmed level flight at 30,000 feet for airspeeds of 300, 400, and 500 mph. Also determine the lift on both the wing and the horizontal tail at level trim for the same altitude and airspeeds. Assume a tail efficiency of 1.0 and use the simplified analysis that was presented in Sec. 4.3.

4.17. For the small jet transport described in problem 4.16, including the effects of downwash, plot the elevator deflection required to trim the airplane at 400 mph as a function of CG location measured relative to the quarter chord of the main wing. Neglect the effects of the thrust deck and fuselage.

4.18. The small jet transport described in problem 4.16 has a fuselage that is 78 feet long. The point of maximum cross-section is located 37 feet aft of the nose, and the center of gravity is also 37 feet aft of the nose. The maximum cross-sectional area of the fuselage is 52 ft². Estimate the static margin in the range of linear lift for the complete airframe, including the effect of the fuselage.

4.19. Two high-bypass-ratio turbofan engines power the small jet transport that is described in problem 4.16. When the airplane is flying at 400 mph, the engines have an ideal propulsive efficiency of 0.64. The engine inlets are 9.4 feet aft of the center of gravity. Estimate the static margin in the linear lift range for the complete airplane, including the effect of the fuselage and thrust deck.

4.20. For the jet transport described in problem 4.16, including the effects of wing downwash but neglecting effects of thrust deck and fuselage, find the elevator deflection and the angle of attack required for trimmed level flight at 30,000 feet for airspeeds of 300, 400, and 500 mph. Also determine the lift on the wing and horizontal tail at level trim for the same airspeeds. Assume a tail efficiency of 1.0 and account for the effects of drag and vertical offset as discussed in Sec. 4.7. Compare the results with those obtained in problem 4.16.

4.21. Assume that the wing and the horizontal tail of the small jet transport have the properties listed in problem 4.16. With the controls fixed at the 400-mph trim condition, plot the pitching moment coefficient as a function of angle of attack from −20 to +12 degrees. Include the effects of wing downwash but neglect the effects of the thrust deck and fuselage. Assume that neither the wing nor the horizontal tail stalls. Assume a tail efficiency of 1.0 and account for the effects of drag and vertical offset as discussed in Sec. 4.7. On the same graph plot the pitching moment coefficient as obtained from the simplified analysis presented in Sec. 4.3.

4.22. Estimate the maximum total lift coefficient and the minimum trimmed airspeed for the wing-tail combination of the jet transport described in problem 4.16. The maximum lift coefficient for the main wing is 1.40. The maximum lift coefficient for the horizontal tail varies with elevator deflection according to

$$C_{L_h \max} = 1.4 + 0.955\delta_e$$

where δ_e is in radians. Consider maximum total lift coefficient and minimum airspeed to be that for trimmed flight when either the wing or the horizontal tail carries its respective maximum lift.

4.23. A small jet transport of the wing-canard design weighs 73,000 pounds. The wing and horizontal canard have the following properties:

$$S_w = 950 \text{ ft}^2, \quad b_w = 75 \text{ ft}, \quad C_{L_w,\alpha} = 4.30, \quad \alpha_{L0w} = -2.0°, \quad C_{m_w} = -0.050,$$

$$S_h = 219 \text{ ft}^2, \quad b_h = 36 \text{ ft}, \quad C_{L_h,\alpha} = 4.30, \quad \alpha_{L0h} = -2.0°, \quad C_{m_h0} = -0.050,$$

$$l_w - l_h = 42.5 \text{ ft}, \quad \varepsilon_e = 0.51, \quad C_{m_h,\delta_e} = -0.49$$

Both the main wing and canard have a taper ratio of 0.35 and a quarter-chord sweep angle of 15 degrees. The canard is mounted directly in front of the wing at the same vertical level. The wing and canard are mounted so that when the airplane is in level flight with an airspeed of 400 mph at an altitude of 30,000 feet, trim is attained with no control surface deflection and a zero angle of attack for the fuselage reference line. In the range of linear lift, the static margin for this wing-canard combination is 5 percent. Neglecting any moments associated with the thrust deck and fuselage, determine the location of the center of gravity, the mounting angles for the wing and the canard, and the pitch stability derivative for this wing-canard combination. Use the simplified analysis presented in Sec. 4.6 and assume a canard efficiency of 1.0. Include the effects of the upwash that is induced on the canard by the main wing, but neglect all other vortex interactions between the canard and the wing.

4.24. For the jet transport described in problem 4.23, recompute the mounting angles and CG location, neglecting the upwash induced on the canard by the wing.

4.25. Two high-bypass-ratio turbofan engines power the jet transport that is described in problem 4.23. Assume that the airplane has the following properties:

$$S_w = 950 \text{ ft}^2, \quad b_w = 75 \text{ ft}, \quad R_{T_w} = 0.35, \quad C_{L_w,\alpha} = 4.30, \quad C_{m_w} = -0.050,$$

$$\Lambda_w = 15°, \quad \alpha_{L0w} = -2.0°, \quad \alpha_{0w} = 1.5°, \quad l_w = 8.6 \text{ ft}, \quad h_w = -3.8 \text{ ft},$$

$$C_{D0w} = 0.01, \quad C_{D_0,L_w} = 0.00, \quad e_w = 0.95,$$

$$S_h = 219 \text{ ft}^2, \quad b_h = 36 \text{ ft}, \quad R_{T_h} = 0.35, \quad C_{L_h,\alpha} = 4.30, \quad C_{m_h0} = -0.050,$$

$$\Lambda_h = 15°, \quad \alpha_{L0h} = -2.0°, \quad \alpha_{0h} = 2.0°, \quad l_h = -33.9 \text{ ft}, \quad h_h = -3.8 \text{ ft},$$

$$C_{D0h} = 0.01, \quad C_{D_0,L_h} = 0.00, \quad e_h = 0.95,$$

$$W = 73,000 \text{ lbf}, \quad \varepsilon_e = 0.51, \quad C_{m_h,\delta_e} = -0.49, \quad l_j = 9.4 \text{ ft}, \quad h_j = 2.4 \text{ ft},$$

$$C_{D_{L0}} = 0.020, \quad C_{D_0,L} = 0.0, \quad e = 0.67$$

Including the effects of wing upwash but neglecting effects of the thrust deck and fuselage, find the control surface deflection and angle of attack required for trimmed level flight at 30,000 feet for airspeeds of 300, 400, and 500 mph. Also determine the lift on both the wing and the canard at level trim for the same altitude and airspeeds. Assume a canard efficiency of 1.0 and use the simplified analysis that was presented in Sec. 4.6.

4.26. For the small jet transport described in problem 4.25, including the effects of upwash, plot the control surface deflection required to trim the airplane at 400 mph as a function of CG location measured relative to the quarter chord of the main wing. Neglect the effects of the thrust deck and fuselage.

4.27. The small jet transport described in problem 4.25 has a fuselage that is 78 feet long. The point of maximum cross-section is located 37 feet aft of the nose and the center of gravity is also 37 feet aft of the nose. The maximum cross-sectional area of the fuselage is 52 ft². Estimate the static margin in the range of linear lift for the complete airframe, including the effect of the fuselage.

4.28. Two high-bypass-ratio turbofan engines power the small jet transport that is described in problem 4.25. When the airplane is flying at 400 mph the engines have an ideal propulsive efficiency of 0.64. The engine inlets are 9.4 feet aft of the center of gravity. Estimate the static margin in the linear lift range for the complete airplane, including the effect of the fuselage and thrust deck.

4.29. For the jet transport described in problem 4.25, including the effects of wing upwash but neglecting the thrust deck and fuselage, find the control surface deflection and the angle of attack required for trimmed level flight at 30,000 feet for airspeeds of 300, 400, and 500 mph. Also determine the lift on the wing and canard at level trim for the same airspeeds. Assume a canard efficiency of 1.0 and account for the effects of drag and vertical offset as discussed in Sec. 4.7. Compare the results with those obtained in problem 4.25.

4.30. Assume that the wing and canard of the small jet transport have the properties that are listed in problem 4.25. With the controls fixed at the 400-mph trim condition, plot the pitching moment coefficient as a function of angle of attack from –20 to +12 degrees. Include effects of wing upwash but neglect the effects of the thrust deck and fuselage. Assume that neither the wing nor the canard stalls. Assume a canard efficiency of 1.0 and account for the effects of drag and vertical offset as discussed in Sec. 4.7. On the same graph plot the pitching moment coefficient as obtained from the simplified analysis presented in Sec. 4.6.

4.31. Estimate the maximum total lift coefficient and the minimum trimmed airspeed for the wing-canard combination of the jet transport described in problem 4.25. The maximum lift coefficient for the main wing is 1.40. The maximum lift coefficient for the canard varies with control surface deflection according to

$$C_{L_h \max} = 1.4 + 0.955\,\delta_e$$

where δ_e is in radians. Consider maximum total lift coefficient and minimum airspeed to be that for trimmed flight when either the wing or the canard carries its respective maximum lift.

4.32. The turboprop transport shown in Fig. 4.12.3 has the following properties:

$$S_w = 1,745 \text{ ft}^2, \quad b_w = 132.6 \text{ ft}, \quad R_{T_w} = 0.57, \quad C_{L_w,\alpha} = 5.16, \quad C_{m_w} = -0.050,$$
$$\Lambda_w = 0°, \quad \alpha_{L0w} = -2.0°, \quad \alpha_{0w} = 0.0°, \quad l_w = -2.5 \text{ ft}, \quad h_w = 7.4 \text{ ft},$$
$$C_{D0w} = 0.01, \quad C_{D_0,L_w} = 0.00, \quad e_w = 0.96,$$
$$S_h = 528 \text{ ft}^2, \quad b_h = 52.7 \text{ ft}, \quad R_{T_h} = 0.35, \quad C_{L_h,\alpha} = 4.52, \quad C_{m_h0} = 0.000,$$
$$\Lambda_h = 7°, \quad \alpha_{L0h} = 0.0°, \quad \alpha_{0h} = -1.5°, \quad l_h = 44.5 \text{ ft}, \quad h_h = 7.4 \text{ ft},$$
$$C_{D0h} = 0.01, \quad C_{D_0,L_h} = 0.00, \quad e_h = 0.97,$$
$$W = 155,000 \text{ lbf}, \quad \varepsilon_e = 0.58, \quad C_{m_h,\delta_e} = -0.49, \quad l_p = -11.9 \text{ ft}, \quad h_p = 6.2 \text{ ft},$$
$$C_{D_{L0}} = 0.031, \quad C_{D_0,L} = 0.0, \quad e = 0.65$$

Including the effects of wing downwash but neglecting effects of the propellers and fuselage, find the elevator deflection and the angle of attack required for trimmed level flight at sea level for airspeeds of 200, 300, and 350 mph. Also determine the lift on both the wing and the horizontal tail at level trim for the same altitude and airspeeds. Assume a tail efficiency of 1.0 and use the simplified analysis that was presented in Sec. 4.3.

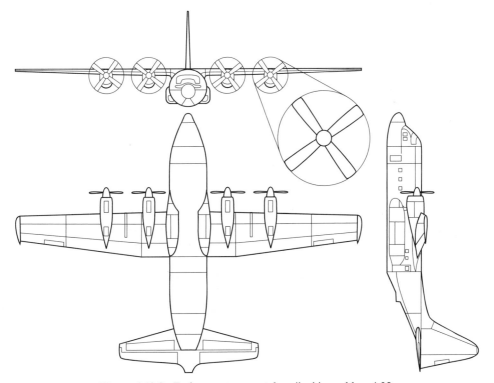

Figure 4.12.3. Turboprop transport described in problem 4.32.

4.33.   For the turboprop transport described in problem 4.32, including the effects of downwash, plot the elevator deflection required to trim the airplane at 200 mph as a function of CG location measured relative to the quarter chord of the main wing. Neglect the effects of the propellers, nacelles, and fuselage.

4.34.   The turboprop transport described in problem 4.32 has a fuselage that is 87.9 feet long. The point of maximum cross-section is located 41 feet aft of the nose and the center of gravity is also 41 feet aft of the nose. The maximum cross-sectional area of the fuselage is 173 ft^2. Estimate the static margin in the range of linear lift for the complete airframe, including the effect of the fuselage.

4.35.   Four 4,910-horsepower constant-speed turboprop engines turning at 13,820 rpm power the transport airplane described in problem 4.32. Each engine turns a 162-inch propeller through a gear reduction assembly having a total reduction ratio of 13.54 to 1. The engines operate at constant rotational speed, independent of flight speed. Changes in power requirements are met by changing the fuel flow and propeller pitch. As the fuel flow is changed, an automatic control system very precisely adjusts the propeller pitch to maintain constant engine rpm. For level flight at sea level with airspeeds of 200, 300, and 350 mph, the change in normal force coefficient with propeller angle of attack for these propellers and operating conditions are 0.277, 0.721, and 1.025, respectively. The propellers are mounted 11 feet 11 inches forward of the center of gravity. At all three of the above airspeeds, estimate the static margin in the linear lift range for the complete airplane, including the effect of the fuselage and propellers.

4.36.   For the turboprop transport described in problem 4.32, including the effects of wing downwash but neglecting effects of the fuselage, nacelles, and propellers, find the elevator deflection and the angle of attack required for trimmed level flight at sea level for airspeeds of 200, 300, and 350 mph. Also determine the lift on the wing and horizontal tail at level trim for the same airspeeds. Assume a tail efficiency of 1.0 and account for the effects of drag and vertical offset as discussed in Sec. 4.7. Compare the results with those obtained in problem 4.32.

4.37.   Assume that the wing and the horizontal tail of the turboprop transport have the properties listed in problem 4.32. With the controls fixed at the 300-mph trim condition, plot the pitching moment coefficient as a function of angle of attack from −20 to +12 degrees. Include the effects of wing downwash but neglect effects of the fuselage, nacelles, and propellers. Assume that neither the wing nor the horizontal tail stalls. Assume a tail efficiency of 1.0 and account for the effects of drag and vertical offset as discussed in Sec. 4.7. On the same graph plot the pitching moment coefficient as obtained from the simplified analysis presented in Sec. 4.3.

Chapter 5
Lateral Static Stability and Trim

5.1. Introduction

In Chapter 4 our consideration of static aircraft stability and trim was limited to what is commonly called *longitudinal motion*, i.e., motion confined to the aircraft's plane of symmetry. For such motion, the total aerodynamic force can be resolved into a lift and drag force, L and D, or a normal and axial force, N and A, all of which fall in the aircraft plane of symmetry. The only aerodynamic moment associated with longitudinal motion is the pitching moment, m. The velocity vector associated with longitudinal motion is completely defined by its magnitude, V, and angle of attack, α. For general three-dimensional motion, there may be an additional component of velocity normal to the aircraft plane of symmetry, which is called *sideslip*, and the aircraft may experience an aerodynamic side force with moments in both roll and yaw as well as in pitch.

Longitudinal motion involves three degrees of freedom: axial translation, normal translation, and rotation in pitch. The other three degrees of freedom are commonly called the *lateral degrees of freedom*. These are sideslip translation, rotation in roll, and rotation in yaw. The *longitudinal degrees of freedom* constitute planar motion, which has the great simplification of being two-dimensional. However, the lateral degrees of freedom do not compose a planar motion and must be treated as three-dimensional.

There is inconsistency in the literature regarding notation for the rolling moment. Both L and l are commonly used. In this text, lowercase script ℓ is used for the rolling moment to avoid confusion with either lift or a distance measured aft of the CG. The yawing moment is commonly given the symbol N or n. Here, we use n for the yawing moment, to avoid confusion with the normal force. Consistent with tradition, an aerodynamic side force to the right is denoted by Y. With this notation, all forces are represented with uppercase letters and lowercase letters are used to denote all moments. By standard notation the sideslip angle is given the symbol β and the bank angle is denoted as ϕ. The sign convention for the moments and angles is shown in Fig. 5.1.1.

Traditionally, for a complete airplane, the aerodynamic coefficients for the rolling and yawing moments as well as the side force are, respectively, defined as

$$C_\ell \equiv \frac{\ell}{\frac{1}{2}\rho V_\infty^2 S_w b_w} \tag{5.1.1}$$

$$C_n \equiv \frac{n}{\frac{1}{2}\rho V_\infty^2 S_w b_w} \tag{5.1.2}$$

$$C_Y \equiv \frac{Y}{\frac{1}{2}\rho V_\infty^2 S_w} \tag{5.1.3}$$

Notice that the traditional reference length used for the rolling and yawing moments is the wingspan, b_w, while that used for the pitching moment is the mean wing chord, $\bar{c}_w$.

497

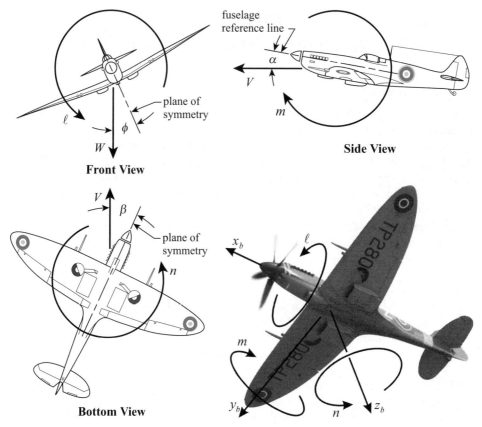

Figure 5.1.1. Sign convention for the aerodynamic angles and the moments acting on an airplane in free flight. (Photograph by Barry Santana)

At this point the reader should memorize the sign convention shown in Fig. 5.1.1. In our study of longitudinal stability and trim we were concerned with one angle and one moment: the angle of attack, α, and the pitching moment, m. It was quite easy to remember that angle of attack is positive with the nose up, in the direction of increasing lift, and the pitching moment is positive in the direction of increasing angle of attack. For our study of lateral stability and trim, we have now introduced two more angles and two more moments: the bank angle, ϕ; the sideslip angle, β; the rolling moment, ℓ; and the yawing moment, n. There would seem to be no preferred direction for roll and yaw on which to base a sign convention. In fact, the choice of sign convention is arbitrary as long as we are consistent. The need to be consistent is the primary reason for defining and memorizing any sign convention. Students are sometimes confused by the sign convention for β in Fig. 5.1.1. This confusion stems from the fact that the pitching moment is positive in the direction of increasing angle of attack, the rolling moment is positive in the direction of increasing bank angle, but the yawing moment is positive in the direction of decreasing sideslip angle.

The traditional way to remember the sign convention for angles or moments is to use the *right-hand rule* relative to some right-hand coordinate system. In this case the coordinate system is the (x_b, y_b, z_b) system shown in Fig. 5.1.1. This coordinate system is fixed to the body of the aircraft with the origin at the CG. The x_b-z_b plane is the aircraft plane of symmetry. The x_b-axis points forward and the z_b-axis points downward. The y_b-axis must then point out the right-hand side of the airplane to complete the right-hand coordinate system. Relative to this coordinate system, the aerodynamic moments ℓ, m, and n are right-hand moments about the x_b, y_b, and z_b axes, respectively.

Students are often confused by trying to apply a similar right-hand rule to the angles ϕ, α, and β. The bank angle, ϕ, in Fig. 5.1.1 is a right-hand rotation of the airplane about the roll axis, x_b. A positive increase in angle of attack could be thought of as a right-hand rotation of the airplane about the pitch axis, y_b. However, to attain a positive sideslip angle, a left-hand rotation of the airplane about the yaw axis, z_b, is required. This could be viewed as an inconsistency in the sign convention shown in Fig. 5.1.1 and it can lead to confusion. To resolve this, we must realize that the bank angle, ϕ, is fundamentally different from the angle of attack, α, and the sideslip angle, β. Both α and β are called *aerodynamic angles* because they are defined from the airplane's velocity vector relative to the surrounding air. In contrast, ϕ is called an *orientation angle*, because it defines one component of the airplane's orientation relative to the earth and is independent of the airplane's velocity. We shall later define two additional orientation angles. Because of the fundamental difference between aerodynamic angles and orientation angles, we should not expect the sign convention for β and ϕ to be defined in exactly the same way. This still does not explain the difference between the sign convention for α and β when viewed as rotations about the y_b and z_b axes. To resolve the confusion, we must think of sideslip not in terms of sideslip angle, but rather in terms of sideslip velocity. **The aircraft is defined to have positive sideslip when the y_b-component of aircraft velocity relative to the surrounding air is positive.** This is consistent with the sign convention for α, since a positive z_b-component of aircraft velocity relative to the surrounding air gives a positive angle of attack. In a similar manner, **the side force, Y, is simply defined to be the y_b-component of the resultant aerodynamic force.**

For equilibrium longitudinal motion, the net side force, rolling moment, and yawing moment must all be zero. All three of these equilibrium requirements can be naturally satisfied with spanwise airplane symmetry. Furthermore, in level flight the gravity vector is in the plane of symmetry and has no spanwise component. Hence, for a completely spanwise symmetric airplane, there is no "fundamental" level flight trim problem in roll or yaw. Level flight of a symmetric airplane is pure longitudinal motion. However, no airplane can always be perfectly symmetric. Asymmetric loading, asymmetric thrust, propeller rotation, or simply an asymmetric distribution of bugs on the wings can cause either aerodynamic or inertial asymmetry. Thus, even for level flight, some provision must be made for trimming the airplane in roll and yaw. Furthermore, if equilibrium longitudinal motion is to be statically stable, the airplane must exhibit lateral static stability as well as longitudinal static stability. This means that any small disturbance in roll or yaw must produce an aerodynamic moment that will return the airplane to its original longitudinal equilibrium. We shall now turn our attention to these aspects of static stability and trim.

5.2. Yaw Stability and Trim

Static yaw stability is provided primarily by the vertical tail and is defined as rotational stability about the yaw axis, as shown in the bottom view of Fig. 5.1.1. This is also commonly referred to as *directional stability*. In the early days of aviation, yaw stability was called *weathercock stability*, in reference to the weathercocks that adorned the roof of almost every barn in rural America. A weathercock points into the wind as a result of the same aerodynamic principles that make an airplane stable in yaw. However, since the word *weathercock* has little meaning in the experience of modern-day college students, the more descriptive terminology *yaw stability* is used in this text.

If an airplane is to be statically stable in yaw, then a small disturbance in yaw must produce a yawing moment that will tend to return the airplane to its original equilibrium state. From Fig. 5.1.1, we see that a yaw disturbance resulting in a positive sideslip angle requires a positive yawing moment to restore the disturbance to zero. Conversely, a negative sideslip angle requires a negative restoring moment. Thus, in mathematical terms, static stability in yaw requires that

$$\frac{\partial n}{\partial \beta} > 0$$

In dimensionless form, this gives **the general criterion for static yaw stability**,

$$\frac{\partial C_n}{\partial \beta} \equiv C_{n,\beta} > 0 \qquad (5.2.1)$$

The aerodynamic derivative, $C_{n,\beta}$, is the *yaw stability derivative* and is often called the *yaw stiffness*. This is analogous to the pitch stability derivative, $C_{m,\alpha}$. The obvious difference between Eqs. (5.2.1) and (4.2.8) is the sign. This difference in sign is simply a result of the sign convention shown in Fig. 5.1.1.

It is very difficult to specify a general yaw stability criterion in terms of the magnitude of $C_{n,\beta}$. **For typical airplane configurations, good handling qualities are normally found with $C_{n,\beta}$ in the range between about 0.06 and 0.15 per radian, with 0.03 being a reasonable lower limit. These values need to be increased for airplanes with very high wing loading and low span. There is no direct upper limit for the magnitude of $C_{n,\beta}$. Usually, the more static stability an airplane has in yaw, the better it will handle.** This does not mean that there is no limit to the allowable size of the vertical tail. The vertical tail also affects the handling qualities of an airplane through certain dynamic characteristics that will be discussed later in this text. The size of the vertical tail is not usually fixed by consideration of static stability. Fortunately, when the vertical tail is sized based on control and dynamic handling quality requirements, which will be discussed later, sufficient static stability is normally provided.

Static yaw stability is very similar to static pitch stability. When the center of pressure acting normal to the plane of symmetry is aft of the center of gravity, any sideslip angle will produce a pressure imbalance with a resultant force that acts aft of the center of gravity. This causes the airplane to "weathervane" into the relative wind.

An aft vertical tail provides static yaw stability in the same way that an aft horizontal tail provides static pitch stability.

The yaw stability derivative, $C_{n,\beta}$, is estimated in much the same way as the pitch stability derivative, $C_{m,\alpha}$. It is synthesized by combining the contributions made by the various components of the airplane. The major contributions are those made by the fuselage, propeller, and vertical tail. The contributions from the fuselage and propeller are typically destabilizing, but these are small and easily countered by the stabilizing effect of an aft vertical tail. In contrast with pitch stability, an unswept wing has little direct effect on static yaw stability and is usually neglected in preliminary calculations.

Effect of the Vertical Tail
When an airplane is flying with a positive sideslip angle as shown in the bottom view of Fig. 5.1.1, the vertical tail is at an angle of attack to the freestream and develops lift, as is shown in Fig. 5.2.1. The lift developed on the vertical tail as a result of positive sideslip produces a side force from right to left, as also shown in Fig. 5.2.1. Since the vertical tail is aft of the CG, this lift produces a positive yawing moment about the center of gravity. This is a restoring moment, which tends to point the airplane into the relative wind and return the sideslip angle to zero.

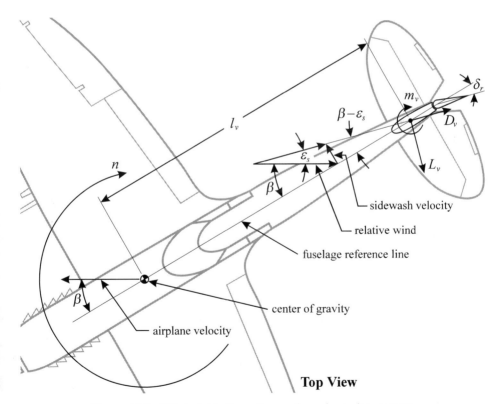

Figure 5.2.1. Effect of sideslip on the aerodynamic yawing moment.

If the vertical tail were isolated in a uniform flow field, the angle of attack for this lifting surface would be equivalent to the sideslip angle, β. However, when installed on an airplane, the airflow relative to the vertical tail is modified in both magnitude and direction. For example, the magnitude of the airflow relative to the vertical tail can be decreased if the surface is in the wake of the wing or the fuselage. This magnitude can be increased if the vertical tail is in the slipstream of a propeller or jet engine. The angle of attack for the vertical tail can be modified by the slipstream of a propeller or by the vorticity shed from the main wing.

As discussed in Chapter 2, the air in the slipstream of a propeller rotates as it moves downstream. Since the vertical tail is usually above the axis of the propeller slipstream, rotation of the slipstream will produce *sidewash* on the vertical tail, as is shown in Fig. 5.2.2. The effect of this sidewash on the vertical tail is analogous to the effect of downwash on the horizontal tail. **Sidewash is defined to be positive from left to right** as shown in Fig. 5.2.1. Thus, if the propeller is rotating counterclockwise when viewed from the front, the propeller slipstream will generate positive sidewash on a vertical tail mounted above the slipstream axis. This produces a positive *sidewash angle*, ε_s, which reduces the sideslip angle felt by the vertical tail as shown in Fig. 5.2.1. The sidewash angle induced on a centered vertical tail by the propeller of a single engine aircraft is maximum when the sideslip angle is zero and decreases when β becomes either positive or negative. This affects the aircraft trim in yaw but has no effect on yaw stability.

The wingtip vortices can also induce sidewash on the vertical tail when the airplane is flying with a sideslip angle. Since the wingtip vortices trail downstream with the relative wind, when the airplane is in sideslip one vortex is closer to the tail than the other. This is shown in Fig. 5.2.3. If the vertical tail is mounted above the wing, inboard sidewash is induced by each wingtip vortex. When an airplane has zero sideslip angle, the vertical tail is midway between the two wingtip vortices and the sidewash induced by the right wing exactly cancels that induced by the left and the resulting sidewash angle is zero. When the sideslip angle is positive, the vertical tail is closer to the right wingtip vortex than it is to the left, as shown in Fig. 5.2.3. Thus, the sidewash moving inboard from the right is greater than that from the left and a negative sidewash angle is induced by the wingtip vortices. Conversely, the sidewash angle induced on a high vertical tail by the wingtip vortices is positive when the sideslip angle is negative. Since ε_s is subtracted from β as shown in Fig. 5.2.1, the wingtip vortices add to the effect of sideslip and are stabilizing, when the vertical tail is above the wing.

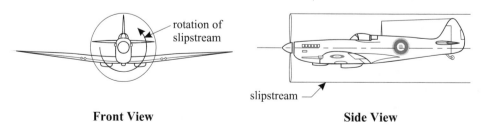

Front View   **Side View**

Figure 5.2.2. Sidewash produced on a vertical tail by the propeller slipstream.

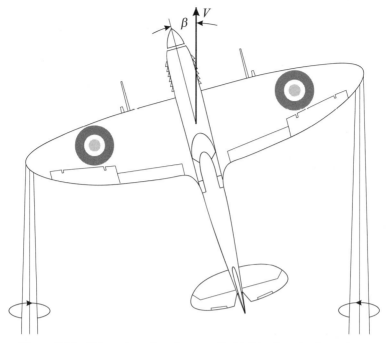

Figure 5.2.3. Sidewash produced on a vertical tail by the wingtip vortices.

Multiple counterrotating propellers produce sidewash on the vertical tail in much the same way as the wingtip vortices. If the engines that are mounted on the right rotate clockwise, when viewed from the front, while the engines mounted on the left rotate counterclockwise, then the propeller slipstream effect is similar to the wingtip vortex effect. That is, the propeller slipstreams have no effect on trim but have a stabilizing effect on an aft vertical tail mounted above the propeller axis. If these propellers were to each rotate in the opposite direction, then there would still be no effect on trim, but the slipstreams would be destabilizing for a high vertical tail.

For small sideslip angles, the sidewash angle on a vertical tail can be considered to be a linear function of β,

$$\varepsilon_s = \varepsilon_{s0} + \varepsilon_{s,\beta}\beta \tag{5.2.2}$$

where ε_{s0} is the sidewash angle at zero sideslip and $\varepsilon_{s,\beta}$ is the sidewash gradient,

$$\varepsilon_{s,\beta} \equiv \frac{\partial \varepsilon_s}{\partial \beta} \tag{5.2.3}$$

For small angles of attack and rudder deflection, the lift developed on the vertical tail is linear with both angle of attack and rudder deflection angle, δ_r. By usual convention, **a leftward deflection of the rudder is defined to be positive**, as shown in Fig. 5.2.1.

Thus, positive deflection of an aft rudder produces a rightward increment in the lift developed on the vertical tail and a negative increment in the aerodynamic yawing moment for the airplane.

Using standard sign convention, the contribution of the vertical tail to the yawing moment is

$$(\Delta n)_v = \tfrac{1}{2}\rho V_v^2 S_v \left[l_v C_{L_v,\alpha}(\beta - \varepsilon_s - \varepsilon_r \delta_r) + \bar{c}_v C_{m_v,\delta_r} \delta_r \right] \tag{5.2.4}$$

where V_v is the local magnitude of the relative airspeed at the position of vertical tail, $C_{L_v,\alpha}$ is the lift slope for the vertical tail, ε_r is the rudder effectiveness, $\bar{c}_v$ is the mean chord length of the vertical tail, and C_{m_v,δ_r} is the change in the moment coefficient for the vertical tail with respect to rudder deflection. Applying Eq. (5.2.2) and nondimensionalizing Eq. (5.2.4) by utilizing the definition in Eq. (5.1.2), **the contribution of the vertical tail to the yawing moment coefficient** can be estimated from

$$(\Delta C_n)_v = \eta_v \frac{S_v l_v}{S_w b_w} \left[C_{L_v,\alpha}(1 - \varepsilon_{s,\beta})_v \beta - C_{L_v,\alpha} \varepsilon_{s0} - \left(\varepsilon_r C_{L_v,\alpha} - \frac{\bar{c}_v}{l_v} C_{m_v,\delta_r} \right) \delta_r \right] \tag{5.2.5}$$

where η_v is the dynamic pressure ratio for the vertical tail, which is analogous to that for the horizontal tail and may be handled in the same way. The ratio $S_v l_v / S_w b_w$ is called the *vertical tail volume ratio*,

$$\mathcal{V}_v \equiv \frac{S_v l_v}{S_w b_w} \tag{5.2.6}$$

and is analogous to the horizontal tail volume ratio.

From Eq. (5.2.5), the contribution of the vertical tail to the yaw stability derivative or yaw stiffness is

$$(\Delta C_{n,\beta})_v = \eta_v \frac{S_v l_v}{S_w b_w} C_{L_v,\alpha}(1 - \varepsilon_{s,\beta})_v \tag{5.2.7}$$

As long as the vertical tail is aft of the center of gravity, l_v and this contribution to the yaw stability derivative are both positive. Thus, an aft vertical tail is always stabilizing.

Equation (5.2.5) also provides us with the contribution that the rudder makes to the yaw control derivative,

$$C_{n,\delta_r} \equiv \frac{\partial C_n}{\partial \delta_r} = -\eta_v \frac{S_v l_v}{S_w b_w} \left(\varepsilon_r C_{L_v,\alpha} - \frac{\bar{c}_v}{l_v} C_{m_v,\delta_r} \right) \tag{5.2.8}$$

The aerodynamic derivative C_{m_v,δ_r} is always negative. Thus, the change in the airplane yawing moment coefficient with rudder deflection is always negative for an aft rudder, when leftward deflection of the rudder is considered to be positive.

The sidewash gradient, $\varepsilon_{s,\beta}$, is typically negative and thus increases the stabilizing effect of the vertical tail. The sidewash gradient resulting from multiple propellers is difficult to estimate with any degree of accuracy and is usually neglected in preliminary design. The sidewash gradient produced by the wingtip vortices can be estimated using the same vortex model that was used in Chapter 4 to estimate the downwash gradient for the horizontal tail.

The lift slope for the vertical tail, $C_{L_v,\alpha}$, is not as easy to estimate as one might at first expect. We have previously used the closed-form solution to Prandtl's lifting-line theory to estimate the lift slope for the wing and the horizontal surface. However, the fuselage and the horizontal tail have a significant effect on the apparent aspect ratio of the vertical tail. These surfaces of the airplane act like winglets for the vertical tail and inhibit vortex shedding, thereby increasing the lift produced at a given angle of attack. The lift slope for a vertical tail in combination with a horizontal tail can be accurately estimated using the numerical lifting-line method or from a three-dimensional panel code. However, prior to such calculations, a very rough estimate for this lift slope can be made from the closed-form solution to Prandtl's lifting-line theory by using an effective aspect ratio that is between about 1.4 and 1.6 times the actual geometric aspect ratio. The lift slope for the vertical tail is sensitive to the geometry of the horizontal tail, so this rough estimate should be used with caution.

Effect of the Fuselage, Nacelles, and External Stores

The contribution of the fuselage, nacelles, and external stores to the static yaw stability of an airplane is very similar to their contribution to pitch stability and is typically destabilizing. In some cases this destabilizing effect can be quite significant. Thus, the fuselage, nacelles, and external stores should be included in the yaw stability analysis during all phases of the airplane design process.

To accurately estimate the yawing moment produced on the fuselage, nacelles, and external stores, we must make use of computer simulations and/or wind tunnel tests. However, as a first approximation for preliminary design, the effect of the fuselage, nacelles, and external stores on yaw stability can be estimated using the method presented in Sec. 4.9. From this approximate experimental correlation, a rough estimate of the fuselage yawing moment coefficient about the airplane center of gravity is given by

$$C_{n_f} \equiv \frac{n_{cg_f}}{\frac{1}{2}\rho V_\infty^2 S_f c_f} = 2\frac{l_f}{c_f}\left[1 - 1.76\left(\frac{d_f}{c_f}\right)^{3/2}\right]\beta \tag{5.2.9}$$

where again S_f is the maximum cross-sectional area of the fuselage, c_f is the chord length of the fuselage, l_f is the distance that the fuselage center of pressure is aft of the airplane center of gravity, and d_f is the diameter of a circle having the same area as the maximum cross-sectional area of the fuselage. The observed difference in sign between Eqs. (5.2.9) and (4.9.1) is simply a function of the sign convention defined in Fig. 5.1.1.

The reference area for the complete airplane is the wing planform area and the reference length for yaw is the wingspan. Thus, the contribution of the fuselage to the total airplane yawing moment coefficient can be approximated by multiplying the result from Eq. (5.2.9) by the product $S_f c_f$ and dividing by the product $S_w b_w$. This gives a result very similar to Eq. (4.9.3),

$$\left(\Delta C_n\right)_f \equiv \frac{n_{cg_f}}{\frac{1}{2}\rho V_\infty^2 S_w b_w} = \frac{S_f c_f}{S_w b_w} C_{n_f} = \left(\Delta C_{n,\beta}\right)_f \beta \tag{5.2.10}$$

$$\left(\Delta C_{n,\beta}\right)_f = 2\frac{S_f l_f}{S_w b_w}\left[1 - 1.76\left(\frac{d_f}{c_f}\right)^{3/2}\right] \tag{5.2.11}$$

By comparing Eq. (5.2.11) with Eq. (4.9.6), we can see that this approximation predicts a fuselage contribution to the yaw stability derivative that is related to its contribution to the pitch stability derivative according to

$$\left(\Delta C_{n,\beta}\right)_f \cong -\left(\Delta C_{m,\alpha}\right)_f \bar{c}_w/b_w \tag{5.2.12}$$

The contribution of the nacelles and external stores to the yaw stability derivative can be approximated in the same manner. Since the fuselage and nacelles are normally symmetric in the spanwise direction, they do not usually contribute to aircraft trim in yaw.

Because the fuselage, nacelles, and external stores are not necessarily axisymmetric and the flow induced by the wing is far from axisymmetric, Eq. (5.2.12) should be used only as a rough approximation. The aerodynamic forces and moments generated on the fuselage, nacelles, and external stores are complex and significantly affected by aerodynamic interactions with the wing and tail. **It is vital to remember that Eq. (5.2.12) gives only a rough estimate of the contribution to the yaw stability derivative generated by the fuselage, nacelles and external stores**. It should only be used as a first approximation for preliminary design. Wind tunnel tests and/or computer simulations should always be used to evaluate this derivative in the later phases of design. Although nacelles and external stores are typically much smaller than the fuselage, their effect on yaw stability can be large. For example, maneuvering of fighters is often limited with external fuel tanks due to their detrimental effect on yaw stiffness.

Effect of Rotating Propellers
A propeller can affect both the trim and static stability of an airplane in yaw. As shown in Fig. 5.2.4, a rotating propeller advancing through the air so that the axis of rotation makes some angle with the freestream can produce both a side force, Y_p, and a yawing moment, n_p, in addition to the thrust, T. From the geometry shown in Fig. 5.2.4, the yawing moment that the propeller produces about the airplane's center of gravity is

$$n_{cg_p} = n_p - y_{bp}T - l_p Y_p \tag{5.2.13}$$

where y_{bp} is the spanwise distance from the CG to the propeller axis, measured positive to the right. For consistency, we are defining l_p as the axial distance that the propeller is mounted aft of the CG. For the airplane in Fig. 5.2.4, l_p is negative.

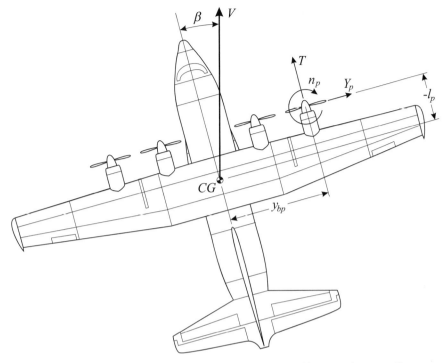

Figure 5.2.4. Side forces and yawing moment generated by a rotating propeller.

The yawing moment computed for each propeller from Eq. (5.2.13) can be added to the yawing moment obtained for all of the other airplane components to determine the yawing moment for the complete airplane. Thus, the contribution of each propeller to the total yawing moment coefficient is found by nondimensionalizing Eq. (5.2.13),

$$
\left(\Delta C_n\right)_p = \frac{n_{cg_p}}{\frac{1}{2}\rho V_\infty^2 S_w b_w} = \frac{n_p}{\frac{1}{2}\rho V_\infty^2 S_w b_w} - \frac{y_{bp}}{b_w}\frac{T}{\frac{1}{2}\rho V_\infty^2 S_w} - \frac{l_p}{b_w}\frac{Y_p}{\frac{1}{2}\rho V_\infty^2 S_w} \tag{5.2.14}
$$

With all engines operational, the total thrust will usually be symmetrically distributed in the spanwise direction and the net contribution of the thrust term in Eq. (5.2.14), for all propellers combined, will typically be zero. However, this term is retained here to account for the possibility of engine failure. For small thrust angles the total equilibrium thrust from all propellers combined is nearly equal to the drag. Thus, the contribution of each propeller can be closely approximated from Eq. (5.2.14) as

$$
\left(\Delta C_n\right)_p = \frac{n_p}{\frac{1}{2}\rho V_\infty^2 S_w b_w} - \frac{y_{bp}}{b_w} f_T C_D - \frac{l_p}{b_w}\frac{Y_p}{\frac{1}{2}\rho V_\infty^2 S_w} \tag{5.2.15}
$$

where f_T is the thrust fraction provided by the propeller.

Following the material presented in Chapter 2, the propeller yawing moment and side force are conventionally expressed in terms of the dimensionless aerodynamic coefficients,

$$C_{n_p} \equiv \frac{n_p}{\rho(\omega/2\pi)^2 d_p^5} \tag{5.2.16}$$

$$C_{Y_p} \equiv \frac{Y_p}{\rho(\omega/2\pi)^2 d_p^4} \tag{5.2.17}$$

where d_p is the propeller diameter and ω is the angular velocity. The yawing moment coefficient is linearly proportional to the propeller angle of attack, α_p,

$$C_{n_p} = C_{n_p,\alpha} \alpha_p \tag{5.2.18}$$

and the side-force coefficient is linearly proportional to the propeller sideslip angle, β_p,

$$C_{Y_p} = C_{Y_p,\beta} \beta_p \tag{5.2.19}$$

The coefficients $C_{n_p,\alpha}$ and $C_{Y_p,\beta}$ depend on propeller geometry and advance ratio, J,

$$J \equiv \frac{2\pi V_\infty}{\omega d_p} \tag{5.2.20}$$

Because the propeller is axisymmetric, the side force is related to the sideslip angle in the same way that the normal force is related to the angle of attack,

$$C_{Y_p,\beta} = -C_{N_p,\alpha} \tag{5.2.21}$$

Here again the negative sign in Eq. (5.2.21) is simply a result of the sign convention shown in Fig. 5.1.1. The reader should examine Fig. 5.1.1 carefully and be sure that the sign in Eq. (5.2.21) can be justified.

Using Eqs. (5.2.16) through (5.2.21) in Eq. (5.2.15), the direct contribution of the propeller to the total airplane yawing moment coefficient is

$$\left(\Delta C_n\right)_p = \frac{2 d_p^3}{S_w b_w} \frac{C_{n_p,\alpha}}{J^2} \alpha_p - \frac{y_{bp}}{b_w} f_T C_D + \frac{2 d_p^2 l_p}{S_w b_w} \frac{C_{N_p,\alpha}}{J^2} \beta_p \tag{5.2.22}$$

The dimensionless yawing moment gradient and the dimensionless normal force gradient, which appear in Eq. (5.2.22), depend on the propeller geometry as well as the advance ratio. For known propeller geometry, these coefficients can be determined as a function of advance ratio using the method presented in Chapter 2. In addition to the direct contribution of the propeller, the interaction of the propeller slipstream with the vertical tail can produce a significant yawing moment, as shown in Fig. 5.2.2.

Both the propeller angle of attack and sideslip angle in Eq. (5.2.22) can be affected by the airflow around the wing. Recall from Eq. (4.10.10) that the angle of attack that the propeller axis makes with the local relative airflow can be written as

$$\alpha_p = \alpha - \varepsilon_{dp} + \alpha_{0p} \tag{5.2.23}$$

where α is the airplane angle of attack relative to the fuselage reference line, ε_{dp} is the local downwash angle at the position of the propeller, and α_{0p} is the angle that the propeller axis makes with the fuselage reference line. Likewise, as was the case for the vertical tail, the propeller sideslip angle depends on the sidewash generated by the wing,

$$\beta_p = \beta - \varepsilon_{sp} \tag{5.2.24}$$

where β is the airplane sideslip angle relative to the fuselage reference line and ε_{sp} is the local sidewash angle at the position of the propeller.

Using Eqs. (5.2.23) and (5.2.24) with Eq. (5.2.22), the direct contribution of the propeller to the total airplane yawing moment coefficient can be written as

$$\left(\Delta C_n\right)_p = \left(\Delta C_{n0}\right)_p + \left(\Delta C_{n,\alpha}\right)_p \alpha + \left(\Delta C_{n,\beta}\right)_p \beta \tag{5.2.25}$$

where

$$\left(\Delta C_{n0}\right)_p = \frac{2 d_p^3}{S_w b_w} \frac{C_{n_p,\alpha}}{J^2} \left(\alpha_{0p} - \varepsilon_{d0p}\right) - \frac{y_{bp}}{b_w} f_T C_D \tag{5.2.26}$$

$$\left(\Delta C_{n,\alpha}\right)_p = \frac{2 d_p^3}{S_w b_w} \left(1 - \varepsilon_{d,\alpha}\right)_p \frac{C_{n_p,\alpha}}{J^2} \tag{5.2.27}$$

$$\left(\Delta C_{n,\beta}\right)_p = \frac{2 d_p^2 l_p}{S_w b_w} \left(1 - \varepsilon_{s,\beta}\right)_p \frac{C_{N_p,\alpha}}{J^2} \tag{5.2.28}$$

ε_{d0p} is the propeller downwash angle with the fuselage reference line at zero angle of attack, $(\varepsilon_{d,\alpha})_p$ is the downwash gradient at the position of the propeller, and $(\varepsilon_{s,\beta})_p$ is the sidewash gradient at the position of the propeller. **For a propeller that is mounted forward of the wing, the sidewash gradient is small and can usually be ignored. However, when the propeller is mounted aft of the wing, the effects of sidewash should be considered.**

Equation (5.2.28) provides the propeller's direct contribution to the total airplane yaw stability derivative. Since the change in the normal force coefficient with angle of attack is positive, if the propeller is forward of the center of gravity, l_p is negative and the propeller contribution to the yaw stability derivative is destabilizing. If the propeller is aft of the center of gravity, l_p is positive and the propeller contribution to the yaw stability derivative is stabilizing.

Notice from Eq. (5.2.25) that the yawing moment contributed by a rotating propeller depends on angle of attack as well as sideslip angle. Thus, rotating propellers can affect trim in yaw as well as yaw stability. The direct contribution of a rotating propeller to the airplane yawing moment coefficient at trim is found from Eq. (5.2.25) by setting the sideslip angle to zero and the angle of attack to the equilibrium angle of attack, α_o. This gives

$$\left(\Delta C_{n_{\text{trim}}}\right)_p = \frac{2 d_p^3}{S_w b_w} \frac{C_{n_p,\alpha}}{J^2} [\alpha_{0p} - \varepsilon_{d0p} + (1 - \varepsilon_{d,\alpha})_p \alpha_o] - \frac{y_{bp}}{b_w} f_T C_D \quad (5.2.29)$$

The slipstream effects shown in Fig. 5.2.2 can add significantly to the trim requirement specified by Eq. (5.2.29).

The change in propeller yawing moment coefficient with propeller angle of attack is negative for right-hand turning propellers and positive for left-hand turning propellers. Since y_{bp} is positive for propellers mounted to the right of center and negative for propellers mounted to the left of center, **for a multiengine airplane the trim requirement in yaw can be eliminated by using counterrotating propellers**. If the engine locations are symmetric in the spanwise direction and those on the left rotate in one direction while those on the right rotate in the opposite direction, the net propeller trim requirement specified by Eq. (5.2.29) will be zero. **Since the change in propeller normal force coefficient with propeller angle of attack is positive for both right- and left-hand turning propellers, the use of counterrotating propellers does not affect yaw stability.** For a multiengine aircraft, the most significant trim requirement predicted by Eq. (5.2.29) occurs under conditions of engine failure. The need to be able to trim a multiengine aircraft with one or more engines inoperative usually provides the most restrictive constraint on rudder size.

For a single-engine aircraft with the propeller axis in the plane of symmetry, there is a yawing moment generated by the propeller at zero sideslip if the propeller angle of attack is not zero. Some rudder deflection is required to trim the airplane against this yawing moment. For an airplane with low thrust-to-weight ratio, this trim requirement is small. However, for a very high-powered propeller-driven airplane, such as the fighter aircraft of World War II, the rudder deflection needed to trim the airplane against the propeller yawing moment can be quite substantial at high angles of attack.

> **EXAMPLE 5.2.1.** The airplane described in Examples 4.4.1, 4.9.1, and 4.10.1 has a vertical tail area of 11.9 ft^2 and the aerodynamic center of the vertical tail is 14.81 feet aft of the CG. For the operating conditions in Example 4.10.1, estimate the yaw stability derivative for this airplane, including the effects of the fuselage and propeller. Assume that the dynamic pressure ratio for the vertical tail is 1.0, the sidewash gradient is -0.10, and the lift slope for the vertical tail is 3.40.
>
> **Solution.** From Example 4.10.1 and the additional information presented here,
>
> $$S_w = 180 \text{ ft}^2, \quad b_w = 33 \text{ ft}, \quad \bar{c}_w = 5.455 \text{ ft}, \quad C_{L_w,\alpha} = 4.44, \quad l_w = -0.71 \text{ ft},$$
> $$S_h = 36 \text{ ft}^2, \quad C_{L_h,\alpha} = 3.97, \quad l_h = 14.29 \text{ ft}, \quad \eta_h = 1.0, \quad (\varepsilon_{d,\alpha})_h = 0.44,$$

$$\left(\Delta C_{m,\alpha}\right)_f = 0.1216, \quad d_p = 74 \text{ in}, \quad l_p = -9 \text{ ft}, \quad C_{N_p,\alpha} = 0.04, \quad J = 0.4858,$$

$$S_v = 11.9 \text{ ft}^2, \quad C_{L_v,\alpha} = 3.40, \quad l_v = 14.81 \text{ ft}, \quad \eta_v = 1.0, \quad (\varepsilon_{s,\beta})_v = -0.10$$

From Eq. (5.2.7), the contribution of the vertical tail to the total yaw stability derivative is

$$\left(\Delta C_{n,\beta}\right)_v = \eta_v \frac{S_v l_v}{S_w b_w} C_{L_v,\alpha}(1 - \varepsilon_{s,\beta})_v$$

$$= 1.0 \frac{11.9(14.81)}{180(33)} 3.40[1 - (-0.10)] = 0.1110$$

and from Eq. (5.2.12), the contribution of the fuselage is approximated as

$$\left(\Delta C_{n,\beta}\right)_f \cong -\left(\Delta C_{m,\alpha}\right)_f \frac{\bar{c}_w}{b_w} = -0.1216 \frac{5.455}{33} = -0.0201$$

Since the propeller is forward of the wing, we can neglect the sidewash gradient, and from Eq. (5.2.28), the propeller contribution to the yaw stability derivative is

$$\left(\Delta C_{n,\beta}\right)_p = \frac{2 d_p^2 l_p}{S_w b_w}(1 - \varepsilon_{s,\beta})_p \frac{C_{N_p,\alpha}}{J^2}$$

$$= \frac{2(74/12)^2(-9.0)}{(180)(33)}(1 - 0.0) \frac{0.04}{0.4858^2} = -0.0195$$

Thus, a reasonable first estimate for the yaw stability derivative of this airplane is

$$C_{n,\beta} \cong \left(\Delta C_{n,\beta}\right)_v + \left(\Delta C_{n,\beta}\right)_f + \left(\Delta C_{n,\beta}\right)_p$$

$$= 0.1110 - 0.0201 - 0.0195 = \underline{0.071}$$

EXAMPLE 5.2.2. The vertical tail of the airplane in Example 5.2.1 has a mean chord of 2.0 feet, the rudder effectiveness is 0.56, and the change in the moment coefficient for the vertical tail with respect to rudder deflection is −0.51. The change in the propeller yawing moment coefficient with respect to propeller angle of attack is −0.013. Assuming that the propeller is mounted with the axis of rotation aligned with the airplane's zero-lift axis, estimate the rudder deflection required to trim the airplane for zero sideslip against the yawing moment produced directly by the propeller in steady level flight at sea level and 80 mph. If no rudder deflection is used, estimate the equilibrium sideslip angle.

Solution. For this airplane we have

$$S_w = 180 \text{ ft}^2, \quad b_w = 33 \text{ ft}, \quad \bar{c}_w = 5.455 \text{ ft}, \quad C_{L_w,\alpha} = 4.44, \quad l_w = -0.71 \text{ ft},$$

$$S_h = 36 \text{ ft}^2, \quad C_{L_h,\alpha} = 3.97, \quad l_h = 14.29 \text{ ft}, \quad \eta_h = 1.0, \quad (\varepsilon_{d,\alpha})_h = 0.44,$$
$$d_p = 74/12 \text{ ft}, \quad l_p = -9 \text{ ft}, \quad (\varepsilon_{d,\alpha})_p = -0.165, \quad C_{n_p,\alpha} = -0.013, \quad J = 0.4858,$$
$$S_v = 11.9 \text{ ft}^2, \quad C_{L_v,\alpha} = 3.40, \quad l_v = 14.81 \text{ ft}, \quad \eta_v = 1.0, \quad (\varepsilon_{s,\beta})_v = -0.10,$$
$$\bar{c}_v = 2.0 \text{ ft}, \quad \varepsilon_r = 0.56, \quad C_{m_v,\delta_r} = -0.51, \quad V = 80 \text{ mph}, \quad W = 2,700 \text{ lbf}$$

From Eq. (5.2.8), the yaw control derivative for this airplane is

$$
\begin{aligned}
C_{n,\delta_r} &= -\eta_v \frac{S_v l_v}{S_w b_w}\left(\varepsilon_r C_{L_v,\alpha} - \frac{\bar{c}_v}{l_v} C_{m_v,\delta_r} \right) \\
&= -1.0 \frac{11.9(14.81)}{180(33)}\left(0.56(3.40) - \frac{2.0}{14.81}(-0.51) \right) = -0.059
\end{aligned}
$$

The rotating propeller produces the only change in the airplane's yawing moment coefficient with respect to angle of attack. From Eq. (5.2.27) we have

$$
\begin{aligned}
C_{n,\alpha} &= \left(\Delta C_{n,\alpha} \right)_p = \frac{2 d_p^3}{S_w b_w}(1 - \varepsilon_{d,\alpha})_p \frac{C_{n_p,\alpha}}{J^2} \\
&= \frac{2(74/12)^3}{180(33)}[1-(-0.165)]\frac{-0.013}{0.4858^2} = -0.0051
\end{aligned}
$$

The lift slope for the complete airplane can be estimated as

$$
\begin{aligned}
C_{L,\alpha} &= C_{L_w,\alpha} + \frac{S_h}{S_w}\eta_h C_{L_h,\alpha}(1-\varepsilon_{d,\alpha})_h \\
&= 4.44 + \frac{36}{180}1.0(3.97)(1-0.44) = 4.88
\end{aligned}
$$

With a level-flight airspeed of 80 mph at sea level, the lift coefficient is

$$
C_L = \frac{W}{\frac{1}{2}\rho V^2 S_w} = \frac{2,700}{\frac{1}{2}0.0023769(80\times 5,280/3,600)^2 180} = 0.9168
$$

and the angle of attack relative to the zero-lift line is closely approximated as

$$
\alpha = \frac{C_L}{C_{L,\alpha}} = \frac{0.9168}{4.88} = 0.1879 = 10.76°
$$

Since the propeller rotation axis is aligned with the zero-lift line, a zero yawing moment with zero sideslip requires

$$
C_{n,\alpha}\,\alpha + C_{n,\delta_r}\,\delta_r = 0
$$

or

$$\delta_r = -\frac{C_{n,\alpha}}{C_{n,\delta_r}}\alpha = -\frac{-0.0051}{-0.059}10.76° = \underline{-0.93°}$$

In a similar manner, if the rudder deflection is zero, the propeller yawing moment must be balanced with sideslip,

$$C_{n,\alpha}\alpha + C_{n,\beta}\beta = 0$$

From Example 5.2.1, $C_{n,\beta} = 0.071$ and

$$\beta = -\frac{C_{n,\alpha}}{C_{n,\beta}}\alpha = -\frac{-0.0051}{0.071}10.76° = \underline{0.77°}$$

From Example 5.2.2, we see that negative rudder deflection is needed to trim the airplane for straight and level flight against the yawing moment produced by the conventional right-hand turning propeller. From the sign convention shown in Fig. 5.2.1, we see that negative rudder deflection corresponds to a deflection of the rudder to the right. This is commonly referred to as *right rudder*, not specifically because the rudder is deflected to the right, but because this is the rudder deflection required to yaw the nose of the airplane to the pilot's right.

In Example 5.2.2, we included only the direct yawing moment that was produced by the propeller. The propeller slipstream can also interact with the vertical tail to produce an additional yawing moment. As can be seen in Fig. 5.2.2, the rotation of the slipstream can induce sidewash on the vertical tail. For the conventional right-hand turning propeller shown in this figure, the sidewash on a conventional high vertical tail is from left to right. This changes the angle of attack for the vertical tail and produces a side force to the right. Since the vertical tail is aft of the CG, this produces a moment that tends to yaw the airplane to the left. This adds to the yawing moment produced directly by the propeller and increases the amount of right rudder needed to trim the airplane. In some cases, this slipstream effect can be as large as or larger than the direct effect from the propeller itself, so a few degrees of right rudder can be required to trim a single-engine propeller-driven airplane for straight and level flight at low airspeeds.

Most single-engine propeller-driven airplanes with a conventional right-hand turning propeller will require some right rudder for trim. If this rudder deflection is not applied, the airplane will yaw to the left. Even more right rudder is required during takeoff, when the engine is at full power and the propeller advance ratio is low. This effect is more pronounced in a tail dragger because during the first phase of the ground roll, the propeller angle of attack is quite high, as seen in Fig. 5.2.5. Once the minimum ground control speed is reached, the pilot can raise the tail, reducing the propeller angle of attack and the required rudder deflection. However, even if the tail is raised enough to reduce the propeller angle of attack to zero, some right rudder is still required to offset the slipstream effect.

Figure 5.2.5. Tail-dragger landing gear typically results in a large propeller angle of attack on takeoff. (Photograph by Barry Santana)

For a tail dragger with very high power-to-weight ratio, such as the P-51 Mustang shown in Fig. 5.2.5 or the Spitfire Mark XVIII shown in Fig. 4.12.2, the yawing moment produced by the propeller can limit the power that can be used on takeoff. Even with full rudder deflection, it is not possible to hold the Spitfire Mark XVIII on the runway if full power is applied before the tail is raised. Most pilots did not apply full power until the spitfire was in the air.

Floatplanes also require a great deal of right rudder during the first phase of the takeoff run, because as full power is applied, the tail is held down with full up elevator to keep spray away from the propeller, as shown in Fig. 5.2.6. This results in a very high propeller angle of attack, which for a conventional right-hand propeller produces a large yawing moment to the left. As the speed increases, the airplane "comes up on the step," as shown in Fig. 5.2.7. This reduces the propeller angle of attack and the required rudder deflection.

Figure 5.2.6. A floatplane requires ample right rudder on takeoff. (Photograph by Lona Santana)

Figure 5.2.7. A floatplane requires less right rudder once it comes up on the step and the propeller angle of attack is reduced. (Photograph by Lona Santana)

For any propeller-driven airplane that does not have balanced counterrotating propellers, the yawing moment produced by the propeller or propellers increases with increasing angle of attack. Thus, the rudder deflection required to trim the airplane against the propeller yawing moment will vary with angle of attack. Since the angle of attack for an airplane in free flight depends on airspeed, rudder deflection at trim is a function of airspeed as well.

Effect of the Wing and Horizontal Tail
As mentioned earlier, an unswept wing has little direct effect on the static yaw stability of an airplane. Generally, the wing's contribution to yaw stability is small and can be neglected in the preliminary design phase. However, if the wing and/or horizontal tail are highly swept, they can contribute to the yaw stability derivative.

When an airplane with swept-back wings yaws to the left with respect to the velocity vector, each section of the right wing is moved farther from the line of flight that passes through the center of gravity. Conversely, each section of the left wing is moved closer to this line of flight. This can be seen in Fig. 5.2.8. If the wing has a quarter-chord sweep angle of Λ, the distance measured parallel to the wing quarter chord from the CG to the wing section located at the spanwise coordinate, y_b, can be expressed as $y_b/\cos(\Lambda)$. Since drag always acts parallel with the relative wind, the yawing moment arm for the section drag acting on any particular wing section is the perpendicular distance from the line of flight passing through the CG to the given wing section. The yawing moment arm for the drag is then $y_b \cos(\Lambda - \beta)/\cos(\Lambda)$ on an arbitrary section of the right wing and $y_b \cos(\Lambda + \beta)/\cos(\Lambda)$ on an arbitrary section of the left wing. Thus, the wing's contribution to the airplane yawing moment coefficient can be written as

$$\left(\Delta C_n\right)_w = \int\limits_{y_b=-b_w/2}^{0} c_w \tilde{C}_{D_w} \frac{y_b \cos(\Lambda + \beta)}{S_w b_w \cos(\Lambda)} dy_b + \int\limits_{y_b=0}^{b_w/2} c_w \tilde{C}_{D_w} \frac{y_b \cos(\Lambda - \beta)}{S_w b_w \cos(\Lambda)} dy_b \qquad (5.2.30)$$

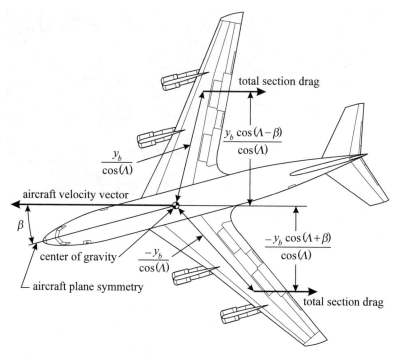

Figure 5.2.8. Wing sweep can affect the static yaw stability of an airplane.

where the section drag coefficient in Eq. (5.2.30) includes both parasitic and induced drag. For small angles of sideslip, we can use the small-angle approximations,

$$\cos(\Lambda + \beta) = \cos\Lambda\cos\beta - \sin\Lambda\sin\beta \cong \cos\Lambda - \beta\sin\Lambda$$
$$\cos(\Lambda - \beta) = \cos\Lambda\cos\beta + \sin\Lambda\sin\beta \cong \cos\Lambda + \beta\sin\Lambda$$

and Eq. (5.2.30) can be closely approximated as

$$\left(\Delta C_n\right)_w = \frac{1}{S_w b_w}\int_{y_b=-b_w/2}^{b_w/2} y_b c_w \widetilde{C}_{D_w}\,dy_b + \frac{\beta\tan\Lambda}{S_w b_w}\int_{y_b=-b_w/2}^{b_w/2}|y_b|c_w\widetilde{C}_{D_w}\,dy_b \qquad (5.2.31)$$

Typically, both the wing chord and section drag coefficient will vary with y_b.

The contribution of the wing to the aircraft yaw stability derivative is found by differentiating Eq. (5.2.31) with respect to β. At $\beta = 0$, this gives

$$\left(\Delta C_{n,\beta}\right)_w = \frac{1}{S_w b_w}\int_{y_b=-b_w/2}^{b_w/2} y_b c_w \frac{\partial\widetilde{C}_{D_w}}{\partial\beta}\,dy_b + \frac{\tan\Lambda}{S_w b_w}\int_{y_b=-b_w/2}^{b_w/2}|y_b|c_w\widetilde{C}_{D_w}\,dy_b \qquad (5.2.32)$$

The total section drag coefficient and its derivative with sideslip angle are complex functions of the spanwise coordinate, y_b, and depend on the wing planform shape, sweep, and dihedral, as well as the aerodynamic interaction with the fuselage. Since there is no closed-form solution for the spanwise distribution of induced drag acting on a swept wing, neither of the two integrals on the right-hand side of Eq. (5.2.32) can be evaluated analytically. Computer simulations or wind tunnel tests must be employed to accurately determine the effect of the wing and horizontal tail on the yaw stability derivative. However, by examining Eq. (5.2.32), we can determine the sign of this contribution.

Wing sweep is defined to be positive when the wings are swept back, placing the wingtips aft of the wing root. The second term on the right-hand side of Eq. (5.2.32) will always have the same sign as $\tan\Lambda$. Thus, this term is stabilizing when the wings are swept back and destabilizing when the wings are swept forward.

The sign of the first term on the right-hand side of Eq. (5.2.32) depends on the sign for the change in section drag coefficient with respect to sideslip angle. For small sideslip angles, the section parasitic drag is nearly independent of sideslip angle, so the first term on the right-hand side of Eq. (5.2.32) is dominated by induced drag. Recall that the local lift at any point on a lifting surface is proportional to the cross product of the local velocity and vorticity vectors. The vorticity on the wing is roughly parallel to the wing quarter chord and the local velocity is roughly parallel to the freestream. From Fig. 5.2.8, we see that as an airplane with positive sweep yaws to the left (positive β), the right wing becomes more perpendicular to the velocity vector and the left wing becomes less perpendicular to the velocity vector. Since the cross product is largest when the two vectors are at 90 degrees to each other, a positive sideslip angle will increase the lift and induced drag on the right wing and reduce the lift and induced drag on the left wing. Thus, for a wing that is swept back, the change in section drag coefficient with respect to sideslip angle is positive on the right wing (positive y_b) and negative on the left wing (negative y_b). So, for a wing with positive sweep, the product of the spanwise coordinate and the change in section drag coefficient with sideslip angle is always positive. For a wing with negative sweep, this product is always negative. As a result, the first term on the right-hand side of Eq. (5.2.32) is also stabilizing when the wings are swept back and destabilizing when the wings are swept forward.

Since both terms on the right-hand side of Eq. (5.2.32) have the same sign as the sweep angle, positive sweep is stabilizing in yaw and negative sweep is destabilizing. For most conventional airplanes with an aft tail, static yaw stability is dominated by the vertical tail. However, for a tailless aircraft, wing sweep is about the only parameter that the designer can vary to provide passive yaw stability. For this reason the wings of a tailless aircraft are almost always swept back. Because of the associated symmetry, sweep in the wing and horizontal tail of a conventional airplane has no effect on the airplane trim in yaw.

Dihedral in the main wing and horizontal tail can also affect the static yaw stability of an airplane. However, these effects are quite complex and intimately coupled with the effect of wing sweep and side flow about the fuselage. Here again, computer simulations or wind tunnel tests must be employed to accurately determine the effect of wing and tail dihedral on the airplane yaw stability derivative. Fortunately, dihedral effects on yaw stability are small and dihedral has no effect on airplane trim in yaw. Dihedral affects roll stability to a much greater extent than it affects yaw stability and will be discussed in greater detail under that heading.

5.3. Estimating the Sidewash Gradient on a Vertical Tail

The sidewash induced on the vertical tail by the wingtip vortices from the main wing can have a significant effect on the static yaw stability of an airplane. For a vertical tail mounted above the main wing, the sidewash gradient is negative and has a stabilizing effect on the airplane. The sidewash gradient produced by the wingtip vortices can be estimated using the same vortex model that was used in Chapter 4 to estimate the downwash gradient for the horizontal tail. Figure 5.3.1 shows this vortex model in relation to a vertical tail mounted above the main wing.

Relative to the aerodynamic coordinate system shown in Fig. 5.3.1, the z-component of velocity induced by this pair of wingtip vortices, at the arbitrary point in space (x,y,z), is readily found from the Biot-Savart law to be

$$
V_z = \frac{\Gamma_{wt}}{4\pi} \left\{ \frac{y}{y^2 + (z + \frac{1}{2}b')^2} \left[1 + \frac{x - \frac{1}{2}b'\tan\Lambda}{\sqrt{(x - \frac{1}{2}b'\tan\Lambda)^2 + y^2 + (z + \frac{1}{2}b')^2}} \right] \right.
$$
$$
\left. - \frac{y}{y^2 + (z - \frac{1}{2}b')^2} \left[1 + \frac{x - \frac{1}{2}b'\tan\Lambda}{\sqrt{(x - \frac{1}{2}b'\tan\Lambda)^2 + y^2 + (z - \frac{1}{2}b')^2}} \right] \right\}
\tag{5.3.1}
$$

From the vortex lifting law, the wingtip vortex strength is proportional to the product of the wing lift coefficient and airspeed. The vortex strength, Γ_{wt}, and spacing, b', can be evaluated from Prandtl's lifting-line theory.

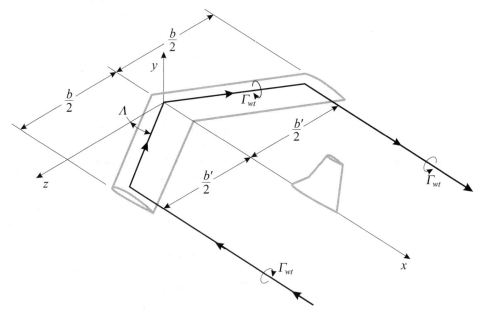

Figure 5.3.1. Vortex model used to estimate the sidewash gradient for the vertical tail.

Since we are using the sign convention that sidewash is positive from left to right, combining Eq. (5.3.1) with Prandtl's lifting-line theory and applying the small-angle approximation, the sidewash angle can be written as

$$
\begin{aligned}
\varepsilon_s &= -\frac{V_z}{V_\infty} \\
&= \frac{C_{L_w}\kappa_v}{\pi^2 R_{A_w}} \left\{ \frac{\bar{y}}{\bar{y}^2 + (\bar{z} - \kappa_b)^2} \left[1 + \frac{\bar{x} - \kappa_b \tan \Lambda}{\sqrt{(\bar{x} - \kappa_b \tan \Lambda)^2 + \bar{y}^2 + (\bar{z} - \kappa_b)^2}} \right] \right. \\
&\left. \quad - \frac{\bar{y}}{\bar{y}^2 + (\bar{z} + \kappa_b)^2} \left[1 + \frac{\bar{x} - \kappa_b \tan \Lambda}{\sqrt{(\bar{x} - \kappa_b \tan \Lambda)^2 + \bar{y}^2 + (\bar{z} + \kappa_b)^2}} \right] \right\}
\end{aligned}
\tag{5.3.2}
$$

where C_{L_w} and R_{A_w} are the lift coefficient and aspect ratio for the wing and

$$
\bar{x} = \frac{x}{b_w/2}, \quad \bar{y} = \frac{y}{b_w/2}, \quad \bar{z} = \frac{z}{b_w/2}
$$

The parameters κ_v and κ_b depend on the planform shape of the wing. The vortex strength factor, κ_v, is a ratio of the wingtip vortex strength to that generated by an elliptic wing having the same lift coefficient and aspect ratio. The vortex span factor, κ_b, is defined as the spacing between the wingtip vortices divided by the wingspan.

For a wing with no sweep or dihedral, both κ_v and κ_b can be determined analytically from the classical series solution to Prandtl's lifting-line equation. For an elliptic wing with no sweep, dihedral, or twist, κ_v is 1.0 and κ_b is $\pi/4$. For an unswept tapered wing with no dihedral or twist, κ_v and κ_b are related to the aspect ratio and taper ratio as shown in Figs. 4.5.2 and 4.5.3. A rough first approximation for swept wings can be made by assuming that κ_v and κ_b are independent of sweep.

Equation (5.3.2) can be directly applied to determine the sidewash angle only when the sideslip angle, β, is zero and the aircraft plane of symmetry is midway between the two trailing vortices. When the airplane has some component of sideslip, the wingtip vortices are displaced relative to the position of the vertical tail, as shown in Fig. 5.3.2. For small sideslip angles, Eq. (5.3.2) gives a very close approximation for the sidewash angle, if the z-coordinate, measured from the aircraft plane of symmetry, is replaced with the z'-coordinate, measured from the centerline midway between the two wingtip vortices, as shown in Fig. 5.3.2. Using the small-angle approximation, the z'-coordinate is related to the z-coordinate according to

$$
z'(\beta) = z\cos\beta - (x - \tfrac{1}{2}b'\tan\Lambda)\sin\beta \cong z - (x - \tfrac{1}{2}b'\tan\Lambda)\beta
$$

or

$$
\bar{z}'(\beta) \cong \bar{z} - (\bar{x} - \kappa_b \tan\Lambda)\beta
\tag{5.3.3}
$$

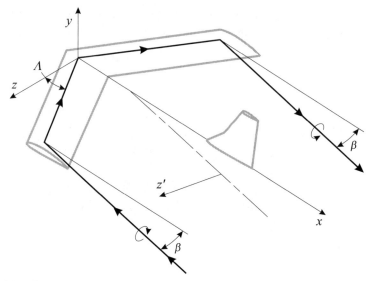

Figure 5.3.2. Effect of sideslip on the position of the wingtip vortices relative to the vertical tail.

Within this small-angle approximation, the sidewash gradient can be written as

$$\frac{\partial \varepsilon_s}{\partial \beta} = \frac{\partial \varepsilon_s}{\partial \bar{z}'} \frac{\partial \bar{z}'}{\partial \beta} \tag{5.3.4}$$

where, from Eq. (5.3.2),

$$
\frac{\partial \varepsilon_s}{\partial \bar{z}'} = \frac{C_{L_w} \kappa_v}{\pi^2 R_{A_w}} \left\{ \frac{-2\bar{y}(\bar{z}-\kappa_b)}{[\bar{y}^2 + (\bar{z}-\kappa_b)^2]^2} \left[1 + \frac{\bar{x} - \kappa_b \tan \Lambda}{\sqrt{(\bar{x}-\kappa_b \tan \Lambda)^2 + \bar{y}^2 + (\bar{z}-\kappa_b)^2}} \right] \right.
$$
$$
+ \frac{2\bar{y}(\bar{z}+\kappa_b)}{[\bar{y}^2 + (\bar{z}+\kappa_b)^2]^2} \left[1 + \frac{\bar{x} - \kappa_b \tan \Lambda}{\sqrt{(\bar{x}-\kappa_b \tan \Lambda)^2 + \bar{y}^2 + (\bar{z}+\kappa_b)^2}} \right]
$$
$$
- \frac{\bar{y}}{\bar{y}^2 + (\bar{z}-\kappa_b)^2} \left[\frac{(\bar{x}-\kappa_b \tan \Lambda)(\bar{z}-\kappa_b)}{[(\bar{x}-\kappa_b \tan \Lambda)^2 + \bar{y}^2 + (\bar{z}-\kappa_b)^2]^{3/2}} \right]
$$
$$
\left. + \frac{\bar{y}}{\bar{y}^2 + (\bar{z}+\kappa_b)^2} \left[\frac{(\bar{x}-\kappa_b \tan \Lambda)(\bar{z}+\kappa_b)}{[(\bar{x}-\kappa_b \tan \Lambda)^2 + \bar{y}^2 + (\bar{z}+\kappa_b)^2]^{3/2}} \right] \right\} \tag{5.3.5}
$$

and from Eq. (5.3.3),

$$\frac{\partial \bar{z}'}{\partial \beta} = -(\bar{x} - \kappa_b \tan \Lambda) \tag{5.3.6}$$

The sidewash gradient induced at an arbitrary point in space can be estimated by using Eqs. (5.3.5) and (5.3.6) in Eq. (5.3.4). In the plane of symmetry this reduces to

$$\frac{\partial \varepsilon_s}{\partial \beta}(\bar{x}, \bar{y}, 0) = -\frac{\kappa_v \kappa_\beta}{\kappa_b} \frac{C_{L_w}}{R_{A_w}} \tag{5.3.7}$$

where

$$\kappa_\beta(\bar{x}, \bar{y}, 0) = \frac{4\bar{y}(\bar{x} - \kappa_b \tan \Lambda)\kappa_b^2}{\pi^2 (\bar{y}^2 + \kappa_b^2)^2} \left[1 + \frac{\bar{x} - \kappa_b \tan \Lambda}{\sqrt{(\bar{x} - \kappa_b \tan \Lambda)^2 + \bar{y}^2 + \kappa_b^2}} \right]$$

$$+ \frac{2\bar{y}\kappa_b}{\pi^2 (\bar{y}^2 + \kappa_b^2)} \left[\frac{(\bar{x} - \kappa_b \tan \Lambda)^2 \kappa_b}{[(\bar{x} - \kappa_b \tan \Lambda)^2 + \bar{y}^2 + \kappa_b^2]^{3/2}} \right] \tag{5.3.8}$$

The tail sidewash factor, κ_β, depends on the planform shape of the wing and the position of the tail relative to the wing. However, the planform shape of the wing affects the value of κ_β only through Λ and κ_b. Thus, κ_β is a unique function of $\bar{x}/\kappa_b - \tan \Lambda$ and $\bar{y}/\kappa_b$, as shown in Fig. 5.3.3.

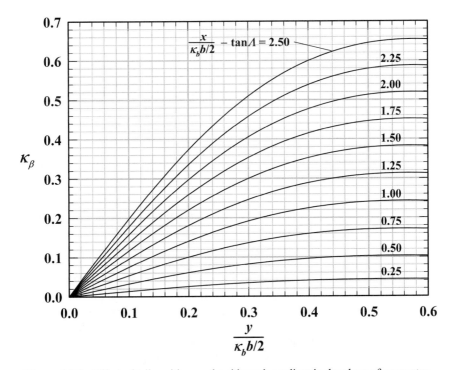

Figure 5.3.3. Effect of tail position on the sidewash gradient in the plane of symmetry.

EXAMPLE 5.3.1. Consider the general aviation aircraft, which was described in Examples 4.3.1 and 4.5.1. If the vertical tail is mounted with the aerodynamic center 5 ft above the wing, using the approximate method presented in this section, estimate the sidewash gradient for an airspeed of 80 mph at sea level.

Solution. For this wing-tail combination we have

$$S_w = 180 \text{ ft}^2, \quad b_w = 33 \text{ ft}, \quad R_{A_w} = 33^2/180 = 6.05, \quad C_{L_w,\alpha} = 4.44, \quad R_{T_w} = 0.40,$$
$$\Lambda = 10°, \quad l_t - l_w = 15 \text{ ft}, \quad \bar{x} = 15/(33/2) = 0.909, \quad \bar{y} = 5/(33/2) = 0.303$$

From Example 4.5.1,

$$\kappa_v = 1.035$$

$$\kappa_b = 0.759$$

From Eq. (5.3.8) or using the results plotted in Fig. 5.3.3, the sidewash factor is found to be

$$\kappa_\beta = 0.230$$

From Example 4.3.1, the lift on the wing for 80 mph at sea level is 2,625.9 lbf and the lift coefficient for the wing is

$$C_{L_w} = \frac{L_w}{\frac{1}{2}\rho V_\infty^2 S_w} = \frac{2,625.9}{\frac{1}{2}0.0023769(80 \times 5,280/3,600)^2 180} = 0.8916$$

From Eq. (5.3.7), the sidewash gradient at the position of the vertical tail is

$$\varepsilon_{s,\beta} \equiv \frac{\partial \varepsilon_s}{\partial \beta} = -\frac{\kappa_v \kappa_\beta}{\kappa_b} \frac{C_{L_w}}{R_{A_w}} = -\frac{1.035(0.230)}{0.759} \frac{0.8916}{6.05} = \underline{-0.046}$$

The predicted sidewash gradient on the vertical tail of the airplane in Example 5.3.1 is small. For this airplane with ample wingspan in cruise configuration, the net effect of sidewash is to increase the yaw stability by about 5 percent. However, the reader should recognize that the sidewash gradient goes up very rapidly with decreasing wingspan. Thus, aircraft having wings of smaller span produce much larger sidewash gradients. For example, if the wingspan for the airplane in Example 5.3.1 were cut in half, while the wing area and all other parameters remained the same, the magnitude of the predicted sidewash gradient would be increased by nearly an order of magnitude, to –0.435. In addition, since the sidewash gradient is proportional to the wing lift coefficient, slow flight with high-lift devices deployed can also result in a significant increase in the sidewash gradient induced on the vertical tail.

It is important to remember the orientation of the coordinate system used in the development of Eqs. (5.3.5) and (5.3.8). As shown in Figs. 5.3.1 and 5.3.2, the x-z plane of this coordinate system coincides with the plane of the trailing wingtip vortices and the x-axis is aligned with the equilibrium relative wind. Thus, the y-coordinate is the vertical distance measured above the plane of the trailing wingtip vortices. When computing the downwash from this vortex model according to the relations presented in Sec. 4.5, the y-coordinate was found to have only a second-order effect on downwash. However, as can be seen from Eqs. (5.3.5) and (5.3.8), the y-coordinate has a first-order effect on the sidewash. Thus, the orientation of the x-axis within the aircraft plane of symmetry is critical for predicting the sidewash gradient. This orientation changes with angle of attack as shown in Fig. 5.3.4. The y-coordinate of any point on the aft vertical stabilizer decreases with increasing angle of attack. From Eq. (5.3.5), we have seen that the sidewash on the vertical stabilizer is proportional to the product of the wing lift coefficient and the dimensionless y-coordinate, $\bar{y}$. As angle of attack is increased, the wing lift coefficient increases but $\bar{y}$ decreases. Thus, for a given flap setting, the magnitude of the sidewash gradient does not increase linearly with lift coefficient. In fact, as the angle of attack increases from zero lift, the magnitude of the sidewash gradient will at some point begin to decrease with increasing angle of attack. Ultimately, the sidewash gradient becomes positive and destabilizing, when the angle of attack is large enough to put the vertical stabilizer below the wingtips. Normally, the wing would stall well before this angle of attack was reached. However, this can produce an important destabilizing effect in post-stall maneuvers.

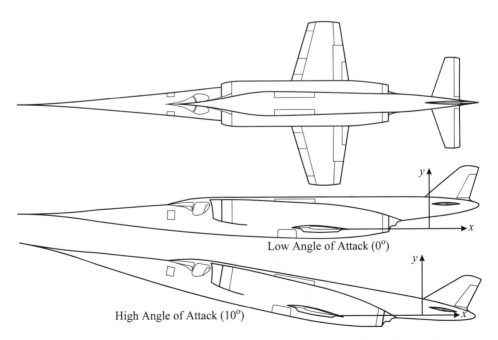

Figure 5.3.4. Effect of aircraft angle of attack on the y-position of the vertical stabilizer.

It is also important to recognize that the sidewash gradient on the vertical stabilizer increases rapidly with its distance aft of the wingtips. From Eq. (5.2.7), we have seen that the yaw stability derivative increases linearly with the product of the area of the vertical stabilizer, S_v, and its distance aft of the center of gravity, l_v. One important consideration in tail design is the trade-off between changing either S_v or l_v to attain the desired level of yaw stability. As the tail is moved aft, the required size of the vertical stabilizer is decreased, which reduces aircraft weight and drag. However, moving the tail aft increases the weight and drag associated with the structure supporting the tail. Thus, there is an optimum in the trade-off between tail area and length. Since the sidewash gradient on a high vertical tail becomes more stabilizing with increasing tail length, sidewash moves this optimum in the direction of smaller tails placed further aft of the CG. As the wingspan is decreased, this effect becomes more important. However, moving the vertical stabilizer farther aft of the wingtips will aggravate the destabilizing effect of sidewash in post-stall maneuvers.

Airplanes are sometimes designed with twin vertical stabilizers symmetrically offset from the midspan. Two examples of aircraft that have used this design feature are the P-38 Lightning and the F-14 Tomcat, both shown in Fig. 5.3.5. One advantage of this tail design is very obvious in this front view of the Lightning. The twin vertical stabilizers are directly aft of the twin counterrotating propellers. This keeps the rudders in the propeller slipstreams and improves rudder control power, especially at low airspeed. A similar but perhaps less obvious advantage is produced by the jet engines of the Tomcat. As shown in Fig. 5.3.6, the entrainment caused by turbulent mixing in the shear layer between the high-speed exhaust jet and the surrounding atmosphere draws increased

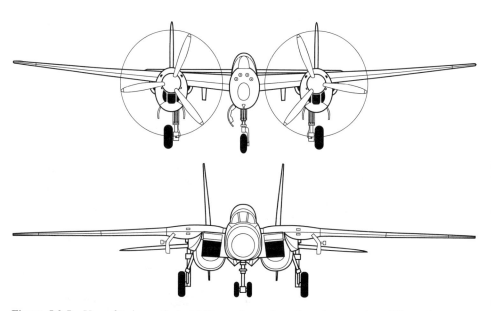

Figure 5.3.5. Use of twin vertical stabilizers above the wing places each stabilizer closer to a wingtip vortex, which increases the sidewash gradient and total yaw stability.

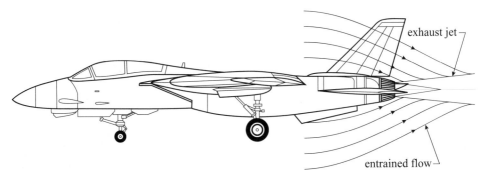

Figure 5.3.6. Increased airflow across the vertical stabilizer caused by exhaust jet entrainment.

airflow over the vertical stabilizers. In the case of the Tomcat, placing the twin vertical stabilizers directly over the exhaust nozzles of the twin turbojet engines maximizes this effect.

Improved rudder control power is not the only advantage of using twin vertical stabilizers. With this tail design, because each of the two vertical stabilizers is closer to one of the wingtip vortices, the sidewash gradient for each stabilizer is greater than that in the aircraft plane of symmetry. For the conventional high tail, this gives the aircraft greater yaw stability than would be attained with a single vertical stabilizer of the same total area, placed the same distance aft of the wing midspan. To estimate the sidewash gradient for this tail design, Eq. (5.3.4) can be used together with Eqs. (5.3.5) and (5.3.6).

5.4. Estimating the Lift Slope for a Vertical Tail

As briefly mentioned in Sec. 5.2, the lift slope for the vertical stabilizer is significantly affected by aerodynamic interactions with the horizontal stabilizer. For either the conventional tail or T-tail configuration, the horizontal stabilizer acts like a very large winglet for the vertical stabilizer. This inhibits vortex shedding and increases the lift slope for the vertical stabilizer over that which would be observed for the same vertical surface isolated in the freestream. Because of the very large number of variations in tail geometry that are possible, it would be extremely difficult to develop a general closed-form expression for estimating the in situ lift slope for the vertical stabilizer as a function of horizontal and vertical tail geometry.

The easiest way to accurately estimate the in situ lift slope for the vertical stabilizer in combination with the horizontal stabilizer is to use the numerical lifting-line method for the actual tail geometry being used. This method gives results that agree almost exactly with results obtained from the more complex vortex panel method. Since fluid viscosity has very little effect on aerodynamic lift in nonseparated flow, these inviscid flow calculations are also in close agreement with experimental data at aerodynamic angles below stall.

To estimate the in situ lift slope for the vertical stabilizer using the numerical lifting-line method, the complete tail geometry is included in the vortex model. The side-force coefficient is determined at a small positive angle of sideslip and again at a small negative angle of sideslip. The in situ lift slope for the vertical stabilizer is estimated as the difference between the two side-force coefficients divided by the difference in sideslip angle in radians.

To give the reader insight into the aerodynamic interactions between the horizontal and vertical stabilizers, Fig. 5.4.1 shows the results of numerical lifting-line computations for a specific conventional tail geometry. The vectors along the quarter-chord line show the magnitude and direction of the spanwise lift distribution, and the vectors along the trailing edge indicate the magnitude and direction of shed vorticity. In this example, both the horizontal and vertical surfaces have symmetric airfoil sections with no geometric or aerodynamic twist. The quarter-chord sweep for each surface is zero and the horizontal surface has no dihedral. The aerodynamic forces shown in Fig. 5.4.1 were computed for zero angle of attack relative to the chord line of the horizontal surface. All aerodynamic forces shown in this figure are the result of a 4-degree sideslip angle. Notice that the lift on the vertical stabilizer goes to zero at the upper tip, and a strong wingtip vortex is shed in this vicinity. If the vertical stabilizer were isolated in the freestream without the horizontal stabilizer, the lift would also go to zero at the lower extremity, and a similar wingtip vortex would be shed in this region. However, the horizontal stabilizer acts like large winglets at the base of the vertical stabilizer and inhibits vortex shedding in this vicinity. The vorticity generated on the vertical stabilizer is spread out over both sides of the horizontal stabilizer. This prevents the lift on the vertical stabilizer from going to zero at the root and generates additional lift on the horizontal stabilizer. The lift developed on the left semispan of the horizontal stabilizer is upward, while that on the right semispan is downward. Thus, in addition to increasing the side force developed on the vertical stabilizer, the aerodynamic interaction produces a net rolling moment as a result of lift developed on the horizontal stabilizer. Since the vertical stabilizer is usually above the center of gravity, the side force developed on this surface as a result of positive sideslip produces a negative rolling moment. The positive rolling moment produced on the horizontal stabilizer will partially counter this effect.

Another way to look at the origin of the forces acting on the horizontal stabilizer shown in Fig. 5.4.1 is to realize that the low pressure generated on the left side of the vertical stabilizer will create positive lift on the left side of the horizontal stabilizer.

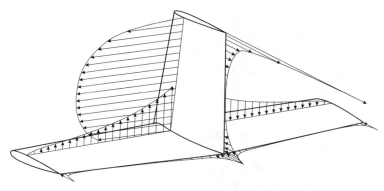

Figure 5.4.1. Aerodynamic interactions between the horizontal and vertical stabilizers in sideslip.

Similarly, the high pressure to the right of the vertical stabilizer generates negative lift on the right semispan of the horizontal stabilizer.

As can be seen from Fig. 5.4.1, the aerodynamic forces produced on the tail of an airplane as a result of sideslip are quite complex, even for this very simple tail geometry. Furthermore, the distribution and magnitude of these forces are affected by the geometry of both the horizontal and vertical stabilizers. When one considers all of the many possible variations in tail geometry, it is not practical to present results that show how the lift slope for the vertical stabilizer varies with all attributes of tail geometry. However, aspect ratio plays a major role in the variation of lift slope with tail geometry.

Figure 5.4.2 shows how the effective aspect ratio of the vertical stabilizer varies with the true aspect ratio of both the horizontal and vertical stabilizers for one particular tail configuration. The tail geometry used to generate the data plotted in Fig. 5.4.2 was a conventional tail similar to that shown in Fig. 5.4.1. Both the horizontal and vertical stabilizers had unswept rectangular planforms with the same chord and the horizontal stabilizer had no dihedral. The true aspect ratio of the vertical stabilizer is defined to be the square of the distance from root to tip divided by the planform area. The effective aspect ratio for the vertical stabilizer was defined to be the aspect ratio that when used in the analytical solution to Prandtl's lifting-line equation for the isolated vertical stabilizer, gives the same lift slope as that predicted by the numerical lifting-line solution for the complete tail. Strictly speaking, Fig. 5.4.2 applies only to the particular tail geometry for which it was obtained. However, in preliminary design the exact tail geometry is not typically known, and Fig. 5.4.2 can be used as a guideline.

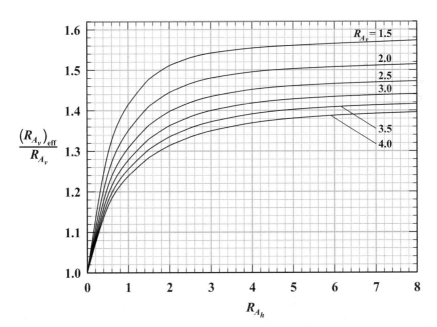

Figure 5.4.2. Effective vertical aspect ratio for a conventional tail with both the horizontal and vertical stabilizers having unswept rectangular planforms.

EXAMPLE 5.4.1. For the preliminary design of an airplane we shall assume that the conventional aft tail is constructed from unswept rectangular surfaces having thin airfoil sections. The size of the tail is not yet known, but the horizontal stabilizer is assumed to have an aspect ratio of 5. Assuming that the vertical stabilizer will have a root-to-tip height of twice the chord, estimate the lift slope for the vertical stabilizer from the results plotted in Fig. 5.4.2.

Solution. For this tail geometry we have

$$R_{A_h} = 5.0, \quad R_{A_v} = 2.0$$

From Fig. 5.4.2,

$$\frac{(R_{A_v})_{\text{eff}}}{R_{A_v}} \cong 1.5$$

and

$$(R_{A_v})_{\text{eff}} - 1.5 R_{A_v} - 1.5(2.0) - 3.0$$

Using this aspect ratio and a taper ratio of 1.0 in the series solution to Prandtl's lifting-line equation presented in Chapter 1, the lift slope factor for the vertical stabilizer is found to be

$$\kappa_L \cong 0.04$$

Using this result with the effective aspect ratio of 3.0 and using the theoretical section lift slope for thin airfoils, 2π, the lift slope for the vertical stabilizer is estimated to be

$$C_{L_v,\alpha} = \frac{\tilde{C}_{L,\alpha}}{[1 + \tilde{C}_{L,\alpha}/\pi(R_{A_v})_{\text{eff}}](1 + \kappa_L)} = \frac{2\pi}{[1 + 2\pi/\pi(3.0)](1 + 0.04)} = \underline{3.62}$$

If we had ignored the vortex interactions between the horizontal stabilizer and the vertical stabilizer in Example 5.4.1, a smaller lift slope would have been predicted. Using the true aspect ratio of 2.0, the lift slope for the isolated vertical stabilizer is predicted from Prandtl's lifting-line theory to be about 3.02. Hence, lifting-line theory predicts that the presence of the horizontal stabilizer increases the lift slope for this vertical stabilizer by about 20 percent.

The aerodynamic interaction between the vertical stabilizer and fuselage can also increase the lift slope for the vertical stabilizer somewhat. This is typically more significant for the T-tail configuration. The vertical lift slope is usually slightly higher for a T-tail than it is for a conventional tail with the same aspect ratio. This is because the horizontal stabilizer inhibits vortex shedding at the top and the fuselage inhibits vortex shedding at the bottom.

5.5. Effects of Tail Dihedral on Yaw Stability

Tail dihedral, such as that shown in Fig. 5.5.1, can significantly influence the yaw stability derivative for an airplane. The dihedral angle for a tail surface is commonly given the symbol Γ and traditionally defined to be positive when the tip is above the root, as is the case for both the right and left semispans of the aft tail in the upper configuration of Fig. 5.5.1. Negative dihedral, as shown in the lower configuration of Fig. 5.5.1, is commonly called *anhedral*. With this usual sign convention, a traditional upright vertical stabilizer has a dihedral angle of 90 degrees and the inverted vertical stabilizers shown in Fig. 5.5.1 are said to have a dihedral angle of −90 degrees, which could also be described as 90 degrees of anhedral

With proper design, either positive or negative dihedral in an aft tail can increase the yaw stability of an airplane. In fact, if sufficient dihedral or anhedral is added to a conventional horizontal stabilizer, the vertical stabilizer can be eliminated completely. Such a configuration, which is commonly called a V-tail, is shown in Fig. 5.5.2. The two semispans of this V-tail provide both pitch and yaw stability for the aircraft. As a first approximation, the pitch and yaw stability derivatives for aft tail configurations such as those shown in Figs. 5.5.1 and 5.5.2 are sometimes estimated from Eqs. (4.3.30)

Figure 5.5.1. Examples of tail dihedral shown in two tail configurations used on different models of the Predator UAV. (NASA and U.S. Air Force photographs)

Figure 5.5.2. Example of a V-tail used on the Global Hawk UAV. (U.S. Air Force photograph)

and (5.2.7) by computing the horizontal and vertical tail volume ratios using the horizontal and vertical projections of the total tail planform area, respectively. Such estimates should be used with caution, because they do not account for the aerodynamic interactions between the different lifting surfaces of the tail and these interactions can substantially alter tail performance.

The spanwise distribution of lift that is produced on a typical V-tail in pure sideslip is shown in Fig. 5.5.3. This lift distribution, which is the result of sideslip to the right with no angle of attack, has positive lift generated on the right semispan and negative lift generated on the left semispan. The vertical components of lift on the two semispans

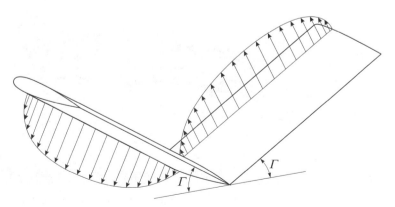

Figure 5.5.3. Spanwise lift distribution on a V-tail in pure sideslip.

approximately cancel, producing little or no vertical force. Conversely, the horizontal components of lift on these two semispans combine to produce a net side force to the left that opposes the sideslip and is proportional to the sideslip angle. The reader may also note from Fig. 5.5.3 that sideslip produces a rolling moment on this V-tail. Consideration of this rolling moment will be covered in Sec. 5.6.

When this same V-tail is operating at a positive angle of attack with no sideslip, it produces the lift distribution shown in Fig. 5.5.4. At this operating condition, the horizontal components of lift on the two semispans cancel and the vertical components combine to produce a net upward force that is proportional to the angle of attack. Thus, when this geometry is used as an aft tail, as shown in Fig. 5.5.2, it provides a stabilizing contribution to both the pitching and yawing moments for the complete aircraft. The magnitudes of the pitch and yaw stability derivatives can be controlled by adjusting the total area and dihedral angle.

An isolated V-tail operating at a positive angle of attack with no sideslip feels an upward component of relative wind equal to the airspeed multiplied by the sine of the angle of attack, which for small angles of attack is approximated as $V\sin\alpha \cong V\alpha$. Similarly, when the isolated V-tail is operating with positive sideslip and no angle of attack, it feels a leftward component of relative wind equal to the airspeed multiplied by the sine of the sideslip angle, which for small sideslip angles can be approximated as $V\sin\beta \cong V\beta$. Thus, for small aerodynamic angles but an arbitrary dihedral angle, the upward normal components of relative wind on the right and left semispans of a V-tail are closely approximated as

$$\left(V_{\text{normal}}\right)_{\text{right}} \cong V\alpha\cos\Gamma + V\beta\sin\Gamma$$

$$\left(V_{\text{normal}}\right)_{\text{left}} \cong V\alpha\cos\Gamma - V\beta\sin\Gamma$$

Accordingly, if the aerodynamic angles α and β are small, the freestream angles of attack normal to the right and left semispans of a V-tail with a dihedral angle of Γ are closely approximated as

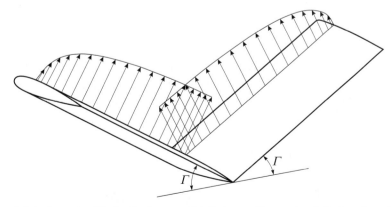

Figure 5.5.4. Spanwise lift distribution on a V-tail at a positive angle of attack and no sideslip.

$$\alpha_{\text{right}} \cong \frac{(V_{\text{normal}})_{\text{right}}}{V} \cong \alpha \cos \Gamma + \beta \sin \Gamma \tag{5.5.1}$$

$$\alpha_{\text{left}} \cong \frac{(V_{\text{normal}})_{\text{left}}}{V} \cong \alpha \cos \Gamma - \beta \sin \Gamma \tag{5.5.2}$$

Referring to Fig. 5.5.4, the vertical lift produced on each semispan of a V-tail is the lift produced normal to that semispan multiplied by $\cos\Gamma$. Similarly, the side force produced on the left semispan is the lift normal to that semispan multiplied by $\sin\Gamma$ and the side force on the right semispan is the normal lift multiplied by $-\sin\Gamma$. Thus, with no control surface deflection and ignoring any vortex interactions between the two semispans, the contribution that the right semispan of a V-tail makes to the vertical lift and side force could be approximated as

$$L_{\text{right}} \cong \frac{1}{2} \rho V^2 S_{\text{right}} (C_{L,\alpha})_{\Gamma=0} (\alpha_{\text{right}} + \alpha_0 - \alpha_{L0})(\cos \Gamma) \tag{5.5.3}$$

$$Y_{\text{right}} \cong \frac{1}{2} \rho V^2 S_{\text{right}} (C_{L,\alpha})_{\Gamma=0} (\alpha_{\text{right}} + \alpha_0 - \alpha_{L0})(-\sin \Gamma) \tag{5.5.4}$$

where the lift slope evaluated at $\Gamma=0$ is simply the isolated lift slope for a horizontal stabilizer having the same semispan planform as the V-tail but with no dihedral. Similarly, the contributions from the left semispan could be approximated as

$$L_{\text{left}} \cong \frac{1}{2} \rho V^2 S_{\text{left}} (C_{L,\alpha})_{\Gamma=0} (\alpha_{\text{left}} + \alpha_0 - \alpha_{L0})(\cos \Gamma) \tag{5.5.5}$$

$$Y_{\text{left}} \cong \frac{1}{2} \rho V^2 S_{\text{left}} (C_{L,\alpha})_{\Gamma=0} (\alpha_{\text{left}} + \alpha_0 - \alpha_{L0})(\sin \Gamma) \tag{5.5.6}$$

Applying Eqs. (5.5.1) and (5.5.2) to Eqs. (5.5.3)–(5.5.6) and adding the contributions from the right and left semispans yields the net lift and side force for the complete V-tail,

$$L_{\text{V-tail}} \cong \frac{1}{2} \rho V^2 S'_V (C_{L,\alpha})_{\Gamma=0} [\alpha \cos^2 \Gamma + (\alpha_0 - \alpha_{L0}) \cos \Gamma] \tag{5.5.7}$$

$$Y_{\text{V-tail}} \cong -\frac{1}{2} \rho V^2 S'_V (C_{L,\alpha})_{\Gamma=0} \beta \sin^2 \Gamma \tag{5.5.8}$$

where S'_V is the combined planform area of both semispans measured parallel with the semispan dihedral, not the horizontal projection. Using this reference area for the lift and side-force coefficients, Eqs. (5.5.7) and (5.5.8) can be written in nondimensional form as

$$(C_L)_{\text{V-tail}} \equiv \frac{L_{\text{V-tail}}}{\frac{1}{2} \rho V^2 S'_V} \cong (C_{L,\alpha})_{\Gamma=0} [\alpha \cos^2 \Gamma + (\alpha_0 - \alpha_{L0}) \cos \Gamma] \tag{5.5.9}$$

$$(C_Y)_{\text{V-tail}} \equiv \frac{Y_{\text{V-tail}}}{\frac{1}{2} \rho V^2 S'_V} \cong -(C_{L,\alpha})_{\Gamma=0} \beta \sin^2 \Gamma \tag{5.5.10}$$

Thus, **neglecting vortex interactions between the two semispans of an isolated V-tail**, the lift slope and side-force derivative are approximated as

$$
\left(C_{L,\alpha}\right)_{\text{V-tail}} \equiv \frac{\partial}{\partial\alpha}\left(C_L\right)_{\text{V-tail}} \cong \left(C_{L,\alpha}\right)_{\Gamma=0} \cos^2\Gamma \qquad (5.5.11)
$$

$$
\left(C_{Y,\beta}\right)_{\text{V-tail}} \equiv \frac{\partial}{\partial\beta}\left(C_Y\right)_{\text{V-tail}} \cong -\left(C_{L,\alpha}\right)_{\Gamma=0} \sin^2\Gamma \qquad (5.5.12)
$$

We may expect that **neglecting vortex interactions between the two semispans of a V-tail is only reasonable for small dihedral angles.** Comparison of Figs. 5.5.3 and 5.5.4 shows that a V-tail does not produce a side force in response to sideslip as efficiently as it produces a vertical force in response to an increase in angle of attack. In Fig. 5.5.3, we see that the lift generated on the two semispans of a V-tail in response to pure sideslip goes to zero at the root as well as at the tips. For this untwisted V-tail in pure sideslip to the right, the aerodynamic angle of attack relative to the freestream has a constant positive value over the right semispan and a constant negative value on the left semispan. This would tend to produce constant positive lift on the right semispan and constant negative lift on the left, if it were not for the downwash induced by the shed vorticity. The reduction in lift at the tips results in the usual manner from the tip vortices. Similarly, the reduction in lift at the root results from vorticity shed near the root.

The simplest method that can be used for accurately estimating the aerodynamic derivatives associated with tail configurations similar to those shown in Figs. 5.5.1 and 5.5.2 is the numerical lifting-line method, which is described in Sec. 1.9. For example, to estimate the change in side-force coefficient with respect to sideslip angle, the side-force coefficient is determined at a small positive sideslip angle and again at a small negative sideslip angle. The change in side-force coefficient with respect to sideslip angle is estimated as the difference between the two side-force coefficients divided by the difference in sideslip angle in radians. This is best demonstrated by example.

EXAMPLE 5.5.1. Consider an inverted V-tail similar to that shown in the lower configuration of Fig. 5.5.1, but with no vertical stabilizer between the two semispans. Each semispan of the V-tail has a constant chord of 2.0 feet and is 6.0 feet from root to tip with a dihedral angle of -35 degrees. Assuming a thin symmetric airfoil section and neglecting parasitic drag, use the numerical lifting-line method to compute the aerodynamic derivatives $C_{L,\alpha}$ and $C_{Y,\beta}$. Compare these results with the simplified approximations neglecting the vortex interactions between the two semispans, which are given by Eqs. (5.5.11) and (5.5.12).

Solution. For this V-tail the reference planform areas is

$$
S_V' = 2(2.0\ \text{ft} \times 6.0\ \text{ft}) = 24.0\ \text{ft}^2
$$

Using this reference area, an airfoil section lift slope of 2π, and 50 nodes per semispan with cosine clustering, the solutions obtained at 3.0 degrees angle of attack with no sideslip and 3.0 degrees sideslip with no angle of attack are shown in Fig. 5.5.5. In this figure, the vectors displayed along the quarter-chord lines show the magnitude and direction of the local section lift, and the lines along the trailing edges indicate the magnitude of the local shed vorticity per unit span. From such double-precision computations, the net lift coefficients at ±1.0 degree angle of attack are found to be

$$\alpha = 1.0°, \ \beta = 0.0° \Rightarrow C_L \equiv \frac{L}{\frac{1}{2}\rho V_\infty^2 S_V'} = 5.36041413258 \times 10^{-2}$$

$$\alpha = -1.0°, \ \beta = 0.0° \Rightarrow C_L \equiv \frac{L}{\frac{1}{2}\rho V_\infty^2 S_V'} = -5.41322891900 \times 10^{-2}$$

Thus, the lift slope for this tail configuration is estimated to be

$$C_{L,\alpha} \cong \frac{\Delta C_L}{\Delta \alpha} = \frac{5.36041413258 \times 10^{-2} - (-5.41322891900 \times 10^{-2})}{2°(\pi/180°)} = 3.086$$

Likewise, the change in side-force coefficient with respect to sideslip angle is estimated from

$$\alpha = 0.0°, \ \beta = 1.0° \Rightarrow C_Y \equiv \frac{Y}{\frac{1}{2}\rho V_\infty^2 S_V'} = -1.86543468320 \times 10^{-2}$$

$$\alpha = 0.0°, \ \beta = -1.0° \Rightarrow C_Y \equiv \frac{Y}{\frac{1}{2}\rho V_\infty^2 S_V'} = 1.86543468320 \times 10^{-2}$$

$$C_{Y,\beta} \cong \frac{\Delta C_Y}{\Delta \beta} = \frac{-1.86543468320 \times 10^{-2} - 1.86543468320 \times 10^{-2}}{2°(\pi/180°)} = -1.069$$

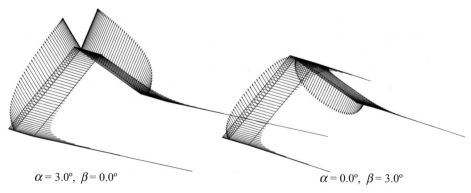

$\alpha = 3.0°, \ \beta = 0.0°$ $\alpha = 0.0°, \ \beta = 3.0°$

Figure 5.5.5. Spanwise lift distributions for the V-tail in Example 5.5.1, as predicted from the numerical lifting-line method.

The isolated lift slope for a horizontal stabilizer having the same semispan planform as this V-tail but with no dihedral can be evaluated from the infinite series solution to Prandtl's lifting-line equation. For this lifting surface the aspect ratio is $R_A=6.0$ and the taper ratio is $R_T=1.0$. Using these values in Fig. 1.8.13 yields $\kappa_L=0.04$ and the lift slope as computed from Eq. (1.8.11) is

$$\left(C_{L,\alpha}\right)_{\Gamma=0} = \frac{\tilde{C}_{L,a}}{(1+\tilde{C}_{L,a}/\pi R_A)(1+\kappa_L)} = \frac{2\pi}{(1+2\pi/\pi 6.0)(1+0.04)} = 4.531$$

Thus, **neglecting vortex interactions between the two semispans of this V-tail**, the lift slope and side-force derivative from Eqs. (5.5.11) and (5.5.12) are

$$C_{L,\alpha} \cong \left(C_{L,\alpha}\right)_{\Gamma=0}\cos^2\Gamma = 4.531\cos^2(-35°) = 3.040$$

$$C_{Y,\beta} \cong -\left(C_{L,\alpha}\right)_{\Gamma=0}\sin^2\Gamma = -4.531\sin^2(-35°) = -1.491$$

Notice that for this V-tail, the approximate lift slope from Eq. (5.5.11) agrees with the numerical lifting-line solution to within less than 2 percent. However, the approximate side-force derivative determined from Eq. (5.5.12) is overestimated by nearly 40 percent. This is because Eq. (5.5.12) neglects the vorticity shed near the root of a V-tail in sideslip, which is shown in Fig. 5.5.5.

While the results obtained in the previous example are typical, the lift slope and side-force derivative for a V-tail depend on its planform and dihedral angle. The numerical lifting-line method is capable of accurately predicting this dependence. For example, Fig. 5.5.6 shows a comparison between the results of numerical lifting-line

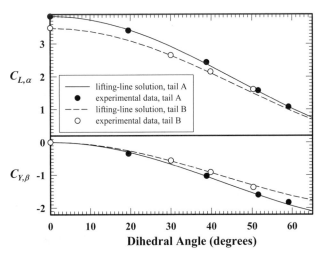

Figure 5.5.6. Lift slope and side-force derivative as predicted from the numerical lifting-line method compared with experimental data for two V-tails with variable dihedral.

computations and experimental data for two different V-tail planforms. The data in this figure are those presented by Purser and Campbell (1945) and Schade (1947). Tail A has an aspect ratio of 5.55 and a taper ratio of 0.39, whereas tail B has an aspect ratio of 3.70 and a taper ratio of 0.56.

The dependence of the lift slope and side-force derivative on planform and dihedral for V-tails with no twist is shown in Figs. 5.5.7 and 5.5.8. Figure 5.5.7 shows the ratio of the lift slope predicted from lifting-line theory to that predicted from Eq. (5.5.11) as a function of dihedral angle, aspect ratio, and taper ratio. Similar results for the side-force derivative are shown in Fig. 5.5.8. **The V-tail aspect ratio used in these two figures is defined to be the square of twice the distance from root to tip divided by the total area of both semispans combined (not the horizontal projection).**

The results shown in Figs. 5.5.7 and 5.5.8 were obtained from numerical lifting-line calculations similar to those used for Example 5.5.1. Although these figures were generated for upright V-tails with positive dihedral, they may also be applied to inverted V-tails with anhedral.

In the preliminary design of an aircraft, we may wish to study some of the trade-offs between using a conventional tail or a V-tail. Such trade studies require a comparison between two tail configurations that provide the same level of static stability for both pitch and yaw. Figures 5.5.7 and 5.5.8 can be used to aid in making an initial estimate for the V-tail geometry that will provide the required stability. This is best demonstrated by example.

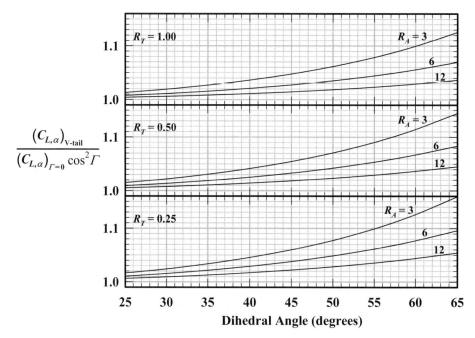

Figure 5.5.7. Ratio of the lift slope for a V-tail as predicted from the numerical lifting-line method to that predicted from Eq. (5.5.11).

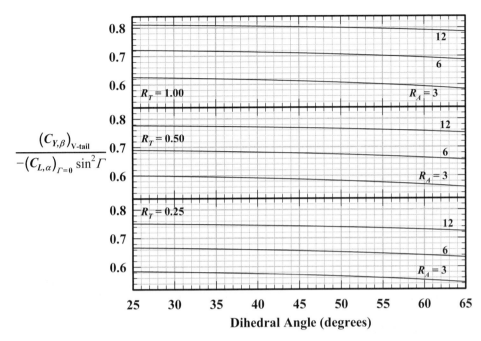

$$\frac{\left(C_{Y,\beta}\right)_{\text{V-tail}}}{-\left(C_{L,\alpha}\right)_{\Gamma=0}\sin^2\Gamma}$$

Figure 5.5.8. Ratio of the side-force derivative for a V-tail as predicted from the numerical lifting-line method to that predicted from Eq. (5.5.12).

EXAMPLE 5.5.2. An airplane has a conventional tail with a rectangular horizontal stabilizer of aspect ratio 4 and a planform area of 36 ft². The vertical stabilizer is also rectangular with the same chord and a planform area of 15 ft². We wish to examine the possibility of replacing this conventional tail with a V-tail. Using the numerical lifting-line method estimate the geometry of a V-tail that would provide the same level of static stability in both pitch and yaw as the original conventional tail. Compare the total planform area of this V-tail to that of the original tail.

Solution. For the original conventional tail we have

$$S_h = 36.0 \text{ ft}^2, \quad b_h = 12.0 \text{ ft}, \quad c_h = 3.0 \text{ ft}, \quad S_v = 15.0 \text{ ft}^2, \quad c_v = 3.0 \text{ ft}$$

To simplify attachment of the replacement tail, we shall keep the V-tail chord the same as the original tail. A good starting point for such designs is to assume the total planform area of the V-tail to be equal to that of the conventional tail,

$$S_V' \cong S_h + S_v = 51 \text{ ft}^2$$

Noting that the lift-slope ratio plotted in Fig. 5.5.7 is greater than 1.0 and that the V-tail aspect ratio will be larger than that for the original horizontal stabilizer, a good starting point for the dihedral angle is typically found from the relation

$$S_V' \cos^2 \Gamma \cong 0.95 S_h$$

which yields

$$\Gamma \cong \cos^{-1}\left(\sqrt{0.95 S_h/S_V'}\right) = \cos^{-1}\left(\sqrt{0.95(36)/51}\right) \cong 35°$$

To maintain a fair comparison between the V-tail and the conventional tail, we should use the same taper ratio for both, in this case $R_T = 1.0$. The V-tail aspect ratio for our initial estimate is $R_A = 5.67$. From Figs. 5.5.7 and 5.5.8, for this taper ratio, aspect ratio, and dihedral angle we obtain

$$\frac{\left(C_{L,\alpha}\right)_{\text{V-tail}}}{\left(C_{L,\alpha}\right)_{\Gamma=0} \cos^2 \Gamma} \cong 1.02 \text{ and } \frac{\left(C_{Y,\beta}\right)_{\text{V-tail}}}{-\left(C_{L,\alpha}\right)_{\Gamma=0} \sin^2 \Gamma} \cong 0.73$$

If the V-tail is to provide the same level of static pitch stability as the original conventional tail, the product of the lift slope and reference area must be the same for both tails. Likewise, to maintain the same yaw stability, the product of the side-force derivative and reference area must be the same for both tails. From the results obtained from Figs. 5.5.7 and 5.5.8 this requires

$$S_V'\left(C_{L,\alpha}\right)_{\text{V-tail}} \cong 1.02 S_V'\left(C_{L,\alpha}\right)_{\Gamma=0} \cos^2 \Gamma = \left(S_h C_{L,\alpha}\right)_{\text{conventional}}$$

$$S_V'\left(C_{Y,\beta}\right)_{\text{V-tail}} \cong -0.73 S_V'\left(C_{L,\alpha}\right)_{\Gamma=0} \sin^2 \Gamma = \left(S_v C_{Y,\beta}\right)_{\text{conventional}}$$

Adding the first relation divided by 1.02 to the second divided by -0.73 yields

$$S_V'\left(C_{L,\alpha}\right)_{\Gamma=0} = \frac{\left(S_h C_{L,\alpha}\right)_{\text{conventional}}}{1.02} - \frac{\left(S_v C_{Y,\beta}\right)_{\text{conventional}}}{0.73}$$

Using an airfoil section lift slope of 2π and 50 nodes per semispan with cosine clustering, the numerical lifting-line method applied to the original conventional tail gives

$$\left(S_h C_{L,\alpha}\right)_{\text{conventional}} = 144.97 \text{ ft}^2 \text{ and } \left(S_v C_{Y,\beta}\right)_{\text{conventional}} = -50.50 \text{ ft}^2$$

The isolated lift slope for the V-tail at $\Gamma=0$ can be evaluated from the infinite series solution to Prandtl's lifting-line equation. From our initial estimate we have $R_A = 5.67$ and $R_T = 1.0$. Using these values in Fig. 1.8.13 yields $\kappa_L = 0.04$ and the lift slope as computed from Eq. (1.8.11) is

$$\left(C_{L,\alpha}\right)_{\Gamma=0} = \frac{\tilde{C}_{L,\alpha}}{(1+\tilde{C}_{L,\alpha}/\pi R_A)(1+\kappa_L)} = \frac{2\pi}{(1+2\pi/\pi 5.67)(1+0.04)} = 4.466$$

Thus, a revised estimate for the required V-tail area is

$$
\begin{aligned}
S'_V &= \frac{\left(S_h C_{L,\alpha}\right)_{\text{conventional}}}{1.02\left(C_{L,\alpha}\right)_{\Gamma=0}} - \frac{\left(S_v C_{Y,\beta}\right)_{\text{conventional}}}{0.73\left(C_{L,\alpha}\right)_{\Gamma=0}} \\
&= \frac{144.97 \text{ ft}^2}{1.02(4.466)} - \frac{-50.50 \text{ ft}^2}{0.73(4.466)} = 47.31 \text{ ft}^2
\end{aligned}
$$

which gives an aspect ratio of 5.26 and an isolated lift slope for the V-tail at $\Gamma=0$ of 4.377. A revised estimate for the dihedral angle is then obtained from

$$
\cos^2\Gamma = \frac{\left(S_h C_{L,\alpha}\right)_{\text{conventional}}}{1.02 S_{V\text{-tail}}\left(C_{L,\alpha}\right)_{\Gamma=0}} = \frac{144.97}{1.02(47.31)(4.377)} = 0.68635 \implies \Gamma = 34.06°
$$

Using an airfoil section lift slope of 2π and 50 nodes per semispan with cosine clustering, the numerical lifting-line method applied to this V-tail geometry gives

$$
\left(SC_{L,\alpha}\right)_{V\text{-tail}} = 144.34 \text{ ft}^2 \quad \text{and} \quad \left(SC_{Y,\beta}\right)_{V\text{-tail}} = -45.39 \text{ ft}^2
$$

We wish to find the planform area and dihedral angle for a V-tail that will satisfy both of the relations

$$
\left(SC_{L,\alpha}\right)_{V\text{-tail}} = \left(S_h C_{L,\alpha}\right)_{\text{conventional}} \quad \text{and} \quad \left(SC_{Y,\beta}\right)_{V\text{-tail}} = \left(S_v C_{Y,\beta}\right)_{\text{conventional}}
$$

Thus, it is convenient to define the following residuals, which depend on the area, $S \equiv S'_V$, and dihedral angle, Γ, of the V-tail,

$$
\begin{aligned}
R_L(S,\Gamma) &\equiv \left(SC_{L,\alpha}\right)_{V\text{-tail}} - \left(S_h C_{L,\alpha}\right)_{\text{conventional}} \\
R_Y(S,\Gamma) &\equiv \left(SC_{Y,\beta}\right)_{V\text{-tail}} - \left(S_v C_{Y,\beta}\right)_{\text{conventional}}
\end{aligned}
$$

We wish to find the area and the dihedral angle that will make both of these residuals go to zero. For our current estimate we have

$$
\begin{aligned}
R_L(47.31 \text{ ft}^2, 34.06°) &= 144.34 \text{ ft}^2 - 144.97 \text{ ft}^2 = -0.63 \text{ ft}^2 \\
R_Y(47.31 \text{ ft}^2, 34.06°) &= -45.39 \text{ ft}^2 - (-50.50 \text{ ft}^2) = 5.11 \text{ ft}^2
\end{aligned}
$$

which is obviously not a solution. However, this geometry is readily corrected using Newton's method. This requires solving the linear system

$$
\begin{bmatrix} \dfrac{\partial R_L}{\partial S} & \dfrac{\partial R_L}{\partial \Gamma} \\[2mm] \dfrac{\partial R_Y}{\partial S} & \dfrac{\partial R_Y}{\partial \Gamma} \end{bmatrix} \begin{Bmatrix} \Delta S \\ \Delta \Gamma \end{Bmatrix} = \begin{Bmatrix} -R_L \\ -R_Y \end{Bmatrix}
$$

These derivatives can be estimated numerically by using the lifting-line method. Increasing the planform area of the V-tail by 1 ft², while holding the dihedral angle constant, results in

$$R_L = 3.23 \text{ ft}^2 \text{ and } R_Y = 3.68 \text{ ft}^2$$

$$\frac{\partial R_L}{\partial S} \cong \frac{\Delta R_L}{\Delta S} = \frac{3.23 \text{ ft}^2 - (-0.63 \text{ ft}^2)}{1.0 \text{ ft}^2} = 3.86$$

$$\frac{\partial R_Y}{\partial S} \cong \frac{\Delta R_Y}{\Delta S} = \frac{3.68 \text{ ft}^2 - (5.11 \text{ ft}^2)}{1.0 \text{ ft}^2} = -1.43$$

Similarly, increasing the dihedral angle by 1 degree, while holding the planform area constant, yields

$$R_L = -3.91 \text{ ft}^2 \text{ and } R_Y = 2.78 \text{ ft}^2$$

$$\frac{\partial R_L}{\partial \Gamma} \cong \frac{\Delta R_L}{\Delta \Gamma} = \frac{-3.91 \text{ ft}^2 - (-0.63 \text{ ft}^2)}{1.0°} = -3.28 \text{ ft}^2/°$$

$$\frac{\partial R_Y}{\partial \Gamma} \cong \frac{\Delta R_Y}{\Delta \Gamma} = \frac{2.78 \text{ ft}^2 - (5.11 \text{ ft}^2)}{1.0°} = -2.33 \text{ ft}^2/°$$

Thus, the corrections for Newton's method are computed from

$$\left\{ \begin{matrix} \Delta S \\ \Delta \Gamma \end{matrix} \right\} = \begin{bmatrix} 3.86 & -3.28 \text{ ft}^2/° \\ -1.43 & -2.33 \text{ ft}^2/° \end{bmatrix}^{-1} \left\{ \begin{matrix} 0.63 \text{ ft}^2 \\ -5.11 \text{ ft}^2 \end{matrix} \right\} = \left\{ \begin{matrix} 1.33 \text{ ft}^2 \\ 1.38° \end{matrix} \right\}$$

and for our next estimate we use

$$S = 47.31 \text{ ft}^2 + 1.33 \text{ ft}^2 = 48.64 \text{ ft}^2$$

$$\Gamma = 34.06° + 1.38° = 35.44°$$

Applying the numerical lifting-line method to this V-tail geometry gives

$$\left(SC_{L,\alpha}\right)_{\text{V-tail}} = 144.767 \text{ ft}^2 \text{ and } \left(SC_{Y,\beta}\right)_{\text{V-tail}} = -50.634 \text{ ft}^2$$

and the new residuals are

$$R_L = -0.203 \text{ ft}^2 \text{ and } R_Y = -0.134 \text{ ft}^2$$

A second iteration using Newton's method yields

$$S = 48.642 \text{ ft}^2 \text{ and } \Gamma = 35.384°$$

$$\left(SC_{L,\alpha}\right)_{\text{V-tail}} = 144.97 \text{ ft}^2 \text{ and } \left(SC_{Y,\beta}\right)_{\text{V-tail}} = -50.50 \text{ ft}^2$$

From this numerical solution, we estimate that a V-tail having a total area of approximately 48.6 ft^2 and a dihedral angle of about 35.4 degrees would provide the same level of static stability as the original conventional tail, which has a total area of 51 ft^2. Thus, using this V-tail in place of the original tail has the potential for reducing tail weight by about 5 percent, based on required area alone. On the other hand, the control mechanism for a V-tail is more complex than that for a conventional tail and this added complexity could add more weight than is saved by the reduction in tail area. Control surfaces for V-tails are covered in Chapter 6. Another advantage of the V-tail is a small reduction in drag due to the fact that a V-tail requires only two fuselage-tail junctions rather than three, which are required for a conventional tail.

We have now seen that an aft V-tail with no vertical stabilizer can be used to provide both pitch and yaw stability for an airplane. However, the aerodynamic interactions between the two semispans of a V-tail reduce its effectiveness in producing a side force in response to sideslip. When a V-tail is used in combination with a vertical stabilizer, as shown in Fig. 5.5.1, the aerodynamic interactions between the three lifting surfaces of the tail become more complex and sometimes produce results that are unexpected.

To give the reader some insight into the performance of tail configurations like those shown in Fig. 5.5.1, we start by reexamining the lift distribution on a conventional tail. Figure 5.5.9 shows a typical lift distribution for a conventional tail in pure sideslip, as predicted from numerical lifting-line computations. For those who might wish to duplicate these computations, this particular tail has a horizontal aspect ratio of 4.0 and a horizontal planform area of 36 ft^2. The vertical stabilizer has the same constant chord with a planform area of 12 ft^2. In this example, both the horizontal and vertical surfaces have thin symmetric airfoil sections and no twist. The sweep for both surfaces is zero and the horizontal surface has no dihedral. At zero angle of attack, lifting-line computations for this geometry predict a side-force derivative, $C_{Y,\beta}$, of -3.078, based on the vertical reference area. The aerodynamic section forces shown in Fig. 5.5.9 were computed at zero angle of attack. All of the forces shown in this figure result from a 5-degree sideslip angle. The aerodynamic forces on the horizontal stabilizer result entirely

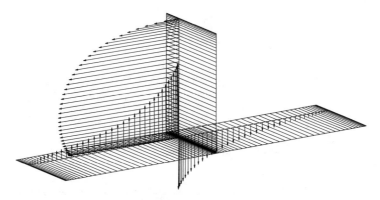

Figure 5.5.9. The lift distribution on a conventional tail in pure sideslip.

from vortex interactions with the vertical stabilizer. These aerodynamic interactions were discussed in Sec. 5.4. If the reader does not understand how these forces are generated, he or she should reread Sec. 5.4 before proceeding further with this section.

When a small amount of positive dihedral is added to the horizontal stabilizer of a conventional tail, the lift distribution shown in Fig. 5.5.9 is changed in two important ways. First, positive dihedral in the horizontal stabilizer combined with sideslip to the right adds positive lift to the right semispan and negative lift to the left semispan, similar to that shown on the V-tail in Fig. 5.5.3. With the added positive dihedral, this change tends to increase the side force to the left. The second change to the lift distribution shown in Fig. 5.5.9, which arises from adding positive dihedral to the horizontal stabilizer, is simply a tilting of the lift vectors that are shown on the horizontal stabilizer in this figure. The positive lift vectors on the left semispan are tilted to the right and the negative lift vectors on the right semispan are also tilted to the right. This change tends to decrease the side force to the left. Figure 5.5.10 shows the net effect of these two changes when 10 degrees of dihedral are added to the horizontal stabilizer of the conventional tail shown in Fig. 5.5.9. Notice that the lift vectors near the root of the horizontal stabilizer in Fig. 5.5.10 contribute to a decrease in the side force to the left, whereas the lift vectors on the outboard sections contribute to an increase in the side force to the left. For this example, the net effect of adding 10 degrees dihedral to the horizontal stabilizer is a 7 percent reduction in the total magnitude of the side force, in spite of a 52 percent increase in the vertical projection of the total planform area.

Figure 5.5.11 shows the effect of adding 10 degrees of anhedral to the horizontal stabilizer of the conventional tail shown in Fig. 5.5.9. In this case the lift generated on the horizontal stabilizer as a result of vortex interactions with the vertical stabilizer is in the same direction as the lift generated by the direct effects of the horizontal-stabilizer anhedral. Thus, both effects contribute to increasing the side force to the left. For this particular example, the net effect of the 10 degrees anhedral is a 24 percent increase in the magnitude of the side force.

Figure 5.5.12 shows how the side-force derivative for the conventional tail shown in Figs. 5.5.9 varies with horizontal-stabilizer dihedral angle from −45 to +45 degrees. The solid lines plotted in this figure were obtained from numerical lifting-line computations

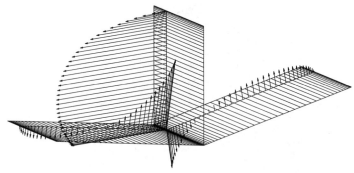

Figure 5.5.10. The effect of horizontal-stabilizer dihedral on the lift distribution for a conventional tail in pure sideslip.

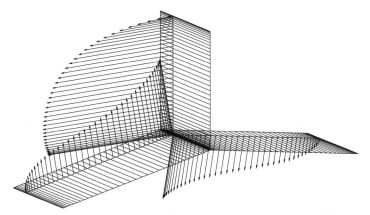

Figure 5.5.11. The effect of horizontal-stabilizer anhedral on the lift distribution for a conventional tail in pure sideslip.

by varying only the dihedral angle of the horizontal stabilizer while keeping the chord and total area constant. For comparison, the square and circular markers display results obtained for the same geometry using a numerical panel code. These results are plotted in terms of what is being called the *total side-force derivative* in square feet. This is simply the product of the dimensionless side-force derivative, $C_{Y,\beta}$, and the corresponding reference area, S_v. This total side-force derivative is also equal to the change in total side force with respect to sideslip angle divided by the freestream dynamic pressure. Similar results for the change in lift with respect to angle of attack are also shown in Fig. 5.5.12.

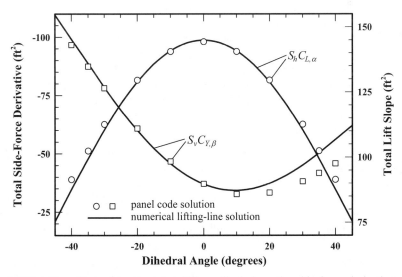

Figure 5.5.12. The effect of horizontal-stabilizer dihedral on the side-force derivative and lift slope, for the conventional tail shown in Fig. 5.5.9.

From the results shown in Fig. 5.5.12, we see that adding a small amount of anhedral to the horizontal stabilizer of a conventional tail produces a significant increase in its ability to provide yaw stability for an airplane, while only slightly reducing its ability to provide pitch stability. On the other hand, adding the same amount of positive dihedral to the horizontal stabilizer of the same conventional tail will actually reduce the tail's effectiveness in providing both pitch and yaw stability.

The results plotted in Fig. 5.5.12 also show that when larger dihedral angles are used in tail configurations like those shown in Fig. 5.5.1, the tail will provide greater yaw stability when the dihedral in the horizontal stabilizer has the opposite sign from that of the vertical stabilizer. Specifically, **a conventional tail having an upright vertical stabilizer mounted above a horizontal stabilizer with a vertical-stabilizer dihedral angle of +90 degrees works best when anhedral is used in the horizontal stabilizer.** Conversely, **an inverted vertical stabilizer with −90 degrees dihedral, or a traditional T-tail, works best when positive dihedral is used in the horizontal stabilizer.** For example, with the geometry used for Fig. 5.5.12, the configuration with 35 degrees of anhedral provides over 75 percent more yaw stability than a similar configuration with 35 degrees of positive dihedral. The reason for the difference can be seen by comparing the two lift distributions for pure sideslip, which are shown in Fig. 5.5.13.

Although the lift distributions shown in Fig. 5.5.13 were obtained using the vortex theory of lift, the lifting forces are transmitted to the surfaces through pressure variations in the air flowing over the tail. The generalized vortex lifting law, which makes three-dimensional lifting-line computations possible, greatly simplifies the mathematical description of lift. On the other hand, it can be difficult to visualize the origin of lift distributions, such as those shown in Fig. 5.5.13, from vortex theory. Another way to look at such lift distributions is through our knowledge of the pressure distributions about lifting surfaces.

Consider, for example, the three lifting surfaces shown in the upper configuration of Fig. 5.5.13. If each of these lifting surfaces were isolated in a freestream and subjected to positive sideslip, high pressure would develop to the right of the surface and low pressure would develop on the left in all three cases. However, when these three surfaces are combined in the freestream and subjected to positive sideslip, there will be some conflicting forces at work in the region where the surfaces are joined. In the triangular region to the left of the vertical surface and to the right of the left-hand surface, the vertical surface is attempting to produce low pressure while the left-hand surface is attempting to produce high pressure. Near the root where the two surfaces are in close proximity, the vertical surface will prevail in this competition, because its orientation relative to the sideslip velocity component produces a greater freestream angle of attack on the surface. Thus, low pressure in the region immediately above and to the left of the root of the vertical stabilizer contributes to a leftward force on the vertical surface and a rightward force on the left-hand surface, as shown in Fig. 5.5.13. In a similar manner, high pressure in the region just above and to the right of the root of the vertical stabilizer will contribute to a leftward force on the vertical surface and a rightward force on the right-hand surface. By similar reasoning, complementary pressure distributions can be used to justify the more favorable lift distribution that is shown on the lower configuration of Fig. 5.5.13.

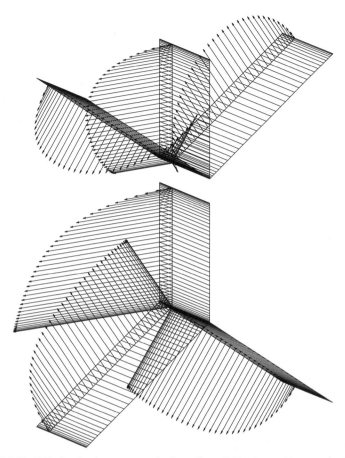

Figure 5.5.13. Lift distributions on two similar tail configurations with opposite dihedral.

The results shown in Fig. 5.5.12 suggest that the weight and parasitic drag for a conventional tail could be reduced somewhat, without sacrificing pitch or yaw stability, by adding anhedral and area to the horizontal stabilizer while decreasing the area of the vertical stabilizer. For the conventional tail geometry shown in Fig. 5.5.9, the results plotted in Fig. 5.5.14 show how the area of the horizontal and vertical stabilizers can be varied with horizontal-stabilizer dihedral angle to maintain constant pitch and yaw stability. The data plotted in this figure were obtained from numerical lifting-line computations. For each dihedral angle, the chord was held constant and the root-to-tip lengths for the horizontal and vertical stabilizers were adjusted iteratively, to maintain the same total lift slope and side-force derivative as the original conventional tail with no dihedral. As either positive or negative dihedral is added to this conventional tail, the planform area of the horizontal stabilizer is increased to maintain constant total lift slope. When negative dihedral is added to the horizontal stabilizer, the area of the vertical stabilizer can be reduced, because a portion of the required side force is supported by the horizontal stabilizer. Thus, as anhedral is added to the horizontal stabilizer, its area

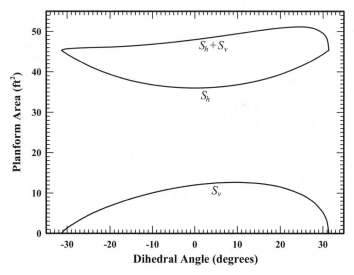

Figure 5.5.14. The effect of horizontal-stabilizer dihedral on the total tail area required to maintain the same total lift slope and side-force derivative as those for the conventional tail that is shown in Fig. 5.5.9.

increases but the area of the vertical stabilizer can be decreased. When the anhedral in the horizontal stabilizer becomes large enough, the required vertical-stabilizer area goes to zero, which yields a V-tail with the same total lift slope and side-force derivative as the original conventional tail.

With all other things being equal, tail weight and parasitic drag are approximately proportional to the total area of the horizontal and vertical stabilizers combined. This total combined area is also plotted in Fig. 5.5.14. Notice that for the conventional tail geometry shown in Fig. 5.5.9, the total required area continues to decrease as anhedral is added to the horizontal stabilizers, until the vertical-stabilizer area goes to zero. Thus, a V-tail provides the minimum-area solution in this particular case and results in a total area reduction of more than 5 percent, relative to the base conventional tail with no horizontal-stabilizer dihedral. This is typical for applications where the required vertical planform is significantly less than the required horizontal planform. In this particular case the vertical planform of the base conventional tail is only one third of the horizontal planform.

When the ratio of the required vertical planform to the required horizontal planform is larger, a V-tail may not provide the minimum-area solution. For example, Fig. 5.5.15 shows results similar to those plotted in Fig. 5.5.14, but for a tail geometry having a base vertical-stabilizer area that is two thirds of that for the horizontal stabilizer. The base conventional tail used for this figure has a horizontal aspect ratio of 4.0 and a horizontal planform area of 36 ft². The vertical stabilizer has the same root chord with a planform area of 24 ft². Both the horizontal and vertical surfaces have taper ratios of 0.5, thin symmetric airfoil sections, and no twist. The quarter-chord sweep for both surfaces is zero and the horizontal surface has no dihedral. At zero angle of attack, lifting-line

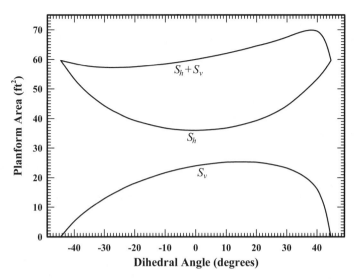

Figure 5.5.15. The effect of horizontal-stabilizer dihedral on the total tail area required to maintain constant total lift slope and side-force derivative for a tail with larger vertical planform requirements than that shown in Fig. 5.5.9.

computations for this geometry predict a total side-force derivative, $S_{\text{ref}}C_{Y,\beta}$, of -96.68 ft^2 and a total lift slope, $S_{\text{ref}}C_{L,\alpha}$, of 148.90 ft^2.

For this base tail geometry, numerical lifting-line computations predict a minimum-area solution at a horizontal-stabilizer dihedral angle of approximately -27 degrees with a vertical-stabilizer planform of just over 14.7 ft^2. This minimum-area solution has a total area of about 57.2 ft^2, which is nearly 5 percent less than the total area of the base conventional tail and more than 4 percent less than that required for an equivalent V-tail. Figure 5.5.15 also shows that adding positive dihedral to this base conventional tail can increase the total required area by as much as 15 percent.

In summary, horizontal-stabilizer dihedral can significantly influence the static yaw stability of an airplane. The vortex interactions between the lifting surfaces of an aft tail substantially alter the side force developed on the tail as a result of sideslip. In the case of a V-tail, these vortex interactions slightly increase the change in lift with respect to angle of attack and substantially decrease the change in side force with respect to sideslip angle. Nevertheless, for applications where the required vertical planform is significantly less than the required horizontal planform, using a V-tail in place of a conventional tail with no horizontal-stabilizer dihedral has the potential for reducing total tail size by about 5 percent. When horizontal-stabilizer dihedral is used in combination with a vertical stabilizer as shown in Fig. 5.5.1, the tail will provide substantially greater yaw stability when the dihedral in the horizontal stabilizer has the opposite sign from that of the vertical stabilizer. If the ratio of the required vertical planform to the required horizontal planform is large enough, properly utilizing horizontal-stabilizer dihedral can result in a minimum total tail size that is less than that for either a V-tail or a conventional tail with no horizontal-stabilizer dihedral.

5.6. Roll Stability and Dihedral Effect

An airplane is said to be statically stable in roll if a restoring rolling moment develops as a result of a disturbance in bank angle. At first thought, this may seem quite similar to the definitions of static pitch and yaw stability. Static pitch stability requires that a restoring pitching moment result from a disturbance in angle of attack, and static yaw stability necessitates that a restoring yawing moment develop from a disturbance in sideslip angle. Static roll stability is fundamentally different from static pitch and yaw stability, because the bank angle is fundamentally different from the angle of attack and sideslip angle. As shown in Fig. 5.1.1, the bank angle, ϕ, is defined in terms of the airplane's orientation with respect to the Earth, while the angle of attack, α, and sideslip angle, β, are defined in terms of the airplane's orientation with respect to the relative wind. The bank angle is completely independent of the relative wind.

The aerodynamic forces and moments acting on an airplane depend only on the orientation of the airplane with respect to the relative wind. Since the gravitational force acts through the center of gravity, the only moments produced about the center of gravity are the aerodynamic moments, including thrust. Thus, there can be no moment about the center of gravity that depends directly on the bank angle. As a result, static roll stability is based on an indirect effect of bank angle rather than a direct effect.

A disturbance in pure roll, with no change in angle of attack, sideslip angle, or forward airspeed, produces no direct change in the aerodynamic forces and moments relative to the airplane. As can be seen in Fig. 5.6.1a, a bank angle disturbance with no change in α or β produce a force imbalance in both the normal and spanwise directions. If the airplane was originally trimmed for straight and level flight, the lift is equal to the weight and there is no aerodynamic side force. After the bank angle disturbance, the lift force remains in the plane of symmetry, and the gravitational force acts at an angle, ϕ, to the plane of symmetry. The net upward normal force is now $W(1-\cos\phi)$ and the net side force is $W\sin\phi$. If there is no change in the control inputs, this force imbalance produces changes in both normal velocity and sideslip. The change in normal velocity decreases the angle of attack until the lift balances the normal component of weight and the sideslip will increase until the side force balances the spanwise component of weight.

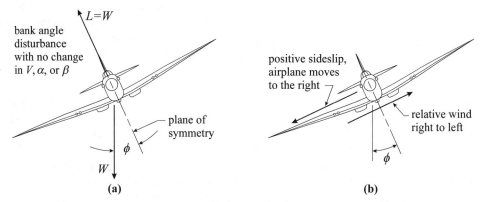

Figure 5.6.1. Force imbalance and resulting sideslip produced by a disturbance in roll.

However, the change in α and β required to restore the force balance can also produce a change in the aerodynamic moments. For a perfectly symmetric airplane, the change in angle of attack does not affect the rolling moment, but the change in sideslip angle can produce a rolling moment that is either positive or negative, depending on the airplane design.

Comparing Fig. 5.6.1b with the sign convention defined in Fig. 5.1.1, we see that a positive disturbance in bank angle results in a positive sideslip velocity component, which was defined to correspond with a positive sideslip angle. Also from the sign convention in Fig. 5.1.1, we see that a negative rolling moment is required to restore a positive disturbance in bank angle. Thus, static roll stability requires that a positive sideslip angle must produce a negative rolling moment,

$$\frac{\partial \ell}{\partial \beta} < 0$$

In dimensionless form, this gives **the general criterion for static roll stability**,

$$\frac{\partial C_\ell}{\partial \beta} \equiv C_{\ell,\beta} < 0 \qquad\qquad (5.6.1)$$

The aerodynamic derivative, $C_{\ell,\beta}$, is the *roll stability derivative*. If $C_{\ell,\beta}$ is negative, the airplane is said to be statically stable in roll, and if $C_{\ell,\beta}$ is positive, the airplane is statically unstable in roll. When $C_{\ell,\beta}$ is zero, the airplane is neutrally stable in roll.

Roll stability gives an airplane the inherent tendency to fly with the wings level when the airplane is trimmed for zero sideslip. An airplane that is neutrally stable or even unstable in roll can still be flown but requires greater attention by the pilot to keep the wings level. The desired magnitude of roll stability depends significantly on the pilot and aircraft application. Very proficient pilots typically prefer less roll stability than inexperienced pilots do. Airplanes that execute rapid rolling maneuvers, such as fighter aircraft and stunt planes, will be disadvantaged by a high degree of roll stability. Transports and other aircraft that use more gradual rolling maneuvers usually have better handling qualities with somewhat higher levels of roll stability.

For any aircraft and pilot, **too much roll stability will be detrimental.** Most of the reasons for this can be traced to the coupling between roll and yaw. When roll stability is excessive, an airplane becomes very sensitive in roll to the minor changes in sideslip caused by very slight deflections of the rudder. This necessitates very precise coordination between the rudder and ailerons, which can be annoying to a proficient pilot and unacceptable to the less experienced. Excessive roll stability can be particularly problematic in a crosswind landing that requires flying the airplane in sideslip with the wings nearly level. In addition, too much roll stability can result in increased passenger discomfort because the airplane becomes very sensitive in roll to changes in sideslip caused by atmospheric turbulence. The roll stability derivative, $C_{\ell,\beta}$, also affects certain dynamic handling characteristics of an airplane, which will be considered later in this textbook.

Because roll stability affects so many different aspects of an aircraft's handling characteristics, it is difficult to define a general roll stability criterion in terms of $C_{\ell,\beta}$. However, **to give the student a place to start, typical airplane configurations normally display good handling qualities with $C_{\ell,\beta}$ in the range from about 0.00 to −0.10 per radian.** This should not be treated as an absolute rule but simply as a starting point. About the only thing that can be said concerning roll stability in general is that a little roll stability is advantageous, but too much can be worse than none at all.

The roll stability derivative is more difficult to estimate than either the pitch or yaw stability derivative. This is because the roll stability derivative is significantly affected by complex interactions between the fuselage, the wing, and the tail. Furthermore, these interactions can change significantly with power and flap settings. Design changes to vary roll stability are commonly required after wind tunnel testing and sometimes after actual flight testing of the airplane. Nevertheless, the designer must have at least a qualitative understanding of all design considerations that affect roll stability and should be able to predict the effects of the design changes that will be used to make any final adjustments in roll stability.

Effect of Wing Dihedral
The change in rolling moment due to sideslip is heavily influenced by the dihedral angle of the wing, which is shown in Fig. 5.6.2. The dihedral angle is commonly given the symbol Γ and is defined to be positive when the wingtips are above the wing root. Negative dihedral is commonly called *anhedral*. Recall that the symbol Γ is also commonly used to represent circulation and vortex strength. Because the circulation theory of lift is used to predict certain dihedral effects, this can be confusing if we are not careful. Here an effort is made to avoid confusion in this regard, but the reader should be forewarned of this potential difficulty.

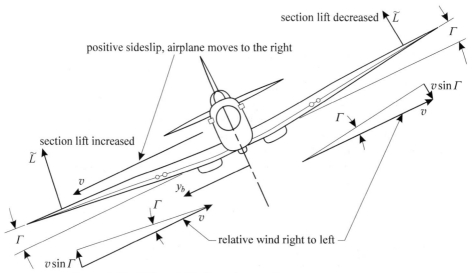

Figure 5.6.2. Effect of sideslip and wing dihedral on local section lift.

Wing dihedral affects the roll stability derivative of an airplane because it causes the lift on the right and left semispans to respond differently to sideslip. As shown in Fig. 5.6.2, on a wing with positive dihedral sideslip produces an increase in angle of attack on the windward semispan and a decrease in angle of attack on the lee-side semispan. The sideslip component of relative wind is denoted in Fig. 5.5.2 as v. When this is resolved into components parallel with and normal to each semispan, as shown in Fig. 5.6.2, the upward normal component is $v \sin \Gamma$ on the right semispan and $-v \sin \Gamma$ on the left semispan. For small aerodynamic angles, the sideslip velocity can be closely approximated as the product of the airspeed and sideslip angle, $v \cong V\beta$. Likewise, the semispan angle of attack is closely approximated as the upward normal component of relative wind divided by the airspeed. Thus, sideslip produces a change in angle of attack that can be closely approximated for the right and left semispans, respectively, as

$$(\Delta\alpha)_{\text{right}} = \frac{v \sin \Gamma}{V} = \frac{V\beta \sin \Gamma}{V} = \beta \sin \Gamma \tag{5.6.2}$$

$$(\Delta\alpha)_{\text{left}} = \frac{-v \sin \Gamma}{V} = \frac{-V\beta \sin \Gamma}{V} = -\beta \sin \Gamma \tag{5.6.3}$$

The differential in angle of attack between the right and left semispans creates a differential in lift that produces a rolling moment. From Fig. 5.1.1, we see that positive lift on the left semispan ($y_b < 0$) produces a positive rolling moment, while positive lift on the right semispan ($y_b > 0$) produces a negative rolling moment. Thus, the rolling moment arm about the junction of the two seimspans for the lift on any wing section is $-y_b/\cos \Gamma$ and the differential length measured perpendicular to the wing sections is $dy_b/\cos \Gamma$. Using the approximation for small sideslip angles given by Eqs. (5.6.2) and (5.6.3), the net rolling moment that results from sideslip combined with wing dihedral can be written

$$\begin{aligned}
(\Delta C_\ell)_{\Gamma w} &= \int_{y_b=-b_w/2}^{0} \frac{\widetilde{C}_{L_i,\alpha} c_w}{S_w b_w}(-\beta \sin \Gamma)\frac{-y_b}{\cos \Gamma}\frac{dy_b}{\cos \Gamma} + \int_{y_b=0}^{b_w/2} \frac{\widetilde{C}_{L_i,\alpha} c_w}{S_w b_w}(\beta \sin \Gamma)\frac{-y_b}{\cos \Gamma}\frac{dy_b}{\cos \Gamma} \\
&= -\frac{\beta \sin \Gamma}{S_w b_w \cos^2 \Gamma}\int_{y_b=-b_w/2}^{b_w/2} \widetilde{C}_{L_i,\alpha} c_w |y_b| dy_b
\end{aligned} \tag{5.6.4}$$

where S_w and b_w are the reference area and reference length, respectively, and $\widetilde{C}_{L_i,\alpha}$ is the *local in situ section lift slope*, including the effects of local induced downwash.

The spanwise variation of local induced downwash is affected by the sideslip. Thus, the derivative of Eq. (5.6.4) with respect to β must be written as

$$\frac{\partial}{\partial\beta}(\Delta C_\ell)_{\Gamma w} = -\frac{\sin \Gamma}{S_w b_w \cos^2 \Gamma}\left(\int_{y_b=-b_w/2}^{b_w/2} \widetilde{C}_{L_i,\alpha} c_w |y_b| dy_b + \beta \int_{y_b=-b_w/2}^{b_w/2} \frac{\partial \widetilde{C}_{L_i,\alpha}}{\partial\beta} c_w |y_b| dy_b\right)$$

However, the roll stability derivative, $C_{\ell,\beta}$, is evaluated at $\beta = 0$, and the contribution of wing dihedral to this aerodynamic derivative can be written

$$\left(\Delta C_{\ell,\beta}\right)_{\Gamma w} = -\frac{\sin \Gamma}{S_w b_w \cos^2 \Gamma} \left[\int_{y_b=-b_w/2}^{b_w/2} \widetilde{C}_{L_i,\alpha} \, c_w \left|y_b\right| dy_b\right]_{\beta=0} \tag{5.6.5}$$

Since the integral in Eq. (5.6.5) is evaluated at $\beta = 0$, both the flow and the wing are symmetric about the midspan. Thus, the integration from $-b_w/2$ to 0 is equivalent to that from 0 to $b_w/2$, and Eq. (5.6.5) can also be written as

$$\left(\Delta C_{\ell,\beta}\right)_{\Gamma w} = -\frac{2\sin \Gamma}{S_w b_w \cos^2 \Gamma} \left[\int_{y_b=0}^{b_w/2} \widetilde{C}_{L_i,\alpha} \, c_w \, y_b \, dy_b\right]_{\beta=0} = -\frac{2\sin \Gamma}{S_w b_w} \left[\int_{y_b'=0}^{b_w'/2} \widetilde{C}_{L_i,\alpha} \, c_w \, y_b' \, dy_b'\right]_{\beta=0} \tag{5.6.6}$$

where $y_b' \equiv y_b/\cos \Gamma$ is a spanwise coordinate measured parallel with the wing dihedral and $b_w' \equiv b_w/\cos \Gamma$ is twice the root-to-tip semispan length.

For low Mach number, Prandtl's lifting-line theory can be used to determine the spanwise variation of in situ section lift slope, which is needed to evaluate the integral on the right-hand side of Eq. (5.6.6). In general, it is necessary to use a numerical lifting-line solution, because the classical infinite series solution applies only to wings with no sweep or dihedral. However, for unswept wings with small dihedral angles, we can begin by **neglecting the spanwise variation of the in situ section lift slope caused by wing dihedral**. The integral in Eq. (5.6.6) can then be evaluated from the classical solution to Prandtl's lifting-line equation. For an unswept wing of arbitrary planform and twist, this solution predicts that the change in local section lift coefficient with respect to the semispan angle of attack is given by

$$\widetilde{C}_{L_i,\alpha} = \frac{4b_w'}{c_w} \sum_{n=1}^{\infty} a_n \sin(n\theta) \tag{5.6.7}$$

where a_n denotes the planform coefficients in the infinite series solution to the lifting-line equation, which is described in Sec. 1.8, and θ is defined by the change of variables

$$y_b' \equiv \tfrac{1}{2} b_w' \cos \theta \tag{5.6.8}$$

$$dy_b' = -\tfrac{1}{2} b_w' \sin \theta \, d\theta \tag{5.6.9}$$

Using Eqs. (5.6.7) through (5.6.9) in Eq. (5.6.6) results in

$$\left(\Delta C_{\ell,\beta}\right)_{\Gamma w} \cong -\frac{2\sin \Gamma}{S_w b_w} \int_{\theta=\pi/2}^{0} \left(\frac{4b_w'}{c_w} \sum_{n=1}^{\infty} a_n \sin(n\theta)\right) c_w \left(\tfrac{1}{2} b_w' \cos \theta\right)\left(-\tfrac{1}{2} b_w' \sin \theta \, d\theta\right)$$

After using the trigonometric identity, $\sin(2\theta) = 2\sin \theta \cos \theta$, this reduces to

$$\left(\Delta C_{\ell,\beta}\right)_{\Gamma w} \cong -\frac{b_w'^3 \sin \Gamma}{S_w b_w} \sum_{n=1}^{\infty} a_n \int_{\theta=0}^{\pi/2} \sin(n\theta)\sin(2\theta)\,d\theta$$

$$= -\frac{b_w'^3 \sin \Gamma}{S_w b_w} \sum_{n=1}^{\infty} a_n \begin{cases} 2\sin(n\pi/2)/(4-n^2), & n \neq 2 \\ \pi/4, & n = 2 \end{cases} \qquad (5.6.10)$$

$$= -\frac{b_w'^3 \sin \Gamma}{S_w b_w}\left[\frac{2}{3}a_1 + \frac{\pi}{4}a_2 + \sum_{n=3}^{\infty}\frac{2\sin(n\pi/2)}{4-n^2}a_n\right]$$

For small dihedral angles, a result identical to Eq. (5.5.11) can be developed for the wing. Using this relation with Eq. (1.8.26) from the classical lifting-line solution yields

$$C_{Lw,\alpha} \cong \left(C_{Lw,\alpha}\right)_{\Gamma=0}\cos^2\Gamma = \pi(b_w'^2/S_w')a_1\cos^2\Gamma$$

where S_w' is twice the semispan planform area measured parallel with the wing dihedral. Thus, if S_w is the horizontal projection of S_w', Eq. (5.6.10) can be rearranged as

$$\left(\Delta C_{\ell,\beta}\right)_{\Gamma w} \cong -\frac{2 S_w' b_w' \sin \Gamma}{3\pi S_w b_w}\kappa_\ell \left(C_{Lw,\alpha}\right)_{\Gamma=0} = -\frac{2\sin \Gamma}{3\pi \cos^4 \Gamma}\kappa_\ell C_{Lw,\alpha} \quad (5.6.11)$$

$$\kappa_\ell = 1 + \frac{3\pi}{8}\frac{a_2}{a_1} + \sum_{n=3}^{\infty}\frac{3\sin(n\pi/2)}{4-n^2}\frac{a_n}{a_1} \qquad (5.6.12)$$

For elliptic wings κ_ℓ is 1.0 and Fig. 5.6.3 shows how κ_ℓ varies with aspect ratio and taper ratio for tapered wings. Since the series solution to Prandtl's lifting-line equation applies to unswept wings with any spanwise variation in section geometry, **the solution can be used for unswept wings of any planform with arbitrary geometric and aerodynamic twist.** An additional correction factor is applied to Eq. (5.6.11) **to account for the change of the in situ section lift slope caused by wing dihedral.** Thus, the wing dihedral contribution to the aircraft's roll stability derivative is

$$\boxed{\left(\Delta C_{\ell,\beta}\right)_{\Gamma w} = -\frac{2 S_w' b_w' \sin \Gamma}{3\pi S_w b_w}\kappa_\Gamma \kappa_\ell \left(C_{Lw,\alpha}\right)_{\Gamma=0} = -\frac{2\sin \Gamma}{3\pi \cos^4 \Gamma}\kappa_\Gamma \kappa_\ell C_{Lw,\alpha} \quad (5.6.13)}$$

For small dihedral angles, the dihedral factor, κ_Γ, is independent of Γ. For wings with linear taper and small dihedral angles, Fig. 5.6.3 shows how κ_Γ varies with aspect ratio and taper ratio as predicted from numerical lifting-line computations. For elliptic wings κ_Γ may be accurately estimated from Fig. 5.6.3 using an effective taper ratio of 0.4.

The development of Eq. (5.6.13) is based on a rolling moment about the junction of the two semispans. If this junction is not aligned vertically with the center of gravity, an additional rolling moment is produced by the side force acting through a vertical moment arm. However, from a development similar to that used to obtain Eq. (5.5.12), the side force on the wing is found to be proportional to $\sin^2\Gamma$. Because the wing dihedral angle is typically small, this contribution to the rolling moment is usually neglected.

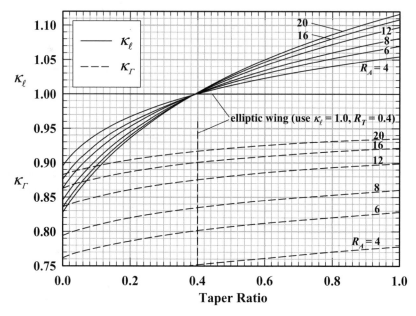

Figure 5.6.3. The small-angle dihedral factors for wings with linear taper.

Strictly speaking, Eqs. (5.6.12) and (5.6.13) apply only at low Mach number to wings with no sweep. However, the contribution of wing dihedral to the roll stability derivative results directly from the lift developed on the wing. Thus, higher subsonic Mach numbers and small sweep angles should affect both lift and wing dihedral contribution to the roll stability derivative in nearly the same proportion. With this line of reasoning, Eqs. (5.6.12) and (5.6.13) could be applied with caution to higher Mach numbers and wings with some sweep, provided that the actual lift slope for the swept wing at the actual flight Mach number is used in Eq. (5.6.13).

One important value in the development that led to Eq. (5.6.13) is the disclosure of a linear relationship between $C_{\ell,\beta}$ and Γ for small values of Γ. Applying the usual small-angle approximation for Γ yields $S'_w b'_w \sin\Gamma / S_w b_w \cong \Gamma$ and we see that Eq. (5.6.13) becomes linear in Γ for small dihedral angles. From Eq. (5.6.6), we see that the linear relation holds for small dihedral angles, regardless of how the in situ section lift slope varies in the spanwise direction. Thus, we should expect the roll stability derivative to change linearly with small dihedral angles for any wing geometry and Mach number.

Effect of Wing Sweep
We now see that wing sweep has an indirect effect on the roll stability derivative, because it affects the lift slope for the wing. This effect is included in our estimate for the contribution of wing dihedral if Eq. (5.6.13) is used. However, wing sweep also has a direct effect on the roll stability derivative, which is not accounted for in Eq. (5.6.13). This direct effect is closely related to the same phenomenon that makes a swept-back wing stabilizing in yaw. This was discussed in Sec. 5.2.

To help the reader better understand the relation between wing sweep and roll stability, we shall again turn to Prandtl's lifting-line theory. In this well-validated model, all of the bound vorticity generated on a lifting wing is represented as a single vortex filament that passes through the aerodynamic centers of the airfoil sections that define the wing (i.e., the wing quarter-chord line for thin airfoil sections in subsonic flow). From the generalized vortex lifting law, for any potential flow containing vortex filaments, a net lift force per unit filament length is exerted by the fluid, which is equal to the fluid density multiplied by the cross product of the local fluid velocity vector and the local circulation vector. Thus, for Prandtl's model of a finite wing, the local lift force per unit length measured along the quarter-chord line is given by

$$\mathbf{f}_L = \rho \mathbf{V} \times \mathbf{\Gamma} \tag{5.6.14}$$

where $\mathbf{V}$ and $\mathbf{\Gamma}$ are, respectively, the local relative wind vector and the local section circulation vector. **Note that the force per unit length computed from Eq. (5.6.14) is not the section lift, because section lift is defined as the lift force per unit span, not the lift force per unit length along the quarter-chord line.**

The local relative wind is the vector sum of the freestream velocity and the velocity induced by the entire vortex system generated by the wing. The induced velocity is very important in evaluating the induced drag component of the aerodynamic force. However, because the induced velocity is much less than the freestream velocity, for determining the magnitude of lift, the local velocity in Eq. (5.6.14) can be closely approximated with the freestream relative wind.

The magnitude of the cross product of any two vectors is the product of the vector magnitudes multiplied by the sine of the angle from the first vector to the second. With no quarter-chord sweep and no sideslip, the circulation vector aligned with the wing quarter chord makes an angle of 90 degrees ($\pi/2$ radians) with the relative wind. As can be seen from the geometry in Fig. 5.6.4, positive sweep for a lifting wing decreases this angle on the right semispan and increases it on the left semispan. Positive sideslip, on the other hand, increases this angle on both semispans. Since a distance measured along the quarter-chord line is the spanwise distance divided by the cosine of the quarter-chord sweep angle, from Eq. (5.6.14) and the geometry in Fig. 5.6.4, the magnitudes for section lift on the right and left semispans are, respectively,

$$(\tilde{L})_{\text{right}} \cong \rho V_\infty \Gamma \frac{\sin[(\pi/2) - \Lambda + \beta]}{\cos \Lambda} \tag{5.6.15}$$

and

$$(\tilde{L})_{\text{left}} \cong \rho V_\infty \Gamma \frac{\sin[(\pi/2) + \Lambda + \beta]}{\cos \Lambda} \tag{5.6.16}$$

Once again the reader is cautioned that the symbol Γ is commonly used to represent both circulation and wing dihedral. **In Eqs. (5.6.15) and (5.6.16), Γ denotes wing section circulation, not wing dihedral.**

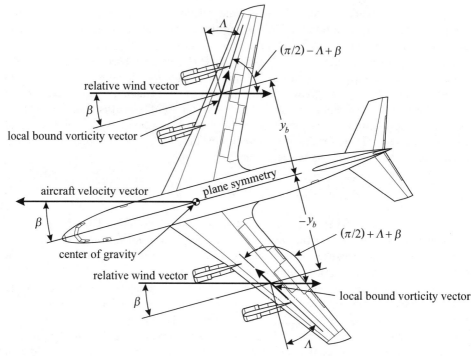

Figure 5.6.4. Effect of wing sweep on the cross product of the relative wind vector with the bound vorticity vector.

Using common trigonometric identities and applying the small-angle approximation for β, it can be shown that

$$\sin[(\pi/2) - \Lambda + \beta] = \cos\Lambda\cos\beta + \sin\Lambda\sin\beta \cong \cos\Lambda + \beta\sin\Lambda \quad (5.6.17)$$

$$\sin[(\pi/2) + \Lambda + \beta] = \cos\Lambda\cos\beta - \sin\Lambda\sin\beta \cong \cos\Lambda - \beta\sin\Lambda \quad (5.6.18)$$

From the sign convention shown in Fig. 5.1.1, positive lift on the left semispan ($y_b < 0$) will produce a positive rolling moment, while positive lift on the right semispan ($y_b > 0$) produces a negative rolling moment. Thus the rolling moment arm for section lift is $-y_b$, and after using Eqs. (5.6.17) and (5.6.18) in Eqs. (5.6.15) and (5.6.16), the contribution to the rolling moment that results from wing sweep is

$$(\Delta\ell)_{\Lambda w} = \int\limits_{y_b=-b_w/2}^{0} \rho V_\infty \Gamma(1 - \beta\tan\Lambda)(-y_b)\,dy_b + \int\limits_{y_b=0}^{b_w/2} \rho V_\infty \Gamma(1 + \beta\tan\Lambda)(-y_b)\,dy_b$$

After rearranging and nondimensionalizing this result according to Eq. (5.1.1), the contribution of wing sweep to the rolling moment coefficient is

$$\left(\Delta C_\ell\right)_{Aw} = -\frac{2}{V_\infty S_w b_w}\left[\int_{y_b=-b_w/2}^{b_w/2}\Gamma\, y_b\, dy_b + \beta \tan\Lambda \int_{y_b=-b_w/2}^{b_w/2}\Gamma\, |y_b|\, dy_b\right] \qquad (5.6.19)$$

Differentiating Eq. (5.6.19) with respect to β gives

$$\frac{\partial}{\partial\beta}\left(\Delta C_\ell\right)_{Aw}$$

$$= -\frac{2}{V_\infty S_w b_w}\left[\int_{y_b=-b_w/2}^{b_w/2}\frac{\partial\Gamma}{\partial\beta}\, y_b\, dy_b + \tan\Lambda \int_{y_b=-b_w/2}^{b_w/2}\Gamma\, |y_b|\, dy_b + \beta\tan\Lambda \int_{y_b=-b_w/2}^{b_w/2}\frac{\partial\Gamma}{\partial\beta}\, |y_b|\, dy_b\right]$$

Evaluating this result at $\beta=0$, the contribution of wing sweep to $C_{\ell,\beta}$ is

$$\left(\Delta C_{\ell,\beta}\right)_{Aw} = -\frac{2}{V_\infty S_w b_w}\left[\int_{y_b=-b_w/2}^{b_w/2}\frac{\partial\Gamma}{\partial\beta}\, y_b\, dy_b + \tan\Lambda \int_{y_b=-b_w/2}^{b_w/2}\Gamma\, |y_b|\, dy_b\right]_{\beta=0} \qquad (5.6.20)$$

The wing section circulation, Γ, varies with the spanwise coordinate, y_b, as a result of vorticity shed from the wing, which rolls up to form the wingtip vortices. The variation of Γ with y_b can be determined from the solution to the lifting-line equation. Since no closed-form solution is available for swept wings, a numerical solution must be used and the integration in Eq. (5.6.20) must be carried out numerically. For swept wings, the spanwise variation in Γ changes with angle of attack. Thus, the contribution of wing sweep to the roll stability derivative depends on airspeed.

The sign of the wing sweep contribution to the roll stability derivative is readily ascertained from Eq. (5.6.20). For small sideslip angles, the change in wing section circulation with sideslip angle is small and the second integral in Eq. (5.6.20) dominates the result. For a wing producing positive lift, the circulation is positive, and the bracketed term in Eq. (5.6.20) will have the same sign as tanΛ. Thus, **positive sweep (sweep back) is stabilizing in roll and negative sweep (sweep forward) is destabilizing.** This is a direct result of the fact that as an airplane with swept-back wings (positive Λ) yaws to the left (positive β), the vorticity generated on the right wing becomes more perpendicular to the freestream, while that generated on the left wing becomes less perpendicular to the freestream. This change in relative wind, which is shown in Fig. 5.6.4, increases lift on the right, decreases lift on the left, and produces a rolling moment to the left (negative ℓ).

Influence of the Fuselage, Nacelles, and External Stores

When an airplane experiences sideslip, the cross-flow component of relative wind must flow around the fuselage, as shown in Fig. 5.6.5. This flow field interacts with the wing and can produce a net change in the roll stability derivative. As the airplane slips to the right, the cross-flow above the fuselage center moves up on the right and back down on the left. Below the center of the fuselage, this flow pattern is inverted.

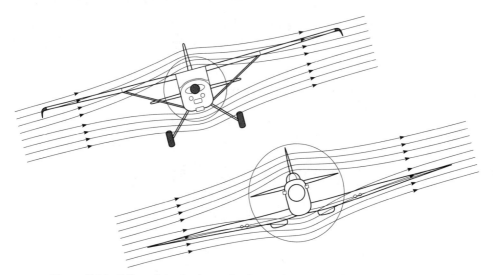

Figure 5.6.5. Effect of the fuselage-wing interaction on the roll stability derivative.

Notice from Fig. 5.6.5 that for a high-wing airplane, the cross-flow around the fuselage creates upwash on the windward semispan and downwash on the lee-side semispan. When the airplane slips to the right, producing a relative wind from right to left (positive β), this increases the angle of attack and lift on the right semispan and decreases the angle of attack and lift on the left, creating a negative and stabilizing contribution to the rolling moment.

For a low-wing airplane, Fig. 5.6.5 shows that flow around the fuselage has the opposite effect on wing lift. When the wing is below the center of the fuselage and the airplane slips to the right, the angle of attack and lift are decreased on the right and increased on the left, as a result of flow around the fuselage. This creates a destabilizing contribution to the rolling moment.

Hence we see that the fuselage contribution to the roll stability derivative of an airplane depends on the vertical position of the wing relative to the fuselage. **A high wing is stabilizing and a low wing is destabilizing.** The interaction between the wing and the fuselage is very complex and depends on the geometry of both the fuselage and the wing. Detailed computer simulations or wind tunnel tests are required to account for this effect accurately. However, **as a very rough initial estimate, a high wing typically has a stabilizing effect equivalent to about 2 to 3 degrees of wing dihedral, and a low wing has a destabilizing effect equivalent to about 3 to 4 degrees of anhedral.** For a mid-wing airplane, the contribution of the fuselage to static roll stability is negligible.

The importance of considering the influence of the fuselage on the roll stability of an airplane becomes very apparent when one examines a variety of airplanes with both high and low wings. The next time that you are near an airport where a large number of general aviation airplanes are tied down, look closely at these airplanes from the front. You will notice that all of the low-wing airplanes have considerably more dihedral than the high-wing airplanes do. The low wings will typically have 4 degrees of dihedral or more. The high wings, on the other hand, will usually have little or no dihedral. Most of the roll stability for a high-wing airplane is commonly provided by the fuselage-wing interaction. Sometimes a small amount of anhedral is even required in a high wing, to partially counter the stabilizing effect of the fuselage-wing interaction.

Nacelles and external stores, such as fuel tanks or armament, interact with the wing in a manner similar to the interaction with the fuselage. As shown in Fig. 5.6.6, when a nacelle or external store is mounted below the wing, any cross-flow creates upwash on the windward side and downwash on the lee side. If the mounting point is sufficiently inboard from the wingtip, this contributes to a stabilizing rolling moment, because the outboard sections of the wing have a greater moment arm for roll than the inboard sections. Conversely, if inboard nacelles or external stores are mounted above the wing, the cross-flow interactions with the wing will be destabilizing.

When fuel tanks or other external stores are mounted at or very near the wingtips, the dihedral effect is different, because only the inboard portion of the cross-flow interacts with the wing. When wingtip tanks are mounted beneath the wing, they create downwash on the upwind semispan and upwash on the downwind semispan. This decreases the angle of attack and lift on the windward side and increases the angle of attack and lift on the lee side, creating a destabilizing contribution to the rolling moment. On the other hand, wingtip tanks or payloads mounted above the wing are stabilizing in roll. Although the nacelles and external stores typically have a greater effect on yaw stability than on roll stability, even small wingtip stores can have a profound effect on

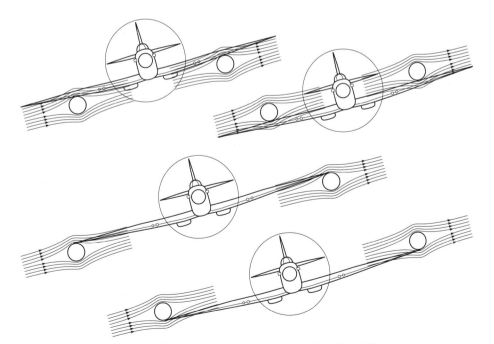

Figure 5.6.6. Effect of nacelles and external stores on the roll stability derivative.

the roll stability of an airplane, because the associated cross-flow interactions are greatest near the wingtips where the moment arm for roll is largest. It would not be extraordinary to have a pair of wingtip fuel tanks change the total dihedral effect for an airplane by 50 percent or more. Although the exact level of dihedral effect is not usually critical to pilots, for some airplane designs that use dihedral to augment roll control through the rudder, wingtip tanks or payloads mounted below the wing can be very detrimental.

Influence of the Tail

As discussed in Sec. 5.2 and shown in Fig.5.2.1, sideslip results in a side force acting on the vertical stabilizer. This side force, applied aft of the CG, generates a stabilizing moment in yaw. When the aerodynamic center of the vertical stabilizer is offset from the axis of roll, as shown in Fig. 5.6.7, this side force can also create a significant rolling moment. When the airplane slips to the right (positive β), the relative wind and side force on the vertical stabilizer are from right to left. If the aerodynamic center of the vertical stabilizer is offset a distance, h_v, above the center of gravity, a negative rolling moment is produced by this side force.

In a manner similar to that used to obtain the yawing moment contribution given by Eq. (5.2.5), the vertical tail's contribution to the roll moment coefficient is found to be

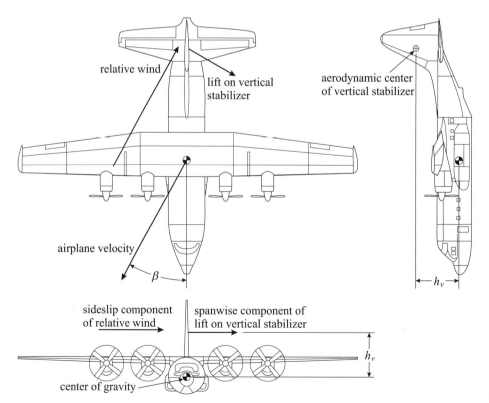

Figure 5.6.7. Effect of the vertical stabilizer on the roll stability derivative.

$$\left(\Delta C_\ell\right)_v = -\eta_v \frac{S_v h_v}{S_w b_w} C_{L_v,\alpha}\left[(1-\varepsilon_{s,\beta})_v \beta - \varepsilon_{s0} - \varepsilon_r \delta_r\right] \tag{5.6.21}$$

and the contribution to the roll stability derivative is

$$\left(\Delta C_{\ell,\beta}\right)_v = -\eta_v \frac{S_v h_v}{S_w b_w}(1-\varepsilon_{s,\beta})_v C_{L_v,\alpha} \tag{5.6.22}$$

From Eq. (5.6.22), we see that when the aerodynamic center of the vertical tail is above the center of gravity ($h_v > 0$), which is typically the case, the vertical stabilizer makes a stabilizing contribution to the roll stability derivative.

Equation (5.6.22) can be used to estimate the direct contribution to the roll stability derivative from lift generated on the vertical tail. However, lift generated on the vertical stabilizer, produces a rolling moment on the horizontal stabilizer, as shown in Fig. 5.4.1. Although the contribution of the horizontal stabilizer to the roll stability derivative has usually been neglected, it can be quite significant. For typical conventional tail geometry, where the horizontal stabilizer is below the vertical stabilizer, the stabilizing contribution that is predicted by Eq. (5.6.22) will be reduced by about 25 to 35 percent as a result of vortex interaction with the horizontal stabilizer. With a T-tail configuration, the horizontal stabilizer mounted above the vertical stabilizer has the opposite effect and will typically increase the roll stability provided by the vertical tail by 25 to 35 percent over that predicted by Eq. (5.6.22). The aerodynamic interaction between the horizontal and vertical stabilizers can be quite accurately determined from a numerical lifting-line solution. As a first approximation for typical tail geometry, a rough estimate for the rolling moment induced on the horizontal stabilizer can be obtained from the relation

$$\left(\Delta C_{\ell,\beta}\right)_h = \pm 0.08 \eta_v \frac{S_v b_h}{S_w b_w}(1-\varepsilon_{s,\beta})_v C_{L_v,\alpha} \tag{5.6.23}$$

where the positive sign is used for horizontal stabilizers mounted below the vertical stabilizer and the negative sign is used for a horizontal stabilizer mounted above the vertical stabilizer. **Equation (5.6.23) should be used as a rough estimate only for tails with no horizontal-stabilizer dihedral when the exact tail geometry is unknown.** A numerical lifting-line solution will give better results if the tail geometry is known.

When dihedral is used in the horizontal stabilizer of an aft tail, it influences roll stability in much the same way as dihedral in the main wing does. For example, when a V-tail like that shown in Fig. 5.5.2 is subjected to sideslip, it produces a significant rolling moment as well as a side force. This can be seen in Fig. 5.5.3. A relation similar to Eq. (5.6.13) could be applied directly to a V-tail. However, because the rolling moment computed from Eq. (5.6.13) is that taken about the junction of the two semispans, an additional contribution must be added to account for the rolling moment produced by the tail side force acting through any vertical offset. This contribution was neglected in Eq. (5.6.13) because the dihedral angle and associated side force for the wing are typically small. For the larger dihedral angles used in a V-tail, the side force contribution to the roll stability derivative is included by using the relation

$$\left(\Delta C_{\ell,\beta}\right)_V = -\eta_V \frac{S_V'}{S_w b_w} \kappa_r \left(\frac{2\kappa_\ell}{3\pi} b_V' + h_{root} \sin \Gamma \right) \sin \Gamma (1 - \varepsilon_{s,\beta})_V \left(C_{L_V,\alpha}\right)_{\Gamma=0} \quad (5.6.24)$$

where b_V' is twice the root-to-tip semispan length, κ_ℓ is obtained either from Eq. (5.6.12) or Fig. 5.6.3, h_{root} is the vertical distance that the semispan junction is above the center of gravity, and S_V' is twice the semispan planform area measured parallel with the tail dihedral, not the horizontal projection. The dihedral factor, κ_r, accounts for the vortex interactions between the semispans of the V-tail and may vary with the dihedral angle.

Although the vortex interactions between the different lifting surfaces of the tail are commonly neglected in the early phases of aircraft design, this practice will significantly overestimate the roll stability contribution of a V-tail. For example, Fig. 5.6.8 shows a comparison between the results of numerical lifting-line computations, Eq. (5.6.24) with $\kappa_r = 1.0$, and experimental data for two different V-tail planforms. The data in this figure are those presented by Purser and Campbell (1945) and Schade (1947). Tails A and B have aspect ratios of 5.55 and 3.70 with taper ratios of 0.39 and 0.56, respectively.

The results that are presented in Fig. 5.6.8 show that while numerical lifting-line computations will accurately predict the roll stability contribution of a V-tail, neglecting vortex interactions between the two semispans overestimates this contribution by 25 to 35 percent. A reasonable first-order correction is obtained by using the small-angle value for κ_r, which can be evaluated from Fig. 5.6.3. However, such results strictly apply only for small dihedral angles. For the larger dihedral angles commonly used in a V-tail, κ_r depends somewhat on the tail dihedral as well as the semispan planform. For lifting surfaces with linear taper, Fig. 5.6.9 shows how κ_r varies with dihedral angle, aspect ratio, and taper ratio, as predicted from numerical lifting-line computations.

When horizontal-stabilizer dihedral is used in combination with a vertical stabilizer as shown in Fig. 5.5.1, Eq. (5.6.24) can be used to estimate the roll stability contribution

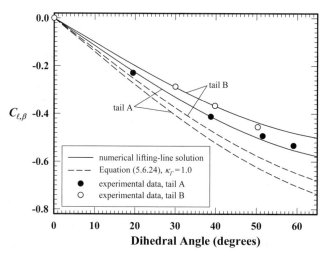

Figure 5.6.8. The roll stability derivative as predicted from the numerical lifting-line method and Eq. (5.6.24) with $\kappa_r = 1.0$, compared with experimental data for two V-tails with variable dihedral.

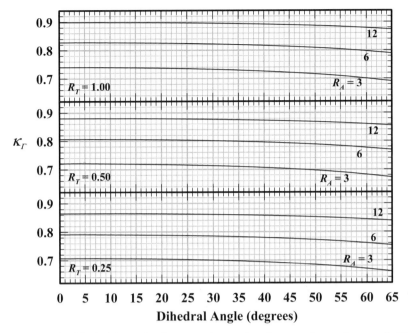

Figure 5.6.9. The dihedral factor for the vortex interactions between the semispans of a V-tail.

resulting from the horizontal-stabilizer dihedral. **The contribution that is predicted from Eq. (5.6.24) must be added to the contribution for a tail with the same lifting-surface geometry but with no horizontal-stabilizer dihedral.**

Although adding positive dihedral to the horizontal stabilizer of a conventional tail will increase the roll stability of an airplane, this is not a particularly efficient way to provide roll stability. There are two primary reasons for this. First, because the span of the horizontal stabilizer is typically much less than that for the wing, the moment arm available for generating a rolling moment on the horizontal stabilizer is small. In addition, a small amount of positive dihedral in the horizontal stabilizer of a conventional tail will significantly decrease the yaw stability provided by the tail. On the other hand, adding positive dihedral to the horizontal stabilizer of a T-tail increases both roll and yaw stability for the airplane.

Effect of Rotating Propellers
As discussed in Sec. 5.2, a rotating propeller can have a significant effect on the trim and static stability of an airplane in yaw. In a similar manner, a propeller will also affect the trim and static stability in roll. One of the largest contributions that a rotating propeller makes to the rolling moment is from the engine torque required to turn the propeller. For a conventional right-hand-turning propeller, the engine must apply a right-hand moment to the propeller. To support this torque, the engine must also exert an equal and opposite moment on the airframe. Thus, a conventional right-hand-turning propeller produces a left-hand rolling moment (negative ℓ).

The rolling moment that results directly from propeller torque is partially countered by the effects of slipstream rotation. The torque applied to the propeller causes the air in the slipstream to rotate in the same direction as the propeller. As shown in Fig. 5.6.10 for a conventional right-hand propeller, slipstream rotation produces sidewash on the vertical stabilizer, downwash on the right semispans of the wing and horizontal stabilizer, and upwash on the left semispans of the wing and horizontal stabilizer. The interactions between the lifting surfaces of the airplane and the rotation of the slipstream produce a rolling moment that is opposite to that produced directly by the propeller (positive ℓ for a right-hand propeller). In some cases, the effects of slipstream rotation will nearly balance the rolling moment produced directly by the propeller. In any case, these moments primarily affect the aircraft trim and have little effect on roll stability.

The axial velocity of the propeller slipstream can also influence the rolling moment acting on an airplane. The slipstream increases the dynamic pressure and lift over the inboard sections of the wing. This does not affect the rolling moment when there is no sideslip. However, as shown in Fig. 5.6.10, sideslip causes the slipstream to impinge on one semispan more than on the other. When the airplane slips to the right (positive β), the slipstream is displaced to the left. This causes the increase in dynamic pressure from the slipstream to affect the left semispan more than the right, which results in more lift produced on the left. The positive increment to the rolling moment produced by this slipstream interaction as a result of positive sideslip is destabilizing. The reduction in roll stability caused by this slipstream effect is appreciably increased with the application of full power at low airspeed, when the ratio of slipstream velocity to forward airspeed is greatest. Flaps and other high-lift devices are normally installed in the inboard sections of the wing where the slipstream influence is high. Thus, slipstream effects are most significant in takeoff and landing configurations with high-lift devices engaged.

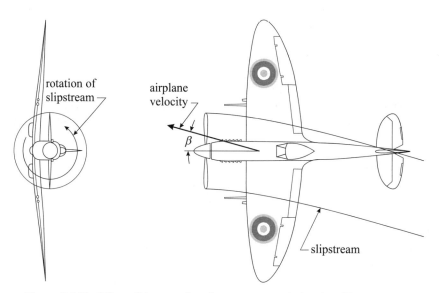

Figure 5.6.10. Effect of the propeller slipstream on an airplane's rolling moment.

The effect of the propeller slipstream on the roll stability derivative is very complex and almost impossible to predict analytically. This effect is usually ignored in the early phases of airplane design.

Summary
We have seen that the roll stability derivative for an airplane is significantly affected by some very complex interactions among the fuselage, wing, vertical stabilizer, horizontal stabilizer, and propulsion system. Furthermore, some of these effects vary with angle of attack as well as flap and power settings. As a result, analytical calculations for the roll stability derivative are not accurate and should be used only as an indication of the expected magnitude of $C_{\ell,\beta}$.

Fortunately, the roll stability derivative is easily adjusted in the later phases of the airplane design process by making small changes to the wing dihedral. The dihedral angle of the wing has such a profound effect on the roll stability derivative that this aerodynamic derivative and its various components are commonly referred to as the *dihedral effect*. For example, the fuselage-wing interaction shown in Fig. 5.6.5 is usually called the *fuselage dihedral effect*.

It is also fortunate that small changes in wing dihedral do not typically have a large effect on other aerodynamic derivatives that also affect the aircraft handling qualities. In a tailless airplane, the roll and yaw stability derivatives are primarily controlled with wing dihedral and sweep. Dihedral has the greatest effect on roll stability, and sweep has the greatest effect on yaw stability. However, the effects of sweep and dihedral are coupled and some iteration is usually required with a tailless aircraft to bring the roll and yaw stability derivatives to the desired magnitude. For an airplane with a conventional aft tail, the roll and yaw stability derivatives are not so closely coupled and a single adjustment of the wing dihedral, after wind tunnel and/or flight tests will usually be sufficient.

> **EXAMPLE 5.6.1.** The airplane described in Examples 5.2.1 and 5.2.2 has a high wing of rectangular planform with no sweep. The tail is the conventional design, with the horizontal stabilizer mounted below the vertical stabilizer. The horizontal stabilizer has no dihedral and the aerodynamic center of the vertical stabilizer is 3.0 feet above the CG. For this airplane in cruise configuration, estimate the wing dihedral required to provide a roll stability derivative of about -0.05.
>
> ***Solution.*** For this airplane we have
>
> $$S_w = 180 \text{ ft}^2, \quad b_w = 33 \text{ ft}, \quad C_{L_w,\alpha} = 4.44, \quad R_{A_w} = 6.05, \quad R_{T_w} = 1.0,$$
> $$S_v = 11.9 \text{ ft}^2, \quad C_{L_v,\alpha} = 3.40, \quad \eta_v = 1.0, \quad (\varepsilon_{s,\beta})_v = -0.10, \quad h_v = 3.0 \text{ ft},$$
> $$S_h = 36 \text{ ft}^2, \quad C_{L_h,\alpha} = 3.97, \quad b_h = 12 \text{ ft}, \quad \eta_h = 1.0, \quad (\varepsilon_{d,\alpha})_h = 0.44$$
>
> From Fig. 5.6.3, for an aspect ratio of about 6.0 and a taper ratio of 1.0, the dihedral factors for the wing are approximately $\kappa_\ell \cong 1.07$ and $\kappa_r \cong 0.83$. Using the small-angle approximation in Eq. (5.6.13), the wing dihedral effect is given by

$$\left(\Delta C_{\ell,\beta}\right)_{\Gamma w} = -\frac{2\,\kappa_\Gamma\,\kappa_\ell\,C_{Lw,\alpha}\sin\Gamma}{3\pi\cos^4\Gamma} \cong -\frac{2(1.07)(0.83)4.44}{3\pi}\Gamma = -0.84\Gamma$$

where Γ is the dihedral angle in radians. Since this is a high-wing airplane and no other information is available, we shall estimate the fuselage dihedral effect as being equivalent to about +2.5 degrees of wing dihedral,

$$\left(\Delta C_{\ell,\beta}\right)_f = -0.84\,\Gamma_f = -0.84\left(2.5\frac{\pi}{180}\right) = -0.037$$

From Eq. (5.6.22), the dihedral effect of the vertical stabilizer is

$$\left(\Delta C_{\ell,\beta}\right)_v = -\eta_v\frac{S_v h_v}{S_w b_w}(1-\varepsilon_{s,\beta})_v\,C_{Lv,\alpha} = -1.0\frac{11.9(3.0)}{180(33)}(1+0.10)3.40$$
$$= -0.022$$

Because this airplane has a conventional tail with the horizontal stabilizer below the vertical stabilizer, the vortex interaction between the lifting surfaces of the tail will produce a destabilizing rolling moment on the horizontal stabilizer. Since the exact geometry of the tail is unknown, we will assume typical conventional tail geometry and apply Eq. (5.6.23),

$$\left(\Delta C_{\ell,\beta}\right)_h = +0.08\,\eta_v\frac{S_v b_h}{S_w b_w}(1-\varepsilon_{s,\beta})_v\,C_{Lv,\alpha}$$
$$= +0.08(1.0)\frac{11.9(12.0)}{180(33)}(1+0.10)3.40 = +0.007$$

Because the airplane is in cruise configuration, the effect of power on the roll stability derivative is probably small. Since no other information is available, we will ignore the propeller slipstream effects.

The net roll stability derivative can now be estimated as

$$C_{\ell,\beta} = \left(\Delta C_{\ell,\beta}\right)_{\Gamma w} + \left(\Delta C_{\ell,\beta}\right)_f + \left(\Delta C_{\ell,\beta}\right)_v + \left(\Delta C_{\ell,\beta}\right)_h$$
$$= -0.84\,\Gamma - 0.037 - 0.022 + 0.007 = -0.05$$

Solving this relation for the wing dihedral gives

$$\Gamma = \frac{0.05 - 0.037 - 0.022 + 0.007}{0.84} = -0.002 = \underline{-0.1°}$$

Thus, the tail and the aerodynamic interaction between the fuselage and the wing provide all of the desired dihedral effect. These approximate calculations predict that a slight amount of wing anhedral is needed.

5.7. Roll Control and Trim Requirements

It is possible to design an airplane that can be flown with only elevator and rudder controls. However, the handling qualities of such an airplane are severely compromised at best. In order to obtain good handling qualities and to trim the airplane in roll over a range of operating conditions, it is necessary to provide the pilot with some means of controlling the aircraft's rolling moment. This is typically accomplished with *ailerons*, which are small trailing-edge flaps located in the outboard sections of the wing. The ailerons are deflected asymmetrically to change the rolling moment. One aileron is deflected downward and the other is deflected upward, as shown in Fig. 5.7.1. This increases lift on the semispan with the downward-deflected aileron and decreases lift on the semispan with the upward-deflected aileron.

The sign convention that is most commonly used for aileron deflection is shown in Fig. 5.7.1. The aileron deflection is assumed positive when the right aileron is deflected downward and the left aileron is deflected upward. With this sign convention, a positive aileron deflection produces a rolling moment to the left (negative ℓ). The two ailerons are not necessarily deflected the same. To help prevent tip stall, the downward-deflected aileron is sometimes deflected less than the upward-deflected aileron. The aileron angle, δ_a, is defined to be the average of the two angular deflections,

$$\delta_a = \frac{(\delta_a)_{\text{right}} + (\delta_a)_{\text{left}}}{2} \tag{5.7.1}$$

At first thought, it may seem quite simple to predict the rolling moment produced by a known deflection of the ailerons from our knowledge of section flap effectiveness. The change in section lift coefficient that results from deflection of a trailing-edge flap was presented in Sec. 1.7 and can be expressed for the ailerons as

$$\left(\Delta \tilde{C}_{L_w}\right)_{\delta_a} = \pm \tilde{C}_{L_w,\alpha}\, \tilde{\varepsilon}_a\, \delta_a \tag{5.7.2}$$

where $\tilde{\varepsilon}_a$ is the local section flap effectiveness and the positive sign corresponds to the right aileron while the negative sign applies to the left.

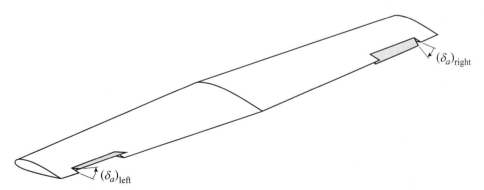

Figure 5.7.1. Ailerons are typically used to control an airplane's rolling moment.

A very simple method that is often recommended for estimating the rolling moment produced by the ailerons is called *strip theory*. This amounts to multiplying the local section lift increment by the local rolling moment arm and integrating over the wing sections containing the ailerons. From Eq. (5.7.2) and the geometry shown in Fig. 5.7.2, the result is

$$
(\Delta \ell)_{\delta_a} = \frac{1}{2}\rho V_\infty^2 \left[(\delta_a)_{\text{right}} \int_{y_i}^{y_o} (-y_b)\tilde{C}_{Lw,\alpha}\, \tilde{\varepsilon}_a\, c_w\, dy_b - (\delta_a)_{\text{left}} \int_{-y_o}^{-y_i} (-y_b)\tilde{C}_{Lw,\alpha}\, \tilde{\varepsilon}_a\, c_w\, dy_b \right]
$$

After nondimensionalizing and differentiating with respect to the aileron angle, for a symmetric wing with constant section lift slope we have

$$
C_{\ell,\delta_a} \equiv \frac{\partial C_\ell}{\partial \delta_a} = \frac{(\Delta \ell)_{\delta_a}}{\frac{1}{2}\rho V_\infty^2 S_w b_w \delta_a} = -\frac{(\delta_a)_{\text{right}} + (\delta_a)_{\text{left}}}{S_w b_w \delta_a} \tilde{C}_{Lw,\alpha} \int_{y_i}^{y_o} y_b\, \tilde{\varepsilon}_a\, c_w\, dy_b
$$

or in view of Eq. (5.7.1),

$$
C_{\ell,\delta_a} = -\frac{2\tilde{C}_{Lw,\alpha}}{S_w b_w} \int_{y_i}^{y_o} y_b\, \tilde{\varepsilon}_a\, c_w\, dy_b
\tag{5.7.3}
$$

The variation in local section flap effectiveness with the spanwise coordinate, y_b, depends on the local ratio of flap chord to wing chord. Thus, the integrand in Eq. (5.7.3) is a function of wing and aileron planforms and the integration is easily carried out. Unfortunately, **the physical model of aileron performance on which Eq. (5.7.3) is based has little correspondence with reality.** This model totally ignores the change in vorticity that must accompany any change in lift. As shown in Fig. 5.7.2, the model assumes that the only change in lift produced by a deflection of the ailerons occurs over those wing sections that include the ailerons. This is not realistic. **Equation (5.7.3) gives only a very rough estimate of aileron performance.** However, it has been recommended in the literature and the student should be aware of its shortcomings.

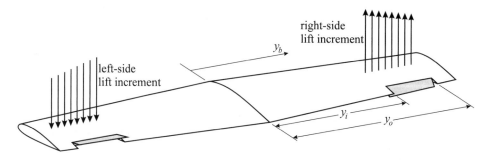

Figure 5.7.2. Lift increment resulting from aileron deflection, as predicted by strip theory.

A more realistic lift distribution is predicted by lifting-line theory. For unswept wings, the series solution presented in Sec. 1.8 can be used to obtain the rolling moment produced directly on the wing by aileron deflection. In general, this can be used as an initial estimate. However, the change in vorticity that results from aileron deflection also produces a significant rolling moment on the tail. To account for the vortex interaction between the wing and the tail, the lifting-line equation can be solved numerically. Results of a numerical lifting-line solution for a particular wing and tail combination are shown in Fig. 5.7.3. In this figure, the vectors displayed along the quarter-chord lines show the magnitude and direction of the local section lift increment, and the lines along the trailing edges indicate the magnitude of the increment in local shed vorticity.

From Fig. 5.7.3, we see that a deflection of the ailerons changes the lift distribution on the entire airplane, not just on those wing sections that contain the ailerons. The changes in lift brought about by the aileron deflection produce a change in the wing vorticity distribution, which in turn modifies the lift distribution on all lifting surfaces. Lift is increased on the entire wing semispan with the downward-deflected aileron and decreased on the entire wing semispan with the upward-deflected aileron. In addition, changes in vorticity shed from the wing modify the distribution of aerodynamic forces acting on the tail. Most notably, a deflection of the ailerons creates sidewash on the vertical stabilizer, which gives rise to a rolling moment that opposes that produced directly by the ailerons.

In addition to producing a rolling moment, aileron deflection gives rise to a change in vorticity that modifies wing downwash and the associated induced drag. The induced drag is increased on the semispan with the downward-deflected aileron and decreased on the semispan with the upward-deflected aileron. This produces a yawing moment in the

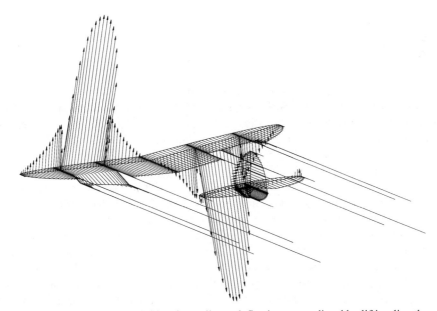

Figure 5.7.3. Lift increment resulting from aileron deflection, as predicted by lifting-line theory.

direction of the downward-deflected aileron. For positive aileron deflection (right aileron down), the rolling moment is to the left (negative ℓ), whereas the yawing moment is to the right (positive n). Because the primary reason for applying positive aileron deflection is to initiate a roll to the left, we see that the yawing moment produced by aileron deflection is in the wrong direction. Accordingly, this aileron-induced yawing moment is commonly called *adverse yaw*. In airplanes having wings with very high aspect ratio, aileron adverse yaw can cause significant lateral control problems. Some potential solutions to this problem are discussed in Chapter 6, which treats aircraft controls and maneuverability in greater detail.

Rudder deflection will also generate a moment about the roll axis of most airplanes. The rolling moment contributed directly by the lift developed on the vertical stabilizer can be estimated using Eq. (5.6.21). Differentiating this result with respect to rudder deflection gives

$$\left(\Delta C_{\ell,\delta_r} \right)_v = \eta_v \frac{S_v h_v}{S_w b_w} \varepsilon_r C_{L_v,\alpha} \tag{5.7.4}$$

As discussed previously, the vorticity associated with lift developed on the vertical stabilizer will modify the distribution of lift developed on the horizontal stabilizer as shown in Fig. 5.4.1. The aerodynamic interaction between the horizontal and vertical stabilizers produces a rolling moment, which occurs whether the lift on the vertical stabilizer is the result of sideslip or rudder deflection. For typical tail geometry, the rolling moment predicted by Eq. (5.7.4) will change by 25 to 35 percent as a result of vortex interactions with the horizontal stabilizer. For typical tail geometry, a rough estimate for the rolling moment induced on the horizontal stabilizer as a result of rudder deflection can be obtained from

$$\left(\Delta C_{\ell,\delta_r} \right)_h = \pm 0.08 \eta_v \frac{S_v b_h}{S_w b_w} \varepsilon_r C_{L_v,\alpha} \tag{5.7.5}$$

where **the negative sign is used when the horizontal stabilizer is mounted below the vertical stabilizer, and the positive sign is used for a horizontal stabilizer mounted above the vertical stabilizer.** Equation (5.7.5) should only be used as a rough estimate when the exact tail geometry is unknown. Here again, a numerical lifting-line solution for the specific tail geometry being used will give a more realistic result.

Four lateral control derivatives, C_{ℓ,δ_a}, C_{n,δ_a}, C_{ℓ,δ_r}, and C_{n,δ_r}, have been discussed. While the ailerons are intended to provide the pilot with roll control and the rudder is meant for yaw control, very seldom are these two controls uncoupled. Aileron deflection usually produces a yawing moment, and rudder deflection will typically generate a rolling moment. This means that trim requirements for roll and yaw are coupled as well. For example, as discussed in Sec. 5.2, a conventional right-hand-turning propeller produces a yawing moment to the left, at positive angles of attack. A right-hand propeller also produces a rolling moment to the left. To trim the airplane against the propeller-induced yawing moment, some right rudder is required. For a conventional high tail, right rudder generates a rolling moment to the left, which adds to that produced by the propeller. Right aileron is needed to compensate for the rolling moment that results from propeller rotation and rudder deflection. However, the adverse yaw associated with right aileron deflection also adds to the right rudder needed to trim the airplane. In addition, deflection of the lateral controls will usually produce a side force through the control derivatives C_{Y,δ_a} and C_{Y,δ_r}. Typically, this side force is balanced by maintaining either a small amount of sideslip or a slight bank angle. These lateral trim requirements are best illustrated by example.

EXAMPLE 5.7.1. The P-38 Lightning shown in Fig. 5.7.4 is powered by two 1,425-horsepower Allison V-1710 engines turning 138-inch, three-blade, counter-rotating Curtiss Electric propellers with 2:1 gear reduction. The propeller on the pilot's right has right-hand rotation, and the propeller on the left has left-hand rotation. Each propeller axis is offset from the plane of symmetry by 8 feet and is aligned with the direction of flight at the maximum airspeed of 400 mph at sea level. With the left engine disabled and the prop feathered, the airplane is flying with an airspeed of 240 mph at 25,000 feet. We wish to estimate the rudder and aileron deflection required to trim the airplane for steady flight at constant altitude with no sideslip. Use the following for this airplane and operating condition:

$$S_w = 327.5 \text{ ft}^2, \quad b_w = 52 \text{ ft}, \quad C_{L,\alpha} = 5.01, \quad V_\infty = 352 \text{ ft/sec}, \quad W = 15,480 \text{ lbf},$$

$$d_p = 138/12 \text{ ft}, \quad h_p = 0.0 \text{ ft}, \quad (\omega/2\pi)_p = 1,350 \text{ rpm}, \quad (\varepsilon_{d,\alpha})_p = -0.175,$$

$$C_{n_p,\alpha} = -0.046, \quad C_D = 0.0466, \quad C_{Y,\delta_a} \cong 0.0, \quad C_{Y,\delta_r} = 0.115,$$

$$C_{\ell,\delta_a} = -0.131, \quad C_{n,\delta_a} = 0.013, \quad C_{\ell,\delta_r} = 0.006, \quad C_{n,\delta_r} = -0.052$$

Solution. For this estimate we shall neglect compressibility and the effects of slipstream rotation. We will also assume a typical propulsive efficiency for the propeller of 80 percent and will ignore any yawing moment produced by the added drag of the feathered propeller on the left.

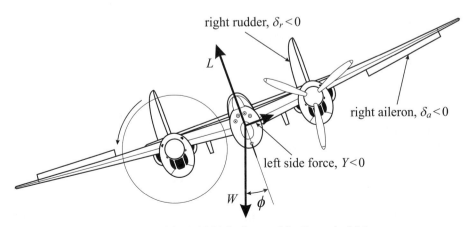

Figure 5.7.4. The P-38 Lightning used for Example 5.7.1.

The lift coefficient in level flight for the maximum airspeed at sea level is

$$(C_L)_{400} \equiv \frac{L}{\frac{1}{2}\rho V_\infty^2 S_w} = \frac{W}{\frac{1}{2}\rho V_\infty^2 S_w} = \frac{15{,}480}{\frac{1}{2}0.0023769(400\times5{,}280/3{,}600)^2 327.5}$$
$$= 0.116$$

Assuming the bank angle is small, at the actual operating condition we have

$$(C_L)_{240} \equiv \frac{L}{\frac{1}{2}\rho V_\infty^2 S_w} = \frac{W\cos\phi}{\frac{1}{2}\rho V_\infty^2 S_w} \cong \frac{W}{\frac{1}{2}\rho V_\infty^2 S_w} = \frac{15{,}480}{\frac{1}{2}0.0010663(352)^2 327.5}$$
$$= 0.716$$

Since the propeller is aligned with the flight direction for 400 mph at sea level, for 240 mph at 25,000 feet the angle between the propeller and the flight velocity is

$$\alpha_{0p} = \left[(C_L)_{240} - (C_L)_{400}\right]/C_{L,\alpha} = (0.716-0.116)/5.01 = 0.1198 = 6.86°$$

The downwash angle in the plane of the propeller is the angle of attack relative to the zero-lift line multiplied by the downwash gradient,

$$\varepsilon_{d0p} = \frac{(C_L)_{240}}{C_{L,\alpha}}(\varepsilon_{d,\alpha})_p = \frac{0.716}{5.01}(-0.175) = -0.0250 = -1.43°$$

For steady flight at constant altitude the thrust is nearly equal to the drag,

$$T \cong D = \frac{1}{2}\rho V_\infty^2 S_w C_D = \frac{1}{2}0.0010663(352)^2 327.5(0.0466) = 1{,}008\ \text{lbf}$$

With the left engine disabled, the thrust vector is offset 8 feet to the right of center and the total yawing moment increment contributed by the propeller is

$$\Delta n_p = -y_{bp} T + \rho(\omega/2\pi)^2 d_p^5 C_{n_p,\alpha}(\alpha_{0p} - \varepsilon_{d0p}) = -8.0(1{,}008)$$
$$+ 0.0010663(1{,}350/60)^2 (138/12)^5(-0.046)[0.1198 - (-0.0250)]$$
$$= -8{,}787\ \text{ft·lbf}$$

Thus, the contribution that the propeller makes to the airplane's yawing moment coefficient is

$$(\Delta C_n)_p \equiv \frac{\Delta n_p}{\frac{1}{2}\rho V_\infty^2 S_w b_w} = \frac{-8{,}787}{\frac{1}{2}0.0010663(352)^2 327.5(52)} = -0.00781$$

The rolling moment increment contributed by the right-hand-turning propeller can be estimated from the thrust, airspeed, and assumed propulsive efficiency,

$$\Delta \ell_p = -\frac{P_b}{\omega_p} = -\frac{TV_\infty/\eta_p}{\omega_p} = -\frac{1,008(352)/0.80}{2\pi(1,350/60)} = -3,137 \text{ ft·lbf}$$

Thus, the running propeller's contribution to the rolling moment coefficient for the complete airplane is estimated to be

$$(\Delta C_\ell)_p \equiv \frac{\Delta \ell_p}{\frac{1}{2}\rho V_\infty^2 S_w b_w} = \frac{-3,137}{\frac{1}{2}0.0010663(352)^2 327.5(52)} = -0.00279$$

To maintain steady flight at constant altitude with no sideslip, the moments produced by the propeller must be balanced with rudder and aileron deflection. Since the total rolling and yawing moments must be zero, this requires

$$\begin{bmatrix} C_{\ell,\delta_a} & C_{\ell,\delta_r} \\ C_{n,\delta_a} & C_{n,\delta_r} \end{bmatrix} \begin{Bmatrix} \delta_a \\ \delta_r \end{Bmatrix} = \begin{Bmatrix} -(\Delta C_\ell)_p \\ -(\Delta C_n)_p \end{Bmatrix}$$

or after substitution of the numerical values

$$\begin{bmatrix} -0.131 & 0.006 \\ 0.013 & -0.052 \end{bmatrix} \begin{Bmatrix} \delta_a \\ \delta_r \end{Bmatrix} = \begin{Bmatrix} 0.00279 \\ 0.00781 \end{Bmatrix}$$

This gives

$$\begin{Bmatrix} \delta_a \\ \delta_r \end{Bmatrix} = \begin{Bmatrix} -0.0285 \\ -0.1573 \end{Bmatrix} = \begin{Bmatrix} -1.6° \\ -9.0° \end{Bmatrix}$$

We can now assess the validity of our original assumption that the bank angle is small. For steady flight, the summation of axial, normal, and side forces acting on the aircraft must all be zero. Within the small-angle approximation, the normal and axial forces were balanced for this solution by setting the lift equal to the weight and the thrust equal to the drag. However, we have not yet enforced Newton's second law in the spanwise direction. For steady flight, this requires that the sum of the aerodynamic side force and the spanwise component of aircraft weight must be zero. Within the small-angle approximation this gives

$$Y + W\sin\phi \cong Y + W\phi = 0.0 \quad \text{or} \quad \phi = -Y/W$$

Because the problem statement requires zero sideslip, and for this airplane the change in side force with aileron deflection is zero, the only aerodynamic side force is that resulting from rudder deflection,

$$Y = \frac{1}{2}\rho V_\infty^2 S_w C_{Y,\delta_r}\delta_r = \frac{1}{2}0.0010663(352)^2 327.5(0.115)(-0.1573) = -391 \text{ lbf}$$

Thus, the required bank angle is

$$\phi = -Y/W = -(-391)/15,480 = 0.0253 = 1.4°$$

which shows that use of the small-angle approximation is well justified.

From this solution we estimate that in the absence of slipstream effects, approximately 9 degrees of right rudder and less than 2 degrees of right aileron are required to trim the airplane for this engine-out flight condition with no sideslip. Slipstream rotation could significantly increase the required rudder deflection while decreasing the required aileron deflection. At lower airspeeds the slipstream effects and required rudder deflection would be larger.

Since wind tunnel data for the P-38 Lightning were not available to the author, the aerodynamic derivatives used in Example 5.7.1 are only approximate and the predicted result is probably not accurate. However, the example introduces what is typically the critical constraint on rudder size for multiengine airplanes. That is, the need to provide the yawing moment necessary to trim the airplane for the most critical allowable engine-out flight configuration. The reader should also note that while some aileron deflection is required for trim, this is quite small. Static trim requirements are not normally a significant consideration in sizing the ailerons of a conventional airplane. When the ailerons are sized on the basis of dynamic considerations, which will be discussed later, they will usually be capable of providing any rolling moments needed for static trim.

5.8. The Generalized Small-Angle Lateral Trim Requirements

In Example 5.7.1, rudder and aileron deflection were used to trim an airplane against asymmetric thrust with no sideslip. While the zero sideslip constraint, which was used in that example, produces minimum drag and greatest passenger comfort, there are many situations in which it is desirable to fly with nonzero sideslip. The most common of these is a crosswind landing. Less frequently, sideslip is used in tight formation flying to move an airplane laterally relative to the flight leader while keeping the wings parallel with those of the leader. Such maneuvers are typically used during aerial refueling, sailplane towing, and precision acrobatics. There are also some rare missions in which sideslip is used to turn an airplane without banking, for example, when some sensor must be kept level with the horizon. Sideslip can also be used to improve the roll performance of a sluggish airplane. Some radio-controlled airplanes have no ailerons and roll is achieved by deflecting the rudder to generate sideslip. The sideslip acts through dihedral effect to generate a rolling moment. The same process is used by many tactical jet fighters, which lose aileron control power at high angles of attack, but can be rolled readily with the rudder. In flight testing, an airplane is sometimes trimmed in steady sideslip at constant airspeed, heading, and altitude, as a means of quantitatively evaluating an airplane's lateral stability derivatives and control powers.

In general, static lateral trim can be achieved with nonzero values for both the sideslip angle, β, and the bank angle, ϕ. The general requirements for lateral trim at constant airspeed, heading, and altitude are simply zero summations for the total side force and two lateral moments. The aerodynamic side force produced on the skin of the aircraft must balance the spanwise component of aircraft weight as well as any side force produced by the aircraft's propulsion system. Similarly, the aerodynamic rolling and yawing moments produced on the aircraft's skin must balance any lateral moments produced by the propulsion system. For small angles, the side force and both lateral moments are linear functions of the sideslip angle and lateral control surface deflections. Also for small elevation and bank angles, the spanwise component of the aircraft's weight can be approximated as a linear function of the bank angle, i.e., $W\cos\theta\sin\phi \cong W\phi$ (see Figs. 4.7.1 and 5.7.4 for the general trigonometric relations). Thus, **the generalized small-angle lateral trim requirements for flight at constant airspeed, heading, and altitude** can be written in dimensionless form as

$$
\begin{bmatrix} C_{Y,\beta} & C_{Y,\delta_a} & C_{Y,\delta_r} \\ C_{\ell,\beta} & C_{\ell,\delta_a} & C_{\ell,\delta_r} \\ C_{n,\beta} & C_{n,\delta_a} & C_{n,\delta_r} \end{bmatrix} \begin{Bmatrix} \beta \\ \delta_a \\ \delta_r \end{Bmatrix} + \begin{Bmatrix} (\Delta C_Y)_p \\ (\Delta C_\ell)_p \\ (\Delta C_n)_p \end{Bmatrix} + \begin{Bmatrix} C_W \\ 0 \\ 0 \end{Bmatrix} \phi = \begin{Bmatrix} 0 \\ 0 \\ 0 \end{Bmatrix}
\tag{5.8.1}
$$

where the subscript p indicates a contribution from the aircraft's propulsion system at zero sideslip and C_W is the weight coefficient,

$$
(\Delta C_Y)_p \equiv \frac{\Delta Y_p}{\frac{1}{2}\rho V_\infty^2 S_w}, \quad (\Delta C_\ell)_p \equiv \frac{\Delta \ell_p}{\frac{1}{2}\rho V_\infty^2 S_w b_w}, \quad (\Delta C_n)_p \equiv \frac{\Delta n_p}{\frac{1}{2}\rho V_\infty^2 S_w b_w}, \quad C_W \equiv \frac{W}{\frac{1}{2}\rho V_\infty^2 S_w}
$$

Examination of Eq. (5.8.1) reveals that even with the weight, dynamic pressure, and all contributions from the propulsion system known, these lateral trim relations are underdetermined without some additional constraint on the four variables, β, δ_a, δ_r, and ϕ. This means that static lateral trim can be attained with many different combinations of sideslip and bank angle. The lateral trim solution that was obtained in Example 5.7.1 is only one of many possible solutions that could be used for the specific engine-out flight condition described in that example. That particular solution was obtained using the constraint $\beta=0$, which typically results in maximum range and endurance for cruise operation and provides the greatest level of comfort for both crew and passengers. However, drag and comfort are not always the highest priorities in all flight scenarios. In some cases, best performance relative to mission objectives is attained when the aircraft is trimmed with some sideslip. In the following example, we shall examine a range of lateral trim solutions that could be used for the same aircraft and engine-out operating condition described in Example 5.7.1.

EXAMPLE 5.8.1. For the P-38 Lightning described in Example 5.7.1, use the small-angle lateral trim relations given by Eq. (5.8.1) to plot the required rudder and aileron deflections as well as the sideslip angle as a function of bank angle for lateral trim at the engine-out operating condition described in that example. In addition to the operating condition and aerodynamic derivatives that are given in Example 5.7.1, use the sideslip derivatives

$$
C_{Y,\beta} = -0.512, \quad C_{\ell,\beta} = -0.025, \quad C_{n,\beta} = 0.083
$$

Solution. From Example 5.7.1 we have

$$C_{Y,\delta_a} = 0.0, \quad C_{\ell,\delta_a} = -0.131, \quad C_{n,\delta_a} = 0.013,$$
$$C_{Y,\delta_r} = 0.115, \quad C_{\ell,\delta_r} = 0.006, \quad C_{n,\delta_r} = -0.052,$$
$$C_W = 0.716, \quad (\Delta C_Y)_p = 0.0, \quad (\Delta C_\ell)_p = -0.00279, \quad (\Delta C_n)_p = -0.00781$$

Thus, Eq. (5.8.1) can be rearranged to yield

$$
\begin{Bmatrix} \beta \\ \delta_a \\ \delta_r \end{Bmatrix} =
\begin{bmatrix}
-0.512 & 0.0 & 0.115 \\
-0.025 & -0.131 & 0.006 \\
0.083 & 0.013 & -0.052
\end{bmatrix}^{-1}
\begin{Bmatrix} -0.716\phi \\ 0.00279 \\ 0.00781 \end{Bmatrix}
$$

For a bank angle of –5 degrees ($\phi=0.0873$) this gives

$$
\begin{Bmatrix} \beta \\ \delta_a \\ \delta_r \end{Bmatrix} =
\begin{Bmatrix} -0.2428 \\ 0.0004 \\ -0.5376 \end{Bmatrix} =
\begin{Bmatrix} -13.91° \\ 0.02° \\ -30.80° \end{Bmatrix}
$$

Repeating similar computations for a range of bank angles produces the linear results that are plotted in Fig. 5.8.1. The zero-sideslip solution that was obtained in Example 5.7.1 is indicated by the vertical dashed line.

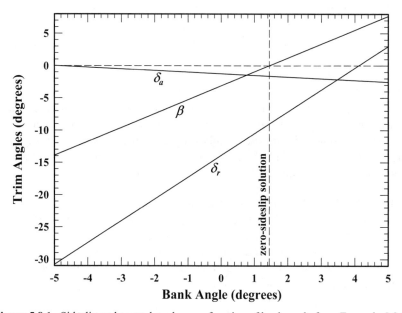

Figure 5.8.1. Sideslip and control angles as a function of bank angle from Example 5.8.1.

For the engine-out operating condition used in the previous example, the results plotted in Fig. 5.8.1 show that the right rudder required to statically trim the airplane against the asymmetric thrust from the right engine can be reduced from that needed for the zero-sideslip solution by slightly increasing the bank angle to the right and allowing the airplane to sideslip to the right. When the airplane is statically trimmed with some sideslip to the right, a yawing moment to the right is generated through the yaw stiffness, $C_{n,\beta}$. This yawing moment helps balance the propulsive yawing moment to the left, thereby reducing the right rudder needed to statically trim the airplane. For this particular airplane and operating condition, using a bank angle of about 4.1 degrees produces sufficient sideslip to totally eliminate the need for rudder deflection at static trim. Larger bank angles require crossed controls, i.e., right aileron and left rudder.

5.9. Steady-Heading Sideslip

The aircraft maneuver associated with maintaining steady sideslip and bank angle at constant airspeed, heading, and altitude is typically referred to as a *steady-heading sideslip* (SHSS). Even with symmetric power, this static maneuver is of significant utility, because it is used in crosswind landings and in flight testing to quantitatively evaluate certain lateral aerodynamic derivatives. In some ways lateral static analysis of steady-heading sideslip is similar to the analysis of longitudinal static trim, because a component of the aircraft's weight is balanced against an aerodynamic force and any propulsive moments are balanced by the aerodynamic moments.

For the special case of symmetric power, the small-angle lateral trim requirements expressed in Eq. (5.8.1) can be rearranged to give

$$
\begin{bmatrix} C_{Y,\delta_a} & C_{Y,\delta_r} & C_W \\ C_{\ell,\delta_a} & C_{\ell,\delta_r} & 0 \\ C_{n,\delta_a} & C_{n,\delta_r} & 0 \end{bmatrix} \begin{Bmatrix} \delta_a \\ \delta_r \\ \phi \end{Bmatrix} = - \begin{Bmatrix} C_{Y,\beta} \\ C_{\ell,\beta} \\ C_{n,\beta} \end{Bmatrix} \beta
\tag{5.9.1}
$$

This linear system is readily solved by Gauss elimination to provide expressions for the aileron deflection, rudder deflection, and bank angle as linear functions of the sideslip angle. Thus, **the small-angle lateral trim requirements for steady-heading sideslip with symmetric power result in the linear relations**

$$
\begin{Bmatrix} \delta_a \\ \delta_r \\ \phi \end{Bmatrix} = \begin{Bmatrix} \left(\dfrac{\partial \delta_a}{\partial \beta} \right)_{\text{SHSS}} \\ \left(\dfrac{\partial \delta_r}{\partial \beta} \right)_{\text{SHSS}} \\ \left(\dfrac{\partial \phi}{\partial \beta} \right)_{\text{SHSS}} \end{Bmatrix} \beta
\tag{5.9.2}
$$

where the SHSS gradients of aileron deflection, rudder deflection, and bank angle with respect to sideslip angle are given by

$$\left(\frac{\partial \delta_a}{\partial \beta}\right)_{SHSS} = \frac{C_{\ell,\delta_r} C_{n,\beta} - C_{\ell,\beta} C_{n,\delta_r}}{C_{\ell,\delta_a} C_{n,\delta_r} - C_{\ell,\delta_r} C_{n,\delta_a}} \qquad (5.9.3)$$

$$\left(\frac{\partial \delta_r}{\partial \beta}\right)_{SHSS} = \frac{C_{\ell,\beta} C_{n,\delta_a} - C_{\ell,\delta_a} C_{n,\beta}}{C_{\ell,\delta_a} C_{n,\delta_r} - C_{\ell,\delta_r} C_{n,\delta_a}} \qquad (5.9.4)$$

$$\left(\frac{\partial \phi}{\partial \beta}\right)_{SHSS} = -\left[C_{Y,\delta_a}\left(\frac{\partial \delta_a}{\partial \beta}\right)_{SHSS} + C_{Y,\delta_r}\left(\frac{\partial \delta_r}{\partial \beta}\right)_{SHSS} + C_{Y,\beta}\right]\bigg/ C_W \qquad (5.9.5)$$

While these relations were developed for the special case of spanwise symmetric power, **if control deflections are measured relative to level trim with no sideslip, Eqs. (5.9.2) through (5.9.5) may be used for aircraft with slightly asymmetric power, such as a propeller-driven airplane without balanced counterrotating propellers.**

EXAMPLE 5.9.1. The P-38 Lightning described in Examples 5.7.1 and 5.8.1 is landing at sea level in a 15-knot crosswind from the right, with both engines in operation. The airplane's airspeed at touchdown is 150 ft/sec. Using the steady-heading sideslip trim relations given by Eqs. (5.9.2) through (5.9.5), we wish to estimate the aileron and rudder deflection that would be required to keep the airplane's heading aligned with the runway at touchdown. Assuming that the pilot touches down in a perfect steady-heading sideslip and that the runway is perfectly level, how far above the runway is the high wheel when the low wheel first makes contact with the runway? The spanwise wheelbase for the P-38 Lightning's main gear is 16 feet. Even though the aerodynamic derivatives may be significantly different at this airspeed with the airplane in landing configuration, for these example computations we shall use the same aerodynamic derivatives given in Examples 5.7.1 and 5.8.1.

Solution. Using the gross weight and wing area that are given in Example 5.7.1, the weight coefficient at touchdown is

$$C_W \equiv \frac{W}{\frac{1}{2}\rho V_\infty^2 S_w} = \frac{15,480}{\frac{1}{2}0.0023769(150)^2 327.5} = 1.768$$

The sideslip angle required to keep the airplane's heading aligned with the runway at touchdown is

$$\beta = \sin^{-1}\left(\frac{V_{crosswind}}{V_{airspeed}}\right) = \sin^{-1}\left(\frac{15 \times 6,076/3,600}{150}\right) = 0.1696 = 9.7°$$

Using the aerodynamic derivatives given in Examples 5.7.1 and 5.8.1, the steady-heading sideslip trim relations given by Eqs. (5.9.2) through (5.9.5) result in

$$\left(\frac{\partial \delta_a}{\partial \beta}\right)_{SHSS} = \frac{C_{\ell,\delta_r} C_{n,\beta} - C_{\ell,\beta} C_{n,\delta_r}}{C_{\ell,\delta_a} C_{n,\delta_r} - C_{\ell,\delta_r} C_{n,\delta_a}}$$

$$= \frac{(0.006)(0.083) - (-0.025)(-0.052)}{(-0.131)(-0.052) - (0.006)(0.013)} = -0.1191$$

$$\delta_a = \left(\frac{\partial \delta_a}{\partial \beta}\right)_{SHSS} \beta = (-0.1191)(0.1696) = -0.0202 = \underline{-1.2°}$$

$$\left(\frac{\partial \delta_r}{\partial \beta}\right)_{SHSS} = \frac{C_{\ell,\beta} C_{n,\delta_a} - C_{\ell,\delta_a} C_{n,\beta}}{C_{\ell,\delta_a} C_{n,\delta_r} - C_{\ell,\delta_r} C_{n,\delta_a}}$$

$$= \frac{(-0.025)(0.013) - (-0.131)(0.083)}{(-0.131)(-0.052) - (0.006)(0.013)} = 1.5664$$

$$\delta_r = \left(\frac{\partial \delta_r}{\partial \beta}\right)_{SHSS} \beta = (1.5664)(0.1696) = 0.2657 = \underline{15.2°}$$

$$\left(\frac{\partial \phi}{\partial \beta}\right)_{SHSS} = -\left[C_{Y,\delta_a}\left(\frac{\partial \delta_a}{\partial \beta}\right)_{SHSS} + C_{Y,\delta_r}\left(\frac{\partial \delta_r}{\partial \beta}\right)_{SHSS} + C_{Y,\beta}\right]\bigg/C_W$$

$$= -\frac{(0.0)(-0.1191) + (0.115)(1.5664) + (-0.512)}{1.768} = 0.1877$$

$$\phi = \left(\frac{\partial \phi}{\partial \beta}\right)_{SHSS} \beta = (0.1877)(0.1696) = 0.0318 = 1.8°$$

Using this bank angle and the 16-foot wheelbase, we obtain the distance that the high wheel is above the runway when the low wheel first makes contact,

$$\text{height} = (\text{wheelbase})\sin\phi = (16\,\text{ft})\sin(0.0318) = \underline{0.51\,\text{ft}} = 6.1\,\text{in}$$

From the results obtained in Example 5.9.1, we see that stabilizing an airplane in a sideslip, such as for a crosswind landing, entails holding the upwind wing down slightly with aileron deflection and balancing the yawing moment with opposite rudder. For a positive sideslip (wind from the pilot's right), the right wing is held down in a positive bank by using right aileron (negative deflection). Left rudder (positive deflection) is used to counter the yawing moment due to directional stability. The opposite crossed-controls inputs are required for a negative sideslip (wind from the pilot's left).

For small sideslip angles with the particular airplane and operating condition used in Example 5.9.1, each degree of positive sideslip requires -0.12 degree of aileron deflection, 1.57 degrees of rudder deflection, and 0.19 degree of bank angle. Because the SHSS gradient of rudder deflection with sideslip angle is much larger than the other SHSS gradients, rudder deflection will most likely limit the maximum sideslip that can be supported with symmetric power, although this is not typically the critical constraint on rudder size. The large difference between the rudder and aileron gradients can be attributed to the large efficient ailerons used on the P-38 Lightning.

Because crossed controls are required to trim an airplane in steady-heading sideslip with symmetric power, the adverse yaw produced by conventional ailerons is beneficial in this static maneuver. Adverse yaw helps support the sideslip and reduces the rudder deflection required for lateral trim. Modern airplanes sometimes use what are called *spoilers* for roll control. Spoilers will be discussed in greater detail in Chapter 6. At this juncture it is sufficient to point out that spoilers produce proverse yaw ($C_{n,\delta_a} < 0$) rather than the adverse yaw exhibited by conventional ailerons. While proverse yaw improves the handling characteristics of an airplane in normal turning flight, it is detrimental to performance in a steady-heading sideslip with symmetric power.

EXAMPLE 5.9.2. (*Contributed by Captain Robert J. Niewoehner, Ph.D., Aerospace Engineering Department, United States Naval Academy*) The F-14 Tomcat shown in Fig. 5.9.1 is a twin-engine supersonic strike fighter designed for carrier-based operations with the U.S. Navy. Its aerodynamic performance across a broad flight envelope is due to its variable geometry design, permitting both very high-speed supersonic operation at aft wingsweep, and excellent low-speed handling and performance with the wings forward. The flight controls include all-moving stabilators for pitch and highspeed roll control, and a conventional rudder on each of the two vertical tails. Low-speed roll control is provided principally by half-span spoilers, supplemented by differential stabilator. (Stabilators and spoilers are discussed in greater detail in Chapter 6.) For the Tomcat in power approach configuration with gear and flaps down, find the SHSS gradients from Eqs. (5.9.3) through (5.9.5) at an approach speed of 130 knots. Use the following approximate data for these computations:

Figure 5.9.1. The F-14 Tomcat described by Captain Niewoehner in Example 5.9.2. Following his experience flying F-14s, Dr. Niewoehner served as the Navy's lead test pilot on the Super Hornet program, where he logged 296 missions and over 450 hours. He has contributed a great deal to the second edition of this textbook, particularly in this and the following sections. (U.S. Navy photo)

$$S_w = 565 \text{ ft}^2, \quad b_w = 64.12 \text{ ft}, \quad W = 54,000 \text{ lbf},$$
$$C_{Y,\beta} = -1.249, \quad C_{\ell,\beta} = -0.138, \quad C_{n,\beta} = 0.120,$$
$$C_{Y,\delta_a} = 0.144, \quad C_{\ell,\delta_a} = -0.103, \quad C_{n,\delta_a} = -0.022,$$
$$C_{Y,\delta_r} = 0.269, \quad C_{\ell,\delta_r} = 0.014, \quad C_{n,\delta_r} = -0.099$$

For these example computations, the aileron control derivatives are defined in terms of an *equivalent aileron deflection*, which is actually provided by a combination of differential stabilator and spoiler deflections.

Solution. From the given gross weight and wing area, the weight coefficient at the approach speed is

$$C_W \equiv \frac{W}{\frac{1}{2}\rho V_\infty^2 S_w} = \frac{54,000}{\frac{1}{2}0.0023769(130 \times 6,076/3,600)^2 565} = 1.67$$

From Eqs. (5.9.3) through (5.9.5), the SHSS gradients are

$$\left(\frac{\partial \delta_a}{\partial \beta}\right)_{\text{SHSS}} = \frac{(0.014)(0.120) - (-0.138)(-.099)}{(-0.103)(-0.099) - (0.014)(-0.022)} = \underline{-1.14}$$

$$\left(\frac{\partial \delta_r}{\partial \beta}\right)_{\text{SHSS}} = \frac{(-0.138)(-0.022) - (-0.103)(0.120)}{(-0.103)(-0.099) - (0.014)(-0.022)} = \underline{1.47}$$

$$\left(\frac{\partial \phi}{\partial \beta}\right)_{\text{SHSS}} = -\frac{(0.144)(-1.14) + (0.269)(1.47) + (-1.249)}{1.67} = \underline{0.61}$$

These three SHSS gradients are dimensionless and can be interpreted as either being in radians per radian or degrees per degree. Over the range of sideslip angles for which linear behavior can be reasonably expected, each degree of positive sideslip requires −1.14 degrees of equivalent aileron deflection (right aileron), +1.47 degrees of rudder deflection (left rudder), and +0.61 degree of bank angle (to the right). Negative sideslip would require the opposite inputs. Notice that for this airplane and operating conditions, the SHSS aileron gradient is nearly as large as the rudder gradient, which is not unusual when roll control is provided by spoilers. For such aircraft, either δ_a or δ_r constraints could limit the maximum sideslip that can be supported with symmetric power.

The three aerodynamic gradients for steady-heading sideslip, which are expressed in Eqs. (5.9.3) through (5.9.5), can be readily determined from flight testing. After trimming for steady level flight, the test aircraft is trimmed in a steady-heading sideslip for several different rudder settings from maximum right rudder to maximum left rudder. Measured values for the aileron and rudder displacements as well as the bank angle are plotted against the measured sideslip angles to discern the SHSS gradients. Knowledge of these static gradients provides quantitative information relating an airplane's directional stability and dihedral effect to the lateral control derivatives. This information can be combined with the results of certain dynamic flight tests and knowledge of the aircraft's geometry and mass properties to directly evaluate the lateral static stability derivatives and control powers.

5.10. Engine Failure and Minimum-Control Airspeed

The critical constraint on rudder size for multiengine airplanes is typically based on the rudder deflection required to trim the airplane against asymmetric thrust with a critical engine inoperative and the remaining engine or engines at full power. The rudder deflection required for such operational states increases as airspeed is decreased. This is because maximum thrust and the associated yawing moment typically decrease or remain constant with increasing airspeed, whereas the aerodynamic moments available from the controls are proportional to dynamic pressure. As the airplane slows down and dynamic pressure decreases, the aerodynamic controls eventually lose the moment-generating capacity to hold against the asymmetric thrust. The airspeed at which either the rudder or aileron reaches its physical limit, called *saturation*, is typically referred to as the *static minimum-control airspeed* and is commonly given the symbol V_{mc}. Usually, the rudder provides the critical constraint on V_{mc}. However, for some airplanes with large dihedral effect and/or minimal aileron control power, the minimum-control airspeed could be dictated by aileron saturation.

For aircraft powered by jet engines, which generate only slight inherent yawing moments, the critical engine used for computing the minimum-control airspeed is either of the two outboard engines. For propeller-driven aircraft, yawing and rolling moments are generated directly by the propellers and from the asymmetric slipstream flow over the fuselage and/or the wing and tail. For such aircraft, more careful thought is required to discern which of the two outboard engines is the critical one. The result of such analysis depends on the direction of propeller rotation.

Minimum-control airspeed is always a concern for either takeoff or landing, because of the possibility of engine failure. Of course, an airplane is not landed at full power, so engine failure on landing does not immediately produce the critical yawing moment. However, the possible need to abort a landing and proceed with a go-around precludes the option of landing at airspeeds below V_{mc}. If the rudder is inadequately sized, the minimum-control airspeed will set the lower limit for both takeoff and landing speeds. Even if the wing with high-lift devices deployed is capable of flying at much lower airspeeds, pilots should avoid flying a multiengine airplane at airspeeds below V_{mc}. To prevent any effect on takeoff or landing performance the rudder should be sized so that minimum-control airspeed is less than 1.1 times the stall speed in takeoff configuration and less than 1.15 times the stall speed in landing configuration.

Operationally, the difficulties associated with maintaining control under conditions of engine failure at low airspeeds have killed pilots in airplanes that otherwise had excellent reputations, notably the P-38 Lightning and F-14 Tomcat. Consequently, V_{mc} is the fundamental reason that regulatory agencies require a separate pilot's license for multiengine airplanes.

In addition to concerns related to engine failure on takeoff and landing, minimum-control airspeed may also be a consideration for some airplanes that routinely shut down one or more engines in cruise flight to improve either range or endurance. This is a common practice for the U.S. Navy's maritime patrol airplane, the turbo-prop P-3C Orion shown in Fig. 5.10.1.

From the results obtained in Example 5.8.1, we have seen that the rudder deflection required to statically trim an airplane against a yawing moment from asymmetric thrust can be reduced by banking the airplane into the operating engine and allowing it to sideslip in the direction of that engine. However, there is a limit to how large a bank angle can be comfortably used for this purpose, particularly in close proximity to the ground. For the purpose of design and certification of conventional piloted airplanes, the magnitude of the bank angle is typically limited to no more than 5 degrees.

For analytically evaluating the minimum-control airspeed, the generalized small-angle lateral trim requirements given by Eq. (5.8.1) can be conveniently rearranged as

$$
\begin{bmatrix} C_{Y,\beta} & C_{Y,\delta_a} & C_{Y,\delta_r} \\ C_{\ell,\beta} & C_{\ell,\delta_a} & C_{\ell,\delta_r} \\ C_{n,\beta} & C_{n,\delta_a} & C_{n,\delta_r} \end{bmatrix} \begin{Bmatrix} \beta \\ \delta_a \\ \delta_r \end{Bmatrix} = - \begin{Bmatrix} (\Delta C_Y)_p + C_W \phi \\ (\Delta C_\ell)_p \\ (\Delta C_n)_p \end{Bmatrix}
\tag{5.10.1}
$$

Assuming that the lateral force and moments produced by the propulsion system are known, and fixing the bank angle at any specified value, this linear system of equations is readily solved for the three unknown angles, β, δ_a, and δ_r divided by the airplane's weight coefficient, C_W. Introducing this change of variables gives

$$
\begin{Bmatrix} \beta/C_W \\ \delta_a/C_W \\ \delta_r/C_W \end{Bmatrix} = \begin{bmatrix} C_{Y,\beta} & C_{Y,\delta_a} & C_{Y,\delta_r} \\ C_{\ell,\beta} & C_{\ell,\delta_a} & C_{\ell,\delta_r} \\ C_{n,\beta} & C_{n,\delta_a} & C_{n,\delta_r} \end{bmatrix}^{-1} \begin{Bmatrix} -\Delta Y_p/W - \phi \\ -\Delta \ell_p/(b_w W) \\ -\Delta n_p/(b_w W) \end{Bmatrix}
\tag{5.10.2}
$$

Figure 5.10.1. To conserve fuel, the P-3C Orion is commonly flown without all four engines in operation. (U.S. Navy photo)

For multiengine airplanes with one engine inoperative, the sign for the bank angle giving the smallest rudder deflection is that for a bank in the direction of the operating engine. Note that with the aerodynamic coefficients, weight, and propulsive contributions fixed, the trim angles per unit weight coefficient are independent of the unknown airspeed.

Because the minimum-control airspeed could be constrained by saturation of either the ailerons or rudder, both possibilities must be considered. Once the trim angles per unit weight coefficient have been determined from Eq. (5.10.2), the aileron- and rudder-limited minimum-control airspeeds for the specified bank angle can be evaluated from the weight coefficient definition and the known control saturation angles,

$$
\left(V_{mc}\right)_{\text{aileron}} = \sqrt{\frac{2W(\delta_a/C_W)}{\rho S_w \delta_{a_{\text{sat}}}}} \tag{5.10.3}
$$

$$
\left(V_{mc}\right)_{\text{rudder}} = \sqrt{\frac{2W(\delta_r/C_W)}{\rho S_w \delta_{r_{\text{sat}}}}} \tag{5.10.4}
$$

To avoid nonphysical imaginary roots, the control saturation angles used in Eqs. (5.10.3) and (5.10.4) must be those with the same sign as the corresponding trim angle per unit weight coefficient. The limiting minimum-control airspeed for the specified bank angle is the larger of the two real solutions. For engine-out operation of typical multiengine airplanes, static V_{mc} is rudder limited.

The reader is cautioned that the above formulation is based on linear relations between control moments and control deflections. While this is reasonable for control deflections having a magnitude less than about 10 degrees, linearity commonly breaks down prior to saturation. Because control power typically decreases as saturation is approached, this formulation will underestimate the minimum-control airspeed if the actual maximum control surface deflection angles are used for saturation. Improved results can be obtained if effective control surface saturation angles are used. These effective angles are defined so that when they are multiplied by the linear control power, the product is equal to the actual moment coefficient at saturation.

For the purpose of evaluating static V_{mc}, writing the lateral trim relations in the form of Eq. (5.10.2) has the advantage of making the dependent variables independent of the unknown airspeed. However, because weight appears on both sides of Eq. (5.10.2), it is less than obvious how minimum-control airspeed is affected by aircraft gross weight. To expose this weight dependence more clearly, Eq. (5.10.2) can be rearranged as

$$
\begin{Bmatrix} \beta \\ \delta_a \\ \delta_r \end{Bmatrix} = \begin{bmatrix} C_{Y,\beta} & C_{Y,\delta_a} & C_{Y,\delta_r} \\ C_{\ell,\beta} & C_{\ell,\delta_a} & C_{\ell,\delta_r} \\ C_{n,\beta} & C_{n,\delta_a} & C_{n,\delta_r} \end{bmatrix}^{-1} \left(-\begin{Bmatrix} \Delta Y_p \\ \Delta \ell_p/b_w \\ \Delta n_p/b_w \end{Bmatrix} - \begin{Bmatrix} W \\ 0 \\ 0 \end{Bmatrix} \phi \right) \frac{1}{\frac{1}{2}\rho V_\infty^2 S_w} \tag{5.10.5}
$$

Writing the lateral trim relations in the form of Eq. (5.10.5) provides considerable insight into the minimum-control airspeed problem. First, it readily displays the previously observed facts that the control deflections required for trim are inversely proportional to dynamic pressure and vary linearly with bank angle. Equation (5.10.5) also shows that for fixed airspeed and thrust contributions, the control deflections required for lateral trim at zero bank are independent of aircraft weight. On the other hand, the derivatives of these control deflections with respect to bank angle are directly proportional to gross weight. This means that using a small bank to reduce the rudder deflection required to trim an airplane against asymmetric thrust is less effective at low gross weights.

Because some bank angle is typically used to establish V_{mc}, the insight gleaned from Eq. (5.10.5) exposes an important and nonintuitive feature of the V_{mc} problem. Whereas many performance metrics, such as takeoff distance and climb, suggest that high gross weight is the most demanding operating condition, for the minimum-control airspeed problem, it is light gross weights that are most dangerous. **Even for a fixed approach speed, when an airplane is banked at a small fixed angle away from a failed engine, the rudder deflection needed for lateral trim increases with decreasing gross weight.** The result is that an airplane is at more risk for loss of control at light weights than at heavy weights. Therefore, the most taxing maneuver is executing a *waveoff* or missed approach at the end of a mission. This is due to the much lower gross weight at the end of the mission, when most of the fuel carried on takeoff has been burned off. For this reason, operating handbooks for many multiengine airplanes recommend landing 10 to 15 knots faster if one engine is out. For an airplane accustomed to landing at 100 knots, an extra 10 knots provides 21 percent greater yawing moment from the rudder.

EXAMPLE 5.10.1. (*Contributed by Captain Robert J. Niewoehner, Ph.D., Aerospace Engineering Department, United States Naval Academy*) Using the data given in Example 5.9.2, estimate static V_{mc} in approach configuration with a 5-degree bank for the F-14B Tomcat with F-110 engines. Each engine develops 16,100 lbf of thrust at maximum dry power and 27,000 lbf at maximum afterburner. Each engine is offset 4.5 feet from the airplane's centerline and contributes a pure yawing moment. Assume that the effective saturation angles for the aileron and rudder are 15 and 25 degrees, respectively. Also plot the change in β, δ_a, and δ_r with respect to bank angle as a function of gross weight from 42,000 to 75,000 pounds using an approach speed of 130 knots.

Solution. Because these jet engines produce no appreciable side force or rolling moment, either of the two engines can be used as the critical one. With the left engine out and the right engine at full power, the appropriate bank angle is positive 5 degrees and the operating engine on the pilot's right produces a negative yawing moment. Thus, from Eq. (5.10.2), at maximum dry power we obtain

$$
\begin{Bmatrix} \beta/C_W \\ \delta_a/C_W \\ \delta_r/C_W \end{Bmatrix} = \begin{bmatrix} C_{Y,\beta} & C_{Y,\delta_a} & C_{Y,\delta_r} \\ C_{\ell,\beta} & C_{\ell,\delta_a} & C_{\ell,\delta_r} \\ C_{n,\beta} & C_{n,\delta_a} & C_{n,\delta_r} \end{bmatrix}^{-1} \begin{Bmatrix} -\phi \\ 0 \\ -(-y_{bp}T_{\max})/(b_w W) \end{Bmatrix}
$$

$$
= \begin{bmatrix} -1.249 & 0.144 & 0.269 \\ -0.138 & -0.103 & 0.014 \\ 0.120 & -0.022 & -0.099 \end{bmatrix}^{-1} \begin{Bmatrix} -5(\pi/180) \\ 0 \\ (4.5 \times 16,100)/(64.12 \times 54,000) \end{Bmatrix} = \begin{Bmatrix} 0.0275 \\ -0.0593 \\ -0.1648 \end{Bmatrix}
$$

Using standard density at sea level in Eqs. (5.10.3) and (5.10.4), the aileron- and rudder-limited minimum-control airspeeds for 5 degrees of bank are

$$\left(V_{mc}\right)_{\text{aileron}} = \sqrt{\frac{2W(\delta_a/C_W)}{\rho S_w \delta_{a_{\text{sat}}}}} = \sqrt{\frac{2(54,000)(-0.0593)}{(0.0023769)(565)(-15 \times \pi/180)}} = 135 \text{ ft/sec}$$

$$= 80 \text{ knots}$$

$$\left(V_{mc}\right)_{\text{rudder}} = \sqrt{\frac{2W(\delta_r/C_W)}{\rho S_w \delta_{r_{\text{sat}}}}} = \sqrt{\frac{2(54,000)(-0.1648)}{(0.0023769)(565)(-25 \times \pi/180)}} = 174 \text{ ft/sec}$$

$$= 103 \text{ knots}$$

Thus, with 5 degrees of bank into the operating engine at maximum dry power, the minimum-control airspeed is rudder limited at 103 knots, which is well below the airplane's typical approach speed of 125 to 135 knots. Repeating similar computations both with and without bank and with and without afterburner yields the following results:

$$\phi = 5°, \ T_{\text{max}} = 16,100 \text{ lbf} \ \Rightarrow \ V_{mc} = 174 \text{ ft/sec} = 103 \text{ knots}$$
$$\phi = 0°, \ T_{\text{max}} = 16,100 \text{ lbf} \ \Rightarrow \ V_{mc} = 231 \text{ ft/sec} = 137 \text{ knots}$$
$$\phi = 5°, \ T_{\text{max}} = 27,000 \text{ lbf} \ \Rightarrow \ V_{mc} = 258 \text{ ft/sec} = 153 \text{ knots}$$
$$\phi = 0°, \ T_{\text{max}} = 27,000 \text{ lbf} \ \Rightarrow \ V_{mc} = 299 \text{ ft/sec} = 177 \text{ knots}$$

The V_{mc} of 153 knots for maximum afterburner with 5 degrees of bank is well above the typical approach speed and nearly equal to the range used for takeoff. It is for this reason that the B and D models of the F-14, equipped with F-110 engines, are prohibited from using afterburners with the landing gear down.

The reader should notice the substantial reduction in V_{mc} that results from a small bank angle. The minimum-control airspeed is very sensitive to bank angle, in this case nearly 7 knots/degree at maximum dry power with a gross weight of 54,000 pounds. For this reason both civil and military specifications allow for a small bank, but prohibit trying to balance the airplane with a more uncomfortable attitude. This bank angle restriction applies to design, specification compliance, and demonstration for certification. **It does not apply to emergency operation, in which the pilot is free to fly the airplane by whatever means he or she deems necessary.** If control cannot be maintained with full rudder, a pilot has little to lose by increasing the bank angle away from a failed engine.

The effect of bank on minimum-control airspeed and its dependence on gross weight are observed by differentiating Eq. (5.10.5) with respect to bank angle,

$$\frac{\partial}{\partial \phi} \left\{\begin{array}{c} \beta \\ \delta_a \\ \delta_r \end{array}\right\} = \begin{bmatrix} C_{Y,\beta} & C_{Y,\delta_a} & C_{Y,\delta_r} \\ C_{\ell,\beta} & C_{\ell,\delta_a} & C_{\ell,\delta_r} \\ C_{n,\beta} & C_{n,\delta_a} & C_{n,\delta_r} \end{bmatrix}^{-1} \left\{\begin{array}{c} 1 \\ 0 \\ 0 \end{array}\right\} \frac{-W}{\frac{1}{2}\rho V_\infty^2 S_w}$$

For a gross weight of 54,000 lbf at sea level ($\rho=0.0023769$ slug/ft^3) and a speed of 130 knots (219.4 ft/sec), the weight coefficient is $C_W=W/(\frac{1}{2}\rho V_\infty^2 S_w)=1.67$ and

$$\frac{\partial}{\partial\phi}\begin{Bmatrix}\beta\\\delta_a\\\delta_r\end{Bmatrix}=\begin{bmatrix}-1.249 & 0.144 & 0.269\\-0.138 & -0.103 & 0.014\\0.120 & -0.022 & -0.099\end{bmatrix}^{-1}\begin{Bmatrix}1\\0\\0\end{Bmatrix}1.67=\begin{Bmatrix}1.64\\-1.87\\2.40\end{Bmatrix}$$

Similar results are plotted in Fig. 5.10.2 as a function of gross weight from the empty weight of 42,000 pounds to the maximum takeoff weight of 75,000 pounds.

Notice that, at this approach speed with a gross weight of 42,000 pounds, each degree of right bank increases the positive sideslip angle by about 1.3 degrees and decrease the magnitude of the negative rudder deflection by nearly 1.9 degrees. However, at 75,000 pounds, each degree of right bank decreases the right rudder by more than 3.3 degrees. The yawing moment generated from the sideslip helps to balance the engine's yawing moment to the left and reduces the required rudder deflection to the right. The constraint here is balancing the force and moments with the available rudder deflection, even if that means flying with sideslip. The goal is control, not minimum drag or passenger comfort. Consequently, this is unlike the engine-out cruise problem in Example 5.7.1, where minimum drag constrained the sideslip to be zero, and the controls were deflected accordingly.

To further examine how and why static V_{mc} depends on bank angle for the engine-out operation of a multiengine airplane, first consider Fig. 5.10.3, which shows how aileron- and rudder-limited minimum-control airspeeds vary with bank angle for the F-14B with symmetric thrust. Results plotted in Fig. 5.10.3 are from Eqs. (5.10.2) through (5.10.4) using the same aerodynamic derivatives and control surface saturation angles that were used for Example 5.10.1, but with the propulsive lateral force and moments all set to zero. Lateral trim at constant airspeed, heading, and altitude can be maintained only for

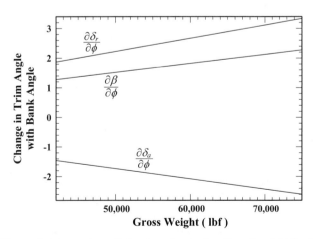

Figure 5.10.2. Change in trim angles with bank angle as a function of weight for Example 5.10.1.

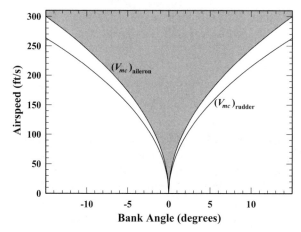

Figure 5.10.3. Aileron- and rudder-limited minimum-control airspeeds as a function of bank angle for the F-14B Tomcat with symmetric thrust.

combinations of airspeed and bank angle that lie within the shaded region above both the aileron- and rudder-limited minimum-control airspeeds. For this airplane with symmetric power, we see that steady-heading sideslip is aileron limited, and as should be expected, V_{mc} is zero with no bank. By contrast, Fig. 5.10.4 shows how aileron- and rudder-limited V_{mc} change with the yawing moment resulting from failure of an F-110 engine at maximum dry power. Note that both the aileron- and rudder-limited curves are shifted to the right, but the rudder-limited curve is shifted farther, due to the large propulsive yawing moment. The small shift in the aileron-limited curve results from roll-yaw coupling.

The trim point in Fig. 5.10.4 corresponding to V_{mc} with no bank is at 231 ft/s and requires crossed controls with $\delta_a=3.3°$, $\delta_r=-25.0°$, and $\beta=-5.0°$. As the bank angle is

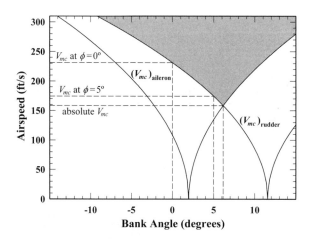

Figure 5.10.4. Aileron- and rudder-limited minimum-control airspeeds as a function of bank angle for the F-14B Tomcat with the right engine at maximum dry power and the left engine inoperative.

increased to the right, static V_{mc}, the positive aileron deflection, and the negative sideslip are all decreased. At a bank angle of 2.0°, V_{mc} is 211 ft/s with δ_a=0.0°, δ_r=−25.0°, and β=−2.5°. Increasing the bank angle farther to the right requires right aileron as well as right rudder. The zero-sideslip solution is obtained at ϕ=3.4° and gives a V_{mc} of 194 ft/s with δ_a=−3.4° and δ_r=−25.0°. With the bank angle at the regulatory limit of 5.0°, V_{mc} is reduced to 174 ft/s with δ_a=−9.0°, δ_r=−25.0°, and β=4.2°. Note that increasing the bank angle somewhat beyond 5 degrees would reduce V_{mc} even further. As the bank angle to the right continues to increase, the sideslip angle to the right also increases and static V_{mc} is reduced, until either the ailerons become saturated or the vertical stabilizers stall. In this case, the ailerons saturate to the right at ϕ=6.1°, giving an *absolute minimum-control airspeed* of 158 ft/s with δ_a=−15.0°, δ_r=−25.0°, and β=8.7°. Any further increase in the bank angle increases V_{mc} as shown in Fig. 5.10.4. At these larger bank angles to the right, V_{mc} remains aileron-limited and the required right rudder is reduced. At a bank angle of 11.6°, V_{mc} is increased to 240 ft/s with δ_a=−15.0°, δ_r=0.0°, and β=11.2°. Trimming the airplane for an even larger bank to the right requires left rudder and more airspeed. The results shown in Fig. 5.10.4 do not account for the possibility of vertical stabilizer stall.

The previous example illustrates how static V_{mc} is affected by a propulsive yawing moment resulting from the engine-out operation of a multiengine airplane. For the jet-powered F-14, the only lateral propulsive contribution is a pure yawing moment. For propeller-driven aircraft, a propulsive rolling moment also contributes to static V_{mc}. To see how aileron- and rudder-limited V_{mc} depend on bank angle for a propeller-driven multiengine airplane, consider the P-38 Lightning shown in Fig. 5.7.4 and described in Examples 5.7.1 and 5.8.1. With the left engine out and full power applied to the right engine, the operating propeller has right-hand rotation at 1,500 rpm and produces rolling and yawing moments of −4,990 and −22,500 ft-lbf, respectively. Results are plotted in Fig. 5.10.5 using the aerodynamic derivatives that were used for Example 5.8.1 combined with aileron and rudder saturation angles of ±20 and ±30 degrees, respectively.

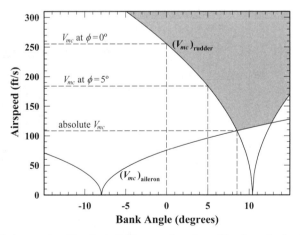

Figure 5.10.5. Aileron- and rudder-limited V_{mc} as a function of bank angle for the P-38 with the right engine producing rolling and yawing moments of −4,990 and −22,500 ft-lbf, respectively.

Important observations can be made by comparing Figs. 5.10.4 and 5.10.5. First, we see that the greater aileron control power of the P-38 produces a flatter aileron-limited curve than that for the F-14. Conversely, the rudder-limited curve is somewhat steeper for the P-38, because its rudder control power is less than that for the F-14. The reader should also note that the aileron-limited curve for the P-38 is shifted well to the left of that for the F-14. This shift results from propeller torque, which produces a propulsive rolling moment to the left. The negative propulsive yawing moment shifts the rudder-limited curve to the right, whereas the negative propulsive rolling moment shifts the aileron-limited curve to the left. This is because $C_{n,\beta}$ is positive while $C_{\ell,\beta}$ is negative.

The direction of propeller rotation can significantly affect static V_{mc}. For example, if the propeller rotation used for Fig. 5.10.5 is reversed to produce a positive propulsive rolling moment of 4,990 ft-lbf while keeping the propulsive yawing moment and all other parameters fixed, the results shown in Fig. 5.10.6 are obtained. Notice that this change in the direction of propeller rotation reduces static V_{mc} with a 5-degree bank from 184 ft/s to 176 ft/s and reduces the absolute static V_{mc} to well below stall. It is interesting to note that the XP-38 prototype was actually built with this propeller rotation.

The difference between the results shown in Figs. 5.10.5 and 5.10.6 was produced by simply changing the sign of the propulsive rolling moment with no change in the propulsive yawing moment. In reality, changing the direction of propeller rotation also affects the propulsive yawing moment as described in Secs. 2.5 and 5.2. At a positive angle of attack, a right-hand-turning propeller produces an inherent yawing moment to the left and a left-hand-turning propeller produces an inherent yawing moment to the right. Thus, because the propeller angle of attack is typically positive at low airspeeds, the magnitude of the net propulsive yawing moment is reduced when the working propeller on the pilot's right has left-hand rotation. When this effect is added to the results shown in Fig. 5.10.6, the net propulsive yawing moment to the left is reduced by about 1,100 ft-lbf and static V_{mc} with a 5-degree bank is further reduced to 167 ft/s.

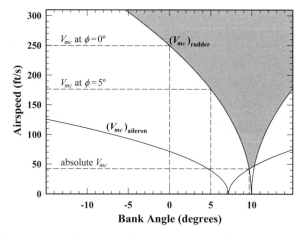

Figure 5.10.6. Aileron- and rudder-limited V_{mc} as a function of bank angle for the P-38 with the right engine producing rolling and yawing moments of $+4,990$ and $-22,500$ ft-lbf, respectively.

The *absolute minimum-control airspeed*, which is shown in Figs. 5.10.4 through 5.10.6, is that for which the aileron and rudder are both saturated simultaneously. For the purpose of evaluating this absolute V_{mc}, Eq. (5.10.1) can be rearranged as

$$
\begin{bmatrix} C_{Y,\beta} & \Delta Y_p + W\phi \\ C_{\ell,\beta} & \Delta \ell_p / b_w \\ C_{n,\beta} & \Delta n_p / b_w \end{bmatrix} \left\{ \begin{array}{c} \beta \\ 1/(\frac{1}{2}\rho V_{mc}^2 S_w) \end{array} \right\} = - \left\{ \begin{array}{c} C_{Y,\delta_a}\delta_a + C_{Y,\delta_r}\delta_r \\ C_{\ell,\delta_a}\delta_a + C_{\ell,\delta_r}\delta_r \\ C_{n,\delta_a}\delta_a + C_{n,\delta_r}\delta_r \end{array} \right\} \qquad (5.10.6)
$$

With the lateral propulsive force and moments known, and both control deflections fixed at saturation, the second and third of these relations are solved for the airspeed and sideslip angle. These results are then used in the first relation to obtain the bank angle. Thus, **the absolute minimum-control airspeed without bank angle restriction is given by**

$$
V_{mc} = \sqrt{\frac{2(C_{\ell,\beta}\Delta n_p - C_{n,\beta}\Delta \ell_p)}{\rho S_w b_w [(C_{n,\beta}C_{\ell,\delta_a} - C_{\ell,\beta}C_{n,\delta_a})\delta_{a_{\text{sat}}} + (C_{n,\beta}C_{\ell,\delta_r} - C_{\ell,\beta}C_{n,\delta_r})\delta_{r_{\text{sat}}}]}} \qquad (5.10.7)
$$

$$
\beta = -\frac{(\Delta C_n)_p + C_{n,\delta_a}\delta_{a_{\text{sat}}} + C_{n,\delta_r}\delta_{r_{\text{sat}}}}{C_{n,\beta}} = -\frac{(\Delta C_\ell)_p + C_{\ell,\delta_a}\delta_{a_{\text{sat}}} + C_{\ell,\delta_r}\delta_{r_{\text{sat}}}}{C_{\ell,\beta}} \qquad (5.10.8)
$$

$$
\phi = -\frac{(\Delta C_Y)_p + C_{Y,\beta}\beta + C_{Y,\delta_a}\delta_{a_{\text{sat}}} + C_{Y,\delta_r}\delta_{r_{\text{sat}}}}{C_W} \qquad (5.10.9)
$$

When the signs are considered, there are four possible cases with both lateral controls at saturation. Absolute static V_{mc} is the smaller of the two real solutions to Eq. (5.10.7).

If the bank angle from Eq. (5.10.9) is less than the limit in the applicable regulation, static V_{mc} can be estimated from Eq. (5.10.7). This is typically the case when V_{mc} from Eqs. (5.10.2) through (5.10.4) is aileron limited at the maximum allowable bank. Notice that for fixed propulsive moments, absolute static V_{mc} is independent of gross weight. However, as the weight decreases, a larger bank angle results from the saturated controls.

The fact that absolute static V_{mc} is independent of gross weight does not eliminate the added danger of control loss at low gross weight, because a larger bank angle is needed to stabilize the airplane. Furthermore, most multiengine airplanes adjust their approach speed to operate at a specified angle of attack and lift coefficient. Thus, lower gross weights permit operation at lower airspeeds. This means that as gross weight is reduced, the allowable approach speed becomes closer to the absolute minimum-control airspeed, because absolute V_{mc} does not decrease with weight as the stall speed does.

Using Eqs. (5.10.7) through (5.10.9) for the F-14B described in Examples 5.9.2 and 5.10.1, at a gross weight of 54,000 pounds with maximum afterburner, the absolute minimum-control airspeed is estimated at 122 knots, with 8.7 degrees of sideslip and 10.3 degrees of bank. Thus, we would predict that even with full afterburner, this airplane would be marginally controllable at airspeeds below the predicted static V_{mc} of 153 knots. It could not be controlled with only 5 degrees of bank. It most likely would not be easy or comfortable, but when an excellent pilot is free to fly the airplane by whatever means deemed necessary, the airplane probably could be controlled at speeds below 153 knots.

Aircraft developers, regulators, and operators have long recognized the importance of static V_{mc} as a critical constraint associated with the asymmetric thrust resulting from engine failure in multiengine airplanes. However, absolute static V_{mc}, which is defined by Eq. (5.10.7), has not typically been a critical concern in the design or operation of conventional piloted aircraft. Concerns for the limitations of human pilots have generally placed it outside the bounds of practical aircraft design space, as is shown in Figs. 5.10.4 through 5.10.6. Nevertheless, absolute static V_{mc} can be important in the design of modern unmanned aerial vehicles (UAV), because autonomous computer-controlled flight systems do not impose the same constraints as those imposed by human pilots.

A special case, where Eq. (5.10.7) is always useful, is in the design of what is called a *three-channel* UAV. Such aircraft have only three controls. Most commonly these are throttle, elevator, and rudder (no ailerons). The only roll control is provided indirectly by the rudder's interaction with wing dihedral. The absence of control surfaces in the wings makes this three-channel design particularly attractive for a small UAV that must be stored compactly and assembled quickly. For such aircraft, V_{mc} can be a critical constraint even for single-engine designs with centerline thrust, due to the contribution of propeller torque. Static V_{mc} is always aileron limited in such three-channel airplanes, because aileron saturation occurs at $\delta_{a_{\text{sat}}} = 0$. In this case, Eqs. (5.10.7) through (5.10.9) must be used to estimate static V_{mc}, because Eqs. (5.10.2) through (5.10.4) are in an indeterminate form for an aircraft without aileron control. This results from the fact that with no aileron control and the propulsive lateral contributions fixed at any given airspeed, only one combination of rudder deflection, sideslip, and bank angle can be used to maintain lateral trim at constant airspeed, heading, and altitude.

Another approach to the three-channel UAV design is where only throttle, aileron, and elevator controls are used (e.g., AeroVironment's *DragonEye*). Obviously in this case, static V_{mc} is rudder limited, because rudder saturation occurs at $\delta_{r_{\text{sat}}} = 0$. Here again, Eqs. (5.10.2) through (5.10.4) are indeterminate and Eqs. (5.10.7) through (5.10.9) must be used to estimate static V_{mc}. For further details on V_{mc} considerations for such aircraft, see Phillips and Niewoehner (2006).

Rudder Sizing

Because minimum-control airspeed is typically the critical constraint on rudder size for multiengine airplanes, in the design of such airplanes we commonly need to estimate the rudder control power that is required to provide a specified static V_{mc}. In solving this problem, we can use the fact that the three rudder control derivatives C_{Y,δ_r}, C_{ℓ,δ_r}, and C_{n,δ_r} are related. When a conventional rudder is deflected, a side-force increment, ΔY_r, is generated on the vertical stabilizer with an aerodynamic center that is aft of the center of gravity by some distance, say l_r. This generates an increment in the yawing moment, $\Delta n_r = -l_r \Delta Y_r$. In a similar manner, the rolling moment increment is $\Delta \ell_r = h_r \Delta Y_r$, where h_r is the distance that the aerodynamic center of the side-force increment is above the center of gravity. Thus, in nondimensional form, increments in the lateral moment coefficients caused by rudder deflection can be related to the increment in the side-force coefficient through simple geometry ratios, $\Delta C_n = -(l_r/b_w)\Delta C_Y$ and $\Delta C_\ell = (h_r/b_w)\Delta C_Y$. From these relations, the change is side-force and rolling moment coefficients with respect to rudder deflection can be written in terms of the rudder control power C_{n,δ_r},

$$C_{Y,\delta_r} = -(b_w/l_r)C_{n,\delta_r} \tag{5.10.10}$$

$$C_{\ell,\delta_r} = -(h_r/l_r)C_{n,\delta_r} \tag{5.10.11}$$

Applying Eqs. (5.10.10) and (5.10.11) to Eq. (5.10.1) yields the trim requirements,

$$\begin{bmatrix} C_{Y,\beta} & C_{Y,\delta_a} & -(b_w/l_r)C_{n,\delta_r} \\ C_{\ell,\beta} & C_{\ell,\delta_a} & -(h_r/l_r)C_{n,\delta_r} \\ C_{n,\beta} & C_{n,\delta_a} & C_{n,\delta_r} \end{bmatrix} \begin{Bmatrix} \beta \\ \delta_a \\ \delta_r \end{Bmatrix} = -\begin{Bmatrix} (\Delta C_Y)_p + C_W\phi \\ (\Delta C_\ell)_p \\ (\Delta C_n)_p \end{Bmatrix} \tag{5.10.12}$$

Assuming that the lateral force and moments produced by the propulsion system are all known, fixing the bank angle at its limiting value, and setting the rudder deflection equal to the desired saturation value, Eq. (5.10.12) can be rearranged to give

$$\begin{Bmatrix} \beta \\ \delta_a \\ C_{n,\delta_r} \end{Bmatrix} = \begin{bmatrix} C_{Y,\beta} & C_{Y,\delta_a} & -(b_w/l_r)\delta_{r\text{sat}} \\ C_{\ell,\beta} & C_{\ell,\delta_a} & -(h_r/l_r)\delta_{r\text{sat}} \\ C_{n,\beta} & C_{n,\delta_a} & \delta_{r\text{sat}} \end{bmatrix}^{-1} \begin{Bmatrix} -(\Delta C_Y)_p - C_W\phi \\ -(\Delta C_\ell)_p \\ -(\Delta C_n)_p \end{Bmatrix} \tag{5.10.13}$$

This linear system can be used to estimate the sideslip angle, aileron deflection, and rudder control power needed to provide a specified static minimum-control airspeed with specified values for the rudder saturation and bank angle limit.

EXAMPLE 5.10.2. Using the same P-38 data given in Examples 5.7.1 and 5.8.1, estimate the rudder control power, C_{n,δ_r}, needed to provide a static minimum-control airspeed at sea level of 100 mph, using no more than 25 degrees of rudder and 5 degrees of bank. At full power and this airspeed, assume that the right-hand propeller is turning at 1,500 rpm, the propulsive efficiency is 65 percent, and $C_{n_p,\alpha} = -0.0089$. Use the relations $C_{Y,\delta_r} = -2.21C_{n,\delta_r}$ and $C_{\ell,\delta_r} = -0.115C_{n,\delta_r}$.

Solution. Following Example 5.7.1, with the right engine at full power and using an airspeed of 100 mph (147 ft/sec) at sea level, we obtain

$$C_L \cong C_W = \frac{W}{\frac{1}{2}\rho V_{mc}^2 S_w} = \frac{15,480}{\frac{1}{2}0.0023769(147)^2 327.5} = 1.841$$

$$\alpha_{0p} = [C_L - (C_L)_{400}]/C_{L,\alpha} = (1.841-0.116)/5.01 = 0.3443 = 19.7°$$

$$\varepsilon_{d0p} = (C_L/C_{L,\alpha})(\varepsilon_{d,\alpha})_p = (1.841/5.01)(-0.175) = -0.0643 = -3.7°$$

$$T_{\text{max}} = \eta_p P_b/V_{mc} = 0.65(1,425\times550)/147 = 3,466 \text{ lbf}$$

$$\Delta n_p = -y_{bp}T_{\text{max}} + \rho(\omega/2\pi)^2 d_p^5 C_{n_p,\alpha}(\alpha_{0p} - \varepsilon_{d0p}) = -8.0(3,466)$$
$$+0.0023769(1,500/60)^2(138/12)^5(-0.0089)[0.3443-(-0.0643)] = -28,815 \text{ ft·lbf}$$

$$\left(\Delta C_n\right)_p \equiv \frac{\Delta n_p}{\frac{1}{2}\rho V_{mc}^2 S_w b_w} = \frac{-28{,}815}{\frac{1}{2}0.0023769(147)^2 327.5(52)} = -0.0659$$

$$\Delta \ell_p = -P_b/\omega_p = -1{,}425\times550/[2\pi(1{,}500/60)] = -4{,}990 \text{ ft}\cdot\text{lbf}$$

$$\left(\Delta C_\ell\right)_p \equiv \frac{\Delta \ell_p}{\frac{1}{2}\rho V_{mc}^2 S_w b_w} = \frac{-4{,}990}{\frac{1}{2}0.0023769(147)^2 327.5(52)} = -0.0114$$

From Eq. (5.10.13) with $\delta_{r_{sat}} = -25° = -0.4363$ and $\phi = 5° = 0.0873$, we obtain

$$\begin{Bmatrix} \beta \\ \delta_a \\ C_{n,\delta_r} \end{Bmatrix} = \begin{bmatrix} C_{Y,\beta} & C_{Y,\delta_a} & -(b_w/l_r)\delta_{r_{sat}} \\ C_{\ell,\beta} & C_{\ell,\delta_a} & -(h_r/l_r)\delta_{r_{sat}} \\ C_{n,\beta} & C_{n,\delta_a} & \delta_{r_{sat}} \end{bmatrix}^{-1} \begin{Bmatrix} -\left(\Delta C_Y\right)_p - C_W\phi \\ -\left(\Delta C_\ell\right)_p \\ -\left(\Delta C_n\right)_p \end{Bmatrix}$$

$$= \begin{bmatrix} -0.512 & 0.0 & -2.21(-0.4363) \\ -0.025 & -0.131 & -0.115(-0.4363) \\ 0.083 & 0.013 & -0.4363 \end{bmatrix}^{-1} \begin{Bmatrix} 0.0-1.841(0.0873) \\ -(-0.0114) \\ -(-0.0659) \end{Bmatrix} = \begin{Bmatrix} 0.033 \\ -0.150 \\ -0.149 \end{Bmatrix}$$

With aileron deflection of −8.6 degrees, this solution is rudder limited. However, a rudder control power of −0.149 is well beyond what was available to P-38 pilots.

Although data used in this example are approximate, the result is reasonable. According to the P-38 *Pilot's Flight Operating Instructions*, on failure of one engine, the pilot must "hold 125 mph or more (at least 160 mph preferred)." If one engine fails below 120 mph the pilot is to "close both throttles and land straight ahead, retracting the landing gear if it is not possible to land on the runway." Historical references indicate that some of the best pilots would not apply full power to the P-38 in single-engine operation at speeds below 150 mph. In 1942, many pilots transitioned to the Lightning with no previous twin-engine experience. Occasionally, pilots were killed when one engine failed on takeoff and the airplane rolled over on its back and into the ground. While otherwise the P-38 had an excellent reputation, some of the top Lightning pilots used a takeoff technique that reflected their distrust in the Allison engines. They would keep the airplane on the ground to 150 mph, even though the manual recommended liftoff at 90 to 100 mph. Usually, no attempt was made to keep the P-38 above V_{mc} on landing with one engine out, because a touchdown speed above 100 mph simply required too much runway. Using the approach procedure recommended for single-engine operation, once below 500 feet the P-38 had to be landed, because the airspeed was too low to make a go-around on one engine.

For conventional piloted aircraft, V_{mc} is governed primarily by the rudder deflection required to trim an airplane against the propulsive yawing moment resulting from engine failure in multiengine airplanes. Still, the reader should remember that a propulsive rolling moment also contributes to V_{mc}. For propeller-driven aircraft with high thrust-to-weight ratio, propeller torque can contribute significantly to V_{mc}. If propeller torque is large enough, V_{mc} can be a consideration even in the design of conventional single-engine aircraft with centerline thrust. This can become more important in the design of small UAVs. Limits on the thrust that can be developed by a propeller are proportional to the propeller disk area and propeller radius. Thus, if s is the length scale for an airplane, these thrust limits are proportional to s^2 and s. Because weight is proportional to s^3, thrust-to-weight ratio limits for propeller-driven aircraft are proportional to s^{-1} and s^{-2}. As airplanes are scaled down in size, very large propeller thrust-to-weight ratios become possible and torque-related V_{mc} issues can become critical.

For any propeller-driven airplane without balanced counterrotating propellers, a rolling moment must be generated by the controls to trim the airplane against the propulsive rolling moment resulting from propeller torque. If propeller torque is large enough to become the primary factor contributing to V_{mc}, then roll control rather than yaw control becomes the critical constraint and V_{mc} is typically aileron limited. For such cases, Eq. (5.10.1) can be rearranged to provide a 3×3 system of equations useful for aileron sizing. This is left as an exercise for the student.

Static V_{mc} discussed in this section is only one of several minimum-control airspeeds. Federal Aviation Regulation (FAR) 23.149 stipulates that V_{mc} may not exceed 1.2 times the stall speed at the maximum takeoff weight. The definition used in the FAR describes a dynamic maneuver from maximum power in which a throttle chop to idle is performed on the critical engine. The pilot must then be capable of neutralizing any resultant rates and returning the airplane to a steady-heading sideslip with no more than 5 degrees of bank angle (though 5 degrees may be exceeded during the dynamic portion of the recovery), no more than 150 pounds of rudder pedal force, and no more than 20 degrees of heading change. Flight test literature describes this as *dynamic* V_{mc} as opposed to the static V_{mc} previously described. The dynamic V_{mc} is more demanding than static V_{mc}, because the rudder must overcome not only the thrust asymmetry, but also any angular rates that may have developed as a result of the pilot's reaction time (2 seconds is assumed in flight testing). Consequently, dynamic V_{mc} should be expected to be a higher airspeed than the static V_{mc}. Dynamic V_{mc} prediction requires high-fidelity man-in-the-loop simulation, since pilot interaction with both the displays and controls is an integral element in the airplane's resultant behavior. The reader must clearly understand which V_{mc} is intended for any specification. For example, the FARs clearly intend dynamic V_{mc} by the included definition, although this terminology is not specifically used.

The roll-control mechanization exerts surprising influence on the dynamic V_{mc} problem. During an engine failure at high power, the resultant roll rate may actually mask the underlying yaw rate, causing the pilot to respond with aileron, rather than rudder. If the ailerons exhibit significant adverse yaw, countering the roll rate with aileron may actually erode control, because the adverse yaw will add to the yawing moment from asymmetric power. For example, dynamic V_{mc} for late-model F/A-18 Hornets was lowered during development by decreasing the aileron authority available to the pilot at certain low-speed conditions, mitigating severe adverse yaw. For further details the reader is referred to Marks, Heller, Traven, and Etz (2000).

Some airplanes also have a ground-minimum-control airspeed, usually denoted V_{mcg}. This is the minimum airspeed at which the rudder has adequate control power to maintain runway centerline within some tolerance, in the event of an engine failure during the takeoff ground roll. While the thrust asymmetry may be eliminated by aborting the takeoff and pulling all engines to idle, there are some situations (dependent on weight, configuration, weather, runway length, field elevation, etc.) in which it might be safer to continue the takeoff and climb to obstacle clearance altitude, followed by an engine-out landing, rather than attempt a high-speed abort. Analytical prediction of V_{mcg} is beyond our scope, due to the complexities of nose-wheel steering, gear dynamics, etc. As with dynamic V_{mc}, prediction of V_{mcg} depends upon high-fidelity simulation, followed by flight-test validation.

Minimum-control airspeed testing is among the most hazardous of flight-test operations because the test team must probe the boundaries of controllability. Due to the strong sensitivity of thrust to density altitude, V_{mc} testing must be done at low altitude with little margin for surprises.

5.11. Longitudinal-Lateral Coupling

In this textbook, consideration of aircraft static stability and trim has been divided into two distinct categories, those associated with longitudinal motion and those associated with lateral motion. The study of most topics within the field of flight mechanics has traditionally been divided along these lines. Longitudinal and lateral motion are usually considered separately and longitudinal-lateral coupling is commonly ignored. Typically, the coupling between the longitudinal and lateral degrees of freedom is very small, and neglecting this coupling is well justified. However, the designer should never lose sight of the fact that longitudinal-lateral coupling can sometimes be very important.

In this chapter we have observed one form of longitudinal-lateral coupling. A rotating propeller produces a yawing moment that varies with angle of attack. This means that lateral trim requirements in yaw can depend on longitudinal trim requirements in pitch. This is only one of many forms of longitudinal-lateral coupling that will be considered as we continue with our study of flight mechanics.

Some forms of longitudinal-lateral coupling were completely unknown prior to the development of high-powered aircraft during and following World War II. On December 10, 1944, a German prototype He-162 jet fighter broke up during a high-speed roll. Later, a similar incident with the British Fairey Delta aircraft resulted in complete control loss but no severe damage to the aircraft. Despite the fact that both incidents were well documented, the aircraft behavior could not be explained on the basis of flight mechanics knowledge available at that time. In both cases, a form of longitudinal-lateral coupling that was not considered in the aircraft design process was at fault.

In the following chapters our study of flight mechanics will continue to be divided along the traditional lines of longitudinal and lateral motion. Initially, longitudinal-lateral coupling will be neglected to simplify the presentation of fundamentals and help the student grasp the underlying concepts more easily. However, **this approach should not be taken as an indication that longitudinal-lateral coupling is never important.**

Most pilots who have flown an airplane with the tendency to yaw to the left at low airspeeds and high angles of attack are aware of at least some aspects of longitudinal-lateral coupling. On the other hand, the effects of longitudinal-lateral coupling have often been neglected by the designer, sometimes at great cost.

5.12. Control Surface Sign Conventions

The control surface sign convention used in this textbook is based on the right-hand rule for control surface rotations. The elevator sign convention is shown in Fig. 4.3.3, that for the rudder is shown in Fig. 5.2.1, and that for the ailerons is shown in Fig. 5.7.1. Downward deflection of the elevator is assumed positive, deflection of the rudder to the left is assumed positive, and aileron deflection is assumed positive when the right aileron is deflected down and the left aileron is deflected up. When this control surface sign convention is combined with the moment sign convention shown in Fig. 5.1.1, the three primary control derivatives (C_{ℓ,δ_a}, C_{m,δ_e}, and C_{n,δ_r}) are all negative. **While the control surface sign convention presented here is probably the most widely used, it is by no means universal.**

A sign convention opposite to that presented in this text is sometimes used for control surface deflections. The argument for using this alternative sign convention is specifically to make the three primary control derivatives positive. For example, since the ailerons are intended to create a rolling moment, it is argued that positive aileron deflection should produce a positive rolling moment. Similar arguments are made for the elevator and rudder. While this argument makes sense, sign conventions are clearly arbitrary and often dictated more by history than by logic.

Any aeronautical engineer will almost assuredly encounter the use of different control surface sign conventions. The author has actually seen different control surface sign conventions used for two different airplanes within the same textbook. With only a little thought, the student should not find this inconsistency too confusing. By simply examining the signs of the primary control derivatives, it is easy to tell what sign convention has been used.

Because there is no universally accepted control surface sign convention, control surface deflections are commonly described as being right or left and up or down. *Right aileron* is always understood to mean an aileron deflection that will tend to bank the airplane into a right turn. *Left aileron* will bank the airplane to the left. Similarly, *right rudder* is always understood to be a rudder deflection that tends to yaw the airplane's nose to the right, and *left rudder* will yaw the nose to the left. *Up elevator* is the elevator deflection that will pitch the nose up, and *down elevator* will pitch the nose down. If you wish to avoid any possibility of confusion, this terminology can be used.

5.13. Problems

5.1. The airplane described in Example 5.2.1 is to be modified to increase the yaw stability derivative to 0.10. As much as possible we would like to leave the other characteristics of the airplane unchanged. What would you suggest as the least expensive design change that would produce the desired result? Your suggestion should be quantitative.

5.2. The vertical stabilizer of the airplane in problem 4.2 is to be sized to give a contribution to the yaw stability derivative of 0.10. Suggest a preliminary design and size for the vertical stabilizer. Use the aircraft dimensions and operating conditions specified in problem 4.1 and clearly state all assumptions.

5.3. During its useful life the British Spitfire underwent many design changes. The first Mark I was powered by a 1,030-horsepower Merlin II engine. Later, a 2,375-horsepower Griffon 67 engine powered the Mark XVIII. Along the way, several changes in the tail were made to accommodate this growth in power. For the Mark XVIII shown in Figs. 4.12.1 and 4.12.2, estimate the vertical tail volume ratio that would be required to give the airplane a yaw stability derivative of 0.10. Use the data given in problems 4.8, 4.10, and 4.11. Assume a lift slope for the vertical stabilizer of 3.50 and clearly state all other assumptions.

5.4. The 2,375-horsepower Griffon engine that powers the Spitfire Mark XVIII turns the 132-inch, five-blade, constant-speed Rotol propeller shown in Fig. 5.13.1. At full power, the propeller turns 1,400 rpm. While taxiing with the tail wheel on the ground, the propeller angle of attack is 11 degrees relative to the freestream. Assume that the pilot uses full power for a takeoff run and raises the tail at an airspeed of 55 mph. At this operating condition, the **magnitude** for the change in the propeller yawing moment coefficient with respect to propeller angle of attack is 0.0785. Estimate the rudder deflection required to trim the airplane for no sideslip at the instant that the tail wheel leaves the ground. Use the data given in problems 4.8, 4.10, and 4.11; neglect the slipstream effects; and assume the following properties for the vertical tail:

Figure 5.13.1. The five-blade, 132-inch, constant-speed propeller on the Spitfire Mark XVIII. (Photograph by Barry Santana)

$$S_v = 21 \text{ ft}^2, \quad C_{L_v,\alpha} = 3.50, \quad l_v = 18.16 \text{ ft}, \quad \eta_v = 1.0, \quad (\varepsilon_{s,\beta})_v = 0.0,$$

$$\bar{c}_v = 3.8 \text{ ft}, \quad \varepsilon_r = 0.60, \quad C_{m_v,\delta_r} = -0.58$$

Based on the photograph shown in Fig. 5.13.1, is right rudder or left rudder required? Explain how you arrived at your conclusion. Do you expect your estimate for rudder deflection to be low or high? What conclusions can you draw from the result predicted?

5.5. For the wing of the Spitfire Mark XVIII described in problem 4.8, use the method presented in Sec. 5.3 to estimate the sidewash gradient in the aircraft plane of symmetry, 3.0 feet above the plane of the wingtip vortices, and 18.0 feet aft of the wing quarter chord.

5.6. Repeat the calculations in problem 5.5 for an unswept rectangular wing of the same span and planform area. Explain any difference that you observe.

5.7. One way to increase the yaw stability of an airplane is to lengthen the tail boom. In addition to increasing the moment arm for the vertical stabilizer, this also changes the sidewash gradient. Repeat the calculations in problem 5.5 for an increase of 25 percent in the distance aft of the wing quarter chord. Explain how this would affect the trade-off between tail boom length and vertical-stabilizer area. How would this trade-off change for a rectangular wing? How would this trade-off change if the height of the vertical stabilizer were increased?

5.8. Airplanes are sometimes designed with twin vertical stabilizers symmetrically offset from the midspan, as shown in Fig. 5.3.5. The sidewash gradient for such a design is significantly different from that of a single vertical stabilizer placed directly aft of the wing midspan. Repeat the calculations in problem 5.5 for a spanwise position that is 30 percent of the distance between the midspan and the wingtip. How does this affect the total vertical-stabilizer area needed to stabilize the airplane, all else being equal?

5.9. Repeat the calculations in problem 5.5 for the case where the wingspan is reduced by 25 percent while the wing area and all other parameters remain unchanged. How would this affect the yaw stability of the airplane if the vertical-stabilizer area and tail boom length were unchanged?

5.10. Consider a conventional tail with the horizontal stabilizer below the vertical stabilizer. Both stabilizers are unswept, rectangular, and have the same chord. The horizontal and vertical stabilizers have aspect ratios of 4.0 and 2.0, respectively. Using Fig. 5.4.2, estimate the lift slope for the vertical stabilizer.

5.11. Use a numerical lifting-line solution to estimate the lift slope for the vertical stabilizer in problem 5.10, with and without the horizontal stabilizer present.

5.12. The conventional aft tail described in problem 5.10 is modified to include taper, sweep, and dihedral. The taper ratio for both stabilizers is changed to 0.50. The quarter-chord sweep is changed so that the trailing edges remain unswept. The horizontal stabilizer is given 3.0 degrees of dihedral. The aspect ratio and planform area of each stabilizer remain unchanged. Use a numerical lifting-line solution to estimate the lift slope for the vertical stabilizer, with and without the horizontal stabilizer present.

5.13. Using Eq. (5.3.7) to approximate the sidewash gradient, estimate the total yaw stability derivative for the Spitfire Mark XVIII shown in Fig. 4.12.1 and described in problems 4.8, 4.10, and 4.11. Assume an airspeed of 200 mph at standard sea level, include the effects of the fuselage and propeller, and use the following properties for the vertical stabilizer:

$$S_v = 21 \, \text{ft}^2, \quad \bar{c}_v = 3.8 \, \text{ft}, \quad C_{L_v,\alpha} = 3.50, \quad l_v = 18.16 \, \text{ft}, \quad h_v = 3.0 \, \text{ft}, \quad \eta_v = 1.0$$

5.14. As a first approximation for the contribution of a swept wing to the yaw stability derivative of an airplane, we can assume that the wing section drag coefficient is constant. Using this assumption in Eq. (5.2.32), obtain an expression for the contribution of a swept tapered wing to the yaw stability derivative.

5.15. Using the result obtained in problem 5.14, estimate the wing contribution to the yaw stability derivative of the small jet transport described in problem 4.16. Assume a constant section drag coefficient of 0.02.

5.16. Using Eq. (5.3.7) to approximate the sidewash gradient, estimate the total yaw stability derivative for the small jet transport described in problems 4.16, 4.18, 4.19, and 5.15. Assume an airspeed of 400 mph at 30,000 feet, include the effects of the fuselage and thrust deck, and use the following properties for the vertical stabilizer:

$$S_v = 138 \, \text{ft}^2, \quad \bar{c}_v = 6.9 \, \text{ft}, \quad C_{L_v,\alpha} = 3.40, \quad l_v = 43.2 \, \text{ft}, \quad h_v = 7.0 \, \text{ft}, \quad \eta_v = 1.0$$

5.17. Using Eq. (5.3.7) to approximate the sidewash gradient, estimate the total yaw stability derivative for the turboprop transport shown in Fig. 4.12.3 and described in problems 4.32, 4.34, and 4.35. Assume an airspeed of 300 mph at standard sea level, include the effects of the fuselage, nacelles, and propeller, and use the following properties for the vertical stabilizer:

$$S_v = 279 \, \text{ft}^2, \quad \bar{c}_v = 12.0 \, \text{ft}, \quad C_{L_v,\alpha} = 3.50, \quad l_v = 43.3 \, \text{ft}, \quad h_v = 15.1 \, \text{ft}, \quad \eta_v = 1.0$$

5.18. Using Eq. (5.3.7) to approximate the sidewash gradient, estimate the total roll stability derivative for the Spitfire Mark XVIII shown in Fig. 4.12.1 and described in problems 4.8, 4.10, 4.11, and 5.13. Assume a wing dihedral angle of 4.8 degrees.

5.19. As a first approximation for the contribution of wing sweep to the roll stability derivative of an airplane, we could assume that the wing section circulation does not change with either β or Λ. With this assumption in Eq. (5.6.20), use the classical infinite series solution to Prandtl's lifting-line equation to obtain an expression for the contribution of a swept tapered wing with no dihedral to the roll stability derivative.

5.20. Using the result from problem 5.19, estimate the wing sweep contribution to the roll stability derivative for a 45-degree swept wing having linear taper with a root chord of 20 feet, a tip chord of 8 feet, and a span of 70 feet. The wing has a thin symmetric airfoil, no twist, and is operating at a lift coefficient of 0.50.

5.21. Using a numerical lifting-line solution, estimate the wing sweep contribution to the roll stability derivative for the wing described in problem 5.20. Compare this result with that obtained from the approximate solution in problem 5.20.

5.22. Using the result from problem 5.19, estimate the wing sweep contribution to the roll stability derivative for a 45-degree swept wing having linear taper with a root chord of 14 feet, a tip chord of 14 feet, and a span of 70 feet. The wing has a thin symmetric airfoil, no twist, and is operating at a lift coefficient of 0.50.

5.23. Using a numerical lifting-line solution, estimate the wing sweep contribution to the roll stability derivative for the wing described in problem 5.22. Compare this result with that obtained from the approximate solution in problem 5.22.

5.24. Using Eq. (5.3.7) to approximate the sidewash gradient, estimate the total roll stability derivative for the small jet transport described in problems 4.16, 4.18, 4.19, and 5.16. Assume a T-tail and a low wing with 4.2 degrees of dihedral.

5.25. Using Eq. (5.3.7) to approximate the sidewash gradient, estimate the total roll stability derivative for the turboprop transport shown in Fig. 4.12.3 and described in problems 4.32, 4.34, 4.35, and 5.17. Assume no wing dihedral.

5.26. The XP-38 prototype for the P-38 Lightning had propellers that turned opposite to those of the later production models. Repeat the computations in Example 5.7.1 for the case where the airplane's propellers turn in the opposite direction and all other conditions remain unchanged. Based on a comparison between the results of these two computations and the photograph of the production model that is shown in Fig. 5.13.2, can you suggest any reason why the direction of propeller rotation may have been changed in going from the prototype to the production model?

5.27. The P-38 Lightning described in Examples 5.7.1 and 5.8.1 is landing at sea level with an approach speed of 200 ft/sec and the right engine inoperative. Assuming that full power is applied to the left engine to abort the landing, use the small-

Figure 5.13.2. A production model P-38 Lightning, showing the right-hand-turning propeller on the pilot's right and the left-hand-turning propeller on the pilot's left.

angle lateral trim relations to estimate the aileron and rudder deflections as well as the sideslip angle required to maintain heading with the wings level. Use the same aerodynamic derivatives given in Examples 5.7.1 and 5.8.1. At full power and this airspeed the left-hand-turning propeller is rotating at 1,500 rpm, the propulsive efficiency is 70 percent, and $C_{n_p,\alpha} = 0.0089$.

5.28. The P-38 described in Examples 5.7.1 and 5.8.1 is landing at sea level with an approach speed of 200 ft/sec and the right engine inoperative. If full power is applied to the left engine, use the small-angle lateral trim relations to estimate the aileron and rudder deflections as well as the sideslip angle required to maintain heading with a 5-degree bank to the left. Use the same aerodynamic derivatives given in Examples 5.7.1 and 5.8.1. At full power and this airspeed the left-hand-turning propeller is rotating at 1,500 rpm, the propulsive efficiency is 70 percent, and $C_{n_p,\alpha} = 0.0089$. Repeat the computations for opposite propeller rotation.

5.29. Estimate sea-level static V_{mc} for the P-38 described in Examples 5.7.1 and 5.8.1. Use the data and aerodynamic derivatives given in Examples 5.7.1 and 5.8.1 and assume that at full power the operating engine contributes a total yawing moment of 22,500 ft-lbf, a rolling moment of 4,990 ft-lbf, and no side force. Use a bank angle limit of 5 degrees and effective saturation angles for the aileron and rudder of 20 and 30 degrees, respectively. Also determine the change in β, δ_a, and δ_r with respect to bank angle at the minimum-control airspeed.

5.30. Repeat the calculation of sea-level static V_{mc} in problem 5.29 for a case where the ailerons are partially jammed. Use an effective saturation angle for the ailerons of 5 degrees, with all other data the same as those used for problem 5.29.

5.31. For the F-14B described in Examples 5.9.2 and 5.10.1, at a gross weight of 54,000 lbf estimate the absolute V_{mc} at sea level with no bank angle restriction. Use the data and aerodynamic derivatives given in Examples 5.9.2 and 5.10.1. Also determine the required sideslip and bank angles.

5.32. (*Contributed by Captain Robert J. Niewoehner, Ph.D., Aerospace Engineering Department, United States Naval Academy*) The F-14A is equipped with TF-30 engines. Each engine develops 19,500 lbf of thrust with maximum afterburner at sea level. Estimate the rudder control power, C_{n,δ_r}, required to trim the airplane for single-engine operation with maximum afterburner at sea level, using no more than 25 degrees of rudder and 5 degrees of bank. Use a gross weight of 54,000 pounds, an airspeed of 135 knots, and the other data and aerodynamic derivatives given in Example 5.9.2. Also determine the change in β, δ_a, and δ_r with bank angle. Assume that $C_{Y,\delta_r} = -2.7 C_{n,\delta_r}$ and $C_{\ell,\delta_r} = -0.14 C_{n,\delta_r}$.

5.33. Develop a closed-form analytical expression that can be used to estimate the rudder control power, C_{n,δ_r}, required to provide a specified absolute minimum-control airspeed with specified values for the aileron and rudder saturation angles. Use the result to estimate the rudder control power needed to provide an absolute minimum-control airspeed of 100 knots at sea level for the F-14B Tomcat described in Examples 5.9.2 and 5.10.1, using no more than 25 degrees of rudder and 15 degrees of aileron at a gross weight of 54,000 lbf with maximum afterburner. Assume that $C_{Y,\delta_r} = -2.7 C_{n,\delta_r}$ and $C_{\ell,\delta_r} = -0.14 C_{n,\delta_r}$.

5.34. A small three-channel UAV has no ailerons. Roll control is provided by deflecting the rudder to generate sideslip, which interacts with wing dihedral to produce roll. Neglect slipstream effects and any yawing moment produced by the single centered propeller and use the following data to estimate static V_{mc} at sea level.

$$S_w = 5.88 \text{ ft}^2, \quad b_w = 7.43 \text{ ft}, \quad C_{L,\alpha} = 5.58, \quad W = 9.0 \text{ lbf},$$
$$d_p = 20/12 \text{ ft}, \quad (\omega/2\pi)_p = 3,100 \text{ rpm}, \quad P_b = 0.95 \text{ hp},$$
$$C_{Y,\beta} = -0.720, \quad C_{\ell,\beta} = -0.065, \quad C_{n,\beta} = 0.177,$$
$$C_{Y,\delta_r} = 0.231, \quad C_{\ell,\delta_r} = 0.014, \quad C_{n,\delta_r} = -0.074, \quad \delta_{r_{sat}} = \pm 25°$$

5.35. For the small three-channel UAV in problem 5.34 estimate static V_{mc} at sea level, including the effects of the propeller yawing moment. The single propeller axis is centered and aligned with the direction of flight for 100 ft/sec at sea level. At low airspeeds, the propeller's yawing moment coefficient varies linearly with propeller advance ratio. Use $C_{n_p,\alpha} = -0.035 J$ and $(\varepsilon_{d,\alpha})_p = -0.085$.

5.36. If ailerons are added to the UAV in problem 5.35, estimate the aileron power, C_{ℓ,δ_a}, needed to reduce static V_{mc} at sea level to 30 ft/sec, using no more than 15 degrees aileron and 5 degrees bank. Use $C_{Y,\delta_a} = 0.45 C_{\ell,\delta_a}$ and $C_{n,\delta_a} = -0.11 C_{\ell,\delta_a}$. Plot the resulting aileron- and rudder-limited V_{mc} as a function of bank angle.

5.37. Consider the conventional tail shown in Fig. 5.5.9. The horizontal stabilizer is rectangular with an aspect ratio of 4.0 and a planform area of 36 ft^2. The vertical stabilizer has the same constant chord with a planform area of 12 ft^2. Both the horizontal and vertical stabilizers have thin symmetric airfoil sections with no twist. The sweep for both surfaces is zero and the horizontal stabilizer has no dihedral. Use the numerical lifting-line method described in Sec. 1.9 to plot the horizontal and vertical stabilizer planform areas required to maintain constant total lift slope and side-force derivative as a function of horizontal-stabilizer dihedral angle. Also plot the total planform area of both stabilizers combined. Hold the chord constant and vary only the root-to-tip length for each surface. Compare the results with Fig. 5.5.14.

5.38. Consider a conventional tail having a horizontal stabilizer of aspect ratio 4.0 with a planform area of 36 ft^2. The vertical stabilizer has the same root chord with a planform area of 24 ft^2. Both the horizontal and vertical stabilizers have taper ratios of 0.5, thin symmetric airfoil sections, and no twist. The quarter-chord sweep for both stabilizers is zero and the horizontal stabilizer has no dihedral. Use the numerical lifting-line method described in Sec. 1.9 to plot the horizontal and vertical stabilizer planform areas required to maintain constant total lift slope and side-force derivative as a function of horizontal-stabilizer dihedral angle. Also plot the total planform area of both stabilizers combined. Hold the root chord and taper ratio constant and vary only the root-to-tip length for each surface. Compare the results with Fig. 5.5.15.

5.39. The wingtip vortices trailing behind a large airplane can cause loss of control or structural damage to another airplane flying into this wake. Near a busy airport, this can present a serious safety hazard. For this reason, the FAA has developed separation requirements for aircraft of different size. Another situation that presents a similar hazard is formation flying. Pilots engaged in formation flying must take care to avoid flying directly through the core of a wingtip vortex shed from one of the companion aircraft. On the other hand, these wingtip vortices can be used to advantage in formation flying if proper position is maintained. While downwash is induced between the wingtip vortices, upwash is induced outboard from the wingtips of any airplane in flight. Another airplane flying slightly aft and outboard from the wingtips of a lead airplane will be flying in a region of constant upwash. This upwash decreases induced drag and can significantly reduce the fuel consumed during a long flight for a large formation of airplanes. Migrating ducks and geese use this same strategy. Use the vortex model presented in Sec. 4.5 to develop an expression for the upwash angle that is induced outboard from the wingtip vortex shed from one semispan of an airplane's wing. Express this upwash angle as a function of the distance aft of the wing, the distance outboard from the center of shed vorticity, and any other parameters that you find to be important.

Chapter 6
Aircraft Controls and Maneuverability

6.1. Longitudinal Control and Maneuverability

From the material presented in the previous chapters, the student should now have a good understanding of the trim and static stability requirements for nonaccelerating flight. Both trim and static stability were defined only for an equilibrium flight condition. While the operation of an airplane in equilibrium flight is a very important consideration in aircraft design, it is also important to understand how aircraft controls are used to accelerate an airplane from one equilibrium state to another. The force that provides nearly all of the acceleration typically used to maneuver an airplane is the lift. For the most part, the aerodynamic controls of an airplane are designed to change the magnitude and direction of this force. Angle of attack governs the magnitude of the lift, and the elevator controls the angle of attack. The ailerons are used primarily to control the direction of the lift vector and the rudder is normally used only to coordinate the maneuver and eliminate the sideslip. Since the elevator provides pitch control, it controls the magnitude of the lift and is used in almost all airplane maneuvers.

In Chapter 4 we learned how the elevator is used to trim an airplane. To determine the elevator angle required for trim, we set the summation of normal forces and pitching moments both equal to zero. These two coupled equations are then solved for the equilibrium angle of attack and elevator deflection. Clearly, the elevator must be large enough to trim the airplane for any possible operating condition, with or without flaps deployed. However, simply being able to trim the airplane is not sufficient. The elevator must also be large enough to maneuver the airplane adequately over the full range of operating conditions as well.

While the elevator trim analysis was based on static equilibrium, for maneuverability analysis we consider accelerating flight. The elevator is used to accelerate the airplane in a direction normal to the flight path. Consider, for example, the constant-speed pull-up maneuver that is shown in Fig. 6.1.1. For this maneuver, the airspeed, V, is held constant but the airplane has an acceleration, a_N, that is normal to the flight path. This normal acceleration requires lift in excess of the weight. From Newton's second law we can write

$$\frac{W}{g} a_N = L - W \tag{6.1.1}$$

or

$$a_N = \left(\frac{L}{W} - 1\right) g \tag{6.1.2}$$

The ratio of the lift generated during an accelerating maneuver to the weight of the airplane is the load factor that was defined in Chapter 3. That is,

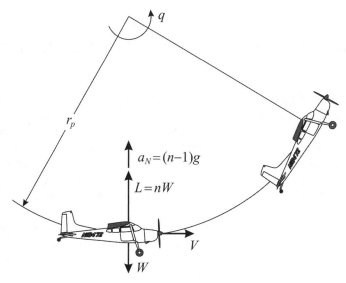

Figure 6.1.1. Constant-speed pull-up maneuver.

$$n \equiv \frac{L}{W} \tag{6.1.3}$$

Recall that although the load factor is dimensionless, it is usually stated in terms of g's. For example, a maneuver that produces a lift force that is five times the aircraft weight is said to be a "5-g maneuver." Using the definition of load factor with Eq. (6.1.2), the normal acceleration for the pull-up maneuver can be written

$$a_N = (n-1)g \tag{6.1.4}$$

This normal acceleration could also be called *centripetal acceleration*, because it acts in the direction normal to the flight path and toward the flight-path center of curvature.

The centripetal acceleration can also be expressed in terms of the radius of curvature of the flight path and the airplane's forward airspeed,

$$a_N = \frac{V^2}{r_p} \tag{6.1.5}$$

where r_p is the radius of curvature for the pull-up maneuver. Using Eq. (6.1.4) to express the centripetal acceleration in terms of the load factor, Eq. (6.1.5) is readily solved for the flight-path radius of curvature,

$$r_p = \frac{V^2}{a_N} = \frac{V^2}{(n-1)g} \tag{6.1.6}$$

which can also be related to the airplane's forward airspeed and angular velocity.

As the airplane travels this curved path it rotates in pitch, as can be seen in Fig. 6.1.1. This is where the elevator comes in. Elevator deflection is required to initiate and sustain this rotation. The rate of pitching rotation is called the *pitch rate* and is commonly given the symbol, q. If the airplane were to travel a curved path of radius r_p at constant speed for one complete revolution, the total rotation angle would be 2π radians. The distance traveled would be $2\pi r_p$ and the time of rotation would be $2\pi r_p/V$. Since the rotation rate is the rotation angle divided by the rotation time, the pitch rate is the forward velocity divided by the radius of curvature,

$$q = \frac{2\pi}{2\pi r_p/V} = \frac{V}{r_p} \qquad (6.1.7)$$

Using Eq. (6.1.6) to eliminate the radius of curvature from Eq. (6.1.7), the pitch rate for a constant-speed pull-up maneuver is expressed in terms of the airspeed and load factor,

$$q = \frac{a_N}{V} = \frac{(n-1)g}{V} \qquad (6.1.8)$$

As the airplane negotiates the pull-up maneuver, the center of gravity has a translational velocity, V, and the airplane is rotating about its center of gravity with an angular velocity, q. Since the airplane is traveling a curved path through the air, the relative airflow past the airplane is also curved, as shown in Fig. 6.1.2. This curvature changes the pressure distribution and the aerodynamic forces and moments acting on the airplane. As can be seen in Fig. 6.1.2, all points aft of the center of gravity experience an increase in angle of attack, while all points forward of the center of gravity experience a decrease in angle of attack. While this alters the pressure distribution over the entire airplane, by far the most significant effects on the aerodynamic forces and moments result from the change in the angle of attack for the horizontal surface and the wing. From Fig. 6.1.2, the change in angle of attack for the horizontal surface can be written as

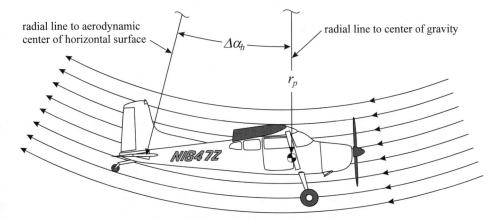

Figure 6.1.2. Airflow relative to the airplane during a pull-up maneuver.

$$\Delta\alpha_h = \tan^{-1}\left(\frac{l_h}{r_p}\right) \cong \frac{l_h}{r_p} \tag{6.1.9}$$

where again l_h is the distance that the aerodynamic center of the horizontal surface is aft of the center of gravity and r_p is the radius of curvature. For an aft tail, this change in angle of attack is positive, and for a forward canard, the change is negative. From Eq. (6.1.7), the radius of curvature is the forward velocity divided by the pitch rate. Thus, Eq. (6.1.9) can be written

$$\Delta\alpha_h = \frac{l_h q}{V} \tag{6.1.10}$$

Unlike the angle of attack, the pitch rate, q, is not dimensionless. Thus, in order to maintain all variables in dimensionless form, it has been standard convention to define the **traditional dimensionless pitch rate** to be

$$\bar{q} \equiv \frac{q\,\bar{c}_w/2}{V} \tag{6.1.11}$$

The choice of $\bar{c}_w/2$ or $\bar{c}_{mac}/2$ as the reference length is the usual convention. However, this is arbitrary and has no real physical significance. Other reference lengths could and have been used. Using Eq. (6.1.11) in Eq. (6.1.10), we can write

$$\Delta\alpha_h = \frac{2l_h}{\bar{c}_w}\bar{q} \tag{6.1.12}$$

Similarly, for the wing we can write

$$\Delta\alpha_w = \frac{2l_w}{\bar{c}_w}\bar{q} \tag{6.1.13}$$

These changes in angle of attack affect both the lift and the pitching moment for the airplane. In the range of linear lift, the lift coefficients for the wing and the horizontal surface are, respectively,

$$C_{L_w} = C_{L_w,\alpha}(\alpha + \Delta\alpha_w + \alpha_{0w} - \alpha_{L0w}) \tag{6.1.14}$$

$$C_{L_h} = C_{L_h,\alpha}(\alpha + \Delta\alpha_h + \alpha_{0h} - \alpha_{L0h} - \varepsilon_d + \varepsilon_e\delta_e) \tag{6.1.15}$$

Thus, including the effects of the pitch rate from Eqs. (6.1.12) and (6.1.13), the total lift coefficient for the two surfaces combined is

$$C_L = C_{L_w,\alpha}\left(\alpha + \frac{2l_w}{\bar{c}_w}\bar{q} + \alpha_{0w} - \alpha_{L0w}\right)$$
$$+ \frac{S_h}{S_w}\eta_h C_{L_h,\alpha}\left(\alpha + \frac{2l_h}{\bar{c}_w}\bar{q} + \alpha_{0h} - \alpha_{L0h} - \varepsilon_d + \varepsilon_e\delta_e\right) \tag{6.1.16}$$

and **assuming no vertical offsets**, the total moment coefficient is approximated as

$$
\begin{aligned}
C_m = {} & C_{m_w} - \frac{l_w}{\bar{c}_w} C_{L_w,\alpha} \left(\alpha + \frac{2l_w}{\bar{c}_w} \bar{q} + \alpha_{0w} - \alpha_{L0w} \right) \\
& + \frac{S_h \bar{c}_h}{S_w \bar{c}_w} \eta_h (C_{m_h 0} + C_{m_h,\delta_e} \delta_e) \\
& - \frac{S_h l_h}{S_w \bar{c}_w} \eta_h C_{L_h,\alpha} \left(\alpha + \frac{2l_h}{\bar{c}_w} \bar{q} + \alpha_{0h} - \alpha_{L0h} - \varepsilon_d + \varepsilon_e \delta_e \right)
\end{aligned}
\tag{6.1.17}
$$

These two equations can be written as

$$
C_L = C_{L0} + C_{L,\alpha} \, \alpha + C_{L,\bar{q}} \, \bar{q} + C_{L,\delta_e} \, \delta_e
\tag{6.1.18}
$$

$$
C_m = C_{m0} + C_{m,\alpha} \, \alpha + C_{m,\bar{q}} \, \bar{q} + C_{m,\delta_e} \, \delta_e
\tag{6.1.19}
$$

where

$$
C_{L0} = C_{L_w,\alpha} (\alpha_{0w} - \alpha_{L0w}) + \frac{S_h}{S_w} \eta_h C_{L_h,\alpha} (\alpha_{0h} - \alpha_{L0h} - \varepsilon_{d0})
\tag{6.1.20}
$$

$$
C_{L,\alpha} = C_{L_w,\alpha} + \frac{S_h}{S_w} \eta_h C_{L_h,\alpha} (1 - \varepsilon_{d,\alpha})
\tag{6.1.21}
$$

$$
C_{L,\bar{q}} = 2 \frac{l_w}{\bar{c}_w} C_{L_w,\alpha} + 2 \frac{S_h l_h}{S_w \bar{c}_w} \eta_h C_{L_h,\alpha}
\tag{6.1.22}
$$

$$
C_{L,\delta_e} = \frac{S_h}{S_w} \eta_h C_{L_h,\alpha} \varepsilon_e
\tag{6.1.23}
$$

$$
\begin{aligned}
C_{m0} = {} & C_{m_w} - \frac{l_w}{\bar{c}_w} C_{L_w,\alpha} (\alpha_{0w} - \alpha_{L0w}) + \frac{S_h \bar{c}_h}{S_w \bar{c}_w} \eta_h C_{m_h 0} \\
& - \frac{S_h l_h}{S_w \bar{c}_w} \eta_h C_{L_h,\alpha} (\alpha_{0h} - \alpha_{L0h} - \varepsilon_{d0})
\end{aligned}
\tag{6.1.24}
$$

$$
C_{m,\alpha} = -\frac{l_w}{\bar{c}_w} C_{L_w,\alpha} - \frac{S_h l_h}{S_w \bar{c}_w} \eta_h C_{L_h,\alpha} (1 - \varepsilon_{d,\alpha})
\tag{6.1.25}
$$

$$
C_{m,\bar{q}} = -2 \frac{l_w^2}{\bar{c}_w^2} C_{L_w,\alpha} - 2 \frac{S_h l_h^2}{S_w \bar{c}_w^2} \eta_h C_{L_h,\alpha}
\tag{6.1.26}
$$

$$
C_{m,\delta_e} = \frac{S_h \bar{c}_h}{S_w \bar{c}_w} \eta_h C_{m_h,\delta_e} - \frac{S_h l_h}{S_w \bar{c}_w} \eta_h C_{L_h,\alpha} \varepsilon_e
\tag{6.1.27}
$$

The coefficient $C_{m,\bar{q}}$ is the change in the pitching moment coefficient with respect to the traditional dimensionless pitch rate. This is commonly referred to as the *pitch-damping derivative*. Notice that in unstalled flight, this derivative is always negative. In the absence of stall, the only parameters in Eq. (6.1.26) that can ever be negative are l_w and l_h. Since both of these parameters are squared in Eq. (6.1.26), the pitch-damping derivative is negative. This means that the contribution to the pitching moment that results from the pitch rate always opposes the pitching motion. A positive pitch rate will produce a negative pitching moment, and a negative pitch rate will produce a positive pitching moment.

When an airplane is flying in trimmed flight, the pitch rate is zero. As the pilot first introduces an elevator deflection to initiate a pull-up maneuver, a nonzero pitching moment is generated about the CG. This causes acceleration in pitch, which in turn produces an increasing pitch rate. As the pitch rate increases, the pitching moment decreases as a result of pitch damping. The acceleration in pitch quickly results in a pitch rate that returns the pitching moment to zero. At this point, the negative pitching moment that results from the positive pitch rate exactly balances the positive pitching moment produced by the added elevator deflection. Since there is no longer an imbalance in the total pitching moment, the pitch rate cannot increase farther without additional elevator deflection.

As a measure of the airplane's maneuverability in pitch, we wish to predict the maximum normal acceleration that could be sustained in a constant-speed pull-up maneuver with a given elevator deflection. To this end, we apply the definition of the traditional dimensionless pitch rate from Eq. (6.1.11) to the pitch rate relation expressed in Eq. (6.1.8). This gives

$$\bar{q} = (n-1)\frac{g\bar{c}_w}{2V^2} \tag{6.1.28}$$

This equation relates the maximum dimensionless pitch rate to the load factor for a constant-speed pull-up maneuver. From the definition of load factor, the lift coefficient for the maneuver is

$$C_L \equiv \frac{L}{\frac{1}{2}\rho V^2 S_w} = \frac{L}{W}\frac{W}{\frac{1}{2}\rho V^2 S_w} = n C_W \tag{6.1.29}$$

where C_W is called the *weight coefficient* and is defined as

$$C_W \equiv \frac{W}{\frac{1}{2}\rho V^2 S_w} \tag{6.1.30}$$

Maximum normal acceleration for any given elevator deflection occurs at maximum pitch rate, when the net pitching moment is zero. Thus, using Eqs. (6.1.28) and (6.1.29) in Eqs. (6.1.18) and (6.1.19) while setting the total pitching moment coefficient to zero gives

$$C_{L,\alpha}\,\alpha + C_{L,\delta_e}\,\delta_e = nC_W - C_{L0} - C_{L,\bar{q}}\,(n-1)\frac{g\bar{c}_w}{2V^2} \tag{6.1.31}$$

$$C_{m,\alpha}\,\alpha + C_{m,\delta_e}\,\delta_e = -C_{m0} - C_{m,\bar{q}}\,(n-1)\frac{g\bar{c}_w}{2V^2} \tag{6.1.32}$$

These two equations are readily solved for the **angle of attack and elevator deflection required for any given load factor in the absence of vertical offsets,**

$$\alpha = \frac{(nC_W - C_{L0})C_{m,\delta_e} + C_{L,\delta_e}\,C_{m0} + (C_{L,\delta_e}\,C_{m,\bar{q}} - C_{L,\bar{q}}\,C_{m,\delta_e})(n-1)(g\bar{c}_w/2V^2)}{C_{L,\alpha}\,C_{m,\delta_e} - C_{L,\delta_e}\,C_{m,\alpha}}$$

$$\tag{6.1.33}$$

$$\delta_e = -\frac{(nC_W - C_{L0})C_{m,\alpha} + C_{L,\alpha}\,C_{m0} + (C_{L,\alpha}\,C_{m,\bar{q}} - C_{L,\bar{q}}\,C_{m,\alpha})(n-1)(g\bar{c}_w/2V^2)}{C_{L,\alpha}\,C_{m,\delta_e} - C_{L,\delta_e}\,C_{m,\alpha}}$$

$$\tag{6.1.34}$$

Since trimmed flight can be viewed as a 1-g maneuver, the elevator deflection required for trim can be found as a special case of Eq. (6.1.34) with the load factor set to 1.0.

An important measure of maneuverability for an airplane is the increase in elevator deflection that is required to produce a given level of normal acceleration. This is typically specified as the elevator deflection that is required to produce a unit increase in load factor. This can be found by differentiating Eq. (6.1.34) with respect to n. Thus, **with no vertical offsets, the change in elevator angle per g of normal acceleration is**

$$\frac{\partial \delta_e}{\partial n} = -\frac{C_W C_{m,\alpha} + (C_{L,\alpha}\,C_{m,\bar{q}} - C_{L,\bar{q}}\,C_{m,\alpha})(g\bar{c}_w/2V^2)}{C_{L,\alpha}\,C_{m,\delta_e} - C_{L,\delta_e}\,C_{m,\alpha}} \tag{6.1.35}$$

This derivative is commonly referred to as the *elevator angle per g*, because it specifies the elevator deflection that is required to sustain each additional g of normal acceleration, beyond the 1-g load that is required to sustain trimmed flight. It provides a measure of airplane maneuverability. The smaller this derivative is, the more maneuverable the airplane. In other words, when this derivative is small, large accelerations are produced with small elevator deflections. While maneuverability is important, particularly for aerobatics and combat, if the elevator angle per g is too small, the airplane becomes overly sensitive to elevator deflection and difficult to fly.

We have seen previously that the location of the center of gravity has a profound effect on airplane stability. In a similar manner, the location of the CG also affects maneuverability. Because we are **neglecting vertical offsets**, from Eq. (4.4.13), the pitch stability derivative can be written in terms of the lift slope and the static margin,

$$C_{m,\alpha} \cong -\frac{l_{np}}{\bar{c}_w}C_{L,\alpha} \tag{6.1.36}$$

Using Eq. (6.1.36) to eliminate the pitch stability derivative, Eq. (6.1.35) can be written

$$\frac{\partial \delta_e}{\partial n} = -\frac{C_W - C_{L,\bar{q}}\,(g\bar{c}_w/2V^2)}{C_{m,\delta_e} + C_{L,\delta_e}\,(l_{np}/\bar{c}_w)}\left[\frac{C_{m,\bar{q}}\,(g\bar{c}_w/2V^2)}{C_W - C_{L,\bar{q}}\,(g\bar{c}_w/2V^2)} - \frac{l_{np}}{\bar{c}_w}\right] \tag{6.1.37}$$

Following the development presented in Example 4.4.2, we let x denote the *aerodynamic coordinate* measured aft of some fixed point on the airplane. (Remember, the x-axis points aft and the x_b-axis points forward; see Figs. 1.1.1, 1.1.3, and 5.1.1.) We then have

$$l_{np} = x_{np} - x_{CG} \tag{6.1.38}$$

Using this relation together with Eq. (6.1.11) in Eq. (6.1.37) gives

$$\frac{\partial \delta_e}{\partial n} = -\frac{C_W - C_{L,q}\,(g/V)}{C_{m,\delta_e} + C_{L,\delta_e}\,[(x_{np} - x_{CG})/\bar{c}_w]}\left[\frac{C_{m,q}\,(g/V)}{C_W - C_{L,q}\,(g/V)} - \frac{x_{np}}{\bar{c}_w} + \frac{x_{CG}}{\bar{c}_w}\right] \tag{6.1.39}$$

By examining the three terms in the brackets on the right-hand side of Eq. (6.1.39), we see that there is some particular location for the center of gravity that will cause the elevator angle per g to go to zero. With the CG located at this point, Eq. (6.1.39) predicts that an infinite acceleration results from an infinitesimal elevator deflection. Obviously, this small-angle solution is not valid for an infinite normal acceleration, because the wing will stall or fail structurally at a finite load factor. Nevertheless, having the CG at this location would make it impossible to control the airplane. The CG location that causes the elevator angle per g to go to zero is commonly called the *stick-fixed maneuver point* and its axial position will be denoted here as x_{mp}. From Eq. (6.1.39) we see that

$$x_{mp} = x_{np} - \frac{\bar{c}_w\,C_{m,q}\,(g/V)}{C_W - C_{L,q}\,(g/V)} = x_{np} - \frac{\bar{c}_w\,C_{m,\bar{q}}\,(g\bar{c}_w/2V^2)}{C_W - C_{L,\bar{q}}\,(g\bar{c}_w/2V^2)} \tag{6.1.40}$$

where $C_{m,q}$ and $C_{L,q}$ are evaluated with the center of gravity located at x_{mp}. Similar to the stick-fixed static margin, the distance aft from the CG to the stick-fixed maneuver point, expressed as a fraction of the arbitrary reference chord length, is usually referred to as the *stick-fixed maneuver margin*,

$$\frac{l_{mp}}{\bar{c}_w} \equiv \frac{x_{mp} - x_{CG}}{\bar{c}_w} = \frac{l_{np}}{\bar{c}_w} - \frac{C_{m,\bar{q}}\,(g\bar{c}_w/2V^2)}{C_W - C_{L,\bar{q}}\,(g\bar{c}_w/2V^2)} = \frac{l_{np}}{\bar{c}_w} - \frac{C_{m,q}\,(g/V)}{C_W - C_{L,q}\,(g/V)} \tag{6.1.41}$$

Because both $C_{m,q}$ and $C_{L,q}$ depend on the location of the CG, finding the stick-fixed maneuver point and maneuver margin from Eqs. (6.1.40) and (6.1.41) is not quite as simple as it looks. However, we can deduce the relative position of the maneuver point by examining the signs and magnitudes of the various terms in Eq. (6.1.40).

The weight coefficient, C_W, is positive and the term that includes the change in lift coefficient with respect to pitching rate, $C_{L,q}\,(g/V)$, is typically very small compared with the weight coefficient. So, in general,

$$C_{L,\bar{q}}\left(g\bar{c}_w/2V^2\right) = C_{L,q}\left(g/V\right) \ll C_W$$

The pitch-damping derivative, $C_{m,\bar{q}}$, is always negative in unstalled flight. Thus,

$$\frac{\bar{c}_w\, C_{m,\bar{q}}\left(g\bar{c}_w/2V^2\right)}{C_W - C_{L,\bar{q}}\left(g\bar{c}_w/2V^2\right)} = \frac{\bar{c}_w\, C_{m,q}\left(g/V\right)}{C_W - C_{L,q}\left(g/V\right)} \cong \frac{\bar{c}_w\, C_{m,q}\left(g/V\right)}{C_W} < 0$$

Using this result with Eq. (6.1.40), we see that **the stick-fixed maneuver point is always aft of the stick-fixed neutral point**. Thus, the maneuver margin is always greater than the static margin, and if the airplane is statically stable, the maneuver margin will always be positive. Therefore, **it is not possible for a statically stable airplane to have a zero stick-fixed maneuver margin.**

Although a positive static margin will always assure a positive maneuver margin, many airplanes have been designed, built, and flown with a negative static margin. In fact, the 1903 Wright Flyer had a static margin that was slightly negative. While a positive static margin is typically desirable for good human handling qualities, it is possible for a human pilot to control an airplane with a static margin that is somewhat negative. Furthermore, to improve maneuverability, most modern fighters have a negative static margin. Such airplanes are typically fitted with computer control systems, which are called *stability augmentation systems*. A computer control system has a much faster reaction time than any human pilot. Thus, a computer can control an airplane that would be much too unstable for a human pilot. However, even a computer cannot control an airplane with a zero maneuver margin. For this reason, **if an airplane is to be operated with a negative static margin, great care must be taken to assure that the center of gravity is always kept forward of the stick-fixed maneuver point.**

Because C_{m,δ_e}, $C_{m,\bar{q}}$, and $C_{L,\bar{q}}$ all depend on the position of the center of gravity, it is not at all obvious from Eq. (6.1.39) exactly how the elevator angle per g varies with the location of the center of gravity. This is examined in the following example.

EXAMPLE 6.1.1. For the wing-tail combination described in Example 4.4.2, plot the elevator angle per g at 80 mph and sea level as a function of the location of the CG measured relative to the quarter chord of the main wing.

Solution. Let x denote the distance measured aft of the quarter chord of the main wing. For this airplane we have

$$S_w = 180\ \text{ft}^2, \quad b_w = 33\ \text{ft}, \quad C_{L_w,\alpha} = 4.44, \quad C_{m_w} = -0.053, \quad W = 2{,}700\ \text{lbf},$$
$$S_h = 36\ \text{ft}^2, \quad b_h = 12\ \text{ft}, \quad C_{L_h,\alpha} = 3.97, \quad C_{m_h 0} = 0.0, \quad C_{m_h,\delta_e} = -0.55,$$
$$\varepsilon_e = 0.60, \quad x_h = 15\ \text{ft}, \quad \eta_h = 1.0$$

From Example 4.4.2,

$$C_{L,\delta_e} = \frac{S_h}{S_w}\eta_h C_{L_h,\alpha}\,\varepsilon_e = 0.4764$$

The weight coefficient at 80 mph is

$$C_W = \frac{W}{\frac{1}{2}\rho V^2 S_w} = 0.9168$$

Also from Example 4.4.2, the change in pitching moment coefficient with respect to elevator deflection for this wing-tail combination is

$$C_{m,\delta_e} = \frac{S_h \bar{c}_h}{S_w \bar{c}_w}\eta_h C_{m_h,\delta_e} - \frac{S_h(x_h - x_{CG})}{S_w \bar{c}_w}\eta_h C_{L_h,\alpha}\varepsilon_e$$
$$= 0.4764\,\hat{x}_{CG} - 1.3706$$

where $\hat{x}_{CG} \equiv x_{CG}/\bar{c}_w$. The position of the stick-fixed neutral point was found to be

$$\frac{x_{np}}{\bar{c}_w} = \frac{l_{np}}{\bar{c}_w} - \frac{l_w}{\bar{c}_w} = 0.2504$$

At 80 mph,

$$\frac{g\bar{c}_w}{2V^2} = \frac{32.2(180/33)}{2(80\times 5{,}280/3{,}600)^2} = 0.006379$$

From Eqs. (6.1.22) and (6.1.26),

$$C_{L,\bar{q}}\frac{g\bar{c}_w}{2V^2} = \left[2\frac{-x_{CG}}{\bar{c}_w}C_{L_w,\alpha} + 2\frac{S_h(x_h - x_{CG})}{S_w\bar{c}_w}\eta_h C_{L_h,\alpha}\right]\frac{g\bar{c}_w}{2V^2}$$
$$= 0.0279 - 0.0668\,\hat{x}_{CG}$$

$$C_{m,\bar{q}}\frac{g\bar{c}_w}{2V^2} = \left[-2\frac{(-x_{CG})^2}{\bar{c}_w^2}C_{L_w,\alpha} - 2\frac{S_h(x_h - x_{CG})^2}{S_w\bar{c}_w^2}\eta_h C_{L_h,\alpha}\right]\frac{g\bar{c}_w}{2V^2}$$
$$= -0.0766 + 0.0557\,\hat{x}_{CG} - 0.0668\,\hat{x}_{CG}^2$$

Using these results in Eq. (6.1.39), the elevator angle per g can be written in terms of the position of the center of gravity relative to the main wing,

$$\frac{\partial \delta_e}{\partial n} = -\frac{C_W - C_{L,\bar{q}}(g\bar{c}_w/2V^2)}{C_{m,\delta_e} + C_{L,\delta_e}[(x_{np} - x_{CG})/\bar{c}_w]}\left[\frac{C_{m,\bar{q}}(g\bar{c}_w/2V^2)}{C_W - C_{L,\bar{q}}(g\bar{c}_w/2V^2)} - \frac{x_{np}}{\bar{c}_w} + \frac{x_{CG}}{\bar{c}_w}\right]$$
$$= -\frac{0.9168 - (0.0279 - 0.0668\,\hat{x}_{CG})}{(0.4764\,\hat{x}_{CG} - 1.3706) + 0.4764(0.2504 - \hat{x}_{CG})}$$
$$\times \left[\frac{(-0.0766 + 0.0557\,\hat{x}_{CG} - 0.0668\,\hat{x}_{CG}^2)}{0.9168 - (0.0279 - 0.0668\,\hat{x}_{CG})} - 0.2504 + \hat{x}_{CG}\right]$$

or after simplifying,

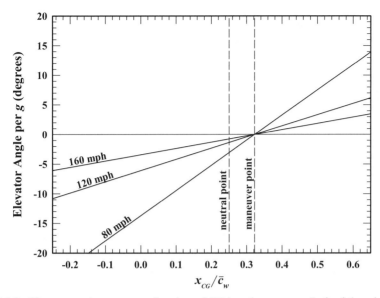

Figure 6.1.3. Elevator angle per g as a function of CG location measured aft of the wing quarter chord for the airplane in Example 6.1.1.

$$\frac{\partial \delta_e}{\partial n} = -\frac{0.8889 + 0.0668\,\hat{x}_{CG}}{-1.2513}\left(\frac{-0.0766 + 0.0557\,\hat{x}_{CG} - 0.0668\,\hat{x}_{CG}^2}{0.8889 + 0.0668\,\hat{x}_{CG}} - 0.2504 + \hat{x}_{CG}\right)$$

$$= \frac{-0.0766 + 0.0557\,\hat{x}_{CG} - 0.0668\,\hat{x}_{CG}^2 + (0.8889 + 0.0668\,\hat{x}_{CG})(-0.2504 + \hat{x}_{CG})}{1.2513}$$

$$= \frac{-0.2992 + 0.9279\,\hat{x}_{CG}}{1.2513} = \underline{-0.2391 + 0.7415\,\hat{x}_{CG}}$$

So we see that within the approximations used for this example, the elevator angle per g is a linear function of CG location. This is not obvious from Eq. (6.1.39). Results for the particular wing-tail combination used in the present example are shown in Fig. 6.1.3 for airspeeds of 80, 120, and 160 mph.

Dynamic Margin and Aft CG Limit
From Fig. 6.1.3 we see that as the CG is moved aft, the elevator control becomes more responsive, so the airplane becomes more maneuverable. That is, smaller elevator deflections are required to produce a given normal acceleration. With the CG located at the stick-fixed maneuver point, this small-angle solution predicts that the elevator control becomes so sensitive that an infinitesimal elevator deflection would result in an infinite normal acceleration. Of course, the small-angle solution is not valid for an infinite normal acceleration. The wing will stall or fail structurally at a finite load factor. When the CG is located aft of the stick-fixed maneuver point, Fig. 6.1.3 shows that the elevator deflection required to produce a constant normal acceleration is reversed.

The axial position of the stick-fixed maneuver point is given by Eq. (6.1.40). This equation suggests an alternative definition for a dimensionless pitch rate, which will be referred to here as the *dynamic pitch rate*

$$\breve{q} \equiv \frac{Vq}{g} = \frac{2V^2}{g\bar{c}_w}\bar{q} \tag{6.1.42}$$

From Eq. (6.1.8) it can be seen that the product of the airspeed and the pitch rate is the centripetal acceleration for the maneuver. Thus, the dimensionless dynamic pitch rate defined by Eq. (6.1.42) is the ratio of the centripetal acceleration to the gravitational acceleration. This is an important dimensionless ratio for accelerating flight, which occurs naturally in this small-angle solution to the governing equations of motion. By contrast, the traditional dimensionless pitch rate defined by Eq. (6.1.11) has no significant basis in the physics of accelerating flight.

Multiplying the numerator and denominator on the far right-hand side of Eq. (6.1.40) by the product of dynamic pressure and wing area, and using Eq. (6.1.36) to express the location of the neutral point in terms of pitch stability, Eq. (6.1.40) is easily rearranged as

$$x_{mp} = x_{np} - \frac{m_{,q}\,(g/V)}{W - L_{,q}\,(g/V)} = x_{np} - \frac{m_{,\breve{q}}}{W - L_{,\breve{q}}} = x_{CG} - \frac{m_{,\alpha}}{L_{,\alpha}} - \frac{m_{,\breve{q}}}{W - L_{,\breve{q}}} \tag{6.1.43}$$

Using the definitions given by Eqs. (6.1.11) and (6.1.42), the aerodynamic derivatives $L_{,\breve{q}}$ and $m_{,\breve{q}}$ can be expressed in terms of the traditional dimensionless aerodynamic derivatives $C_{L,\bar{q}}$ and $C_{m,\bar{q}}$, which are commonly determined from wind tunnel testing or approximated from relations like Eqs. (6.1.20) through (6.1.27). Thus, we obtain

$$L_{,\alpha} \equiv \tfrac{1}{2}\rho V^2 S_w C_{L,\alpha} \cong \tfrac{1}{2}\rho V^2 [S_w C_{L_w,\alpha} + S_h \eta_h C_{L_h,\alpha}(1 - \varepsilon_{d,\alpha})] \tag{6.1.44}$$

$$L_{,\breve{q}} \equiv \tfrac{1}{4}\rho g S_w \bar{c}_w C_{L,\bar{q}} \cong \tfrac{1}{2}\rho g (S_w l_w C_{L_w,\alpha} + S_h l_h \eta_h C_{L_h,\alpha}) \tag{6.1.45}$$

$$m_{,\alpha} \equiv \tfrac{1}{2}\rho V^2 S_w \bar{c}_w C_{m,\alpha} \cong -\tfrac{1}{2}\rho V^2 [S_w l_w C_{L_w,\alpha} + S_h l_h \eta_h C_{L_h,\alpha}(1 - \varepsilon_{d,\alpha})] \tag{6.1.46}$$

$$m_{,\breve{q}} \equiv \tfrac{1}{4}\rho g S_w \bar{c}_w^2 C_{m,\bar{q}} \cong -\tfrac{1}{2}\rho g (S_w l_w^2 C_{L_w,\alpha} + S_h l_h^2 \eta_h C_{L_h,\alpha}) \tag{6.1.47}$$

From Eqs. (6.1.45) and (6.1.47) we see that the derivatives $L_{,\breve{q}}$ and $m_{,\breve{q}}$ are independent of airspeed. With this knowledge, Eq. (6.1.43) displays the important fact that the position of the maneuver point is independent of airspeed, which is also shown in Fig. 6.1.3.

It is also important to notice from Eq. (6.1.43) that although the stick-fixed maneuver point was defined in terms of the elevator angle per g, its location is completely independent of the elevator power. Even if the elevator were completely fixed and unmovable, the location of the stick-fixed maneuver point would not change.

As we will see, the location of the stick-fixed maneuver point has greater significance than is suggested by its relation to the elevator angle per g. From Example 6.1.1 and Eqs. (6.1.43) through (6.1.47) we see that within the approximations used, **the location of the stick-fixed maneuver point is independent of airspeed, elevator power, location of the CG, the definition of the reference chord length, and our choice of a datum.**

Subtracting x_{CG} from both sides of Eq. (6.1.40) or (6.1.43), the stick-fixed maneuver point can be equivalently located relative to the center of gravity from the relation

$$l_{mp} = -\frac{\overline{c}_w\, C_{m,\alpha}}{C_{L,\alpha}} - \frac{\overline{c}_w\, C_{m,\breve{q}}}{C_W - C_{L,\breve{q}}} = -\frac{m_{,\alpha}}{L_{,\alpha}} - \frac{m_{,\breve{q}}}{W - L_{,\breve{q}}} \tag{6.1.48}$$

Example 6.1.1 reveals that the term $C_{L,\breve{q}}(g\overline{c}_w/2V^2) = C_{L,q}(g/V) = C_{L,\breve{q}}$ is small compared with the weight coefficient. **For all practical purposes, the location of the stick-fixed maneuver point can be accurately estimated from the simplified relation**

$$l_{mp} = -\frac{\overline{c}_w\, C_{m,\alpha}}{C_{L,\alpha}} - \frac{\overline{c}_w\, C_{m,\breve{q}}}{C_W} = -\frac{m_{,\alpha}}{L_{,\alpha}} - \frac{m_{,\breve{q}}}{W} \tag{6.1.49}$$

Extensive experimental investigation with human-piloted airplanes has shown that to maintain satisfactory handling qualities for a human pilot, the center of gravity must be kept forward of the stick-fixed maneuver point by a distance that can be expressed in terms of what will be called the **dynamic margin**,

$$\frac{l_{mp}}{r_{yy_b}} = -\frac{C_{\tilde{m},\alpha}}{C_{L,\alpha}} - \frac{C_{\tilde{m},\breve{q}}}{C_W} = -\frac{m_{,\alpha}}{L_{,\alpha}\, r_{yy_b}} - \frac{m_{,\breve{q}}}{W r_{yy_b}} \ge \frac{\dot{q}_c}{g/r_{yy_b}} \tag{6.1.50}$$

where r_{yy_b} is the **pitch radius of gyration** defined from the pitch moment of inertia, I_{yy_b}, and the mass of the aircraft, W/g,

$$r_{yy_b} \equiv \sqrt{g I_{yy_b}/W}, \quad I_{yy_b} \equiv (1/g)\iiint_W (x_b^2 + z_b^2)\, dW \tag{6.1.51}$$

and $C_{\tilde{m}}$ is a **dynamic moment coefficient**

$$C_{\tilde{m}} \equiv \frac{m}{\frac{1}{2}\rho V^2 S_w\, r_{yy_b}} \equiv \frac{\overline{c}_w}{r_{yy_b}} C_m \tag{6.1.52}$$

For establishing the value of the dimensionless ratio on the right-hand side of Eq. (6.1.50), *the parameter $\dot{q}_c$ is an experimentally determined constraint having the dimensions of angular acceleration, which accounts for the combined effects of pilot limitations and flight phase requirements.* Extensive experimental research suggests that the following minimum values should be used for all classes of airplanes, **to obtain adequate handling qualities for human pilots under normal flight conditions:**

$\dot{q}_c = 0.28\ \text{s}^{-2}$ for flight phases that require rapid maneuvering and precise flight-path control (i.e., combat, aerobatics, close-formation flying, etc.).

$\dot{q}_c = 0.15\ \text{s}^{-2}$ for flight phases that require accurate flight-path control but are usually accomplished with gradual maneuvers (i.e., takeoff, landing, approach, etc.).

$\dot{q}_c = 0.085\ \text{s}^{-2}$ for flight phases that are accomplished with gradual maneuvers and do not require precision tracking (i.e., climb, cruise, decent, etc.).

The **dynamic margin** *is formally defined here to be the distance that the stick-fixed maneuver point lies aft of the center of gravity, expressed as a fraction of the airplane's pitch radius of gyration.* The minimum acceptable dynamic margin for a human pilot depends on the pilot's ability to sense and correct for disturbances in pitch. Thus, it might be more enlightening to think of $\dot{q}_c$ as *an experimental constraint resulting from the pilot's **pitch-acceleration-sensitivity limit*** *as well as the flight phase requirements.*

The dynamically scaled derivatives in Eq. (6.1.50) are easily related to the traditional aerodynamic derivatives, which are commonly obtained from wind tunnel tests. This requires knowing only the airspeed, acceleration of gravity, radius of gyration, and the arbitrary reference length used to report the wind tunnel results. From the definitions in Eqs. (6.1.11), (6.1.42), and (6.1.52), it is easily shown that

$$C_{\tilde{m},\alpha} = \frac{\bar{c}_w}{r_{yy_b}} C_{m,\alpha} \qquad (6.1.53)$$

$$C_{\tilde{m},\bar{q}} = \frac{g\bar{c}_w^2}{2 r_{yy_b} V^2} C_{m,\bar{q}} \qquad (6.1.54)$$

With these relations, Eq. (6.1.50) provides a simple but important fundamental constraint on pitch stability that can be used in the early phases of airplane design. This provides a significant improvement over any constraint that is based on static margin alone. The pitch radius of gyration can be estimated in the early phases of design from the mass properties data that are presented in Appendix C.

As seen in Eq. (6.1.43), the stick-fixed maneuver point for any airplane is aft of the stick-fixed neutral point by a distance that is proportional to the pitch damping. Any pitch stability constraint that is based on static margin alone does not account for the fact that both pitch stability and pitch damping affect the handling qualities of a piloted airplane. When pitch damping is high, it is easier for a pilot to cope with less pitch stability. Using the dynamic-margin constraint given by Eq. (6.1.50) accounts for this effect. This relation predicts that if the pitch damping were sufficiently large, a human pilot could fly an airplane having a zero or slightly negative static margin, even without the aid of a computer-controlled stability augmentation system. This could explain the success of the 1903 Wright Flyer despite a slightly negative static margin.

The reader should note that the aft CG limit specified by Eq. (6.1.50) is significantly affected by gross weight, through its dependence on the weight coefficient and the pitch radius of gyration. Airplanes can have fuel/cargo fractions as large as 50% or more. Because the variable fraction of the gross weight (fuel, stores, and cargo) is usually concentrated close to the dry CG, the radius of gyration defined by Eq. (6.1.51) will typically increase as fuel, stores, and cargo are removed. Hence, the critical aft CG limit determined from the dynamic-margin constraint specified by Eq. (6.1.50) will typically be that for the minimum landing weight, not that for the maximum takeoff weight.

The extensive research on aircraft handling qualities, which forms the foundation for Eq. (6.1.50), is reported in the U.S. military specifications MIL-F-8785C (1980) and MIL-STD-1797A (1995). For a detailed developed of Eq. (6.1.50) see Phillips and Niewoehner (2009). This material is also covered in Chapter 10 of this textbook.

EXAMPLE 6.1.2. For a Boeing 747-100 landing at sea level with an airspeed of 221 ft/s, consider three different landing weights; for case I, $W = 491,400$ lbf, $I_{yy_b} = 31,500,000$ slug·ft^2; for case II, $W = 564,000$ lbf, $I_{yy_b} = 32,300,000$ slug·ft^2; and for case III, $W = 636,600$ lbf, $I_{yy_b} = 33,100,000$ slug·ft^2. The aerodynamic derivatives for the airplane at this Mach number as determined from wind tunnel test are $C_{L,\alpha} = 5.8760$, $C_{m,\alpha} = -1.4325$, and $C_{m,\bar{q}} = -21.184$. These traditional aerodynamic derivatives are based on the wing planform area, $S_w = 5,500$ ft^2. The pitching moment coefficient was measured about a fixed reference origin and nondimensionalized with respect to a reference chord length, $c_{ref} = 27.31$ feet. For each weight, find the location of the stick-fixed neural point and the stick-fixed maneuver point as distances measured aft of the wind tunnel reference origin. For each weight, compute the aft CG limit based on Eq. (6.1.50) and compute the static margin corresponding to this aft CG limit. Neglect the change in lift with respect to pitch rate and any change in r_{yy_b} with CG location.

Solution. Let x denote a distance measured aft of the wind tunnel reference origin that was used to determine the given aerodynamic derivatives. The location of the stick-fixed neutral point relative to this origin, as obtained from Eq. (6.1.36), is

$$x_{np} = -\frac{c_{ref} C_{m,\alpha}}{C_{L,\alpha}} = -\frac{(27.31 \text{ ft})(-1.4325)}{5.8760} = \underline{6.658 \text{ ft}}$$

which does not change with landing weight. From Eq. (6.1.47),

$$m_{,\bar{q}} = \tfrac{1}{4}\rho g S_w \bar{c}_w^2 C_{m,\bar{q}}$$

$$= \tfrac{1}{4}(0.0023769)(32.2)(5,500)(27.31^2)(-21.184) = -1,662,727 \text{ ft·lbf}$$

Neglecting the change in lift with respect to pitch rate, at the lightest of the three landing weights Eq. (6.1.43) yields

$$x_{mp} = x_{np} - \frac{m_{,\bar{q}}}{W} = 6.658 \text{ ft} - \frac{-1,662,727 \text{ ft·lbf}}{491,400 \text{ lbf}} = \underline{10.042 \text{ ft}}$$

At the same landing weight, the airplane's pitch radius of gyration computed from Eq. (6.1.51) is

$$r_{yy_b} \equiv \sqrt{g I_{yy_b}/W} = \sqrt{(32.2)(31,500,000)/491,400} = 45.432 \text{ ft}$$

For the landing flight phase, the pitch-acceleration-sensitivity limit is $\dot{q}_c = 0.15$ s^{-2} and the dynamic margin at the aft CG limit computed from Eq. (6.1.50) is

$$\left(\frac{l_{mp}}{r_{yy_b}}\right)_{min} = \frac{\dot{q}_c}{g/r_{yy_b}} = \frac{0.15}{32.2/45.432} = 0.2116$$

The CG location that results in this dynamic margin is obtained from the definition

$$(l_{mp}/r_{yy_b})_{\min} \equiv [x_{mp} - (x_{CG})_{\max}]/r_{yy_b} = 0.2116$$

Neglecting any change in r_{yy_b}, this is easily rearranged to give

$$(x_{CG})_{\max} = x_{mp} - 0.2116\, r_{yy_b} = 10.042 \text{ ft} - 0.2116\,(45.432 \text{ ft}) = \underline{0.426 \text{ ft}}$$

The static margin that corresponds to this CG location is

$$(l_{np}/c_{\text{ref}})_{\min} \equiv [x_{np} - (x_{CG})_{\max}]/c_{\text{ref}} = (6.658 \text{ ft} - 0.426 \text{ ft})/27.31 \text{ ft} = \underline{0.228}$$

Repeating these computations for the other two landing weights yields

	Case I	Case II	Case III
W (lbf)	491,400	564,000	636,600
I_{yy_b} (slug·ft^2)	31,500,000	32,300,000	33,100,000
r_{yy_b} (ft)	45.432	42.943	40.917
$m_{,\bar{q}}$ (ft·lbf)	−1,662,727	−1,662,727	−1,662,727
x_{np} (ft)	6.658	6.658	6.658
x_{mp} (ft)	10.042	9.606	9.270
$(l_{mp}/r_{yy_b})_{\min}$	0.2116	0.2000	0.1906
$(x_{CG})_{\max}$ (ft)	0.426	1.016	1.470
$(l_{np}/c_{\text{ref}})_{\min}$	0.228	0.207	0.190

Note that the critical aft-center-of-gravity limit for this example occurs at minimum weight. As weight is increased, the maneuver point is moved forward as a result of the last term on the right-hand side of Eq. (6.1.43). This increase in weight also decreases the radius of gyration, so there is often little change in the ratio l_{mp}/r_{yy} for a fixed CG location. However, the decrease in radius of gyration caused by an increase in weight also decreases the minimum-dynamic-margin constraint on the right-hand side of Eq. (6.1.50). Thus, the minimum-weight case is typically constraining.

The previous example demonstrates one possible candidate for the aft CG limit of an airplane. When the CG is aft of the limit specified by Eq. (6.1.50), the airplane becomes overly sensitive to control inputs or disturbances in pitch and is difficult to fly. For large airplanes, the dynamic margin constraint expressed in Eq. (6.1.50) usually provides the most restrictive aft CG limit. Notice that for the Boeing 747 in Example 6.1.2, the static margin is in excess of 20 percent when the CG is at its aft limit with respect to dynamic margin. The data for this example are based on Heffley and Jewell (1972) and are typical of large transports. It should be noted here that Eq. (6.1.50) does not typically provide the aft CG limit for general aviation airplanes, because FAR Part 23 requirements specify pitch stability limits in terms of the control force and these requirements are typically more restrictive than Eq. (6.1.50) for light airplanes. The effect of CG location on the pitch control force is covered in Sec. 6.3.

Elevator Power and Forward CG Limit

In Example 6.1.1 we saw that within the approximations used, the location of the stick-fixed maneuver point is independent of airspeed. We also saw that for a stable aircraft, the elevator deflection required to produce a given normal acceleration increases as the airspeed is decreased. Moreover, as the CG is moved forward, the elevator deflection required for a given maneuver is further increased. At low airspeed with the CG forward of the wing quarter chord, large elevator deflections are required to maneuver the airplane. In Fig. 4.4.3 we saw that large negative elevator deflections are required just to trim the airplane under such conditions. Furthermore, with flaps fully deployed in landing configuration, this condition is further aggravated.

> **EXAMPLE 6.1.3.** (*Suggested by Captain Robert J. Niewoehner, Ph.D., Aerospace Engineering Department, United States Naval Academy*) For the wing-tail combination described in Example 6.1.1 with the CG located 0.70 feet aft of the wing quarter chord, plot the load factor as a function of airspeed at sea level for several fixed values of elevator angles. Overlay these plots on a traditional *V-n* diagram, which is discussed in Sec. 3.9. In addition to the geometric and aerodynamic properties that are used for Example 6.1.1, assume that the stall limits are fixed by $C_{L_{max}} = 1.5$ and $C_{L_{min}} = -0.7$. For the load limits use $n_{max} = 4$ and $n_{min} = -2$.

Solution. After applying the definition of weight coefficient from Eq. (6.1.30) and rearranging, Eq. (6.1.34) can be readily solved for the load factor as a function of elevator angle and airspeed,

$$
n = \frac{C_{L,\alpha}\,C_{m,\bar{q}} - C_{L,\bar{q}}\,C_{m,\alpha}}{C_{m,\alpha}\,R_W + C_{L,\alpha}\,C_{m,\bar{q}} - C_{L,\bar{q}}\,C_{m,\alpha}}
$$
$$
+ \frac{C_{m,\alpha}\,C_{L0} - C_{L,\alpha}\,C_{m0} - (C_{L,\alpha}\,C_{m,\delta_e} - C_{L,\delta_e}\,C_{m,\alpha})\,\delta_e}{C_{m,\alpha}\,C_W\,(2V^2/g\bar{c}_w) + C_{L,\alpha}\,C_{m,\bar{q}} - C_{L,\bar{q}}\,C_{m,\alpha}}\,\frac{2V^2}{g\bar{c}_w}
\tag{6.1.55}
$$

where, for this particular airplane at standard sea level,

$$
C_W\,\frac{2V^2}{g\bar{c}_w} = \frac{W}{\frac{1}{2}\rho V^2 S_w}\,\frac{2V^2}{g\bar{c}_w} = \frac{4W}{\rho g S_w \bar{c}_w} = 143.7
$$

Thus we see that with both the CG location and the elevator angle fixed, the load factor is a parabolic function of airspeed. From Examples 4.4.2 and 6.1.1, the following aerodynamic coefficients are independent of CG location:

$$
C_{L0} = 0.4075, \quad C_{L,\alpha} = 4.8844, \quad C_{L,\delta_e} = 0.4764
$$

With the CG located 0.70 feet aft of the wing quarter chord, the dimensionless CG location is $\hat{x}_{CG} \equiv x_{CG}/\bar{c}_w = 0.1283$, and following Examples 4.4.2 and 6.1.1, we obtain the aerodynamic coefficients that depend on CG location

$$
C_{L,\bar{q}} = 3.0236, \quad C_{m0} = -0.0007, \quad C_{m,\alpha} = -0.5959,
$$

$$C_{m,\bar{q}} = -11.0608, \quad C_{m,\delta_e} = -1.3095$$

For an airspeed of 80 mph, we obtain

$$2V^2/g\bar{c}_w = 156.77$$

Using these values for the aerodynamic derivatives in Eq. (6.1.55) with an elevator deflection angle of −25 degrees yields

$$n = 3.684$$

Repeating these computations for other values of airspeed and elevator deflection produces the results shown in Fig. 6.1.4. Notice that for this CG location, only 10 degrees of up elevator are required to produce a 4-g load factor at the maneuver speed.

 As the CG is moved forward, the elevator deflection required to produce a given maneuver is increased. For example, from the results shown in Fig. 6.1.5, we see that 25 degrees of up elevator are required to produce a 4-g load factor at the maneuver speed, when the CG is 0.48 foot forward of the wing quarter chord. Notice that with this CG location, the allowable flight envelope is nicely covered by elevator angles in the range $-25° \leq \delta_e \leq +17°$. Hence, 0.48 foot forward of the wing quarter chord would be a candidate forward CG limit. With the CG forward of this point, more elevator power would be required to produce the allowable 4-g load factor at the maneuver speed.

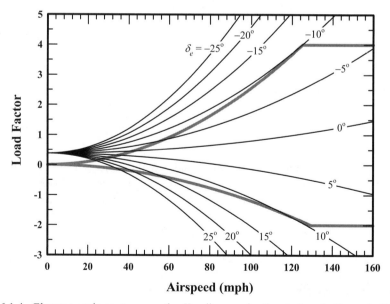

Figure 6.1.4. Elevator-angle contours on the V-n diagram for the airplane in Example 6.1.3 with the CG located 0.70 foot aft of the wing quarter chord.

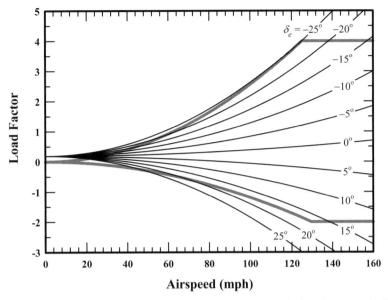

Figure 6.1.5. Elevator-angle contours on the *V-n* diagram for the airplane in Example 6.1.3 with the CG located 0.48 foot forward of the wing quarter chord.

6.2. Effects of Structural Flexibility

To this point we have considered the airplane to be a rigid body. In reality, no structure is completely rigid. This is particularly true of an airplane. By necessity, airplanes are designed to be as light as possible, and elastic flexibility is a natural characteristic of lightweight structures. We have seen that even very small changes in airplane geometry (particularly angular changes) can have very significant effects on the aerodynamic forces and moments produced on the airplane. Furthermore, because of unavoidable

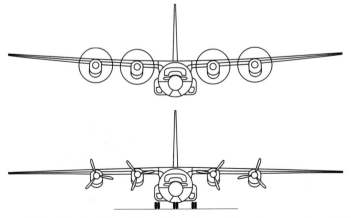

Figure 6.2.1. Elastic deformation of the wings resulting from aerodynamic lift.

elastic flexibility, the geometry of the airplane is not fixed but changes as the aerodynamic forces and moments are applied to the structure. For example, when an airplane is at rest on the ground, the fuselage supports the weight of the wings. When the airplane is in the air, the wings support the weight of the fuselage. The lift produced by the wings results in significant elastic deformation, as shown in Fig. 6.2.1. In a large airplane, the wingtips can deflect several feet as a result of the applied lift. This deformation in turn affects the forces and moments acting on the airplane. This is only one of many ways in which an airplane can deform under aerodynamic loading.

The study of the coupling between the aerodynamic forces and moment and the elastic deformation of the airplane is a separate branch of aeronautical engineering known as *aeroelasticity*. Complete textbooks have been written on aeroelasticity and no attempt is made here to cover this topic in detail. However, to give the student an appreciation for the importance of the topic, a very simple example is considered. The serious student is encouraged to study the topic further, either through formal coursework or independent study.

Like the wings of an airplane, the structure supporting the tail is subject to elastic deformation. As shown schematically in Fig. 6.2.2, the stiffness of a typical tail boom is not constant but varies along the length of the boom, from the center of gravity to the aerodynamic center of the horizontal surface. Typically, the fuselage is quite stiff near the CG. However, the aft portion near the tail is usually much more flexible. The aerodynamic forces and moments produced on the tail cause the tail boom to deflect. This deflection changes the angle of attack for the horizontal surface, which in turn affects the aerodynamic load. For example, when the elevator is deflected downward, the lift on the tail is increased and the pitching moment becomes more negative. Thus, both the force and the moment resulting from a downward elevator deflection tend to

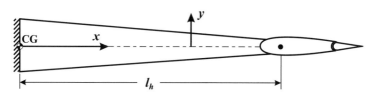

Figure 6.2.2. Schematic diagram of the structure supporting the aft tail of an airplane.

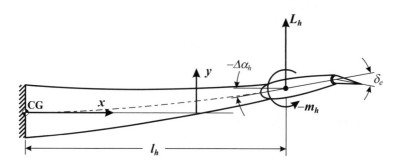

Figure 6.2.3. Deflection of the aft tail as a result of elevator deflection.

deflect the tail boom upward and rotate the tail to a more negative angle of attack, as shown in Fig. 6.2.3. The reduction in tail angle of attack, which results from this elastic deformation, reduces the lift produced by the tail. Thus, flexibility in the tail boom reduces the effectiveness of the elevator to produce the pitching moments necessary to trim and maneuver the airplane.

The elastic deflection of a beam can be found by relating the local curvature to the local bending moment and the local stiffness. The governing differential equation is

$$\frac{d^2 y}{dx^2} = \frac{M_b}{EI} \tag{6.2.1}$$

where M_b is the local bending moment, E is the material modulus of elasticity, and I is the local area moment of inertia. In terms of the lift and pitching moment acting on the tail, this can be written as

$$\frac{d^2 y}{dx^2} = \frac{L_h (l_h - x) - m_h}{EI} \tag{6.2.2}$$

The change in tail angle of attack that results from bending of the tail boom is found by integrating this equation from the center of gravity to the aerodynamic center of the tail. This gives

$$\Delta \alpha_h = -\frac{dy}{dx}\bigg|_{x=l_h} = m_h \int_{x=0}^{l_h} \frac{1}{EI} dx - L_h \int_{x=0}^{l_h} \frac{l_h - x}{EI} dx \tag{6.2.3}$$

The lift and pitching moment acting on the horizontal tail can be written as

$$L_h = \tfrac{1}{2} \rho V^2 S_h \, \eta_h C_{L_h,\alpha} [(1 - \varepsilon_{d,\alpha})\alpha + \Delta \alpha_h + \alpha_{0h} - \alpha_{L0h} - \varepsilon_{d0} + \varepsilon_e \delta_e] \tag{6.2.4}$$

$$m_h = \tfrac{1}{2} \rho V^2 S_h \bar{c}_h \, \eta_h (C_{m_h 0} + C_{m_h, \delta_e} \, \delta_e) \tag{6.2.5}$$

So the change in the tail angle of attack must satisfy the equation

$$\Delta \alpha_h = \left(\tfrac{1}{2} \rho V^2 S_h \bar{c}_h \, \eta_h \int_{x=0}^{l_h} \frac{1}{EI} dx \right) (C_{m_h 0} + C_{m_h, \delta_e} \, \delta_e)$$
$$- \left(\tfrac{1}{2} \rho V^2 S_h \, \eta_h \int_{x=0}^{l_h} \frac{l_h - x}{EI} dx \right) C_{L_h,\alpha} [(1 - \varepsilon_{d,\alpha})\alpha + \Delta \alpha_h + \alpha_{0h} - \alpha_{L0h} - \varepsilon_{d0} + \varepsilon_e \delta_e] \tag{6.2.6}$$

or

$$\Delta \alpha_h = K_m (C_{m_h 0} + C_{m_h, \delta_e} \, \delta_e)$$
$$- K_L C_{L_h,\alpha} [(1 - \varepsilon_{d,\alpha})\alpha + \Delta \alpha_h + \alpha_{0h} - \alpha_{L0h} - \varepsilon_{d0} + \varepsilon_e \delta_e] \tag{6.2.7}$$

where

$$K_m \equiv \frac{1}{2}\rho V^2 S_h \bar{c}_h \, \eta_h \int\limits_{x=0}^{l_h} \frac{1}{EI} dx \tag{6.2.8}$$

$$K_L \equiv \frac{1}{2}\rho V^2 S_h \, \eta_h \int\limits_{x=0}^{l_h} \frac{l_h - x}{EI} dx \tag{6.2.9}$$

The elastic deflection angle for the horizontal tail appears on both sides of Eq. (6.2.7). This is readily solved for the elastic deflection angle to give

$$\Delta\alpha_h = -\frac{K_L\,C_{L_h,\alpha}[(1-\varepsilon_{d,\alpha})\alpha + \alpha_{0h} - \alpha_{L0h} - \varepsilon_{d0} + \varepsilon_e\delta_e] - K_m(C_{m_h0} + C_{m_h,\delta_e}\,\delta_e)}{1 + K_L\,C_{L_h,\alpha}} \tag{6.2.10}$$

Thus, the change in the elastic deflection angle with respect to elevator angle is

$$\frac{\partial\Delta\alpha_h}{\partial\delta_e} = -\frac{K_L\,C_{L_h,\alpha}\,\varepsilon_e - K_m\,C_{m_h,\delta_e}}{1 + K_L\,C_{L_h,\alpha}} \tag{6.2.11}$$

The contribution that the aft horizontal stabilizer makes to the aerodynamic pitching moment about the airplane center of gravity is given by

$$\left(\Delta m_{CG}\right)_h = m_h - l_h\,L_h \tag{6.2.12}$$

Using Eqs. (6.2.4) and (6.2.5), this can be written as

$$\begin{aligned}\left(\Delta m_{CG}\right)_h = {}&\frac{1}{2}\rho V^2 S_h \bar{c}_h \, \eta_h(C_{m_h0} + C_{m_h,\delta_e}\,\delta_e) \\ &- \frac{1}{2}\rho V^2 S_h l_h \, \eta_h C_{L_h,\alpha}[(1-\varepsilon_{d,\alpha})\alpha + \Delta\alpha_h + \alpha_{0h} - \alpha_{L0h} - \varepsilon_{d0} + \varepsilon_e\delta_e]\end{aligned} \tag{6.2.13}$$

In dimensionless form, this gives

$$\begin{aligned}\left(\Delta C_m\right)_h = {}&\frac{S_h\bar{c}_h}{S_w\bar{c}_w}\eta_h(C_{m_h0} + C_{m_h,\delta_e}\,\delta_e) \\ &- \frac{S_h l_h}{S_w\bar{c}_w}\eta_h C_{L_h,\alpha}[(1-\varepsilon_{d,\alpha})\alpha + \Delta\alpha_h + \alpha_{0h} - \alpha_{L0h} - \varepsilon_{d0} + \varepsilon_e\delta_e]\end{aligned} \tag{6.2.14}$$

The total change in the pitching moment coefficient with respect to elevator deflection, including the effect of elastic deformation of the fuselage, is found by differentiating Eq. (6.2.14) with respect to δ_e. This gives

$$C_{m,\delta_e} \equiv \frac{\partial C_m}{\partial\delta_e} = \frac{S_h\bar{c}_h}{S_w\bar{c}_w}\eta_h C_{m_h,\delta_e} - \frac{S_h l_h}{S_w\bar{c}_w}\eta_h C_{L_h,\alpha}\left(\frac{\partial\Delta\alpha_h}{\partial\delta_e} + \varepsilon_e\right) \tag{6.2.15}$$

or after using Eq. (6.2.11),

$$
C_{m,\delta_e} = \frac{S_h \bar{c}_h}{S_w \bar{c}_w} \eta_h C_{m_h,\delta_e} - \frac{S_h l_h}{S_w \bar{c}_w} \eta_h C_{L_h,\alpha} \left(\varepsilon_e - \frac{K_L C_{L_h,\alpha} \varepsilon_e - K_m C_{m_h,\delta_e}}{1 + K_L C_{L_h,\alpha}} \right) \quad (6.2.16)
$$

In Eq. (6.2.16), ε_e denotes the rigid-body elevator effectiveness. From Eqs. (6.2.8) and (6.2.9), we see that both K_m and K_L are zero for a rigid airframe. However, for an airframe of finite stiffness, both K_m and K_L increase with the square of the airspeed. Thus, as the airspeed is increased, the actual elevator effectiveness is reduced, due to the flexibility of the fuselage. This reduces the capacity of the elevator to control the pitching moment for the airplane. From Eq. (6.2.16), the reduction in the elevator control derivative that results from fuselage bending can be expressed as

$$
\left(\Delta C_{m,\delta_e} \right)_{fb} = \frac{S_h l_h}{S_w \bar{c}_w} \eta_h C_{L_h,\alpha} \left(\frac{K_L C_{L_h,\alpha} \varepsilon_e - K_m C_{m_h,\delta_e}}{1 + K_L C_{L_h,\alpha}} \right) \quad (6.2.17)
$$

In a similar manner, bending of the fuselage also reduces the stability of the airplane. The contribution that the aft tail makes to the pitch stability derivative is found by differentiating Eq. (6.2.14) with respect to the angle of attack. This gives

$$
\left(\Delta C_{m,\alpha} \right)_h = -\frac{S_h l_h}{S_w \bar{c}_w} \eta_h C_{L_h,\alpha} \left(1 - \varepsilon_{d,\alpha} + \frac{\partial \Delta \alpha_h}{\partial \alpha} \right) \quad (6.2.18)
$$

Differentiating Eq. (6.2.10) with respect to angle of attack, we have

$$
\frac{\partial \Delta \alpha_h}{\partial \alpha} = -\frac{K_L C_{L_h,\alpha} (1 - \varepsilon_{d,\alpha})}{1 + K_L C_{L_h,\alpha}} \quad (6.2.19)
$$

Using Eq. (6.2.19) in Eq. (6.2.18), the net contribution that the aft tail makes to the pitch stability derivative, including the effect of fuselage bending, is found to be

$$
\left(\Delta C_{m,\alpha} \right)_h = -\frac{S_h l_h}{S_w \bar{c}_w} \eta_h C_{L_h,\alpha} (1 - \varepsilon_{d,\alpha}) \left(1 - \frac{K_L C_{L_h,\alpha}}{1 + K_L C_{L_h,\alpha}} \right) \quad (6.2.20)
$$

Thus, we find that the contribution to the pitch stability derivative that results from fuselage bending is

$$
\left(\Delta C_{m,\alpha} \right)_{fb} = \frac{S_h l_h}{S_w \bar{c}_w} \eta_h C_{L_h,\alpha} (1 - \varepsilon_{d,\alpha}) \left(\frac{K_L C_{L_h,\alpha}}{1 + K_L C_{L_h,\alpha}} \right) \quad (6.2.21)
$$

From this result, we see that fuselage bending makes a positive contribution to the pitch stability derivative and is therefore destabilizing.

From Eq. (4.4.13), the static margin can be written in terms of the pitch stability derivative and the lift slope,

$$\frac{l_{np}}{\bar{c}_w} = -\frac{C_{m,\alpha}}{C_{L,\alpha}} \qquad (6.2.22)$$

Since the tail has only a small effect on the net lift slope for the complete airplane, combining Eq. (6.2.21) with Eq. (6.2.22), the effect of fuselage bending on the static margin is approximated as

$$\left(\frac{\Delta l_{np}}{\bar{c}_w}\right)_{fb} \cong -\frac{S_h l_h}{S_w \bar{c}_w} \eta_h \frac{C_{L_h,\alpha}}{C_{L,\alpha}} (1-\varepsilon_{d,\alpha}) \left(\frac{K_L \, C_{L_h,\alpha}}{1+K_L \, C_{L_h,\alpha}}\right) \qquad (6.2.23)$$

From Eq. (6.2.23), we see that fuselage bending causes the stick-fixed neutral point to move forward, making the airplane less stable. Both the decrease in elevator effectiveness and the decrease in stability become more severe at higher airspeeds.

EXAMPLE 6.2.1. To demonstrate the effects of insufficient tail boom stiffness, consider the wing-tail combination that is described in Examples 4.3.1 through 4.4.2. Assume that the tail is attached to the wing with a thin-walled aluminum tube ($E = 10 \times 10^6$ psi) that extends from the tail to the center of gravity, which is located at the wing quarter chord. The aluminum tube has a constant cross-section of 4-inch diameter and 0.125-inch wall thickness. Neglecting all effects of compressibility but accounting for tail boom bending, plot C_{m,δ_e} and $C_{m,\alpha}$ as a function of the airspeed at sea level from 80 to 600 mph.

Solution. For this airplane we have

$$S_w = 180 \text{ ft}^2, \quad b_w = 33 \text{ ft}, \quad l_w = 0.0 \text{ ft}, \quad C_{L_w,\alpha} = 4.44, \quad C_{m_w} = -0.053,$$
$$S_h = 36 \text{ ft}^2, \quad b_h = 12 \text{ ft}, \quad l_h = 15.0 \text{ ft}, \quad C_{L_h,\alpha} = 3.97, \quad C_{m_h 0} = 0.0,$$
$$\varepsilon_e = 0.60, \quad C_{m_h,\delta_e} = -0.55, \quad \varepsilon_{d,\alpha} = 0.44, \quad \eta_h = 1.0$$

Following Example 4.3.1 and assuming a rigid tail boom, we find that

$$\left(C_{m,\alpha}\right)_{\text{rigid}} = -\frac{l_w}{\bar{c}_w} C_{L_w,\alpha} - \frac{S_h l_h}{S_w \bar{c}_w} \eta_h C_{L_h,\alpha} (1-\varepsilon_{d,\alpha}) = -1.22276$$

$$\left(C_{m,\delta_e}\right)_{\text{rigid}} = \frac{S_h \bar{c}_h}{S_w \bar{c}_w} \eta_h C_{m_h,\delta_e} - \frac{S_h l_h}{S_w \bar{c}_w} \eta_h C_{L_h,\alpha} \varepsilon_e = -1.37060$$

For this wing-tail combination, the tail volume ratio multiplied by the tail lift slope is

$$\frac{S_h l_h}{S_w \bar{c}_w} \eta_h C_{L_h,\alpha} = \frac{36(15.0)}{180(180/33)}1.0(3.97) = 2.1835$$

The moment of inertia for a thin-walled tube of radius, r, and wall thickness, t, is

$$I = \pi r^3 t = \pi (2/12)^3 0.125/12 = 0.0001515\,\text{ft}^4$$

Thus, the tail boom stiffness is

$$EI = (10\times144\times10^6\,\text{lbf/ft}^2)0.0001515\,\text{ft}^4 = 218,160\,\text{lbf}\cdot\text{ft}^2$$

From Eqs. (6.2.8) and (6.2.9),

$$K_m \equiv \frac{1}{2}\rho V^2 S_h \bar{c}_h \eta_h \int_{x=0}^{l_h} \frac{1}{EI}\,dx$$

$$K_L \equiv \frac{1}{2}\rho V^2 S_h \eta_h \int_{x=0}^{l_h} \frac{l_h - x}{EI}\,dx$$

For a tail boom of constant cross-section

$$\int_{x=0}^{l_h} \frac{1}{EI}\,dx = \frac{l_h}{EI}$$

$$\int_{x=0}^{l_h} \frac{l_h - x}{EI}\,dx = \frac{l_h^2}{2EI}$$

Thus,

$$K_m = \frac{\rho V^2 S_h \bar{c}_h l_h \eta_h}{2EI} = \frac{\rho V^2 (36\,\text{ft}^2)(3\,\text{ft})(15.0\,\text{ft})1.0}{2(218,160\,\text{lbf}\cdot\text{ft}^2)} = \frac{\rho V^2}{269.33\,\text{lbf/ft}^2}$$

$$K_L = \frac{\rho V^2 S_h l_h^2 \eta_h}{4EI} = \frac{\rho V^2 (36\,\text{ft}^2)(15.0\,\text{ft})^2 1.0}{4(218,160\,\text{lbf}\cdot\text{ft}^2)} = \frac{\rho V^2}{107.73\,\text{lbf/ft}^2}$$

For example, at 200 mph,

$$K_m = \frac{\rho V^2}{269.33\,\text{lbf/ft}^2} = \frac{0.0023769(200\times5,280/3,600)^2}{269.33} = 0.75935$$

$$K_L = \frac{\rho V^2}{107.73 \text{ lbf/ft}^2} = \frac{0.0023769(200 \times 5{,}280/3{,}600)^2}{107.73} = 1.89838$$

Thus, from Eqs. (6.2.17) and (6.2.21),

$$
\begin{aligned}
C_{m,\delta_e} &= \left(C_{m,\delta_e}\right)_{\text{rigid}} + \left(\Delta C_{m,\delta_e}\right)_{fb} \\
&= \left(C_{m,\delta_e}\right)_{\text{rigid}} + \frac{S_h l_h}{S_w \bar{c}_w} \eta_h C_{L_h,\alpha} \left(\frac{K_L C_{L_h,\alpha}\, \varepsilon_e - K_m\, C_{m_h,\delta_e}}{1 + K_L C_{L_h,\alpha}}\right) \\
&= -1.37060 + 2.1835\left(\frac{1.89838(3.97)(0.60) - 0.75935(-0.55)}{1 + 1.89838(3.97)}\right) = \underline{-0.107}
\end{aligned}
$$

$$
\begin{aligned}
C_{m,\alpha} &= \left(C_{m,\alpha}\right)_{\text{rigid}} + \left(\Delta C_{m,\alpha}\right)_{fb} \\
&= \left(C_{m,\alpha}\right)_{\text{rigid}} + \frac{S_h l_h}{S_w \bar{c}_w} \eta_h C_{L_h,\alpha}\, (1 - \varepsilon_{d,\alpha})\left(\frac{K_L C_{L_h,\alpha}}{1 + K_L C_{L_h,\alpha}}\right) \\
&= -1.22276 + 2.1835(1 - 0.44)\left(\frac{1.89838(3.97)}{1 + 1.89838(3.97)}\right) = \underline{-0.143}
\end{aligned}
$$

Repeating these calculations for airspeeds from 80 to 600 mph, the results shown in Fig. 6.2.4 are obtained.

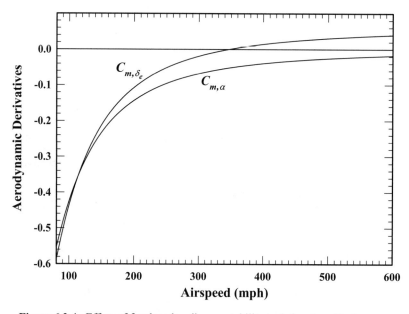

Figure 6.2.4. Effect of fuselage bending on stability and elevator effectiveness.

From Example 6.2.1, excluding compressibility effects, we have seen that both pitch stability and elevator control can be reduced drastically as a result of fuselage bending. Furthermore, from Fig. 6.2.4 we see that these effects are rapidly amplified as airspeed is increased. This could cause a very serious problem for an aircraft that enters a steep dive. The airplane might have adequate pitch stability and elevator power for normal flight conditions. However, if the tail boom stiffness is inadequate and the airplane enters a steep dive, as airspeed is quickly increased, both pitch stability and elevator power are reduced. This can cause the airplane to nose over or "tuck," which places the aircraft in an even steeper dive. With the ensuing reduction in elevator power, the airplane could become locked in the dive, and even with maximum elevator deflection, it may be impossible for the pilot to recover.

Compressibility effects at high subsonic airspeeds can amplify the tendency for an airplane to tuck under in a dive as a result of aeroelastic bending. When an airplane exceeds some critical Mach number, shock waves form on the upper surface of the wing. Such shock waves can induce premature flow separation, which increases the drag and decreases the lift. Since the downwash on an aft tail is directly related to the lift developed by the main wing, compressibility effects at high subsonic speeds can drastically reduce the downwash on an aft tail. This decrease in downwash causes an increase in the angle of attack for the tail and creates a nose-down pitching moment, which in turn aggravates any tendency for the airplane to tuck under in a dive.

Early versions of the Lockheed P-38 Lightning, shown in Fig. 6.2.5, were victims of such phenomena. Several pilots, including Ralph Virden, one of Lockheed's top test

Figure 6.2.5. Lockheed P-38 Lightning. (Photograph by Terry Gleason)

pilots, were killed as a result of the Lightning's inability to recover from a steep dive. Enemy pilots soon learned of this flaw in the P-38 design and used the trait to their advantage. The problem was solved with the addition of dive recovery flaps on the lower surface of the main wing and a small fillet at the juncture between the wing leading edge and the canopy. When the dive recovery flaps were engaged in a high-speed dive, more lift was developed on the main wing, and as a result, the downwash was increased on the tail plane. This decreased the angle of attack for the horizontal tail, making it possible for the pilot to recover from the dive.

As shown in Fig. 6.2.4, under extreme conditions aeroelastic bending of the tail boom can result in elevator control reversal. As can be seen from Eqs. (6.2.8) and (6.2.9), both K_m and K_L are directly proportional to the airspeed squared and inversely proportional to the tail boom stiffness. From this fact combined with Eq. (6.2.16), we see that elevator control reversal occurs at an airspeed that is proportional to the square root of the tail boom stiffness. **The tail boom must be stiff enough so that the elevator control reversal speed exceeds the maximum airspeed that could ever be attained by the airplane.** A conservative rule for fixing the minimum tail boom stiffness is to allow no more than 10 percent control loss at the maximum allowable airspeed.

6.3. Control Force and Trim Tabs

We have now seen that an airplane's elevator must be large enough to trim and maneuver the craft in pitch, throughout the allowable flight envelope. We have also seen that the fuselage must be stiff enough to avoid substantial aeroelastic control loss at high airspeeds. These are not, however, the only constraints that must be placed on the pitch control system of an airplane. If we are to provide good handling qualities for a human pilot, consideration should also be given to the forces that must be exerted by the pilot to initiate and maintain the required elevator deflections.

An airplane's elevator is actuated with either a *stick* or a *wheel* that pilots refer to as a *control yoke* or *control column*. The stick or control yoke is pushed forward to produce a nose-down pitching moment, and the pilot pulls back on the control to provide a nose-up pitching moment. Obviously, the force required to move the stick or control yoke must fall within acceptable limits throughout the entire flight envelope. FAR Part 23 allows a maximum temporary elevator-actuation force of 60 lbf on a stick, 50 lbf for one hand on a wheel, and 75 lbf for two hands on a wheel. For prolonged application, the elevator-actuation force cannot exceed 10 pounds for either a stick or a wheel.

In addition to limiting the maximum force required to provide the needed elevator deflection, the variation in elevator control force with airplane trim speed must be sufficient to provide the pilot with a good "feel" for the trim attitude of the airplane. As we have seen, elevator deflection is used to control angle of attack and trim the airplane for level flight at any desired airspeed. To decrease the angle of attack and increase the trim speed, the pilot must push forward on the control. The control force should increase smoothly as the trim speed is increased beyond that associated with the neutral position of the control. Likewise, the pull force required to decrease the trim speed should increase smoothly as the trim speed is decreased. The change in elevator control force with respect to aircraft trim speed is commonly called the *control force gradient*.

For longitudinal maneuvers, like the pull-up or push-over maneuver, good handling qualities require a smooth increase in the stick force with increasing load factor. This gives the pilot a good feel for the severity of the maneuver being executed. In transport aircraft, the large elevator deflection needed to induce a rapid pull-up maneuver, near the structural limit of the aircraft, should require a fairly large stick force, near the FAR Part 23 limit. To provide good handling qualities for combat or stunt aircraft, a much lower stick force is needed. In combat, a *longitudinal stick force per g* between about 4 and 10 pounds per *g* is usually acceptable. Experienced pilots can easily adapt to lower stick force gradients, which make an airplane less tiring to fly in combat.

For light aircraft, the power required to move the control surfaces is often supplied entirely by the pilot through a reversible system of mechanical cables, pulleys, gears, rods, and levers. The range of options available for mechanically connecting the cockpit controls to the aerodynamic control surfaces of an airplane is very great, and a discussion of actual mechanisms is not appropriate for this text. For our purpose it is sufficient to understand that **with *reversible mechanical controls*, a pilot-supplied control force is required that is directly proportional to the aerodynamic hinge moment needed to deflect the control surface.** Such mechanical mechanisms are called *reversible controls*, because if the cockpit control is left free, the mechanism can be operated in reverse from the tail of the aircraft. In other words, a force applied to the cockpit control will move the elevator or a force applied to the elevator will move the cockpit control.

Modern transport and fighter aircraft typically have irreversible hydraulic or electric systems for actuating the control surfaces. In these aircraft, the stick forces are provided by a system commonly called a *control loader*. There are many ways to provide such artificial feedback to the pilot, and the details of control loader design are not covered here. **With *irreversible controls*, a motor provides the power required to move each control surface, and the control force felt by the pilot can be uncoupled from the aerodynamic hinge moment needed to deflect the control surface.** Such mechanisms are called *irreversible controls*, because although a force applied to the cockpit control will move the elevator, a force applied to the elevator will not move the cockpit control.

Whether the cockpit controls are linked to the aerodynamic control surfaces with a reversible mechanical linkage or an irreversible hydraulic or "fly-by-wire" control system, knowledge of the control surface hinge moment is needed to properly design the aircraft control system. The moment that is required to rotate an airplane control surface about its hinge line is primarily a function of the aerodynamic pressure difference between the upper and lower surface. For example, consider the symmetric airfoil section of the horizontal tail that is shown schematically in Fig. 6.3.1. With no angle of attack and no elevator deflection, the aerodynamic pressure acting on the upper surface exactly balances that acting on the lower surface and no aerodynamic moment is produced about the hinge line. If this tail plane is given a positive angle of attack, the pressure on the upper surface is decreased and that on the lower surface is increased. This produces an aerodynamic hinge moment that must be countered by the control system to prevent the elevator from rotating upward. A negative angle of attack will produce an aerodynamic hinge moment in the opposite direction. Similarly, when the elevator is deflected downward, the pressure on the upper surface becomes less and the pressure on the lower surface becomes greater. This produces an aerodynamic hinge moment that opposes the

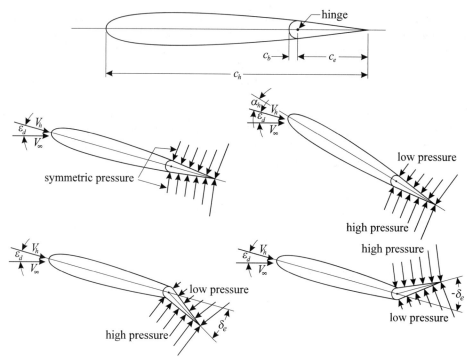

Figure 6.3.1. Aerodynamic hinge moment acting on the elevator of an aft tail.

downward rotation of the elevator. Likewise, an upward rotation of the elevator produces a downward-acting hinge moment, which also opposes the rotation of the control surface. At any given angle of attack there is one elevator deflection angle that will result in a zero aerodynamic hinge moment. Elevator deflection in either direction from this neutral elevator position will produce an opposing aerodynamic hinge moment.

Since elevator deflection controls the airplane's angle of attack and trim speed, any airspeed that requires an elevator deflection that differs from the neutral position of the elevator will result in a nonzero elevator hinge moment. If the cockpit controls have a reversible mechanical connection to the elevator, the pilot must supply a control force proportional to this hinge moment. Suppose, for example, that an airplane with an aft tail is flying at some airspeed that produces steady level flight with no force on the control (i.e., neutral elevator deflection). If the pilot wishes to increase the airspeed to a faster level trim setting, she or he must increase the power in order to balance the thrust with the drag at this higher airspeed. However, if the angle of attack remains unchanged, the increased airspeed will produce lift in excess of the weight and the airplane will begin to climb. To maintain constant altitude, the pilot must also push forward on the control to deflect the aft elevator downward and rotate the airplane to a lower angle of attack. With the proper elevator deflection, the balance between lift and weight is reestablished at the higher airspeed. To balance the aerodynamic hinge moment produced on the elevator at this new trim speed, if no other trim adjustment is supplied, the pilot will need to

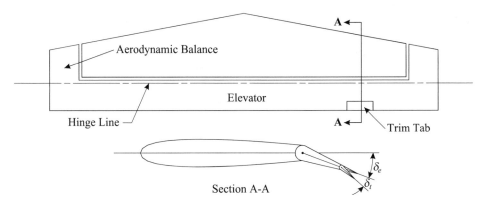

Figure 6.3.2. Elevator trim tab on an aft tail.

maintain a continuous push force to hold the control at the new trim position. When flying at constant airspeed for a long period of time, it becomes uncomfortable for a pilot to maintain even a small force on the stick or control yoke. If the cockpit controls are connected to the elevator through a reversible mechanical linkage, a trim tab similar to that shown in Fig. 6.3.2 can be used to relieve the pilot of this load.

A trim tab is a small secondary control surface located somewhere along the trailing edge of the primary control surface. Since the planform area of the trim tab is very small, it usually has only a negligible effect on the lift produced by the aerodynamic surface to which it is attached. However, because there is a fairly large moment arm between the aerodynamic center of the trim tab and the hinge line of the primary control surface, deflection of the trim tab does significantly alter the aerodynamic hinge moment for the primary control surface. A downward deflection of the trim tab produces a change in the hinge moment that tends to deflect the primary control surface upward. Conversely, when the trim tab is deflected upward, a more downward-deflecting hinge moment is produced on the primary control surface.

The aerodynamic moment produced about the hinge line of an airplane's elevator depends on the local angle of attack for the horizontal surface, the deflection angle of the elevator itself, and the deflection angle of any trim tab. An elevator hinge moment coefficient is commonly defined as

$$C_{H_e} \equiv \frac{H_e}{\frac{1}{2}\rho V_h^2 S_e \bar{c}_e} \tag{6.3.1}$$

where, according to common notation, H_e is the aerodynamic elevator hinge moment, S_e is the planform area of that portion of the elevator that is aft of the hinge line, and $\bar{c}_e$ is the mean chord for that same portion of the elevator, aft of the hinge line. Since our sign convention defines a downward rotation of the elevator to be positive, a hinge moment that tends to rotate the elevator downward is also defined as positive. According to this convention, a positive elevator deflection usually tends to produce a negative hinge moment, and a negative elevator deflection will normally produce a positive hinge moment. In other words, the change in elevator hinge moment with respect to elevator deflection angle is typically negative.

Aerodynamic theory predicts that for small angles of attack and small control surface deflections, in both subsonic and supersonic flight, the hinge moment coefficient is very nearly a linear function of the local angle of attack for the horizontal surface, the elevator deflection angle, and the trim tab deflection angle,

$$C_{H_e} = C_{H_e0} + C_{H_e,\alpha_h}\alpha_h + C_{H_e,\delta_e}\delta_e + C_{H_e,\delta_t}\delta_t \tag{6.3.2}$$

where

$$C_{H_e,\alpha_h} \equiv \frac{\partial C_{H_e}}{\partial \alpha_h}$$

$$C_{H_e,\delta_e} \equiv \frac{\partial C_{H_e}}{\partial \delta_e}$$

$$C_{H_e,\delta_t} \equiv \frac{\partial C_{H_e}}{\partial \delta_t}$$

and C_{H_e0} is the elevator hinge moment coefficient with the local angle of attack, the elevator deflection angle, and the trim tab deflection angle all set at zero. In nearly all cases, an aft tail will have a symmetric airfoil section. Thus, C_{H_e0} is usually zero for an aft tail. For a canard, C_{H_e0} is typically negative.

While experiments confirm the linear relation given by Eq. (6.3.2) for small angles, the actual magnitudes of the coefficients cannot be accurately determined by analytical means. These constants depend in a very complex way on the section and planform geometry of the elevator, the horizontal surface, and the hinge. The coefficients are very sensitive to local boundary layer separation near the hinge and the state of the boundary layer near the trailing edge. These factors vary with Reynolds number. Typically, the coefficients in Eq. (6.3.2) are determined from wind tunnel tests.

From Eq. (6.3.2) we can show how a trim tab is used to zero the hinge moment at any desired elevator deflection angle and eliminate the pilot-supplied control force at any desired trim speed. Setting the hinge moment coefficient in Eq. (6.3.2) to zero and solving for the elevator deflection angle, we obtain

$$(\delta_e)_{H_e=0} = -\frac{C_{H_e0}}{C_{H_e,\delta_e}} - \frac{C_{H_e,\alpha_h}}{C_{H_e,\delta_e}}\alpha_h - \frac{C_{H_e,\delta_t}}{C_{H_e,\delta_e}}\delta_t \tag{6.3.3}$$

From Eq. (6.3.3) we have

$$\left(\frac{\partial \delta_e}{\partial \delta_t}\right)_{H_e=0} = -\frac{C_{H_e,\delta_t}}{C_{H_e,\delta_e}} \tag{6.3.4}$$

Since C_{H_e,δ_e} and C_{H_e,δ_t} are both negative, we see that the change in the neutral elevator position with respect to the trim tab deflection angle is negative. In other words, a downward deflection of the trim tab will move the neutral elevator position upward, and an upward deflection of the tab will move the neutral elevator position downward.

The trim tab on a reversible mechanical control system is typically held in position by an irreversible mechanism, such as a worm gear or jackscrew that can be adjusted by the pilot but requires no pilot input to maintain. By adjusting the trim tab deflection angle, the pilot can change the neutral position of the elevator at any angle of attack. Thus, the pilot is able to zero out the elevator control force that is required to trim the airplane for any desired airspeed.

While a trim tab can be used to zero the reversible mechanical stick force at trim, it does not significantly alter the force needed to deflect the elevator away from its neutral position. The magnitude of the force needed for such elevator deflection is also important to the handling qualities of an airplane, since the pilot will need to make temporary changes in elevator deflection to maneuver the airplane or simply to correct for disturbances caused by turbulence.

As the size of an airplane is increased, the size of the elevator and the power needed to deflect it from its neutral position, against the aerodynamic hinge moment, will typically increase as well. Within limits, the designer has some control over the aerodynamic hinge moment produced on an elevator of a given size. While that portion of the elevator aft of the hinge line produces a negative change in hinge moment with respect to deflection angle, any part of the elevator that is forward of the hinge line will tend to make this derivative more positive. The power required to deflect the elevator from neutral can be decreased by increasing the fraction of the total elevator area that is forward of the hinge line. When a significant portion of the elevator is forward of the hinge line, as shown in Fig. 6.3.2, it is commonly called an *aerodynamic balance*.

Two important control force derivatives are commonly used to quantify the stick force needed to deflect the elevator from its neutral position with a reversible mechanical linkage. The first of these derivatives is the *control force gradient*, which is defined to be the change in elevator control force with respect to airplane trim speed for a constant trim tab setting. The second of these important control force derivatives is the change in elevator control force with respect to load factor for the constant-speed pull-up maneuver. This derivative is commonly referred to as the *longitudinal control force per g*. To gain a better understanding of these stick force derivatives, we shall now examine in more detail the longitudinal control force required for a reversible mechanical control linkage.

When the cockpit controls are connected to the aerodynamic control surfaces with a reversible mechanical linkage, the pitch control force provided by the pilot, F_p, is proportional to the aerodynamic hinge moment produced on the elevator,

$$F_p = H_e/l_g = \tfrac{1}{2}\rho V_h^2 S_e \bar{c}_e C_{H_e}/l_g \qquad (6.3.5)$$

where l_g is the length of the effective lever arm associated with the mechanism or *gearing* that links the cockpit control to the elevator. Using Eq. (6.3.2) in Eq. (6.3.5), the pitch control force can be written as

$$F_p = \tfrac{1}{2}\rho V^2 \eta_h S_e \frac{\bar{c}_e}{l_g}(C_{H_e 0} + C_{H_e,\alpha_h}\alpha_h + C_{H_e,\delta_e}\delta_e + C_{H_e,\delta_t}\delta_t) \qquad (6.3.6)$$

where η_h is the usual dynamic pressure ratio for the horizontal surface and V is the airplane's true airspeed.

Including the effect of pitch rate given by Eq. (6.1.12), the local angle of attack for the horizontal surface can be written as

$$\alpha_h = \alpha_{0h} - \varepsilon_{d0} + (1 - \varepsilon_{d,\alpha})\alpha + \frac{2l_h}{\bar{c}_w}\bar{q} \tag{6.3.7}$$

where α_{0h} is the angle that the horizontal surface makes with the fuselage reference line, ε_{d0} is the downwash angle when the airplane angle of attack relative to the fuselage reference line is zero, $\varepsilon_{d,\alpha}$ is the downwash gradient, α is the airplane angle of attack relative to the fuselage reference line, and $\bar{q}$ is the dimensionless pitch rate that was defined in Eq. (6.1.11). Using Eq. (6.3.7) in Eq. (6.3.6), the pilot-supplied pitch control force required with a reversible mechanical control linkage is

$$
F_p = \frac{\frac{1}{2}\rho V^2 \eta_h S_e \bar{c}_e}{l_g}
$$

$$
\times \left\{ C_{H_e 0} + C_{H_e,\alpha_h}\left[\alpha_{0h} - \varepsilon_{d0} + (1 - \varepsilon_{d,\alpha})\alpha + \frac{2l_h}{\bar{c}_w}\bar{q}\right] + C_{H_e,\delta_e}\delta_e + C_{H_e,\delta_t}\delta_t \right\} \tag{6.3.8}
$$

The elevator deflection affects both the airplane angle of attack and the pitch rate. Recall from Sec. 6.1 that for constant airspeed and constant normal acceleration, the angle of attack, α, the elevator deflection angle, δ_e, and the dimensionless pitch rates, $\bar{q}$ and $\breve{q}$, can all be related to the airspeed, V, and the load factor, n. From Eqs. (6.1.33), (6.1.34), (6.1.11), (6.1.28), and (6.1.42), we have

$$
\alpha = \frac{(nC_W - C_{L0})C_{m,\delta_e} + C_{L,\delta_e}C_{m0} + (C_{L,\delta_e}C_{m,\bar{q}} - C_{L,\bar{q}}C_{m,\delta_e})(n-1)(g\bar{c}_w/2V^2)}{C_{L,\alpha}C_{m,\delta_e} - C_{L,\delta_e}C_{m,\alpha}} \tag{6.3.9}
$$

$$
\delta_e = -\frac{(nC_W - C_{L0})C_{m,\alpha} + C_{L,\alpha}C_{m0} + (C_{L,\alpha}C_{m,\bar{q}} - C_{L,\bar{q}}C_{m,\alpha})(n-1)(g\bar{c}_w/2V^2)}{C_{L,\alpha}C_{m,\delta_e} - C_{L,\delta_e}C_{m,\alpha}} \tag{6.3.10}
$$

$$
\bar{q} \equiv \frac{\bar{c}_w q}{2V} = (n-1)\frac{g\bar{c}_w}{2V^2} \quad \text{and} \quad \breve{q} \equiv \frac{Vq}{g} = \frac{2V^2}{g\bar{c}_w}\bar{q} = n-1 \tag{6.3.11}
$$

Here it is important to remember that the weight coefficient, C_W, is a function of the airspeed,

$$
C_W \equiv \frac{W}{\frac{1}{2}\rho V^2 S_w} \tag{6.3.12}
$$

Using Eqs. (6.3.9) through (6.3.12) in Eq. (6.3.8), after some algebraic manipulation, the pitch control force for a reversible mechanical linkage can be written as

$$F_p = \eta_h \frac{S_e \bar{c}_e}{S_w l_g} \left[C_1 n W + C_2 (n-1) \frac{\rho g S_w \bar{c}_w}{4} + (C_3 + C_{H_e,\delta_t} \delta_t)\left(\tfrac{1}{2}\rho V^2 S_w\right) \right] \quad (6.3.13)$$

where

$$C_1 = \frac{C_{m,\delta_e} C_{H_e,\alpha_h}(1 - \varepsilon_{d,\alpha}) - C_{m,\alpha} C_{H_e,\delta_e}}{C_{L,\alpha} C_{m,\delta_e} - C_{L,\delta_e} C_{m,\alpha}} \quad (6.3.14)$$

$$C_2 = \frac{2 l_h C_{H_e,\alpha_h}}{\bar{c}_w}$$
$$+ \frac{(C_{L,\delta_e} C_{m,\bar{q}} - C_{L,\bar{q}} C_{m,\delta_e}) C_{H_e,\alpha_h}(1 - \varepsilon_{d,\alpha}) - (C_{L,\alpha} C_{m,\bar{q}} - C_{L,\bar{q}} C_{m,\alpha}) C_{H_e,\delta_e}}{C_{L,\alpha} C_{m,\delta_e} - C_{L,\delta_e} C_{m,\alpha}} \quad (6.3.15)$$

and

$$C_3 = C_{H_e 0} + C_{H_e,\alpha_h}(\alpha_{0h} - \varepsilon_{d0})$$
$$+ \frac{(C_{m,\alpha} C_{L0} - C_{L,\alpha} C_{m0}) C_{H_e,\delta_e} - (C_{m,\delta_e} C_{L0} - C_{L,\delta_e} C_{m0}) C_{H_e,\alpha_h}(1 - \varepsilon_{d,\alpha})}{C_{L,\alpha} C_{m,\delta_e} - C_{L,\delta_e} C_{m,\alpha}} \quad (6.3.16)$$

While the coefficients C_1, C_2, and C_3 are not affected by the cockpit controls, the trim tab setting can be used to adjust the value of δ_t so that the reversible mechanical control force is zero at some desired airspeed. At level trim, the load factor is unity. Thus, if we let V_{trim} denote the airspeed that requires no pitch control force for level trimmed flight with a trim tab setting of δ_t, then Eq. (6.3.13) requires

$$0 = \eta_h \frac{S_e \bar{c}_e}{S_w l_g} \left[C_1 W + (C_3 + C_{H_e,\delta_t} \delta_t)\left(\tfrac{1}{2}\rho V_{\text{trim}}^2 S_w\right) \right]$$

or

$$C_3 + C_{H_e,\delta_t} \delta_t = -\frac{W}{\tfrac{1}{2}\rho V_{\text{trim}}^2 S_w} C_1 \quad (6.3.17)$$

Using this result in Eq. (6.3.13), the **pitch control force required for constant airspeed and constant normal acceleration with a reversible mechanical linkage** is

$$F_p = \eta_h \frac{S_e \bar{c}_e}{S_w l_g} \left[C_1 \left(n - \frac{V^2}{V_{\text{trim}}^2} \right) W + C_2 (n-1) \frac{\rho g S_w \bar{c}_w}{4} \right] \quad (6.3.18)$$

In the absence of Reynolds number and Mach number effects, the **control force gradient for a reversible mechanical linkage** is obtained by differentiating Eq. (6.3.18) with respect to airspeed at constant load factor and evaluating the result at $V = V_{trim}$,

$$\left(\frac{\partial F_p}{\partial V}\right)_{\substack{n=\text{constant} \\ V=V_{trim}}} = -2\eta_h \frac{S_e \bar{c}_e}{S_w l_g} C_1 \frac{W}{V_{trim}} \qquad (6.3.19)$$

Similarly, the **longitudinal control force per g for a reversible mechanical linkage** is found by differentiating Eq. (6.3.18) with respect to load factor at constant airspeed,

$$\left(\frac{\partial F_p}{\partial n}\right)_{V=\text{constant}} = \eta_h \frac{S_e \bar{c}_e}{S_w l_g}\left(C_1 W + C_2 \frac{\rho g S_w \bar{c}_w}{4}\right) \qquad (6.3.20)$$

From Eq. (6.3.19) we see that for a reversible mechanical linkage, the magnitude of the control force gradient is proportional to aircraft gross weight, inversely proportional to trim speed, and independent of altitude. Thus, reversible mechanical controls become heavier when gross weight is increased or trim speed is decreased. From Eq. (6.3.20) we see that for a reversible mechanical linkage, the longitudinal control force per g also increases linearly with gross weight. However, apart from Mach number and Reynolds number effects, the control force per g for a reversible mechanical linkage is independent of airspeed. Altitude affects the control force per g only through the second term in Eq. (6.3.20). Comparing the two equations, we find that the change in the two control force derivatives with gross weight have opposite signs. The coefficient C_1 is positive, so the control force gradient decreases with gross weight, whereas the control force per g increases with gross weight. This is simply a result of our sign convention for the control force. This sign convention, which was implicitly stated in Eq. (6.3.5), specifies that a pull force is positive and a push force is negative. The difference in sign between Eqs. (6.3.19) and (6.3.20) results from the fact that increasing the trim speed requires a push force, whereas increasing the load factor requires a pull force.

Note that both control force derivatives are proportional to the ratio $S_e\bar{c}_e/S_w l_g$. For geometrically similar airplanes of different size, the ratio of elevator area to wing area does not change appreciably. Thus, in addition to the effect of weight, we see that aircraft size increases the control force derivatives in direct proportion to the elevator chord length, which is roughly proportional to the linear size of the airplane. When the effects of weight and linear size are combined, we see that the magnitudes of the control force derivatives for a reversible mechanical linkage increase very rapidly with increasing aircraft size.

Increasing the length of the lever arm associated with the mechanical gearing can decrease the magnitude of the control force derivatives. However, this is severely limited by the range of motion of the pilot's arms. When the aerodynamic control surfaces are connected to the cockpit controls with a reversible mechanical linkage, the control force can be reduced by changing the gearing. However, this does not decrease the mechanical work required for a given control surface deflection. For example, if the control force were cut in half by doubling the mechanical gear ratio, the required displacement of the pilot's control would be doubled. The pilot can comfortably move the stick or control yoke only through a limited range. Thus, the designer's ability to vary the control force with mechanical gearing is very limited. Because the reversible mechanical control force increases rapidly with aircraft size, large airplanes typically have irreversible controls, which utilize hydraulic or electric power boost or complete fly-by-wire control systems.

Because several of the aerodynamic derivatives used in the definitions of C_1 and C_2 depend on the position of the center of gravity, the pitch control force for a reversible mechanical linkage is also CG dependent. The nature of this dependence is not obvious from Eq. (6.3.18). This is examined in the following example.

EXAMPLE 6.3.1. For the wing-tail combination that is described in Examples 4.4.2 and 6.1.1 with a reversible mechanical control linkage, plot the control force gradient and longitudinal control force per g at sea level and 80 mph as a function of CG location measured relative to the quarter chord of the main wing. For this elevator and stabilizer the change in the hinge moment coefficient with angle of attack is -0.63 and the change in the hinge moment coefficient with elevator deflection is -0.97, based on a 12-ft^2 elevator with a mean chord of 1.1 feet. Assume that the reversible mechanical linkage between the stick and the elevator has an effective lever arm of 2.0 feet.

Solution. Again we let x denote the distance measured aft of the wing quarter chord. For this airplane we have

$$S_w = 180 \text{ ft}^2, \quad b_w = 33 \text{ ft}, \quad C_{L_w,\alpha} = 4.44, \quad C_{m_w} = -0.053, \quad W = 2,700 \text{ lbf},$$

$$S_h = 36 \text{ ft}^2, \quad b_h = 12 \text{ ft}, \quad C_{L_h,\alpha} = 3.97, \quad C_{m_h0} = 0.0, \quad C_{m_h,\delta_e} = -0.55,$$

$$S_e = 12 \text{ ft}^2, \quad \bar{c}_e = 1.1 \text{ ft}, \quad \varepsilon_e = 0.60, \quad x_h = 15 \text{ ft}, \quad \eta_h = 1.0, \quad \varepsilon_{d,\alpha} = 0.44,$$

$$C_{H_e,\alpha_h} = -0.63, \quad C_{H_e,\delta_e} = -0.97, \quad l_g = 2.0 \text{ ft}$$

By definition

$$l_w = x_w - x_{CG} = -x_{CG}$$

$$l_h = x_h - x_{CG}$$

Using results obtained in Examples 4.3.1, 4.4.1, 4.4.2, and 6.1.1,

$$C_{L,\alpha} = C_{L_w,\alpha} + \frac{S_h}{S_w}\eta_h C_{L_h,\alpha}(1 - \varepsilon_{d,\alpha}) = 4.8846$$

$$C_{L,\bar{q}} = 2\frac{-x_{CG}}{\bar{c}_w}C_{L_w,\alpha} + 2\eta_t \frac{S_t(x_t - x_{CG})}{S_w\bar{c}_w}C_{L_t,\alpha}$$

$$= 4.367 - 10.468\hat{x}_{CG}$$

$$C_{L,\delta_e} = \frac{S_h}{S_w}\eta_h C_{L_h,\alpha}\varepsilon_e = 0.4764$$

$$C_{m,\alpha} = -\frac{-x_{CG}}{\overline{c}_w}C_{L_w,\alpha} - \eta_t\frac{S_t(x_t - x_{CG})}{S_w\overline{c}_w}C_{L_t,\alpha}(1 - \varepsilon_{d,\alpha})$$

$$= 4.8846\,\hat{x}_{CG} - 1.2228$$

$$C_{m,\overline{q}} = -2\frac{(-x_{CG})^2}{\overline{c}_w^2}C_{L_w,\alpha} - 2\eta_h\frac{S_h(x_h - x_{CG})^2}{S_w\overline{c}_w^2}C_{L_h,\alpha}$$

$$= -12.009 + 8.734\,\hat{x}_{CG} - 10.468\,\hat{x}_{CG}^2$$

$$C_{m,\delta_e} = \frac{S_h\overline{c}_h}{S_w\overline{c}_w}\eta_h C_{m_h,\delta_e} - \frac{S_h(x_h - x_{CG})}{S_w\overline{c}_w}\eta_h C_{L_h,\alpha}\varepsilon_e$$

$$= 0.4764\,\hat{x}_{CG} - 1.3706$$

where $\hat{x}_{CG} = x_{CG}/\overline{c}_w$. Using these results in Eq. (6.3.14), we obtain

$$
\begin{aligned}
C_1 &= \frac{-0.3528(C_{m,\delta_e}) + 0.97(C_{m,\alpha})}{4.88464(C_{m,\delta_e}) - 0.4764(C_{m,\alpha})} \\
&= \frac{-0.3528(0.4764\,\hat{x}_{CG} - 1.3706) + 0.97(4.8846\,\hat{x}_{CG} - 1.2228)}{4.8846(0.4764\,\hat{x}_{CG} - 1.3706) - 0.4764(4.8846\,\hat{x}_{CG} - 1.2228)} \\
&= 0.11494 - 0.74767\,\hat{x}_{CG}
\end{aligned}
$$

Similarly, using Eq. (6.3.15) with considerable algebraic manipulation gives

$$C_2 = 5.01202 - 0.80192\,\hat{x}_{CG}$$

For level trimmed flight at 80 mph, Eq. (6.3.19) results in

$$\left(\frac{\partial F_p}{\partial V}\right)_{\substack{n=\text{constant}\\V=V_{\text{trim}}}} = -2\eta_h\frac{S_e\overline{c}_e}{S_w l_g}C_1\frac{W}{V_{\text{trim}}}$$

$$= \underline{(-0.1940 + 1.262\,\hat{x}_{CG})}\ \text{lbf}\cdot\text{s/ft}$$

and at sea level, Eq. (6.3.20) reduces to

$$\left(\frac{\partial F_p}{\partial n}\right)_{V=\text{constant}} = \eta_h\frac{S_e\overline{c}_e}{S_w l_g}\left(C_1 W + C_2\frac{\rho g S_w\overline{c}_w}{4}\right)$$

$$= \underline{(14.83 - 74.57\,\hat{x}_{CG})}\ \text{lbf/g}$$

Thus, with the approximations used in the present example, the control force derivatives for a reversible mechanical linkage vary linearly with the axial position of the CG. Results for the particular wing-tail combination used in this example are shown in Figs. 6.3.3 and 6.3.4 for trim speeds of 80, 120, and 160 mph.

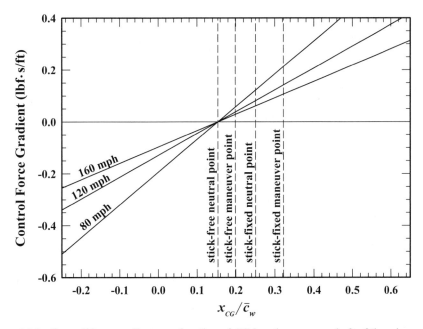

Figure 6.3.3. Control force gradient as a function of CG location measured aft of the wing quarter chord for the airplane and reversible mechanical control linkage in Example 6.3.1.

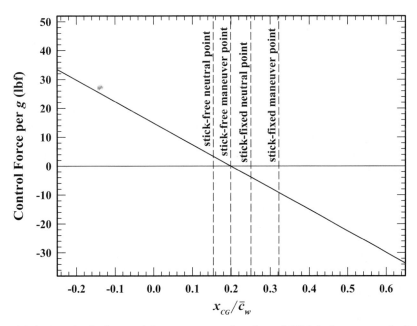

Figure 6.3.4. Longitudinal control force per g as a function of CG location measured aft of the wing quarter chord for the airplane and reversible mechanical control linkage in Example 6.3.1.

The exact linear nature of the relations found in Example 6.3.1 is a direct result of neglecting the vertical offsets. If the vertical offsets were accounted for as described in Sec. 4.7, then the relations would no longer be exactly linear. However, for a reversible mechanical control linkage, the variation in pitch control force with the location of the center of gravity is very nearly linear for most airplanes.

For the airplane in Example 6.3.1, with the CG located near the wing quarter chord, a stick force of about 15 pounds per g is required for the constant-speed pull-up maneuver. This means that executing a 4-g pull-up, which is near the structural limit of most general aviation airplanes, would require a stick force of about 60 pounds, which corresponds to the FAR Part 23 limit for a stick. With the CG so located, even at speeds above the maneuver speed, the pilot would need to exert considerable effort to initiate a pull-up that would damage the aircraft. It is not likely that such a maneuver could be initiated unintentionally.

In all cases, as the center of gravity is moved forward, reversible mechanical controls becomes heavier, and as the CG is moved aft, reversible mechanical controls becomes lighter. Notice that the control force gradient goes to zero for a CG location that is forward of the stick-fixed neutral point. The CG location that produces a zero reversible mechanical control force gradient is called the *stick-free neutral point*. The CG location that results in a zero control force per g is aft of the stick-free neutral point but forward of the stick-fixed maneuver point. The CG location that gives a zero reversible mechanical control force per g is normally referred to as the *stick-free maneuver point*.

6.4. Stick-Free Neutral and Maneuver Points

The physical significance of the stick-free neutral point and stick-free maneuver point has often been misinterpreted. For this reason, some space has been devoted here to examining these two parameters. First, we discuss how the stick-free neutral point gets its name.

Consider a hypothetical elevator that has no mass and rotates about a frictionless hinge. Assume that this elevator is coupled to the cockpit control with a reversible mechanical linkage that also has no mass and no friction. If no force were applied to the cockpit control of this hypothetical elevator, then the elevator deflection would always instantaneously take the value that would make the hinge moment go to zero. That is, from Eqs. (6.3.3) and (6.3.7),

$$\left(\delta_e\right)_{H_e=0} = -\frac{C_{H_e 0}}{C_{H_e,\delta_e}} - \frac{C_{H_e,\alpha_h}}{C_{H_e,\delta_e}}\left[\alpha_{0h} - \varepsilon_{d0} + (1-\varepsilon_{d,\alpha})\alpha + \frac{2l_h}{\bar{c}_w}\bar{q}\right] - \frac{C_{H_e,\delta_t}}{C_{H_e,\delta_e}}\delta_t \quad (6.4.1)$$

What is normally referred to as the lift slope for a complete airplane is defined to be the change in lift coefficient for the complete airplane differentiated with respect to angle of attack at constant elevator deflection. That is,

$$C_{L,\alpha} \equiv \left(\frac{\partial C_L}{\partial \alpha}\right)_{\delta_e = \text{constant}} \quad (6.4.2)$$

Similarly, what is commonly referred to as the *pitch stability derivative* is defined to be the change in pitching moment coefficient with respect to angle of attack at constant elevator deflection,

$$C_{m,\alpha} \equiv \left(\frac{\partial C_m}{\partial \alpha}\right)_{\delta_e=\text{constant}} \tag{6.4.3}$$

For our hypothetical massless and frictionless elevator, we can define a stick-free lift slope to be the change in lift coefficient with respect to angle of attack at zero hinge moment. From Eqs. (6.4.1) and (6.4.2), this gives

$$C'_{L,\alpha} \equiv \left(\frac{\partial C_L}{\partial \alpha}\right)_{H_e=0} = \left(\frac{\partial C_L}{\partial \alpha}\right)_{\delta_e=\text{constant}} + \left(\frac{\partial C_L}{\partial \delta_e}\right)_{\alpha=\text{constant}} \left(\frac{\partial \delta_e}{\partial \alpha}\right)_{H_e=0}$$
$$= C_{L,\alpha} - \frac{C_{L,\delta_e} C_{H_e,\alpha_h}(1-\varepsilon_{d,\alpha})}{C_{H_e,\delta_e}} \tag{6.4.4}$$

Similarly, from Eqs. (6.4.1) and (6.4.3), we define the stick-free pitch stability derivative for this massless and frictionless elevator as

$$C'_{m,\alpha} \equiv \left(\frac{\partial C_m}{\partial \alpha}\right)_{H_e=0} = \left(\frac{\partial C_m}{\partial \alpha}\right)_{\delta_e=\text{constant}} + \left(\frac{\partial C_m}{\partial \delta_e}\right)_{\alpha=\text{constant}} \left(\frac{\partial \delta_e}{\partial \alpha}\right)_{H_e=0}$$
$$= C_{m,\alpha} - \frac{C_{m,\delta_e} C_{H_e,\alpha_h}(1-\varepsilon_{d,\alpha})}{C_{H_e,\delta_e}} \tag{6.4.5}$$

If we neglect the effects of vertical offsets, the stick-fixed neutral point is related to the lift slope and the pitch stability derivative according to Eq. (4.4.13),

$$\frac{l_{np}}{\bar{c}_w} = -\frac{C_{m,\alpha}}{C_{L,\alpha}} \tag{6.4.6}$$

For our hypothetical massless and frictionless elevator, we can define a stick-free neutral point as

$$\frac{l'_{np}}{\bar{c}_w} = -\frac{C'_{m,\alpha}}{C'_{L,\alpha}} \tag{6.4.7}$$

This is exactly the CG location where the longitudinal control force gradient goes to zero, as was shown in Fig. 6.3.3, hence the name.

The stick-free neutral point is the aerodynamic center that an airplane would have with no pilot control force if the elevator and its reversible mechanical linkage to the cockpit control had no mass or friction. The mass and friction associated with any real elevator and reversible mechanical control linkage will move the actual stick-free aerodynamic center of an airplane considerably aft of the point given by Eq. (6.4.7).

Thus, the name *stick-free neutral point* is somewhat misleading. **More appropriately, the stick-free neutral point should be thought of simply as the CG location for which the stick force gradient goes to zero for a reversible mechanical control linkage.**

The stick-fixed maneuver point for an airplane is that CG location for which the elevator angle per g goes to zero. In Sec. 6.1 it was stated that the elevator deflection required to produce constant normal acceleration is reversed when the center of gravity is located aft of the stick-fixed maneuver point. In that context, control reversal was implicitly defined in terms of elevator deflection. Under normal conditions, with the CG located forward of the stick-fixed maneuver point, downward rotation of an aft elevator, which occurs when the pilot moves the control stick forward, will result in a downward increment in the normal acceleration. Conversely, when the pilot moves the control stick back to rotate the aft elevator upward, the resulting increment in normal acceleration is upward. If the CG were located aft of the stick-fixed maneuver point, this relationship between control stick displacement and steady normal acceleration would be reversed.

With reversible mechanical controls and typical CG location, the pilot exerts a pull force to move the stick back and initiate upward acceleration from trim. Conversely, a push force is needed to move the stick forward and initiate downward acceleration. From the result shown in Fig. 6.3.4, one might conclude that when the CG is located aft of the stick-free maneuver point, this normal relation between control force and the initiation of normal acceleration is reversed. This is also misleading. We must remember that the control force per g, which is plotted in Fig. 6.3.4, is defined in terms of the *steady-state load factor*. This is the control force required when the pitch damping moment exactly balances the pitching moment produced by the elevator deflection. This says nothing about the stick force required to initiate a normal acceleration from trimmed flight. **Even when the CG is located at the stick-free maneuver point, a mechanical pull force is typically required to initiate upward acceleration and a push force is required to initiate downward acceleration.** However, when the CG is located near the stick-free maneuver point, these forces become very small with reversible mechanical controls.

In summary, the stick-free neutral point is defined to be the CG location for which the equilibrium stick force gradient goes to zero for a reversible mechanical control system. The stick-free maneuver point is defined as the CG location for which the equilibrium stick force per g goes to zero for a reversible mechanical control system. **Any interpretation of these parameters beyond those by which they were defined should be made with extreme caution.** The stick-free neutral and maneuver points for an airplane are of little or no importance in the design of an irreversible fly-by-wire control system.

6.5. Ground Effect, Elevator Sizing, and CG Limits

During takeoff and landing an airplane must briefly fly in very close proximity to the ground. Under such conditions the flow around the airplane is significantly modified by the presence of the ground. The resulting changes in lift and drag are commonly referred to as *ground effect*. The vorticity trailing aft of a lifting wing is substantially diminished in ground effect. This reduces the downwash that is induced on the wing and other components of the aircraft. The reduction in downwash increases the lift slope, decreases the induced drag, and shifts the stick-fixed neutral point aft.

The shift in the stick-fixed neutral point that results from ground effect has an important influence on aircraft controls. For a conventional airplane with an aft tail, there are three primary factors that contribute to this shift in the neutral point: an increase in lift slope for the wing, an increase in lift slope for the tail, and a reduction in the downwash induced on the tail by the main wing. All three of these factors arise directly from a reduction in downwash. The increase in wing lift slope will move the stick-fixed neutral point toward the aerodynamic center of the wing, and the increase in lift slope for the horizontal stabilizer will move the neutral point toward the tail. These two effects tend to cancel. As can be seen from Eq. (4.4.3), if the lift slopes for the wing and horizontal stabilizer are increased in the same proportion, there is no net shift in the neutral point resulting directly from these lift slope changes. However, for most conventional airplanes, ground effect increases the lift slope for the wing by more than that for the horizontal stabilizer. Thus, lift slope changes resulting from ground effect will normally produce a slight forward shift in the neutral point. On the other hand, ground effect also reduces the downwash induced on the horizontal stabilizer by the main wing, and since this downwash is destabilizing, its reduction produces an aft shift in the neutral point. Typically, the aft shift caused by the decrease in $\varepsilon_{d,\alpha}$ is much greater than the forward shift caused by the change in lift slope. As a result, ground effect usually produces a substantial rearward shift in the stick-fixed neutral point for a conventional airplane.

Moving the stick-fixed neutral point farther aft affects the airplane controls in much the same way as moving the center of gravity forward. Thus, ground effect increases the amount of up elevator required to trim the airplane at low airspeeds. In a similar manner, ground effect also increases the magnitude of the control force gradient for a reversible mechanical linkage. For airplanes with reversible all-mechanical control systems, this causes the control to become heavier as the airplane approaches the ground.

The fundamental analytical methods presented in this and the earlier chapters apply to airplanes in ground effect as well as in flight remote from the ground. The only significant difference is that some of the aerodynamic derivatives change as the airplane approaches the ground. Thus, the problem of evaluating stability and control for aircraft flying close to the ground is resolved into that of estimating the aerodynamic derivatives in ground effect. For the dimensions and airspeeds encountered in most atmospheric flight, ground effect is essentially an inviscid phenomenon. Thus, a potential flow solution will normally provide a satisfactory estimate for the aerodynamic changes produced by ground effect.

One convenient way to model an aircraft in ground effect for use with a potential flow algorithm is to replace the surface of the ground with an *image* of the aircraft, positioned and oriented as though it were reflected in the surface of the ground. Such a model is shown in Fig. 6.5.1. By design, the flow around this aircraft combined with its mirror image is symmetric across the plane of reflection. At any point on this plane of symmetry, the downwash generated by the aircraft will be exactly offset with upwash generated by its mirror image. Thus, there can be no net flow normal to the plane of reflection, which is accordingly a stream surface for the flow. This means that potential flow about the aircraft combined with its mirror image is identical to potential flow about the aircraft combined with a flat solid surface representing the ground. Since the two flows are identical, the associated aerodynamic forces will be identical as well.

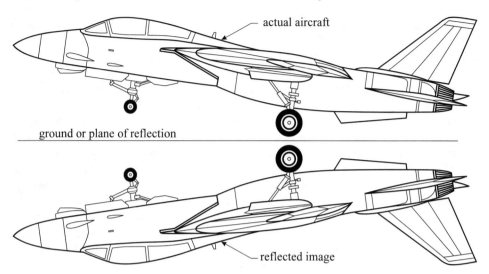

Figure 6.5.1. Mirror image model used to simulate ground effect.

The model shown in Fig. 6.5.1 can be used with either the vortex panel method or the vortex lifting-line method to obtain a detailed numerical estimate for the important aerodynamic derivatives, which affect an airplane's controls in ground effect. However, since lift slope effects for the wing and horizontal stabilizer tend to cancel, a reasonable first estimate can be obtained by considering only the reduction in downwash gradient induced on the tail by the main wing. This procedure is best demonstrated by example.

EXAMPLE 6.5.1. For the wing-tail combination that is described in Examples 4.3.1, 4.4.1, 4.4.2, and 4.5.1, estimate the downwash gradient induced on the horizontal stabilizer when the airplane is flying in ground effect with the main wing 4.0 feet above the ground. Also estimate the static margin.

Solution. For this estimate we will ignore changes in lift slope and will model the downwash using the vortex system described in Sec. 4.5. To account for ground effect, we combine this vortex system with an image of the vortex system that is positioned and oriented as though it were reflected in the surface of the ground, much like the image shown in Fig. 6.5.1. From Example 4.5.1, the geometry associated with the primary vortex system gives

$$S_w = 180 \text{ ft}^2, \quad b_w = 33 \text{ ft}, \quad R_{A_w} = 33^2/180 = 6.05, \quad C_{L_w,\alpha} = 4.44, \quad R_{T_w} = 0.40,$$
$$\Lambda = 10°, \quad l_h - l_w = 15 \text{ ft}, \quad \bar{x} = 15/(33/2) = 0.909, \quad \bar{y} = 3.4/(33/2) = 0.206$$

which was found to yield

$$\kappa_v = 1.035, \quad \kappa_b = 0.759, \quad \kappa_p = 0.433, \quad \kappa_s = 1.015$$

From Eq. (4.5.11), the downwash induced on the horizontal stabilizer by the primary vortex system was found to be

$$(\varepsilon_d)_{\text{primary}} = \left(\frac{\kappa_v \kappa_p \kappa_s}{\kappa_b}\right)_{\text{primary}} \frac{C_{L_w}}{R_{A_w}} = \frac{1.035(0.433)1.015}{0.759} \frac{C_{L_w}}{R_{A_w}} = 0.599 \frac{C_{L_w}}{R_{A_w}}$$

As shown in Fig. 6.5.2, the mirror image of this vortex system is identical to the primary system, except for the direction of rotation and its position relative to the horizontal stabilizer of the actual airplane. The image vortex system is located a distance below the ground equal to the height of the wing above the ground. Thus, the height of the horizontal stabilizer above the image vortex system is twice the distance from the wing to the ground plus the distance that the horizontal stabilizer is above the primary vortex system. Therefore, relative to the image vortex system, we have

$$\bar{y} = (2 \times 4.0 + 3.4)/(33/2) = 0.691$$

From Eqs. (4.5.6) and (4.5.12), the position factor and sweep factor for the image vortex system are found to be

$$(\kappa_p)_{\text{image}} = 0.244, \quad (\kappa_s)_{\text{image}} = 1.001$$

The span factor, κ_b, for the image vortex system is the same as that for the primary system. Since the image vortex rotates in the opposite direction from the primary vortex, the strength factor, κ_v, for the image vortex is equal in magnitude but opposite in sign to that of the primary vortex. Thus, from Eq. (4.5.11), the downwash induced on the horizontal stabilizer by the image vortex system is

$$(\varepsilon_d)_{\text{image}} = \left(\frac{\kappa_v \kappa_p \kappa_s}{\kappa_b}\right)_{\text{image}} \frac{C_{L_w}}{R_{A_w}} = \frac{(-1.035)(0.244)1.001}{0.759} \frac{C_{L_w}}{R_{A_w}} = -0.333 \frac{C_{L_w}}{R_{A_w}}$$

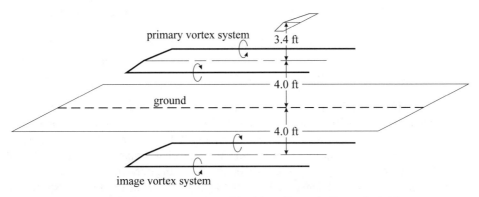

Figure 6.5.2. Vortex system and its mirror image for Example 6.5.1.

The net downwash in ground effect is then

$$\varepsilon_d = (\varepsilon_d)_{\text{primary}} + (\varepsilon_d)_{\text{image}} = (0.599 - 0.333)\frac{C_{L_w}}{R_{A_w}} = 0.266\frac{C_{L_w}}{R_{A_w}}$$

For this first approximation we are neglecting the change in lift slope caused by ground effect. Thus, the downwash gradient in ground effect is estimated to be

$$\varepsilon_{d,\alpha} = 0.266\frac{C_{L_w,\alpha}}{R_{A_w}} = 0.266\frac{4.44}{6.05} = \underline{0.195}$$

From Example 4.4.1, the center of gravity is located 0.71 foot aft of the aerodynamic center of the wing. Thus we have

$$S_w = 180 \text{ ft}^2, \quad b_w = 33 \text{ ft}, \quad C_{L_w,\alpha} = 4.44, \quad l_w = -0.71 \text{ ft},$$

$$S_h = 36 \text{ ft}^2, \quad b_h = 12 \text{ ft}, \quad C_{L_h,\alpha} = 3.97, \quad l_h = 14.29 \text{ ft}, \quad \eta_h = 1.0$$

Using these values combined with the downwash gradient that was predicted for the horizontal stabilizer in ground effect and following Example 4.4.1, the static margin in ground effect is estimated to be

$$\frac{l_{np}}{\bar{c}_w} = \frac{l_w C_{L_w,\alpha} + \dfrac{S_h l_h}{S_w}\eta_h C_{L_h,\alpha}(1-\varepsilon_{d,\alpha})}{\bar{c}_w\left[C_{L_w,\alpha} + \dfrac{S_h}{S_w}\eta_h C_{L_h,\alpha}(1-\varepsilon_{d,\alpha})\right]} = 0.2158 = \underline{22\%}$$

For the particular geometry and flight condition that were used in Example 6.5.1, we see that ground effect reduces the downwash on the horizontal stabilizer by more than 50 percent and nearly doubles the static margin. Since the computations used in this example were made neglecting changes in lift slope, the estimated changes in downwash and static margin are somewhat greater than would be observed for this airplane in actual flight.

The decrease in downwash that results from ground effect increases the local angle of attack for the horizontal stabilizer. This increases the lift on the aft stabilizer and produces a negative pitching moment about the center of gravity. To counter this nose-down pitching moment and maintain trim, some additional up elevator is required in ground effect. Again this is best demonstrated by example.

EXAMPLE 6.5.2. We wish to examine the near-ground trim requirements for the wing-tail combination described in Example 4.4.2. For the ground-effect operating condition described in Example 6.5.1, plot the elevator deflection required to trim the airplane for 80 mph at sea level as a function of the location of the center of gravity measured aft of the wing quarter chord. Compare these results with those obtained in Example 4.4.2 for flight at the same airspeed and altitude but out of ground effect.

Solution. Again we let x denote the distance measured aft of the quarter chord of the main wing and we shall continue to ignore changes in lift slope caused by ground effect. For this wing-tail combination we have

$$S_w = 180 \text{ ft}^2, \quad b_w = 33 \text{ ft}, \quad C_{L_w,\alpha} = 4.44, \quad C_{m_w} = -0.053, \quad W = 2,700 \text{ lbf},$$

$$S_h = 36 \text{ ft}^2, \quad b_h = 12 \text{ ft}, \quad C_{L_h,\alpha} = 3.97, \quad C_{m_h,\delta_e} = -0.55,$$

$$\varepsilon_e = 0.60, \quad x_h = 15 \text{ ft}, \quad \eta_h = 1.0,$$

$$\alpha_{L0w} = -2.20°, \quad \alpha_{0w} = 3.1°, \quad \alpha_{L0h} = 0.0°, \quad \alpha_{0h} = 2.3°$$

For this geometry, the lift coefficient for the main wing at zero fuselage angle of attack is

$$C_{L_w0} = C_{L_w,\alpha}(\alpha_{0w} - \alpha_{L0w}) = 0.4107$$

Thus, from Example 6.5.1 for the specified ground-effect flight condition, the downwash angle for the horizontal stabilizer at zero fuselage angle of attack is

$$\varepsilon_{d0} = 0.266 \frac{C_{L_w0}}{R_{A_w}} = 0.0181 = 1.03°$$

Following Example 4.4.2, we obtain

$$C_{L0} = C_{L_w,\alpha}(\alpha_{0w} - \alpha_{L0w}) + \frac{S_h}{S_w} \eta_h C_{L_h,\alpha}(\alpha_{0h} - \varepsilon_{d0}) = 0.4282$$

$$C_{L,\delta_e} = \frac{S_h}{S_w} \eta_h C_{L_h,\alpha} \varepsilon_e = 0.4764$$

$$C_L = \frac{W}{\frac{1}{2}\rho V_\infty^2 S_w} = 0.9168$$

For these sample calculations we continue to follow Example 4.4.2 for the case where the center of gravity is located at $x_{CG}/\bar{c}_w = 0.1$. In ground effect, we obtain

$$C_{m0} = C_{m_w} + \frac{x_{CG}}{\bar{c}_w} C_{L_w,\alpha}(\alpha_{0w} - \alpha_{L0w}) - \frac{S_h(x_h - x_{CG})}{S_w \bar{c}_w} \eta_h C_{L_h,\alpha}(\alpha_{0h} - \varepsilon_{d0})$$

$$= -0.05840$$

$$C_{m,\delta_e} = \frac{S_h \bar{c}_h}{S_w \bar{c}_w} \eta_h C_{m_h,\delta_e} - \frac{S_h(x_h - x_{CG})}{S_w \bar{c}_w} \eta_h C_{L_h,\alpha} \varepsilon_e = -1.3230$$

$$\frac{x_{np}}{\bar{c}_w} = \frac{x_w C_{L_w,\alpha} + \frac{S_h x_h}{S_w} \eta_h C_{L_h,\alpha}(1 - \varepsilon_{d,\alpha})}{\bar{c}_w \left[C_{L_w,\alpha} + \frac{S_h}{S_w} \eta_h C_{L_h,\alpha}(1 - \varepsilon_{d,\alpha}) \right]} = 0.3460$$

$$(\delta_e)_{\text{trim}} = -\frac{C_{m0} - (C_L - C_{L0})[(x_{np} - x_{CG})/\bar{c}_w]}{C_{m,\delta_e} + C_{L,\delta_e}[(x_{np} - x_{CG})/\bar{c}_w]} = -0.1481 = \underline{-8.5°}$$

This compares to an elevator deflection of about −4 degrees for the same operating condition outside of ground effect.

Repeating the previous computations for several different values of $x_{CG}/\bar{c}_w$, we obtain the result shown in Fig. 6.5.3. It should be remembered that these computations are based on constant elevator effectiveness. In reality, elevator effectiveness decreases when the deflection angle exceeds about 10 degrees. Thus, even larger elevator deflections would be required for the most forward CG locations shown in Fig. 6.5.3.

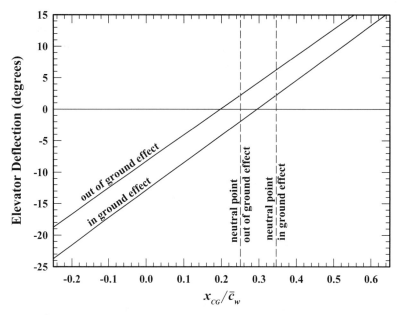

Figure 6.5.3. Elevator angle required for level trim at 80 mph and sea level as a function of CG location measured aft of the wing quarter chord for the airplane in Example 6.5.2.

From Example 6.5.2 we see that trimming a conventional airplane in ground effect requires more up elevator than is required for the same operating condition remote from the ground. Additionally, from the results obtained in Example 4.4.2 and plotted in Fig. 4.4.3, we have seen that reducing airspeed also requires more up elevator for trim. Because it is desirable to land an airplane as slowly as possible, the combination of low airspeed and ground effect often make the landing case a critical constraint on elevator size and forward CG limit. In a perfect *full-stall landing*, the airplane will stall just as the wheels touch the ground. Thus, the elevator must be sized and the forward CG limit set such that the elevator angle required to trim the airplane in ground effect for a lift coefficient of C_{Lmax} does not exceed full up elevator. If the forward CG limit has been specified by some other means, this constraint can be used to size the elevator. On the other hand, for a given elevator geometry, the constraint could be used to specify one possible candidate for the forward CG limit.

In Sec. 6.1 we learned that another possible candidate for the forward-center-of-gravity limit is the location that requires full up elevator to produce the maximum allowable load factor at the maneuver speed. For the same wing-tail combination used for Examples 6.1.1, 6.1.3, 6.5.1 and 6.5.2, we found that the forward CG limit based on a 4-g load factor at the maneuver speed was 0.48 foot forward of the wing quarter chord. This was demonstrated in Fig. 6.1.5. Repeating the computations that were used to obtain Fig. 6.1.5 with the same CG location but using the ground-effect downwash gradient obtained from Example 6.5.1 produces the results shown in Fig. 6.5.4. By comparing Figs. 6.1.5 and 6.5.4, we see that ground effect substantially reduces the load factor that can be produced with maximum elevator deflection. In fact, Fig. 6.5.4 shows that this airplane cannot even be trimmed in ground effect for the 1-g stall speed with maximum up elevator when the CG is 0.48 foot forward of the wing quarter chord. This means that for this airplane, the forward-center-of-gravity limit based on landing is more restrictive than that based on in-flight maneuvering.

Just being able to trim the airplane in ground effect for a full-stall landing may not be sufficient if a go around becomes necessary. The pull-up maneuver required to abort a landing in ground effect is stall limited, because airplanes are not normally landed at airspeeds more than 20 percent above stall. Thus, another candidate for the forward CG limit is the location that requires full up elevator to produce a stall-limited pull-up

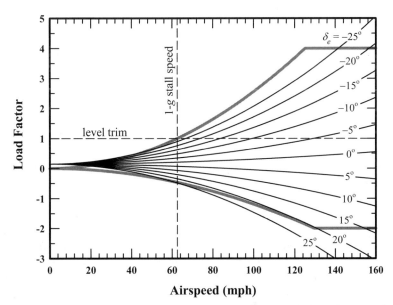

Figure 6.5.4. Elevator-angle contours on the V-n diagram for the airplane in Example 6.1.3 in ground effect with the CG located 0.48 foot forward of the wing quarter chord.

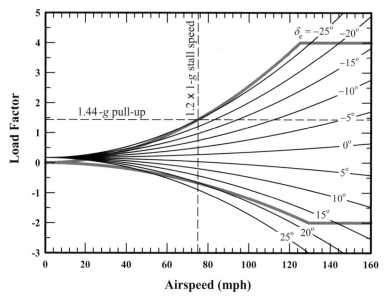

Figure 6.5.5. Elevator-angle contours on the *V-n* diagram for the airplane in Example 6.1.3 in ground effect with the CG located 0.07 foot forward of the wing quarter chord.

maneuver at 1.2 times the 1-*g* stall speed (i.e., a load factor of 1.2 squared, or 1.44). This condition is shown in Fig. 6.5.5 for the same airplane and ground-effect landing conditions as those used for Fig. 6.5.4, except that for the results shown in Fig. 6.5.5 the CG was moved aft to a point 0.07 foot forward of the wing quarter chord.

When a pilot sets up for landing, the airplane's trim setting is usually adjusted to zero the control force at the approach speed, which is below cruise but somewhat above landing speed. As the airplane is slowed from approach speed to landing speed, the pilot must apply a continuously increasing pull force to move a reversible mechanically linked control and provide the additional up elevator needed to maintain trim. Obviously, the control force required to land the airplane from the approach speed should not exceed that which can be applied comfortably by a human pilot. If the control force gradient were constant, the control force required to land would simply be the control force gradient multiplied by the difference between the approach speed and the landing speed. However, as an airplane approaches the ground, the control force gradient for a reversible mechanical linkage is increased by ground effect. This is readily demonstrated by repeating the computations in Example 6.3.1 for the ground-effect operating condition described in Example 6.5.1. The result of these computations is shown in Fig. 6.5.6. Notice that the stick-free neutral point is moved aft as a result of ground effect. When the center of gravity is forward of the stick-free neutral point, the magnitude of the control force gradient is increased in ground effect. This means that for a reversible mechanically linked control, a greater pull force is required to reduce the trim speed in ground effect than would be required to achieve a similar speed reduction when the airplane is flying far from the ground.

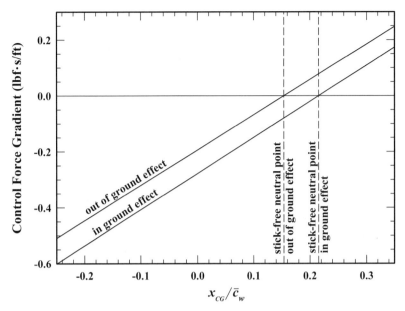

Figure 6.5.6. Control force gradient at 80 mph as a function of CG location measured aft of the wing quarter chord for the airplane and reversible mechanical control linkage in Example 6.5.2.

EXAMPLE 6.5.3. For the wing-tail combination described in Example 6.3.1, we wish to estimate the reversible mechanical control force required for a full-stall landing from an approach speed of 80 mph, with the CG located 15 inches forward of the aerodynamic center of the wing. We will assume that the touchdown wing position relative to the ground is that described in Example 6.5.1. Using the trim tab setting, the control force is zeroed at the approach speed outside of ground effect, and the maximum lift coefficient for the wing-tail combination is 1.3.

Solution. For this wing-tail combination and CG location, we have

$$S_w = 180 \text{ ft}^2, \quad b_w = 33 \text{ ft}, \quad C_{L_w,\alpha} = 4.44, \quad C_{m_w} = -0.053, \quad W = 2{,}700 \text{ lbf},$$

$$S_h = 36 \text{ ft}^2, \quad b_h = 12 \text{ ft}, \quad C_{L_h,\alpha} = 3.97, \quad C_{m_h 0} = 0.0, \quad C_{m_h,\delta_e} = -0.55,$$

$$\alpha_{L0w} = -2.20°, \quad \alpha_{0w} = 3.1°, \quad \alpha_{L0h} = 0.0°, \quad \alpha_{0h} = 2.3°,$$

$$S_e = 12 \text{ ft}^2, \quad \bar{c}_e = 1.1 \text{ ft}, \quad \varepsilon_e = 0.60, \quad l_w = 1.25 \text{ ft}, \quad l_h = 16.25 \text{ ft}, \quad \eta_h = 1.0,$$

$$C_{H_e 0} = 0.00, \quad C_{H_e,\alpha_h} = -0.63, \quad C_{H_e,\delta_e} = -0.97, \quad l_g = 2.0 \text{ ft}$$

From Eq. (6.3.13) evaluated for a load factor of 1.0, the pitch control force required for a reversible mechanical control linkage at level trim is

$$F_p = \eta_h \frac{S_e \bar{c}_e}{S_w l_g} \left[C_1 W + (C_3 + C_{H_e,\delta_t} \delta_t) \left(\tfrac{1}{2} \rho V^2 S_w \right) \right] \qquad (6.5.1)$$

where

$$C_1 = \frac{C_{m,\delta_e} C_{H_e,\alpha_h} (1 - \varepsilon_{d,\alpha}) - C_{m,\alpha} C_{H_e,\delta_e}}{C_{L,\alpha} C_{m,\delta_e} - C_{L,\delta_e} C_{m,\alpha}}$$

$$C_3 = C_{H_e 0} + C_{H_e,\alpha_h} (\alpha_{0h} - \varepsilon_{d0})$$
$$+ \frac{(C_{m,\alpha} C_{L0} - C_{L,\alpha} C_{m0}) C_{H_e,\delta_e} - (C_{m,\delta_e} C_{L0} - C_{L,\delta_e} C_{m0}) C_{H_e,\alpha_h} (1 - \varepsilon_{d,\alpha})}{C_{L,\alpha} C_{m,\delta_e} - C_{L,\delta_e} C_{m,\alpha}}$$

Since both C_1 and C_3 depend on the downwash induced on the horizontal stabilizer, these two coefficients change when the airplane enters ground effect. From Example 6.5.1, outside of ground effect the downwash angle for the horizontal stabilizer is given by

$$\varepsilon_d = 0.599 \frac{C_{L_w}}{R_{A_w}}$$

and we have

$$C_{L_w 0} = C_{L_w,\alpha} (\alpha_{0w} - \alpha_{L0w}) = 0.4107$$

$$\varepsilon_{d0} = 0.599 \frac{C_{L_w 0}}{R_{A_w}} = 0.0407 = 2.33°$$

$$\varepsilon_{d,\alpha} = 0.599 \frac{C_{L_w,\alpha}}{R_{A_w}} = 0.4396$$

$$C_{L0} = C_{L_w,\alpha} (\alpha_{0w} - \alpha_{L0w}) + \frac{S_h}{S_w} \eta_h C_{L_h,\alpha} (\alpha_{0h} - \alpha_{L0h} - \varepsilon_{d0}) = 0.4103$$

$$C_{L,\alpha} = C_{L_w,\alpha} + \frac{S_h}{S_w} \eta_h C_{L_h,\alpha} (1 - \varepsilon_{d,\alpha}) = 4.8850$$

$$C_{L,\delta_e} = \frac{S_h}{S_w} \eta_h C_{L_h,\alpha} \varepsilon_e = 0.4764$$

$$C_{m0} = C_{m_w} - \frac{l_w}{\bar{c}_w} C_{L_w,\alpha} (\alpha_{0w} - \alpha_{L0w}) + \frac{S_h \bar{c}_h}{S_w \bar{c}_w} \eta_h C_{m_h 0}$$
$$- \frac{S_h l_h}{S_w \bar{c}_w} \eta_h C_{L_h,\alpha} (\alpha_{0h} - \alpha_{L0h} - \varepsilon_{d0}) = -0.1459$$

$$C_{m,\alpha} = -\frac{l_w}{\bar{c}_w} C_{L_w,\alpha} - \frac{S_h l_h}{S_w \bar{c}_w} \eta_h C_{L_h,\alpha} (1 - \varepsilon_{d,\alpha}) = -2.3431$$

$$C_{m,\delta_e} = \frac{S_h \bar{c}_h}{S_w \bar{c}_w} \eta_h C_{m_h,\delta_e} - \frac{S_h l_h}{S_w \bar{c}_w} \eta_h C_{L_h,\alpha} \varepsilon_e = -1.4798$$

$$C_1 = \frac{C_{m,\delta_e} C_{H_e,\alpha_h}(1-\varepsilon_{d,\alpha}) - C_{m,\alpha} C_{H_e,\delta_e}}{C_{L,\alpha} C_{m,\delta_e} - C_{L,\delta_e} C_{m,\alpha}} = 0.2864$$

$$C_3 = C_{H_e 0} + C_{H_e,\alpha_h}(\alpha_{0h} - \varepsilon_{d0})$$
$$+ \frac{(C_{m,\alpha} C_{L0} - C_{L,\alpha} C_{m0}) C_{H_e,\delta_e} - (C_{m,\delta_e} C_{L0} - C_{L,\delta_e} C_{m0}) C_{H_e,\alpha_h}(1-\varepsilon_{d,\alpha})}{C_{L,\alpha} C_{m,\delta_e} - C_{L,\delta_e} C_{m,\alpha}}$$
$$= -0.0081$$

Since the pitch control force is set to zero when the airplane is trimmed for 80 mph out of ground effect, from Eq. (6.5.1) we have

$$C_{H_e,\delta_t} \delta_t = -C_1 \frac{W}{\frac{1}{2}\rho V^2 S_w} - C_3 = -0.2544$$

For the full-stall landing, the touchdown airspeed is

$$V_{\text{stall}} = \sqrt{\frac{2W}{\rho S_w C_{L\max}}} = 98.53 \text{ ft/sec}$$

From Example 6.5.1, in ground effect the downwash angle for the horizontal stabilizer is given by

$$\varepsilon_d = 0.266 \frac{C_{L_w}}{R_{A_w}}$$

Repeating the previous computations using the airspeed at stall and the downwash in ground effect gives

$$C_1 = 0.3402$$

$$C_3 = -0.0071$$

and from Eq. (6.5.1), the control force at touchdown is

$$F_p = \eta_h \frac{S_e \bar{c}_e}{S_w l_g}\left[C_1 W + (C_3 + C_{H_e,\delta_t}\delta_t)\left(\frac{1}{2}\rho V^2 S_w\right)\right] = 13.8 \text{ lbf}$$

If these same computations are carried out excluding the reduction in downwash produced by ground effect, we find that a control force of only 8.4 lbf is predicted to stall the airplane outside of ground effect.

In most cases, the increase in pitch control force that results from ground effect with a reversible mechanically linked control system is not a critical concern to the designer. In fact, the slightly heavier pitch control gives the pilot increased sensory feedback at a time in the flight when maintaining precise pitch attitude is very critical. Since pilot effort and concentration is quite naturally increased as the airplane approaches the ground, many pilots are not aware of the increasing control force gradient.

Since requirements for landing often provide the critical constraint on elevator size and forward CG limit, it is important to understand how the airplane's landing configuration affects trim. Because the landing gear is always below the center of gravity, drag on the gear produces a nose-down pitching moment about the CG. Thus, when the landing gear is down, some additional up elevator is needed to compensate for the negative increment in pitching moment produced on the gear. Flaps and/or other high-lift devices are also commonly deployed for landing. The deployment of wing flaps gives rise to four significant aerodynamic changes: a positive increment in maximum lift coefficient, a negative increment in zero-lift angle of attack, a negative increment in wing pitching moment, and a positive increment in wing drag. The elevator angle required to trim the airplane for any given lift coefficient is significantly affected by the zero-lift angle of attack as well as the pitching moment produced on wing and the landing gear.

From Eqs. (6.1.33) and (6.1.34) evaluated for a load factor of 1.0 combined with the relation for static margin given by Eq. (4.4.13), the **fuselage angle of attack and elevator angle at level trim can be estimated as**

$$(\alpha)_{\text{trim}} = \frac{(C_W - C_{L0})C_{m,\delta_e} + C_{L,\delta_e}C_{m0}}{C_{L,\alpha}[C_{m,\delta_e} + C_{L,\delta_e}(l_{np}/\overline{c}_w)]} \tag{6.5.2}$$

$$(\delta_e)_{\text{trim}} = \frac{(C_W - C_{L0})(l_{np}/\overline{c}_w) - C_{m0}}{C_{m,\delta_e} + C_{L,\delta_e}(l_{np}/\overline{c}_w)} \tag{6.5.3}$$

When flaps are deployed for landing, C_{L0} is increased while the flaps and gear contribute to a negative increment in C_{m0}. The drag increment that results from deploying flaps on a high wing contributes to a positive increment in C_{m0}, but the negative increment in wing pitching moment will override this positive increment. If the flap deflection is not too large and the airspeed is held constant, deploying flaps and gear does not appreciably change the remaining terms in Eqs. (6.5.2) and (6.5.3). Thus, the increments in fuselage angle of attack and elevator angle at trim, which result from deploying flaps and landing gear at constant airspeed, are, respectively,

$$(\Delta\alpha)_{fg} = \frac{-C_{m,\delta_e}(\Delta C_{L0})_{fg} + C_{L,\delta_e}(\Delta C_{m0})_{fg}}{C_{L,\alpha}[C_{m,\delta_e} + C_{L,\delta_e}(l_{np}/\overline{c}_w)]} \tag{6.5.4}$$

$$(\Delta\delta_e)_{fg} = \frac{-(l_{np}/\overline{c}_w)(\Delta C_{L0})_{fg} - (\Delta C_{m0})_{fg}}{C_{m,\delta_e} + C_{L,\delta_e}(l_{np}/\overline{c}_w)} \tag{6.5.5}$$

where $(\Delta C_{L0})_{fg}$ and $(\Delta C_{m0})_{fg}$ are, respectively, the increments in C_{L0} and C_{m0} that result from deploying the flaps and gear at constant airspeed.

The change in pitching moment coefficient with elevator deflection is negative and larger in magnitude than the change in lift coefficient with elevator deflection, which is positive. Thus, the denominators in Eqs. (6.5.4) and (6.5.5) are both negative. With this knowledge, inspection of the right-hand side of Eq. (6.5.4) reveals that the positive increment in C_{L0} produced by the flaps results in a decrease in fuselage angle of attack at constant airspeed. On the other hand, the negative increment in C_{m0} will tend to increase the fuselage angle of attack at constant airspeed. Because the change in lift coefficient with elevator deflection is typically small, the first term in the numerator of Eq. (6.5.4) dominates and the fuselage angle of attack at any given airspeed is significantly reduced in landing configuration. From a pilot's point of view, these constant-airspeed relations can be misleading, because as flaps are deployed for landing, airspeed is reduced. Flap deflection is typically used to attain lower airspeed, not lower attitude.

In a similar manner, inspection of the right-hand side of Eq. (6.5.5) shows that when the CG is forward of the neutral point, the positive increment in C_{L0} produced by the flap deflection requires a positive increment in elevator deflection to maintain trim. However, the negative increment in C_{m0} requires a negative increment in elevator deflection and this term dominates. The magnitude of the first term in the numerator of Eq. (6.5.5) increases as the CG is moved forward, while that of the second term remains constant. The second term in the numerator of Eq. (6.5.5) dominates and additional up elevator is needed to compensate for the nose-down pitching moment produced by deploying flaps and gear. As the CG is moved forward, the magnitude of the first term in the numerator of Eq. (6.5.5) increases and less up elevator is needed to compensate for flaps and gear at constant airspeed. **This does not mean that a down-elevator increment is required to compensate for forward movement of the CG.** Equation (6.5.3) clearly shows that the net effect of forward CG movement always requires up elevator for $C_W > C_{L0}$.

Landing often provides a critical constraint on elevator size. To provide slower landing speed, airplanes are typically landed with flaps. However, at some point it may be necessary to land the airplane without flaps. Thus, the elevator should be sized large enough to trim the airplane at C_{Lmax}, in ground effect, with the CG at its forward limit, and the landing gear down, both with and without flaps. Depending on the forward CG limit, either of these two cases could provide the critical constraint on elevator size. Demonstration of this point is left as an exercise for the student.

Summary of Common Candidate CG Limits
As we have seen, the stability and maneuverability of an airplane can be dramatically affected by the way in which the airplane is loaded. The difference between good and poor handling qualities can be a matter of a shift in the center of gravity of only a few inches. Because airplanes can have fuel/cargo fractions of 50 percent or more, such a shift in the CG could easily result from insufficient care in loading the airplane. The designer must correctly specify and communicate the allowable CG range for an airplane to the end user. However, from that point on it is the responsibility of the pilot and crew to see that the airplane is properly loaded so as to keep the CG within design limits. As we have seen, there are several possible constraints that could provide the critical CG limits. Because of their crucial importance, some of the most common constraints on the CG limits for a conventional airplane will be reviewed here.

Aft CG Limit, Candidate 1, Dynamic Margin: For airplanes without stability augmentation, the CG should remain forward of the stick-fixed maneuver point by a distance sufficient to provide a dynamic margin that will satisfy Eq. (6.1.50). With the CG aft of this limit, an airplane becomes overly sensitive to control inputs or disturbances in pitch and is difficult to fly. For airplanes with stability augmentation or when flying in a failure state, this constraint can be significantly relaxed, but the center of gravity must always remain forward of the stick-fixed maneuver point. Further details on the foundation for the constraint specified by Eq. (6.1.50) will be covered in Chapters 8 and 10.

Aft CG Limit, Candidate 2, Control Force per g: For airplanes with reversible mechanical controls, the CG should remain forward of the stick-free maneuver point by a distance that will ensure a control force per g sufficient to prevent the pilot from inadvertently overstressing the airplane during maneuvers. The local control force per g should not be less than 3 lbf/g on a stick or 4 lbf/g on a wheel throughout the required range of load factor and airspeed. To meet FAR Part 23 requirements, the elevator control force needed to achieve the positive limit load factor may not be less than; for a wheel, $W_{max}/100$ or 20 lbf whichever is greater, except that it need not exceed 50 lbf; or for a stick, $W_{max}/140$ or 15 lbf whichever is greater, except that it need not exceed 35 lbf.

Aft CG Limit, Candidate 3, Control Force Gradient: For airplanes with reversible mechanical controls, the CG should remain forward of the stick-free neutral point by a distance sufficient to provide a control force that increases with departure from the trim speed. No quantitative values are specified for the control force gradient in FAR Part 23. The regulations merely require that the airplane must have "*sufficient change in control force, as it is displaced from the trimmed condition, to produce suitable control feel for safe operation.*" A pull force must be required to fly slower than trim, a push force must be required to fly faster than trim, and the gradient should be "*clearly perceptible to the pilot*" at any allowable airspeed.

Forward CG Limit, Candidate 1, In-Flight Elevator Angle per g: To ensure good maneuverability when the airplane is flying outside of ground effect, the CG should be kept aft of the point that requires full up elevator to produce the maximum required load factor at the maneuver speed. This is not an absolute requirement for certification or safe operation of an airplane. However, with the center of gravity located forward of this point, the maneuverability of the airplane will be somewhat compromised, which could be problematic, particularly for aerobatic or combat aircraft.

Forward CG Limit, Candidate 2, Elevator Angle per g in Ground Effect: For best takeoff and landing performance, the CG should be kept aft of the point that requires full up elevator to produce a 1.44-g load factor at 1.2 times the 1-g stall speed. Here again, this is not an absolute requirement for certification or safe operation of an airplane. However, adhering to this constraint will normally ensure that the airplane's takeoff and landing performance is not compromised by insufficient elevator power.

Forward CG Limit, Candidate 3, Elevator Angle to Trim in Ground Effect: For best landing performance, the CG should be kept aft of the point that requires full up elevator to trim the airplane for level flight in ground effect with a lift coefficient of C_{Lmax} and the landing gear deployed, both with and without the flaps in landing configuration. Although the airplane could be landed safely with the CG located forward of this point, the landing airspeed and ground roll distance would be significantly increased.

Forward CG Limit, Candidate 4, Control Force in Ground Effect: For airplanes with reversible mechanical controls, the CG must remain sufficiently aft to ensure that the airplane does not encounter an elevator control force in excess of what the pilot can apply with one hand while completing a safe landing following an approach to land. The trim control is set to minimize or zero the control force at the approach speed, and no further trim adjustment is allowed. To meet FAR Part 23 requirements, the temporary one-hand control force required of the pilot may not exceed 60 lbf on a stick or 50 lbf on a wheel.

The critical CG limits that are specified for an airplane are the most restrictive of all possible candidates. The candidate CG limits that are summarized here are not the only possible constraints that could be placed on the CG limits required for the safe operation of an airplane. However, if these candidates are considered in the early phases of design, fewer problems will typically arise in the later phases. Because regulations can and do change, **designers and operators should always refer to the current version of the appropriate regulation when formulating engineering requirements and test plans.**

6.6. Stall Recovery

Most of the stability and control analysis that we have considered to this point is based on a linear relation between angle of attack and lift. Furthermore, we have assumed a positive lift slope for all lifting surfaces. These assumptions are reasonable when the angle of attack is sufficiently below stall. However, as the angle of attack approaches stall, the variation in lift with angle of attack becomes nonlinear, and at angles of attack beyond stall, the lift slope becomes negative. Consequently, the post-stall stability and control characteristics of an aircraft are quite different from those at low angles of attack.

Fighter and stunt airplanes commonly experience angles of attack well beyond stall. The ability of a modern fighter to perform post-stall maneuvers, like the one shown in Fig. 3.9.7, is critical and missiles frequently operate at angles of attack that produce separated flow. For these reasons and many others, high angle of attack aerodynamics, stability, and control are of great interest, particularly to military aircraft designers.

Although most transports and general aviation airplanes are not routinely flown at angles of attack beyond stall, the possibility of stall always exists. Hence, any airplane should be designed so that recovery from stall is at least possible. For most civilian airplanes, it is usually preferable if the airplane is designed to provide stall recovery with little or no control input from the pilot. In the most forgiving designs, the airplane will naturally recover from a stall if the pilot simply releases the controls.

At angles of attack beyond stall the airflow about an airplane is extremely complex and the aerodynamic forces and moments become highly nonlinear. At high angles of attack the drag forces acting on the wing and tail will contribute significantly to the aircraft's pitching moment, which adds to the nonlinear variation of pitching moment coefficient with angle of attack. Figure 6.6.1 shows how airplane lift, drag, and pitching moment coefficients vary with angle of attack for one particular geometry. The airplane used to generate this figure had a rectangular wing with no sweep and no geometric or aerodynamic twist. The airplane had a conventional aft tail and was trimmed for steady level flight with a fuselage angle of attack of about 4 degrees. The CG was located to provide a static margin of about 7 percent.

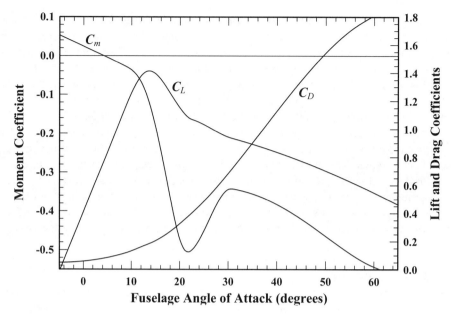

Figure 6.6.1. Lift, drag, and pitching moment coefficients for an airplane with an unswept rectangular wing and conventional aft tail.

Notice in Fig. 6.6.1 that for the region just to the right of C_{Lmax}, the change in pitching moment coefficient with angle of attack becomes significantly more negative. This is primarily the result of two contributing factors. First of all, for this aircraft the CG was located somewhat aft of the aerodynamic center of the main wing, which makes the wing destabilizing when the lift slope is positive and the effects of drag are small. When the main wing stalls, its lift slope becomes negative and the contribution of wing lift becomes stabilizing (i.e., $C_{m,\alpha}$ becomes more negative).

The aft horizontal stabilizer also contributes to the more negative pitch stability derivative, which is seen in Fig. 6.6.1 for the region between 10 and 20 degrees angle of attack. As is often the case for an airplane with a conventional aft tail, the main wing of this airplane stalls before the horizontal stabilizer. Because downwash is induced on the aft tail by the main wing, the local angle of attack for the horizontal stabilizer is less than that for the main wing and stall of the stabilizer is postponed. We have seen that when the lift slope for the main wing is positive, this downwash reduces the stabilizing contribution that an aft tail makes to the pitch stability derivative. In a similar manner, when the lift slope for the main wing is negative, downwash increases the stabilizing contribution of an unstalled aft tail. When the main wing stalls, the decrease in lift reduces the downwash aft of the wing and increases the local angle of attack for the horizontal stabilizer. This increases the lift developed on the unstalled stabilizer and produces a greater nose-down pitching moment, which tends to reduce the angle of attack and return the airplane to unstalled flight. This increased contribution to pitch stability continues only as long as the horizontal stabilizer does not stall.

At the larger angles of attack shown in Fig. 6.6.1, both the wing and the horizontal stabilizer are stalled. From Eq. (4.7.6), for the special case where the vertical offset, h_h, is zero, the contribution of the horizontal stabilizer to the airplane pitching moment coefficient is

$$\left(\Delta C_m\right)_h = \frac{S_h \bar{c}_h}{S_w \bar{c}_w} \eta_h C_{m_h} - \frac{S_h l_h}{S_w \bar{c}_w} \eta_h \left[\cos(\alpha - \varepsilon_d) C_{L_h} + \sin(\alpha - \varepsilon_d) C_{D_h}\right] \quad (6.6.1)$$

When both the wing and the horizontal stabilizer are stalled, the downwash angle and lift coefficient decrease with increasing angle of attack, while the drag coefficient increases with angle of attack. With these facts in mind, from Eq. (6.6.1) we see that lift developed on this stalled aft stabilizer produces a nose-down pitching moment that decreases with increasing angle of attack. Drag, on the other hand, produces a nose-down pitching moment that increases with angle of attack. Thus, the lift developed on this stalled horizontal surface has a destabilizing effect on the moment slope, whereas the drag has a stabilizing effect. At angles of attack just slightly beyond the stabilizer stall point, the effects of lift are greater than the effects of drag and the aft horizontal surface is destabilizing in pitch. At somewhat higher angles of attack the drag term dominates and the horizontal surface again becomes stabilizing. This is the reason for the two changes in slope that are seen in the pitching moment curve of Fig. 6.6.1.

While the derivative of the pitching moment coefficient with respect to angle of attack is positive over a portion of the range shown in Fig. 6.6.1, the pitching moment coefficient is always negative at angles of attack beyond stall. This means that with the controls in this position, the stalled aircraft will always experience the nose-down pitching moment needed to recover from stall and return the airplane to trimmed flight. While post-stall stability characteristics similar to those shown in Fig. 6.6.1 are exhibited by many well-designed conventional airplanes, this is not necessarily the case for all airplanes and loading conditions.

The vertical position of the aircraft center of gravity relative to the wing and tail can have a significant effect on the post-stall stability characteristics of an airplane. This is demonstrated in Fig. 6.6.2. This figure shows the lift, drag, and pitching moment coefficients for the same airplane geometry that was used to generate Fig. 6.6.1. The only difference between the airplanes used for these two figures is the location of the center of gravity. For the results plotted in Fig. 6.6.2, the CG was lowered by a distance equal to one-half of the wing chord length, and the trim setting was adjusted to maintain the same level trim airspeed as that in Fig. 6.6.1. Notice that for the lower center of gravity, the destabilizing effect of the stalled aft tail is greater and extends over a larger range of angle of attack.

Wing sweep also affects the post-stall stability of an airplane. Figure 6.6.3 shows the lift, drag, and pitching moment coefficients for one particular airplane with a swept-back wing. The wing has a tapered planform with no geometric or aerodynamic twist. The airplane has the same conventional aft tail as that used for Fig. 6.6.1 and was trimmed for steady level flight at the same fuselage angle of attack. The center of gravity was located to provide the same static margin of about 7 percent, and the wing, horizontal stabilizer, and center of gravity were all placed on the fuselage reference line.

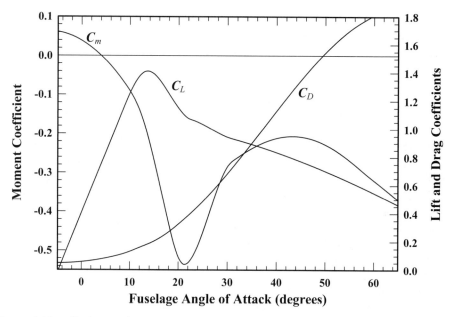

Figure 6.6.2. Lift, drag, and pitching moment coefficients for an airplane with an unswept rectangular wing, a conventional aft tail, and a low center of gravity.

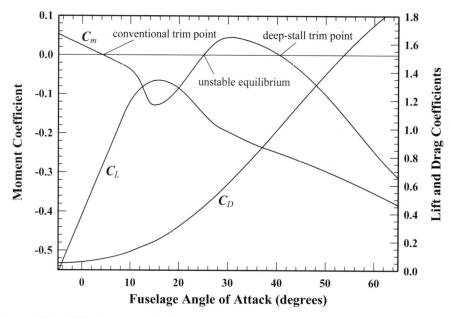

Figure 6.6.3. Lift, drag, and pitching moment coefficients for an airplane with a swept-back tapered wing and conventional aft tail.

Figure 6.6.3 demonstrates what can be a very significant problem for an airplane with a swept-back tapered wing. On a wing of this planform shape with no geometric or aerodynamic twist, the outboard wing sections will tend to stall before the sections near the root. As the angle of attack is increased from unstalled flight, the wingtips will stall first and the stall will progress inward and forward toward the root of the swept wing, as shown schematically in Fig. 6.6.4. Since positive sweep places the wingtips aft of the root, losing lift on the aft wingtips while retaining lift on the forward wing root produces a nose-up increment in the pitching moment, which tends to further increase the angle of attack and deepen the stall. This effect, combined with the destabilizing effects of a stalled aft tail, can cause the total airplane pitching moment to become positive over some range of angle of attack beyond stall. At even larger angles of attack, the pitching moment will again become negative, due to the stabilizing effects of drag. This behavior can be seen in Fig. 6.6.3.

Notice that for the pitching moment coefficient curve shown in Fig. 6.6.3, there are two stable trim points. The first occurs at an angle of attack of about 4 degrees and is the conventional trim point for unstalled flight. For this airplane, a second stable trim point occurs at about 40 degrees angle of attack. Since this point is well past stall, it is frequently called a *deep-stall trim point* or sometimes simply *deep stall*. An airplane

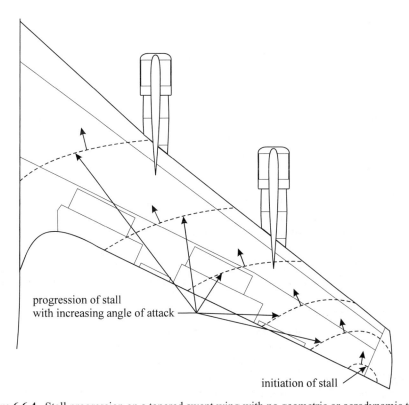

Figure 6.6.4. Stall progression on a tapered swept wing with no geometric or aerodynamic twist.

with the stall characteristics shown in Fig. 6.6.3 will recover from a shallow stall in much the same way as those shown in Figs. 6.6.1 and 6.6.2. For this particular case, as long as the angle of attack remains less than about 25 degrees, a nose-down pitching moment is created with the control in this position, and the airplane will recover from the shallow stall. However, if the angle of attack should ever exceed that of the 25-degree unstable equilibrium point with the control held in the same position that would result in recovery from a shallow stall, the airplane would quickly rotate to the deep-stall trim point. It can be difficult or even impossible to recover from this deep stall with conventional controls, because control surfaces may be immersed in the stalled wake and become ineffective. The T-tail can be particularly problematic in this regard.

One way to avoid the stall characteristic shown in Fig. 6.6.3 is through the use of washout in the swept wing. By decreasing the geometric angle of attack at the wingtips relative to that at the root, wingtip stall can be postponed to a higher angle of attack. If sufficient washout is used, the wing root will stall prior to the wingtips, and the onset of stall will produce a nose-down pitching moment that counters the destabilizing effect of the lift developed on a stalled aft tail. Sometimes it is not possible to avoid the stall characteristic shown in Fig. 6.6.3 and still maintain the other characteristics needed to provide mission effectiveness. In such cases, some means must be provided for the pilot to recover from deep stall. Alternative controls such as movable canard surfaces, flaperons, and thrust vectoring are sometimes used for this purpose.

6.7. Lateral Control and Maneuverability

In Sec. 6.1 we examined how the elevator is used to control the normal acceleration of an aircraft in longitudinal motion. The elevator is used to control the angle of attack at which the pitching moment is zero. Thus, the elevator indirectly controls the magnitude of lift developed by the wing. If the lift does not equal the component of weight normal to the flight path, a normal acceleration will occur. Up to this point in the chapter, we have limited our study of aircraft maneuverability to the case where the normal acceleration is confined to the longitudinal plane.

To maneuver an aircraft outside the longitudinal plane, the ailerons and rudder are typically used in combination with the elevator to control not only the magnitude of the lift but also the direction in which the lift vector acts, thereby controlling both the magnitude and direction of acceleration for the maneuver. Consider, for example, the *steady coordinated turn* shown in Fig. 6.7.1. The term *steady* requires that both the airspeed and angular velocity are held constant. By definition, *a coordinated maneuver is one in which the controls are coordinated so that the vector sum of the airplane's acceleration and the acceleration of gravity produces an apparent body force that falls in the aircraft's plane of symmetry.* In other words, the apparent body force has no component in the spanwise direction. By definition, the turning axis for the steady coordinated turn is vertical, but the airplane may be climbing or descending during the turn. The climb angle, γ, is defined to be the angle that the airplane's velocity vector makes with the horizontal. For convenience, the body axis is chosen to align with the direction of flight, and the bank angle, ϕ, is the angle that the airplane is rolled about this velocity vector, as measured with respect to the upright "wings-level" orientation.

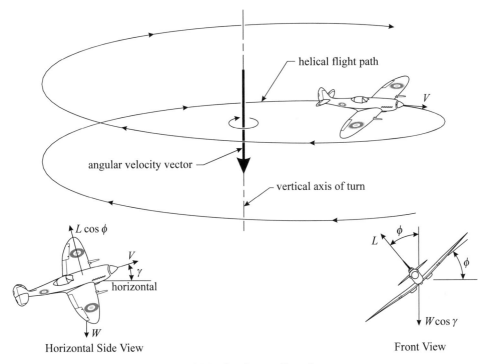

angular velocity vector

helical flight path

V

vertical axis of turn

$L \cos \phi$

V

γ

horizontal

W

Horizontal Side View

L

ϕ

ϕ

$W \cos \gamma$

Front View

Figure 6.7.1. Steady coordinated turn.

We will use the (x_b, y_b, z_b) coordinate system shown in Fig. 5.1.1. The origin of this coordinate system is located at the aircraft center of gravity. For convenience, the x_b-axis is in the aircraft plane of symmetry aligned with the direction of flight and the z_b-axis points downward in the aircraft plane of symmetry. To complete the right-hand coordinate system, the y_b-axis points in the spanwise direction to the pilot's right.

From Fig. 6.7.1, we see that the component of aircraft weight in the direction of flight is $-W \sin \gamma$ and the component of weight in the plane normal to the direction of flight is $W \cos \gamma$. The normal component of weight is easily resolved into a spanwise component, $W \cos \gamma \sin \phi$, and a component in the plane of symmetry, $W \cos \gamma \cos \phi$. Thus, the components of weight expressed in the (x_b, y_b, z_b) coordinate system are

$$\mathbf{W} = \begin{Bmatrix} -W \sin \gamma \\ W \cos \gamma \sin \phi \\ W \cos \gamma \cos \phi \end{Bmatrix} \tag{6.7.1}$$

The angular velocity vector for this steady coordinated turn is constant and parallel to the weight vector. Thus, using the same geometric relations that were used for the weight vector, the components of the airplane's angular velocity vector in (x_b, y_b, z_b) coordinates are

$$\Omega = \begin{Bmatrix} -\Omega \sin \gamma \\ \Omega \cos \gamma \sin \phi \\ \Omega \cos \gamma \cos \phi \end{Bmatrix} \tag{6.7.2}$$

where Ω is the angular velocity magnitude. The (x_b, y_b, z_b) coordinates were chosen to align the x_b-axis with the velocity component in the x_b-z_b plane. Thus, for small sideslip angles, the components of translational velocity in the (x_b, y_b, z_b) coordinate system are

$$\mathbf{V} = \begin{Bmatrix} V \\ V\beta \\ 0 \end{Bmatrix} \tag{6.7.3}$$

Because neither the translational velocity nor the angular velocity is changing with time, the only acceleration associated with the steady coordinated turn is the centripetal acceleration resulting from the changing direction of flight. Thus, from elementary rigid-body dynamics, the airplane's acceleration vector in the steady coordinated turn can be written as the cross product of the angular and translational velocity vectors. Moreover, because β is very small at the angular rates typically encountered in a coordinated turn, we can neglect the $\Omega\beta$ product and the acceleration vector is accurately approximated as

$$\frac{d\mathbf{V}}{dt} = \Omega \times \mathbf{V} = \begin{Bmatrix} 0 \\ \Omega V \cos \gamma \cos \phi \\ -\Omega V \cos \gamma \sin \phi \end{Bmatrix} \tag{6.7.4}$$

This acceleration vector includes the restriction that the turn is steady. However, an additional restriction is needed to account for the fact that the turn is also coordinated. The total apparent body force for this accelerating coordinate frame is the vector sum of that due to gravity and that due to the acceleration of the coordinate frame. Since an upward acceleration adds to the downward force of gravity, the apparent body force is

$$\mathbf{B} \equiv \mathbf{W} - \frac{W}{g}\frac{d\mathbf{V}}{dt} \tag{6.7.5}$$

Using Eqs. (6.7.1) and (6.7.4) in Eq. (6.7.5), the (x_b, y_b, z_b) components are

$$\mathbf{B} = W\begin{Bmatrix} -\sin \gamma \\ \cos \gamma \sin \phi \\ \cos \gamma \cos \phi \end{Bmatrix} - \frac{W}{g}\Omega V \begin{Bmatrix} 0 \\ \cos \gamma \cos \phi \\ -\cos \gamma \sin \phi \end{Bmatrix} \tag{6.7.6}$$

The definition of a coordinated maneuver requires that the spanwise component of this apparent body force must be zero, i.e.,

$$W \cos \gamma \sin \phi - (W\Omega V/g)\cos \gamma \cos \phi = 0 \tag{6.7.7}$$

This imposes a relation between the turning rate and the bank angle,

$$\Omega = \frac{g \tan \phi}{V} \tag{6.7.8}$$

Using Eq. (6.7.8) in Eq. (6.7.4), the acceleration for this steady coordinated turn is

$$\frac{d\mathbf{V}}{dt} = \begin{Bmatrix} 0 \\ g \cos \gamma \sin \phi \\ -g \cos \gamma \sin^2 \phi / \cos \phi \end{Bmatrix} \tag{6.7.9}$$

The total force acting on the aircraft is the vector sum of the thrust force, the aerodynamic force, and the weight. Drag acts in the negative x_b-direction and lift acts in the negative z_b-direction. For this analysis we assume a thrust vector aligned with the x_b-axis and will let Y represent the side force acting in the positive y_b-direction. Thus, using Eqs. (6.7.1) and (6.7.9), Newton's second law requires that

$$\begin{Bmatrix} T - D \\ Y \\ -L \end{Bmatrix} + \begin{Bmatrix} -W \sin \gamma \\ W \cos \gamma \sin \phi \\ W \cos \gamma \cos \phi \end{Bmatrix} = \frac{W}{g} \begin{Bmatrix} 0 \\ g \cos \gamma \sin \phi \\ -g \cos \gamma \sin^2 \phi / \cos \phi \end{Bmatrix} \tag{6.7.10}$$

or after rearranging and applying the trigonometric identity $\sin^2 \phi + \cos^2 \phi = 1$,

$$\begin{Bmatrix} T - D \\ Y \\ L \end{Bmatrix} = \begin{Bmatrix} W \sin \gamma \\ 0 \\ W \cos \gamma / \cos \phi \end{Bmatrix} \tag{6.7.11}$$

Because the turn is steady, the angular rates are constant. Furthermore, at the angular rates encountered in a coordinated turn, second-order angular momentum effects are negligible and the moments acting on the airplane are very close to zero,

$$\begin{Bmatrix} \ell \\ m \\ n \end{Bmatrix} = \begin{Bmatrix} 0 \\ 0 \\ 0 \end{Bmatrix} \tag{6.7.12}$$

Newton's second law, as expressed in Eqs. (6.7.11) and (6.7.12), provides six scalar equations that can be solved for the thrust, angle of attack, sideslip angle, and the three control surface deflections. From a conceptual point of view, this problem is identical to the longitudinal control problem considered in Sec. 6.1. Since longitudinal motion is two-dimensional, the longitudinal control problem involves only three degrees of freedom and three unknowns: the thrust, the angle of attack, and the elevator deflection. Because lateral motion is three-dimensional, the present problem involves six degrees of freedom and six unknowns, but is otherwise identical.

The first component of Eq. (6.7.11) allows us to determine the thrust required to maintain constant airspeed at a specified climb angle, or if the thrust were specified, the climb angle could be determined from this relation. The remaining five components of

Eqs. (6.7.11) and (6.7.12) will be used to determine the aerodynamic angles and control surface deflections. In traditional dimensionless form, these five scalar equations are

$$
\begin{Bmatrix} C_L \\ C_Y \\ C_\ell \\ C_m \\ C_n \end{Bmatrix} = \begin{Bmatrix} C_W \cos\gamma/\cos\phi \\ 0 \\ 0 \\ 0 \\ 0 \end{Bmatrix} \tag{6.7.13}
$$

From Eqs. (6.1.11), (6.1.18), and (6.1.42), the lift coefficient can be written as

$$
C_L = C_{L0} + C_{L,\alpha}\,\alpha + C_{L,\breve{q}}\,\breve{q} + C_{L,\delta_e}\,\delta_e \tag{6.7.14}
$$

where $\breve{q}$ is the dimensionless dynamic pitch rate defined in Eq. (6.1.42). The pitching moment coefficient for pure longitudinal motion was given by Eq. (6.1.19). However, lateral motion allows sideslip and we have seen that a rotating propeller produces a pitching moment that is proportional to the sideslip angle. Thus, for turning flight, Eqs. (6.1.11), (6.1.19), and (6.1.42), modified to account for the change in the pitching moment coefficient with respect to sideslip angle, yield

$$
C_m = C_{m0} + C_{m,\alpha}\,\alpha + C_{m,\beta}\,\beta + C_{m,\breve{q}}\,\breve{q} + C_{m,\delta_e}\,\delta_e \tag{6.7.15}
$$

Just as the longitudinal forces and moment depend on the pitch rate, the lateral force and moments depend on the roll and yaw rates. By standard convention, the rolling and yawing rates are given the symbols p and r, respectively, and they are traditionally nondimensionalized with respect to the forward velocity divided by the wing semispan,

$$
\begin{Bmatrix} \bar{p} \\ \bar{q} \\ \bar{r} \end{Bmatrix} \equiv \begin{Bmatrix} p b_w/2V \\ q \bar{c}_w/2V \\ r b_w/2V \end{Bmatrix} \tag{6.7.16}
$$

However, following Eq. (6.1.42), we will use the dimensionless **dynamic angular rates**

$$
\begin{Bmatrix} \breve{p} \\ \breve{q} \\ \breve{r} \end{Bmatrix} \equiv \frac{V}{g} \begin{Bmatrix} p \\ q \\ r \end{Bmatrix} \tag{6.7.17}
$$

With these definitions, the lateral aerodynamic force and moments are written

$$
C_Y = C_{Y,\beta}\,\beta + C_{Y,\breve{p}}\,\breve{p} + C_{Y,\breve{r}}\,\breve{r} + C_{Y,\delta_r}\,\delta_r \tag{6.7.18}
$$

$$
C_\ell = C_{\ell,\beta}\,\beta + C_{\ell,\breve{p}}\,\breve{p} + C_{\ell,\breve{r}}\,\breve{r} + C_{\ell,\delta_a}\,\delta_a + C_{\ell,\delta_r}\,\delta_r \tag{6.7.19}
$$

$$
C_n = C_{n,\alpha}\,\alpha + C_{n,\beta}\,\beta + C_{n,\breve{p}}\,\breve{p} + C_{n,\breve{r}}\,\breve{r} + C_{n,\delta_a}\,\delta_a + C_{n,\delta_r}\,\delta_r \tag{6.7.20}
$$

Using Eqs. (6.7.14) through (6.7.20) in Eq. (6.7.13), we have

$$
\begin{bmatrix}
C_{L,\alpha} & 0 & 0 & C_{L,\delta_e} & 0 \\
0 & C_{Y,\beta} & 0 & 0 & C_{Y,\delta_r} \\
0 & C_{\ell,\beta} & C_{\ell,\delta_a} & 0 & C_{\ell,\delta_r} \\
C_{m,\alpha} & C_{m,\beta} & 0 & C_{m,\delta_e} & 0 \\
C_{n,\alpha} & C_{n,\beta} & C_{n,\delta_a} & 0 & C_{n,\delta_r}
\end{bmatrix}
\begin{Bmatrix}
\alpha \\ \beta \\ \delta_a \\ \delta_e \\ \delta_r
\end{Bmatrix}
$$

$$
= \begin{Bmatrix}
C_W \cos\gamma / \cos\phi - C_{L0} \\
0 \\
0 \\
-C_{m0} \\
0
\end{Bmatrix}
- \begin{bmatrix}
0 & C_{L,\bar{q}} & 0 \\
C_{Y,\bar{p}} & 0 & C_{Y,\bar{r}} \\
C_{\ell,\bar{p}} & 0 & C_{\ell,\bar{r}} \\
0 & C_{m,\bar{q}} & 0 \\
C_{n,\bar{p}} & 0 & C_{n,\bar{r}}
\end{bmatrix}
\begin{Bmatrix}
\breve{p} \\ \breve{q} \\ \breve{r}
\end{Bmatrix}
\tag{6.7.21}
$$

The aerodynamic derivatives with respect to the dynamic angular rates are easily related to the traditional aerodynamic derivatives, which are commonly obtained from wind tunnel tests. From Eqs. (6.7.16) and (6.7.17), it is easily shown that

$$
\begin{Bmatrix}
\partial/\partial\breve{p} \\
\partial/\partial\breve{q} \\
\partial/\partial\breve{r}
\end{Bmatrix}
\equiv \frac{g}{V}
\begin{Bmatrix}
\partial/\partial p \\
\partial/\partial q \\
\partial/\partial r
\end{Bmatrix}
\equiv
\begin{Bmatrix}
(gb_w/2V^2)\partial/\partial\bar{p} \\
(g\bar{c}_w/2V^2)\partial/\partial\bar{q} \\
(gb_w/2V^2)\partial/\partial\bar{r}
\end{Bmatrix}
\tag{6.7.22}
$$

From Eqs. (6.7.2), (6.7.8), and (6.7.17), we obtain

$$
\begin{Bmatrix}
\breve{p} \\ \breve{q} \\ \breve{r}
\end{Bmatrix}
= \begin{Bmatrix}
-\sin\gamma \tan\phi \\
\cos\gamma \sin\phi \tan\phi \\
\cos\gamma \sin\phi
\end{Bmatrix}
\tag{6.7.23}
$$

Using Eq. (6.7.23) in Eq. (6.7.21) gives

$$
\begin{bmatrix}
C_{L,\alpha} & 0 & 0 & C_{L,\delta_e} & 0 \\
0 & C_{Y,\beta} & 0 & 0 & C_{Y,\delta_r} \\
0 & C_{\ell,\beta} & C_{\ell,\delta_a} & 0 & C_{\ell,\delta_r} \\
C_{m,\alpha} & C_{m,\beta} & 0 & C_{m,\delta_e} & 0 \\
C_{n,\alpha} & C_{n,\beta} & C_{n,\delta_a} & 0 & C_{n,\delta_r}
\end{bmatrix}
\begin{Bmatrix}
\alpha \\ \beta \\ \delta_a \\ \delta_e \\ \delta_r
\end{Bmatrix}
$$

$$
= \begin{Bmatrix}
C_W \cos\gamma / \cos\phi - C_{L0} - C_{L,\bar{q}} \cos\gamma \sin\phi \tan\phi \\
C_{Y,\bar{p}} \sin\gamma \tan\phi - C_{Y,\bar{r}} \cos\gamma \sin\phi \\
C_{\ell,\bar{p}} \sin\gamma \tan\phi - C_{\ell,\bar{r}} \cos\gamma \sin\phi \\
-C_{m0} - C_{m,\bar{q}} \cos\gamma \sin\phi \tan\phi \\
C_{n,\bar{p}} \sin\gamma \tan\phi - C_{n,\bar{r}} \cos\gamma \sin\phi
\end{Bmatrix}
\tag{6.7.24}
$$

If the aerodynamic derivatives are all known, Eq. (6.7.24) can be solved for the control surface deflection angles required to maintain a steady coordinated turn with airspeed, V, climb angle, γ, and bank angle, ϕ. In this formulation, we have used the bank angle to characterize the severity of the turn. Other important measures that are commonly used to characterize a turn include the turning radius, the turning rate, and the load factor. In the steady coordinated turn, these four quantities are all related. If any one is known together with the climb angle and airspeed, the other three can easily be determined. From Eq. (3.9.6), the bank angle can be related to the turning radius, R,

$$\phi = \tan^{-1}\left(\frac{V^2 \cos\gamma}{Rg}\right) \tag{6.7.25}$$

Similarly, from Eq. (6.7.8), the bank angle can be computed from the turning rate, Ω,

$$\phi = \tan^{-1}\left(\frac{\Omega V}{g}\right) \tag{6.7.26}$$

and from Eq. (6.7.11), the bank angle can also be related to the load factor, L/W,

$$\phi = \cos^{-1}\left(\frac{\cos\gamma}{L/W}\right) \tag{6.7.27}$$

Here, the reader may recall that the traditional symbol used to denote load factor is n, which is also the symbol traditionally used to signify the yawing moment. This was a very unfortunate choice of notation, but it is well established in the literature and we must learn to live with it. Both load factor and yawing moment are important parameters in the analysis of turning flight and it is common to see the symbol, n, used to represent both of these parameters in the same formulation. It is usually left up to the reader to determine where n denotes load factor and where it denotes yawing moment. In this text, tradition will not be carried to that extreme. Whenever load factor and yawing moment are used in the same formulation, notation will be used that distinguishes between these two variables. Here the explicit notation, L/W, is used for the load factor. However, the reader should be forewarned that this is not the norm.

EXAMPLE 6.7.1. Consider a level coordinated turn for the airplane described in Examples 5.2.1, 5.2.2, 5.6.1, and 6.1.1. All control surfaces are trimmed for steady level flight at 80 mph and sea level. Ignoring any possibility of stall, we wish to plot the change in aileron, elevator, and rudder deflection angles required to maintain the level coordinated turn, at the same airspeed and altitude, as a function of bank angle. For this airplane and operating condition we will use

$$S_w = 180 \text{ ft}^2, \quad b_w = 33 \text{ ft}, \quad W = 2{,}700 \text{ lbf}, \quad V = 80 \text{ mph}, \quad C_W = 0.9168,$$
$$C_{L,\alpha} = 4.88, \quad C_{L,\delta_e} = 0.476, \quad C_{L,\bar{q}} = 3.00,$$
$$C_{Y,\beta} = -0.508, \quad C_{Y,\delta_r} = 0.131, \quad C_{Y,\bar{p}} = 0.00, \quad C_{Y,\bar{r}} = 0.142,$$

$$C_{\ell,\beta} = -0.050, \quad C_{\ell,\delta_a} = -0.134, \quad C_{\ell,\delta_r} = 0.012, \quad C_{\ell,\bar{p}} = -0.411,$$

$$C_{m,\alpha} = -0.587, \quad C_{m,\beta} = -0.031, \quad C_{m,\delta_e} = -1.31, \quad C_{m,\bar{q}} = -11.05,$$

$$C_{n,\alpha} = -0.0051, \quad C_{n,\beta} = 0.071, \quad C_{n,\delta_a} = 0.0035, \quad C_{n,\delta_r} = -0.059, \quad C_{n,\bar{p}} = -0.057,$$

$$C_{\ell,\bar{r}} = 0.020 + 0.25 C_L, \quad C_{n,\bar{r}} = -0.114 - 0.024 C_L^2,$$

$$g\bar{c}_w/2V^2 = 0.00637, \quad g b_w/2V^2 = 0.03856$$

Solution. Let the aerodynamic angles and control surface deflections for the steady level coordinated turn be expressed as

$$\begin{Bmatrix} \alpha \\ \beta \\ \delta_a \\ \delta_e \\ \delta_r \end{Bmatrix} = \begin{Bmatrix} \alpha \\ \beta \\ \delta_a \\ \delta_e \\ \delta_r \end{Bmatrix}_{\text{trim}} + \begin{Bmatrix} \Delta\alpha \\ \Delta\beta \\ \Delta\delta_a \\ \Delta\delta_e \\ \Delta\delta_r \end{Bmatrix} \tag{6.7.28}$$

where the trim condition is evaluated for steady level flight at the same airspeed and altitude. Steady level flight is a special case of the steady coordinated turn where both the bank angle and climb angle are zero. Thus, the trim condition must satisfy Eq. (6.7.24) with both ϕ and γ set to zero.

$$\begin{bmatrix} C_{L,\alpha} & 0 & 0 & C_{L,\delta_e} & 0 \\ 0 & C_{Y,\beta} & 0 & 0 & C_{Y,\delta_r} \\ 0 & C_{\ell,\beta} & C_{\ell,\delta_a} & 0 & C_{\ell,\delta_r} \\ C_{m,\alpha} & C_{m,\beta} & 0 & C_{m,\delta_e} & 0 \\ C_{n,\alpha} & C_{n,\beta} & C_{n,\delta_a} & 0 & C_{n,\delta_r} \end{bmatrix} \begin{Bmatrix} \alpha \\ \beta \\ \delta_a \\ \delta_e \\ \delta_r \end{Bmatrix}_{\text{trim}} = \begin{Bmatrix} C_W - C_{L0} \\ 0 \\ 0 \\ -C_{m0} \\ 0 \end{Bmatrix} \tag{6.7.29}$$

Using Eq. (6.7.28) in Eq. (6.7.24) and applying Eq. (6.7.29) gives

$$\begin{bmatrix} C_{L,\alpha} & 0 & 0 & C_{L,\delta_e} & 0 \\ 0 & C_{Y,\beta} & 0 & 0 & C_{Y,\delta_r} \\ 0 & C_{\ell,\beta} & C_{\ell,\delta_a} & 0 & C_{\ell,\delta_r} \\ C_{m,\alpha} & C_{m,\beta} & 0 & C_{m,\delta_e} & 0 \\ C_{n,\alpha} & C_{n,\beta} & C_{n,\delta_a} & 0 & C_{n,\delta_r} \end{bmatrix} \begin{Bmatrix} \Delta\alpha \\ \Delta\beta \\ \Delta\delta_a \\ \Delta\delta_e \\ \Delta\delta_r \end{Bmatrix}$$

$$= \begin{Bmatrix} C_W[(\cos\gamma/\cos\phi) - 1] - C_{L,\bar{q}}\cos\gamma\sin\phi\tan\phi \\ C_{Y,\bar{p}}\sin\gamma\tan\phi - C_{Y,\bar{r}}\cos\gamma\sin\phi \\ C_{\ell,\bar{p}}\sin\gamma\tan\phi - C_{\ell,\bar{r}}\cos\gamma\sin\phi \\ -C_{m,\bar{q}}\cos\gamma\sin\phi\tan\phi \\ C_{n,\bar{p}}\sin\gamma\tan\phi - C_{n,\bar{r}}\cos\gamma\sin\phi \end{Bmatrix} \tag{6.7.30}$$

Using the aerodynamic derivatives for this airplane, the 5×5 matrix on the left-hand side of Eq. (6.7.30) is

$$
[\mathbf{A}] \equiv
\begin{bmatrix}
C_{L,\alpha} & 0 & 0 & C_{L,\delta_e} & 0 \\
0 & C_{Y,\beta} & 0 & 0 & C_{Y,\delta_r} \\
0 & C_{\ell,\beta} & C_{\ell,\delta_a} & 0 & C_{\ell,\delta_r} \\
C_{m,\alpha} & C_{m,\beta} & 0 & C_{m,\delta_e} & 0 \\
C_{n,\alpha} & C_{n,\beta} & C_{n,\delta_a} & 0 & C_{n,\delta_r}
\end{bmatrix}
$$

$$
=
\begin{bmatrix}
4.88 & 0 & 0 & 0.476 & 0 \\
0 & -0.508 & 0 & 0 & 0.131 \\
0 & -0.050 & -0.134 & 0 & 0.012 \\
-0.587 & -0.031 & 0 & -1.31 & 0 \\
-0.0051 & 0.071 & 0.0035 & 0 & -0.059
\end{bmatrix}
$$

Since this matrix is independent of both climb angle and bank angle, the system of equations does not need to be solved separately for each value of bank angle to be used in the plot. The matrix need only be inverted once. The matrix can be inverted numerically to give

$$
[\mathbf{A}]^{-1} =
\begin{bmatrix}
0.21427 & -0.00685 & -0.00040 & 0.07786 & -0.01529 \\
-0.00692 & -2.83723 & -0.16542 & -0.00251 & -6.33325 \\
0.00018 & 0.75698 & -7.45841 & 0.00007 & 0.16380 \\
-0.09585 & 0.07021 & 0.00409 & -0.79819 & 0.15672 \\
-0.02684 & -3.36879 & -0.64148 & -0.00975 & -24.55949
\end{bmatrix}
$$

Choosing an example bank angle of 30 degrees and applying Eqs. (6.7.11) and (6.7.22), the coordinated turn with $\gamma = 0$ at 80 mph and sea level gives

$$
C_L = C_W \cos\gamma / \cos\phi = 1.05861
$$

$$
C_{\ell,\bar{r}} = 0.020 + 0.25 C_L = 0.28465
$$

$$
C_{n,\bar{r}} = -0.114 - 0.024 C_L^2 = -0.14090
$$

$$
\left\{
\begin{array}{l}
C_W[(1/\cos\phi) - 1] - (g\bar{c}_w/2V^2)C_{L,\bar{q}} \sin\phi \tan\phi \\
-(gb_w/2V^2)C_{Y,\bar{r}} \sin\phi \\
-(gb_w/2V^2)C_{\ell,\bar{r}} \sin\phi \\
-(g\bar{c}_w/2V^2)C_{m,\bar{q}} \sin\phi \tan\phi \\
-(gb_w/2V^2)C_{n,\bar{r}} \sin\phi
\end{array}
\right\}
=
\left\{
\begin{array}{r}
0.13631 \\
-0.00274 \\
-0.00549 \\
0.02033 \\
0.00272
\end{array}
\right\}
$$

The changes in the aerodynamic angles and control surface deflections for this particular bank angle can now be found by multiplying this vector by the inverted matrix $[\mathbf{A}]^{-1}$,

$$
\begin{Bmatrix} \Delta\alpha \\ \Delta\beta \\ \Delta\delta_a \\ \Delta\delta_e \\ \Delta\delta_r \end{Bmatrix}_{\phi=30°} = [\mathbf{A}]^{-1} \begin{Bmatrix} 0.13631 \\ -0.00274 \\ -0.00549 \\ 0.02033 \\ 0.00272 \end{Bmatrix} = \begin{Bmatrix} 0.03077 \\ -0.00952 \\ 0.03933 \\ -0.02908 \\ -0.05782 \end{Bmatrix} = \begin{Bmatrix} 1.76° \\ -0.55° \\ 2.25° \\ -1.67° \\ -3.31° \end{Bmatrix}
$$

Similarly, a bank angle of −30 degrees requires

$$
\begin{Bmatrix} \Delta\alpha \\ \Delta\beta \\ \Delta\delta_a \\ \Delta\delta_e \\ \Delta\delta_r \end{Bmatrix}_{\phi=-30°} = [\mathbf{A}]^{-1} \begin{Bmatrix} 0.13631 \\ 0.00274 \\ 0.00549 \\ 0.02033 \\ -0.00272 \end{Bmatrix} = \begin{Bmatrix} 0.03081 \\ 0.00753 \\ -0.03928 \\ -0.02950 \\ 0.05011 \end{Bmatrix} = \begin{Bmatrix} 1.77° \\ 0.43° \\ -2.25° \\ -1.69° \\ 2.87° \end{Bmatrix}
$$

Repeating these computations for a range of bank angles while neglecting the possibility of stall, we obtain the results that are plotted as the thin black lines in Fig. 6.7.2.

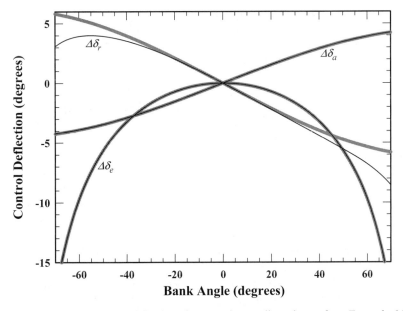

Figure 6.7.2. Control surface deflections for a steady coordinated turn, from Example 6.7.1.

Notice that for the airplane in Example 6.7.1, the rudder deflection is not symmetric with respect to right and left turns. This is a direct result of the yawing moment produced by the airplane's propeller. When the airplane is turned either to the right or to the left, additional lift is needed to produce the turning acceleration. As seen in Fig. 6.7.2, an increment of up elevator (negative $\Delta\delta_e$) must be applied to increase the angle of attack and generate this added lift. However, the increase in angle of attack also produces an increased propeller yawing moment to the left. This must be countered with an increment of right rudder (negative $\Delta\delta_r$). If the airplane is being turned to the right (positive ϕ), the right rudder needed to compensate for the propeller yawing moment adds to the right rudder needed for the turn. If the airplane is being turned to the left, the right rudder needed to compensate for the propeller yawing moment decreases the left rudder needed for the turn. For comparison, the thick gray lines in Fig. 6.7.2 show results obtained by repeating the computations in Example 6.7.1 with $C_{m,\beta} = C_{n,\alpha} = 0$. Note that the elevator and aileron deflections are not significantly affected by $C_{m,\beta}$ and $C_{n,\alpha}$.

Also note from Fig. 6.7.2 that left aileron (positive $\Delta\delta_a$) is needed to maintain a right turn (positive ϕ). This may seem counterintuitive. From the sign convention presented in Sec. 5.7, it is clear that right aileron (negative $\Delta\delta_a$) is needed to initiate a roll to the right. So why was left aileron predicted for the right turn in Example 6.7.1? To answer this question we must recognize that in a steady level coordinated turn, the bank angle is held constant and the airplane is not rolling. To initiate a right turn the pilot does need to provide right aileron input. This will start the airplane rolling to the right, increasing the bank angle. However, if the pilot continues to apply right aileron, the airplane will continue to roll and the bank angle will continue to increase. To stop the roll when the desired bank angle is reached, it seems reasonable that the pilot may need to briefly apply a little left aileron. Nevertheless, this still does not explain why left aileron should be needed to hold the bank angle constant in a coordinated right turn. This can be better understood by examining the reason that the rudder is needed to coordinate the turn.

The amount of right rudder required to sustain a right turn depends primarily on the change in the yawing moment coefficient with respect to the yawing rate, $C_{n,\bar{r}}$, which is called the *yaw-damping derivative*. As the airplane negotiates a right turn, the center of gravity follows a curved path and the airflow relative to the airplane is curved as shown in Fig. 6.7.3. This curvature of the relative wind has a significant effect on the yawing moment acting on the airplane. As seen in Fig. 6.7.3, this curvature produces sidewash on the vertical stabilizer, increases the airspeed over the outside wing semispan, and decreases the airspeed over the inside wing semispan. Thus, a right turn generates a side force to the right on the aft vertical stabilizer, increases the lift and drag on the left wing semispan, and decreases the lift and drag on the right wing semispan. For a conventional airplane with an aft vertical stabilizer having an aerodynamic center above the airplane's center of gravity, all of these effects tend to yaw the airplane to the left and roll it to the right. To counter the yawing moment produced by flight-path curvature in a right turn, some right rudder is required. Additionally, a left increment in the rolling moment is needed to compensate for the effects of flight-path curvature on the rolling moment produced in a right turn. Because of the roll-yaw coupling, right rudder will usually produce some left increment in the rolling moment, but this is almost never sufficient to counter the effects of flight-path curvature, and some left aileron is also typically needed.

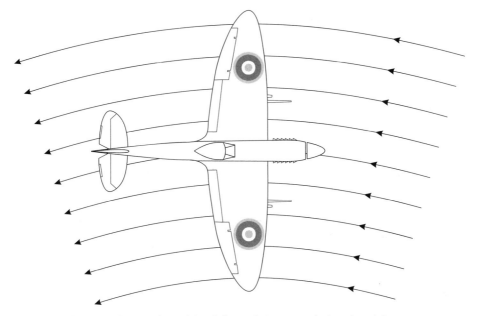

Figure 6.7.3. Top view of the airflow relative to an airplane in a right turn.

Example 6.7.1 was confined to the special case of a level turn (i.e., $\gamma = 0$). In a steady level coordinated turn the rolling rate is zero. This is not the case for a climbing or diving turn. Even though the bank angle remains constant in any steady turn, the airplane will experience a constant rolling rate when climbing or descending during a turn. As can be seen in Eq. (6.7.2), when climbing or descending while turning about the vertical axis, an airplane has a rolling rate that is proportional to the turning rate multiplied by the sine of the climb angle, $p = -\Omega \sin \gamma$. This rolling rate can have a significant effect on the control surface deflections required to sustain the turn, because the rolling rate substantially affects the aerodynamic moments. The investigation of these effects is left as an exercise for the student.

Elevator Angle to Turn
In Example 6.7.1 we saw that the elevator deflection required to support a steady constant-altitude coordinated turn is not significantly affected by the asymmetric power produced by a typical single-engine propeller-driven airplane. For all practical purposes, the elevator deflection in a steady coordinated turn at constant altitude can be evaluated from the first and fourth components of Eq. (6.7.30) with γ and $C_{m,\beta}$ set to zero,

$$\begin{bmatrix} C_{L,\alpha} & C_{L,\delta_e} \\ C_{m,\alpha} & C_{m,\delta_e} \end{bmatrix} \begin{Bmatrix} \Delta\alpha \\ \Delta\delta_e \end{Bmatrix} = \begin{Bmatrix} C_W[(1/\cos\phi)-1] - C_{L,\bar{q}} \sin\phi \tan\phi \\ -C_{m,\bar{q}} \sin\phi \tan\phi \end{Bmatrix} \tag{6.7.31}$$

where $\Delta\alpha$ and $\Delta\delta_e$ are relative to steady level flight at the same airspeed and altitude.

From the third component of Eq. (6.7.11), the load factor for the steady coordinated turn at constant altitude is

$$n \equiv \frac{L}{W} = \frac{1}{\cos\phi} \tag{6.7.32}$$

The reader should be careful to note that in this subsection we are using n to denote **load factor, not the yawing moment**. From the second component of Eqs. (6.7.16), (6.7.17), and (6.7.23), the dynamic pitch rate for the steady coordinated turn at constant altitude is

$$\breve{q} \equiv \frac{Vq}{g} = \frac{2V^2}{g\bar{c}_w}\bar{q} = \sin\phi\tan\phi = \frac{\sin^2\phi}{\cos\phi} = \frac{1-\cos^2\phi}{\cos\phi} = \frac{1}{\cos\phi} - \cos\phi \tag{6.7.33}$$

or after applying Eq. (6.7.32), **the dynamic pitch rate for a steady coordinated turn at constant altitude** can be related directly to the load factor, n,

$$\breve{q} = \sin\phi\tan\phi = n - 1/n \tag{6.7.34}$$

Using Eqs. (6.7.32) and (6.7.34) in Eq. (6.7.31) yields

$$\begin{bmatrix} C_{L,\alpha} & C_{L,\delta_e} \\ C_{m,\alpha} & C_{m,\delta_e} \end{bmatrix} \begin{Bmatrix} \Delta\alpha \\ \Delta\delta_e \end{Bmatrix} = \begin{Bmatrix} C_W(n-1) - C_{L,\breve{q}}(n-1/n) \\ -C_{m,\breve{q}}(n-1/n) \end{Bmatrix} \tag{6.7.35}$$

Equation (6.7.35) is readily solved for the **increments in angle of attack and elevator deflection relative to steady level flight, which are required to support a coordinated turn at constant airspeed and altitude,**

$$\begin{Bmatrix} \Delta\alpha \\ \Delta\delta_e \end{Bmatrix}_{turn} = \begin{Bmatrix} \dfrac{C_W C_{m,\delta_e}(n-1) + (C_{L,\delta_e}C_{m,\breve{q}} - C_{L,\breve{q}}C_{m,\delta_e})(n-1/n)}{C_{L,\alpha}C_{m,\delta_e} - C_{L,\delta_e}C_{m,\alpha}} \\[2ex] -\dfrac{C_W C_{m,\alpha}(n-1) + (C_{L,\alpha}C_{m,\breve{q}} - C_{L,\breve{q}}C_{m,\alpha})(n-1/n)}{C_{L,\alpha}C_{m,\delta_e} - C_{L,\delta_e}C_{m,\alpha}} \end{Bmatrix} \tag{6.7.36}$$

Equation (6.7.22) applied to Eqs. (6.1.33) and (6.1.34) produces similar results for the **increments in angle of attack and elevator deflection relative to steady level flight, which are required to support the constant-speed pull-up maneuver,**

$$\begin{Bmatrix} \Delta\alpha \\ \Delta\delta_e \end{Bmatrix}_{pull-up} = \begin{Bmatrix} \dfrac{C_W C_{m,\delta_e}(n-1) + (C_{L,\delta_e}C_{m,\breve{q}} - C_{L,\breve{q}}C_{m,\delta_e})(n-1)}{C_{L,\alpha}C_{m,\delta_e} - C_{L,\delta_e}C_{m,\alpha}} \\[2ex] -\dfrac{C_W C_{m,\alpha}(n-1) + (C_{L,\alpha}C_{m,\breve{q}} - C_{L,\breve{q}}C_{m,\alpha})(n-1)}{C_{L,\alpha}C_{m,\delta_e} - C_{L,\delta_e}C_{m,\alpha}} \end{Bmatrix} \tag{6.7.37}$$

From a comparison of Eqs. (6.7.36) and (6.7.37) we see that these relations are similar but not identical. The angle of attack and elevator deflection required to support a given load factor are not the same for a coordinated turn as they are for the pull-up maneuver.

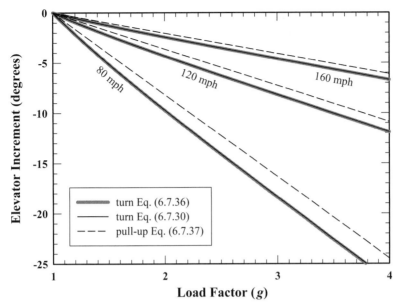

Figure 6.7.4. Elevator increment relative to steady level flight as a function of load factor for the airplane in Example 6.7.1.

For the airplane in Example 6.7.1, Fig. 6.7.4 shows the elevator increment relative to steady level flight, which is required to support a given load factor in a steady constant-altitude coordinated turn. The thick gray lines were obtained from Eq. (6.7.36), which neglects the effects of asymmetric power. The thin black lines include the effects of asymmetric power as predicted from Eq. (6.7.30). For comparison, the dashed lines show similar results for the constant-speed pull-up maneuver. The larger elevator angle for the coordinated turn is required to support a larger pitch rate. From Eq. (6.1.8), the dynamic pitch rate for the pull-up maneuver is $\bar{q} \equiv Vq/g = n-1$, whereas Eq. (6.7.34) gives the dynamic pitch rate for the coordinated turn as $\bar{q} = n-1/n$. This translates to a 50 percent increase in pitch rate for a 2-g coordinated turn relative to that for a 2-g pull-up maneuver.

6.8. Aileron Reversal

In Sec. 6.2 we have seen that the lift developed on the horizontal stabilizer at a given fuselage angle of attack and the pitching moment created as the result of a given elevator deflection are reduced by aeroelastic bending of the fuselage. The aerodynamic forces and moments produced on the wing are also affected by deformation, which is caused by the aerodynamic loading. Aeroelastic bending of the wing causes a change in wing dihedral, as was shown in Fig. 6.2.1. However, the wing's bending stiffness typically has little effect on the rolling moment produced by the ailerons. The torsional stiffness of the wing, on the other hand, can significantly alter aileron effectiveness.

Flap deflection produces a pitching moment about the aerodynamic center of an airfoil section. Thus, when the right aileron is deflected down and the left aileron is deflected up, the outboard wing sections on the right experience a negative increment in pitching moment while those on the left experience a positive increment in pitching moment. This differential pitching moment between the outboard sections of right and left semispans produces a twisting deformation of the wing, as shown in Fig. 6.8.1. The wing twist increases the angle of attack on the left and decreases it on the right, thereby reducing the rolling moment produced by the ailerons.

For small deflection angles, the aerodynamic twisting moment generated by aileron deflection is linearly proportional to the product of the dynamic pressure and the aileron deflection angle,

$$\Delta m_t = \tfrac{1}{2}\rho V^2 S_w \bar{c}_w C_{m_t,\delta_a}\delta_a \tag{6.8.1}$$

where C_{m_t,δ_a} can be simply viewed as an aerodynamic proportionality constant. From a structural viewpoint, the twisting moment on the wing must be linearly proportional to the twist deformation angle, τ,

$$\Delta m_t = k_t \tau \tag{6.8.2}$$

where k_t is a torsional stiffness coefficient. For small deflection angles, the net change in the rolling moment coefficient is linearly proportional to both the aileron deflection angle and the twist deformation angle,

$$\Delta C_\ell = \left(C_{\ell,\delta_a}\right)_{\text{rigid}}\delta_a - C_{\ell,\tau}\tau \tag{6.8.3}$$

Combining Eqs. (6.8.1) through (6.8.3), we obtain

$$\Delta C_\ell - \left[\left(C_{\ell,\delta_a}\right)_{\text{rigid}} - \tfrac{1}{2}\rho V^2 S_w \bar{c}_w\, C_{\ell,\tau}C_{m_t,\delta_a}/k_t\right]\delta_a \tag{6.8.4}$$

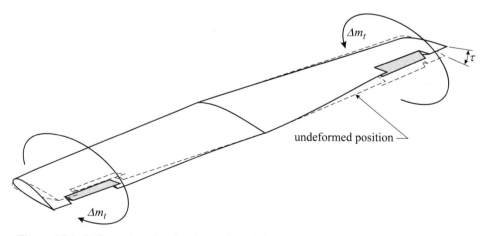

Figure 6.8.1. Deformation of a wing due to the twisting moment produced by aileron deflection.

From Eq. (6.8.4), we see that the net lateral control derivative, including the effect of wing twist, is given by

$$C_{\ell,\delta_a} = \left(C_{\ell,\delta_a}\right)_{\text{rigid}} - \frac{\rho V^2 S_w \bar{c}_w C_{\ell,\tau} C_{m_t,\delta_a}}{2k_t} \tag{6.8.5}$$

which shows that aileron effectiveness is decreased in direct proportion to the airspeed squared and in inverse proportion to the torsional stiffness of the wing. From this result we see that there is an aileron reversal speed that is proportional to the square root of the wing's torsional stiffness,

$$V_{ar} = \sqrt{\frac{2k_t \left(C_{\ell,\delta_a}\right)_{\text{rigid}}}{\rho S_w \bar{c}_w C_{\ell,\tau} C_{m_t,\delta_a}}} \tag{6.8.6}$$

At this airspeed, aileron deflection produces no rolling moment. At higher airspeeds, aileron control is reversed.

Because of the complex cross-sections used to construct the wing of an airplane, its torsional stiffness can only be determined experimentally or through the use of finite element analysis on a digital computer. Accurate estimation of the aerodynamic loading resulting from aileron deflection and wing twist also requires computer simulation. Since wing twist depends on aerodynamic loading and aerodynamic loading depends on wing twist, an iterative procedure must be used to estimate the change in rolling moment with respect to aileron deflection for a nonrigid wing. We can start by computing the aerodynamic loading for the undeformed or rigid wing. This loading is then used in a finite element structural analysis to determine the wing deformation. The aerodynamic computations are then repeated using the deformed geometry to obtain an improved estimate for the aerodynamic loading. The process is repeated until no further changes are observed.

All *fixed-wing* aircraft can be classified in one of two very broad categories, *rigid-wing* and *flex-wing* aircraft. Since aeroelastic deformation can never be eliminated in the lightweight structures required for flight, one might think that no airplane could ever be classified as a rigid-wing. However, this is not how the two categories are defined. **A rigid-wing aircraft is one in which the moments that result from the elastic deformation of the wing are less than those that result from control surface deflections.** On the other hand, **a flex-wing aircraft is one that derives the control moment primarily from elastic deformation of the wing.** When all things that fly are considered, most could be categorized as flex-wings. However, when we consider only those aircraft designed and built by humans, most are rigid-wings. **In a rigid-wing airplane, the wing should have sufficient torsional stiffness so that the aileron control reversal speed exceeds the maximum airspeed that could ever be attained by the airplane.**

When a wing is very flexible in torsion, as it is on birds and some ultralight aircraft such as hang gliders, the aileron control reversal speed for the wing could actually be less than the wing stall speed, and ailerons would not work at all in the conventional manner.

On such a wing, ailerons would work very much like trim tabs. A downward deflection of an aileron would twist the outboard sections of the wing semispan to sufficiently reduce the angle of attack so that a rolling moment would be produced in the same direction as the aileron deflection. This could still work as an aircraft control. However, the control derivative would be positive instead of negative and the ailerons would need to be linked to the stick or control yoke accordingly.

6.9. Other Control Surface Configurations

To this point we have treated trim, control, and maneuverability as though the airplane had fixed horizontal and vertical stabilizers combined with a control set composed of movable ailerons, elevator, and rudder, all with adjustable trim tabs. Alternative control surface configurations are commonly used. However, with suitable variable definitions, results obtained for these other configurations are very similar to those obtained for the conventional controls that have previously been discussed.

Three different control surface configurations that have been used for pitch control are shown in Fig. 6.9.1. The first is the conventional fixed stabilizer with movable elevator, which has previously been treated. The second is the *simple stabilator*, which is often referred to as an *all-moving* or *all-flying tail*. This could be thought of as a special case of the stabilizer-elevator configuration, which has an elevator chord fraction of 1.0.

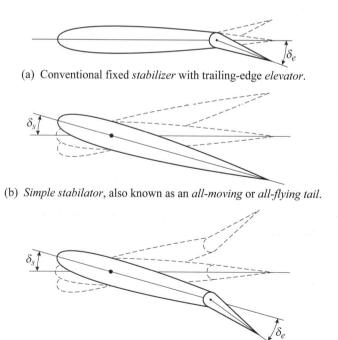

(a) Conventional fixed *stabilizer* with trailing-edge *elevator*.

(b) *Simple stabilator*, also known as an *all-moving* or *all-flying tail*.

(c) *Compound stabilator* with trailing-edge flap called an *anti-servo tab*.

Figure 6.9.1. Three different aft-tail configurations, which have been used for pitch control.

The last configuration shown in Fig. 6.9.1 is referred to here as a *compound stabilator*. With this configuration, the entire horizontal surface is movable like the simple stabilator, but there is also an aft trailing-edge flap, sometimes called an *anti-servo tab*, that moves relative to the forward portion of this control surface.

The Compound Stabilator

The compound stabilator is the least common of the three configurations that are shown in Fig. 6.9.1. However, since both the conventional stabilizer-elevator configuration and the simple stabilator can be viewed as special cases of the compound stabilator, this configuration is the most general of the three. From an inanimate visual inspection, most people could not distinguish between a compound stabilator and a conventional stabilizer-elevator configuration. The external geometry is essentially the same. The difference is that there are two degrees of freedom associated with the deflection of a compound stabilator, as shown in Fig. 6.9.2. The entire horizontal surface can be moved through a deflection angle that shall be denoted here as δ_s. Additionally, the trailing-edge flap can be deflected relative to the forward portion of the surface through a deflection angle indicated here by δ_e. The sign convention that will be used for both of these angles is positive with the leading edge up and the trailing edge down, as shown in Fig. 6.9.1.

While the compound stabilator has two deflection angles that could be controlled independently, the pilot has only one pitch control. To induce a nose-down pitching moment the pilot pushes the stick or control yoke forward, and pulling the control back creates a nose-up pitching moment. To maintain the conventional cockpit controls, the two stabilator deflection angles are linked together, either mechanically or electronically. The primary control surface deflection angle, δ_s, is linked directly to the pilot's stick or control yoke and the trailing-edge flap deflection, δ_e, is linked to the primary control surface deflection so that

$$\delta_e = k_{es}\delta_s + \delta_{et} \tag{6.9.1}$$

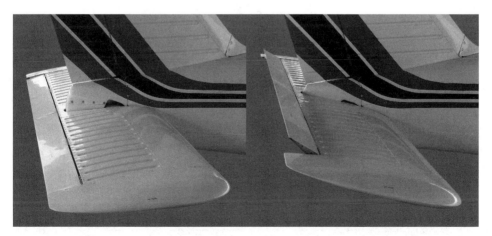

Figure 6.9.2. The compound stabilator with anti-servo tab used on the Piper Cherokee.

where the constant k_{es} can be varied by the designer but not usually by the pilot. In a mechanical system, k_{es} is fixed by the mechanical linkage between the primary control surface deflection and the trailing-edge flap deflection. The offset, δ_{et}, can usually be adjusted by the pilot as a trim setting. This is done to zero the control force at trim for some desired airspeed, much as a trim tab is used on a conventional elevator.

Following a procedure similar to that used to obtain Eqs. (6.1.16) and (6.1.17), the lift and pitching moment coefficients for a wing combined with a compound stabilator are

$$
\begin{aligned}
C_L = {} & C_{L_w,\alpha}\left(\alpha + \frac{2l_w}{\bar{c}_w}\bar{q} + \alpha_{0w} - \alpha_{L0w}\right) \\
& + \frac{S_h}{S_w}\eta_h C_{L_h,\alpha}\left(\alpha + \frac{2l_h}{\bar{c}_w}\bar{q} + \alpha_{0h} - \alpha_{L0h} - \varepsilon_d + \delta_s + \varepsilon_e\delta_e\right)
\end{aligned}
\tag{6.9.2}
$$

$$
\begin{aligned}
C_m = {} & C_{m_w} - \frac{l_w}{\bar{c}_w}C_{L_w,\alpha}\left(\alpha + \frac{2l_w}{\bar{c}_w}\bar{q} + \alpha_{0w} - \alpha_{L0w}\right) \\
& + \frac{S_h\bar{c}_h}{S_w\bar{c}_w}\eta_h(C_{m_h0} + C_{m_h,\delta_e}\delta_e) \\
& - \frac{S_h l_h}{S_w\bar{c}_w}\eta_h C_{L_h,\alpha}\left(\alpha + \frac{2l_h}{\bar{c}_w}\bar{q} + \alpha_{0h} - \alpha_{L0h} - \varepsilon_d + \delta_s + \varepsilon_e\delta_e\right)
\end{aligned}
\tag{6.9.3}
$$

where the remaining notation used here is the same as that defined for the stabilizer-elevator configuration. Using Eq. (6.9.1), these two equations can be written as

$$
C_L = C_{L0} + C_{L,\delta_e}\delta_{et} + C_{L,\alpha}\alpha + C_{L,\bar{q}}\bar{q} + C_{L,\delta_s}\delta_s
\tag{6.9.4}
$$

$$
C_m = C_{m0} + C_{m,\delta_e}\delta_{et} + C_{m,\alpha}\alpha + C_{m,\bar{q}}\bar{q} + C_{m,\delta_s}\delta_s
\tag{6.9.5}
$$

where

$$
C_{L,\delta_s} = \frac{S_h}{S_w}\eta_h C_{L_h,\alpha}(1 + \varepsilon_e k_{es})
\tag{6.9.6}
$$

$$
C_{m,\delta_s} = \frac{S_h\bar{c}_h}{S_w\bar{c}_w}\eta_h C_{m_h,\delta_e}k_{es} - \frac{S_h l_h}{S_w\bar{c}_w}\eta_h C_{L_h,\alpha}(1 + \varepsilon_e k_{es})
\tag{6.9.7}
$$

and all of the other aerodynamic coefficients are the same as those for the stabilizer-elevator configuration as defined in Eqs. (6.1.20) through (6.1.27). Continuing to follow a development similar to that used for the stabilizer-elevator configuration, the fuselage angle of attack at level trim for the compound stabilator configuration is found to be

$$
(\alpha)_{\text{trim}} = \frac{(C_W - C_{L0} - C_{L,\delta_e}\delta_{et})C_{m,\delta_s} + (C_{m0} + C_{m,\delta_e}\delta_{et})C_{L,\delta_s}}{C_{L,\alpha}C_{m,\delta_s} - C_{L,\delta_s}C_{m,\alpha}}
\tag{6.9.8}
$$

The stabilator angle at level trim is

$$(\delta_s)_{\text{trim}} = -\frac{(C_W - C_{L0} - C_{L,\delta_e}\delta_{et})C_{m,\alpha} + (C_{m0} + C_{m,\delta_e}\delta_{et})C_{L,\alpha}}{C_{L,\alpha}C_{m,\delta_s} - C_{L,\delta_s}C_{m,\alpha}} \tag{6.9.9}$$

Similarly, the stabilator angle per g is

$$\frac{\partial \delta_s}{\partial n} = -\frac{C_W C_{m,\alpha} + (C_{L,\alpha}C_{m,\bar{q}} - C_{L,\bar{q}}C_{m,\alpha})(g\bar{c}_w/2V^2)}{C_{L,\alpha}C_{m,\delta_s} - C_{L,\delta_s}C_{m,\alpha}} \tag{6.9.10}$$

Because the primary stabilator hinge is used to rotate the entire horizontal surface, the stabilator hinge moment coefficient is defined as

$$C_{H_s} \equiv \frac{H_s}{\frac{1}{2}\rho V_h^2 S_h \bar{c}_h} \tag{6.9.11}$$

where H_s is the hinge moment for the primary stabilator hinge. With this definition, the stabilator hinge moment coefficient can be written as

$$C_{H_s} = C_{m_h 0} + C_{m_h,\delta_e}\delta_e \\ - \frac{l_{Hs}}{\bar{c}_h}C_{L_h,\alpha}\left(\alpha + \frac{2l_h}{\bar{c}_w}\bar{q} + \alpha_{0h} - \alpha_{L0h} - \varepsilon_d + \delta_s + \varepsilon_e\delta_e\right) \tag{6.9.12}$$

where l_{Hs} is the distance that the aerodynamic center of the entire stabilator surface is aft of the primary hinge. The designer can vary this length, but it is typically set close to zero so that the pitch control force can be kept within allowable limits. Applying Eq. (6.9.1), Eq. (6.9.12) can be rearranged as

$$C_{H_s} = C_{H_s 0} + C_{H_s,\alpha_h}\left[\alpha_{0h} - \varepsilon_{d0} + (1 - \varepsilon_{d,\alpha})\alpha + \frac{2l_h}{\bar{c}_w}\bar{q}\right] \\ + C_{H_s,\delta_s}\delta_s + C_{H_s,\delta_e}\delta_{et} \tag{6.9.13}$$

where

$$C_{H_s 0} = C_{m_h 0} + \frac{l_{Hs}}{\bar{c}_h}C_{L_h,\alpha}\alpha_{L0h} \tag{6.9.14}$$

$$C_{H_s,\alpha_h} = -\frac{l_{Hs}}{\bar{c}_h}C_{L_h,\alpha} \tag{6.9.15}$$

$$C_{H_s,\delta_s} = C_{m_h,\delta_e}k_{es} - \frac{l_{Hs}}{\bar{c}_h}C_{L_h,\alpha}(1 + \varepsilon_e k_{es}) \tag{6.9.16}$$

$$C_{H_s,\delta_e} = C_{m_h,\delta_e} - \frac{l_{Hs}}{\bar{c}_h} C_{L_h,\alpha} \varepsilon_e \tag{6.9.17}$$

The pitch control force that is exerted by the pilot for a compound stabilator operated by a reversible all-mechanical control system would then be

$$F_p = \frac{H_s}{l_g} = \frac{\frac{1}{2}\rho V^2 \eta_h S_h \bar{c}_h}{l_g}$$

$$\times \left\{ C_{H_s0} + C_{H_s,\alpha_h} [\alpha_{0h} - \varepsilon_{d0} + (1 - \varepsilon_{d,\alpha})\alpha + \frac{2l_h}{\bar{c}_w}\bar{q}] + C_{H_s,\delta_s}\delta_s + C_{H_s,\delta_e}\delta_{et} \right\} \tag{6.9.18}$$

Notice that this has exactly the same form as Eq. (6.3.8). The only difference between Eq. (6.3.8) and Eq. (6.9.18) is in the hinge moment derivatives.

Continuing to follow the development in Sec. 6.3, when pitch control is provided by a compound stabilator, the pitch control force required for constant airspeed and constant normal acceleration is

$$F_p = \eta_h \frac{S_h \bar{c}_h}{S_w l_g} \left[C_1 \left(n - \frac{V^2}{V_{\text{trim}}^2} \right) W + C_2(n-1)\frac{\rho g S_w \bar{c}_w}{4} \right] \tag{6.9.19}$$

where

$$C_1 = \frac{C_{m,\delta_s} C_{H_s,\alpha_h}(1 - \varepsilon_{d,\alpha}) - C_{m,\alpha} C_{H_s,\delta_s}}{C_{L,\alpha} C_{m,\delta_s} - C_{L,\delta_s} C_{m,\alpha}} \tag{6.9.20}$$

$$C_2 = \frac{2l_h C_{H_s,\alpha_h}}{\bar{c}_w}$$

$$+ \frac{(C_{L,\delta_s} C_{m,\bar{q}} - C_{L,\bar{q}} C_{m,\delta_s}) C_{H_s,\alpha_h}(1 - \varepsilon_{d,\alpha}) - (C_{L,\alpha} C_{m,\bar{q}} - C_{L,\bar{q}} C_{m,\alpha}) C_{H_s,\delta_s}}{C_{L,\alpha} C_{m,\delta_s} - C_{L,\delta_s} C_{m,\alpha}} \tag{6.9.21}$$

The reversible control force gradient and longitudinal control force per g for a compound stabilator are obtained from Eq. (6.9.19) in the same way that Eqs. (6.3.19) and (6.3.20) were obtained from Eq. (6.3.18).

For subsonic airplanes, the primary advantage of the compound stabilator over the stabilizer-elevator configuration is the extra flexibility afforded the designer in adjusting the control force derivatives for a reversible all-mechanical control system. As discussed in Sec. 6.3, the reversible mechanical control force for a stabilizer-elevator configuration increases rapidly with aircraft size. The added versatility of the compound stabilator allows the designer the possibility of using reversible mechanical controls on larger airplanes. Investigation of this potential is left as a homework exercise for the student. A disadvantage of the compound stabilator is added complexity. An additional mechanism is required to link the primary hinge to the trailing-edge flap hinge. With all else being equal, increased complexity always translates to reduced reliability.

The Simple Stabilator

The simple stabilator or all-flying tail, which is also shown in Fig. 6.9.1, is hinged like the primary hinge of the compound stabilator, but it has no movable trailing-edge flap. For the purpose of analysis, the simple stabilator can be treated like an elevator having a flap chord fraction of 1.0, which translates to an elevator with $\varepsilon_e = 1.0$ and $C_{m_h,\delta_e} = 0.0$. On the other hand, it could also be treated as a special case of the compound stabilator, having both k_{es} and δ_{et} set to zero. For these reasons, no additional analysis is needed for this tail configuration. **In its general form, the stabilator is defined by the geometry shown in Figs. 6.9.1c and 6.9.2. The geometry shown in Figs. 6.9.1b and 6.9.3 is simply a special case of the more general stabilator configuration, which was treated in the preceding subsection.** This is true withstanding the fact that military terminology commonly considers the stabilator to include only the geometry shown in Fig. 6.9.3.

The simple stabilator is most commonly used on supersonic fighters like the F/A-18 Super Hornet shown in Fig. 6.9.3. The primary motivation for using an all-movable tail on a supersonic aircraft is the loss of elevator effectiveness at transonic and supersonic airspeeds. With the exception of delta wings, most modern supersonic aircraft use the simple stabilator, because at transonic airspeeds, a shockwave can form at the hinge of a trailing-edge flap, rendering an elevator ineffective as an aerodynamic control surface.

Figure 6.9.3. The simple stabilator used on the F/A-18F supersonic fighter. (U.S. Navy photos)

Furthermore, at supersonic airspeeds, camber has little effect on the production of lift, for the reasons discussed in Sec. 1.14. For these reasons, neither a conventional stabilizer-elevator configuration nor a compound stabilator is practical for providing pitch control on a supersonic aircraft.

The simple stabilator was first investigated as a pitch control mechanism during the British World War II Miles M-52 project. When the project was canceled, the research results were passed to the U.S. supersonic project and the simple stabilator was used on the experimental Bell X-1. The North American Aviation F-86 Sabre, which was the first U.S. Airforce aircraft to exceed the speed of sound in a shallow dive, was introduced with a conventional stabilizer-elevator tail configuration. This was eventually replaced with a stabilator.

Although the primary motivation for the supersonic stabilator is the loss of elevator effectiveness at transonic and supersonic speeds, there are some advantages to the simple stabilator in subsonic flight as well. Like the compound stabilator, the simple stabilator can provide an advantage for designs where it is difficult to keep the pitch control force within allowable limits and a reversible mechanical control system is desired. If a symmetric airfoil is used and the hinge is located at the aerodynamic center, then the only moment needed to rotate the horizontal surface is that needed to overcome friction. With this design, the required control force and trim adjustment could be provided by an adjustable spring mechanism. This allows a reversible mechanical control system to be used on larger airplanes.

The simple stabilator on military aircraft has the same overcontrol problem as on general aviation aircraft. In older jet fighters, a resisting stick force was generated within the control linkage rather than by an external anti-servo tab. For example, in the F-100 Super Sabre, springs were attached to the control stick to provide increasing resistance to pilot input. In modern jet fighters, the stick forces are generated within the irreversible fly-by-wire control system. There is no direct connection between the pilot's stick and the stabilator.

A disadvantage of the simple stabilator configuration, which is also shared by the compound stabilator, is that the hinge must carry all of the lift developed on the surface. The design of a freely moving hinge with adequate strength and stiffness for this purpose can be challenging but is not impossible. The problem is amplified by the fact that failure of the hinge would be catastrophic.

The Stabilizer-Elevator Configuration with All-Movable Trim Adjustment

Most modern airliners and many general aviation airplanes adjust the incidence angle of the tailplane to trim the aircraft during flight as fuel is burned and the weight coefficient changes. These adjustments are handled by an adjustable horizontal stabilizer referred to here as the *all-movable trim adjustment*. However, such an adjustable stabilizer is not the same as a stabilator. A stabilator is controlled by the pilot's stick or control yoke, and an adjustable stabilizer is controlled by the pilot's trim control. This common variation of the conventional stabilizer-elevator configuration has exactly the same external geometry as the compound stabilator, which is shown in Fig. 6.9.1c. The only difference between these two configurations is in how the pilot controls the two deflection angles. For this stabilizer-elevator configuration, the pilot's cockpit pitch control is linked to only the elevator deflection angle, δ_e, just as it is in any stabilizer-elevator configuration. However, instead of using a trim tab, the trim adjustment is linked to the stabilizer deflection angle, δ_s. This provides a broader range of trim adjustment than can be attained with a trim tab. With this type of trim adjustment the control force can be zeroed over a wider range airspeed, gross weight, and CG location. In addition to increased manufacturing cost, the price for this increased flexibility is linked to the possibility of failure. While the mechanism used to adjust the trim setting by changing the incidence angle of the stabilizer is probably no more likely to fail than a trim tab mechanism, the consequences of failure are very different. If the pilot looses the ability to adjust a trim tab, the result will be increased control forces over some portions of the flight. However, if the jackscrew controlling an airplane's stabilizer incidence angle were to fail, the result would likely be complete control loss. For instance, failure of the jackscrew on a McDonnell Douglas MD-83 resulted in the death of all passengers and crew members in the crash of Alaska Airlines Flight 261 off the coast of California on January 31, 2000.

The V-Tail

A variation of the conventional aft tail with separate horizontal and vertical stabilizers is the V-tail, shown in Figs. 5.5.2 and 6.9.4. With this configuration, the vertical stabilizer is absent and the "horizontal" stabilizer has substantial dihedral, usually close to 45 degrees. The normal and spanwise components of lift developed on this V-shaped surface can be used to provide stability and control in both pitch and yaw.

Unlike the conventional elevator, the control surface on each side of a V-tail can be deflected separately. Each of these control surfaces is linked to both the stick or control

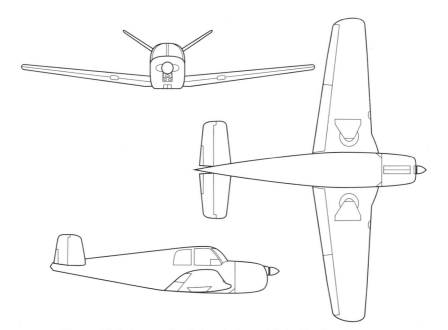

Figure 6.9.4. A general aviation airplane with the V-tail empennage.

yoke and to the rudder pedals. The position of the stick or control yoke determines the sum of the two deflection angles, and the position of the rudder pedals determines the difference. With the rudder pedals in the neutral position and the stick or control yoke pushed forward, both control surfaces are deflected downward through the same deflection angle, as shown in Fig. 6.9.5a. With the pitch control in the neutral position and the left rudder pedal depressed, the control surface on the left is deflected downward and that on the right is deflected upward through the same deflection angle, as shown in Fig. 6.9.5b. In addition to generating a yawing moment to the left, the control surface deflections that are shown in Fig. 6.9.5b will also produce a substantial rolling moment to the right.

Estimating the aerodynamic forces and moments developed by a V-tail as a function of angle of attack, sideslip angle, and control surface deflection angles is not a simple matter of using the results for a horizontal stabilizer multiplied by the appropriate sines and cosines. Lift developed on one side of the V-tail generates vorticity that produces downwash on the other side. This downwash affects the lift distribution over the entire empennage, and its effect depends significantly on the dihedral angle. The vortex interactions between the two sides of a V-tail are similar to those between the horizontal and vertical stabilizers of a conventional tail. This was discussed in Sec. 5.5 and shown schematically in Figs. 5.5.3, 5.5.4, and 5.5.5. The classical infinite series solution to Prandtl's lifting-line theory cannot be used to estimate the forces and moments created by a V-tail, because this solution is valid only for lifting surfaces with no sweep or dihedral. In Sec. 5.5 a method is presented for predicting the effects of tail dihedral and sideslip on the yawing moment coefficient. Similarly, Sec. 5.6 presents a method for predicting the effects of dihedral and sideslip on the rolling moment coefficient. The control derivatives for a V-tail can be estimated in a similar fashion. The simplest analytical method that can be used to accurately estimate the aerodynamic forces and moments developed by a V-tail is the numerical lifting-line method.

The V-tail empennage has not been widely used. Perhaps the best known example of the V-tail design is the Beech Bonanza, which was the model for Fig. 6.9.4. The earliest versions of this airplane were manufactured in the late 1950s and early 1960s. Although many of these V-tail Bonanzas are still flying today, in later models the V-tail was replaced with a conventional aft tail. One possible reason for the dearth of V-tails is the inefficiency associated with producing control moments in this manner. When the

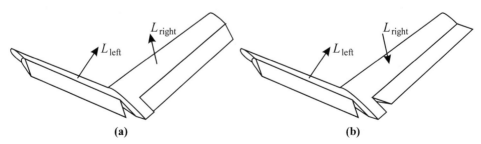

(a) **(b)**

Figure 6.9.5. Control surface deflections for the V-tail empennage showing (a) down elevator equivalent and (b) left rudder equivalent.

control surfaces are deflected as shown in Fig. 6.9.5, only about 70 percent of the lift developed on the empennage is useful in producing the desired moment. When the control surfaces are used to generate a pitching moment as shown in Fig. 6.9.5a, the horizontal components of lift on the right and left sides of the V-tail cancel. Only the vertical components of lift contribute to the pitching moment about the center of gravity. Similarly, when the control surfaces are used as shown in Fig. 6.9.5b to produce the equivalent of rudder deflection, only the horizontal components of lift contribute to the desired yawing moment and the vertical components cancel. The need to generate opposing components of lift on the two sides of a V-tail increases induced drag and reduces the maximum moment that can be produced by an empennage of any given size.

Ailerons, Spoilers, and Adverse Yaw
In Sec. 5.7 we learned that when conventional ailerons are used for roll control, an adverse yawing moment is generated as a result of the change in induced drag produced by aileron deflection. This tends to yaw the airplane in the direction opposite to the intended roll. The Frise aileron, shown in Fig. 6.9.6, can be used to reduce or eliminate this adverse yaw. These ailerons are shaped and hinged so that on the upward deflected aileron only, the leading edge projects out into the airflow. This increases the parasitic drag on the semispan with up aileron, producing a favorable yawing moment that counters the adverse yawing moment generated by the change in induced drag.

Aileron adverse yaw can also be avoided by using spoilers for roll control. A spoiler is a panel on the upper surface of a wing that can be raised into the airflow as shown in Fig. 6.9.7. The purpose of raising the spoiler is to induce flow separation, which destroys or "spoils" the lift on the wing sections involved. Raising a spoiler on the outboard sections of the left wing semispan produces a rolling moment to the left (negative ℓ). The raised left spoiler also increases the drag on the left semispan and produces a favorable yawing moment (negative n).

Spoilers could be used as the sole means of roll control, or they could be used in conjunction ailerons, as shown in Fig. 6.9.7. Since conventional ailerons produce adverse yaw and spoilers generate favorable yaw, aileron and spoiler deflection could be coordinated to produce a pure rolling moment. This could help to relieve the pilot of the need for careful aileron-rudder coordination in situations where maintaining precise aircraft attitude is critical.

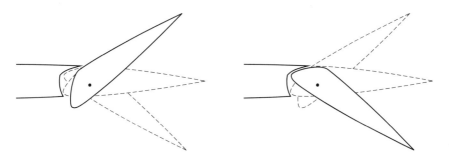

Figure 6.9.6. Frise ailerons used to eliminate adverse yaw.

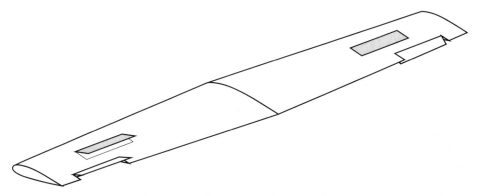

Figure 6.9.7. Spoilers used for roll control.

There are some objections to the use of spoilers as the primary source of roll control. For one, spoilers become ineffective at high angles of attack. Additionally, spoilers produce a nonlinear rolling moment, and they introduce hysteresis in the roll control because the boundary layer does not reattach as quickly as it separates. Another objection, which applies only to reversible mechanical control systems, is the nonlinear hinge moment associated with spoiler deflection. Nevertheless, with proper attention to design detail, these difficulties can be circumvented, and spoilers are becoming more popular as a means of roll control.

With a modern fly-by-wire control system, it is possible to electronically reduce or eliminate coupling between the roll and yaw controls. This can be accomplished by "mixing" the electronic input signals. Output signals to determine aileron and rudder control surface deflections could be computed from input signals obtained from the pilot's cockpit controls according to the relation

$$
\left\{ \begin{matrix} \delta_a \\ \delta_r \end{matrix} \right\}_{\substack{\text{control} \\ \text{surface}}} = \begin{bmatrix} k_{aa} & k_{ar} \\ k_{ra} & k_{rr} \end{bmatrix} \left\{ \begin{matrix} \delta_a \\ \delta_r \end{matrix} \right\}_{\substack{\text{pilot} \\ \text{input}}}
\tag{6.9.22}
$$

where k_{aa}, k_{ar}, k_{ra}, and k_{rr} are control system gain coefficients. For traditional "unmixed" control, both k_{ar} and k_{ra} would be zero. For small deflection angles, the contributions of control surface deflections to the rolling and yawing moment coefficients are

$$
\begin{aligned}
\left\{ \begin{matrix} \Delta C_\ell \\ \Delta C_n \end{matrix} \right\}_{\text{control}} &= \begin{bmatrix} C_{\ell,\delta_a} & C_{\ell,\delta_r} \\ C_{n,\delta_a} & C_{n,\delta_r} \end{bmatrix} \left\{ \begin{matrix} \delta_a \\ \delta_r \end{matrix} \right\}_{\substack{\text{control} \\ \text{surface}}} \\[2mm]
&= \begin{bmatrix} C_{\ell,\delta_a} & C_{\ell,\delta_r} \\ C_{n,\delta_a} & C_{n,\delta_r} \end{bmatrix} \begin{bmatrix} k_{aa} & k_{ar} \\ k_{ra} & k_{rr} \end{bmatrix} \left\{ \begin{matrix} \delta_a \\ \delta_r \end{matrix} \right\}_{\substack{\text{pilot} \\ \text{input}}} \\[2mm]
&= \begin{bmatrix} (C_{\ell,\delta_a}k_{aa} + C_{\ell,\delta_r}k_{ra}) & (C_{\ell,\delta_a}k_{ar} + C_{\ell,\delta_r}k_{rr}) \\ (C_{n,\delta_a}k_{aa} + C_{n,\delta_r}k_{ra}) & (C_{n,\delta_a}k_{ar} + C_{n,\delta_r}k_{rr}) \end{bmatrix} \left\{ \begin{matrix} \delta_a \\ \delta_r \end{matrix} \right\}_{\substack{\text{pilot} \\ \text{input}}}
\end{aligned}
\tag{6.9.23}
$$

By choosing suitable values for the off-diagonal gain coefficients, the airplane could be made to "feel" as though the roll and yaw controls were uncoupled. Setting the off-diagonal terms on the far right-hand side of Eq. (6.9.23) to zero, this requires that

$$
\begin{Bmatrix} k_{ar} \\ k_{ra} \end{Bmatrix} = \begin{Bmatrix} -(C_{\ell,\delta_r}/C_{\ell,\delta_a})k_{rr} \\ -(C_{n,\delta_a}/C_{n,\delta_r})k_{aa} \end{Bmatrix}
\tag{6.9.24}
$$

The diagonal gain coefficients can be adjusted to give the desired sensitivity to the cockpit controls. One significant advantage of this approach to adverse yaw abatement is that the gain coefficients are easily modified to provide improved handling qualities, even after final design and flight testing. In fact, these gain coefficients could be varied with airspeed, altitude, and flap settings; or they could be adjusted by the pilot to provide improved handling qualities for a range of operating conditions and pilot preference.

6.10. Airplane Spin

In Sec. 6.6, airplane stall was treated as longitudinal motion. In reality, an airplane almost always has sufficient asymmetry to cause one wing to stall before the other. When one wing stalls, the asymmetric loss of lift produces a moment that rolls the airplane in the direction of the stalled wing. At stall, the fuselage angle of attack is typically quite high. When an airplane at a high angle of attack rolls about the fuselage axis, angle of attack is converted to sideslip. Since the airplane is stable in yaw, this sideslip will produce a moment that tends to yaw the airplane in the direction of the stalled wing. Because the wing typically stalls before the horizontal stabilizer, loss of lift on the wing also results in a nose-down pitching moment. These moments, combined with the force imbalance between lift and weight, cause the airplane to roll and yaw in the direction of the stalled wing while pitching nose down as it falls in response to the lost lift. This motion clearly involves all six degrees of freedom.

The yawing rate that is typically generated in an asymmetric stall can induce what is commonly called a *spin*. Airplane spin is a stalled motion in which the airplane rapidly descends while rotating about a nearly vertical axis that passes fairly close to the center of gravity. During the spinning descent, the airplane's center of gravity follows a helical path centered about the spin axis. As shown in Fig. 6.10.1, the spin can occur with the airplane in a rather steep nose-down attitude, or it can occur with the wings and fuselage closer to horizontal. The later case is called a *flat spin*. As shown in Fig. 6.10.1, the axis of a flat spin is closer to the center of gravity than that of a steep spin. In a flat spin, the wings of the airplane have a motion that is somewhat similar to that of the rotating blades of a helicopter, and in the limiting case, the spin axis may actually pass through the center of gravity. Because of the high descent rate, the wings of an airplane remain completely stalled in a spin. Both steep and flat spins can occur in either direction, right or left.

A spin can develop into stable motion. Spin rates on the order of 90 to 180 degrees per second can occur in the fully developed spin of a light aircraft. It can be difficult for the pilot to recover from a developed spin if the airplane is not specifically designed for spin recovery or spin resistance. Whether or not an airplane develops a stable spin depends on the airplane's inertia properties and aerodynamic moments.

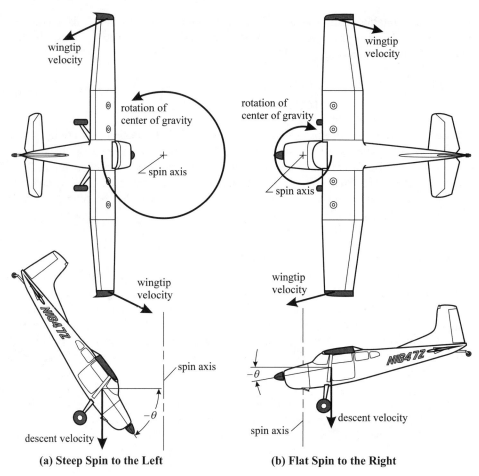

(a) Steep Spin to the Left   (b) Flat Spin to the Right

Figure 6.10.1. Top and side views of airplane pitch attitude and motion in steep and flat spins.

Airplane spin was encountered very early in the history of aviation. In the early days of flight, airplane spin was the cause of many fatal accidents. Even today, spin is one of the primary causes of accidents in general aviation. Since spin is always preceded by stall, the first solution to the spin problem was stall avoidance. Today, thanks to years of experimental investigation, most general aviation airplanes will recover from a spin if the pilot simply relinquishes control of the airplane. This does not mean that spins are not dangerous. FAR Part 23 requires that a single-engine airplane be able to recover from a one-turn spin in no more than one additional turn. Even so, because the airplane is descending very rapidly (100 to 200 ft/sec), a great deal of altitude is lost during any spin recovery. With insufficient altitude, the initiation of any spin will almost certainly be disastrous.

The aerodynamic forces and moments produced by the separated flow around a spinning airplane are highly nonlinear and quite complex. Thus, it is very difficult to accurately predict the unsteady motion of a spinning airplane from theoretical analysis alone. Most of what is known about designing airplanes for spin resistance and spin recovery has been obtained from experimental investigation. Nevertheless, some insight into the physics of airplane spin can be gained by examining the equations of motion for a fully developed steady spin.

Development of the equations of motion for steady spin about a vertical axis is similar to that for the steady coordinated turn, which was presented in Sec. 6.7. In both cases the airplane's center of gravity traces a helical path centered about a vertical axis. The primary differences are the orientation of the aircraft and the radius of the helix. During the steady coordinated turn depicted in Fig. 6.7.1, the fuselage axis is nearly tangent to the flight path, and the radius of the helix is large compared to the wingspan. As shown schematically in Fig. 6.10.1, in a spin, the velocity vector for the airplane's center of gravity makes a rather large angle with the fuselage axis, and the radius of the helix is less than the wingspan. For simplicity, Fig. 6.10.1 was drawn with the bank angle set to zero. Although the bank angle can be small in a spin, it is not necessarily zero. Figure 6.10.2 shows a schematic of a spinning airplane with nonzero bank angle. In this figure, the spin radius has been somewhat enlarged for the sake of clarity. If the radius of the helix in Fig. 6.10.2 had been drawn to scale, the spin axis would pass through or very near the airplane, as was shown in Fig. 6.10.1.

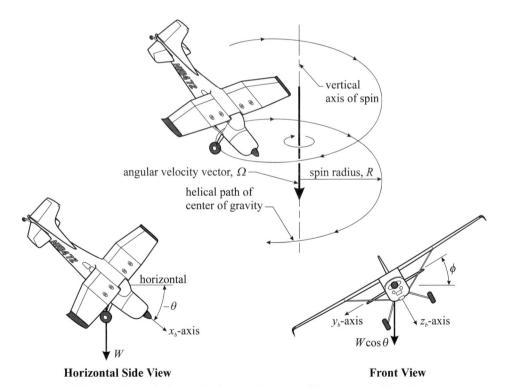

Figure 6.10.2. Airplane spin, showing the spin radius exaggerated for clarity.

Here again we shall use the (x_b, y_b, z_b) coordinate system that is shown in Fig. 5.1.1. For convenience in simplifying the angular momentum equations, we will choose this coordinate system to coincide with the primary inertial axes of the airplane. The angle that the fuselage axis makes with the horizontal, commonly called the *elevation angle*, is usually given the symbol θ. According to standard convention, θ is positive with the nose of the airplane above the horizontal. In a spin, the airplane has a nose-down attitude, which as shown in Fig. 6.10.2 corresponds to a negative elevation angle. As can be seen in Fig. 6.10.2, the airplane's weight can be resolved into components parallel with the x_b-axis, $W\sin(-\theta)$, and normal to the x_b-axis, $W\cos(-\theta)$. The component of weight normal to the x_b-axis can be resolved into components parallel with the y_b-axis, $W\cos(-\theta)\sin(\phi)$, and parallel with the z_b-axis, $W\cos(-\theta)\cos(\phi)$. Thus, the components of the weight vector expressed in our chosen coordinate system are

$$\mathbf{W} = \begin{Bmatrix} -W\sin\theta \\ W\cos\theta\sin\phi \\ W\cos\theta\cos\phi \end{Bmatrix} \tag{6.10.1}$$

Since the angular velocity vector is parallel with the weight vector, the airplane's angular velocity components can similarly be written in the same coordinate system as

$$\mathbf{\Omega} \equiv \begin{Bmatrix} p \\ q \\ r \end{Bmatrix} = \begin{Bmatrix} -\Omega\sin\theta \\ \Omega\cos\theta\sin\phi \\ \Omega\cos\theta\cos\phi \end{Bmatrix} \tag{6.10.2}$$

Likewise, the airplane's descent velocity, $\mathbf{V}_d$, is parallel to the weight vector and

$$\mathbf{V}_d = \begin{Bmatrix} -V_d\sin\theta \\ V_d\cos\theta\sin\phi \\ V_d\cos\theta\cos\phi \end{Bmatrix} \tag{6.10.3}$$

The total velocity of the airplane's center of gravity is the vector sum of the descent velocity and the circumferential velocity. By definition, the spin axis is vertical and the circumferential velocity is in the horizontal plane. In Fig. 6.10.1, the fuselage axis was drawn in the radial plane of the spin helix. We have no way of knowing a priori that this will be the case. In general, we must allow the fuselage axis to be rotated relative to the radial plane through some angle, σ, as shown in the top view of Fig. 6.10.3. The magnitude of the circumferential velocity is $R\Omega$, and as can be seen from Fig. 6.10.3, its x_b-, y_b-, and z_b-components are

$$\mathbf{V}_c = \begin{Bmatrix} -R\Omega\sin\sigma\cos\theta \\ R\Omega(-\cos\sigma\cos\phi - \sin\sigma\sin\theta\sin\phi) \\ R\Omega(\cos\sigma\sin\phi - \sin\sigma\sin\theta\cos\phi) \end{Bmatrix} \tag{6.10.4}$$

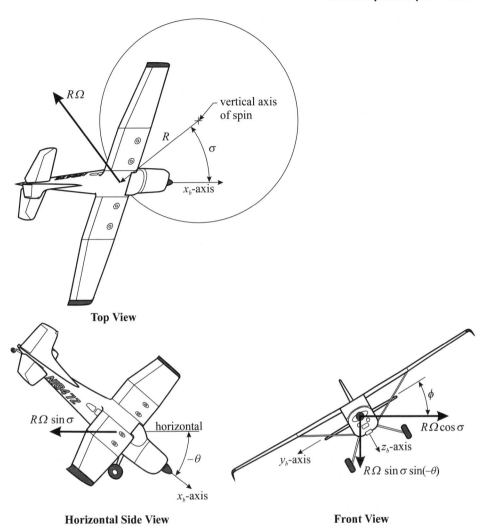

Figure 6.10.3. Top, side, and front views of an airplane's circumferential velocity in a spin.

Combining the components of Eq. (6.10.3) with those from Eq. (6.10.4), the total velocity of the airplane's center of gravity in our body-fixed coordinate system is

$$\mathbf{V} \equiv \begin{Bmatrix} u \\ v \\ w \end{Bmatrix} = \begin{Bmatrix} -V_d \sin\theta - R\Omega\sin\sigma\cos\theta \\ V_d \cos\theta\sin\phi - R\Omega(\cos\sigma\cos\phi + \sin\sigma\sin\theta\sin\phi) \\ V_d \cos\theta\cos\phi + R\Omega(\cos\sigma\sin\phi - \sin\sigma\sin\theta\cos\phi) \end{Bmatrix} \quad (6.10.5)$$

It is sometimes assumed that both ϕ and σ are zero in the developed spin of a conventional airplane. However, this cannot be justified in general.

Because we have chosen our coordinate system to coincide with the primary inertial axes of the airplane, the angular momentum vector can be written from Eq. (6.10.2) as

$$
\mathbf{H} = \left\{ \begin{array}{c} -I_{xx_b} \Omega \sin\theta \\ I_{yy_b} \Omega \cos\theta \sin\phi \\ I_{zz_b} \Omega \cos\theta \cos\phi \end{array} \right\} \tag{6.10.6}
$$

where I_{xx_b}, I_{yy_b}, and I_{zz_b} are, respectively, the x_b, y_b, and z_b moments of inertia.

By definition, we are considering the case of a fully developed steady spin. Thus, the translational velocity components in our body-fixed coordinate system are not changing with time. The only translational acceleration is the centripetal acceleration associated with the rotating coordinate system. From Eqs. (6.10.2) and (6.10.5), we have

$$
\frac{d\mathbf{V}}{dt} = \Omega \times \mathbf{V} = R\Omega^2 \left\{ \begin{array}{c} \cos\sigma \cos\theta \\ \cos\sigma \sin\theta \sin\phi - \sin\sigma \cos\phi \\ \cos\sigma \sin\theta \cos\phi + \sin\sigma \sin\phi \end{array} \right\} \tag{6.10.7}
$$

Similarly, the airplane's angular velocity components in this body-fixed coordinate system are constant and the only angular acceleration is that associated with coordinate system rotation. Using Eqs. (6.10.2) and (6.10.6) gives

$$
\frac{d\mathbf{H}}{dt} = \Omega \times \mathbf{H} = \Omega^2 \left\{ \begin{array}{c} (I_{zz_b} - I_{yy_b}) \cos^2\theta \cos\phi \sin\phi \\ (I_{zz_b} - I_{xx_b}) \cos\theta \sin\theta \cos\phi \\ (I_{xx_b} - I_{yy_b}) \cos\theta \sin\theta \sin\phi \end{array} \right\} \tag{6.10.8}
$$

Applying Eqs. (6.10.1) and (6.10.7), the body-fixed components of the translational momentum equation are written as

$$
\left\{ \begin{array}{c} T+X \\ Y \\ Z \end{array} \right\} + W \left\{ \begin{array}{c} -\sin\theta \\ \cos\theta \sin\phi \\ \cos\theta \cos\phi \end{array} \right\} = \frac{WR\Omega^2}{g} \left\{ \begin{array}{c} \cos\sigma \cos\theta \\ \cos\sigma \sin\theta \sin\phi - \sin\sigma \cos\phi \\ \cos\sigma \sin\theta \cos\phi + \sin\sigma \sin\phi \end{array} \right\} \tag{6.10.9}
$$

where X, Y, and Z are, respectively, the x_b-, y_b-, and z_b-components of the aerodynamic force acting on the airplane and T is the thrust developed by the engine or engines. Here, we are assuming that the thrust vector is aligned with the x_b-axis. Similarly, assuming that the thrust vector passes through the airplane's center of gravity, using Eq. (6.10.8) the angular momentum equation requires that

$$
\left\{ \begin{array}{c} \ell \\ m \\ n \end{array} \right\} = \Omega^2 \left\{ \begin{array}{c} (I_{zz_b} - I_{yy_b}) \cos^2\theta \cos\phi \sin\phi \\ (I_{zz_b} - I_{xx_b}) \cos\theta \sin\theta \cos\phi \\ (I_{xx_b} - I_{yy_b}) \cos\theta \sin\theta \sin\phi \end{array} \right\} \tag{6.10.10}
$$

Equations (6.10.9) and (6.10.10) provide six scalar equations that relate the forces, moments, and mass properties of the airplane to the dynamic parameters of the spin. If we could express the forces and moments in terms of the translational and rotational velocity components, we would have six equations in six unknowns, ϕ, θ, σ, Ω, R, and V_d. The aerodynamic forces and moments depend on the relative wind. At any point on the airplane, the relative wind can be expressed in terms of the position relative to the center of gravity and the translational and rotational velocities. If we let $\mathbf{r}$ represent the position vector from the center of gravity to some arbitrary point on the airplane, using the definitions in Eqs. (6.10.2) and (6.10.5), the local relative wind can be expressed as

$$\mathbf{V}_r \;=\; -\mathbf{V}+\mathbf{r}\times\mathbf{\Omega} \;=\; -\left\{\begin{matrix} u \\ v \\ w \end{matrix}\right\} + \left\{\begin{matrix} y_b r - z_b q \\ z_b p - x_b r \\ x_b q - y_b p \end{matrix}\right\} \tag{6.10.11}$$

Since the wing is stalled, the force on the wing's surface dominates the aerodynamic force on the airplane. Furthermore, the force acting on a stalled surface is dominated by pressure, which acts normal to the surface. Thus, the total longitudinal force is closely aligned with the negative z_b-direction and is proportional to the square of the normal velocity, V_N, which is the negative z_b-component of relative wind,

$$V_N^2 \;=\; w^2 + 2y_b pw + y_b^2 p^2 - 2x_b qw - 2x_b y_b pq + x_b^2 q^2 \tag{6.10.12}$$

For example, the force and associated moments produced on the wing are then given by

$$\left\{\begin{matrix} Z_w \\ \ell_w \\ m_w \end{matrix}\right\} = \tfrac{1}{2}\rho\tilde{C}_w \int\limits_{-b_w/2}^{b_w/2} \left\{\begin{matrix} -V_N^2 \\ -y_b V_N^2 \\ x_b V_N^2 \end{matrix}\right\} c_w dy_b$$

$$= \tfrac{1}{2}\rho S_w\tilde{C}_w \left\{\begin{matrix} -w^2 - \overline{y_w^2}p^2 + 2\overline{x_w}qw - \overline{x_w^2}q^2 \\ -2\overline{y_w^2}pw + 2x_w\overline{y_w^2}pq \\ \overline{x_w}w^2 + x_w\overline{y_w^2}p^2 - 2\overline{x_w^2}qw + \overline{x_w^3}q^2 \end{matrix}\right\} \tag{6.10.13}$$

where $\tilde{C}_w$ is the section normal force coefficient for the stalled wing (i.e., about 2.0 for a thin airfoil section), S_w is the planform area of the wing, and

$$\overline{f_w^k} \;\equiv\; \frac{1}{S_w}\int\limits_{-b_w/2}^{b_w/2} f^k c_w dy_b \tag{6.10.14}$$

Similar results are obtained for the other surfaces of the airplane, with the side force and yawing moment being proportional to the square of the sideslip velocity.

As discussed previously, a rotating propeller produces a yawing moment that is proportional to the normal component of relative wind. Because an airplane's normal velocity component is typically quite large in a spin, a rotating propeller can contribute significantly to the yawing moment needed to initiate and sustain the spin. If the propeller's rotational velocity is sufficiently large compared to the relative wind, the propeller yawing moment can be approximated according to Eq. (2.5.28). Using the negative z_b-component of Eq. (6.10.11) for the normal component of relative wind, this gives

$$n_p = \frac{\partial n_p}{\partial V_N} V_N \cong -\frac{T V_N}{\omega_p} = \frac{\partial n_p}{\partial V_N}(w + y_p p - x_p q) \cong -\frac{T}{\omega_p}(w + y_p p - x_p q) \quad (6.10.15)$$

where x_p and y_p are the x_b- and y_b-coordinate of the propeller axis and the propeller's angular velocity, ω_p, is positive for a conventional right-hand propeller and negative for a left-hand propeller. Notice from Eq. (6.10.15) that even balanced counterrotating propellers will produce a net yawing moment in a spin. For counterrotating propellers mounted equal distance on opposite sides of the CG, contributions from the first and third terms on the right-hand side of Eq. (6.10.15) will cancel, since ω_p will be positive for one propeller and negative for the other. However, the contributions from the second term on the right of Eq. (6.10.15) will add, because both ω_p and y_p have opposite signs.

Assuming the thrust to be a linear function of the forward velocity and combining contributions from the wing, horizontal stabilizer, vertical stabilizer, fuselage, propellers, and rudder, the net forces and moments are approximated as

$$\begin{Bmatrix} T+X \\ Y \\ Z \end{Bmatrix} = \frac{1}{2}\rho S_w \begin{Bmatrix} V_{T0}^2 + V_{T1}u + C_X u^2 \\ (C_{Y1}v^2 + 2C_{n1}b_w rv + C_{Y2}b_w^2 r^2 - 2C_{Y3}b_w pv \\ \qquad -2C_{n3}b_w^2 pr + C_{Y4}b_w^2 p^2)\Omega/|\Omega| \\ -C_{N1}w^2 - C_{N2}b_w^2 p^2 + 2C_{m1}b_w qw - C_{N3}b_w^2 q^2 \end{Bmatrix} \quad (6.10.16)$$

$$\begin{Bmatrix} \ell \\ m \\ n \end{Bmatrix} = \frac{1}{2}\rho S_w b_w \begin{Bmatrix} -2C_{N2}b_w pw + 2C_{m2}b_w^2 pq \\ C_{m1}w^2 + C_{m2}b_w^2 p^2 - 2C_{N3}b_w qw + C_{m3}b_w^2 q^2 \\ [(C_{n1}v^2 + 2C_{Y2}b_w rv + C_{n2}b_w^2 r^2 - 2C_{n3}b_w pv - 2C_{n4}b_w^2 pr \\ + C_{n5}b_w^2 p^2)\Omega/|\Omega| - V_{p1}w - V_{p2}b_w p + V_{p3}b_w q + (\Delta C_n)_r u^2] \end{Bmatrix} \quad (6.10.17)$$

where

$$V_{T0}^2 \equiv \frac{2T_{u=0}}{\rho S_w}, \quad V_{T1} \equiv \frac{2}{\rho S_w}\frac{\partial T}{\partial u}, \quad C_X \equiv \frac{X}{\frac{1}{2}\rho S_w u^2}, \quad (\Delta C_n)_r \equiv \frac{(\Delta n)_r}{\frac{1}{2}\rho S_w b_w u^2},$$

$$C_{N1} \equiv \frac{S_w \tilde{C}_w + S_h \tilde{C}_h + S_{fh}\tilde{C}_f}{S_w}, \quad C_{N2} \equiv \frac{S_w \tilde{C}_w \overline{y_w^2} + S_h \tilde{C}_h \overline{y_h^2} + S_{fh}\tilde{C}_f \overline{y_{fh}^2}}{S_w b_w^2},$$

$$C_{N3} \equiv \frac{S_w \tilde{C}_w \overline{x_w^2} + S_h \tilde{C}_h \overline{x_h^2} + S_{fh}\tilde{C}_f \overline{x_{fh}^2}}{S_w b_w^2}, \quad C_{Y1} \equiv \frac{\eta_v S_v \tilde{C}_v + S_{fv}\tilde{C}_f}{S_w},$$

$$C_{Y2} \equiv \frac{\eta_v S_v \widetilde{C}_v \overline{x_v^2} + S_{fv} \widetilde{C}_f \overline{x_{fv}^2}}{S_w b_w^2} , \quad C_{Y3} \equiv \frac{\eta_v S_v \widetilde{C}_v \overline{z_v} + S_{fv} \widetilde{C}_f \overline{z_{fv}}}{S_w b_w} ,$$

$$C_{Y4} \equiv \frac{\eta_v S_v \widetilde{C}_v \overline{z_v^2} + S_{fv} \widetilde{C}_f \overline{z_{fv}^2}}{S_w b_w^2} , \quad C_{m1} \equiv \frac{S_w \widetilde{C}_w \overline{x_w} + S_h \widetilde{C}_h \overline{x_h} + S_{fh} \widetilde{C}_f \overline{x_{fh}}}{S_w b_w} ,$$

$$C_{m2} \equiv \frac{S_w \widetilde{C}_w \overline{x_w y_w^2} + S_h \widetilde{C}_h \overline{x_h y_h^2} + S_{fh} \widetilde{C}_f \overline{x_{fh} y_{fh}^2}}{S_w b_w^3} ,$$

$$C_{m3} \equiv \frac{S_w \widetilde{C}_w \overline{x_w^3} + S_h \widetilde{C}_h \overline{x_h^3} + S_{fh} \widetilde{C}_f \overline{x_{fh}^3}}{S_w b_w^3} , \quad C_{n1} \equiv \frac{\eta_v S_v \widetilde{C}_v \overline{x_v} + S_{fv} \widetilde{C}_f \overline{x_{fv}}}{S_w b_w} ,$$

$$C_{n2} \equiv \frac{\eta_v S_v \widetilde{C}_v \overline{x_v^3} + S_{fv} \widetilde{C}_f \overline{x_{fv}^3}}{S_w b_w^3} , \quad C_{n3} \equiv \frac{\eta_v S_v \widetilde{C}_v \overline{x_v z_v} + S_{fv} \widetilde{C}_f \overline{x_{fv} z_{fv}}}{S_w b_w^2} ,$$

$$C_{n4} \equiv \frac{\eta_v S_v \widetilde{C}_v \overline{x_v^2 z_v} + S_{fv} \widetilde{C}_f \overline{x_{fv}^2 z_{fv}}}{S_w b_w^3} , \quad C_{n5} \equiv \frac{\eta_v S_v \widetilde{C}_v \overline{x_v z_v^2} + S_{fv} \widetilde{C}_f \overline{x_{fv} z_{fv}^2}}{S_w b_w^3} ,$$

$$V_{p1} = -\sum \frac{2}{\rho S_w b_w} \frac{\partial n_p}{\partial V_N} , \quad V_{p2} = -\sum \frac{2 y_p}{\rho S_w b_w^2} \frac{\partial n_p}{\partial V_N} , \quad V_{p3} = -\sum \frac{2 x_p}{\rho S_w b_w^2} \frac{\partial n_p}{\partial V_N}$$

The vertical stabilizer efficiency factor, η_v, is less than unity to account for the reduced yaw stability resulting from the vertical stabilizer being partly shadowed by the wake of the horizontal stabilizer as shown in Fig. 6.10.4. The vertical stabilizer efficiency factor can be approximated as the ratio of the unshadowed stabilizer area to the total stabilizer area. The 15-degree expansion angle shown in Fig. 6.10.4 is typical but depends somewhat on the section geometry of the horizontal stabilizer.

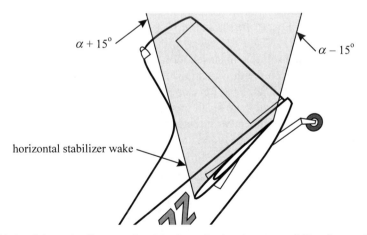

Figure 6.10.4. Schematic diagram showing the reduction in yaw stability due to the vertical stabilizer being partially shadowed by the wake of the horizontal stabilizer.

Starting with the definitions in Eqs. (6.10.2) and (6.10.5), then using Eqs. (6.10.16) and (6.10.17) in Eqs. (6.10.9) and (6.10.10), we have a complete formulation for the steady developed spin of a stalled airplane,

$$
\begin{Bmatrix} u \\ v \\ w \end{Bmatrix} \equiv \begin{Bmatrix} -V_d \sin\theta - R\Omega \sin\sigma \cos\theta \\ V_d \cos\theta \sin\phi - R\Omega(\cos\sigma \cos\phi + \sin\sigma \sin\theta \sin\phi) \\ V_d \cos\theta \cos\phi + R\Omega(\cos\sigma \sin\phi - \sin\sigma \sin\theta \cos\phi) \end{Bmatrix}, \quad \begin{Bmatrix} p \\ q \\ r \end{Bmatrix} \equiv \begin{Bmatrix} -\Omega\sin\theta \\ \Omega\cos\theta\sin\phi \\ \Omega\cos\theta\cos\phi \end{Bmatrix}
$$

$$
\tfrac{1}{2}\rho S_w \begin{Bmatrix} V_{T0}^2 + V_{T1}u + C_X u^2 \\ (C_{Y1}v^2 + 2C_{m1}b_w rv + C_{Y2}b_w^2 r^2 - 2C_{Y3}b_w pv \\ \quad - 2C_{n3}b_w^2 pr + C_{Y4}b_w^2 p^2)\Omega/|\Omega| \\ -C_{N1}w^2 - C_{N2}b_w^2 p^2 + 2C_{m1}b_w qw - C_{N3}b_w^2 q^2 \end{Bmatrix} + W \begin{Bmatrix} -\sin\theta \\ \cos\theta\sin\phi \\ \cos\theta\cos\phi \end{Bmatrix}
$$

$$
= \frac{WR\Omega^2}{g} \begin{Bmatrix} \cos\sigma\cos\theta \\ \cos\sigma\sin\theta\sin\phi - \sin\sigma\cos\phi \\ \cos\sigma\sin\theta\cos\phi + \sin\sigma\sin\phi \end{Bmatrix} \tag{6.10.18}
$$

$$
\tfrac{1}{2}\rho S_w b_w \begin{Bmatrix} -2C_{N2}b_w pw + 2C_{m2}b_w^2 pq \\ C_{m1}w^2 + C_{m2}b_w^2 p^2 - 2C_{N3}b_w qw + C_{m3}b_w^2 q^2 \\ [(C_{n1}v^2 + 2C_{Y2}b_w rv + C_{n2}b_w^2 r^2 - 2C_{n3}b_w pv - 2C_{n4}b_w^2 pr \\ + C_{n5}b_w^2 p^2)\Omega/|\Omega| - V_{p1}w - V_{p2}b_w p + V_{p3}b_w q + (\Delta C_n)_r u^2] \end{Bmatrix}
$$

$$
= \Omega^2 \begin{Bmatrix} (I_{zz_b} - I_{yy_b})\cos^2\theta\cos\phi\sin\phi \\ (I_{zz_b} - I_{xx_b})\cos\theta\sin\theta\cos\phi \\ (I_{xx_b} - I_{yy_b})\cos\theta\sin\theta\sin\phi \end{Bmatrix} \tag{6.10.19}
$$

The six components of Eqs. (6.10.18) and (6.10.19) can be solved numerically for the six unknowns ϕ, θ, σ, Ω, R, and V_d. However, with some further assumptions it is possible to obtain an approximate solution in closed form.

The difference between the thrust and the axial component of drag is typically quite small compared to the normal force developed on the stalled wing. Furthermore, it is often assumed that both σ and ϕ are small in a spin. If we also assume that the wing and tail are in close vertical alignment with the center of gravity ($z_b \cong 0$), a reasonable first approximation for the aerodynamic forces and moments is obtained by using

$$
V_{T0}^2 + V_{T1}u + C_X u^2 \cong C_{Y3} \cong C_{Y4} \cong C_{n3} \cong C_{n4} \cong C_{n5} \cong 0
$$

$$
\begin{Bmatrix} u \\ v \\ w \end{Bmatrix} \cong \begin{Bmatrix} -V_d \sin\theta \\ -R\Omega \\ V_d \cos\theta \end{Bmatrix}, \quad \begin{Bmatrix} p \\ q \\ r \end{Bmatrix} \cong \begin{Bmatrix} -\Omega\sin\theta \\ 0 \\ \Omega\cos\theta \end{Bmatrix}
$$

Also applying the small-angle approximation for ϕ and σ, the formulation reduces to

$$
\frac{1}{2}\rho S_w \left\{
\begin{array}{c}
0 \\
(C_{Y1}R^2 - 2C_{n1}b_w R \cos\theta + C_{Y2}b_w^2 \cos^2\theta)\Omega\,|\Omega| \\
- C_{N1}V_d^2 \cos^2\theta - C_{N2}b_w^2\Omega^2 \sin^2\theta
\end{array}
\right\}
$$

$$
+ W \left\{
\begin{array}{c}
-\sin\theta \\
\phi\cos\theta \\
\cos\theta
\end{array}
\right\}
= \frac{WR\Omega^2}{g}
\left\{
\begin{array}{c}
\cos\theta \\
\phi\sin\theta - \sigma \\
\sin\theta
\end{array}
\right\}
\tag{6.10.20}
$$

$$
\frac{1}{2}\rho S_w b_w \left\{
\begin{array}{c}
2C_{N2}b_w\Omega V_d \cos\theta\sin\theta \\
C_{m1}V_d^2 \cos^2\theta + C_{m2}b_w^2\Omega^2 \sin^2\theta \\
[(C_{n1}R^2 - 2C_{Y2}b_w R\cos\theta + C_{n2}b_w^2 \cos^2\theta)\Omega\,|\Omega| \\
- V_{p1}V_d \cos\theta + V_{p2}b_w\Omega\sin\theta + (\Delta C_n)_r V_d^2 \sin^2\theta]
\end{array}
\right\}
$$

$$
= \Omega^2 \left\{
\begin{array}{c}
(I_{zz_b} - I_{yy_b})\phi\cos^2\theta \\
(I_{zz_b} - I_{xx_b})\cos\theta\sin\theta \\
(I_{xx_b} - I_{yy_b})\phi\cos\theta\sin\theta
\end{array}
\right\}
\tag{6.10.21}
$$

The yawing moment is very critical to the physics of airplane spin. Both spin initiation and spin recovery depend primarily on the magnitude and sign of the yawing moment. For positive spin ($\Omega > 0$), a positive yawing moment supports spin and a negative yawing moment opposes spin. From the definition of C_{Y2} it can be seen that the second yaw term on the left-hand side of Eq. (6.10.21) always opposes spin, while the definitions of C_{n1} and C_{n2} show that the first and third terms can either support or oppose spin, depending on airplane geometry. Since the vertical stabilizer is aft of the CG ($x_b < 0$), the aerodynamic yawing moment generated by the vertical stabilizer opposes spin. However, the aerodynamic center of the fuselage is typically forward of the CG and the yawing moment generated on the fuselage usually supports spin. If the vertical stabilizer is sufficiently shadowed by the wake of the horizontal stabilizer, the fuselage can dominate to produce a net yawing moment in support of the spin.

The fourth and fifth yaw terms on the left-hand side of Eq. (6.10.21) result from the propeller or propellers. For a single-engine airplane with the propeller axis in the plane of symmetry, only the first of these terms is nonzero and the sign of the propeller yawing moment is independent of the direction of airplane spin. In such configurations a conventional right-hand propeller supports negative spin, while a left-hand propeller supports positive spin. For a twin-engine configuration with counterrotating propellers, the first of the two propeller terms that appear in Eq. (6.10.21) is zero and the propeller-induced yawing moment is proportional to Ω. If the right-hand propeller has right-hand rotation and the left-hand propeller has left-hand rotation, the resulting yawing moment always opposes the spin (remember $\sin\theta < 0$), which makes the design more spin resistant. If each of these twin propellers were to rotate in the opposite direction, the propeller-induced yawing moment would support spin in either direction.

With algebraic manipulation, the six components of Eqs. (6.10.20) and (6.10.21) can be rearranged to eliminate ϕ and σ. Thus, the formulation can be reduced to a system of four nonlinear equations that can be directly solved for the spin rate, Ω, the spin radius, R, the sink rate, V_d, and the required yaw control coefficient, $(\Delta C_n)_r$, all as a function of the elevation angle, θ,

$$\Omega^2 = \frac{2W}{\rho S_w C_{N_1} \cos\theta \sin^2\theta}\left[\frac{2(I_{zz_b} - I_{xx_b})}{\rho S_w b_w C_{m_1} \tan\theta} - \frac{C_{m_2} b_w^2}{C_{m_1}} + \frac{C_{N_2} b_w^2}{C_{N_1}}\right]^{-1} \quad (6.10.22)$$

$$R = -\frac{g \tan\theta}{\Omega^2} \quad (6.10.23)$$

$$V_d^2 = \frac{2W}{\rho S_w C_{N_1} \cos^3\theta} - \frac{C_{N_2} b_w^2 \Omega^2 \tan^2\theta}{C_{N_1}} \quad (6.10.24)$$

$$(\Delta C_n)_r = \frac{2C_{N_2}(I_{xx_b} - I_{yy_b})b_w\Omega}{(I_{zz_b} - I_{yy_b})V_d} + \frac{V_{p_1}\cos\theta}{V_d\sin^2\theta} - \frac{V_{p_2}b_w\Omega}{V_d^2\sin\theta}$$
$$- \left(\frac{C_{n_1}R^2}{\sin^2\theta} - \frac{2C_{Y_2}b_wR}{\sin\theta\tan\theta} + \frac{C_{n_2}b_w^2}{\tan^2\theta}\right)\frac{\Omega|\Omega|}{V_d^2} \quad (6.10.25)$$

Typically, the elevation angle required to support a fully developed steady spin is not known a priori. Instead, the yaw control coefficient might be known for a given rudder deflection. For example, $(\Delta C_n)_r$ would be zero with no rudder deflection. After using Eqs. (6.10.22) through (6.10.24) to eliminate Ω, R, and V_d, Eq. (6.10.25) provides a single nonlinear equation for the unknown elevation angle, θ. The roots of this equation, if they exist, are the elevation angles that will support a fully developed steady spin, with the assumption that ϕ and σ are small.

Figure 6.10.5 shows an example of spin characteristics that were predicted from Eqs. (6.10.22) through (6.10.25) for a light aircraft having similar moments of inertia for roll and pitch. For these predictions, the normal force coefficient for the wing and both surfaces of the tail were assumed to be 2.0 and that for the fuselage was taken as 1.0. Several important observations can be made from Fig. 6.10.5. First of all, notice that the spin radius for a flat spin is much smaller than that for a steep spin. For a very flat spin, the spin radius approaches zero. For the more common steeper spins, which typically occur with the airplane's nose about 40 to 60 degrees below horizontal, the predicted spin radius is about 25 to 30 percent of the wingspan. Also notice that the predicted sink rate is much larger for a steep spin than for a flat spin. For the common elevation angle range between –40 and –60 degrees, the current analytical model predicts sink rates ranging from about 100 to 200 ft/sec, which is in good agreement with experimental data (see, for example, Stough, DiCarlo, and Patton 1985). For this same range of elevation angles, the predicted spin rate only varies from about 110 to 140 deg/sec. This is also in good agreement with experimental observation.

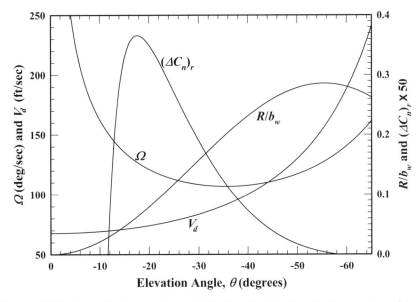

Figure 6.10.5. Example of spin characteristics for a light airplane that is not spin resistant.

For the airplane used to generate Fig. 6.10.5, the analytical model predicts that there are two elevation angles that will support a steady developed spin with no rudder deflection, i.e., $(\Delta C_n)_r = 0$. These correspond to a flat spin at about -12 degrees and a steep spin at about -58 degrees. The flat spin would most likely be difficult to initiate, because the airplane's nose usually drops very rapidly when the wing stalls. However, if this flat spin were to develop, it could be difficult or impossible for the pilot to recover, because the rudder may be completely shadowed by the horizontal stabilizer unless the airplane is specifically designed for this contingency.

Today, unlike the airplane used to generate Fig. 6.10.5, most general aviation airplanes will not support a steady spin without maintained rudder deflection. In such spin-resistant airplanes a spin can usually be initiated by applying full rudder just as the wing stalls. If some rudder deflection is maintained, the airplane may develop a steady spin. However, when the controls are returned to neutral, the airplane will usually recover from the spin in less than one complete revolution. The spin characteristics predicted from Eqs. (6.10.22) through (6.10.25) for a typical general aviation airplane are shown in Fig. 6.10.6. Notice that a very large yaw control input would be required to maintain a flat spin. This is usually significantly more than can be achieved with a conventional rudder. In a steeper nose-down attitude there would be sufficient rudder power to stabilize the spin. However, the predicted rudder control moment required to support the spin only goes to zero in the limit as the elevation angle for the spin approaches -90 degrees. Such a steep maneuver is not a spin at all but a spiral dive, and the present analytical model does not apply because the angle of attack would be below stall. Furthermore, the present model does not account for compressibility effects, which become important in very steep spins and dives.

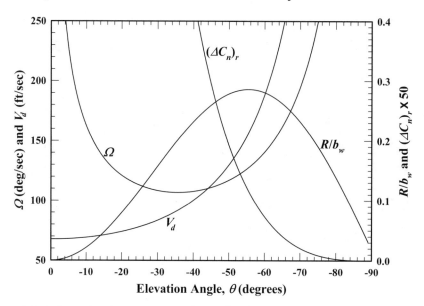

Figure 6.10.6. Example of spin characteristics predicted for a typical spin-resistant light airplane.

It should be reemphasized that Eqs. (6.10.22) through (6.10.25) are based on an approximation to the steady equations of motion for airplane spin. In reality, the net axial force will not be exactly zero in a spin. However, compared to the normal force generated on the stalled wing, this force is small. Furthermore, the angles σ and ϕ are not necessarily small in a spin. Thus, the model does not apply to all spins and all airplanes. The model is based on the assumption of large angle of attack and thus does not apply to extremely steep spins and spiral dives because the angle of attack becomes small in such cases. Also, the model does not account for compressibility effects, which become important in very steep spins and dives. Nevertheless, this approximate model can provide some valuable insight into the physics of airplane spin.

6.11. Problems

6.1.   Consider the British Spitfire Mark XVIII that is described in problem 4.8. The wing and horizontal tail have the following properties:

$$S_w = 244 \text{ ft}^2, \quad b_w = 36.83 \text{ ft}, \quad C_{L_w,\alpha} = 4.62, \quad C_{m_w} = -0.053, \quad W = 8{,}375 \text{ lbf},$$

$$S_h = 31 \text{ ft}^2, \quad b_h = 10.64 \text{ ft}, \quad C_{L_h,\alpha} = 4.06, \quad C_{m_h 0} = 0.00, \quad C_{m_h,\delta_e} = -0.55,$$

$$\varepsilon_e = 0.60, \quad l_h - l_w = 18.16 \text{ ft}, \quad \eta_h = 1.0$$

Including the effects of wing downwash but neglecting all effects of the propeller and fuselage, plot the elevator angle per g as a function of CG location relative to the wing quarter chord. Use an airspeed of 100 mph at sea level.

6.2. The small jet transport that is described in problem 4.16 has the following properties:

$$S_w = 950 \text{ ft}^2, \quad b_w = 75 \text{ ft}, \quad C_{L_w,\alpha} = 4.30, \quad C_{m_w} = -0.050, \quad W = 73{,}000 \text{ lbf},$$
$$S_h = 207 \text{ ft}^2, \quad b_h = 31 \text{ ft}, \quad C_{L_h,\alpha} = 4.00, \quad C_{m_h0} = 0.00, \quad C_{m_h,\delta_e} = -0.49,$$
$$\varepsilon_e = 0.51, \quad l_h - l_w = 42.5 \text{ ft}, \quad \eta_h = 1.0$$

Including the effects of wing downwash but neglecting effects of the thrust deck and fuselage, plot the elevator angle per g as a function of CG location relative to the wing quarter chord. Use an airspeed of 300 mph at sea level.

6.3. Consider the small jet transport that is described in problem 4.25. The wing and canard have the following properties:

$$S_w = 950 \text{ ft}^2, \quad b_w = 75 \text{ ft}, \quad C_{L_w,\alpha} = 4.30, \quad C_{m_w} = -0.050, \quad W = 73{,}000 \text{ lbf},$$
$$S_h = 219 \text{ ft}^2, \quad b_h = 36 \text{ ft}, \quad C_{L_h,\alpha} = 4.30, \quad C_{m_h0} = -0.050, \quad C_{m_h,\delta_e} = -0.49,$$
$$\varepsilon_e = 0.51, \quad l_h - l_w = -42.5 \text{ ft}, \quad \eta_h = 1.0$$

Including the effects of wing downwash but neglecting effects of the thrust deck and fuselage, plot the elevator angle per g as a function of CG location relative to the wing quarter chord. Use an airspeed of 300 mph at sea level.

6.4. The turboprop transport that is described in problem 4.32 has the following properties:

$$S_w = 1{,}745 \text{ ft}^2, \quad b_w = 132.6 \text{ ft}, \quad C_{L_w,\alpha} = 5.16, \quad C_{m_w} = -0.050, \quad W = 155{,}000 \text{ lbf},$$
$$S_h = 528 \text{ ft}^2, \quad b_h = 52.7 \text{ ft}, \quad C_{L_h,\alpha} = 4.52, \quad C_{m_h0} = 0.00, \quad C_{m_h,\delta_e} = -0.49,$$
$$\varepsilon_e = 0.58, \quad l_h - l_w = 47.0 \text{ ft}, \quad \eta_h = 1.0$$

Including the effects of wing downwash but neglecting effects of the propellers and fuselage, plot the elevator angle per g as a function of CG location relative to the wing quarter chord. Use an airspeed of 200 mph at sea level.

6.5. Consider the wing-tail combination that is described in Example 6.2.1. Assume that the tail is attached to the wing with a thin-walled aluminum tube that extends from the tail to the center of gravity, which is located at the wing quarter chord. The aluminum tube has a constant circular cross-section and constant wall thickness. To avoid buckling, the ratio of tube diameter to wall thickness is kept less than or equal to 32. Neglecting all effects of compressibility but accounting for tail boom bending, what are the diameter, wall thickness, and weight of the lightest aluminum tube that would result in no more than 10 percent control loss with an airspeed of 400 mph at sea level?

6.6.   The diameter of the thin-walled aluminum tube connecting the wing and tail in problem 6.5 is allowed to vary linearly along the length of the tube. To avoid buckling, the ratio of tube diameter to wall thickness is to be kept equal to 32 at every cross-section. Neglecting all effects of compressibility but accounting for tail boom bending, what are the maximum diameter, maximum wall thickness, taper ratio, and weight of the lightest tapered aluminum tube that would result in no more than 10 percent control loss with an airspeed of 400 mph at sea level?

6.7.   For the Spitfire in problem 6.1, neglecting the effects of propeller and fuselage, plot the control force gradient at sea level and 100 mph as a function of CG location relative to the wing quarter chord. The change in elevator hinge moment coefficient with angle of attack is –0.59 and the change in hinge moment coefficient with elevator deflection is –0.90, based on a 14-ft² elevator with a 1.4-foot mean chord. The reversible mechanical linkage between the stick and elevator has an effective lever arm of 2.0 feet. Where is the stick-free neutral point?

6.8.   For the British Spitfire Mark XVIII and operating conditions in problem 6.7, neglecting the effects of the propeller and fuselage, plot the longitudinal control force per *g* as a function of the location of the CG measured relative to the quarter chord of the main wing. Also find the stick-free maneuver point.

6.9.   For the small jet transport in problem 6.2, neglecting the effects of the thrust deck and fuselage, plot the control force gradient at sea level and 300 mph as a function of the location of the CG measured relative to the quarter chord of the main wing. For this tail the change in the hinge moment coefficient with angle of attack is –0.63 and the change in the hinge moment coefficient with elevator deflection is –0.97, based on a 69-ft² elevator with a mean chord of 2.5 feet. Assume a reversible mechanical linkage between the control yoke and elevator that has an effective lever arm of 2.0 feet. Also find the stick-free neutral point. Do you think that a reversible mechanical control is feasible for this airplane?

6.10.   For the small jet transport and operating conditions in problem 6.9, neglecting the effects of the thrust deck and fuselage, plot the longitudinal control force per *g* as a function of the location of the CG measured relative to the quarter chord of the main wing. Also find the stick-free maneuver point.

6.11.   For the small jet transport in problem 6.3, neglecting the effects of the thrust deck and fuselage, plot the control force gradient at sea level and 300 mph as a function of CG location measured relative to the quarter chord of the main wing. For the canard control surface, the change in hinge moment coefficient with angle of attack is –0.63 and the change in hinge moment coefficient with elevator deflection is –0.97, based on a 69-ft² control surface with a mean chord of 2.5 feet. Assume a reversible mechanical linkage between the control yoke and the control surface with an effective lever arm of 2.0 feet. Also find the stick-free neutral point. Do you think that a reversible mechanical control is feasible for this airplane?

6.12. For the small jet transport and operating conditions in problem 6.11, neglecting the effects of the thrust deck and fuselage, plot the longitudinal control force per *g* as a function of the location of the CG measured relative to the quarter chord of the main wing. Also find the stick-free maneuver point.

6.13. For the large turboprop transport described in problem 6.4, neglecting the effects of the propellers and fuselage, plot the control force gradient at sea level and 200 mph as a function of the location of the CG measured relative to the quarter chord of the main wing. For this elevator and stabilizer, the change in the hinge moment coefficient with angle of attack is –0.63 and the change in the hinge moment coefficient with elevator deflection is –0.97, based on a 176-ft^2 elevator with a mean chord of 3.4 feet. Assume a reversible mechanical linkage between the control yoke and the elevator that has an effective lever arm of 2.0 feet. Also find the stick-free neutral point. Do you think that a reversible mechanical control is feasible for this airplane?

6.14. For the turboprop transport and operating conditions in problem 6.13, neglecting the effects of the propellers and fuselage, plot the longitudinal control force per *g* as a function of the location of the CG measured relative to the quarter chord of the main wing. Also find the stick-free maneuver point.

6.15. For the British Spitfire Mark XVIII that is described in problems 4.3 and 4.5, estimate the downwash gradient induced on the aft horizontal stabilizer when the airplane is flying in ground effect with the tips of the main wing 4.0 feet above the ground.

6.16. For the small jet transport that is described in problem 4.14, estimate the downwash gradient induced on the horizontal stabilizer when the airplane is flying in ground effect with the main wing 8.0 feet above the ground.

6.17. For the turboprop transport that is described in problem 4.32, estimate the downwash gradient induced on the horizontal stabilizer when the airplane is flying in ground effect with the main wing 18.3 feet above the ground.

6.18. For the British Spitfire and the ground effect operating condition described in problem 6.15, plot the elevator deflection required to trim the airplane at 100 mph and sea level as a function of the location of the center of gravity measured aft of the wing quarter chord. Compare this result with that obtained in problem 4.9 for flight at the same airspeed and altitude but out of ground effect.

6.19. For the turboprop transport and the ground effect operating condition described in problem 6.17, plot the elevator deflection required to trim the airplane for 200 mph at sea level as a function of the location of the center of gravity measured aft of the wing quarter chord. Compare this result with that obtained in problem 4.33 for flight at the same airspeed and altitude but out of ground effect.

6.20. For the British Spitfire Mark XVIII in problem 6.7, estimate the control force required for a full stall landing from an approach speed of 120 mph, with the CG located at the aerodynamic center of the wing. Assume that the touchdown wing position relative to the ground is that described in problem 6.15. Using the trim tab setting, the control force is zeroed at the approach speed outside of ground effect and the maximum lift coefficient for the wing-tail combination is 1.4.

6.21. Rework Example 6.7.1 for the case of a climbing coordinated turn. Use a climb angle of 20 degrees. Assume that the airplane's engine has sufficient power to support the climbing turn. Ignore any possibility of stall and hold all parameters, except the climb angle, the same as those used in Example 6.7.1.

6.22. Rework Example 6.7.1 for the case of a descending coordinated turn. Use a climb angle of −20 degrees. Assume that the airplane's drag is sufficient to maintain the desired airspeed simply by throttling back the engine. Ignore any possibility of stall and hold all parameters, except the climb angle, the same as those used in Example 6.7.1.

6.23. Consider a level coordinated turn for the British Spitfire described in problem 6.1. All control surfaces are trimmed for steady level flight at 300 mph and sea level. Ignoring any possibility of stall, plot the change in aileron, elevator, and rudder deflection angles required to maintain the level coordinated turn, at the same airspeed and altitude, as a function bank angle. For this airplane and operating condition use

$$S_w = 244 \text{ ft}^2, \quad b_w = 36.83 \text{ ft}, \quad W = 8{,}375 \text{ lbf}, \quad V = 300 \text{ mph},$$
$$C_{L,\alpha} = 4.88, \quad C_{L,\delta_e} = 0.305, \quad C_{L,\bar{q}} = 2.83,$$
$$C_{Y,\beta} = -0.212, \quad C_{Y,\delta_r} = 0.209, \quad C_{Y,\bar{p}} = 0.045, \quad C_{Y,\bar{r}} = 0.282,$$
$$C_{\ell,\beta} = 0.00, \quad C_{\ell,\delta_a} = -0.201, \quad C_{\ell,\delta_r} = 0.019, \quad C_{\ell,\bar{p}} = -0.889,$$
$$C_{m,\alpha} = -0.469, \quad C_{m,\beta} = 0.0085, \quad C_{m,\delta_e} = -0.879, \quad C_{m,\bar{q}} = -7.75,$$
$$C_{n,\alpha} = 0.0018, \quad C_{n,\beta} = 0.095, \quad C_{n,\delta_a} = 0.0053, \quad C_{n,\delta_r} = -0.094,$$
$$C_{n,\bar{p}} = -0.0186, \quad C_{\ell,\bar{r}} = 0.017 + 0.148 C_L, \quad C_{n,\bar{r}} = -0.137 - 0.0274 C_L^2$$

Discuss the differences that you observe between the results predicted for this airplane and the one in Example 6.7.1.

6.24. A later version of the British Spitfire eliminated the propeller torque problem by using twin coaxial-contrarotating propellers. Repeat problem 6.23 assuming that all parameters remain the same except that $C_{n,\alpha} = C_{m,\beta} = 0$.

6.25. Following a procedure similar to that used in Sec. 6.1 for the stabilizer-elevator configuration, show that a compound stabilator configuration leads to the results given in Eqs. (6.9.8), (6.9.9), and (6.9.10).

6.26. Following a procedure similar to that used in Sec. 6.3 for the stabilizer-elevator configuration, show that a compound stabilator configuration leads to the results given in Eqs. (6.9.19), (6.9.20), and (6.9.21).

6.27. Following a procedure similar to that used in Sec. 6.3 for the stabilizer-elevator configuration, show that the trim setting, δ_{et}, for a compound stabilator configuration varies with level flight trim speed according to the relation

$$\delta_{et} = -[C_1(W/\tfrac{1}{2}\rho V^2_{\text{trim}} S_w) + C_3]/C_4$$

where C_1 is given by Eq. (6.9.20) and

$$C_3 = C_{H_s,0} + C_{H_s,\alpha_h}(\alpha_{0h} - \varepsilon_{d0})$$
$$+ \frac{(C_{m,\alpha}C_{L0} - C_{L,\alpha}C_{m0})C_{H_s,\delta_s} - (C_{m,\delta_s}C_{L0} - C_{L,\delta_s}C_{m0})C_{H_s,\alpha_h}(1 - \varepsilon_{d,\alpha})}{C_{L,\alpha}C_{m,\delta_s} - C_{L,\delta_s}C_{m,\alpha}}$$

$$C_4 = C_{H_s,\delta_e}$$
$$+ \frac{(C_{m,\alpha}C_{L,\delta_e} - C_{L,\alpha}C_{m,\delta_e})C_{H_s,\delta_s} - (C_{m,\delta_s}C_{L,\delta_e} - C_{L,\delta_s}C_{m,\delta_e})C_{H_s,\alpha_h}(1 - \varepsilon_{d,\alpha})}{C_{L,\alpha}C_{m,\delta_s} - C_{L,\delta_s}C_{m,\alpha}}$$

6.28. Consider replacing the aft stabilizer and elevator of the airplane described in Example 6.3.1 with a compound stabilator having exactly the same geometry. Assume that the primary hinge of the stabilator will pass through the aerodynamic center of this horizontal surface. The trailing-edge flap will be mechanically linked to the stabilator so that k_{es} will be 0.50. Assume a reversible mechanical linkage between the stick and the stabilator having an effective lever arm of 2.0 feet. For this stabilator, repeat the computations that were carried out in Example 6.3.1 for the stabilizer-elevator configuration and compare the results.

6.29 One advantage of using the compound stabilator configuration for pitch control is that the hinge location, l_{Hs}, and the linkage coefficient between the stabilator and the trailing-edge flap, k_{es}, can be varied by the designer to adjust the control forces. In the absence of Reynolds number and Mach number effects, show that a desired control force gradient, $F_{p,V}$, can be attained at a specified trim speed, V_{trim}, by enforcing the relation

$$\frac{l_{Hs}}{\bar{c}_h} = \frac{C_1(C_{L,\alpha}C_{m,\delta_s} - C_{L,\delta_s}C_{m,\alpha}) + C_{m_h,\delta_e}C_{m,\alpha}k_{es}}{C_{L_h,\alpha}[C_{m,\alpha}(1 + \varepsilon_e k_{es}) - C_{m,\delta_s}(1 - \varepsilon_{d,\alpha})]}$$

where

$$C_1 = -\frac{S_w l_g F_{p,V} V_{\text{trim}}}{2\eta_h S_h \bar{c}_h W}$$

6.30. For the airplane and compound stabilator that is described in problem 6.28, use the result obtained in problem 6.29 to estimate the stabilator hinge location that would be required to provide a control force gradient of –0.20 lbf·sec/ft at a trim speed of 80 mph. Assume that the trailing-edge flap is mechanically linked to the stabilator so that k_{es} is 0.50, and that the CG is located 10 percent forward of the wing quarter chord.

6.31. The designer can also adjust the control forces for an airplane with a reversible mechanical compound stabilator by changing the gear ratio for the linkage between the main hinge of the stabilator and the trailing-edge flap. Starting with the result obtained in problem 6.29, develop a relation for the linkage coefficient, k_{es}, that is required to attain a desired control force gradient at a specified trim speed for a fixed location of the main stabilator hinge.

6.32. For the airplane and compound stabilator described in problem 6.28, use the result obtained in problem 6.31 to estimate the linkage coefficient, k_{es}, that would be required to provide a control force gradient of –0.20 lbf·sec/ft at a trim speed of 80 mph. Assume that the primary hinge of the stabilator passes through the aerodynamic center of the horizontal surface, and that the CG is located 10 percent forward of the wing quarter chord.

6.33. For a light airplane with the properties listed below, use the approximate model for airplane spin that is given by Eqs. (6.10.22) through (6.10.25) to plot the fully developed spin rate, spin radius, and sink rate as a function of elevation angle. Assume standard sea-level air density and neglect the yawing moment produced by the airplane's propeller.

$$I_{xx_b} = 1{,}000 \text{ slug·ft}^2, \quad I_{yy_b} = 1{,}000 \text{ slug·ft}^2, \quad I_{zz_b} = 3{,}300 \text{ slug·ft}^2,$$

$$S_w = 180 \text{ ft}^2, \quad b_w = 33 \text{ ft}, \quad \tilde{C}_w = 2.0, \quad x_w = 0.71 \text{ ft}, \quad W = 2{,}700 \text{ lbf},$$

$$S_h = 36 \text{ ft}^2, \quad b_h = 12 \text{ ft}, \quad \tilde{C}_h = 2.0, \quad x_h = -14.29 \text{ ft},$$

$$S_v = 11.9 \text{ ft}^2, \quad \tilde{C}_v = 2.0, \quad x_v = -14.81 \text{ ft}, \quad \eta_v = 0.1,$$

$$S_{fh} = 63.8 \text{ ft}^2, \quad C_f = 1.0, \quad \overline{x_{fh}} = 3.5 \text{ ft},$$

$$S_{fv} = 80.2 \text{ ft}^2, \quad \overline{x_{fv}} = 3.5 \text{ ft}, \quad \overline{x_{fv}^2} = 22.3 \text{ ft}^2, \quad \overline{x_{fv}^3} = 149 \text{ ft}^3$$

6.34. For the steady coordinated turn analysis presented in Sec. 6.7, it was assumed that the angular rate was sufficiently small so that the angular acceleration associated with the rotating coordinate system could be ignored. Because the angular rate is much higher in a spin, this assumption was not made for the spin analysis presented in Sec. 6.10. Assume that the body-fixed coordinate system coincides with the primary inertial axes of the airplane and follow a procedure similar to that used in Sec. 6.10 to add the angular acceleration terms to the steady coordinated turn analysis presented in Sec. 6.7.

6.35. Using the more general formulation developed in problem 6.34, repeat the computations in problem 6.22, accounting for the angular acceleration terms. Use the following moments of inertia,

$$I_{xx_b} = 1,000 \text{ slug}\cdot\text{ft}^2, \quad I_{yy_b} = 3,000 \text{ slug}\cdot\text{ft}^2, \quad I_{zz_b} = 3,500 \text{ slug}\cdot\text{ft}^2$$

Compare your results with those obtained in problem 6.22. Do you think that angular acceleration would ever be significant in a typical steady coordinated turn?

6.36. The XP-38 prototype for the P-38 Lightning had propellers that turned opposite to those of the later production models. Based on spin considerations and the photograph of the production model that is shown in Fig. 5.13.2, can you suggest a reason why the direction of propeller rotation may have been changed in going from the prototype to the production model?

6.37. Supersonic flow is characteristically very different from subsonic flow. Recall that camber does not contribute to the section lift developed on a thin airfoil at supersonic airspeeds. As a result, control surface effectiveness is reduced at Mach numbers greater than 1. Consider, for example, the symmetric diamond-wedge airfoil section with trailing-edge flap shown in Fig. 6.11.1. The thickness distribution increases linearly from zero at the leading edge to the point of maximum thickness located at the half-chord. The thickness then decreases linearly from the half-chord to the sharp trailing edge. Recall that the pressure coefficient at any point in the flow over a thin airfoil at supersonic airspeeds depends only on the freestream Mach number and the slope of the local streamline relative to the freestream, i.e.,

$$C_p \equiv \frac{p - p_\infty}{\frac{1}{2}\rho_\infty V_\infty^2} = \pm \frac{2}{\sqrt{M_\infty^2 - 1}}\left(\frac{\partial y}{\partial x}\right)_{\text{streamline}}$$

where the positive sign applies above the airfoil and the negative sign applies below. With the maximum thickness denoted as t and assuming that all angles are small, start with this result and develop an expression for the ideal section flap effectiveness.

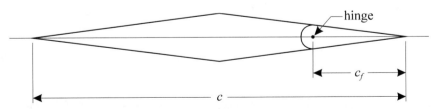

Figure 6.11.1. Diamond-wedge supersonic airfoil and trailing-edge flap used in problem 6.37.

6.38. For the supersonic airfoil section that was described in problem 6.37, develop a small-angle expression for the half-chord section moment coefficient as a function of the flap deflection angle and the angle of attack measured relative to the undeflected chord line.

6.39. For the supersonic airfoil section that was described in problem 6.37, develop a small-angle expression for the section wave drag coefficient as a function of the flap deflection angle and the angle of attack measured relative to the undeflected chord line.

6.40. A trim tab with chord length c_t is added to the supersonic airfoil section that was described in problem 6.37 (see Fig. 6.11.2). Develop a small-angle expression for the section hinge moment coefficient as a function of the flap deflection angle, trim tab deflection angle, and angle of attack measured relative to the undeflected chord line.

6.41. (*Contributed by Captain Robert J. Niewoehner, Ph.D., Aerospace Engineering Department, United States Naval Academy*) Consider the British Spitfire Mark XVIII described in problems 4.8 and 6.1. Including the effects of wing down-wash on the aft tail but neglecting all effects of the propeller, the fuselage, and any vertical offsets, plot the load factor as a function of airspeed at standard sea level for several fixed values of elevator angles with the CG located at the wing quarter chord. Overlay these plots on a traditional *V-n* diagram, which is discussed in Sec. 3.9. In addition to the geometric and aerodynamic properties used for problems 4.8 and 6.1, assume that the stall limits are fixed by $C_{L_{max}} = 1.5$ and $C_{L_{min}} = -0.7$. For the load limits use $n_{max} = 7$ and $n_{min} = -3$. Also find the CG location that requires 25 degrees of up elevator to produce a 7-g load factor at the maneuver speed. Repeat the overlay plots on the *V-n* diagram for this forward CG location.

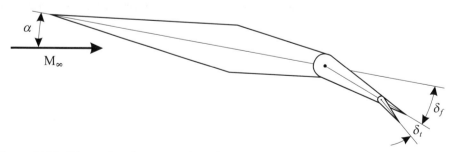

Figure 6.11.2. Diamond-wedge supersonic airfoil section with trailing-edge flap and trim tab used in problem 6.40.

Chapter 7
Aircraft Equations of Motion

7.1. Introduction

The process involved in carrying out any engineering analysis can be divided into two phases, problem formulation and problem solution. The formulation phase requires understanding the physics of an engineering problem and converting that physics into a consistent set of mathematical equations. The solution phase involves applying the rules of mathematics and/or numerical methods to obtain a solution to the system of equations that was developed in the formulation phase. Both phases are equally important. A system of equations is of little direct use in engineering analysis unless it can be accurately solved. Likewise, a mathematical solution is of no use unless the formulation from which it came provides a reasonable description for the physics of the original engineering problem.

In Chapters 3 through 6 of this text, the formulation and solution for many flight mechanics problems were obtained starting from fundamental principles of Newtonian mechanics. In each case we started with the physics of a particular engineering problem and applied all appropriate simplifications as the governing mathematical equations were being developed. This approach minimizes mathematical complexity and emphasizes understanding the physics of each engineering problem. This approach also allows us to begin with simple problems, like those associated with steady level flight, and progress toward more complex problems, such as the analysis of turns and spins. However, as the reader has probably observed, there is a great deal of repetition involved in formulating different but related problems in this manner. For most students, this repetition and the progression from the simple to more complex are helpful in building an understanding of new material.

Once an engineer has developed a good understanding of the physics associated with a particular discipline, there is another approach that can be used for the formulation of engineering problems, that is, starting with a very general formulation and obtaining the formulation for more restricted problems by direct mathematical reduction. For example, had we begun our study of flight mechanics with analysis of the steady coordinated turn, the formulation for steady level flight could have been obtained from the more general formulation by simply setting the climb angle and bank angle to zero. The obvious advantage of this method is reduced formulation time once the more general formulation has been developed. There is a less obvious disadvantage. **No formulation has ever been completely general.** Human limitations being what they are, we all have a tendency to forget the restrictions that apply to a "general" formulation. Sooner or later most engineers have used a so-called "general formulation" for a problem to which it does not apply. Nevertheless, if the engineer exercises appropriate care, the advantage of using this method of formulation outweighs the disadvantage.

In this chapter we develop a *more general* formulation of the rigid-body equations of motion. Since the chapter could serve as a starting point for a second-semester course in flight mechanics, it includes a review of coordinate systems and mathematical notation that were introduced gradually throughout Chapters 3 through 6. This chapter was written to be more or less independent of the presentation in Chapters 3 through 6. Thus, a student or practicing engineer who is already familiar with the principles of static stability and control could start with this chapter to begin a study of aircraft dynamics.

Before the equations of motion for a physical system can be formulated mathematically, a coordinate system must be chosen. Furthermore, this choice of coordinate system can have a profound effect on the complexity of the ensuing mathematical formulation. Thus, before developing the more general rigid-body equations of motion for an aircraft in flight, it is worth spending some time considering the advantages and disadvantages of the coordinate systems that could be used. Here we discuss the advantages and disadvantages of some common coordinate systems and settle on three different coordinate systems to be used in our development of the aircraft equations of motion.

Study of aerodynamics typically begins with the consideration of two-dimensional flows, such as the flow around airfoils and long cylinders. When formulating any two-dimensional problem in Cartesian coordinates, it is conventional to use an x-y coordinate system. When considering flow over an airfoil, it was quite convenient to have the x-axis aligned with the chord, pointing in the general direction of flow with the origin at the leading edge. The y-axis was quite naturally chosen normal to the x-axis in the upward direction. It was also quite natural to choose the angle of attack to be positive with the leading edge high, since increasing the angle of attack in this direction increases the lift produced by the airfoil. Once the sign convention for angle of attack was established, both the pitching moment and the pitching rate were naturally defined to be positive in the direction of increasing angle of attack.

When the study of aerodynamics is extended to include three-dimensional flow around finite wings, retaining the original x-y coordinate system and choosing a conventional right-handed system requires the z-axis to be pointing in the spanwise direction from right to left, as shown in Fig. 7.1.1. Here, this coordinate system is referred to as *aerodynamic coordinates*. When using aerodynamic coordinates it is

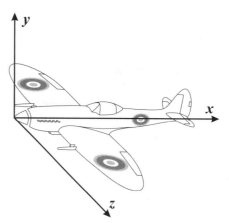

Figure 7.1.1. The aerodynamic coordinate system.

conventional to choose the origin to be at the leading edge of the object of study, in this case the nose of an airplane, with the x-axis lying along a somewhat arbitrary fuselage reference line.

There are two primary objections to continuing with aerodynamic coordinates as the sole coordinate system to be used in our study of aircraft dynamics. First, because the aerodynamic coordinate system is fixed to the aircraft, the position and orientation of the aircraft cannot be conveniently described in terms of aerodynamic coordinates. Second, the sign convention for the moment and angular velocity components used with aerodynamic coordinates does not follow the conventional right-hand rule.

To describe the position and orientation of an aircraft relative to the Earth, we normally use a coordinate system fixed to the Earth. Position relative to the Earth is usually described in terms of latitude, longitude, and altitude above mean sea level. However, because this coordinate system is not a Cartesian system, it introduces unnecessary complexities to the dynamic formulation. **Since the radius of the Earth is large compared to the distance traveled by an airplane in the relatively short time periods that characterize aircraft motion, the formulation can be greatly simplified by using a Cartesian coordinate system fixed to the Earth. For our purposes here, such a Cartesian system can be considered to be an inertial coordinate system. Of course, this so-called "flat-Earth" approximation is not appropriate for tracking the position of an aircraft over long distances.**

While an Earth-fixed coordinate system allows us to conveniently describe aircraft position and orientation, the components of the inertia tensor become time dependent in Earth-fixed coordinates. The complexity associated with carrying a time-dependent inertia tensor in the equations of motion is almost prohibitive. The mathematical description of the inertia tensor can be greatly simplified by formulating the angular momentum equation in terms of a coordinate system that is fixed to the aircraft with the origin located at the aircraft's center of gravity.

The aerodynamic forces and moments acting on an aircraft depend not on the velocity of the aircraft relative to the ground, but rather, on the velocity relative to the surrounding air. Thus, these forces and moments are most conveniently described in terms of a coordinate system fixed to the atmosphere. The atmosphere can have motion relative to the Earth, i.e., wind. **In our development of the aircraft equations of motion we shall consider the atmosphere in the immediate vicinity of the aircraft to be moving at a constant velocity relative to the Earth-fixed coordinate system. Thus, within this approximation, an atmosphere-fixed coordinate system can also be considered to be an inertial coordinate system.**

In summary, position and orientation are best described in terms of an Earth-fixed coordinate system; the inertia tensor is most conveniently described in terms of a body-fixed coordinate system; and forces and moments are more easily described in terms of an atmosphere-fixed coordinate system. For this reason, these three different coordinate systems will all be used in developing the aircraft equations of motion. Furthermore, it is desirable to define the body-fixed system such that the usual sign conventions for the moment and angular velocity components follow the conventional right-hand rule. The Earth-fixed and body-fixed Cartesian coordinate systems commonly used for this purpose are shown in Fig. 7.1.2.

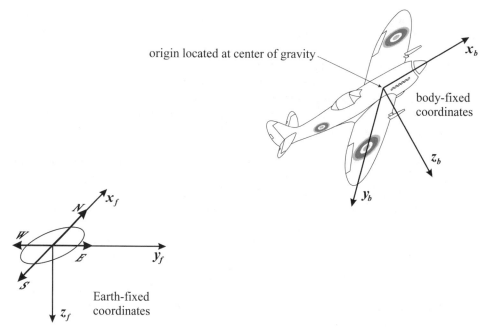

origin located at center of gravity

body-fixed
coordinates

Earth-fixed
coordinates

Figure 7.1.2. The body-fixed and Earth-fixed coordinate systems.

The **body-fixed coordinate system** has its origin located at the aircraft center of gravity. The x_b-axis points forward along some convenient axis of the fuselage in the aircraft's plane of symmetry. The y_b-axis is normal to the plane of symmetry pointing in the direction of the right wing. The z_b-axis then points downward in the aircraft plane of symmetry, completing the right-handed Cartesian system. In the present text, (x_b, y_b, z_b) will always denote this particular body-fixed coordinate system, which is often referred to simply as *body coordinates*. In this coordinate system, a positive angle of attack corresponds to a positive z_b-component of airspeed, and a positive sideslip angle results from a positive y_b-component of airspeed. The sign conventions for roll, pitch, and yaw follow the right-hand rule relative to the x_b-, y_b-, and z_b-axes, respectively. In Chapters 1 through 6, a distance aft of the CG was denoted as l, where in aerodynamic coordinates $l \equiv x - x_{CG}$. Note that in body-fixed coordinates l is simply given by $l \equiv -x_b$.

For the **Earth-fixed coordinate system** shown in Fig. 7.1.2, the x_f-y_f plane is normal to the local gravitational vector with the x_f-axis pointing north and the y_f-axis pointing east. The z_f-axis then points down, completing the right-handed Cartesian system. In the present text, this coordinate system is designated (x_f, y_f, z_f) and is often referred to simply as *fixed coordinates*. While the student may find a downward-pointing z-axis to be somewhat unconventional, when the aircraft heading or azimuth angle is described in terms of a conventional right-handed angular rotation about the z_f-axis, the heading angles conform to conventional compass headings, that is, a heading angle measured clockwise from north with east being 90 degrees, south being 180 degrees, west being 270 degrees, and north being either 0 or 360 degrees.

The **atmosphere-fixed coordinate system** is denoted (x_a, y_a, z_a) and defined such that all three axes are always parallel to those of the Earth-fixed coordinate system. However, the atmosphere-fixed system moves at a **constant velocity** relative to the Earth-fixed system. The atmosphere-fixed and Earth-fixed systems are defined to be coincident at time equal to zero. The atmosphere-fixed coordinate system will be used only temporarily, in the development of the aircraft equations of motion. The final equations will be expressed in terms of only Earth-fixed and body-fixed coordinates.

The vector $\mathbf{V}_g$ will denote the velocity of the body-fixed coordinate frame relative to the Earth-fixed coordinate frame. The magnitude of this velocity vector is commonly termed the *ground speed*, which is simply the speed of the aircraft relative to the ground. The velocity of the body-fixed coordinate frame relative to the atmosphere-fixed coordinate frame is denoted here simply as $\mathbf{V}$, and its magnitude is commonly termed the *airspeed*. Ground speed and airspeed are related through the simple vector equation

$$\mathbf{V}_g = \mathbf{V} + \mathbf{V}_w \qquad (7.1.1)$$

where $\mathbf{V}_w$ is the velocity of the atmosphere relative to the Earth, which is commonly called the *wind*.

The aerodynamic forces and moments acting on an aircraft could be expressed as functions of the translational and rotational velocity components relative to the surrounding air. However, in the study of aerodynamic forces and moments, it is more convenient to express the translational dependence in terms of total airspeed, angle of attack, and sideslip angle. Of course, the velocity vector relative to the surrounding air can be equivalently described either in terms of three velocity components or in terms of a magnitude and any two independent aerodynamic angles. The total airspeed is readily expressed in terms of the body-fixed components according to the relation

$$V \equiv \sqrt{V_{x_b}^2 + V_{y_b}^2 + V_{z_b}^2} \qquad (7.1.2)$$

As standard convention, angle of attack has always traditionally been defined according to the mathematical relation

$$\alpha \equiv \tan^{-1}\left(\frac{V_{z_b}}{V_{x_b}}\right) \qquad (7.1.3)$$

Unfortunately, there are two different mathematical definitions for the sideslip angle that are sometimes used in the determination of aerodynamic forces.

In Chapter 5, the side force and yawing moment contributed by the vertical stabilizer were estimated from the sideslip angle and section properties of the vertical stabilizer in the same way that the lift and pitching moment were estimated from the angle of attack and section properties for the wing and horizontal stabilizer. This method of estimating the aerodynamic forces and moments tacitly assumes that the sideslip angle is related to a vertical surface in the same way that the angle of attack is related to a horizontal surface. This implies a mathematical definition of sideslip angle analogous to Eq. (7.1.3), i.e.,

$$\beta_a \equiv \tan^{-1}\left(\frac{V_{y_b}}{V_{x_b}}\right) \tag{7.1.4}$$

Here the subscript a is used with this definition of sideslip angle because the definition is most convenient for the *analytical* estimation of aerodynamic forces and moments.

For the experimental determination of aerodynamic forces and moments, another definition of sideslip angle is used almost exclusively. This definition comes about as a result of the way in which data are gathered in a wind tunnel. Typically, the rotational mechanism used to adjust the angle of attack is mounted on a turntable, which is used to set the sideslip angle. When this gimbal mechanism is used to orient a model in a wind tunnel, one of two gimbal angles is exactly the angle of attack as defined by Eq. (7.1.3). However, the other gimbal angle is not that defined by Eq. (7.1.4). The angle set by the turntable of such a mechanism is related to the body-fixed velocity components by

$$\beta_e \equiv \sin^{-1}\left(\frac{V_{y_b}}{\sqrt{V_{x_b}^2 + V_{y_b}^2 + V_{z_b}^2}}\right) \tag{7.1.5}$$

where the subscript e is used here to indicate the definition that is most convenient for *experimental* determination of aerodynamic forces and moments. The definitions given by Eqs. (7.1.4) and (7.1.5) are identical only when the angle of attack is zero. Of course, these two definitions are essentially equivalent for small angles of attack, but as the angle of attack becomes large, the difference becomes significant.

Clearly, both the aerodynamic angle defined by Eq. (7.1.4) and that defined by Eq. (7.1.5) are independent of the angle of attack defined by Eq. (7.1.3). Thus, either of these two sideslip definitions could be combined with the angle of attack and the total airspeed to uniquely describe the velocity vector relative to the air. However, the aerodynamic angle defined by Eq. (7.1.4) has the same geometric relationship to a vertical surface as the angle of attack has to a horizontal surface. This cannot be said of the aerodynamic angle defined by Eq. (7.1.5). Most textbooks define the sideslip angle only in terms of Eq. (7.1.5), but for analytical predictions this angle is often used as if it were defined according to Eq. (7.1.4).

The sideslip angle defined by Eq. (7.1.5) depends on the z_b-velocity component, while the angle of attack as defined from Eq. (7.1.3) does not depend on the y_b-velocity component. This asymmetry in the two definitions is not at all attractive from a physical point of view, especially when considering large angles of attack. For example, consider an aircraft that is symmetric about both the x_b-y_b and the x_b-z_b planes, such as a rocket with four identical tail fins. Let the velocity of this aircraft be such that the x_b-, y_b-, and z_b-components of velocity are all equal. From Eqs. (7.1.3), (7.1.4), and (7.1.5) we have $\alpha = 45°$, $\beta_a = 45°$, and $\beta_e = 35.26°$. If the sideslip angle were used to obtain the side force in the same way that the angle of attack is used to obtain the normal force, the aerodynamic angles defined by Eqs. (7.1.3) and (7.1.5) would result in an asymmetric aerodynamic loading produced by this completely symmetric flight condition. This is clearly not consistent with reality.

One reason commonly given for defining the sideslip angle according to Eq. (7.1.5) is that this definition gives a sideslip angle that is completely independent of the orientation of the x_b-axis within the x_b-z_b plane of symmetry. Clearly, the side force should not depend on our choice for the direction of the x_b-axis within this plane. However, the sweep angle that any vertical surface makes with the z_b-axis does depend on the x_b-axis orientation. Thus, the sideslip angle should also vary with the x_b-axis orientation, such that the combined effects of sweep and sideslip make the airflow relative to a vertical surface independent of the x_b-axis orientation within the x_b-z_b plane of symmetry. The definition of sideslip angle that is given by Eq. (7.1.4) satisfies this requirement, while that given by Eq. (7.1.5) does not. Thus, the definition in Eq. (7.1.5) is not nearly as convenient for analytical prediction as that in Eq. (7.1.4).

Either of these two definitions for sideslip angle may be used in a formulation of the rigid-body equations of motion for aircraft dynamics. **The important thing to remember is to be consistent.** Whenever a relation for some particular aerodynamic coefficient has been obtained using one definition of sideslip angle, it should be applied using that same definition. It is easy to violate this requirement unintentionally if some aerodynamic coefficients are estimated analytically from airfoil section properties, while others are taken from experimental data.

During most of the flight of a conventional airplane the aerodynamic angles remain relatively small and the difference between the two definitions for sideslip angle is insignificant. However, **for flight conditions that may involve large aerodynamic angles, care should be taken to avoid the common pitfall of using one definition for sideslip angle in a relation that was developed using the other**.

The transformation equations given by Eqs. (7.1.2) through (7.1.4) can readily be rearranged to yield an inverse transformation, which can be used to compute the three body-fixed velocity components from known values of V, α, and β_a. We start by solving Eq. (7.1.2) for the x_b-velocity component. This gives

$$V_{x_b} = \sqrt{V^2 - V_{y_b}^2 - V_{z_b}^2} \tag{7.1.6}$$

Solving Eq. (7.1.3) for the z_b-velocity component yields

$$V_{z_b} = V_{x_b} \tan\alpha = V_{x_b} \frac{\sin\alpha}{\cos\alpha} \tag{7.1.7}$$

and solving Eq. (7.1.4) for the y_b-velocity component results in

$$V_{y_b} = V_{x_b} \tan\beta_a = V_{x_b} \frac{\sin\beta_a}{\cos\beta_a} \tag{7.1.8}$$

Using Eqs. (7.1.7) and (7.1.8) in Eq. (7.1.6), we obtain

$$V_{x_b} = \sqrt{V^2 - V_{x_b}^2 \tan^2\beta_a - V_{x_b}^2 \tan^2\alpha} = \sqrt{V^2 - V_{x_b}^2 \frac{\sin^2\beta_a}{\cos^2\beta_a} - V_{x_b}^2 \frac{\sin^2\alpha}{\cos^2\alpha}} \tag{7.1.9}$$

Solving Eq. (7.1.9) for the x_b-component of velocity and using the result in Eqs. (7.1.7) and (7.1.8), we obtain the inverse transformation equations

$$V_{x_b} = V \frac{1}{\sqrt{1+\tan^2\alpha+\tan^2\beta_a}} = V \frac{\cos\alpha\cos\beta_a}{\sqrt{1-\sin^2\alpha\,\sin^2\beta_a}} \qquad (7.1.10)$$

$$V_{y_b} = V \frac{\tan\beta_a}{\sqrt{1+\tan^2\alpha+\tan^2\beta_a}} = V \frac{\cos\alpha\sin\beta_a}{\sqrt{1-\sin^2\alpha\,\sin^2\beta_a}} \qquad (7.1.11)$$

$$V_{z_b} = V \frac{\tan\alpha}{\sqrt{1+\tan^2\alpha+\tan^2\beta_a}} = V \frac{\sin\alpha\cos\beta_a}{\sqrt{1-\sin^2\alpha\,\sin^2\beta_a}} \qquad (7.1.12)$$

The relations on the far right of Eqs. (7.1.10) through (7.1.12) are preferred for large-angle computations where either of the aerodynamic angles may pass through 90 degrees. If Eq. (7.1.5) is used for the sideslip angle, the inverse transformation is

$$V_{x_b} = V\cos\alpha\cos\beta_e \qquad (7.1.13)$$

$$V_{y_b} = V\sin\beta_e \qquad (7.1.14)$$

$$V_{z_b} = V\sin\alpha\cos\beta_e \qquad (7.1.15)$$

Notice the symmetry that exists between Eqs. (7.1.11) and (7.1.12). There is no such symmetry between Eqs. (7.1.14) and (7.1.15).

EXAMPLE 7.1.1. A rocket is flying at an airspeed of 1,000 ft/sec. The angle of attack is 30 degrees and the sideslip angle, β_a, is 20 degrees. Determine the axial, sideslip, and normal velocity components.

Solution. From Eqs. (7.1.10), (7.1.11), and (7.1.12), the velocity components are

$$V_{x_b} = \frac{V}{\sqrt{1+\tan^2\alpha+\tan^2\beta_a}} = \frac{1{,}000 \text{ ft/sec}}{\sqrt{1+(0.577350)^2+(0.363970)^2}}$$
$$= 825.96 \text{ ft/sec}$$

$$V_{y_b} = \frac{V\tan\beta_a}{\sqrt{1+\tan^2\alpha+\tan^2\beta_a}} = \frac{1{,}000 \text{ ft/sec}(0.363970)}{\sqrt{1+(0.577350)^2+(0.363970)^2}}$$
$$= 300.63 \text{ ft/sec}$$

$$V_{z_b} = \frac{V\tan\alpha}{\sqrt{1+\tan^2\alpha+\tan^2\beta_a}} = \frac{1{,}000 \text{ ft/sec}(0.577350)}{\sqrt{1+(0.577350)^2+(0.363970)^2}}$$
$$= 476.87 \text{ ft/sec}$$

EXAMPLE 7.1.2. A rocket is flying at an airspeed of 1,000 ft/sec. The angle of attack is 30 degrees and the sideslip angle, β_e, is 20 degrees. Determine the axial, sideslip, and normal velocity components.

Solution. From Eqs. (7.1.13), (7.1.14), and (7.1.15), the three body-fixed velocity components are

$$V_{x_b} = V \cos\alpha \cos\beta_e = 1{,}000 \text{ ft/sec}(0.866025)(0.939693) = \underline{813.80 \text{ ft/sec}}$$

$$V_{y_b} = V \sin\beta_e = 1{,}000 \text{ ft/sec}(0.342020) = \underline{342.02 \text{ ft/sec}}$$

$$V_{z_b} = V \sin\alpha \cos\beta_e = 1{,}000 \text{ ft/sec}(0.500000)(0.939693) = \underline{469.85 \text{ ft/sec}}$$

In certain situations it may be necessary to estimate some aerodynamic coefficients using airfoil section properties and the definition given by Eq. (7.1.4), while others may be taken from experimental results that were obtained using the definition in Eq. (7.1.5). In such cases, it is convenient to be able to transform an aerodynamic derivative with respect to sideslip from one definition of sideslip angle to the other. The mathematical relation between β_a and β_e can be readily established by using Eqs. (7.1.13) and (7.1.14) in Eq. (7.1.4). This results in

$$\tan\beta_a = \frac{\tan\beta_e}{\cos\alpha} \tag{7.1.16}$$

Now consider some aerodynamic coefficient, C, that is somehow known as a function of V, α, and β_e. Suppose that we also know the partial derivative of C with respect to β_e at constant V and α. The partial derivative of C with respect to β_a at constant V and α can then be expressed as

$$\frac{\partial C}{\partial\beta_a} = \frac{\partial C}{\partial\beta_e}\frac{\partial\beta_e}{\partial\beta_a} \tag{7.1.17}$$

Differentiating both sides of Eq. (7.1.16) with respect to β_a and rearranging, it is readily shown that

$$\frac{\partial\beta_e}{\partial\beta_a} = \frac{\cos\alpha \cos^2\beta_e}{\cos^2\beta_a}$$

After applying the trigonometric identity, $\cos^2\beta_e = 1/(1 + \tan^2\beta_e)$, and using Eq. (7.1.16) to eliminate $\tan\beta_e$, we have

$$\frac{\partial\beta_e}{\partial\beta_a} = \frac{\cos\alpha}{\cos^2\beta_a + \cos^2\alpha\sin^2\beta_a} \tag{7.1.18}$$

Using Eq. (7.1.18) in Eq. (7.1.17) results in

$$\frac{\partial C}{\partial \beta_a} = \frac{\cos\alpha}{\cos^2\beta_a + \cos^2\alpha \sin^2\beta_a}\frac{\partial C}{\partial \beta_e} \tag{7.1.19}$$

In a similar manner, it can be shown that

$$\frac{\partial C}{\partial \beta_e} = \frac{\cos\alpha}{\sin^2\beta_e + \cos^2\alpha \cos^2\beta_e}\frac{\partial C}{\partial \beta_a} \tag{7.1.20}$$

EXAMPLE 7.1.3. An airplane model is mounted in a wind tunnel with the angle of attack set at 12.0 degrees and the sideslip angle, β_e, set to 8.0 degrees. By varying β_e slightly to both sides from this orientation, the change in the yawing moment coefficient with respect to β_e is estimated to be 0.250. Find the angle, β_a, and estimate the change in the yawing moment coefficient with respect to β_a.

Solution. From Eq. (7.1.16),

$$\beta_a = \tan^{-1}\left(\frac{\tan\beta_e}{\cos\alpha}\right) = \tan^{-1}\left(\frac{0.140541}{0.978148}\right) = 0.142704 = \underline{8.2°}$$

Using this in Eq. (7.1.19) yields

$$\frac{\partial C}{\partial \beta_a} = \frac{\cos\alpha}{\cos^2\beta_a + \cos^2\alpha \sin^2\beta_a}\frac{\partial C}{\partial \beta_e}$$

$$= \frac{0.978148}{(0.989835)^2 + (0.978148)^2(0.142220)^2}0.250 = \underline{0.245}$$

EXAMPLE 7.1.4. An airplane model is mounted in a wind tunnel with the angle of attack set at 40 degrees and the sideslip angle, β_e, set to 30 degrees. By varying β_e slightly to both sides from this orientation, the change in the yawing moment coefficient with respect to β_e is estimated to be 0.250. Find the angle, β_a, and estimate the change in the yawing moment coefficient with respect to β_a.

Solution. From Eq. (7.1.16),

$$\beta_a = \tan^{-1}\left(\frac{\tan\beta_e}{\cos\alpha}\right) = \tan^{-1}\left(\frac{0.577350}{0.766044}\right) = 0.645850 = \underline{37°}$$

Using this in Eq. (7.1.19) yields

$$\frac{\partial C}{\partial \beta_a} = \frac{\cos\alpha}{\cos^2\beta_a + \cos^2\alpha \sin^2\beta_a}\frac{\partial C}{\partial \beta_e}$$

$$= \frac{0.766044}{(0.798588)^2 + (0.766044)^2(0.601878)^2}0.250 = \underline{0.225}$$

From the results of Example 7.1.3 it can be seen that at angles of attack below stall, there is no significant difference between our two definitions for sideslip angle. Even for the relatively high angles used in Example 7.1.4, the difference between C_{n,β_e} and C_{n,β_a} is only about 10 percent. This should not be taken as a general license for carelessness with regard to sideslip angle definition. There are important applications where angles of attack can approach or even exceed 90 degrees. For such cases, the difference between the definitions in Eqs. (7.1.4) and (7.1.5) is quite large. For instance, when the angle of attack is 80 degrees, a value for β_e of 30 degrees corresponds to a value of β_a of approximately 73 degrees. Nevertheless, in the vast majority of all flight, the angle of attack remains small. For analysis of such flight the subscript on β will be dropped and no distinction will be made.

7.2. Newton's Second Law for Rigid-Body Dynamics

The foundation for the formulation of most problems in flight mechanics is Newton's second law (often stated as $F = ma$). Newton's second law applied to all points in a rigid body requires that the summation of forces must equal the time rate of change of translational momentum, and the summation of moments must equal the time rate of change of angular momentum. Application of this principle is not always as simple as it sounds. For the present formulation, we will be using a body-fixed coordinate system and allowing both translational and angular acceleration. At this point you may wish to review the chapter on accelerating coordinate systems in your introductory dynamics textbook.

Newton's second law for rigid-body motion can be expressed in the following vector equations, which are written in terms of a body-fixed coordinate system (see, for example, Hibbeler 1998 or Beer and Johnston 1997),

$$\mathbf{F}_S + \mathbf{W} = \frac{d}{dt}(m\mathbf{V}) + \omega \times (m\mathbf{V}) \tag{7.2.1}$$

$$\mathbf{M}_S = \frac{d}{dt}([\mathbf{I}]\omega) + \omega \times ([\mathbf{I}]\omega) \tag{7.2.2}$$

where $\mathbf{F}_S$ is the net surface force vector, $\mathbf{W}$ is the weight vector, $\mathbf{M}_S$ is the net moment vector about the body-fixed origin (i.e., the CG), m is the mass, $[\mathbf{I}]$ is the inertia tensor,

$$[\mathbf{I}] = \begin{bmatrix} I_{xx_b} & -I_{xy_b} & -I_{xz_b} \\ -I_{yx_b} & I_{yy_b} & -I_{yz_b} \\ -I_{zx_b} & -I_{zy_b} & I_{zz_b} \end{bmatrix} \tag{7.2.3}$$

$$I_{xx_b} = \iiint_m (y_b^2 + z_b^2)\,dm \tag{7.2.4}$$

$$I_{yy_b} = \iiint_m (x_b^2 + z_b^2)\,dm \tag{7.2.5}$$

$$I_{zz_b} = \iiint_m (x_b^2 + y_b^2)\,dm \tag{7.2.6}$$

$$I_{xy_b} = I_{yx_b} = \iiint_m x_b y_b\,dm \tag{7.2.7}$$

$$I_{yz_b} = I_{zy_b} = \iiint_m y_b z_b\,dm \tag{7.2.8}$$

$$I_{xz_b} = I_{zx_b} = \iiint_m x_b z_b\,dm \tag{7.2.9}$$

and $\mathbf{V}$ and ω are, respectively, the translational and rotational velocities of the body-fixed coordinate frame relative to some inertial reference frame. For convenience in describing the aerodynamic forces and moments, the atmosphere-fixed coordinate system is selected as our inertial reference frame.

Mass is a scalar, so the second term on the right-hand side of Eq. (7.2.1) is written

$$\omega \times (m\mathbf{V}) = m\begin{vmatrix} \mathbf{i}_{x_b} & \mathbf{i}_{y_b} & \mathbf{i}_{z_b} \\ \omega_{x_b} & \omega_{y_b} & \omega_{z_b} \\ V_{x_b} & V_{y_b} & V_{z_b} \end{vmatrix} = m\begin{Bmatrix} \omega_{y_b}V_{z_b} - \omega_{z_b}V_{y_b} \\ \omega_{z_b}V_{x_b} - \omega_{x_b}V_{z_b} \\ \omega_{x_b}V_{y_b} - \omega_{y_b}V_{x_b} \end{Bmatrix} \tag{7.2.10}$$

Equation (7.2.10) can be used in Eq. (7.2.1) to yield the body-fixed Cartesian components of the translational momentum equation,

$$\frac{d}{dt}\left(m\begin{Bmatrix} V_{x_b} \\ V_{y_b} \\ V_{z_b} \end{Bmatrix} \right) = \begin{Bmatrix} F_{Sx_b} + W_{x_b} - m\omega_{y_b}V_{z_b} + m\omega_{z_b}V_{y_b} \\ F_{Sy_b} + W_{y_b} - m\omega_{z_b}V_{x_b} + m\omega_{x_b}V_{z_b} \\ F_{Sz_b} + W_{z_b} - m\omega_{x_b}V_{y_b} + m\omega_{y_b}V_{x_b} \end{Bmatrix} \tag{7.2.11}$$

The total surface force, $\mathbf{F}_S$, is simply the vector sum of all pressure and shear forces acting on the surface of the control volume enclosing the aircraft.

Aircraft mass typically changes with time as a result of fuel burn. Furthermore, fuel weight can be a very significant fraction of total aircraft weight, and the mass of an aircraft can change very rapidly with time. For example, a rocket may burn off a large fraction of its total weight in fuel within just a few seconds. During midair refueling, the mass of both airplanes can change quite rapidly. Also, bombers, crop dusters, and fire-fighting aircraft can drop a large fraction of their total mass in a very short time.

When mass exits an aircraft, it carries momentum with it. Any mass that leaves an aircraft (or enters, in the case of an aircraft being refueled) could have velocity different from that of the aircraft. With a jet or rocket, the fuel mass leaves at a velocity that is significantly higher than the velocity of the aircraft. However, the difference between the exit velocity and the aircraft velocity is commonly included in the thrust force. Care must be taken to see that this momentum flux is not counted twice when formulating Newton's second law. Thus, including mass transfer, Eq. (7.2.11) is written as

$$
m\frac{\partial}{\partial t}\begin{Bmatrix} V_{x_b} \\ V_{y_b} \\ V_{z_b} \end{Bmatrix} + \begin{Bmatrix} V_{x_b} \\ V_{y_b} \\ V_{z_b} \end{Bmatrix}\frac{\partial m}{\partial t} + \dot{m}_e\begin{Bmatrix} V_{ex_b} \\ V_{ey_b} \\ V_{ez_b} \end{Bmatrix} = \begin{Bmatrix} F_{Sx_b} + W_{x_b} - m\omega_{y_b}V_{z_b} + m\omega_{z_b}V_{y_b} \\ F_{Sy_b} + W_{y_b} - m\omega_{z_b}V_{x_b} + m\omega_{x_b}V_{z_b} \\ F_{Sz_b} + W_{z_b} - m\omega_{x_b}V_{y_b} + m\omega_{y_b}V_{x_b} \end{Bmatrix} \quad (7.2.12)
$$

where $\dot{m}_e$ is the mass flow rate leaving the aircraft and $\mathbf{V}_e$ is the velocity of the mass leaving or entering the aircraft, as measured relative to the inertial coordinate system. The mass flow rate, $\dot{m}_e$, is negative for an aircraft being refueled.

In addition to Newton's second law, we must also assure that mass is conserved. The continuity equation requires that

$$
\frac{\partial m}{\partial t} + \dot{m}_e = 0 \quad (7.2.13)
$$

Using Eq. (7.2.13) in Eq. (7.2.12), the body-fixed Cartesian components of the translational momentum equation can be written

$$
m\begin{Bmatrix} \dot{V}_{x_b} \\ \dot{V}_{y_b} \\ \dot{V}_{z_b} \end{Bmatrix} = \begin{Bmatrix} F_{x_b} + W_{x_b} - m\omega_{y_b}V_{z_b} + m\omega_{z_b}V_{y_b} \\ F_{y_b} + W_{y_b} - m\omega_{z_b}V_{x_b} + m\omega_{x_b}V_{z_b} \\ F_{z_b} + W_{z_b} - m\omega_{x_b}V_{y_b} + m\omega_{y_b}V_{x_b} \end{Bmatrix} \quad (7.2.14)
$$

where $\mathbf{F}$ is a pseudo aerodynamic force including thrust, $\mathbf{F} = \mathbf{F}_S - \dot{m}(\mathbf{V}_e - \mathbf{V})$, and the dot indicates a time derivative. **It is important to recognize that Eq. (7.2.14) is not based on the assumption that aircraft mass does not change with time. This equation is valid regardless of how rapidly the aircraft mass is changing, provided that the pseudo thrust force accounts for any difference between the exit velocity and the velocity of the aircraft.**

As discussed in the preceding section, the Cartesian body-fixed coordinate system, which is commonly used in aircraft dynamics, has the x_b-axis pointing forward from the center of mass along some convenient body-fixed axis in the aircraft plane of symmetry. The y_b-axis is normal to the plane of symmetry pointing out the right-hand side of the aircraft, and the z_b-axis points downward in the aircraft plane of symmetry. Here the designation (x_b, y_b, z_b) always refers to this body-fixed coordinate system. Since the coordinate system was chosen so that the airplane is symmetric in y_b, we have

$$
I_{xy_b} = I_{yx_b} = I_{yz_b} = I_{zy_b} = 0 \quad (7.2.15)
$$

and

$$
[\mathbf{I}] = \begin{bmatrix} I_{xx_b} & 0 & -I_{xz_b} \\ 0 & I_{yy_b} & 0 \\ -I_{zx_b} & 0 & I_{zz_b} \end{bmatrix} \quad (7.2.16)
$$

The product of inertia terms in Eq. (7.2.15) could be carried through the analysis. However, virtually all aircraft possess this symmetry, which simplifies the linearized equations of motion. With this symmetry, the angular momentum vector, $\mathbf{H}$, is

$$\mathbf{H} \equiv [\mathbf{I}]\boldsymbol{\omega} = \begin{bmatrix} I_{xx_b} & 0 & -I_{xz_b} \\ 0 & I_{yy_b} & 0 \\ -I_{zx_b} & 0 & I_{zz_b} \end{bmatrix} \begin{Bmatrix} \omega_{x_b} \\ \omega_{y_b} \\ \omega_{z_b} \end{Bmatrix} = \begin{Bmatrix} I_{xx_b}\omega_{x_b} - I_{xz_b}\omega_{z_b} \\ I_{yy_b}\omega_{y_b} \\ I_{zz_b}\omega_{z_b} - I_{xz_b}\omega_{x_b} \end{Bmatrix}$$

$$\boldsymbol{\omega} \times ([\mathbf{I}]\boldsymbol{\omega}) = \boldsymbol{\omega} \times \mathbf{H}$$

$$= \begin{vmatrix} \mathbf{i}_{x_b} & \mathbf{i}_{y_b} & \mathbf{i}_{z_b} \\ \omega_{x_b} & \omega_{y_b} & \omega_{z_b} \\ H_{x_b} & H_{y_b} & H_{z_b} \end{vmatrix} = \begin{Bmatrix} \omega_{y_b}H_{z_b} - \omega_{z_b}H_{y_b} \\ \omega_{z_b}H_{x_b} - \omega_{x_b}H_{z_b} \\ \omega_{x_b}H_{y_b} - \omega_{y_b}H_{x_b} \end{Bmatrix} = \begin{Bmatrix} (I_{zz_b} - I_{yy_b})\omega_{y_b}\omega_{z_b} - I_{xz_b}\omega_{x_b}\omega_{y_b} \\ (I_{xx_b} - I_{zz_b})\omega_{x_b}\omega_{z_b} + I_{xz_b}(\omega_{x_b}^2 - \omega_{z_b}^2) \\ (I_{yy_b} - I_{xx_b})\omega_{x_b}\omega_{y_b} + I_{xz_b}\omega_{y_b}\omega_{z_b} \end{Bmatrix}$$

$$(7.2.17)$$

Using Eq. (7.2.17), the body-fixed Cartesian components of Eq. (7.2.2) are

$$\frac{d}{dt}\left(\begin{bmatrix} I_{xx_b} & 0 & -I_{xz_b} \\ 0 & I_{yy_b} & 0 \\ -I_{zx_b} & 0 & I_{zz_b} \end{bmatrix} \begin{Bmatrix} \omega_{x_b} \\ \omega_{y_b} \\ \omega_{z_b} \end{Bmatrix}\right) = \begin{Bmatrix} M_{Sx_b} + (I_{yy_b} - I_{zz_b})\omega_{y_b}\omega_{z_b} + I_{xz_b}\omega_{x_b}\omega_{y_b} \\ M_{Sy_b} + (I_{zz_b} - I_{xx_b})\omega_{x_b}\omega_{z_b} + I_{xz_b}(\omega_{z_b}^2 - \omega_{x_b}^2) \\ M_{Sz_b} + (I_{xx_b} - I_{yy_b})\omega_{x_b}\omega_{y_b} - I_{xz_b}\omega_{y_b}\omega_{z_b} \end{Bmatrix}$$

$$(7.2.18)$$

As aircraft mass changes with time, so does the moment of inertia. However, when mass is expelled from an aircraft, it can carry angular momentum with it. Thus, including the effects of mass transfer and applying conservation of mass in a manner similar to that used for translational momentum, Eq. (7.2.18) can be written as

$$\begin{bmatrix} I_{xx_b} & 0 & -I_{xz_b} \\ 0 & I_{yy_b} & 0 \\ -I_{zx_b} & 0 & I_{zz_b} \end{bmatrix} \begin{Bmatrix} \dot{\omega}_{x_b} \\ \dot{\omega}_{y_b} \\ \dot{\omega}_{z_b} \end{Bmatrix} = \begin{Bmatrix} M_{x_b} + (I_{yy_b} - I_{zz_b})\omega_{y_b}\omega_{z_b} + I_{xz_b}\omega_{x_b}\omega_{y_b} \\ M_{y_b} + (I_{zz_b} - I_{xx_b})\omega_{x_b}\omega_{z_b} + I_{xz_b}(\omega_{z_b}^2 - \omega_{x_b}^2) \\ M_{z_b} + (I_{xx_b} - I_{yy_b})\omega_{x_b}\omega_{y_b} - I_{xz_b}\omega_{y_b}\omega_{z_b} \end{Bmatrix}$$

$$(7.2.19)$$

where **M** is a pseudo aerodynamic moment including any and all effects of thrust. **This pseudo moment accounts for all the effects of mass leaving or entering the aircraft. Equation (7.2.19) does not require constant moment of inertia.**

The explicit notation $(V_{x_b}, V_{y_b}, V_{z_b})$ and $(\omega_{x_b}, \omega_{y_b}, \omega_{z_b})$, which is commonly used in classical dynamics for the components of the translational and angular velocity vectors in body-fixed coordinates, is not normally used in the field of aircraft flight dynamics. The notation that is used almost exclusively by the aircraft community is

$$\begin{Bmatrix} u \\ v \\ w \end{Bmatrix} \equiv \begin{Bmatrix} V_{x_b} \\ V_{y_b} \\ V_{z_b} \end{Bmatrix} = \begin{Bmatrix} \text{the axial velocity} \\ \text{the sideslip velocity} \\ \text{the normal velocity} \end{Bmatrix}, \quad \begin{Bmatrix} p \\ q \\ r \end{Bmatrix} \equiv \begin{Bmatrix} \omega_{x_b} \\ \omega_{y_b} \\ \omega_{z_b} \end{Bmatrix} = \begin{Bmatrix} \text{the rolling rate} \\ \text{the pitching rate} \\ \text{the yawing rate} \end{Bmatrix}$$

This notation has always had the potential for causing some confusion with new students. However, the notation is well established in the literature and it should be memorized. **Be careful not to confuse the rolling rate with the pressure and remember that p does not stand for pitch and r does not stand for roll.**

As a memory aid, notice that p, q, and r are in alphabetical order, as are the rotation axes x_b, y_b, and z_b. If we think of the rotation triad "roll, pitch, and yaw," in that order, the symbols used for the angular velocity components and the axes of rotation form alphabetic triads in the same order.

There is another potential confusion associated with using lowercase m for mass, as is commonly done in classical dynamics. This is because in the field of aerodynamics, lowercase m is normally used to signify the aerodynamic pitching moment. To avoid this confusion, the x_b-, y_b-, and z_b-components of the aerodynamic moment vector (excluding the effects of thrust) are denoted ℓ, m, and n, whereas the mass is designated as W/g, with W denoting the weight and g signifying the acceleration of gravity.

The flight mechanics community sometimes uses the symbols L, M, and N to designate the x_b-, y_b-, and z_b-components of the total moment vector, including any moment produced by the thrust vector, which in some cases may not act through the center of gravity. This has been extremely confusing to new students, since L is also used to signify lift and N is commonly used to denote the normal force. In at least one textbook, the author has seen the symbol L used twice in the same equation, once to represent lift and once to represent the rolling moment. In the present text, we shall not carry tradition quite that far. Here we shall continue to use M_{x_b}, M_{y_b}, and M_{z_b} to designate the body-fixed components of the total moment vector.

In the remainder of our development of the aircraft equations of motion we shall use the following notation:

u	$\equiv$	x_b-component of aircraft translational velocity relative to surrounding air.
v	$\equiv$	y_b-component of aircraft translational velocity relative to surrounding air.
w	$\equiv$	z_b-component of aircraft translational velocity relative to surrounding air.
p	$\equiv$	x_b-component of aircraft rotational velocity (rolling rate).
q	$\equiv$	y_b-component of aircraft rotational velocity (pitching rate).
r	$\equiv$	z_b-component of aircraft rotational velocity (yawing rate).
F_{x_b}	$\equiv$	x_b-component of total pseudo aerodynamic force vector, including thrust.
F_{y_b}	$\equiv$	y_b-component of total pseudo aerodynamic force vector, including thrust.
F_{z_b}	$\equiv$	z_b-component of total pseudo aerodynamic force vector, including thrust.
M_{x_b}	$\equiv$	x_b-component of total pseudo aerodynamic moment vector, including thrust.
M_{y_b}	$\equiv$	y_b-component of total pseudo aerodynamic moment vector, including thrust.
M_{z_b}	$\equiv$	z_b-component of total pseudo aerodynamic moment vector, including thrust.
X	$\equiv$	x_b-component of aerodynamic force, excluding thrust (axial force).
Y	$\equiv$	y_b-component of aerodynamic force, excluding thrust (side force).
Z	$\equiv$	z_b-component of aerodynamic force, excluding thrust (normal force).
ℓ	$\equiv$	x_b-component of aerodynamic moment, excluding thrust (rolling moment).
m	$\equiv$	y_b-component of aerodynamic moment, excluding thrust (pitching moment).
n	$\equiv$	z_b-component of aerodynamic moment, excluding thrust (yawing moment).

Students should memorize this notation before proceeding with their study of this and the following chapters.

Substituting the aircraft notation just defined into Eqs. (7.2.14) and (7.2.19), we have a system of six first-order differential equations for the six components of aircraft velocity relative to the surrounding atmosphere. These are the velocity components associated with the six degrees of freedom for the aircraft treated as a rigid body. This is our statement of **Newton's second law,**

$$
\begin{bmatrix} W/g & 0 & 0 \\ 0 & W/g & 0 \\ 0 & 0 & W/g \end{bmatrix} \begin{Bmatrix} \dot{u} \\ \dot{v} \\ \dot{w} \end{Bmatrix} = \begin{Bmatrix} F_{x_b} + W_{x_b} + (rv - qw)W/g \\ F_{y_b} + W_{y_b} + (pw - ru)W/g \\ F_{z_b} + W_{z_b} + (qu - pv)W/g \end{Bmatrix}
$$
$$
\begin{bmatrix} I_{xx_b} & 0 & -I_{xz_b} \\ 0 & I_{yy_b} & 0 \\ -I_{zx_b} & 0 & I_{zz_b} \end{bmatrix} \begin{Bmatrix} \dot{p} \\ \dot{q} \\ \dot{r} \end{Bmatrix} = \begin{Bmatrix} M_{x_b} + (I_{yy_b} - I_{zz_b})qr + I_{xz_b}pq \\ M_{y_b} + (I_{zz_b} - I_{xx_b})pr + I_{xz_b}(r^2 - p^2) \\ M_{z_b} + (I_{xx_b} - I_{yy_b})pq - I_{xz_b}qr \end{Bmatrix}
$$

(7.2.20)

where W_{x_b}, W_{y_b}, and W_{z_b} are body-fixed components of aircraft weight. It is important to keep in mind that the forces and moments listed in Eq. (7.2.20) represent the total summation of forces and moments, including those due to thrust and gravity. Thrust can contribute to both the net force and the net moment if the thrust vector does not act through the center of gravity. The gravitational force can never contribute to the net moment, since by definition it always acts through the center of gravity.

When an aircraft is operating in steady flight, with some equilibrium airspeed, V_o, some equilibrium angle of attack, α_o, no sideslip velocity, and no angular velocity, the summation of forces and moments will vanish and the terms on the right-hand side of Eq. (7.2.20) will be zero. For example, in steady level flight with the thrust vector aligned with the direction of flight, the lift is equal to the weight, the thrust is equal to the drag, and the summation of moments must be zero. If any of the parameters on the right deviate from the equilibrium values, the aircraft will accelerate and the translational and rotational velocities will change. The terms on the right may be thought of as disturbances to steady equilibrium flight. These disturbances are usually divided into two categories. The disturbances that affect the motion only in the x_b-z_b plane are called the *longitudinal disturbances,*

$$ u - V_o \cos(\alpha_o), \ w - V_o \sin(\alpha_o), \ q, \ F_{x_b} + W_{x_b}, \ F_{z_b} + W_{z_b}, \ M_{y_b} $$

The remaining disturbances are called the *lateral disturbances,*

$$ v, \ p, \ r, \ F_{y_b} + W_{y_b}, \ M_{x_b}, \ M_{z_b} $$

If all of the lateral disturbances are zero,

$$ v = p = r = F_{y_b} + W_{y_b} = M_{x_b} = M_{z_b} = 0 $$

(7.2.21)

and Eq. (7.2.20) becomes

$$
\begin{bmatrix}
W/g & 0 & 0 & 0 & 0 & 0 \\
0 & W/g & 0 & 0 & 0 & 0 \\
0 & 0 & W/g & 0 & 0 & 0 \\
0 & 0 & 0 & I_{xx_b} & 0 & -I_{xz_b} \\
0 & 0 & 0 & 0 & I_{yy_b} & 0 \\
0 & 0 & 0 & -I_{xz_b} & 0 & I_{zz_b}
\end{bmatrix}
\begin{Bmatrix}
\dot{u} \\ \dot{v} \\ \dot{w} \\ \dot{p} \\ \dot{q} \\ \dot{r}
\end{Bmatrix}
=
\begin{Bmatrix}
F_{x_b} + W_{x_b} - qwW/g \\
0 \\
F_{z_b} + W_{z_b} + quW/g \\
0 \\
M_{y_b} \\
0
\end{Bmatrix}
\tag{7.2.22}
$$

From Eq. (7.2.22), we see that the longitudinal disturbances affect only the longitudinal velocity components, u, w, and q. The longitudinal disturbances produce no change in the lateral velocity components, v, p, and r. This results from the assumed symmetry of the aircraft about the x_b-z_b plane. Eliminating the trivial results, $\dot{v} = \dot{p} = \dot{r} = 0$, which are easily obtained from the second, fourth, and sixth rows of Eq. (7.2.22), the longitudinal equations of motion for the aircraft reduce to

$$
\begin{bmatrix}
W/g & 0 & 0 \\
0 & W/g & 0 \\
0 & 0 & I_{yy_b}
\end{bmatrix}
\begin{Bmatrix}
\dot{u} \\ \dot{w} \\ \dot{q}
\end{Bmatrix}
=
\begin{Bmatrix}
F_{x_b} + W_{x_b} - qwW/g \\
F_{z_b} + W_{z_b} + quW/g \\
M_{y_b}
\end{Bmatrix}
\tag{7.2.23}
$$

If, on the other hand, all of the longitudinal disturbances are zero,

$$
u - V_o \cos(\alpha_o) = w - V_o \sin(\alpha_o) = q
$$
$$
= F_{x_b} + W_{x_b} = F_{z_b} + W_{z_b} = M_{y_b} = 0
\tag{7.2.24}
$$

and Eq. (7.2.20) becomes

$$
\begin{bmatrix}
W/g & 0 & 0 & 0 & 0 & 0 \\
0 & W/g & 0 & 0 & 0 & 0 \\
0 & 0 & W/g & 0 & 0 & 0 \\
0 & 0 & 0 & I_{xx_b} & 0 & -I_{xz_b} \\
0 & 0 & 0 & 0 & I_{yy_b} & 0 \\
0 & 0 & 0 & -I_{xz_b} & 0 & I_{zz_b}
\end{bmatrix}
\begin{Bmatrix}
\dot{u} \\ \dot{v} \\ \dot{w} \\ \dot{p} \\ \dot{q} \\ \dot{r}
\end{Bmatrix}
\tag{7.2.25}
$$
$$
=
\begin{Bmatrix}
rvW/g \\
F_{y_b} + W_{y_b} + [pV_o \sin(\alpha_o) - rV_o \cos(\alpha_o)]W/g \\
- pvW/g \\
M_{x_b} \\
(I_{zz_b} - I_{xx_b})pr + I_{xz_b}(r^2 - p^2) \\
M_{z_b}
\end{Bmatrix}
$$

Thus we see that while the longitudinal disturbances affect only the longitudinal velocity components, the lateral disturbances have an effect on both the lateral and the longitudinal velocity components. It should be noted that longitudinal motion (axial translation, normal translation, and pitching rotation) is simple planar motion confined to the x_b-z_b plane. On the other hand, lateral motion (sideslip translation, rolling rotation, and yawing rotation) is not a simple planar motion at all. Roll is rotation in the y_b-z_b plane and yaw is rotation in the x_b-y_b plane. The longitudinal velocity components are directly affected by the lateral velocity components, in part because of the Coriolis accelerations. For example, sideslip combined with a yawing rate results in a Coriolis acceleration in the axial direction. Similarly, sideslip combined with a rolling rate produces a normal Coriolis acceleration. Rolling and yawing rates also produce a pitching acceleration that results from the asymmetry of the aircraft in the x_b-z_b plane. The latter coupling term is usually referred to as the *inertial cross coupling*.

EXAMPLE 7.2.1. Consider an object moving through space with no forces of any kind acting on the body. The body is initially moving relative to an inertial coordinate system with a velocity of V_0 and is rotating about an axis normal to the velocity vector with an angular velocity of ω_0. We wish to use Newton's second law to determine the translational and rotational velocity of this object as a function of time. This problem is, of course, trivial to solve using an inertial coordinate system. However, to gain some familiarity and confidence in working with body-fixed coordinates, we wish to solve this simple problem using only body-fixed coordinates.

Solution. Because the rotation vector is normal to the translation vector, this is planar motion. Choosing the body-fixed coordinate system so that at time $t = 0$ the x_b-axis coincides with the direction of motion and the y_b-axis coincides with the axis of rotation, the motion is planar longitudinal motion. Newton's second law is given by Eq. (7.2.23) with all forces and moments set to zero,

$$
\begin{bmatrix} W/g & 0 & 0 \\ 0 & W/g & 0 \\ 0 & 0 & I_{yy_b} \end{bmatrix} \begin{Bmatrix} \dot{u} \\ \dot{w} \\ \dot{q} \end{Bmatrix} = \begin{Bmatrix} -qwW/g \\ quW/g \\ 0 \end{Bmatrix} \tag{7.2.26}
$$

The initial conditions for this first-order system of differential equations is

$$
\begin{Bmatrix} u \\ w \\ q \end{Bmatrix}_{t=0} = \begin{Bmatrix} V_0 \\ 0 \\ \omega_0 \end{Bmatrix} \tag{7.2.27}
$$

The third equation in Eq. (7.2.26) is not coupled to the other two. This equation is easily integrated subject to the third initial condition in Eq. (7.2.27) to give

$$
q = \omega_0
$$

Using this result in the remaining two equations from Eq. (7.2.26), we have

$$\begin{bmatrix} 1 & 0 \\ 0 & 1 \end{bmatrix} \begin{Bmatrix} \dot{u} \\ \dot{w} \end{Bmatrix} + \begin{bmatrix} 0 & \omega_0 \\ -\omega_0 & 0 \end{bmatrix} \begin{Bmatrix} u \\ w \end{Bmatrix} = \begin{Bmatrix} 0 \\ 0 \end{Bmatrix}$$

This is a homogeneous system of first-order linear differential equations. At this point students may wish to review the chapter on systems of linear differential equations from their introductory course on ordinary differential equations. Rewriting this system of equations in the usual differential operator notation, we have

$$\begin{bmatrix} D & \omega_0 \\ -\omega_0 & D \end{bmatrix} \begin{Bmatrix} u \\ w \end{Bmatrix} = \begin{Bmatrix} 0 \\ 0 \end{Bmatrix} \qquad (7.2.28)$$

where D the differential operator. Differentiating the first equation and multiplying the second equation by ω_0 gives

$$\begin{bmatrix} D^2 & \omega_0 D \\ -\omega_0^2 & \omega_0 D \end{bmatrix} \begin{Bmatrix} u \\ w \end{Bmatrix} = \begin{Bmatrix} 0 \\ 0 \end{Bmatrix}$$

Subtracting the second equation from the first, we have

$$\begin{bmatrix} (D^2 + \omega_0^2) & 0 \\ -\omega_0^2 & \omega_0 D \end{bmatrix} \begin{Bmatrix} u \\ w \end{Bmatrix} = \begin{Bmatrix} 0 \\ 0 \end{Bmatrix}$$

In a similar manner the system can be further reduced to yield

$$\begin{bmatrix} (D^2 + \omega_0^2) & 0 \\ 0 & (D^2 + \omega_0^2) \end{bmatrix} \begin{Bmatrix} u \\ w \end{Bmatrix} = \begin{Bmatrix} 0 \\ 0 \end{Bmatrix}$$

We have now reduced a system of two coupled first-order equations to a system of two uncoupled second-order equations. Each of these is easily solved separately to give

$$\begin{Bmatrix} u \\ w \end{Bmatrix} = \begin{Bmatrix} C_1 \sin(\omega_0 t) + C_2 \cos(\omega_0 t) \\ C_3 \sin(\omega_0 t) + C_4 \cos(\omega_0 t) \end{Bmatrix} \qquad (7.2.29)$$

Using this result in Eq. (7.2.28), we have

$$\begin{Bmatrix} C_1 \omega_0 \cos(\omega_0 t) - C_2 \omega_0 \sin(\omega_0 t) + C_3 \omega_0 \sin(\omega_0 t) + C_4 \omega_0 \cos(\omega_0 t) \\ -C_1 \omega_0 \sin(\omega_0 t) - C_2 \omega_0 \cos(\omega_0 t) + C_3 \omega_0 \cos(\omega_0 t) - C_4 \omega_0 \sin(\omega_0 t) \end{Bmatrix} = \begin{Bmatrix} 0 \\ 0 \end{Bmatrix}$$

which requires that

$$\begin{Bmatrix} C_3 \\ C_4 \end{Bmatrix} = \begin{Bmatrix} C_2 \\ -C_1 \end{Bmatrix} \tag{7.2.30}$$

Using Eq. (7.2.30) in Eq. (7.2.29) gives

$$\begin{Bmatrix} u \\ w \end{Bmatrix} = \begin{Bmatrix} C_1 \sin(\omega_0 t) + C_2 \cos(\omega_0 t) \\ C_2 \sin(\omega_0 t) - C_1 \cos(\omega_0 t) \end{Bmatrix} \tag{7.2.31}$$

Applying the initial conditions,

$$\begin{Bmatrix} u(0) \\ w(0) \end{Bmatrix} = \begin{Bmatrix} V_0 \\ 0 \end{Bmatrix}$$

we obtain

$$\begin{Bmatrix} C_1 \\ C_2 \end{Bmatrix} = \begin{Bmatrix} 0 \\ V_0 \end{Bmatrix} \tag{7.2.32}$$

Using Eq. (7.2.32) in Eq. (7.2.31) and including the result obtained for the angular velocity gives the complete solution,

$$\begin{Bmatrix} u \\ w \\ q \end{Bmatrix} = \begin{Bmatrix} V_0 \cos(\omega_0 t) \\ V_0 \sin(\omega_0 t) \\ \omega_0 \end{Bmatrix}$$

This problem is, of course, trivial to solve relative to an inertial coordinate system. For an inertial coordinate system, Eq. (7.2.26) reduces to the completely uncoupled system,

$$\begin{bmatrix} W/g & 0 & 0 \\ 0 & W/g & 0 \\ 0 & 0 & I_{yy_b} \end{bmatrix} \begin{Bmatrix} \dot{V}_{x_f} \\ \dot{V}_{z_f} \\ \dot{\omega}_{y_f} \end{Bmatrix} = \begin{Bmatrix} 0 \\ 0 \\ 0 \end{Bmatrix}$$

which can be directly integrated subject to the initial conditions to give

$$\begin{Bmatrix} V_{x_f} \\ V_{z_f} \\ \omega_{y_f} \end{Bmatrix} = \begin{Bmatrix} V_0 \\ 0 \\ \omega_0 \end{Bmatrix}$$

Thus, to a fixed observer in the inertial coordinate system, the velocity of the object appears to be constant with time. On the other hand, to an observer in the body-fixed coordinate system, the velocity appears to be changing sinusoidal with time. These are simply two equivalent ways to describe exactly the same motion.

The advantage of using the body-fixed coordinate system is not at all obvious from Example 7.2.1. This very simple problem is much easier to solve using the inertial coordinate system. We do see, however, that Newton's second law can be stated and solved either in terms of inertial coordinates or in terms of body-fixed coordinates. The advantage of using a body-fixed coordinate system will become more apparent in the following sections.

7.3. Position and Orientation: The Euler Angle Formulation

In the preceding section, Newton's second law was written in terms of a coordinate system fixed to the aircraft. For complex geometry and/or complex motion, this coordinate system can simplify the equations of motion by making the inertia tensor independent of time. Furthermore, onboard instruments report flight parameters only as viewed from a body-fixed coordinate system. Thus, the body-fixed coordinate system extends considerable utility to the study of aircraft dynamics. Unfortunately, this moving coordinate system cannot easily be used for describing the position and orientation of the airplane. The aircraft position and orientation are most conveniently described in terms of a coordinate system fixed to the Earth.

The position of an aircraft relative to the Earth is normally described in terms of latitude, longitude, and altitude. However, over short time periods, the distance traveled is small compared to the radius of the Earth and we can describe position in terms of a Cartesian system. As previously discussed, the Cartesian system commonly used has the x-axis pointing north, the y-axis pointing east, and the z-axis pointing down. This Earth-fixed coordinate system has been designated (x_f, y_f, z_f). The position of the aircraft is specified by the location of the origin of the body-fixed coordinate frame relative to the Earth-fixed coordinate frame.

The orientation of an aircraft relative to the Earth can be described in terms of what are called *Euler angles*. The orientation of the body-fixed coordinate frame, (x_b, y_b, z_b), relative to the Earth-fixed coordinate frame, (x_f, y_f, z_f), is described in terms of three consecutive rotations through three Euler angles, in the specific order as follows:

(a) Rotate the (x_f, y_f, z_f) coordinate system about the z_f-axis through an angle ψ to the coordinate system (x_1, y_1, z_1), as shown in Fig. 7.3.1a.

(b) Rotate the (x_1, y_1, z_1) coordinate system about the y_1-axis through an angle θ to the coordinate system (x_2, y_2, z_2), as shown in Fig. 7.3.1b.

(c) Rotate the (x_2, y_2, z_2) coordinate system about the x_2-axis through an angle ϕ to the coordinate system (x_b, y_b, z_b), as shown in Fig. 7.3.1c.

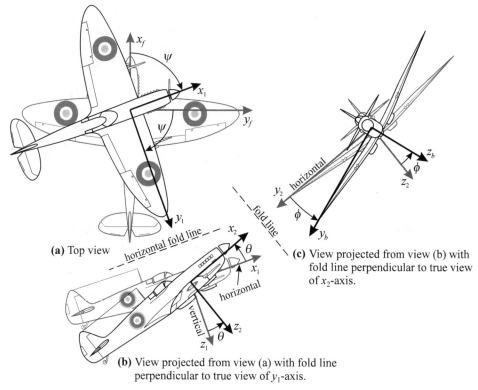

(a) Top view

(b) View projected from view (a) with fold line perpendicular to true view of y_1-axis.

(c) View projected from view (b) with fold line perpendicular to true view of x_2-axis.

Figure 7.3.1. True views of the three Euler angles shown following the standard conventions of engineering graphics and descriptive geometry.

An isometric view of these three rotations is shown in Fig. 7.3.2. The order of rotation is very important. In general, if these three rotations were to take place in a different order, the final orientation would be different.

The three Euler angles, ϕ, θ, and ψ are, respectively, called the bank angle, the elevation angle, and the azimuth angle or heading. These three angles are sometimes incorrectly referred to as the roll, pitch, and yaw. There is a subtle but very important difference between the Euler angles and roll, pitch, and yaw. The azimuth angle is a rotation about the z_f-axis and yaw is a rotation about the z_b-axis. Similarly, the elevation angle is a rotation about the y_1-axis and pitch is a rotation about the y_b-axis. Roll, pitch, and yaw are orthogonal. The Euler angles are not.

The Euler angles can be related to roll, pitch, and yaw, but this is not a one-to-one relationship. For example, it is possible to change the bank angle of an aircraft without using any roll input whatsoever. To demonstrate this, consider an airplane in a level attitude with all three Euler angles at zero, as shown in Fig. 7.3.3. Now let the airplane move through a pitch angle of 90 degrees, followed by a yaw angle of 90 degrees and a pitch angle of −90 degrees. The airplane now has a bank angle of 90 degrees, but there was no roll input throughout the process. Try it with a model.

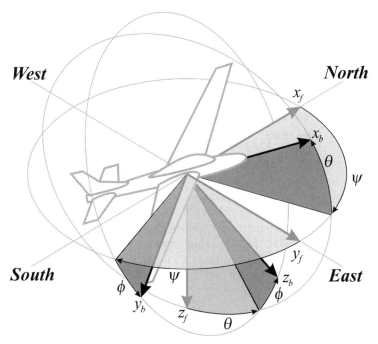

Figure 7.3.2. Euler angles shown in an isometric view.

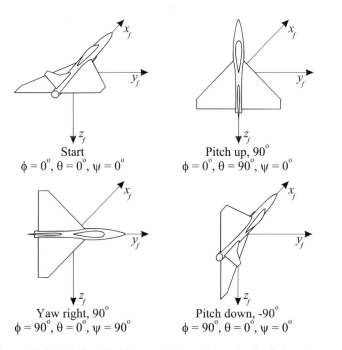

Start
$\phi = 0°, \theta = 0°, \psi = 0°$

Pitch up, 90°
$\phi = 0°, \theta = 90°, \psi = 0°$

Yaw right, 90°
$\phi = 90°, \theta = 0°, \psi = 90°$

Pitch down, -90°
$\phi = 90°, \theta = 0°, \psi = 0°$

Figure 7.3.3. Relationship between Euler angles and roll, pitch, and yaw.

Any set of three Euler angles describes a unique orientation of the aircraft (i.e., no two different orientations have the same three Euler angles). A specific set of Euler angles specifies an instantaneous orientation that does not depend in any way on how the aircraft came to be so oriented. While a given set of Euler angles specifies a unique orientation for the aircraft, orientation is a periodic function of the Euler angles. Thus, any orientation of an aircraft could be described by more than one set of Euler angles. Clearly, adding 360 degrees to any one of the Euler angles will result in a different but equivalent set of Euler angles. If we are to define the Euler angles so that any given orientation has a unique set of Euler angles, then we must restrict each Euler angle to fall within some range including no more than 360 degrees. However, even this is not sufficient. This is demonstrated in Fig. 7.3.4. In this figure we start with an airplane pointed due north with both the wing span and the fuselage reference line parallel to the ground. Normally, we would specify this orientation as having all three Euler angles equal to zero. From this starting point, shown in Fig. 7.3.4a, we let the airplane move through a pitch angle of 180 degrees to the orientation shown in Fig. 7.3.4b. Clearly, this orientation could be described as having a bank angle of zero, an elevation angle of 180 degrees, and azimuth angle of zero. The airplane could be placed in this same

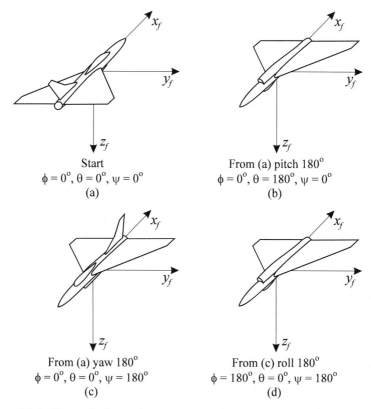

Figure 7.3.4. Two equivalent Euler angle descriptions for the same aircraft orientation.

orientation by starting with all Euler angles at zero and rotating the airplane through a yaw angle of 180 degrees to the orientation shown in Fig. 7.3.4c. From this point the airplane is rolled through 180 degrees to the orientation shown in Fig. 7.3.4d. This final orientation could be described as having a bank angle of 180 degrees, an elevation angle of zero, and an azimuth angle of 180 degrees. Since the orientation shown in Fig. 7.3.4b is identical to that shown in Fig. 7.3.4d, the Euler angle set $\phi = 0°$, $\theta = 180°$, and $\psi = 0°$ is obviously equivalent to the set $\phi = 180°$, $\theta = 0°$, and $\psi = 180°$. To avoid this duality and make the Euler angle definitions unique, the Euler angles are by standard convention usually restricted to the range $-180° < \phi \leq 180°$, $-90° \leq \theta \leq 90°$, and $0° \leq \psi < 360°$.

To obtain the transformations between coordinate systems (x_b, y_b, z_b) and (x_f, y_f, z_f), we now review some principles from calculus and analytic geometry. In terms of the three Euler angle rotations shown in Fig. 7.3.1, we wish to develop two transformation matrices, one to transform from body-fixed to Earth-fixed coordinates and another to transform from Earth-fixed to body-fixed coordinates. We first consider each of the three rotations shown in Fig. 7.3.1 as a separate transformation and then combine the three transformations into one transformation matrix.

Consider first an arbitrary vector $\mathbf{v}$ having components $(v_{x_1}, v_{y_1}, v_{z_1})$ in coordinate system (x_1, y_1, z_1) and having components $(v_{x_f}, v_{y_f}, v_{z_f})$ in the (x_f, y_f, z_f) coordinate system. From the geometry in Fig. 7.3.1a, we see that the x-component of the vector $\mathbf{v}$ in coordinate system 1 can be broken up into two components in coordinate system f. Thus we have

$$v_{x_1}\mathbf{i}_{x_1} = v_{x_1}\cos(\psi)\mathbf{i}_{x_f} + v_{x_1}\sin(\psi)\mathbf{i}_{y_f} \tag{7.3.1}$$

Similarly, the y-component transforms as

$$v_{y_1}\mathbf{i}_{y_1} = -v_{y_1}\sin(\psi)\mathbf{i}_{x_f} + v_{y_1}\cos(\psi)\mathbf{i}_{y_f} \tag{7.3.2}$$

and the z-component transforms as

$$v_{z_1}\mathbf{i}_{z_1} = v_{z_1}\mathbf{i}_{z_f} \tag{7.3.3}$$

Combining Eqs. (7.3.1) through (7.3.3), we obtain the transformation equation

$$\begin{Bmatrix} v_{x_f} \\ v_{y_f} \\ v_{z_f} \end{Bmatrix} = \begin{bmatrix} \cos(\psi) & -\sin(\psi) & 0 \\ \sin(\psi) & \cos(\psi) & 0 \\ 0 & 0 & 1 \end{bmatrix} \begin{Bmatrix} v_{x_1} \\ v_{y_1} \\ v_{z_1} \end{Bmatrix} \tag{7.3.4}$$

In a similar manner, the inverse transformation can be written

$$\begin{Bmatrix} v_{x_1} \\ v_{y_1} \\ v_{z_1} \end{Bmatrix} = \begin{bmatrix} \cos(\psi) & \sin(\psi) & 0 \\ -\sin(\psi) & \cos(\psi) & 0 \\ 0 & 0 & 1 \end{bmatrix} \begin{Bmatrix} v_{x_f} \\ v_{y_f} \\ v_{z_f} \end{Bmatrix} \tag{7.3.5}$$

Likewise, from Fig. 7.3.1b, $(v_{x_1}, v_{y_1}, v_{z_1})$ can be expressed in terms of $(v_{x_2}, v_{y_2}, v_{z_2})$,

$$\begin{Bmatrix} v_{x_1} \\ v_{y_1} \\ v_{z_1} \end{Bmatrix} = \begin{bmatrix} \cos(\theta) & 0 & \sin(\theta) \\ 0 & 1 & 0 \\ -\sin(\theta) & 0 & \cos(\theta) \end{bmatrix} \begin{Bmatrix} v_{x_2} \\ v_{y_2} \\ v_{z_2} \end{Bmatrix} \tag{7.3.6}$$

and the inverse transformation is found to be

$$\begin{Bmatrix} v_{x_2} \\ v_{y_2} \\ v_{z_2} \end{Bmatrix} = \begin{bmatrix} \cos(\theta) & 0 & -\sin(\theta) \\ 0 & 1 & 0 \\ \sin(\theta) & 0 & \cos(\theta) \end{bmatrix} \begin{Bmatrix} v_{x_1} \\ v_{y_1} \\ v_{z_1} \end{Bmatrix} \tag{7.3.7}$$

From Fig. 7.3.1c we obtain

$$\begin{Bmatrix} v_{x_2} \\ v_{y_2} \\ v_{z_2} \end{Bmatrix} = \begin{bmatrix} 1 & 0 & 0 \\ 0 & \cos(\phi) & -\sin(\phi) \\ 0 & \sin(\phi) & \cos(\phi) \end{bmatrix} \begin{Bmatrix} v_{x_b} \\ v_{y_b} \\ v_{z_b} \end{Bmatrix} \tag{7.3.8}$$

and

$$\begin{Bmatrix} v_{x_b} \\ v_{y_b} \\ v_{z_b} \end{Bmatrix} = \begin{bmatrix} 1 & 0 & 0 \\ 0 & \cos(\phi) & \sin(\phi) \\ 0 & -\sin(\phi) & \cos(\phi) \end{bmatrix} \begin{Bmatrix} v_{x_2} \\ v_{y_2} \\ v_{z_2} \end{Bmatrix} \tag{7.3.9}$$

Combining Eqs. (7.3.4), (7.3.6), and (7.3.8), we have

$$\begin{Bmatrix} v_{x_f} \\ v_{y_f} \\ v_{z_f} \end{Bmatrix} = \begin{bmatrix} \cos(\psi) & -\sin(\psi) & 0 \\ \sin(\psi) & \cos(\psi) & 0 \\ 0 & 0 & 1 \end{bmatrix} \begin{bmatrix} \cos(\theta) & 0 & \sin(\theta) \\ 0 & 1 & 0 \\ -\sin(\theta) & 0 & \cos(\theta) \end{bmatrix} \begin{bmatrix} 1 & 0 & 0 \\ 0 & \cos(\phi) & -\sin(\phi) \\ 0 & \sin(\phi) & \cos(\phi) \end{bmatrix} \begin{Bmatrix} v_{x_b} \\ v_{y_b} \\ v_{z_b} \end{Bmatrix} \tag{7.3.10}$$

and combining Eqs. (7.3.5), (7.3.7), and (7.3.9) gives the inverse transformation,

$$\begin{Bmatrix} v_{x_b} \\ v_{y_b} \\ v_{z_b} \end{Bmatrix} = \begin{bmatrix} 1 & 0 & 0 \\ 0 & \cos(\phi) & \sin(\phi) \\ 0 & -\sin(\phi) & \cos(\phi) \end{bmatrix} \begin{bmatrix} \cos(\theta) & 0 & -\sin(\theta) \\ 0 & 1 & 0 \\ \sin(\theta) & 0 & \cos(\theta) \end{bmatrix} \begin{bmatrix} \cos(\psi) & \sin(\psi) & 0 \\ -\sin(\psi) & \cos(\psi) & 0 \\ 0 & 0 & 1 \end{bmatrix} \begin{Bmatrix} v_{x_f} \\ v_{y_f} \\ v_{z_f} \end{Bmatrix} \tag{7.3.11}$$

The individual transformation matrices in Eq. (7.3.10) and those in Eq. (7.3.11) may be multiplied together to obtain a single transformation matrix for each equation. Using the shorthand notation $S_\phi = \sin(\phi)$, $C_\phi = \cos(\phi)$, $S_\theta = \sin(\theta)$, $C_\theta = \cos(\theta)$, $S_\psi = \sin(\psi)$, and $C_\psi = \cos(\psi)$, from Eq. (7.3.10), we have the transformation equation,

$$
\begin{Bmatrix} v_{x_f} \\ v_{y_f} \\ v_{z_f} \end{Bmatrix} = \begin{bmatrix} C_\theta C_\psi & S_\phi S_\theta C_\psi - C_\phi S_\psi & C_\phi S_\theta C_\psi + S_\phi S_\psi \\ C_\theta S_\psi & S_\phi S_\theta S_\psi + C_\phi C_\psi & C_\phi S_\theta S_\psi - S_\phi C_\psi \\ -S_\theta & S_\phi C_\theta & C_\phi C_\theta \end{bmatrix} \begin{Bmatrix} v_{x_b} \\ v_{y_b} \\ v_{z_b} \end{Bmatrix} \tag{7.3.12}
$$

and from Eq. (7.3.11) we obtain the inverse transformation equation,

$$
\begin{Bmatrix} v_{x_b} \\ v_{y_b} \\ v_{z_b} \end{Bmatrix} = \begin{bmatrix} C_\theta C_\psi & C_\theta S_\psi & -S_\theta \\ S_\phi S_\theta C_\psi - C_\phi S_\psi & S_\phi S_\theta S_\psi + C_\phi C_\psi & S_\phi C_\theta \\ C_\phi S_\theta C_\psi + S_\phi S_\psi & C_\phi S_\theta S_\psi - S_\phi C_\psi & C_\phi C_\theta \end{bmatrix} \begin{Bmatrix} v_{x_f} \\ v_{y_f} \\ v_{z_f} \end{Bmatrix} \tag{7.3.13}
$$

Equation (7.3.12) can be used to transform the components of any vector from body-fixed to Earth-fixed coordinates, and Eq. (7.3.13) can be used to transform vector components from Earth-fixed to body-fixed coordinates. **It should be noted that the inverse of the transformation matrix in Eq. (7.3.12) is its transpose.**

In body-fixed coordinates, the velocity of the aircraft relative to the surrounding air has been designated (u, v, w), and in Earth-fixed coordinates, the velocity of the aircraft relative to the Earth is the time rate of change of the position vector $(\dot{x}_f, \dot{y}_f, \dot{z}_f)$. The ground speed is related to the airspeed and wind through Eq. (7.1.1). Thus, combining Eqs. (7.1.1) and (7.3.12), we can write

$$
\begin{Bmatrix} \dot{x}_f \\ \dot{y}_f \\ \dot{z}_f \end{Bmatrix} = \begin{bmatrix} C_\theta C_\psi & S_\phi S_\theta C_\psi - C_\phi S_\psi & C_\phi S_\theta C_\psi + S_\phi S_\psi \\ C_\theta S_\psi & S_\phi S_\theta S_\psi + C_\phi C_\psi & C_\phi S_\theta S_\psi - S_\phi C_\psi \\ -S_\theta & S_\phi C_\theta & C_\phi C_\theta \end{bmatrix} \begin{Bmatrix} u \\ v \\ w \end{Bmatrix} + \begin{Bmatrix} V_{wx_f} \\ V_{wy_f} \\ V_{wz_f} \end{Bmatrix} \tag{7.3.14}
$$

where V_{wx_f}, V_{wy_f}, and V_{wz_f} are the components of the constant wind vector in Earth-fixed coordinates. Integration of this equation will yield the airplane's position as a function of time, relative to the Earth-fixed coordinate system.

Since we are using Newton's second law in body-fixed coordinates, we must express the gravitational force in body-fixed coordinates as well. In Earth-fixed coordinates, the representation of the gravitational force vector is $(0, 0, W)$. Thus, from Eq. (7.3.13), the **gravitational force vector in body-fixed coordinates** is represented as

$$
\begin{Bmatrix} W_{x_b} \\ W_{y_b} \\ W_{z_b} \end{Bmatrix} = W \begin{Bmatrix} -\sin(\theta) \\ \sin(\phi)\cos(\theta) \\ \cos(\phi)\cos(\theta) \end{Bmatrix} \tag{7.3.15}
$$

In a similar manner, the time derivatives of the Euler angles, $(\dot{\phi}, \dot{\theta}, \dot{\psi})$, are related to the body-fixed components of the angular velocity vector, (p, q, r). However, we must remember that each of the Euler angles is defined relative to a different coordinate system. The bank angle, ϕ, is defined relative to the coordinate system (x_2, y_2, z_2), the elevation angle, θ, is defined relative to (x_1, y_1, z_1), and the azimuth angle, ψ, is defined relative to (x_f, y_f, z_f). Thus, using Eqs. (7.3.5), (7.3.7), and (7.3.9), we can write

$$
\begin{Bmatrix} p \\ q \\ r \end{Bmatrix} = \begin{bmatrix} 1 & 0 & 0 \\ 0 & C_\phi & S_\phi \\ 0 & -S_\phi & C_\phi \end{bmatrix} \begin{Bmatrix} \dot\phi \\ 0 \\ 0 \end{Bmatrix} + \begin{bmatrix} 1 & 0 & 0 \\ 0 & C_\phi & S_\phi \\ 0 & -S_\phi & C_\phi \end{bmatrix} \begin{bmatrix} C_\theta & 0 & -S_\theta \\ 0 & 1 & 0 \\ S_\theta & 0 & C_\theta \end{bmatrix} \begin{Bmatrix} 0 \\ \dot\theta \\ 0 \end{Bmatrix}
$$
$$
+ \begin{bmatrix} 1 & 0 & 0 \\ 0 & C_\phi & S_\phi \\ 0 & -S_\phi & C_\phi \end{bmatrix} \begin{bmatrix} C_\theta & 0 & -S_\theta \\ 0 & 1 & 0 \\ S_\theta & 0 & C_\theta \end{bmatrix} \begin{bmatrix} C_\psi & S_\psi & 0 \\ -S_\psi & C_\psi & 0 \\ 0 & 0 & 1 \end{bmatrix} \begin{Bmatrix} 0 \\ 0 \\ \dot\psi \end{Bmatrix}
$$

(7.3.16)

or after combining terms,

$$
\begin{Bmatrix} p \\ q \\ r \end{Bmatrix} = \begin{bmatrix} 1 & 0 & -S_\theta \\ 0 & C_\phi & S_\phi C_\theta \\ 0 & -S_\phi & C_\phi C_\theta \end{bmatrix} \begin{Bmatrix} \dot\phi \\ \dot\theta \\ \dot\psi \end{Bmatrix}
$$

(7.3.17)

Inverting Eq. (7.3.17), we get the Euler rates in terms of the roll, pitch, and yaw rates,

$$
\begin{Bmatrix} \dot\phi \\ \dot\theta \\ \dot\psi \end{Bmatrix} = \begin{bmatrix} 1 & S_\phi S_\theta / C_\theta & C_\phi S_\theta / C_\theta \\ 0 & C_\phi & -S_\phi \\ 0 & S_\phi / C_\theta & C_\phi / C_\theta \end{bmatrix} \begin{Bmatrix} p \\ q \\ r \end{Bmatrix}
$$

(7.3.18)

Here, we see that **the inverse of the matrix in Eq. (7.3.17) is not its transpose.** Integration of this equation will yield the airplane's orientation as a function of time in terms of the Euler angles ϕ, θ, and ψ.

Combining Eqs. (7.3.14) and (7.3.18), we have a system of six first-order differential equations that relate the time rate of change of the position and orientation of the aircraft in Earth-fixed coordinates to the translational and rotational velocity components, relative to the surrounding air, in body-fixed coordinates,

$$
\begin{Bmatrix} \dot x_f \\ \dot y_f \\ \dot z_f \end{Bmatrix} = \begin{bmatrix} C_\theta C_\psi & S_\phi S_\theta C_\psi - C_\phi S_\psi & C_\phi S_\theta C_\psi + S_\phi S_\psi \\ C_\theta S_\psi & S_\phi S_\theta S_\psi + C_\phi C_\psi & C_\phi S_\theta S_\psi - S_\phi C_\psi \\ -S_\theta & S_\phi C_\theta & C_\phi C_\theta \end{bmatrix} \begin{Bmatrix} u \\ v \\ w \end{Bmatrix} + \begin{Bmatrix} V_{wx_f} \\ V_{wy_f} \\ V_{wz_f} \end{Bmatrix}
$$
$$
\begin{Bmatrix} \dot\phi \\ \dot\theta \\ \dot\psi \end{Bmatrix} = \begin{bmatrix} 1 & S_\phi S_\theta / C_\theta & C_\phi S_\theta / C_\theta \\ 0 & C_\phi & -S_\phi \\ 0 & S_\phi / C_\theta & C_\phi / C_\theta \end{bmatrix} \begin{Bmatrix} p \\ q \\ r \end{Bmatrix}
$$

(7.3.19)

These six equations are called the **kinematic transformation equations**, in terms of Euler angles. This transformation contains a singularity. This singularity can be seen in the last two terms in both the fourth and sixth equations in Eq. (7.3.19). Clearly, these four terms become singular when the Euler angle, θ, is $\pm90°$ and $\cos\theta$ becomes zero. At these points the transformation breaks down, and integration of the Euler angles becomes indeterminate. This singularity in the integration of the Euler angles is commonly known as *gimbal lock*. Due to its existence, virtually no commercial or widely used six-degree-of-freedom (6-DOF) computer code employs the transformation given by Eq. (7.3.19) for determination of the Euler angles. Rather, other methods are generally employed, which avoids these singularities through the use of either a *quaternion transformation* or a *direction cosine transformation*. However, the Euler angle transformation is widely used in the linearized equations of motion. Clearly, under normal operation, most airplanes stay within the range of motion that can be described by Eq. (7.3.19). Except for vertical launch vehicles, fighter aircraft, and stunt planes that might pass through these attitudes, the Euler angle singularities at ±90 degrees are not a problem. The main disadvantage of using the quaternion formulation for aircraft attitude is the increased complexity of the physical interpretation. Thus, the student should have a good understanding of the Euler angles discussed in this section before attempting to understand the quaternion formulation to be presented later in this text.

For pure longitudinal motion, the bank angle is typically zero and the azimuth angle must remain constant. For simplicity, we could align the x_f-axis of the Earth-fixed coordinate system with the x_b-z_b plane of the body-fixed coordinate system and make the assumption of no wind. Thus, we can write

$$v = p = r = \phi = \psi = 0 \tag{7.3.20}$$

For this pure longitudinal motion, from Eq. (7.3.15), the gravitational force described in body-fixed coordinates reduces to

$$\begin{Bmatrix} W_{x_b} \\ W_{y_b} \\ W_{z_b} \end{Bmatrix} = W \begin{Bmatrix} -\sin(\theta) \\ 0 \\ \cos(\theta) \end{Bmatrix} \tag{7.3.21}$$

and Eq. (7.3.19) becomes

$$\begin{Bmatrix} \dot{x}_f \\ \dot{y}_f \\ \dot{z}_f \\ \dot{\phi} \\ \dot{\theta} \\ \dot{\psi} \end{Bmatrix} = \begin{bmatrix} C_\theta & 0 & S_\theta & 0 & 0 & 0 \\ 0 & 1 & 0 & 0 & 0 & 0 \\ -S_\theta & 0 & C_\theta & 0 & 0 & 0 \\ 0 & 0 & 0 & 1 & 0 & S_\theta / C_\theta \\ 0 & 0 & 0 & 0 & 1 & 0 \\ 0 & 0 & 0 & 0 & 0 & 1/C_\theta \end{bmatrix} \begin{Bmatrix} u \\ 0 \\ w \\ 0 \\ q \\ 0 \end{Bmatrix} \tag{7.3.22}$$

Here again we see that longitudinal motion has no tendency to introduce lateral motion. Eliminating the trivial results, $W_{y_b} = \dot{y}_f = \dot{\phi} = \dot{\psi} = 0$, longitudinal motion requires that

$$\begin{Bmatrix} W_{x_b} \\ W_{z_b} \end{Bmatrix} = W \begin{Bmatrix} -\sin(\theta) \\ \cos(\theta) \end{Bmatrix} \tag{7.3.23}$$

$$\begin{Bmatrix} \dot{x}_f \\ \dot{z}_f \\ \dot{\theta} \end{Bmatrix} = \begin{bmatrix} \cos(\theta) & \sin(\theta) & 0 \\ -\sin(\theta) & \cos(\theta) & 0 \\ 0 & 0 & 1 \end{bmatrix} \begin{Bmatrix} u \\ w \\ q \end{Bmatrix} \tag{7.3.24}$$

For pure lateral motion, the axial velocity, normal velocity, and elevation angle must remain constant and the y_b-component of the angular velocity must be zero. For simplicity, we shall neglect wind and consider the special case of lateral motion relative to an equilibrium flight path where the elevation angle is zero. Thus, we can write

$$u - V_o \cos(\alpha_o) = w - V_o \sin(\alpha_o) = q = \theta = 0 \tag{7.3.25}$$

Thus, for this pure lateral motion, Eqs. (7.3.15) and (7.3.19) become

$$\begin{Bmatrix} W_{x_b} \\ W_{y_b} \\ W_{z_b} \end{Bmatrix} = W \begin{Bmatrix} 0 \\ \sin(\phi) \\ \cos(\phi) \end{Bmatrix} \tag{7.3.26}$$

$$\begin{Bmatrix} \dot{x}_f \\ \dot{y}_f \\ \dot{z}_f \\ \dot{\phi} \\ \dot{\theta} \\ \dot{\psi} \end{Bmatrix} = \begin{bmatrix} C_\psi & -C_\phi S_\psi & S_\phi S_\psi & 0 & 0 & 0 \\ S_\psi & C_\phi C_\psi & -S_\phi C_\psi & 0 & 0 & 0 \\ 0 & S_\phi & C_\phi & 0 & 0 & 0 \\ 0 & 0 & 0 & 1 & 0 & 0 \\ 0 & 0 & 0 & 0 & C_\phi & -S_\phi \\ 0 & 0 & 0 & 0 & S_\phi & C_\phi \end{bmatrix} \begin{Bmatrix} V_o \cos(\alpha_o) \\ V_{y_b} \\ V_o \sin(\alpha_o) \\ p \\ 0 \\ r \end{Bmatrix} \tag{7.3.27}$$

Here again we find that while the longitudinal disturbances affect only the longitudinal motion, the lateral disturbances have an effect on both the lateral and the longitudinal motion.

EXAMPLE 7.3.1.   Determine the position and orientation of the object in Example 7.2.1 as a function of time. Starting with the body-fixed solution to Example 7.2.1, describe the position in terms of Earth-fixed coordinates, and the orientation in terms of Euler angles.

Solution. From the solution to Example 7.2.1,

$$\begin{Bmatrix} u \\ w \\ q \end{Bmatrix} = \begin{Bmatrix} V_0 \cos(\omega_0 t) \\ V_0 \sin(\omega_0 t) \\ \omega_0 \end{Bmatrix}$$

Since this is longitudinal motion, we can apply this result to Eq. (7.3.24):

$$\begin{Bmatrix} \dot{x}_f \\ \dot{z}_f \\ \dot{\theta} \end{Bmatrix} = \begin{bmatrix} \cos(\theta) & \sin(\theta) & 0 \\ -\sin(\theta) & \cos(\theta) & 0 \\ 0 & 0 & 1 \end{bmatrix} \begin{Bmatrix} V_0 \cos(\omega_0 t) \\ V_0 \sin(\omega_0 t) \\ \omega_0 \end{Bmatrix} \tag{7.3.28}$$

The initial conditions are

$$\begin{Bmatrix} x_f \\ z_f \\ \theta \end{Bmatrix}_{t=0} = \begin{Bmatrix} 0 \\ 0 \\ 0 \end{Bmatrix} \tag{7.3.29}$$

The third equation in Eq. (7.3.28) is easily integrated subject to the third initial condition in Eq. (7.3.29) to yield

$$\theta = \omega_0 t$$

Using this result in the first two equations in Eq. (7.3.28), we have

$$\begin{Bmatrix} \dot{x}_f \\ \dot{z}_f \end{Bmatrix} = \begin{bmatrix} \cos(\omega_0 t) & \sin(\omega_0 t) \\ -\sin(\omega_0 t) & \cos(\omega_0 t) \end{bmatrix} \begin{Bmatrix} V_0 \cos(\omega_0 t) \\ V_0 \sin(\omega_0 t) \end{Bmatrix}$$

Performing the indicated multiplication gives

$$\begin{Bmatrix} \dot{x}_f \\ \dot{z}_f \end{Bmatrix} = \begin{Bmatrix} V_0 \cos^2(\omega_0 t) + V_0 \sin^2(\omega_0 t) \\ -V_0 \cos(\omega_0 t)\sin(\omega_0 t) + V_0 \sin(\omega_0 t)\cos(\omega_0 t) \end{Bmatrix} = \begin{Bmatrix} V_0 \\ 0 \end{Bmatrix}$$

Integrating this result subject to Eq. (7.3.29), the complete solution is

$$\begin{Bmatrix} x_f \\ z_f \\ \theta \end{Bmatrix} = \begin{Bmatrix} V_0 t \\ 0 \\ \omega_0 t \end{Bmatrix}$$

EXAMPLE 7.3.2. The rocket in Example 7.1.1 has a bank angle of 40 degrees, an elevation angle of 20 degrees, and an azimuth angle of 70 degrees. Assuming no wind, what is its velocity in Earth-fixed coordinates?

Solution. From Example 7.1.1,

$$\begin{Bmatrix} V_{x_b} \\ V_{y_b} \\ V_{z_b} \end{Bmatrix} \equiv \begin{Bmatrix} u \\ v \\ w \end{Bmatrix} = \begin{Bmatrix} 825.96 \text{ ft/sec} \\ 300.63 \text{ ft/sec} \\ 476.87 \text{ ft/sec} \end{Bmatrix}$$

From Eq. (7.3.19),

$$\begin{Bmatrix} V_{x_f} \\ V_{y_f} \\ V_{z_f} \end{Bmatrix} = \begin{bmatrix} C_\theta C_\psi & S_\phi S_\theta C_\psi - C_\phi S_\psi & C_\phi S_\theta C_\psi + S_\phi S_\psi \\ C_\theta S_\psi & S_\phi S_\theta S_\psi + C_\phi C_\psi & C_\phi S_\theta S_\psi - S_\phi C_\psi \\ -S_\theta & S_\phi C_\theta & C_\phi C_\theta \end{bmatrix} \begin{Bmatrix} u \\ v \\ w \end{Bmatrix}$$

For this example,

$$\begin{aligned}
C_\phi &= \cos 40° = 0.766044, & S_\phi &= \sin 40° = 0.642788, \\
C_\theta &= \cos 20° = 0.939693, & S_\theta &= \sin 20° = 0.342020, \\
C_\psi &= \cos 70° = 0.342020, & S_\psi &= \sin 70° = 0.939693
\end{aligned}$$

Substituting, we have

$$\begin{Bmatrix} V_{x_f} \\ V_{y_f} \\ V_{z_f} \end{Bmatrix} = \begin{bmatrix} 0.321394 & -0.644654 & 0.693633 \\ 0.883022 & 0.468591 & 0.026356 \\ -0.342020 & 0.604023 & 0.719846 \end{bmatrix} \begin{Bmatrix} 825.96 \text{ ft/sec} \\ 300.63 \text{ ft/sec} \\ 476.87 \text{ ft/sec} \end{Bmatrix}$$

and the matrix multiplication gives

$$\begin{Bmatrix} V_{x_f} \\ V_{y_f} \\ V_{z_f} \end{Bmatrix} = \begin{Bmatrix} 402.43 \text{ ft/sec} \\ 882.78 \text{ ft/sec} \\ 242.37 \text{ ft/sec} \end{Bmatrix}$$

As a cross-check on our math we can compute the magnitude of this vector. It should be the same as the magnitude in body-fixed coordinates, 1,000 ft/sec.

$$V = \sqrt{V_{x_f}^2 + V_{y_f}^2 + V_{z_f}^2} = \sqrt{402.43^2 + 882.78^2 + 242.37^2} \text{ ft/sec} = 1,000.00 \text{ ft/sec}$$

EXAMPLE 7.3.3. The flight of a stable projectile, such as an arrow or a stable bomb, can be approximated as pure longitudinal motion. If the projectile is stable enough, the angle of attack remains small and we can approximate the longitudinal aerodynamic forces and moment as

$$X \cong -\tfrac{1}{2}\rho u^2 S_r \left(k_0 + k_1 \alpha^2 \right) \tag{7.3.30}$$

$$Z \cong -\tfrac{1}{2}\rho u^2 S_r k_2 \alpha \tag{7.3.31}$$

$$m \cong -\tfrac{1}{2}\rho u^2 S_r l_r \left(k_3 \alpha + k_4 l_r \, q/u \right) \tag{7.3.32}$$

where X is the x_b-component of the aerodynamic force, Z is the z_b-component of the aerodynamic force, m is the aerodynamic pitching moment, k_0, k_1, k_2, k_3, and k_4 are dimensionless constants, S_r is a reference area, l_r is a reference length, and α is the angle of attack, which can be approximated as

$$\alpha \cong \frac{w}{u} \tag{7.3.33}$$

Using these approximations and assuming no wind, we wish to determine the position and orientation of the projectile as a function of time. Formulate the problem in a manner suitable for solution by fourth-order Runge-Kutta. Assume that at time equal zero, the position and orientation are (x_0, z_0, θ_0), the magnitude of the velocity is V_0, while the angle of attack and angular velocity are both zero.

Solution. Fourth-order Runge-Kutta requires a system of first-order differential equations and initial conditions in the form

$$\dot{y}_i = f_i(y_1, y_2, y_3, ..., y_n, t) \quad , \quad i = 1, 2, 3, ..., n$$
$$y_i(0) = Y_i$$

Using Eq. (7.3.33) in Eqs. (7.3.30) through (7.3.32), we can write

$$\begin{Bmatrix} X \\ Z \\ m \end{Bmatrix} = -\frac{1}{2} \rho S_r \begin{Bmatrix} k_0 u^2 + k_1 w^2 \\ k_2 u w \\ k_3 l_r u w + k_4 l_r^2 q u \end{Bmatrix}$$

The total force is the aerodynamic force plus the gravitational force. Since this is longitudinal motion, we can apply this result along with Eq. (7.3.23) to Eq. (7.2.23),

$$\begin{bmatrix} W/g & 0 & 0 \\ 0 & W/g & 0 \\ 0 & 0 & I_{yy_b} \end{bmatrix} \begin{Bmatrix} \dot{u} \\ \dot{w} \\ \dot{q} \end{Bmatrix}$$

$$= -\frac{1}{2} \rho S_r \begin{Bmatrix} k_0 u^2 + k_1 w^2 \\ k_2 u w \\ k_3 l_r u w + k_4 l_r^2 q u \end{Bmatrix} + \begin{Bmatrix} -W \sin(\theta) \\ W \cos(\theta) \\ 0 \end{Bmatrix} + \begin{Bmatrix} -q w W/g \\ q u W/g \\ 0 \end{Bmatrix}$$

Solving for the time derivatives, we have

$$\begin{Bmatrix} \dot{u} \\ \dot{w} \\ \dot{q} \end{Bmatrix} = \begin{Bmatrix} -K_0 u^2 - K_1 w^2 - g \sin(\theta) - q w \\ -K_2 u w + g \cos(\theta) + q u \\ -K_3 u w - K_4 q u \end{Bmatrix} \tag{7.3.34}$$

where

$$
\begin{Bmatrix} K_0 \\ K_1 \\ K_2 \\ K_3 \\ K_4 \end{Bmatrix} \equiv \begin{Bmatrix} \frac{1}{2}\rho S_r k_0 g/W \\ \frac{1}{2}\rho S_r k_1 g/W \\ \frac{1}{2}\rho S_r k_2 g/W \\ \frac{1}{2}\rho S_r l_r k_3 / I_{yy_b} \\ \frac{1}{2}\rho S_r l_r^2 k_4 / I_{yy_b} \end{Bmatrix}
$$

Combining Eq. (7.3.24) with Eq. (7.3.34), we obtain a complete system of first-order differential equations in a form suitable for solution by fourth-order Runge-Kutta,

$$
\begin{Bmatrix} \dot{u} \\ \dot{w} \\ \dot{q} \\ \dot{x}_f \\ \dot{z}_f \\ \dot{\theta} \end{Bmatrix} = \begin{Bmatrix} -K_0 u^2 - K_1 w^2 - g\sin\theta - qw \\ -K_2 uw + g\cos\theta + qu \\ -K_3 uw - K_4 qu \\ u\cos\theta + w\sin\theta \\ w\cos\theta - u\sin\theta \\ q \end{Bmatrix} \tag{7.3.35}
$$

The appropriate initial conditions are

$$
\begin{Bmatrix} u(0) \\ w(0) \\ q(0) \\ x_f(0) \\ z_f(0) \\ \theta(0) \end{Bmatrix} = \begin{Bmatrix} V_0 \\ 0 \\ 0 \\ x_0 \\ z_0 \\ \theta_0 \end{Bmatrix} \tag{7.3.36}
$$

Starting with the initial values given in Eq. (7.3.36), Eq. (7.3.35) can be integrated forward in time using fourth-order Runge-Kutta. After the first four steps the computation time can be decreased with no loss of accuracy by switching to the fourth-order Adams-Moulton method. The details of the fourth-order Runge-Kutta and the fourth-order Adams-Moulton numerical methods can be found in any undergraduate text on numerical methods (see, for example, Hoffman 1992). The details of obtaining this solution are left as an exercise for the student.

EXAMPLE 7.3.4. An arrow is shot from a bow with a velocity of 210 ft/sec and an elevation angle of 5.0 degrees. Using the formulation developed in Example 7.3.3, we wish to obtain the position, orientation, and airspeed of the arrow 100 yards downrange. For this arrow the aerodynamic coefficients that were defined following Eq. (7.3.34) in Example 7.3.3 are $K_0 = 0.00061$ ft^{-1}, $K_1 = 0.14$ ft^{-1},

$K_2 = 0.00059$ ft^{-1}, $K_3 = 0.0016$ ft^{-2}, and $K_4 = 0.0064$ ft^{-1}. Also plot the elevation of the arrow in inches above the starting point, its airspeed in feet per second, its elevation angle in degrees, and its angle of attack in degrees as a function of horizontal distance downrange in yards, from zero to 100 yards.

Solution. The details of obtaining a Runge-Kutta solution are straightforward and left as an exercise for the student. However, the formulation given by Eq. (7.3.35) is written with time as the independent variable and we wish to obtain a solution for a specified downrange distance of 100 yards. We have no way of knowing, a priori, the length of time required for the arrow to travel 100 yards. Since x_f is a monotonically increasing function of time, rather than using the formulation with time as the independent variable, which requires stepping forward in time, it would be convenient to have x_f as the independent variable, so that we could step forward in range. To facilitate this change of variables, we recognize that

$$\frac{d\mathscr{F}}{dx_f} = \frac{d\mathscr{F}}{dt} \bigg/ \frac{dx_f}{dt}$$

where $\mathscr{F}$ could be any one of the dependent variables. The change in x_f with respect to time is given by the fourth equation in Eq. (7.3.35). Thus we can write

$$\frac{d\mathscr{F}}{dx_f} = \frac{d\mathscr{F}}{dt} \bigg/ (u\cos\theta + w\sin\theta) \tag{7.3.37}$$

Using this result in Eq. (7.3.35) gives

$$\begin{Bmatrix} u' \\ w' \\ q' \\ x'_f \\ z'_f \\ \theta' \end{Bmatrix} = \left(\frac{1}{u\cos\theta + w\sin\theta}\right) \begin{Bmatrix} -K_0\, u^2 - K_1\, w^2 - g\sin\theta - q\,w \\ -K_2\, uw + g\cos\theta + q\,u \\ -K_3\, uw - K_4\, q\,u \\ u\cos\theta + w\sin\theta \\ w\cos\theta - u\sin\theta \\ q \end{Bmatrix}$$

where the prime indicates differentiation with respect to x_f. With this change of variables, the fourth equation becomes trivial, and if we do not care about time, we can reduce the order of the system from six to five. However, since computation time is of no concern with this very simple system, we could as well compute time as a dependent variable. Using Eq. (7.3.37) for $\mathscr{F} = t$, we have

$$\frac{dt}{dx_f} = \frac{dt}{dt} \bigg/ \frac{dx_f}{dt} = \frac{1}{u\cos\theta + w\sin\theta}$$

Replacing the fourth equation with this result, we obtain the complete system of first-order differential equations with x_f as the independent variable,

$$
\begin{Bmatrix} u' \\ w' \\ q' \\ t' \\ z'_f \\ \theta' \end{Bmatrix} = \left(\frac{1}{u\cos\theta + w\sin\theta} \right) \begin{Bmatrix} -K_0 u^2 - K_1 w^2 - g\sin\theta - qw \\ -K_2 uw + g\cos\theta + qu \\ -K_3 uw - K_4 qu \\ 1 \\ w\cos\theta - u\sin\theta \\ q \end{Bmatrix} \qquad (7.3.38)
$$

The appropriate initial conditions are

$$
\begin{Bmatrix} u(0) \\ w(0) \\ q(0) \\ t(0) \\ z_f(0) \\ \theta(0) \end{Bmatrix} = \begin{Bmatrix} V_0 \\ 0 \\ 0 \\ 0 \\ 0 \\ \theta_0 \end{Bmatrix} \qquad (7.3.39)
$$

We can now use fourth-order Runge-Kutta to step the solution in horizontal position rather than time. Using a specified change in horizontal position, say δx_f, we can integrate numerically from zero to some specified downrange distance, say x_{max}, as demonstrated in the following algorithm:

$$g = 32.2$$
$$u = V_0$$
$$w = 0.0$$
$$q = 0.0$$
$$t = 0.0$$
$$z_f = 0.0$$
$$\theta = \theta_0$$
$$\alpha = \tan^{-1}(w/u)$$
$$V = \sqrt{u^2 + w^2}$$
print t, $x_f/3.$, $-12.z_f$, V, 57.2957795θ, 57.2957795α
do $x_f = 0.0$, x_{max}, δx_f
$\qquad$ call Runge-Kutta for range step δx_f to find u, w, q, t, z_f, θ
$$\qquad \alpha = \tan^{-1}(w/u)$$
$$\qquad V = \sqrt{u^2 + w^2}$$
$\qquad$ print t, $x_f/3.$, $-12.z_f$, V, 57.2957795θ, 57.2957795α
end do

Using this algorithm with $\delta x_f = 3$ ft and $x_{max} = 300$ ft, we obtain the following numerical results, which are plotted in Figs. 7.3.5 and 7.3.6. For comparison, Fig. 7.3.5 also shows the elevation and airspeed of the projectile for the classical solution corresponding to the case of no drag.

Time (sec)	Range (yd)	Elevation (in.)	Speed (ft/sec)	θ (deg)	α (deg)
0.000000	0.000	0.000	210.000	5.000000	0.000000
0.014353	1.000	3.110	209.575	4.999698	0.125461
0.028733	2.000	6.140	209.150	4.997596	0.249642
⋮	⋮	⋮	⋮	⋮	⋮
1.543669	98.000	−125.007	176.654	−10.014533	−0.126531
1.560924	99.000	−131.340	176.422	−10.156230	−0.090636
1.578210	100.000	−137.788	176.192	−10.294110	−0.050468

Example 7.3.4 demonstrates how to compute the free trajectory of an aerodynamically stable object moving through a fluid. The object, in this case an arrow, is given some initial velocity in some initial direction. In free trajectory problems, the only forces acting on the moving object are the gravitational force and the aerodynamic force. The force of gravity tends to give the object a constant acceleration in the positive z_f-direction. If there were no aerodynamic force, the trajectory would be a parabola, as is shown in Fig. 7.3.5. Since the aerodynamic force on a streamlined body such as an arrow is quite small compared to the weight, the trajectory does not deviate

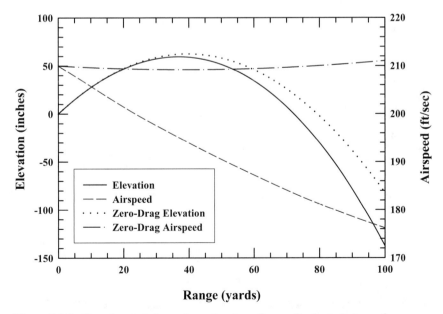

Figure 7.3.5. Elevation and airspeed as a function of range for the trajectory of an arrow.

dramatically from the zero-drag parabola, but the deviation is significant. Most readers will probably find nothing surprising in the results that are shown in Fig. 7.3.5.

This problem would be very difficult to solve using only Earth-fixed coordinates. Because the object is aerodynamically stable, it rotates as it moves, weathervaning into the relative wind as the force of gravity curves its trajectory. This change in orientation causes the inertia tensor to change with time in Earth-fixed coordinates. In addition, the aerodynamic force and moment components were much easier to describe in body-fixed coordinates than they would be in Earth-fixed coordinates. The change in orientation of the arrow as it moves over the ground is shown in Fig. 7.3.6.

This figure also shows something that the reader may not have expected. The angle of attack oscillates as the arrow moves through the air. This is our first look at a phenomenon that can affect any object in stable flight. The reader may ask, why should the angle of attack oscillate? The answer is linked directly to the same aerodynamic forces and moments that make the arrow stable in flight. When the angle of attack deviates from zero, an aerodynamic lift force is generated on the vanes of the arrow. Since the vanes are well aft of the center of gravity, the center of pressure is aft of the center of gravity, and this lift produces a moment that tends to restore the angle of attack toward zero. This restoring moment acts like an *aerodynamic torsional spring*, since the restoring moment increases with angle of attack. Whenever a torsional spring force is applied to an object with a finite mass moment of inertia, there is always the possibility of oscillating angular motion. Similar oscillating motion can occur in any aircraft and is discussed in greater detail in later chapters.

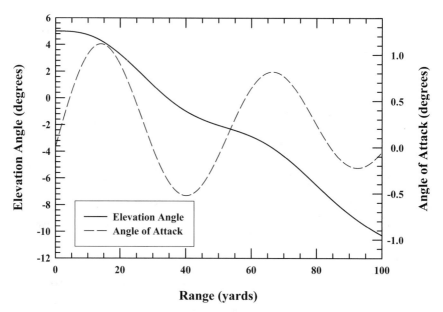

Figure 7.3.6. Elevation angle and angle of attack as a function of range for the trajectory of an arrow.

7.4. Rigid-Body 6-DOF Equations of Motion

Starting at a time when we know the airplane's position, orientation, translational velocity, and rotational velocity, we can determine the airspeed and the aerodynamic angles from the translational velocity components,

$$V = \sqrt{u^2 + v^2 + w^2} \qquad (7.4.1)$$

$$\alpha = \tan^{-1}\left(\frac{w}{u}\right) \qquad (7.4.2)$$

$$\beta_a = \tan^{-1}\left(\frac{v}{u}\right) \text{ and/or } \beta_e = \sin^{-1}\left(v\big/\sqrt{u^2 + v^2 + w^2}\right) \qquad (7.4.3)$$

From the known airspeed, aerodynamic angles, and angular velocity components, we can compute the aerodynamic forces and moments, including thrust. We can then compute the time derivatives from Newton's second law and the kinematic transformation equations. For a spanwise symmetric aircraft, the **Euler angle formulation** is

$$\begin{Bmatrix} \dot{u} \\ \dot{v} \\ \dot{w} \end{Bmatrix} = \frac{g}{W}\begin{Bmatrix} F_{x_b} \\ F_{y_b} \\ F_{z_b} \end{Bmatrix} + g\begin{Bmatrix} -S_\theta \\ S_\phi C_\theta \\ C_\phi C_\theta \end{Bmatrix} + \begin{Bmatrix} rv - qw \\ pw - ru \\ qu - pv \end{Bmatrix} \qquad (7.4.4)$$

$$\begin{Bmatrix} \dot{p} \\ \dot{q} \\ \dot{r} \end{Bmatrix} = \begin{bmatrix} I_{xx_b} & 0 & -I_{xz_b} \\ 0 & I_{yy_b} & 0 \\ -I_{zx_b} & 0 & I_{zz_b} \end{bmatrix}^{-1} \begin{Bmatrix} M_{x_b} + (I_{yy_b} - I_{zz_b})qr + I_{xz_b}pq \\ M_{y_b} + (I_{zz_b} - I_{xx_b})pr + I_{xz_b}(r^2 - p^2) \\ M_{z_b} + (I_{xx_b} - I_{yy_b})pq - I_{xz_b}qr \end{Bmatrix} \quad (7.4.5)$$

$$\begin{Bmatrix} \dot{x}_f \\ \dot{y}_f \\ \dot{z}_f \end{Bmatrix} = \begin{bmatrix} C_\theta C_\psi & S_\phi S_\theta C_\psi - C_\phi S_\psi & C_\phi S_\theta C_\psi + S_\phi S_\psi \\ C_\theta S_\psi & S_\phi S_\theta S_\psi + C_\phi C_\psi & C_\phi S_\theta S_\psi - S_\phi C_\psi \\ -S_\theta & S_\phi C_\theta & C_\phi C_\theta \end{bmatrix} \begin{Bmatrix} u \\ v \\ w \end{Bmatrix} + \begin{Bmatrix} V_{wx_f} \\ V_{wy_f} \\ V_{wz_f} \end{Bmatrix} \qquad (7.4.6)$$

$$\begin{Bmatrix} \dot{\phi} \\ \dot{\theta} \\ \dot{\psi} \end{Bmatrix} = \begin{bmatrix} 1 & S_\phi S_\theta / C_\theta & C_\phi S_\theta / C_\theta \\ 0 & C_\phi & -S_\phi \\ 0 & S_\phi / C_\theta & C_\phi / C_\theta \end{bmatrix} \begin{Bmatrix} p \\ q \\ r \end{Bmatrix} \qquad (7.4.7)$$

where F_{x_b}, F_{y_b}, F_{z_b}, M_{x_b}, M_{y_b}, and M_{z_b} are the body-fixed components of the aerodynamic force and moment vectors, including thrust, and V_{wx_f}, V_{wy_f}, and V_{wz_f} are the Earth-fixed components of the constant wind vector. This system of 12 equations can be integrated numerically to yield the position vector, Euler angles, translational velocity, and rotational velocity of the aircraft as a function of time.

7.5. Linearized Equations of Motion

The rigid-body 6-DOF equations of motion, summarized in Sec. 7.4, are nonlinear. Solutions can only be obtained numerically, for all but the most trivial conditions. However, the dynamics of aircraft motion depend on many aircraft design and operating parameters, and the nature of this dependence is not easily observable from a numerical solution. For this reason, closed-form solutions that accurately describe the essential features of aircraft dynamics are desirable. In addition, closed-form solutions are useful for the optimization of aircraft control systems. In most cases, closed-form solutions can be obtained only for linear systems of equations.

In this section we simplify the rigid-body equations of motion by applying a very powerful analytical tool, which is widely used in almost every field of engineering and science. Since this analytical technique, known as *small-disturbance theory*, has broad application outside the field of flight mechanics, the student should strive to understand the method at a fundamental level. To help facilitate this understanding, we shall begin by examining how small-disturbance theory is applied to a simple one-dimensional problem.

For example, consider some function of time, $x(t)$, that is governed by a nonlinear second-order differential equation of the form

$$\ddot{x} = f(\dot{x}, x, t)$$

Now suppose that there is a known particular solution to this problem, $x_o(t)$. Expanding the differential equation in a Taylor series about this particular solution gives

$$\ddot{x} = f(\dot{x}_o, x_o, t) + \frac{\partial f}{\partial \dot{x}}\bigg|_{\substack{\dot{x}=\dot{x}_o \\ x=x_o}} (\dot{x} - \dot{x}_o) + \frac{\partial f}{\partial x}\bigg|_{\substack{\dot{x}=\dot{x}_o \\ x=x_o}} (x - x_o) + \cdots$$

If we assume that the general solution does not deviate greatly from the particular solution, then the higher-order terms in the Taylor series can be ignored. Since the particular solution satisfies $\ddot{x}_o = f(\dot{x}_o, x_o, t)$, any small disturbances from the particular solution must satisfy the differential equation

$$\Delta \ddot{x} = \frac{\partial f}{\partial \dot{x}}\bigg|_{\substack{\dot{x}=\dot{x}_o \\ x=x_o}} \Delta \dot{x} + \frac{\partial f}{\partial x}\bigg|_{\substack{\dot{x}=\dot{x}_o \\ x=x_o}} \Delta x$$

where Δx is the deviation from the particular solution, $\Delta x = x - x_o$. Since the derivatives of f with respect to $\dot{x}$ and x in this approximate differential equation are evaluated for the known particular solution, the approximate equation is linear in Δx.

EXAMPLE 7.5.1. Consider a tractor pulling a trailer of mass m. The trailer is attached to the tractor with an elastic tongue having a spring constant k. The tractor is moving along a straight and level roadway with constant velocity V_o. Assuming that the total resistance force on the trailer is proportional to velocity squared, obtain a linearized differential equation for the position of the trailer.

Solution. Newton's second law governs the motion of the trailer. This requires that the mass of the trailer multiplied by its acceleration must equal the net force acting on the trailer. There are two forces acting along the axis of motion, the force applied to the tongue by the tractor and the resistance force between the trailer and its surroundings. Because of the elastic tongue, the force between the tractor and trailer varies linearly with the distance between them. Since the tractor is moving with a constant velocity, its position is changing linearly with time. Letting x denote the position of the trailer relative to some convenient origin, you should be able to draw a free-body diagram of the trailer and show that the governing differential equation is

$$m\ddot{x} = k(V_o t - x) - C\dot{x}^2$$

where C is the proportionality constant between the resistance force acting on the trailer and its velocity squared.

Physics dictates that there should be a steady solution, $x_o(t)$, for which the trailer velocity is constant and equal to the tractor velocity. From the governing differential equation this requires that

$$0 = k(V_o t - x_o) - CV_o^2$$

or

$$x_o = V_o t - \frac{CV_o^2}{k}$$

which requires that

$$\dot{x}_o = V_o$$

We now define the position disturbance to be the difference between the general solution, $x(t)$, and this static equilibrium solution, $x_o(t)$,

$$\Delta x \equiv x - x_o = x - V_o t + \frac{CV_o^2}{k}$$

Since the trailer is attached to the tractor by the elastic tongue, its position cannot deviate far from the equilibrium position. Thus, assuming the disturbance to be small, the governing differential equation is expanded in a Taylor series with all but the first-order terms neglected. This gives

$$m\Delta\ddot{x} = \frac{\partial}{\partial\dot{x}}\left[k(V_o t - x) - C\dot{x}^2\right]\Bigg|_{\substack{\dot{x}=\dot{x}_o \\ x=x_o}} \Delta\dot{x} + \frac{\partial}{\partial x}\left[k(V_o t - x) - C\dot{x}^2\right]\Bigg|_{\substack{\dot{x}=\dot{x}_o \\ x=x_o}} \Delta x$$

or

$$m\Delta\ddot{x} = -2CV_o\Delta\dot{x} - k\Delta x$$

This is the linearized second-order differential equation that governs the small disturbance motion of the trailer relative to its equilibrium position behind the moving tractor.

Small-disturbance theory can also be applied to systems of differential equations. The rigid-body 6-DOF equations of motion, written in terms of Euler angles, consist of 12 coupled nonlinear differential equations in 12 unknowns. Using small-disturbance theory, these equations can be linearized. The small-disturbance approximation requires that the motion of the aircraft be confined to small deviations about some steady equilibrium flight condition. Under normal operating conditions, the forces and moments acting on an aircraft change slowly and do not deviate far from equilibrium. Thus, small-disturbance theory can be applied to aircraft motion for many commonly encountered flight conditions.

In the application of small-disturbance theory to aircraft dynamics, we write each variable in the equations of motion as its value at the equilibrium reference state plus a small disturbance from that equilibrium state,

$$
\begin{array}{lll}
u = u_o + \Delta u & v = v_o + \Delta v & w = w_o + \Delta w \\
p = p_o + \Delta p & q = q_o + \Delta q & r = r_o + \Delta r \\
x_f = x_o + \Delta x_f & y_f = y_o + \Delta y_f & z_f = z_o + \Delta z_f \\
\phi = \phi_o + \Delta\phi & \theta = \theta_o + \Delta\theta & \psi = \psi_o + \Delta\psi \\
F_{x_b} = F_{x_bo} + \Delta F_{x_b} & F_{y_b} = F_{y_bo} + \Delta F_{y_b} & F_{z_b} = F_{z_bo} + \Delta F_{z_b} \quad (7.5.1) \\
W_{x_b} = W_{x_bo} + \Delta W_{x_b} & W_{y_b} = W_{y_bo} + \Delta W_{y_b} & W_{z_b} = W_{z_bo} + \Delta W_{z_b} \\
M_{x_b} = M_{x_bo} + \Delta M_{x_b} & M_{y_b} = M_{y_bo} + \Delta M_{y_b} & M_{z_b} = M_{z_bo} + \Delta M_{z_b} \\
\delta_a = \delta_{ao} + \Delta\delta_a & \delta_e = \delta_{eo} + \Delta\delta_e & \delta_r = \delta_{ro} + \Delta\delta_r
\end{array}
$$

The first 12 variables in Eq. (7.5.1) are commonly known as the *state variables*, because they define the instantaneous state of the aircraft in terms of velocity, position, and orientation. Since there are only 12 scalar components in Eqs. (7.4.4) through (7.4.7), the remaining 12 variables must somehow be known as functions of the state variables and pilot control inputs.

Here we consider the reference condition to be steady flight, at some equilibrium airspeed, V_o, with no sideslip, no bank angle, and no angular velocity. Since the alignment of the axes is arbitrary, we shall align the x_b-axis with the aircraft's equilibrium direction of flight and place the x_f-axis so that the equilibrium azimuth angle is zero. Furthermore, since the summation of forces is zero in steady flight, we have

$$u_o = V_o, \quad x_o = (V_{wx_f} + V_o\cos\theta_o)t, \quad y_o = V_{wy_f}t, \quad z_o = (V_{wz_f} - V_o\sin\theta_o)t,$$

$$F_{x_bo} = -W_{x_bo}, \quad F_{z_bo} = -W_{z_bo}, \quad v_o = w_o = p_o = q_o = r_o = \phi_o = \psi_o = F_{y_bo} = W_{y_bo} = \mathbf{M}_o = 0$$

$$(7.5.2)$$

By substituting the results from Eq. (7.5.2) for the equilibrium values in Eq. (7.5.1), the system variables are expanded in terms of the disturbances from this equilibrium flight condition. This gives

$$
\begin{aligned}
u &= V_o + \Delta u & v &= \Delta v & w &= \Delta w \\
p &= \Delta p & q &= \Delta q & r &= \Delta r \\
x_f &= (V_{wx_f} + V_o \cos\theta_o)t + \Delta x_f & y_f &= V_{wy_f}t + \Delta y_f & z_f &= (V_{wz_f} - V_o \sin\theta_o)t + \Delta z_f \\
\phi &= \Delta\phi & \theta &= \theta_o + \Delta\theta & \psi &= \Delta\psi \qquad (7.5.3) \\
F_{x_b} &= -W_{x_b o} + \Delta F_{x_b} & F_{y_b} &= \Delta F_{y_b} & F_{z_b} &= -W_{z_b o} + \Delta F_{z_b} \\
W_{x_b} &= W_{x_b o} + \Delta W_{x_b} & W_{y_b} &= \Delta W_{y_b} & W_{z_b} &= W_{z_b o} + \Delta W_{z_b} \\
M_{x_b} &= \Delta M_{x_b} & M_{y_b} &= \Delta M_{y_b} & M_{z_b} &= \Delta M_{z_b} \\
\delta_a &= \delta_{ao} + \Delta\delta_a & \delta_r &= \delta_{ro} + \Delta\delta_r & \delta_e &= \delta_{eo} + \Delta\delta_e
\end{aligned}
$$

Newton's second law of motion can be written in terms of the disturbance variables by using Eq. (7.5.3) in Eq. (7.2.20). Since the disturbances are assumed small, we can neglect the product of disturbances, and Newton's second law can be approximated as

$$
\begin{bmatrix}
W/g & 0 & 0 & 0 & 0 & 0 \\
0 & W/g & 0 & 0 & 0 & 0 \\
0 & 0 & W/g & 0 & 0 & 0 \\
0 & 0 & 0 & I_{xx_b} & 0 & -I_{xz_b} \\
0 & 0 & 0 & 0 & I_{yy_b} & 0 \\
0 & 0 & 0 & -I_{xz_b} & 0 & I_{zz_b}
\end{bmatrix}
\begin{Bmatrix}
\Delta\dot{u} \\ \Delta\dot{v} \\ \Delta\dot{w} \\ \Delta\dot{p} \\ \Delta\dot{q} \\ \Delta\dot{r}
\end{Bmatrix}
=
\begin{Bmatrix}
\Delta F_{x_b} + \Delta W_{x_b} \\
\Delta F_{y_b} + \Delta W_{y_b} - \Delta r V_o W/g \\
\Delta F_{z_b} + \Delta W_{z_b} + \Delta q V_o W/g \\
\Delta M_{x_b} \\
\Delta M_{y_b} \\
\Delta M_{z_b}
\end{Bmatrix}
\quad (7.5.4)
$$

The aerodynamic forces and moments are functions of the translational and rotational velocities, the translational acceleration, and the deflection of the control surfaces. Thus, for small-disturbance theory, the aerodynamic force is approximated as

$$
\begin{Bmatrix}
\Delta F_{x_b} \\ \Delta F_{y_b} \\ \Delta F_{z_b}
\end{Bmatrix}
=
\begin{bmatrix}
\dfrac{\partial F_{x_b}}{\partial u} & \dfrac{\partial F_{x_b}}{\partial v} & \dfrac{\partial F_{x_b}}{\partial w} \\[8pt]
\dfrac{\partial F_{y_b}}{\partial u} & \dfrac{\partial F_{y_b}}{\partial v} & \dfrac{\partial F_{y_b}}{\partial w} \\[8pt]
\dfrac{\partial F_{z_b}}{\partial u} & \dfrac{\partial F_{z_b}}{\partial v} & \dfrac{\partial F_{z_b}}{\partial w}
\end{bmatrix}
\begin{Bmatrix}
\Delta u \\ \Delta v \\ \Delta w
\end{Bmatrix}
+
\begin{bmatrix}
\dfrac{\partial F_{x_b}}{\partial p} & \dfrac{\partial F_{x_b}}{\partial q} & \dfrac{\partial F_{x_b}}{\partial r} \\[8pt]
\dfrac{\partial F_{y_b}}{\partial p} & \dfrac{\partial F_{y_b}}{\partial q} & \dfrac{\partial F_{y_b}}{\partial r} \\[8pt]
\dfrac{\partial F_{z_b}}{\partial p} & \dfrac{\partial F_{z_b}}{\partial q} & \dfrac{\partial F_{z_b}}{\partial r}
\end{bmatrix}
\begin{Bmatrix}
\Delta p \\ \Delta q \\ \Delta r
\end{Bmatrix}
$$

$$
+
\begin{bmatrix}
\dfrac{\partial F_{x_b}}{\partial \dot{u}} & \dfrac{\partial F_{x_b}}{\partial \dot{v}} & \dfrac{\partial F_{x_b}}{\partial \dot{w}} \\[8pt]
\dfrac{\partial F_{y_b}}{\partial \dot{u}} & \dfrac{\partial F_{y_b}}{\partial \dot{v}} & \dfrac{\partial F_{y_b}}{\partial \dot{w}} \\[8pt]
\dfrac{\partial F_{z_b}}{\partial \dot{u}} & \dfrac{\partial F_{z_b}}{\partial \dot{v}} & \dfrac{\partial F_{z_b}}{\partial \dot{w}}
\end{bmatrix}
\begin{Bmatrix}
\Delta\dot{u} \\ \Delta\dot{v} \\ \Delta\dot{w}
\end{Bmatrix}
+
\begin{bmatrix}
\dfrac{\partial F_{x_b}}{\partial \delta_a} & \dfrac{\partial F_{x_b}}{\partial \delta_e} & \dfrac{\partial F_{x_b}}{\partial \delta_r} \\[8pt]
\dfrac{\partial F_{y_b}}{\partial \delta_a} & \dfrac{\partial F_{y_b}}{\partial \delta_e} & \dfrac{\partial F_{y_b}}{\partial \delta_r} \\[8pt]
\dfrac{\partial F_{z_b}}{\partial \delta_a} & \dfrac{\partial F_{z_b}}{\partial \delta_e} & \dfrac{\partial F_{z_b}}{\partial \delta_r}
\end{bmatrix}
\begin{Bmatrix}
\Delta\delta_a \\ \Delta\delta_e \\ \Delta\delta_r
\end{Bmatrix}
\quad (7.5.5)
$$

and for the aerodynamic moment we use the approximation

$$
\begin{Bmatrix} \Delta M_{x_b} \\ \Delta M_{y_b} \\ \Delta M_{z_b} \end{Bmatrix} =
\begin{bmatrix}
\dfrac{\partial M_{x_b}}{\partial u} & \dfrac{\partial M_{x_b}}{\partial v} & \dfrac{\partial M_{x_b}}{\partial w} \\[2ex]
\dfrac{\partial M_{y_b}}{\partial u} & \dfrac{\partial M_{y_b}}{\partial v} & \dfrac{\partial M_{y_b}}{\partial w} \\[2ex]
\dfrac{\partial M_{z_b}}{\partial u} & \dfrac{\partial M_{z_b}}{\partial v} & \dfrac{\partial M_{z_b}}{\partial w}
\end{bmatrix}
\begin{Bmatrix} \Delta u \\ \Delta v \\ \Delta w \end{Bmatrix} +
\begin{bmatrix}
\dfrac{\partial M_{x_b}}{\partial p} & \dfrac{\partial M_{x_b}}{\partial q} & \dfrac{\partial M_{x_b}}{\partial r} \\[2ex]
\dfrac{\partial M_{y_b}}{\partial p} & \dfrac{\partial M_{y_b}}{\partial q} & \dfrac{\partial M_{y_b}}{\partial r} \\[2ex]
\dfrac{\partial M_{z_b}}{\partial p} & \dfrac{\partial M_{z_b}}{\partial q} & \dfrac{\partial M_{z_b}}{\partial r}
\end{bmatrix}
\begin{Bmatrix} \Delta p \\ \Delta q \\ \Delta r \end{Bmatrix}
$$

$$
+
\begin{bmatrix}
\dfrac{\partial M_{x_b}}{\partial \dot{u}} & \dfrac{\partial M_{x_b}}{\partial \dot{v}} & \dfrac{\partial M_{x_b}}{\partial \dot{w}} \\[2ex]
\dfrac{\partial M_{y_b}}{\partial \dot{u}} & \dfrac{\partial M_{y_b}}{\partial \dot{v}} & \dfrac{\partial M_{y_b}}{\partial \dot{w}} \\[2ex]
\dfrac{\partial M_{z_b}}{\partial \dot{u}} & \dfrac{\partial M_{z_b}}{\partial \dot{v}} & \dfrac{\partial M_{z_b}}{\partial \dot{w}}
\end{bmatrix}
\begin{Bmatrix} \Delta \dot{u} \\ \Delta \dot{v} \\ \Delta \dot{w} \end{Bmatrix} +
\begin{bmatrix}
\dfrac{\partial M_{x_b}}{\partial \delta_a} & \dfrac{\partial M_{x_b}}{\partial \delta_e} & \dfrac{\partial M_{x_b}}{\partial \delta_r} \\[2ex]
\dfrac{\partial M_{y_b}}{\partial \delta_a} & \dfrac{\partial M_{y_b}}{\partial \delta_e} & \dfrac{\partial M_{y_b}}{\partial \delta_r} \\[2ex]
\dfrac{\partial M_{z_b}}{\partial \delta_a} & \dfrac{\partial M_{z_b}}{\partial \delta_e} & \dfrac{\partial M_{z_b}}{\partial \delta_r}
\end{bmatrix}
\begin{Bmatrix} \Delta \delta_a \\ \Delta \delta_e \\ \Delta \delta_r \end{Bmatrix}
\tag{7.5.6}
$$

where all derivatives are evaluated at the equilibrium flight condition.

The body-fixed aerodynamic force and moment components, which are expressed in Eqs. (7.5.5) and (7.5.6), are shown as independent of the Euler angles. The aerodynamic force and moment components in body-fixed coordinates do not depend on the aircraft orientation relative to the Earth, but only on the motion of the aircraft relative to the surrounding air. This is why atmosphere-fixed coordinates were chosen as the inertial coordinate system in the formulation of Newton's second law.

In Eqs. (7.5.5) and (7.5.6) we have assumed that the force and moment components depend on the translational acceleration components. The forces and moments depend on the translational accelerations because the vorticity generated by the wing requires a finite time to travel from the wing to the tail. As the angle of attack changes, the lift and vorticity generated by the wing also change. This change in vorticity alters the downwash at the tail and hence alters the forces and moments produced by the tail. However, since the vorticity travels from the wing to the tail at the speed of the aircraft, it takes a finite time for the altered downwash to reach the tail. Since sideslip has little effect on lift or vorticity, the sideslip acceleration terms are not significant. In addition, several of the remaining terms can be eliminated on the basis of symmetry. Thus, for the linearized equations of motion we shall use

$$
\frac{\partial F_{y_b}}{\partial \dot{u}} = \frac{\partial F_{x_b}}{\partial \dot{v}} = \frac{\partial F_{y_b}}{\partial \dot{v}} = \frac{\partial F_{z_b}}{\partial \dot{v}} = \frac{\partial F_{y_b}}{\partial \dot{w}} = 0
$$

$$
\frac{\partial M_{x_b}}{\partial \dot{u}} = \frac{\partial M_{z_b}}{\partial \dot{u}} = \frac{\partial M_{x_b}}{\partial \dot{v}} = \frac{\partial M_{y_b}}{\partial \dot{v}} = \frac{\partial M_{z_b}}{\partial \dot{v}} = \frac{\partial M_{x_b}}{\partial \dot{w}} = \frac{\partial M_{z_b}}{\partial \dot{w}} = 0
\tag{7.5.7}
$$

Several more of the derivatives in Eqs. (7.5.5) and (7.5.6) can be eliminated on the basis of symmetry. For example, at the equilibrium flight condition, the change in axial force with respect to the sideslip velocity must be zero, because the aircraft is assumed symmetric about the x_b-z_b plane and the equilibrium sideslip velocity is zero. Keep in mind that the derivatives in Eqs. (7.5.5) and (7.5.6) are all evaluated at the equilibrium flight condition. Clearly, the axial force is not totally independent of sideslip. However, from the assumed symmetry, a positive sideslip should have the same effect on the axial force as a negative sideslip. Thus, the axial force must be a symmetric function of sideslip, and the change in axial force with respect to sideslip velocity must be zero, when the sideslip velocity is zero. By similar reasoning we can eliminate all of the following derivatives evaluated at the reference flight condition:

$$
\frac{\partial F_{x_b}}{\partial v} = \frac{\partial F_{y_b}}{\partial u} = \frac{\partial F_{y_b}}{\partial w} = \frac{\partial F_{z_b}}{\partial v} = 0
$$

$$
\frac{\partial M_{x_b}}{\partial u} = \frac{\partial M_{x_b}}{\partial w} = \frac{\partial M_{y_b}}{\partial v} = \frac{\partial M_{z_b}}{\partial u} = \frac{\partial M_{z_b}}{\partial w} = 0
$$

$$
\frac{\partial F_{x_b}}{\partial p} = \frac{\partial F_{x_b}}{\partial r} = \frac{\partial F_{y_b}}{\partial q} = \frac{\partial F_{z_b}}{\partial p} = \frac{\partial F_{z_b}}{\partial r} = 0
$$

$$
\frac{\partial M_{x_b}}{\partial q} = \frac{\partial M_{y_b}}{\partial p} = \frac{\partial M_{y_b}}{\partial r} = \frac{\partial M_{z_b}}{\partial q} = 0 \tag{7.5.8}
$$

$$
\frac{\partial F_{x_b}}{\partial \delta_a} = \frac{\partial F_{x_b}}{\partial \delta_r} = \frac{\partial F_{y_b}}{\partial \delta_e} = \frac{\partial F_{z_b}}{\partial \delta_a} = \frac{\partial F_{z_b}}{\partial \delta_r} = 0
$$

$$
\frac{\partial M_{x_b}}{\partial \delta_e} = \frac{\partial M_{y_b}}{\partial \delta_a} = \frac{\partial M_{y_b}}{\partial \delta_r} = \frac{\partial M_{z_b}}{\partial \delta_e} = 0
$$

The body-fixed components of the gravitational force vector change with aircraft orientation. Since the body-fixed gravitational components only depend on the Euler angles, from Eq. (7.3.15) we have

$$
\begin{Bmatrix} W_{x_b} \\ W_{y_b} \\ W_{z_b} \end{Bmatrix} = W \begin{Bmatrix} -\sin(\theta) \\ \sin(\phi)\cos(\theta) \\ \cos(\phi)\cos(\theta) \end{Bmatrix} \tag{7.5.9}
$$

which at the equilibrium flight condition gives

$$
\begin{Bmatrix} \Delta W_{x_b} \\ \Delta W_{y_b} \\ \Delta W_{z_b} \end{Bmatrix} = \begin{bmatrix} \dfrac{\partial W_{x_b}}{\partial \phi} & \dfrac{\partial W_{x_b}}{\partial \theta} & \dfrac{\partial W_{x_b}}{\partial \psi} \\ \dfrac{\partial W_{y_b}}{\partial \phi} & \dfrac{\partial W_{y_b}}{\partial \theta} & \dfrac{\partial W_{y_b}}{\partial \psi} \\ \dfrac{\partial W_{z_b}}{\partial \phi} & \dfrac{\partial W_{z_b}}{\partial \theta} & \dfrac{\partial W_{z_b}}{\partial \psi} \end{bmatrix} \begin{Bmatrix} \Delta\phi \\ \Delta\theta \\ \Delta\psi \end{Bmatrix} = \begin{bmatrix} 0 & -W\cos\theta_0 & 0 \\ W\cos\theta_0 & 0 & 0 \\ 0 & -W\sin\theta_0 & 0 \end{bmatrix} \begin{Bmatrix} \Delta\phi \\ \Delta\theta \\ \Delta\psi \end{Bmatrix} \tag{7.5.10}
$$

By definition, the gravitational force does not produce a moment about the CG.

Using Eqs. (7.5.7) and (7.5.8) in Eqs. (7.5.5) and (7.5.6) and adding Eq. (7.5.10), the small-disturbance force approximations are given by

$$
\begin{Bmatrix} \Delta F_{x_b} + \Delta W_{x_b} \\ \Delta F_{y_b} + \Delta W_{y_b} \\ \Delta F_{z_b} + \Delta W_{z_b} \end{Bmatrix} = \begin{bmatrix} F_{x_b,u} & 0 & F_{x_b,w} \\ 0 & F_{y_b,v} & 0 \\ F_{z_b,u} & 0 & F_{z_b,w} \end{bmatrix} \begin{Bmatrix} \Delta u \\ \Delta v \\ \Delta w \end{Bmatrix} + \begin{bmatrix} 0 & F_{x_b,q} & 0 \\ F_{y_b,p} & 0 & F_{y_b,r} \\ 0 & F_{z_b,q} & 0 \end{bmatrix} \begin{Bmatrix} \Delta p \\ \Delta q \\ \Delta r \end{Bmatrix}
$$

$$
+ \begin{bmatrix} F_{x_b,\dot{u}} & 0 & F_{x_b,\dot{w}} \\ 0 & 0 & 0 \\ F_{z_b,\dot{u}} & 0 & F_{z_b,\dot{w}} \end{bmatrix} \begin{Bmatrix} \Delta \dot{u} \\ \Delta \dot{v} \\ \Delta \dot{w} \end{Bmatrix} + \begin{bmatrix} 0 & F_{x_b,\delta_e} & 0 \\ F_{y_b,\delta_a} & 0 & F_{y_b,\delta_r} \\ 0 & F_{z_b,\delta_e} & 0 \end{bmatrix} \begin{Bmatrix} \Delta \delta_a \\ \Delta \delta_e \\ \Delta \delta_r \end{Bmatrix} \quad (7.5.11)
$$

$$
+ \begin{bmatrix} 0 & -W\cos\theta_0 & 0 \\ W\cos\theta_0 & 0 & 0 \\ 0 & -W\sin\theta_0 & 0 \end{bmatrix} \begin{Bmatrix} \Delta\phi \\ \Delta\theta \\ \Delta\psi \end{Bmatrix}
$$

and the small-disturbance moment approximations become

$$
\begin{Bmatrix} \Delta M_{x_b} \\ \Delta M_{y_b} \\ \Delta M_{z_b} \end{Bmatrix} = \begin{bmatrix} 0 & M_{x_b,v} & 0 \\ M_{y_b,u} & 0 & M_{y_b,w} \\ 0 & M_{z_b,v} & 0 \end{bmatrix} \begin{Bmatrix} \Delta u \\ \Delta v \\ \Delta w \end{Bmatrix} + \begin{bmatrix} M_{x_b,p} & 0 & M_{x_b,r} \\ 0 & M_{y_b,q} & 0 \\ M_{z_b,p} & 0 & M_{z_b,r} \end{bmatrix} \begin{Bmatrix} \Delta p \\ \Delta q \\ \Delta r \end{Bmatrix}
$$

$$
+ \begin{bmatrix} 0 & 0 & 0 \\ M_{y_b,\dot{u}} & 0 & M_{y_b,\dot{w}} \\ 0 & 0 & 0 \end{bmatrix} \begin{Bmatrix} \Delta \dot{u} \\ \Delta \dot{v} \\ \Delta \dot{w} \end{Bmatrix} + \begin{bmatrix} M_{x_b,\delta_a} & 0 & M_{x_b,\delta_r} \\ 0 & M_{y_b,\delta_e} & 0 \\ M_{z_b,\delta_a} & 0 & M_{z_b,\delta_r} \end{bmatrix} \begin{Bmatrix} \Delta \delta_a \\ \Delta \delta_e \\ \Delta \delta_r \end{Bmatrix} \quad (7.5.12)
$$

Notice that within the accuracy of small-disturbance theory, ΔF_{x_b}, ΔF_{z_b}, ΔW_{x_b}, ΔW_{z_b}, and ΔM_{y_b} depend only on $\Delta\dot{u}$, $\Delta\dot{w}$, Δu, Δw, Δq, $\Delta\delta_e$, and $\Delta\theta$, while ΔF_{y_b}, ΔW_{y_b}, ΔM_{x_b}, and ΔM_{z_b} depend only on Δv, Δp, Δr, $\Delta\delta_a$, $\Delta\delta_r$, and $\Delta\phi$.

Rearranging Eqs. (7.5.11) and (7.5.12), we have

$$
\begin{Bmatrix} \Delta F_{x_b} + \Delta W_{x_b} \\ \Delta F_{z_b} + \Delta W_{z_b} \\ \Delta M_{y_b} \end{Bmatrix} = \begin{bmatrix} F_{x_b,\dot{u}} & F_{x_b,\dot{w}} & 0 \\ F_{z_b,\dot{u}} & F_{z_b,\dot{w}} & 0 \\ M_{y_b,\dot{u}} & M_{y_b,\dot{w}} & 0 \end{bmatrix} \begin{Bmatrix} \Delta \dot{u} \\ \Delta \dot{w} \\ \Delta \dot{q} \end{Bmatrix} + \begin{bmatrix} F_{x_b,u} & F_{x_b,w} & F_{x_b,q} \\ F_{z_b,u} & F_{z_b,w} & F_{z_b,q} \\ M_{y_b,u} & M_{y_b,w} & M_{y_b,q} \end{bmatrix} \begin{Bmatrix} \Delta u \\ \Delta w \\ \Delta q \end{Bmatrix}
$$

$$
+ \begin{Bmatrix} F_{x_b,\delta_e} \\ F_{z_b,\delta_e} \\ M_{y_b,\delta_e} \end{Bmatrix} \Delta\delta_e + \begin{Bmatrix} -W\cos(\theta_0) \\ -W\sin(\theta_0) \\ 0 \end{Bmatrix} \Delta\theta \quad (7.5.13)
$$

$$
\begin{Bmatrix} \Delta F_{y_b} + \Delta W_{y_b} \\ \Delta M_{x_b} \\ \Delta M_{z_b} \end{Bmatrix} = \begin{bmatrix} F_{y_b,v} & F_{y_b,p} & F_{y_b,r} \\ M_{x_b,v} & M_{x_b,p} & M_{x_b,r} \\ M_{z_b,v} & M_{z_b,p} & M_{z_b,r} \end{bmatrix} \begin{Bmatrix} \Delta v \\ \Delta p \\ \Delta r \end{Bmatrix} + \begin{bmatrix} F_{y_b,\delta_a} & F_{y_b,\delta_r} \\ M_{x_b,\delta_a} & M_{x_b,\delta_r} \\ M_{z_b,\delta_a} & M_{z_b,\delta_r} \end{bmatrix} \begin{Bmatrix} \Delta\delta_a \\ \Delta\delta_r \end{Bmatrix}
$$

$$
+ \begin{Bmatrix} W\cos(\theta_0) \\ 0 \\ 0 \end{Bmatrix} \Delta\phi \quad (7.5.14)
$$

Applying these results, the system of six equations in Eq. (7.5.4) may be separated into two uncoupled systems, each containing three equations,

$$
\begin{bmatrix} W/g & 0 & 0 \\ 0 & W/g & 0 \\ 0 & 0 & I_{yy_b} \end{bmatrix} \begin{Bmatrix} \Delta\dot{u} \\ \Delta\dot{w} \\ \Delta\dot{q} \end{Bmatrix} = \begin{Bmatrix} \Delta F_{x_b} + \Delta W_{x_b} \\ \Delta F_{z_b} + \Delta W_{z_b} + \Delta q V_o W/g \\ \Delta M_{y_b} \end{Bmatrix} \tag{7.5.15}
$$

$$
\begin{bmatrix} W/g & 0 & 0 \\ 0 & I_{xx_b} & -I_{xz_b} \\ 0 & -I_{xz_b} & I_{zz_b} \end{bmatrix} \begin{Bmatrix} \Delta\dot{v} \\ \Delta\dot{p} \\ \Delta\dot{r} \end{Bmatrix} = \begin{Bmatrix} \Delta F_{y_b} + \Delta W_{y_b} - \Delta r V_o W/g \\ \Delta M_{x_b} \\ \Delta M_{z_b} \end{Bmatrix} \tag{7.5.16}
$$

or after using Eqs. (7.5.13) and (7.5.14),

$$
\begin{bmatrix} W/g - F_{x_b,\dot{u}} & -F_{x_b,\dot{w}} & 0 \\ -F_{z_b,\dot{u}} & W/g - F_{z_b,\dot{w}} & 0 \\ -M_{y_b,\dot{u}} & -M_{y_b,\dot{w}} & I_{yy_b} \end{bmatrix} \begin{Bmatrix} \Delta\dot{u} \\ \Delta\dot{w} \\ \Delta\dot{q} \end{Bmatrix}
$$
$$
= \begin{bmatrix} F_{x_b,u} & F_{x_b,w} & F_{x_b,q} \\ F_{z_b,u} & F_{z_b,w} & F_{z_b,q} + V_o W/g \\ M_{y_b,u} & M_{y_b,w} & M_{y_b,q} \end{bmatrix} \begin{Bmatrix} \Delta u \\ \Delta w \\ \Delta q \end{Bmatrix} + \begin{Bmatrix} F_{x_b,\delta_e} \\ F_{z_b,\delta_e} \\ M_{y_b,\delta_e} \end{Bmatrix} \Delta\delta_e + \begin{Bmatrix} -W\cos(\theta_0) \\ -W\sin(\theta_0) \\ 0 \end{Bmatrix} \Delta\theta \tag{7.5.17}
$$

$$
\begin{bmatrix} W/g & 0 & 0 \\ 0 & I_{xx_b} & -I_{xz_b} \\ 0 & -I_{xz_b} & I_{zz_b} \end{bmatrix} \begin{Bmatrix} \Delta\dot{v} \\ \Delta\dot{p} \\ \Delta\dot{r} \end{Bmatrix}
$$
$$
= \begin{bmatrix} F_{y_b,v} & F_{y_b,p} & F_{y_b,r} - V_o W/g \\ M_{x_b,v} & M_{x_b,p} & M_{x_b,r} \\ M_{z_b,v} & M_{z_b,p} & M_{z_b,r} \end{bmatrix} \begin{Bmatrix} \Delta v \\ \Delta p \\ \Delta r \end{Bmatrix} + \begin{bmatrix} F_{y_b,\delta_a} & F_{y_b,\delta_r} \\ M_{x_b,\delta_a} & M_{x_b,\delta_r} \\ M_{z_b,\delta_a} & M_{z_b,\delta_r} \end{bmatrix} \begin{Bmatrix} \Delta\delta_a \\ \Delta\delta_r \end{Bmatrix} \tag{7.5.18}
$$
$$
+ \begin{Bmatrix} W\cos(\theta_0) \\ 0 \\ 0 \end{Bmatrix} \Delta\phi
$$

Using the small-disturbance variables from Eq. (7.5.3), applying the small-angle approximations,

$$
\begin{aligned}
\sin(\phi) &\cong \Delta\phi & \sin(\theta) &\cong \sin(\theta_o) + \cos(\theta_o)\Delta\theta & \sin(\psi) &\cong \Delta\psi \\
\cos(\phi) &\cong 1 & \cos(\theta) &\cong \cos(\theta_o) - \sin(\theta_o)\Delta\theta & \cos(\psi) &\cong 1
\end{aligned} \tag{7.5.19}
$$

and neglecting second-order terms in the kinematic transformation matrix, Eq. (7.3.19) reduces to

$$
\begin{Bmatrix} V_{wx_f} + V_o C_{\theta_o} + \Delta \dot{x}_f \\ V_{wy_f} + \Delta \dot{y}_f \\ V_{wz_f} - V_o S_{\theta_o} + \Delta \dot{z}_f \end{Bmatrix}
$$

$$
= \begin{bmatrix} C_{\theta_o} - S_{\theta_o}\Delta\theta & S_{\theta_o}\Delta\phi - \Delta\psi & S_{\theta_o} + C_{\theta_o}\Delta\theta \\ C_{\theta_o}\Delta\psi & 1 & S_{\theta_o}\Delta\psi - \Delta\phi \\ -S_{\theta_o} - C_{\theta_o}\Delta\theta & C_{\theta_o}\Delta\phi & C_{\theta_o} - S_{\theta_o}\Delta\theta \end{bmatrix} \begin{Bmatrix} V_o + \Delta u \\ \Delta v \\ \Delta w \end{Bmatrix} + \begin{Bmatrix} V_{wx_f} \\ V_{wy_f} \\ V_{wz_f} \end{Bmatrix} \tag{7.5.20}
$$

$$
\begin{Bmatrix} \Delta\dot{\phi} \\ \Delta\dot{\theta} \\ \Delta\dot{\psi} \end{Bmatrix} = \begin{bmatrix} 1 & S_{\theta_o}\Delta\phi/C_{\theta_o} & S_{\theta_o}/C_{\theta_o} + \Delta\theta/C_{\theta_o}^2 \\ 0 & 1 & -\Delta\phi \\ 0 & \Delta\phi/C_{\theta_o} & 1/C_{\theta_o} + S_{\theta_o}\Delta\theta/C_{\theta_o}^2 \end{bmatrix} \begin{Bmatrix} \Delta p \\ \Delta q \\ \Delta r \end{Bmatrix} \tag{7.5.21}
$$

or, after eliminating the remaining second-order terms, we have

$$
\begin{Bmatrix} \Delta\dot{x}_f \\ \Delta\dot{y}_f \\ \Delta\dot{z}_f \\ \Delta\dot{\phi} \\ \Delta\dot{\theta} \\ \Delta\dot{\psi} \end{Bmatrix} = \begin{bmatrix} \cos(\theta_o) & 0 & \sin(\theta_o) & 0 & 0 & 0 \\ 0 & 1 & 0 & 0 & 0 & 0 \\ -\sin(\theta_o) & 0 & \cos(\theta_o) & 0 & 0 & 0 \\ 0 & 0 & 0 & 1 & 0 & \tan(\theta_o) \\ 0 & 0 & 0 & 0 & 1 & 0 \\ 0 & 0 & 0 & 0 & 0 & \sec(\theta_o) \end{bmatrix} \begin{Bmatrix} \Delta u \\ \Delta v \\ \Delta w \\ \Delta p \\ \Delta q \\ \Delta r \end{Bmatrix}
$$

$$
+ \begin{bmatrix} 0 & -V_o \sin(\theta_o) & 0 \\ 0 & 0 & V_o \cos(\theta_o) \\ 0 & -V_o \cos(\theta_o) & 0 \\ 0 & 0 & 0 \\ 0 & 0 & 0 \\ 0 & 0 & 0 \end{bmatrix} \begin{Bmatrix} \Delta\phi \\ \Delta\theta \\ \Delta\psi \end{Bmatrix} \tag{7.5.22}
$$

Equations (7.5.17), (7.5.18), and (7.5.22) provide 12 first-order linearized differential equations for the velocity and position associated with each of the six degrees of freedom that are possible in rigid-body flight. These include three translational degrees of freedom (axial, normal, and transverse) and three rotational degrees of freedom (roll, pitch, and yaw). We could solve this system of equations as a fully coupled system of 12 equations in 12 unknowns. However, this is not a fully coupled system of equations. The subsystem given by Eq. (7.5.17) provides three equations, which include only four of the 12 unknowns, Δu, Δw, Δq, and $\Delta\theta$. The subsystem in Eq. (7.5.18) provides three more equations in four entirely different unknowns, Δv, Δp, Δr, and $\Delta\phi$. Combining Eq. (7.5.17) with the first, third, and fifth equations from Eq. (7.5.22), we have a complete subsystem that can be solved for six of the 12 unknowns, independent of the other six. Similarly, by combining Eq. (7.5.18) with the second, fourth, and sixth equations from Eq. (7.5.22), we can solve for the remaining six of the 12 unknowns.

Thus, rearranging Eqs. (7.5.17), (7.5.18), and (7.5.22), we obtain two separate systems of six equations in six unknowns:

the **linearized longitudinal equations,**

$$
\begin{bmatrix}
W/g - F_{x_b,\dot{u}} & -F_{x_b,\dot{w}} & 0 & 0 & 0 & 0 \\
-F_{z_b,\dot{u}} & W/g - F_{z_b,\dot{w}} & 0 & 0 & 0 & 0 \\
-M_{y_b,\dot{u}} & -M_{y_b,\dot{w}} & I_{yy_b} & 0 & 0 & 0 \\
0 & 0 & 0 & 1 & 0 & 0 \\
0 & 0 & 0 & 0 & 1 & 0 \\
0 & 0 & 0 & 0 & 0 & 1
\end{bmatrix}
\begin{Bmatrix}
\Delta\dot{u} \\ \Delta\dot{w} \\ \Delta\dot{q} \\ \Delta\dot{x}_f \\ \Delta\dot{z}_f \\ \Delta\dot{\theta}
\end{Bmatrix}
=
\begin{Bmatrix}
F_{x_b,\delta_e} \\ F_{z_b,\delta_e} \\ M_{y_b,\delta_e} \\ 0 \\ 0 \\ 0
\end{Bmatrix}
\Delta\delta_e
$$

$$
+
\begin{bmatrix}
F_{x_b,u} & F_{x_b,w} & F_{x_b,q} & 0 & 0 & -W\cos\theta_0 \\
F_{z_b,u} & F_{z_b,w} & F_{z_b,q}+V_oW/g & 0 & 0 & -W\sin\theta_0 \\
M_{y_b,u} & M_{y_b,w} & M_{y_b,q} & 0 & 0 & 0 \\
\cos\theta_o & \sin\theta_o & 0 & 0 & 0 & -V_o\sin\theta_o \\
-\sin\theta_o & \cos\theta_o & 0 & 0 & 0 & -V_o\cos\theta_o \\
0 & 0 & 1 & 0 & 0 & 0
\end{bmatrix}
\begin{Bmatrix}
\Delta u \\ \Delta w \\ \Delta q \\ \Delta x_f \\ \Delta z_f \\ \Delta\theta
\end{Bmatrix}
\tag{7.5.23}
$$

and the **linearized lateral equations,**

$$
\begin{bmatrix}
W/g & 0 & 0 & 0 & 0 & 0 \\
0 & I_{xx_b} & -I_{xz_b} & 0 & 0 & 0 \\
0 & -I_{xz_b} & I_{zz_b} & 0 & 0 & 0 \\
0 & 0 & 0 & 1 & 0 & 0 \\
0 & 0 & 0 & 0 & 1 & 0 \\
0 & 0 & 0 & 0 & 0 & 1
\end{bmatrix}
\begin{Bmatrix}
\Delta\dot{v} \\ \Delta\dot{p} \\ \Delta\dot{r} \\ \Delta\dot{y}_f \\ \Delta\dot{\phi} \\ \Delta\dot{\psi}
\end{Bmatrix}
=
\begin{bmatrix}
F_{y_b,\delta_a} & F_{y_b,\delta_r} \\
M_{x_b,\delta_a} & M_{x_b,\delta_r} \\
M_{z_b,\delta_a} & M_{z_b,\delta_r} \\
0 & 0 \\
0 & 0 \\
0 & 0
\end{bmatrix}
\begin{Bmatrix}
\Delta\delta_a \\ \Delta\delta_r
\end{Bmatrix}
$$

$$
+
\begin{bmatrix}
F_{y_b,v} & F_{y_b,p} & F_{y_b,r}-V_oW/g & 0 & W\cos\theta_0 & 0 \\
M_{x_b,v} & M_{x_b,p} & M_{x_b,r} & 0 & 0 & 0 \\
M_{z_b,v} & M_{z_b,p} & M_{z_b,r} & 0 & 0 & 0 \\
1 & 0 & 0 & 0 & 0 & V_o\cos\theta_o \\
0 & 1 & \tan\theta_o & 0 & 0 & 0 \\
0 & 0 & \sec\theta_o & 0 & 0 & 0
\end{bmatrix}
\begin{Bmatrix}
\Delta v \\ \Delta p \\ \Delta r \\ \Delta y_f \\ \Delta\phi \\ \Delta\psi
\end{Bmatrix}
\tag{7.5.24}
$$

Equations (7.5.23) and (7.5.24) are the linearized 6-DOF equations of motion. They provide an important simplification over the more general nonlinear 6-DOF equations in Sec. 7.4. However, we must remember that these equations are valid only if the motion of the aircraft is confined to small deviations about some steady equilibrium flight condition. Larger deviations will introduce coupling between the lateral and longitudinal motions and nonlinearity in the aerodynamic forces and moments.

With this said, it should be pointed out that this linearized theory yields considerable information and many valuable insights with comparatively little effort. Furthermore, it gives good engineering accuracy over a surprisingly wide range of application. The small-disturbance model is actually valid for disturbance magnitudes that would seem quite violent to most passengers. In fact, after experiencing flight disturbances of a magnitude that is near the limit of applicability for small-disturbance theory, most individuals would not care to investigate a large-disturbance theory experimentally. Small-disturbance theory is valid for most commonly encountered flying situations. Nearly all of the flying time logged by commercial and private pilots is spent within the boundaries of the constraints imposed by small-disturbance theory. There are, of course, limitations. Small-disturbance theory is not applicable to spinning motion, stall recovery, or any other application that involves rapid maneuvers or large-amplitude oscillations.

EXAMPLE 7.5.2. Over a short distance, the flight of an arrow or other such projectile does not deviate a great deal from motion along a straight line at constant speed. Small-disturbance theory can be used to approximate such motion. Using small-disturbance theory, linearize the system of equations that was developed in Example 7.3.3. Linearize the equations relative to motion at constant speed, V_o, along a horizontal line. After the full system of equations has been linearized, solve the linear system for the special case where K_2 is zero and K_3 is infinity. Assume that initially the projectile starts from $x_f = 0$ and $z_f = 0$ with a speed of V_o and a small elevation angle of $\Delta\theta_0$.

Solution. From the solution to Example 7.3.3,

$$
\begin{Bmatrix}
\dot{u} \\
\dot{w} \\
\dot{q} \\
\dot{x}_f \\
\dot{z}_f \\
\dot{\theta}
\end{Bmatrix}
=
\begin{Bmatrix}
-K_0 u^2 - K_1 w^2 - g\sin\theta - qw \\
-K_2 uw + g\cos\theta + qu \\
-K_3 uw - K_4 qu \\
u\cos\theta + w\sin\theta \\
w\cos\theta - u\sin\theta \\
q
\end{Bmatrix}
\tag{7.5.25}
$$

Expanding relative to horizontal motion at constant speed, we use

$$
\begin{Bmatrix}
u(t) \\
w(t) \\
q(t) \\
x_f(t) \\
z_f(t) \\
\theta(t)
\end{Bmatrix}
=
\begin{Bmatrix}
V_o + \Delta u(t) \\
\Delta w(t) \\
\Delta q(t) \\
V_o t + \Delta x_f(t) \\
\Delta z_f(t) \\
\Delta\theta(t)
\end{Bmatrix}
\tag{7.5.26}
$$

Using these definitions in Eq. (7.5.25) gives

$$
\begin{Bmatrix}
\Delta \dot{u} \\
\Delta \dot{w} \\
\Delta \dot{q} \\
V_o + \Delta \dot{x}_f \\
\Delta \dot{z}_f \\
\Delta \dot{\theta}
\end{Bmatrix}
=
\begin{Bmatrix}
-K_0(V_o + \Delta u)^2 - K_1(\Delta w)^2 - g\sin(\Delta\theta) - \Delta q\,\Delta w \\
-K_2(V_o + \Delta u)\Delta w + g\cos(\Delta\theta) + \Delta q(V_o + \Delta u) \\
-K_3(V_o + \Delta u)\Delta w - K_4\Delta q(V_o + \Delta u) \\
(V_o + \Delta u)\cos(\Delta\theta) + \Delta w\sin(\Delta\theta) \\
\Delta w\cos(\Delta\theta) - (V_o + \Delta u)\sin(\Delta\theta) \\
\Delta q
\end{Bmatrix}
$$

Retaining only first-order terms, we have

$$
\begin{Bmatrix}
\Delta \dot{u} \\
\Delta \dot{w} \\
\Delta \dot{q} \\
V_o + \Delta \dot{x}_f \\
\Delta \dot{z}_f \\
\Delta \dot{\theta}
\end{Bmatrix}
=
\begin{Bmatrix}
-K_0(V_o^2 + 2V_o\Delta u) - g\Delta\theta \\
-K_2V_o\Delta w + g + V_o\Delta q \\
-K_3V_o\Delta w - K_4V_o\Delta q \\
V_o + \Delta u \\
\Delta w - V_o\Delta\theta \\
\Delta q
\end{Bmatrix}
$$

or

$$
\begin{Bmatrix}
\Delta \dot{u} \\
\Delta \dot{w} \\
\Delta \dot{q} \\
\Delta \dot{x}_f \\
\Delta \dot{z}_f \\
\Delta \dot{\theta}
\end{Bmatrix}
=
\begin{bmatrix}
-2K_0V_o & 0 & 0 & 0 & 0 & -g \\
0 & -K_2V_o & V_o & 0 & 0 & 0 \\
0 & -K_3V_o & -K_4V_o & 0 & 0 & 0 \\
1 & 0 & 0 & 0 & 0 & 0 \\
0 & 1 & 0 & 0 & 0 & -V_o \\
0 & 0 & 1 & 0 & 0 & 0
\end{bmatrix}
\begin{Bmatrix}
\Delta u \\
\Delta w \\
\Delta q \\
\Delta x_f \\
\Delta z_f \\
\Delta\theta
\end{Bmatrix}
+
\begin{Bmatrix}
-K_0V_o^2 \\
g \\
0 \\
0 \\
0 \\
0
\end{Bmatrix}
$$

This can be written in differential operator notation as

$$
\begin{bmatrix}
(D + 2K_0V_o) & 0 & 0 & 0 & 0 & g \\
0 & (D + K_2V_o) & -V_o & 0 & 0 & 0 \\
0 & K_3V_o & (D + K_4V_o) & 0 & 0 & 0 \\
-1 & 0 & 0 & D & 0 & 0 \\
0 & -1 & 0 & 0 & D & V_o \\
0 & 0 & -1 & 0 & 0 & D
\end{bmatrix}
\begin{Bmatrix}
\Delta u \\
\Delta w \\
\Delta q \\
\Delta x_f \\
\Delta z_f \\
\Delta\theta
\end{Bmatrix}
=
\begin{Bmatrix}
-K_0V_o^2 \\
g \\
0 \\
0 \\
0 \\
0
\end{Bmatrix}
$$

where D denotes differentiation with respect to time.

This is the linearized system of differential equations that applies to the motion of a stable projectile having small deviations from motion along a horizontal line at constant speed. For a projectile such as an arrow that has a large static margin but does not have a large lifting surface, the lift force has little effect on the motion, and the change in aerodynamic moment with respect to angle of attack is quite large. For such projectiles, we can assume that K_2 is very small and K_3 is very large.

Setting K_2 to zero and K_3 to infinity, this system becomes

$$\begin{bmatrix} (D+2K_0V_o) & 0 & 0 & 0 & 0 & g \\ 0 & D & -V_o & 0 & 0 & 0 \\ 0 & 1 & 0 & 0 & 0 & 0 \\ -1 & 0 & 0 & D & 0 & 0 \\ 0 & -1 & 0 & 0 & D & V_o \\ 0 & 0 & -1 & 0 & 0 & D \end{bmatrix} \begin{Bmatrix} \Delta u \\ \Delta w \\ \Delta q \\ \Delta x_f \\ \Delta z_f \\ \Delta \theta \end{Bmatrix} = \begin{Bmatrix} -K_0V_o^2 \\ g \\ 0 \\ 0 \\ 0 \\ 0 \end{Bmatrix} \qquad (7.5.27)$$

The appropriate initial conditions are

$$\begin{Bmatrix} \Delta u(0) \\ \Delta x_f(0) \\ \Delta z_f(0) \\ \Delta \theta(0) \end{Bmatrix} = \begin{Bmatrix} 0 \\ 0 \\ 0 \\ \Delta \theta_0 \end{Bmatrix} \qquad (7.5.28)$$

From the third equation in Eq. (7.5.27),

$$\Delta w = 0 \qquad (7.5.29)$$

Using this result in the second equation in Eq. (7.5.27), we have

$$\Delta q = -\frac{g}{V_o} \qquad (7.5.30)$$

Using this with the last equation in Eq. (7.5.27) gives

$$D\Delta \theta = -\frac{g}{V_o}$$

Integrating and applying the fourth initial condition from Eq. (7.5.28) yields

$$\Delta \theta = -\frac{g}{V_o}t + C_1$$

$$\Delta \theta = \Delta \theta_0 - \frac{g}{V_o}t \qquad (7.5.31)$$

Using Eqs. (7.5.29) and (7.5.31) in the fifth equation in Eq. (7.5.27) results in

$$D\Delta z_f = gt - V_o \Delta\theta_0$$

Integrating and applying the third initial condition from Eq. (7.5.28) yields

$$\Delta z_f = \frac{g}{2}t^2 - V_o \Delta\theta_0 t \tag{7.5.32}$$

Using Eq. (7.5.31) in the first equation in Eq. (7.5.27) gives

$$(D + 2K_0 V_o)\Delta u = \frac{g^2}{V_o}t - K_0 V_o^2 - g\Delta\theta_0$$

Integrating this subject to the first initial condition from Eq. (7.5.28) yields

$$\Delta u = \left(\frac{V_o}{2} + \frac{g\Delta\theta_0}{2K_0 V_o} + \frac{g^2}{4K_0^2 V_o^3}\right)\left(e^{-2K_0 V_o t} - 1\right) + \frac{g^2 t}{2K_0 V_o^2} \tag{7.5.33}$$

Using this result in the fourth equation in Eq. (7.5.27), we have

$$D\Delta x_f = \left(\frac{V_o}{2} + \frac{g\Delta\theta_0}{2K_0 V_o} + \frac{g^2}{4K_0^2 V_o^3}\right)\left(e^{-2K_0 V_o t} - 1\right) + \frac{g^2 t}{2K_0 V_o^2}$$

Integrating subject to the second initial condition from Eq. (7.5.28) gives

$$\Delta x_f = \left(\frac{V_o}{2} + \frac{g\Delta\theta_0}{2K_0 V_o} + \frac{g^2}{4K_0^2 V_o^3}\right)\left(\frac{1 - e^{-2K_0 V_o t}}{2K_0 V_o} - t\right) + \frac{g^2 t^2}{4K_0 V_o^2} \tag{7.5.34}$$

Applying Eqs. (7.5.29) through (7.5.34) to Eq. (7.5.26), the full solution is

$$
\begin{Bmatrix} u(t) \\ w(t) \\ q(t) \\ x_f(t) \\ z_f(t) \\ \theta(t) \end{Bmatrix} = \begin{Bmatrix} V_o + \left(\dfrac{V_o}{2} + \dfrac{g\Delta\theta_0}{2K_0 V_o} + \dfrac{g^2}{4K_0^2 V_o^3}\right)\left(e^{-2K_0 V_o t} - 1\right) + \dfrac{g^2 t}{2K_0 V_o^2} \\[4pt] 0 \\[4pt] -\dfrac{g}{V_o} \\[4pt] V_o t + \left(\dfrac{V_o}{2} + \dfrac{g\Delta\theta_0}{2K_0 V_o} + \dfrac{g^2}{4K_0^2 V_o^3}\right)\left(\dfrac{1 - e^{-2K_0 V_o t}}{2K_0 V_o} - t\right) + \dfrac{g^2 t^2}{4K_0 V_o^2} \\[4pt] \dfrac{g}{2}t^2 - V_o \Delta\theta_0 t \\[4pt] \Delta\theta_0 - \dfrac{g}{V_o}t \end{Bmatrix}
$$

7.6. Force and Moment Derivatives

While computer simulations or wind tunnel tests are required for accurate determination of aerodynamic derivatives, some first-order approximations can be made analytically. Formulation of the linearized equations of motion requires evaluation of the derivatives of the aerodynamic forces and moments with respect to the translational velocity components. However, in our study of aerodynamic forces and moments, we found it convenient to express these forces and moments in terms of total airspeed,

$$V = \sqrt{u^2 + v^2 + w^2} \tag{7.6.1}$$

angle of attack,

$$\alpha = \tan^{-1}\left(\frac{w}{u}\right) \cong \frac{w}{u} \tag{7.6.2}$$

and sideslip angle,

$$\beta \cong \beta_a = \tan^{-1}\left(\frac{v}{u}\right) \cong \beta_e = \sin^{-1}\left(\frac{v}{\sqrt{u^2 + v^2 + w^2}}\right) \cong \frac{v}{u} \tag{7.6.3}$$

The change in any function, $\mathcal{F}(V, \alpha, \beta)$, with respect to axial velocity component is

$$\frac{\partial \mathcal{F}}{\partial u} = \frac{\partial \mathcal{F}}{\partial V}\frac{\partial V}{\partial u} + \frac{\partial \mathcal{F}}{\partial \alpha}\frac{\partial \alpha}{\partial u} + \frac{\partial \mathcal{F}}{\partial \beta}\frac{\partial \beta}{\partial u} \tag{7.6.4}$$

Since the angle of attack is not a function of the sideslip velocity, a change with respect to the sideslip velocity component is

$$\frac{\partial \mathcal{F}}{\partial v} = \frac{\partial \mathcal{F}}{\partial V}\frac{\partial V}{\partial v} + \frac{\partial \mathcal{F}}{\partial \beta}\frac{\partial \beta}{\partial v} \tag{7.6.5}$$

Likewise, within the approximation of Eq. (7.6.3), the sideslip angle is not a function of normal velocity, so a change with respect to normal velocity component is given by

$$\frac{\partial \mathcal{F}}{\partial w} = \frac{\partial \mathcal{F}}{\partial V}\frac{\partial V}{\partial w} + \frac{\partial \mathcal{F}}{\partial \alpha}\frac{\partial \alpha}{\partial w} \tag{7.6.6}$$

From Eq. (7.6.1),

$$\frac{\partial V}{\partial u} = \frac{u}{\sqrt{u^2 + v^2 + w^2}} \tag{7.6.7}$$

$$\frac{\partial V}{\partial v} = \frac{v}{\sqrt{u^2 + v^2 + w^2}} \tag{7.6.8}$$

$$\frac{\partial V}{\partial w} = \frac{w}{\sqrt{u^2 + v^2 + w^2}} \tag{7.6.9}$$

For small angles, Eq. (7.6.2) gives

$$\frac{\partial \alpha}{\partial u} \cong -\frac{w}{u^2} \tag{7.6.10}$$

$$\frac{\partial \alpha}{\partial w} \cong \frac{1}{u} \tag{7.6.11}$$

and Eq. (7.6.3) results in

$$\frac{\partial \beta}{\partial u} \cong -\frac{v}{u^2} \tag{7.6.12}$$

$$\frac{\partial \beta}{\partial v} \cong \frac{1}{u} \tag{7.6.13}$$

The particular body-fixed coordinate system that we are using was chosen so that at the equilibrium reference state,

$$\left\{ \begin{array}{c} u \\ v \\ w \end{array} \right\} = \left\{ \begin{array}{c} V_o \\ 0 \\ 0 \end{array} \right\} \tag{7.6.14}$$

The result given by Eq. (7.6.14) is valid only for this particular body-fixed coordinate system, which is normally referred to as the *stability axes*. This particular coordinate system was chosen specifically to make the angle of attack zero at the reference flight condition. This choice of coordinate system simplifies the results considerably, just as it did for the static analysis.

Using Eq. (7.6.14) in Eqs. (7.6.7) through (7.6.13), at the reference state, we obtain the results

$$\frac{\partial V}{\partial u} = 1 \tag{7.6.15}$$

$$\frac{\partial \alpha}{\partial w} \cong \frac{\partial \beta}{\partial v} \cong \frac{1}{V_o} \tag{7.6.16}$$

$$\frac{\partial V}{\partial v} = \frac{\partial V}{\partial w} = \frac{\partial \alpha}{\partial u} = \frac{\partial \beta}{\partial u} = 0 \tag{7.6.17}$$

Using Eqs. (7.6.15) through (7.6.17) in Eqs. (7.6.4) through (7.6.6) results in

$$\frac{\partial \mathcal{F}}{\partial u} = \frac{\partial \mathcal{F}}{\partial V} \tag{7.6.18}$$

$$\frac{\partial \mathcal{F}}{\partial v} = \frac{1}{V_o}\frac{\partial \mathcal{F}}{\partial \beta} \tag{7.6.19}$$

and

$$\frac{\partial \mathcal{F}}{\partial w} = \frac{1}{V_o}\frac{\partial \mathcal{F}}{\partial \alpha} \tag{7.6.20}$$

The forces and moments acting on an aircraft are the sum of those due to gravity, aerodynamic loading, and thrust. Figure 7.6.1a shows the body-fixed components of the force due to gravity. Figure 7.6.1b shows the body-fixed components of the aerodynamic forces and moments. Finally, the thrust force, which acts in the plane of symmetry, is shown in Fig. 7.6.2 along with the longitudinal components of the aerodynamic and gravitational forces. Combining these, the total force and moment in body-fixed coordinates are given by

$$\bar{\mathbf{F}} + \bar{\mathbf{W}} = \frac{1}{2}\rho V^2 S_w \begin{Bmatrix} C_X \\ C_Y \\ C_Z \end{Bmatrix} + T \begin{Bmatrix} \cos(\alpha_{T0}) \\ 0 \\ -\sin(\alpha_{T0}) \end{Bmatrix} + W \begin{Bmatrix} -\sin(\theta) \\ \sin(\phi)\cos(\theta) \\ \cos(\phi)\cos(\theta) \end{Bmatrix} \tag{7.6.21}$$

$$\bar{\mathbf{M}} = \frac{1}{2}\rho V^2 S_w \begin{Bmatrix} b_w C_\ell \\ \bar{c}_w C_m \\ b_w C_n \end{Bmatrix} + T \begin{Bmatrix} 0 \\ z_T \cos(\alpha_{T0}) + x_T \sin(\alpha_{T0}) \\ 0 \end{Bmatrix} \tag{7.6.22}$$

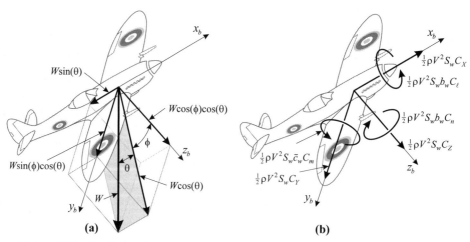

Figure 7.6.1. Gravitational and aerodynamic forces and moments in body-fixed coordinates.

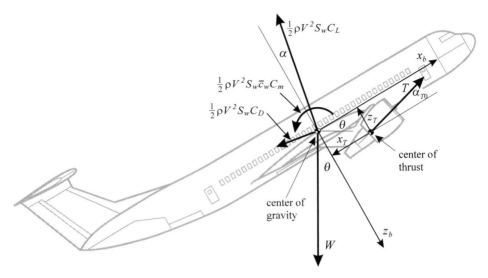

Figure 7.6.2. Longitudinal forces and moments acting in the aircraft plane of symmetry.

It is sometimes convenient to express the x_b- and z_b-components of the aerodynamic force in terms of lift and drag. As shown in Fig. 7.6.2, these two equivalent descriptions of the aerodynamic force are related through the angle of attack measured relative to the stability axes,

$$\tfrac{1}{2}\rho V^2 S_w \begin{Bmatrix} C_X \\ C_Y \\ C_Z \end{Bmatrix} = \tfrac{1}{2}\rho V^2 S_w \begin{Bmatrix} C_L \sin(\alpha) - C_D \cos(\alpha) \\ C_Y \\ -C_L \cos(\alpha) - C_D \sin(\alpha) \end{Bmatrix} \tag{7.6.23}$$

For the nonaccelerating reference state, the lift, drag, and aerodynamic pitching moment can be related to the weight and thrust. The fuselage reference line was defined so that α is zero at the equilibrium reference state. Thus at equilibrium, the drag force must balance the x_b-components of thrust and weight; the lift force must balance the z_b-components of thrust and weight; the aerodynamic pitching moment must balance the moment due to thrust; and the remaining aerodynamic force and moment components must vanish. Therefore, at the equilibrium reference state, we have

$$\tfrac{1}{2}\rho V_o^2 S_w \begin{Bmatrix} C_X \\ C_Y \\ C_Z \\ b_w C_\ell \\ \overline{c}_w C_m \\ b_w C_n \end{Bmatrix} = \tfrac{1}{2}\rho V_o^2 S_w \begin{Bmatrix} -C_D \\ 0 \\ -C_L \\ 0 \\ \overline{c}_w C_m \\ 0 \end{Bmatrix} = \begin{Bmatrix} -T\cos(\alpha_{T0}) + W\sin(\theta_o) \\ 0 \\ T\sin(\alpha_{T0}) - W\cos(\theta_o) \\ 0 \\ -z_T T\cos(\alpha_{T0}) - x_T T\sin(\alpha_{T0}) \\ 0 \end{Bmatrix} \tag{7.6.24}$$

An important and often used special case of this equilibrium force and moment balance occurs for level flight, with the thrust vector aligned with the center of gravity and the direction of flight. In this case, for nonaccelerating flight, the thrust must equal the drag; the lift must equal the weight; and the aerodynamic pitching moment must vanish along with the remaining force and moment components. The lift, weight, thrust, and drag are sometimes referred to as the *four forces of flight* and the relations *lift equals weight* and *thrust equals drag* are often used. However, we must keep in mind that these are only for the special case of steady level flight when the thrust vector is aligned with the direction of flight. A more general relation is provided by Eq. (7.6.24).

The force and moment derivatives with respect to total airspeed are obtained by differentiating Eqs. (7.6.21) and (7.6.22) with respect to V at constant angle of attack. Using Eq. (7.6.23) to replace the axial and normal coefficients, this gives

$$
\begin{aligned}
\frac{\partial F_{x_b}}{\partial V} &= \frac{\partial}{\partial V}\left[\frac{1}{2}\rho V^2 S_w C_X + T\cos(\alpha_{T0}) - W\sin(\theta)\right] \\
&= \rho V S_w\left(C_X + \frac{V}{2}\frac{\partial C_X}{\partial V}\right) + \frac{\partial T}{\partial V}\cos(\alpha_{T0}) \\
&= \rho V S_w\left[\left(C_L + \frac{V}{2}\frac{\partial C_L}{\partial V}\right)\sin(\alpha) - \left(C_D + \frac{V}{2}\frac{\partial C_D}{\partial V}\right)\cos(\alpha)\right] + \frac{\partial T}{\partial V}\cos(\alpha_{T0})
\end{aligned}
\tag{7.6.25}
$$

$$
\frac{\partial F_{y_b}}{\partial V} = \frac{\partial}{\partial V}\left[\frac{1}{2}\rho V^2 S_w C_Y + W\sin(\phi)\cos(\theta)\right] = \rho V S_w\left(C_Y + \frac{V}{2}\frac{\partial C_Y}{\partial V}\right)
\tag{7.6.26}
$$

$$
\begin{aligned}
\frac{\partial F_{z_b}}{\partial V} &= \frac{\partial}{\partial V}\left[\frac{1}{2}\rho V^2 S_w C_Z - T\sin(\alpha_{T0}) + W\cos(\phi)\cos(\theta)\right] \\
&= \rho V S_w\left(C_Z + \frac{V}{2}\frac{\partial C_Z}{\partial V}\right) - \frac{\partial T}{\partial V}\sin(\alpha_{T0}) \\
&= \rho V S_w\left[-\left(C_L + \frac{V}{2}\frac{\partial C_L}{\partial V}\right)\cos(\alpha) - \left(C_D + \frac{V}{2}\frac{\partial C_D}{\partial V}\right)\sin(\alpha)\right] - \frac{\partial T}{\partial V}\sin(\alpha_{T0})
\end{aligned}
\tag{7.6.27}
$$

$$
\frac{\partial M_{x_b}}{\partial V} = \frac{\partial}{\partial V}\left(\frac{1}{2}\rho V^2 S_w b_w C_\ell\right) = \rho V S_w b_w\left(C_\ell + \frac{V}{2}\frac{\partial C_\ell}{\partial V}\right)
\tag{7.6.28}
$$

$$
\begin{aligned}
\frac{\partial M_{y_b}}{\partial V} &= \frac{\partial}{\partial V}\left[\frac{1}{2}\rho V^2 S_w \bar{c}_w C_m + z_T T\cos(\alpha_{T0}) + x_T T\sin(\alpha_{T0})\right] \\
&= \rho V S_w \bar{c}_w\left(C_m + \frac{V}{2}\frac{\partial C_m}{\partial V}\right) + \frac{\partial T}{\partial V}\left[z_T\cos(\alpha_{T0}) + x_T\sin(\alpha_{T0})\right]
\end{aligned}
\tag{7.6.29}
$$

$$
\frac{\partial M_{z_b}}{\partial V} = \frac{\partial}{\partial V}\left(\frac{1}{2}\rho V^2 S_w b_w C_n\right) = \rho V S_w b_w\left(C_n + \frac{V}{2}\frac{\partial C_n}{\partial V}\right)
\tag{7.6.30}
$$

Remember, these derivatives are all evaluated at the reference flight condition.

The aerodynamic coefficients are generally functions of geometry, Reynolds number, and Mach number. Both the Reynolds number and the Mach number depend on the flight speed. In certain flow regimes, the effect of Reynolds number on the aerodynamic coefficients can be quite pronounced. The coefficients can change dramatically over the Reynolds number region that encompasses the transition from laminar to turbulent boundary layers. However, in other flow regimes, the effect of the Reynolds number is much less. As a result, aerodynamic coefficients are often considered to be rather weak functions of Reynolds number, R, and strong functions of the Mach number, M, and the aerodynamic angles, α and β. Furthermore, the drag coefficient, the lift coefficient, and the pitching moment coefficient can be considered functions only of R, M, and α, while the side force coefficient, the rolling moment coefficient, and the yawing moment coefficient can be taken as functions only of R, M, and β.

Aerodynamic data are generally computed analytically or collected in the wind tunnel as functions of Mach number and Reynolds number rather than the flight velocity. Thus, derivatives of the aerodynamic coefficients with respect to the flight velocity are not as convenient as derivatives with respect to Mach number and Reynolds number. However, if the Mach number and Reynolds number derivatives are known, the derivative of any aerodynamic coefficient with respect to flight velocity can easily be computed from the definition of Mach number and Reynolds number,

$$ M \equiv \frac{V}{a} $$

$$ R \equiv \frac{V \, l_{\text{ref}}}{v} $$

where M is the Mach number, a is the speed of sound, R is the Reynolds number, l_{ref} is the reference length, and v is the kinematic viscosity. From these definitions we have

$$ V \frac{\partial C}{\partial V} = V \left(\frac{\partial M}{\partial V} \frac{\partial C}{\partial M} + \frac{\partial R}{\partial V} \frac{\partial C}{\partial R} \right) = V \left(\frac{1}{a} \frac{\partial C}{\partial M} + \frac{l_{\text{ref}}}{v} \frac{\partial C}{\partial R} \right) = M \frac{\partial C}{\partial M} + R \frac{\partial C}{\partial R} \quad (7.6.31) $$

where C could be any one of the aerodynamic coefficients. In general, the derivatives of the aerodynamic coefficients with respect to Mach number and Reynolds number could be determined from wind tunnel tests or computer simulations. The results could then be used with Eq. (7.6.31) in Eqs. (7.6.25) through (7.6.30) to obtain the required derivatives with respect to forward velocity.

For applications with very low Mach number, the flow can be considered to be incompressible and the aerodynamic coefficients are independent of Mach number but may be strongly dependent on the Reynolds number. However, for most applications, the Reynolds numbers encountered in normal flight are quite large, and for large Reynolds numbers, the aerodynamic coefficients are nearly independent of Reynolds number. Thus, for most atmospheric flight we can neglect derivatives with respect to Reynolds number, and Eq. (7.6.31) can be approximated as

$$\frac{V}{2}\frac{\partial C}{\partial V} \cong \frac{M}{2}\frac{\partial C}{\partial M} \tag{7.6.32}$$

Applying Eq. (7.6.18) with Eqs. (7.6.25) through (7.6.30) and Eq. (13.7.32), **for high Reynolds number at the equilibrium reference state, the derivatives with respect to forward velocity** can be approximated as

$$
\begin{aligned}
F_{x_b,u} &\equiv \frac{\partial F_{x_b}}{\partial u} = \frac{\partial F_{x_b}}{\partial V} = \rho V_o S_w\left(C_X + \frac{M}{2}C_{X,M}\right) + \frac{\partial T}{\partial V}\cos(\alpha_{T0}) \\
&= -\tfrac{1}{2}\rho V_o^2 S_w\left(2\frac{C_D}{V_o} + \frac{C_{D,M}}{a}\right) + \frac{\partial T}{\partial V}\cos(\alpha_{T0}) \\[4pt]
F_{y_b,u} &\equiv \frac{\partial F_{y_b}}{\partial u} = \frac{\partial F_{y_b}}{\partial V} = \rho V_o S_w\left(C_Y + \frac{M}{2}C_{Y,M}\right) = 0 \\[4pt]
F_{z_b,u} &\equiv \frac{\partial F_{z_b}}{\partial u} = \frac{\partial F_{z_b}}{\partial V} = \rho V_o S_w\left(C_Z + \frac{M}{2}C_{Z,M}\right) - \frac{\partial T}{\partial V}\sin(\alpha_{T0}) \\
&= -\tfrac{1}{2}\rho V_o^2 S_w\left(2\frac{C_L}{V_o} + \frac{C_{L,M}}{a}\right) - \frac{\partial T}{\partial V}\sin(\alpha_{T0}) \quad (7.6.33) \\[4pt]
M_{x_b,u} &\equiv \frac{\partial M_{x_b}}{\partial u} = \frac{\partial M_{x_b}}{\partial V} = \rho V_o S_w b_w\left(C_\ell + \frac{M}{2}C_{\ell,M}\right) = 0 \\[4pt]
M_{y_b,u} &\equiv \frac{\partial M_{y_b}}{\partial u} = \frac{\partial M_{y_b}}{\partial V} = \rho V_o S_w \bar{c}_w\left(C_m + \frac{M}{2}C_{m,M}\right) + z_{T0}\frac{\partial T}{\partial V} \\
&= \tfrac{1}{2}\rho V_o^2 S_w \bar{c}_w\left(2\frac{C_m}{V_o} + \frac{C_{m,M}}{a}\right) + z_{T0}\frac{\partial T}{\partial V} \\[4pt]
M_{z_b,u} &\equiv \frac{\partial M_{z_b}}{\partial u} = \frac{\partial M_{z_b}}{\partial V} = \rho V_o S_w b_w\left(C_n + \frac{M}{2}C_{n,M}\right) = 0
\end{aligned}
$$

where z_{T0} is the perpendicular offset between the thrust vector and the CG,

$$z_{T0} \equiv z_T \cos(\alpha_{T0}) + x_T \sin(\alpha_{T0})$$

The derivatives of the aerodynamic coefficients with respect to Mach number can be determined from wind tunnel data, which are generally gathered as a function of the Mach number. The derivatives are estimated from the data using finite difference. However, the required wind tunnel tests are time consuming and expensive. Thus, for preliminary design, it is convenient to have a first-order approximation for these derivatives that can be used to estimate their values prior to wind tunnel testing.

When the flow around an airfoil section remains subsonic, thin airfoil theory with the Prandtl-Glauert compressibility correction predicts a section lift coefficient given by

$$\tilde{C}_L = \frac{2\pi(\alpha - \alpha_{L0})}{\sqrt{1 - M^2}}$$

Within the limitations of thin airfoil theory, the section lift coefficient for a subsonic airfoil depends only on angle of attack, Mach number, and camber line shape.

At supersonic flight speeds, Ackeret's theory predicts that the section lift coefficient for thin airfoils at low angles of attack is determined from

$$\tilde{C}_L = \frac{4\alpha}{\sqrt{M^2 - 1}}$$

Within the accuracy of this theory, the supersonic section lift coefficient is found to be a function only of angle of attack and Mach number. Camber is predicted to have no effect on the supersonic section lift.

These theoretical results for thin airfoils in both subsonic and supersonic flow are plotted in Fig. 7.6.3. From this figure we see that the variation in lift coefficient with Mach number is negligible only for Mach numbers less than about 0.3. In the transonic region near Mach 1, both of these theories break down. There is no simple theory that is capable of predicting the variation in section lift coefficient with Mach number in the transonic region. At Mach numbers near 1, the flow around an airfoil is extremely complex and very sensitive to section shape. In the transonic region we must rely on experimental data.

At subsonic speeds, a first approximation for the variation in airplane lift coefficient with Mach number can be obtained by assuming that the Prandtl-Glauert compressibility correction for airfoils can be applied to the lift coefficient for a complete airplane. This approximation results in

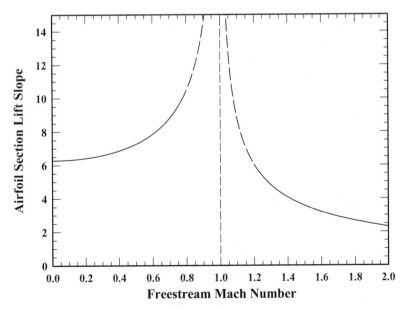

Figure 7.6.3. Theoretical lift slope for a thin airfoil as a function of freestream Mach number.

$$C_L \cong \frac{C_{L_{M=0}}}{\sqrt{1-M^2}} \qquad (7.6.34)$$

where $C_{L_{M=0}}$ is the lift coefficient for incompressible flow. Equation (7.6.34) can easily be differentiated with respect to the Mach number to yield

$$C_{L,M} \cong \frac{M}{1-M^2} C_L \qquad (7.6.35)$$

Equation (7.6.35) provides a reasonable first approximation for the change in lift coefficient with Mach number, which can be used in the absence of experimental data. However, it should be used with caution. Remember that Eq. (7.6.35) applies only for Mach numbers less than the critical Mach number, where sonic flow is first attained at some point around the airfoil. Furthermore, even below the critical Mach number, experimental observations show that Eq. (7.6.35) underestimates $C_{L,M}$ for some airfoil sections and overestimates it for others.

In a similar manner, subsonic thin airfoil theory predicts that the Prandtl-Glauert compressibility correction, which was used for the section lift coefficient, also applies directly to the section moment coefficient. Thus, by reasoning similar to that used for the lift coefficient, a first-order approximation for the subcritical derivative of the pitching moment coefficient with Mach number is given by

$$C_{m,M} \cong \frac{M}{1-M^2} C_m \qquad (7.6.36)$$

Here again, this should be used with caution and only in absence of experimental data.

Unlike the lift and moment coefficients, which increase with subsonic Mach number, the zero-lift drag coefficient remains relatively constant with Mach number, up to the critical value at which sonic flow is first encountered at some location around the aircraft. The total drag is the zero-lift drag plus the drag due to lift. Using Eq. (7.6.34), the drag coefficient for subsonic flow can be approximated as

$$C_D = C_{D_{L0}} + C_{D_i} \cong C_{D_{L0}} + \frac{C_L^2}{\pi e R_A} \cong C_{D_{L0}} + \frac{C_{L_{M=0}}^2}{\pi e R_A (1-M^2)} \qquad (7.6.37)$$

Differentiating this result with respect to Mach number gives a first approximation for the change in drag coefficient with Mach number, which can be used with caution for subcritical Mach numbers in the absence of experimental data,

$$C_{D,M} \cong \frac{2M C_{L_{M=0}}^2}{\pi e R_A (1-M^2)^2} = \frac{2M}{1-M^2} C_{D_i} \qquad (7.6.38)$$

At some Mach number slightly above the critical, *drag divergence* is encountered and Eq. (7.6.38) becomes very inaccurate.

Using Eqs. (7.6.35), (7.6.36), and (7.6.38) together with the equilibrium requirement given by Eq. (7.6.24) in Eq. (7.6.33), we obtain a **first-order approximation for the derivatives with respect to forward velocity in subsonic flight,**

$$F_{x_b,u} \cong -\frac{1}{2}\rho V_o^2 S_w \left(2 + \frac{2M^2}{1-M^2}\frac{C_{D_i}}{C_D}\right)\frac{C_D}{V_o} + \frac{\partial T}{\partial V}\cos(\alpha_{T0})$$

$$= -\left(2 + \frac{2M^2}{1-M^2}\frac{C_{D_i}}{C_D}\right)\frac{T\cos(\alpha_{T0}) - W\sin(\theta_o)}{V_o} + \frac{\partial T}{\partial V}\cos(\alpha_{T0})$$

$$F_{z_b,u} \cong -\frac{1}{2}\rho V_o^2 S_w \frac{2-M^2}{1-M^2}\frac{C_L}{V_o} - \frac{\partial T}{\partial V}\sin(\alpha_{T0})$$

$$= -\frac{2-M^2}{1-M^2}\frac{W\cos(\theta_o) - T\sin(\alpha_{T0})}{V_o} - \frac{\partial T}{\partial V}\sin(\alpha_{T0}) \qquad (7.6.39)$$

$$M_{y_b,u} \cong \frac{1}{2}\rho V_o^2 S_w \bar{c}_w \frac{2-M^2}{1-M^2}\frac{C_m}{V_o} + \frac{\partial T}{\partial V}\left[z_T\cos(\alpha_{T0}) + x_T\sin(\alpha_{T0})\right]$$

$$= \left(-\frac{2-M^2}{1-M^2}\frac{T}{V_o} + \frac{\partial T}{\partial V}\right)\left[z_T\cos(\alpha_{T0}) + x_T\sin(\alpha_{T0})\right]$$

$$F_{y_b,u} = M_{x_b,u} = M_{z_b,u} = 0$$

These approximations should only be used for subsonic flight below the critical Mach number and only for preliminary results in the absence of wind tunnel data.

In a similar manner, the force and moment derivatives with respect to angle of attack are obtained by differentiating Eqs. (7.6.21) and (7.6.22) with respect to α. Since only the axial force, the normal force, and the pitching moment depend on the angle of attack, we have

$$\frac{\partial F_{x_b}}{\partial \alpha} = \frac{\partial}{\partial \alpha}\left[\frac{1}{2}\rho V^2 S_w C_X + T\cos(\alpha_{T0}) - W\sin(\theta)\right] = \frac{1}{2}\rho V^2 S_w \frac{\partial C_X}{\partial \alpha}$$

$$= \frac{1}{2}\rho V^2 S_w\left[\frac{\partial C_L}{\partial \alpha}\sin(\alpha) + C_L\cos(\alpha) - \frac{\partial C_D}{\partial \alpha}\cos(\alpha) + C_D\sin(\alpha)\right] \qquad (7.6.40)$$

$$\frac{\partial F_{z_b}}{\partial \alpha} = \frac{\partial}{\partial \alpha}\left[\frac{1}{2}\rho V^2 S_w C_Z - T\sin(\alpha_{T0}) + W\cos(\phi)\cos(\theta)\right] = \frac{1}{2}\rho V^2 S_w \frac{\partial C_Z}{\partial \alpha}$$

$$= \frac{1}{2}\rho V^2 S_w\left[-\frac{\partial C_L}{\partial \alpha}\cos(\alpha) + C_L\sin(\alpha) - \frac{\partial C_D}{\partial \alpha}\sin(\alpha) - C_D\cos(\alpha)\right] \qquad (7.6.41)$$

$$\frac{\partial M_{y_b}}{\partial \alpha} = \frac{\partial}{\partial \alpha}\left[\frac{1}{2}\rho V^2 S_w \bar{c}_w C_m + z_T T\cos(\alpha_{T0}) + x_T T\sin(\alpha_{T0})\right]$$

$$= \frac{1}{2}\rho V^2 S_w \bar{c}_w \frac{\partial C_m}{\partial \alpha} \qquad (7.6.42)$$

The side force and the rolling and yawing moments do not depend on angle of attack. Thus, the derivatives of the remaining force and moment components with angle of attack are zero,

$$\frac{\partial F_{y_b}}{\partial \alpha} = \frac{\partial M_{x_b}}{\partial \alpha} = \frac{\partial M_{z_b}}{\partial \alpha} = 0$$

Using Eqs. (7.6.20), (7.6.24), and (7.6.40) through (7.6.42), the **derivatives with respect to normal velocity** evaluated at the reference flight condition are

$$
\begin{aligned}
F_{x_b,w} &\equiv \frac{\partial F_{x_b}}{\partial w} = \frac{1}{V_o}\frac{\partial F_{x_b}}{\partial \alpha} = \frac{\rho V_o S_w}{2}\frac{\partial C_X}{\partial \alpha} = \frac{\rho V_o S_w}{2}\left(C_L - \frac{\partial C_D}{\partial \alpha}\right) \\
&= \frac{W\cos(\theta_o) - T\sin(\alpha_T)}{V_o} - \frac{\rho V_o S_w}{2}\frac{\partial C_D}{\partial \alpha} \\
F_{z_b,w} &\equiv \frac{\partial F_{z_b}}{\partial w} = \frac{1}{V_o}\frac{\partial F_{z_b}}{\partial \alpha} = \frac{\rho V_o S_w}{2}\frac{\partial C_Z}{\partial \alpha} = -\frac{\rho V_o S_w}{2}\left(\frac{\partial C_L}{\partial \alpha} + C_D\right) \\
&= -\frac{\rho V_o S_w}{2}\frac{\partial C_L}{\partial \alpha} - \frac{T\cos(\alpha_T) - W\sin(\theta_o)}{V_o} \\
M_{y_b,w} &\equiv \frac{\partial M_{y_b}}{\partial w} = \frac{1}{V_o}\frac{\partial M_{y_b}}{\partial \alpha} = \frac{\rho V_o S_w \bar{c}_w}{2}\frac{\partial C_m}{\partial \alpha} \\
F_{y_b,w} &= M_{z_b,w} = M_{x_b,w} = 0
\end{aligned}
\qquad (7.6.43)
$$

Likewise, at the reference state, the **derivatives with respect to sideslip velocity** are

$$
\begin{aligned}
F_{y_b,v} &\equiv \frac{\partial F_{y_b}}{\partial v} = \frac{1}{V_o}\frac{\partial F_{y_b}}{\partial \beta} = \frac{\rho V_o S_w}{2}\frac{\partial C_Y}{\partial \beta} \\
M_{x_b,v} &\equiv \frac{\partial M_{x_b}}{\partial v} = \frac{1}{V_o}\frac{\partial M_{x_b}}{\partial \beta} = \frac{\rho V_o S_w b_w}{2}\frac{\partial C_\ell}{\partial \beta} \\
M_{z_b,v} &\equiv \frac{\partial M_{z_b}}{\partial v} = \frac{1}{V_o}\frac{\partial M_{z_b}}{\partial \beta} = \frac{\rho V_o S_w b_w}{2}\frac{\partial C_n}{\partial \beta} \\
F_{x_b,v} &= F_{z_b,v} = M_{y_b,v} = 0
\end{aligned}
\qquad (7.6.44)
$$

As we have seen previously, the aerodynamic forces and moments depend on the angular velocity components as well as the translational velocity components. As an aircraft rotates in roll, one wing moves down and experiences an increase in the z_b-component of velocity, while the other wing moves up, experiencing a decrease in the z_b-component of velocity. This results in a differential angle of attack between the two wings, as shown in Fig. 7.6.4. The rolling motion increasing the lift on the falling wing, decreases the lift on the rising wing, and creates a rolling moment that opposes the rolling rate. The change in the rolling moment with respect to the rolling rate is called the *roll-damping derivative*.

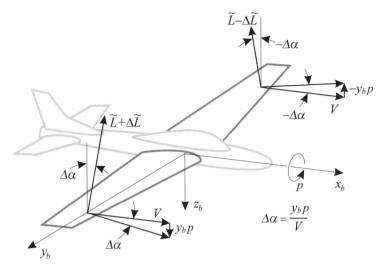

Figure 7.6.4. Effects of rolling rate on the local angle of attack and section lift.

The rolling rate also affects the yawing moment. The primary contribution to this yawing moment comes from the wing. The wing contribution arises from the increase in angle of attack on the falling wing and the decrease in angle of attack on the rising wing. Since the lift vector at any section of the wing must be perpendicular to the local freestream, the section lift vector on the falling wing $\widetilde{L} + \Delta\widetilde{L}$ is tilted forward and the section lift vector on the rising wing $\widetilde{L} - \Delta\widetilde{L}$ is tilted back, as is shown in Fig. 7.6.4. The result is the generation of a negative yawing moment from a positive rolling rate.

Using Prandtl's lifting-line theory, the change in the rolling and yawing moment coefficients with respect to rolling rate can be approximated as

$$\left(\frac{\partial C_\ell}{\partial p}\right)_{\text{wing}} \cong -\frac{\kappa_{\ell\overline{p}} b_w}{16 V_o} \frac{\partial C_{L_w}}{\partial \alpha} \tag{7.6.45}$$

$$\left(\frac{\partial C_n}{\partial p}\right)_{\text{wing}} \cong -\left(1 - \frac{3\kappa_{\ell\overline{p}}}{\pi R_{A_w}} \frac{\partial C_{L_w}}{\partial \alpha}\right) \frac{b_w C_{L_w}}{16 V_o} \tag{7.6.46}$$

where b_w, R_{A_w}, and C_{L_w} are, respectively, the wingspan, aspect ratio, and lift coefficient for the wing at the equilibrium reference state. The coefficient $\kappa_{\ell\overline{p}}$ depends on wing planform and aspect ratio as described in Sec. 1.8 and shown in Fig. 1.8.25. Both the horizontal and vertical tail also contribute to the change in the rolling and yawing moments with respect to rolling rate. However, the tail contribution is small compared to that of the wing and is usually neglected in a first analysis.

If the wing has dihedral, then the lift differential between the right wing and the left wing that is caused by the roll rate will also produce a side force. However, this term is small and can be neglected as a first approximation.

Thus, for a first approximation, the **derivatives with respect to roll rate** evaluated at the reference flight condition are

$$
F_{y_b,p} \equiv \frac{\partial F_{y_b}}{\partial p} = \tfrac{1}{2}\rho V_o^2 S_w \frac{\partial C_Y}{\partial p} \cong 0
$$

$$
M_{x_b,p} \equiv \frac{\partial M_{x_b}}{\partial p} = \tfrac{1}{2}\rho V_o^2 S_w b_w \frac{\partial C_\ell}{\partial p} \cong -\rho V_o S_w b_w^2 \frac{\kappa_{\ell\overline{p}}}{32} \frac{\partial C_{L_w}}{\partial \alpha}
$$

$$
M_{z_b,p} \equiv \frac{\partial M_{z_b}}{\partial p} = \tfrac{1}{2}\rho V_o^2 S_w b_w \frac{\partial C_n}{\partial p} \cong -\rho V_o S_w b_w^2 \left(1 - \frac{3\kappa_{\ell\overline{p}}}{\pi R_{A_w}} \frac{\partial C_{L_w}}{\partial \alpha}\right) \frac{C_{L_w}}{32}
$$

$$
F_{x_b,p} = F_{z_b,p} = M_{y_b,p} = 0
$$

(7.6.47)

The pitching rate affects the forces and moments on the aircraft primarily through the horizontal tail. As the aircraft rotates in pitch about the center of mass, an aft tail experiences an increase in the z_b-component of velocity that is equal to the product of the pitching rate and the distance the tail is aft of the center of mass (see Fig. 7.6.5). This increases the effective angle of attack for the horizontal tail, which produces a change in the lift force and the associated induced drag, as well as a pitching moment that opposes the pitching rate. The change in the pitching moment with respect to the pitching rate is called the *pitch-damping derivative*. The wing and the fuselage can also affect the pitch-damping derivative, but these effects are small and are often neglected in a first analysis.

The lift produced by the horizontal tail of a pitching aircraft is proportional to the local angle of attack, which is the sum of the geometric angle of attack and the angle of attack that is induced by the pitching rate $(-x_{bh}\,q/V_o)$. Thus, we can write

$$
\left(\frac{\partial C_Z}{\partial q}\right)_{\text{tail}} \cong \eta_h \frac{S_h x_{bh}}{S_w V_o} \frac{\partial C_{L_h}}{\partial \alpha}
$$

(7.6.48)

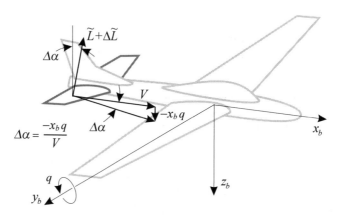

Figure 7.6.5. Effects of pitching rate on the local angle of attack at the tail.

and

$$
\left(\frac{\partial C_m}{\partial q} \right)_{\text{tail}} \cong -\eta_h \frac{S_h x_{bh}^2}{S_w V_o \bar{c}_w} \frac{\partial C_{L_h}}{\partial \alpha}
\tag{7.6.49}
$$

where η_h, S_h, and x_{bh} are, respectively, the efficiency factor, area, and x_b-coordinate of the aerodynamic center for the horizontal tail. Because the pitching rate changes the lift on the horizontal tail, it also changes the induced drag. However, this effect is small and is usually neglected in a first analysis. The wing is included in the same manner.

As a first approximation the **derivatives with respect to pitch rate** evaluated at the reference flight condition are then given by

$$
\begin{aligned}
F_{x_b,q} &\equiv \frac{\partial F_{x_b}}{\partial q} = -\frac{1}{2}\rho V_o^2 S_w \frac{\partial C_D}{\partial q} \cong 0 \\[2mm]
F_{z_b,q} &\equiv \frac{\partial F_{z_b}}{\partial q} = -\frac{1}{2}\rho V_o^2 S_w \frac{\partial C_L}{\partial q} \cong \frac{1}{2}\rho V_o \left(S_w x_{bw} \frac{\partial C_{L_w}}{\partial \alpha} + \eta_h S_h x_{bh} \frac{\partial C_{L_h}}{\partial \alpha} \right) \\[2mm]
M_{y_b,q} &\equiv \frac{\partial M_{y_b}}{\partial q} = \frac{1}{2}\rho V_o^2 S_w \bar{c}_w \frac{\partial C_m}{\partial q} \cong -\frac{1}{2}\rho V_o \left(S_w x_{bw}^2 \frac{\partial C_{L_w}}{\partial \alpha} + \eta_h S_h x_{bh}^2 \frac{\partial C_{L_h}}{\partial \alpha} \right) \\[2mm]
F_{y_b,q} &= M_{x_b,q} = M_{z_b,q} = 0
\end{aligned}
\tag{7.6.50}
$$

In a similar manner, the yawing rate affects the side force, the rolling moment, and the yawing moment by changing the effective sideslip angle for the vertical tail, as shown in Fig. 7.6.6. Thus we have

$$
\left(\frac{\partial C_Y}{\partial r} \right)_{\text{tail}} \cong -\eta_v \frac{S_v x_{bv}}{S_w V_o} \frac{\partial C_{L_v}}{\partial \beta}
\tag{7.6.51}
$$

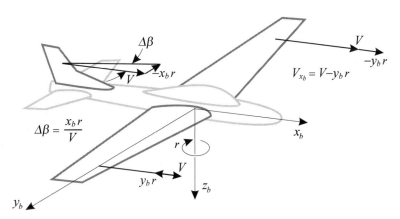

Figure 7.6.6. Effects of yawing rate on the sideslip angle at the tail and the local forward velocity of the wings.

$$\left(\frac{\partial C_\ell}{\partial r}\right)_{\text{tail}} \cong \eta_v \frac{S_v x_{bv} z_{bv}}{S_w b_w V_o} \frac{\partial C_{L_v}}{\partial \beta} \tag{7.6.52}$$

and

$$\left(\frac{\partial C_n}{\partial r}\right)_{\text{tail}} \cong -\eta_v \frac{S_v x_{bv}^2}{S_w b_w V_o} \frac{\partial C_{L_v}}{\partial \beta} \tag{7.6.53}$$

where η_v, S_v, x_{bv}, and z_{bv} are, respectively, the efficiency factor, area, x_b-coordinate, and z_b-coordinate of the aerodynamic center for the vertical tail. The change in the yawing moment with respect to the yawing rate is called the *yaw-damping derivative*.

The main wing also makes a significant contribution to the change in the rolling moment with respect to the yawing rate. If the airplane is rotating with a positive yawing rate, the forward speed of the left wing is increased and the forward speed of the right wing is decreased, as shown is in Fig. 7.6.6. This increases the lift on the left wing and decreases the lift on the right wing, creating a positive rolling moment. Again using Prandtl's lifting-line theory, the change in the wing's rolling moment coefficients with respect to yawing rate can be approximated as

$$\left(\frac{\partial C_\ell}{\partial r}\right)_{\text{wing}} \cong -2\left(\frac{\partial C_n}{\partial p}\right)_{\text{wing}} \cong \left(1 - \frac{3\kappa_{\ell\overline{p}}}{\pi R_{A_w}}\frac{\partial C_{L_w}}{\partial \alpha}\right)\frac{b_w C_{L_w}}{8 V_o} \tag{7.6.54}$$

Thus, at the reference state, **derivatives with respect to yaw rate** are approximated as

$$
\begin{aligned}
F_{y_b,r} &\equiv \frac{\partial F_{y_b}}{\partial r} = \tfrac{1}{2}\rho V_o^2 S_w \frac{\partial C_Y}{\partial r} \cong -\tfrac{1}{2}\rho V_o \eta_v S_v x_{bv}\frac{\partial C_{L_v}}{\partial \beta} \\[2mm]
M_{x_b,r} &\equiv \frac{\partial M_{x_b}}{\partial r} = \tfrac{1}{2}\rho V_o^2 S_w b_w \frac{\partial C_\ell}{\partial r} \\[1mm]
&\cong \tfrac{1}{2}\rho V_o\left[\left(1 - \frac{3\kappa_{\ell\overline{p}}}{\pi R_{A_w}}\frac{\partial C_{L_w}}{\partial \alpha}\right)\frac{S_w b_w^2 C_{L_w}}{8} + \eta_v S_v x_{bv} z_{bv}\frac{\partial C_{L_v}}{\partial \beta}\right] \\[2mm]
M_{z_b,r} &\equiv \frac{\partial M_{z_b}}{\partial r} = \tfrac{1}{2}\rho V_o^2 S_w b_w \frac{\partial C_n}{\partial r} \cong -\tfrac{1}{2}\rho V_o \eta_v S_v x_{bv}^2 \frac{\partial C_{L_v}}{\partial \beta} \\[2mm]
F_{x_b,r} &= F_{z_b,r} = M_{y_b,r} = 0
\end{aligned}
\tag{7.6.55}
$$

As discussed previously, the longitudinal forces and the pitching moment also vary with translational acceleration, because a change in the vorticity generated by the wing requires a finite time to alter the downwash on the tail and fuselage. The aerodynamic derivatives with respect to the translational acceleration components are quite complex, depending on the shape and position of the wing, the tail, and the fuselage. In general, these derivatives would need to be determined from an unsteady fluid flow analysis. However, for a first estimate, the following quasi-steady approximations may be used with some caution.

From Prandtl's lifting-line theory, the downwash angle behind an elliptic wing in steady flight can be approximated as

$$\varepsilon_d \cong \frac{V_d}{V_o} \cong \frac{2S_w}{\pi b_w^2} C_{L_w} \tag{7.6.56}$$

The time required for the downwash generated by the wing to reach the tail is the streamwise distance from the wingtip to the tail, l_{wt}, divided by the forward velocity, V (see Fig. 7.6.7). During unsteady flight, this time lag can be approximated as $\Delta t = l_{wt}/V_o$ and the downwash angle at the tail can be approximated as

$$\varepsilon_d \cong \frac{2S_w}{\pi b_w^2}\left(C_{L_w} - \frac{\partial C_{L_w}}{\partial t}\Delta t\right) = \frac{2S_w}{\pi b_w^2}\left(C_{L_w} - \frac{\partial C_{L_w}}{\partial t}\frac{l_{wt}}{V_o}\right)$$
$$= \frac{2S_w}{\pi b_w^2}\left(C_{L_w} - \frac{\partial C_{L_w}}{\partial \alpha}\frac{\partial \alpha}{\partial t}\frac{l_{wt}}{V_o}\right) \tag{7.6.57}$$

From Eq. (7.6.2),

$$\frac{\partial \alpha}{\partial t} = \frac{\partial \alpha}{\partial w}\frac{\partial w}{\partial t} + \frac{\partial \alpha}{\partial u}\frac{\partial u}{\partial t} \tag{7.6.58}$$

or using Eqs. (7.6.16) and (7.6.17),

$$\frac{\partial \alpha}{\partial t} = \frac{1}{V_o}\frac{\partial w}{\partial t} \equiv \frac{1}{V_o}\dot{w} \tag{7.6.59}$$

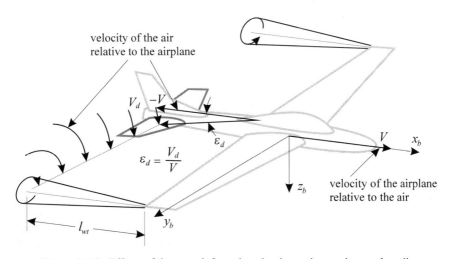

Figure 7.6.7. Effects of downwash from the wingtip vortices acting on the tail.

Using Eq. (7.6.59) in Eq. (7.6.57), we have

$$\varepsilon_d = \frac{2S_w}{\pi b_w^2}\left(C_{L_w} - \frac{\partial C_{L_w}}{\partial \alpha}\frac{l_{wt}}{V_o^2}\dot{w}\right) \tag{7.6.60}$$

The lift produced by the horizontal tail is proportional to its local angle of attack, which is the geometric angle of attack less the downwash angle,

$$(L)_{tail} = \eta_h\left(\tfrac{1}{2}\rho V_o^2 S_h\right)\frac{\partial C_{L_h}}{\partial \alpha}(\alpha_t - \varepsilon_d) \tag{7.6.61}$$

or in view of Eq. (7.6.60),

$$(L)_{tail} = \eta_h\left(\tfrac{1}{2}\rho V_o^2 S_h\right)\frac{\partial C_{L_h}}{\partial \alpha}\left[\alpha_t - \frac{2S_w}{\pi b_w^2}\left(C_{L_w} - \frac{\partial C_{L_w}}{\partial \alpha}\frac{l_{wt}}{V_o^2}\dot{w}\right)\right] \tag{7.6.62}$$

The pitching moment produced by the tail is the product of the lift developed on the horizontal tail and the moment arm between the aerodynamic center of the horizontal tail and the aircraft center of gravity,

$$\begin{aligned}\left(M_{y_b}\right)_{tail} &= x_{bh}(L)_{tail}\\ &= \eta_h\left(\tfrac{1}{2}\rho V_o^2 S_h\right)\frac{\partial C_{L_h}}{\partial \alpha}\left[\alpha_t - \frac{2S_w}{\pi b_w^2}\left(C_{L_w} - \frac{\partial C_{L_w}}{\partial \alpha}\frac{l_{wt}}{V_o^2}\dot{w}\right)\right]x_{bh}\end{aligned} \tag{7.6.63}$$

Differentiating Eqs. (7.6.62) and (7.6.63) with respect to the normal acceleration, we have an approximation for the contribution that the horizontal tail makes to the change in normal force with respect to normal acceleration,

$$\left(\frac{\partial F_{z_b}}{\partial \dot{w}}\right)_{tail} = -\left(\frac{\partial L}{\partial \dot{w}}\right)_{tail} = -\eta_h\frac{\rho S_w S_h l_{wt}}{\pi b_w^2}\frac{\partial C_{L_w}}{\partial \alpha}\frac{\partial C_{L_h}}{\partial \alpha} \tag{7.6.64}$$

and the change in pitching moment with respect to normal acceleration,

$$\left(\frac{\partial M_{y_b}}{\partial \dot{w}}\right)_{tail} = \eta_h\frac{\rho S_w S_h l_{wt} x_{bh}}{\pi b_w^2}\frac{\partial C_{L_w}}{\partial \alpha}\frac{\partial C_{L_h}}{\partial \alpha} \tag{7.6.65}$$

To obtain an estimate of these derivatives for the complete aircraft, including the effects of the wing-fuselage interaction, it is recommended that these values be increased by 10 percent for an airplane with an aft tail and set to zero for an airplane with a forward canard. As a first approximation, the remaining longitudinal derivatives with respect to the translational acceleration components are usually assumed to be zero.

Thus, as a first estimate, the **longitudinal derivatives with respect to translational acceleration** can be approximated as

$$
\boxed{
\begin{aligned}
F_{x_b,\dot{u}} &\cong F_{z_b,\dot{u}} \cong M_{y_b,\dot{u}} \cong F_{x_b,\dot{w}} \cong 0 \\[6pt]
F_{z_b,\dot{w}} &\equiv \frac{\partial F_{z_b}}{\partial \dot{w}} = \frac{1}{V_o}\frac{\partial F_{z_b}}{\partial \dot{\alpha}} = -\frac{\tfrac{1}{2}\rho V_o^2 S_w}{V_o}\frac{\partial C_L}{\partial \dot{\alpha}} \\[6pt]
&\cong -\eta_h \frac{\rho S_w S_h l_{wt}}{\pi b_w^2}\frac{\partial C_{L_w}}{\partial \alpha}\frac{\partial C_{L_h}}{\partial \alpha} \\[10pt]
M_{y_b,\dot{w}} &\equiv \frac{\partial M_{y_b}}{\partial \dot{w}} = \frac{1}{V_o}\frac{\partial M_{y_b}}{\partial \dot{\alpha}} = \frac{\tfrac{1}{2}\rho V_o^2 S_w \bar{c}_w}{V_o}\frac{\partial C_m}{\partial \dot{\alpha}} \\[6pt]
&\cong \eta_h \frac{\rho S_w S_h l_{wt} x_{bh}}{\pi b_w^2}\frac{\partial C_{L_w}}{\partial \alpha}\frac{\partial C_{L_h}}{\partial \alpha} \\[10pt]
l_{wt} &= \begin{cases} 1.1\left(x_{b\,\text{wingtip}} - x_{bh}\right), & x_{b\,\text{wingtip}} > x_{bh} \\ 0.0, & x_{bh} > x_{b\,\text{wingtip}} \end{cases}
\end{aligned}
}
\tag{7.6.66}
$$

As mentioned previously, all of the **lateral derivatives with respect to translational acceleration are either zero or negligible.**

All of the aerodynamic derivatives that appear in the linearized equations of motion are complex functions of the shape and placement of the wing, tail, fuselage, and engines. For any given aircraft, these derivatives also vary with airspeed and other operating conditions. In general, these derivatives will need to be determined either from computer simulations or from wind tunnel tests. However, as a first estimate, the approximations given in this section may be used.

EXAMPLE 7.6.1. A large jet transport is flying at 30,000 feet with a steady level airspeed of 450 mph, which is the minimum drag airspeed for this altitude. This airplane has the following properties in cruise configuration:

$$
S_w = 5{,}500 \text{ ft}^2, \quad b_w = 196 \text{ ft}, \quad \bar{c}_w = 28 \text{ ft}, \quad x_{bw} = 0, \quad C_{L_w,\alpha} = 4.67,
$$

$$
S_h = 1{,}300 \text{ ft}^2, \quad x_{bh} = -100 \text{ ft}, \quad x_{b\,\text{wingtip}} = -50 \text{ ft}, \quad l_{wt} = 55 \text{ ft}, \quad C_{L_h,\alpha} = 3.50,
$$

$$
W = 636{,}600 \text{ lbf}, \quad C_{L,\alpha} = 5.50, \quad C_{m,\alpha} = -1.26
$$

The maximum lift-to-drag ratio is 14. Assuming constant thrust aligned with the direction of flight and the center of gravity, estimate all dimensional longitudinal aerodynamic derivatives that appear in the homogeneous form of Eq. (7.5.23).

Solution. For this airplane the airspeed and the Mach number are

$$
V_o = 450\frac{5{,}280}{3{,}600} \text{ ft/sec} = 660 \text{ ft/sec}
$$

$$\mathrm{M} = \frac{V_o}{a} = \frac{660 \text{ ft/sec}}{994.85 \text{ ft/sec}} = 0.663$$

At the minimum drag airspeed the lift-to-drag ratio is equal to the maximum lift-to-drag ratio, and since the thrust is aligned with direction of flight, we have

$$\alpha_T = 0, \qquad L = W,$$

$$T = D = \frac{L}{L/D} = \frac{W}{L/D} = \frac{636,600 \text{ lbf}}{14} = 45,470 \text{ lbf}$$

At the minimum drag airspeed, the drag due to lift is approximately equal to the zero-lift drag. Thus, the drag due to lift is about half the total drag. Since we are assuming constant thrust, from Eq. (7.6.39), this level flight reference state gives

$$F_{x_b,u} \cong -\left(2 + \frac{2\mathrm{M}^2}{1-\mathrm{M}^2} \frac{C_{D_i}}{C_D}\right)\frac{T}{V_o} = -\left(2 + \frac{2(0.663)^2}{1-(0.663)^2} \frac{1}{2}\right)\frac{45,470 \text{ lbf}}{660 \text{ ft/sec}}$$

$$= \underline{-192 \text{ lbf·sec/ft}}$$

$$F_{z_b,u} \cong -\frac{2-\mathrm{M}^2}{1-\mathrm{M}^2}\frac{W}{V_o} = -\left(\frac{2-(0.663)^2}{1-(0.663)^2}\right)\frac{636,600 \text{ lbf}}{660 \text{ ft/sec}} = \underline{-2,686 \text{ lbf·sec/ft}}$$

Since the thrust is aligned with the direction of flight and the center of gravity, also from Eq. (7.6.39),

$$\alpha_{T0} = 0, \qquad z_T = 0,$$

$$M_{y_b,u} \cong -\frac{2-\mathrm{M}^2}{1-\mathrm{M}^2}\frac{T}{V_o}\left[z_T\cos(\alpha_{T0}) + x_T\sin(\alpha_{T0})\right] = \underline{0.0}$$

Since the total drag is the zero-lift drag plus the drag due to lift, the drag coefficient for subsonic flow can be approximated as

$$C_D = C_{D_{L0}} + C_{D_i} \cong C_{D_{L0}} + \frac{C_L^2}{\pi e R_A}$$

and the change in C_D with respect to angle of attack is approximated as

$$\frac{\partial C_D}{\partial \alpha} \cong \frac{2C_L}{\pi e R_A}\frac{\partial C_L}{\partial \alpha} = \frac{2C_{D_i}}{C_L}\frac{\partial C_L}{\partial \alpha}$$

At the minimum drag airspeed, the drag due to lift is approximately equal to the zero-lift drag. Thus, the drag due to lift is about half the total drag and

$$\frac{\partial C_D}{\partial \alpha} \cong \frac{2C_{D_i}}{C_L} \frac{\partial C_L}{\partial \alpha} \cong \frac{C_D}{C_L} \frac{\partial C_L}{\partial \alpha} = \frac{5.50}{14} = 0.393$$

Thus, from Eq. (7.6.43),

$$
\begin{aligned}
F_{x_b, w} &= \frac{W}{V_o} - \frac{\rho V_o S_w}{2} \frac{\partial C_D}{\partial \alpha} \\
&= \frac{636{,}600 \text{ lbf}}{660 \text{ ft/sec}} - \frac{0.00089068 \text{ slug/ft}^3 (660 \text{ ft/sec})(5{,}500 \text{ ft}^2)}{2} 0.393 \\
&= 329 \text{ lbf·sec/ft}
\end{aligned}
$$

$$
\begin{aligned}
F_{z_b, w} &= -\frac{\rho V_o S_w}{2} \frac{\partial C_L}{\partial \alpha} - \frac{T}{V_o} \\
&= -\frac{0.00089068 \text{ slug/ft}^3 (660 \text{ ft/sec})(5{,}500 \text{ ft}^2)}{2} 5.50 - \frac{45{,}470 \text{ lbf}}{660 \text{ ft/sec}} \\
&= -8{,}960 \text{ lbf·sec/ft}
\end{aligned}
$$

$$
\begin{aligned}
M_{y_b, w} &= \frac{\rho V_o S_w \bar{c}_w}{2} \frac{\partial C_m}{\partial \alpha} \\
&= \frac{0.00089068 \text{ slug/ft}^3 (660 \text{ ft/sec})(5{,}500 \text{ ft}^2)(28 \text{ ft})}{2} (-1.26) \\
&= -57{,}033 \text{ lbf·sec}
\end{aligned}
$$

From Eq. (7.6.50), assuming $\eta_h = 1.0$, the derivatives with pitching rate are

$$F_{x_b, q} \cong \underline{0.0}$$

$$
\begin{aligned}
F_{z_b, q} &\cong \frac{1}{2} \rho V_o \left(S_w x_{bw} \frac{\partial C_{L_w}}{\partial \alpha} + \eta_h S_h x_{bh} \frac{\partial C_{L_h}}{\partial \alpha} \right) \\
&= \frac{0.00089068 \text{ slug/ft}^3 (660 \text{ ft/sec})}{2} \left[0.0 + 1.0(1{,}300 \text{ ft}^2)(-100 \text{ ft})3.50 \right] \\
&= -133{,}700 \text{ lbf·sec}
\end{aligned}
$$

$$
\begin{aligned}
M_{y_b, q} &\cong -\frac{1}{2} \rho V_o \left(S_w x_{bw}^2 \frac{\partial C_{L_w}}{\partial \alpha} + \eta_h S_h x_{bh}^2 \frac{\partial C_{L_h}}{\partial \alpha} \right) \\
&= -\frac{0.00089068 \text{ slug/ft}^3 (660 \text{ ft/sec})}{2} \left[0.0 + 1.0(1{,}300 \text{ ft}^2)(-100 \text{ ft})^2 3.50 \right] \\
&= -13{,}370{,}000 \text{ ft·lbf·sec}
\end{aligned}
$$

From Eq. (7.6.66), the derivatives with respect to translational acceleration are

$$F_{x_b,\dot{u}} \cong F_{z_b,\dot{u}} \cong M_{y_b,\dot{u}} \cong F_{x_b,\dot{w}} \cong \underline{0.0}$$

$$F_{z_b,\dot{w}} \cong -\eta_h \frac{\rho S_w S_h l_{wt}}{\pi b_w^2} \frac{\partial C_{L_w}}{\partial \alpha} \frac{\partial C_{L_h}}{\partial \alpha}$$

$$= -1.0 \frac{0.00089068 \text{ slug/ft}^3 (5{,}500 \text{ ft}^2)(1{,}300 \text{ ft}^2)(55 \text{ ft})}{\pi (196 \text{ ft})^2}(4.67)(3.50)$$

$$= \underline{-47.4 \text{ slugs}}$$

$$M_{y_b,\dot{w}} \cong \eta_h \frac{\rho S_w S_h l_{wt} x_{bh}}{\pi b_w^2} \frac{\partial C_{L_w}}{\partial \alpha} \frac{\partial C_{L_h}}{\partial \alpha}$$

$$= 1.0 \frac{0.00089068 \text{ slug/ft}^3 (5{,}500 \text{ ft}^2)(1{,}300 \text{ ft}^2)(55 \text{ ft})(-100 \text{ ft})}{\pi (196 \text{ ft})^2}(4.67)(3.50)$$

$$= \underline{-4{,}740 \text{ slugs} \cdot \text{ft}}$$

Notice that for this example, the effect of the translational acceleration derivative on the z_b-component of the translational momentum equation is essentially equivalent to adding about 1,500 pounds of weight to the aircraft. This is hardly significant for an airplane that weights more than 300 tons. On the other hand, its effect on the pitching component of the angular momentum equation could be significant.

7.7. Nondimensional Linearized Equations of Motion

We can see from the results presented in Sec. 7.6 that even within the approximation of linear aerodynamics, most of the force and moment derivatives needed to fill out the matrices in Eqs. (7.5.23) and (7.5.24) depend on airspeed. It is sometimes convenient to rearrange the equations of motion in a nondimensional form that makes many of the nondimensional force and moment derivatives independent of airspeed, at least within the accuracy of the approximations used for linear aerodynamics.

To nondimensionalize the independent and dependent variables, we need references for length, velocity, and time. One half the mean wing chord length is traditionally used as the reference length for longitudinal motion, and the wing semispan is the traditional reference length for lateral motion. The equilibrium airspeed is the reference velocity for both longitudinal and lateral motion, and the reference lengths divided by this reference velocity provide the traditional reference time scales. Although these reference lengths and time scales are the traditional choices, they are also somewhat arbitrary and have no crucial physical significances. Thus, we should not expect the dynamic characteristics of airplanes to necessarily scale with these reference lengths and time scales.

The three components of velocity disturbance are nondimensionalized with respect to the reference velocity. Within the approximation of small disturbance theory, the normal velocity disturbance divided by the reference velocity is simply the disturbance in angle of attack and the sideslip velocity disturbance divided by the reference velocity is the sideslip angle disturbance. Thus, nondimensionalizing the three velocity components in this manner introduces only one new nondimensional variable, and we have

$$\Delta\mu \equiv \frac{\Delta u}{V_o}$$

$$\Delta\beta \cong \frac{\Delta v}{V_o} \qquad (7.7.1)$$

$$\Delta\alpha \cong \frac{\Delta w}{V_o}$$

Angular velocity has dimensions of inverse time and is traditionally nondimensionalized with respect to the reference velocity divided by the reference length. Since we are using different reference lengths for longitudinal and lateral motion, we use

$$\Delta\hat{\alpha} \equiv \frac{\Delta\dot{\alpha}\,\bar{c}_w}{2V_o}$$

$$\Delta\bar{p} \equiv \frac{\Delta p\,b_w}{2V_o}$$

$$\Delta\bar{q} \equiv \frac{\Delta q\,\bar{c}_w}{2V_o} \qquad (7.7.2)$$

$$\Delta\bar{r} \equiv \frac{\Delta r\,b_w}{2V_o}$$

In a similar manner, the disturbances in aircraft position are nondimensionalized as

$$\Delta\xi_x \equiv \frac{2\Delta x_f}{\bar{c}_w}$$

$$\Delta\xi_y \equiv \frac{2\Delta y_f}{b_w} \qquad (7.7.3)$$

$$\Delta\xi_z \equiv \frac{2\Delta z_f}{\bar{c}_w}$$

The Euler angles are already dimensionless, so no change in variables is required for describing the aircraft orientation.

Time is traditionally nondimensionalized with respect to reference length divided by reference velocity. Again, since we are using different reference lengths for longitudinal and lateral motion, the traditional dimensionless time for longitudinal motion, τ_x, is not the same as the traditional dimensionless time for lateral motion, τ_y,

$$\tau_x \equiv \frac{2V_o t}{\bar{c}_w}$$

$$\tau_y \equiv \frac{2V_o t}{b_w} \qquad (7.7.4)$$

Using the dimensionless variables that were defined in Eqs. (7.7.1) through (7.7.4), the linearized equations of motion that were given by Eqs. (7.5.23) and (7.5.24) may be written as

the **nondimensional linearized longitudinal equations,**

$$
\begin{bmatrix}
(1-R_{x,\hat{\mu}}) & -R_{x,\hat{\alpha}} & 0 & 0 & 0 & 0 \\
-R_{z,\hat{\mu}} & (1-R_{z,\hat{\alpha}}) & 0 & 0 & 0 & 0 \\
-R_{m,\hat{\mu}} & -R_{m,\hat{\alpha}} & 1 & 0 & 0 & 0 \\
0 & 0 & 0 & 1 & 0 & 0 \\
0 & 0 & 0 & 0 & 1 & 0 \\
0 & 0 & 0 & 0 & 0 & 1
\end{bmatrix}
\begin{Bmatrix}
\Delta\hat{\mu} \\
\Delta\hat{\alpha} \\
\Delta\hat{q} \\
\Delta\hat{\xi}_x \\
\Delta\hat{\xi}_z \\
\Delta\hat{\theta}
\end{Bmatrix}
$$

$$
=
\begin{bmatrix}
R_{x,\mu} & R_{x,\alpha} & R_{x,\bar{q}} & 0 & 0 & -R_{gx}\cos\theta_0 \\
R_{z,\mu} & R_{z,\alpha} & (R_{z,\bar{q}}+1) & 0 & 0 & -R_{gx}\sin\theta_0 \\
R_{m,\mu} & R_{m,\alpha} & R_{m,\bar{q}} & 0 & 0 & 0 \\
\cos\theta_o & \sin\theta_o & 0 & 0 & 0 & -\sin\theta_o \\
-\sin\theta_o & \cos\theta_o & 0 & 0 & 0 & -\cos\theta_o \\
0 & 0 & 1 & 0 & 0 & 0
\end{bmatrix}
\begin{Bmatrix}
\Delta\mu \\
\Delta\alpha \\
\Delta\bar{q} \\
\Delta\xi_x \\
\Delta\xi_z \\
\Delta\theta
\end{Bmatrix}
+
\begin{Bmatrix}
R_{x,\delta_e} \\
R_{z,\delta_e} \\
R_{m,\delta_e} \\
0 \\
0 \\
0
\end{Bmatrix}
\Delta\delta_e
$$

(7.7.5)

and the **nondimensional linearized lateral equations,**

$$
\begin{bmatrix}
1 & 0 & 0 & 0 & 0 & 0 \\
0 & 1 & -\iota_{xz} & 0 & 0 & 0 \\
0 & -\iota_{zx} & 1 & 0 & 0 & 0 \\
0 & 0 & 0 & 1 & 0 & 0 \\
0 & 0 & 0 & 0 & 1 & 0 \\
0 & 0 & 0 & 0 & 0 & 1
\end{bmatrix}
\begin{Bmatrix}
\Delta\hat{\beta} \\
\Delta\hat{p} \\
\Delta\hat{r} \\
\Delta\hat{\xi}_y \\
\Delta\hat{\phi} \\
\Delta\hat{\psi}
\end{Bmatrix}
$$

$$
=
\begin{bmatrix}
R_{y,\beta} & R_{y,\bar{p}} & (R_{y,\bar{r}}-1) & 0 & R_{gy}\cos\theta_0 & 0 \\
R_{\ell,\beta} & R_{\ell,\bar{p}} & R_{\ell,\bar{r}} & 0 & 0 & 0 \\
R_{n,\beta} & R_{n,\bar{p}} & R_{n,\bar{r}} & 0 & 0 & 0 \\
1 & 0 & 0 & 0 & 0 & \cos\theta_o \\
0 & 1 & \tan\theta_o & 0 & 0 & 0 \\
0 & 0 & \sec\theta_o & 0 & 0 & 0
\end{bmatrix}
\begin{Bmatrix}
\Delta\beta \\
\Delta\bar{p} \\
\Delta\bar{r} \\
\Delta\xi_y \\
\Delta\phi \\
\Delta\psi
\end{Bmatrix}
+
\begin{bmatrix}
R_{y,\delta_a} & R_{y,\delta_r} \\
R_{\ell,\delta_a} & R_{\ell,\delta_r} \\
R_{n,\delta_a} & R_{n,\delta_r} \\
0 & 0 \\
0 & 0 \\
0 & 0
\end{bmatrix}
\begin{Bmatrix}
\Delta\delta_a \\
\Delta\delta_r
\end{Bmatrix}
$$

(7.7.6)

where we are using the notation for both longitudinal and lateral motion,

$$
\hat{f} \equiv \frac{\partial f}{\partial \tau_x} \text{ for longitudinal motion and } \hat{f} \equiv \frac{\partial f}{\partial \tau_y} \text{ for lateral motion} \tag{7.7.7}
$$

Here the nondimensional components of the coefficient matrices are those that arise naturally as a result of nondimensionalizing each of the 12 equations using the dimensionless variables defined in Eqs. (7.7.1) through (7.7.4). Thus we define

$$
\iota_{xz} \equiv \frac{I_{xz_b}}{I_{xx_b}} \qquad \iota_{zx} \equiv \frac{I_{xz_b}}{I_{zz_b}} \qquad R_{gx} \equiv \frac{g\bar{c}_w}{2V_o^2} \qquad R_{gy} \equiv \frac{gb_w}{2V_o^2}
$$

$$
R_{x,\hat{\mu}} \equiv \frac{\rho S_w \bar{c}_w}{4W/g} C_{X,\hat{\mu}} \qquad R_{z,\hat{\mu}} \equiv \frac{\rho S_w \bar{c}_w}{4W/g} C_{Z,\hat{\mu}} \qquad R_{m,\hat{\mu}} \equiv \frac{\rho S_w \bar{c}_w^3}{8I_{yy_b}} C_{m,\hat{\mu}}
$$

$$
R_{x,\hat{\alpha}} \equiv \frac{\rho S_w \bar{c}_w}{4W/g} C_{X,\hat{\alpha}} \qquad R_{z,\hat{\alpha}} \equiv \frac{\rho S_w \bar{c}_w}{4W/g} C_{Z,\hat{\alpha}} \qquad R_{m,\hat{\alpha}} \equiv \frac{\rho S_w \bar{c}_w^3}{8I_{yy_b}} C_{m,\hat{\alpha}}
$$

$$
R_{x,\mu} \equiv \frac{\rho S_w \bar{c}_w}{4W/g} \left(2C_X + C_{X,\mu} + \frac{T_V \cos(\alpha_{T0})}{\frac{1}{2}\rho V_o S_w} \right)
$$

$$
R_{z,\mu} \equiv \frac{\rho S_w \bar{c}_w}{4W/g} \left(2C_Z + C_{Z,\mu} - \frac{T_V \sin(\alpha_{T0})}{\frac{1}{2}\rho V_o S_w} \right)
$$

$$
R_{m,\mu} \equiv \frac{\rho S_w \bar{c}_w^3}{8I_{yy_b}} \left(2C_m + C_{m,\mu} + \frac{T_V [z_T \cos(\alpha_{T0}) + x_T \sin(\alpha_{T0})]}{\frac{1}{2}\rho V_o S_w \bar{c}_w} \right)
$$

$$
R_{x,\alpha} \equiv \frac{\rho S_w \bar{c}_w}{4W/g} C_{X,\alpha} \qquad R_{z,\alpha} \equiv \frac{\rho S_w \bar{c}_w}{4W/g} C_{Z,\alpha} \qquad R_{m,\alpha} \equiv \frac{\rho S_w \bar{c}_w^3}{8I_{yy_b}} C_{m,\alpha}
$$

$$
R_{y,\beta} \equiv \frac{\rho S_w b_w}{4W/g} C_{Y,\beta} \qquad R_{\ell,\beta} \equiv \frac{\rho S_w b_w^3}{8I_{xx_b}} C_{\ell,\beta} \qquad R_{n,\beta} \equiv \frac{\rho S_w b_w^3}{8I_{zz_b}} C_{n,\beta} \qquad (7.7.8)
$$

$$
R_{y,\bar{p}} \equiv \frac{\rho S_w b_w}{4W/g} C_{Y,\bar{p}} \qquad R_{\ell,\bar{p}} \equiv \frac{\rho S_w b_w^3}{8I_{xx_b}} C_{\ell,\bar{p}} \qquad R_{n,\bar{p}} \equiv \frac{\rho S_w b_w^3}{8I_{zz_b}} C_{n,\bar{p}}
$$

$$
R_{x,\bar{q}} \equiv \frac{\rho S_w \bar{c}_w}{4W/g} C_{X,\bar{q}} \qquad R_{z,\bar{q}} \equiv \frac{\rho S_w \bar{c}_w}{4W/g} C_{Z,\bar{q}} \qquad R_{m,\bar{q}} \equiv \frac{\rho S_w \bar{c}_w^3}{8I_{yy_b}} C_{m,\bar{q}}
$$

$$
R_{y,\bar{r}} \equiv \frac{\rho S_w b_w}{4W/g} C_{Y,\bar{r}} \qquad R_{\ell,\bar{r}} \equiv \frac{\rho S_w b_w^3}{8I_{xx_b}} C_{\ell,\bar{r}} \qquad R_{n,\bar{r}} \equiv \frac{\rho S_w b_w^3}{8I_{zz_b}} C_{n,\bar{r}}
$$

$$
R_{y,\delta_a} \equiv \frac{\rho S_w b_w}{4W/g} C_{Y,\delta_a} \qquad R_{\ell,\delta_a} \equiv \frac{\rho S_w b_w^3}{8I_{xx_b}} C_{\ell,\delta_a} \qquad R_{n,\delta_a} \equiv \frac{\rho S_w b_w^3}{8I_{zz_b}} C_{n,\delta_a}
$$

$$
R_{x,\delta_e} \equiv \frac{\rho S_w \bar{c}_w}{4W/g} C_{X,\delta_e} \qquad R_{z,\delta_e} \equiv \frac{\rho S_w \bar{c}_w}{4W/g} C_{Z,\delta_e} \qquad R_{m,\delta_e} \equiv \frac{\rho S_w \bar{c}_w^3}{8I_{yy_b}} C_{m,\delta_e}
$$

$$
R_{y,\delta_r} \equiv \frac{\rho S_w b_w}{4W/g} C_{Y,\delta_r} \qquad R_{\ell,\delta_r} \equiv \frac{\rho S_w b_w^3}{8I_{xx_b}} C_{\ell,\delta_r} \qquad R_{n,\delta_r} \equiv \frac{\rho S_w b_w^3}{8I_{zz_b}} C_{n,\delta_r}
$$

All of the dimensionless force and moment derivatives are expressed in terms of the usual force and moment coefficients and their derivatives with respect to the previously defined dimensionless variables. Derivatives with respect to normal acceleration have been replaced by derivatives with respect to the traditional dimensionless rate of change in angle of attack. The normal and sideslip velocity derivatives are replaced with derivatives with respect to the aerodynamic angles, and the derivatives with respect to the angular rates are replaced with their traditional nondimensional counterparts.

In this nondimensional form, many of the aerodynamic derivatives no longer depend on the airspeed. From Eq. (7.6.66), the **traditional nondimensional longitudinal derivatives with respect to translational acceleration** are approximated as

$$C_{X,\hat{\mu}} \equiv \frac{\partial C_X}{\partial \hat{\mu}} \cong C_{Z,\hat{\mu}} \equiv \frac{\partial C_Z}{\partial \hat{\mu}} \cong C_{m,\hat{\mu}} \equiv \frac{\partial C_m}{\partial \hat{\mu}} \cong C_{X,\hat{\alpha}} \equiv \frac{\partial C_X}{\partial \hat{\alpha}} \cong 0$$

$$C_{Z,\hat{\alpha}} \equiv \frac{\partial C_Z}{\partial \hat{\alpha}} = -\frac{\partial C_L}{\partial \hat{\alpha}} = -\frac{2V_o}{\overline{c}_w}\frac{\partial C_L}{\partial \dot{\alpha}} \cong -\eta_h \frac{4 S_h l_{wt}}{\pi b_w^2 \overline{c}_w}\frac{\partial C_{L_w}}{\partial \alpha}\frac{\partial C_{L_h}}{\partial \alpha}$$

$$C_{m,\hat{\alpha}} \equiv \frac{\partial C_m}{\partial \hat{\alpha}} = \frac{2V_o}{\overline{c}_w}\frac{\partial C_m}{\partial \dot{\alpha}} \cong \eta_h \frac{4 S_h l_{wt} x_{bh}}{\pi b_w^2 \overline{c}_w^2}\frac{\partial C_{L_w}}{\partial \alpha}\frac{\partial C_{L_h}}{\partial \alpha} \tag{7.7.9}$$

$$l_{wt} = \begin{cases} 1.1(x_{b\,\text{wingtip}} - x_{bh}), & x_{b\,\text{wingtip}} > x_{bh} \\ 0.0, & x_{bh} > x_{b\,\text{wingtip}} \end{cases}$$

In general, the aerodynamic forces and moments as well as the thrust are functions of forward velocity. However, for equilibrium flight, the change in side force with respect to forward velocity is zero. In addition, the derivatives of the rolling and yawing moments with forward velocity are also zero at equilibrium. Thus, from Eqs. (7.6.33) and (7.6.39), the **nondimensional derivatives with respect to forward velocity** are

$$2C_X + C_{X,\mu} + \frac{T_{,V}\cos(\alpha_{T0})}{\frac{1}{2}\rho V_o S_w} = -2C_D - M C_{D,M} + \frac{T_{,V}\cos(\alpha_{T0})}{\frac{1}{2}\rho V_o S_w}$$

$$\cong -\left(2 + \frac{2M^2}{1-M^2}\frac{C_{D_i}}{C_D}\right)C_D + \frac{T_{,V}\cos(\alpha_{T0})}{\frac{1}{2}\rho V_o S_w}$$

$$2C_Z + C_{Z,\mu} - \frac{T_{,V}\sin(\alpha_{T0})}{\frac{1}{2}\rho V_o S_w} = -2C_L - M C_{L,M} - \frac{T_{,V}\sin(\alpha_{T0})}{\frac{1}{2}\rho V_o S_w}$$

$$\cong -\left(\frac{2-M^2}{1-M^2}\right)C_L - \frac{T_{,V}\sin(\alpha_{T0})}{\frac{1}{2}\rho V_o S_w} \tag{7.7.10}$$

$$2C_m + C_{m,\mu} + \frac{T_{,V}[z_T\cos(\alpha_{T0}) + x_T\sin(\alpha_{T0})]}{\frac{1}{2}\rho V_o S_w \overline{c}_w}$$

$$= 2C_m + M C_{m,M} + \frac{T_{,V}[z_T\cos(\alpha_{T0}) + x_T\sin(\alpha_{T0})]}{\frac{1}{2}\rho V_o S_w \overline{c}_w}$$

$$\cong \left(\frac{2-M^2}{1-M^2}\right)C_m + \frac{T_{,V}[z_T\cos(\alpha_{T0}) + x_T\sin(\alpha_{T0})]}{\frac{1}{2}\rho V_o S_w \overline{c}_w}$$

The stability derivatives with respect to angle of attack and sideslip angle were discussed in Chapters 4 and 5. From Eq. (7.6.43), the **nondimensional derivatives with respect to angle of attack** are

$$
\begin{aligned}
C_{X,\alpha} &\equiv \frac{\partial C_X}{\partial \alpha} = C_L - \frac{\partial C_D}{\partial \alpha} \\
C_{Z,\alpha} &\equiv \frac{\partial C_Z}{\partial \alpha} = -\frac{\partial C_L}{\partial \alpha} - C_D \\
C_{m,\alpha} &\equiv \frac{\partial C_m}{\partial \alpha}
\end{aligned}
\tag{7.7.11}
$$

The pitch stability derivative, $C_{m,\alpha}$, can be estimated from the material presented in Chapter 4. From Eq. (7.6.44), the **nondimensional derivatives with respect to sideslip angle** are

$$
\begin{aligned}
C_{Y,\beta} &\equiv \frac{\partial C_Y}{\partial \beta} \\
C_{\ell,\beta} &\equiv \frac{\partial C_\ell}{\partial \beta} \\
C_{n,\beta} &\equiv \frac{\partial C_n}{\partial \beta}
\end{aligned}
\tag{7.7.12}
$$

These lateral stability derivatives can be estimated from the material that was presented in Chapter 5. From Eq. (7.6.47), the **traditional nondimensional derivatives with respect to roll rate** are approximated as

$$
\begin{aligned}
C_{Y,\overline{p}} &\equiv \frac{\partial C_Y}{\partial \overline{p}} = \frac{2V_o}{b_w}\frac{\partial C_Y}{\partial p} \cong 0 \\
C_{\ell,\overline{p}} &\equiv \frac{\partial C_\ell}{\partial \overline{p}} = \frac{2V_o}{b_w}\frac{\partial C_\ell}{\partial p} \cong -\frac{\kappa_{\ell\overline{p}}}{8}\frac{\partial C_{L_w}}{\partial \alpha} \\
C_{n,\overline{p}} &\equiv \frac{\partial C_n}{\partial \overline{p}} = \frac{2V_o}{b_w}\frac{\partial C_n}{\partial p} \cong -\left(1-\frac{3\kappa_{\ell\overline{p}}}{\pi R_{A_w}}\frac{\partial C_{L_w}}{\partial \alpha}\right)\frac{C_{L_w}}{8}
\end{aligned}
\tag{7.7.13}
$$

From Eq. (7.6.50), the **traditional nondimensional derivatives with respect to pitch rate** are approximated as

$$
\begin{aligned}
C_{X,\overline{q}} &\equiv \frac{\partial C_X}{\partial \overline{q}} = -\frac{\partial C_D}{\partial \overline{q}} = -\frac{2V_o}{\overline{c}_w}\frac{\partial C_D}{\partial q} \cong 0 \\
C_{Z,\overline{q}} &\equiv \frac{\partial C_Z}{\partial \overline{q}} = -\frac{\partial C_L}{\partial \overline{q}} = -\frac{2V_o}{\overline{c}_w}\frac{\partial C_L}{\partial q} \cong \frac{2x_{bw}}{\overline{c}_w}\frac{\partial C_{L_w}}{\partial \alpha} + \eta_h\frac{2S_h x_{bh}}{S_w \overline{c}_w}\frac{\partial C_{L_h}}{\partial \alpha} \\
C_{m,\overline{q}} &\equiv \frac{\partial C_m}{\partial \overline{q}} = \frac{2V_o}{\overline{c}_w}\frac{\partial C_m}{\partial q} \cong -\frac{2x_{bw}^2}{\overline{c}_w^2}\frac{\partial C_{L_w}}{\partial \alpha} - \eta_h\frac{2S_h x_{bh}^2}{S_w \overline{c}_w^2}\frac{\partial C_{L_h}}{\partial \alpha}
\end{aligned}
\tag{7.7.14}
$$

From Eq. (7.6.55), the **traditional nondimensional derivatives with respect to yaw rate** are approximated as

$$
\begin{aligned}
C_{Y,\bar{r}} &\equiv \frac{\partial C_Y}{\partial \bar{r}} = \frac{2V_o}{b_w}\frac{\partial C_Y}{\partial r} \cong -\eta_v \frac{2S_v x_{bv}}{S_w b_w}\frac{\partial C_{L_v}}{\partial \beta} \\
C_{\ell,\bar{r}} &\equiv \frac{\partial C_\ell}{\partial \bar{r}} = \frac{2V_o}{b_w}\frac{\partial C_\ell}{\partial r} \cong \left(1 - \frac{3\kappa_{\bar{p}}}{\pi R_{A_w}}\frac{\partial C_{L_w}}{\partial \alpha}\right)\frac{C_{L_w}}{4} + \eta_v \frac{2S_v x_{bv} z_{bv}}{S_w b_w^2}\frac{\partial C_{L_v}}{\partial \beta} \quad (7.7.15) \\
C_{n,\bar{r}} &\equiv \frac{\partial C_n}{\partial \bar{r}} = \frac{2V_o}{b_w}\frac{\partial C_n}{\partial r} \cong -\eta_v \frac{2S_v x_{bv}^2}{S_w b_w^2}\frac{\partial C_{L_v}}{\partial \beta}
\end{aligned}
$$

Equations (7.7.5) and (7.7.6) are both nonhomogeneous systems of differential equations with constant coefficients. The general solution to such a system of equations is the sum of the homogeneous solution and a particular solution. The homogeneous solution is the solution to the system with all of the control inputs set to zero. The particular solution is any set of values for the state variables that when substituted into the nonhomogeneous system will satisfy all equations.

EXAMPLE 7.7.1. The large jet transport described in Example 7.6.1 has a pitching moment of inertia of 3.31×10^7 slug·ft^2. Using the same operating conditions and assumptions that were used in Example 7.6.1, fill out the nondimensional longitudinal coefficient matrices associated with the homogeneous form of Eq. (7.7.5).

Solution. From Example 7.6.1, for this airplane and flight condition we have

$$
S_w = 5{,}500 \text{ ft}^2, \quad b_w = 196 \text{ ft}, \quad \bar{c}_w = 28 \text{ ft}, \quad x_{bw} = 0, \quad C_{L_w,\alpha} = 4.67,
$$
$$
W = 636{,}600 \text{ lbf}, \quad C_{L,\alpha} = 5.50, \quad C_{m,\alpha} = -1.26, \quad C_{D,\alpha} = 0.393,
$$
$$
S_h = 1{,}300 \text{ ft}^2, \quad x_{bh} = -100 \text{ ft}, \quad l_{wt} = 55 \text{ ft}, \quad C_{L_h,\alpha} = 3.50,
$$
$$
V_o = 660 \text{ ft/sec}, \quad M = 0.663, \quad C_L/C_D = 14, \quad C_{D_i}/C_D = 0.5,
$$
$$
I_{yy_b} = 3.31 \times 10^7 \text{ slug·ft}^2
$$

At the minimum drag airspeed the lift-to-drag ratio is equal to the maximum lift-to-drag ratio, and since the thrust is aligned with direction of flight, we have

$$
\alpha_T = 0, \quad L = W
$$

$$
C_L \equiv \frac{L}{\frac{1}{2}\rho V_o^2 S_w} = \frac{W}{\frac{1}{2}\rho V_o^2 S_w} = \frac{636{,}600 \text{ lbf}}{\frac{1}{2}(0.00089068 \text{ slug/ft}^3)(660 \text{ ft/sec})^2 (5{,}500 \text{ ft}^2)}
$$
$$
= 0.597
$$

$$
C_D = \frac{C_L}{C_L/C_D} = \frac{0.597}{14} = 0.0426
$$

Thus, from Eq. (7.7.9), the approximate nondimensional derivatives with respect to translational acceleration are

$$C_{X,\hat{\mu}} \cong C_{Z,\hat{\mu}} \cong C_{m,\hat{\mu}} \cong C_{X,\hat{\alpha}} \cong 0$$

$$C_{Z,\hat{\alpha}} \cong -\eta_h \frac{4S_h l_{wt}}{\pi b_w^2 \bar{c}_w} \frac{\partial C_{L_w}}{\partial \alpha} \frac{\partial C_{L_h}}{\partial \alpha} = -1.0 \frac{4(1{,}300 \text{ ft}^2)(55 \text{ ft})}{\pi (196 \text{ ft})^2 (28 \text{ ft})}(4.67)3.50 = -1.38$$

$$C_{m,\hat{\alpha}} \cong \eta_h \frac{4S_h l_{wt} x_{bh}}{\pi b_w^2 \bar{c}_w^2} \frac{\partial C_{L_w}}{\partial \alpha} \frac{\partial C_{L_h}}{\partial \alpha} = 1.0 \frac{4(1{,}300 \text{ ft}^2)(55 \text{ ft})(-100 \text{ ft})}{\pi (196 \text{ ft})^2 (28 \text{ ft})^2}(4.67)3.50$$

$$= -4.94$$

From Eq. (7.7.10), the approximate nondimensional derivatives with respect to forward velocity are

$$2C_X + C_{X,\mu} + \frac{T_{,V} \cos(\alpha_{T0})}{\frac{1}{2}\rho V_o S_w} = -2C_D - M C_{D,M} \cong -\left(2 + \frac{2M^2}{1-M^2} \frac{C_{D_i}}{C_D}\right)C_D$$

$$= -\left(2 + \frac{2(0.663)^2}{1-0.663^2} 0.5\right)0.0426 = -0.1186$$

$$2C_Z + C_{Z,\mu} - \frac{T_{,V} \sin(\alpha_{T0})}{\frac{1}{2}\rho V_o S_w} = -2C_L - M C_{L,M} \cong -\left(\frac{2-M^2}{1-M^2}\right)C_L$$

$$= -\left(\frac{2-0.663^2}{1-0.663^2}\right)0.597 = -1.662$$

and since the thrust is aligned with the center of gravity, the aerodynamic moment for the reference state must be zero:

$$2C_m + C_{m,\mu} + \frac{T_{,V}[z_T \cos(\alpha_{T0}) + x_T \sin(\alpha_{T0})]}{\frac{1}{2}\rho V_o S_w \bar{c}_w} \cong \left(\frac{2-M^2}{1-M^2}\right)C_m = 0.0$$

From Eq. (7.7.11), the nondimensional derivatives with respect to angle of attack are

$$C_{X,\alpha} = C_L - C_{D,\alpha} = 0.597 - 0.393 = 0.204$$

$$C_{Z,\alpha} = -C_{L,\alpha} - C_D = -5.50 - 0.0426 = -5.54$$

$$C_{m,\alpha} = -1.26$$

From Eq. (7.7.14), the nondimensional derivatives with respect to pitch rate are

$$C_{X,\bar{q}} \cong 0$$

$$C_{Z,\bar{q}} \cong \eta_h \frac{2S_h x_{bh}}{S_w \bar{c}_w} \frac{\partial C_{L_h}}{\partial \alpha} = 1.0 \frac{2(1,300 \text{ ft}^2)(-100 \text{ ft})}{5,500 \text{ ft}^2 (28 \text{ ft})} 3.50 = -5.91$$

$$C_{m,\bar{q}} \cong -\eta_h \frac{2S_h x_{bh}^2}{S_w \bar{c}_w^2} \frac{\partial C_{L_h}}{\partial \alpha} = -1.0 \frac{2(1,300 \text{ ft}^2)(-100 \text{ ft})^2}{5,500 \text{ ft}^2 (28 \text{ ft})^2} 3.50 = -21.10$$

From Eq. (7.7.8), we have

$$\frac{\rho S_w \bar{c}_w}{4W/g} = \frac{0.00089068 \text{ slug/ft}^3 (5,500 \text{ ft}^2)(28 \text{ ft})}{4(636,600/32.17) \text{ slug}} = 0.001733$$

$$\frac{\rho S_w \bar{c}_w^3}{8I_{yy_b}} = \frac{0.00089068 \text{ slug/ft}^3 (5,500 \text{ ft}^2)(28 \text{ ft})^3}{8(3.31 \times 10^7 \text{ slug} \cdot \text{ft}^2)} = 0.0004061$$

and

$$R_{gx} \equiv \frac{g\bar{c}_w}{2V_o^2} = \frac{32.17 \text{ ft/sec}^2 (28 \text{ ft})}{2(660 \text{ ft/sec})^2} = 0.00103$$

$$R_{x,\hat{\mu}} \equiv \frac{\rho S_w \bar{c}_w}{4W/g} C_{X,\hat{\mu}} \cong 0.0$$

$$R_{z,\hat{\mu}} \equiv \frac{\rho S_w \bar{c}_w}{4W/g} C_{Z,\hat{\mu}} \cong 0.0$$

$$R_{m,\hat{\mu}} \equiv \frac{\rho S_w \bar{c}_w^3}{8I_{yy_b}} C_{m,\hat{\mu}} \cong 0.0$$

$$R_{x,\hat{\alpha}} \equiv \frac{\rho S_w \bar{c}_w}{4W/g} C_{X,\hat{\alpha}} \cong 0.0$$

$$R_{z,\hat{\alpha}} \equiv \frac{\rho S_w \bar{c}_w}{4W/g} C_{Z,\hat{\alpha}} = 0.001733(-1.38) = -0.00239$$

$$R_{m,\hat{\alpha}} \equiv \frac{\rho S_w \bar{c}_w^3}{8I_{yy_b}} C_{m,\hat{\alpha}} = 0.0004061(-4.94) = -0.00201$$

$$R_{x,\mu} \equiv \frac{\rho S_w \bar{c}_w}{4W/g} \left(2C_X + C_{X,\mu} + \frac{T_V \cos(\alpha_{T0})}{\frac{1}{2}\rho V_o S_w} \right) = 0.001733(-0.1186)$$

$$= -0.000206$$

$$R_{z,\mu} \equiv \frac{\rho S_w \bar{c}_w}{4W/g} \left(2C_Z + C_{Z,\mu} - \frac{T_{,V} \sin(\alpha_{T0})}{\frac{1}{2}\rho V_o S_w} \right) = 0.001733(-1.662)$$

$$= -0.00288$$

$$R_{m,\mu} \equiv \frac{\rho S_w \bar{c}_w^3}{8I_{yy_b}} \left(2C_m + C_{m,\mu} + \frac{T_{,V}[z_T \cos(\alpha_{T0}) + x_T \sin(\alpha_{T0})]}{\frac{1}{2}\rho V_o S_w \bar{c}_w} \right) = 0.0$$

$$R_{x,\alpha} \equiv \frac{\rho S_w \bar{c}_w}{4W/g} C_{X,\alpha} = 0.001733(0.204) = 0.000354$$

$$R_{z,\alpha} \equiv \frac{\rho S_w \bar{c}_w}{4W/g} C_{Z,\alpha} = 0.001733(-5.54) = -0.00960$$

$$R_{m,\alpha} \equiv \frac{\rho S_w \bar{c}_w^3}{8I_{yy_b}} C_{m,\alpha} = 0.0004061(-1.26) = -0.000512$$

$$R_{x,\bar{q}} \equiv \frac{\rho S_w \bar{c}_w}{4W/g} C_{X,\bar{q}} = 0.0$$

$$R_{z,\bar{q}} \equiv \frac{\rho S_w \bar{c}_w}{4W/g} C_{Z,\bar{q}} = 0.001733(-5.91) = -0.0102$$

$$R_{m,\bar{q}} \equiv \frac{\rho S_w \bar{c}_w^3}{8I_{yy_b}} C_{m,\bar{q}} = 0.0004061(-21.10) = -0.00857$$

Using these results in Eq. (7.7.5), the nondimensional form of the homogeneous linearized system of equations for longitudinal motion is

$$\begin{bmatrix} 1 & 0 & 0 & 0 & 0 & 0 \\ 0 & 1.00239 & 0 & 0 & 0 & 0 \\ 0 & 0.00201 & 1 & 0 & 0 & 0 \\ 0 & 0 & 0 & 1 & 0 & 0 \\ 0 & 0 & 0 & 0 & 1 & 0 \\ 0 & 0 & 0 & 0 & 0 & 1 \end{bmatrix} \begin{Bmatrix} \Delta\hat{\mu} \\ \Delta\hat{\alpha} \\ \Delta\hat{\bar{q}} \\ \Delta\hat{\xi}_x \\ \Delta\hat{\xi}_z \\ \Delta\hat{\theta} \end{Bmatrix}$$

$$= \begin{bmatrix} -0.000206 & 0.000354 & 0 & 0 & 0 & -0.00103 \\ -0.00288 & -0.00960 & 0.9898 & 0 & 0 & 0 \\ 0 & -0.000512 & -0.00857 & 0 & 0 & 0 \\ 1 & 0 & 0 & 0 & 0 & 0 \\ 0 & 1 & 0 & 0 & 0 & -1 \\ 0 & 0 & 1 & 0 & 0 & 0 \end{bmatrix} \begin{Bmatrix} \Delta\mu \\ \Delta\alpha \\ \Delta\bar{q} \\ \Delta\xi_x \\ \Delta\xi_z \\ \Delta\theta \end{Bmatrix}$$

7.8. Transformation of Stability Axes

In our original definition of the body-fixed coordinate system, the origin was fixed to the aircraft center of gravity; the x_b-axis was defined to point "forward" along "some convenient line" in the aircraft's plane of symmetry; and the z_b-axis was defined to point "downward" in the plane of symmetry. Thus, the y_b-axis is normal to the plane of symmetry pointing in the direction of the right wing. When developing the linearized equations of motion, we found that the most convenient orientation for the x_b-axis was pointing in the direction of the aircraft's equilibrium velocity relative to the surrounding air. This particular body-fixed coordinate system is usually called the *stability axes*.

Aligning our body-fixed coordinate system with the stability axes significantly simplifies the linearized equations of motion. However, since angle of attack is a function of aircraft operating parameters such as airspeed and elevation angle, the orientation of the stability axes within the plane of symmetry depends on the equilibrium operating conditions. The inertia tensor and the aerodynamic derivatives are measured relative to some reference body-fixed coordinate system, which may or may not be aligned with the stability axes for a given equilibrium operating condition. Thus, it is convenient to be able to transform the inertia tensor and the aerodynamic derivatives from one body-fixed coordinate system to another within the aircraft plane of symmetry.

Consider the two different body-fixed coordinate systems shown in Fig. 7.8.1. For both of these coordinate systems the x-z plane is coincident with the plane of symmetry. Thus, transformation from one of these coordinate systems to the other is a simple rotation about the y_b-axis. For an arbitrary vector, **v**, we can write

$$\begin{Bmatrix} v_{x_b} \\ v_{y_b} \\ v_{z_b} \end{Bmatrix} = \begin{bmatrix} \cos\varphi & 0 & \sin\varphi \\ 0 & 1 & 0 \\ -\sin\varphi & 0 & \cos\varphi \end{bmatrix} \begin{Bmatrix} v_{x_b'} \\ v_{y_b'} \\ v_{z_b'} \end{Bmatrix} \tag{7.8.1}$$

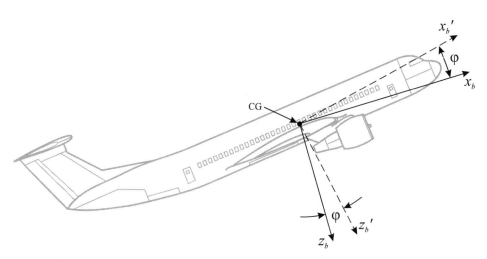

Figure 7.8.1. Two choices for the body-fixed coordinate system.

and

$$\begin{Bmatrix} v_{x_b{'}} \\ v_{y_b{'}} \\ v_{z_b{'}} \end{Bmatrix} = \begin{bmatrix} \cos\varphi & 0 & -\sin\varphi \\ 0 & 1 & 0 \\ \sin\varphi & 0 & \cos\varphi \end{bmatrix} \begin{Bmatrix} v_{x_b} \\ v_{y_b} \\ v_{z_b} \end{Bmatrix} \tag{7.8.2}$$

The angular momentum in the body-fixed coordinate system (x_b, y_b, z_b) is

$$\begin{Bmatrix} H_{x_b} \\ H_{y_b} \\ H_{z_b} \end{Bmatrix} = \begin{bmatrix} I_{xx_b} & 0 & -I_{xz_b} \\ 0 & I_{yy_b} & 0 \\ -I_{zx_b} & 0 & I_{zz_b} \end{bmatrix} \begin{Bmatrix} p \\ q \\ r \end{Bmatrix} \tag{7.8.3}$$

Transforming both sides of Eq. (7.8.3) using Eq. (7.8.2), we have

$$\begin{Bmatrix} H_{x_b{'}} \\ H_{y_b{'}} \\ H_{z_b{'}} \end{Bmatrix} = \begin{bmatrix} \cos\varphi & 0 & -\sin\varphi \\ 0 & 1 & 0 \\ \sin\varphi & 0 & \cos\varphi \end{bmatrix} \begin{bmatrix} I_{xx_b} & 0 & -I_{xz_b} \\ 0 & I_{yy_b} & 0 \\ -I_{zx_b} & 0 & I_{zz_b} \end{bmatrix} \begin{Bmatrix} p \\ q \\ r \end{Bmatrix} \tag{7.8.4}$$

The angular velocity vector on the right-hand side of Eq. (7.8.4) can be written in terms of the body-fixed coordinate system $(x_b{'}, y_b{'}, z_b{'})$ by using Eq. (7.8.1),

$$\begin{Bmatrix} H_{x_b{'}} \\ H_{y_b{'}} \\ H_{z_b{'}} \end{Bmatrix} = \begin{bmatrix} \cos\varphi & 0 & -\sin\varphi \\ 0 & 1 & 0 \\ \sin\varphi & 0 & \cos\varphi \end{bmatrix} \begin{bmatrix} I_{xx_b} & 0 & -I_{xz_b} \\ 0 & I_{yy_b} & 0 \\ -I_{zx_b} & 0 & I_{zz_b} \end{bmatrix} \begin{bmatrix} \cos\varphi & 0 & \sin\varphi \\ 0 & 1 & 0 \\ -\sin\varphi & 0 & \cos\varphi \end{bmatrix} \begin{Bmatrix} p{'} \\ q{'} \\ r{'} \end{Bmatrix} \tag{7.8.5}$$

The angular momentum in the body-fixed coordinate system $(x_b{'}, y_b{'}, z_b{'})$ can also be written in terms of the inertia tensor written in the same coordinate system,

$$\begin{Bmatrix} H_{x_b{'}} \\ H_{y_b{'}} \\ H_{z_b{'}} \end{Bmatrix} = [\mathbf{I}]{'} \begin{Bmatrix} p{'} \\ q{'} \\ r{'} \end{Bmatrix} = \begin{bmatrix} I_{xx_b} & 0 & -I_{xz_b} \\ 0 & I_{yy_b} & 0 \\ -I_{zx_b} & 0 & I_{zz_b} \end{bmatrix}{'} \begin{Bmatrix} p{'} \\ q{'} \\ r{'} \end{Bmatrix} \tag{7.8.6}$$

Combining Eqs. (7.8.5) and (7.8.6), we see that the inertia tensor written in the body-fixed coordinate system $(x_b{'}, y_b{'}, z_b{'})$ can be found from

$$[\mathbf{I}]{'} = \begin{bmatrix} \cos\varphi & 0 & -\sin\varphi \\ 0 & 1 & 0 \\ \sin\varphi & 0 & \cos\varphi \end{bmatrix} \begin{bmatrix} I_{xx_b} & 0 & -I_{xz_b} \\ 0 & I_{yy_b} & 0 \\ -I_{zx_b} & 0 & I_{zz_b} \end{bmatrix} \begin{bmatrix} \cos\varphi & 0 & \sin\varphi \\ 0 & 1 & 0 \\ -\sin\varphi & 0 & \cos\varphi \end{bmatrix} \tag{7.8.7}$$

Equation (7.8.7) can be used to transform the inertia tensor from one body-fixed coordinate system to another whenever the two coordinate systems are related to each other through a simple rotation of φ about the y_b-axis from x_b to x_b'. Multiplying the three matrices on the right-hand side of Eq. (7.8.7), we obtain the component transformation equations

$$
\begin{aligned}
\left(I_{xx_b}\right)' &= I_{xx_b}\cos^2\varphi + 2I_{xz_b}\cos\varphi\sin\varphi + I_{zz_b}\sin^2\varphi \\
\left(I_{yy_b}\right)' &= I_{yy_b} \\
\left(I_{zz_b}\right)' &= I_{zz_b}\cos^2\varphi - 2I_{xz_b}\cos\varphi\sin\varphi + I_{xx_b}\sin^2\varphi \\
\left(I_{xz_b}\right)' &= I_{xz_b}(\cos^2\varphi - \sin^2\varphi) + (I_{zz_b} - I_{xx_b})\cos\varphi\sin\varphi
\end{aligned}
\tag{7.8.8}
$$

The aerodynamic force and moment derivatives for the airplane can be transformed in a similar manner. In the linearized equations of motion, the aerodynamic force disturbances are written as

$$
\begin{Bmatrix} \Delta F_{x_b} \\ \Delta F_{y_b} \\ \Delta F_{z_b} \end{Bmatrix} =
\begin{bmatrix} F_{x_b,u} & 0 & F_{x_b,w} \\ 0 & F_{y_b,v} & 0 \\ F_{z_b,u} & 0 & F_{z_b,w} \end{bmatrix}
\begin{Bmatrix} \Delta u \\ \Delta v \\ \Delta w \end{Bmatrix} +
\begin{bmatrix} 0 & F_{x_b,q} & 0 \\ F_{y_b,p} & 0 & F_{y_b,r} \\ 0 & F_{z_b,q} & 0 \end{bmatrix}
\begin{Bmatrix} \Delta p \\ \Delta q \\ \Delta r \end{Bmatrix}
$$

$$
+ \begin{bmatrix} F_{x_b,\dot{u}} & 0 & F_{x_b,\dot{w}} \\ 0 & 0 & 0 \\ F_{z_b,\dot{u}} & 0 & F_{z_b,\dot{w}} \end{bmatrix}
\begin{Bmatrix} \Delta\dot{u} \\ \Delta\dot{v} \\ \Delta\dot{w} \end{Bmatrix} +
\begin{bmatrix} 0 & F_{x_b,\delta_e} & 0 \\ F_{y_b,\delta_a} & 0 & F_{y_b,\delta_r} \\ 0 & F_{z_b,\delta_e} & 0 \end{bmatrix}
\begin{Bmatrix} \Delta\delta_a \\ \Delta\delta_e \\ \Delta\delta_r \end{Bmatrix}
\tag{7.8.9}
$$

and the aerodynamic moment disturbances are

$$
\begin{Bmatrix} \Delta M_{x_b} \\ \Delta M_{y_b} \\ \Delta M_{z_b} \end{Bmatrix} =
\begin{bmatrix} 0 & M_{x_b,v} & 0 \\ M_{y_b,u} & 0 & M_{y_b,w} \\ 0 & M_{z_b,v} & 0 \end{bmatrix}
\begin{Bmatrix} \Delta u \\ \Delta v \\ \Delta w \end{Bmatrix} +
\begin{bmatrix} M_{x_b,p} & 0 & M_{x_b,r} \\ 0 & M_{y_b,q} & 0 \\ M_{z_b,p} & 0 & M_{z_b,r} \end{bmatrix}
\begin{Bmatrix} \Delta p \\ \Delta q \\ \Delta r \end{Bmatrix}
$$

$$
+ \begin{bmatrix} 0 & 0 & 0 \\ M_{y_b,\dot{u}} & 0 & M_{y_b,\dot{w}} \\ 0 & 0 & 0 \end{bmatrix}
\begin{Bmatrix} \Delta\dot{u} \\ \Delta\dot{v} \\ \Delta\dot{w} \end{Bmatrix} +
\begin{bmatrix} M_{x_b,\delta_a} & 0 & M_{x_b,\delta_r} \\ 0 & M_{y_b,\delta_e} & 0 \\ M_{z_b,\delta_a} & 0 & M_{z_b,\delta_r} \end{bmatrix}
\begin{Bmatrix} \Delta\delta_a \\ \Delta\delta_e \\ \Delta\delta_r \end{Bmatrix}
\tag{7.8.10}
$$

Each of the eight matrices in Eqs. (7.8.9) and (7.8.10) can be transformed using Eqs. (7.8.1) and (7.8.2) in a manner similar to that used to obtain Eq. (7.8.7). For example,

$$
\begin{bmatrix} F_{x_b,\dot{u}} & 0 & F_{x_b,\dot{w}} \\ 0 & 0 & 0 \\ F_{z_b,\dot{u}} & 0 & F_{z_b,\dot{w}} \end{bmatrix}' =
\begin{bmatrix} \cos\varphi & 0 & -\sin\varphi \\ 0 & 1 & 0 \\ \sin\varphi & 0 & \cos\varphi \end{bmatrix}
\begin{bmatrix} F_{x_b,\dot{u}} & 0 & F_{x_b,\dot{w}} \\ 0 & 0 & 0 \\ F_{z_b,\dot{u}} & 0 & F_{z_b,\dot{w}} \end{bmatrix}
\begin{bmatrix} \cos\varphi & 0 & \sin\varphi \\ 0 & 1 & 0 \\ -\sin\varphi & 0 & \cos\varphi \end{bmatrix}
\tag{7.8.11}
$$

and

$$
\begin{bmatrix} 0 & F_{x_b,\delta_e} & 0 \\ F_{y_b,\delta_a} & 0 & F_{y_b,\delta_r} \\ 0 & F_{z_b,\delta_e} & 0 \end{bmatrix}' = \begin{bmatrix} \cos\varphi & 0 & -\sin\varphi \\ 0 & 1 & 0 \\ \sin\varphi & 0 & \cos\varphi \end{bmatrix} \begin{bmatrix} 0 & F_{x_b,\delta_e} & 0 \\ F_{y_b,\delta_a} & 0 & F_{y_b,\delta_r} \\ 0 & F_{z_b,\delta_e} & 0 \end{bmatrix} \quad (7.8.12)
$$

When the matrix of aerodynamic derivatives in Eq. (7.8.12) is transformed, it is only premultiplied by the transformation matrix from Eq. (7.8.2). It is not postmultiplied by the transformation matrix from Eq. (7.8.1), because only the output vector for the last term in Eq. (7.8.9) depends on the coordinate system. The input vector of control surface deflections is the same in both coordinate systems. Similarly, we can also write

$$
\begin{bmatrix} F_{x_b,u} & 0 & F_{x_b,w} \\ 0 & F_{y_b,v} & 0 \\ F_{z_b,u} & 0 & F_{z_b,w} \end{bmatrix}'
$$
$$
= \begin{bmatrix} \cos\varphi & 0 & -\sin\varphi \\ 0 & 1 & 0 \\ \sin\varphi & 0 & \cos\varphi \end{bmatrix} \begin{bmatrix} F_{x_b,u} & 0 & F_{x_b,w} \\ 0 & F_{y_b,v} & 0 \\ F_{z_b,u} & 0 & F_{z_b,w} \end{bmatrix} \begin{bmatrix} \cos\varphi & 0 & \sin\varphi \\ 0 & 1 & 0 \\ -\sin\varphi & 0 & \cos\varphi \end{bmatrix}
$$

$$
\begin{bmatrix} 0 & F_{x_b,q} & 0 \\ F_{y_b,p} & 0 & F_{y_b,r} \\ 0 & F_{z_b,q} & 0 \end{bmatrix}'
$$
$$
= \begin{bmatrix} \cos\varphi & 0 & -\sin\varphi \\ 0 & 1 & 0 \\ \sin\varphi & 0 & \cos\varphi \end{bmatrix} \begin{bmatrix} 0 & F_{x_b,q} & 0 \\ F_{y_b,p} & 0 & F_{y_b,r} \\ 0 & F_{z_b,q} & 0 \end{bmatrix} \begin{bmatrix} \cos\varphi & 0 & \sin\varphi \\ 0 & 1 & 0 \\ -\sin\varphi & 0 & \cos\varphi \end{bmatrix}
$$

$$
\begin{bmatrix} 0 & M_{x_b,v} & 0 \\ M_{y_b,u} & 0 & M_{y_b,w} \\ 0 & M_{z_b,v} & 0 \end{bmatrix}'
$$
$$
= \begin{bmatrix} \cos\varphi & 0 & -\sin\varphi \\ 0 & 1 & 0 \\ \sin\varphi & 0 & \cos\varphi \end{bmatrix} \begin{bmatrix} 0 & M_{x_b,v} & 0 \\ M_{y_b,u} & 0 & M_{y_b,w} \\ 0 & M_{z_b,v} & 0 \end{bmatrix} \begin{bmatrix} \cos\varphi & 0 & \sin\varphi \\ 0 & 1 & 0 \\ -\sin\varphi & 0 & \cos\varphi \end{bmatrix}
$$

$$
\begin{bmatrix} M_{x_b,\delta_a} & 0 & M_{x_b,\delta_r} \\ 0 & M_{y_b,\delta_e} & 0 \\ M_{z_b,\delta_a} & 0 & M_{z_b,\delta_r} \end{bmatrix}' = \begin{bmatrix} \cos\varphi & 0 & -\sin\varphi \\ 0 & 1 & 0 \\ \sin\varphi & 0 & \cos\varphi \end{bmatrix} \begin{bmatrix} M_{x_b,\delta_a} & 0 & M_{x_b,\delta_r} \\ 0 & M_{y_b,\delta_e} & 0 \\ M_{z_b,\delta_a} & 0 & M_{z_b,\delta_r} \end{bmatrix}
$$

and so on.

By transforming each of the eight individual matrices in Eqs. (7.8.9) and (7.8.10), we obtain the component transformation equations for the **longitudinal aerodynamic derivatives**:

$$\left(F_{x_b,u}\right)' = F_{x_b,u} \cos^2\varphi - \left(F_{x_b,w} + F_{z_b,u}\right)\cos\varphi\sin\varphi + F_{z_b,w} \sin^2\varphi$$

$$\left(F_{x_b,w}\right)' = F_{x_b,w} \cos^2\varphi + \left(F_{x_b,u} - F_{z_b,w}\right)\cos\varphi\sin\varphi - F_{z_b,u} \sin^2\varphi$$

$$\left(F_{x_b,\dot{u}}\right)' = F_{x_b,\dot{u}} \cos^2\varphi - \left(F_{x_b,\dot{w}} + F_{z_b,\dot{u}}\right)\cos\varphi\sin\varphi + F_{z_b,\dot{w}} \sin^2\varphi$$

$$\left(F_{x_b,\dot{w}}\right)' = F_{x_b,\dot{w}} \cos^2\varphi + \left(F_{x_b,\dot{u}} - F_{z_b,\dot{w}}\right)\cos\varphi\sin\varphi - F_{z_b,\dot{u}} \sin^2\varphi$$

$$\left(F_{x_b,q}\right)' = F_{x_b,q} \cos\varphi - F_{z_b,q} \sin\varphi$$

$$\left(F_{x_b,\delta_e}\right)' = F_{x_b,\delta_e} \cos\varphi - F_{z_b,\delta_e} \sin\varphi$$

$$\left(F_{z_b,u}\right)' = F_{z_b,u} \cos^2\varphi + \left(F_{x_b,u} - F_{z_b,w}\right)\cos\varphi\sin\varphi - F_{x_b,w} \sin^2\varphi$$

$$\left(F_{z_b,w}\right)' = F_{z_b,w} \cos^2\varphi + \left(F_{x_b,w} + F_{z_b,u}\right)\cos\varphi\sin\varphi + F_{x_b,u} \sin^2\varphi$$

$$\left(F_{z_b,\dot{u}}\right)' = F_{z_b,\dot{u}} \cos^2\varphi + \left(F_{x_b,\dot{u}} - F_{z_b,\dot{w}}\right)\cos\varphi\sin\varphi - F_{x_b,\dot{w}} \sin^2\varphi \qquad (7.8.13)$$

$$\left(F_{z_b,\dot{w}}\right)' = F_{z_b,\dot{w}} \cos^2\varphi + \left(F_{x_b,\dot{w}} + F_{z_b,\dot{u}}\right)\cos\varphi\sin\varphi + F_{x_b,\dot{u}} \sin^2\varphi$$

$$\left(F_{z_b,q}\right)' = F_{z_b,q} \cos\varphi + F_{x_b,q} \sin\varphi$$

$$\left(F_{z_b,\delta_e}\right)' = F_{z_b,\delta_e} \cos\varphi + F_{x_b,\delta_e} \sin\varphi$$

$$\left(M_{y_b,u}\right)' = M_{y_b,u} \cos\varphi - M_{y_b,w} \sin\varphi$$

$$\left(M_{y_b,w}\right)' = M_{y_b,w} \cos\varphi + M_{y_b,u} \sin\varphi$$

$$\left(M_{y_b,\dot{u}}\right)' = M_{y_b,\dot{u}} \cos\varphi - M_{y_b,\dot{w}} \sin\varphi$$

$$\left(M_{y_b,\dot{w}}\right)' = M_{y_b,\dot{w}} \cos\varphi + M_{y_b,\dot{u}} \sin\varphi$$

$$\left(M_{y_b,q}\right)' = M_{y_b,q}$$

$$\left(M_{y_b,\delta_e}\right)' = M_{y_b,\delta_e}$$

and the component transformation equations for the **lateral aerodynamic derivatives**:

$$\left(F_{y_b,v}\right)' = F_{y_b,v}$$

$$\left(F_{y_b,p}\right)' = F_{y_b,p}\cos\varphi - F_{y_b,r}\sin\varphi$$

$$\left(F_{y_b,r}\right)' = F_{y_b,r}\cos\varphi + F_{y_b,p}\sin\varphi$$

$$\left(F_{y_b,\delta_a}\right)' = F_{y_b,\delta_a}$$

$$\left(F_{y_b,\delta_r}\right)' = F_{y_b,\delta_r}$$

$$\left(M_{x_b,v}\right)' = M_{x_b,v}\cos\varphi - M_{z_b,v}\sin\varphi$$

$$\left(M_{x_b,p}\right)' = M_{x_b,p}\cos^2\varphi - \left(M_{x_b,r} + M_{z_b,p}\right)\cos\varphi\sin\varphi + M_{z_b,r}\sin^2\varphi$$

$$\left(M_{x_b,r}\right)' = M_{x_b,r}\cos^2\varphi + \left(M_{x_b,p} - M_{z_b,r}\right)\cos\varphi\sin\varphi - M_{z_b,p}\sin^2\varphi \qquad (7.8.14)$$

$$\left(M_{x_b,\delta_a}\right)' = M_{x_b,\delta_a}\cos\varphi - M_{z_b,\delta_a}\sin\varphi$$

$$\left(M_{x_b,\delta_r}\right)' = M_{x_b,\delta_r}\cos\varphi - M_{z_b,\delta_r}\sin\varphi$$

$$\left(M_{z_b,v}\right)' = M_{z_b,v}\cos\varphi + M_{x_b,v}\sin\varphi$$

$$\left(M_{z_b,p}\right)' = M_{z_b,p}\cos^2\varphi + \left(M_{x_b,p} - M_{z_b,r}\right)\cos\varphi\sin\varphi - M_{x_b,r}\sin^2\varphi$$

$$\left(M_{z_b,r}\right)' = M_{z_b,r}\cos^2\varphi + \left(M_{x_b,r} + M_{z_b,p}\right)\cos\varphi\sin\varphi + M_{x_b,p}\sin^2\varphi$$

$$\left(M_{z_b,\delta_a}\right)' = M_{z_b,\delta_a}\cos\varphi + M_{x_b,\delta_a}\sin\varphi$$

$$\left(M_{z_b,\delta_r}\right)' = M_{z_b,\delta_r}\cos\varphi + M_{x_b,\delta_r}\sin\varphi$$

The orientation of the stability axes for any given operating condition depends on some of the aerodynamic derivatives. Thus, we cannot know the exact orientation of the stability axes prior to determining these derivatives. For this reason, it is not convenient to directly measure the aerodynamic derivatives relative to the stability axes. However, the aerodynamic derivatives and inertia tensor can be measured relative to another convenient reference system, and the results can be transformed to the stability axes using Eqs. (7.8.8), (7.8.13), and (7.8.14).

EXAMPLE 7.8.1. The large jet transport described in Examples 7.6.1 and 7.7.1 has the following moments of inertia and product of inertia relative to the zero-lift axes of the airplane:

$$I_{xx_b} = 1.888 \times 10^7 \text{ slug} \cdot \text{ft}^2$$
$$I_{yy_b} = 3.310 \times 10^7 \text{ slug} \cdot \text{ft}^2$$
$$I_{zz_b} = 4.902 \times 10^7 \text{ slug} \cdot \text{ft}^2$$
$$I_{xz_b} = 4.075 \times 10^6 \text{ slug} \cdot \text{ft}^2$$

For the same flight condition used in Examples 7.6.1 and 7.7.1, estimate the components of the inertia tensor relative to the stability axes.

Solution. From Example 7.7.1, for this airplane and flight condition we have

$$C_L = 0.597, \quad C_{L,\alpha} = 5.50$$

Assuming the lift to be linear with angle of attack, we can approximate the equilibrium angle of attack relative to the zero-lift axes as

$$\alpha_o = \frac{C_L}{C_{L,\alpha}} = \frac{0.597}{5.50} = 0.1085 \text{ rad} = 6.22°$$

Since the equilibrium angle of attack is zero relative to the stability axes, the rotation about the y_b-axis from the zero-lift axes to the stability axes is

$$\varphi = -\alpha_o = -6.22°$$

Using Eq. (7.8.8), the four nonzero components of the inertia tensor relative to the stability axes are

$$\left(I_{xx_b}\right)_s = \left(I_{xx_b}\right)_{L0} \cos^2 \varphi + 2\left(I_{xz_b}\right)_{L0} \cos\varphi \sin\varphi + \left(I_{zz_b}\right)_{L0} \sin^2 \varphi$$
$$= 1.888 \times 10^7 \cos^2 \varphi + 2\left(4.075 \times 10^6\right)\cos\varphi \sin\varphi + 4.902 \times 10^7 \sin^2 \varphi$$
$$= \underline{1.836 \times 10^7 \text{ slug} \cdot \text{ft}^2}$$

$$\left(I_{yy_b}\right)_s = \left(I_{yy_b}\right)_{L0} = \underline{3.310 \times 10^7 \text{ slug} \cdot \text{ft}^2}$$

$$\left(I_{zz_b}\right)_s = \left(I_{zz_b}\right)_{L0} \cos^2 \varphi - 2\left(I_{xz_b}\right)_{L0} \cos\varphi \sin\varphi + \left(I_{xx_b}\right)_{L0} \sin^2 \varphi$$
$$= 4.902 \times 10^7 \cos^2 \varphi - 2\left(4.075 \times 10^6\right)\cos\varphi \sin\varphi + 1.888 \times 10^7 \sin^2 \varphi$$
$$= \underline{4.954 \times 10^7 \text{ slug} \cdot \text{ft}^2}$$

$$
\begin{aligned}
\left(I_{xz_b}\right)_s &= \left(I_{xz_b}\right)_{L0}(\cos^2\varphi - \sin^2\varphi) + \left[\left(I_{zz_b}\right)_{L0} - \left(I_{xx_b}\right)_{L0}\right]\,\cos\varphi\sin\varphi \\
&= \left(4.075\times10^6\right)\!\left(\cos^2\varphi - \sin^2\varphi\right) + \left(4.902\times10^7 - 1.888\times10^7\right)\!\cos\varphi\sin\varphi \\
&= \underline{7.330\times10^5 \text{ slug}\cdot\text{ft}^2}
\end{aligned}
$$

Notice that this transformation has little effect on the moments of inertia but has a large effect on the product of inertia.

7.9. Inertial and Gyroscopic Coupling

In our previous development of the rigid-body equations of motion, we assumed that the aircraft was symmetric about the x_b-z_b plane. It was this assumed symmetry that allowed us to uncouple the longitudinal equations from the lateral equations after applying the small-disturbance approximation. While most aircraft possess this symmetry, it is not unusual for an aircraft to be flown in an asymmetrical configuration. For example, a military airplane might be flown with a missile under only one wing, and any airplane could be flown with more fuel in one wing than in the other. If the symmetry assumption is relaxed, the longitudinal and lateral equations become coupled, even within the approximation of small-disturbance theory.

There is also another form of cross coupling that arises from the spinning rotors associated with aircraft power plants, i.e., turbines, compressors, and propellers. In treating the aircraft as a rigid body, we tacitly neglected the angular momentum of these rotors relative to the body-fixed coordinates. If we relax the rigid-body assumption to the extent that we allow some parts of the aircraft to be rotating relative to the body-fixed coordinate system, then additional terms appear on the right-hand side of the angular momentum equation, Eq. (7.2.2). These terms are called the *gyroscopic couples*.

Relaxing both the symmetry assumption and the rigid-body assumption as mentioned above, the angular momentum equation can be written as

$$
\mathbf{M} = \frac{d\mathbf{H}}{dt} + \boldsymbol{\omega}\times\mathbf{H} \tag{7.9.1}
$$

where

$$
\mathbf{H} = [\mathbf{I}]\boldsymbol{\omega} + \mathbf{h} \tag{7.9.2}
$$

$$
[\mathbf{I}] = \begin{bmatrix} I_{xx_b} & -I_{xy_b} & -I_{xz_b} \\ -I_{yx_b} & I_{yy_b} & -I_{yz_b} \\ -I_{zx_b} & -I_{zy_b} & I_{zz_b} \end{bmatrix} \tag{7.9.3}
$$

and $\mathbf{h}$ is the total angular momentum of all spinning rotors relative to the body-fixed coordinate system. If we assume all rotors to be spinning with a constant angular velocity relative to the aircraft, Eq. (7.9.1) can be written as

$$\mathbf{M} = \frac{d}{dt}\big([\mathbf{I}]\boldsymbol{\omega}\big) + \boldsymbol{\omega} \times \big([\mathbf{I}]\boldsymbol{\omega} + \mathbf{h}\big) \tag{7.9.4}$$

or after expanding and rearranging,

$$
\begin{bmatrix} I_{xx_b} & -I_{xy_b} & -I_{xz_b} \\ -I_{xy_b} & I_{yy_b} & -I_{yz_b} \\ -I_{xz_b} & -I_{yz_b} & I_{zz_b} \end{bmatrix}
\begin{Bmatrix} \dot{p} \\ \dot{q} \\ \dot{r} \end{Bmatrix}
=
\begin{Bmatrix} M_{x_b} \\ M_{y_b} \\ M_{z_b} \end{Bmatrix}
+
\begin{bmatrix} 0 & -h_{z_b} & h_{y_b} \\ h_{z_b} & 0 & -h_{x_b} \\ -h_{y_b} & h_{x_b} & 0 \end{bmatrix}
\begin{Bmatrix} p \\ q \\ r \end{Bmatrix}
$$
$$
+
\begin{Bmatrix}
(I_{yy_b} - I_{zz_b})qr + I_{yz_b}(q^2 - r^2) + I_{xz_b}pq - I_{xy_b}pr \\
(I_{zz_b} - I_{xx_b})pr + I_{xz_b}(r^2 - p^2) + I_{xy_b}qr - I_{yz_b}pq \\
(I_{xx_b} - I_{yy_b})pq + I_{xy_b}(p^2 - q^2) + I_{yz_b}pr - I_{xz_b}qr
\end{Bmatrix}
\tag{7.9.5}
$$

From Eq. (7.9.5) we see that for this more general case, there is both linear and nonlinear coupling between the longitudinal and lateral degrees of freedom. These effects can be accounted for in the Euler angle formulation of the nonlinear 6-DOF equations of motion simply by replacing Eq. (7.4.5) with Eq. (7.9.5). This adds no significant complexity to the nonlinear 6-DOF equations.

In steady linear flight, the first-order coupling between the longitudinal and lateral equations of motion results only from asymmetry about the x_b-z_b plane and/or from the gyroscopic effects of the spinning rotors. From Eq. (7.9.5), we see that both the rolling and yawing rates affect the pitching equation through the terms $-I_{xy_b}\dot{p}$ and $-I_{yz_b}\dot{r}$. Likewise, the pitching rate affects both the rolling and yawing equations through the terms $-I_{xy_b}\dot{q}$ and $-I_{yz_b}\dot{q}$. In conventional airplanes the spinning rotors are usually aligned very closely with the x_b-axis. In this case the gyroscopic effects result primarily in a coupling between pitch and yaw, through the terms $h_{x_b}q$ and $-h_{x_b}r$. For helicopters, the primary rotors are aligned more closely with the z_b-axis, which produces a roll-pitch coupling from the terms $h_{z_b}p$ and $-h_{z_b}q$.

During very rapid maneuvers, even the second-order terms in Eq. (7.9.5) can become significant. While pitching and yawing rates are not usually large, fighter aircraft routinely execute maneuvers with very high rolling rates. Such maneuvers can result in significant longitudinal-lateral coupling. Most notably, long slender aircraft have a moment of inertia about the roll axis that is much less than that about either the pitch or yaw axis. When such an aircraft executes a rapid rolling maneuver, the second-order term $(I_{zz_b} - I_{xx_b})pr$ can produce a pitching motion large enough to cause loss of control or even structural failure. Similarly, an aircraft with significant asymmetry will have significant product of inertia terms. A rapid rolling maneuver in this type of aircraft will also produce significant pitching motion as a result of the terms $I_{xz_b}(r^2 - p^2)$ and $-I_{yz_b}pq$. Such coupling between rolling and pitching motion is particularly important, since rolling rates can be very large in fighter aircraft and because pitching motion can result in rapid changes in angle of attack. These changes in angle of attack can have a profound and sometimes catastrophic effect on the aerodynamic forces. Fortunately, the angular rates in most flying situations are small and the second-order terms in Eq. (7.9.5) are not significant.

7.10. Problems

7.1. A rocket is flying with an axial velocity component of 1,000 ft/sec, a sideslip velocity component of 500 ft/sec, and a normal velocity component of 2,000 ft/sec. Determine the total airspeed in miles per hour and the angle of attack and sideslip angles β_a and β_e in degrees.

7.2. An aircraft has an airspeed of 400 ft/sec. The angle of attack is 20 degrees and the sideslip angle, β_a, is 10 degrees. Determine the axial velocity component, the sideslip velocity component, and the normal velocity component relative to the same body-fixed coordinate system.

7.3. An aircraft has an airspeed of 400 ft/sec. The angle of attack is 20 degrees and the sideslip angle, β_e, is 10 degrees. Determine the axial velocity component, the sideslip velocity component, and the normal velocity component relative to the same body-fixed coordinate system.

7.4. An aircraft has an airspeed of 800 ft/sec. The angle of attack is 75 degrees and the sideslip angle, β_a, is 60 degrees. Determine the axial velocity component, the sideslip velocity component, and the normal velocity component.

7.5. An aircraft has an airspeed of 800 ft/sec. The angle of attack is 75 degrees and the sideslip angle, β_e, is 60 degrees. Determine the axial velocity component, the sideslip velocity component, and the normal velocity component.

7.6. Consider some aerodynamic coefficient, C, that is somehow known as a function of V, α, and β_a. Using the definitions in Eqs. (7.1.2) through (7.1.4), obtain an expression for the change in C with respect to sideslip velocity in terms of the derivatives with respect to V, α, and β_a.

7.7. Consider some aerodynamic coefficient, C, that is somehow known as a function of V, α, and β_e. Using the definitions in Eqs. (7.1.2), (7.1.3), and (7.1.5), obtain an expression for the change in C with respect to sideslip velocity in terms of the derivatives with respect to V, α, and β_e.

7.8. Using Eqs. (7.1.16), (7.1.19), and (7.1.20), show that the results obtained in problems 7.6 and 7.7 are equivalent.

7.9. A rectangular solid of density ρ has length L_x in the x_b-direction, length L_y in the y_b-direction, and length L_z in the z_b-direction. Starting with Eqs. (7.2.3) through (7.2.9), obtain an expression for the inertia tensor for this object. Use a body-fixed coordinate system aligned with the primary axes of the body.

7.10. Starting with Eqs. (7.2.3) through (7.2.9), obtain an expression for the inertia tensor for a cylinder having density ρ, radius R, and length L. Use a body-fixed coordinate system aligned with the primary axes of the cylinder.

7.11. If the rocket in problem 7.1 has a bank angle of -30 degrees, an elevation angle of 60 degrees, and an azimuth angle of 230 degrees, what is its velocity in Earth-fixed coordinates?

7.12. If the aircraft in problem 7.2 has a bank angle of 40 degrees, an elevation angle of -20 degrees, and an azimuth angle of 290 degrees, what is its velocity in Earth-fixed coordinates?

7.13. An airplane has an angle of attack of 10 degrees and the sideslip angle, β_a, is 5 degrees. If the airplane has a bank angle of 60 degrees, an elevation angle of -30 degrees, and an azimuth angle of 0 degrees, what direction is the nose of the airplane pointing, and in what direction is the airplane moving?

7.14. An airplane has an angle of attack of 10 degrees and the sideslip angle, β_e, is 5 degrees. If the airplane has a bank angle of 60 degrees, an elevation angle of -30 degrees, and an azimuth angle of 0 degrees, what direction is the nose of the airplane pointing, and in what direction is the airplane moving?

7.15. During a post-stall maneuver, an airplane's angle of attack is 80 degrees and the sideslip angle, β_a, is 30 degrees. If the bank angle is 90 degrees, the elevation angle is 10 degrees, and the azimuth angle is 180 degrees, what direction is the nose of the airplane pointing, and in what direction is the airplane moving?

7.16. During a post-stall maneuver, an airplane's angle of attack is 80 degrees and the sideslip angle, β_e, is 30 degrees. If the bank angle is 90 degrees, the elevation angle is 10 degrees, and the azimuth angle is 180 degrees, what direction is the nose of the airplane pointing, and in what direction is the airplane moving?

7.17. Write a computer subroutine that uses fourth-order Runge-Kutta to solve the projectile equations developed in Example 7.3.4. Start with Eq. (7.3.38).

7.18. Using the subroutine from problem 7.17, obtain the Earth-fixed coordinates and airspeed at 25 and 100 yards downrange for an arrow having $K_0 = 0.00061$ ft^{-1}, $K_1 = 0.14$ ft^{-1}, $K_2 = 0.00059$ ft^{-1}, $K_3 = 0.0016$ ft^{-2}, and $K_4 = 0.0064$ ft^{-1}. The arrow starts at the Earth-fixed origin with a speed of 210 ft/sec, no angle of attack, no angular velocity, and an elevation angle of 0.0 degree. How far is the arrow below the initial line of motion at 25 and 100 yards downrange?

7.19. Using the subroutine from problem 7.17, obtain the Earth-fixed coordinates and airspeed at 25 and 100 yards downrange for an arrow having $K_0 = 0.00061$ ft^{-1}, $K_1 = 0.14$ ft^{-1}, $K_2 = 0.00059$ ft^{-1}, $K_3 = 0.0016$ ft^{-2}, and $K_4 = 0.0064$ ft^{-1}. The arrow starts at the Earth-fixed origin with a speed of 210 ft/sec, no angle of attack, no angular velocity, and an elevation angle of 45 degrees. How far is the arrow below the initial line of motion at 25 and 100 yards downrange? Consider the downrange distance to be measured along the initial 45-degree flight line.

7.20. Modify the computer subroutine written for problem 7.17 so that time is used as the independent variable, i.e., use Eq. (7.3.35) instead of Eq. (7.3.38). Also account for the variation in air density with altitude.

7.21. A projectile is dropped from an aircraft flying at 10,000 feet with an airspeed of 500 ft/sec. At sea level, the projectile has $K_0 = 0.002$ ft^{-1}, $K_1 = 0.2$ ft^{-1}, $K_2 = 0.001$ ft^{-1}, $K_3 = 0.001$ ft^{-2}, and $K_4 = 0.01$ ft^{-1}. Use the subroutine written for problem 7.20 to determine the horizontal distance traveled by the projectile from the time it is dropped from the aircraft until it falls to sea level. What are the elevation angle and the angle of attack for the projectile when it reaches sea level? Assume that the projectile leaves the aircraft with no angular velocity and no angle of attack. Use a time step of 0.05 second and assume no wind.

7.22. Using the subroutine that was written for problem 7.20, plot the trajectory of the projectile in problem 7.21 from 10,000 feet to sea level.

7.23. Using the subroutine that was written for problem 7.20, plot both the horizontal and vertical components of velocity as a function of time for the projectile in problem 7.21.

7.24. Using the subroutine written for problem 7.20, plot the angle of attack as a function of time for the projectile in problem 7.21.

7.25. A spherical projectile 24 inches in diameter and weighing 50 pounds is dropped from an aircraft flying at 10,000 feet and 500 ft/sec. Use the subroutine written for problem 7.20 to determine the horizontal distance traveled by the projectile from the time it is dropped until it falls to sea level. Assume a constant drag coefficient of 1.0. Use a time step of 0.05 second and assume no wind.

7.26. Modify the computer subroutine written for problem 7.20 to account for the effects of wind. Assume constant headwind and no crosswind.

7.27. Using the subroutine written for problem 7.26, plot the trajectory of the projectile in problem 7.21 from 10,000 feet to sea level for the case of a constant headwind of 75 ft/sec.

7.28. Using the subroutine that was written for problem 7.26, plot the horizontal and vertical components of ground speed as a function of time for the projectile in problem 7.27.

7.29. For the arrow described in problem 7.18, use the fourth-order Runge-Kutta integration method with the linearized system of equations developed in Example 7.5.2 to plot the elevation and airspeed as a function of range from zero to 100 yards. Use a time step of 0.015 second. Compare your results to those obtained in problem 7.18.

7.30. In Example 7.5.2, the projectile equations developed in Example 7.3.3 were linearized relative to motion along a horizontal line at constant speed. We wish to develop a more general system of linearized projectile equations. Linearize the equations developed in Example 7.3.3 relative to constant-speed motion along a straight line having an elevation angle of θ_o.

7.31. For the arrow described in problem 7.19, use the linearized system of equations developed in problem 7.30 to plot the elevation and airspeed as a function of range from zero to 100 yards. Use a time step of 0.015 second. Compare your results to those obtained in problem 7.19.

7.32. For the airplane and flight conditions in Example 7.6.1, estimate all of the dimensional lateral aerodynamic derivatives that appear in the homogeneous form of Eq. (7.5.24). In addition to the properties given in Example 7.6.1, this airplane has the following properties:

$$(c_t/c_r)_w = 0.286, \quad S_v = 875 \text{ ft}^2, \quad x_{bv} = -96 \text{ ft}, \quad z_{bv} = -23 \text{ ft},$$
$$C_{L_v,\beta} = 3.30, \quad C_{Y,\beta} = -0.890, \quad C_{\ell,\beta} = -0.144, \quad C_{n,\beta} = 0.182$$

7.33. Repeat Example 7.6.1, for a thrust vector that is inclined upward at an angle of 9.0 degrees to the direction of flight and passes through a point that is 9.0 ft below the center of gravity. Assume that all other parameters, including the lift-to-drag ratio, remain unchanged.

7.34. For the airplane and flight conditions in problem 7.32, fill out the nondimensional lateral coefficient matrices associated with the homogeneous form of Eq. (7.7.6). For this airplane, the components of the inertia tensor relative to the stability axes were computed in Example 7.8.1.

7.35. For the airplane described in Example 7.8.1, what rotation angle about the y_b-axis is required to move from the zero-lift axes to the primary axes?

7.36. The Spitfire Mark XVIII that is shown in Figs. 4.12.1 and 4.12.2 is in level flight at sea level. Assume the following properties and operating conditions:

$$S_w = 244 \text{ ft}^2, \quad b_w = 36.83 \text{ ft}, \quad C_{L_w,\alpha} = 4.62, \quad C_{m_w} = -0.053,$$
$$A_w = 0.0°, \quad \alpha_{L0w} = -2.2°, \quad \alpha_{0w} = 3.0°, \quad l_w = 0.0 \text{ ft}, \quad h_w = -2.6 \text{ ft},$$
$$C_{D0w} = 0.01, \quad C_{D0,L_w} = 0.00, \quad e_w = 1.00,$$
$$S_h = 31 \text{ ft}^2, \quad b_h = 10.64 \text{ ft}, \quad C_{L_h,\alpha} = 4.06, \quad C_{m_h0} = 0.000,$$
$$A_h = 0.0°, \quad \alpha_{L0h} = 0.0°, \quad \alpha_{0h} = 0.0°, \quad l_h = 18.16 \text{ ft}, \quad h_h = 0.0 \text{ ft},$$
$$C_{D0h} = 0.01, \quad C_{D0,L_h} = 0.00, \quad e_h = 1.00, \quad \eta_h = 1.0,$$
$$S_f = 15 \text{ ft}^2, \quad c_f = 32.67 \text{ ft}, \quad \alpha_{0f} = 0.0°, \quad l_f = -3.0 \text{ ft},$$

$$d_p = 132/12 \text{ ft}, \quad \omega/2\pi = 1350 \text{ rpm}, \quad C_{N_p,\alpha} = 0.390,$$

$$T_V = -2.40 \text{ lbf·sec/ft}, \quad \alpha_{0p} = 0.0°, \quad l_p = -9.33 \text{ ft}, \quad h_p = 0.0 \text{ ft},$$

$$W = 8,375 \text{ lbf}, \quad I_{yy_b} = 10,060 \text{ slug·ft}^2, \quad V_o = 300 \text{ mph},$$

$$C_{L,\alpha} = 4.884, \quad C_{D_{L0}} = 0.0173, \quad C_{D_0,L} = 0.0, \quad e = 0.80, \quad C_{m,\alpha} = -0.3786$$

Estimate all of the dimensionless longitudinal aerodynamic derivatives and fill out the nondimensional longitudinal coefficient matrices associated with the homogeneous form of Eq. (7.7.5).

7.37. Two high-bypass-ratio turbofan engines power a small jet transport of the wing-canard design. The airplane is in level flight at 30,000 feet. Assume the following properties and operating conditions:

$$S_w = 950 \text{ ft}^2, \quad b_w = 75 \text{ ft}, \quad R_{T_w} = 0.35, \quad C_{L_w,\alpha} = 4.30, \quad C_{m_w} = -0.050,$$

$$\Lambda_w = 15°, \quad \alpha_{L0w} = -2.0°, \quad \alpha_{0w} = 1.5°, \quad l_w = 8.6 \text{ ft}, \quad h_w = -3.8 \text{ ft},$$

$$C_{D_{0w}} = 0.01, \quad C_{D_0,L_w} = 0.00, \quad e_w = 0.95,$$

$$S_h = 219 \text{ ft}^2, \quad b_h = 36 \text{ ft}, \quad R_{T_h} = 0.35, \quad C_{L_h,\alpha} = 4.30, \quad C_{m_h0} = -0.050,$$

$$\Lambda_h = 15°, \quad \alpha_{L0h} = -2.0°, \quad \alpha_{0h} = 2.0°, \quad l_h = -33.9 \text{ ft}, \quad h_h = -3.8 \text{ ft},$$

$$C_{D_{0h}} = 0.01, \quad C_{D_0,L_h} = 0.00, \quad e_h = 0.95, \quad \eta_h = 1.0,$$

$$S_f = 52 \text{ ft}^2, \quad c_f = 78 \text{ ft}, \quad \alpha_{0f} = 0.0°, \quad l_f = -18.5 \text{ ft},$$

$$T_V = -10.5 \text{ lbf·sec/ft}, \quad \alpha_{0j} = 0.0°, \quad \eta_{p_i} = 0.64, \quad l_j = 9.4 \text{ ft}, \quad h_j = 2.4 \text{ ft},$$

$$W = 73,000 \text{ lbf}, \quad I_{yy_b} = 451,500 \text{ slug·ft}^2, \quad V_o = 400 \text{ mph},$$

$$C_{L,\alpha} = 5.327, \quad C_{D_{L0}} = 0.020, \quad C_{D_0,L} = 0.0, \quad e = 0.67, \quad C_{m,\alpha} = -0.122$$

Estimate all of the dimensionless longitudinal aerodynamic derivatives and fill out the nondimensional longitudinal coefficient matrices associated with the homogeneous form of Eq. (7.7.5).

7.38. The turboprop transport that is shown in Fig. 4.12.3 is in level flight at sea level. Assume the following properties and operating conditions:

$$S_w = 1,745 \text{ ft}^2, \quad b_w = 132.6 \text{ ft}, \quad R_{T_w} = 0.57, \quad C_{L_w,\alpha} = 5.16, \quad C_{m_w} = -0.050,$$

$$\Lambda_w = 0.0°, \quad \alpha_{L0w} = -2.0°, \quad \alpha_{0w} = 0.0°, \quad l_w = -2.5 \text{ ft}, \quad h_w = 7.4 \text{ ft},$$

$$C_{D_{0w}} = 0.01, \quad C_{D_0,L_w} = 0.00, \quad e_w = 0.96,$$

$$S_h = 528 \text{ ft}^2, \quad b_h = 52.7 \text{ ft}, \quad R_{T_h} = 0.35, \quad C_{L_h,\alpha} = 4.52, \quad C_{m_h0} = 0.000,$$

$$\Lambda_h = 7°, \quad \alpha_{L0h} = 0.0°, \quad \alpha_{0h} = -1.5°, \quad l_h = 44.5 \text{ ft}, \quad h_h = 7.4 \text{ ft},$$

$$C_{D_{0h}} = 0.01, \quad C_{D_0,L_h} = 0.00, \quad e_h = 0.97, \quad \eta_h = 1.0,$$

$$S_f = 173 \text{ ft}^2, \quad c_f = 87.9 \text{ ft}, \quad \alpha_{0f} = 0.0°, \quad l_f = -20.5 \text{ ft},$$

$$S_n = 28.7 \text{ ft}^2, \quad c_n = 23 \text{ ft}, \quad \alpha_{0n} = 0.0°, \quad l_n = -9.2 \text{ ft},$$
$$d_p = 162/12 \text{ ft}, \quad \omega/2\pi = 1020 \text{ rpm}, \quad C_{N_p,\alpha} = 0.721,$$
$$T_{,V} = -40.4 \text{ lbf·sec/ft}, \quad \alpha_{0p} = 0.0°, \quad l_p = -11.9 \text{ ft}, \quad h_p = 6.2 \text{ ft},$$
$$W = 155,000 \text{ lbf}, \quad I_{yy_b} = 894,700 \text{ slug·ft}^2, \quad V_o = 300 \text{ mph},$$
$$C_{L,\alpha} = 6.087, \quad C_{D_{L0}} = 0.031, \quad C_{D_0,L} = 0.0, \quad e = 0.65, \quad C_{m,\alpha} = -1.645$$

Estimate all of the dimensionless longitudinal aerodynamic derivatives and fill out the nondimensional longitudinal coefficient matrices associated with the homogeneous form of Eq. (7.7.5).

7.39. For the Spitfire Mark XVIII that is described in problem 7.36, fill out the nondimensional lateral coefficient matrices associated with the homogeneous form of Eq. (7.7.6). In addition to the properties and operating conditions listed in problem 7.36, assume the following:

$$I_{xx_b} = 1,500 \text{ slug·ft}^2, \quad I_{zz_b} = 11,700 \text{ slug·ft}^2, \quad I_{xz_b} = 100 \text{ slug·ft}^2,$$
$$C_{Y,\beta} = -0.212, \quad C_{\ell,\beta} = 0.000, \quad C_{n,\beta} = 0.095,$$
$$C_{Y,\bar{p}} = 0.000, \quad C_{\ell,\bar{p}} = -0.8885, \quad C_{n,\bar{p}} = -0.0186,$$
$$C_{Y,\bar{r}} = 0.282, \quad C_{\ell,\bar{r}} = 0.037, \quad C_{n,\bar{r}} = -0.139$$

7.40. For the small jet transport that is described in problem 7.37, fill out the nondimensional lateral coefficient matrices associated with the homogeneous form of Eq. (7.7.6). In addition to the properties and operating conditions listed in problem 7.37, assume the following:

$$I_{xx_b} = 393,700 \text{ slug·ft}^2, \quad I_{zz_b} = 808,200 \text{ slug·ft}^2, \quad I_{xz_b} = 75,000 \text{ slug·ft}^2,$$
$$C_{Y,\beta} = -0.735, \quad C_{\ell,\beta} = -0.220, \quad C_{n,\beta} = 0.135,$$
$$C_{Y,\bar{p}} = 0.000, \quad C_{\ell,\bar{p}} = -0.385, \quad C_{n,\bar{p}} = -0.049,$$
$$C_{Y,\bar{r}} = 0.092, \quad C_{\ell,\bar{r}} = 0.143, \quad C_{n,\bar{r}} = -0.165$$

7.41. For the turboprop transport that is described in problem 7.38, fill out the nondimensional lateral coefficient matrices associated with the homogeneous form of Eq. (7.7.6). In addition to the properties and operating conditions listed in problem 7.38, assume the following:

$$I_{xx_b} = 1,136,000 \text{ slug·ft}^2, \quad I_{zz_b} = 1,860,000 \text{ slug·ft}^2, \quad I_{xz_b} = 1,100 \text{ slug·ft}^2,$$
$$C_{Y,\beta} = -0.362, \quad C_{\ell,\beta} = -0.125, \quad C_{n,\beta} = 0.101,$$
$$C_{Y,\bar{p}} = 0.000, \quad C_{\ell,\bar{p}} = -0.530, \quad C_{n,\bar{p}} = -0.037,$$
$$C_{Y,\bar{r}} = 0.082, \quad C_{\ell,\bar{r}} = 0.113, \quad C_{n,\bar{r}} = -0.171$$

Chapter 8
Linearized Longitudinal Dynamics

8.1. Fundamentals of Dynamics: Eigenproblems

The linearized 6-DOF equations of motion developed in Chapter 7 consist of 12 first-order linear differential equations with constant coefficients. These 12 first-order equations are equivalent to a system of six second-order equations for the six degrees of freedom associated with rigid-body flight. Similar systems of second-order equations can be used to model many other physical phenomena. Systems of second-order differential equations with constant coefficients are frequently encountered in every branch of engineering. For this reason, such systems of equations have been studied at great length and their solutions are well understood. Because an understanding of second-order systems is prerequisite to an understanding of aircraft dynamics, here we review the characteristics of such systems in general before considering their application to dynamic aircraft stability and control.

First consider a one-degree-of-freedom system, modeled with a single second-order differential equation with constant coefficients. While several electrical and mechanical systems can be modeled with a single equation of this type, here we illustrate the problem by examining the motion of a one-degree-of-freedom spring-mass system with linearized viscous damping (see Fig. 8.1.1). The equation of motion is developed by summing forces acting on the mass to obtain the second-order differential equation

$$m\frac{d^2x}{dt^2} + c\frac{dx}{dt} + k(x - x_o) = F(t) \tag{8.1.1}$$

where m is the mass, c is the damping coefficient, k is the spring constant, x_o is the equilibrium value of the x-position that results in no spring force, and $F(t)$ is a forcing function that may vary with time, t. The general solution to this equation, referred to as the *forced response*, is the sum of the homogeneous solution and a particular solution. The homogeneous solution, or *free response*, is the solution with the forcing function set

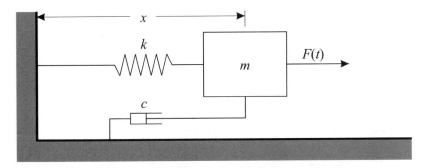

Figure 8.1.1. Damped spring-mass system with one degree of freedom.

to zero, and the particular solution is any solution that will satisfy the nonhomogeneous equation. Because system dynamics are characterized primarily by the free response, we start our discussion with a consideration of the homogeneous solution.

Equation (8.1.1) can be rewritten as

$$m\frac{d^2\Delta x}{dt^2} + c\frac{d\Delta x}{dt} + k\Delta x = F(t) \tag{8.1.2}$$

where Δx is the displacement from equilibrium, $\Delta x = x - x_o$. The reader should recognize the homogeneous solution to Eq. (8.1.2) from his or her introductory class on ordinary differential equations. The only existing solutions are given by

$$\Delta x(t) = C_1 e^{\lambda_1 t} + C_2 e^{\lambda_2 t} \tag{8.1.3}$$

where C_1 and C_2 are arbitrary constants, to be determined from the initial conditions, and λ_1 and λ_2 are the *eigenvalues*, which are the roots of the characteristic equation,

$$\lambda_{1,2} = \frac{-c \pm \sqrt{c^2 - 4mk}}{2m} = -\frac{c}{2m} \pm \sqrt{\left(\frac{c}{2m}\right)^2 - \frac{k}{m}} \tag{8.1.4}$$

The motion that occurs when the system is disturbed from equilibrium and then released will depend on the eigenvalues λ_1 and λ_2. The reader should be familiar with three types of motion that are shown in Fig. 8.1.2. **Overdamped motion** occurs when the eigenvalues are real, negative, and distinct,

$$\left(\frac{c}{2m}\right)^2 > \frac{k}{m} \tag{8.1.5}$$

In this case any disturbance from equilibrium will result in a smooth exponential return to the equilibrium position with no oscillation or overshoot. **Underdamped motion** occurs when the eigenvalues are complex,

$$\left(\frac{c}{2m}\right)^2 < \frac{k}{m} \tag{8.1.6}$$

In this case, disturbance from equilibrium results in damped sinusoidal motion with a *damped natural frequency* equal to the magnitude of the eigenvalue's imaginary part,

$$\text{damped natural frequency} \equiv \omega_d = |\text{imag}(\lambda)| = \sqrt{\frac{k}{m} - \left(\frac{c}{2m}\right)^2} \tag{8.1.7}$$

and a *damping rate* equal to the negative of the real part of the eigenvalue,

$$\text{damping rate} \equiv \sigma = -\text{real}(\lambda) = \frac{c}{2m} \tag{8.1.8}$$

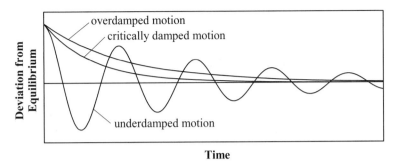

Figure 8.1.2. Overdamped, critically damped, and underdamped motion.

Critically damped motion occurs when the eigenvalues are real, negative, and identical,

$$\frac{c}{2m} = \sqrt{\frac{k}{m}} \tag{8.1.9}$$

This case is the boundary between overdamped exponential motion and underdamped sinusoidal motion. Critically damped motion provides the most rapid exponential return to equilibrium that is possible without inducing sinusoidal oscillations.

For an underdamped mode, the eigenvalues form a complex pair. This complex pair and the nature of the associated motion can be specified in terms of the damped natural frequency, ω_d, and the damping rate, σ, defined by Eqs. (8.1.7) and (8.1.8),

$$\lambda_{1,2} = -\sigma \pm \omega_d i \tag{8.1.10}$$

where $i = \sqrt{-1}$. For underdamped motion, both the damped natural frequency and the damping rate have rather obvious physical meanings. However, for overdamped or critically damped motion, the concept of damped natural frequency is meaningless. For this reason, it is often more convenient to describe an eigenvalue pair in terms of what are called the *undamped natural frequency* and the *damping ratio*. The undamped natural frequency is the frequency that the mode would exhibit if the damping were identically zero. This undamped natural frequency is related to the eigenvalue pair according to

$$\text{undamped natural frequency} \equiv \omega_n = \sqrt{\lambda_1 \lambda_2} \tag{8.1.11}$$

The damping ratio is the actual damping rate divided by the damping rate that would be required to make the mode critically damped. In general, to determine the damping ratio from the eigenvalue pair, we use

$$\text{damping ratio} \equiv \zeta = -\frac{\lambda_1 + \lambda_2}{2\sqrt{\lambda_1 \lambda_2}} \tag{8.1.12}$$

Using Eq. (8.1.10) in Eqs. (8.1.11) and (8.1.12), for an underdamped mode, the undamped natural frequency and the damping ratio can be related to the damped natural frequency and the damping rate. This gives

$$\omega_n = \sqrt{(-\sigma + \omega_d\, i)(-\sigma - \omega_d\, i)} = \sqrt{\sigma^2 + \omega_d^2} \tag{8.1.13}$$

$$\zeta = -\frac{-\sigma + \omega_d\, i - \sigma - \omega_d\, i}{2\sqrt{(-\sigma + \omega_d\, i)(-\sigma - \omega_d\, i)}} = \frac{\sigma}{\sqrt{\sigma^2 + \omega_d^2}} = \frac{\sigma}{\omega_n} \tag{8.1.14}$$

While Eqs. (8.1.13) and (8.1.14) can only be applied to a mode that is underdamped, Eqs. (8.1.11) and (8.1.12) can be applied to underdamped, overdamped, and critically damped modes. For the damped spring-mass system that is shown in Fig. 8.1.1, using Eq. (8.1.4) in Eq. (8.1.11) gives

$$\omega_n = \sqrt{k/m} \tag{8.1.15}$$

and using Eq. (8.1.4) in Eq. (8.1.12) results in

$$\zeta = c/\left(2\sqrt{km}\right) \tag{8.1.16}$$

Combining Eqs. (8.1.11) and (8.1.12), the eigenvalue pair can be expressed in terms of the undamped natural frequency and the damping ratio,

$$\lambda_{1,2} = -\zeta\,\omega_n \pm \omega_n\sqrt{\zeta^2 - 1} \tag{8.1.17}$$

If the damping ratio is less than 1.0, comparing Eq. (8.1.17) with Eq. (8.1.10), we see that the damping rate and the damped natural frequency are related to the damping ratio and the undamped natural frequency according to

$$\sigma = \zeta\,\omega_n \tag{8.1.18}$$

$$\omega_d = \omega_n\sqrt{1 - \zeta^2} \tag{8.1.19}$$

When the eigenvalues are complex, using Eq. (8.1.10), the homogeneous solution given by Eq. (8.1.3) can also be written in terms of sinusoidal functions,

$$\Delta x(t) = e^{-\sigma t}[C_3 \sin(\omega_d t) + C_4 \cos(\omega_d t)] \tag{8.1.20}$$

where C_3 and C_4 are arbitrary constants, different from C_1 and C_2. In this form it is easy to see that the homogeneous solution is a sinusoidal oscillation with period $2\pi/\omega_d$. We also see that the amplitude of oscillation decreases exponentially with time. The rate at which the oscillations decay is sometimes specified by the *time constant*, which is $1/\sigma$ or the time required for the amplitude to be reduced by about 63 percent. Another common measure of the decay rate is the *time to half amplitude*, which is $-\ln(0.5)/\sigma$ or about 69 percent of the time constant.

   In summary, the nature of the homogeneous solution to Eq. (8.1.2) depends on two constants, the eigenvalues λ_1 and λ_2, which can be either real or complex. In general, these eigenvalues can be written in terms of the damping ratio, ζ, and the undamped natural frequency, ω_n, according to Eq. (8.1.17). If the damping ratio is less than 1.0, the eigenvalues are complex and can be written in terms of the damping rate, σ, and the damped natural frequency, ω_d, in accordance with Eq. (8.1.10). The homogeneous solution obtained from these eigenvalues is called the *free response*.

System motion resulting from the application of a time varying force is called the *forced response*. For example, consider the case where the forcing function, $F(t)$, is sinusoidal with frequency, ω_f, and amplitude, F_a. In this case Eq. (8.1.2) becomes

$$m\frac{d^2\Delta x}{dt^2} + c\frac{d\Delta x}{dt} + k\Delta x = F_a \sin(\omega_f t) \qquad (8.1.21)$$

The general solution to this equation is the homogeneous solution in Eq. (8.1.3), plus a particular solution, which can be found using the method of undetermined coefficients,

$$\Delta x(t) = C_1 e^{\lambda_1 t} + C_2 e^{\lambda_2 t} + A_f \frac{F_a}{k}\sin(\omega_f t - \Psi) \qquad (8.1.22)$$

where A_f is the amplification factor shown in Fig. 8.1.3 and given by

$$A_f = \frac{1}{\sqrt{\left[1-(\omega_f/\omega_n)^2\right]^2 + \left[2\zeta(\omega_f/\omega_n)\right]^2}} \qquad (8.1.23)$$

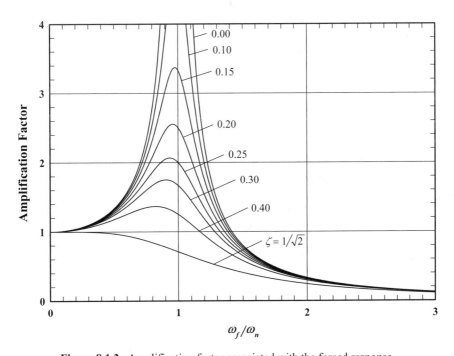

Figure 8.1.3. Amplification factor associated with the forced response.

and Ψ is the phase angle given by

$$\Psi = \frac{2\zeta(\omega_f/\omega_n)}{1-(\omega_f/\omega_n)^2} \tag{8.1.24}$$

Figure 8.1.3 shows the amplification factor, given by Eq. (8.1.23), plotted as a function of the forcing frequency, ω_f, divided by the undamped natural frequency, ω_n. In the limiting case of zero damping, the amplitude of the forced vibration is seen to approach infinity as the forcing frequency approaches the undamped natural frequency. This phenomenon is called *resonance*. The response amplitude of the forced vibration is maximum when the forcing frequency is somewhat less than the undamped natural frequency. The forcing frequency that results in maximum response amplification is found by differentiating the right-hand side of Eq. (8.1.23) with respect to ω_f and setting the result to zero. This gives

$$\text{maximum response frequency} \equiv \omega_m = \omega_n\sqrt{1-2\zeta^2} \tag{8.1.25}$$

Clearly, from Eq. (8.1.25) we see that the frequency of maximum response, ω_m, is less than the undamped natural frequency, ω_n. Furthermore, after comparing Eq. (8.1.25) with Eq. (8.1.19), we see that the frequency resulting in maximum response is also less than the damped natural frequency, ω_d.

For lightly damped systems when the forcing frequency is close to ω_m, the amplification factor is much greater than 1; when the forcing frequency is much less than ω_m, the amplification factor approaches 1; and for a forcing frequency much greater than ω_m, the amplification factor approaches zero. As the damping ratio is increased from zero, the amplification due to resonance is reduced. For damping ratios that are greater than $\zeta = \sqrt{1/2}$, all resonance amplification is eliminated.

Because the spring-mass system having one degree of freedom is modeled with a single second-order differential equation, the eigenvalues can be found analytically. However, this is not the case for systems with many degrees of freedom. For second-order systems having more than two degrees of freedom, the eigenvalues must be found numerically. To facilitate the numerical solution, it is convenient to express each second-order differential equation as two first-order equations. To demonstrate, we shall first apply this procedure to Eq. (8.1.2).

Defining a second dependent variable to be the time rate of change of the position disturbance, i.e., the velocity disturbance,

$$\Delta V \equiv \frac{d\Delta x}{dt} \tag{8.1.26}$$

we can rewrite the homogeneous form of Eq. (8.1.2) as

$$\begin{bmatrix} m & 0 \\ 0 & 1 \end{bmatrix}\begin{Bmatrix} \Delta\dot{V} \\ \Delta\dot{x} \end{Bmatrix} = \begin{bmatrix} -c & -k \\ 1 & 0 \end{bmatrix}\begin{Bmatrix} \Delta V \\ \Delta x \end{Bmatrix} \tag{8.1.27}$$

where the dot indicates a time derivative. Equation (8.1.27) is of the form

$$[\mathbf{B}]\{\dot{\mathbf{X}}\} = [\mathbf{A}]\{\mathbf{X}\} \tag{8.1.28}$$

where

$$\{\mathbf{X}\} = \begin{Bmatrix} \Delta V \\ \Delta x \end{Bmatrix} \tag{8.1.29}$$

$$[\mathbf{A}] = \begin{bmatrix} -c & -k \\ 1 & 0 \end{bmatrix} \tag{8.1.30}$$

$$[\mathbf{B}] = \begin{bmatrix} m & 0 \\ 0 & 1 \end{bmatrix} \tag{8.1.31}$$

The solutions to Eq. (8.1.28) are of the form

$$\{\mathbf{X}\} = C\{\mathcal{X}\}e^{\lambda t} \tag{8.1.32}$$

where C is an arbitrary scalar constant, $\{\mathcal{X}\}$ is a vector with explicit constant components, and λ is an explicit scalar constant. Using Eq. (8.1.32) in Eq. (8.1.28), we have

$$C\lambda[\mathbf{B}]\{\mathcal{X}\}e^{\lambda t} = C[\mathbf{A}]\{\mathcal{X}\}e^{\lambda t} \tag{8.1.33}$$

Equation (8.1.33) can be written in the alternative form

$$\Big([\mathbf{A}] - \lambda\,[\mathbf{B}]\Big)\{\mathcal{X}\} = 0 \tag{8.1.34}$$

Equation (8.1.34) represents a homogeneous system of linear algebraic equations. This is the *generalized eigenproblem*. Nontrivial solutions to this system of equation will exist only if the determinant of $[\mathbf{A}] - \lambda[\mathbf{B}]$ is zero. The values of λ that will force this determinant to be zero are the eigenvalues. There exists a pair of eigenvalues for each degree of freedom associated with the problem. A one-degree-of-freedom problem has two eigenvalues; a two-degree-of-freedom problem has four eigenvalues; and so on. For each eigenvalue there exists a nontrivial solution for $\{\mathcal{X}\}$ that will satisfy Eq. (8.1.34). These solution vectors are called *eigenvectors*. While the eigenvectors are nontrivial, they are not unique. Clearly, because Eq. (8.1.34) is homogeneous, any multiple of an eigenvector will also be an eigenvector.

In summary, the eigenvalues are found by obtaining the roots to the characteristic equation obtained by setting the determinant of $[\mathbf{A}] - \lambda[\mathbf{B}]$ to zero,

$$\big|[\mathbf{A}] - \lambda_i[\mathbf{B}]\big| = 0 \tag{8.1.35}$$

Each set of eigenvectors is found by using the corresponding eigenvalue in the system of equations given by

$$([\mathbf{A}] - \lambda_i[\mathbf{B}])\{\mathcal{X}\}_i = 0 \tag{8.1.36}$$

Because the eigenvectors are unique only to within some arbitrary constant multiplier, they are commonly defined so that each eigenvector has a magnitude of unity, with the largest component being real,

$$\sqrt{\sum_{j=1}^{N}\left|\mathcal{X}_j\right|_i^2} = 1, \quad \mathrm{imag}\big(\mathrm{max}(\mathcal{X}_j)\big)_i = 0 \tag{8.1.37}$$

The *absolute value* or *magnitude* of any component of an eigenvector is defined to be the square root of the sum of the squares of its real and imaginary parts. Once any form of an eigenvector is found, it can easily be normalized according to Eq. (8.1.37). First, the component having the largest magnitude is found. All components of the eigenvector are then multiplied by the complex conjugate of this largest component. This makes the largest component real. Next, the magnitude of the entire eigenvector is found by taking the square root of the sum of the squares of the magnitudes of all components. Finally, all components of the eigenvector are divided by this magnitude, giving the final form of the eigenvector a magnitude of unity.

The two eigenvalues for a one-degree-of-freedom problem can easily be obtained analytically. Using Eqs. (8.1.30) and (8.1.31) in Eq. (8.1.35) and evaluating the determinate, we have

$$\left|[\mathbf{A}] - \lambda[\mathbf{B}]\right| = \left|\begin{bmatrix} -c & -k \\ 1 & 0 \end{bmatrix} - \lambda\begin{bmatrix} m & 0 \\ 0 & 1 \end{bmatrix}\right| = \left|\begin{matrix} -(c+m\lambda) & -k \\ 1 & -\lambda \end{matrix}\right| \tag{8.1.38}$$

$$= m\lambda^2 + c\lambda + k = 0$$

which gives the well-known solutions expressed in Eq. (8.1.4).

The eigenvalues and eigenvectors for systems with multiple degrees of freedom can readily be obtained numerically using a digital computer. Because of the importance of eigenproblems in all fields of engineering and science, algorithms for computing eigenvalues and eigenvectors are widely available. Most modern computer facilities will have subroutines available for determining eigenvalues and eigenvectors. In fact, today, most handheld scientific calculators have preprogrammed algorithms for obtaining numerical solutions to eigenproblems.

The details associated with obtaining numerical solutions to eigenproblems can be found in any undergraduate text on engineering numerical methods (see, for example, Hoffman 1992 or Press, Flannery, Teukolsky, and Vetterling 1986). Because an introductory course in numerical methods is assumed to be a prerequisite to the material presented in this text, the details associated with implementing numerical solutions to the eigenproblem will not be repeated here.

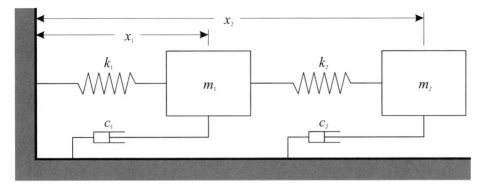

Figure 8.1.4. Damped spring-mass system with two degrees of freedom.

As an example of an eigenproblem with multiple degrees of freedom, consider the free response of the damped spring-mass system illustrated in Fig. 8.1.4. This system has two degrees of freedom, one associated with the independent translation of each mass. Applying Newton's second law of motion to this system, we have

$$m_1 \frac{d^2 x_1}{dt^2} = -c_1 \frac{dx_1}{dt} - k_1(x_1 - x_{1o}) + k_2[(x_2 - x_1) - (x_{2o} - x_{1o})] \tag{8.1.39}$$

$$m_2 \frac{d^2 x_2}{dt^2} = -c_2 \frac{dx_2}{dt} - k_2[(x_2 - x_1) - (x_{2o} - x_{1o})] \tag{8.1.40}$$

Equations (8.1.39) and (8.1.40) can be rewritten in terms of the axial displacement of each mass from equilibrium, $\Delta x_1 = x_1 - x_{1o}$ and $\Delta x_2 = x_2 - x_{2o}$,

$$m_1 \frac{d^2 \Delta x_1}{dt^2} + c_1 \frac{d\Delta x_1}{dt} + (k_1 + k_2)\Delta x_1 - k_2 \Delta x_2 = 0 \tag{8.1.41}$$

$$m_2 \frac{d^2 \Delta x_2}{dt^2} + c_2 \frac{d\Delta x_2}{dt} - k_2 \Delta x_1 + k_2 \Delta x_2 = 0 \tag{8.1.42}$$

Defining the dependent variables,

$$\Delta V_1 \equiv \frac{d\Delta x_1}{dt} \tag{8.1.43}$$

$$\Delta V_2 \equiv \frac{d\Delta x_2}{dt} \tag{8.1.44}$$

we can rewrite Eqs. (8.1.41) and (8.1.42) as

$$
\begin{bmatrix} m_1 & 0 & 0 & 0 \\ 0 & m_2 & 0 & 0 \\ 0 & 0 & 1 & 0 \\ 0 & 0 & 0 & 1 \end{bmatrix} \begin{Bmatrix} \Delta\dot{V}_1 \\ \Delta\dot{V}_2 \\ \Delta\dot{x}_1 \\ \Delta\dot{x}_2 \end{Bmatrix} = \begin{bmatrix} -c_1 & 0 & -(k_1+k_2) & k_2 \\ 0 & -c_2 & k_2 & -k_2 \\ 1 & 0 & 0 & 0 \\ 0 & 1 & 0 & 0 \end{bmatrix} \begin{Bmatrix} \Delta V_1 \\ \Delta V_2 \\ \Delta x_1 \\ \Delta x_2 \end{Bmatrix} \qquad (8.1.45)
$$

In general, the solutions to the homogeneous system of equations given by Eq. (8.1.45) are all of the form

$$
\begin{Bmatrix} \Delta V_1 \\ \Delta V_2 \\ \Delta x_1 \\ \Delta x_2 \end{Bmatrix} = C \begin{Bmatrix} \chi_{V_1} \\ \chi_{V_2} \\ \chi_{x_1} \\ \chi_{x_2} \end{Bmatrix} e^{\lambda t} \qquad (8.1.46)
$$

Using Eq. (8.1.46) in Eq. (8.1.45), we have

$$
C\lambda \begin{bmatrix} m_1 & 0 & 0 & 0 \\ 0 & m_2 & 0 & 0 \\ 0 & 0 & 1 & 0 \\ 0 & 0 & 0 & 1 \end{bmatrix} \begin{Bmatrix} \chi_{V_1} \\ \chi_{V_2} \\ \chi_{x_1} \\ \chi_{x_2} \end{Bmatrix} e^{\lambda t} = C \begin{bmatrix} -c_1 & 0 & -(k_1+k_2) & k_2 \\ 0 & -c_2 & k_2 & -k_2 \\ 1 & 0 & 0 & 0 \\ 0 & 1 & 0 & 0 \end{bmatrix} \begin{Bmatrix} \chi_{V_1} \\ \chi_{V_2} \\ \chi_{x_1} \\ \chi_{x_2} \end{Bmatrix} e^{\lambda t} \qquad (8.1.47)
$$

or after rearranging,

$$
\left(\begin{bmatrix} -c_1 & 0 & -(k_1+k_2) & k_2 \\ 0 & -c_2 & k_2 & -k_2 \\ 1 & 0 & 0 & 0 \\ 0 & 1 & 0 & 0 \end{bmatrix} - \lambda \begin{bmatrix} m_1 & 0 & 0 & 0 \\ 0 & m_2 & 0 & 0 \\ 0 & 0 & 1 & 0 \\ 0 & 0 & 0 & 1 \end{bmatrix} \right) \begin{Bmatrix} \chi_{V_1} \\ \chi_{V_2} \\ \chi_{x_1} \\ \chi_{x_2} \end{Bmatrix} = \begin{Bmatrix} 0 \\ 0 \\ 0 \\ 0 \end{Bmatrix} \qquad (8.1.48)
$$

This system of linear algebraic equations is in the form of the generalized eigenproblem given by Eq. (8.1.34), where

$$
[\mathbf{A}] = \begin{bmatrix} -c_1 & 0 & -(k_1+k_2) & k_2 \\ 0 & -c_2 & k_2 & -k_2 \\ 1 & 0 & 0 & 0 \\ 0 & 1 & 0 & 0 \end{bmatrix} \qquad (8.1.49)
$$

$$
[\mathbf{B}] = \begin{bmatrix} m_1 & 0 & 0 & 0 \\ 0 & m_2 & 0 & 0 \\ 0 & 0 & 1 & 0 \\ 0 & 0 & 0 & 1 \end{bmatrix} \qquad (8.1.50)
$$

If the masses, damping coefficients, and spring constants in Eq. (8.1.48) are all known, the eigenvalues and eigenvectors associated with the motion of this system can be found by using these matrices in any generalized eigensolver.

The character of each mode of motion associated with the free response of this system can be identified as being overdamped, underdamped, or critically damped, by examining the eigenvalues, just as was done for the one-degree-of-freedom problem. Any eigenvalue that is real, distinct, and negative corresponds to an overdamped mode, with the damping rate being equal to the negative of this real eigenvalue. A pair of eigenvalues that are real, identical, and negative corresponds to a critically damped mode, having a damping rate that is also equal to the negative of the real eigenvalue. When complex eigenvalues are found, they are always found in complex pairs. Each complex pair of eigenvalues corresponds to a mode that is underdamped. Such modes result in sinusoidal motion having a frequency equal to the magnitude of the imaginary part of the eigenvalue and a damping rate equal to the negative of the real part of the eigenvalue.

Most readers should be quite familiar with the foregoing interpretation of the eigenvalues, from their study of elementary dynamics with one degree of freedom. However, many readers may not be as familiar with the physical interpretation of the eigenvectors. Since a real number is a special case of a complex number, we shall first describe the interpretation of an eigenvector that is associated with a complex eigenvalue. The components of the eigenvectors associated with a complex pair of eigenvalues also form complex pairs. To demonstrate the physical meaning of such eigenvectors, consider the eigenvectors associated with a complex pair of eigenvalues that are solutions to Eq. (8.1.48). Let the eigenvalues be

$$\lambda_1, \lambda_2 = -\sigma \pm \omega_d i \tag{8.1.51}$$

and let the eigenvectors associated with these two eigenvalues be

$$\begin{Bmatrix} \chi_{V_1} \\ \chi_{V_2} \\ \chi_{x_1} \\ \chi_{x_2} \end{Bmatrix}_1, \begin{Bmatrix} \chi_{V_1} \\ \chi_{V_2} \\ \chi_{x_1} \\ \chi_{x_2} \end{Bmatrix}_2 = \begin{Bmatrix} R_{V_1} \pm I_{V_1} i \\ R_{V_2} \pm I_{V_2} i \\ R_{x_1} \pm I_{x_1} i \\ R_{x_2} \pm I_{x_2} i \end{Bmatrix} \tag{8.1.52}$$

where σ is the damping rate and ω_d is the damped natural frequency.

Recall that a complex number can be written in exponential form in terms of its amplitude and phase. Thus, Eq. (8.1.52) can be written as

$$\begin{Bmatrix} \chi_{V_1} \\ \chi_{V_2} \\ \chi_{x_1} \\ \chi_{x_2} \end{Bmatrix}_1, \begin{Bmatrix} \chi_{V_1} \\ \chi_{V_2} \\ \chi_{x_1} \\ \chi_{x_2} \end{Bmatrix}_2 = \begin{Bmatrix} A_{V_1} e^{\pm\Theta_{v_1} i} \\ A_{V_2} e^{\pm\Theta_{v_2} i} \\ A_{x_1} e^{\pm\Theta_{x_1} i} \\ A_{x_2} e^{\pm\Theta_{x_2} i} \end{Bmatrix} \tag{8.1.53}$$

where for each component of this complex pair of eigenvectors, the amplitude, A, and the phase angle, Θ, are related to the real part, R, and the imaginary part, I, according to the relations

$$A = \sqrt{R^2 + I^2} \tag{8.1.54}$$

and

$$\Theta = \tan^{-1}(I/R) \tag{8.1.55}$$

From Eq. (8.1.46), the general solution corresponding to this pair of eigenvalues can be written as

$$
\begin{Bmatrix} \Delta V_1 \\ \Delta V_2 \\ \Delta x_1 \\ \Delta x_2 \end{Bmatrix} = C_1 \begin{Bmatrix} A_{V_1} e^{\Theta_{v_1} i} \\ A_{V_2} e^{\Theta_{v_2} i} \\ A_{x_1} e^{\Theta_{x_1} i} \\ A_{x_2} e^{\Theta_{x_2} i} \end{Bmatrix} e^{(-\sigma + \omega_d i)t} + C_2 \begin{Bmatrix} A_{V_1} e^{-\Theta_{v_1} i} \\ A_{V_2} e^{-\Theta_{v_2} i} \\ A_{x_1} e^{-\Theta_{x_1} i} \\ A_{x_2} e^{-\Theta_{x_2} i} \end{Bmatrix} e^{(-\sigma - \omega_d i)t} \tag{8.1.56}
$$

or after rearranging,

$$
\begin{Bmatrix} \Delta V_1 \\ \Delta V_2 \\ \Delta x_1 \\ \Delta x_2 \end{Bmatrix} = e^{-\sigma t} \left(C_1 \begin{Bmatrix} A_{V_1} e^{i(\omega_d t + \Theta_{v_1})} \\ A_{V_2} e^{i(\omega_d t + \Theta_{v_2})} \\ A_{x_1} e^{i(\omega_d t + \Theta_{x_1})} \\ A_{x_2} e^{i(\omega_d t + \Theta_{x_2})} \end{Bmatrix} + C_2 \begin{Bmatrix} A_{V_1} e^{-i(\omega_d t + \Theta_{v_1})} \\ A_{V_2} e^{-i(\omega_d t + \Theta_{v_2})} \\ A_{x_1} e^{-i(\omega_d t + \Theta_{x_1})} \\ A_{x_2} e^{-i(\omega_d t + \Theta_{x_2})} \end{Bmatrix} \right) \tag{8.1.57}
$$

To facilitate writing the solution in terms of circular functions, Eq. (8.1.57) can be further rearranged to give

$$
\begin{Bmatrix} \Delta V_1 \\ \Delta V_2 \\ \Delta x_1 \\ \Delta x_2 \end{Bmatrix} = e^{-\sigma t} \left(\frac{C_1 + C_2}{2} \begin{Bmatrix} A_{V_1} e^{i(\omega_d t + \Theta_{v_1})} \\ A_{V_2} e^{i(\omega_d t + \Theta_{v_2})} \\ A_{x_1} e^{i(\omega_d t + \Theta_{x_1})} \\ A_{x_2} e^{i(\omega_d t + \Theta_{x_2})} \end{Bmatrix} + \frac{C_1 + C_2}{2} \begin{Bmatrix} A_{V_1} e^{-i(\omega_d t + \Theta_{v_1})} \\ A_{V_2} e^{-i(\omega_d t + \Theta_{v_2})} \\ A_{x_1} e^{-i(\omega_d t + \Theta_{x_1})} \\ A_{x_2} e^{-i(\omega_d t + \Theta_{x_2})} \end{Bmatrix} \right)
$$

$$
+ e^{-\sigma t} \left(\frac{C_1 - C_2}{2} \begin{Bmatrix} A_{V_1} e^{i(\omega_d t + \Theta_{v_1})} \\ A_{V_2} e^{i(\omega_d t + \Theta_{v_2})} \\ A_{x_1} e^{i(\omega_d t + \Theta_{x_1})} \\ A_{x_2} e^{i(\omega_d t + \Theta_{x_2})} \end{Bmatrix} - \frac{C_1 - C_2}{2} \begin{Bmatrix} A_{V_1} e^{-i(\omega_d t + \Theta_{v_1})} \\ A_{V_2} e^{-i(\omega_d t + \Theta_{v_2})} \\ A_{x_1} e^{-i(\omega_d t + \Theta_{x_1})} \\ A_{x_2} e^{-i(\omega_d t + \Theta_{x_2})} \end{Bmatrix} \right) \tag{8.1.58}
$$

Now, recalling the exponential definitions of the circular functions,

$$\cos x = \frac{e^{ix} + e^{-ix}}{2} \tag{8.1.59}$$

$$\sin x = \frac{e^{ix} - e^{-ix}}{2i} \tag{8.1.60}$$

Eq. (8.1.58) becomes

$$
\begin{Bmatrix} \Delta V_1 \\ \Delta V_2 \\ \Delta x_1 \\ \Delta x_2 \end{Bmatrix} = (C_1 + C_2)e^{-\sigma t} \begin{Bmatrix} A_{V_1}\cos(\omega_d t + \Theta_{V_1}) \\ A_{V_2}\cos(\omega_d t + \Theta_{V_2}) \\ A_{x_1}\cos(\omega_d t + \Theta_{x_1}) \\ A_{x_2}\cos(\omega_d t + \Theta_{x_2}) \end{Bmatrix}
$$
$$
+ (C_1 - C_2)e^{-\sigma t} \begin{Bmatrix} A_{V_1}\sin(\omega_d t + \Theta_{V_1}) \\ A_{V_2}\sin(\omega_d t + \Theta_{V_2}) \\ A_{x_1}\sin(\omega_d t + \Theta_{x_1}) \\ A_{x_2}\sin(\omega_d t + \Theta_{x_2}) \end{Bmatrix} i
\tag{8.1.61}
$$

Here we see that this solution corresponds to damped sinusoidal oscillation. Each of the four dependent variables oscillates with exactly the same frequency. However, each dependent variable has a different amplitude and a different phase shift. The eigenvalues give us the damped natural frequency and the damping rate for all of the oscillation, while the components of the eigenvectors give us the amplitudes and the phase shifts. Since the sinusoidal terms are multiplied by arbitrary constants, the eigenvectors describe only relative amplitudes and phase shifts. If the eigenvector has been normalized according to Eq. (8.1.37), making the largest component real, from Eq. (8.1.55) we see that the phase angle for this component will be zero. Thus, an eigenvector in this form describes the phase of each component relative to the component having the largest magnitude.

Eigenvectors corresponding to purely real or purely imaginary eigenvalues can be viewed as special cases of the complex eigenvalue. Purely imaginary eigenvalues simply describe undamped sinusoidal motion, and the eigenvectors are interpreted exactly the same as those corresponding to complex eigenvalues. A purely real eigenvalue could be thought of as describing damped sinusoidal motion with a frequency of zero. All components of the eigenvector corresponding to a purely real eigenvalue are also real. Thus, the magnitude of each component of such an eigenvector is simply equal to its absolute value and the phase angle is either zero or 180 degrees. So we see that for a purely real eigenvalue, the components of the eigenvector also describe the relative amplitudes associated with the temporal variations of the dependent variables. The only difference between the results obtained from real eigenvalues and those obtained from complex eigenvalues is that the temporal variations associated with the real eigenvalues are exponential, while those associated with the complex eigenvalues represent damped sinusoidal motion.

Before proceeding with some example calculations, it is worth spending a few words talking about the *special eigenproblem*. The generalized eigenproblem specified by Eq. (8.1.34) contains two $n \times n$ matrices, $[A]$ and $[B]$. No special requirements are placed on either $[A]$ or $[B]$. In general, both of these matrices could have n^2 nonzero terms. However, many physical problems can readily be formulated in terms of a somewhat simpler system of equations called the *special eigenproblem*. For example, consider the spring-mass system shown in Fig. 8.1.4, which yields the eigenproblem specified by Eq. (8.1.48):

$$\left(\begin{bmatrix} -c_1 & 0 & -(k_1+k_2) & k_2 \\ 0 & -c_2 & k_2 & -k_2 \\ 1 & 0 & 0 & 0 \\ 0 & 1 & 0 & 0 \end{bmatrix} - \lambda \begin{bmatrix} m_1 & 0 & 0 & 0 \\ 0 & m_2 & 0 & 0 \\ 0 & 0 & 1 & 0 \\ 0 & 0 & 0 & 1 \end{bmatrix} \right) \begin{Bmatrix} \chi_{V_1} \\ \chi_{V_2} \\ \chi_{x_1} \\ \chi_{x_2} \end{Bmatrix} = \begin{Bmatrix} 0 \\ 0 \\ 0 \\ 0 \end{Bmatrix}$$

This eigenproblem can readily be rearranged to the form of the special eigenproblem. Dividing the first row by m_1 and the second row by m_2, we have

$$\left(\begin{bmatrix} -c_1/m_1 & 0 & -(k_1+k_2)/m_1 & k_2/m_1 \\ 0 & -c_2/m_2 & k_2/m_2 & -k_2/m_2 \\ 1 & 0 & 0 & 0 \\ 0 & 1 & 0 & 0 \end{bmatrix} \right.$$

$$\left. - \lambda \begin{bmatrix} 1 & 0 & 0 & 0 \\ 0 & 1 & 0 & 0 \\ 0 & 0 & 1 & 0 \\ 0 & 0 & 0 & 1 \end{bmatrix} \right) \begin{Bmatrix} \chi_{V_1} \\ \chi_{V_2} \\ \chi_{x_1} \\ \chi_{x_2} \end{Bmatrix} = \begin{Bmatrix} 0 \\ 0 \\ 0 \\ 0 \end{Bmatrix}$$

(8.1.62)

This is exactly the form of the **special eigenproblem**, which is a special case of the generalized eigenproblem where the matrix $[B]$ is exactly equal to the identity matrix $[i]$,

$$\left([A] - \lambda [i] \right) \{\chi\} = 0 \tag{8.1.63}$$

Eigensolvers for the special eigenproblem are more widely available than are those for the generalized eigenproblem. If you have an eigensolver that requires only one matrix as input, it is more than likely solving for the eigenvalues and eigenvectors associated with Eq. (8.1.63).

The specifications for a subroutine intended to solve the special eigenproblem might say that the subroutine finds the eigenvalues and eigenvectors of the matrix $[A]$. Strictly speaking, a single matrix $[A]$ by itself does not have eigenvalues. Only when the matrix $[A]$ is combined with a second matrix $[B]$ and used in Eq. (8.1.34) to form a system of equations can we define the eigenvalues. Strictly speaking, it is the system of equations that has eigenvalues, not the matrix. However, in the usual manner of speaking, whenever the second matrix $[\mathbf{B}]$ is not mentioned, it is assumed to be the identity matrix and it is assumed that we are solving the special eigenproblem given by Eq. (8.1.63).

The generalized eigenproblem can be reduced to the special eigenproblem. In the special case of Eq. (8.1.48) this was particularly easy because $[\mathbf{B}]$ was a diagonal matrix. In the more general case, a special eigenproblem can be obtained from a generalized eigenproblem by multiplying by the inverse of the matrix $[\mathbf{B}]$. From Eq. (8.1.34) we can write

$$\left([\mathbf{B}]^{-1}[\mathbf{A}] - \lambda [\mathbf{B}]^{-1}[\mathbf{B}]\right)\{\chi\} = 0 \tag{8.1.64}$$

Since any matrix multiplied by its inverse is equal to the identity matrix, Eq. (8.1.64) is in the form of the special eigenproblem. The only difficulty comes in finding the inverse of $[\mathbf{B}]$. If the matrix $[\mathbf{B}]$ is of a form that easily allows us to evaluate its inverse, then the generalized eigenproblem is easily reduced to the special eigenproblem. For example, if $[\mathbf{B}]$ is tridiagonal, the Thomas algorithm can be used to invert $[\mathbf{B}]$. If $[\mathbf{B}]$ is diagonally dominant and otherwise quite sparse, it may be very simple to manually reduce the generalized eigenproblem to the special eigenproblem by using Gauss elimination. However, if a good generalized eigensolver is available, it is usually easier to leave the eigenproblem in the generalized form and let the computer do the work.

EXAMPLE 8.1.1. For the damped spring-mass system that is shown in Fig. 8.1.4, find the eigenvalues, the eigenvectors, the damping rate, the 99 percent damping time, and the damped natural frequency and period if underdamped, for each mode associated with the case:

$$m_1 = 20.0 \text{ slugs}, \qquad c_1 = 30.0 \text{ lbf} \cdot \text{sec}/\text{ft}, \qquad k_1 = 2.0 \text{ lbf}/\text{ft},$$
$$m_2 = 20.0 \text{ slugs}, \qquad c_2 = 15.0 \text{ lbf} \cdot \text{sec}/\text{ft}, \qquad k_2 = 100.0 \text{ lbf}/\text{ft}$$

For each pair of eigenvalues, compute the undamped natural frequency and the damping ratio. For each oscillatory mode, compute the amplitude and phase angle for each component of the eigenvector.

Solution. Using these values for the masses, damping coefficients, and spring constants in Eq. (8.1.48), the generalized eigenproblem for this system is

$$\left(\begin{bmatrix} -30.0 & 0 & -102.0 & 100.0 \\ 0 & -15.0 & 100.0 & -100.0 \\ 1 & 0 & 0 & 0 \\ 0 & 1 & 0 & 0 \end{bmatrix} - \lambda \begin{bmatrix} 20.0 & 0 & 0 & 0 \\ 0 & 20.0 & 0 & 0 \\ 0 & 0 & 1 & 0 \\ 0 & 0 & 0 & 1 \end{bmatrix} \right) \begin{Bmatrix} \chi_{V_1} \\ \chi_{V_2} \\ \chi_{x_1} \\ \chi_{x_2} \end{Bmatrix} = \begin{Bmatrix} 0 \\ 0 \\ 0 \\ 0 \end{Bmatrix}$$

The solution to this characteristic equation can be obtained numerically using a generalized eigensolver. This system of equations is also easily reduced to the special eigenproblem

$$
\left(
\begin{bmatrix}
-1.5 & 0 & -5.1 & 5.0 \\
0 & -0.75 & 5.0 & -5.0 \\
1 & 0 & 0 & 0 \\
0 & 1 & 0 & 0
\end{bmatrix}
- \lambda
\begin{bmatrix}
1 & 0 & 0 & 0 \\
0 & 1 & 0 & 0 \\
0 & 0 & 1 & 0 \\
0 & 0 & 0 & 1
\end{bmatrix}
\right)
\begin{Bmatrix}
\chi_{V_1} \\
\chi_{V_2} \\
\chi_{x_1} \\
\chi_{x_2}
\end{Bmatrix}
=
\begin{Bmatrix}
0 \\
0 \\
0 \\
0
\end{Bmatrix}
$$

The solution to either of these systems yields the eigenvalues and eigenvectors that correspond to the following modes:

Mode 1 $(\lambda_{1,2} = -0.556387 \ \text{sec}^{-1} \pm 3.098306 \ i \ \text{sec}^{-1})$
From the eigenvectors associated with this mode, the corresponding homogeneous solutions are

$$
\begin{Bmatrix}
\Delta V_1 \\
\Delta V_2 \\
\Delta x_1 \\
\Delta x_2
\end{Bmatrix}
= C_{1,2}
\begin{Bmatrix}
(-0.644717 \mp 0.153945 \ i) \ \text{ft/sec} \\
0.684817 \ \text{ft/sec} \\
(-0.011934 \pm 0.210230 \ i) \ \text{ft} \\
(-0.038452 \mp 0.214124 \ i) \ \text{ft}
\end{Bmatrix}
$$
$$
\times \exp[(-0.556387 \pm 3.098306 \ i) \ t/\text{sec}]
$$

The damping rate is

$$
\sigma = -\text{real}(\lambda) = 0.556387 \ \text{sec}^{-1}
$$

The 99 percent damping time is the time required for the amplitude of the oscillations to be reduced by 99 percent, in other words, the time required for the amplitude to be reduced to 1 percent of its original value. Thus, we can write

$$
\frac{\exp(-\sigma t_{99})}{\exp(-\sigma 0.0)} = \exp(-\sigma t_{99}) = 0.01
$$

$$
99\% \ \text{damping time} = \frac{\ln(0.01)}{-\sigma} = \frac{\ln(0.01)}{-0.556387 \ \text{sec}^{-1}} = 8.277 \ \text{sec}
$$

The damped natural frequency and period for this motion are

$$
\omega_d = |\text{imag}(\lambda)| = 3.098306 \ \text{rad/sec}
$$

$$
\text{period} = \frac{2\pi}{\omega_d} = \frac{2\pi}{3.098306 \ \text{sec}^{-1}} = 2.028 \ \text{sec}
$$

For the pair of complex roots associated with this mode, the undamped natural frequency is

$$\omega_n = \sqrt{\lambda_1 \lambda_2} = \sqrt{(-0.556387 + 3.098306\,i)(-0.556387 - 3.098306\,i)}$$

$$= \sqrt{0.556387^2 + 3.098306^2} = 3.147867 \text{ rad/sec}$$

and the damping ratio is

$$\zeta = -\frac{\lambda_1 + \lambda_2}{2\sqrt{\lambda_1 \lambda_2}} = -\frac{-0.556387 + 3.098306\,i - 0.556387 - 3.098306\,i}{2\sqrt{(-0.556387 + 3.098306\,i)(-0.556387 - 3.098306\,i)}}$$

$$= \frac{0.556387}{\sqrt{0.556387^2 + 3.098306^2}} = 0.176750$$

Notice that for this mode, the frequency is decreased by less than 2 percent as a result of damping.

Mode 2 ($\lambda_3 = -1.090975 \text{ sec}^{-1}$)
From the eigenvector associated with this mode, the homogeneous solution that results from this real eigenvalue is

$$\begin{Bmatrix} \Delta V_1 \\ \Delta V_2 \\ \Delta x_1 \\ \Delta x_2 \end{Bmatrix} = C_3 \begin{Bmatrix} 0.539609 \text{ ft/sec} \\ 0.502243 \text{ ft/sec} \\ -0.494612 \text{ ft} \\ -0.460362 \text{ ft} \end{Bmatrix} \exp(-1.090975\,t/\text{sec})$$

Since the eigenvalue is real and negative, this is overdamped motion having a damping rate and 99 percent damping time of

$$\sigma = -\text{real}(\lambda) = 1.090975 \text{ sec}^{-1}$$

$$99\% \text{ damping time} = \frac{\ln(0.01)}{-\sigma} = \frac{\ln(0.01)}{-1.090975 \text{ sec}^{-1}} = 4.221 \text{ sec}$$

Mode 3 ($\lambda_4 = -0.046251 \text{ sec}^{-1}$)
From the associated eigenvector, the homogeneous solution for this overdamped mode is

$$\begin{Bmatrix} \Delta V_1 \\ \Delta V_2 \\ \Delta x_1 \\ \Delta x_2 \end{Bmatrix} = C_4 \begin{Bmatrix} -0.032563 \text{ ft/sec} \\ -0.032776 \text{ ft/sec} \\ 0.704041 \text{ ft} \\ 0.708654 \text{ ft} \end{Bmatrix} \exp(-0.046251\,t/\text{sec})$$

The damping rate and 99 percent damping time are

$$\sigma = -\text{real}(\lambda) = 0.046251 \, \text{sec}^{-1}$$

$$99\% \text{ damping time} = \frac{\ln(0.01)}{-\sigma} = \frac{\ln(0.01)}{-0.046251 \, \text{sec}^{-1}} = 99.569 \, \text{sec}$$

For the pair of roots that are associated with modes 2 and 3, the undamped natural frequency is

$$\omega_n = \sqrt{\lambda_3 \lambda_4} = \sqrt{(-1.090975)(-0.046251)} = 0.224630 \, \text{rad/sec}$$

and the damping ratio is

$$\zeta = -\frac{\lambda_3 + \lambda_4}{2\sqrt{\lambda_3 \lambda_4}} = -\frac{-1.090975 - 0.046251}{2\sqrt{(-1.090975)(-0.046251)}} = 2.531330$$

For this eigenproblem, mode 1 is underdamped, while modes 2 and 3 are overdamped. Thus, mode 1 is the only oscillatory mode. The complex components of the eigenvectors associated with any complex pair of eigenvalues can be written in terms of amplitude and phase. The amplitude of each component is simply the magnitude of that complex number. Thus, for component χ_i,

$$\text{amplitude} = \sqrt{[\text{real}(\chi_i)]^2 + [\text{imag}(\chi_i)]^2}$$

The phase angle associated with each eigenvector component is the arctangent of the imaginary part divided by the real part. Thus, for component χ_i,

$$\text{phase} = \text{atan2}[\text{imag}(\chi_i), \text{real}(\chi_i)]$$

For mode 1, the amplitude and phase of the eigenvectors associated with the positive and negative frequencies are, respectively,

$$\text{amplitude} \begin{Bmatrix} \Delta V_1 \\ \Delta V_2 \\ \Delta x_1 \\ \Delta x_2 \end{Bmatrix} = C_{1,2} \begin{Bmatrix} 0.662842 \, \text{ft/sec} \\ 0.684817 \, \text{ft/sec} \\ 0.210569 \, \text{ft} \\ 0.217549 \, \text{ft} \end{Bmatrix}$$

$$\text{phase} \begin{Bmatrix} \Delta V_1 \\ \Delta V_2 \\ \Delta x_1 \\ \Delta x_2 \end{Bmatrix} = \pm \begin{Bmatrix} -2.9039 \\ 0.0000 \\ 1.6256 \\ -1.7466 \end{Bmatrix} = \pm \begin{Bmatrix} -166.38° \\ 0.00° \\ 93.14° \\ -100.07° \end{Bmatrix}$$

Thus, the homogeneous solutions associated with this pair of eigenvalues can be written as

$$\begin{Bmatrix} \Delta V_1 \\ \Delta V_2 \\ \Delta x_1 \\ \Delta x_2 \end{Bmatrix} = C_{1,2} \begin{Bmatrix} 0.662842 \ \exp[i(\mp 2.9039 \pm 3.098306 \ t/\text{sec})] \ \text{ft/sec} \\ 0.684817 \ \exp[i(\pm 3.098306 \ t/\text{sec})] \ \text{ft/sec} \\ 0.210568 \ \exp[i(\pm 1.6256 \pm 3.098306 \ t/\text{sec})] \ \text{ft} \\ 0.217549 \ \exp[i(\mp 1.7466 \pm 3.098306 \ t/\text{sec})] \ \text{ft} \end{Bmatrix}$$

$$\times \exp(-0.556387 \ t/\text{sec})$$

The damped spring-mass system in Example 8.1.1 is typical of systems that result in underdamped, overdamped, and/or critically damped motion, when disturbed from equilibrium. Underdamped modes produce sinusoidal oscillations in velocity and displacement for all degrees of freedom. The oscillations for each component of velocity and displacement can have a different amplitude and phase. Notice in Example 8.1.1 that the oscillations in Δx_1 and Δx_2 have similar amplitudes but are nearly 180 degrees out of phase. In general, it is also possible for second-order systems to result in modes that are not underdamped, overdamped, or critically damped. One such mode, called a *rigid-body displacement mode*, is demonstrated in the following example.

EXAMPLE 8.1.2. Repeat Example 8.1.1 for the case

$$m_1 = 20.0 \text{ slugs}, \qquad c_1 = 30.0 \text{ lbf} \cdot \text{sec/ft}, \qquad k_1 = 0.0 \text{ lbf/ft},$$
$$m_2 = 20.0 \text{ slugs}, \qquad c_2 = 15.0 \text{ lbf} \cdot \text{sec/ft}, \qquad k_2 = 100.0 \text{ lbf/ft}$$

Solution. The generalized eigenproblem for this system is

$$\left(\begin{bmatrix} -30.0 & 0 & -100.0 & 100.0 \\ 0 & -15.0 & 100.0 & -100.0 \\ 1 & 0 & 0 & 0 \\ 0 & 1 & 0 & 0 \end{bmatrix} - \lambda \begin{bmatrix} 20.0 & 0 & 0 & 0 \\ 0 & 20.0 & 0 & 0 \\ 0 & 0 & 1 & 0 \\ 0 & 0 & 0 & 1 \end{bmatrix} \right) \begin{Bmatrix} \chi_{V_1} \\ \chi_{V_2} \\ \chi_{x_1} \\ \chi_{x_2} \end{Bmatrix} = \begin{Bmatrix} 0 \\ 0 \\ 0 \\ 0 \end{Bmatrix}$$

The numerical solution to this characteristic equation yields the eigenvalues and eigenvectors corresponding to the following modes:

Mode 1 ($\lambda_{1,2} = -0.554492 \ \text{sec}^{-1} \pm 3.090658 \ i \ \text{sec}^{-1}$)

$$\begin{Bmatrix} \Delta V_1 \\ \Delta V_2 \\ \Delta x_1 \\ \Delta x_2 \end{Bmatrix} = C_{1,2} \begin{Bmatrix} (-0.641297 \mp 0.152667 \ i) \ \text{ft/sec} \\ 0.688002 \ \text{ft/sec} \\ (-0.011790 \pm 0.209611 \ i) \ \text{ft} \\ (-0.038692 \mp 0.215665 \ i) \ \text{ft} \end{Bmatrix}$$

$$\times \exp[(-0.554492 \pm 3.090658 \ i) \ t/\text{sec}]$$

$$\sigma = -\text{real}(\lambda) = 0.554492 \text{ sec}^{-1}$$

$$99\% \text{ damping time} = \frac{\ln(0.01)}{-\sigma} = \frac{\ln(0.01)}{-0.554492 \text{ sec}^{-1}} = 8.305 \text{ sec}$$

$$\omega_d = |\text{imag}(\lambda)| = 3.090658 \text{ rad/sec}$$

$$\text{period} = \frac{2\pi}{\omega_d} = \frac{2\pi}{3.090658 \text{ sec}^{-1}} = 2.033 \text{ sec}$$

Following Example 8.1.1, the amplitudes for the four components of this pair of eigenvectors are

$$\text{amplitude} \begin{Bmatrix} \Delta V_1 \\ \Delta V_2 \\ \Delta x_1 \\ \Delta x_2 \end{Bmatrix} = C_{1,2} \begin{Bmatrix} 0.659219 \text{ ft/sec} \\ 0.688002 \text{ ft/sec} \\ 0.209942 \text{ ft} \\ 0.219109 \text{ ft} \end{Bmatrix}$$

and the phase shifts relative to ΔV_2 are

$$\text{phase} \begin{Bmatrix} \Delta V_1 \\ \Delta V_2 \\ \Delta x_1 \\ \Delta x_2 \end{Bmatrix} = \pm \begin{Bmatrix} -2.9045 \\ 0.0000 \\ 1.6251 \\ -1.7464 \end{Bmatrix} = \pm \begin{Bmatrix} -166.42° \\ 0.00° \\ 93.11° \\ -100.06° \end{Bmatrix}$$

Notice that this mode has changed very little from mode 1 in Example 8.1.1. This is because the mode primarily involves the vibration of only the second spring, which has not changed.

The undamped natural frequency for this mode is

$$\omega_n = \sqrt{\lambda_1 \lambda_2} = \sqrt{(-0.554492 + 3.090658 \, i)(-0.554492 - 3.090658 \, i)}$$

$$= \sqrt{0.554492^2 + 3.090658^2} = 3.140004 \text{ rad/sec}$$

and the damping ratio is

$$\zeta = -\frac{\lambda_1 + \lambda_2}{2\sqrt{\lambda_1 \lambda_2}} = -\frac{-0.554492 + 3.090658 \, i - 0.554492 - 3.090658 \, i}{2\sqrt{(-0.554492 + 3.090658 \, i)(-0.554492 - 3.090658 \, i)}}$$

$$= \frac{0.556387}{\sqrt{0.554492^2 + 3.090658^2}} = 0.176590$$

Mode 2 ($\lambda_3 = -1.141016 \text{ sec}^{-1}$)
This overdamped mode corresponds to the homogeneous solution

$$
\begin{Bmatrix} \Delta V_1 \\ \Delta V_2 \\ \Delta x_1 \\ \Delta x_2 \end{Bmatrix} = C_3 \begin{bmatrix} 0.553987 \text{ ft/sec} \\ 0.508604 \text{ ft/sec} \\ -0.485521 \text{ ft} \\ -0.445746 \text{ ft} \end{bmatrix} \exp(-1.141016\, t/\text{sec})
$$

$$
\sigma = -\text{real}(\lambda) = 1.141016 \text{ sec}^{-1}
$$

$$
99\% \text{ damping time} = \frac{\ln(0.01)}{-\sigma} = \frac{\ln(0.01)}{-1.141016 \text{ sec}^{-1}} = 4.036 \text{ sec}
$$

Mode 3 ($\lambda_4 = 0.0 \text{ sec}^{-1}$)
This trivial mode results in

$$
\begin{Bmatrix} \Delta V_1 \\ \Delta V_2 \\ \Delta x_1 \\ \Delta x_2 \end{Bmatrix} = C_4 \begin{Bmatrix} 0.0 \text{ ft/sec} \\ 0.0 \text{ ft/sec} \\ 0.707107 \text{ ft} \\ 0.707107 \text{ ft} \end{Bmatrix} \exp(0.0\, t/\text{sec})
$$

For the pair of roots associated with modes 2 and 3, the undamped natural frequency is

$$
\omega_n = \sqrt{\lambda_3 \lambda_4} = \sqrt{(-1.141016)(0.0)} = 0.0 \text{ rad/sec}
$$

and the damping ratio is

$$
\zeta = -\frac{\lambda_3 + \lambda_4}{2\sqrt{\lambda_3 \lambda_4}} = -\frac{-1.141016 + 0.0}{2\sqrt{(-1.141016)(0.0)}} = \infty
$$

In Example 8.1.2, the eigenvalue for mode 3 is identically zero. This results in a constant solution with no velocity components and equal but arbitrary values for both displacements. This rather trivial mode arises from the fact that the spring constant, k_1, in Eq. (8.1.39) is zero. This is equivalent to removing the left-hand spring that is shown in Fig. 8.1.4. With this spring missing, there is no force that depends on the absolute position of the center of mass. The forces exerted on the components of the system depend only on the distance between the two masses and their respective velocities. Thus, the center of mass can be placed at any arbitrary position along the x-axis without changing the characteristics of the system response. This is simply a *rigid-body displacement* of the complete spring-mass system.

Many other physical systems exhibit one or more modes with eigenvalues of zero. Modes that have zero eigenvalues are called *rigid-body displacement modes*, since they correspond to simple rigid-body displacements of the system. Such a mode occurs whenever there is no force that depends on the displacement associated with a particular degree of freedom. The linearized rigid-body equations of motion for an aircraft exhibit four rigid-body displacement modes. In the linearized equations, there are no forces that depend on the translational displacements. Within the small-displacement approximation, we neglect changes in air density and all forces are independent of longitude, latitude, and altitude. The aerodynamic forces depend only on the velocity components, and the gravitational forces depend only on orientation. Furthermore, while components of the gravitational force in body-fixed coordinates depend on the elevation angle and the bank angle, they do not depend on the azimuth angle. Thus, in the homogeneous solution to the linearized equations of motion for an aircraft, we should expect to find rigid-body displacement modes associated with four of the six degrees of freedom, x_f, y_f, z_f, and ψ.

There is yet another type of motion that can be associated with the free response of physical systems. This type of motion normally does not occur in simple spring-mass systems. Linearized viscous damping coefficients associated with conventional spring-mass systems are always positive or zero. As a result, the real part of each eigenvalue associated with such spring-mass systems will be either negative or zero. However, the analogous terms in systems of equations describing other physical phenomena, including aircraft dynamics, can result in eigenvalues with positive real parts. The motion associated with an eigenvalue having a positive real part is called a *divergent mode*. When a system having a divergent mode is disturbed from equilibrium, the disturbance will grow with time rather than decaying, as is shown in Fig. 8.1.5. An aircraft can exhibit some divergent modes. In fact, divergence in some modes can be acceptable, provided that the time required to double the amplitude of a disturbance is not too short.

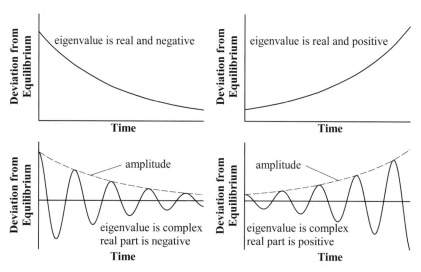

Figure 8.1.5. Types of motion associated with the free response.

In summary, the nature of the motion associated with the free response of a physical system, which is described by an eigenproblem, can be characterized by examining the eigenvalues and eigenvectors. The **eigenvalues** describe the form of the motion as shown in Fig. 8.1.5.

A **purely real eigenvalue** is associated with a **nonoscillatory mode**, having a damping rate equal to the negative of the eigenvalue.

If a **real eigenvalue is zero**, the solution corresponds to a **rigid-body mode**, describing a simple rigid-body displacement.

If a **real eigenvalue is negative**, the solution corresponds to a **convergent mode**, describing an exponential return to equilibrium following any disturbance.

If a **real eigenvalue is positive**, the solution corresponds to a **divergent mode**, describing an exponential deviation from equilibrium following any disturbance.

A **complex pair of eigenvalues** is associated with an **oscillatory mode**, having a damping rate equal to the negative of the real part of the eigenvalue and a damped natural frequency equal to the magnitude of the imaginary part of the eigenvalue.

If the **real part of the eigenvalue is zero**, the solution corresponds to an **undamped mode**, describing sinusoidal motion having constant amplitude.

If the **real part of the eigenvalue is negative**, the solution corresponds to a **damped mode**, describing sinusoidal motion having an amplitude that decreases exponentially with time.

If the **real part of the eigenvalue is positive**, the solution corresponds to a **divergent mode**, describing sinusoidal motion having an amplitude that increases exponentially with time.

The components of the eigenvector that is associated with any particular eigenvalue describe the relative participation of each dependent variable in the motion that is described by that eigenvalue. The **eigenvectors** are interpreted as follows:

A **purely real eigenvalue** always has a **purely real eigenvector**, whose components describe the relative amplitudes associated with the exponential motion.

For a real eigenvalue, the amplitude of the exponential variation in each dependent variable is proportional to the real component of the eigenvector that is associated with that dependent variable.

In general, a **complex eigenvalue** will result in an **eigenvector with complex components**, which describe the relative amplitudes and phase shifts associated with the sinusoidal motion.

> For a complex eigenvalue, the amplitude of the sinusoidal variation in each dependent variable is proportional to the magnitude of the complex eigenvector component that is associated with that dependent variable.

> For a complex eigenvalue, the phase of the sinusoidal variation in each dependent variable is equal to the arctangent of the imaginary part divided by the real part of the complex eigenvector component that is associated with that dependent variable. If the eigenvector is normalized to make the largest component real, then this phase is measured relative to the phase of the dependent variable having the largest amplitude.

In this section we have talked about three separate frequencies that can all be associated with any oscillatory mode, which represents a solution to an eigenproblem. These are the *undamped natural frequency*, ω_n, the *damped natural frequency*, ω_d, and the *maximum response frequency*, ω_m. In structural vibration problems these three frequencies are often used interchangeably, because structural damping is typically very light. With extremely light damping these three frequencies are, for all practical purposes, identical. However, **the modes encountered in aircraft dynamics are not all lightly damped. For this reason, we cannot use these three frequencies interchangeably** and the reader must be sure that he or she understands which of these three frequencies is being used in each case. Throughout this text an effort has been made to keep this distinction clear.

8.2. Longitudinal Motion: The Linearized Coupled Equations

The dynamic response of any real aircraft involves an extremely large number of degrees of freedom. An aircraft in flight has three translational degrees of freedom, three rotational degrees of freedom, and more elastic degrees of freedom than one would care to count. The equations of motion developed in Chapter 7 assume the aircraft to be a rigid body. This reduces the problem to one of six degrees of freedom. While the elastic degrees of freedom are important and cannot be completely neglected in aircraft design, much can be learned from the study of rigid-body aircraft dynamics. Thus, in this chapter we shall restrict our discussion entirely to the six degrees of freedom associated with rigid-body flight.

As we have seen, the general 6-DOF equations of motion are highly nonlinear. However, when the motion of the aircraft is confined to small deviations from steady equilibrium flight, the equations can be linearized. Furthermore, the linearized 6-DOF equations of motion can be separated into two subsystems, each with three degrees of freedom. These have been called the *linearized longitudinal equations* and the *linearized lateral equations*. For small deviations from equilibrium, there is no coupling between the longitudinal motion and the lateral motion. Thus, each of these three-degree-of-freedom subsystems can be solved independently. In this section we examine the stick-fixed free response associated with linearized, longitudinal rigid-body motion. These results provide a reasonable description of longitudinal aircraft dynamics provided that the aircraft is not undergoing large-amplitude oscillations or rapid maneuvers.

The longitudinal motion comprises two translational degrees of freedom (axial and normal) and one rotational degree of freedom (pitch). Since this motion has three degrees of freedom, it is characterized by six eigenvalues. The eigenvalues associated with the linearized longitudinal motion of a typical airplane contain two complex pairs, corresponding to two oscillatory modes. One of these modes is lightly damped with a period on the order of 30 seconds or more and is called the *long-period* or *phugoid mode*. The other oscillatory mode is heavily damped with a much shorter period and is simply called the *short-period mode*. The remaining two longitudinal eigenvalues are always zero, corresponding to simple rigid-body displacement modes.

The free response of the longitudinal system is obtained from the homogeneous equations, with the control deviations set to zero. As indicated in Eq. (7.7.9), for all practical purposes, four of the derivatives with respect to the translational accelerations are zero,

$$R_{x,\hat{\mu}} \cong 0$$

$$R_{z,\hat{\mu}} \cong 0$$

$$R_{m,\hat{\mu}} \cong 0$$

$$R_{x,\hat{\alpha}} \cong 0$$

From Eq. (7.7.5), the homogeneous linearized longitudinal equations can then be written in the form

$$
\begin{bmatrix}
1 & 0 & 0 & 0 & 0 & 0 \\
0 & (1-R_{z,\hat{\alpha}}) & 0 & 0 & 0 & 0 \\
0 & -R_{m,\hat{\alpha}} & 1 & 0 & 0 & 0 \\
0 & 0 & 0 & 1 & 0 & 0 \\
0 & 0 & 0 & 0 & 1 & 0 \\
0 & 0 & 0 & 0 & 0 & 1
\end{bmatrix}
\begin{Bmatrix}
\Delta\hat{\mu} \\
\Delta\hat{\alpha} \\
\Delta\hat{q} \\
\Delta\hat{\xi}_x \\
\Delta\hat{\xi}_z \\
\Delta\hat{\theta}
\end{Bmatrix}
$$

$$
-
\begin{bmatrix}
R_{x,\mu} & R_{x,\alpha} & R_{x,\overline{q}} & 0 & 0 & -R_{gx}\cos\theta_0 \\
R_{z,\mu} & R_{z,\alpha} & (1+R_{z,\overline{q}}) & 0 & 0 & -R_{gx}\sin\theta_0 \\
R_{m,\mu} & R_{m,\alpha} & R_{m,\overline{q}} & 0 & 0 & 0 \\
\cos\theta_o & \sin\theta_o & 0 & 0 & 0 & -\sin\theta_o \\
-\sin\theta_o & \cos\theta_o & 0 & 0 & 0 & -\cos\theta_o \\
0 & 0 & 1 & 0 & 0 & 0
\end{bmatrix}
\begin{Bmatrix}
\Delta\mu \\
\Delta\alpha \\
\Delta q \\
\Delta\xi_x \\
\Delta\xi_z \\
\Delta\theta
\end{Bmatrix}
=
\begin{Bmatrix}
0 \\
0 \\
0 \\
0 \\
0 \\
0
\end{Bmatrix}
$$

(8.2.1)

Equation (8.2.1) exactly fits the form of the eigenvalue problem given by Eq. (8.1.28). Thus, the **generalized eigenproblem for longitudinal motion is**

$$
\left(
\begin{bmatrix}
R_{x,\mu} & R_{x,\alpha} & R_{x,\overline{q}} & 0 & 0 & -R_{gx}\cos\theta_0 \\
R_{z,\mu} & R_{z,\alpha} & \left(1+R_{z,\overline{q}}\right) & 0 & 0 & -R_{gx}\sin\theta_0 \\
R_{m,\mu} & R_{m,\alpha} & R_{m,\overline{q}} & 0 & 0 & 0 \\
\cos\theta_o & \sin\theta_o & 0 & 0 & 0 & -\sin\theta_o \\
-\sin\theta_o & \cos\theta_o & 0 & 0 & 0 & -\cos\theta_o \\
0 & 0 & 1 & 0 & 0 & 0
\end{bmatrix}
\right.
$$

$$
\left.
-\lambda
\begin{bmatrix}
1 & 0 & 0 & 0 & 0 & 0 \\
0 & \left(1-R_{z,\hat{\alpha}}\right) & 0 & 0 & 0 & 0 \\
0 & -R_{m,\hat{\alpha}} & 1 & 0 & 0 & 0 \\
0 & 0 & 0 & 1 & 0 & 0 \\
0 & 0 & 0 & 0 & 1 & 0 \\
0 & 0 & 0 & 0 & 0 & 1
\end{bmatrix}
\right)
\begin{Bmatrix}
\chi_\mu \\ \chi_\alpha \\ \chi_{\overline{q}} \\ \chi_{\xi_x} \\ \chi_{\xi_z} \\ \chi_\theta
\end{Bmatrix}
=
\begin{Bmatrix}
0 \\ 0 \\ 0 \\ 0 \\ 0 \\ 0
\end{Bmatrix}
\tag{8.2.2}
$$

The eigenvalues and eigenvectors associated with longitudinal motion can be found by solving Eq. (8.2.2) using a generalized eigensolver. If a generalized eigensolver is not available, this system of equations is easily rearranged to give the **special eigenproblem for longitudinal motion:**

$$
\left(
\begin{bmatrix}
R_{x,\mu} & R_{x,\alpha} & R_{x,\overline{q}} & 0 & 0 & -R_{gx}\cos\theta_0 \\[2mm]
\dfrac{R_{z,\mu}}{1-R_{z,\hat{\alpha}}} & \dfrac{R_{z,\alpha}}{1-R_{z,\hat{\alpha}}} & \dfrac{1+R_{z,\overline{q}}}{1-R_{z,\hat{\alpha}}} & 0 & 0 & -\dfrac{R_{gx}\sin\theta_0}{1-R_{z,\hat{\alpha}}} \\[2mm]
R_{m,\mu}+\dfrac{R_{m,\hat{\alpha}}R_{z,\mu}}{1-R_{z,\hat{\alpha}}} & R_{m,\alpha}+\dfrac{R_{m,\hat{\alpha}}R_{z,\alpha}}{1-R_{z,\hat{\alpha}}} & R_{m,\overline{q}}+\dfrac{R_{m,\hat{\alpha}}\left(1+R_{z,\overline{q}}\right)}{1-R_{z,\hat{\alpha}}} & 0 & 0 & -\dfrac{R_{m,\hat{\alpha}}R_{gx}\sin\theta_0}{1-R_{z,\hat{\alpha}}} \\[2mm]
\cos\theta_o & \sin\theta_o & 0 & 0 & 0 & -\sin\theta_o \\[1mm]
-\sin\theta_o & \cos\theta_o & 0 & 0 & 0 & -\cos\theta_o \\[1mm]
0 & 0 & 1 & 0 & 0 & 0
\end{bmatrix}
\right.
$$

$$
\left.
-\lambda
\begin{bmatrix}
1 & 0 & 0 & 0 & 0 & 0 \\
0 & 1 & 0 & 0 & 0 & 0 \\
0 & 0 & 1 & 0 & 0 & 0 \\
0 & 0 & 0 & 1 & 0 & 0 \\
0 & 0 & 0 & 0 & 1 & 0 \\
0 & 0 & 0 & 0 & 0 & 1
\end{bmatrix}
\right)
\begin{Bmatrix}
\chi_\mu \\ \chi_\alpha \\ \chi_{\overline{q}} \\ \chi_{\xi_x} \\ \chi_{\xi_z} \\ \chi_\theta
\end{Bmatrix}
=
\begin{Bmatrix}
0 \\ 0 \\ 0 \\ 0 \\ 0 \\ 0
\end{Bmatrix}
\tag{8.2.3}
$$

A variety of different numerical techniques are commonly used for finding eigenvalues and eigenvectors. The most popular of these techniques uses the method of Householder (1964) to transform the system to upper Hessenberg. The transformed system is then solved using the QR method. See Wilkinson (1965) for a discussion of the QR method.

From Eq. (8.2.2) or Eq. (8.2.3) we obtain dimensionless eigenvalues. To obtain the dimensional eigenvalues in terms of time, the dimensionless eigenvalues must be divided by the reference time, $\bar{c}_w/2V_o$.

Thus, the **damping rate for a longitudinal mode** is

$$\sigma = -\text{real}(\lambda)\frac{2V_o}{\bar{c}_w} \tag{8.2.4}$$

and the **damped natural frequency for a longitudinal mode** is

$$\omega_d = |\text{imag}(\lambda)|\frac{2V_o}{\bar{c}_w} \tag{8.2.5}$$

The **damping ratio** is

$$\zeta = -\frac{\lambda_1 + \lambda_2}{2\sqrt{\lambda_1\lambda_2}} \tag{8.2.6}$$

and the **undamped natural frequency for a longitudinal mode** is

$$\omega_n = \sqrt{\lambda_1\lambda_2}\,\frac{2V_o}{\bar{c}_w} \tag{8.2.7}$$

EXAMPLE 8.2.1. For steady level flight at sea level, find the dimensionless eigenvalues and eigenvectors, the damping rate, the 99 percent damping time, and the damped natural frequency and period, if underdamped, for each longitudinal mode of the typical general aviation airplane with the characteristics listed below. For each pair of eigenvalues, find the undamped natural frequency and the damping ratio. For each oscillatory mode, compute the amplitude and phase for each component of the eigenvector associated with the positive frequency. Neglect the effects of compressibility and assume constant thrust aligned with the flight path and center of gravity.

$$S_w = 185 \text{ ft}^2, \quad b_w = 33 \text{ ft}, \quad W = 2{,}800 \text{ lbf}, \quad V_o = 180 \text{ ft/sec}, \quad C_{Do} = 0.05,$$

$$I_{xx_b} = 1{,}000 \text{ slug·ft}^2, \quad I_{yy_b} = 3{,}000 \text{ slug·ft}^2, \quad I_{zz_b} = 3{,}500 \text{ slug·ft}^2,$$

$$I_{xz_b} = 30 \text{ slug·ft}^2,$$

$$C_{L,\alpha} = 4.40, \quad C_{D,\alpha} = 0.35, \quad C_{m,\alpha} = -0.68, \quad C_{L,\hat{\alpha}} = 1.60, \quad C_{m,\hat{\alpha}} = -4.35,$$
$$C_{Y,\beta} = -0.560, \quad C_{\ell,\beta} = -0.075, \quad C_{n,\beta} = 0.070,$$
$$C_{D,\bar{q}} \cong 0.0, \quad C_{L,\bar{q}} = 3.80, \quad C_{m,\bar{q}} = -9.95,$$
$$C_{Y,\bar{p}} \cong 0.0, \quad C_{\ell,\bar{p}} = -0.410, \quad C_{n,\bar{p}} = -0.0575,$$
$$C_{Y,\bar{r}} = 0.240, \quad C_{\ell,\bar{r}} = 0.105, \quad C_{n,\bar{r}} = -0.125$$

Solution. For this aircraft we have

$$W/g = 2{,}800/32.17 = 87.04 \text{ slugs}$$

$$\bar{c}_w = S_w/b_w = 185/33 = 5.606 \text{ ft}$$

$$\frac{\rho S_w \bar{c}_w}{4W/g} = \frac{(0.0023769)(185)(5.606)}{4(87.04)} = 0.007080$$

$$\frac{\rho S_w \bar{c}_w^3}{8I_{yy_b}} = \frac{(0.0023769)(185)(5.606)^3}{8(3{,}000)} - 0.003228$$

$$R_{gx} \equiv \frac{g\bar{c}_w}{2V_o^2} = \frac{32.17(5.606)}{2(180)^2} = 0.00278$$

$$C_{Lo} = \frac{W\cos\theta_o}{\frac{1}{2}\rho V_o^2 S_w} = \frac{2{,}800(1.0)}{(0.5)(0.0023769)(180)^2(185)} = 0.393$$

$$R_{z,\hat{\alpha}} \equiv \frac{\rho S_w \bar{c}_w}{4W/g} C_{Z,\hat{\alpha}} = \frac{\rho S_w \bar{c}_w}{4W/g}\left(-C_{L,\hat{\alpha}}\right) = 0.007080(-1.60) = -0.01133$$

$$R_{m,\hat{\alpha}} \equiv \frac{\rho S_w \bar{c}_w^3}{8I_{yy_b}} C_{m,\hat{\alpha}} = 0.003228(-4.35) = -0.01404$$

$$R_{x,\mu} \equiv \frac{\rho S_w \bar{c}_w}{4W/g}\left(2C_{Xo}\right) = \frac{\rho S_w \bar{c}_w}{4W/g}\left(-2C_{Do}\right) = 0.007080(-2(0.05))$$
$$= -0.00071$$

$$R_{z,\mu} \equiv \frac{\rho S_w \bar{c}_w}{4W/g}\left(2C_{Zo}\right) = \frac{\rho S_w \bar{c}_w}{4W/g}\left(-2C_{Lo}\right) = 0.007080(-2(0.393))$$
$$= -0.00556$$

$$R_{m,\mu} \equiv \frac{\rho S_w \bar{c}_w^3}{4I_{yy_b}}\left(\frac{-2T_o + V_o T_{,V}}{\rho V_o^2 S_w}\right)\frac{[z_T \cos(\alpha_{To}) + x_T \sin(\alpha_{To})]}{\bar{c}_w} = 0.0$$

$$R_{x,\alpha} \equiv \frac{\rho S_w \bar{c}_w}{4W/g} C_{X,\alpha} = \frac{\rho S_w \bar{c}_w}{4W/g}\left(C_{Lo} - C_{D,\alpha}\right) = 0.007080(0.393 - 0.35)$$
$$= 0.00030$$

$$R_{z,\alpha} \equiv \frac{\rho S_w \bar{c}_w}{4W/g} C_{Z,\alpha} = \frac{\rho S_w \bar{c}_w}{4W/g}\left(-C_{L,\alpha} - C_{Do}\right) = 0.007080(-4.40 - 0.05)$$
$$= -0.03151$$

$$R_{m,\alpha} \equiv \frac{\rho S_w \bar{c}_w^3}{8I_{yy_b}} C_{m,\alpha} = 0.003228(-0.68) = -0.00220$$

$$R_{x,\bar{q}} \equiv \frac{\rho S_w \bar{c}_w}{4W/g} C_{X,\bar{q}} = \frac{\rho S_w \bar{c}_w}{4W/g}\left(-C_{D,\bar{q}}\right) = 0.007080(0.0) = 0.0$$

$$R_{z,\bar{q}} \equiv \frac{\rho S_w \bar{c}_w}{4W/g} C_{Z,\bar{q}} = \frac{\rho S_w \bar{c}_w}{4W/g}\left(-C_{L,\bar{q}}\right) = 0.007080(-3.80) = -0.02691$$

$$R_{m,\bar{q}} \equiv \frac{\rho S_w \bar{c}_w^3}{8I_{yy_b}} C_{m,\bar{q}} = 0.003228(-9.95) = -0.03212$$

For this level-flight equilibrium condition, the characteristic equation given by Eq. (8.2.2) then becomes

$$\left(\begin{bmatrix} -0.00071 & 0.00030 & 0 & 0 & 0 & -0.00278 \\ -0.00556 & -0.03151 & 0.97309 & 0 & 0 & 0 \\ 0 & -0.00220 & -0.03212 & 0 & 0 & 0 \\ 1 & 0 & 0 & 0 & 0 & 0 \\ 0 & 1 & 0 & 0 & 0 & -1 \\ 0 & 0 & 1 & 0 & 0 & 0 \end{bmatrix}\right.$$

$$\left. -\lambda \begin{bmatrix} 1 & 0 & 0 & 0 & 0 & 0 \\ 0 & 1.01133 & 0 & 0 & 0 & 0 \\ 0 & 0.01404 & 1 & 0 & 0 & 0 \\ 0 & 0 & 0 & 1 & 0 & 0 \\ 0 & 0 & 0 & 0 & 1 & 0 \\ 0 & 0 & 0 & 0 & 0 & 1 \end{bmatrix}\right) \begin{Bmatrix} \chi_\mu \\ \chi_\alpha \\ \chi_{\bar{q}} \\ \chi_{\xi_x} \\ \chi_{\xi_z} \\ \chi_\theta \end{Bmatrix} = \begin{Bmatrix} 0 \\ 0 \\ 0 \\ 0 \\ 0 \\ 0 \end{Bmatrix}$$

Using a generalized numerical eigensolver, the dimensionless eigenvalues and eigenvectors for this eigenproblem are easily obtained. The solution to the characteristic equation results in two complex pairs of eigenvalues and two eigenvalues that are identically zero. The two complex pairs describe the two oscillatory modes (the short-period mode and the long-period or phugoid mode) that are associated with the longitudinal free response of an aircraft in flight. The two zero eigenvalues correspond to two longitudinal rigid-body displacement modes (horizontal displacement and vertical displacement). From the numerical solution, these longitudinal modes are:

The short-period mode $(\lambda = -0.038484 \pm 0.040513\,i)$
The eigenvalue pair having the largest magnitude will always correspond to the short-period mode. In this case we have

$$
\begin{Bmatrix} \Delta\mu \\ \Delta\alpha \\ \Delta\overline{q} \\ \Delta\xi_x \\ \Delta\xi_z \\ \Delta\theta \end{Bmatrix} = C_{1,2}
\begin{Bmatrix} -0.000147 \pm 0.003381\,i \\ 0.010268 \pm 0.098873\,i \\ -0.004242 \mp 0.000301\,i \\ 0.045683 \mp 0.039762\,i \\ 0.990267 \\ 0.048378 \pm 0.058755\,i \end{Bmatrix} \exp[(-0.038484 \pm 0.040513\,i)\,\tau_x]
$$

$$
\sigma = -\operatorname{real}(\lambda)\frac{2V_o}{\overline{c}_w} = 0.038484\frac{2(180)}{5.606} = 2.471324 \text{ sec}^{-1}
$$

$$
99\% \text{ damping time} = \frac{\ln(0.01)}{-\sigma} = \frac{\ln(0.01)}{-2.471324 \text{ sec}^{-1}} = 1.86 \text{ sec}
$$

$$
\zeta = -\frac{\lambda_1 + \lambda_2}{2\sqrt{\lambda_1\lambda_2}} = -\frac{-0.038484 + 0.040513\,i - 0.038484 - 0.040513\,i}{2\sqrt{(-0.038484 + 0.040513\,i)(-0.038484 - 0.040513\,i)}}
$$

$$
= \frac{0.038484}{\sqrt{0.038484^2 + 0.040513^2}} = 0.688718
$$

$$
\omega_d = |\operatorname{imag}(\lambda)|\frac{2V_o}{\overline{c}_w} = 0.040513\frac{2(180)}{5.606} = 2.601620 \text{ rad/sec}
$$

$$
\text{period} = \frac{2\pi}{\omega_d} = \frac{2\pi}{2.601620 \text{ sec}^{-1}} = 2.42 \text{ sec}
$$

$$
\omega_n = \sqrt{\lambda_1\lambda_2}\frac{2V_o}{\overline{c}_w}
$$

$$
= \sqrt{(-0.038484 + 0.040513\,i)(-0.038484 - 0.040513\,i)}\frac{2(180)}{5.606}
$$

$$
= \sqrt{0.038484^2 + 0.040513^2}\frac{2(180)}{5.606} = 3.588296 \text{ rad/sec}
$$

The long-period (phugoid) mode $(\lambda = -0.000264 \pm 0.003272\,i)$
The second largest eigenvalue pair will always be associated with the phugoid.
For the current example,

$$
\begin{Bmatrix} \Delta\mu \\ \Delta\alpha \\ \Delta\overline{q} \\ \Delta\xi_x \\ \Delta\xi_z \\ \Delta\theta \end{Bmatrix} = C_{3,4} \begin{Bmatrix} 0.002090 \pm 0.000366\,i \\ -0.000118 \mp 0.000023\,i \\ 0.000008 \pm 0.000001\,i \\ 0.059947 \mp 0.643541\,i \\ 0.763053 \\ 0.000083 \mp 0.002520\,i \end{Bmatrix} \exp[(-0.000264 \pm 0.003272\,i)\,\tau_x]
$$

$$
\sigma = -\text{real}(\lambda)\frac{2V_o}{\overline{c}_w} = 0.000264\frac{2(180)}{5.606} = 0.016953\ \text{sec}^{-1}
$$

$$
99\%\ \text{damping time} = \frac{\ln(0.01)}{-\sigma} = \frac{\ln(0.01)}{-0.016953\ \text{sec}^{-1}} = 271.6\ \text{sec}
$$

$$
\zeta = -\frac{\lambda_1 + \lambda_2}{2\sqrt{\lambda_1\lambda_2}} = -\frac{-0.000264 + 0.003272\,i - 0.000264 - 0.003272\,i}{2\sqrt{(-0.000264 + 0.003272\,i)(-0.000264 - 0.003272\,i)}}
$$

$$
= \frac{0.000264}{\sqrt{0.000264^2 + 0.003272^2}} = 0.080423
$$

$$
\omega_d = |\text{imag}(\lambda)|\frac{2V_o}{\overline{c}_w} = 0.003272\frac{2(180)}{5.606} = 0.210118\ \text{rad/sec}
$$

$$
\text{period} = \frac{2\pi}{\omega_d} = \frac{2\pi}{0.210118\ \text{sec}^{-1}} = 29.9\ \text{sec}
$$

$$
\omega_n = \sqrt{\lambda_1\lambda_2}\frac{2V_o}{\overline{c}_w}
$$

$$
= \sqrt{(-0.000264 + 0.003272\,i)(-0.000264 - 0.003272\,i)}\frac{2(180)}{5.606}
$$

$$
= \sqrt{0.000264^2 + 0.003272^2}\frac{2(180)}{5.606} = 0.210801\ \text{rad/sec}
$$

The amplitudes and phase shifts for both the short-period and long-period eigenvectors are listed in Table 8.2.1. The amplitudes in this table are normalized. In this case, for both short-period and long-period motion, the largest amplitude is associated with the oscillation in vertical displacement. Thus, all of the phase shifts listed in Table 8.2.1 are reported relative to the phase for the vertical displacement.

Eigenvector Component	Short-Period		Long-Period	
	Amplitude	**Phase**	**Amplitude**	**Phase**
$\Delta\mu$	0.003384	92.39°	0.002122	9.93°
$\Delta\alpha$	0.099405	83.97°	0.000121	−168.60°
$\Delta\overline{q}$	0.004253	−175.74°	0.000008	6.52°
$\Delta\xi_x$	0.060564	−40.99°	0.646327	−84.58°
$\Delta\xi_z$	0.990267	0.00°	0.763054	0.00°
$\Delta\theta$	0.076108	50.47°	0.002521	−88.00°

Table 8.2.1. Longitudinal eigenvectors for a typical general aviation airplane.

The rigid-body modes $(\lambda = 0.0, 0.0)$

$$
\begin{Bmatrix} \Delta\mu \\ \Delta\alpha \\ \Delta\overline{q} \\ \Delta\xi_x \\ \Delta\xi_z \\ \Delta\theta \end{Bmatrix} = C_5 \begin{Bmatrix} 0 \\ 0 \\ 0 \\ 1 \\ 0 \\ 0 \end{Bmatrix} \exp(0.0\,\tau_x), \; C_6 \begin{Bmatrix} 0 \\ 0 \\ 0 \\ 0 \\ 1 \\ 0 \end{Bmatrix} \exp(0.0\,\tau_x)
$$

These trivial rigid-body modes simply state the obvious fact that steady level flight can be maintained at different latitudes and/or different altitudes.

The two oscillatory longitudinal modes observed in Example 8.2.1 are illustrated in Figs. 8.2.1 through 8.2.3. These oscillatory modes are typical of the small-amplitude longitudinal modes associated with any airplane in flight. While the period and damping will vary with aircraft design and operating conditions, all airplanes exhibit these two modes of longitudinal motion.

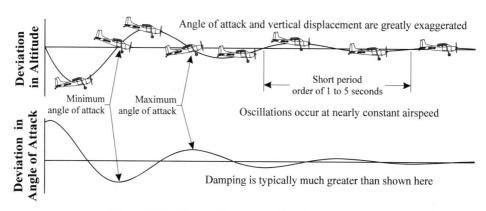

Figure 8.2.1. Short-period mode with reduced damping.

From examination of Example 8.2.1, several observations can be made. From this example we see that short-period motion is a heavily damped oscillatory mode that is composed of pitching motion with superimposed vertical oscillations. From the short-period eigenvector, we see that the amplitude for the dimensionless axial velocity deviation is small compared to that for the deviation in angle of attack. Thus, this mode is characterized by rapid changes in angle of attack and vertical position, with little change in axial velocity. Note that the deviation in angle of attack leads the deviation in vertical displacement by about 90 degrees in phase. This can be seen in Fig. 8.2.1.

The motion depicted in Fig. 8.2.1 is shown starting from the left with the airplane moving downward from the equilibrium altitude to the point of maximum positive displacement in z_f (i.e., minimum altitude). At this point, the vertical velocity component is zero, but the angle of attack is somewhat greater than the equilibrium value. Thus, the lift is greater than the weight and the airplane is accelerated upward, back toward the equilibrium altitude. However, at the minimum altitude, the angle of attack is rapidly decreasing with time and the angular momentum of the aircraft causes the angle of attack to overshoot the equilibrium value. Once the angle of attack becomes less than the equilibrium value, the lift becomes less than the weight and the vertical velocity of the airplane begins to slow. As the airplane reaches its equilibrium altitude, the angle of attack approaches its minimum but the vertical velocity is still quite large. The vertical momentum of the airplane carries it above the equilibrium altitude as the angle of attack begins to increase. As the airplane reaches maximum altitude, the angle of attack is still less than the equilibrium value. So, the weight is still greater than the lift and the airplane accelerates back toward the equilibrium altitude once again. This oscillatory interchange between rotational kinetic energy and potential energy continues until it is damped out, primarily as a result of pitch damping.

In a well-designed airplane, the short-period damping is much greater than shown in Fig. 8.2.1. In this figure the damping was deliberately reduced to better show the physics of the oscillatory motion. Short-period motion with more typical damping is shown in Fig. 8.2.2. Because short-period motion is heavily damped in most airplanes, it is not normally objectionable and most pilots are not even aware of its existence.

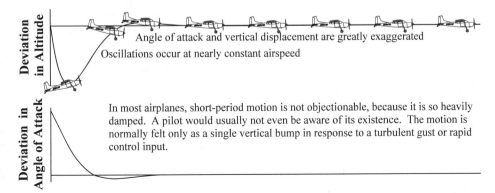

Figure 8.2.2. Short-period mode with typical damping.

While the short-period mode is a heavily damped oscillatory interchange between rotational kinetic energy and potential energy, the long-period or phugoid mode is a very lightly damped oscillatory interchange between translational kinetic energy and potential energy. From the long-period eigenvector, we see that the amplitude for the deviation in angle of attack is small compared to that for the dimensionless axial velocity deviation. This mode is then characterized by gradual changes in airspeed, altitude, and elevation angle, with very small changes in angle of attack. Notice that the axial velocity deviation is approximately 90 degrees out of phase with the deviation in elevation angle, with the velocity leading the elevation angle. Also notice that the velocity deviation leads the deviation in vertical displacement by only a few degrees. This can be seen in Fig. 8.2.3.

The phugoid mode is, at best, very lightly damped, and some modern airplanes can have negative phugoid damping at certain airspeeds. Even so, this motion is usually not objectionable to VFR[*] pilots, because the period of the phugoid is so long. When a pilot is controlling an airplane using visual references, his or her response time is very short compared to the phugoid period and the pilot will normally correct for this motion without ever being cognizant of its existence. On the other hand, when flying an airplane solely from the references provided by instruments, the pilot's response to the gradual changes associated with phugoid motion is not nearly so rapid. Thus, IFR[†] pilots often find low phugoid damping to be very objectionable.

Once the aerodynamic derivatives for an aircraft are determined from wind tunnel tests or other means, the frequency and damping for both the short-period and phugoid modes can be evaluated by numerically determining the eigenvalues of Eq. (8.2.2). However, these eigenvalues depend on many aircraft design and operating parameters, and the nature of this dependence is not easily observable from a numerical solution. Thus, a closed-form solution that describes the essential features of these modes is desirable. In addition, closed-form solutions have always been useful for application to aircraft control systems. For this reason, we shall now proceed to the development of some closed-form approximations for the two modes of longitudinal motion.

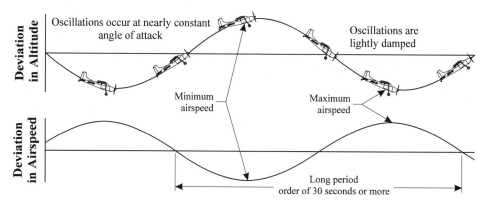

Figure 8.2.3. Phugoid or long-period mode.

---

[*]Visual flight rules.   [†]Instrument flight rules.

8.3. Short-Period Approximation

For the short-period mode, the amplitude of the dimensionless axial velocity deviation is small compared to that for the deviation in angle of attack. Thus, an approximation for short-period motion can be obtained by neglecting changes in axial velocity and ignoring the x-momentum equation. For motion relative to steady level flight, assuming that

$$\Delta\mu = 0 \tag{8.3.1}$$

and replacing the x-momentum equation from Eq. (8.2.1) with Eq. (8.3.1) gives

$$\begin{bmatrix} 0 & 0 & 0 & 0 & 0 & 0 \\ 0 & (1-R_{z,\hat{\alpha}}) & 0 & 0 & 0 & 0 \\ 0 & -R_{m,\hat{\alpha}} & 1 & 0 & 0 & 0 \\ 0 & 0 & 0 & 1 & 0 & 0 \\ 0 & 0 & 0 & 0 & 1 & 0 \\ 0 & 0 & 0 & 0 & 0 & 1 \end{bmatrix} \begin{Bmatrix} \Delta\hat{\mu} \\ \Delta\hat{\alpha} \\ \Delta\hat{q} \\ \Delta\hat{\xi}_x \\ \Delta\hat{\xi}_z \\ \Delta\hat{\theta} \end{Bmatrix} = \begin{bmatrix} 1 & 0 & 0 & 0 & 0 & 0 \\ 0 & R_{z,\alpha} & (R_{z,\bar{q}}+1) & 0 & 0 & 0 \\ 0 & R_{m,\alpha} & R_{m,\bar{q}} & 0 & 0 & 0 \\ 0 & 0 & 0 & 0 & 0 & 0 \\ 0 & 1 & 0 & 0 & 0 & -1 \\ 0 & 0 & 1 & 0 & 0 & 0 \end{bmatrix} \begin{Bmatrix} \Delta\mu \\ \Delta\alpha \\ \Delta q \\ \Delta\xi_x \\ \Delta\xi_z \\ \Delta\theta \end{Bmatrix} \tag{8.3.2}$$

The z-force ratios, $R_{z,\hat{\alpha}}$ and $R_{z,\bar{q}}$, are much less than 1 and usually neglected, to give

$$\begin{bmatrix} 0 & 0 & 0 & 0 & 0 & 0 \\ 0 & 1 & 0 & 0 & 0 & 0 \\ 0 & -R_{m,\hat{\alpha}} & 1 & 0 & 0 & 0 \\ 0 & 0 & 0 & 1 & 0 & 0 \\ 0 & 0 & 0 & 0 & 1 & 0 \\ 0 & 0 & 0 & 0 & 0 & 1 \end{bmatrix} \begin{Bmatrix} \Delta\hat{\mu} \\ \Delta\hat{\alpha} \\ \Delta\hat{q} \\ \Delta\hat{\xi}_x \\ \Delta\hat{\xi}_z \\ \Delta\hat{\theta} \end{Bmatrix} = \begin{bmatrix} 1 & 0 & 0 & 0 & 0 & 0 \\ 0 & R_{z,\alpha} & 1 & 0 & 0 & 0 \\ 0 & R_{m,\alpha} & R_{m,\bar{q}} & 0 & 0 & 0 \\ 0 & 0 & 0 & 0 & 0 & 0 \\ 0 & 1 & 0 & 0 & 0 & -1 \\ 0 & 0 & 1 & 0 & 0 & 0 \end{bmatrix} \begin{Bmatrix} \Delta\mu \\ \Delta\alpha \\ \Delta q \\ \Delta\xi_x \\ \Delta\xi_z \\ \Delta\theta \end{Bmatrix} \tag{8.3.3}$$

The second and third equations in Eq. (8.3.3) are not coupled to any of the other four equations. Thus, Eq. (8.3.3) can be rearranged and separated to yield

$$\begin{bmatrix} 1 & 0 \\ -R_{m,\hat{\alpha}} & 1 \end{bmatrix} \begin{Bmatrix} \lambda\Delta\alpha \\ \lambda\Delta\bar{q} \end{Bmatrix} = \begin{bmatrix} R_{z,\alpha} & 1 \\ R_{m,\alpha} & R_{m,\bar{q}} \end{bmatrix} \begin{Bmatrix} \Delta\alpha \\ \Delta\bar{q} \end{Bmatrix} \tag{8.3.4}$$

and

$$\begin{Bmatrix} \Delta\mu \\ \Delta\alpha \\ \Delta\bar{q} \\ \Delta\xi_x \\ \Delta\theta \end{Bmatrix} = \frac{\lambda}{R_{z,\alpha}} \begin{Bmatrix} 0 \\ \lambda \\ \lambda(\lambda - R_{z,\alpha}) \\ 0 \\ \lambda - R_{z,\alpha} \end{Bmatrix} \Delta\xi_z \tag{8.3.5}$$

The eigenvalues are obtained from Eq. (8.3.4) and the eigenvectors from Eq. (8.3.5).

The eigenproblem associated with Eq. (8.3.4) is

$$
\left(
\begin{bmatrix} R_{z,\alpha} & 1 \\ R_{m,\alpha} & R_{m,\bar{q}} \end{bmatrix}
- \lambda
\begin{bmatrix} 1 & 0 \\ -R_{m,\hat{\alpha}} & 1 \end{bmatrix}
\right)
\begin{Bmatrix} \chi_\alpha \\ \chi_{\bar{q}} \end{Bmatrix}
=
\begin{Bmatrix} 0 \\ 0 \end{Bmatrix}
\tag{8.3.6}
$$

The eigenvalues are obtained from the roots of the characteristic equation, which is found by setting the determinate of the coefficient matrix on the left-hand side of Eq. (8.3.6) equal to zero,

$$
\begin{vmatrix}
\left(R_{z,\alpha} - \lambda \right) & 1 \\
\left(R_{m,\alpha} + R_{m,\hat{\alpha}}\,\lambda \right) & \left(R_{m,\bar{q}} - \lambda \right)
\end{vmatrix}
= 0
\tag{8.3.7}
$$

Equation (8.3.7) can be expanded to yield

$$
\lambda^2 - \left(R_{z,\alpha} + R_{m,\bar{q}} + R_{m,\hat{\alpha}} \right)\lambda + \left(R_{z,\alpha} R_{m,\bar{q}} - R_{m,\alpha} \right) = 0
\tag{8.3.8}
$$

From Eq. (8.3.8), the **dimensionless eigenvalues for the short-period approximation** are easily found to be

$$
\lambda_{sp} = \left(\frac{R_{z,\alpha} + R_{m,\bar{q}} + R_{m,\hat{\alpha}}}{2} \right) \pm \sqrt{ \left(\frac{R_{z,\alpha} + R_{m,\bar{q}} + R_{m,\hat{\alpha}}}{2} \right)^2 - \left(R_{z,\alpha} R_{m,\bar{q}} - R_{m,\alpha} \right) }
\tag{8.3.9}
$$

which gives the approximate **short-period damped natural frequency**,

$$
\omega_{d_{sp}} = \frac{2V_o}{\bar{c}_w} \left| \mathrm{imag}(\lambda_{sp}) \right| = \frac{2V_o}{\bar{c}_w} \sqrt{ \left(R_{z,\alpha} R_{m,\bar{q}} - R_{m,\alpha} \right) - \left(\frac{R_{z,\alpha} + R_{m,\bar{q}} + R_{m,\hat{\alpha}}}{2} \right)^2 }
\tag{8.3.10}
$$

and the approximate **short-period damping rate**,

$$
\sigma_{sp} = -\frac{2V_o}{\bar{c}_w}\,\mathrm{real}(\lambda_{sp}) = -\frac{V_o}{\bar{c}_w}\left(R_{z,\alpha} + R_{m,\bar{q}} + R_{m,\hat{\alpha}} \right)
\tag{8.3.11}
$$

The short-period approximation, as given by Eq. (8.3.9), agrees very closely with the exact solution for the short-period mode obtained from Eq. (8.2.2). This is demonstrated in the following example.

EXAMPLE 8.3.1. Using the short-period approximation given by Eq. (8.3.9), find the dimensionless eigenvalues and eigenvectors, the damping rate, and the damped natural frequency for the short-period mode of the typical general aviation airplane described in Example 8.2.1. Compare these approximate results with the fully coupled short-period solution that was obtained in Example 8.2.1.

Solution. From Example 8.2.1, for this general aviation airplane and operating conditions, we have

$$V_o = 180 \text{ ft/sec}, \qquad \bar{c}_w = 5.606 \text{ ft},$$

$$R_{z,\alpha} = -0.03151, \quad R_{m,\alpha} = -0.00220, \quad R_{m,\hat{\alpha}} = -0.01404, \quad R_{m,\bar{q}} = -0.03212$$

Using these values in Eq. (8.3.9), the short-period approximation gives the eigenvalues

$$\lambda_{sp} = \left(\frac{-0.03151 - 0.03212 - 0.01404}{2} \right)$$

$$\pm \sqrt{\left(\frac{-0.03151 - 0.03212 - 0.01404}{2} \right)^2 - \left[(-0.03151)(-0.03212) - (-0.00220) \right]}$$

$$= -0.038835 \pm 0.041279 \, i$$

The damping rate is

$$\sigma = -\text{real}(\lambda)\frac{2V_o}{\bar{c}} = 0.038835\frac{2(180)}{5.606} = 2.493864 \text{ sec}^{-1}$$

and the damped natural frequency is

$$\omega_d = |\text{imag}(\lambda)|\frac{2V_o}{\bar{c}} = 0.041279\frac{2(180)}{5.606} = 2.650801 \text{ rad/sec}$$

The eigenvectors for the short-period approximation are found by using the complex eigenvalues in Eq. (8.3.5),

$$\begin{Bmatrix} \Delta\mu \\ \Delta\alpha \\ \Delta q \\ \Delta\xi_x \\ \Delta\theta \end{Bmatrix} = \frac{\lambda}{R_{z,\alpha}} \begin{Bmatrix} 0 \\ \lambda \\ \lambda(\lambda - R_{z,\alpha}) \\ 0 \\ \lambda - R_{z,\alpha} \end{Bmatrix} \Delta\xi_z$$

The only requirement for the six components of this eigenvector is that they must satisfy these five equations. Thus, we can only determine the eigenvector to within an arbitrary constant. We can find one possible form for the eigenvector by setting $\Delta\xi_z$ equal to some arbitrary constant and then using the relations above to find the other five components. For example, in this particular case we can simplify the complex arithmetic by using

$$\Delta\xi_z = R_{z,\alpha}/\lambda$$

This gives the eigenvector

$$
\begin{Bmatrix} \chi_\mu \\ \chi_\alpha \\ \chi_{\bar{q}} \\ \chi_{\xi_x} \\ \chi_{\xi_z} \\ \chi_\theta \end{Bmatrix} = \begin{Bmatrix} 0 \\ \lambda \\ \lambda(\lambda - R_{z,\alpha}) \\ 0 \\ R_{z,\alpha}/\lambda \\ \lambda - R_{z,\alpha} \end{Bmatrix}
$$

or

$$
\begin{Bmatrix} \chi_\mu \\ \chi_\alpha \\ \chi_{\bar{q}} \\ \chi_{\xi_x} \\ \chi_{\xi_z} \\ \chi_\theta \end{Bmatrix} = \begin{Bmatrix} 0 \\ -0.038835 \pm 0.041279\,i \\ (-0.038835 \pm 0.041279\,i)(-0.038835 \pm 0.041279\,i + 0.03151) \\ 0 \\ -0.03151/(-0.038835 \pm 0.041279\,i) \\ -0.038835 \pm 0.041279\,i + 0.03151 \end{Bmatrix}
$$

After performing the complex arithmetic, this yields

$$
\begin{Bmatrix} \chi_\mu \\ \chi_\alpha \\ \chi_{\bar{q}} \\ \chi_{\xi_x} \\ \chi_{\xi_z} \\ \chi_\theta \end{Bmatrix} = \begin{Bmatrix} 0 \\ -0.038835 \pm 0.041279\,i \\ -0.001419 \mp 0.001905\,i \\ 0 \\ 0.380963 \pm 0.404936\,i \\ -0.007325 \pm 0.041279\,i \end{Bmatrix}
$$

To make the largest component real we multiply the entire eigenvector by the complex conjugate of the largest component,

$$
\begin{Bmatrix} \chi_\mu \\ \chi_\alpha \\ \chi_{\bar{q}} \\ \chi_{\xi_x} \\ \chi_{\xi_z} \\ \chi_\theta \end{Bmatrix} = (0.380963 \mp 0.404936\,i) \begin{Bmatrix} 0 \\ -0.038835 \pm 0.041279\,i \\ -0.001419 \mp 0.001905\,i \\ 0 \\ 0.380963 \pm 0.404936\,i \\ -0.007325 \pm 0.041279\,i \end{Bmatrix}
$$

This gives

$$
\begin{Bmatrix}
\chi_\mu \\
\chi_\alpha \\
\chi_{\overline{q}} \\
\chi_{\xi_x} \\
\chi_{\xi_z} \\
\chi_\theta
\end{Bmatrix}
=
\begin{Bmatrix}
0 \\
0.001921 \pm 0.031451\,i \\
-0.001312 \mp 0.000151\,i \\
0 \\
0.309106 \\
0.013925 \pm 0.018692\,i
\end{Bmatrix}
$$

This eigenvector has a magnitude

$$
\sqrt{\sum_{j=1}^{N}\left|\chi_j\right|_i^2} = 0.311584
$$

Normalizing the eigenvector, we have

$$
\begin{Bmatrix}
\chi_\mu \\
\chi_\alpha \\
\chi_{\overline{q}} \\
\chi_{\xi_x} \\
\chi_{\xi_z} \\
\chi_\theta
\end{Bmatrix}
=
\frac{1}{0.311584}
\begin{Bmatrix}
0 \\
0.001921 \pm 0.031451\,i \\
-0.001312 \mp 0.000151\,i \\
0 \\
0.309106 \\
0.013925 \pm 0.018692\,i
\end{Bmatrix}
=
\begin{Bmatrix}
0 \\
0.006164 \pm 0.100940\,i \\
-0.004212 \mp 0.000485\,i \\
0 \\
0.992048 \\
0.044690 \pm 0.059990\,i
\end{Bmatrix}
$$

This gives

$$
\begin{Bmatrix}
\Delta\mu \\
\Delta\alpha \\
\Delta\overline{q} \\
\Delta\xi_x \\
\Delta\xi_z \\
\Delta\theta
\end{Bmatrix}
= C_{1,2}
\begin{Bmatrix}
0.000000 \\
0.006164 \pm 0.100940\,i \\
-0.004212 \mp 0.000485\,i \\
0.000000 \\
0.992048 \\
0.044690 \pm 0.059990\,i
\end{Bmatrix}
\exp[(-0.038835 \pm 0.041279\,i)\,\tau_x]
$$

The exact solution from Example 8.2.1 is

$$
\begin{Bmatrix}
\Delta\mu \\
\Delta\alpha \\
\Delta\overline{q} \\
\Delta\xi_x \\
\Delta\xi_z \\
\Delta\theta
\end{Bmatrix}
= C_{1,2}
\begin{Bmatrix}
-0.000147 \pm 0.003381\,i \\
0.010268 \pm 0.098873\,i \\
-0.004242 \mp 0.000301\,i \\
0.045683 \mp 0.039762\,i \\
0.990267 \\
0.048378 \pm 0.058755\,i
\end{Bmatrix}
\exp[(-0.038484 \pm 0.040513\,i)\,\tau_x]
$$

For the airplane and operating conditions used in Example 8.3.1, the eigenvalues computed from the short-period approximation agree with the exact solution to within about 2 percent. This result is quite typical of what can be expected from this short-period approximation.

Because the short-period mode could be overdamped, it is convenient to express the short-period approximation given by Eq. (8.3.9) in terms of the undamped natural frequency and damping ratio. After using Eqs. (8.3.10) and (8.3.11) in Eq. (8.1.13) and applying the definitions from Eq. (7.7.8), it is easily shown that the undamped natural frequency for this short-period approximation can be rearranged to yield

$$\omega_{n_{sp}} = \sqrt{\frac{\rho V_o^2 S_w C_{Z,\alpha}}{2W} \frac{\rho V_o^2 S_w \bar{c}_w C_{m,\bar{q}}}{2I_{yy_b}} \frac{g\bar{c}_w}{2V_o^2} - \frac{\rho V_o^2 S_w \bar{c}_w C_{m,\alpha}}{2I_{yy_b}}} = \sqrt{\frac{Z_{,\alpha}}{W} \frac{m_{,\bar{q}}}{I_{yy_b}} \frac{g\bar{c}_w}{2V_o^2} - \frac{m_{,\alpha}}{I_{yy_b}}}$$

which in view of Eq. (7.7.2) reduces to

$$\omega_{n_{sp}} = \sqrt{\frac{Z_{,\alpha}}{W} \frac{m_{,q}}{I_{yy_b}} \frac{g}{V_o} - \frac{m_{,\alpha}}{I_{yy_b}}} \tag{8.3.12}$$

In a similar manner, using Eqs. (7.7.2) and (7.7.8) in Eq. (8.3.11), the damping rate for this short-period approximation can be written as

$$\sigma_{sp} = -\frac{1}{2}\left(\frac{Z_{,\alpha}}{W} \frac{g}{V_o} + \frac{m_{,q}}{I_{yy_b}} + \frac{m_{,\dot{\alpha}}}{I_{yy_b}}\right) \tag{8.3.13}$$

Using Eq. (8.3.13) in Eq. (8.1.14), the damping ratio for this short-period approximation can be expressed as

$$\zeta_{sp} = -\frac{1}{2}\left(\frac{Z_{,\alpha}}{W} \frac{g}{V_o} + \frac{m_{,q}}{I_{yy_b}} + \frac{m_{,\dot{\alpha}}}{I_{yy_b}}\right)\Big/\omega_{n_{sp}} \tag{8.3.14}$$

The normal force can be expressed in terms of lift and drag, but drag contributes little to the normal force. Thus, using the small-angle approximation and neglecting the contribution from drag yields $Z = -L\cos\alpha - D\sin\alpha \cong -L$. Using this approximation in Eqs. (8.3.12) and (8.3.14) produces **an accurate approximation for the short-period undamped natural frequency and damping ratio**

$$\omega_{n_{sp}} = \sqrt{-\frac{L_{,\alpha}}{W} \frac{m_{,q}}{I_{yy_b}} \frac{g}{V_o} - \frac{m_{,\alpha}}{I_{yy_b}}} \tag{8.3.15}$$

$$\zeta_{sp} = \frac{1}{2}\left(\frac{L_{,\alpha}}{W} \frac{g}{V_o} - \frac{m_{,q}}{I_{yy_b}} - \frac{m_{,\dot{\alpha}}}{I_{yy_b}}\right)\Big/\omega_{n_{sp}} \tag{8.3.16}$$

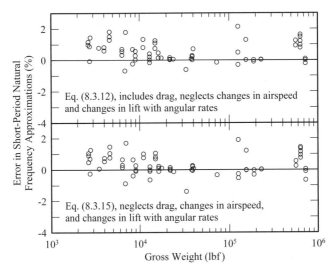

Figure 8.3.1. Short-period approximation errors. (From Phillips and Niewoehner 2009)

The accuracy of the short-period approximation can be quantitatively assessed by comparing results obtained from the approximation with more exact numerical solutions obtained from the full linearized longitudinal equations of motion. Errors associated with such comparisons are presented in Fig. 8.3.1 for 70 different cases encompassing widely different airplanes and operating conditions. For these 70 cases, the maximum positive error for Eq. (8.3.15) is +1.90 percent, the maximum negative error is −1.38 percent, the mean error is +0.31 percent, and the root-mean-square error is 0.69 percent. For these same cases, the maximum positive error for Eq. (8.3.12) is +2.13 percent, the maximum negative error is −0.69 percent, the mean error is +0.52 percent, and the root-mean-square error is 0.81 percent. The results presented in Fig. 8.3.1 clearly show that including the contribution of drag does not improve the accuracy of the short-period approximation.

An alternative useful form of the short-period approximation given by Eq. (8.3.15) is obtained by applying the definitions of the dimensionless dynamic pitch rate given by Eq. (6.1.42), $\tilde{q} \equiv V_o q / g$, and the pitch radius of gyration from Eq. (6.1.51), $r_{yy_b}^2 \equiv g I_{yy_b} / W$,

$$\omega_{n_{sp}} = \sqrt{\frac{g L_{,\alpha}}{r_{yy_b}^2 W} \left(-\frac{m_{,\alpha}}{L_{,\alpha}} - \frac{m_{,\tilde{q}}}{W} \right)} \qquad (8.3.17)$$

Note from Eq. (6.1.49) that the collection of terms in parentheses on the right-hand side of Eq. (8.3.17) is equal to the distance aft from the center of gravity to the stick-fixed maneuver point. Thus, **the short-period undamped natural frequency is proportional to the square root of the dynamic margin**, which is defined in Eq. (6.1.50):

$$\omega_{n_{sp}} = \sqrt{g l_{mp} L_{,\alpha} / (r_{yy_b}^2 W)} \qquad (8.3.18)$$

We see from Eq. (8.3.18) that the short-period natural frequency varies with the position of the center of gravity. Moving the CG forward increases the short-period natural frequency, and as the CG is moved aft, the short-period frequency decreases. As the CG approaches the stick-fixed maneuver point the natural frequency goes to zero.

8.4. Long-Period Approximation

The first closed-form solution for phugoid motion was developed by Lanchester (1908). In his original solution, Lanchester assumed no change in angle of attack and no change in the net axial force. With these assumptions, Lanchester obtained an approximation for the frequency of phugoid motion. However, this approximation predicts completely undamped sinusoidal motion and gives no information about the phugoid damping. A variation of Lanchester's solution, which does include an approximation for phugoid damping, has been widely used. In this approximation, Lanchester's original assumption of no change in axial force is dropped, but the assumption of no change in angle of attack is retained,

$$\Delta \alpha = 0 \qquad (8.4.1)$$

Restricting the analysis to deviations about level flight and neglecting the force derivatives with respect to rate of change of angle of attack and pitching rate, the linearized longitudinal equations from Eq. (8.2.1) can be written as

$$
\begin{bmatrix}
1 & 0 & 0 & 0 & 0 & 0 \\
0 & 1 & 0 & 0 & 0 & 0 \\
0 & -R_{m,\hat{\alpha}} & 1 & 0 & 0 & 0 \\
0 & 0 & 0 & 1 & 0 & 0 \\
0 & 0 & 0 & 0 & 1 & 0 \\
0 & 0 & 0 & 0 & 0 & 1
\end{bmatrix}
\begin{Bmatrix}
\Delta\hat{\mu} \\
\Delta\hat{\alpha} \\
\Delta\hat{q} \\
\Delta\hat{\xi}_x \\
\Delta\hat{\xi}_z \\
\Delta\hat{\theta}
\end{Bmatrix}
=
\begin{bmatrix}
R_{x,\mu} & R_{x,\alpha} & 0 & 0 & 0 & -R_{gx} \\
R_{z,\mu} & R_{z,\alpha} & 1 & 0 & 0 & 0 \\
R_{m,\mu} & R_{m,\alpha} & R_{m,\overline{q}} & 0 & 0 & 0 \\
1 & 0 & 0 & 0 & 0 & 0 \\
0 & 1 & 0 & 0 & 0 & -1 \\
0 & 0 & 1 & 0 & 0 & 0
\end{bmatrix}
\begin{Bmatrix}
\Delta\mu \\
\Delta\alpha \\
\Delta\overline{q} \\
\Delta\xi_x \\
\Delta\xi_z \\
\Delta\theta
\end{Bmatrix}
\qquad (8.4.2)
$$

Applying Eq. (8.4.1) to Eq. (8.4.2) and then replacing the angular momentum equation with Eq. (8.4.1) yields

$$
\begin{bmatrix}
1 & 0 & 0 & 0 & 0 & 0 \\
0 & 0 & 0 & 0 & 0 & 0 \\
0 & 0 & 0 & 0 & 0 & 0 \\
0 & 0 & 0 & 1 & 0 & 0 \\
0 & 0 & 0 & 0 & 1 & 0 \\
0 & 0 & 0 & 0 & 0 & 1
\end{bmatrix}
\begin{Bmatrix}
\Delta\hat{\mu} \\
\Delta\hat{\alpha} \\
\Delta\hat{q} \\
\Delta\hat{\xi}_x \\
\Delta\hat{\xi}_z \\
\Delta\hat{\theta}
\end{Bmatrix}
=
\begin{bmatrix}
R_{x,\mu} & 0 & 0 & 0 & 0 & -R_{gx} \\
R_{z,\mu} & 0 & 1 & 0 & 0 & 0 \\
0 & 1 & 0 & 0 & 0 & 0 \\
1 & 0 & 0 & 0 & 0 & 0 \\
0 & 0 & 0 & 0 & 0 & -1 \\
0 & 0 & 1 & 0 & 0 & 0
\end{bmatrix}
\begin{Bmatrix}
\Delta\mu \\
\Delta\alpha \\
\Delta\overline{q} \\
\Delta\xi_x \\
\Delta\xi_z \\
\Delta\theta
\end{Bmatrix}
\qquad (8.4.3)
$$

Notice that within this approximation, the second and third equations in Eq. (8.4.3) do not contain any derivative terms, and are thus algebraic.

Rewriting Eq. (8.4.3) after subtracting the second of these equations from the last gives the eigenproblem

$$
\begin{bmatrix}
1 & 0 & 0 & 0 & 0 & 0 \\
0 & 0 & 0 & 0 & 0 & 0 \\
0 & 0 & 0 & 0 & 0 & 0 \\
0 & 0 & 0 & 1 & 0 & 0 \\
0 & 0 & 0 & 0 & 1 & 0 \\
0 & 0 & 0 & 0 & 0 & 1
\end{bmatrix}
\begin{Bmatrix}
\Delta\hat{\mu} \\
\Delta\hat{\alpha} \\
\Delta\hat{q} \\
\Delta\hat{\xi}_x \\
\Delta\hat{\xi}_z \\
\Delta\hat{\theta}
\end{Bmatrix}
=
\begin{bmatrix}
R_{x,\mu} & 0 & 0 & 0 & 0 & -R_{gx} \\
R_{z,\mu} & 0 & 1 & 0 & 0 & 0 \\
0 & 1 & 0 & 0 & 0 & 0 \\
1 & 0 & 0 & 0 & 0 & 0 \\
0 & 0 & 0 & 0 & 0 & -1 \\
-R_{z,\mu} & 0 & 0 & 0 & 0 & 0
\end{bmatrix}
\begin{Bmatrix}
\Delta\mu \\
\Delta\alpha \\
\Delta\overline{q} \\
\Delta\xi_x \\
\Delta\xi_z \\
\Delta\theta
\end{Bmatrix}
\tag{8.4.4}
$$

The first and last equations in Eq. (8.4.4) are not coupled to any of the other four equations. Thus, Eq. (8.4.4) can be rearranged and separated to yield

$$
\begin{bmatrix}
\left(R_{x,\mu}-\lambda\right) & -R_{gx} \\
-R_{z,\mu} & -\lambda
\end{bmatrix}
\begin{Bmatrix}
\Delta\mu \\
\Delta\theta
\end{Bmatrix}
=
\begin{Bmatrix}
0 \\
0
\end{Bmatrix}
\tag{8.4.5}
$$

$$
\begin{Bmatrix}
\Delta\mu \\
\Delta\alpha \\
\Delta\overline{q} \\
\Delta\xi_z \\
\Delta\theta
\end{Bmatrix}
=
\begin{Bmatrix}
\lambda \\
0 \\
-R_{z,\mu}\lambda \\
R_{z,\mu}/\lambda \\
-R_{z,\mu}
\end{Bmatrix}
\Delta\xi_x
\tag{8.4.6}
$$

The eigenvalues for Eq. (8.4.5) are obtained from the roots of the characteristic equation, which is readily expanded to give

$$
\lambda^2 - R_{x,\mu}\lambda - R_{gx}R_{z,\mu} = 0
\tag{8.4.7}
$$

Thus, from Eq. (8.4.7), the dimensionless eigenvalues for this phugoid approximation are given by

$$
\lambda_p = \frac{R_{x,\mu}}{2} \pm \sqrt{\left(\frac{R_{x,\mu}}{2}\right)^2 + R_{gx}R_{z,\mu}}
\tag{8.4.8}
$$

which gives the approximate phugoid damping rate,

$$
\sigma_p = -\frac{2V_o}{\overline{c}_w}\,\mathrm{real}(\lambda_p) = -\frac{2V_o}{\overline{c}_w}\left(\frac{R_{x,\mu}}{2}\right) = -\frac{V_o}{\overline{c}_w}R_{x,\mu}
\tag{8.4.9}
$$

and the approximate phugoid damped natural frequency,

$$\omega_{d_p} = \frac{2V_o}{\bar{c}_w}\left|\text{imag}(\lambda_p)\right|$$

$$= \frac{2V_o}{\bar{c}_w}\sqrt{-R_{gx}R_{z,\mu} - \left(\frac{R_{x,\mu}}{2}\right)^2} = \frac{2V_o}{\bar{c}_w}\sqrt{-R_{gx}R_{z,\mu}}\sqrt{1 + \frac{R_{x,\mu}^2}{4R_{gx}R_{z,\mu}}} \qquad (8.4.10)$$

If we also neglect compressibility and assume that the thrust is constant and aligned with the direction of flight, then from Eqs. (7.7.8) and (7.710), the equilibrium lift force must equal the weight and we can write

$$R_{z,\mu} = \frac{\rho S_w \bar{c}_w}{2W/g} C_{Zo} = -\frac{\rho S_w \bar{c}_w}{2W/g} C_{Lo}$$

$$= -\frac{\rho S_w \bar{c}_w}{2W/g}\left(\frac{W}{\frac{1}{2}\rho V_o^2 S_w}\right) = -\frac{g\bar{c}_w}{V_o^2} = -2R_{gx} \qquad (8.4.11)$$

and

$$R_{x,\mu} = \frac{\rho S_w \bar{c}_w}{2W/g} C_{Xo} = -\frac{\rho S_w \bar{c}_w}{2W/g} C_{Do} = -\frac{\rho S_w \bar{c}_w}{2W/g} C_{Lo}\frac{C_{Do}}{C_{Lo}} = -2R_{gx}\frac{C_{Do}}{C_{Lo}} \qquad (8.4.12)$$

Using Eqs. (8.4.11) and (8.4.12) in Eq. (8.4.8), the dimensionless eigenvalues become

$$\lambda_p = R_{gx}\left[-\frac{C_{Do}}{C_{Lo}} \pm i\sqrt{2 - \left(\frac{C_{Do}}{C_{Lo}}\right)^2}\right] = \frac{g\bar{c}_w}{2V_o^2}\left[-\frac{C_{Do}}{C_{Lo}} \pm i\sqrt{2 - \left(\frac{C_{Do}}{C_{Lo}}\right)^2}\right] \qquad (8.4.13)$$

The approximate phugoid damping rate then reduces to

$$\sigma_p = \frac{2V_o}{\bar{c}_w}\left(R_{gx}\frac{C_{Do}}{C_{Lo}}\right) = \frac{2V_o}{\bar{c}_w}\left(\frac{g\bar{c}_w}{2V_o^2}\right)\frac{C_{Do}}{C_{Lo}} = \frac{g}{V_o}\frac{C_{Do}}{C_{Lo}} \qquad (8.4.14)$$

and the approximate phugoid damped natural frequency is

$$\omega_{d_p} = \frac{2V_o}{\bar{c}_w}\left(\sqrt{2}R_{gx}\sqrt{1 - \frac{1}{2}\left(\frac{C_{Do}}{C_{Lo}}\right)^2}\right) = \frac{2V_o}{\bar{c}_w}\left(\sqrt{2}\frac{g\bar{c}_w}{2V_o^2}\right)\sqrt{1 - \frac{1}{2}\left(\frac{C_{Do}}{C_{Lo}}\right)^2}$$

$$= \sqrt{2}\frac{g}{V_o}\sqrt{1 - \frac{1}{2}\left(\frac{C_{Do}}{C_{Lo}}\right)^2} \qquad (8.4.15)$$

This approximation predicts the same undamped natural frequency as Lanchester's original approximation. This very simple result gives a phugoid frequency that depends only on airspeed and not at all on the airplane or its altitude. The phugoid damping as predicted by Eq. (8.4.14) is also quite simple, depending only on the airspeed and the lift-to-drag ratio for the airplane.

The phugoid approximation, given by Eq. (8.4.13), is not nearly as accurate as the short-period approximation, given by Eq. (8.3.9). Furthermore, many of the fundamental characteristics of phugoid motion are not captured in Eq. (8.4.13) and the approximation can be quite misleading in some regards. This approximation has, however, been used extensively for many years as a first approximation for the frequency and damping associated with phugoid motion. The inaccuracy of this approximation is demonstrated in the following example.

EXAMPLE 8.4.1. Using the long-period approximation given by Eq. (8.4.13), find the dimensionless eigenvalues and eigenvectors, the damping rate, and the damped natural frequency for the phugoid mode of the typical general aviation airplane described in Example 8.2.1. Compare these approximate results with the long-period solution that was obtained in Example 8.2.1 from the fully coupled longitudinal equations.

Solution. From Example 8.2.1 we have

$$V_o = 180 \text{ ft/sec}, \qquad \bar{c}_w = 5.606 \text{ ft,}$$
$$C_{Lo} = 0.393, \qquad C_{Do} = 0.050$$

Using these values in Eq. (8.4.13), the long-period approximation gives

$$
\lambda_p = \frac{g\bar{c}_w}{2V_o^2}\left[-\frac{C_{Do}}{C_{Lo}} \pm i\sqrt{2-\left(\frac{C_{Do}}{C_{Lo}}\right)^2}\right]
$$

$$
= \frac{(32.17 \text{ ft/sec}^2)5.606 \text{ ft}}{2(180 \text{ ft/sec})^2}\left[-\frac{0.050}{0.393} \pm i\sqrt{2-\left(\frac{0.050}{0.393}\right)^2}\right]
$$

$$
= -0.000354 \pm 0.003920\,i
$$

$$
\sigma = -\text{real}(\lambda)\frac{2V_o}{\bar{c}_w} = \frac{C_{Do}}{C_{Lo}}\frac{g}{V_o} = \frac{0.050}{0.393}\frac{32.17}{180} = 0.022738 \text{ sec}^{-1}
$$

$$
\omega_d = |\text{imag}(\lambda)|\frac{2V_o}{\bar{c}_w} = \frac{g}{V_o}\sqrt{2-\left(C_{Do}/C_{Lo}\right)^2}
$$

$$
= \frac{32.17}{180}\sqrt{2-\left(0.050/0.393\right)^2} = 0.251727 \text{ rad/sec}
$$

The long-period eigenvectors for this first-order approximation, as are found from Eq. (8.4.6), give

$$
\begin{Bmatrix} \Delta\mu \\ \Delta\alpha \\ \Delta\overline{q} \\ \Delta\xi_x \\ \Delta\xi_z \\ \Delta\theta \end{Bmatrix} = C_{1,2} \begin{Bmatrix} 0.002237 \pm 0.000408\,i \\ 0.000000 \\ 0.000012 \pm 0.000002\,i \\ 0.051978 \mp 0.575432\,i \\ 0.816187 \\ 0.000289 \mp 0.003199\,i \end{Bmatrix} \exp[(-0.000354 \pm 0.003920\,i)\,\tau_x]
$$

The exact long-period solution from Example 8.2.1 is

$$
\begin{Bmatrix} \Delta\mu \\ \Delta\alpha \\ \Delta\overline{q} \\ \Delta\xi_x \\ \Delta\xi_z \\ \Delta\theta \end{Bmatrix} = C_{3,4} \begin{Bmatrix} 0.002090 \pm 0.000366\,i \\ -0.000118 \mp 0.000023\,i \\ 0.000008 \pm 0.000001\,i \\ 0.059947 \mp 0.643541\,i \\ 0.763053 \\ 0.000083 \mp 0.002520\,i \end{Bmatrix} \exp[(-0.000264 \pm 0.003272\,i)\,\tau_x]
$$

From this example we see that the phugoid eigenvalues, computed from Eq. (8.4.13), are significantly in error. For this example, the frequency from Eq. (8.4.13) is about 20 percent high and the damping is overpredicted by more than 33 percent. Furthermore, this type of aircraft, with low aspect ratio and high drag, is particularly flattering to this approximation. For high-performance aircraft, the results predicted from Eq. (8.4.13) are much worse. In fact, under certain conditions, this approximation does not even predict the correct sign for the phugoid damping. Since the gravitational acceleration, airspeed, lift coefficient, and drag coefficient are always positive, Eq. (8.4.14) will always predict a convergent phugoid mode. In reality, however, modern airplanes can exhibit a divergent phugoid at certain airspeeds.

The most restrictive approximation associated with the development of Eq. (8.4.13) comes from the assumption that all changes in angle of attack are exactly zero. Clearly, we have seen in Example 8.2.1 that phugoid motion does produce changes in angle of attack that are small compared to changes in the dimensionless axial velocity. On the other hand, we have also seen that even very small changes in angle of attack can have a significant effect on aircraft motion. In fact, angle of attack has such a profound effect on aircraft dynamics that one is usually ill advised to neglect changes in angle of attack under almost any circumstance.

Clearly, the approximation given by Eq. (8.4.13) does not accurately describe the phugoid motion under many conditions. Another closed-form approximation, which was first presented by Bairstow (1939), gives a much more accurate prediction of the phugoid frequency. However, this approximation was never widely used, since in most cases the damping predicted by Bairstow's approximation is less accurate than that predicted by Eq. (8.4.13). Here we shall use a closed-form approximation, which was developed by Phillips (2000a) and more accurately describes both the period and the damping for the long-period mode.

To this end, we recognize that phugoid motion is always underdamped. Thus, the eigenvalues describing the phugoid will always form a complex pair, $\lambda = \lambda_r \pm i\lambda_i$, where the imaginary part specifies the phugoid frequency and the real part specifies the damping. With this recognition, the eigenproblem associated with Eq. (8.4.2) could be written as

$$
\begin{Bmatrix}
(\lambda_r \pm i\lambda_i)\Delta\mu \\
(\lambda_r \pm i\lambda_i)\Delta\alpha \\
(\lambda_r \pm i\lambda_i)(\Delta\bar{q} - R_{m,\hat{\alpha}}\Delta\alpha) \\
(\lambda_r \pm i\lambda_i)\Delta\xi_x \\
(\lambda_r \pm i\lambda_i)\Delta\xi_z \\
(\lambda_r \pm i\lambda_i)\Delta\theta
\end{Bmatrix}
=
\begin{bmatrix}
R_{x,\mu} & R_{x,\alpha} & 0 & 0 & 0 & -R_{gx} \\
R_{z,\mu} & R_{z,\alpha} & 1 & 0 & 0 & 0 \\
R_{m,\mu} & R_{m,\alpha} & R_{m,\bar{q}} & 0 & 0 & 0 \\
1 & 0 & 0 & 0 & 0 & 0 \\
0 & 1 & 0 & 0 & 0 & -1 \\
0 & 0 & 1 & 0 & 0 & 0
\end{bmatrix}
\begin{Bmatrix}
\Delta\mu \\
\Delta\alpha \\
\Delta\bar{q} \\
\Delta\xi_x \\
\Delta\xi_z \\
\Delta\theta
\end{Bmatrix}
\quad (8.4.16)
$$

Since phugoid motion is lightly damped, the real part of the phugoid eigenvalue pair is very small. Furthermore, while both the angle of attack and the pitching rate change significantly with time, the amplitudes for these changes are also very small. Thus, the products $\lambda_r\Delta\alpha$ and $\lambda_r\Delta\bar{q}$ are extremely small and can be ignored. This means that we can approximate the variation in angle of attack and pitching rate as undamped sinusoidal motion,

$$
(\lambda_r \pm i\lambda_i)\Delta\alpha \cong \pm i\varpi_n\Delta\alpha \quad (8.4.17)
$$

$$
(\lambda_r \pm i\lambda_i)\Delta\bar{q} \cong \pm i\varpi_n\Delta\bar{q} \quad (8.4.18)
$$

where ϖ_n is the dimensionless undamped natural frequency. Because the rate of change of angle of attack is very small, we also neglect the change in pitching moment with respect to the time rate of change of angle of attack. Furthermore, we shall assume that the thrust force is aligned with the center of gravity so that the equilibrium aerodynamic pitching moment is zero. With these approximations, Eqs. (7.6.24) and (7.7.8) require that $R_{m,\mu} = 0$ and Eq. (8.4.16) reduces to

$$
\begin{Bmatrix}
\lambda\Delta\mu \\
\pm i\varpi_n\Delta\alpha \\
\pm i\varpi_n\Delta\bar{q} \\
\lambda\Delta\xi_x \\
\lambda\Delta\xi_z \\
\lambda\Delta\theta
\end{Bmatrix}
=
\begin{bmatrix}
R_{x,\mu} & R_{x,\alpha} & 0 & 0 & 0 & -R_{gx} \\
R_{z,\mu} & R_{z,\alpha} & 1 & 0 & 0 & 0 \\
0 & R_{m,\alpha} & R_{m,\bar{q}} & 0 & 0 & 0 \\
1 & 0 & 0 & 0 & 0 & 0 \\
0 & 1 & 0 & 0 & 0 & -1 \\
0 & 0 & 1 & 0 & 0 & 0
\end{bmatrix}
\begin{Bmatrix}
\Delta\mu \\
\Delta\alpha \\
\Delta\bar{q} \\
\Delta\xi_x \\
\Delta\xi_z \\
\Delta\theta
\end{Bmatrix}
\quad (8.4.19)
$$

The second and third equations in Eq. (8.4.19) provide an algebraic system that can be used to relate the deviation in angle of attack and pitching rate to the deviation in forward velocity,

$$
\begin{bmatrix} (R_{z,\alpha} \mp i\varpi_n) & 1 \\ R_{m,\alpha} & (R_{m,\bar{q}} \mp i\varpi_n) \end{bmatrix} \begin{Bmatrix} \Delta\alpha \\ \Delta\bar{q} \end{Bmatrix} = \begin{Bmatrix} -R_{z,\mu}\Delta\mu \\ 0 \end{Bmatrix}
\tag{8.4.20}
$$

The remaining four equations form the eigenproblem,

$$
\begin{Bmatrix} \lambda\Delta\mu \\ \lambda\Delta\xi_x \\ \lambda\Delta\xi_z \\ \lambda\Delta\theta \end{Bmatrix} = \begin{bmatrix} R_{x,\mu} & R_{x,\alpha} & 0 & -R_{gx} \\ 1 & 0 & 0 & 0 \\ 0 & 1 & 0 & -1 \\ 0 & 0 & 1 & 0 \end{bmatrix} \begin{Bmatrix} \Delta\mu \\ \Delta\alpha \\ \Delta\bar{q} \\ \Delta\theta \end{Bmatrix}
\tag{8.4.21}
$$

The algebraic solution to Eq. (8.4.20) is rather straightforward and gives

$$
\begin{Bmatrix} \Delta\alpha \\ \Delta\bar{q} \end{Bmatrix} = \begin{Bmatrix} R_{xc} \\ R_{zc} \end{Bmatrix} \Delta\mu
\tag{8.4.22}
$$

where the complex coefficients R_{xc} and R_{zc} are defined as

$$
R_{xc} \equiv \frac{R_{z,\mu}(R_{m,\bar{q}} \mp i\varpi_n)}{R_{m,\alpha} - R_{z,\alpha}R_{m,\bar{q}} + \varpi_n^2 \pm i\varpi_n(R_{z,\alpha} + R_{m,\bar{q}})}
\tag{8.4.23}
$$

and

$$
R_{zc} \equiv \frac{-R_{z,\mu}R_{m,\alpha}}{R_{m,\alpha} - R_{z,\alpha}R_{m,\bar{q}} + \varpi_n^2 \pm i\varpi_n(R_{z,\alpha} + R_{m,\bar{q}})}
\tag{8.4.24}
$$

Equation (8.4.22) describes both the amplitude and phase relationships between the angle of attack, pitching rate, and dimensionless axial velocity deviation. Using Eq. (8.4.22) in Eq. (8.4.21) to eliminate the angle of attack from the first equation and the pitching rate from the last equation, we obtain

$$
\begin{Bmatrix} \lambda\Delta\mu \\ \lambda\Delta\xi_x \\ \lambda\Delta\xi_z \\ \lambda\Delta\theta \end{Bmatrix} = \begin{bmatrix} (R_{x,\mu} + R_{x,\alpha}R_{xc}) & 0 & 0 & -R_{gx} \\ 1 & 0 & 0 & 0 \\ 0 & 1 & 0 & -1 \\ R_{zc} & 0 & 0 & 0 \end{bmatrix} \begin{Bmatrix} \Delta\mu \\ \Delta\alpha \\ \Delta\bar{q} \\ \Delta\theta \end{Bmatrix}
\tag{8.4.25}
$$

The first and fourth equations in Eq. (8.4.25) contain only the forward velocity deviation and the elevation angle. Thus, these two equations can be separated from the other two equations to give

$$
\begin{bmatrix} (R_{x,\mu} + R_{x,\alpha}R_{xc} - \lambda) & -R_{gx} \\ R_{zc} & -\lambda \end{bmatrix} \begin{Bmatrix} \Delta\mu \\ \Delta\theta \end{Bmatrix} = \begin{Bmatrix} 0 \\ 0 \end{Bmatrix}
\tag{8.4.26}
$$

Combining Eqs. (8.4.22) and (8.4.25) and rearranging, we also have

$$
\begin{Bmatrix} \Delta\mu \\ \Delta\alpha \\ \Delta\overline{q} \\ \Delta\xi_z \\ \Delta\theta \end{Bmatrix} = \begin{Bmatrix} \lambda \\ R_{xc}\lambda \\ R_{zc}\lambda \\ R_{xc} - R_{zc}/\lambda \\ R_{zc} \end{Bmatrix} \Delta\xi_x
\tag{8.4.27}
$$

Equation (8.4.26) can be used to solve for the phugoid eigenvalues, and the resulting complex eigenvalues can then be used in Eq. (8.4.27) to obtain the eigenvectors. The characteristic equation associated with Eq. (8.4.26) is

$$
\lambda^2 - \left(R_{x,\mu} + R_{x,\alpha} R_{xc} \right)\lambda + R_{gx} R_{zc} = 0
\tag{8.4.28}
$$

and the eigenvalues are the roots of this quadratic equation,

$$
\lambda = \frac{R_{x,\mu} + R_{x,\alpha} R_{xc}}{2} \pm \sqrt{\left(\frac{R_{x,\mu} + R_{x,\alpha} R_{xc}}{2} \right)^2 - R_{gx} R_{zc}}
\tag{8.4.29}
$$

Because the phugoid frequency is very low, we can neglect terms in Eqs. (8.4.23) and (8.4.24) that contain the phugoid frequency raised to any power greater than 1. Thus, these complex coefficients can be approximated as

$$
R_{xc} \cong \frac{R_{z,\mu} R_{m,\overline{q}}}{R_{m,\alpha} - R_{z,\alpha} R_{m,\overline{q}}} \left(1 \mp i \frac{\varpi_n}{R_{m,\overline{q}}} \right) \left(1 \mp i \frac{\varpi_n \left(R_{z,\alpha} + R_{m,\overline{q}} \right)}{R_{m,\alpha} - R_{z,\alpha} R_{m,\overline{q}}} \right) \cong R_{xa} \left(1 \mp i \varpi_n R_{xp} \right) \tag{8.4.30}
$$

$$
R_{zc} \cong \frac{- R_{z,\mu} R_{m,\alpha}}{R_{m,\alpha} - R_{z,\alpha} R_{m,\overline{q}}} \left(1 \mp i \frac{\varpi_n \left(R_{z,\alpha} + R_{m,\overline{q}} \right)}{R_{m,\alpha} - R_{z,\alpha} R_{m,\overline{q}}} \right) \cong R_{za} \left(1 \mp i \varpi_n R_{zp} \right) \tag{8.4.31}
$$

where R_{xa}, R_{za}, R_{xp}, and R_{zp} are defined,

$$
R_{xa} \equiv \frac{R_{z,\mu} R_{m,\overline{q}}}{R_{m,\alpha} - R_{z,\alpha} R_{m,\overline{q}}}
\tag{8.4.32}
$$

$$
R_{za} \equiv \frac{-R_{z,\mu} R_{m,\alpha}}{R_{m,\alpha} - R_{z,\alpha} R_{m,\overline{q}}}
\tag{8.4.33}
$$

$$
R_{xp} \equiv \frac{R_{m,\alpha} + R_{m,\overline{q}}^2}{R_{m,\overline{q}} \left(R_{m,\alpha} - R_{z,\alpha} R_{m,\overline{q}} \right)}
\tag{8.4.34}
$$

$$R_{zp} \equiv \frac{R_{z,\alpha} + R_{m,\bar{q}}}{R_{m,\alpha} - R_{z,\alpha} R_{m,\bar{q}}} \tag{8.4.35}$$

Using the approximations that are given by Eqs. (8.4.30) and (8.4.31) in Eq. (8.4.29), the phugoid eigenvalues can be approximated as

$$\lambda \cong \frac{R_{x,\mu} + R_{x,\alpha} R_{xa}\left(1 \mp i\varpi_n R_{xp}\right)}{2}$$
$$\pm \sqrt{\left(\frac{R_{x,\mu} + R_{x,\alpha} R_{xa}\left(1 \mp i\varpi_n R_{xp}\right)}{2}\right)^2 - R_{gx}R_{za}\left(1 \mp i\varpi_n R_{zp}\right)} \tag{8.4.36}$$

The imaginary components of the complex coefficients R_{xc} and R_{zc} are very small. The result of this very small phase shift is to transfer a small fraction of the imaginary part of the eigenvalue to the real part and to transfer a small fraction of the real part to the imaginary part. Because phugoid motion is so lightly damped, the small fraction of the real part that is phase-shifted to the imaginary part can be neglected. However, when a very small fraction of the much larger imaginary part is phase-shifted to the real part, it becomes significant.

Thus, ignoring the phase shift in R_{xc}, Eq. (8.4.36) can be written

$$\lambda \cong \frac{R_{x,\mu} + R_{x,\alpha} R_{xa}}{2} \pm \sqrt{\left(\frac{R_{x,\mu} + R_{x,\alpha} R_{xa}}{2}\right)^2 - R_{gx}R_{za}\left(1 \mp i\varpi_n R_{zp}\right)} \tag{8.4.37}$$

or

$$\lambda \cong \frac{R_{x,\mu} + R_{x,\alpha} R_{xa}}{2} \pm i\sqrt{R_{gx}R_{za}} \sqrt{1 \mp i\varpi_n R_{zp} - \frac{\left(R_{x,\mu} + R_{x,\alpha} R_{xa}\right)^2}{4 R_{gx}R_{za}}} \tag{8.4.38}$$

Since both the phase shift and the square of the damping term are very small, we can further approximate Eq. (8.4.38) as

$$\lambda \cong \frac{R_{x,\mu} + R_{x,\alpha} R_{xa}}{2} \pm i\sqrt{R_{gx}R_{za}}\left[1 \mp i\frac{\varpi_n R_{zp}}{2} - \frac{\left(R_{x,\mu} + R_{x,\alpha} R_{xa}\right)^2}{8 R_{gx}R_{za}}\right] \tag{8.4.39}$$

which can be rearranged to give

$$\lambda \cong \frac{R_{x,\mu} + R_{x,\alpha} R_{xa}}{2} + (\pm i)(\mp i)\frac{\varpi_n R_{zp}\sqrt{R_{gx}R_{za}}}{2} \pm i\sqrt{R_{gx}R_{za}}\left[1 - \frac{\left(R_{x,\mu} + R_{x,\alpha} R_{xa}\right)^2}{8 R_{gx}R_{za}}\right] \tag{8.4.40}$$

or

$$\lambda \cong \frac{R_{x,\mu} + R_{x,\alpha} R_{xa}}{2} + \frac{\varpi_n R_{zp} \sqrt{R_{gx} R_{za}}}{2} \pm i \sqrt{R_{gx} R_{za} - \left(\frac{R_{x,\mu} + R_{x,\alpha} R_{xa}}{2}\right)^2} \qquad (8.4.41)$$

From Eq. (8.4.41) it is clearly seen that there is a component of phugoid damping that is related to the phugoid frequency. This will be called *phase damping*, because it results from the phase shift between angle of attack, pitching rate, and forward velocity. As will be shown, the phase damping contribution to the real component of the eigenvalue is always positive and tends to reduce the total phugoid damping. In fact, under certain circumstances, the phase damping can result in a large enough positive contribution to render the phugoid divergent.

The dimensionless undamped natural frequency, ϖ_n, has to this point been unknown. We can now evaluate this frequency by neglecting the damping terms in Eq. (8.4.41). This gives

$$\varpi_n = \sqrt{R_{gx} R_{za}} \qquad (8.4.42)$$

Using Eq. (8.4.42) in Eq. (8.4.41), the phugoid eigenvalues can be expressed as

$$\lambda \cong \frac{R_{x,\mu} + R_{x,\alpha} R_{xa}}{2} + \frac{R_{zp} R_{gx} R_{za}}{2} \pm i \sqrt{R_{gx} R_{za} - \left(\frac{R_{x,\mu} + R_{x,\alpha} R_{xa}}{2}\right)^2} \qquad (8.4.43)$$

Thus, for this approximation, the dimensionless **phugoid eigenvalues** can be written as

$$\lambda_p \cong \frac{R_{z,\mu}}{2}\left[\left(\frac{R_{x,\mu}}{R_{z,\mu}} + R_d - R_p\right) \pm i \sqrt{-\frac{4R_{gx}}{R_{z,\mu}} R_s - \left(\frac{R_{x,\mu}}{R_{z,\mu}} + R_d\right)^2}\right] \qquad (8.4.44)$$

where we define the **phugoid stability ratio**,

$$R_s \equiv -\frac{R_{za}}{R_{z,\mu}} = \frac{R_{m,\alpha}}{R_{m,\alpha} - R_{z,\alpha} R_{m,\bar{q}}} \qquad (8.4.45)$$

the **phugoid pitch-damping ratio**,

$$R_d \equiv \frac{R_{x,\alpha} R_{xa}}{R_{z,\mu}} = \frac{R_{x,\alpha} R_{m,\bar{q}}}{R_{m,\alpha} - R_{z,\alpha} R_{m,\bar{q}}} \qquad (8.4.46)$$

and the **phugoid phase-divergence ratio**,

$$R_p \equiv -\frac{R_{zp} R_{gx} R_{za}}{R_z} = R_{gx} R_s \left(\frac{R_{z,\alpha} + R_{m,\bar{q}}}{R_{m,\alpha} - R_{z,\alpha} R_{m,\bar{q}}}\right) \qquad (8.4.47)$$

This gives the approximate **phugoid damped natural frequency,**

$$
\omega_{d_p} = \frac{2V_o}{\bar{c}_w}\left|\text{imag}(\lambda_p)\right| \cong \frac{2V_o}{\bar{c}_w}\left[\frac{R_{z,\mu}}{2}\sqrt{-\frac{4R_{gx}}{R_{z,\mu}}R_s - \left(\frac{R_{x,\mu}}{R_{z,\mu}} + R_d\right)^2}\right] \tag{8.4.48}
$$

and the approximate **phugoid damping rate,**

$$
\sigma_p = -\frac{2V_o}{\bar{c}_w}\text{real}(\lambda_p) \cong -\frac{2V_o}{\bar{c}_w}\left[\frac{R_{z,\mu}}{2}\left(\frac{R_{x,\mu}}{R_{z,\mu}} + R_d - R_p\right)\right] \tag{8.4.49}
$$

If the flow is incompressible and the thrust is independent of airspeed and aligned with the direction of flight, the approximation can be somewhat simplified by invoking Eqs. (8.4.11) and (8.4.12). From these equations we can write

$$
\frac{R_{z,\mu}}{2} = -R_{gx} = -\frac{g\bar{c}_w}{2V_o^2} \tag{8.4.50}
$$

$$
\frac{R_{x,\mu}}{R_{z,\mu}} = \frac{C_{Do}}{C_{Lo}} \tag{8.4.51}
$$

Thus, for **incompressible flow and constant thrust aligned with the direction of flight,** the approximate dimensionless **phugoid eigenvalues** can be written as

$$
\lambda_p \cong \frac{g\bar{c}_w}{2V_o^2}\left[\left(-\frac{C_{Do}}{C_{Lo}} - R_d + R_p\right) \pm i\sqrt{2R_s - \left(\frac{C_{Do}}{C_{Lo}} + R_d\right)^2}\right] \tag{8.4.52}
$$

This gives the approximate **phugoid damped natural frequency,**

$$
\omega_{d_p} \cong \frac{2V_o}{\bar{c}_w}\left[\frac{g\bar{c}_w}{2V_o^2}\sqrt{2R_s - \left(\frac{C_{Do}}{C_{Lo}} + R_d\right)^2}\right] = \sqrt{2}\frac{g}{V_o}\sqrt{R_s - \frac{1}{2}\left(\frac{C_{Do}}{C_{Lo}} + R_d\right)^2} \tag{8.4.53}
$$

and the approximate **phugoid damping rate,**

$$
\sigma_p \cong -\frac{2V_o}{\bar{c}_w}\left[\frac{g\bar{c}}{2V_o^2}\left(-\frac{C_{Do}}{C_{Lo}} - R_d + R_p\right)\right] = \frac{g}{V_o}\left(\frac{C_{Do}}{C_{Lo}} + R_d - R_p\right) \tag{8.4.54}
$$

The eigenvalues and eigenvectors predicted using Eq. (8.4.44) or Eq. (8.4.52) are greatly improved over those predicted from Eq. (8.4.13). This is demonstrated in the following example.

EXAMPLE 8.4.2. Repeat Example 8.4.1 using Eq. (8.4.52). Again, compare the results obtained from this approximation with the long-period solution that was obtained in Example 8.2.1.

Solution. From Example 8.2.1, we have

$$V_o = 180 \text{ ft/sec}, \quad \bar{c}_w = 5.606 \text{ ft}, \quad R_{gx} = 0.00278, \quad C_{Lo} = 0.393, \quad C_{Do} = 0.050,$$
$$R_{x,\alpha} = 0.00030, \quad R_{z,\alpha} = -0.03151, \quad R_{m,\alpha} = -0.00220, \quad R_{m,\bar{q}} = -0.03212$$

From Eq. (8.4.45), the pitch-stability ratio is

$$R_s \equiv \frac{R_{m,\alpha}}{R_{m,\alpha} - R_{z,\alpha} R_{m,\bar{q}}} = \frac{-0.00220}{-0.00220 - (-0.03151)(-0.03212)} = 0.684910$$

From Eq. (8.4.46), the pitch-damping ratio is

$$R_d \equiv \frac{R_{x,\alpha} R_{m,\bar{q}}}{R_{m,\alpha} - R_{z,\alpha} R_{m,\bar{q}}} = \frac{0.00030(-0.03212)}{-0.00220 - (-0.03151)(-0.03212)} = 0.003000$$

From Eq. (8.4.47), the phase-divergence ratio is

$$R_p \equiv R_{gx} R_s \left(\frac{R_{z,\alpha} + R_{m,\bar{q}}}{R_{m,\alpha} - R_{z,\alpha} R_{m,\bar{q}}} \right)$$

$$= 0.00278(0.684910) \left(\frac{-0.03151 - 0.03212}{-0.00220 - (-0.03151)(-0.03212)} \right) = 0.037718$$

From Eq. (8.4.52), we have

$$\lambda_p \cong R_{gx} \left[\left(-\frac{C_{Do}}{C_{Lo}} - R_d + R_p \right) \pm i \sqrt{2R_s - \left(\frac{C_{Do}}{C_{Lo}} + R_d \right)^2} \right]$$

$$= 0.00278 \left[\left(-\frac{0.050}{0.393} - 0.003000 + 0.037718 \right) \right.$$

$$\left. \pm i \sqrt{2(0.684910) - \left(\frac{0.050}{0.393} + 0.003000 \right)^2} \right] = -0.000257 \pm 0.003233 \, i$$

$$\sigma = -\text{real}(\lambda) \frac{2V_o}{\bar{c}_w} = 0.000257 \frac{2(180)}{5.606} = 0.016504 \text{ sec}^{-1}$$

$$\omega_d = |\text{imag}(\lambda)| \frac{2V_o}{\bar{c}_w} = 0.003233 \frac{2(180)}{5.606} = 0.207613 \text{ rad/sec}$$

From Eqs. (8.4.32) through (8.4.35),

$$R_{xa} \equiv \frac{R_{z,\mu} R_{m,\bar{q}}}{R_{m,\alpha} - R_{z,\alpha} R_{m,\bar{q}}} = -0.055598$$

$$R_{za} \equiv \frac{-R_{z,\mu} R_{m,\alpha}}{R_{m,\alpha} - R_{z,\alpha} R_{m,\bar{q}}} = 0.003808$$

$$R_{xp} \equiv \frac{R_{m,\alpha} + R_{m,\bar{q}}^2}{R_{m,\bar{q}}(R_{m,\alpha} - R_{z,\alpha} R_{m,\bar{q}})} = -11.323787$$

$$R_{zp} \equiv \frac{R_{z,\alpha} + R_{m,\bar{q}}}{R_{m,\alpha} - R_{z,\alpha} R_{m,\bar{q}}} = 19.809463$$

From Eqs. (8.4.30), (8.4.31), and (8.4.42),

$$\varpi_n = \sqrt{R_{gx} R_{za}} = \sqrt{0.00278\,(0.003808)} = 0.003254$$

$$R_{xc} \cong R_{xa}(1 \mp i\varpi_n R_{xp}) = -0.055598 \mp i0.002049$$

$$R_{zc} \cong R_{za}(1 \mp i\varpi_n R_{zp}) = 0.003808 \mp i0.000245$$

Using these results in Eq. (8.4.27), the eigenvectors for this long-period approximation are

$$\begin{Bmatrix} \Delta\mu \\ \Delta\alpha \\ \Delta\bar{q} \\ \Delta\xi_x \\ \Delta\xi_z \\ \Delta\theta \end{Bmatrix} = C_{1,2} \begin{Bmatrix} 0.002077 \pm 0.000370\,i \\ -0.000115 \mp 0.000025\,i \\ 0.000008 \pm 0.000001\,i \\ 0.062894 \mp 0.647399\,i \\ 0.759545 \\ 0.000081 \mp 0.002481\,i \end{Bmatrix} \exp[(-0.000257 \pm 0.003233\,i)\,\tau_x]$$

The exact long-period solution from Example 8.2.1 is

$$\begin{Bmatrix} \Delta\mu \\ \Delta\alpha \\ \Delta\bar{q} \\ \Delta\xi_x \\ \Delta\xi_z \\ \Delta\theta \end{Bmatrix} = C_{3,4} \begin{Bmatrix} 0.002090 \pm 0.000366\,i \\ -0.000118 \mp 0.000023\,i \\ 0.000008 \pm 0.000001\,i \\ 0.059947 \mp 0.643541\,i \\ 0.763053 \\ 0.000083 \mp 0.002520\,i \end{Bmatrix} \exp[(-0.000264 \pm 0.003272\,i)\,\tau_x]$$

From Example 8.4.2 we see that the long-period eigenvalues that are computed from Eq. (8.4.52) are in excellent agreement with the fully coupled solution. For this example, the frequency prediction from Eq. (8.4.52) is accurate to within about 1 percent and the damping error is less than 3 percent. Furthermore, all components of the eigenvectors predicted from this approximation are in good agreement with the exact solution. Figures 8.4.1 through 8.4.3 show how this approximation compares with Eq. (8.4.13) and the exact solution over a broad range of aerodynamic parameters.

Here it should be noted that as the pitch-stability derivative approaches infinity, the phugoid stability ratio approaches unity, while both the phugoid pitch-damping ratio and the phugoid phase-divergence ratio approach zero. So we see that in the limit of infinite pitch stability, the result given by Eq. (8.4.52) reduces exactly to Eq. (8.4.13). Thus, Eq. (8.4.13) represents an asymptotic solution for very large pitch stability. This can be

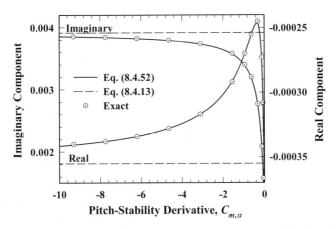

Figure 8.4.1. Effect of pitch stability on the dimensionless phugoid eigenvalues.

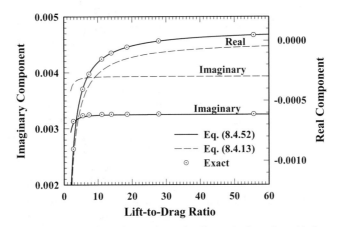

Figure 8.4.2. Effect of lift-to-drag ratio on the dimensionless phugoid eigenvalues.

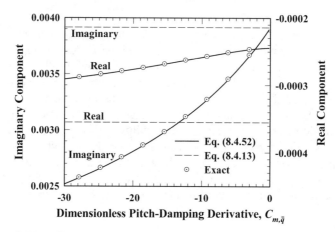

Figure 8.4.3. Effect of pitch damping on the dimensionless phugoid eigenvalues.

seen graphically in Fig. 8.4.1. Notice that as the pitch stability is increased, the result predicted by Eq. (8.4.52) approaches that predicted by Eq. (8.4.13). In this figure all parameters except the pitch stability ratio have been held constant at those values given in Example 8.2.1. In Fig. 8.4.2, a comparison between Eq. (8.4.13), Eq. (8.4.52), and the exact solution is shown for a broad range of lift-to-drag ratio. A similar comparison showing the effect of pitch damping is displayed in Fig. 8.4.3.

This approximate closed-form solution allows us to more easily examine how the aerodynamic coefficients and stability derivatives affect phugoid motion. As we have seen, damping has very little effect on the phugoid frequency. Thus, neglecting the damping in Eq. (8.4.53) and applying the definition of the phugoid stability ratio from Eq. (8.4.45), the undamped natural frequency for phugoid motion is

$$
\omega_n = \sqrt{2}\,\frac{g}{V_o}\sqrt{R_s} = \sqrt{2}\,\frac{g}{V_o}\sqrt{\frac{R_{m,\alpha}}{R_{m,\alpha}-R_{z,\alpha}R_{m,\bar q}}} = \sqrt{2}\,\frac{g}{V_o}\sqrt{\frac{\dfrac{4W/g}{\rho S_w \bar c_w}C_{m,\alpha}}{\dfrac{4W/g}{\rho S_w \bar c_w}C_{m,\alpha}-C_{Z,\alpha}C_{m,\bar q}}}
$$

$$(8.4.55)$$

From Eq. (8.4.55), we first notice the well-known fact that the phugoid frequency is inversely proportional to forward speed. In addition, there are four other parameters that affect the phugoid frequency: the dimensionless mass, the change in pitching moment with respect to angle of attack, the change in z-force with respect to angle of attack, and the change in pitching moment with respect to pitching rate. For a stable aircraft, all of these parameters except the dimensionless mass are negative. Thus, the phugoid stability ratio is always between zero and unity. From this term we see that the phugoid frequency increases with dimensionless aircraft mass and pitch stability, while it decreases with lift slope and pitch damping. As the product of dimensionless mass and pitch stability becomes large compared to the product of lift slope and pitch damping, the phugoid stability ratio approaches unity and the undamped phugoid frequency predicted from Eq. (8.4.55) reduces to that predicted by Eq. (8.4.15). Conversely, as the product of pitch stability and dimensionless mass approaches zero, so do the phugoid stability ratio and phugoid frequency.

From Eq. (8.4.54) we see that there are three distinct components to phugoid damping. Here these are called the *drag damping*, the *pitch damping*, and the *phase damping*. In most cases the drag damping is the largest of these three components. From Eq. (8.4.54), we define

$$\text{drag damping} \equiv \frac{g}{V_o} \frac{C_{Do}}{C_{Lo}} \tag{8.4.56}$$

Clearly, since the velocity, drag coefficient, and lift coefficient are all positive, the drag damping is always positive. The total drag is the sum of the parasitic drag and the induced drag. The parasitic drag coefficient is very nearly constant. The induced drag coefficient is proportional to the lift coefficient squared, and the lift coefficient is inversely proportional to the velocity squared. Thus, we can write

$$\text{drag damping} \cong \frac{g}{V_o} \frac{C_{D_{L0}} + C_{Lo}^2/\pi e R_A}{C_{Lo}} = \frac{g}{V_o}\left(\frac{C_{D_{L0}}}{C_{Lo}} + \frac{C_{Lo}}{\pi e R_A}\right)$$

$$= \frac{g}{V_o}\left(\frac{C_{D_{L0}}(\frac{1}{2}\rho V_o^2 S_w)}{W} + \frac{W}{\pi e R_A(\frac{1}{2}\rho V_o^2 S_w)}\right) \tag{8.4.57}$$

At very high airspeeds the parasitic drag dominates, the drag-to-lift ratio is proportional to velocity squared, and the phugoid drag damping increases linearly with forward velocity. At very low airspeeds the induced drag dominates, the drag-to-lift ratio is inversely proportional to velocity squared, and the phugoid drag damping varies inversely with the forward velocity cubed. At some intermediate airspeed, phugoid drag damping will exhibit a minimum. The airspeed that results in minimum phugoid drag damping is somewhat faster than the minimum drag airspeed.

The phugoid pitch damping is defined from Eqs. (8.4.54) and (8.4.46) to be

$$\text{pitch damping} \equiv \frac{g}{V_o}R_d = \frac{g}{V_o}\left(\frac{R_{x,\alpha}R_{m,\bar{q}}}{R_{m,\alpha} - R_{z,\alpha}R_{m,\bar{q}}}\right) = \frac{g}{V_o}\left(\frac{C_{X,\alpha}C_{m,\bar{q}}}{\frac{4W/g}{\rho S_w \bar{c}_w}C_{m,\alpha} - C_{Z,\alpha}C_{m,\bar{q}}}\right)$$

$$\tag{8.4.58}$$

To determine the sign of the phugoid pitch damping, we note that the dimensionless mass is positive, while the pitch-stability derivative, the z-force derivative, and the pitch-damping derivative are all negative. Thus, the phugoid pitch damping will have the same sign as the change in axial force with respect to angle of attack. Depending on the airplane design, this axial force derivative can be either positive or negative.

The change in the x-force coefficient with respect to angle of attack can be written in terms of the lift and drag coefficients,

$$
\begin{aligned}
C_{X,\alpha} &= \frac{\partial}{\partial\alpha}(C_L \sin\alpha - C_D \cos\alpha) \\
&= C_L \cos\alpha + C_{L,\alpha}\sin\alpha + C_D \sin\alpha - C_{D,\alpha}\cos\alpha \cong C_{Lo} - C_{D,\alpha}
\end{aligned}
\tag{8.4.59}
$$

The change in drag coefficient with angle of attack can be approximated as

$$
C_{D,\alpha} \cong \frac{\partial C_D}{\partial C_L}\frac{\partial C_L}{\partial\alpha} = \frac{\partial}{\partial C_L}\left(C_{D_{Lo}} + \frac{C_L^2}{\pi e R_A}\right)C_{L,\alpha} = \frac{2C_{L,\alpha}}{\pi e R_A}C_L
\tag{8.4.60}
$$

Using Eq. (8.4.60) in Eq. (8.4.59), we have

$$
C_{X,\alpha} \cong C_{Lo}\left(1 - \frac{2C_{L,\alpha}}{\pi e R_A}\right) = \frac{W}{\frac{1}{2}\rho V_o^2 S_w}\left(1 - \frac{2C_{L,\alpha}}{\pi e R_A}\right)
\tag{8.4.61}
$$

Using Eq. (8.4.61) in Eq. (8.4.58), the phugoid pitch damping can be expressed as

$$
\text{pitch damping} \cong \frac{gW}{\frac{1}{2}\rho V_o^3 S_w}\left(1 - \frac{2C_{L,\alpha}}{\pi e R_A}\right)\left(\frac{C_{m,\bar{q}}}{\frac{4W/g}{\rho S_w \bar{c}_w}C_{m,\alpha} - C_{Z,\alpha}C_{m,\bar{q}}}\right)
\tag{8.4.62}
$$

From Eq. (8.4.62), we see that phugoid pitch damping is inversely proportional to the cube of the forward velocity; it increases with the pitch-damping derivative; and it decreases with increasing pitch stability. We also see that phugoid pitch damping can be either positive or negative. Furthermore, the pitch damping is more negative for planes with low aspect ratio and low Oswald efficiency, and it is more positive for planes with high aspect ratio and Oswald efficiency. This may seem counterintuitive. However, the total phugoid damping is the sum of the drag damping, the pitch damping, and the phase damping. Lowering the aspect ratio or Oswald efficiency factor will increase the drag damping more than it will decrease the pitch damping. Thus, total phugoid damping will increase with decreasing aerodynamic efficiency, as expected.

Because lift slope decreases with decreasing aspect ratio, phugoid pitch damping can be negative only for airplanes with very low aspect ratio. For example, Prandtl's lifting-line theory combined with thin airfoil theory predicts that for an elliptic wing,

$$
\frac{2C_{L,\alpha}}{\pi e R_A} \cong \frac{4}{R_A + 2}
\tag{8.4.63}
$$

If we were to use this result in Eq. (8.4.62), it would predict that phugoid pitch damping would be negative only for an airplane having an aspect ratio less than 2. Of course,

Prandtl's lifting-line theory does not apply to wings having an aspect ratio as low as 2. However, Eq. (8.4.63) does show that phugoid pitch damping will be positive for typical subsonic airplanes, which usually have aspect ratios of 5 or above.

Using Eq. (8.4.47) in Eq. (8.4.54), the phugoid phase damping is defined as

$$
\text{phase damping} \equiv -\frac{g}{V_o} R_p = -\frac{g}{V_o} R_{gx} R_s \left(\frac{R_{z,\alpha} + R_{m,\bar{q}}}{R_{m,\alpha} - R_{z,\alpha} R_{m,\bar{q}}} \right)
$$

$$
= -\frac{g^2 \bar{c}_w}{2V_o^3} R_s \left(\frac{\dfrac{8I_{yy_b}}{\rho S_w \bar{c}_w^3} C_{Z,\alpha} + \dfrac{4W/g}{\rho S_w \bar{c}_w} C_{m,\bar{q}}}{\dfrac{4W/g}{\rho S_w \bar{c}_w} C_{m,\alpha} - C_{Z,\alpha} C_{m,\bar{q}}} \right) \tag{8.4.64}
$$

We have already seen that the phugoid stability ratio can vary from zero to positive unity. The dimensionless mass and moment of inertia are always positive, while the pitch-stability derivative, the pitch-damping derivative, and the *z*-force derivative are all negative. Thus, for a stable aircraft, the phugoid phase damping is always negative, tending to decrease the total phugoid damping. In fact, under certain circumstances, it is possible for this negative phase damping to overpower the drag and pitch damping, which makes the phugoid divergent. Since the phugoid phase damping is inversely proportional to the forward velocity cubed, this condition is aggravated at low airspeed.

The phugoid approximation given by Eq. (8.4.13) predicts that only by increasing drag can we increase phugoid damping. Increasing aircraft drag to improve phugoid damping is obviously not a particularly desirable solution. From the present improved approximation, we see that phugoid damping can in fact be increased without increasing aircraft drag. This can be done by either increasing the phugoid pitch damping or by decreasing the magnitude of the negative phugoid phase damping.

8.5. Pure Pitching Motion

In addition to the short-period and long-period modes associated with free flight, there is a third type of oscillatory longitudinal motion that does not occur in free flight but which has implications for wind tunnel testing. While it is not practical to perform free flight tests in a wind tunnel, it is possible to obtain much information about the free-flight characteristics of an aircraft from wind tunnel tests.

Here we shall consider the case in which the airplane's center of gravity is fixed in the wind tunnel. The airplane is constrained so that the lateral motions (sideslip, roll, and yaw) are all maintained at zero. In addition, the airplane's center of gravity cannot move axially or vertically within the wind tunnel. The aircraft is, however, free to rotate in pitch about the center of gravity. A model constrained in this manner has only one degree of freedom, rotation in pitch. Figure 8.5.1 shows a schematic drawing of a wind tunnel model so constrained. The oscillatory motion that can develop in such a wind tunnel test is *pure pitching motion*.

The equations of motion for this model can be obtained by applying the following constraints to the general longitudinal equations:

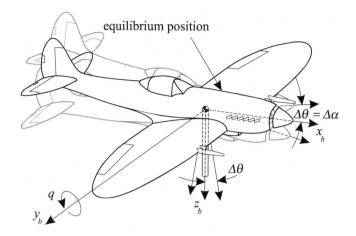

Figure 8.5.1. Wind tunnel model constrained to pure pitching motion.

$$\Delta\mu = \Delta\xi_x = \Delta\xi_z = 0 \tag{8.5.1}$$

$$\Delta\theta = \Delta\alpha \tag{8.5.2}$$

Because the axial and vertical motion is constrained, we can disregard the axial and vertical momentum equations. Thus, applying Eqs. (8.5.1) and (8.5.2) to Eq. (8.2.1) and eliminating the trivial equations, we have

$$\begin{bmatrix} -R_{m,\hat{\alpha}} & 1 \\ 1 & 0 \end{bmatrix} \begin{Bmatrix} \Delta\hat{\alpha} \\ \Delta\hat{\bar{q}} \end{Bmatrix} = \begin{bmatrix} R_{m,\alpha} & R_{m,\bar{q}} \\ 0 & 1 \end{bmatrix} \begin{Bmatrix} \Delta\alpha \\ \Delta\bar{q} \end{Bmatrix} \tag{8.5.3}$$

The eigenproblem associated with Eq. (8.5.3) is

$$\left(\begin{bmatrix} R_{m,\alpha} & R_{m,\bar{q}} \\ 0 & 1 \end{bmatrix} - \lambda \begin{bmatrix} -R_{m,\hat{\alpha}} & 1 \\ 1 & 0 \end{bmatrix} \right) \begin{Bmatrix} \chi_\alpha \\ \chi_{\bar{q}} \end{Bmatrix} = \begin{Bmatrix} 0 \\ 0 \end{Bmatrix} \tag{8.5.4}$$

The characteristic equation is

$$\begin{vmatrix} (R_{m,\alpha} + R_{m,\hat{\alpha}}\lambda) & (R_{m,\bar{q}} - \lambda) \\ -\lambda & 1 \end{vmatrix} = 0 \tag{8.5.5}$$

or

$$\lambda^2 - \left(R_{m,\bar{q}} + R_{m,\hat{\alpha}} \right)\lambda - R_{m,\alpha} = 0 \tag{8.5.6}$$

The **eigenvalues for pure pitching motion**, as obtained from Eq. (8.5.6), are then

$$
\lambda = \frac{R_{m,\bar{q}} + R_{m,\hat{\alpha}}}{2} \pm \sqrt{\left(\frac{R_{m,\bar{q}} + R_{m,\hat{\alpha}}}{2}\right)^2 + R_{m,\alpha}} \tag{8.5.7}
$$

Equation (8.5.7) looks very similar to Eq. (8.3.9), which expresses an approximate relationship for the eigenvalues associated with short-period motion in free flight. In fact, if we allow the mass to approach infinity, the z-force ratio in Eq. (8.3.9) goes to zero and Eq. (8.3.9) reduces to Eq. (8.5.7). Thus we see that as the dimensionless mass becomes large compared to the dimensionless moment of inertia, the short-period motion associated with free flight becomes pure pitching motion.

By performing wind tunnel tests on an aircraft model that is constrained to pure pitching motion, we could determine the eigenvalues experimentally. The model's moment of inertia should be adjusted to produce underdamped pitching motion. In the wind tunnel, the model is displaced in pitch from the equilibrium angle of attack and then released. The resulting oscillatory pitching motion is then recorded as a function of time. By measuring the period of oscillation and the time required to reduce the amplitude of oscillation by one-half, we could determine the sum of the two pitch-damping derivatives as well as the pitch-stability derivative.

From Eq. (8.5.7), the 50 percent damping time and the period for this pure pitching motion are given by

$$
50\% \text{ damping time} \equiv t_{50} = \frac{\bar{c}_w}{2V_o} \frac{\ln(0.50)}{\text{real}(\lambda)} = \frac{\bar{c}_w}{2V_o} \frac{\ln(0.50)}{\frac{1}{2}(R_{m,\bar{q}} + R_{m,\hat{\alpha}})}
$$

$$
= \frac{\bar{c}_w}{2V_o} \frac{\ln(0.50)}{\frac{1}{2}\frac{\rho S_w \bar{c}_w^3}{8I_{yy_b}}(C_{m,\bar{q}} + C_{m,\hat{\alpha}})} = \left(\frac{8I_{yy_b}}{\rho S_w \bar{c}_w^2 V_o}\right)\frac{\ln(0.50)}{C_{m,\bar{q}} + C_{m,\hat{\alpha}}} \tag{8.5.8}
$$

$$
\text{period} \equiv t_p = \frac{\bar{c}_w}{2V_o} \frac{2\pi}{|\text{imag}(\lambda)|} = \frac{\pi \bar{c}_w / V_o}{\sqrt{-R_{m,\alpha} - \left(\frac{R_{m,\bar{q}} + R_{m,\hat{\alpha}}}{2}\right)^2}}
$$

$$
= \frac{\pi \bar{c}_w / V_o}{\sqrt{-\left(\frac{\rho S_w \bar{c}^3}{8I_{yy_b}}\right)C_{m,\alpha} - \left(\frac{\rho S_w \bar{c}^3}{8I_{yy_b}} \frac{C_{m,\bar{q}} + C_{m,\hat{\alpha}}}{2}\right)^2}} \tag{8.5.9}
$$

$$
= \frac{\frac{16\pi I_{yy_b}}{\rho S_w \bar{c}_w^2 V_o}}{\sqrt{-\left(\frac{32I_{yy_b}}{\rho S_w \bar{c}_w^3}\right)C_{m,\alpha} - (C_{m,\bar{q}} + C_{m,\hat{\alpha}})^2}}
$$

Rearranging Eqs. (8.5.8) and (8.5.9), we have

$$C_{m,\bar{q}} + C_{m,\dot{\alpha}} = \frac{8 I_{yy_b} \ln(0.50)}{\rho S_w \bar{c}^2 V_o t_{50}} \tag{8.5.10}$$

and

$$C_{m,\alpha} = -\frac{2 I_{yy_b}}{\rho S_w \bar{c} V_o^2} \left[\left(\frac{2\pi}{t_p} \right)^2 + \left(\frac{\ln(0.50)}{t_{50}} \right)^2 \right] \tag{8.5.11}$$

Using Eqs. (8.5.10) and (8.5.11), we can determine the sum of the two pitch-damping derivatives as well as the pitch-stability derivative from the measured period and damping time, provided that we know the wind tunnel airspeed and density, along with the geometry and moment of inertia for the aircraft model. While the airflow and geometric parameters are readily determined, the mass moment of inertia is more easily resolved from a separate but related experiment.

Components of the mass moment of inertia tensor for any object can be determined by hanging the object to create a *trifilar* torsional pendulum as shown in Fig. 8.5.2. The object is suspended from three long parallel cables or strings of lengths s_1, s_2, and s_3. One end of each string is attached to the object and the other end is attached to a rigid supporting member. **It is critical to locate the attachments points so that the strings are parallel.** The vertical axis passing through the center of gravity is located between the strings a distance r_1 from string 1, a distance r_2 from string 2, and a distance r_3 from string 3. As the object rotates through a small angle θ about the vertical axis passing through the center of gravity, the lower attachment points move through small arcs. At the beginning of the arcs, all three strings are vertical. At the end of the arcs, the strings make angles ϕ_1, ϕ_2, and ϕ_3 with the vertical. It can be shown that the restoring torque is given by

$$T = \left(\frac{r_1^2 C_1}{s_1 \cos \phi_1} + \frac{r_2^2 C_2}{s_2 \cos \phi_2} + \frac{r_3^2 C_3}{s_3 \cos \phi_3} \right) W \sin \theta \tag{8.5.12}$$

where C_1, C_2, and C_3 are the vertical components of tension in each string divided by the weight, W. The development of Eq. (8.5.12) is left as an exercise for the student.

If the strings are long and the rotation angle is small, the vertical components of acceleration can be neglected and the vertical components of string tension support only the weight. If the weight supported by each string is experimentally determined, the position of the center of gravity can be related to the positions of the strings

$$\begin{Bmatrix} x_{cg} \\ z_{cg} \end{Bmatrix} = \begin{Bmatrix} x_1 C_1 + x_2 C_2 + x_3 C_3 \\ z_1 C_1 + z_2 C_2 + z_3 C_3 \end{Bmatrix} \tag{8.5.13}$$

With the weight supported by each string and the position of the center of gravity known, the distance from each string to the center of gravity is easily determined from

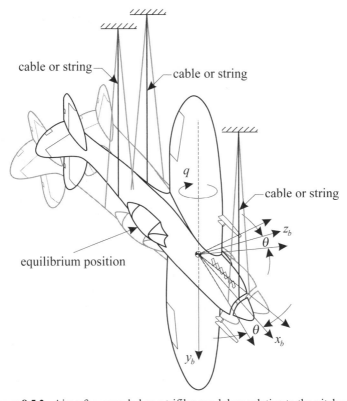

Figure 8.5.2. Aircraft suspended as a trifilar pendulum relative to the pitch axis.

$$\begin{Bmatrix} r_1^2 \\ r_2^2 \\ r_3^2 \end{Bmatrix} = \begin{Bmatrix} (x_1 - x_{cg})^2 + (z_1 - z_{cg})^2 \\ (x_2 - x_{cg})^2 + (z_2 - z_{cg})^2 \\ (x_3 - x_{cg})^2 + (z_3 - z_{cg})^2 \end{Bmatrix} \tag{8.5.14}$$

If the strings are long and the rotation angle is small, we can apply the usual small angle approximations, $\sin \theta \cong \theta$ and $\cos \phi_1 \cong \cos \phi_2 \cong \cos \phi_3 \cong 1$. From Eq. (8.5.12), this yields

$$T = \left(\frac{r_1^2 C_1}{s_1} + \frac{r_2^2 C_2}{s_2} + \frac{r_3^2 C_3}{s_3} \right) W\theta \tag{8.5.15}$$

When this trifilar pendulum is given a small angular displacement and released, it will oscillate in rotation about the vertical axis passing through the center of gravity. Neglecting all aerodynamic forces, Newton's second law for this trifilar pendulum requires

$$I_{yy_b} \frac{d^2\theta}{dt^2} + \left(\frac{r_1^2 C_1}{s_1} + \frac{r_2^2 C_2}{s_2} + \frac{r_3^2 C_3}{s_3} \right) W\theta = 0 \tag{8.5.16}$$

The eigenvalues for this second-order differential equation predict undamped oscillations with a natural frequency of

$$\omega_n = \sqrt{\frac{W}{I_{yy_b}}\left(\frac{r_1^2 C_1}{s_1} + \frac{r_2^2 C_2}{s_2} + \frac{r_3^2 C_3}{s_3}\right)} \tag{8.5.17}$$

Of course, in any real experiment there will be damping. From Eq. (8.5.17) we see that long strings will produce low-frequency oscillations, which result in light damping. By recording the angular position, velocity, or acceleration as a function of time, we can determine both the damped frequency, ω_d, and damping rate, σ. The undamped natural frequency and damping ratio can then be determined from Eqs. (8.1.13) and (8.1.14):

$$\omega_n = \sqrt{\omega_d^2 + \sigma^2}, \qquad \zeta = \sigma/\omega_n \tag{8.5.18}$$

These relations assume linear damping. However, if the strings are long enough so that the damping ratio is much less than unity, the nonlinearities in the damping will have no significant effect on the frequency. Thus, the mass moment of inertia can be accurately evaluated from the experimentally determined natural frequency. The data can be fit to a damped sinusoid and from Eqs. (8.5.17) and (8.5.18) we can use the relation

$$I_{yy_b} = \frac{W}{\omega_n^2}\left(\frac{r_1^2 C_1}{s_1} + \frac{r_2^2 C_2}{s_2} + \frac{r_3^2 C_3}{s_3}\right) = \frac{W}{\omega_d^2 + \sigma^2}\left(\frac{r_1^2 C_1}{s_1} + \frac{r_2^2 C_2}{s_2} + \frac{r_3^2 C_3}{s_3}\right) \tag{8.5.19}$$

and the radius of gyration can be obtained from

$$r_{yy_b} \equiv \sqrt{\frac{g I_{yy_b}}{W}} = \sqrt{\frac{g}{\omega_d^2 + \sigma^2}\left(\frac{r_1^2 C_1}{s_1} + \frac{r_2^2 C_2}{s_2} + \frac{r_3^2 C_3}{s_3}\right)} \tag{8.5.20}$$

To determine the moment of inertia and radius of gyration for pitch, an aircraft is suspended so that the pitch axis is vertical as shown in Fig. 8.5.2. The pitching moment of inertia so determined can be used in Eqs. (8.5.10) and (8.5.11) to obtain the related aerodynamic derivatives. Other components of the moment of inertia tensor can be obtained in a similar manner, by suspending the aircraft in different orientations.

8.6. Summary

In this chapter we have examined the stick-fixed longitudinal dynamics of an aircraft in free flight. Our analysis was based on the assumptions of rigid-body motion and small deviations from equilibrium. We found that the same aerodynamic forces and moment that make an airplane statically stable in trimmed flight will tend to produce oscillations in the aircraft operating parameters, such as airspeed and angle of attack, whenever the aircraft is disturbed from equilibrium. Furthermore, we found that under certain conditions it is possible for these oscillations to be dynamically unstable (i.e., divergent), even though the aircraft is statically stable in all regards. For a divergent mode, the amplitude of oscillation grows exponentially with time, if uncontrolled.

Two longitudinal modes of oscillation were found for a rigid airplane in free flight. These are the short-period mode and the phugoid or long-period mode. The short-period mode is composed of relatively high-frequency oscillations primarily in angle of attack and vertical displacement, with little change in airspeed. The short-period mode is typically heavily damped, with a period and 99 percent damping time both on the order of 1 to 5 seconds. The phugoid, on the other hand, is a low-frequency oscillation composed primarily of changes in airspeed and altitude, with little variation in angle of attack. The phugoid is usually very lightly damped and may even be divergent for some airplanes at certain airspeeds. The period for the phugoid is typically on the order of 30 seconds or more, and even when the mode is convergent, the 99 percent damping time is usually on the order of several minutes. The frequency and damping for both the short-period mode and the phugoid are determined from the eigenvalues, which are obtained from the linearized longitudinal equations of motion. The eigenvectors obtained from this system of equations tell us the relative amplitudes and phase shifts for the oscillations in each of the longitudinal flight variables.

These natural modes of oscillation for an aircraft in free flight are like the natural frequencies of vibration for a spring-mass system. If periodic flight disturbances are encountered of a frequency close to either that of the short-period mode or the phugoid, the oscillations associated with the affected mode can be excited in a manner similar to what is called resonance in a spring-mass system. With an aircraft, as with a spring-mass system, there are two ways to avoid resonance. One way is to make sure that the natural frequency of oscillation is not close to the frequency of any periodic disturbances that may be encountered. The other way is to provide sufficient damping in the oscillatory mode so that resonance does not produce significant amplification. For the longitudinal modes of a typical airplane, resonance is not usually a problem, because the short-period mode is heavily damped and the phugoid frequency is far below the frequency of any disturbances that are normally encountered in flight.

The material presented in this chapter is based on the assumption that the aircraft is a rigid body. Of course, in reality, airplanes are not rigid and have many modes of elastic vibration in addition to the two longitudinal flight modes discussed in this chapter. The flight modes are not significantly affected by the elastic modes provided that frequencies are not too closely aligned. The study of the effect of elastic deflection on the dynamics and aerodynamics of flight is a branch of aeronautical engineering called *aeroelasticity*. This is normally taught as a complete course, separate from flight mechanics, and many books have been written on the subject. The coupling between the elastic vibration modes of an airplane and the dynamic flight modes is beyond the scope of this textbook and is not covered here in any detail.

The reader should never lose sight of the fact that the material presented in this chapter is based on the assumption of small deviations from equilibrium flight. This approximation is quite good for the low-angle-of-attack conditions that are encountered in nearly all flight of commercial and general aviation aircraft. However, fighter aircraft and stunt airplanes routinely execute maneuvers that involve high angles of attack, large angular rates, and substantial deviation from equilibrium flight. Dynamic analysis of such aircraft motion requires a much more sophisticated analysis than the application of linear theory that is presented in this chapter.

8.7. Problems

8.1. Use a numerical eigensolver to find the eigenvalues and eigenvectors for the spring-mass system described in Example 8.1.1. Verify that the results obtained from your solver agree with the results presented in Example 8.1.1. The solution may be obtained using either a generalized or a special eigensolver.

8.2. Use a numerical eigensolver to find the eigenvalues and eigenvectors for the spring-mass system described in Example 8.1.2. Verify that the results obtained from your solver agree with the results presented in Example 8.1.2.

8.3. Use a numerical eigensolver to find the eigenvalues and eigenvectors associated with the longitudinal modes of the typical general aviation airplane described in Example 8.2.1. Verify that the results obtained from your solver agree with the results presented in Example 8.2.1.

8.4. For the large jet transport and flight conditions in Examples 7.6.1 and 7.7.1, find the eigenvalues, the eigenvectors, the 99 percent damping time, and the period, if underdamped, for each longitudinal mode. For each oscillatory mode, compute the amplitude and phase for each component of the eigenvector associated with the positive frequency.

8.5. For the large jet transport and flight conditions in Examples 7.6.1 and 7.7.1, using the short-period eigenvalue approximation given by Eq. (8.3.9), find the eigenvalues, the 99 percent damping time, and the period for the short-period mode. Compare these approximate results with the fully coupled short-period solution that was obtained in problem 8.4.

8.6. For the large jet transport and flight conditions in Examples 7.6.1 and 7.7.1, using the short-period eigenvector approximation given by Eq. (8.3.5), find the eigenvectors for the short-period mode. Normalize these eigenvectors according to Eq. (8.1.37). Compare these approximate eigenvectors with the fully coupled short-period solution that was obtained in problem 8.4.

8.7. For the large jet transport and flight conditions in Examples 7.6.1 and 7.7.1, using the phugoid eigenvalue approximation given by Eq. (8.4.8), find the dimensionless phugoid eigenvalues, the 99 percent damping time, and the period for the long-period mode. Compare these approximate results with the fully coupled long-period solution that was obtained in problem 8.4.

8.8. For the large jet transport and flight conditions in Examples 7.6.1 and 7.7.1, using the phugoid eigenvector approximation given by Eq. (8.4.6), find the dimensionless phugoid eigenvectors for the long-period mode. Normalize these eigenvectors according to Eq. (8.1.37). Compare these approximate eigenvectors with those obtained from the fully coupled long-period solution that was obtained in problem 8.4.

8.9. For the large jet transport and flight conditions in Examples 7.6.1 and 7.7.1, using the phugoid eigenvalue approximation given by Eq. (8.4.44), find the dimensionless phugoid eigenvalues, the 99 percent damping time, and the period for the long-period mode. Compare these approximate results with the fully coupled long-period solution that was obtained in problem 8.4.

8.10. For the large jet transport and flight conditions in Examples 7.6.1 and 7.7.1, using the phugoid eigenvector approximation given by Eq. (8.4.27), find the dimensionless phugoid eigenvectors for the long-period mode. Normalize these eigenvectors according to Eq. (8.1.37). Compare these approximate eigenvectors with the fully coupled long-period solution that was obtained in problem 8.4.

8.11. For the large jet transport in Examples 7.6.1 and 7.7.1, the level flight airspeed is changed from 450 mph to 350 mph. All other operating conditions remain unchanged. Find the eigenvalues, the eigenvectors, the 99 percent damping time, and the period, if underdamped, for each longitudinal mode. For each oscillatory mode, compute the amplitude and phase for each component of the eigenvector associated with the positive frequency. Compare these results with those that were obtained for an airspeed of 450 mph in problem 8.4.

8.12. For the large jet transport in Examples 7.6.1 and 7.7.1, the airspeed is maintained at 450 mph while the aircraft is climbing at 2,000 ft/min. All other operating conditions remain unchanged. Find the eigenvalues, the eigenvectors, the 99 percent damping time, and the period, if underdamped, for each longitudinal mode. For each oscillatory mode, compute the amplitude and phase for each component of the eigenvector associated with the positive frequency. Compare the results with those obtained for level flight at 450 mph in problem 8.4.

8.13. For the Spitfire Mark XVIII and flight conditions described in problem 7.36, find the eigenvalues, the eigenvectors, the 99 percent damping time, and the period, if underdamped, for each longitudinal mode. For each oscillatory mode, compute the amplitude and phase for each component of the eigenvector associated with the positive frequency.

8.14. For the Spitfire Mark XVIII and flight conditions described in problem 7.36, using the short-period eigenvalue approximation given by Eq. (8.3.9), find the eigenvalues, the 99 percent damping time, and the period for the short-period mode. Compare these approximate results with the fully coupled short-period solution that was obtained in problem 8.13.

8.15. For the Spitfire Mark XVIII and flight conditions described in problem 7.36, using the short-period eigenvector approximation given by Eq. (8.3.5), find the eigenvectors for the short-period mode. Normalize these eigenvectors according to Eq. (8.1.37). Compare these approximate eigenvectors with the fully coupled short-period solution that was obtained in problem 8.13.

8.16. For the Spitfire Mark XVIII and flight conditions described in problem 7.36, using the phugoid eigenvalue approximation given by Eq. (8.4.8), find the dimensionless phugoid eigenvalues, the 99 percent damping time, and the period for the long-period mode. Compare these approximate results with the fully coupled long-period solution that was obtained in problem 8.13.

8.17. For the Spitfire Mark XVIII and flight conditions described in problem 7.36, using the phugoid eigenvector approximation given by Eq. (8.4.6), find the dimensionless phugoid eigenvectors for the long-period mode. Normalize these eigenvectors according to Eq. (8.1.37). Compare these approximate eigenvectors with those obtained from the fully coupled long-period solution that was obtained in problem 8.13.

8.18. For the Spitfire Mark XVIII and flight conditions described in problem 7.36, using the phugoid eigenvalue approximation given by Eq. (8.4.44), find the dimensionless phugoid eigenvalues, the 99 percent damping time, and the period for the long-period mode. Compare these approximate results with the fully coupled long-period solution that was obtained in problem 8.13.

8.19. For the Spitfire Mark XVIII and flight conditions described in problem 7.36, using the phugoid eigenvector approximation given by Eq. (8.4.27), find the dimensionless phugoid eigenvectors for the long-period mode. Normalize these eigenvectors according to Eq. (8.1.37). Compare these approximate eigenvectors with the fully coupled long-period solution that was obtained in problem 8.13.

8.20. For the Spitfire Mark XVIII described in problem 7.36, the level-flight airspeed is changed from 300 mph to 200 mph. All other operating conditions remain unchanged. Find the eigenvalues, the eigenvectors, the 99 percent damping time, and the period, if underdamped, for each longitudinal mode. For each oscillatory mode, compute the amplitude and phase for each component of the eigenvector associated with the positive frequency. Compare these results with those that were obtained for an airspeed of 300 mph in problem 8.13.

8.21. For the Spitfire Mark XVIII described in problem 7.36, the airspeed is maintained at 300 mph while the aircraft is climbing at 5,000 ft/min. All other operating conditions remain unchanged. Find the eigenvalues, the eigenvectors, the 99 percent damping time, and the period, if underdamped, for each longitudinal mode. For each oscillatory mode, compute the amplitude and phase for each component of the eigenvector associated with the positive frequency. Compare the results with those obtained for level flight at 300 mph in problem 8.13.

8.22. For the jet transport and flight conditions that are described in problem 7.37, find the eigenvalues, the eigenvectors, the 99 percent damping time, and the period, if underdamped, for each longitudinal mode. For each oscillatory mode, compute the amplitude and phase for each component of the eigenvector associated with the positive frequency.

8.23. For the jet transport and flight conditions that are described in problem 7.37, using the short-period eigenvalue approximation given by Eq. (8.3.9), find the eigenvalues, the 99 percent damping time, and the period for the short-period mode. Compare these approximate results with those obtained from the fully coupled short-period solution that was obtained in problem 8.22.

8.24. For the jet transport and flight conditions that are described in problem 7.37, using the short-period eigenvector approximation given by Eq. (8.3.5), find the eigenvectors for the short-period mode. Normalize these eigenvectors according to Eq. (8.1.37). Compare these approximate eigenvectors with the fully coupled short-period solution that was obtained in problem 8.22.

8.25. For the jet transport and flight conditions that are described in problem 7.37, using the phugoid eigenvalue approximation given by Eq. (8.4.8), find the dimensionless phugoid eigenvalues, the 99 percent damping time, and the period for the long-period mode. Compare these approximate results with the fully coupled long-period solution that was obtained in problem 8.22.

8.26. For the jet transport and flight conditions that are described in problem 7.37, using the phugoid eigenvector approximation given by Eq. (8.4.6), find the dimensionless phugoid eigenvectors for the long-period mode. Normalize these eigenvectors according to Eq. (8.1.37). Compare these approximate eigenvectors with those obtained from the fully coupled long-period solution that was obtained in problem 8.22.

8.27. For the jet transport and flight conditions that are described in problem 7.37, using the phugoid eigenvalue approximation given by Eq. (8.4.44), find the dimensionless phugoid eigenvalues, the 99 percent damping time, and the period for the long-period mode. Compare these approximate results with the fully coupled long-period solution that was obtained in problem 8.22.

8.28. For the jet transport and flight conditions that are described in problem 7.37, using the phugoid eigenvector approximation given by Eq. (8.4.27), find the dimensionless phugoid eigenvectors for the long-period mode. Normalize these eigenvectors according to Eq. (8.1.37). Compare these approximate eigenvectors with those obtained from the fully coupled long-period solution that was obtained in problem 8.22.

8.29. For the jet transport described in problem 7.37, assume that the pitch-stability and pitch-damping derivatives are doubled while all other parameters remain unchanged. Find the eigenvalues, the eigenvectors, the 99 percent damping time, and the period, if underdamped, for each longitudinal mode. For each oscillatory mode, compute the amplitude and phase for each component of the eigenvector associated with the positive frequency. Compare these results with those that were obtained in problem 8.22.

8.30. The jet transport described in problem 7.37 is climbing at 2,000 ft/min. Assume that the airspeed and all aerodynamic derivatives remain unchanged. Find the eigenvalues, the eigenvectors, the 99 percent damping time, and the period, if underdamped, for each longitudinal mode. For each oscillatory mode, compute the amplitude and phase for each component of the eigenvector associated with the positive frequency. Compare these results with those that were obtained for level flight in problem 8.22.

8.31. For the turboprop transport and flight conditions described in problem 7.38, find the eigenvalues, the eigenvectors, the 99 percent damping time, and the period, if underdamped, for each longitudinal mode. For each oscillatory mode, compute the amplitude and phase for each component of the eigenvector associated with the positive frequency.

8.32. For the turboprop transport and flight conditions described in problem 7.38, using the short-period eigenvalue approximation given by Eq. (8.3.9), find the eigenvalues, the 99 percent damping time, and the period for the short-period mode. Compare these approximate results with the fully coupled short-period solution that was obtained in problem 8.31.

8.33. For the turboprop transport and flight conditions described in problem 7.38, using the short-period eigenvector approximation given by Eq. (8.3.5), find the eigenvectors for the short-period mode. Normalize these eigenvectors according to Eq. (8.1.37). Compare these approximate eigenvectors with those from the exact solution for the fully coupled short-period mode, which was obtained in problem 8.31.

8.34. For the turboprop transport and flight conditions described in problem 7.38, using the phugoid eigenvalue approximation given by Eq. (8.4.8), find the dimensionless phugoid eigenvalues, the 99 percent damping time, and the period for the long-period mode. Compare these approximate results with the fully coupled long-period solution that was obtained in problem 8.31.

8.35. For the turboprop transport and flight conditions described in problem 7.38, using the phugoid eigenvector approximation given by Eq. (8.4.6), find the dimensionless phugoid eigenvectors for the long-period mode. Normalize these eigenvectors according to Eq. (8.1.37). Compare these approximate eigenvectors with those from the fully coupled long-period solution, which was obtained in problem 8.31.

8.36. For the turboprop transport and flight conditions described in problem 7.38, using the phugoid eigenvalue approximation given by Eq. (8.4.44), find the dimensionless phugoid eigenvalues, the 99 percent damping time, and the period for the long-period mode. Compare these approximate results with the fully coupled long-period solution that was obtained in problem 8.31.

8.37. For the turboprop transport and flight conditions described in problem 7.38, using the phugoid eigenvector approximation given by Eq. (8.4.27), find the dimensionless phugoid eigenvectors for the long-period mode. Normalize these eigenvectors according to Eq. (8.1.37). Compare these approximate eigenvectors with those from the fully coupled long-period solution, which was obtained in problem 8.31.

8.38. For the turboprop transport described in problem 7.38, assume that $R_{m,\mu}$ is doubled while all other aerodynamic derivatives and operating conditions remain unchanged. Find the eigenvalues, the eigenvectors, the 99 percent damping time, and the period, if underdamped, for each longitudinal mode. For each oscillatory mode, compute the amplitude and phase for each component of the eigenvector associated with the positive frequency. Compare these results with those that were obtained in problem 8.31.

8.39. The turboprop transport described in problem 7.38 is climbing at 5,000 ft/min. Assume that the airspeed and all aerodynamic derivatives remain unchanged. Find the eigenvalues, the eigenvectors, the 99 percent damping time, and the period, if underdamped, for each longitudinal mode. For each oscillatory mode, compute the amplitude and phase for each component of the eigenvector associated with the positive frequency. Compare these results with those obtained for level flight in problem 8.31.

8.40. Consider an object such as the airplane shown in Fig. 8.7.1 that is hung from two long parallel strings of lengths s_1 and s_2, which are separated by a distance Δ. The vertical axis passing through the center of gravity is located directly between the strings a distance r_1 from string 1 and a distance r_2 from string 2. As the object rotates through an angle θ about the vertical axis passing through the center of gravity, the lower attachment points for strings 1 and 2 move through arcs with horizontal chord-length projections of c_1 and c_2, respectively. At the beginning of the arcs, both strings are vertical. At the end of the arcs, the strings make small angles ϕ_1 and ϕ_2 with the vertical. Show that the restoring torque is

$$T = \frac{r_1 r_2 W}{\Delta}\left(\frac{r_1}{s_1 \cos \phi_1} + \frac{r_2}{s_2 \cos \phi_2}\right)\sin \theta = \frac{\Delta^2 W_1 W_2}{W^2}\left(\frac{W_2}{s_1 \cos \phi_1} + \frac{W_1}{s_2 \cos \phi_2}\right)\sin \theta$$

where W, W_1, and W_2 are the total weight and the vertical components of tension in each string, respectively. Also show that for long strings and small rotational oscillations, the natural frequency is

$$\omega_n = \sqrt{\bar{r} W / I_{yy_b}}$$

where

$$\bar{r} \equiv \frac{r_1 r_2}{\Delta}\left(\frac{r_1}{s_1} + \frac{r_2}{s_2}\right) = \frac{\Delta^2 W_1 W_2}{W^3}\left(\frac{W_2}{s_1} + \frac{W_1}{s_2}\right)$$

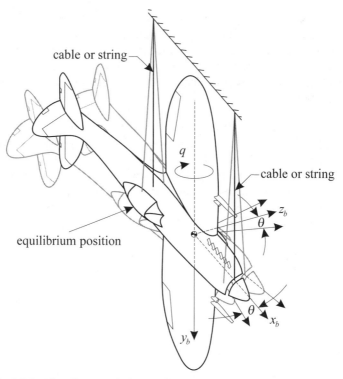

Fig. 8.7.1. Aircraft suspended as a bifilar pendulum relative to the pitch axis.

8.41. Consider the airplane shown in Fig. 8.5.2 that is hung from three parallel cables or strings of lengths s_1, s_2, and s_3. The vertical axis passing through the center of gravity is located between the strings a distance r_1 from string 1, a distance r_2 from string 2, and a distance r_3 from string 3. As the object rotates through an angle θ about the vertical axis passing through the center of gravity, the lower attachment points for strings 1, 2, and 3 move through arcs with horizontal chord-length projections of c_1, c_2, and c_3, respectively. At the beginning of the arcs, all three strings are vertical. At the end of the arcs, the strings make angles ϕ_1, ϕ_2, and ϕ_3 with the vertical. Show that the restoring torque is given by

$$T = \left(\frac{r_1^2 C_1}{s_1 \cos \phi_1} + \frac{r_2^2 C_2}{s_2 \cos \phi_2} + \frac{r_3^2 C_3}{s_3 \cos \phi_3} \right) W \sin \theta$$

where C_1, C_2, and C_3 are the vertical components of tension in each string divided by the weight, W. For long strings and small rotation angles, develop relations for C_1, C_2, and C_3 as a function of the coordinates of the string attachment points and center of gravity. Also for long strings and small rotation angles, develop relations for the coordinates of the center of gravity as a function of C_1, C_2, C_3, and the coordinates of the string attachment points.

Chapter 9
Linearized Lateral Dynamics

9.1. Introduction

If one understands the material presented in Chapter 8, it should be fairly easy to understand the material presented in this chapter, at least from an analytical point of view. Mathematically speaking, there is little difference between linearized lateral dynamics and linearized longitudinal dynamics. In each case we are dealing with an eigenproblem associated with a homogeneous system of six first-order linear differential equations. In both cases the six first-order equations were obtained from a second-order system representing uncontrolled motion with three degrees of freedom. It is a matter of little wonder that the homogeneous solutions for lateral aircraft dynamics have much in common with those for longitudinal dynamics. There are, however, some significant differences between the uncontrolled dynamic response of an aircraft to a lateral disturbance and its response to a similar longitudinal disturbance. In studying the material in this chapter, the student should strive to understand both the similarities and the differences between longitudinal and lateral aircraft dynamics.

A significant difference between lateral and longitudinal motion is that longitudinal motion is true two-dimensional motion, involving two translational degrees of freedom and one rotational degree of freedom. Lateral motion, however, is three-dimensional, involving two rotational degrees of freedom and one translational degree of freedom. Pure longitudinal motion can exist in free flight, even outside the approximation of small-disturbance theory. Because longitudinal motion is two-dimensional, there is no tendency for a longitudinal disturbance to induce lateral motion in a perfectly symmetric aircraft, even if the disturbance is large. On the other hand, recall that the lateral equations of motion were separated from the longitudinal equations only as a result of the small-disturbance approximation. Thus, lateral motion is only an approximation that can never truly exist in the world of finite disturbance. The actual response of an aircraft to a finite lateral disturbance will always involve all six degrees of freedom. Nevertheless, the lateral approximation to be studied in this chapter is very close to the response of a real aircraft for the disturbance magnitudes that are encountered in normal flight.

9.2. Lateral Motion: The Linearized Coupled Equations

The lateral dynamic response of an aircraft disturbed from equilibrium flight involves a complex combination of sideslip, roll, and yaw. These three degrees of freedom are very closely coupled. As we have already seen, sideslip produces both rolling and yawing moments, which in turn produce rolling and yawing rates. The rolling rate produces a significant yawing moment in addition to the rolling moment, and the yawing rate generates a side force and a rolling moment as well as a yawing moment. In general, the sideslip, roll, and yaw are also coupled to the longitudinal and elastic degrees of freedom. However, in this chapter we are restricting our discussion to the six degrees of freedom associated with rigid-body flight and we are confining the aircraft

885

motion to small deviations from steady equilibrium flight. For small deviations from equilibrium, the lateral motion is not coupled to the longitudinal motion and the lateral equations can be solved independently. In this section we examine the stick-fixed free response associated with linearized, lateral, rigid-body motion. Here again, these results provide a reasonable description of lateral aircraft dynamics provided that the aircraft is not undergoing large-amplitude oscillations or rapid maneuvers. We should also keep in mind that the separation of the longitudinal and lateral equations of motion depended on our neglecting the angular momentum of the spinning rotors in the aircraft's engines. In conventional airplanes, such gyroscopic effects can sometimes cause significant yaw-pitch coupling.

Since rigid-body lateral motion has three degrees of freedom (sideslip, roll, and yaw), it is characterized by six eigenvalues. The eigenvalues associated with the linearized lateral motion of a typical airplane contain two distinct real roots, one complex pair, and two identical zero roots. The real and distinct roots correspond to two nonoscillatory modes, one heavily damped motion called the *roll mode* and one lightly damped or divergent motion called the *spiral mode*. The complex pair describes a damped oscillatory motion that is called *Dutch roll*. The remaining two zero eigenvalues correspond to simple rigid-body displacement modes. The roll mode is so heavily damped that it is of no concern to pilots or passengers. As we shall see, the nonoscillatory spiral mode can be divergent even in an aircraft with a high degree of directional stability. However, in most aircraft, the spiral mode is convergent or so slowly divergent that it does not create significant problems for either the pilot or the designer. The Dutch roll mode is an out-of-phase combination of sideslipping, rolling, and yawing oscillations that can be very objectionable. In this section we examine these coupled lateral motions in detail.

The free response of a conventional airplane to small lateral disturbances is obtained from the linearized homogeneous system of lateral equations, with both lateral control inputs set to zero. From Eq. (7.7.6), the homogeneous linearized lateral equations can be written as

$$
\begin{bmatrix}
1 & 0 & 0 & 0 & 0 & 0 \\
0 & 1 & -\iota_{xz} & 0 & 0 & 0 \\
0 & -\iota_{zx} & 1 & 0 & 0 & 0 \\
0 & 0 & 0 & 1 & 0 & 0 \\
0 & 0 & 0 & 0 & 1 & 0 \\
0 & 0 & 0 & 0 & 0 & 1
\end{bmatrix}
\begin{Bmatrix}
\Delta\hat{\beta} \\
\Delta\hat{p} \\
\Delta\hat{r} \\
\Delta\hat{\xi}_y \\
\Delta\hat{\phi} \\
\Delta\hat{\psi}
\end{Bmatrix}
$$

$$
-\begin{bmatrix}
R_{y,\beta} & R_{y,\bar{p}} & (R_{y,\bar{r}}-1) & 0 & R_{gy}\cos\theta_0 & 0 \\
R_{\ell,\beta} & R_{\ell,\bar{p}} & R_{\ell,\bar{r}} & 0 & 0 & 0 \\
R_{n,\beta} & R_{n,\bar{p}} & R_{n,\bar{r}} & 0 & 0 & 0 \\
1 & 0 & 0 & 0 & 0 & \cos\theta_o \\
0 & 1 & \tan\theta_o & 0 & 0 & 0 \\
0 & 0 & \sec\theta_o & 0 & 0 & 0
\end{bmatrix}
\begin{Bmatrix}
\Delta\beta \\
\Delta p \\
\Delta r \\
\Delta\xi_y \\
\Delta\phi \\
\Delta\psi
\end{Bmatrix}
=
\begin{Bmatrix}
0 \\
0 \\
0 \\
0 \\
0 \\
0
\end{Bmatrix}
\tag{9.2.1}
$$

Here again, Eq. (9.2.1) exactly fits the form of the eigenvalue problem that was given by Eq. (8.1.28), and the **generalized eigenproblem for lateral motion** is

$$
\left(
\begin{bmatrix}
R_{y,\beta} & R_{y,\bar{p}} & (R_{y,\bar{r}}-1) & 0 & R_{gy}\cos\theta_0 & 0 \\
R_{\ell,\beta} & R_{\ell,\bar{p}} & R_{\ell,\bar{r}} & 0 & 0 & 0 \\
R_{n,\beta} & R_{n,\bar{p}} & R_{n,\bar{r}} & 0 & 0 & 0 \\
1 & 0 & 0 & 0 & 0 & \cos\theta_0 \\
0 & 1 & \tan\theta_0 & 0 & 0 & 0 \\
0 & 0 & \sec\theta_0 & 0 & 0 & 0
\end{bmatrix}
\right.
$$

$$
\left.
-\lambda
\begin{bmatrix}
1 & 0 & 0 & 0 & 0 & 0 \\
0 & 1 & -\iota_{xz} & 0 & 0 & 0 \\
0 & -\iota_{zx} & 1 & 0 & 0 & 0 \\
0 & 0 & 0 & 1 & 0 & 0 \\
0 & 0 & 0 & 0 & 1 & 0 \\
0 & 0 & 0 & 0 & 0 & 1
\end{bmatrix}
\right)
\begin{Bmatrix}
\chi_\beta \\
\chi_{\bar{p}} \\
\chi_{\bar{r}} \\
\chi_{\xi_y} \\
\chi_\phi \\
\chi_\psi
\end{Bmatrix}
=
\begin{Bmatrix}
0 \\ 0 \\ 0 \\ 0 \\ 0 \\ 0
\end{Bmatrix}
\tag{9.2.2}
$$

The eigenvalues and eigenvectors associated with linearized lateral aircraft motion can be found by solving Eq. (9.2.2) using a generalized eigensolver. If a generalized eigensolver is not available, the system can easily be rearranged to give the **special eigenproblem for lateral motion:**

$$
\left(
\begin{bmatrix}
R_{y,\beta} & R_{y,\bar{p}} & (R_{y,\bar{r}}-1) & 0 & R_{gy}\cos\theta_0 & 0 \\[2mm]
\dfrac{R_{\ell,\beta}+\iota_{xz}R_{n,\beta}}{1-\iota_{xz}\iota_{zx}} & \dfrac{R_{\ell,\bar{p}}+\iota_{xz}R_{n,\bar{p}}}{1-\iota_{xz}\iota_{zx}} & \dfrac{R_{\ell,\bar{r}}+\iota_{xz}R_{n,\bar{r}}}{1-\iota_{xz}\iota_{zx}} & 0 & 0 & 0 \\[2mm]
\dfrac{R_{n,\beta}+\iota_{zx}R_{\ell,\beta}}{1-\iota_{xz}\iota_{zx}} & \dfrac{R_{n,\bar{p}}+\iota_{zx}R_{\ell,\bar{p}}}{1-\iota_{xz}\iota_{zx}} & \dfrac{R_{n,\bar{r}}+\iota_{zx}R_{\ell,\bar{r}}}{1-\iota_{xz}\iota_{zx}} & 0 & 0 & 0 \\[2mm]
1 & 0 & 0 & 0 & 0 & \cos\theta_0 \\
0 & 1 & \tan\theta_0 & 0 & 0 & 0 \\
0 & 0 & \sec\theta_0 & 0 & 0 & 0
\end{bmatrix}
\right.
$$

$$
\left.
-\lambda
\begin{bmatrix}
1 & 0 & 0 & 0 & 0 & 0 \\
0 & 1 & 0 & 0 & 0 & 0 \\
0 & 0 & 1 & 0 & 0 & 0 \\
0 & 0 & 0 & 1 & 0 & 0 \\
0 & 0 & 0 & 0 & 1 & 0 \\
0 & 0 & 0 & 0 & 0 & 1
\end{bmatrix}
\right)
\begin{Bmatrix}
\chi_\beta \\
\chi_{\bar{p}} \\
\chi_{\bar{r}} \\
\chi_{\xi_y} \\
\chi_\phi \\
\chi_\psi
\end{Bmatrix}
=
\begin{Bmatrix}
0 \\ 0 \\ 0 \\ 0 \\ 0 \\ 0
\end{Bmatrix}
\tag{9.2.3}
$$

From Eq. (9.2.2) or Eq. (9.2.3), we obtain dimensionless eigenvalues. To obtain the dimensional eigenvalues in terms of time, the dimensionless eigenvalues must be divided by the reference time, $b_w/2V_o$.

Thus, the **damping rate for a lateral mode** is

$$\sigma = -\text{real}(\lambda)\frac{2V_o}{b_w} \tag{9.2.4}$$

and the **damped natural frequency for a lateral mode** is

$$\omega_d = \left|\text{imag}(\lambda)\right|\frac{2V_o}{b_w} \tag{9.2.5}$$

The **damping ratio** is

$$\zeta = -\frac{\lambda_1 + \lambda_2}{2\sqrt{\lambda_1\lambda_2}} \tag{9.2.6}$$

and the **undamped natural frequency for a lateral mode** is

$$\omega_n = \sqrt{\lambda_1\lambda_2}\,\frac{2V_o}{b_w} \tag{9.2.7}$$

EXAMPLE 9.2.1. For steady level flight at sea level, find the dimensionless eigenvalues and eigenvectors, the damping rate, the 99 percent damping time, and the damped natural frequency and period, if underdamped, for each lateral mode of the typical general aviation airplane described in Example 8.2.1. For each pair of eigenvalues, find the undamped natural frequency and the damping ratio. For the oscillatory mode, compute the amplitude and phase. The characteristics of this airplane are again listed below.

$$S_w = 185\,\text{ft}^2, \quad b_w = 33\,\text{ft}, \quad W = 2{,}800\,\text{lbf}, \quad V_o = 180\,\text{ft/sec}, \quad C_{Do} = 0.05,$$

$$I_{xx_b} = 1{,}000\,\text{slug·ft}^2, \quad I_{yy_b} = 3{,}000\,\text{slug·ft}^2, \quad I_{zz_b} = 3{,}500\,\text{slug·ft}^2,$$

$$I_{xz_b} = 30\,\text{slug·ft}^2,$$

$$C_{L,\alpha} = 4.40, \quad C_{D,\alpha} = 0.35, \quad C_{m,\alpha} = -0.68, \quad C_{L,\hat{\alpha}} = 1.60, \quad C_{m,\hat{\alpha}} = -4.35,$$

$$C_{Y,\beta} = -0.560, \quad C_{\ell,\beta} = -0.075, \quad C_{n,\beta} = 0.070,$$

$$C_{D,\bar{q}} \cong 0.0, \quad C_{L,\bar{q}} = 3.80, \quad C_{m,\bar{q}} = -9.95,$$

$$C_{Y,\bar{p}} \cong 0.0, \quad C_{\ell,\bar{p}} = -0.410, \quad C_{n,\bar{p}} = -0.0575,$$

$$C_{Y,\bar{r}} = 0.240, \quad C_{\ell,\bar{r}} = 0.105, \quad C_{n,\bar{r}} = -0.125$$

Solution. The additional parameters that affect the lateral eigenvalues and eigenvectors for this aircraft are:

$$\iota_{xz} \equiv I_{xz_b}/I_{xx_b} = 0.0300, \qquad \iota_{zx} \equiv I_{xz_b}/I_{zz_b} = 0.0086,$$

$$\frac{\rho S_w b_w}{4W/g} = 0.041679, \qquad \frac{\rho S_w b_w^3}{8I_{xx_b}} = 1.975306, \qquad \frac{\rho S_w b_w^3}{8I_{zz_b}} = 0.564373,$$

$$R_{y,\beta} \equiv \frac{\rho S_w b_w}{4W/g} C_{Y,\beta} = -0.0233, \qquad R_{\ell,\beta} \equiv \frac{\rho S_w b_w^3}{8I_{xx_b}} C_{\ell,\beta} = -0.1481,$$

$$R_{n,\beta} \equiv \frac{\rho S_w b_w^3}{8I_{zz_b}} C_{n,\beta} = 0.0395, \qquad R_{y,\bar{p}} \equiv \frac{\rho S_w b_w}{4W/g} C_{Y,\bar{p}} = 0.0,$$

$$R_{\ell,\bar{p}} \equiv \frac{\rho S_w b_w^3}{8I_{xx_b}} C_{\ell,\bar{p}} = -0.8099, \qquad R_{n,\bar{p}} \equiv \frac{\rho S_w b_w^3}{8I_{zz_b}} C_{n,\bar{p}} = -0.0325,$$

$$R_{y,\bar{r}} \equiv \frac{\rho S_w b_w}{4W/g} C_{Y,\bar{r}} = 0.0100, \qquad R_{\ell,\bar{r}} \equiv \frac{\rho S_w b_w^3}{8I_{xx_b}} C_{\ell,\bar{r}} = 0.2074,$$

$$R_{n,\bar{r}} \equiv \frac{\rho S_w b_w^3}{8I_{zz_b}} C_{n,\bar{r}} = -0.0705, \qquad R_{gy} \equiv \frac{g b_w}{2V_o^2} = 0.0164$$

For this level-flight equilibrium condition, the cosine of the equilibrium elevation angle is unity while the tangent is zero. Thus, for this aircraft and flight condition, the eigenproblem given by Eq. (9.2.2) becomes

$$\left(\begin{bmatrix} -0.0233 & 0 & -0.9900 & 0 & 0.0164 & 0 \\ -0.1481 & -0.8099 & 0.2074 & 0 & 0 & 0 \\ 0.0395 & -0.0325 & -0.0705 & 0 & 0 & 0 \\ 1 & 0 & 0 & 0 & 0 & 1 \\ 0 & 1 & 0 & 0 & 0 & 0 \\ 0 & 0 & 1 & 0 & 0 & 0 \end{bmatrix} \right.$$

$$\left. -\lambda \begin{bmatrix} 1 & 0 & 0 & 0 & 0 & 0 \\ 0 & 1 & -0.0300 & 0 & 0 & 0 \\ 0 & -0.0086 & 1 & 0 & 0 & 0 \\ 0 & 0 & 0 & 1 & 0 & 0 \\ 0 & 0 & 0 & 0 & 1 & 0 \\ 0 & 0 & 0 & 0 & 0 & 1 \end{bmatrix}\right) \begin{Bmatrix} \chi_\beta \\ \chi_{\bar{p}} \\ \chi_{\bar{r}} \\ \chi_y \\ \chi_\phi \\ \chi_\psi \end{Bmatrix} = \begin{Bmatrix} 0 \\ 0 \\ 0 \\ 0 \\ 0 \\ 0 \end{Bmatrix}$$

The solution to this characteristic equation yields one pair of distinct real eigenvalues, one complex pair of eigenvalues, and two eigenvalues that are identically zero. These roots describe two nonoscillatory modes, one oscillatory mode, and two rigid-body displacement modes that are associated with lateral motion. These modes are:

The roll mode $(\lambda_1 = -0.813797)$

$$
\begin{Bmatrix} \Delta\beta \\ \Delta p \\ \Delta \bar{r} \\ \Delta \xi_y \\ \Delta\phi \\ \Delta\psi \end{Bmatrix} = C_1 \begin{Bmatrix} -0.054317 \\ -0.629391 \\ -0.030559 \\ 0.020602 \\ 0.773401 \\ 0.037551 \end{Bmatrix} \exp(-0.813797\,\tau_y)
$$

The damping rate is

$$
\sigma = -\operatorname{real}(\lambda)\frac{2V_o}{b_w} = 0.813797\frac{2(180)}{33} = 8.877785 \text{ sec}^{-1}
$$

and the 99 percent damping time is

$$
99\% \text{ damping time} = \frac{\ln(0.01)}{-\sigma} = \frac{\ln(0.01)}{-8.877785 \text{ sec}^{-1}} = 0.519 \text{ sec}
$$

The spiral mode $(\lambda_2 = -0.000918)$

$$
\begin{Bmatrix} \Delta\beta \\ \Delta p \\ \Delta \bar{r} \\ \Delta \xi_y \\ \Delta\phi \\ \Delta\psi \end{Bmatrix} = C_2 \begin{Bmatrix} 0.000001 \\ 0.000000 \\ 0.000001 \\ 0.999999 \\ 0.000053 \\ -0.000920 \end{Bmatrix} \exp(-0.000918\,\tau_y)
$$

The damping rate and the 99 percent damping time are

$$
\sigma = -\operatorname{real}(\lambda)\frac{2V_o}{b_w} = 0.000918\frac{2(180)}{33} = 0.010015 \text{ sec}^{-1}
$$

$$
99\% \text{ damping time} = \frac{\ln(0.01)}{-\sigma} = \frac{\ln(0.01)}{-0.010015 \text{ sec}^{-1}} = 460 \text{ sec}
$$

The eigenvalues for the roll and spiral modes form a mathematical pair. For this roll-spiral pair, the undamped natural frequency is

$$\omega_n = \sqrt{\lambda_1 \lambda_2} \frac{2V_o}{b_w} = \sqrt{(-0.813797)(-0.000918)} \frac{2(180)}{33} = 0.298173 \text{ rad/sec}$$

and the damping ratio is

$$\zeta = -\frac{\lambda_1 + \lambda_2}{2\sqrt{\lambda_1 \lambda_2}} = -\frac{-0.813797 - 0.000918}{2\sqrt{(-0.813797)(-0.000918)}} = 14.9038$$

The Dutch roll mode $(\lambda_{3,4} = -0.044202 \pm 0.217908\,i)$

$$
\begin{Bmatrix} \Delta\beta \\ \Delta\bar{p} \\ \Delta\bar{r} \\ \Delta\xi_y \\ \Delta\phi \\ \Delta\psi \end{Bmatrix} = C_{3,4}
\begin{Bmatrix} 0.589580 \\ -0.109648 \mp 0.001442\,i \\ 0.013967 \mp 0.121744\,i \\ 0.172251 \mp 0.220690\,i \\ 0.091681 \pm 0.484586\,i \\ -0.549104 \pm 0.047290\,i \end{Bmatrix} \exp[(-0.044202 \pm 0.217908\,i)\,\tau_y]
$$

The damping rate for the Dutch roll mode is

$$\sigma = -\text{real}(\lambda)\frac{2V_o}{b_w} = 0.044202 \frac{2(180)}{33} = 0.482204 \text{ sec}^{-1}$$

The 99 percent damping time is

$$99\% \text{ damping time} = \frac{\ln(0.01)}{-\sigma} = \frac{\ln(0.01)}{-0.482204} - = 9.55 \text{ sec}$$

and the damping ratio for this mode is

$$\zeta = -\frac{\lambda_3 + \lambda_4}{2\sqrt{\lambda_3 \lambda_4}} = -\frac{-0.044202 + 0.217908\,i - 0.044202 - 0.217908\,i}{2\sqrt{(-0.044202 + 0.217908\,i)(-0.044202 - 0.217908\,i)}}$$

$$= \frac{0.044202}{\sqrt{0.044202^2 + 0.217908^2}} = 0.198798$$

The damped natural frequency for the Dutch roll oscillations is

$$\omega_d = |\text{imag}(\lambda)|\frac{2V_o}{b_w} = 0.217908 \frac{2(180)}{33} = 2.377178 \text{ rad/sec}$$

and the period is

$$\text{period} = \frac{2\pi}{\omega_d} = \frac{2\pi}{2.377178} = 2.64 \text{ sec}$$

The undamped natural frequency for the Dutch roll mode is

$$\omega_n = \sqrt{\lambda_3 \lambda_4} \frac{2V_o}{b_w}$$

$$= \sqrt{(-0.044202 + 0.217908\, i)(-0.044202 - 0.217908\, i)} \frac{2(180)}{33}$$

$$= \sqrt{0.044202^{\,2} + 0.217908^{\,2}} \frac{2(180)}{33} = 2.425592 \text{ rad/sec}$$

Notice that for this airplane and these operating conditions, damping reduces the Dutch roll frequency by about 2 percent.

The rigid-body modes $(\lambda_{5,6} = 0.0, 0.0)$

$$\begin{Bmatrix} \Delta\beta \\ \Delta\overline{p} \\ \Delta\overline{r} \\ \Delta\xi_y \\ \Delta\phi \\ \Delta\psi \end{Bmatrix} = C_5 \begin{Bmatrix} 0 \\ 0 \\ 0 \\ 1 \\ 0 \\ 0 \end{Bmatrix} \exp(0.0\, \tau_y), \; C_6 \begin{Bmatrix} 0 \\ 0 \\ 0 \\ 0 \\ 0 \\ 1 \end{Bmatrix} \exp(0.0\, \tau_y)$$

The trivial rigid-body displacement modes simply state the fact that steady level flight can be maintained, with no change in dynamic characteristics, at different longitudes and/or at different headings. These modes arise from the fact that the forces and moments are independent of ξ_y and ψ. The eigenvectors for the nontrivial modes are summarized in Table 9.2.1.

Eigenvector Component	Roll Mode Amplitude	Spiral Mode Amplitude	Dutch Roll Amplitude	Dutch Roll Phase
$\Delta\beta$	0.054317	0.000001	0.589580	0.00°
$\Delta\overline{p}$	0.629391	0.000000	0.109657	−179.04°
$\Delta\overline{r}$	0.030559	0.000001	0.122543	−83.36°
$\Delta\xi_y$	0.020602	0.999999	0.279954	−51.97°
$\Delta\phi$	0.773401	0.000053	0.493182	79.20°
$\Delta\psi$	0.037551	0.000920	0.551136	174.88°

Table 9.2.1. Lateral eigenvectors for a typical general aviation airplane.

The lateral modes observed in Example 9.2.1 are quite typical of the small-amplitude lateral motions associated with any conventional airplane in flight. While the damping associated with all of the nontrivial modes and the period associated with Dutch roll can vary with aircraft design and operating conditions, all conventional airplanes exhibit these three modes of lateral motion.

In this example we see that the roll mode is a heavily overdamped motion, which has a rolling rate that is much greater than the yawing rate. The significance of the roll mode is best explained in connection with the airplane's response to aileron deflection. Recall that the solutions given in Example 9.2.1 are homogeneous solutions, obtained from the linearized lateral equations with all control inputs set to zero. The aileron deflection enters the lateral equations of motion as a nonhomogeneity. The solution to any nonhomogeneous linear system of equations is obtained by adding a particular solution to the homogeneous solution. The particular solution is a steady-state solution. If the motion is convergent, after some time the homogeneous solution approaches zero and the nonhomogeneous solution approaches the particular solution. The roll response of a typical airplane to a step change in aileron deflection is shown in Fig. 9.2.1. In this figure the initial bank angle and initial rolling rate are both zero. At time $t = 0$, a step change in deflection is applied to the ailerons. Notice that as time increases, the rolling rate approaches a constant value. The time required for the rolling motion to approach a constant rate is determined by the roll mode eigenvalue. Since the roll mode for typical fixed-wing aircraft is overdamped and highly convergent, transients in roll are very quickly damped out. Thus, when an airplane responds to a change in aileron input, the transient time is very short and the airplane quickly approaches a steady rolling rate.

The spiral mode is characterized primarily by changes in heading and direction of travel. For conventional airplanes, this root is real but is usually only slowly convergent or even somewhat divergent. A slightly divergent spiral mode is acceptable provided that the time required to double the amplitude is not too short. The uncontrolled motion resulting from a divergent spiral mode is shown in Fig. 9.2.2. Since the lateral motion is uncoupled from the longitudinal motion only for small deviations from equilibrium,

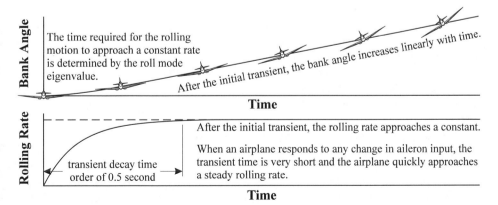

Figure 9.2.1. Roll response to a step change in aileron deflection.

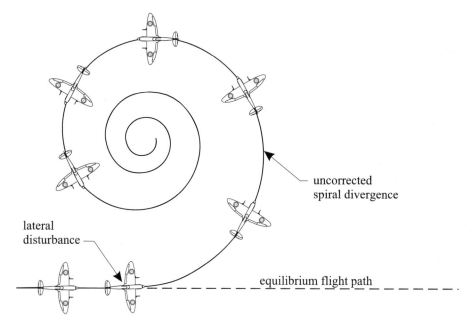

Figure 9.2.2. Spiral divergence.

as the spiral deviation becomes large, longitudinal motion is also induced. When an airplane with a divergent spiral mode is disturbed laterally from trimmed flight, the airplane will begin a slow spiral in the direction of the disturbance. If uncorrected, this spiral will continue to tighten as the airplane gradually enters an ever-tightening high-speed spiral dive. However, since the spiral mode is normally only slowly divergent, the time required to double the amplitude is long compared to the reaction time of either a VFR or an IFR pilot. Thus, even though the airplane is unstable, the pilot will usually not be aware of this instability and will instinctively provide the control inputs necessary to maintain straight and level flight.

Although an airplane with spiral instability is not considered dangerous in the hands of a qualified pilot, spiral divergence can result in extreme attitude changes and even structural failure if for some reason pilot control is absent for a few minutes. If a pilot without proper working instruments and the ability to use them should fly into a cloud, he or she would lose all visual reference with which to control the airplane. In this situation spiral divergence will prove catastrophic if the airplane does not emerge from the cloud very quickly.

While the spiral mode and the roll mode seem very different in character, they are mathematically quite closely related. Since the equations of motion are second order in each of the six degrees of freedom, the eigenvalues always occur in pairs. The eigenvalues for the roll and spiral modes together form one of these pairs. Because complex eigenvalues can occur only in complex pairs, if one of these roots is real, they both must be real. The large wing area of conventional airplanes inherently produces ample roll damping, causing the eigenvalues for both the roll and spiral modes to be real. However, some modern aircraft have very little wing area and the roll damping for such aircraft can be quite low. For aircraft of this type, it is occasionally possible for the eigenvalues associated with the usual roll and spiral modes to combine, forming a complex pair. In this case, the aircraft will exhibit a long-period lateral oscillation similar to the phugoid mode associated with longitudinal dynamics.

Dutch roll is the lateral counterpart to the short-period mode associated with longitudinal dynamics. As shown in Fig. 9.2.3, Dutch roll is a damped oscillatory motion that is characterized by a combination of rolling, yawing, and sideslip. This yawing and rolling motion is similar to the rolling and twisting gait of an ice skater. Hence, the mode derives its name from the country well known for ice-skating. Here again, the lateral equations become coupled to the longitudinal equations for all but very small lateral oscillations. Thus, when the Dutch roll amplitude becomes large, longitudinal motions are also induced. Since the period for this motion is typically on the order of a few seconds, Dutch roll motion can be very annoying to both passengers and pilots. For well-designed airplanes, Dutch roll is not objectionable because the damping is large. However, this motion can be quite noticeable in flight and has been very objectionable in some aircraft, because of light damping. When flying an airplane with a lightly damped Dutch roll mode, the pilot will experience an unpleasant feel to the aircraft. Because of the large yaw component in the Dutch roll oscillations, this feel sometimes gives the impression that the tail of the airplane is trying to pass the nose. If the Dutch roll damping is extremely light, the airplane can be quite difficult to handle and very uncomfortable for passengers.

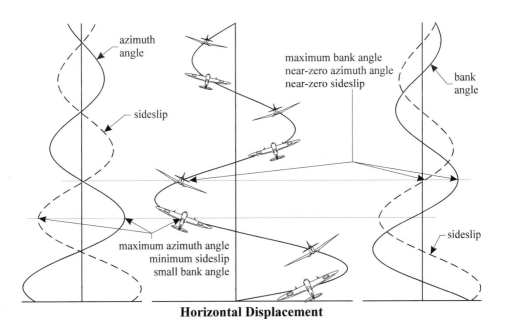

Horizontal Displacement

Figure 9.2.3. Dutch roll mode.

9.3. Roll Approximation

For the roll mode, the sideslip angle and yawing rate are small compared to the rolling rate. Thus, an approximation for the roll mode can be obtained by neglecting the rolling and yawing moments that result from the sideslip and yawing rate compared to those generated from the rolling rate. Since the yawing rate is much less than the rolling rate, we can also neglect the x-z product of inertia in the x-component of the angular momentum equation. The side force derivatives with respect to sideslip and yawing rate are also neglected, since they have little effect on the roll motion. Using these approximations for motion relative to steady level flight, Eq. (9.2.1) can be written

$$
\begin{bmatrix}
1 & 0 & 0 & 0 & 0 & 0 \\
0 & 1 & 0 & 0 & 0 & 0 \\
0 & -\iota_{zx} & 1 & 0 & 0 & 0 \\
0 & 0 & 0 & 1 & 0 & 0 \\
0 & 0 & 0 & 0 & 1 & 0 \\
0 & 0 & 0 & 0 & 0 & 1
\end{bmatrix}
\begin{Bmatrix}
\Delta\hat{\beta} \\
\Delta\hat{p} \\
\Delta\hat{r} \\
\Delta\hat{\xi}_y \\
\Delta\hat{\phi} \\
\Delta\hat{\psi}
\end{Bmatrix}
=
\begin{bmatrix}
0 & R_{y,\bar{p}} & -1 & 0 & R_{gy} & 0 \\
0 & R_{\ell,\bar{p}} & 0 & 0 & 0 & 0 \\
0 & R_{n,\bar{p}} & 0 & 0 & 0 & 0 \\
1 & 0 & 0 & 0 & 0 & 1 \\
0 & 1 & 0 & 0 & 0 & 0 \\
0 & 0 & 1 & 0 & 0 & 0
\end{bmatrix}
\begin{Bmatrix}
\Delta\beta \\
\Delta\bar{p} \\
\Delta\bar{r} \\
\Delta\xi_y \\
\Delta\phi \\
\Delta\psi
\end{Bmatrix}
\tag{9.3.1}
$$

The second equation in Eq. (9.3.1) is uncoupled from the remaining five,

$$
\Delta\hat{p} = R_{\ell,\bar{p}}\Delta\bar{p}
\tag{9.3.2}
$$

and can be solved directly to give

the dimensionless **eigenvalue for the roll mode approximation,**

$$
\lambda_r \cong R_{\ell,\bar{p}}
\tag{9.3.3}
$$

and the approximate **roll mode damping rate,**

$$
\sigma_r \cong -\frac{2V_o}{b_w}R_{\ell,\bar{p}} = -\frac{2V_o}{b_w}\left(\frac{\rho S_w b_w^3}{8 I_{xx_b}}C_{\ell,\bar{p}}\right) = -\frac{\rho S_w b_w^2 V_o}{4 I_{xx_b}}C_{\ell,\bar{p}}
\tag{9.3.4}
$$

Since the eigenvalue is now known, the remaining five equations in Eq. (9.3.1) can be written in algebraic form,

$$
\begin{Bmatrix}
\lambda\Delta\beta \\
\lambda\Delta\bar{r} - \iota_{zx}\lambda\Delta\bar{p} \\
\lambda\Delta\xi_y \\
\lambda\Delta\phi \\
\lambda\Delta\psi
\end{Bmatrix}
=
\begin{bmatrix}
0 & R_{y,\bar{p}} & -1 & R_{gy} & 0 \\
0 & R_{n,\bar{p}} & 0 & 0 & 0 \\
1 & 0 & 0 & 0 & 1 \\
0 & 1 & 0 & 0 & 0 \\
0 & 0 & 1 & 0 & 0
\end{bmatrix}
\begin{Bmatrix}
\Delta\beta \\
\Delta\bar{p} \\
\Delta\bar{r} \\
\Delta\phi \\
\Delta\psi
\end{Bmatrix}
\tag{9.3.5}
$$

Equation (9.3.5) can be rearranged to yield

$$
\left\{\begin{array}{c} \Delta\beta \\ \Delta\bar{p} \\ \Delta\bar{r} \\ \Delta\xi_y \\ \Delta\phi \\ \Delta\psi \end{array}\right\} = \left\{\begin{array}{c} \left[R_{y,\bar{p}} - I_{zx} + \left(R_{gy} - R_{n,\bar{p}}\right)/\lambda\right] \\ \lambda \\ I_{zx}\lambda + R_{n,\bar{p}} \\ R_{y,\bar{p}}/\lambda + R_{gy}/\lambda^2 \\ I_{zx} + R_{n,\bar{p}}/\lambda \end{array}\right\} \Delta\phi \qquad (9.3.6)
$$

The roll mode eigenvalue is determined from Eq. (9.3.3), and the eigenvector is found from Eq. (9.3.6).

The roll approximation given by Eq. (9.3.3) agrees closely with the exact solution for the roll mode obtained from Eq. (9.2.2). For the typical general aviation airplane in Example 9.2.1, this roll mode approximation results in

$$
\left\{\begin{array}{c} \Delta\beta \\ \Delta\bar{p} \\ \Delta\bar{r} \\ \Delta\xi_y \\ \Delta\phi \\ \Delta\psi \end{array}\right\} = C \left\{\begin{array}{c} -0.053453 \\ -0.627612 \\ -0.030583 \\ 0.019375 \\ 0.774925 \\ 0.037761 \end{array}\right\} \exp(-0.809900\,\tau_y)
$$

The details of obtaining this result are left as an exercise for the reader. The exact roll mode solution obtained in Example 9.2.1 is

$$
\left\{\begin{array}{c} \Delta\beta \\ \Delta\bar{p} \\ \Delta\bar{r} \\ \Delta\xi_y \\ \Delta\phi \\ \Delta\psi \end{array}\right\} = C_1 \left\{\begin{array}{c} -0.054317 \\ -0.629391 \\ -0.030559 \\ 0.020602 \\ 0.773401 \\ 0.037551 \end{array}\right\} \exp(-0.813797\,\tau_y)
$$

For this airplane, the approximate roll mode eigenvalue is accurate to within one-half of 1 percent, which is typical of what can be expected for most conventional airplanes.

9.4. Spiral Approximation

The spiral mode is characterized primarily by changes in heading and direction of travel. This mode is either slowly convergent or slowly divergent. Because the magnitude of the eigenvalue is so small, the change in heading occurs very slowly, with very little sideslip. The rolling and yawing rates also remain very low during spiral motion. The usual approximation for the spiral mode is obtained by neglecting the product of inertia, rolling rate, bank angle, and sideslip momentum equation. With these approximations, Eq. (9.2.1) reduces to

$$
\begin{bmatrix}
0 & 0 & 0 & 0 & 0 & 0 \\
0 & 1 & 0 & 0 & 0 & 0 \\
0 & 0 & 1 & 0 & 0 & 0 \\
0 & 0 & 0 & 1 & 0 & 0 \\
0 & 0 & 0 & 0 & 1 & 0 \\
0 & 0 & 0 & 0 & 0 & 1
\end{bmatrix}
\begin{Bmatrix}
0 \\
0 \\
\Delta\hat{\bar{r}} \\
\Delta\hat{\xi}_y \\
0 \\
\Delta\hat{\psi}
\end{Bmatrix}
=
\begin{bmatrix}
0 & 1 & 0 & 0 & 0 & 0 \\
R_{\ell,\beta} & 0 & R_{\ell,\bar{r}} & 0 & 0 & 0 \\
R_{n,\beta} & 0 & R_{n,\bar{r}} & 0 & 0 & 0 \\
1 & 0 & 0 & 0 & 0 & 1 \\
0 & 0 & 0 & 0 & 1 & 0 \\
0 & 0 & 1 & 0 & 0 & 0
\end{bmatrix}
\begin{Bmatrix}
\Delta\beta \\
\Delta\bar{p} \\
\Delta\bar{r} \\
\Delta\xi_y \\
\Delta\phi \\
\Delta\psi
\end{Bmatrix}
\tag{9.4.1}
$$

Using the second equation in Eq. (9.4.1) to eliminate the sideslip angle from the third equations, Eq. (9.4.1) becomes

$$
\begin{Bmatrix}
0 \\
0 \\
\Delta\hat{\bar{r}} \\
\Delta\hat{\xi}_y \\
0 \\
\Delta\hat{\psi}
\end{Bmatrix}
=
\begin{bmatrix}
0 & 1 & 0 & 0 & 0 & 0 \\
R_{\ell,\beta} & 0 & R_{\ell,\bar{r}} & 0 & 0 & 0 \\
0 & 0 & \left(\dfrac{R_{\ell,\beta}R_{n,\bar{r}} - R_{\ell,\bar{r}}R_{n,\beta}}{R_{\ell,\beta}}\right) & 0 & 0 & 0 \\
1 & 0 & 0 & 0 & 0 & 1 \\
0 & 0 & 0 & 0 & 1 & 0 \\
0 & 0 & 1 & 0 & 0 & 0
\end{bmatrix}
\begin{Bmatrix}
\Delta\beta \\
\Delta\bar{p} \\
\Delta\bar{r} \\
\Delta\xi_y \\
\Delta\phi \\
\Delta\psi
\end{Bmatrix}
\tag{9.4.2}
$$

The third equation in Eq. (9.4.2) can be solved directly for the characteristic root,

$$
\lambda_s \cong \frac{R_{\ell,\beta}R_{n,\bar{r}} - R_{\ell,\bar{r}}R_{n,\beta}}{R_{\ell,\beta}}
\tag{9.4.3}
$$

The remaining equations in Eq. (9.4.2) can be combined with Eq. (9.4.3) to obtain the spiral mode eigenvector for this approximation. Rearranging Eq. (9.4.2), we have

$$
\begin{Bmatrix}
\Delta\beta \\
\Delta\bar{p} \\
\Delta\bar{r} \\
\Delta\xi_y \\
\Delta\phi
\end{Bmatrix}
=
\begin{Bmatrix}
-R_{\ell,\bar{r}}\lambda/R_{\ell,\beta} \\
0 \\
\lambda \\
1/\lambda - R_{\ell,\bar{r}}/R_{\ell,\beta} \\
0
\end{Bmatrix}
\Delta\psi
\tag{9.4.4}
$$

Using the definitions from Eq. (7.7.8) in Eq. (9.4.3), the dimensionless spiral mode eigenvalue for this approximation can be written

$$
\lambda_s \cong \frac{\rho S_w b_w^3}{8 I_{zz_b}}\left(\frac{C_{\ell,\beta}C_{n,\bar{r}} - C_{\ell,\bar{r}}C_{n,\beta}}{C_{\ell,\beta}}\right)
\tag{9.4.5}
$$

and the approximate spiral mode damping rate is

$$\sigma_s \cong -\frac{2V_o}{b_w}\lambda_s = \frac{\rho S_w b_w^2 V_o}{4I_{zz_b}}\left(\frac{C_{\ell,\bar{r}}C_{n,\beta} - C_{\ell,\beta}C_{n,\bar{r}}}{C_{\ell,\beta}}\right) \tag{9.4.6}$$

Equation (9.4.5) has been widely used as a spiral mode approximation. However, the accuracy of this approximation is very poor. The eigenvalue and some components of the eigenvector obtained from this approximation can be in error by more than two orders of magnitude. Furthermore, Eq. (9.4.6) predicts a damping rate that is directly proportional to airspeed, when in fact, the damping rate for spiral motion decreases with airspeed. A more realistic closed-form approximation for the spiral mode can be obtained by retaining the sideslip momentum equation and including the effects of rolling rate and bank angle.

While changes in rolling rate and bank angle are significant for spiral motion, the damping rate is very low, so all changes occur very slowly. Thus, for the spiral mode approximation, we neglect all acceleration terms in the lateral equations of motion,

$$\Delta\hat{\beta} \cong \Delta\hat{p} \cong \Delta\hat{r} \cong 0 \tag{9.4.7}$$

The side force derivatives with respect to sideslip, rolling rate, and yawing rate are also neglected in the spiral approximation, since they have little effect on spiral motion. Using these approximations for lateral motion relative to steady level flight, Eq. (9.2.1) reduces to

$$\begin{bmatrix} 1 & 0 & 0 & 0 & 0 & 0 \\ 0 & 1 & -l_{xz} & 0 & 0 & 0 \\ 0 & -l_{zx} & 1 & 0 & 0 & 0 \\ 0 & 0 & 0 & 1 & 0 & 0 \\ 0 & 0 & 0 & 0 & 1 & 0 \\ 0 & 0 & 0 & 0 & 0 & 1 \end{bmatrix}\begin{Bmatrix} 0 \\ 0 \\ 0 \\ \Delta\hat{\xi}_y \\ \Delta\hat{\phi} \\ \Delta\hat{\psi} \end{Bmatrix} = \begin{bmatrix} 0 & 0 & -1 & 0 & R_{gy} & 0 \\ R_{\ell,\beta} & R_{\ell,\bar{p}} & R_{\ell,\bar{r}} & 0 & 0 & 0 \\ R_{n,\beta} & R_{n,\bar{p}} & R_{n,\bar{r}} & 0 & 0 & 0 \\ 1 & 0 & 0 & 0 & 0 & 1 \\ 0 & 1 & 0 & 0 & 0 & 0 \\ 0 & 0 & 1 & 0 & 0 & 0 \end{bmatrix}\begin{Bmatrix} \Delta\beta \\ \Delta\bar{p} \\ \Delta\bar{r} \\ \Delta\xi_y \\ \Delta\phi \\ \Delta\psi \end{Bmatrix} \tag{9.4.8}$$

These approximate lateral equations contain only three differential equations. The remaining three equations are algebraic. These three algebraic equations can be combined to express the rolling rate in terms of the bank angle. This result can be used in the right-hand side of the fifth equation in Eq. (9.4.8) to obtain a single first-order differential equation containing only one dependent variable, the bank angle,

$$\Delta\hat{\phi} = -R_{gy}\left(\frac{R_{\ell,\beta}R_{n,\bar{r}} - R_{\ell,\bar{r}}R_{n,\beta}}{R_{\ell,\beta}R_{n,\bar{p}} - R_{\ell,\bar{p}}R_{n,\beta}}\right)\Delta\phi \tag{9.4.9}$$

Equation (9.4.9) can be used to obtain the spiral mode eigenvalue and the remaining five equations in Eq. (9.4.8) can be rearranged to obtain the eigenvector,

$$\begin{Bmatrix} \Delta\beta \\ \Delta\bar{p} \\ \Delta\bar{r} \\ \Delta\xi_y \\ \Delta\psi \end{Bmatrix} = \begin{Bmatrix} -\left(R_{n,\bar{p}}\lambda + R_{n,\bar{r}}R_{gy}\right)/R_{n,\beta} \\ \lambda \\ R_{gy} \\ R_{gy}/\lambda^2 - \left(R_{n,\bar{p}}\lambda + R_{n,\bar{r}}R_{gy}\right)/\left(R_{n,\beta}\lambda\right) \\ R_{gy}/\lambda \end{Bmatrix} \Delta\phi \qquad (9.4.10)$$

From Eq. (9.4.9) the **approximate dimensionless spiral mode eigenvalue** is

$$\lambda_s \cong -R_{gy}\left(\frac{R_{\ell,\beta}R_{n,\bar{r}} - R_{\ell,\bar{r}}R_{n,\beta}}{R_{\ell,\beta}R_{n,\bar{p}} - R_{\ell,\bar{p}}R_{n,\beta}}\right) = -\frac{gb_w}{2V_o^2}\left(\frac{C_{\ell,\beta}C_{n,\bar{r}} - C_{\ell,\bar{r}}C_{n,\beta}}{C_{\ell,\beta}C_{n,\bar{p}} - C_{\ell,\bar{p}}C_{n,\beta}}\right) \qquad (9.4.11)$$

and the **approximate spiral mode damping rate** is

$$\sigma_s \cong -\frac{2V_o}{b_w}\lambda_s = \frac{g}{V_o}\left(\frac{C_{\ell,\beta}C_{n,\bar{r}} - C_{\ell,\bar{r}}C_{n,\beta}}{C_{\ell,\beta}C_{n,\bar{p}} - C_{\ell,\bar{p}}C_{n,\beta}}\right) \qquad (9.4.12)$$

The spiral mode approximation given by Eq. (9.4.11) agrees with the exact solution obtained from Eq. (9.2.2) to within about 10 percent. For the typical general aviation airplane in Example 9.2.1, obtaining the eigenvalue from Eq. (9.4.11) and using the result in Eq. (9.4.10) to obtain the eigenvector results in

$$\begin{Bmatrix} \Delta\beta \\ \Delta\bar{p} \\ \Delta\bar{r} \\ \Delta\xi_y \\ \Delta\phi \\ \Delta\psi \end{Bmatrix} = C\begin{Bmatrix} 0.000002 \\ 0.000000 \\ 0.000001 \\ 0.999999 \\ 0.000061 \\ -0.001004 \end{Bmatrix} \exp(-0.001002\,\tau_y)$$

The exact solution for the spiral mode from Example 9.2.1 is

$$\begin{Bmatrix} \Delta\beta \\ \Delta\bar{p} \\ \Delta\bar{r} \\ \Delta\xi_y \\ \Delta\phi \\ \Delta\psi \end{Bmatrix} = C_2\begin{Bmatrix} 0.000001 \\ 0.000000 \\ 0.000001 \\ 0.999999 \\ 0.000053 \\ -0.000920 \end{Bmatrix} \exp(-0.000918\,\tau_y)$$

For this same airplane, the spiral mode approximation that is given by Eq. (9.4.5) yields the very different result

$$
\begin{Bmatrix} \Delta\beta \\ \Delta\bar{p} \\ \Delta\bar{r} \\ \Delta\xi_y \\ \Delta\phi \\ \Delta\psi \end{Bmatrix} = C \begin{Bmatrix} 0.000330 \\ 0.000000 \\ 0.000236 \\ 0.999880 \\ 0.000000 \\ -0.015512 \end{Bmatrix} \exp(-0.015184\,\tau_y)
$$

The details of this numerical comparison are left as an exercise for the reader.

Clearly, the spiral mode approximation expressed in Eq. (9.4.11) represents a tremendous improvement over Eq. (9.4.5). The most serious inaccuracy in the approximations associated with obtaining Eq. (9.4.5) is neglecting the effects of bank angle. The bank angle component of the eigenvector for spiral motion is one to two orders of magnitude larger than either the sideslip angle or the dimensionless yawing rate, both of which were retained in the earlier approximation. In fact, the change in bank angle provides the primary driving force associated with spiral motion.

Comparing spiral mode damping rate from Eq. (9.4.6) with that from Eq. (9.4.12) shows some similarity in dependence on stability and damping derivatives. However, these two approximations show a very different dependence on other aircraft design and operating parameters. The spiral approximation given by Eq. (9.4.6) predicts a damping rate that is proportional to airspeed, whereas Eq. (9.4.12) shows an inverse relation between airspeed and spiral mode damping rate. Equation (9.4.6) predicts a spiral mode damping rate that is directly proportional to air density, wing area, and the square of the wing span, while being inversely proportional to the yawing moment of inertia. On the other hand, Eq. (9.4.12) predicts no dependence on any of these parameters. Results obtained from Eq. (9.4.12) are much more consistent with reality.

EXAMPLE 9.4.1. The general aviation airplane described in Example 9.2.1 is modified so that the directional stability is doubled,

$$
C_{n,\beta} = +0.140
$$

All other parameters affecting lateral motion are unchanged. For this modified airplane, use Eq. (9.2.2) to find the eigenvalue and eigenvector corresponding to the spiral mode. Compare this exact solution with the spiral mode approximation given by Eq. (9.4.11).

Solution. For the modified airplane, we have

$$
R_{\ell,\beta} \equiv \frac{\rho S_w b_w^3}{8 I_{xx_b}} C_{\ell,\beta} = -0.1481, \qquad R_{n,\beta} \equiv \frac{\rho S_w b_w^3}{8 I_{zz_b}} C_{n,\beta} = 0.0790,
$$

$$
R_{\ell,\bar{p}} \equiv \frac{\rho S_w b_w^3}{8 I_{xx_b}} C_{\ell,\bar{p}} = -0.8099, \qquad R_{n,\bar{p}} \equiv \frac{\rho S_w b_w^3}{8 I_{zz_b}} C_{n,\bar{p}} = -0.0325,
$$

$$R_{\ell,\bar{r}} \equiv \frac{\rho S_w b_w^3}{8 I_{xx_b}} C_{\ell,\bar{r}} = 0.2074, \qquad R_{n,\bar{r}} \equiv \frac{\rho S_w b_w^3}{8 I_{zz_b}} C_{n,\bar{r}} = -0.0705,$$

$$R_{gy} \equiv \frac{g b_w}{2 V_o^2} = 0.0164$$

With this design change, the eigenproblem given by Eq. (9.2.2) is the same as that in Example 9.2.1, with the exception that the first term in the third row of the first matrix is doubled,

$$\left(\begin{bmatrix} -0.0233 & 0 & -0.9900 & 0 & 0.0164 & 0 \\ -0.1481 & -0.8099 & 0.2074 & 0 & 0 & 0 \\ 0.0790 & -0.0325 & -0.0705 & 0 & 0 & 0 \\ 1 & 0 & 0 & 0 & 0 & 1 \\ 0 & 1 & 0 & 0 & 0 & 0 \\ 0 & 0 & 1 & 0 & 0 & 0 \end{bmatrix}\right.$$

$$\left. - \lambda \begin{bmatrix} 1 & 0 & 0 & 0 & 0 & 0 \\ 0 & 1 & -0.0300 & 0 & 0 & 0 \\ 0 & -0.0086 & 1 & 0 & 0 & 0 \\ 0 & 0 & 0 & 1 & 0 & 0 \\ 0 & 0 & 0 & 0 & 1 & 0 \\ 0 & 0 & 0 & 0 & 0 & 1 \end{bmatrix}\right) \begin{Bmatrix} \chi_\beta \\ \chi_{\bar{p}} \\ \chi_{\bar{r}} \\ \chi_{\xi_y} \\ \chi_\phi \\ \chi_\psi \end{Bmatrix} = \begin{Bmatrix} 0 \\ 0 \\ 0 \\ 0 \\ 0 \\ 0 \end{Bmatrix}$$

and the spiral mode solution is

$$\begin{Bmatrix} \Delta\beta \\ \Delta\bar{p} \\ \Delta\bar{r} \\ \Delta\xi_y \\ \Delta\phi \\ \Delta\psi \end{Bmatrix} = C_2 \begin{Bmatrix} 0.000002 \\ 0.000000 \\ 0.000002 \\ 0.999999 \\ 0.000112 \\ 0.001346 \end{Bmatrix} \exp(+0.001348\,\tau_y)$$

Using the parameters for the modified aircraft in Eq. (9.4.11), the approximate spiral mode eigenvalue is

$$\lambda_s = -R_{gy} \left(\frac{R_{\ell,\beta} R_{n,\bar{r}} - R_{\ell,\bar{r}} R_{n,\beta}}{R_{\ell,\beta} R_{n,\bar{p}} - R_{\ell,\bar{p}} R_{n,\beta}} \right)$$

$$= -0.0164 \left(\frac{(-0.1481)(-0.0705) - (0.2074)(0.0790)}{(-0.1481)(-0.0325) - (-0.8099)(0.0790)} \right) = +0.001417$$

Applying this result to Eq. (9.4.10), we have

$$
\begin{Bmatrix} \Delta\beta \\ \Delta\overline{p} \\ \Delta\overline{r} \\ \Delta\xi_y \\ \Delta\psi \end{Bmatrix} = \begin{Bmatrix} -\left(R_{n,\overline{p}}\lambda + R_{n,\overline{r}}R_{gy}\right)/R_{n,\beta} \\ \lambda \\ R_{gy} \\ R_{gy}/\lambda^2 - \left(R_{n,\overline{p}}\lambda + R_{n,\overline{r}}R_{gy}\right)/\left(R_{n,\beta}\lambda\right) \\ R_{gy}/\lambda \end{Bmatrix} \Delta\phi
$$

$$
= \begin{Bmatrix} -\left[-0.0325(0.001417)+0.0164(-0.0705)\right]/0.0790 \\ 0.001417 \\ 0.0164 \\ 0.0164/(0.001417)^2 + 0.015218/0.001417 \\ 0.0164/0.001417 \end{Bmatrix} \Delta\phi = \begin{Bmatrix} 0.015218 \\ 0.001417 \\ 0.016400 \\ 8178.5220 \\ 11.573747 \end{Bmatrix} \Delta\phi
$$

which gives the normalized solution,

$$
\begin{Bmatrix} \Delta\beta \\ \Delta\overline{p} \\ \Delta\overline{r} \\ \Delta\xi_y \\ \Delta\phi \\ \Delta\psi \end{Bmatrix} = C \begin{Bmatrix} 0.000002 \\ 0.000000 \\ 0.000002 \\ 0.999999 \\ 0.000122 \\ 0.001415 \end{Bmatrix} \exp(+0.001417\,\tau_y)
$$

For comparison, using Eq. (9.4.5) with Eq. (9.4.4) gives the very different result

$$
\begin{Bmatrix} \Delta\beta \\ \Delta\overline{p} \\ \Delta\overline{r} \\ \Delta\xi_y \\ \Delta\phi \\ \Delta\psi \end{Bmatrix} = C \begin{Bmatrix} 0.002134 \\ 0.000000 \\ 0.001524 \\ 0.999276 \\ 0.000000 \\ 0.037969 \end{Bmatrix} \exp(+0.040132\,\tau_y)
$$

From Example 9.4.1, we see that doubling the directional stability of this particular aircraft will change the spiral motion from slowly convergent to slowly divergent. This may seem counterintuitive at first thought. Because divergent spiral motion is characterized primarily by changes in heading and direction of travel, with a seemingly insignificant rolling rate, it could be thought of as some type of directional instability. With this line of reasoning, it may seem natural to think that increasing the directional stability should improve the convergence of the spiral mode. To understand why this is not so, one must understand that spiral motion is not caused directly by a disturbance in yaw, but rather by a disturbance in roll. This is not to say that a disturbance in yaw will not induce spiral motion. Quite the contrary, because of the roll-yaw coupling, any disturbance in yaw will induce a disturbance in roll, and even the slightest roll disturbance can induce spiral motion.

To continue with our explanation of what was observed in Example 9.4.1, let us consider exactly how a disturbance in roll induces spiral motion. Even the smallest disturbance in roll will tend to induce some sideslip, because it results in a component of gravity in the y_b-direction. If the airplane is stable in both roll and yaw, this sideslip will tend to return the bank angle to zero, but it will also tend to yaw the aircraft in the direction of sideslip, in an attempt to return the sideslip angle to zero. If the directional stability is large enough, or the roll stability is low enough, the yaw correction will occur faster than the roll correction, and the airplane's heading will change continuously in the direction of the original roll disturbance, unless corrected with control input. Under such conditions the spiral motion is divergent. If the directional stability is low enough compared to the roll stability, then the roll correction will occur faster than the yaw correction and the spiral mode will be convergent. Thus, we see why increasing directional stability makes the spiral mode more divergent. A comparison between directional divergence and spiral divergence for a conventional airplane is shown in Fig. 9.4.1.

From this description of spiral motion, we also see why the commonly used spiral approximation given by Eq. (9.4.5) is so poor. This approximation totally neglects all changes in rolling rate and bank angle. In so doing, the approximation neglects the very essence of the spiral motion that it is attempting to capture. This can be seen more dramatically by considering an aircraft with neutral roll stability. From Eq. (9.4.5) we see that if the change in rolling moment with respect to yawing rate has the usual positive sign and the aircraft is stable in yaw, then this approximation predicts infinite spiral convergence for neutral roll stability. This is not consistent with reality. In fact, neutral roll stability normally produces spiral divergence.

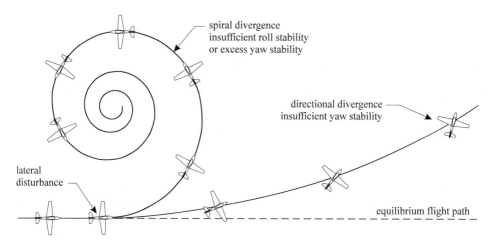

Figure 9.4.1. Comparison between directional divergence and spiral divergence.

If the aircraft in Example 9.2.1 is modified so that the roll stability becomes neutral while all other parameters remain unchanged, we have

$$R_{\ell,\beta} = 0.0$$

and the exact solution for the spiral mode becomes

$$\begin{Bmatrix} \Delta\beta \\ \Delta\bar{p} \\ \Delta\bar{r} \\ \Delta\xi_y \\ \Delta\phi \\ \Delta\psi \end{Bmatrix} = C \begin{Bmatrix} 0.000033 \\ 0.000004 \\ 0.000016 \\ 0.999992 \\ 0.001010 \\ 0.003960 \end{Bmatrix} \exp(+0.003993\,\tau_y)$$

For this case, the approximate spiral mode solution given by Eq. (9.4.11) gives

$$\begin{Bmatrix} \Delta\beta \\ \Delta\bar{p} \\ \Delta\bar{r} \\ \Delta\xi_y \\ \Delta\phi \\ \Delta\psi \end{Bmatrix} = C \begin{Bmatrix} 0.000035 \\ 0.000004 \\ 0.000017 \\ 0.999991 \\ 0.001067 \\ 0.004165 \end{Bmatrix} \exp(+0.004200\,\tau_y)$$

As mentioned previously, with neutral roll stability, the spiral mode approximation from Eq. (9.4.5) gives

$$\begin{Bmatrix} \Delta\beta \\ \Delta\bar{p} \\ \Delta\bar{r} \\ \Delta\xi_y \\ \Delta\phi \\ \Delta\psi \end{Bmatrix} = C\exp(-\infty\,\tau_y) = 0.0$$

Here again we see that the spiral approximation given by Eq. (9.4.11) agrees closely with the exact solution, whereas Eq. (9.4.5) predicts a much different result. In this case, the eigenvalue predicted by Eq. (9.4.5) does not even have the correct sign.

The spiral mode approximation given by Eq. (9.4.5) should never be used. It is based on a mathematical approximation that is not consistent with physical reality and is very misleading. It has been included here for historical reasons only and because the reader may encounter this approximation in the literature.

9.5. Dutch Roll Approximation

As noted previously, Dutch roll is a damped oscillatory motion that is characterized by a combination of sideslip, roll, and yaw. Nevertheless, the usual approximation for Dutch roll is obtained by assuming the motion to consist solely of sideslip and yaw. The rolling rate, bank angle, and rolling momentum equation are completely neglected in the usual Dutch roll approximation. Also, the change in side force with respect to yawing rate is usually neglected, because this is small and has little effect on lateral motion. Using these approximations for motion relative to level flight, Eq. (9.2.1) becomes

$$
\begin{bmatrix}
1 & 0 & 0 & 0 & 0 & 0 \\
0 & 0 & 0 & 0 & 0 & 0 \\
0 & -I_{zx} & 1 & 0 & 0 & 0 \\
0 & 0 & 0 & 1 & 0 & 0 \\
0 & 0 & 0 & 0 & 0 & 0 \\
0 & 0 & 0 & 0 & 0 & 1
\end{bmatrix}
\begin{Bmatrix}
\Delta\hat{\beta} \\
0 \\
\Delta\hat{\bar{r}} \\
\Delta\hat{\xi}_y \\
0 \\
\Delta\hat{\psi}
\end{Bmatrix}
=
\begin{bmatrix}
R_{y,\beta} & 0 & (R_{y,\bar{r}}-1) & 0 & 0 & 0 \\
0 & 1 & 0 & 0 & 0 & 0 \\
R_{n,\beta} & 0 & R_{n,\bar{r}} & 0 & 0 & 0 \\
1 & 0 & 0 & 0 & 0 & 1 \\
0 & 0 & 0 & 0 & 1 & 0 \\
0 & 0 & 1 & 0 & 0 & 0
\end{bmatrix}
\begin{Bmatrix}
\Delta\beta \\
\Delta\bar{p} \\
\Delta\bar{r} \\
\Delta\xi_y \\
\Delta\phi \\
\Delta\psi
\end{Bmatrix}
\tag{9.5.1}
$$

The first and third equations in Eq. (9.5.1) are not coupled to the other four equations. Thus, this system of equations can be rearranged and separated to yield

$$
\begin{bmatrix}
(R_{y,\beta}-\lambda) & (R_{y,\bar{r}}-1) \\
R_{n,\beta} & (R_{n,\bar{r}}-\lambda)
\end{bmatrix}
\begin{Bmatrix}
\Delta\beta \\
\Delta\bar{r}
\end{Bmatrix}
=
\begin{Bmatrix}
0 \\
0
\end{Bmatrix}
\tag{9.5.2}
$$

and

$$
\begin{Bmatrix}
\Delta\bar{p} \\
\Delta\bar{r} \\
\Delta\xi_y \\
\Delta\phi \\
\Delta\psi
\end{Bmatrix}
=
\begin{Bmatrix}
0 \\
R_{n,\beta}/(\lambda-R_{n,\bar{r}}) \\
1/\lambda + \left[R_{n,\beta}/(\lambda-R_{n,\bar{r}})\right]/\lambda^2 \\
0 \\
\left[R_{n,\beta}/(\lambda-R_{n,\bar{r}})\right]/\lambda
\end{Bmatrix}
\Delta\beta
\tag{9.5.3}
$$

The approximate Dutch roll eigenvalues are obtained from the roots of Eq. (9.5.2), and the approximate eigenvectors are obtained by using the result in Eq. (9.5.3).

The characteristic equation for Eq. (9.5.2) is

$$
\lambda^2 - (R_{y,\beta}+R_{n,\bar{r}})\lambda + (1-R_{y,\bar{r}})R_{n,\beta} + R_{y,\beta}R_{n,\bar{r}} = 0
\tag{9.5.4}
$$

From this quadratic equation the dimensionless Dutch roll eigenvalues predicted by this approximation are found to be

$$
\lambda_{Dr} \cong \frac{R_{y,\beta}+R_{n,\bar{r}}}{2} \pm i\sqrt{(1-R_{y,\bar{r}})R_{n,\beta} + R_{y,\beta}R_{n,\bar{r}} - \left(\frac{R_{y,\beta}+R_{n,\bar{r}}}{2}\right)^2}
\tag{9.5.5}
$$

From Eq. (9.5.5), the approximate damped natural frequency and damping rate are

$$\omega_{d_{Dr}} \cong \frac{2V_o}{b_w}\,\mathrm{imag}(\lambda_{Dr}) = \frac{2V_o}{b_w}\sqrt{\left(1-R_{y,\bar{r}}\right)R_{n,\beta}+R_{y,\beta}\,R_{n,\bar{r}}-\left(\frac{R_{y,\beta}+R_{n,\bar{r}}}{2}\right)^2} \quad (9.5.6)$$

$$\sigma_{Dr} \cong -\frac{2V_o}{b_w}\,\mathrm{real}(\lambda_{Dr}) = -\frac{V_o}{b_w}\left(R_{y,\beta}+R_{n,\bar{r}}\right) \quad (9.5.7)$$

Not surprisingly, results obtained from this commonly used approximation do not always give particularly good values for the frequency, damping, or eigenvectors that are associated with Dutch roll. The reason for poor agreement between this uncoupled Dutch roll approximation and the exact solution is that Dutch roll strongly involves all three lateral degrees of freedom, with significant coupling between sideslip, roll, and yaw. As we have seen in Example 9.2.1, the largest Dutch roll oscillations are typically in sideslip. However, the Dutch roll amplitude for rolling rate is nearly as large as that for sideslip and is about the same as that for yawing rate. Furthermore, Dutch roll oscillations in bank angle are typically as large as those in heading. Thus, it is very difficult to justify neglecting any of the three lateral degrees of freedom when attempting to develop a Dutch roll approximation. Here, we consider an improved Dutch roll approximation developed by Phillips (2000b), which includes the effects of roll as well as those of sideslip and yaw.

In this approximation, we neglect the product of inertia and the change in side force with respect to rolling rate because these values are typically small and have little effect on Dutch roll. Using these approximations for motion relative to steady level flight, the lateral equations of motion, given by Eq. (9.2.1), become

$$\begin{bmatrix} 1&0&0&0&0&0\\ 0&1&0&0&0&0\\ 0&0&1&0&0&0\\ 0&0&0&1&0&0\\ 0&0&0&0&1&0\\ 0&0&0&0&0&1 \end{bmatrix}\begin{Bmatrix} \Delta\hat{\beta}\\ \Delta\hat{p}\\ \Delta\hat{r}\\ \Delta\hat{\xi}_y\\ \Delta\hat{\phi}\\ \Delta\hat{\psi} \end{Bmatrix} = \begin{bmatrix} R_{y,\beta}&0&(R_{y,\bar{r}}-1)&0&R_{gy}&0\\ R_{\ell,\beta}&R_{\ell,\bar{p}}&R_{\ell,\bar{r}}&0&0&0\\ R_{n,\beta}&R_{n,\bar{p}}&R_{n,\bar{r}}&0&0&0\\ 1&0&0&0&0&1\\ 0&1&0&0&0&0\\ 0&0&1&0&0&0 \end{bmatrix}\begin{Bmatrix} \Delta\beta\\ \Delta p\\ \Delta r\\ \Delta\xi_y\\ \Delta\phi\\ \Delta\psi \end{Bmatrix} \quad (9.5.8)$$

which results in the eigenproblem

$$\begin{bmatrix} (R_{y,\beta}-\lambda)&0&(R_{y,\bar{r}}-1)&0&R_{gy}&0\\ R_{\ell,\beta}&(R_{\ell,\bar{p}}-\lambda)&R_{\ell,\bar{r}}&0&0&0\\ R_{n,\beta}&R_{n,\bar{p}}&(R_{n,\bar{r}}-\lambda)&0&0&0\\ 1&0&0&-\lambda&0&1\\ 0&1&0&0&-\lambda&0\\ 0&0&1&0&0&-\lambda \end{bmatrix}\begin{Bmatrix} \Delta\beta\\ \Delta p\\ \Delta r\\ \Delta\xi_y\\ \Delta\phi\\ \Delta\psi \end{Bmatrix} = \begin{Bmatrix} 0\\ 0\\ 0\\ 0\\ 0\\ 0 \end{Bmatrix} \quad (9.5.9)$$

Using the second and fifth equations in Eq. (9.5.9), we can eliminate the rolling rate and bank angle from the first and third equations. Thus, Eq. (9.5.9) can be rearranged and separated to yield

$$
\left[
\begin{array}{cc}
\left(R_{y,\beta} - \dfrac{R_{gy}\, R_{\ell,\beta}}{(R_{\ell,\bar{p}} - \lambda)\lambda} - \lambda \right) & -\left(1 - R_{y,\bar{r}} + \dfrac{R_{gy}\, R_{\ell,\bar{r}}}{(R_{\ell,\bar{p}} - \lambda)\lambda} \right) \\[4mm]
\left(R_{n,\beta} - \dfrac{R_{\ell,\beta}\, R_{n,\bar{p}}}{R_{\ell,\bar{p}} - \lambda} \right) & \left(R_{n,\bar{r}} - \dfrac{R_{\ell,\bar{r}}\, R_{n,\bar{p}}}{R_{\ell,\bar{p}} - \lambda} - \lambda \right)
\end{array}
\right]
\left\{ \begin{array}{c} \Delta\beta \\ \Delta\bar{r} \end{array} \right\}
= \left\{ \begin{array}{c} 0 \\ 0 \end{array} \right\}
\tag{9.5.10}
$$

and

$$
\left\{ \begin{array}{c} \Delta\bar{p} \\ \Delta\bar{r} \\ \Delta\xi_y \\ \Delta\phi \\ \Delta\psi \end{array} \right\}
= \left\{ \begin{array}{c} R_{xc} \\ R_{zc} \\ (1 + R_{zc}/\lambda)/\lambda \\ R_{xc}/\lambda \\ R_{zc}/\lambda \end{array} \right\} \Delta\beta,
\tag{9.5.11}
$$

where

$$
R_{xc} \equiv \frac{R_{n,\beta}\, R_{\ell,\bar{r}} - R_{\ell,\beta}\,(R_{n,\bar{r}} - \lambda)}{(R_{\ell,\bar{p}} - \lambda)(R_{n,\bar{r}} - \lambda) - R_{\ell,\bar{r}}\, R_{n,\bar{p}}}
$$

and

$$
R_{zc} \equiv \frac{R_{\ell,\beta}\, R_{n,\bar{p}} - R_{n,\beta}\,(R_{\ell,\bar{p}} - \lambda)}{(R_{\ell,\bar{p}} - \lambda)(R_{n,\bar{r}} - \lambda) - R_{\ell,\bar{r}}\, R_{n,\bar{p}}}
$$

The characteristic equation for Eq. (9.5.10) is

$$
\left|
\begin{array}{cc}
\left(R_{y,\beta} - \dfrac{R_{gy}\, R_{\ell,\beta}}{(R_{\ell,\bar{p}} - \lambda)\lambda} - \lambda \right) & -\left(1 - R_{y,\bar{r}} + \dfrac{R_{gy}\, R_{\ell,\bar{r}}}{(R_{\ell,\bar{p}} - \lambda)\lambda} \right) \\[4mm]
\left(R_{n,\beta} - \dfrac{R_{\ell,\beta}\, R_{n,\bar{p}}}{R_{\ell,\bar{p}} - \lambda} \right) & \left(R_{n,\bar{r}} - \dfrac{R_{\ell,\bar{r}}\, R_{n,\bar{p}}}{R_{\ell,\bar{p}} - \lambda} - \lambda \right)
\end{array}
\right| = 0
\tag{9.5.12}
$$

which can be expanded to give

$$
\lambda^2 - \left(R_{y,\beta} + R_{n,\bar{r}} - \frac{R_{\ell,\bar{r}}\, R_{n,\bar{p}}}{R_{\ell,\bar{p}} - \lambda} \right)\lambda + (1 - R_{y,\bar{r}})\, R_{n,\beta} + R_{y,\beta}\, R_{n,\bar{r}}
$$
$$
+ \frac{R_{\ell,\beta}\left[R_{gy} - (1 - R_{y,\bar{r}})\, R_{n,\bar{p}} \right] - R_{y,\beta}\, R_{\ell,\bar{r}}\, R_{n,\bar{p}}}{R_{\ell,\bar{p}} - \lambda} + \frac{R_{gy}\,(R_{\ell,\bar{r}}\, R_{n,\beta} - R_{\ell,\beta}\, R_{n,\bar{r}})}{(R_{\ell,\bar{p}} - \lambda)\lambda} = 0
\tag{9.5.13}
$$

In our study of the roll mode, we found that roll is typically very heavily damped. Thus, all of those terms in Eq. (9.5.13) that are inversely proportional to the roll-damping ratio are small, but not necessarily totally negligible. This suggests that we might expand Eq. (9.5.13) in terms of a Taylor series in one over the roll-damping ratio. For this purpose we use the expansion

$$\frac{1}{R_{\ell,\bar{p}} - \lambda} = \frac{1}{R_{\ell,\bar{p}}} + \frac{\lambda}{R_{\ell,\bar{p}}^2} + \frac{\lambda^2}{R_{\ell,\bar{p}}^3} + \frac{\lambda^3}{R_{\ell,\bar{p}}^4} + \cdots \qquad (9.5.14)$$

If we define λ_∞ to be the eigenvalue corresponding to an infinite roll-damping ratio,

$$R_{\ell,\bar{p}} \to \infty \qquad (9.5.15)$$

then from Eq. (9.5.13), we have

$$\lambda_\infty^2 - (R_{y,\beta} + R_{n,\bar{r}})\lambda_\infty + (1 - R_{y,\bar{r}})R_{n,\beta} + R_{y,\beta}R_{n,\bar{r}} = 0 \qquad (9.5.16)$$

and

$$\lambda_\infty = \frac{R_{y,\beta} + R_{n,\bar{r}}}{2} \pm i\sqrt{(1 - R_{y,\bar{r}})R_{n,\beta} + R_{y,\beta}R_{n,\bar{r}} - \left(\frac{R_{y,\beta} + R_{n,\bar{r}}}{2}\right)^2} \qquad (9.5.17)$$

At this point it is interesting to note that the eigenvalues obtained from Eq. (9.5.17) are identical to those obtained from Eq. (9.5.5). Thus, we see that this earlier Dutch roll approximation is valid only in the limit as the roll-damping ratio approaches infinity. If the roll-damping ratio were large enough, the rolling motion associated with Dutch roll would be completely suppressed and the motion would indeed consist solely of sideslip and yaw. However, as the name implies, an aircraft seldom has sufficient roll damping to eliminate the rolling motion from Dutch roll. The effects of roll can be included in the Dutch roll approximation by carrying some of the low-order terms from the Taylor series in Eq. (9.5.14).

Because the roll-damping ratio is several times larger than the Dutch roll eigenvalue, all terms beyond the first in Eq. (9.5.14) are small and become smaller with increasing order. Thus, for the purpose of evaluating this Taylor series, we can replace the actual eigenvalue on the right-hand side with the approximate eigenvalue from Eq. (9.5.17),

$$\frac{1}{R_{\ell,\bar{p}} - \lambda} \cong \frac{1}{R_{\ell,\bar{p}}} + \frac{\lambda_\infty}{R_{\ell,\bar{p}}^2} + \frac{\lambda_\infty^2}{R_{\ell,\bar{p}}^3} + \frac{\lambda_\infty^3}{R_{\ell,\bar{p}}^4} + \cdots \qquad (9.5.18)$$

Furthermore, because the Dutch roll damping is an order of magnitude less than the Dutch roll frequency, we can also neglect the damping from Eq. (9.5.17) and use the approximations

$$\frac{1}{R_{\ell,\bar{p}} - \lambda} \cong \frac{1}{R_{\ell,\bar{p}}} \pm i\frac{\varpi_\infty}{R_{\ell,\bar{p}}^2} - \frac{\varpi_\infty^2}{R_{\ell,\bar{p}}^3} \mp i\frac{\varpi_\infty^3}{R_{\ell,\bar{p}}^4} + \cdots \tag{9.5.19}$$

$$\frac{1}{(R_{\ell,\bar{p}} - \lambda)\lambda} \cong \mp i\frac{1}{\varpi_\infty R_{\ell,\bar{p}}} + \frac{1}{R_{\ell,\bar{p}}^2} \pm i\frac{\varpi_\infty}{R_{\ell,\bar{p}}^3} - \frac{\varpi_\infty^2}{R_{\ell,\bar{p}}^4} + \cdots \tag{9.5.20}$$

where

$$\varpi_\infty \equiv \sqrt{(1 - R_{y,\bar{r}})R_{n,\beta} + R_{y,\beta}R_{n,\bar{r}}} \tag{9.5.21}$$

Retaining the first two terms in each of these Taylor series and applying the results to the characteristic equation given by Eq. (9.5.13), we obtain a quadratic equation for the approximate Dutch roll eigenvalues,

$$\lambda^2 - (R_{y,\beta} + R_{n,\bar{r}} + R_{dc})\lambda + (1 - R_{y,\bar{r}})R_{n,\beta} + R_{y,\beta}R_{n,\bar{r}} + R_{fc} = 0 \tag{9.5.22}$$

where the complex coefficients R_{dc} and R_{fc} are defined as

$$R_{dc} \equiv -\frac{R_{\ell,\bar{r}}R_{n,\bar{p}}}{R_{\ell,\bar{p}}} \mp i\frac{\varpi_\infty R_{\ell,\bar{r}}R_{n,\bar{p}}}{R_{\ell,\bar{p}}^2} \tag{9.5.23}$$

and

$$R_{fc} \equiv \left[\frac{R_{\ell,\beta}\left[R_{gy} - (1 - R_{y,\bar{r}})R_{n,\bar{p}}\right] - R_{y,\beta}R_{\ell,\bar{r}}R_{n,\bar{p}}}{R_{\ell,\bar{p}}} + \frac{R_{gy}(R_{\ell,\bar{r}}R_{n,\beta} - R_{\ell,\beta}R_{n,\bar{r}})}{R_{\ell,\bar{p}}^2}\right]$$
$$\pm i\left[\frac{R_{gy}(R_{\ell,\beta}R_{n,\bar{r}} - R_{\ell,\bar{r}}R_{n,\beta})}{\varpi_\infty R_{\ell,\bar{p}}} + \frac{\varpi_\infty\left\{R_{\ell,\beta}\left[R_{gy} - (1 - R_{y,\bar{r}})R_{n,\bar{p}}\right] - R_{y,\beta}R_{\ell,\bar{r}}R_{n,\bar{p}}\right\}}{R_{\ell,\bar{p}}^2}\right] \tag{9.5.24}$$

The eigenvalues obtained from Eq. (9.5.22), with the application of Eq. (9.5.21), are found to be

$$\lambda = \frac{R_{y,\beta} + R_{n,\bar{r}} + R_{dc}}{2} \pm i\sqrt{(1 - R_{y,\bar{r}})R_{n,\beta} + R_{y,\beta}R_{n,\bar{r}} + R_{fc} - \left(\frac{R_{y,\beta} + R_{n,\bar{r}} + R_{dc}}{2}\right)^2}$$

$$= \frac{R_{y,\beta} + R_{n,\bar{r}} + R_{dc}}{2} \pm i\varpi_\infty\sqrt{1 + \frac{R_{fc}}{\varpi_\infty^2} - \left(\frac{R_{y,\beta} + R_{n,\bar{r}} + R_{dc}}{2\varpi_\infty}\right)^2} \tag{9.5.25}$$

Since both R_{fc} and the damping term are small compared to ϖ_∞, Eq. (9.5.25) can be further approximated as

$$\lambda \cong \frac{R_{y,\beta} + R_{n,\bar{r}} + R_{dc}}{2} \pm i\varpi_\infty \left[1 + \frac{\text{real}(R_{fc})}{2\varpi_\infty^2} \pm i\frac{\text{imag}(R_{fc})}{2\varpi_\infty^2} - \frac{(R_{y,\beta} + R_{n,\bar{r}} + R_{dc})^2}{8\varpi_\infty^2} \right]$$

$$\cong \frac{R_{y,\beta} + R_{n,\bar{r}} + R_{dc} - \text{imag}(R_{fc})/\varpi_\infty}{2} \pm i\varpi_\infty \sqrt{1 + \frac{\text{real}(R_{fc})}{\varpi_\infty^2} - \left(\frac{R_{y,\beta} + R_{n,\bar{r}} + R_{dc}}{2\varpi_\infty}\right)^2}$$

$$= \frac{R_{y,\beta} + R_{n,\bar{r}} + R_{dc} - \text{imag}(R_{fc})/\varpi_\infty}{2} \pm i\sqrt{\varpi_\infty^2 + \text{real}(R_{fc}) - \left(\frac{R_{y,\beta} + R_{n,\bar{r}} + R_{dc}}{2}\right)^2} \tag{9.5.26}$$

Because the imaginary component of the Dutch roll eigenvalue is typically an order of magnitude larger than the real component, the second-order terms in R_{dc} and R_{fc} do not have a significant effect on the Dutch roll frequency. However, the second-order terms are crucial to an accurate determination of the much smaller Dutch roll damping rate. For this reason we retain all second-order terms in the real component of the eigenvalue but neglect second-order terms in the imaginary component. Thus, because the imaginary component of R_{dc} is second order, it can be neglected in the Dutch roll approximation. We can also neglect the second-order term in the real part of R_{fc} since it affects only the frequency. On the other hand, we must retain the second term in the imaginary part of R_{fc} because this term is part of the damping.

Thus, for this approximation, the **dimensionless Dutch roll eigenvalues** are

$$\lambda_{Dr} \cong \frac{R_{y,\beta} + R_{n,\bar{r}} - R_{Dc} + R_{Dp}}{2}$$

$$\pm i\sqrt{(1 - R_{y,\bar{r}})R_{n,\beta} + R_{y,\beta}R_{n,\bar{r}} + R_{Ds} - \left(\frac{R_{y,\beta} + R_{n,\bar{r}}}{2}\right)^2} \tag{9.5.27}$$

where we define the **Dutch roll stability ratio**,

$$R_{Ds} \equiv \frac{R_{\ell,\beta}\left[R_{gy} - (1 - R_{y,\bar{r}})R_{n,\bar{p}}\right] - R_{y,\beta}R_{\ell,\bar{r}}R_{n,\bar{p}}}{R_{\ell,\bar{p}}} \tag{9.5.28}$$

the **Dutch roll coupling ratio**,

$$R_{Dc} \equiv \frac{R_{\ell,\bar{r}}R_{n,\bar{p}}}{R_{\ell,\bar{p}}} \tag{9.5.29}$$

and the **Dutch roll phase-divergence ratio**,

$$R_{Dp} \equiv \frac{R_{gy}(R_{\ell,\bar{r}}R_{n,\beta} - R_{\ell,\beta}R_{n,\bar{r}})}{R_{\ell,\bar{p}}(R_{n,\beta} + R_{y,\beta}R_{n,\bar{r}})} - \frac{R_{Ds}}{R_{\ell,\bar{p}}} \tag{9.5.30}$$

This gives the approximate **Dutch roll damped natural frequency,**

$$\omega_{d_{Dr}} \cong \frac{2V_o}{b_w}\sqrt{(1-R_{y,\bar{r}})R_{n,\beta} + R_{y,\beta}R_{n,\bar{r}} + R_{Ds} - \left(\frac{R_{y,\beta}+R_{n,\bar{r}}}{2}\right)^2} \tag{9.5.31}$$

and the approximate **Dutch roll damping rate,**

$$\sigma_{Dr} \cong -\frac{V_o}{b_w}\left(R_{y,\beta} + R_{n,\bar{r}} - R_{Dc} + R_{Dp}\right) \tag{9.5.32}$$

EXAMPLE 9.5.1. Using the Dutch roll approximation given by Eq. (9.5.27), find the eigenvalues, eigenvectors, damping rate, and damped natural frequency for the Dutch roll mode of the airplane described in Example 9.2.1. Compare these approximate results with the exact Dutch roll solution that was obtained in Example 9.2.1 and with the results obtained from Eq. (9.5.5).

Solution. From Example 9.2.1 we have

$$V_o = 180 \text{ ft/sec}, \qquad b_w = 33 \text{ ft}, \qquad R_{gy} = 0.0164,$$
$$R_{y,\beta} = -0.0233, \qquad R_{y,\bar{p}} = 0.0, \qquad R_{y,\bar{r}} = 0.0100,$$
$$R_{\ell,\beta} = -0.1481, \qquad R_{\ell,\bar{p}} = -0.8099, \qquad R_{\ell,\bar{r}} = 0.2074,$$
$$R_{n,\beta} = 0.0395, \qquad R_{n,\bar{p}} = -0.0325, \qquad R_{n,\bar{r}} = -0.0705$$

From Eq. (9.5.28), the Dutch roll stability ratio is

$$\begin{aligned}R_{Ds} &\equiv \frac{R_{\ell,\beta}\left[R_{gy} - (1-R_{y,\bar{r}})R_{n,\bar{p}}\right] - R_{y,\beta}R_{\ell,\bar{r}}R_{n,\bar{p}}}{R_{\ell,\bar{p}}}\\[2mm]
&= \frac{(-0.1481)[0.0164 - (0.9900)(-0.0325)] - (-0.0233)(0.2074)(-0.0325)}{-0.8099}\\[2mm]
&= 0.009076\end{aligned}$$

From Eq. (9.5.29), the Dutch roll coupling ratio is

$$R_{Dc} \equiv \frac{R_{\ell,\bar{r}}R_{n,\bar{p}}}{R_{\ell,\bar{p}}} = \frac{(0.2074)(-0.0325)}{-0.8099} = 0.008323$$

From Eq. (9.5.30), the Dutch roll phase-divergence ratio is

$$\begin{aligned}R_{Dp} &\equiv \frac{R_{gy}(R_{\ell,\bar{r}}R_{n,\beta} - R_{\ell,\beta}R_{n,\bar{r}})}{R_{\ell,\bar{p}}(R_{n,\beta} + R_{y,\beta}R_{n,\bar{r}})} - \frac{R_{Ds}}{R_{\ell,\bar{p}}}\\[2mm]
&= \frac{0.0164[(0.2074)(0.0395) - (-0.1481)(-0.0705)]}{-0.8099[0.0395 + (-0.0233)(-0.0705)]} - \frac{0.009076}{-0.8099} = 0.012313\end{aligned}$$

Using these values in Eq. (9.5.27), the approximate Dutch roll eigenvalues are

$$
\lambda_{Dr} \cong \frac{R_{y,\beta} + R_{n,\bar{r}} - R_{Dc} + R_{Dp}}{2}
$$

$$
\pm i \sqrt{(1 - R_{y,\bar{r}}) R_{n,\beta} + R_{y,\beta} R_{n,\bar{r}} + R_{Ds} - \left(\frac{R_{y,\beta} + R_{n,\bar{r}}}{2}\right)^2}
$$

$$
= \frac{-0.0233 - 0.0705 - 0.008323 + 0.012313}{2}
$$

$$
\pm i \sqrt{0.9900(0.0395) + (-0.0233)(-0.0705) + 0.009076 - \left(\frac{-0.0233 - 0.0705}{2}\right)^2}
$$

$$
= -0.044904 \pm i\, 0.218230
$$

The damping rate and damped natural frequency of oscillation are

$$
\sigma = \text{real}(\lambda)\frac{2V_o}{b_w} = 0.044904\,\frac{2(180)}{33} = 0.489862 \text{ sec}^{-1}
$$

$$
\omega_d = |\text{imag}(\lambda)|\frac{2V_o}{b_w} = 0.218230\,\frac{2(180)}{33} = 2.380691 \text{ rad/sec}
$$

The eigenvectors for this Dutch roll approximation are found by using the complex eigenvalues in Eq. (9.5.11). This gives

$$
\begin{Bmatrix} \Delta\beta \\ \Delta\bar{p} \\ \Delta\bar{r} \\ \Delta\xi_y \\ \Delta\phi \\ \Delta\psi \end{Bmatrix} = C_{1,2} \begin{Bmatrix} 0.588877 \\ -0.110484 \mp 0.001381\,i \\ 0.014438 \mp 0.121348\,i \\ 0.165221 \mp 0.228052\,i \\ 0.093870 \pm 0.486957\,i \\ -0.546529 \pm 0.046297\,i \end{Bmatrix} \exp[(-0.044904 \pm 0.218230\,i)\,\tau_y]
$$

The exact solution from Example 9.2.1 is

$$
\begin{Bmatrix} \Delta\beta \\ \Delta\bar{p} \\ \Delta\bar{r} \\ \Delta\xi_y \\ \Delta\phi \\ \Delta\psi \end{Bmatrix} = C_{3,4} \begin{Bmatrix} 0.589580 \\ -0.109648 \mp 0.001442\,i \\ 0.013967 \mp 0.121744\,i \\ 0.172251 \mp 0.220690\,i \\ 0.091681 \pm 0.484586\,i \\ -0.549104 \pm 0.047290\,i \end{Bmatrix} \exp[(-0.044202 \pm 0.217908\,i)\,\tau_y]
$$

From Eqs. (9.5.3) and (9.5.5), the infinite roll-damping solution is

$$\begin{Bmatrix} \Delta\beta \\ \Delta\bar{p} \\ \Delta\bar{r} \\ \Delta\xi_y \\ \Delta\phi \\ \Delta\psi \end{Bmatrix} = C_{1,2} \begin{Bmatrix} 0.653476 \\ 0.000000 \\ 0.015578 \mp 0.129597\,i \\ 0.344288 \mp 0.138783\,i \\ 0.000000 \\ -0.642375 \pm 0.074105\,i \end{Bmatrix} \exp[(-0.046900 \pm 0.196337\,i)\,\tau_y]$$

Figures 9.5.1 through 9.5.5 show how the Dutch roll approximation that is given by Eq. (9.5.27) compares with that given by Eq. (9.5.5) and the exact solution for a broad range of aerodynamic parameters. Here it should be noted that as the roll-damping derivative approaches infinity, the Dutch roll stability ratio, coupling ratio, and phase-divergence ratio all approach zero. So we see that in the limit of infinite roll damping, the result that is given by Eq. (9.5.27) reduces exactly to that given by Eq. (9.5.5). Thus, as was mentioned previously, Eq. (9.5.5) represents an asymptotic solution for infinite roll damping. This can be seen graphically in Fig. 9.5.1. In this figure, all parameters except the roll-damping ratio have been held constant at those values given in Example 9.5.1. For a conventional airplane, the roll-damping derivative is large compared to the other aerodynamic derivatives. Thus, it may seem reasonable to use an asymptotic solution for large roll damping as an approximation for Dutch roll. However, the roll-damping ratio for conventional airplanes is typically somewhat less than 1.0. Figure 9.5.1 shows that this level of roll damping is not sufficient to produce the asymptotic behavior.

Figures 9.5.2 through 9.5.5 show how other aerodynamic derivatives affect Dutch roll. In Fig. 9.5.2, a comparison between Eqs. (9.5.5), (9.5.27), and the exact solution is shown for a broad range of roll-stability ratio. Figure 9.5.3 shows the same comparison for a broad range of yaw-stability ratio. Similar comparisons, showing the effects of roll and yaw coupling, are displayed in Figs. 9.5.4 and 9.5.5.

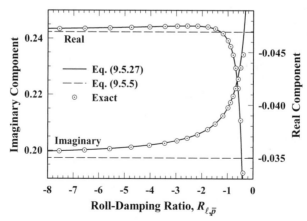

Figure 9.5.1. Effect of roll damping on the dimensionless Dutch roll eigenvalues.

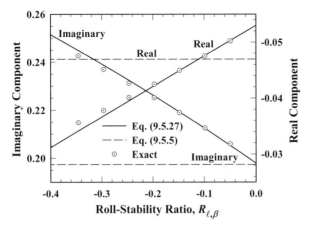

Figure 9.5.2. Effect of roll stability on the dimensionless Dutch roll eigenvalues.

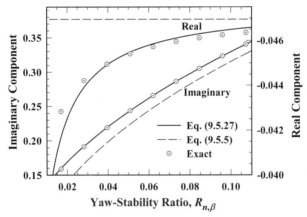

Figure 9.5.3. Effect of yaw stability on the dimensionless Dutch roll eigenvalues.

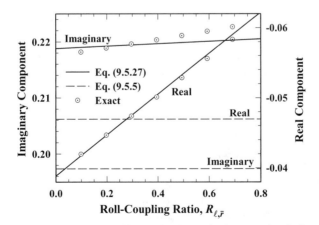

Figure 9.5.4. Effect of roll coupling on the dimensionless Dutch roll eigenvalues.

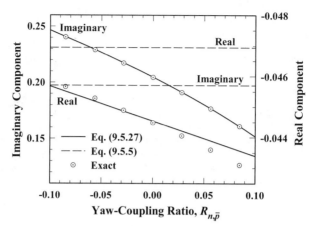

Figure 9.5.5. Effect of yaw coupling on the dimensionless Dutch roll eigenvalues.

The improved closed-form approximation allows us to see more easily how the aerodynamic coefficients and stability derivatives affect the Dutch roll motion. In order to help the student develop some insight into the genesis of Dutch roll, we shall now examine the predictions of this improved approximation in greater detail.

Damping has very little effect on the Dutch roll frequency. Thus, neglecting the damping terms in Eq. (9.5.31) and applying the definition of Dutch roll stability ratio from Eq. (9.5.28), the undamped natural frequency for Dutch roll motion is approximated as

$$
\begin{aligned}
\omega_{n_{Dr}} &\cong \frac{2V_o}{b_w}\sqrt{\omega_\infty^2 + R_{Ds}} \\
&= \frac{2V_o}{b_w}\sqrt{\omega_\infty^2 + \frac{R_{\ell,\beta}\left[R_{gy} - (1 - R_{y,\bar{r}})R_{n,\bar{p}}\right] - R_{y,\beta}R_{\ell,\bar{r}}R_{n,\bar{p}}}{R_{\ell,\bar{p}}}}
\end{aligned}
\tag{9.5.33}
$$

where ω_∞ is the traditional dimensionless undamped natural frequency for infinite roll damping. The roll-damping derivative is always negative, and for a stable aircraft, the roll-stability derivative is also negative. The gravitational acceleration ratio is positive; the dimensionless change in side force with yawing rate is always less than unity; and for a conventional airplane generating positive lift, the change in yawing moment with rolling rate is typically negative. Thus, the Dutch roll stability ratio is positive, and from Eq. (9.5.33) we see that increasing roll stability will increase the Dutch roll frequency while increasing roll damping will decrease the Dutch roll frequency. Furthermore, since the change in side force with sideslip angle is negative and the change in rolling moment with yawing rate is typically positive, Eq. (9.5.33) shows that increasing the roll-yaw coupling will normally increase the Dutch roll frequency.

Using the definitions from Eqs. (9.5.28) through (9.5.30) in Eq. (9.5.32), the Dutch roll damping for the improved approximation can be written as

$$\sigma_{Dr} \cong \sigma_\infty + \frac{V_o}{b_w}(R_{Dc} - R_{Dp})$$

$$= \sigma_\infty + \frac{V_o}{b_w}\left(\frac{R_{\ell,\bar{r}} R_{n,\bar{p}}}{R_{\ell,\bar{p}}} - \frac{R_{gy}(R_{\ell,\bar{r}} R_{n,\beta} - R_{\ell,\beta} R_{n,\bar{r}})}{R_{\ell,\bar{p}}(R_{n,\beta} + R_{y,\beta} R_{n,\bar{r}})} \right. \tag{9.5.34}$$

$$\left. + \frac{R_{\ell,\beta}[R_{gy} - (1 - R_{y,\bar{r}}) R_{n,\bar{p}}] - R_{y,\beta} R_{\ell,\bar{r}} R_{n,\bar{p}}}{R_{\ell,\bar{p}}^2} \right)$$

where σ_∞ is the traditional yaw damping term, which provides the total Dutch roll damping for the case of infinite roll damping. The second term on the right-hand side of Eq. (9.5.34) is the Dutch roll coupling ratio. Since the roll-yaw coupling derivatives typically have opposite signs and the roll-damping derivative is always negative, the Dutch roll coupling ratio will normally increase the Dutch roll damping. However, it is possible for the roll-yaw coupling derivatives to have the same sign, in which case this Dutch roll coupling would tend to decrease the total Dutch roll damping.

The third and fourth terms on the right-hand side of Eq. (9.5.34) compose the *Dutch roll phase-divergence ratio*. This has been called the phase-divergence ratio because it results from the phase shift between the various components that comprise the Dutch roll motion and because it will typically decrease the total Dutch roll damping. For an aircraft that is stable in yaw, the denominator in the third term on the right-hand side of Eq. (9.5.34) is negative, and the denominator in the last term is always positive. Thus, since the roll-stability derivative is negative for a stable aircraft, Eq. (9.5.34) shows that increasing roll stability will decrease Dutch roll damping. In a similar manner, from further examination of Eq. (9.5.34) it can be shown that increasing yaw stability will increase Dutch roll damping.

The design of an airplane having good Dutch roll characteristics while maintaining acceptable roll, yaw, and spiral stability can present somewhat of a dilemma. The aerodynamic derivatives that have greatest influence on the Dutch roll and spiral damping are the roll-stability derivative, the roll-damping derivative, the yaw-stability derivative, and the yaw-damping derivative. From the improved approximations for the Dutch roll and spiral modes, we see how each of these derivatives affects the damping for these two modes. For a conventional airplane that is stable in both roll and yaw, the influence of these derivatives is summarized as follows:

Increasing roll stability will: Increasing roll damping will:
    Decrease Dutch roll damping.   Increase Dutch roll damping.
    Increase spiral damping.   Decrease spiral damping.

Increasing yaw stability will: Increasing yaw damping will:
    Increase Dutch roll damping.   Increase Dutch roll damping.
    Decrease spiral damping.   Increase spiral damping.

While increasing roll stability has a favorable effect on the spiral damping, it has an adverse effect on the Dutch roll damping. On the other hand, roll damping has an adverse effect on spiral stability while improving the Dutch roll damping. Increasing yaw stability improves the damping for Dutch roll but renders the spiral mode more unstable. Only by increasing yaw damping can we have a favorable impact on the damping for both Dutch roll and the spiral mode. Unfortunately, increasing yaw damping without increasing yaw stability is not easy to achieve through geometric design changes, without significantly increasing the drag.

The most obvious way to increase yaw damping is to increase the area of the vertical tail. However, this will add drag and increase yaw stability, which has a destabilizing effect on the spiral mode. By adding a vertical surface forward of the center of gravity, we could increase yaw damping while decreasing yaw stability. By selecting the proper combination of vertical surfaces both forward and aft of the center of gravity, it would be possible to design an airplane to have any desired combination of yaw stability and yaw damping. However, vertical surfaces placed in front of the center of gravity add unnecessarily to the drag and are not normally used. As a better solution to this problem, modern airplanes are often fitted with a feedback control system to provide additional yaw damping without the added drag and yaw stability that are produced by increasing the area of the vertical tail.

An electronic yaw damper can be constructed by using a rate gyro that produces an output signal proportional to the yawing rate of the aircraft. This output signal is multiplied by a gain coefficient and fed back into a control system to produce a rudder deflection proportional to the yawing rate. From the yawing momentum equation given by Eq. (7.7.6), we have

$$\Delta \hat{\dot{r}} - \iota_{zx} \Delta \hat{\dot{p}} - R_{n,\beta} \Delta \beta - R_{n,\bar{p}} \Delta \bar{p} - R_{n,\bar{r}} \Delta \bar{r} \;=\; R_{n,\delta_a} \Delta \delta_a + R_{n,\delta_r} \Delta \delta_r \qquad (9.5.35)$$

With the yaw damper activated, the total rudder deflection is the sum of that imposed by the feedback control system and that imposed by the pilot,

$$\Delta \delta_r \;=\; K \Delta \bar{r} + \left(\Delta \delta_r \right)_{\text{pilot}} \qquad (9.5.36)$$

where K is a dimensionless gain coefficient for the control system. Using Eq. (9.5.36) in Eq. (9.5.35) and rearranging gives

$$\Delta \hat{\dot{r}} - \iota_{zx} \Delta \hat{\dot{p}} - R_{n,\beta} \Delta \beta - R_{n,\bar{p}} \Delta \bar{p} - \left(R_{n,\bar{r}} + K R_{n,\delta_r} \right) \Delta \bar{r}$$
$$= R_{n,\delta_a} \Delta \delta_a + R_{n,\delta_r} \left(\Delta \delta_r \right)_{\text{pilot}} \qquad (9.5.37)$$

Thus, the effective yaw-damping ratio for the aircraft is

$$\left(R_{n,\bar{r}} \right)_{\text{effective}} \;=\; R_{n,\bar{r}} + K R_{n,\delta_r} \qquad (9.5.38)$$

With the sign convention used in this textbook, the sign of the rudder deflection is defined such that a negative yawing moment is produced by positive rudder deflection. Thus, K must be positive to increase the effective yaw damping (i.e., to make the yaw damping derivative more negative). By adjusting the gain on the feedback control system, we can provide any desired amount of yaw damping without affecting the yaw stability. This allows us to control the Dutch roll damping without the adverse effects associated with increasing the vertical tail area.

While the discussion above was aimed at controlling the yaw damping for an airplane, it is possible to control both the yaw damping and yaw stability. This is accomplished by producing a rudder deflection that is proportional to both the sideslip and yawing rate,

$$\Delta\delta_r = K_1\Delta\beta + K_2\Delta\bar{r} + \left(\Delta\delta_r\right)_{\text{pilot}} \tag{9.5.39}$$

Including the effects of this feedback control system, the linearized yawing momentum equation becomes

$$\Delta\hat{\bar{r}} - \iota_{zx}\Delta\hat{\bar{p}} - \left(R_{n,\beta} + K_1 R_{n,\delta_r}\right)\Delta\beta - R_{n,\bar{p}}\Delta\bar{p} - \left(R_{n,\bar{r}} + K_2 R_{n,\delta_r}\right)\Delta\bar{r}$$
$$= R_{n,\delta_a}\left(\Delta\delta_a\right)_{\text{pilot}} + R_{n,\delta_r}\left(\Delta\delta_r\right)_{\text{pilot}} \tag{9.5.40}$$

The dimensionless gains for the feedback control system, K_1 and K_2, can be selected to provide any desired level of yaw damping and yaw stability. With appropriate adjustment of these two gains it is possible to control the Dutch roll damping and frequency independently. Furthermore, these gains can be varied with airspeed and flap setting to provide good Dutch roll characteristics over the full range of operating conditions. However, we must remember that both directional and spiral stability also impose constraints on the yaw stability and damping. The use of feedback control systems for stability augmentation is not a cure-all to be used in place of good geometric design. However, it does provide the designer with an additional degree of freedom, which can be used to ensure good flying qualities over the full range of operating conditions while maintaining satisfactory performance relative to other mission objectives.

9.6. Pure Rolling Motion

While pure rolling motion does not occur in free flight, a wind tunnel model could be constrained to allow motion only in pure roll. In free flight, rolling motion will always produce some sideslip and yaw. However, if a wind tunnel model is constrained as is shown in Fig. 9.6.1, the mounting sting will supply whatever moments are necessary to resist the aerodynamic moments except for those about the roll axis. With a model so constrained, it is possible to obtain information from which certain aerodynamic derivatives can be extracted.

The equation of motion for this model can be obtained from the general lateral equations by applying the constraints

$$\Delta\beta = \Delta\bar{r} = \Delta\delta_r = 0 \tag{9.6.1}$$

Using these constraints in Eq. (7.7.6) with $\Delta\bar{p} = \bar{p}$, the rolling momentum equation becomes

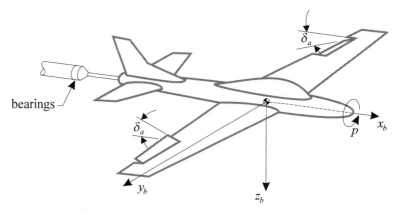

Figure 9.6.1. Wind tunnel model constrained to pure rolling motion.

$$\frac{d\bar{p}}{d\tau_y} - R_{\ell,\bar{p}}\,\bar{p} = R_{\ell,\delta_a}\,\Delta\delta_a \tag{9.6.2}$$

where

$$\bar{p} \equiv \frac{pb_w}{2V_o}, \quad \tau_y \equiv \frac{2V_o t}{b_w}, \quad R_{\ell,\bar{p}} \equiv \frac{\rho S_w b_w^3}{8I_{xx_b}}\,C_{\ell,\bar{p}}, \quad R_{\ell,\delta_a} \equiv \frac{\rho S_w b_w^3}{8I_{xx_b}}\,C_{\ell,\delta_a} \tag{9.6.3}$$

The solution to this first-order differential equation for a step change in aileron deflection is

$$\bar{p} = \frac{R_{\ell,\delta_a}\,\Delta\delta_a}{-R_{\ell,\bar{p}}}\big[1 - \exp(R_{\ell,\bar{p}}\,\tau_y)\big] \tag{9.6.4}$$

After applying the definitions in Eq. (9.6.3), this solution can be written as

$$\frac{pb_w}{2V_o} = \frac{C_{\ell,\delta_a}\,\Delta\delta_a}{-C_{\ell,\bar{p}}}\big[1 - \exp(-\sigma_r t)\big] \tag{9.6.5}$$

where σ_r is the roll-damping rate,

$$\sigma_r = -\frac{\rho S_w b_w^2 V_o}{4I_{xx_b}}\,C_{\ell,\bar{p}} \tag{9.6.6}$$

Because the roll-damping derivative is always negative, the rolling rate will exponentially approach a steady value, as shown in Fig. 9.6.2. From Eq. (9.6.5), the steady rolling rate, which develops after the rolling transient is damped out, must satisfy the equation

$$\frac{pb_w}{2V_o} = \frac{C_{\ell,\delta_a} \Delta\delta_a}{-C_{\ell,\bar{p}}} \tag{9.6.7}$$

Since the roll-damping rate is large, the transient term, $\exp(-\sigma_r t)$, decays to zero very rapidly and the rolling rate quickly approaches a constant value. This large roll-damping rate gives the pilot the impression that aileron deflection controls rolling rate rather than rolling acceleration. For this reason, the maximum rolling rate that can be attained with maximum aileron deflection is an important measure of the lateral control power available from the ailerons. Maximum rolling rate is attained when the rolling moment produced by the aileron deflection is exactly balanced by the rolling moment resulting from the rolling rate. This produces a net rolling moment of zero. The maximum dimensionless rolling rate that is attained at full aileron deflection can be used for sizing the ailerons. The **minimum requirements commonly used for aileron sizing are:**

$$
\begin{array}{ll}
\text{low- to medium- maneuverability aircraft} \\
\text{such as cargo and transport airplanes}
\end{array} \right\} \quad \left(\frac{pb_w}{2V_o}\right)_{\max} \geq 0.07
$$

$$
\begin{array}{ll}
\text{high- maneuverability aircraft} \\
\text{such as fighters and stunt airplanes}
\end{array} \right\} \quad \left(\frac{pb_w}{2V_o}\right)_{\max} \geq 0.09
$$

The rolling rate of the aircraft could be measured as a function of time using a rate gyro. By recording the roll response to a step change in aileron deflection, we could determine both the roll-damping rate and the steady rolling rate that develops after the transient has decayed. This could be done by fitting the measured response to the analytical solution given by Eq. (9.6.5). From the measured damping rate, the dimensionless roll-damping derivative could be determined by rearranging Eq. (9.6.6),

$$C_{\ell,\bar{p}} = -\frac{4I_{xx_b}\sigma_r}{\rho S_w b_w^2 V_o} \tag{9.6.8}$$

Using Eq. (9.6.8) in Eq. (9.6.7), the change in rolling moment with respect to aileron deflection could be evaluated from the measured values of the roll-damping rate and the steady rolling rate, which develops after the damping transient has decayed,

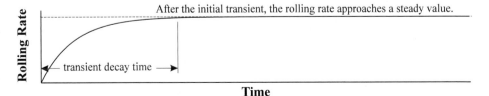

Figure 9.6.2. Roll response to a step change in aileron deflection.

$$C_{\ell,\delta_a} = \frac{2I_{xx_b}\sigma_r\, p_{\text{steady}}}{\rho S_w b_w V_o^2\, \Delta\delta_a} \tag{9.6.9}$$

If the rolling moment of inertia, geometric properties of the model, and wind tunnel operating conditions are all known, Eqs. (9.6.8) and (9.6.9) can be used to determine the aerodynamic derivatives. The rolling moment of inertia for the model could be determined in a manner similar to that described in Sec. 8.5.

9.7. Pure Yawing Motion

In a similar manner, a wind tunnel model could be constrained to pure yawing motion, as shown in Fig. 9.7.1. The equation of motion for this model is obtained from the general lateral equations by applying the constraints

$$\Delta\overline{p} = \Delta\xi_y = \Delta\phi = \Delta\delta_a = 0 \tag{9.7.1}$$

$$\Delta\beta = -\Delta\psi \tag{9.7.2}$$

Using these constraints in Eq. (7.7.6) with $\Delta\overline{r} = \overline{r}$ and $\Delta\psi = \psi$, from the third and sixth rows we have

$$\frac{d\overline{r}}{d\tau_y} = R_{n,\delta_r}\,\Delta\delta_r - R_{n,\beta}\,\dot{\psi} + R_{n,\overline{r}}\,\overline{r} \tag{9.7.3}$$

$$\frac{d\psi}{d\tau_y} = \overline{r} \tag{9.7.4}$$

where

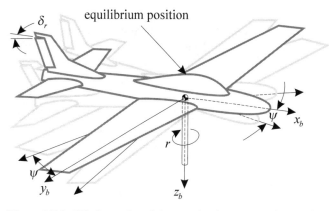

Figure 9.7.1. Wind tunnel model constrained to pure yawing motion.

$$\bar{r} \equiv \frac{rb_w}{2V_o}, \quad \tau_y \equiv \frac{2V_o t}{b_w}, \quad R_{n,\delta_r} \equiv \frac{\rho S_w b_w^3}{8 I_{zz_b}} C_{n,\delta_r},$$

$$R_{n,\beta} \equiv \frac{\rho S_w b_w^3}{8 I_{zz_b}} C_{n,\beta}, \quad R_{n,\bar{r}} \equiv \frac{\rho S_w b_w^3}{8 I_{zz_b}} C_{n,\bar{r}} \tag{9.7.5}$$

Using Eq. (9.7.4) in Eq. (9.7.3) and rearranging, we have

$$\frac{d^2 \psi}{d\tau_y^2} - R_{n,\bar{r}} \frac{d\psi}{d\tau_y} + R_{n,\beta} \psi = R_{n,\delta_r} \Delta \delta_r \tag{9.7.6}$$

The characteristic equation for Eq. (9.7.6) is

$$\lambda^2 - R_{n,\bar{r}} \lambda + R_{n,\beta} = 0 \tag{9.7.7}$$

Thus, the **eigenvalues for pure yawing motion** are

$$\lambda = \frac{R_{n,\bar{r}}}{2} \pm \sqrt{\left(\frac{R_{n,\bar{r}}}{2}\right)^2 - R_{n,\beta}} \tag{9.7.8}$$

By performing wind tunnel tests on an aircraft model that is constrained to pure yawing motion, we could determine the eigenvalues experimentally. The model's moment of inertia should be adjusted to produce underdamped motion. In the wind tunnel, the model is displaced from equilibrium and then released. The resulting oscillatory yawing motion is recorded as a function of time. By measuring the period of oscillation and the time required to reduce the amplitude of oscillation by one-half, we could determine the yaw-damping derivative as well as the yaw-stability derivative for the model.

From Eqs. (9.7.5) and (9.7.8), the 50 percent damping time is given by

$$50\% \text{ damping time} \equiv t_{50} = \frac{b_w}{2V_o} \frac{\ln(0.50)}{\text{real}(\lambda)} = \frac{b_w}{V_o} \frac{\ln(0.50)}{R_{n,\bar{r}}} = \frac{8 I_{zz_b}}{\rho S_w b_w^2 V_o} \frac{\ln(0.50)}{C_{n,\bar{r}}} \tag{9.7.9}$$

and the period is

$$\text{period} \equiv t_p = \frac{b_w}{2V_o} \frac{2\pi}{|\text{imag}(\lambda)|} = \frac{\dfrac{\pi b_w}{V_o}}{\sqrt{R_{n,\beta} - \left(\dfrac{R_{n,\bar{r}}}{2}\right)^2}} = \frac{\dfrac{16\pi I_{zz_b}}{\rho S_w b_w^2 V_o}}{\sqrt{\dfrac{32 I_{zz_b}}{\rho S_w b_w^3} C_{n,\beta} - C_{n,\bar{r}}^2}} \tag{9.7.10}$$

Rearranging Eqs. (9.7.9) and (9.7.10), we have

$$C_{n,\bar{r}} = \frac{8I_{zz_b} \ln(0.50)}{\rho S_w b_w^2 V_o t_{50}} \tag{9.7.11}$$

$$C_{n,\beta} = \frac{2I_{zz_b}}{\rho S_w b_w V_o^2} \left[\left(\frac{2\pi}{t_p} \right)^2 + \left(\frac{\ln(0.50)}{t_{50}} \right)^2 \right] \tag{9.7.12}$$

Using Eqs. (9.7.11) and (9.7.12), we could determine the yaw-damping derivative as well as the yaw-stability derivative from the measured period and 50 percent damping time, provided that we know the wind tunnel airspeed and density, along with the geometry and yawing moment of inertia for the aircraft model. Here again the yawing moment of inertia could be determined in a manner similar to that described in Sec. 8.5.

9.8. Longitudinal-Lateral Coupling

As discussed previously, the lateral equations of motion were uncoupled from the longitudinal equations of motion only as a result of imposing the small-disturbance approximation. The actual response of an aircraft to a finite lateral disturbance will involve all six degrees of freedom. Furthermore, as discussed in Sec. 7.9, inertial and gyroscopic effects can produce coupling between the lateral and longitudinal equations, even within the framework of small disturbance theory. What is normally referred to as *inertial coupling* occurs whenever the aircraft's mass is not completely symmetric about the x_b-z_b plane, and *gyroscopic coupling* occurs as a result of any net angular momentum associated with rotors spinning relative to the body-fixed coordinate system. Aircraft asymmetry can also give rise to aerodynamic coupling, because aerodynamic derivatives such as $C_{\ell,\alpha}$, $C_{n,\alpha}$, and $C_{m,\beta}$ can become nonzero. The occurrence of such first-order coupling is not unusual.

Uncoupling the longitudinal and lateral equations of motion reduces the eigen-problem associated with linearized aircraft dynamics from one 12×12 system to two 6×6 systems. Decades ago, when the theory of linearized aircraft dynamics was developed, this represented a significant computational savings. To reduce computation time even further, each of these 6×6 systems was commonly reduced algebraically to a 4×4 system. The eigenvalues obtained from the 4×4 system are identical to those obtained from the 6×6 system, except that the lower-order system does not yield the trivial rigid-body displacement modes. This represents no loss of information. However, there is a loss of information in the eigenvectors when the 6×6 systems are reduced to 4×4. Two components of each eigenvector are lost. This was a small price to pay for the computational savings that were realized in the mid-twentieth century.

Today, there are no significant computational savings to be gained by uncoupling the linearized longitudinal and lateral equations of motion. Using modern computational facilities, the solution of a 12×12 eigenproblem is trivial. Thus, it is quite simple to account for the effects of longitudinal-lateral coupling in linearized aircraft dynamics.

The Euler angle formulation of the rigid-body 6-DOF equations of motion, which was summarized in Sec. 7.4, can be modified to account for the effects of longitudinal-lateral coupling by replacing Eq. (7.4.5) with Eq. (7.9.5). Thus, **including the effects of gyroscopic and inertial coupling, the 12 first-order differential equations that govern the motion of a rigid aircraft in constant wind** can be written as

$$\frac{W}{g}\begin{Bmatrix} \dot{u} \\ \dot{v} \\ \dot{w} \end{Bmatrix} = \begin{Bmatrix} F_{x_b} \\ F_{y_b} \\ F_{z_b} \end{Bmatrix} + W\begin{Bmatrix} -S_\theta \\ S_\phi C_\theta \\ C_\phi C_\theta \end{Bmatrix} + \frac{W}{g}\begin{Bmatrix} rv - qw \\ pw - ru \\ qu - pv \end{Bmatrix} \tag{9.8.1}$$

$$\begin{bmatrix} I_{xx_b} & -I_{xy_b} & -I_{xz_b} \\ -I_{xy_b} & I_{yy_b} & -I_{yz_b} \\ -I_{xz_b} & -I_{yz_b} & I_{zz_b} \end{bmatrix}\begin{Bmatrix} \dot{p} \\ \dot{q} \\ \dot{r} \end{Bmatrix} = \begin{Bmatrix} M_{x_b} \\ M_{y_b} \\ M_{z_b} \end{Bmatrix} + \begin{bmatrix} 0 & -h_{z_b} & h_{y_b} \\ h_{z_b} & 0 & -h_{x_b} \\ -h_{y_b} & h_{x_b} & 0 \end{bmatrix}\begin{Bmatrix} p \\ q \\ r \end{Bmatrix}$$
$$+\begin{Bmatrix} (I_{yy_b} - I_{zz_b})qr + I_{yz_b}(q^2 - r^2) + I_{xz_b}pq - I_{xy_b}pr \\ (I_{zz_b} - I_{xx_b})pr + I_{xz_b}(r^2 - p^2) + I_{xy_b}qr - I_{yz_b}pq \\ (I_{xx_b} - I_{yy_b})pq + I_{xy_b}(p^2 - q^2) + I_{yz_b}pr - I_{xz_b}qr \end{Bmatrix} \tag{9.8.2}$$

$$\begin{Bmatrix} \dot{x}_f \\ \dot{y}_f \\ \dot{z}_f \end{Bmatrix} = \begin{bmatrix} C_\theta C_\psi & S_\phi S_\theta C_\psi - C_\phi S_\psi & C_\phi S_\theta C_\psi + S_\phi S_\psi \\ C_\theta S_\psi & S_\phi S_\theta S_\psi + C_\phi C_\psi & C_\phi S_\theta S_\psi - S_\phi C_\psi \\ -S_\theta & S_\phi C_\theta & C_\phi C_\theta \end{bmatrix}\begin{Bmatrix} u \\ v \\ w \end{Bmatrix} + \begin{Bmatrix} V_{wx_f} \\ V_{wy_f} \\ V_{wz_f} \end{Bmatrix} \tag{9.8.3}$$

$$\begin{Bmatrix} \dot{\phi} \\ \dot{\theta} \\ \dot{\psi} \end{Bmatrix} = \begin{bmatrix} 1 & S_\phi S_\theta / C_\theta & C_\phi S_\theta / C_\theta \\ 0 & C_\phi & -S_\phi \\ 0 & S_\phi / C_\theta & C_\phi / C_\theta \end{bmatrix}\begin{Bmatrix} p \\ q \\ r \end{Bmatrix} \tag{9.8.4}$$

To obtain the classical dynamic modes from these 12 equations, the system must be linearized relative to some equilibrium flight condition. In Sec. 7.5, the equilibrium flight condition chosen for this purpose was steady climbing flight. This condition was chosen specifically to eliminate the coupling between the longitudinal and lateral modes. However, this is not the most general equilibrium flight condition that can be used to linearize the equations of motion if we are willing to accept longitudinal-lateral coupling in the final result.

Perhaps the most common cause of longitudinal-lateral coupling is simple turning flight. In Sec. 7.5 the equations of motion were linearized relative to a flight path having both the equilibrium bank angle and azimuth angle set to zero. Since there is no force in body-fixed coordinates that depends on azimuth angle, the choice of equilibrium azimuth angle is arbitrary and has no effect on the disturbance equations. On the other hand, allowing for a nonzero equilibrium bank angle will significantly affect the disturbance equations, introducing longitudinal-lateral coupling.

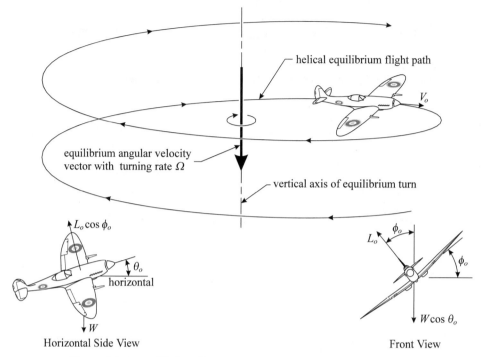

Figure 9.8.1. Equilibrium flight conditions for a steady coordinated turn.

For example, consider the steady coordinated turn shown in Fig. 9.8.1. In a steady coordinated turn the ailerons, elevator, and rudder are coordinated to maintain constant airspeed and angular acceleration while eliminating the side force. The equilibrium angular velocity vector is aligned with the z_f-axis, and by our choice of the body-fixed coordinate system, the equilibrium translational velocity vector is aligned with the x_b-axis. The equilibrium angular velocity vector in body-fixed coordinates can be determined by applying Eq. (7.3.13) to the angular velocity vector in Earth-fixed coordinates,

$$
\begin{Bmatrix} p \\ q \\ r \end{Bmatrix}_o = \begin{bmatrix} C_\theta C_\psi & C_\theta S_\psi & -S_\theta \\ S_\phi S_\theta C_\psi - C_\phi S_\psi & S_\phi S_\theta S_\psi + C_\phi C_\psi & S_\phi C_\theta \\ C_\phi S_\theta C_\psi + S_\phi S_\psi & C_\phi S_\theta S_\psi - S_\phi C_\psi & C_\phi C_\theta \end{bmatrix} \begin{Bmatrix} 0 \\ 0 \\ \Omega \end{Bmatrix} = \Omega \begin{Bmatrix} -S_{\theta_o} \\ S_{\phi_o} C_{\theta_o} \\ C_{\phi_o} C_{\theta_o} \end{Bmatrix} \quad (9.8.5)
$$

The translational velocity vector, in body-fixed coordinates, for this equilibrium flight condition and our choice for the body-fixed coordinate system is

$$
\begin{Bmatrix} u \\ v \\ w \end{Bmatrix}_o = \begin{Bmatrix} V_o \\ 0 \\ 0 \end{Bmatrix} \quad (9.8.6)
$$

From Eq. (9.8.1), constant airspeed and zero side force requires that

$$
\begin{Bmatrix} 0 \\ 0 \\ 0 \end{Bmatrix} = \begin{Bmatrix} F_{x_b} \\ 0 \\ F_{z_b} \end{Bmatrix} + W \begin{Bmatrix} -S_\theta \\ S_\phi C_\theta \\ C_\phi C_\theta \end{Bmatrix} + \frac{W}{g} \begin{Bmatrix} rv - qw \\ pw - ru \\ qu - pv \end{Bmatrix}
\tag{9.8.7}
$$

Applying Eqs. (9.8.5) and (9.8.6) to Eq. (9.8.7), this equilibrium flight condition gives

$$
\begin{Bmatrix} 0 \\ 0 \\ 0 \end{Bmatrix} = \begin{Bmatrix} F_{x_bo} \\ 0 \\ F_{z_bo} \end{Bmatrix} + W \begin{Bmatrix} -S_{\theta_o} \\ S_{\phi_o} C_{\theta_o} \\ C_{\phi_o} C_{\theta_o} \end{Bmatrix} + \frac{W\Omega V_o}{g} \begin{Bmatrix} 0 \\ -C_{\phi_o} C_{\theta_o} \\ S_{\phi_o} C_{\theta_o} \end{Bmatrix}
\tag{9.8.8}
$$

The second equation in Eq. (9.8.8) can be solved for the equilibrium turning rate as a function of airspeed and bank angle,

$$
\Omega = \frac{g \tan \phi_o}{V_o}
\tag{9.8.9}
$$

From Eqs. (9.8.8) and (9.8.9), the net equilibrium aerodynamic force required to maintain a steady coordinated turn is

$$
\begin{Bmatrix} F_{x_b} \\ F_{y_b} \\ F_{z_b} \end{Bmatrix}_o = W \begin{Bmatrix} S_{\theta_o} \\ 0 \\ -C_{\theta_o}/C_{\phi_o} \end{Bmatrix}
\tag{9.8.10}
$$

From Eqs. (9.8.4) and (9.8.5), the equilibrium Euler angle rates are readily found to be

$$
\begin{Bmatrix} \dot\phi \\ \dot\theta \\ \dot\psi \end{Bmatrix}_o = \begin{Bmatrix} 0 \\ 0 \\ \Omega \end{Bmatrix}
\tag{9.8.11}
$$

Integrating Eq. (9.8.11) subject to the initial condition of zero heading at time $t = 0$, we find that the equilibrium Euler angles for a steady coordinated turn are

$$
\begin{Bmatrix} \phi \\ \theta \\ \psi \end{Bmatrix}_o = \begin{Bmatrix} \phi_o \\ \theta_o \\ \Omega t \end{Bmatrix}
\tag{9.8.12}
$$

Similarly, using this result in Eq. (9.8.3) and integrating, the equilibrium position for the steady coordinated turn can be written as

$$\begin{Bmatrix} x_f \\ y_f \\ z_f \end{Bmatrix}_o = \begin{Bmatrix} V_{wx_f}t + V_o C_{\theta_o} S_{\Omega t}/\Omega \\ V_{wy_f}t - V_o C_{\theta_o} C_{\Omega t}/\Omega \\ (V_{wz_f} - V_o S_{\theta_o})t \end{Bmatrix} \tag{9.8.13}$$

Expanding the aircraft flight variables relative to this equilibrium state, we have

$$u = V_o + \Delta u, \qquad\qquad v = \Delta v, \qquad\qquad w = \Delta w,$$

$$p = -gT_{\phi_o}S_{\theta_o}/V_o + \Delta p, \quad q = gT_{\phi_o}S_{\phi_o}C_{\theta_o}/V_o + \Delta q, \quad r = gS_{\phi_o}C_{\theta_o}/V_o + \Delta r,$$

$$F_{x_b} = WS_{\theta_o} + \Delta F_{x_b}, \qquad F_{y_b} = \Delta F_{y_b}, \qquad F_{z_b} = -WC_{\theta_o}/C_{\phi_o} + \Delta F_{z_b},$$

$$M_{x_b} = M_{x_b 0} + \Delta M_{x_b}, \qquad M_{y_b} = M_{y_b 0} + \Delta M_{y_b}, \qquad M_{z_b} = M_{z_b 0} + \Delta M_{z_b},$$

$$x_f = V_{wx_f}t + \frac{V_o^2 C_{\theta_o} S_{\Omega t}}{gT_{\phi_o}} \qquad y_f = V_{wy_f}t - \frac{V_o^2 C_{\theta_o} C_{\Omega t}}{gT_{\phi_o}} \qquad z_f = (V_{wz_f} - V_o S_{\theta_o})t$$
$$+ \Delta x_f, \qquad\qquad\qquad + \Delta y_f, \qquad\qquad\qquad + \Delta z_f,$$

$$\phi = \phi_o + \Delta\phi, \qquad\qquad \theta = \theta_o + \Delta\theta, \qquad\qquad \psi = (g/V_o)T_{\phi_o}t + \Delta\psi$$

where T_θ is shorthand for $\tan\theta$ and the equilibrium moments are obtained by using these values in Eq. (9.8.2) with all disturbances set to zero. Equations (9.8.1) through (9.8.4) can now be linearized following Sec. 7.5 (see Phillips and Santana 2002). This gives **the general linearized 6-DOF equations of motion for aircraft dynamics,**

$$\begin{bmatrix} [\mathbf{B}]_m & [\mathbf{n}] & [\mathbf{n}] & [\mathbf{n}] \\ [\mathbf{B}]_{iv} & [\mathbf{I}] & [\mathbf{n}] & [\mathbf{n}] \\ [\mathbf{n}] & [\mathbf{n}] & [\mathbf{i}] & [\mathbf{n}] \\ [\mathbf{n}] & [\mathbf{n}] & [\mathbf{n}] & [\mathbf{i}] \end{bmatrix} \begin{Bmatrix} \Delta\dot{u} \\ \Delta\dot{v} \\ \Delta\dot{w} \\ \Delta\dot{p} \\ \Delta\dot{q} \\ \Delta\dot{r} \\ \Delta\dot{x}_c \\ \Delta\dot{y}_c \\ \Delta\dot{z}_c \\ \Delta\dot{\phi} \\ \Delta\dot{\theta} \\ \Delta\dot{\psi} \end{Bmatrix} - \begin{bmatrix} [\mathbf{A}]_{fv} & [\mathbf{A}]_{f\omega} & [\mathbf{n}] & [\mathbf{A}]_w \\ [\mathbf{A}]_{mv} & [\mathbf{A}]_{m\omega} & [\mathbf{n}] & [\mathbf{n}] \\ [\mathbf{A}]_{pv} & [\mathbf{n}] & [\mathbf{n}] & [\mathbf{A}]_{pe} \\ [\mathbf{n}] & [\mathbf{A}]_{e\omega} & [\mathbf{n}] & [\mathbf{A}]_{ee} \end{bmatrix} \begin{Bmatrix} \Delta u \\ \Delta v \\ \Delta w \\ \Delta p \\ \Delta q \\ \Delta r \\ \Delta x_c \\ \Delta y_c \\ \Delta z_c \\ \Delta\phi \\ \Delta\theta \\ \Delta\psi \end{Bmatrix}$$

$$= \begin{bmatrix} [\mathbf{C}]_f \\ [\mathbf{C}]_m \\ [\mathbf{n}] \\ [\mathbf{n}] \end{bmatrix} \begin{Bmatrix} \Delta\delta_a \\ \Delta\delta_e \\ \Delta\delta_r \end{Bmatrix} \tag{9.8.14}$$

where

$$[\mathbf{B}]_m = \begin{bmatrix} \dfrac{W}{g} - F_{x_b,\dot{u}} & 0 & -F_{x_b,\dot{w}} \\[2mm] 0 & \dfrac{W}{g} & 0 \\[2mm] -F_{z_b,\dot{u}} & 0 & \dfrac{W}{g} - F_{z_b,\dot{w}} \end{bmatrix}$$

$$[\mathbf{B}]_{iv} = \begin{bmatrix} 0 & 0 & 0 \\ -M_{y_b,\dot{u}} & 0 & -M_{y_b,\dot{w}} \\ 0 & 0 & 0 \end{bmatrix} \qquad [\mathbf{I}] = \begin{bmatrix} I_{xx_b} & -I_{xy_b} & -I_{xz_b} \\ -I_{xy_b} & I_{yy_b} & -I_{yz_b} \\ -I_{xz_b} & -I_{yz_b} & I_{zz_b} \end{bmatrix}$$

$$[\mathbf{A}]_{fv} = \begin{bmatrix} F_{x_b,u} & \dfrac{WS_{\phi_o}C_{\theta_o}}{V_o} & F_{x_b,w} - \dfrac{WT_{\phi_o}S_{\phi_o}C_{\theta_o}}{V_o} \\[3mm] -\dfrac{WS_{\phi_o}C_{\theta_o}}{V_o} & F_{y_b,v} & -\dfrac{WT_{\phi_o}S_{\theta_o}}{V_o} \\[3mm] F_{z_b,u} + \dfrac{WT_{\phi_o}S_{\phi_o}C_{\theta_o}}{V_o} & \dfrac{WT_{\phi_o}S_{\theta_o}}{V_o} & F_{z_b,w} \end{bmatrix}$$

$$[\mathbf{A}]_{f\omega} = \begin{bmatrix} 0 & F_{x_b,q} & 0 \\[2mm] F_{y_b,p} & 0 & F_{y_b,r} - \dfrac{V_oW}{g} \\[2mm] 0 & F_{z_b,q} + \dfrac{V_oW}{g} & 0 \end{bmatrix} \qquad [\mathbf{A}]_w = \begin{bmatrix} 0 & -WC_{\theta_o} & 0 \\ WC_{\phi_o}C_{\theta_o} & -WS_{\phi_o}S_{\theta_o} & 0 \\ -WS_{\phi_o}C_{\theta_o} & -WC_{\phi_o}S_{\theta_o} & 0 \end{bmatrix}$$

$$[\mathbf{A}]_{mv} = \begin{bmatrix} 0 & M_{x_b,v} & M_{x_b,w} \\ M_{y_b,u} & M_{y_b,v} & M_{y_b,w} \\ 0 & M_{z_b,v} & M_{z_b,w} \end{bmatrix} \qquad [\mathbf{A}]_{m\omega} = \begin{bmatrix} A_{\ell p} & A_{\ell q} & A_{\ell r} \\ A_{mp} & A_{mq} & A_{mr} \\ A_{np} & A_{nq} & A_{nr} \end{bmatrix}$$

$$A_{\ell p} = M_{x_b,p} + \frac{g(I_{xz_b}T_{\phi_o}S_{\phi_o}C_{\theta_o} - I_{xy_b}S_{\phi_o}C_{\theta_o})}{V_o}$$

$$A_{\ell q} = -h_{z_b} + \frac{g[(I_{yy_b} - I_{zz_b})S_{\phi_o}C_{\theta_o} + 2I_{yz_b}T_{\phi_o}S_{\phi_o}C_{\theta_o} - I_{xz_b}T_{\phi_o}S_{\theta_o}]}{V_o}$$

$$A_{\ell r} = M_{x_b,r} + h_{y_b} + \frac{g[(I_{yy_b} - I_{zz_b})T_{\phi_o}S_{\phi_o}C_{\theta_o} - 2I_{yz_b}S_{\phi_o}C_{\theta_o} + I_{xy_b}T_{\phi_o}S_{\theta_o}]}{V_o}$$

$$A_{mp} = h_{z_b} + \frac{g[(I_{zz_b} - I_{xx_b})S_{\phi_o}C_{\theta_o} + 2I_{xz_b}T_{\phi_o}S_{\theta_o} - I_{yz_b}T_{\phi_o}S_{\phi_o}C_{\theta_o}]}{V_o}$$

$$A_{mq} = M_{y_b,q} + \frac{g(I_{xy_b}S_{\phi_o}C_{\theta_o} + I_{yz_b}T_{\phi_o}S_{\theta_o})}{V_o}$$

$$A_{mr} = -h_{x_b} + \frac{g[(I_{xx_b} - I_{zz_b})T_{\phi_o}S_{\theta_o} + 2I_{xz_b}S_{\phi_o}C_{\theta_o} + I_{xy_b}T_{\phi_o}S_{\phi_o}C_{\theta_o}]}{V_o}$$

$$A_{np} = M_{z_b,p} - h_{y_b} + \frac{g[(I_{xx_b} - I_{yy_b})T_{\phi_o}S_{\phi_o}C_{\theta_o} - 2I_{xy_b}T_{\phi_o}S_{\theta_o} + I_{yz_b}S_{\phi_o}C_{\theta_o}]}{V_o}$$

$$A_{nq} = h_{x_b} + \frac{g[(I_{yy_b} - I_{xx_b})T_{\phi_o}S_{\theta_o} - 2I_{xy_b}T_{\phi_o}S_{\phi_o}C_{\theta_o} - I_{xz_b}S_{\phi_o}C_{\theta_o}]}{V_o}$$

$$A_{nr} = M_{z_b,r} + \frac{g(-I_{yz_b}T_{\phi_o}S_{\theta_o} - I_{xz_b}T_{\phi_o}S_{\phi_o}C_{\theta_o})}{V_o}$$

$$[\mathbf{A}]_{pv} = \begin{bmatrix} C_{\theta_o} & S_{\phi_o}S_{\theta_o} & C_{\phi_o}S_{\theta_o} \\ 0 & C_{\phi_o} & -S_{\phi_o} \\ -S_{\theta_o} & S_{\phi_o}C_{\theta_o} & C_{\phi_o}C_{\theta_o} \end{bmatrix} \qquad [\mathbf{A}]_{pe} = \begin{bmatrix} 0 & -V_oS_{\theta_o} & 0 \\ 0 & 0 & V_oC_{\theta_o} \\ 0 & -V_oC_{\theta_o} & 0 \end{bmatrix}$$

$$[\mathbf{A}]_{e\omega} = \begin{bmatrix} 1 & S_{\phi_o}T_{\theta_o} & C_{\phi_o}T_{\theta_o} \\ 0 & C_{\phi_o} & -S_{\phi_o} \\ 0 & \dfrac{S_{\phi_o}}{C_{\theta_o}} & \dfrac{C_{\phi_o}}{C_{\theta_o}} \end{bmatrix} \qquad [\mathbf{A}]_{ee} = \begin{bmatrix} 0 & \dfrac{gT_{\phi_o}}{V_oC_{\theta_o}} & 0 \\ -\dfrac{gT_{\phi_o}C_{\theta_o}}{V_o} & 0 & 0 \\ 0 & \dfrac{gT_{\phi_o}T_{\theta_o}}{V_o} & 0 \end{bmatrix}$$

$$[\mathbf{C}]_f = \begin{bmatrix} 0 & F_{x_b,\delta_e} & 0 \\ F_{y_b,\delta_a} & 0 & F_{y_b,\delta_r} \\ 0 & F_{z_b,\delta_e} & 0 \end{bmatrix} \qquad [\mathbf{C}]_m = \begin{bmatrix} M_{x_b,\delta_a} & 0 & M_{x_b,\delta_r} \\ 0 & M_{y_b,\delta_e} & 0 \\ M_{z_b,\delta_a} & 0 & M_{z_b,\delta_r} \end{bmatrix}$$

$$\begin{Bmatrix} \Delta\dot{x}_c \\ \Delta\dot{y}_c \\ \Delta\dot{z}_c \end{Bmatrix} = \begin{bmatrix} C_{\Omega t} & S_{\Omega t} & 0 \\ -S_{\Omega t} & C_{\Omega t} & 0 \\ 0 & 0 & 1 \end{bmatrix} \begin{Bmatrix} \Delta\dot{x}_f \\ \Delta\dot{y}_f \\ \Delta\dot{z}_f \end{Bmatrix} \qquad [\mathbf{i}] = \begin{bmatrix} 1 & 0 & 0 \\ 0 & 1 & 0 \\ 0 & 0 & 1 \end{bmatrix} \qquad [\mathbf{n}] = \begin{bmatrix} 0 & 0 & 0 \\ 0 & 0 & 0 \\ 0 & 0 & 0 \end{bmatrix}$$

Equation (9.8.14) can also be used to study steady climbing flight along a linear flight path. With the equilibrium bank angle, ϕ_o, set to zero, Eq. (9.8.14) provides the linearized equations of motion for linear climbing flight, including the effects of inertial, gyroscopic, and aerodynamic coupling. Steady level flight corresponds to the special case where both ϕ_o and θ_o are zero.

As can be seen from the formulation presented here, a wide range of longitudinal-lateral coupling is introduced by simple turning flight. Gyroscopic coupling enters the linearized equations of motion only through its effect on $[\mathbf{A}]_{m\omega}$, and with no bank angle, inertial asymmetry produces coupling only through the moment of inertia tensor, $[\mathbf{I}]$. A nonzero bank angle, on the other hand, has broader effects on the linearized equations of motion. Turning flight introduces coupling in the translational velocity components as a result of the Coriolis forces seen in $[\mathbf{A}]_{fv}$. Additional translational coupling is introduced through the weight tensor, $[\mathbf{A}]_w$, when the bank angle is nonzero. Turning also produces considerable rotational coupling as a result of the gyroscopic effects seen in $[\mathbf{A}]_{m\omega}$. Finally, turning produces coupling in the position and orientation equations through $[\mathbf{A}]_{pv}$, $[\mathbf{A}]_{e\omega}$, and $[\mathbf{A}]_{ee}$.

To nondimensionalize Eq. (9.8.14), it is convenient to select a single reference length for all longitudinal and lateral variables. Retaining the equilibrium airspeed as the reference velocity, the nondimensional velocity disturbances remain,

$$\Delta\mu \equiv \frac{\Delta u}{V_o}$$

$$\Delta\beta \equiv \frac{\Delta v}{V_o} \tag{9.8.15}$$

$$\Delta\alpha \equiv \frac{\Delta w}{V_o}$$

Denoting the reference length as l_{ref}, the dimensionless angular rate disturbances are defined as

$$\Delta\alpha' \equiv \frac{\Delta\dot\alpha\, l_{\text{ref}}}{V_o} = \frac{2l_{\text{ref}}}{\bar{c}_w}\Delta\hat\alpha$$

$$\Delta\breve{p} \equiv \frac{\Delta p\, l_{\text{ref}}}{V_o} = \frac{2l_{\text{ref}}}{b_w}\Delta\bar{p}$$

$$\Delta\breve{q} \equiv \frac{\Delta q\, l_{\text{ref}}}{V_o} = \frac{2l_{\text{ref}}}{\bar{c}_w}\Delta\bar{q} \tag{9.8.16}$$

$$\Delta\breve{r} \equiv \frac{\Delta r\, l_{\text{ref}}}{V_o} = \frac{2l_{\text{ref}}}{b_w}\Delta\bar{r}$$

Similarly, the disturbances in aircraft position are now nondimensionalized as

$$\Delta\varsigma_x \equiv \frac{\Delta x_c}{l_{\text{ref}}} = \frac{\bar{c}_w}{2l_{\text{ref}}}\Delta\xi_x$$

$$\Delta\varsigma_y \equiv \frac{\Delta y_c}{l_{\text{ref}}} = \frac{b_w}{2l_{\text{ref}}}\Delta\xi_y \tag{9.8.17}$$

$$\Delta\varsigma_z \equiv \frac{\Delta z_c}{l_{\text{ref}}} = \frac{\bar{c}_w}{2l_{\text{ref}}}\Delta\xi_z$$

For dimensionless time we define

$$\tau \equiv \frac{V_o t}{l_{\text{ref}}} = \frac{\bar{c}_w}{2l_{\text{ref}}}\tau_x = \frac{b_w}{2l_{\text{ref}}}\tau_y \tag{9.8.18}$$

and

$$f' \equiv \frac{\partial f}{\partial \tau} = \frac{l_{\text{ref}}}{V_o}\frac{\partial f}{\partial t} = \frac{2l_{\text{ref}}}{\bar{c}_w}\left(\hat{f}\right)_{\text{longitudinal}} = \frac{2l_{\text{ref}}}{b_w}\left(\hat{f}\right)_{\text{lateral}} \tag{9.8.19}$$

With these definitions, **the nondimensional linearized equations of motion** are

$$
\begin{bmatrix}
1-B_{x,\mu'} & 0 & -B_{x,\alpha'} & 0 & 0 & 0 & 0 & 0 & 0 & 0 & 0 & 0\\
0 & 1 & 0 & 0 & 0 & 0 & 0 & 0 & 0 & 0 & 0 & 0\\
-B_{z,\mu'} & 0 & 1-B_{z,\alpha'} & 0 & 0 & 0 & 0 & 0 & 0 & 0 & 0 & 0\\
0 & 0 & 0 & 1 & -t_{xy} & -t_{xz} & 0 & 0 & 0 & 0 & 0 & 0\\
-B_{m,\mu'} & 0 & -B_{m,\alpha'} & -t_{yx} & 1 & -t_{yz} & 0 & 0 & 0 & 0 & 0 & 0\\
0 & 0 & 0 & -t_{zx} & -t_{zy} & 1 & 0 & 0 & 0 & 0 & 0 & 0\\
0 & 0 & 0 & 0 & 0 & 0 & 1 & 0 & 0 & 0 & 0 & 0\\
0 & 0 & 0 & 0 & 0 & 0 & 0 & 1 & 0 & 0 & 0 & 0\\
0 & 0 & 0 & 0 & 0 & 0 & 0 & 0 & 1 & 0 & 0 & 0\\
0 & 0 & 0 & 0 & 0 & 0 & 0 & 0 & 0 & 1 & 0 & 0\\
0 & 0 & 0 & 0 & 0 & 0 & 0 & 0 & 0 & 0 & 1 & 0\\
0 & 0 & 0 & 0 & 0 & 0 & 0 & 0 & 0 & 0 & 0 & 1
\end{bmatrix}
\begin{Bmatrix}
\Delta\mu'\\ \Delta\beta'\\ \Delta\alpha'\\ \Delta\bar{p}'\\ \Delta\bar{q}'\\ \Delta\bar{r}'\\ \Delta\varsigma'_x\\ \Delta\varsigma'_y\\ \Delta\varsigma'_z\\ \Delta\phi'\\ \Delta\theta'\\ \Delta\psi'
\end{Bmatrix}
=
$$

$$
\begin{bmatrix}
A_{x,\mu} & A_g S_{\phi_o} C_{\theta_o} & A_{x,\alpha}-A_g T_{\phi_o} S_{\phi_o} C_{\theta_o} & 0 & A_{x,\bar{q}} & 0 & 0 & 0 & 0 & 0 & -A_g C_{\theta_o} & 0\\
-A_g S_{\phi_o} C_{\theta_o} & A_{y,\beta} & -A_g T_{\phi_o} S_{\theta_o} & A_{y,\bar{p}} & 0 & A_{y,\bar{r}}-1 & 0 & 0 & 0 & A_g C_{\phi_o} C_{\theta_o} & -A_g S_{\phi_o} S_{\theta_o} & 0\\
A_{z,\mu}+A_g T_{\phi_o} S_{\phi_o} C_{\theta_o} & A_g T_{\phi_o} S_{\theta_o} & A_{z,\alpha} & 0 & A_{z,\bar{q}}+1 & 0 & 0 & 0 & 0 & -A_g S_{\phi_o} C_{\theta_o} & -A_g C_{\phi_o} S_{\theta_o} & 0\\
0 & A_{\ell,\beta} & A_{\ell,\alpha} & A_{\ell,\bar{p}}+\eta_{xx} & -\eta_{xy} & A_{\ell,\bar{r}}+\eta_{xz} & 0 & 0 & 0 & 0 & 0 & 0\\
A_{m,\mu} & A_{m,\beta} & A_{m,\alpha} & \eta_{yx} & A_{m,\bar{q}}+\eta_{yy} & -\eta_{yz} & 0 & 0 & 0 & 0 & 0 & 0\\
0 & A_{n,\beta} & A_{n,\alpha} & A_{n,\bar{p}}-\eta_{zx} & \eta_{zy} & A_{n,\bar{r}}+\eta_{zz} & 0 & 0 & 0 & 0 & 0 & 0\\
C_{\theta_o} & S_{\phi_o} S_{\theta_o} & C_{\phi_o} S_{\theta_o} & 0 & 0 & 0 & 0 & 0 & 0 & 0 & -S_{\theta_o} & 0\\
0 & C_{\phi_o} & -S_{\phi_o} & 0 & 0 & 0 & 0 & 0 & 0 & 0 & 0 & C_{\theta_o}\\
-S_{\theta_o} & S_{\phi_o} C_{\theta_o} & C_{\phi_o} C_{\theta_o} & 0 & 0 & 0 & 0 & 0 & 0 & 0 & -C_{\theta_o} & 0\\
0 & 0 & 0 & 1 & S_{\phi_o} T_{\theta_o} & C_{\phi_o} T_{\theta_o} & 0 & 0 & 0 & 0 & A_g T_{\phi_o}/C_{\theta_o} & 0\\
0 & 0 & 0 & 0 & C_{\phi_o} & -S_{\phi_o} & 0 & 0 & 0 & -A_g T_{\phi_o} C_{\theta_o} & -C_{\theta_o} & 0\\
0 & 0 & 0 & 0 & S_{\phi_o}/C_{\theta_o} & C_{\phi_o}/C_{\theta_o} & 0 & 0 & 0 & 0 & A_g T_{\phi_o} T_{\theta_o} & 0
\end{bmatrix}
\begin{Bmatrix}
\Delta\mu\\ \Delta\beta\\ \Delta\alpha\\ \Delta\bar{p}\\ \Delta\bar{q}\\ \Delta\bar{r}\\ \Delta\varsigma_x\\ \Delta\varsigma_y\\ \Delta\varsigma_z\\ \Delta\phi\\ \Delta\theta\\ \Delta\psi
\end{Bmatrix}
$$

$$
+
\begin{bmatrix}
0 & D_{x,\delta_e} & D_{x,\delta_r}\\
D_{y,\delta_a} & 0 & D_{y,\delta_r}\\
0 & D_{z,\delta_e} & 0\\
D_{\ell,\delta_a} & 0 & D_{\ell,\delta_r}\\
0 & D_{m,\delta_e} & 0\\
D_{n,\delta_a} & 0 & D_{n,\delta_r}\\
0 & 0 & 0\\
0 & 0 & 0\\
0 & 0 & 0\\
0 & 0 & 0\\
0 & 0 & 0\\
0 & 0 & 0
\end{bmatrix}
\begin{Bmatrix}
\Delta\delta_a\\ \Delta\delta_e\\ \Delta\delta_r
\end{Bmatrix}
\tag{9.8.20}
$$

where

$$\iota_{xy} \equiv \frac{I_{xy_b}}{I_{xx_b}} \quad \iota_{xz} \equiv \frac{I_{xz_b}}{I_{xx_b}} \quad \iota_{yx} \equiv \frac{I_{xy_b}}{I_{yy_b}} \quad \iota_{yz} \equiv \frac{I_{yz_b}}{I_{yy_b}} \quad \iota_{zx} \equiv \frac{I_{xz_b}}{I_{zz_b}} \quad \iota_{zy} \equiv \frac{I_{yz_b}}{I_{zz_b}}$$

$$B_{x,\mu'} \equiv \frac{\rho S_w \bar{c}_w}{4W/g} C_{X,\hat{\mu}} \quad B_{z,\mu'} \equiv \frac{\rho S_w \bar{c}_w}{4W/g} C_{Z,\hat{\mu}} \quad B_{m,\mu'} \equiv \frac{\rho S_w \bar{c}_w^2 l_{\text{ref}}}{4 I_{yy_b}} C_{m,\hat{\mu}}$$

$$B_{x,\alpha'} \equiv \frac{\rho S_w \bar{c}_w}{4W/g} C_{X,\hat{\alpha}} \quad B_{z,\alpha'} \equiv \frac{\rho S_w \bar{c}_w}{4W/g} C_{Z,\hat{\alpha}} \quad B_{m,\alpha'} \equiv \frac{\rho S_w \bar{c}_w^2 l_{\text{ref}}}{4 I_{yy_b}} C_{m,\hat{\alpha}}$$

$$\eta_{xx} = A_g \frac{I_{xz_b} T_{\phi_o} S_{\phi_o} C_{\theta_o} - I_{xy_b} S_{\phi_o} C_{\theta_o}}{I_{xx_b}}$$

$$\eta_{xy} = \frac{h_{z_b} l_{\text{ref}}}{I_{xx_b} V_o} + A_g \frac{(I_{zz_b} - I_{yy_b}) S_{\phi_o} C_{\theta_o} - 2 I_{yz_b} T_{\phi_o} S_{\phi_o} C_{\theta_o} + I_{xz_b} T_{\phi_o} S_{\theta_o}}{I_{xx_b}}$$

$$\eta_{xz} = \frac{h_{y_b} l_{\text{ref}}}{I_{xx_b} V_o} + A_g \frac{(I_{yy_b} - I_{zz_b}) T_{\phi_o} S_{\phi_o} C_{\theta_o} - 2 I_{yz_b} S_{\phi_o} C_{\theta_o} + I_{xy_b} T_{\phi_o} S_{\theta_o}}{I_{xx_b}}$$

$$\eta_{yx} = \frac{h_{z_b} l_{\text{ref}}}{I_{yy_b} V_o} + A_g \frac{(I_{zz_b} - I_{xx_b}) S_{\phi_o} C_{\theta_o} + 2 I_{xz_b} T_{\phi_o} S_{\theta_o} - I_{yz_b} T_{\phi_o} S_{\phi_o} C_{\theta_o}}{I_{yy_b}}$$

$$\eta_{yy} = A_g \frac{I_{xy_b} S_{\phi_o} C_{\theta_o} + I_{yz_b} T_{\phi_o} S_{\theta_o}}{I_{yy_b}}$$

$$\eta_{yz} = \frac{h_{x_b} l_{\text{ref}}}{I_{yy_b} V_o} + A_g \frac{(I_{zz_b} - I_{xx_b}) T_{\phi_o} S_{\theta_o} - 2 I_{xz_b} S_{\phi_o} C_{\theta_o} - I_{xy_b} T_{\phi_o} S_{\phi_o} C_{\theta_o}}{I_{yy_b}}$$

$$\eta_{zx} = \frac{h_{y_b} l_{\text{ref}}}{I_{zz_b} V_o} + A_g \frac{(I_{yy_b} - I_{xx_b}) T_{\phi_o} S_{\phi_o} C_{\theta_o} + 2 I_{xy_b} T_{\phi_o} S_{\theta_o} - I_{yz_b} S_{\phi_o} C_{\theta_o}}{I_{zz_b}}$$

$$\eta_{zy} = \frac{h_{x_b} l_{\text{ref}}}{I_{zz_b} V_o} + A_g \frac{(I_{yy_b} - I_{xx_b}) T_{\phi_o} S_{\theta_o} - 2 I_{xy_b} T_{\phi_o} S_{\phi_o} C_{\theta_o} - I_{xz_b} S_{\phi_o} C_{\theta_o}}{I_{zz_b}}$$

$$\eta_{zz} = A_g \frac{-I_{yz_b} T_{\phi_o} S_{\theta_o} - I_{xz_b} T_{\phi_o} S_{\phi_o} C_{\theta_o}}{I_{zz_b}}$$

$$A_g \equiv \frac{g l_{\text{ref}}}{V_o^2} \quad A_{x,\mu} \equiv \frac{\rho S_w l_{\text{ref}}}{2W/g} \left(2 C_X + C_{X,\mu} + \frac{T_{,V} \cos(\alpha_{T0})}{\frac{1}{2} \rho V_o S_w} \right)$$

$$A_{z,\mu} \equiv \frac{\rho S_w l_{\text{ref}}}{2W/g} \left(2 C_Z + C_{Z,\mu} - \frac{T_{,V} \sin(\alpha_{T0})}{\frac{1}{2} \rho V_o S_w} \right)$$

$$A_{m,\mu} \equiv \frac{\rho S_w \bar{c}_w l_{\text{ref}}^2}{2 I_{yy_b}} \left(2 C_m + C_{m,\mu} + \frac{T_{,V} [z_T \cos(\alpha_{T0}) + x_T \sin(\alpha_{T0})]}{\frac{1}{2} \rho V_o S_w \bar{c}_w} \right)$$

$$A_{x,\alpha} \equiv \frac{\rho S_w l_{ref}}{2W/g} C_{X,\alpha} \qquad A_{z,\alpha} \equiv \frac{\rho S_w l_{ref}}{2W/g} C_{Z,\alpha} \qquad A_{m,\alpha} \equiv \frac{\rho S_w \bar{c}_w l_{ref}^2}{2 I_{yy_b}} C_{m,\alpha}$$

$$A_{y,\beta} \equiv \frac{\rho S_w l_{ref}}{2W/g} C_{Y,\beta} \qquad A_{\ell,\beta} \equiv \frac{\rho S_w b_w l_{ref}^2}{2 I_{xx_b}} C_{\ell,\beta} \qquad A_{n,\beta} \equiv \frac{\rho S_w b_w l_{ref}^2}{2 I_{zz_b}} C_{n,\beta}$$

$$A_{\ell,\alpha} \equiv \frac{\rho S_w b_w l_{ref}^2}{2 I_{xx_b}} C_{\ell,\alpha} \qquad A_{m,\beta} \equiv \frac{\rho S_w \bar{c}_w l_{ref}^2}{2 I_{yy_b}} C_{m,\beta} \qquad A_{n,\alpha} \equiv \frac{\rho S_w b_w l_{ref}^2}{2 I_{zz_b}} C_{n,\alpha}$$

$$A_{y,\bar{p}} \equiv \frac{\rho S_w b_w}{4W/g} C_{Y,\bar{p}} \qquad A_{\ell,\bar{p}} \equiv \frac{\rho S_w b_w^2 l_{ref}}{4 I_{xx_b}} C_{\ell,\bar{p}} \qquad A_{n,\bar{p}} \equiv \frac{\rho S_w b_w^2 l_{ref}}{4 I_{zz_b}} C_{n,\bar{p}}$$

$$A_{x,\bar{q}} \equiv \frac{\rho S_w \bar{c}_w}{4W/g} C_{X,\bar{q}} \qquad A_{z,\bar{q}} \equiv \frac{\rho S_w \bar{c}_w}{4W/g} C_{Z,\bar{q}} \qquad A_{m,\bar{q}} \equiv \frac{\rho S_w \bar{c}_w^2 l_{ref}}{4 I_{yy_b}} C_{m,\bar{q}}$$

$$A_{y,\bar{r}} \equiv \frac{\rho S_w b_w}{4W/g} C_{Y,\bar{r}} \qquad A_{\ell,\bar{r}} \equiv \frac{\rho S_w b_w^2 l_{ref}}{4 I_{xx_b}} C_{\ell,\bar{r}} \qquad A_{n,\bar{r}} \equiv \frac{\rho S_w b_w^2 l_{ref}}{4 I_{zz_b}} C_{n,\bar{r}}$$

$$D_{y,\delta_a} \equiv \frac{\rho S_w l_{ref}}{2W/g} C_{Y,\delta_a} \qquad D_{\ell,\delta_a} \equiv \frac{\rho S_w b_w l_{ref}^2}{2 I_{xx_b}} C_{\ell,\delta_a} \qquad D_{n,\delta_a} \equiv \frac{\rho S_w b_w l_{ref}^2}{2 I_{zz_b}} C_{n,\delta_a}$$

$$D_{x,\delta_e} \equiv \frac{\rho S_w l_{ref}}{2W/g} C_{X,\delta_e} \qquad D_{z,\delta_e} \equiv \frac{\rho S_w l_{ref}}{2W/g} C_{Z,\delta_e} \qquad D_{m,\delta_e} \equiv \frac{\rho S_w \bar{c}_w l_{ref}^2}{2 I_{yy_b}} C_{m,\delta_e}$$

$$D_{y,\delta_r} \equiv \frac{\rho S_w l_{ref}}{2W/g} C_{Y,\delta_r} \qquad D_{\ell,\delta_r} \equiv \frac{\rho S_w b_w l_{ref}^2}{2 I_{xx_b}} C_{\ell,\delta_r} \qquad D_{n,\delta_r} \equiv \frac{\rho S_w b_w l_{ref}^2}{2 I_{zz_b}} C_{n,\delta_r}$$

As was the case for Eqs. (7.7.5) and (7.7.6), **the reference length and time scale used to nondimensionalize Eq. (9.8.20) are arbitrary and have no physical significances.**

Even for the case of no coupling, Eq. (9.8.20) can be used to obtain the eigenvalues and eigenvectors for both longitudinal and lateral motion. When all coupling terms are zero, results obtained from Eq. (9.8.20) are identical to those obtained from Eqs. (7.7.5) and (7.7.6). The only advantage to be gained from using Eqs. (7.7.5) and (7.7.6) in place of Eq. (9.8.20) is computational efficiency. Two 6×6 eigenproblems can be numerically solved slightly faster than one 12×12 eigenproblem. However, with the computational hardware and software available today, when solving something as simple as a 12×12 eigenproblem, it typically takes longer to display the results than it does to perform the numerical computations. Thus, with problems of this level of complexity, computation time is no longer an issue. With current computational tools, the advantages of being able to account for inertial, gyroscopic, and aerodynamic coupling, as well as turning flight, far outweigh any computational advantage of using the uncoupled equations for longitudinal and lateral motion. Thus, when developing computer code for today's hardware, it is recommended that Eq. (9.8.20) be used instead of Eqs. (7.7.5) and (7.7.6).

When the longitudinal-lateral coupling terms in Eq. (9.8.20) are present but small, all of the traditional longitudinal and lateral modes are still observed in the solution to the associated eigenproblem. The net effect of a small amount of coupling is to produce modest changes in the frequencies and damping rates associated with the traditional modes while introducing some lateral motion into the longitudinal modes, and vice versa. This is the case that most accurately describes the majority of conventional airplanes. At least a small amount of longitudinal-lateral coupling is usually present in any aircraft. If the coupling terms in Eq. (9.8.20) are large enough, the interactions between the longitudinal and lateral degrees of freedom can be sufficient to render the traditional longitudinal and lateral modes unrecognizable.

Aerodynamic coupling between the longitudinal and lateral modes is embodied in the linearized formulation through the inclusion of additional aerodynamic derivatives, which were not included in Eqs. (7.7.5) and (7.7.6). In the present formulation three additional aerodynamic derivatives have been included: the change in rolling moment coefficient with angle of attack, $C_{\ell,\alpha}$; the change in yawing moment coefficient with angle of attack, $C_{n,\alpha}$; and the change in pitching moment coefficient with sideslip angle, $C_{m,\beta}$. These three derivatives were included in the present formulation because they are the coupling derivatives that most commonly become significant when an airplane is flown in an asymmetric configuration. This type of aerodynamic coupling is produced by a rotating propeller and can occur whenever an aircraft is flown with an asymmetric distribution of external stores attached to the wings. This is not to say that no other longitudinal-lateral coupling derivatives can ever be significant. Additional aerodynamic coupling derivatives can easily be added to the formulation presented here. Once the first coupling term is included, adding others does not significantly increase the complexity.

Perhaps the most common cause of aerodynamic coupling between the longitudinal and lateral modes is simply a rotating propeller. No propeller-driven aircraft can be truly symmetric unless it has an even number of counterrotating propellers symmetrically distributed in the spanwise direction. An aircraft with a single rotating propeller will always display some degree of aerodynamic coupling between the longitudinal and lateral modes. Recall from Chapter 2 that a rotating propeller produces a yawing moment that is proportional to the propeller angle of attack. In Eq. (5.2.27) we found that the propeller's contribution to the change in airplane yawing moment coefficient with angle of attack can be expressed as

$$
\left(\Delta C_{n,\alpha}\right)_p = \frac{2d_p^3}{S_w b_w}(1-\varepsilon_{d,\alpha})_p \frac{C_{n_p,\alpha}}{J^2}
\tag{9.8.21}
$$

Similarly, the propeller also produces a change in the airplane's pitching moment coefficient with sideslip angle, which can be expressed as

$$
\left(\Delta C_{m,\beta}\right)_p = \frac{2d_p^3}{S_w \bar{c}_w}(1-\varepsilon_{s,\beta})_p \frac{C_{n_p,\alpha}}{J^2}
\tag{9.8.22}
$$

Before attempting to use Eqs. (9.8.21) and (9.8.22), the reader may wish to review the related material in Chapters 2 and 5. For airplanes with low thrust-to-weight ratio these coupling terms can usually be neglected.

EXAMPLE 9.8.1. The airplane described in Examples 8.2.1 and 9.2.1 is in a steady level coordinated turn at sea level. The airspeed is 180 ft/sec and the bank angle is 60 degrees. Determine the period and damping rate of the dynamic modes for this equilibrium flight condition, neglecting both gyroscopic and aerodynamic coupling. Compare these results with those obtained for steady level flight in Examples 8.2.1 and 9.2.1.

Solution. For this aircraft and a reference state corresponding to steady level flight at sea level, from Example 9.2.1 we have

$$S_w = 185 \text{ ft}^2, \quad b_w = 33 \text{ ft}, \quad W = 2{,}800 \text{ lbf}, \quad V_o = 180 \text{ ft/sec},$$

$$I_{xx_b} = 1{,}000 \text{ slug} \cdot \text{ft}^2, \quad I_{yy_b} = 3{,}000 \text{ slug} \cdot \text{ft}^2, \quad I_{zz_b} = 3{,}500 \text{ slug} \cdot \text{ft}^2,$$

$$I_{xy_b} = 0 \text{ slug} \cdot \text{ft}^2, \quad I_{xz_b} = 30 \text{ slug} \cdot \text{ft}^2, \quad I_{yz_b} = 0 \text{ slug} \cdot \text{ft}^2,$$

$$h_{x_b} = 0 \text{ slug} \cdot \text{ft}^2/\text{sec}, \quad h_{y_b} = 0 \text{ slug} \cdot \text{ft}^2/\text{sec}, \quad h_{z_b} = 0 \text{ slug} \cdot \text{ft}^2/\text{sec},$$

$$C_{Lo} = 0.393, \quad C_{Do} = 0.050, \quad C_{mo} = 0.000,$$

$$C_{L,\hat{\mu}} = 0.000, \quad C_{D,\hat{\mu}} = 0.000, \quad C_{m,\hat{\mu}} = 0.000,$$

$$C_{L,\hat{\alpha}} = 1.600, \quad C_{D,\hat{\alpha}} = 0.000, \quad C_{m,\hat{\alpha}} = -4.350,$$

$$C_{L,M} = 0.0, \quad C_{D,M} = 0.0, \quad C_{m,M} = 0.0,$$

$$C_{L,\alpha} = 4.400, \quad C_{D,\alpha} = 0.350, \quad C_{\ell,\alpha} = 0.0, \quad C_{m,\alpha} = -0.680, \quad C_{n,\alpha} = 0.0,$$

$$C_{Y,\beta} = -0.560, \quad C_{\ell,\beta} = -0.075, \quad C_{m,\beta} = 0.0, \quad C_{n,\beta} = 0.070,$$

$$C_{Y,\bar{p}} = 0.000, \quad C_{\ell,\bar{p}} = -0.410, \quad C_{n,\bar{p}} = -0.0575,$$

$$C_{L,\bar{q}} = 3.800, \quad C_{D,\bar{q}} = 0.000, \quad C_{m,\bar{q}} = -9.950,$$

$$C_{Y,\bar{r}} = 0.240, \quad C_{\ell,\bar{r}} = 0.105, \quad C_{n,\bar{r}} = -0.125$$

At the level-flight reference state, the lift coefficient was found to be

$$\left(C_L\right)_{\text{ref}} = \frac{W}{\frac{1}{2}\rho V_{\text{ref}}^2 S_w} = 0.393$$

From Eq. (9.8.10), the lift coefficient in the coordinated turn is

$$C_L = \frac{W \cos\theta_o}{\frac{1}{2}\rho V_o^2 S_w \cos\phi_o} = 0.786 \tag{9.8.23}$$

The drag coefficient can be closely approximated as a linear function of the lift coefficient squared,

$$C_D = C_1 + C_2 C_L^2 \tag{9.8.24}$$

where C_1 and C_2 are constants. Thus, the change in C_D with α is given by

$$C_{D,\alpha} = 2C_2 C_L C_{L,\alpha} \tag{9.8.25}$$

From the known characteristics at the level-flight reference state, Eqs. (9.8.24) and (9.8.25) can be solved for the constants C_1 and C_2:

$$C_1 = \left(C_D - \frac{C_{D,\alpha} C_L}{2 C_{L,\alpha}}\right)_{\text{ref}} \qquad C_2 = \left(\frac{C_{D,\alpha}}{2 C_L C_{L,\alpha}}\right)_{\text{ref}}$$

Thus, Eqs. (9.8.24) and (9.8.25) can be written as

$$
\begin{aligned}
C_D &= \left(C_D\right)_{\text{ref}} + \left(\frac{C_{D,\alpha} C_L}{2 C_{L,\alpha}}\right)_{\text{ref}} \left[\frac{C_L^2}{(C_L)_{\text{ref}}^2} - 1\right] \\
&= 0.05 + \frac{(0.35)(0.393)}{2(4.40)} \left[\left(\frac{0.786}{0.393}\right)^2 - 1\right] = 0.097
\end{aligned} \tag{9.8.26}
$$

$$C_{D,\alpha} = \left(C_{D,\alpha}\right)_{\text{ref}} \frac{C_L}{(C_L)_{\text{ref}}} = 0.350 \frac{0.786}{0.393} = 0.700 \tag{9.8.27}$$

and

$$C_{X,\alpha} = C_L - C_{D,\alpha} = 0.786 - 0.700 = 0.086$$

$$C_{Z,\alpha} = -C_{L,\alpha} - C_D = -4.400 - 0.097 = -4.497$$

Assuming constant lift slope, the increase in angle of attack between the level-flight reference state and the coordinated turn is

$$\Delta\alpha = \frac{C_L - (C_L)_{\text{ref}}}{C_{L,\alpha}} = \frac{0.786 - 0.393}{4.40} = 0.08932 = 5.118°$$

Using the results presented in Sec. 7.8, the stability derivatives can be transformed from the stability axis for level flight to the stability axis for the coordinated turn. It should be pointed out that the lift and drag do not need to be transformed, since the required transformation is already included in Eqs. (9.8.23), (9.8.26), and (9.8.27). This is due to the fact that lift and drag are defined relative to the direction of motion. For this transformation

$$\varphi = -\Delta\alpha = -0.08932 = -5.118°$$

and after performing the required transformation, the aircraft characteristics for the coordinated turn are found to be

$$I_{xx_b} = 1{,}015 \text{ slug} \cdot \text{ft}^2, \quad I_{yy_b} = 3{,}000 \text{ slug} \cdot \text{ft}^2, \quad I_{zz_b} = 3{,}485 \text{ slug} \cdot \text{ft}^2,$$

$$I_{xy_b} = 0 \text{ slug} \cdot \text{ft}^2, \quad I_{xz_b} = -193 \text{ slug} \cdot \text{ft}^2, \quad I_{yz_b} = 0 \text{ slug} \cdot \text{ft}^2,$$

$$h_{x_b} = 0 \text{ slug} \cdot \text{ft}^2/\text{sec}, \quad h_{y_b} = 0 \text{ slug} \cdot \text{ft}^2/\text{sec}, \quad h_{z_b} = 0 \text{ slug} \cdot \text{ft}^2/\text{sec},$$

$$C_{Xo} = -0.097, \quad C_{Zo} = -0.786, \quad C_{mo} = 0.000,$$

$$C_{X,\hat{\mu}} = -0.013, \quad C_{Z,\hat{\mu}} = -0.142, \quad C_{m,\hat{\mu}} = -0.388,$$

$$C_{X,\hat{\alpha}} = -0.142, \quad C_{Z,\hat{\alpha}} = -1.587, \quad C_{m,\hat{\alpha}} = -4.333,$$

$$C_{X,\mu} = 0.0, \quad C_{Z,\mu} = 0.0, \quad C_{m,\mu} = 0.0,$$

$$C_{X,\alpha} = 0.086, \quad C_{Z,\alpha} = -4.497, \quad C_{\ell,\alpha} = 0.0, \quad C_{m,\alpha} = -0.677, \quad C_{n,\alpha} = 0.0,$$

$$C_{Y,\beta} = -0.560, \quad C_{\ell,\beta} = -0.0685, \quad C_{m,\beta} = 0.0, \quad C_{n,\beta} = 0.0764,$$

$$C_{Y,\bar{p}} = 0.0214, \quad C_{\ell,\bar{p}} = -0.4035, \quad C_{n,\bar{p}} = -0.0326,$$

$$C_{X,\bar{q}} = -0.339, \quad C_{Z,\bar{q}} = -3.785, \quad C_{m,\bar{q}} = -9.950,$$

$$C_{Y,\bar{r}} = 0.239, \quad C_{\ell,\bar{r}} = 0.130, \quad C_{n,\bar{r}} = -0.131$$

Choosing the reference length to be the square root of the wing area, $l_{\text{ref}} = 13.6$ ft, the nondimensional [**A**] and [**B**] matrices are, respectively,

```
-0.0067  0.0117 -0.0173  0.0000 -0.0024  0.0000  0.0000  0.0000  0.0000  0.0000 -0.0135  0.0000
-0.0117 -0.0192  0.0000  0.0009  0.0000 -0.9900  0.0000  0.0000  0.0000  0.0068  0.0000  0.0000
-0.0338  0.0000 -0.1545  0.0000  0.9732  0.0000  0.0000  0.0000  0.0000 -0.0117  0.0000  0.0000
 0.0000 -0.0905  0.0000 -0.6514 -0.0056  0.1988  0.0000  0.0000  0.0000  0.0000  0.0000  0.0000
 0.0000  0.0000 -0.0515  0.0096 -0.1558 -0.0015  0.0000  0.0000  0.0000  0.0000  0.0000  0.0000
 0.0000  0.0294  0.0000 -0.0267  0.0006 -0.0603  0.0000  0.0000  0.0000  0.0000  0.0000  0.0000
 1.0000  0.0000  0.0000  0.0000  0.0000  0.0000  0.0000  0.0000  0.0000  0.0000  0.0000  0.0000
 0.0000  0.5000 -0.8660  0.0000  0.0000  0.0000  0.0000  0.0000  0.0000  0.0000  0.0000  1.0000
 0.0000  0.8660  0.5000  0.0000  0.0000  0.0000  0.0000  0.0000  0.0000  0.0000 -1.0000  0.0000
 0.0000  0.0000  0.0000  1.0000  0.0000  0.0000  0.0000  0.0000  0.0000  0.0000  0.0234  0.0000
 0.0000  0.0000  0.0000  0.0000  0.5000 -0.8660  0.0000  0.0000  0.0000 -0.0234  0.0000  0.0000
 0.0000  0.0000  0.0000  0.0000  0.8660  0.5000  0.0000  0.0000  0.0000  0.0000  0.0000  0.0000
```

```
 1.0001  0.0000  0.0010  0.0000  0.0000  0.0000  0.0000  0.0000  0.0000  0.0000  0.0000  0.0000
 0.0000  1.0000  0.0000  0.0000  0.0000  0.0000  0.0000  0.0000  0.0000  0.0000  0.0000  0.0000
 0.0010  0.0000  1.0112  0.0000  0.0000  0.0000  0.0000  0.0000  0.0000  0.0000  0.0000  0.0000
 0.0000  0.0000  0.0000  1.0000  0.0000  0.1899  0.0000  0.0000  0.0000  0.0000  0.0000  0.0000
 0.0061  0.0000  0.0679  0.0000  1.0000  0.0000  0.0000  0.0000  0.0000  0.0000  0.0000  0.0000
 0.0000  0.0000  0.0000  0.0553  0.0000  1.0000  0.0000  0.0000  0.0000  0.0000  0.0000  0.0000
 0.0000  0.0000  0.0000  0.0000  0.0000  0.0000  1.0000  0.0000  0.0000  0.0000  0.0000  0.0000
 0.0000  0.0000  0.0000  0.0000  0.0000  0.0000  0.0000  1.0000  0.0000  0.0000  0.0000  0.0000
 0.0000  0.0000  0.0000  0.0000  0.0000  0.0000  0.0000  0.0000  1.0000  0.0000  0.0000  0.0000
 0.0000  0.0000  0.0000  0.0000  0.0000  0.0000  0.0000  0.0000  0.0000  1.0000  0.0000  0.0000
 0.0000  0.0000  0.0000  0.0000  0.0000  0.0000  0.0000  0.0000  0.0000  0.0000  1.0000  0.0000
 0.0000  0.0000  0.0000  0.0000  0.0000  0.0000  0.0000  0.0000  0.0000  0.0000  0.0000  1.0000
```

Using a numerical eigensolver to obtain the eigenvalues associated with this generalized eigenproblem, the dimensionless phugoid eigenvalues are found to be $\lambda = -0.000556 \pm 0.029026\,i$. Thus, for the phugoid we have

$$\sigma = -\text{real}(\lambda)\frac{V_o}{l_{ref}} = 0.000556\frac{180}{13.6} = 0.00736\,\text{sec}^{-1}$$

$$\omega_d = |\text{imag}(\lambda)|\frac{V_o}{l_{ref}} = 0.029026\frac{180}{13.6} = 0.384\,\text{sec}^{-1}$$

Similarly, for the other modes

| | 60° Bank Angle | | Level Flight | |
Mode	Period (sec)	Damping Rate (sec^{-1})	Period (sec)	Damping Rate (sec^{-1})
Short-period	2.44	2.48	2.42	2.47
Phugoid	16.4	0.0074	29.9	0.0169
Roll	∞	8.67	∞	8.88
Spiral	∞	0.0488	∞	0.0100
Dutch roll	2.61	0.591	2.64	0.482

From these results we see that turning flight and longitudinal-lateral coupling can have a significant effect on aircraft dynamics. For the conventional airplane used in this example, there is nothing too detrimental in the observed changes. However, for some aircraft configurations, longitudinal-lateral coupling has proven more deleterious.

9.9. Nonlinear Effects

Here again it is emphasized that the linearized analysis used in this and Chapter 8 is based on the assumption of small deviations from equilibrium flight. The linearized aerodynamic model provides a good approximation for most flight conditions when the angular rates and aerodynamic angles are small. However, many of the maneuvers executed by modern fighter aircraft and stunt planes involve high angular rates and/or large angles of attack that sometimes approach 90 degrees. Such maneuvers can produce important dynamic effects that are not predicted by linear theory. A few of these are mentioned here.

Inertial Pitch and Yaw Divergence in Roll Maneuvers
When the angular rates are large, nonlinear inertial coupling is introduced between the longitudinal and lateral degrees of freedom. This nonlinear coupling results from the product terms seen on the right-hand side of Eqs. (9.8.1) and (9.8.2). As mentioned in Sec. 7.9, the most problematic nonlinear inertial coupling results from the terms $(I_{zz_b} - I_{xx_b})pr$ and $(I_{xx_b} - I_{yy_b})pq$, which occur in the pitching and yawing momentum equations, respectively. For long slender aircraft undergoing rapid roll maneuvers, this nonlinear coupling can produce changes in angle of attack and sideslip angle that are

large enough to cause pitch and/or yaw divergence, which can result in complete control loss or even structural failure. This problem is often aggravated by the fact that to improve maneuverability, modern fighters have reduced pitch stability.

The dynamic effects of nonlinear inertial coupling were not understood or even encountered until late in World War II. On December 10, 1944, the German He-162 prototype jet fighter disintegrated during a high-speed roll maneuver just four days after its maiden flight. In spite of the fact that the incident was well recorded on film, the response of the aircraft could not be explained based on the knowledge of flight dynamics available at that time. It was not until several years later that the cause of this and other similar incidents was understood. This behavior cannot be predicted using the linearized theory presented in this chapter. To analyze such motion we must utilize the fully coupled nonlinear equations of motion.

Aerodynamic Yaw Departure at High Angles of Attack

Maneuvers involving large aerodynamic angles produce strong nonlinearity in the aerodynamic forces and moments as a result of flow separation. This flow separation can also produce nonlinear coupling between the longitudinal and lateral degrees of freedom. For example, at high angles of attack, the flow around an airplane's fuselage will separate. Separated flow around a slender body at high angles of attack can result in stable forebody vortices that are shed near the nose. As the angle of attack for a slender body is increased slightly beyond the point of separation, the cross flow generates a pair of stable symmetric vortices, as shown in Fig.9.9.1. As the angle of attack continues to

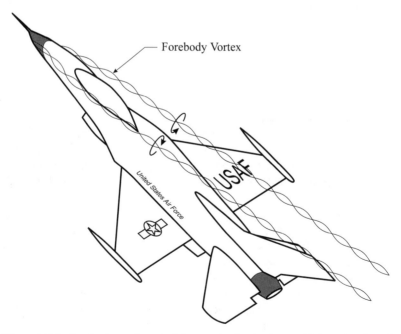

Figure 9.9.1. Forebody vortices shed from a slender fuselage at high angle of attack.

increase somewhat beyond the point of initial separation, the normal force increases nonlinearly as a result of vortex lift. At angles of attack that are only slightly above the separation point, the stable vortex pair remains symmetric and no side force is generated without sideslip. As the angle of attack is increased even further, the axial flow component remains sufficient to maintain stable vortex shedding, but the vortex pair becomes asymmetric. This can produce a large side force, even in the absence of sideslip. Since the vortex pair is shed far forward from the center of gravity, this side force also produces a strong yawing moment. The interaction of these asymmetric vortices with the wing and tail can also produce a rolling moment. At even higher angles of attack, the axial flow component is no longer adequate to maintain the stable vortex pair and the flow again becomes symmetric, as a vortex wake forms over the entire leeward side of the fuselage. At the angles of attack that produce the greatest vortex-induced yawing moment, yaw control can be drastically reduced as well, because the rudder is typically shadowed by the wake of the fuselage and wing. The problem can be reduced or eliminated with nose strakes or other reshaping that promotes symmetric vortex shedding. Clearly, the nonlinear dynamics of such high-angle-of-attack flight cannot be studied using the method presented in this and Chapter 8.

Wing Rock

The nonlinear aerodynamics associated with high-angle-of-attack flight can also produce dynamic oscillations that are not predicted by linear theory. One of the most common phenomena of this type is called *wing rock*. As the name implies, wing rock is an oscillatory motion, predominantly in roll. However, the motion can be very complex and typically involves several degrees of freedom. Any aircraft with highly swept wings and/or a long slender forebody is usually susceptible to wing rock at high angles of attack.

When such an aircraft is exposed to high angles of attack, a very complex vortex flow-field develops, as shown in Fig. 9.9.2. The interactions of the vortex wakes with

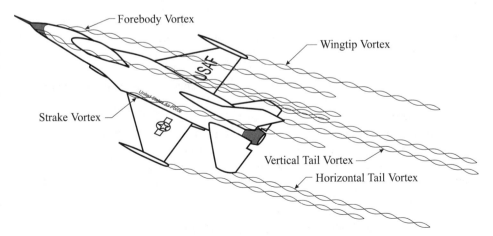

Figure 9.9.2. High-angle-of-attack vortex wakes shed from a typical fighter aircraft.

the wing and tail can cause the aerodynamic derivatives to change very dramatically with angle of attack. For example, Fig. 9.9.3 shows how the roll-damping derivative might vary with angle of attack for a typical jet fighter. Notice that at certain angles of attack, the roll-damping derivative is positive. For angles of attack in this range, a disturbance in rolling rate produces a rolling moment in the same direction as the roll rather than opposing it. This destabilizing phenomenon results from the fact that the lift slope for the wing is negative beyond stall. At higher angles of attack the roll-damping derivative again becomes negative, because the effect of wing drag on roll damping overpowers the effect of wing lift.

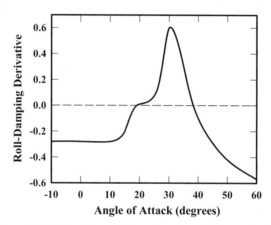

Figure 9.9.3. Loss of roll damping at high angles of attack for a typical fighter aircraft.

The presence of vortex wakes at high angles of attack also introduces aerodynamic hysteresis in the forces and moments produced on the aircraft. For example, the forebody vortices influence the forces and moments developed on the wing and tail. However, since the vortex wake is convected relative to the aircraft by the freestream, the airflow around the tail at any given instant in time depends on the motion history and vortex development as well as the instantaneous airspeed and aerodynamic angles. The vortex-induced forces and moments developed on the vertical tail depend not only on the instantaneous angle of attack, but also on the angle of attack history and the time required for the vortex wake to be transported from the nose to the tail.

The genesis of wing rock is related directly to nonlinearities and hysteresis in the aerodynamic forces and moments acting on an aircraft at high angles of attack. Oscillations that occur as a result of such nonlinear behavior can be quite different in character from those associated with the classical modes of linear dynamics. The amplitude for any classical oscillatory mode will always change exponentially with time if uncontrolled. If the mode is convergent, the amplitude of oscillation will decay exponentially to zero. If the mode is divergent, the uncontrolled oscillation will grow exponentially without limit. However, as the oscillations associated with wing rock begin to develop, the amplitude grows with an increasing rate and the frequency decreases.   After a time, the rate of amplitude growth begins to decrease and the amplitude eventually approaches a steady-state value. From this point on, the oscillations continue at constant amplitude and frequency, unless controlled in some manner. This type of motion, which is shown in Fig. 9.9.4, is called *limit-cycle oscillation*.

Since high angle of attack hysteresis and nonlinearities are the direct cause of wing rock, this and similar phenomena cannot be predicted from linear theory. To study such phenomena, we must turn to experimental methods or numerical flight simulation using the complete nonlinear equations of motion, including a model for the unsteady aerodynamics. Numerical flight simulation is treated in Chapter 11.

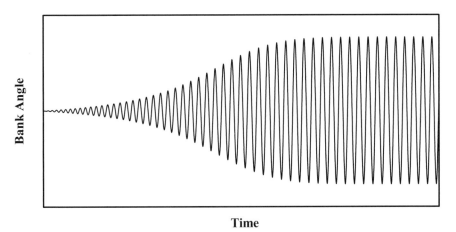

Time

Figure 9.9.4. Limit-cycle oscillations in bank angle associated with wing rock.

9.10. Summary

In this chapter we have examined the stick-fixed lateral dynamics of an aircraft, based on the linearized rigid-body equations of motion. As was the case for longitudinal motion, the linearized lateral equations have four nonzero eigenvalues. These typically consist of one pair of distinct real eigenvalues and one complex pair. The real eigenvalues correspond to two nonoscillatory modes, the roll mode and the spiral mode. The complex pair describes an oscillatory mode called Dutch roll. For a conventional fixed-wing airplane, the roll mode is very heavily damped, with a 99 percent damping time on the order of 1 second. The spiral mode is very slowly convergent or even slightly divergent and is characterized by changes in heading and direction of travel. Even when the spiral mode is convergent, the 99 percent damping time is on the order of several minutes or longer. For an aircraft with very little wing area, the roll and spiral modes can combine to produce a long-period lateral oscillation. Dutch roll is a damped oscillatory motion that involves an out-of-phase combination of sideslip, roll, and yaw. The period for Dutch roll is typically on the order of 2 to 10 seconds, with a 99 percent damping time of 10 seconds or longer. Dutch roll is typically the most troublesome of the dynamic modes for a conventional airplane.

The reason that Dutch roll can sometimes be troublesome is because the quarter-period of oscillation is typically quite close to the response time of a human pilot, and Dutch roll is often lightly damped. The natural response of most pilots to a disturbance that causes one wing to drop is initiation of an aileron input to bring the low wing back up. However, there is always some delay between the time of the disturbance and the time of the control input, because it takes a finite length of time for the pilot to recognize the disturbance and respond. If the disturbance is periodic in nature and the quarter-period of oscillation is equal to or slightly less than the pilot's response time, then the control input to raise the low wing comes not as the wing is falling but as the wing is rising. This means that the pilot's control input will actually be applied in the direction of motion and not in the opposing direction, as the pilot intended. This control input can still feel correct to the pilot, because the aileron input to raise the low wing comes near the time when the wing is at its lowest point and the aircraft seems to respond correctly to the control input as the low wing begins to rise. However, such pilot response will actually amplify the oscillations of the airplane. This is much like pushing a child in a swing. If the push comes at or slightly after the swing reaches the apex of oscillation, the amplitude will increase. On the other hand, if the push comes just prior to the apex, the oscillation of the swing will quickly cease. In both cases the push is in a direction that returns the swing toward its equilibrium position, but the response is very different, depending on the timing.

The suitability of this analogy is demonstrated by an experience that was related to the author by Dr. Frank Redd. In the early 1960s Frank was flying KC-135s for the U.S. Air Force. Early models of the KC-135 had very light Dutch roll damping when the yaw damper was disengaged or not functioning for some other reason. Since the yaw damper could be damaged and become inoperative, it was important for pilots to learn to control the aircraft with the yaw damper disengaged. For safety sake, this was done at altitude. With few exceptions, when pilots first tried to fly the KC-135 with the yaw damper disengaged, they only succeeded in aggravating the Dutch roll. To suppress the Dutch roll oscillations rather than aggravate them, a pilot needs to supply out-of-phase control inputs rather than the in-phase control inputs that result from doing what at first seems natural. One way to achieve the desired result would be to shorten pilot reaction time. However, there are human limits that place constraints on what can be done in this regard. Another option is to keep the reaction time constant, but supply the opposite control input. While this control strategy seemed counterintuitive at first, with a little practice, most pilots were able to reprogram their reflexes and compensate successfully for the disengaged yaw damper.

From this example and the analogy with pushing a child in a swing, we can see that the dynamic response of an airplane can be very closely coupled to the reaction time of the pilot, and human reaction times are widely varied. For example, a Dutch roll mode that would be totally unacceptable to one pilot might be only mildly annoying to a second pilot and go unnoticed by a third.

Because of their tailless design, hang gliders often have light Dutch roll damping. In these tiny aircraft, Dutch roll oscillations have sometimes been referred to as *pilot-induced oscillations* (PIOs). There have been many heated arguments about whether these oscillations are caused by the aircraft or the pilot. For example, Dick flies glider A on several occasions and observes lateral oscillations on each flight. Jane, who has also flown glider A many times, has never observed such oscillations. Jane tells Dick that there is absolutely nothing wrong with the glider and that the oscillations must be entirely pilot induced, since she has never observed these oscillations while flying the same glider. The counterargument made by Dick is that the oscillations cannot possibly be pilot induced since he has flown glider B for many hours and has never observed any lateral oscillations in that glider. Thus, Dick strongly argues that he is not causing these oscillations, so they must be inherent in the aircraft. Both pilots have a strong argument from their own frame of reference, and the conflict is never resolved. The truth is that both pilots are right and both pilots are wrong. The Dutch roll oscillations are inherent in the aircraft and Dick is exciting this natural frequency with the timing of his control inputs. Most likely, Jane has a faster reaction time than Dick and is thus able to instinctively suppress the Dutch roll without even being aware of its existence.

The dynamic interaction between an aircraft and its human pilot can be a very important consideration in aircraft design. In this chapter and in Chapter 8 we confined our study of aircraft dynamics to the stick-fixed dynamic response of an aircraft to small disturbances from equilibrium flight. In so doing we have removed the pilot from the equation. To continue to ignore the pilot's influence on the human-machine system that constitutes a modern airplane would be ill advised. This is the topic of Chapter 10.

9.11 Problems

9.1. For the large jet transport and flight conditions in problem 7.34, find the eigenvalues, the eigenvectors, the damping rate, the 99 percent damping time, and the damped natural frequency and period, if underdamped, for each lateral mode. For each pair of eigenvalues, find the undamped natural frequency and the damping ratio. Compute the amplitude and phase for each component of the eigenvectors associated with the oscillatory mode.

9.2. To improve the Dutch roll characteristics, the large jet transport in problem 9.1 is to be fitted with a feedback control system to provide additional yaw damping. For the flight conditions used in problem 9.1, find the yaw-damping ratio, $R_{n,\bar{r}}$, necessary to make the Dutch roll damping rate equal to 0.35 sec^{-1}. Repeat all calculations in problem 9.1 for the modified airplane.

9.3. For the large jet transport and flight conditions in problem 7.34, using the roll mode eigenvalue approximation given by Eq. (9.3.3), find the eigenvalue and the 99 percent damping time for the roll mode. Compare this approximate result with the fully coupled roll mode solution that was obtained in problem 9.1.

9.4. For the large jet transport and flight conditions in problem 7.34, using the roll mode eigenvector approximation given by Eq. (9.3.6), find the eigenvector for the roll mode. Normalize this eigenvector according to Eq. (8.1.37). Compare this approximate eigenvector with the fully coupled roll mode solution that was obtained in problem 9.1.

9.5. For the large jet transport and flight conditions in problem 7.34, using the spiral mode eigenvalue approximation given by Eq. (9.4.3), find the eigenvalue and the 99 percent damping time for the spiral mode. Compare this approximate result with the fully coupled spiral mode solution that was obtained in problem 9.1.

9.6. For the large jet transport and flight conditions in problem 7.34, using the spiral mode eigenvector approximation given by Eq. (9.4.4), find the eigenvector for the spiral mode. Normalize this eigenvector according to Eq. (8.1.37). Compare this approximate eigenvector with the fully coupled spiral solution that was obtained in problem 9.1.

9.7. For the large jet transport and flight conditions in problem 7.34, using the spiral mode eigenvalue approximation given by Eq. (9.4.11), find the eigenvalue and the 99 percent damping time for the spiral mode. Compare this approximate result with the fully coupled spiral mode solution obtained in problem 9.1.

9.8. For the large jet transport and flight conditions in problem 7.34, using the spiral mode eigenvector approximation given by Eq. (9.4.10), find the eigenvector for the spiral mode. Normalize this eigenvector according to Eq. (8.1.37). Compare this approximate eigenvector with the fully coupled spiral solution that was obtained in problem 9.1.

9.9. For the large jet transport and flight conditions in problem 7.34, using the Dutch roll eigenvalue approximation given by Eq. (9.5.5), find the eigenvalues, the 99 percent damping time, and the period for the Dutch roll mode. Compare these approximate results with the fully coupled Dutch roll solution that was obtained in problem 9.1.

9.10. For the large jet transport and flight conditions in problem 7.34, using the Dutch roll eigenvector approximation given by Eq. (9.5.3), find the eigenvectors for the Dutch roll mode. Normalize these eigenvectors according to Eq. (8.1.37). Compare these approximate eigenvectors with the fully coupled Dutch roll solution that was obtained in problem 9.1.

9.11. For the large jet transport and flight conditions in problem 7.34, using the Dutch roll eigenvalue approximation given by Eq. (9.5.27), find the eigenvalues, the 99 percent damping time, and the period for the Dutch roll mode. Compare these approximate results with the fully coupled Dutch roll solution that was obtained in problem 9.1.

9.12. For the large jet transport and flight conditions in problem 7.34, using the Dutch roll eigenvector approximation given by Eq. (9.5.11), find the eigenvectors for the Dutch roll mode. Normalize these eigenvectors according to Eq. (8.1.37). Compare these approximate eigenvectors with the fully coupled Dutch roll solution that was obtained in problem 9.1.

9.13. For the Spitfire Mark XVIII in problems 7.36 and 7.39, find the eigenvalues, the eigenvectors, the damping rate, the 99 percent damping time, and the damped natural frequency and period, if underdamped, for each lateral mode. For each pair of eigenvalues, find the undamped natural frequency and the damping ratio. Compute the amplitude and phase for each component of the eigenvectors associated with the oscillatory mode.

9.14. For the Spitfire Mark XVIII in problems 7.36 and 7.39, using the roll mode eigenvalue approximation given by Eq. (9.3.3), find the eigenvalue and the 99 percent damping time for the roll mode. Compare this approximate result with the fully coupled roll mode solution that was obtained in problem 9.13.

9.15. For the Spitfire Mark XVIII in problems 7.36 and 7.39, using the roll mode eigenvector approximation given by Eq. (9.3.6), find the eigenvector for the roll mode. Normalize this eigenvector according to Eq. (8.1.37). Compare this approximate eigenvector with the fully coupled roll mode solution that was obtained in problem 9.13.

9.16. For the Spitfire Mark XVIII in problems 7.36 and 7.39, using the spiral mode eigenvalue approximation given by Eq. (9.4.3), find the eigenvalue and the 99 percent damping time for the spiral mode. Compare this approximate result with the fully coupled spiral mode solution that was obtained in problem 9.13.

9.17. For the Spitfire Mark XVIII in problems 7.36 and 7.39, using the spiral mode eigenvector approximation given by Eq. (9.4.4), find the eigenvector for the spiral mode. Normalize this eigenvector according to Eq. (8.1.37). Compare this approximate eigenvector with the fully coupled spiral solution that was obtained in problem 9.13.

9.18. For the Spitfire Mark XVIII in problems 7.36 and 7.39, using the spiral mode eigenvalue approximation given by Eq. (9.4.11), find the eigenvalue and the 99 percent damping time for the spiral mode. Compare this approximate result with the fully coupled spiral mode solution that was obtained in problem 9.13.

9.19. For the Spitfire Mark XVIII in problems 7.36 and 7.39, using the spiral mode eigenvector approximation given by Eq. (9.4.10), find the eigenvector for the spiral mode. Normalize this eigenvector according to Eq. (8.1.37). Compare this approximate eigenvector with the fully coupled spiral solution that was obtained in problem 9.13.

9.20. For the Spitfire Mark XVIII in problems 7.36 and 7.39, using the Dutch roll eigenvalue approximation given by Eq. (9.5.5), find the eigenvalues, the 99 percent damping time, and the period for the Dutch roll mode. Compare these approximate results with the fully coupled Dutch roll solution that was obtained in problem 9.13.

9.21. For the Spitfire Mark XVIII in problems 7.36 and 7.39, using the Dutch roll eigenvector approximation given by Eq. (9.5.3), find the eigenvectors for the Dutch roll mode. Normalize these eigenvectors according to Eq. (8.1.37). Compare these approximate eigenvectors with the fully coupled Dutch roll solution that was obtained in problem 9.13.

9.22. For the Spitfire Mark XVIII in problems 7.36 and 7.39, using the Dutch roll eigenvalue approximation given by Eq. (9.5.27), find the eigenvalues, the 99 percent damping time, and the period for the Dutch roll mode. Compare these approximate results with the fully coupled Dutch roll solution that was obtained in problem 9.13.

9.23. For the Spitfire Mark XVIII in problems 7.36 and 7.39, using the Dutch roll eigenvector approximation given by Eq. (9.5.11), find the eigenvectors for the Dutch roll mode. Normalize these eigenvectors according to Eq. (8.1.37). Compare these approximate eigenvectors with the fully coupled Dutch roll solution that was obtained in problem 9.13.

9.24. For the small jet transport in problems 7.37 and 7.40, find the eigenvalues, the eigenvectors, the damping rate, the 99 percent damping time, and the damped natural frequency and period, if underdamped, for each lateral mode. For each pair of eigenvalues, find the undamped natural frequency and the damping ratio. Compute the amplitude and phase for each component of the eigenvectors associated with the oscillatory mode.

9.25. For the small jet transport in problems 7.37 and 7.40, using the roll mode eigenvalue approximation given by Eq. (9.3.3), find the eigenvalue and the 99 percent damping time for the roll mode. Compare this approximate result with the fully coupled roll mode solution that was obtained in problem 9.24.

9.26. For the small jet transport in problems 7.37 and 7.40, using the roll mode eigenvector approximation given by Eq. (9.3.6), find the eigenvector for the roll mode. Normalize this eigenvector according to Eq. (8.1.37). Compare this approximate eigenvector with the fully coupled roll mode solution that was obtained in problem 9.24.

9.27. For the small jet transport in problems 7.37 and 7.40, using the spiral mode eigenvalue approximation given by Eq. (9.4.3), find the eigenvalue and the 99 percent damping time for the spiral mode. Compare this approximate result with the fully coupled spiral mode solution that was obtained in problem 9.24.

9.28. For the small jet transport in problems 7.37 and 7.40, using the spiral mode eigenvector approximation given by Eq. (9.4.4), find the eigenvector for the spiral mode. Normalize this eigenvector according to Eq. (8.1.37). Compare this approximate eigenvector with the fully coupled spiral solution that was obtained in problem 9.24.

9.29. For the small jet transport in problems 7.37 and 7.40, using the spiral mode eigenvalue approximation given by Eq. (9.4.11), find the eigenvalue and the 99 percent damping time for the spiral mode. Compare this approximate result with the fully coupled spiral mode solution that was obtained in problem 9.24.

9.30. For the small jet transport in problems 7.37 and 7.40, using the spiral mode eigenvector approximation given by Eq. (9.4.10), find the eigenvector for the spiral mode. Normalize this eigenvector according to Eq. (8.1.37). Compare this approximate eigenvector with the fully coupled spiral solution that was obtained in problem 9.24.

9.31. For the small jet transport in problems 7.37 and 7.40, using the Dutch roll eigenvalue approximation given by Eq. (9.5.5), find the eigenvalues, the 99 percent damping time, and the period for the Dutch roll mode. Compare these approximate results with the fully coupled Dutch roll solution that was obtained in problem 9.24.

9.32. For the small jet transport in problems 7.37 and 7.40, using the Dutch roll eigenvector approximation given by Eq. (9.5.3), find the eigenvectors for the Dutch roll mode. Normalize these eigenvectors according to Eq. (8.1.37). Compare these approximate eigenvectors with the fully coupled Dutch roll solution that was obtained in problem 9.24.

9.33. For the small jet transport in problems 7.37 and 7.40, using the Dutch roll eigenvalue approximation given by Eq. (9.5.27), find the eigenvalues, the 99 percent damping time, and the period for the Dutch roll mode. Compare these approximate results with the fully coupled Dutch roll solution that was obtained in problem 9.24.

9.34. For the small jet transport in problems 7.37 and 7.40, using the Dutch roll eigenvector approximation given by Eq. (9.5.11), find the eigenvectors for the Dutch roll mode. Normalize these eigenvectors according to Eq. (8.1.37). Compare these approximate eigenvectors with the fully coupled Dutch roll solution that was obtained in problem 9.24.

9.35. For the turboprop transport in problems 7.38 and 7.41, find the eigenvalues, the eigenvectors, the damping rate, the 99 percent damping time, and the damped natural frequency and period, if underdamped, for each lateral mode. For each pair of eigenvalues, find the undamped natural frequency and the damping ratio. Compute the amplitude and phase for each component of the eigenvectors associated with the oscillatory mode.

9.36. For the turboprop transport in problems 7.38 and 7.41, using the roll mode eigenvalue approximation given by Eq. (9.3.3), find the eigenvalue and the 99 percent damping time for the roll mode. Compare this approximate result with the fully coupled roll mode solution that was obtained in problem 9.35.

9.37. For the turboprop transport in problems 7.38 and 7.41, using the roll mode eigenvector approximation given by Eq. (9.3.6), find the eigenvector for the roll mode. Normalize this eigenvector according to Eq. (8.1.37). Compare this approximate eigenvector with the fully coupled roll mode solution that was obtained in problem 9.35.

9.38. For the turboprop transport in problems 7.38 and 7.41, using the spiral mode eigenvalue approximation given by Eq. (9.4.3), find the eigenvalue and the 99 percent damping time for the spiral mode. Compare this approximate result with the fully coupled spiral mode solution that was obtained in problem 9.35.

9.39. For the turboprop transport in problems 7.38 and 7.41, using the spiral mode eigenvector approximation given by Eq. (9.4.4), find the eigenvector for the spiral mode. Normalize this eigenvector according to Eq. (8.1.37). Compare this approximate eigenvector with the fully coupled spiral solution that was obtained in problem 9.35.

9.40. For the turboprop transport in problems 7.38 and 7.41, using the spiral mode eigenvalue approximation given by Eq. (9.4.11), find the eigenvalue and the 99 percent damping time for the spiral mode. Compare this approximate result with the fully coupled spiral mode solution that was obtained in problem 9.35.

9.41. For the turboprop transport in problems 7.38 and 7.41, using the spiral mode eigenvector approximation given by Eq. (9.4.10), find the eigenvector for the spiral mode. Normalize this eigenvector according to Eq. (8.1.37). Compare this approximate eigenvector with the fully coupled spiral solution that was obtained in problem 9.35.

9.42. For the turboprop transport in problems 7.38 and 7.41, using the Dutch roll eigenvalue approximation given by Eq. (9.5.5), find the eigenvalues, the 99 percent damping time, and the period for the Dutch roll mode. Compare these approximate results with the fully coupled Dutch roll solution that was obtained in problem 9.35.

9.43. For the turboprop transport in problems 7.38 and 7.41, using the Dutch roll eigenvector approximation given by Eq. (9.5.3), find the eigenvectors for the Dutch roll mode. Normalize these eigenvectors according to Eq. (8.1.37). Compare these approximate eigenvectors with the fully coupled Dutch roll solution that was obtained in problem 9.35.

9.44. For the turboprop transport in problems 7.38 and 7.41, using the Dutch roll eigenvalue approximation given by Eq. (9.5.27), find the eigenvalues, the 99 percent damping time, and the period for the Dutch roll mode. Compare these approximate results with the fully coupled Dutch roll solution that was obtained in problem 9.35.

9.45. For the turboprop transport in problems 7.38 and 7.41, using the Dutch roll eigenvector approximation given by Eq. (9.5.11), find the eigenvectors for the Dutch roll mode. Normalize these eigenvectors according to Eq. (8.1.37). Compare these approximate eigenvectors with the fully coupled Dutch roll solution that was obtained in problem 9.35.

9.46. Write a computer program that can be used to solve the 12×12 eigenproblem associated with the homogeneous form of Eq. (9.8.20). The program should import all required data from an input file of the same format as the sample file *airplane.txt*, which can be provided by your instructor. This sample data file is for the airplane and operating conditions in Example 9.8.1. Using only the data obtained from the input file, the program should fill out the two dimensionless 12×12 matrices needed for the generalized eigenproblem. The program should then use a generalized eigensolver to obtain the eigenvalues and eigenvectors from these matrices. A special eigensolver can be used if the system is first numerically reduced to the special eigenproblem. Be sure that all computations are done in double precision. Test your program by comparing results with those presented in Example 9.8.1.

9.47. Use the computer program that was written for problem 9.46 to obtain the solutions to problems 8.4 and 9.1. A formatted data file for this airplane and operation conditions is available as *B747.txt*, which can be provided by your instructor.

9.48. Use the computer program that was written for problem 9.46 to obtain the solutions to problems 8.13 and 9.13. A formatted data file for this airplane and operation conditions is available as *Spitfire.txt*, which can be provided by your instructor.

9.49. Use the computer program that was written for problem 9.46 to obtain the solutions to problems 8.22 and 9.24. A formatted data file for this airplane and operation conditions is available as *CJT001.txt*, which can be provided by your instructor.

9.50. Use the computer program that was written for problem 9.46 to obtain the solutions to problems 8.31 and 9.35. A formatted data file for this airplane and operation conditions is available as *C130H.txt*, which can be provided by your instructor.

9.51. Modify the data file *airplane.txt*, which was used in problem 9.46, to include the gyroscopic effects of the propeller and crankshaft turning at 2,350 rpm. The moment of inertia for the crankshaft and propeller combined is 4.7 slug·ft^2. Use the computer program that was written for problem 9.46 to obtain an improved solution to Example 9.8.1, including the gyroscopic effects of the rotating propeller and crankshaft.

9.52. The 2,375-horsepower, 12-cylinder Rolls-Royce Griffon engine of the Spitfire Mark XVIII in problem 9.48 is turning the five-blade, 132-inch Rotol propeller at 1,350 rpm. The effective mass moment of inertia for the crankshaft, hub, and propeller combined is 95 slug·ft^2, based on propeller speed. With a forward airspeed of 440 ft/sec, the change in propeller yawing moment coefficient with angle of attack is $C_{n_p,\alpha} = +0.296$ (the Griffon engine has left-hand rotation). Use the computer program that was written for problem 9.46 to obtain an improved solution to problem 9.48, including the gyroscopic and aerodynamic effects of the rotating propulsion system. Compare the short-period and Dutch roll eigenvectors both with and without the inclusion of the gyroscopic and aerodynamic coupling terms.

9.53. The Spitfire Mark XVIII in problem 9.48 is executing a level coordinated turn with an airspeed of 440 ft/sec and a bank angle of 60 degrees. Use the computer program that was written for problem 9.46 to obtain a solution to problem 9.48 for this operating condition. Obtain the solution both with and without the gyroscopic and aerodynamic effects of the rotating propulsion system, which were described in problem 9.52.

9.54. Repeat problem 9.52 assuming that mass moment of inertia for the crankshaft, hub, and propeller combined is doubled while all other parameters and operating conditions remain unchanged.

9.55. Four 4,910-horsepower constant-speed turboprop engines, turning at 13,820 rpm, power the transport airplane described in problem 9.50. Each engine turns a 162-inch propeller through a gear reduction assembly having a total reduction ratio of 13.54 to 1. The equivalent moment of inertia for each turbine rotor and gear reduction assembly combined is 6.5 slug·ft^2, based on turbine speed. The moment of inertia for each hub and propeller is 388 slug·ft^2. The change in propeller yawing moment coefficient with angle of attack is $C_{n_p,\alpha} = -0.154$. Assuming that all four engines turn in the same direction, use the computer program that was written for problem 9.46 to obtain a solution to problem 9.50, including the gyroscopic and aerodynamic effects of the rotating propulsion system. Compare the short-period and Dutch roll eigenvectors both with and without the inclusion of the gyroscopic and aerodynamic coupling terms.

9.56. The turboprop transport in problem 9.55 is executing a level coordinated turn with an airspeed of 440 ft/sec and a bank angle of 60 degrees. Use the computer program that was written for problem 9.46 to obtain a solution to problem 9.50 for this operating condition. Include the gyroscopic and aerodynamic effects of all four engines and propellers turning in the same direction.

Chapter 10
Aircraft Handling Qualities and Control Response

10.1. Introduction

The material presented in Chapters 8 and 9 dealt with determining the eigenvalues and eigenvectors associated with the stick-fixed longitudinal and lateral modes of an airplane relative to some equilibrium flight condition. We found that longitudinal motion includes two oscillatory modes, the short-period mode and the long-period or phugoid mode. The lateral motion of a typical airplane includes two nonoscillatory modes, the roll mode and the spiral mode, as well as one oscillatory mode, Dutch roll. The damping rate for each mode and the damped natural frequency for the oscillatory modes are determined from the eigenvalues. We found that the damping rates and frequencies depend on geometric and mass properties of the airplane as well as aerodynamic derivatives and flight speed. The designer has some control over the characteristics of each dynamic mode through his or her selection of airplane geometry. Furthermore, the designer can exercise additional control over the characteristics of these dynamic modes through implementation of an active feedback control system, which is referred to as *stability augmentation*.

Because the airplane designer has some control over the frequencies and damping rates associated with the longitudinal and lateral modes, a simple but important question quite naturally arises. What is considered to be good? The answer to this question is closely linked to the pilot's involvement in the control of the aircraft and has long been an important topic of research. Recent developments, such as fly-by-wire and digitally controlled aircraft, have generated renewed interest in such research, and the final statement relating to this topic has certainly not yet been expressed. It is still not uncommon to have a modern aircraft development project delayed by difficulties associated with dynamic stability and control. Such difficulties are often due, at least in part, to an insufficient understanding of the aircraft handling qualities discipline. This chapter gives only a brief introduction to the topic of aircraft handling qualities, as it applies to the dynamic modes of a typical airplane. For a more in-depth treatment of this topic, the reader is referred to the U.S. Military Specifications MIL-F-8785C (1980) and MIL-STD-1797A (1995), which are summarized by Hodgkinson (1999).

10.2. Pilot Opinion

Ultimately, pilots will assess the handling qualities of any aircraft. For this reason, research on aircraft handling qualities has always centered on pilot opinion. Extensive research has been conducted by the federal government and the aviation industry to relate the static and dynamic stability characteristics of an airplane to the pilot's opinion of the aircraft's handling qualities. Since pilot opinion is subjective, it is difficult to quantify and measure. By contrast, measuring the dynamic characteristics of the aircraft itself is relatively simple.

953

The *Cooper-Harper rating scale* is most commonly used to quantify pilot opinion of aircraft handling qualities. The decision tree used by pilots when rating an aircraft according to the Cooper-Harper scale is shown in Fig. 10.2.1. This scale is dichotomous in that it forces the pilot to make a series of binary decisions, which ultimately lead to a quantitative rating from 1 to 10. A rating of 1 indicates excellent handling qualities, and a rating of 10 represents uncontrollable conditions. Subsequent to each dichotomous decision, the pilot further refines the rating according to the level of *pilot compensation* required. In the Cooper-Harper scale, pilot compensation is defined to be *the measure of additional pilot effort and attention required to maintain a given level of performance in the face of deficient vehicle characteristics.*

A 10-point rating scale is usually considered to be excessive for gathering subjective opinions. Perhaps for this reason, the U.S. Military Specifications for aircraft handling qualities are centered around *levels* of handling qualities, which are defined with reference to the Cooper-Harper scale. Cooper-Harper ratings of 1, 2, and 3 are defined to be *level 1*, while *level 2* consists of ratings 4, 5, and 6. Ratings of 7, 8, and 9 fall in *level 3*. Common practice has unofficially labeled flying qualities having a Cooper-Harper rating of 10 as being *level 4*. Table 10.2.1 shows the relationship between these U.S. Military Specifications for levels of handling qualities and mission effectiveness (see MIL-F-8785C, 1980) as summarized by Hodgkinson (1999).

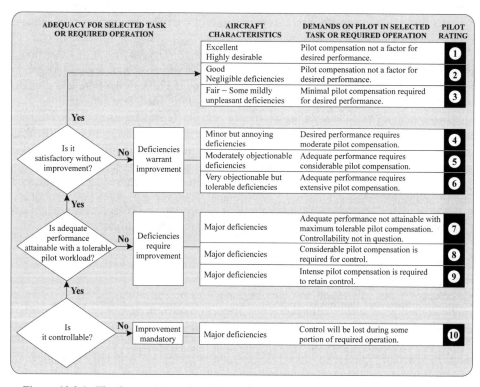

Figure 10.2.1. The Cooper-Harper handling qualities rating scale. (Cooper and Harper 1969)

	Pilot Rating	Handling Characteristics	Pilot Workload and Mission Effectiveness
Level 1	1,2,3	Satisfactory	Flying qualities clearly adequate for the mission flight phase. Desired performance is achievable with no more than minimal pilot compensation.
Level 2	4,5,6	Acceptable	Flying qualities adequate to accomplish the mission flight phase, but some increase in pilot workload or degradation in mission effectiveness, or both, exist.
Level 3	7,8,9	Controllable	Flying qualities such that the aircraft can be controlled safely in the context of the mission flight phase, but pilot workload is excessive or mission effectiveness is inadequate, or both. Category A flight phase can be terminated safely, and category B and C flight phases can be completed.
Level 4	10	Uncontrollable	Flying qualities worse than level 3 (unofficial).

Table 10.2.1. U.S. Military Specifications for levels of handling qualities and their relation to the Cooper-Harper scale.

It is not surprising that a pilot's opinion of an aircraft's handling qualities depends somewhat on the task that the pilot is performing. For those who have never flown an aircraft, this may be more easily understood in reference to driving a car. For example, drivers might rate the handling characteristics of a particular automobile as being excellent when referenced to high-speed highway driving. However, the same automobile may be rated as having very poor handling characteristics when drivers are ask to consider the ease of parking the car in a confined space or driving in heavy traffic. In a similar manner, pilot opinion of aircraft handling qualities must be referenced in some way to piloting task. For example, a pilot may rate an airplane as having very poor handling characteristics during takeoff. Yet the same pilot could rate the same airplane as having good handling characteristics on landing and during the gradual coordinated maneuvers associated with climb, cruise, and descent.

The essence of piloting task definitions can be crucial to the process of gathering reliable pilot opinions on aircraft handling qualities. For many years, piloting tasks have been loosely classified into three *categories*, which are called *flight phase categories*. As reported in MIL-F-8785C (1980) and summarized by Nelson (1998), these categories are defined in Table 10.2.2. Today, piloting tasks are often more narrowly defined in terms of what are called *mission task elements*, which are more precisely defined tasks that can be incorporated into flight tests and flight simulations. While more precisely defined piloting tasks lead to more reliable pilot ratings, the rather loosely defined task categories specified in Table 10.2.2 are still useful for classifying pilot opinion relative to piloting task, particularly in the design phase of an aircraft development program.

Flight Phase	Piloting Task Classification
Category A	Nonterminal flight phases that require rapid maneuvering, precision tracking, or precise flight-path control. Included in the category are air-to-air combat, ground attack, weapon delivery or launch, aerial recovery, reconnaissance, in-flight refueling (receiver), terrain following, antisubmarine search, and close-formation flying.
Category B	Nonterminal flight phases that are normally accomplished using gradual maneuvers and without precision tracking, although accurate flight-path control may be required. Included in the category are climb, cruise, loiter, in-flight refueling (tanker), descent, emergency descent, emergency deceleration, and aerial delivery.
Category C	Terminal flight phases that are normally accomplished using gradual maneuvers and usually require accurate flight-path control. Included in this category are takeoff, catapult takeoff, approach, wave-off/go-around and landing.

Table 10.2.2. Flight phase categories.

The reader should likewise not be surprised to learn that the size and type of aircraft being flown have a significant effect on the handling qualities that are expected by a pilot and on the ratings that result from a particular handling characteristic. Here again we can relate this to the more commonplace experience of driving a car. For example, a handling characteristic that might be considered only mildly unpleasant in a passenger car may be considered totally unacceptable in a racing car. On the other hand, a delivery truck that exhibits exactly the same characteristic could be rated as having excellent handling qualities. Likewise, some handling characteristics that are typically considered to be excellent in a large transport aircraft may be considered to be poor in a fighter or stunt airplane. To help account for the influence of airplane size and type on pilot opinion of aircraft handling qualities, airplanes are usually classified according to size and maneuverability. Table 10.2.3 shows the classification of aircraft that was adopted in the U.S. Military Specification on flying qualities, as reported in MIL-F-8785C (1980) and summarized by Hodgkinson (1999).

Pilot opinion of one handling characteristic can also be influenced significantly by the presence of another. For example, if significant pilot concentration is required to compensate for poor longitudinal handling qualities, the pilot will be less tolerant of deficiencies in the lateral modes. In measuring and specifying handling qualities it is commonly assumed that any degradation in pilot opinion is the result of degrading only one handling characteristic at a time. To estimate what might be expected from the combined effects of more than one poor handling characteristic, the reader is referred to Mitchell, Aponso, and Hoh (1990).

Aircraft Classification	Aircraft Size and Type
Class I	Small, light aircraft such as: Light utility Primary trainer Light observation
Class II	Medium-weight, low- to medium-maneuverability aircraft such as: Heavy utility/search and rescue Light or medium transport/cargo/tanker Early warning/electronic countermeasures/airborne command, control or communications relay Antisubmarine Assault transport Reconnaissance Tactical bomber Heavy attack Trainer for class II
Class III	Large, heavy, low- to medium-maneuverability aircraft such as: Heavy transport/cargo/tanker Heavy bomber Patrol/early warning/electronic countermeasures/airborne command, control or communications relay Trainer for class III
Class IV	High-maneuverability aircraft such as: Fighter-interceptor Attack Tactical reconnaissance Observation Trainer for class IV

Table 10.2.3. Aircraft classification, from the U.S. Military Specification on flying qualities.

Level 1 aircraft handling qualities are not necessarily required for all possible flight circumstances. Handling quality requirements for a particular situation depend to some extent on the probability that the situation will be encountered. The handling qualities that are required for a particular aircraft in a particular situation depend on what are called the *flight envelope* and the *aircraft state*.

Three nested flight envelopes are defined. The **operational flight envelope** encompasses those flight tasks that are encountered in the normal day-to-day operation of the aircraft: for example, takeoff and landing. The **service flight envelope** is defined by the service limits of the airplane, such as the service ceiling and maximum airspeed. The **permissible flight envelope** comprises all flight conditions that are allowable and possible, including stalls, spins, and other such extreme situations.

A **normal state** is defined as any normal operational configuration for the airplane, in terms of flap settings, gross weight, CG location, and so on. A **failure state** is a normal state modified by a malfunction in one or more of the airplane's components. Failure states are categorized according to the probability of occurrence.

In general, level 1 handling qualities are always required in the operational flight envelope for an aircraft in any normal state. In a failure state, the handling qualities are allowed to degrade in accordance with the probability of occurrence for that particular state. For failure states that are very likely to occur, handling qualities must remain at level 1. A failure state that is likely to occur more than once every 100 flights can result in handling qualities no worse than level 2. Handling qualities as poor as level 3 are allowed only if the failure state is likely to occur less than once in every 10,000 flights. In some cases, handling qualities have been allowed to degrade to level 2 in moderate turbulence and to level 3 in the presence of severe turbulence.

10.3. Dynamic Handling Quality Prediction

We have the ability to mathematically estimate the eigenvalues for the longitudinal and lateral modes associated with the dynamic response of a particular airplane design at some particular operating condition, prior to fabrication and flight testing. However, these eigenvalues do not provide a direct measure of aircraft handling qualities. What is needed is some means of predicting aircraft handling qualities from knowledge of the airplane's dynamic response eigenvalues and other design characteristics. This is one of the primary objectives of handling qualities research and is the topic of this section. Here each of the airplane's dynamic modes will be addressed separately. The handling qualities predicted for any mode are based on the assumption of level 1 handling qualities for all other modes.

The Short-Period Mode

As discussed previously, the short-period mode is comprised primarily of high-frequency oscillations in angle of attack and vertical displacement, with little change in forward velocity. Humans are highly sensitive to acceleration but quite insensitive to small changes in position and orientation. From the pilot's viewpoint it is the acceleration associated with short-period motion that is of greatest importance. Since the forward velocity remains nearly constant for the short period, it is primarily the normal acceleration associated with this mode that pilots feel. Oscillations in angle of attack and the resulting oscillations in lift produce this normal acceleration. Thus, it is not surprising that a pilot's opinion of how the airplane's handling qualities are affected by the short-period mode depends on how the airplane's normal acceleration varies with angle of attack. The load factor, n, is the usual measure of an airplane's normal acceleration. Recall that because normal acceleration is generated by lift, the load factor can be expressed as the lift divided by the weight. The change in load factor with respect to angle of attack is called the *acceleration sensitivity*,

$$\text{acceleration sensitivity} \equiv \frac{\partial n}{\partial \alpha} = \frac{1}{W}\frac{\partial L}{\partial \alpha} = \frac{1}{C_W}\frac{\partial C_L}{\partial \alpha} \qquad (10.3.1)$$

Pilot opinion of how the short-period frequency influences handling qualities has been found to vary with the acceleration sensitivity of the airplane. When pilot opinion data are correlated with the short-period frequency and acceleration sensitivity, results similar to those reported in MIL-F-8785C (1980) and shown here in Fig. 10.3.1 are obtained. These results are based on adequate short-period damping.

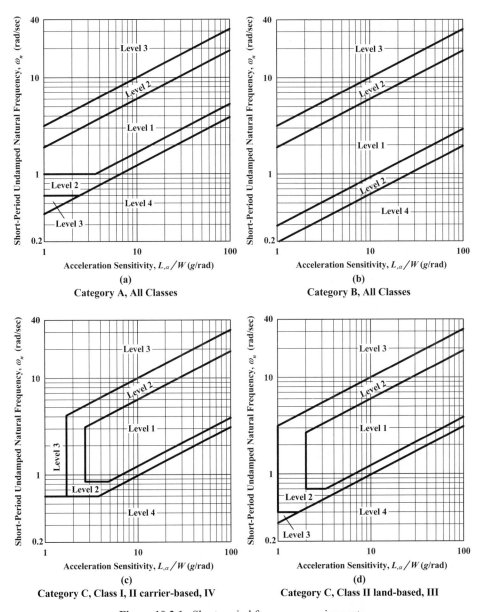

Figure 10.3.1. Short-period frequency requirements.

From Fig. 10.3.1 it can be seen that when the acceleration sensitivity is above some minimum value, the boundaries between the different levels of handling qualities are all defined by lines of constant slope on a log-log plot. Furthermore, the slope for each of these boundary lines is exactly ½. This means that the square of the undamped natural frequency at any point along any one of these boundary lines is proportional to the acceleration sensitivity. Thus, the square of the short-period undamped natural frequency divided by the acceleration sensitivity is an important correlation parameter associated with the short-period frequency requirements. This parameter is commonly called the *control anticipation parameter* (CAP),

$$
\text{CAP} \equiv (\omega_n^2)_{SP} \Big/ \frac{\partial n}{\partial \alpha} = \frac{(\omega_n^2)_{SP}}{L_{,\alpha}/W} = \frac{(\omega_n^2)_{SP}}{C_{L,\alpha}/C_W}
\tag{10.3.2}
$$

For the prediction of short-period frequency requirements, when the airplane's acceleration sensitivity is above a minimum value (3.5 for category A and 5.0 for category C), the boundaries between different levels of handling qualities are lines of constant CAP. For example, the top two sloping lines in Fig. 10.3.1(a)–(d) have constant CAP values of 3.6 and 10.0 s^{-2}. The lower two sloping lines have constant CAP values of 0.15 and 0.28 s^{-2} in Fig. 10.3.1(a), 0.038 and 0.085 s^{-2} in Fig. 10.3.1(b), and 0.096 and 0.15 s^{-2} in Fig. 10.3.1(c) and (d).

From results presented in Sec. 8.3, it has been shown that the short-period undamped natural frequency is proportional to the square root of the maneuver margin, l_{mp}, which is the axial distance aft from the center of gravity to the stick-fixed maneuver point. From Eq. (8.3.18), an accurate approximation for the short-period undamped natural frequency is given by

$$
(\omega_n^2)_{SP} = \frac{g l_{mp} L_{,\alpha}}{r_{yy_b}^2 W}
\tag{10.3.3}
$$

where r_{yy_b} is the pitch radius of gyration defined in Eq. (6.1.51),

$$
r_{yy_b}^2 \equiv \frac{g I_{yy_b}}{W}
\tag{10.3.4}
$$

Analytical relations for the stick-fixed maneuver margin were presented and discussed in Sec. 6.1. Recall that the stick-fixed maneuver point is aft of the stick-fixed neutral point by a distance that is proportional to pitch damping. From Eqs. (6.1.36) and (6.1.49),

$$
l_{mp} = -\frac{m_{,\alpha}}{L_{,\alpha}} - \frac{m_{,\bar{q}}}{W} = l_{np} - \frac{m_{,\bar{q}}}{W}
\tag{10.3.5}
$$

where $\bar{q}$ is the dynamic pitch rate defined in Eq. (6.1.42),

$$
\bar{q} \equiv \frac{Vq}{g} = \frac{2V^2}{g\bar{c}_w} q
\tag{10.3.6}
$$

Using Eq. (10.3.3) in Eq. (10.3.2), the CAP is found to depend only on the acceleration of gravity, the maneuver margin, and the pitch radius of gyration,

$$\text{CAP} \equiv \frac{(\omega_n^2)_{SP}}{L_{,\alpha}/W} = \frac{g l_{mp}}{r_{yy_b}^2} \tag{10.3.7}$$

With the application of Eq. (10.3.7), the short-period natural frequency requirements shown in Fig. 10.3.1 can be written as a function of the defined handling-quality levels and flight-phase categories:

$$
\left\{
\begin{array}{l}
\left(\begin{array}{l} 0.28\ \text{s}^{-2},\ \text{level 1} \\ 0.15\ \text{s}^{-2},\ \text{level 2} \end{array}\right),\ \text{category A} \\[2em]
\left(\begin{array}{l} 0.085\ \text{s}^{-2},\ \text{level 1} \\ 0.038\ \text{s}^{-2},\ \text{level 2} \end{array}\right),\ \text{category B} \\[2em]
\left(\begin{array}{l} 0.15\ \text{s}^{-2},\ \text{level 1} \\ 0.096\ \text{s}^{-2},\ \text{level 2} \end{array}\right),\ \text{category C}
\end{array}
\right\}
\le \frac{g l_{mp}}{r_{yy_b}^2} \le
\left\{
\left(\begin{array}{l} 3.6\ \text{s}^{-2},\ \text{level 1} \\ 10.\ \text{s}^{-2},\ \text{level 2} \end{array}\right),\ \text{all categories}
\right\}
$$

$$\tag{10.3.8}$$

With the Earth's gravitational acceleration fixed, Eq. (10.3.8) shows that compliance with the short-period frequency requirements depends only on an airplane's maneuver margin and pitch radius of gyration. Thus, compliance with these requirements depends on the location of the airplane's center of gravity and the mass distribution within the airplane. The empirical correlation expressed in Eq. (10.3.8) and the extensive flight-testing research that led to its development are the foundation for the dynamic margin constraint on aft CG limit, which is discussed in Sec. 6.1 and expressed in Eq. (6.1.50). At this point the student may find a review of Sec. 6.1 helpful.

As a word of caution, keep in mind that the short-period frequency requirements given by Eq. (10.3.8) are not the only CG limitations imposed by pilot demands. The control forces required to maneuver an airplane are also of great importance to the pilot. The control force per g must be light enough that the pilot can comfortably attain the maximum allowable load factor, and it must be heavy enough to prevent the pilot from inadvertently overstressing the airplane during maneuvering flight. Likewise, the control force gradient must be heavy enough to provide the pilot a suitable control force that increases with departure from trim speed, and it must be light enough to ensure that the pilot does not encounter an elevator control force in excess of what can be applied with one hand when completing a safe landing following an approach to land. For airplanes with reversible mechanical controls, the elevator control force derivatives depend on CG location, and control-force CG limitations are typically more restrictive than those imposed by the short-period frequency requirements given in Eq. (10.3.8).

The short-period damping ratio also has a significant effect on airplane handling qualities as judged by pilot opinion. The short-period damping requirements given in MIL-F-8785C (1980) and summarized by Hodgkinson (1999) are listed in Table 10.3.1.

Level	Category A and C Flight Phases		Category B Flight Phases	
	Minimum ζ	Maximum ζ	Minimum ζ	Maximum ζ
1	0.35	1.30	0.30	2.00
2	0.25	2.00	0.20	2.00
3	0.15	–	0.15	–

Table 10.3.1. Short-period damping ratio requirements for all aircraft classes.

The results presented in Table 10.3.1 are based on the assumption that the short-period frequency falls within the range that would give level 1 handling qualities according to Fig. 10.3.1. The short-period damping ratio can be evaluated from Eq. (8.3.16).

If the acceleration sensitivity is large enough so that the short-period frequency requirements depend only on the CAP, then the short-period frequency and damping requirements can be combined in a single plot with CAP on one axis and damping ratio on the other. For example, from Eq. (10.3.8) we see that level 1 handling qualities for category A flight phases require a control anticipation parameter between 0.28 s^{-2} and 3.6 s^{-2}, while Table 10.3.1 shows a required damping ratio between 0.35 and 1.30. These combined requirements are shown as the inner rectangle in Fig. 10.3.2. The level 2 and 3 handling qualities boundaries are also shown in this figure. Similar results for flight phase categories B and C are shown in Figs. 10.3.3 and 10.3.4.

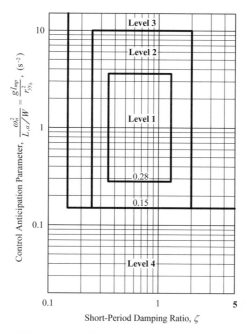

Figure 10.3.2. Short-period frequency and damping requirements for category A flight phases and airplanes of all classes having acceleration sensitivity greater than 3.5 g/rad.

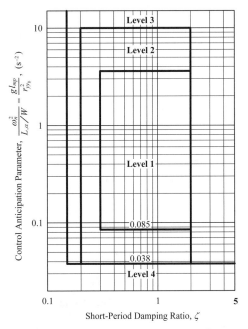

Figure 10.3.3. Short-period frequency and damping requirements for category B flight phases and airplanes of all classes.

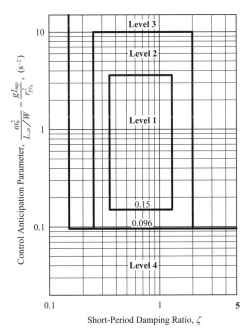

Figure 10.3.4. Short-period frequency and damping requirements for category C flight phases and airplanes of all classes having acceleration sensitivity greater than 5 *g*/rad.

The Phugoid

The period of the phugoid is always long compared to a pilot's response time. Thus, pilots can easily suppress phugoid oscillations with appropriate control inputs. Because of the large difference between the phugoid period and a pilot's response time, the exact value of the phugoid frequency has almost no effect on a pilot's opinion of an airplane's handling qualities. However, if the phugoid damping is too low, suppressing the phugoid requires some pilot attention, and this makes other tasks more difficult. For this reason, aircraft handling qualities do correlate with phugoid damping. The phugoid damping requirements presented in MIL-F-8785C (1980) and summarized by Hodgkinson (1999) are given in Table 10.3.2. From the pilot opinion data currently available, these damping requirements appear to be independent of both aircraft class and flight phase.

Level	All Flight Phases for All Aircraft Categories Minimum Phugoid Damping
1	Damping ratio should be greater than 0.04
2	Damping ratio should be greater than 0.00
3	Time to double amplitude should be greater than 55 sec

Table 10.3.2. Phugoid damping requirements for all aircraft classes.

The Roll Mode

As discussed in Chapter 9, the roll mode is a heavily overdamped motion. Thus, when a typical-fixed wing aircraft responds to changes in aileron input, the time constant is very short and the airplane quickly approaches a steady rolling rate that is proportional to aileron deflection. This gives the pilot the perception that aileron input commands the airplane's rolling rate. The time constant for any mode is defined simply as the inverse of the damping rate. For favorable handling qualities, the time constant for the roll mode is typically on the order of 1 second or slightly longer. If the roll mode time constant is much longer than a second, the transient persists well into the maneuver and the pilot begins to feel as though he or she is commanding roll acceleration instead of rolling rate. This will normally degrade the pilot's opinion of the airplane's handling qualities. Table 10.3.3 lists the minimum damping requirements in terms of maximum time constant for the roll mode from MIL-F-8785C (1980) as summarized by Nelson (1998).

Flight Phase Category	Aircraft Class	Handling Quality		
		Level 1	Level 2	Level 3
A and C	I and IV	1.0	1.4	10
	II and III	1.4	3.0	10
B	All	1.4	3.0	10

Table 10.3.3. Roll mode damping requirements, maximum time constant $(1/\sigma)$ in seconds.

The Spiral Mode

In Chapter 9 we found that the spiral mode is usually either slowly convergent or slowly divergent. The time constant for the spiral mode is typically large compared to the pilot's reaction time. Thus, like the phugoid, this motion is readily controlled with pilot input. Even if the spiral mode is divergent, the pilot can typically control the motion with little or no conscious effort, provided that the time required to double the disturbance amplitude is not too short. However, if the spiral divergence is too rapid, pilot attention is required, which will detract from the performance of other tasks and degrade the pilot's opinion of the aircraft's handling qualities. The spiral damping requirements presented in MIL-F-8785C (1980) and summarized by Nelson (1998) are shown in Table 10.3.4. The damping criteria in this table are based on the minimum allowable doubling time for a divergent mode, i.e., $-\ln(2)/\sigma$.

Flight Phase Category	Aircraft Class	Handling Quality		
		Level 1	Level 2	Level 3
A	I and IV	12	12	4
	II and III	20	12	4
B and C	All	20	12	4

Table 10.3.4. Spiral mode damping requirements, minimum time-to-double-amplitude in seconds.

The Lateral Phugoid

Some modern aircraft have very little wing area and the roll damping for such aircraft can be quite low. For aircraft of this type, it is occasionally possible for the eigenvalues associated with the usual roll and spiral modes to combine, forming a complex pair. In such cases, the aircraft's lateral motion will exhibit a low-frequency oscillatory mode, which is sometimes called the *lateral phugoid*. While this characteristic is not normally associated with good handling qualities, it can be acceptable if the damping is sufficient. The damping requirements for this mode, which are reported in MIL-F-8785C (1980) and summarized by Hodgkinson (1999), are given in Table 10.3.5. The damping criteria in this table are based on the minimum allowable damping rate, σ, which is the product of the undamped natural frequency, ω_n, and the damping ratio, ζ.

Level	All Flight Phases for All Aircraft Categories Minimum Damping for the Oscillatory Roll-Spiral Mode
1	The $\zeta\omega_n$ product should be greater than 0.50 rad/sec
2	The $\zeta\omega_n$ product should be greater than 0.30 rad/sec
3	The $\zeta\omega_n$ product should be greater than 0.15 rad/sec

Table 10.3.5. Total damping requirements for the seldom-encountered low-frequency mode that is sometimes called the lateral phugoid.

Dutch Roll

Dutch roll is often the most troublesome of the dynamic modes associated with the stick-fixed longitudinal and lateral response of a conventional airplane. As discussed in Chapter 9, Dutch roll is a damped oscillatory motion that involves an out-of-phase combination of sideslip, roll, and yaw. Since the period for this mode is typically on the order of 2 to 10 seconds, Dutch roll motion can be very annoying to both pilots and passengers if the damping is light. Because the quarter-period of oscillation for Dutch roll can be quite close to the response time of a human pilot, it can be extremely difficult for a pilot to suppress this motion with control inputs. In fact, as discussed in Chapter 9, the natural Dutch roll characteristics of some aircraft are such that most pilots will actually excite the Dutch roll in an attempt to suppress it. If the period of oscillation is not too short and the damping is not too light, Dutch roll motion does not require significant pilot attention. However, if the Dutch roll damping is extremely light and the period is short, the airplane can be quite difficult to handle.

As might be expected, both the Dutch roll frequency and damping ratio affect the handling qualities of an airplane. The Dutch roll frequency and damping requirements reported in MIL-F-8785C (1980) and summarized by Hodgkinson (1999) are presented in Table 10.3.6. Notice that this table gives a minimum value for the damping ratio, the undamped natural frequency, and the product of the damping ratio and the undamped natural frequency. All of these requirements should be met. The minimum frequency requirement is always based on the undamped natural frequency, and the minimum damping requirement is that which yields the largest damping ratio.

As a word of caution, it is worth reiterating that although the classical Dutch roll approximation given by Eq. (9.5.5) is commonly used in the literature, this is a poor approximation at best. To evaluate compliance with requirements given in Table 10.3.6, either Eq. (9.5.27) or a full numerical solution should be used.

Level	Flight Phase Category	Aircraft Class	Minimum[5] ζ	Minimum[5] $\zeta\omega_n$ (rad/sec)	Minimum[6] ω_n (rad/sec)
1	A	IV-CO[1] and GA[2]	0.40	0.40	1.0
		I and IV-other	0.19	0.35	1.0
		II and III	0.19	0.35	0.4
	B	All	0.08	0.15	0.4
	C	I, II-C[3], and IV	0.08	0.15	1.0
		II-L[4] and III	0.08	0.10	0.4
2	All	All	0.02	0.05	0.4
3	All	All	0.00	–	0.4

[1]CO: combat. [2]GA: ground attack. [3]C: carrier-based. [4]L: land-based.
[5]The minimum damping requirement is that yielding the greatest damping ratio.
[6]The minimum frequency requirement is always based on the undamped natural frequency.

Table 10.3.6. The Dutch roll frequency and damping requirements.

EXAMPLE 10.3.1. For the airplane and operating condition that was described in Examples 8.2.1 and 9.2.1, compare the predicted dynamic response with the frequency and damping requirements presented in this section. What level of handling qualities should be expected for this airplane and operating condition, based on the dynamic response modes? Estimate the aft CG limit for this flight phase based on the level 1 short-period frequency requirement.

Solution. This is a class I aircraft. Since the operating condition is cruise with no special requirement for precision tracking, the flight phase is in category B.

From Example 8.2.1, the short-period damping ratio and undamped natural frequency are

$$\zeta = 0.69, \quad \omega_n = 3.59 \text{ rad/sec}$$

The lift slope and weight coefficient for this airplane and operating condition are

$$C_{L,\alpha} = 4.40, \quad C_W = 0.393$$

Thus, the acceleration sensitivity is

$$\text{acceleration sensitivity} = C_{L,\alpha}/C_W = 4.40/0.393 = 11.2 \text{ g/rad}$$

and the control anticipation parameter is

$$\text{CAP} = \frac{(\omega_n^2)_{SP}}{C_{L,\alpha}/C_W} = \frac{3.59^2}{11.2} = 1.15 \text{ s}^{-2}$$

For this airplane and category B flight phase, the short-period frequency and damping requirements are prescribed in Fig. 10.3.3. The damping ratio of 0.69 and CAP of 1.15 fall well within the limits for which level 1 handling qualities would normally be expected, based on short-period requirements.

Also from Example 8.2.1, the phugoid damping ratio for this airplane and operating condition is 0.080. Based on the requirements given in Table 10.3.2, this is well over the limit needed to provide level 1 handling qualities.

From Example 9.2.1, the predicted damping rate for the roll mode is 8.9 sec^{-1}, so the roll mode time constant is

$$\text{roll mode time constant} = 1/\sigma = 1/8.9 = 0.11 \text{ sec}$$

From Fig. 10.3.7 we see that this is considerably less than the maximum time constant that would normally yield level 1 handling qualities.

A convergent spiral mode was predicted in Example 9.2.1. Thus, from the results presented in Table 10.3.4, we see that the spiral mode should not contribute to reducing the handling qualities below level 1.

For this airplane and operating condition, the Dutch roll damping ratio and undamped natural frequency as predicted in Example 9.2.1 are

$$\zeta = 0.20, \quad \omega_n = 2.43 \text{ rad/sec}, \quad \text{and} \quad \zeta\omega_n = 0.49 \text{ rad/sec}$$

All three of these values satisfy the requirements for level 1 handling qualities as specified in Table 10.3.6.

The frequencies and damping rates for all of the stick-fixed dynamic modes associated with this airplane and operating conditions are well within the limits that would typically provide level 1 handling qualities. Thus, for this operating condition, we would expect that the stick-fixed dynamic response of this airplane is adequate to provide level 1 handling qualities. However, to predict the airplane's handling qualities for all normal operation, these calculations would need to be repeated for the full range of operating conditions within the *operational flight envelope*.

From the data given in Example 8.2.1, the pitch radius of gyration squared is

$$r_{yy_b}^2 \equiv \frac{g I_{yy_b}}{W} = \frac{32.2 \text{ ft/s}^2 (3{,}000 \text{ slug·ft}^2)}{2{,}800 \text{ lbf}} = 34.5 \text{ ft}^2$$

From Eq. (10.3.7), the maneuver margin is

$$l_{mp} = \frac{r_{yy_b}^2 (\text{CAP})}{g} = \frac{34.5 \text{ ft}^2 (1.15 \text{ s}^{-2})}{32.2 \text{ ft/s}^2} = 1.232 \text{ ft}$$

Using Eq. (10.3.8) and assuming that the pitch radius of gyration does not change, the minimum allowable maneuver margin needed to provide level 1 handling qualities for this category B flight phase is

$$(l_{mp})_{\min} = \frac{r_{yy_b}^2 (\text{CAP})_{\min}}{g} = \frac{34.5 \text{ ft}^2 (0.085 \text{ s}^{-2})}{32.2 \text{ ft/s}^2} = 0.091 \text{ ft}$$

Thus, the CG could be moved aft about 13.7 inches without violating the level 1 short-period frequency requirement for the category B flight phase. Requirements for takeoff and landing would likely be more restrictive.

10.4. Response to Control Inputs

Pilot opinion of aircraft handling qualities seems to correlate quite well with the eigenvalues for the traditional stick fixed dynamic modes, which are obtained from aircraft small-disturbance theory with all control inputs set to zero. One might ask the question: Why should results obtained using the case of no control input have such a significant effect on a pilot's ability to control the airplane? The answer is linked to the transient response of the airplane to control inputs.

As we have seen, airplane controls are used to establish trim and maneuver or accelerate an airplane from one trimmed state to another. To this point, our treatment of aircraft control and maneuverability has been mostly confined to the case of constant angular rates, which produce constant acceleration. For example, our analysis of the elevator angle and control force per g in the pull-up maneuver was based on the assumption that the maneuver had progressed to the point where the pitching rate had reached a steady value. Similarly, our analysis of the steady coordinated turn was based on the assumption that the wings were already banked at the correct angle and that the airplane was turning with a constant angular rate.

In a typical flying situation, an airplane might be flying at level trim. If the pilot wishes to climb, descend, or turn, then he or she must change the control inputs. For example, assume that the pilot wishes to change the airplane's flight path from level trimmed flight to climbing flight without changing the throttle setting. To accomplish this, he or she must pull the stick or control yoke back. After the control has been held back for a period of time, the airplane will attain a new trim state, which then has a positive climb angle and lower airspeed. The maneuver that takes the airplane between these two trimmed states is the pull-up maneuver. However, this maneuver starts with zero pitch rate and ends with zero pitch rate. While there may be a period of time during the maneuver when the pitch rate is constant, clearly this entire maneuver cannot take place at constant pitch rate and constant airspeed. Thus, our Chapter 6 analysis for the constant-speed pull-up maneuver cannot possibly tell the whole story in regard to this common airplane maneuver.

When the pilot first initiates a pull-up maneuver from trimmed flight, the pitching moment generated by the elevator deflection produces acceleration in pitch and the pitch rate begins to rise. As the pitch rate increases, an opposing moment develops as a result of pitch damping. If this pitch rate becomes large enough, the pitch damping moment will just balance the moment produced by the elevator deflection, returning the pitch acceleration to zero. A question then naturally arises: How does the airplane respond during the transient period of angular acceleration?

We have at least superficially examined the analogous problem associated with aileron deflection. Because the roll damping is quite large for a conventional airplane, transients associated with aileron deflection are quickly damped out and the airplane rapidly approaches a constant rolling rate. This gives pilots the perception that aileron deflection commands rolling rate. However, the pitch and yaw damping for a typical airplane are much less than that for roll. Thus, the pitch and yaw controls "feel" very different from the roll control. In this section we examine the transients associated with elevator and rudder deflection.

In general, maneuvering flight with substantial translational and rotational acceleration can involve significant inertial, gravitational, and aerodynamic nonlinearities. To accurately predict the response of an airplane during such maneuvers, we must return to the nonlinear equations of motion developed in Chapter 7. These equations would need to be integrated numerically. This is the topic of Chapter 11. However, by continuing somewhat further with small-disturbance theory, we can obtain some useful results that provide valuable insight into the control process. Furthermore, such results are of great value in the design of autopilot control systems.

Transfer Functions

The linearized aircraft equations of motion presented in Sec. 7.5, 7.7, or 9.8 are of the mathematical form of the generalized linear control problem,

$$[B]\{\dot{X}\} = [A]\{X\} + [C]\{D\}$$

where $\{X\}$ is a vector of n state variables, $[A]$ and $[B]$ are constant $n \times n$ matrices, $\{D\}$ is a vector of m control deflections, and $[C]$ is a constant $n \times m$ matrix. A special case of the generalized linear control problem is of particular interest. That is the special linear control problem, where $[B]$ is the identity matrix. The generalized linear control problem is easily converted to the special linear control problem by multiplying both sides of the equation by the inverse of $[B]$,

$$[B]^{-1}[B]\{\dot{X}\} = [B]^{-1}[A]\{X\} + [B]^{-1}[C]\{D\}$$

Thus, any generalized linear control problem is readily converted to the special linear control problem, which has the form

$$\{\dot{X}\} = [A]\{X\} + [C]\{D\} \tag{10.4.1}$$

If the control deflections, $\{D\}$, were known as a function of time, the solution to this linear control problem would allow us to determine the state variables, $\{X\}$, as a function of time.

Linear time-invariant systems of the form given by Eq. (10.4.1) are commonly solved using the method of Laplace transforms. Recall that the Laplace transform of some function of t, say $x(t)$, is defined by

$$\bar{x}(s) \equiv \int_0^\infty x(t) e^{-st} dt \tag{10.4.2}$$

Thus, using integration by parts, the Laplace transform of the derivative of x with respect to t is given by

$$\overline{\frac{dx}{dt}} \equiv \int_0^\infty \frac{dx}{dt} e^{-st} dt = \int_0^\infty e^{-st} dx = x e^{-st} \Big|_0^\infty - \int_0^\infty x d(e^{-st})$$
$$= -x(0) + s \int_0^\infty x e^{-st} dt = s\bar{x} - x(0) \tag{10.4.3}$$

An important concept called the *transfer function* is widely used in the application of Laplace transforms to Eq. (10.4.1). Each element of the transfer-function matrix, $[G]$, is defined to be the ratio of the Laplace transform of the response in one of the state variables to the Laplace transform of one control input, for the special case where the state variable response is zero at time $t = 0$. Specifically, $[G]$ is defined to yield

$$\{\bar{X}\} = [G]\{\bar{D}\} \tag{10.4.4}$$

If the transfer-function matrix is known, the state variable response vector in the Laplace domain can be found for any control deflections input vector by simple matrix multiplication according to Eq. (10.4.4).

The transfer-function matrix, $[\mathbf{G}]$, is readily obtained from Eq. (10.4.1). Taking the Laplace transform of Eq. (10.4.1) and applying Eq. (10.4.3) along with the null initial condition gives

$$s\{\overline{\mathbf{X}}\} = [\mathbf{A}]\{\overline{\mathbf{X}}\} + [\mathbf{C}]\{\overline{\mathbf{D}}\} \tag{10.4.5}$$

or after solving for the Laplace transform of the state variable response vector,

$$\{\overline{\mathbf{X}}\} = \left[s\,[\mathbf{i}] - [\mathbf{A}]\right]^{-1} [\mathbf{C}]\{\overline{\mathbf{D}}\} \tag{10.4.6}$$

where $[\mathbf{i}]$ is the identity matrix. By comparing Eq. (10.4.6) with Eq. (10.4.4), we see that the transfer-function matrix is given by

$$[\mathbf{G}] = \left[s\,[\mathbf{i}] - [\mathbf{A}]\right]^{-1} [\mathbf{C}] \tag{10.4.7}$$

To evaluate the inverse matrix in Eq. (10.4.7), we recall from linear algebra that if the determinant is not zero,

$$\left[s\,[\mathbf{i}] - [\mathbf{A}]\right]^{-1} = \frac{[\mathbf{M}]^T}{\left|s\,[\mathbf{i}] - [\mathbf{A}]\right|} \tag{10.4.8}$$

where $[\mathbf{M}]^T$ is the transpose of the matrix of cofactors. The components of $[\mathbf{M}]$ are defined as

$$M_{ij} = (-1)^{i+j}\, m_{ij} \tag{10.4.9}$$

where m_{ij} is the minor determinate of $s[\mathbf{i}]-[\mathbf{A}]$, obtained from the matrix that remains after row i and column j are removed.

The significance of this mathematics is best demonstrated by example. To help the student relate the mathematics to the physics of commonplace experience, we shall begin with a simple spring-mass system.

EXAMPLE 10.4.1. Consider the spring-mass system described in Example 8.1.1. An independent time-varying control force is applied to each of the two masses as shown in Fig. 10.4.1. Find the transfer-function matrix for this system. For the special case where ΔF_1 is a unit step function and ΔF_2 is zero, find the Laplace transform of the state variable response vector.

$$m_1 = 20.0 \text{ slugs}, \qquad c_1 = 30.0 \text{ lbf·sec/ft}, \qquad k_1 = 2.0 \text{ lbf/ft},$$
$$m_2 = 20.0 \text{ slugs}, \qquad c_2 = 15.0 \text{ lbf·sec/ft}, \qquad k_2 = 100.0 \text{ lbf/ft}$$

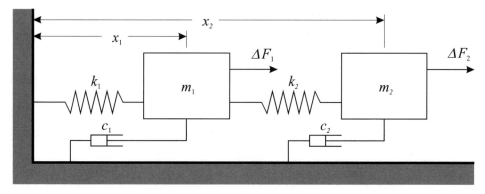

Figure 10.4.1. Damped spring-mass system with two degrees of freedom.

Solution. Following the development that led to Eq. (8.1.45) but including the control forces, we have

$$
\begin{Bmatrix} \Delta \dot{V}_1 \\ \Delta \dot{V}_2 \\ \Delta \dot{x}_1 \\ \Delta \dot{x}_2 \end{Bmatrix} = \begin{bmatrix} -c_1/m_1 & 0 & -(k_1+k_2)/m_1 & k_2/m_1 \\ 0 & -c_2/m_2 & k_2/m_2 & -k_2/m_2 \\ 1 & 0 & 0 & 0 \\ 0 & 1 & 0 & 0 \end{bmatrix} \begin{Bmatrix} \Delta V_1 \\ \Delta V_2 \\ \Delta x_1 \\ \Delta x_2 \end{Bmatrix}
$$

$$
+ \begin{bmatrix} 1/m_1 & 0 \\ 0 & 1/m_2 \\ 0 & 0 \\ 0 & 0 \end{bmatrix} \begin{Bmatrix} \Delta F_1 \\ \Delta F_2 \end{Bmatrix}
$$

After using the specified masses, damping coefficients, and spring constants, the special linear control problem for this system is

$$
\begin{Bmatrix} \Delta \dot{V}_1 \\ \Delta \dot{V}_2 \\ \Delta \dot{x}_1 \\ \Delta \dot{x}_2 \end{Bmatrix} = \begin{bmatrix} -1.5 & 0 & -5.1 & 5.0 \\ 0 & -0.75 & 5.0 & -5.0 \\ 1 & 0 & 0 & 0 \\ 0 & 1 & 0 & 0 \end{bmatrix} \begin{Bmatrix} \Delta V_1 \\ \Delta V_2 \\ \Delta x_1 \\ \Delta x_2 \end{Bmatrix} + \begin{bmatrix} 0.05 & 0 \\ 0 & 0.05 \\ 0 & 0 \\ 0 & 0 \end{bmatrix} \begin{Bmatrix} \Delta F_1 \\ \Delta F_2 \end{Bmatrix}
$$

By direct comparison with Eq. (10.4.1), we have

$$
[s\,[\mathbf{i}]-[\mathbf{A}]] = \begin{bmatrix} s+1.5 & 0 & 5.1 & -5.0 \\ 0 & s+0.75 & -5.0 & 5.0 \\ -1 & 0 & s & 0 \\ 0 & -1 & 0 & s \end{bmatrix} \equiv [\mathbf{E}]
$$

The cofactors of this matrix as determined from Eq. (10.4.9) are

$$M_{11} = s^3 + 0.75s^2 + 5.0s$$
$$M_{12} = 5.0s$$
$$M_{13} = s^2 + 0.75s + 5.0$$
$$M_{14} = 5.0$$
$$M_{21} = 5.0s$$
$$M_{22} = s^3 + 1.5s^2 + 5.1s$$
$$M_{23} = 5.0$$
$$M_{24} = s^2 + 1.5s + 5.1$$
$$M_{31} = -5.1s^2 - 3.825s - 0.5$$
$$M_{32} = 5.0s^2 + 7.5s$$
$$M_{33} = s^3 + 2.25s^2 + 6.125s + 7.5$$
$$M_{34} = 5.0s + 7.5$$
$$M_{41} = 5.0s^2 + 3.75s$$
$$M_{42} = -5.0s^2 - 7.5s - 0.5$$
$$M_{43} = 5.0s + 3.75$$
$$M_{44} = s^3 + 2.25s^2 + 6.225s + 3.825$$

From these cofactors, the determinate is found to be

$$\left| s\,[\mathbf{i}] - [\mathbf{A}] \right| \equiv |\mathbf{E}| = \sum_{i=1}^{4} E_{ij} M_{ij} = \sum_{j=1}^{4} E_{ij} M_{ij}$$
$$= s^4 + 2.25s^3 + 11.225s^2 + 11.325s + 0.5$$

From Eqs. (10.4.7) and (10.4.8), the transfer-function matrix can be determined from the cofactors and the control force coefficient matrix,

$$[\mathbf{G}] = \frac{[\mathbf{M}]^T [\mathbf{C}]}{\left| s\,[\mathbf{i}] - [\mathbf{A}] \right|}$$

where in this case

$$[\mathbf{C}] = \begin{bmatrix} 0.05 & 0 \\ 0 & 0.05 \\ 0 & 0 \\ 0 & 0 \end{bmatrix}$$

After performing the indicated algebra, we have

$$[\mathbf{G}] = \frac{\begin{bmatrix} 0.05s^3 + 0.0375s^2 + 0.25s & 0.25s \\ 0.25s & 0.05s^3 + 0.075s^2 + 0.255s \\ 0.05s^2 + 0.0375s + 0.25 & 0.25 \\ 0.25 & 0.05s^2 + 0.075s + 0.255 \end{bmatrix}}{s^4 + 2.25s^3 + 11.225s^2 + 11.325s + 0.5}$$

Tables of Laplace transforms are widely available, and from any such table we would find that the Laplace transform of a unit step function is $1/s$. Thus, for the special case where ΔF_1 is a unit step function and ΔF_2 is zero,

$$\left\{ \begin{matrix} \overline{\Delta F}_1 \\ \overline{\Delta F}_2 \end{matrix} \right\} = \left\{ \begin{matrix} 1/s \\ 0 \end{matrix} \right\}$$

From Eq. (10.4.4), the Laplace transform of the state variable response vector is found to be

$$\left\{ \begin{matrix} \overline{\Delta V}_1 \\ \overline{\Delta V}_2 \\ \overline{\Delta x}_1 \\ \overline{\Delta x}_2 \end{matrix} \right\} = [\mathbf{G}] \left\{ \begin{matrix} 1/s \\ 0 \end{matrix} \right\} = \left\{ \begin{matrix} \left(\dfrac{0.05s^3 + 0.0375s^2 + 0.25s}{s(s^4 + 2.25s^3 + 11.225s^2 + 11.325s + 0.5)} \right) \\[2ex] \left(\dfrac{0.25s}{s(s^4 + 2.25s^3 + 11.225s^2 + 11.325s + 0.5)} \right) \\[2ex] \left(\dfrac{0.05s^2 + 0.0375s + 0.25}{s(s^4 + 2.25s^3 + 11.225s^2 + 11.325s + 0.5)} \right) \\[2ex] \left(\dfrac{0.25}{s(s^4 + 2.25s^3 + 11.225s^2 + 11.325s + 0.5)} \right) \end{matrix} \right\}$$

Step Response

Results like those obtained in Example 10.4.1 are quite useless unless we are able to transform the solution back to the time domain. As a precursor to doing exactly that, several important observations can be made from Example 10.4.1.

First, notice that each of the transfer functions for the system is the ratio of two polynomials in s. Furthermore, it is seen that all of the transfer functions have exactly the same denominator and the order of the polynomial in the numerator is lower than that in the denominator. These characteristics hold for the solution to Eq. (10.4.1) in general, regardless of the order of the system.

The polynomial in the denominator of each transfer function is the *characteristic polynomial* for the system of equations. From Eq. (10.4.8) we see that the characteristic polynomial comes from the determinate of a characteristic matrix, which is identical to that used to obtain the eigenvalues, i.e., Eq. (8.1.63). Thus, the roots of the characteristic polynomial are the eigenvalues for the homogeneous system.

With the discussion above in mind, Eq. (10.4.8) can be used in Eq. (10.4.7) and the denominator can be factored to give

$$[\mathbf{G}] = \frac{[\mathbf{M}]^T [\mathbf{C}]}{(s - \lambda_1)(s - \lambda_2)(s - \lambda_3) \cdots (s - \lambda_n)} \tag{10.4.10}$$

where n is the order of the system. Since $[\mathbf{M}]^T$ is the transpose of the matrix of cofactors and $[\mathbf{C}]$ is a matrix of constants, each component of the numerator on the right-hand side of Eq. (10.4.10) is a polynomial of order less than n.

When the Laplace transform of some function of t is the ratio of two polynomials in s, the inverse transform can be obtained from the *Heaviside expansion theorem*. For instance, assume that the Laplace transform of $x(t)$ is given by

$$\overline{x}(s) = \frac{N(s)}{(s - r_1)(s - r_2)(s - r_3) \cdots (s - r_n)} \tag{10.4.11}$$

where $N(s)$ is a polynomial in s. In the absence of repeated roots, the inverse transform is then given by

$$x(t) = \sum_{k=1}^{n} \left[\frac{N(r_k)}{\prod_{j=1}^{n} (r_k - r_j)_{j \neq k}} \right] \exp(r_k t) \tag{10.4.12}$$

Again, the significance of the mathematics is best demonstrated by example.

EXAMPLE 10.4.2. Consider the spring-mass system in Example 10.4.1. For the special case where ΔF_1 is a unit step function and ΔF_2 is zero, obtain a time-domain expression for the Δx_1 component of the system response vector.

Solution. From Example 10.4.1,

$$\overline{\Delta x_1}(s) = \frac{0.05s^2 + 0.0375s + 0.25}{s(s^4 + 2.25s^3 + 11.225s^2 + 11.325s + 0.5)}$$

The roots of the polynomial in the denominator of this Laplace transform are the eigenvalues found in Example 8.1.1 combined with the root zero,

$$r_1 = \lambda_1 = -0.556387 + 3.098306\, i$$

$$r_2 = \lambda_2 = -0.556387 - 3.098306\, i$$

$$r_3 = \lambda_3 = -1.090975$$

$$r_4 = \lambda_4 = -0.046251$$

$$r_5 = 0.0$$

Thus, using Eq. (10.4.12), we have

$$
\Delta x_1(t) = \left[\frac{0.05\lambda_1^2 + 0.0375\lambda_1 + 0.25}{\lambda_1(\lambda_1 - \lambda_2)(\lambda_1 - \lambda_3)(\lambda_1 - \lambda_4)} \right] \exp(\lambda_1 t)
$$
$$
+ \left[\frac{0.05\lambda_2^2 + 0.0375\lambda_2 + 0.25}{\lambda_2(\lambda_2 - \lambda_1)(\lambda_2 - \lambda_3)(\lambda_2 - \lambda_4)} \right] \exp(\lambda_2 t)
$$
$$
+ \left[\frac{0.05\lambda_3^2 + 0.0375\lambda_3 + 0.25}{\lambda_3(\lambda_3 - \lambda_1)(\lambda_3 - \lambda_2)(\lambda_3 - \lambda_4)} \right] \exp(\lambda_3 t)
$$
$$
+ \left[\frac{0.05\lambda_4^2 + 0.0375\lambda_4 + 0.25}{\lambda_4(\lambda_4 - \lambda_1)(\lambda_4 - \lambda_2)(\lambda_4 - \lambda_3)} \right] \exp(\lambda_4 t) + \left[\frac{0.25}{(-\lambda_1)(-\lambda_2)(-\lambda_3)(-\lambda_4)} \right]
$$

Using the known eigenvalues and carrying out the algebra gives

$$
\Delta x_1(t) = (-0.001254 - 0.000081\,i)\exp[(-0.556387 + 3.098306\,i)t]
$$
$$
+ (-0.001254 + 0.000081\,i)\exp[(-0.556387 - 3.098306\,i)t]
$$
$$
+ 0.023840\exp(-1.090975t) - 0.521332\exp(-0.046251t) + 0.500000
$$

The first two terms in this solution could be combined and written as a damped sinusoid. The solution for the output response to a step function input that is not of unit amplitude is found simply by multiplying this solution by the amplitude of the input step function.

We see from Example 10.4.2 that the response to a step function control input is a constant plus a weighted sum of the homogeneous solutions, which are obtained directly from the system eigenvalues independent of the control inputs. If the two matrices [A] and [C] in Eq. (10.4.1) are known, the method used in Examples 10.4.1 and 10.4.2 to obtain the system response to a step control input is very systematic and quite suitable for programming on a digital computer.

EXAMPLE 10.4.3. For steady level flight at sea level, determine the longitudinal response to a negative 1-degree step change in elevator deflection for the general aviation airplane, which was described in Example 8.2.1. Using the linearized longitudinal equations of motion, plot the angle of attack and airspeed as a function of time. For this airplane and operating condition, we have

$$
S_w = 185 \text{ ft}^2, \quad b_w = 33 \text{ ft}, \quad \bar{c}_w = 5.606 \text{ ft}, \quad W = 2{,}800 \text{ lbf},
$$
$$
V_o = 180 \text{ ft/sec}, \quad C_{Do} = 0.05,
$$
$$
I_{xx_b} = 1{,}000 \text{ slug·ft}^2, \quad I_{yy_b} = 3{,}000 \text{ slug·ft}^2, \quad I_{zz_b} = 3{,}500 \text{ slug·ft}^2,
$$
$$
I_{xz_b} = 30 \text{ slug·ft}^2,
$$
$$
C_{L,\alpha} = 4.40, \quad C_{D,\alpha} = 0.35, \quad C_{m,\alpha} = -0.68, \quad C_{L,\hat{\alpha}} = 1.60, \quad C_{m,\hat{\alpha}} = -4.35,
$$

$$C_{Y,\beta} = -0.560, \quad C_{\ell,\beta} = -0.075, \quad C_{n,\beta} = 0.070,$$
$$C_{D,\bar{q}} \cong 0.0, \quad C_{L,\bar{q}} = 3.80, \quad C_{m,\bar{q}} = -9.95,$$
$$C_{Y,\bar{p}} \cong 0.0, \quad C_{\ell,\bar{p}} = -0.410, \quad C_{n,\bar{p}} = -0.0575,$$
$$C_{Y,\bar{r}} = 0.240, \quad C_{\ell,\bar{r}} = 0.105, \quad C_{n,\bar{r}} = -0.125,$$
$$C_{L,\delta_e} = 0.350, \quad C_{D,\delta_e} \cong 0.00, \quad C_{m,\delta_e} = -0.920,$$
$$C_{Y,\delta_a} \cong 0.00, \quad C_{\ell,\delta_a} = -0.135, \quad C_{n,\delta_a} = 0.0035,$$
$$C_{Y,\delta_r} = 0.155, \quad C_{\ell,\delta_r} = 0.105, \quad C_{n,\delta_r} = -0.075$$

Solution. From Example 8.2.1 we have

$$\frac{\rho S_w \bar{c}_w}{4W/g} = 0.007080, \qquad \frac{\rho S_w \bar{c}_w^3}{8 I_{yy_b}} = 0.003228$$

From Eq. (7.7.8), this gives

$$R_{x,\delta_e} \equiv \frac{\rho S_w \bar{c}_w}{4W/g} C_{X,\delta_e} = -\frac{\rho S_w \bar{c}_w}{4W/g} C_{D,\delta_e} = 0.00$$

$$R_{z,\delta_e} \equiv \frac{\rho S_w \bar{c}_w}{4W/g} C_{Z,\delta_e} = -\frac{\rho S_w \bar{c}_w}{4W/g} C_{L,\delta_e} = -0.00248$$

$$R_{m,\delta_e} \equiv \frac{\rho S_w \bar{c}_w^3}{8 I_{yy_b}} C_{m,\delta_e} = -0.00297$$

Combining these results with those obtained in Example 8.2.1, the linearized longitudinal equations of motion as given by Eq. (7.7.5) result in a system of equations in the form of the generalized linear control problem,

$$
\begin{bmatrix}
1 & 0 & 0 & 0 & 0 & 0 \\
0 & 1.01133 & 0 & 0 & 0 & 0 \\
0 & 0.01404 & 1 & 0 & 0 & 0 \\
0 & 0 & 0 & 1 & 0 & 0 \\
0 & 0 & 0 & 0 & 1 & 0 \\
0 & 0 & 0 & 0 & 0 & 1
\end{bmatrix}
\begin{Bmatrix}
\Delta\hat{\mu} \\
\Delta\hat{\alpha} \\
\Delta\hat{q} \\
\Delta\hat{\xi}_x \\
\Delta\hat{\xi}_z \\
\Delta\hat{\theta}
\end{Bmatrix}
=
\begin{Bmatrix}
0 \\
-0.00248 \\
-0.00297 \\
0 \\
0 \\
0
\end{Bmatrix}
\Delta\delta_e
$$

$$
+
\begin{bmatrix}
-0.00071 & 0.00030 & 0 & 0 & 0 & -0.00278 \\
-0.00556 & -0.03151 & 0.97309 & 0 & 0 & 0 \\
0 & -0.00220 & -0.03212 & 0 & 0 & 0 \\
1 & 0 & 0 & 0 & 0 & 0 \\
0 & 1 & 0 & 0 & 0 & -1 \\
0 & 0 & 1 & 0 & 0 & 0
\end{bmatrix}
\begin{Bmatrix}
\Delta\mu \\
\Delta\alpha \\
\Delta q \\
\Delta\xi_x \\
\Delta\xi_z \\
\Delta\theta
\end{Bmatrix}
$$

Since here we are only interested in airspeed and angle of attack, and because all coefficients of $\Delta\xi_x$ and $\Delta\xi_z$ are zero, this system can be reduced to fourth order by removing the fourth and fifth rows and columns. In addition, the system is readily reduced to the special linear control problem that is in the form of Eq. (10.4.1). This gives

$$
\begin{Bmatrix} \Delta\hat{\mu} \\ \Delta\hat{\alpha} \\ \Delta\hat{q} \\ \Delta\hat{\theta} \end{Bmatrix} = \begin{bmatrix} -0.00071 & 0.00030 & 0 & -0.00278 \\ -0.00550 & -0.03116 & 0.9622 & 0 \\ 0.00008 & -0.00176 & -0.04563 & 0 \\ 0 & 0 & 1 & 0 \end{bmatrix} \begin{Bmatrix} \Delta\mu \\ \Delta\alpha \\ \Delta q \\ \Delta\theta \end{Bmatrix}
$$
$$
+ \begin{Bmatrix} 0 \\ -0.00245 \\ -0.00294 \\ 0 \end{Bmatrix} \Delta\delta_e
$$

The eigenvalues of this system are the nonzero eigenvalues that were obtained in Example 8.2.1. Only the rigid-body displacement modes are lost as a result of reducing the system to fourth order. The airspeed and angle-of-attack response to a negative 1-degree step change in elevator deflection ($\Delta\delta_e = -0.01745$), as computed from this system of equations, is shown in Fig. 10.4.2.

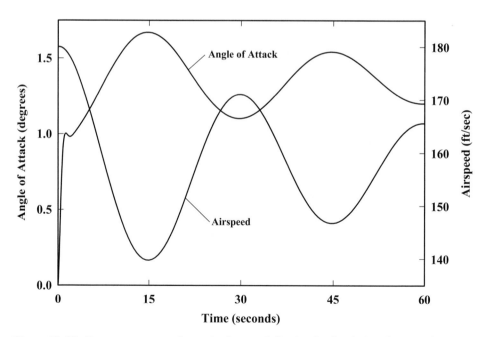

Figure 10.4.2. Response to a step change in elevator deflection for the airplane in Example 10.4.3.

A few observations should be made from the results predicted in Example 10.4.3. Here again we see that system response to a step change in control input has the same characteristic form as the homogeneous solution, which is obtained with the control input set to zero. From Fig. 10.4.2, we clearly see the nature of both the short period and the phugoid in this solution. Because the frequency of the phugoid and short period differ so greatly, for the first second or so after the elevator deflection has been applied, the response of the system depends almost entirely on the short-period characteristics. After a few seconds have passed, damping has removed all traces of the short period from this solution. However, the phugoid oscillations persist for much longer. With the elevator deflection maintained, a new trim state would eventually be reached, corresponding to climbing flight at about 160 ft/sec with an increase in angle of attack of approximately 1.3 degrees. If the elevator deflection were actually held constant, it would take several minutes for the phugoid oscillations to die out. However, if the pilot uses the stick or control yoke to hold the nose of the airplane at the desired position relative to the horizon, the phugoid oscillations will disappear almost immediately. In fact, since pilots always control an airplane by watching the horizon or an instrument, not by watching the stick or control yoke, phugoid oscillations are not normally encountered in a pull-up maneuver. The short-period characteristics, on the other hand, affect the pull-up maneuver significantly, even more so in airplanes which exhibit lower short-period damping. Figure 10.4.3 shows how reduced pitch damping affects the step response to elevator deflection.

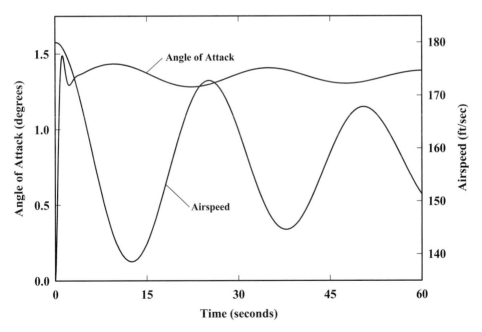

Figure 10.4.3. Response to a step change in elevator deflection for the airplane in Example 10.4.3, with a significant reduction in pitch damping.

EXAMPLE 10.4.4. For steady level flight at sea level, determine the lateral response to a 1-degree step change in rudder deflection for the general aviation airplane that was described in Example 9.2.1. Using the linearized lateral equations of motion, plot the sideslip angle and bank angle as a function of time. For this airplane and operating conditions, we have

$$S_w = 185 \text{ ft}^2, \quad b_w = 33 \text{ ft}, \quad \bar{c}_w = 5.606 \text{ ft}, \quad W = 2,800 \text{ lbf},$$

$$V_o = 180 \text{ ft/sec}, \quad C_{Do} = 0.05,$$

$$I_{xx_b} = 1,000 \text{ slug·ft}^2, \quad I_{yy_b} = 3,000 \text{ slug·ft}^2, \quad I_{zz_b} = 3,500 \text{ slug·ft}^2,$$

$$I_{xz_b} = 30 \text{ slug·ft}^2,$$

$$C_{L,\alpha} = 4.40, \quad C_{D,\alpha} = 0.35, \quad C_{m,\alpha} = -0.68, \quad C_{L,\hat{\alpha}} = 1.60, \quad C_{m,\hat{\alpha}} = -4.35,$$

$$C_{Y,\beta} = -0.560, \quad C_{\ell,\beta} = -0.075, \quad C_{n,\beta} = 0.070,$$

$$C_{D,\bar{q}} \cong 0.0, \quad C_{L,\bar{q}} = 3.80, \quad C_{m,\bar{q}} = -9.95,$$

$$C_{Y,\bar{p}} \cong 0.0, \quad C_{\ell,\bar{p}} = -0.410, \quad C_{n,\bar{p}} = -0.0575,$$

$$C_{Y,\bar{r}} = 0.240, \quad C_{\ell,\bar{r}} = 0.105, \quad C_{n,\bar{r}} = -0.125,$$

$$C_{L,\delta_e} = 0.350, \quad C_{D,\delta_e} \cong 0.00, \quad C_{m,\delta_e} = -0.920,$$

$$C_{Y,\delta_a} \cong 0.00, \quad C_{\ell,\delta_a} = -0.135, \quad C_{n,\delta_a} = 0.0035,$$

$$C_{Y,\delta_r} = 0.155, \quad C_{\ell,\delta_r} = 0.105, \quad C_{n,\delta_r} = -0.075$$

Solution. From Example 9.2.1, we have

$$\frac{\rho S_w b_w}{4W/g} = 0.041679, \quad \frac{\rho S_w b_w^3}{8I_{xx_b}} = 1.975306, \quad \frac{\rho S_w b_w^3}{8I_{zz_b}} = 0.564373$$

From Eq. (7.7.8), this gives

$$R_{y,\delta_a} \equiv \frac{\rho S_w b_w}{4W/g} C_{Y,\delta_a} = 0.00$$

$$R_{\ell,\delta_a} \equiv \frac{\rho S_w b_w^3}{8I_{xx_b}} C_{\ell,\delta_a} = -0.2667$$

$$R_{n,\delta_a} \equiv \frac{\rho S_w b_w^3}{8I_{zz_b}} C_{n,\delta_a} = 0.00198$$

$$R_{y,\delta_r} \equiv \frac{\rho S_w b_w}{4W/g} C_{Y,\delta_r} = 0.00646$$

$$R_{\ell,\delta_r} \equiv \frac{\rho S_w b_w^3}{8I_{xx_b}} C_{\ell,\delta_r} = 0.2074$$

$$R_{n,\delta_r} \equiv \frac{\rho S_w b_w^3}{8I_{zz_b}} C_{n,\delta_r} = -0.04233$$

Combining these results with those obtained in Example 9.2.1, the linearized lateral equations of motion as given by Eq. (7.7.6) result in a generalized linear control problem,

$$
\begin{bmatrix}
1 & 0 & 0 & 0 & 0 & 0 \\
0 & 1 & -0.0300 & 0 & 0 & 0 \\
0 & -0.0086 & 1 & 0 & 0 & 0 \\
0 & 0 & 0 & 1 & 0 & 0 \\
0 & 0 & 0 & 0 & 1 & 0 \\
0 & 0 & 0 & 0 & 0 & 1
\end{bmatrix}
\begin{Bmatrix}
\Delta\hat{\beta} \\
\Delta\hat{p} \\
\Delta\hat{r} \\
\Delta\hat{\xi}_y \\
\Delta\hat{\phi} \\
\Delta\hat{\psi}
\end{Bmatrix}
=
\begin{bmatrix}
0 & 0.00646 \\
-0.2667 & 0.2074 \\
0.00198 & -0.04233 \\
0 & 0 \\
0 & 0 \\
0 & 0
\end{bmatrix}
\begin{Bmatrix}
\Delta\delta_a \\
\Delta\delta_r
\end{Bmatrix}
$$

$$
+
\begin{bmatrix}
-0.0233 & 0 & -0.9900 & 0 & 0.0164 & 0 \\
-0.1481 & -0.8099 & 0.2074 & 0 & 0 & 0 \\
0.0395 & -0.0325 & -0.0705 & 0 & 0 & 0 \\
1 & 0 & 0 & 0 & 0 & 1 \\
0 & 1 & 0 & 0 & 0 & 0 \\
0 & 0 & 1 & 0 & 0 & 0
\end{bmatrix}
\begin{Bmatrix}
\Delta\beta \\
\Delta\overline{p} \\
\Delta\overline{r} \\
\Delta\xi_y \\
\Delta\phi \\
\Delta\psi
\end{Bmatrix}
$$

Similar to Example 10.4.3, this system can also be reduced to fourth order. Notice that rows 1, 2, 3, and 5 are not coupled in any way to rows 4 and 6. Thus, removing rows and columns 4 and 6 reduces the order of the system. The result is readily reduced to the form of Eq. (10.4.1),

$$
\begin{Bmatrix}
\Delta\hat{\beta} \\
\Delta\hat{p} \\
\Delta\hat{r} \\
\Delta\hat{\phi}
\end{Bmatrix}
=
\begin{bmatrix}
-0.0233 & 0 & -0.9900 & 0.0164 \\
-0.1470 & -0.8111 & 0.2053 & 0 \\
0.0382 & -0.0395 & -0.0687 & 0 \\
0 & 1 & 0 & 0
\end{bmatrix}
\begin{Bmatrix}
\Delta\beta \\
\Delta\overline{p} \\
\Delta\overline{r} \\
\Delta\phi
\end{Bmatrix}
$$

$$
+
\begin{bmatrix}
0 & 0.00646 \\
-0.2667 & 0.2062 \\
-0.00031 & -0.04056 \\
0 & 0
\end{bmatrix}
\begin{Bmatrix}
\Delta\delta_a \\
\Delta\delta_r
\end{Bmatrix}
$$

The eigenvalues for this system were already obtained in Example 9.2.1. These are the eigenvalues for the roll, spiral, and Dutch roll modes, which were found in that example. The trivial rigid-body displacement modes are not found in this system of reduced order. Figure 10.4.4 shows the response in bank angle and sideslip angle to a 1-degree step change in rudder deflection, as computed from this linear system.

Despite rather heavy yaw damping, excitation of Dutch roll can clearly be seen in Fig. 10.4.4. To demonstrate how yaw damping affects control response, Fig. 10.4.5 shows similar results, as predicted with reduced yaw damping.

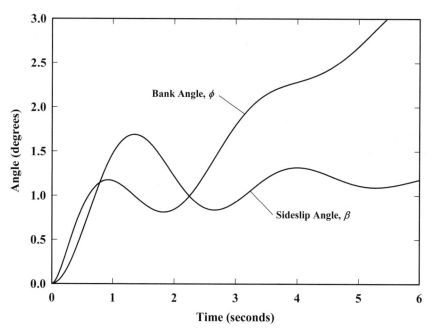

Figure 10.4.4. Response to a step change in rudder deflection for the airplane in Example 10.4.4.

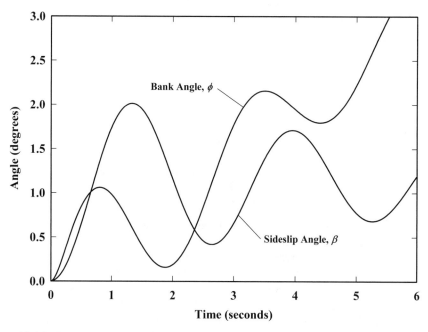

Figure 10.4.5. Response to a step change in rudder deflection for the airplane in Example 10.4.4, with a significant reduction in yaw damping.

In Examples 10.4.3 and 10.4.4, we have seen that an airplane's initial response to a change in control input has the same characteristic form as the stick-fixed dynamic modes, which are obtained from the homogeneous equations of motion with all control inputs set to zero. For this reason, the "feel" of an airplane's controls is significantly affected by the eigenvalues for the stick-fixed dynamic modes. For example, the feel of the elevator control is affected by the short-period eigenvalues, and the feel of the rudder control is affected by the Dutch roll eigenvalues. This explains why pilot opinion of aircraft handling qualities correlates so well with these eigenvalues.

Frequency Response
Another very important type of input that can be analyzed with the aid of the transfer function is the sinusoidal input. Consider a particular state variable that has a transfer function with respect to some particular input parameter, which has the usual form

$$G(s) = \frac{N(s)}{(s-\lambda_1)(s-\lambda_2)(s-\lambda_3)\cdots(s-\lambda_n)} \tag{10.4.13}$$

where $N(s)$ is a polynomial of order less than n. Now let the input be given by

$$\exp(i\omega t) = \cos(\omega t) + i\sin(\omega t) \tag{10.4.14}$$

If we multiply the Laplace transform of this input by the transfer function, the result is the Laplace transform of the response to a unit sinusoid. The real part of the result will be the Laplace transform of the response to a unit cosine input, and the imaginary part of the result will be the Laplace transform of the response to a unit sine input. From any table of Laplace transforms we find that

$$\overline{\exp(i\omega t)} = 1/(s-i\omega) \tag{10.4.15}$$

Thus, combining Eqs. (10.4.13) and (10.4.15), the Laplace transform of the response to a unit sinusoid is

$$\overline{x}(s) = G(s)\,\overline{\exp(i\omega t)} = \frac{N(s)}{(s-i\omega)(s-\lambda_1)(s-\lambda_2)\cdots(s-\lambda_n)} \tag{10.4.16}$$

After applying Eq. (10.4.12) to Eq. (10.4.16), we find that the time-domain response to a unit sinusoidal input is

$$x(t) = \sum_{k=1}^{n}\left[\frac{N(\lambda_k)}{\prod_{j=1}^{n}(\lambda_k-\lambda_j)_{j\neq k}}\right]\frac{\exp(\lambda_k t)}{(\lambda_k - i\omega)}$$
$$+ \frac{N(i\omega)}{(i\omega-\lambda_1)(i\omega-\lambda_2)(i\omega-\lambda_3)\cdots(i\omega-\lambda_n)}\exp(i\omega t) \tag{10.4.17}$$

Some important observations can be made from Eq. (10.4.17). Here again we see that the initial response to the control input includes a weighted sum of the homogeneous solutions that are obtained from the system eigenvalues. The last term in Eq. (10.4.17) describes a sinusoidal oscillation having the same frequency as the input. If all of the system eigenvalues are associated with convergent modes, after some time has passed, each of the homogeneous solutions will go to zero. Thus, the last term in Eq. (10.4.17) is the steady periodic response to a unit sinusoidal input. This is a sinusoidal output having the same frequency as the input but with different amplitude and phase.

Notice by comparing the transfer function in Eq. (10.4.13) with the response predicted by Eq. (10.4.17) that the steady periodic solution can simply be written as

$$x(t) \underset{t \to \infty}{=} G(i\omega) \exp(i\omega t) \qquad (10.4.18)$$

The function $G(i\omega)$ is called the *complex amplitude* and is simply the transfer function evaluated at $s = i\omega$. In general, $G(i\omega)$ is a complex number that can be expressed in terms of amplitude and phase, just as we did with the eigenvectors in Chapter 8. The magnitude of $G(i\omega)$ is called the *gain* and is simply the ratio of the response amplitude to the input amplitude. When the input frequency ω is zero, Eq. (10.4.17) reduces to the step response. Thus, $G(0)$ is called the *static gain* and $|G(i\omega)|/|G(0)|$ is called the *dynamic gain*. Figure 10.4.6 shows how the dynamic gain varies with frequency and damping ratio for a second-order system.

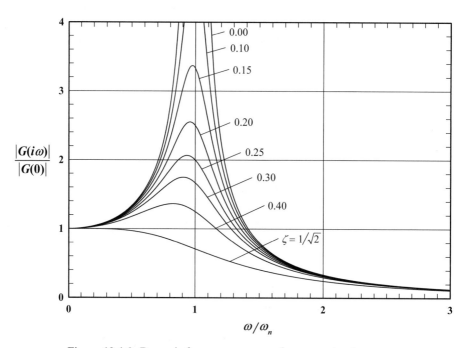

Figure 10.4.6. Dynamic frequency response for a second-order system.

Curves like those shown in Fig. 10.4.6 are commonly called *frequency-response curves*. A second-order linear-invariant system, like that used for Fig. 10.4.6, has only one undamped natural frequency and will exhibit at most one resonance peak. Higher-order linear-invariant systems, such as those associated with aircraft controls, typically have multiple natural frequencies and will commonly exhibit multiple resonance peaks. The examination of frequency-response curves for aircraft controls is left as an exercise for the student.

Autopilots and Automatic Control

The concept of transfer functions and the Laplace transform description of system dynamics are employed extensively in the application of classical automatic control theory to the design of autopilots and aircraft stability augmentation systems. To study autopilot design successfully, the student would need a good background in the fundamentals of automatic control theory as well as flight mechanics. At least one undergraduate engineering course in control theory would typically be needed to provide the minimum essential background. A detailed treatment of automatic control theory is not within the intended scope of this textbook. However, a few words on automatic control may help the interested student to decide whether or not additional coursework along these lines would be in order.

The important and exciting field of automatic control theory is commonly divided into two very broad areas, which are traditionally called *classical control theory* and *modern control theory*. The term "modern" control theory can be misleading, because it seems to imply that classical automatic control theory is obsolete and no longer used. This is not the case. To fully understand the design of autopilots and aircraft stability augmentation systems, an engineer would need to understand both the classical and modern approaches to automatic control.

Classical automatic control theory is based on the frequency-response method and utilizes the Laplace transforms and transfer functions that were briefly introduced in this section. As we have seen in Eq. (10.4.10), knowledge of the transfer functions depends on knowledge of the system's eigenvalues. Thus, the reader may begin to see why closed-form approximations for these eigenvalues would be of value in autopilot design. If the reader has made only a cursory attempt to follow the material that was presented here, he or she may have the false impression that Laplace transforms and transfer functions are very complicated and difficult to understand. However, if an honest attempt is made to follow these mathematical developments, most readers will see that describing system dynamics through the use of Laplace transforms and transfer functions is remarkably simple. In fact, the primary advantage of the classical approach to automatic control theory is the great simplification in the mathematical description of system dynamics that is afforded by the Laplace transform.

Modern automatic control theory, which is sometimes referred to as the *state space approach*, has its origins in the early 1960s with the development of the high-speed digital computer. With this approach, system dynamics is expressed directly in terms of a system of first-order linear differential equations, like those presented in Chapter 7 for aircraft dynamics. One significant advantage of this approach to automatic control has been the application of numerical optimization algorithms to the development of what is commonly called *optimal control theory*. Modern automatic control is mathematically quite similar to digital flight simulation. Thus, an understanding of the material to be covered in Chapter 11 should be helpful in a later study of automatic control theory.

For many years autopilots have been used to relieve human pilots of the monotonous task of maintaining attitude and heading during long periods of steady level flight. Stability augmentation systems have long been used to enhance the handling qualities of airplanes and provide maneuverability that would otherwise be unattainable. Today the application of automatic control theory to the world of aviation goes far beyond these traditional applications. Totally autonomous flight is now an attainable reality. Recent developments of extremely small electromechanical systems are making flight on a very small scale possible and practicable. The next 50 years will surely see some very exciting advancements come out of the marriage between the disciplines of flight mechanics and automatic control theory.

If a student finds the topic of autopilots and aircraft stability augmentation systems to be of particular interest, he or she is encouraged to pursue this interest further with formal coursework on automatic control theory. The starting point would typically be the first undergraduate engineering course in automatic control. If interest continues, this introductory class should be followed with courses on digital and optimal control. These three classes, combined with a thorough background in fluid mechanics, aerodynamics, and flight mechanics, should find the engineering graduate well prepared to contribute to this exciting field.

10.5. Nonlinear Effects and Longitudinal-Lateral Coupling

It should be reemphasized that all of the material on aircraft handling qualities and control response that has been presented in this chapter is based on the linearized equations of motion with negligible coupling between the longitudinal and lateral degrees of freedom. Such results are restricted to low-angle-of-attack flight with relatively small departures from equilibrium. These results are also restricted to aircraft and operating conditions having little or no inertial, gyroscopic, or aerodynamic coupling between the pitch axis and the roll or yaw axes. While these conditions are satisfied for many airplanes and operating conditions, there are also important cases for which these approximations are not valid.

In Sec. 9.8, the longitudinal-lateral coupling resulting from inertial, gyroscopic, and aerodynamic asymmetry was discussed. Since coupling of this type has first-order components, it can affect aircraft handling qualities even during gradual maneuvers at low angles of attack. While these effects are not accounted for in the handling quality predictions presented in this chapter, they are of significant interest, particularly in some military application where substantial aircraft asymmetry can be encountered.

Handling qualities during very rapid maneuvers have also been ignored in this chapter. Even in a completely symmetric airplane, during very rapid maneuvers, second-order inertial coupling between roll and pitch (Secs. 7.9 and 9.8) can significantly alter control response and aircraft handling qualities. The analytical treatment of such phenomena requires numerical solution of the nonlinear equations of motion, which is discussed in Chapter 11.

Today, aircraft control and handling qualities at angles of attack beyond stall are of significant interest. For a combat aircraft, high turning rates are very critical to mission effectiveness. The maximum turning rate that can be achieved in a conventional turn is severely limited by the load factor that can be endured by an aircraft and pilot. Turning rates at low load factor can be increased substantially if the aircraft is capable of controlled flight at angles of attack well beyond stall. The cobra maneuver, which is shown in Fig. 3.9.7, is one example of how high turning rate can be achieved in such aircraft. An aircraft capable of such maneuvers has a decisive tactical advantage over an airplane that is limited to conventional turns. Because conventional airplane control surfaces become ineffective at angles of attack beyond stall, alternative methods of control have been developed for high-angle-of-attack flight. These include thrust vectoring, actuated forebody strakes, and forebody blowing and suction. Because flight dynamics becomes highly nonlinear at angles of attack beyond stall, the methods presented in this chapter do not apply. Here again, the numerical solution techniques presented in Chapter 11 provide a good alternative for the analytical investigation of such phenomena.

10.6. Problems

10.1. For the airplane and operating conditions in problem 9.47, compare the predicted dynamic response with the frequency and damping requirements presented in Sec. 10.3. What level of handling qualities should be expected for this airplane and operating conditions, based on the dynamic response modes?

10.2. For the airplane and operating conditions in problem 9.48, compare the predicted dynamic response with the frequency and damping requirements presented in Sec. 10.3. What level of handling qualities should be expected for this airplane and operating conditions, based on the dynamic response modes?

10.3. For the airplane and operating conditions in problem 9.49, compare the predicted dynamic response with the frequency and damping requirements presented in Sec. 10.3. What level of handling qualities should be expected for this airplane and operating conditions, based on the dynamic response modes?

10.4. For the airplane and operating conditions in problem 9.50, compare the predicted dynamic response with the frequency and damping requirements presented in Sec. 10.3. What level of handling qualities should be expected for this airplane and operating conditions, based on the dynamic response modes?

10.5. For the airplane and operating conditions in Example 10.4.3, find the angle-of-attack transfer function for response to elevator deflection.

10.6. From the transfer function obtained in problem 10.5, find the Laplace transform of the angle-of-attack component of the state variable response vector for a unit step change in elevator deflection.

10.7. From the Laplace transform acquired in problem 10.6, obtain a time-domain expression for the angle-of-attack response to a negative 1-degree step change in elevator deflection. Using this expression, plot the angle of attack as a function of time.

10.8. For the airplane and operating conditions in Example 10.4.4, find the sideslip angle transfer function for response to rudder deflection.

10.9. From the transfer function obtained in problem 10.8, find the Laplace transform of the sideslip angle component of the state variable response vector for a unit step change in rudder deflection.

10.10. From the Laplace transform acquired in problem 10.9, obtain a time-domain expression for the sideslip angle response to a 1-degree step change in rudder deflection. Using this expression, plot the sideslip angle as a function of time.

10.11. From the transfer function obtained in problem 10.5, examine the angle-of-attack frequency response for sinusoidal variation in elevator deflection. Plot the dynamic gain as a function of input frequency.

10.12. From the transfer function obtained in problem 10.8, examine the sideslip angle frequency response for sinusoidal variation in rudder deflection. Plot the dynamic gain as a function of input frequency.

Chapter 11
Aircraft Flight Simulation

11.1. Introduction

In Chapter 7, the six-degree-of-freedom (6-DOF) rigid-body equations of motion for an aircraft in flight were developed in terms of Euler angles. This Euler angle formulation was then linearized, and in Chapters 8, 9, and 10, we examined some aspects of aircraft dynamics as predicted from the linearized formulation. However, there are many applications in the discipline of flight dynamics to which the linearized equations of motion do not apply. For such applications, we must return to a fully coupled nonlinear formulation. For this purpose, the Euler angle formulation developed in Chapter 7 could be used. However, as we shall see, this is not the best alternative.

The Euler angle formulation contains a singularity known as *gimbal lock*, which is encountered when the nose of the airplane is pointed either straight up or straight down. While this singularity is of no concern when simulating the flight of most conventional airplanes under typical operating conditions, it does limit the general utility of a flight simulator based on the Euler angle formulation. In addition, the Euler angle formulation has significant computational disadvantages compared to other formulations that have been developed for flight simulation. In this chapter we compare some of these alternative formulations. Emphasis is placed on what is called the *quaternion formulation*, which has substantial advantages over the Euler angle formulation. For example, the computation time required for vector transformation using the Euler angle formulation is about 11 times that required with the quaternion formulation.

For simulating aircraft motion, Newton's second law is written in terms of a non-inertial coordinate system fixed to the aircraft. Position and orientation are described using a coordinate frame fixed to the Earth, which for all practical purposes is an inertial coordinate system. The location and orientation of the noninertial frame relative to the inertial frame specify the aircraft's position and orientation. The noninertial or *body-fixed* coordinates are designated (x_b, y_b, z_b) and the inertial or *Earth-fixed* coordinates, which have been used to this point in the text, are denoted (x_f, y_f, z_f). The orientations of these two Cartesian coordinate systems relative to the Earth and the aircraft were shown in Fig. 7.1.2. The *flat-Earth* coordinate system (x_f, y_f, z_f) can only be used for relatively short flights that do not traverse a significant portion of the Earth's surface. For longer flights, position is described in terms of latitude, longitude, and altitude.

To develop a complete 6-DOF formulation for flight simulation, some means of transforming the components of a vector between noninertial and inertial coordinates is required. The equations used for this purpose are called the *kinematic transformation equations*. For a complete review of the literature on the mathematical formulations that have been used for aircraft kinematics the reader is referred to Phillips, Hailey, and Gebert (2000 and 2001). Much of the material presented in this chapter was taken directly from these references. The reader is cautioned that the equations in the 2001 *Journal of Aircraft* paper contain numerous typesetting errors, which are not found in the 2000 AIAA meeting paper or in the material presented in this chapter.

11.2. Euler Angle Formulations

The orientation of the noninertial reference frame relative to the inertial reference frame can be described in terms of three consecutive rotations through three *body-referenced Euler angles*. There are 12 possible ways to define three independent body-referenced Euler angles. Starting from the inertial reference frame, three consecutive rotations are performed, each about one of the three body-referenced axes, to arrive at the final noninertial reference frame. The only restriction required to provide three degrees of freedom is that no two consecutive rotations can be about the same axis. With this restriction, there are six symmetric sets of Euler angles,

$$\varphi_x \to \varphi_y \to \varphi_x, \qquad \varphi_x \to \varphi_z \to \varphi_x,$$
$$\varphi_y \to \varphi_x \to \varphi_y, \qquad \varphi_y \to \varphi_z \to \varphi_y,$$
$$\varphi_z \to \varphi_x \to \varphi_z, \qquad \varphi_z \to \varphi_y \to \varphi_z$$

and six asymmetric sets,

$$\varphi_x \to \varphi_y \to \varphi_z, \qquad \varphi_x \to \varphi_z \to \varphi_y,$$
$$\varphi_y \to \varphi_x \to \varphi_z, \qquad \varphi_y \to \varphi_z \to \varphi_x,$$
$$\varphi_z \to \varphi_x \to \varphi_y, \qquad \varphi_z \to \varphi_y \to \varphi_x$$

Tandon (1978) presents the rotation matrix for each of the 12 sets of Euler angles. Any of these 12 Euler angle sets can be used for attitude representation. For example, in the field of orbital mechanics the $\varphi_z \to \varphi_x \to \varphi_z$ sequence is typically used. The aeronautics community commonly uses the last of the rotation sequences above, $\varphi_z \to \varphi_y \to \varphi_x$. These three Euler angles, commonly written ψ, θ, and ϕ, are, respectively, the azimuth angle, elevation angle, and bank angle, which were described in detail in Sec. 7.3 and shown in Figs. 7.3.1 and 7.3.2.

With these common Euler angle definitions, in Sec. 7.3 we considered an arbitrary vector $\mathbf{v}$ having components $(v_{x_f}, v_{y_f}, v_{z_f})$ in the inertial coordinate system and having components $(v_{x_b}, v_{y_b}, v_{z_b})$ in the noninertial coordinate system. Using the shorthand notation $S_\chi = \sin(\chi)$ and $C_\chi = \cos(\chi)$, the transformation of the components of this vector from inertial coordinates to noninertial coordinates was found to be

$$\begin{Bmatrix} v_{x_b} \\ v_{y_b} \\ v_{z_b} \end{Bmatrix} = \begin{bmatrix} 1 & 0 & 0 \\ 0 & C_\phi & S_\phi \\ 0 & -S_\phi & C_\phi \end{bmatrix} \begin{bmatrix} C_\theta & 0 & -S_\theta \\ 0 & 1 & 0 \\ S_\theta & 0 & C_\theta \end{bmatrix} \begin{bmatrix} C_\psi & S_\psi & 0 \\ -S_\psi & C_\psi & 0 \\ 0 & 0 & 1 \end{bmatrix} \begin{Bmatrix} v_{x_f} \\ v_{y_f} \\ v_{z_f} \end{Bmatrix}$$

or

$$\begin{Bmatrix} v_{x_b} \\ v_{y_b} \\ v_{z_b} \end{Bmatrix} = \begin{bmatrix} C_\theta C_\psi & C_\theta S_\psi & -S_\theta \\ S_\phi S_\theta C_\psi - C_\phi S_\psi & S_\phi S_\theta S_\psi + C_\phi C_\psi & S_\phi C_\theta \\ C_\phi S_\theta C_\psi + S_\phi S_\psi & C_\phi S_\theta S_\psi - S_\phi C_\psi & C_\phi C_\theta \end{bmatrix} \begin{Bmatrix} v_{x_f} \\ v_{y_f} \\ v_{z_f} \end{Bmatrix} \qquad (11.2.1)$$

The inverse of the transformation matrix in Eq. (11.2.1) is its transpose, which gives

$$
\begin{Bmatrix} v_{x_f} \\ v_{y_f} \\ v_{z_f} \end{Bmatrix} = \begin{bmatrix} C_\theta C_\psi & S_\phi S_\theta C_\psi - C_\phi S_\psi & C_\phi S_\theta C_\psi + S_\phi S_\psi \\ C_\theta S_\psi & S_\phi S_\theta S_\psi + C_\phi C_\psi & C_\phi S_\theta S_\psi - S_\phi C_\psi \\ -S_\theta & S_\phi C_\theta & C_\phi C_\theta \end{bmatrix} \begin{Bmatrix} v_{x_b} \\ v_{y_b} \\ v_{z_b} \end{Bmatrix} \tag{11.2.2}
$$

Since Newton's second law is written in terms of the noninertial coordinate system, the gravitational force must be expressed in terms of noninertial coordinates. Applying Eq. (11.2.1) to the weight vector, we found that

$$
\begin{Bmatrix} W_{x_b} \\ W_{y_b} \\ W_{z_b} \end{Bmatrix} = \begin{bmatrix} C_\theta C_\psi & C_\theta S_\psi & -S_\theta \\ S_\phi S_\theta C_\psi - C_\phi S_\psi & S_\phi S_\theta S_\psi + C_\phi C_\psi & S_\phi C_\theta \\ C_\phi S_\theta C_\psi + S_\phi S_\psi & C_\phi S_\theta S_\psi - S_\phi C_\psi & C_\phi C_\theta \end{bmatrix} \begin{Bmatrix} 0 \\ 0 \\ W \end{Bmatrix}
$$

or

$$
\begin{Bmatrix} W_{x_b} \\ W_{y_b} \\ W_{z_b} \end{Bmatrix} = W \begin{Bmatrix} -S_\theta \\ S_\phi C_\theta \\ C_\phi C_\theta \end{Bmatrix} = W \begin{Bmatrix} -\sin\theta \\ \sin\phi\cos\theta \\ \cos\phi\cos\theta \end{Bmatrix} \tag{11.2.3}
$$

Similarly, from Eq. (11.2.2), the ground speed was related to the airspeed and wind by

$$
\begin{Bmatrix} \dot{x}_f \\ \dot{y}_f \\ \dot{z}_f \end{Bmatrix} = \begin{Bmatrix} V_{wx_f} \\ V_{wy_f} \\ V_{wz_f} \end{Bmatrix} + \begin{bmatrix} C_\theta C_\psi & S_\phi S_\theta C_\psi - C_\phi S_\psi & C_\phi S_\theta C_\psi + S_\phi S_\psi \\ C_\theta S_\psi & S_\phi S_\theta S_\psi + C_\phi C_\psi & C_\phi S_\theta S_\psi - S_\phi C_\psi \\ -S_\theta & S_\phi C_\theta & C_\phi C_\theta \end{bmatrix} \begin{Bmatrix} u \\ v \\ w \end{Bmatrix} \tag{11.2.4}
$$

Using these same Euler angle definitions, we found that the relationship between the body-fixed angular rates and the time rate of change of the Euler angles is

$$
\begin{Bmatrix} \dot{\phi} \\ \dot{\theta} \\ \dot{\psi} \end{Bmatrix} = \begin{bmatrix} 1 & S_\phi S_\theta / C_\theta & C_\phi S_\theta / C_\theta \\ 0 & C_\phi & -S_\phi \\ 0 & S_\phi / C_\theta & C_\phi / C_\theta \end{bmatrix} \begin{Bmatrix} p \\ q \\ r \end{Bmatrix} \tag{11.2.5}
$$

In terms of this particular set of Euler angles, Eqs. (11.2.4) and (11.2.5) provide the kinematic transformation equations, which are needed to update the position and orientation of the aircraft with time. The gimbal lock singularity is seen in the last two terms of the first and last rows of Eq. (11.2.5). Likewise, a similar set of kinematic transformation equations can be developed for any one of the other 11 Euler angle sets. However, in no case will a singularity-free formulation result. One possible means to avoid the singularity associated with an Euler angle formulation is to use two different Euler angle formulations, which have their singularities located at different orientations.

As the singularity associated with one formulation is approached, the computation algorithm is switched to the other formulation. While this avoids the gimbal lock singularity, for reasons of computational efficiency, it is not the best solution.

11.3. Direction-Cosine Formulation

The *direction-cosine matrix* can be formed from any of the symmetric or asymmetric Euler angle sets. For example, the matrix on the right side of Eq. (11.2.1) is a direction-cosine matrix. One way to avoid the singularity in Eq. (11.2.5) is to treat the nine components of this matrix as a fundamental description of orientation. The elements of this matrix are called the *direction cosines*. If these nine direction cosines are known, the components of an arbitrary vector in body-fixed coordinates are quite simply related to the components of the same vector in inertial coordinates through the definition of the direction-cosine matrix, $[\mathbf{C}]$,

$$\begin{Bmatrix} V_{x_b} \\ V_{y_b} \\ V_{z_b} \end{Bmatrix} \equiv \begin{bmatrix} C_{11} & C_{12} & C_{13} \\ C_{21} & C_{22} & C_{23} \\ C_{31} & C_{32} & C_{33} \end{bmatrix} \begin{Bmatrix} V_{x_f} \\ V_{y_f} \\ V_{z_f} \end{Bmatrix} \tag{11.3.1}$$

and the gravitational force vector, expressed in body-fixed coordinates, is

$$\begin{Bmatrix} W_{x_b} \\ W_{y_b} \\ W_{z_b} \end{Bmatrix} = W \begin{Bmatrix} C_{13} \\ C_{23} \\ C_{33} \end{Bmatrix} \tag{11.3.2}$$

The inverse of the direction-cosine matrix is simply its transpose, so the ground speed is related to the airspeed and wind by

$$\begin{Bmatrix} \dot{x}_f \\ \dot{y}_f \\ \dot{z}_f \end{Bmatrix} = \begin{Bmatrix} V_{wx_f} \\ V_{wy_f} \\ V_{wz_f} \end{Bmatrix} + \begin{bmatrix} C_{11} & C_{21} & C_{31} \\ C_{12} & C_{22} & C_{32} \\ C_{13} & C_{23} & C_{33} \end{bmatrix} \begin{Bmatrix} u \\ v \\ w \end{Bmatrix} \tag{11.3.3}$$

When the body-fixed reference frame is undergoing rotation, the elements of the direction-cosine matrix are functions of time. To complete the kinematic formulation in terms of direction cosines, a set of differential equations relating the temporal derivatives of the nine elements of the direction-cosine matrix to the body-fixed angular rates is required. These nine equations, known as *Poisson's kinematic equations*, can be written in matrix form as

$$\begin{bmatrix} \dot{C}_{11} & \dot{C}_{12} & \dot{C}_{13} \\ \dot{C}_{21} & \dot{C}_{22} & \dot{C}_{23} \\ \dot{C}_{31} & \dot{C}_{32} & \dot{C}_{33} \end{bmatrix} = \begin{bmatrix} 0 & r & -q \\ -r & 0 & p \\ q & -p & 0 \end{bmatrix} \begin{bmatrix} C_{11} & C_{12} & C_{13} \\ C_{21} & C_{22} & C_{23} \\ C_{31} & C_{32} & C_{33} \end{bmatrix} \tag{11.3.4}$$

It should be noted that there is no standard convention for numbering the direction cosines. Here, the numbering in Eq. (11.3.1) has been defined relative to the forward transformation, from Earth-fixed coordinates to body-fixed coordinates. This is the most commonly used notation. However, other authors have numbered the direction cosines relative to the inverse transformation, from body-fixed to Earth-fixed coordinates. One notation is simply the transpose of the other. Nevertheless, the reader should be careful to observe which of these notations is being used when reviewing a publication or simulation code.

There are only three degrees of freedom associated with orientation. However, there are nine components in the direction-cosine matrix. Thus, these nine components cannot be independent. There must be six levels of redundancy associated with this formulation. It can be shown that rigid-body rotation is a *proper orthogonal transformation*. Therefore, the direction-cosine matrix must be proper orthogonal, which means that the inverse of the direction-cosine matrix must be its transpose, $[\mathbf{C}]^{-1} = [\mathbf{C}]^T$. This requires the constraints

$$C_{11}^2 + C_{21}^2 + C_{31}^2 = 1,$$
$$C_{12}^2 + C_{22}^2 + C_{32}^2 = 1,$$
$$C_{13}^2 + C_{23}^2 + C_{33}^2 = 1,$$
$$C_{11}C_{12} + C_{21}C_{22} + C_{31}C_{32} = 0,$$
$$C_{11}C_{13} + C_{21}C_{23} + C_{31}C_{33} = 0,$$
$$C_{12}C_{13} + C_{22}C_{23} + C_{32}C_{33} = 0$$

$$(11.3.5)$$

These six constraints are usually called the *redundancy relations*. Mathematically, the relation in Eq. (11.3.4) preserves the orthogonality of the direction-cosine matrix. However, errors associated with the numerical integration of Eq. (11.3.4) can cause degradation in the orthogonality of the matrix. Furthermore, these orthogonality errors can build up with time during the simulation. A method for avoiding this error buildup with the direction-cosine formulation was first developed by Corbett and Wright (1957) and is still commonly used today.

The direction-cosine formulation contains no singularities and is frequently used by the aeronautics community to avoid the possibility of gimbal lock. However, numerical integration of Eq. (11.3.4) is excessively time-consuming and a computational penalty is paid for its use.

11.4. Euler Axis Formulation

The orientation of the noninertial reference frame relative to the inertial reference frame can also be described in terms of a single rotation through an angle, Θ, about a particular axis, $\mathbf{E}$, which is commonly called the *Euler axis* or the *eigenaxis*. This description of orientation was first introduced by Euler (1775a). A comparison between this Euler axis rotation and the three Euler angle rotations that are commonly used by the aircraft community is shown in Fig. 11.4.1.

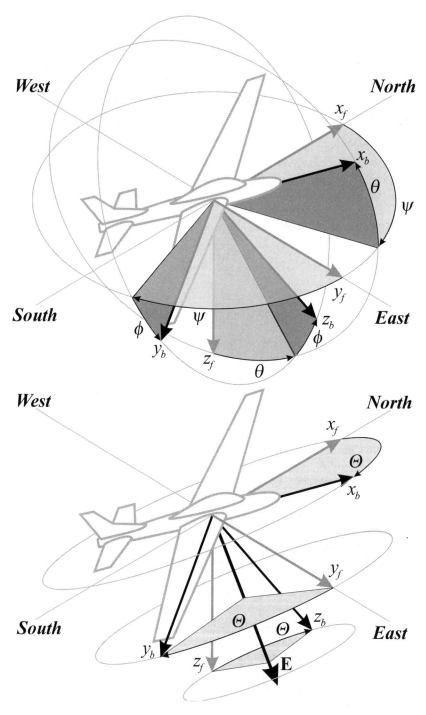

Figure 11.4.1. Comparison between the Euler angle rotations and the Euler axis rotation.

The Euler axis rotation gives rise to a four-component description of orientation. The four components in this description are the total rotation angle, Θ, and the three components of a vector directed along the Euler axis, E_x, E_y, and E_z. Since these four parameters describe an orientation having only three degrees of freedom, there is some redundancy in the Euler axis description. By describing orientation in this manner, a fourth mathematical degree of freedom has been introduced. Clearly, the length of the vector describing the orientation of the Euler axis is arbitrary. To remove this additional mathematical degree of freedom, a constraint must be applied to fix the magnitude of the Euler axis vector. While this constraint is somewhat arbitrary, the usual and most obvious solution is to require the Euler axis vector to be of unit magnitude,

$$E_x^2 + E_y^2 + E_z^2 \equiv 1 \qquad (11.4.1)$$

During the Euler axis rotation, the orientation of the Euler axis is invariant. Thus, the vector, **E**, has the same components in both inertial and noninertial coordinates,

$$\begin{Bmatrix} E_{x_f} \\ E_{y_f} \\ E_{z_f} \end{Bmatrix} = \begin{Bmatrix} E_{x_b} \\ E_{y_b} \\ E_{z_b} \end{Bmatrix} \equiv \begin{Bmatrix} E_x \\ E_y \\ E_z \end{Bmatrix} \qquad (11.4.2)$$

The components of an arbitrary vector, **v**, in body-fixed coordinates are related to the components of the same vector in Earth-fixed coordinates through what is commonly called *Euler's formula*,

$$\begin{Bmatrix} v_{x_b} \\ v_{y_b} \\ v_{z_b} \end{Bmatrix} = \begin{bmatrix} E_{xx} + C_\Theta & E_{xy} + E_z S_\Theta & E_{xz} - E_y S_\Theta \\ E_{xy} - E_z S_\Theta & E_{yy} + C_\Theta & E_{yz} + E_x S_\Theta \\ E_{xz} + E_y S_\Theta & E_{yz} - E_x S_\Theta & E_{zz} + C_\Theta \end{bmatrix} \begin{Bmatrix} v_{x_f} \\ v_{y_f} \\ v_{z_f} \end{Bmatrix} \qquad (11.4.3)$$

where $E_{ij} = E_i E_j (1 - C_\Theta)$. The inverse of the transformation matrix in Eq. (11.4.3) is obtained by simply rotating through the negative of the total rotation angle that is used in the forward transformation (i.e., change the sign of each sine),

$$\begin{Bmatrix} v_{x_f} \\ v_{y_f} \\ v_{z_f} \end{Bmatrix} = \begin{bmatrix} E_{xx} + C_\Theta & E_{xy} - E_z S_\Theta & E_{xz} + E_y S_\Theta \\ E_{xy} + E_z S_\Theta & E_{yy} + C_\Theta & E_{yz} - E_x S_\Theta \\ E_{xz} - E_y S_\Theta & E_{yz} + E_x S_\Theta & E_{zz} + C_\Theta \end{bmatrix} \begin{Bmatrix} v_{x_b} \\ v_{y_b} \\ v_{z_b} \end{Bmatrix} \qquad (11.4.4)$$

From Eq. (11.4.3) the weight vector in noninertial coordinates is

$$\begin{Bmatrix} W_{x_b} \\ W_{y_b} \\ W_{z_b} \end{Bmatrix} = W \begin{Bmatrix} E_{xz} - E_y S_\Theta \\ E_{yz} + E_x S_\Theta \\ E_{zz} + C_\Theta \end{Bmatrix} \qquad (11.4.5)$$

and from Eq. (11.4.4), the ground speed can be related to the airspeed and wind by

$$
\begin{Bmatrix} \dot{x}_f \\ \dot{y}_f \\ \dot{z}_f \end{Bmatrix} = \begin{Bmatrix} V_{wx_f} \\ V_{wy_f} \\ V_{wz_f} \end{Bmatrix} + \begin{bmatrix} E_{xx} + C_\Theta & E_{xy} - E_z S_\Theta & E_{xz} + E_y S_\Theta \\ E_{xy} + E_z S_\Theta & E_{yy} + C_\Theta & E_{yz} - E_x S_\Theta \\ E_{xz} - E_y S_\Theta & E_{yz} + E_x S_\Theta & E_{zz} + C_\Theta \end{bmatrix} \begin{Bmatrix} u \\ v \\ w \end{Bmatrix} \tag{11.4.6}
$$

The relationship between the noninertial angular rates and the rate of change of the Euler axis rotation parameters is given by

$$
\begin{Bmatrix} \dot{\Theta} \\ \dot{E}_x \\ \dot{E}_y \\ \dot{E}_z \end{Bmatrix} = \frac{1}{2} \begin{bmatrix} 2E_x & 2E_y & 2E_z \\ E'_{xx} + C/S & E'_{xy} - E_z & E'_{xz} + E_y \\ E'_{xy} + E_z & E'_{yy} + C/S & E'_{yz} - E_x \\ E'_{xz} - E_y & E'_{yz} + E_x & E'_{zz} + C/S \end{bmatrix} \begin{Bmatrix} p \\ q \\ r \end{Bmatrix} \tag{11.4.7}
$$

where $E'_{ij} = -E_i E_j\, C/S$, $S = \sin(\Theta/2)$, and $C = \cos(\Theta/2)$. Equations (11.4.6) and (11.4.7) are the kinematic transformation equations in terms of the Euler axis rotation parameters. Notice that Eq. (11.4.7) also has a singularity. When the total rotation angle, Θ, is zero or 180 degrees, the integration of these equations is indeterminate.

The singularity in the Euler axis formulation is particularly problematic in airplane flight simulation, because it occurs when the fuselage axis is level with the ground and the aircraft is headed either due north or due south. These aircraft orientations are much more likely to occur in normal flight than the vertical orientations that give rise to the gimbal lock singularity associated with the traditional Euler angle formulation. For this reason, the Euler axis formulation is never directly applied to flight simulation. The primary reason for examining the Euler axis formulation is that it can be used with a change of variables to develop another formulation that is singularity free.

11.5. The Euler-Rodrigues Quaternion Formulation

The Euler-Rodrigues formulation is related to the Euler axis formulation through a simple change of variables. The four parameters in the Euler axis description of orientation are used to define four different parameters, which are significantly more convenient. These four new parameters are defined as

$$
\begin{Bmatrix} e_0 \\ e_x \\ e_y \\ e_z \end{Bmatrix} \equiv \begin{Bmatrix} \cos(\Theta/2) \\ E_x \sin(\Theta/2) \\ E_y \sin(\Theta/2) \\ E_z \sin(\Theta/2) \end{Bmatrix} \tag{11.5.1}
$$

These four parameters, known as the *Euler-Rodrigues symmetric parameters* or the quaternion of finite rotation, form the basis of a very widely used description of orientation and rigid-body rotation. The Euler-Rodrigues formulation for rigid-body kinematics was developed well before 1844, when the well-known mathematician William R. Hamilton first developed the quaternion and the detailed theory of a noncommutative algebraic system known as quaternion algebra (see Hamilton 1844). Much has been written about the life of Hamilton and his moment of enlightenment concerning quaternion formulas while crossing a bridge with his wife (see Graves 1975, Hankins 1980, and O'Donnell 1983). Hamilton, in a letter to his son, noted that he could not resist the impulse to carve the fundamental quaternion formula into the stone on the bridge. However, it is clear that Hamilton did not develop quaternion algebra as a means of describing rotational transformations.

The brief review of the history of the Euler-Rodrigues formulation that is presented here is taken primarily from Altmann (1986 and 1989), Shuster (1993b), and Cheng and Gupta (1989). An early work of Euler (1758) showed that any differential movement of a rigid body can be expressed as a translation and a rotation about some specific axis. Euler's theorem on the motion of a rigid body and Euler's formula were both published in 1775. The first publication of the derivation of the Euler angles is commonly reported as being in 1862. Although the work was published posthumously, the date is probably in error, as Euler died in 1783.

Whether or not Euler knew of the Euler-Rodrigues symmetric parameters is a subject of some debate. Euler (1770) did develop four symmetric parameters for orthogonal transformations (without the use of half angles). Roberson (1968) and Jacobi (1969) argue that Euler had presented a rotation matrix in terms of the so-called Euler-Rodrigues parameters. Shuster (1993b) notes that the symmetric parameters developed by Euler in the 1770 paper contained sign errors and formed an improper orthogonal matrix. It is important to note that Euler viewed a matrix as a table. The matrix as a mathematical object would not evolve until vector space was studied by Grassman (1862 and 1878) and Gibbs (1901). As a side note, Hamilton (1884) worked out the chief properties of vectors in his investigation of quaternion algebra (see Boyer and Merzbach 1989).

Four years before Hamilton began his algebraic study of quaternions, Olinde Rodrigues (1840) published his work on the Euler-Rodrigues symmetric parameters, the rules for the compositions, and a geometrical construction for combining two rotations. Unlike Hamilton, whose accomplishments are well documented, the only detailed published article on Rodrigues did not appear until Gray (1980). Some additional historical details can be found in Van der Waerden (1985), Kline (1972), Crowe (1967), and McDuffee (1946).

There is no universal agreement on the choice of indices for the Euler-Rodrigues symmetric parameters. Most authors use the indices 1 through 4 or 0 through 3. Some authors have chosen indices 1 through 3 for the vector components while using 4 as the scalar index, whereas others have used 1 for the scalar index and 2 through 4 for the vector components. The scalar index has also been chosen as 0, with indices 1 through 3 denoting the vector components. There is not even universal agreement on the order of the vector components. Whereas most authors assign the x-, y-, and z-components in ascending order, some authors have used 4 to denote the x-component, 3 for the y-component, and 2 for the z-component. To avoid confusion in this text, the vector components are explicitly labeled using the subscripts x, y, and z, while a 0 subscript is used to denote the scalar component.

Since the four parameters defined by Eq. (11.5.1) uniquely describe an orientation having only three degrees of freedom, these four parameters must be related in some way. This relation is easily seen by squaring the four components of Eq. (11.5.1) and adding them together. This gives

$$e_0^2 + e_x^2 + e_y^2 + e_z^2 = \cos^2(\Theta/2) + (E_x^2 + E_y^2 + E_z^2)\sin^2(\Theta/2) \tag{11.5.2}$$

Since the Euler axis vector, $\mathbf{E}$, is a unit vector and $\cos^2(\Theta/2) + \sin^2(\Theta/2) = 1$, it follows directly that

$$e_0^2 + e_x^2 + e_y^2 + e_z^2 = 1 \tag{11.5.3}$$

The transformation expressed in Eq. (11.4.3) can be written in terms of half the total rotation angle by applying the trigonometric identities, $\sin(\Theta) = 2\sin(\Theta/2)\cos(\Theta/2)$, $\cos(\Theta) = \cos^2(\Theta/2) - \sin^2(\Theta/2)$, and $1 - \cos(\Theta) = 2\sin^2(\Theta/2)$. Thus, with shorthand notation $S = \sin(\Theta/2)$, $C = \cos(\Theta/2)$, and $E_{ij} = 2E_iE_jS^2$, Eq. (11.4.3) can be rewritten as

$$\begin{Bmatrix} v_{x_b} \\ v_{y_b} \\ v_{z_b} \end{Bmatrix} = \begin{bmatrix} E_{xx} + C^2 - S^2 & E_{xy} + 2E_zSC & E_{xz} - 2E_ySC \\ E_{xy} - 2E_zSC & E_{yy} + C^2 - S^2 & E_{yz} + 2E_xSC \\ E_{xz} + 2E_ySC & E_{yz} - 2E_xSC & E_{zz} + C^2 - S^2 \end{bmatrix} \begin{Bmatrix} v_{x_f} \\ v_{y_f} \\ v_{z_f} \end{Bmatrix} \tag{11.5.4}$$

Recognizing that $E_{ij} = 2e_ie_j$ and $S^2 = (E_x^2 + E_y^2 + E_z^2)S^2 = e_x^2 + e_y^2 + e_z^2$, after applying Eq. (11.5.1) to Eq. (11.5.4), we obtain

$$\begin{Bmatrix} v_{x_b} \\ v_{y_b} \\ v_{z_b} \end{Bmatrix} = \begin{bmatrix} e_x^2 + e_0^2 - e_y^2 - e_z^2 & 2(e_xe_y + e_ze_0) & 2(e_xe_z - e_ye_0) \\ 2(e_xe_y - e_ze_0) & e_y^2 + e_0^2 - e_x^2 - e_z^2 & 2(e_ye_z + e_xe_0) \\ 2(e_xe_z + e_ye_0) & 2(e_ye_z - e_xe_0) & e_z^2 + e_0^2 - e_x^2 - e_y^2 \end{bmatrix} \begin{Bmatrix} v_{x_f} \\ v_{y_f} \\ v_{z_f} \end{Bmatrix} \tag{11.5.5}$$

The inverse of the transformation matrix in Eq. (11.5.5) is

$$\begin{Bmatrix} v_{x_f} \\ v_{y_f} \\ v_{z_f} \end{Bmatrix} = \begin{bmatrix} e_x^2 + e_0^2 - e_y^2 - e_z^2 & 2(e_xe_y - e_ze_0) & 2(e_xe_z + e_ye_0) \\ 2(e_xe_y + e_ze_0) & e_y^2 + e_0^2 - e_x^2 - e_z^2 & 2(e_ye_z - e_xe_0) \\ 2(e_xe_z - e_ye_0) & 2(e_ye_z + e_xe_0) & e_z^2 + e_0^2 - e_x^2 - e_y^2 \end{bmatrix} \begin{Bmatrix} v_{x_b} \\ v_{y_b} \\ v_{z_b} \end{Bmatrix} \tag{11.5.6}$$

The transformation equation given by Eq. (11.5.5) can be used to obtain the gravitational force vector in body-fixed coordinates,

$$\begin{Bmatrix} W_{x_b} \\ W_{y_b} \\ W_{z_b} \end{Bmatrix} = W \begin{Bmatrix} 2(e_xe_z - e_ye_0) \\ 2(e_ye_z + e_xe_0) \\ e_z^2 + e_0^2 - e_x^2 - e_y^2 \end{Bmatrix} \tag{11.5.7}$$

and from Eq. (11.5.6), the ground speed is related to the airspeed and wind by

$$
\left\{\begin{array}{c} \dot{x}_f \\ \dot{y}_f \\ \dot{z}_f \end{array}\right\} = \left\{\begin{array}{c} V_{wx_f} \\ V_{wy_f} \\ V_{wz_f} \end{array}\right\} + \left[\begin{array}{ccc} e_x^2 + e_0^2 - e_y^2 - e_z^2 & 2(e_xe_y - e_ze_0) & 2(e_xe_z + e_ye_0) \\ 2(e_xe_y + e_ze_0) & e_y^2 + e_0^2 - e_x^2 - e_z^2 & 2(e_ye_z - e_xe_0) \\ 2(e_xe_z - e_ye_0) & 2(e_ye_z + e_xe_0) & e_z^2 + e_0^2 - e_x^2 - e_y^2 \end{array}\right] \left\{\begin{array}{c} u \\ v \\ w \end{array}\right\} \quad (11.5.8)
$$

The time rate of change of the Euler-Rodrigues symmetric parameters can be related to the time rate of change of the Euler axis rotation parameters. Simply differentiating Eq. (11.5.1) with respect to time produces

$$
\left\{\begin{array}{c} \dot{e}_0 \\ \dot{e}_x \\ \dot{e}_y \\ \dot{e}_z \end{array}\right\} = \left\{\begin{array}{c} -\sin(\Theta/2) \\ E_x \cos(\Theta/2) \\ E_y \cos(\Theta/2) \\ E_z \cos(\Theta/2) \end{array}\right\} \frac{\dot{\Theta}}{2} + \left\{\begin{array}{c} 0 \\ \dot{E}_x \sin(\Theta/2) \\ \dot{E}_y \sin(\Theta/2) \\ \dot{E}_z \sin(\Theta/2) \end{array}\right\} \quad (11.5.9)
$$

Using Eq. (11.4.7) to express the time rate of change of the Euler axis rotation parameters in terms of the noninertial angular rates, Eq. (11.5.9) can be written as

$$
\left\{\begin{array}{c} \dot{e}_0 \\ \dot{e}_x \\ \dot{e}_y \\ \dot{e}_z \end{array}\right\} = \frac{1}{2} \left[\begin{array}{ccc} -E_xS & -E_yS & -E_zS \\ C & -E_zS & E_yS \\ E_zS & C & -E_xS \\ -E_yS & E_xS & C \end{array}\right] \left\{\begin{array}{c} p \\ q \\ r \end{array}\right\} \quad (11.5.10)
$$

or after applying Eq. (11.5.1),

$$
\left\{\begin{array}{c} \dot{e}_0 \\ \dot{e}_x \\ \dot{e}_y \\ \dot{e}_z \end{array}\right\} = \frac{1}{2} \left[\begin{array}{ccc} -e_x & -e_y & -e_z \\ e_0 & -e_z & e_y \\ e_z & e_0 & -e_x \\ -e_y & e_x & e_0 \end{array}\right] \left\{\begin{array}{c} p \\ q \\ r \end{array}\right\} \quad (11.5.11)
$$

Because Eq. (11.5.11) is linear in both the noninertial angular rates and the Euler-Rodrigues symmetric parameters, it can also be written as

$$
\left\{\begin{array}{c} \dot{e}_0 \\ \dot{e}_x \\ \dot{e}_y \\ \dot{e}_z \end{array}\right\} = \frac{1}{2} \left[\begin{array}{cccc} 0 & -p & -q & -r \\ p & 0 & r & -q \\ q & -r & 0 & p \\ r & q & -p & 0 \end{array}\right] \left\{\begin{array}{c} e_0 \\ e_x \\ e_y \\ e_z \end{array}\right\} \quad (11.5.12)
$$

Equation (11.5.8) and either Eq. (11.5.11) or Eq. (11.5.12) provide the kinematic transformation equations in terms of the Euler-Rodrigues symmetric parameters.

Robinson (1958a) described the advantages of the Euler-Rodrigues quaternion formulation over the Euler angle or the direction-cosine formulation almost 50 years ago. Most of his observations, obtained on analog computers, are still valid in today's digital world. When the Euler elevation angle, θ, is $\pm90°$, the Euler angle integration becomes indeterminate. Despite the singularity, the Euler angle formulation is widely used because the three Euler angles have the simple interpretation of heading, elevation angle, and bank angle. The Euler-Rodrigues formulation is free of singularities. However, the physical interpretation of the quaternion is much less intuitive than that associated with the Euler angles. In a mathematical study, Stuelpnagel (1964) considered parameterization of a general three-parameter rotation group, four-parameter rotation group, and five- and six-parameter groups. He proves that the three-parameter rotation group always leads to nonlinear kinematic equations and that the Euler-Rodrigues symmetric parameters represent the smallest number of parameters (four) with linear kinematic equations. Errors associated with numerical integration of the kinematic equations for attitude have been characterized for both the quaternion and the direction-cosine parameterizations, and the superiority of the quaternion parameterization is well documented. Another advantage of the Euler-Rodrigues formulation is the application of Kalman filtering to quaternion estimation (see, for example, Shuster 1993a).

A very significant advantage of the quaternion formulation over either the Euler angle formulation or the direction-cosine formulation is increased computational speed. Numerical integration of the nine component direction-cosine formulation is excessively time-consuming compared with the four component Euler-Rodrigues formulation. The trigonometric functions in the Euler angle transformation matrix make simulations that use this nonlinear formulation much more computationally intensive than those using the linear equations of the quaternion formulation. Furthermore, the computational advantage of the quaternion formulation can be extended even further through use of Hamilton's quaternion algebra. While the Euler-Rodrigues formulation was originally developed before quaternion algebra was conceived, the most computationally efficient algorithms are in fact based on this mathematical rule set, which was first presented by Hamilton (1844). To demonstrate this computational advantage, a brief introduction to quaternion algebra is presented here.

11.6. Quaternion Algebra

A general quaternion, $\{\mathbf{Q}\}$, is defined as

$$\{\mathbf{Q}\} \equiv Q_0 + Q_x\mathbf{i}_x + Q_y\mathbf{i}_y + Q_z\mathbf{i}_z \qquad (11.6.1)$$

where Q_0, Q_x, Q_y, and Q_z are scalars and $\mathbf{i}_x$, $\mathbf{i}_y$, and $\mathbf{i}_z$ are unit vectors in the Cartesian x-, y-, and z-directions, respectively. A quaternion has both vector and scalar properties. Vectors and scalars can be thought of as special cases of the more general quaternion. A scalar is a quaternion with Q_x, Q_y, and Q_z equal to zero and a vector is a quaternion where Q_0 equals zero.

For quaternion multiplication, a procedure called the *quaternion product* is defined. Since a vector is a special case of a quaternion, care should be taken to distinguish the quaternion product from either the dot product or the cross product. Here, the quaternion product will be indicated using the operator $\otimes$. The quaternion products of the usual Cartesian unit vectors are defined according to the following rule set:

$$
\begin{array}{lll}
\mathbf{i}_x \otimes \mathbf{i}_x \equiv -1, & \mathbf{i}_x \otimes \mathbf{i}_y \equiv \mathbf{i}_z, & \mathbf{i}_x \otimes \mathbf{i}_z \equiv -\mathbf{i}_y \\
\mathbf{i}_y \otimes \mathbf{i}_x \equiv -\mathbf{i}_z, & \mathbf{i}_y \otimes \mathbf{i}_y \equiv -1, & \mathbf{i}_y \otimes \mathbf{i}_z \equiv \mathbf{i}_x, \\
\mathbf{i}_z \otimes \mathbf{i}_x \equiv \mathbf{i}_y, & \mathbf{i}_z \otimes \mathbf{i}_y \equiv -\mathbf{i}_x, & \mathbf{i}_z \otimes \mathbf{i}_z \equiv -1
\end{array}
\tag{11.6.2}
$$

The quaternion product simply follows the distributive law. Thus, the quaternion product of one quaternion, $\{\mathbf{A}\}$, with another quaternion, $\{\mathbf{B}\}$, is

$$
\begin{aligned}
\{\mathbf{A}\} \otimes \{\mathbf{B}\} &= (A_0 + A_x \mathbf{i}_x + A_y \mathbf{i}_y + A_z \mathbf{i}_z) \otimes (B_0 + B_x \mathbf{i}_x + B_y \mathbf{i}_y + B_z \mathbf{i}_z) \\
&= (A_0 B_0 - A_x B_x - A_y B_y - A_z B_z) \\
&\quad + (A_0 B_x + A_x B_0 + A_y B_z - A_z B_y)\mathbf{i}_x \\
&\quad + (A_0 B_y - A_x B_z + A_y B_0 + A_z B_x)\mathbf{i}_y \\
&\quad + (A_0 B_z + A_x B_y - A_y B_x + A_z B_0)\mathbf{i}_z
\end{aligned}
\tag{11.6.3}
$$

The quaternion products defined in Eq. (11.6.2) are nearly like cross products, following the right-hand rule. However, special treatment is given to the quaternion product of a unit vector with itself. It should also be noted that, in general, $\{\mathbf{A}\} \otimes \{\mathbf{B}\} \neq \{\mathbf{B}\} \otimes \{\mathbf{A}\}$, so the quaternion product is not commutative. Also notice that the quaternion product reduces to simple multiplication for the special case when either operand is a scalar. However, it does not reduce to either the dot product or the cross product when both operands are vectors. In general, the quaternion product of two simple vectors is a four-component quaternion, equal to the negative of the dot product added to the cross product,

$$
\begin{aligned}
\mathbf{A} \otimes \mathbf{B} &= -(A_x B_x + A_y B_y + A_z B_z) \\
&\quad + (A_y B_z - A_z B_y)\mathbf{i}_x \\
&\quad + (A_z B_x - A_x B_z)\mathbf{i}_y \\
&\quad + (A_x B_y - A_y B_x)\mathbf{i}_z = -\mathbf{A} \cdot \mathbf{B} + \mathbf{A} \times \mathbf{B}
\end{aligned}
\tag{11.6.4}
$$

Quaternions not only have properties of scalars and vectors, but quaternion algebra also has similarities to complex algebra. The magnitude of a quaternion is defined similar to that of a complex number or a vector,

$$
|\{\mathbf{Q}\}| \equiv \sqrt{Q_0^2 + Q_x^2 + Q_y^2 + Q_z^2}
\tag{11.6.5}
$$

Also, the conjugate of a quaternion is defined as

$$
\{\mathbf{Q}\}^* \equiv Q_0 - Q_x \mathbf{i}_x - Q_y \mathbf{i}_y - Q_z \mathbf{i}_z
\tag{11.6.6}
$$

where the asterisk indicates a quaternion conjugate. Using Eq. (11.6.6) with Eq. (11.6.3) produces

$$\{\mathbf{Q}\} \otimes \{\mathbf{Q}\}^* \;=\; Q_0^2 + Q_x^2 + Q_y^2 + Q_z^2 \;=\; \left|\{\mathbf{Q}\}\right|^2 \tag{11.6.7}$$

Thus, similar to a complex variable, the quaternion product of a quaternion with its conjugate generates a scalar equal to the square of the magnitude of the quaternion.

The Euler-Rodrigues symmetric parameters can be thought of as the components of a particular unit quaternion, $\{\mathbf{e}\}$, having components (e_0, e_x, e_y, e_z), which uniquely specifies the orientation of the noninertial coordinate system relative to the inertial coordinate system. Furthermore, the rotational transformation given by Eq. (11.5.5) can also be expressed in terms of quaternion products. Let $\mathbf{v}$ be an arbitrary vector having quaternion components ($0, v_{x_b}, v_{y_b}, v_{z_b}$) in the body-fixed coordinate system and having quaternion components ($0, v_{x_f}, v_{y_f}, v_{z_f}$) in the Earth-fixed coordinate system. The components of $\mathbf{v}$ in body-fixed coordinates are related to the components of $\mathbf{v}$ in Earth-fixed coordinates through the quaternion transformation

$$\mathbf{v}_b \;=\; \{\mathbf{e}\}^* \otimes \left(\mathbf{v}_f \otimes \{\mathbf{e}\}\right) \tag{11.6.8}$$

Expanding Eq. (11.6.8) using Eq. (11.6.3), the orthogonal transformation can be written as a two-step process using a temporary quaternion, $\{\mathbf{T}\}$,

$$
\begin{aligned}
\{\mathbf{T}\} \;=\; \mathbf{v}_f \otimes \{\mathbf{e}\} \;=\; & (-v_{x_f} e_x - v_{y_f} e_y - v_{z_f} e_z) \\
& + (v_{x_f} e_0 + v_{y_f} e_z - v_{z_f} e_y)\mathbf{i}_{x_f} \\
& + (-v_{x_f} e_z + v_{y_f} e_0 + v_{z_f} e_x)\mathbf{i}_{y_f} \\
& + (v_{x_f} e_y - v_{y_f} e_x + v_{z_f} e_0)\mathbf{i}_{z_f}
\end{aligned}
\tag{11.6.9}
$$

$$
\begin{aligned}
\mathbf{v}_b \;=\; \{\mathbf{e}\}^* \otimes \{\mathbf{T}\} \;=\; & (e_0 T_x - e_x T_0 - e_y T_z + e_z T_y)\mathbf{i}_{x_b} \\
& + (e_0 T_y + e_x T_z - e_y T_0 - e_z T_x)\mathbf{i}_{y_b} \\
& + (e_0 T_z - e_x T_y + e_y T_x - e_z T_0)\mathbf{i}_{z_b}
\end{aligned}
\tag{11.6.10}
$$

From Eq. (11.6.8) and the definitions in Eqs. (11.6.1) and (11.6.6), it is worth noting that **the conjugate of an Euler-Rodrigues quaternion represents the inverse rotation**.

The advantage of Eqs. (11.6.9) and (11.6.10) over Eq. (11.5.5) is a matter of reduced computation time. The rotational transformation as expressed in Eq. (11.5.5) requires 39 multiplications and 21 additions. On the other hand, computing the same transformation from Eqs. (11.6.9) and (11.6.10) requires only 24 multiplications and 17 additions. This translates to a significant computational saving. For typical modern computers, the transformation computed from Eq. (11.5.5) requires about 50 percent greater computation time than does that computed from Eqs. (11.6.9) and (11.6.10). This can be quite significant in a flight simulator, which requires a very large number of such transformations for the visual display.

For aircraft that are not all-attitude vehicles, such as transports, helicopters, and roll-stabilized missiles, Euler angles are often used to compute attitude because the singularity is not encountered. However, the computational advantage of the quaternion transformation is even more impressive when compared with the Euler angle transformation given by Eq. (11.2.1). Even when each trigonometric function in Eq. (11.2.1) is only evaluated once, the transformation computed from Eq. (11.2.1) requires about 11 times as long to evaluate as the quaternion transformation computed from Eqs. (11.6.9) and (11.6.10). Thus, even ignoring the singularity in the Euler angle formulation, the quaternion formulation is far superior to the Euler angle formulation, based on computational efficiency alone.

In the mid-1980s, algorithms were developed, which further reduced the computation time for quaternion multiplication on the hardware available at that time. For example, Dvornychenko (1985), based on the work of Winograd (1970 and 1977), presented two algorithms that were claimed to reduce computation time. The first of these algorithms requires 11 multiplications and 19 additions. The second requires 10 multiplications and 26 additions. This compares to 16 multiplications and 12 additions for the conventional quaternion multiplication described in Eq. (11.6.3). The hardware available at that time required something on the order of five instruction cycles for multiplication and only one cycle for addition. Thus, trading five multiplications for seven additions would result in a significant reduction in computation time, for the hardware of that era. However, typical modern hardware requires only one instruction cycle for multiplication and one instruction cycle for addition. Thus, the algorithms that were developed in the 1980s to speed up quaternion transformations will actually slow down the transformations when used with today's hardware. For simulations run on modern hardware, these old algorithms should be replaced with the more straightforward algorithm specified by Eq. (11.6.3).

The system of differential equations that governs the change in the transformation quaternion with time can also be written in terms of a quaternion product. This differential system, specified by either Eq. (11.5.11) or (11.5.12), can be written as

$$\{\dot{\mathbf{e}}\} = \tfrac{1}{2}\{\mathbf{e}\} \otimes \omega \qquad (11.6.11)$$

where ω is the angular velocity vector in body-fixed coordinates. Writing the quaternion differential equation in this form provides no computational savings over the matrix form given by Eq. (11.5.12). From Eq. (11.6.3), it is observed that the quaternion product in Eq. (11.6.11) requires 16 multiplications and 12 additions. This is exactly equivalent to that required for the matrix multiplication in Eq. (11.5.12). However, the matrix multiplication in Eq. (11.5.11) requires only 12 multiplications and eight additions. Thus, the temporal derivative of the transformation quaternion should always be computed from Eq. (11.5.11).

Aircraft simulations often deal with more than two coordinate systems. For example, a single simulation could use an inertial coordinate system fixed to the center of the Earth, an Earth-fixed coordinate system at the aircraft's local latitude and longitude, an atmosphere-fixed coordinate system, and an aircraft body-fixed coordinate system. In such simulations, it is convenient to be able to efficiently combine a succession of coordinate rotations into a single transformation. This is, of course, readily accomplished when using a direction-cosine transformation by simply multiplying the direction-cosine matrices for each successive rotation. For example, if $[C]_{1-2}$ is the direction-cosine matrix for the transformation from coordinate system 1 to coordinate system 2 and $[C]_{2-3}$ is the direction-cosine matrix for the transformation from coordinate system 2 to coordinate system 3, we can write

$$[C]_{1-3} = [C]_{2-3}[C]_{1-2} \tag{11.6.12}$$

where $[C]_{1-3}$ is the direction-cosine matrix for the transformation from coordinate system 1 to coordinate system 3. Since matrix multiplication is not commutative, the right-to-left order of the multiplication is important.

In an analogous manner, a succession of coordinate rotations can be combined into a single Euler-Rodrigues quaternion by making use of the quaternion product defined in Eq. (11.6.3). If $\{e\}_{1-2}$ is the quaternion for the transformation from coordinate system 1 to coordinate system 2 and $\{e\}_{2-3}$ is the quaternion for the transformation from coordinate system 2 to coordinate system 3, then the Euler-Rodrigues quaternion for the transformation from coordinate system 1 to coordinate system 3 is given by

$$\{e\}_{1-3} = \{e\}_{1-2} \otimes \{e\}_{2-3} \tag{11.6.13}$$

Here again the order of multiplication is important because the quaternion product is not commutative. Notice that this quaternion product requires a left-to-right order, which contrasts with the right-to-left order required for direction-cosine matrix multiplication. The relation given by Eq. (11.6.13) is not obvious from examination of Eq. (11.6.8). Whittaker (1937) demonstrates the validity of this result.

There is some confusion in the literature regarding the order of multiplication required when using the Euler-Rodrigues quaternion for a succession of coordinate rotations. For example, Dvornychenko (1985) presents a right-to-left ordering of the quaternion product, which is opposite to that shown in Eq. (11.6.13). This is because Dvornychenko has also used an unconventional quaternion definition, which has the vector components defined as the negative of those defined in Eq. (11.5.1). The positive signs used in Eq. (11.5.1) correspond to the conventional right-hand Euler-Rodrigues quaternion, which is based on a right-hand rotation about the Euler axis. The negative signs used by Dvornychenko result in a left-hand quaternion, which is not normally used. When the right-hand Euler-Rodrigues quaternion is used, successive coordinate rotations are expressed using the left-to-right quaternion product that is shown in Eq. (11.6.13).

11.7. Relations between the Quaternion and Other Attitude Descriptors

The physical interpretation of the Euler-Rodrigues quaternion is much less intuitive than that associated with the Euler angles. For this reason it is convenient to be able to relate the Euler-Rodrigues quaternion to the Euler angles. Such a relation can be obtained by combining Eqs. (11.2.1) and (11.5.5). These two equations require that

$$
\begin{bmatrix}
e_x^2 + e_0^2 - e_y^2 - e_z^2 & 2(e_x e_y + e_z e_0) & 2(e_x e_z - e_y e_0) \\
2(e_x e_y - e_z e_0) & e_y^2 + e_0^2 - e_x^2 - e_z^2 & 2(e_y e_z + e_x e_0) \\
2(e_x e_z + e_y e_0) & 2(e_y e_z - e_x e_0) & e_z^2 + e_0^2 - e_x^2 - e_y^2
\end{bmatrix}
$$

$$
=
\begin{bmatrix}
C_\theta C_\psi & C_\theta S_\psi & -S_\theta \\
S_\phi S_\theta C_\psi - C_\phi S_\psi & S_\phi S_\theta S_\psi + C_\phi C_\psi & S_\phi C_\theta \\
C_\phi S_\theta C_\psi + S_\phi S_\psi & C_\phi S_\theta S_\psi - S_\phi C_\psi & C_\phi C_\theta
\end{bmatrix}
\tag{11.7.1}
$$

This matrix equation provides nine scalar equations relating the four components of the Euler-Rodrigues quaternion, $\{\mathbf{e}\}$, to the three Euler angles, ϕ, θ, and ψ. However, the nine components of the matrices on both sides of Eq. (11.7.1) each describe a rotation having only three degrees of freedom. Thus, there are six levels of redundancy in Eq. (11.7.1). Additionally, Eq. (11.5.3) must also be satisfied, which simply requires $\{\mathbf{e}\}$ to be a unit quaternion. The three degrees of freedom in Eq. (11.7.1), combined with the requirement in Eq. (11.5.3), provide exactly the four degrees of freedom needed to solve for the four components of the quaternion $\{\mathbf{e}\}$.

Combining the diagonal components of Eq. (11.7.1) with the requirement for a unit quaternion expressed in Eq. (11.5.3), a 4×4 system of algebraic equations for the squares of the quaternion components is obtained,

$$
\begin{bmatrix}
1 & 1 & -1 & -1 \\
1 & -1 & 1 & -1 \\
1 & -1 & -1 & 1 \\
1 & 1 & 1 & 1
\end{bmatrix}
\begin{Bmatrix}
e_0^2 \\
e_x^2 \\
e_y^2 \\
e_z^2
\end{Bmatrix}
=
\begin{Bmatrix}
C_\theta C_\psi \\
S_\phi S_\theta S_\psi + C_\phi C_\psi \\
C_\phi C_\theta \\
1
\end{Bmatrix}
\tag{11.7.2}
$$

This system is readily solved by direct elimination to give

$$
\begin{Bmatrix}
e_0^2 \\
e_x^2 \\
e_y^2 \\
e_z^2
\end{Bmatrix}
= \frac{1}{4}
\begin{Bmatrix}
1 + C_\theta C_\psi + S_\phi S_\theta S_\psi + C_\phi C_\theta + C_\phi C_\psi \\
1 - C_\phi C_\theta - S_\phi S_\theta S_\psi - C_\phi C_\psi + C_\theta C_\psi \\
1 - C_\theta C_\psi - C_\phi C_\theta + S_\phi S_\theta S_\psi + C_\phi C_\psi \\
1 - C_\theta C_\psi + C_\phi C_\theta - S_\phi S_\theta S_\psi - C_\phi C_\psi
\end{Bmatrix}
\tag{11.7.3}
$$

Applying the half-angle identities, Eq. (11.7.3) is written as

$$
\begin{Bmatrix}
e_0^2 \\
e_x^2 \\
e_y^2 \\
e_z^2
\end{Bmatrix}
=
\begin{Bmatrix}
(C_{\phi/2} C_{\theta/2} C_{\psi/2} + S_{\phi/2} S_{\theta/2} S_{\psi/2})^2 \\
(S_{\phi/2} C_{\theta/2} C_{\psi/2} - C_{\phi/2} S_{\theta/2} S_{\psi/2})^2 \\
(C_{\phi/2} S_{\theta/2} C_{\psi/2} + S_{\phi/2} C_{\theta/2} S_{\psi/2})^2 \\
(S_{\phi/2} S_{\theta/2} C_{\psi/2} - C_{\phi/2} C_{\theta/2} S_{\psi/2})^2
\end{Bmatrix}
\tag{11.7.4}
$$

Each of these component equations has two possible solutions, so the signs are indeterminate at this point.

The off-diagonal components of Eq. (11.7.1) may be extracted to produce

$$
\begin{bmatrix}
0 & 0 & 2 & 2 & 0 & 0 \\
0 & 0 & -2 & 2 & 0 & 0 \\
0 & -2 & 0 & 0 & 2 & 0 \\
0 & 2 & 0 & 0 & 2 & 0 \\
2 & 0 & 0 & 0 & 0 & 2 \\
-2 & 0 & 0 & 0 & 0 & 2
\end{bmatrix}
\begin{Bmatrix}
e_0 e_x \\
e_0 e_y \\
e_0 e_z \\
e_x e_y \\
e_x e_z \\
e_y e_z
\end{Bmatrix}
=
\begin{Bmatrix}
C_\theta S_\psi \\
S_\phi S_\theta C_\psi - C_\phi S_\psi \\
-S_\theta \\
C_\phi S_\theta C_\psi + S_\phi S_\psi \\
S_\phi C_\theta \\
C_\phi S_\theta S_\psi - S_\phi C_\psi
\end{Bmatrix}
\tag{11.7.5}
$$

This simple algebraic system is easily solved by adding and subtracting appropriate pairs of equations. Replacing the first and second equations with their sum and their difference and doing likewise with the other two pairs of equations results in

$$
\begin{Bmatrix}
e_0 e_x \\
e_0 e_y \\
e_0 e_z \\
e_x e_y \\
e_x e_z \\
e_y e_z
\end{Bmatrix}
= \frac{1}{4}
\begin{Bmatrix}
S_\phi C_\theta - C_\phi S_\theta S_\psi + S_\phi C_\psi \\
C_\phi S_\theta C_\psi + S_\phi S_\psi + S_\theta \\
C_\theta S_\psi - S_\phi S_\theta C_\psi + C_\phi S_\psi \\
C_\theta S_\psi + S_\phi S_\theta C_\psi - C_\phi S_\psi \\
C_\phi S_\theta C_\psi + S_\phi S_\psi - S_\theta \\
S_\phi C_\theta + C_\phi S_\theta S_\psi - S_\phi C_\psi
\end{Bmatrix}
\tag{11.7.6}
$$

Using Eq. (11.7.4) in Eq. (11.7.6) and applying the half-angle identities, the off-diagonal components of Eq. (11.7.1) reduce to

$$
\begin{Bmatrix}
s_0 s_x (S_\phi C_\theta - C_\phi S_\theta S_\psi + S_\phi C_\psi) \\
s_0 s_y (C_\phi S_\theta C_\psi + S_\phi S_\psi + S_\theta) \\
s_0 s_z (-C_\phi S_\psi + S_\phi S_\theta C_\psi - C_\phi S_\psi) \\
s_x s_y (C_\theta S_\psi + S_\phi S_\theta C_\psi - C_\phi S_\psi) \\
s_x s_z (-C_\phi S_\theta C_\psi - S_\phi S_\psi + S_\theta) \\
s_y s_z (-S_\phi C_\theta - C_\phi S_\theta S_\psi + S_\phi C_\psi)
\end{Bmatrix}
=
\begin{Bmatrix}
S_\phi C_\theta - C_\phi S_\theta S_\psi + S_\phi C_\psi \\
C_\phi S_\theta C_\psi + S_\phi S_\psi + S_\theta \\
C_\theta S_\psi - S_\phi S_\theta C_\psi + C_\phi S_\psi \\
C_\theta S_\psi + S_\phi S_\theta C_\psi - C_\phi S_\psi \\
C_\phi S_\theta C_\psi + S_\phi S_\psi - S_\theta \\
S_\phi C_\theta + C_\phi S_\theta S_\psi - S_\phi C_\psi
\end{Bmatrix}
\tag{11.7.7}
$$

where s_0, s_x, s_y, and s_z are the unknown signs of the quaternion components obtained from Eq. (11.7.4). Thus, the off-diagonal components of Eq. (11.7.1) provide only three additional pieces of information, $s_0 s_x = 1$, $s_0 s_y = 1$, and $s_0 s_z = -1$. Using these signs with Eq. (11.7.4), only two possible solutions to Eq. (11.7.1) exist, which are

$$
\begin{Bmatrix}
e_0 \\
e_x \\
e_y \\
e_z
\end{Bmatrix}
= \pm
\begin{Bmatrix}
C_{\phi/2} C_{\theta/2} C_{\psi/2} + S_{\phi/2} S_{\theta/2} S_{\psi/2} \\
S_{\phi/2} C_{\theta/2} C_{\psi/2} - C_{\phi/2} S_{\theta/2} S_{\psi/2} \\
C_{\phi/2} S_{\theta/2} C_{\psi/2} + S_{\phi/2} C_{\theta/2} S_{\psi/2} \\
C_{\phi/2} C_{\theta/2} S_{\psi/2} - S_{\phi/2} S_{\theta/2} C_{\psi/2}
\end{Bmatrix}
\tag{11.7.8}
$$

Both solutions in Eq. (11.7.8) are valid. Any orientation of one coordinate system relative to another can be described in terms of two right-hand rotations. For example, a right-hand rotation of 90 degrees about the positive x-axis is equivalent to a right-hand rotation of 270 degrees about the negative x-axis. The two solutions in Eq. (11.7.8) represent these two equivalent rotations. Usually, the positive sign is selected.

The inverse of Eq. (11.7.8) is obtained from Eq. (11.7.1) as well. This is rather straightforward and yields

$$\begin{Bmatrix} \phi \\ \theta \\ \psi \end{Bmatrix} = \begin{Bmatrix} \text{atan2}[2(e_0 e_x + e_y e_z), (e_0^2 + e_z^2 - e_x^2 - e_y^2)] \\ \text{asin}[2(e_0 e_y - e_x e_z)] \\ \text{atan2}[2(e_0 e_z + e_x e_y), (e_0^2 + e_x^2 - e_y^2 - e_z^2)] \end{Bmatrix} \qquad (11.7.9)$$

The function atan2 in Eq. (11.7.9) is a two-argument arctangent that returns a result in the proper quadrant, such as the atan2 intrinsic provided in Fortran and C. The two-argument function is not needed for the elevation angle, θ, because this angle is defined only in the range $-\pi/2$ to $\pi/2$.

The gimbal lock singularity appears again in Eq. (11.7.9). For example, consider a due north heading with an elevation angle of -90 degrees and no bank angle, i.e.,

$$\begin{Bmatrix} \phi \\ \theta \\ \psi \end{Bmatrix} = \begin{Bmatrix} 0 \\ -\pi/2 \\ 0 \end{Bmatrix}$$

Using the positive sign in Eq. (11.7.8), the quaternion representation for this attitude is

$$\begin{Bmatrix} e_0 \\ e_x \\ e_y \\ e_z \end{Bmatrix} = \begin{Bmatrix} C_{\pi/4} \\ 0 \\ -S_{\pi/4} \\ 0 \end{Bmatrix} = \begin{Bmatrix} 0.70710678 \\ 0 \\ -0.70710678 \\ 0 \end{Bmatrix}$$

Using these quaternion components in Eq. (11.7.9) to evaluate the Euler angles gives

$$\begin{Bmatrix} \phi \\ \theta \\ \psi \end{Bmatrix} = \begin{Bmatrix} \text{atan2}(0.,0.) \\ \text{asin}(-1.) \\ \text{atan2}(0.,0.) \end{Bmatrix}$$

This returns an elevation angle of -90 degrees. However, both the bank angle and the heading are indeterminate. This results from the fact that when the elevation angle is exactly ± 90 degrees, the Euler angles are not unique. For example, the attitude specified by a due north heading with an elevation angle of -90 degrees and no bank angle can also be specified as a due east heading with an elevation angle of -90 degrees and a bank angle of -90 degrees, i.e.,

$$\begin{Bmatrix} \phi \\ \theta \\ \psi \end{Bmatrix} = \begin{Bmatrix} -\pi/2 \\ -\pi/2 \\ \pi/2 \end{Bmatrix}$$

Again using the positive sign in Eq. (11.7.8), this set of Euler angles also yields the quaternion representation

$$\begin{Bmatrix} e_0 \\ e_x \\ e_y \\ e_z \end{Bmatrix} = \begin{Bmatrix} C_{\pi/4}C_{\pi/4}C_{\pi/4} + S_{\pi/4}S_{\pi/4}S_{\pi/4} \\ -S_{\pi/4}C_{\pi/4}C_{\pi/4} + C_{\pi/4}S_{\pi/4}S_{\pi/4} \\ -C_{\pi/4}S_{\pi/4}C_{\pi/4} - S_{\pi/4}C_{\pi/4}S_{\pi/4} \\ C_{\pi/4}C_{\pi/4}S_{\pi/4} - S_{\pi/4}S_{\pi/4}C_{\pi/4} \end{Bmatrix} = \begin{Bmatrix} 0.70710678 \\ 0 \\ -0.70710678 \\ 0 \end{Bmatrix}$$

When the Euler elevation angle is exactly ±90 degrees, the heading angle is completely arbitrary because the orientation of the x_b-axis is fully specified by the elevation angle alone. Specifying a heading with a ±90-degree elevation angle is like specifying the longitude at the North or South Pole on the Earth's surface. When the latitude is ±90 degrees, there is no need to specify the longitude because each of these two latitudes alone uniquely describes a point on the surface of the Earth. Thus, longitude is arbitrary for latitudes of ±90 degrees.

Because of gimbal lock, the transformation given by Eq. (11.7.9) cannot be used for a completely arbitrary Euler-Rodrigues quaternion. Whenever the Euler elevation angle passes through ±90 degrees, the Euler heading and bank angle will each undergo a step change of 180 degrees. Furthermore, at an elevation angle of exactly ±90 degrees the heading angle is arbitrary and the bank angle can only be expressed as a function of this arbitrary heading. From Eq. (11.7.8), the Euler-Rodrigues quaternion at gimbal lock is found to be

$$\begin{Bmatrix} e_0 \\ e_x \\ e_y \\ e_z \end{Bmatrix} = C_{\pi/4} \begin{Bmatrix} C_{\phi/2}C_{\psi/2} \pm S_{\phi/2}S_{\psi/2} \\ S_{\phi/2}C_{\psi/2} \mp C_{\phi/2}S_{\psi/2} \\ \pm C_{\phi/2}C_{\psi/2} + S_{\phi/2}S_{\psi/2} \\ C_{\phi/2}S_{\psi/2} \mp S_{\phi/2}C_{\psi/2} \end{Bmatrix} = C_{\pi/4} \begin{Bmatrix} \cos(\phi/2 \mp \psi/2) \\ \sin(\phi/2 \mp \psi/2) \\ \pm\cos(\phi/2 \mp \psi/2) \\ \mp\sin(\phi/2 \mp \psi/2) \end{Bmatrix}, \quad \theta = \pm\frac{\pi}{2}$$

Using this result, the Euler angles at gimbal lock can be found from the Euler-Rodrigues quaternion at gimbal lock according to the relation

$$\begin{Bmatrix} \phi \\ \theta \\ \psi \end{Bmatrix} = \begin{Bmatrix} 2\operatorname{asin}[e_x/\cos(\pi/4)] \pm \psi \\ \pm\pi/2 \\ \text{arbitrary} \end{Bmatrix} \tag{11.7.10}$$

Combining Eq. (11.7.9) with Eq. (11.7.10), a generalized algorithm for computing the Euler angles from the components of an arbitrary Euler-Rodrigues quaternion can be written as if $(e_0 e_y - e_x e_z = 0.5)$

$$\begin{Bmatrix} \phi \\ \theta \\ \psi \end{Bmatrix} = \begin{Bmatrix} 2\,\text{asin}[e_x/\cos(\pi/4)] + \psi \\ \pi/2 \\ \text{arbitrary} \end{Bmatrix}$$

if $(e_0 e_y - e_x e_z = -0.5)$

$$\begin{Bmatrix} \phi \\ \theta \\ \psi \end{Bmatrix} = \begin{Bmatrix} 2\,\text{asin}[e_x/\cos(\pi/4)] - \psi \\ -\pi/2 \\ \text{arbitrary} \end{Bmatrix} \qquad (11.7.11)$$

else

$$\begin{Bmatrix} \phi \\ \theta \\ \psi \end{Bmatrix} = \begin{Bmatrix} \text{atan2}[2(e_0 e_x + e_y e_z), (e_0^2 + e_z^2 - e_x^2 - e_y^2)] \\ \text{asin}[2(e_0 e_y - e_x e_z)] \\ \text{atan2}[2(e_0 e_z + e_x e_y), (e_0^2 + e_x^2 - e_y^2 - e_z^2)] \end{Bmatrix}$$

Since the heading at gimbal lock is arbitrary, any convenient value may be used.

The Euler-Rodrigues quaternion is also related to the total rotation angle and the components of the Euler axis vector through the definitions of the Euler-Rodrigues symmetric parameters,

$$\begin{Bmatrix} e_0 \\ e_x \\ e_y \\ e_z \end{Bmatrix} \equiv \begin{Bmatrix} \cos(\Theta/2) \\ E_x \sin(\Theta/2) \\ E_y \sin(\Theta/2) \\ E_z \sin(\Theta/2) \end{Bmatrix} \qquad (11.7.12)$$

The inverse of Eq. (11.7.12) is readily found to be

$$\begin{Bmatrix} \Theta \\ E_x \\ E_y \\ E_z \end{Bmatrix} = \begin{Bmatrix} 2\cos^{-1}(e_0) \\ e_x/\sin(\Theta/2) \\ e_y/\sin(\Theta/2) \\ e_z/\sin(\Theta/2) \end{Bmatrix} \qquad (11.7.13)$$

Thus, from the four components of the Euler-Rodrigues quaternion, the orientation of the noninertial reference frame with respect to the inertial reference frame can be expressed as a single rotation about a known axis through a known angle. For nonzero rotation angles, the three vector components of this unit quaternion compose a vector directed along the Euler axis. The scalar component of this quaternion is equal to the cosine of one half the angle of rotation about this axis, in a direction defined by the right-hand rule. If the scalar component of this quaternion is ±1, the orientation of the Euler axis is indeterminate. However, a scalar component of ±1 would indicate a total rotation angle of zero, making knowledge of the Euler axis unnecessary. It is important to recognize that like the Euler axis vector, $\mathbf{E}$, the four components of the unit quaternion, $\{\mathbf{e}\}$, are the same in both the inertial and the noninertial reference frames. Thus, a reference frame need not be specified when referring to the components of this particular quaternion.

During the 1970s the spacecraft community utilized both the direction cosines and the Euler-Rodrigues quaternion to describe attitude. For example, in space shuttle steering algorithms the command attitude was computed as a direction-cosine matrix, whereas the attitude error was more conveniently described as a quaternion. The quaternion was used to describe the attitude error because of its simpler physical interpretation. The error axis coincides with the vector part of the error quaternion and the total error is readily determined from the scalar part. To compute the error quaternion, the command-attitude quaternion was first extracted from the direction-cosine matrix. Several other space shuttle flight algorithms also required the determination of an attitude quaternion from a direction-cosine matrix. Consequently, a number of algorithms have been developed for extracting the quaternion from the direction cosines. Shepperd (1978) first published the most efficient method. The aircraft and missile community is still using both the direction cosines and the Euler-Rodrigues quaternion as attitude descriptors for aircraft flight simulation.

The direction cosines can be expressed in terms of the four components of the Euler-Rodrigues quaternion by equating the left-hand sides of Eqs. (11.3.1) and (11.5.5),

$$
\begin{bmatrix} C_{11} & C_{12} & C_{13} \\ C_{21} & C_{22} & C_{23} \\ C_{31} & C_{32} & C_{33} \end{bmatrix} = \begin{bmatrix} e_x^2 + e_0^2 - e_y^2 - e_z^2 & 2(e_x e_y + e_z e_0) & 2(e_x e_z - e_y e_0) \\ 2(e_x e_y - e_z e_0) & e_y^2 + e_0^2 - e_x^2 - e_z^2 & 2(e_y e_z + e_x e_0) \\ 2(e_x e_z + e_y e_0) & 2(e_y e_z - e_x e_0) & e_z^2 + e_0^2 - e_x^2 - e_y^2 \end{bmatrix} \quad (11.7.14)
$$

Combining the diagonal components of Eq. (11.7.14) with Eq. (11.5.3) provides a 4×4 system of equations that is readily solved to relate the squares of the quaternion components to the diagonal components of the direction-cosine matrix,

$$
\begin{Bmatrix} e_0^2 \\ e_x^2 \\ e_y^2 \\ e_z^2 \end{Bmatrix} = \frac{1}{4} \begin{Bmatrix} 1 + C_{11} + C_{22} + C_{33} \\ 1 + C_{11} - C_{22} - C_{33} \\ 1 - C_{11} + C_{22} - C_{33} \\ 1 - C_{11} - C_{22} + C_{33} \end{Bmatrix} \quad (11.7.15)
$$

The off-diagonal components of Eq. (11.7.14) can be rearranged to give

$$
\begin{Bmatrix} e_0 e_x \\ e_0 e_y \\ e_0 e_z \\ e_x e_y \\ e_x e_z \\ e_y e_z \end{Bmatrix} = \frac{1}{4} \begin{Bmatrix} C_{23} - C_{32} \\ C_{31} - C_{13} \\ C_{12} - C_{21} \\ C_{12} + C_{21} \\ C_{31} + C_{13} \\ C_{23} + C_{32} \end{Bmatrix} \quad (11.7.16)
$$

The components of the quaternion can be computed, without singularities, from the direction cosines by using Eqs. (11.7.15) and (11.7.16). This is done by first finding the quaternion component of greatest magnitude from Eq. (11.7.15),

$$e_{max}^2 = \max(e_0^2, e_x^2, e_y^2, e_z^2) \tag{11.7.17}$$

Once the component of largest magnitude has been determined, the quaternion is computed from one of the following algorithms:

If $e_0^2 = e_{max}^2$,

$$
\begin{aligned}
e_0 &= \pm\sqrt{1 + C_{11} + C_{22} + C_{33}}\,/2 \\
e_x &= (C_{23} - C_{32})/4e_0 \\
e_y &= (C_{31} - C_{13})/4e_0 \\
e_z &= (C_{12} - C_{21})/4e_0
\end{aligned} \tag{11.7.18}
$$

If $e_x^2 = e_{max}^2$,

$$
\begin{aligned}
e_x &= \pm\sqrt{1 + C_{11} - C_{22} - C_{33}}\,/2 \\
e_0 &= (C_{23} - C_{32})/4e_x \\
e_y &= (C_{12} + C_{21})/4e_x \\
e_z &= (C_{31} + C_{13})/4e_x
\end{aligned} \tag{11.7.19}
$$

If $e_y^2 = e_{max}^2$,

$$
\begin{aligned}
e_y &= \pm\sqrt{1 - C_{11} + C_{22} - C_{33}}\,/2 \\
e_0 &- (C_{31} - C_{13})/4e_y \\
e_x &= (C_{12} + C_{21})/4e_y \\
e_z &= (C_{23} + C_{32})/4e_y
\end{aligned} \tag{11.7.20}
$$

If $e_z^2 = e_{max}^2$,

$$
\begin{aligned}
e_z &= \pm\sqrt{1 - C_{11} - C_{22} + C_{33}}\,/2 \\
e_0 &= (C_{12} - C_{21})/4e_z \\
e_x &= (C_{31} + C_{13})/4e_z \\
e_y &= (C_{23} + C_{32})/4e_z
\end{aligned} \tag{11.7.21}
$$

Notice that as was the case with Eq. (11.7.1), there are two possible quaternions that will satisfy Eq. (11.7.14). These are the same two equivalent quaternions that were discussed following Eq. (11.7.8).

EXAMPLE 11.7.1. For the rocket and flight condition described in Example 7.3.2, find the Euler-Rodrigues quaternion that specifies the rocket's orientation. Using this quaternion and the quaternion product description of rotation, determine the rocket's velocity components in Earth-fixed coordinates.

Solution. From Example 7.3.2, the orientation of the rocket expressed in terms of Euler angles and the rocket's velocity components expressed in terms of body-fixed coordinates are

$$
\begin{Bmatrix} \phi \\ \theta \\ \psi \end{Bmatrix} = \begin{Bmatrix} 40° \\ 20° \\ 70° \end{Bmatrix} \text{ and } \begin{Bmatrix} V_{x_b} \\ V_{y_b} \\ V_{z_b} \end{Bmatrix} \equiv \begin{Bmatrix} u \\ v \\ w \end{Bmatrix} = \begin{Bmatrix} 825.96 \text{ ft/sec} \\ 300.63 \text{ ft/sec} \\ 476.87 \text{ ft/sec} \end{Bmatrix}
$$

Using the positive sign in Eq. (11.7.8), the Euler-Rodrigues quaternion that is equivalent to this set of Euler angles is

$$
\begin{Bmatrix} e_0 \\ e_x \\ e_y \\ e_z \end{Bmatrix} = \begin{Bmatrix} C_{\phi/2}C_{\theta/2}C_{\psi/2} + S_{\phi/2}S_{\theta/2}S_{\psi/2} \\ S_{\phi/2}C_{\theta/2}C_{\psi/2} - C_{\phi/2}S_{\theta/2}S_{\psi/2} \\ C_{\phi/2}S_{\theta/2}C_{\psi/2} + S_{\phi/2}C_{\theta/2}S_{\psi/2} \\ C_{\phi/2}C_{\theta/2}S_{\psi/2} - S_{\phi/2}S_{\theta/2}C_{\psi/2} \end{Bmatrix} = \begin{Bmatrix} 0.79212226 \\ 0.18231628 \\ 0.32686024 \\ 0.48214674 \end{Bmatrix}
$$

Because the conjugate of an Euler-Rodrigues quaternion represents the inverse rotation, from Eq. (11.6.8) we have

$$
\mathbf{V}_f = \{e\} \otimes \left(\mathbf{V}_b \otimes \{e\}^* \right)
$$

or using a temporary quaternion, $\{T\}$,

$$
\begin{Bmatrix} T_0 \\ T_x \\ T_y \\ T_z \end{Bmatrix} = \begin{Bmatrix} 0 \\ u \\ v \\ w \end{Bmatrix} \otimes \begin{Bmatrix} e_0 \\ -e_x \\ -e_y \\ -e_z \end{Bmatrix} = \begin{Bmatrix} +ue_x + ve_y + we_z \\ +ue_0 - ve_z + we_y \\ +ue_z + ve_0 - we_x \\ -ue_y + ve_x + we_0 \end{Bmatrix} = \begin{Bmatrix} 478.7713 \text{ ft/sec} \\ 665.1834 \text{ ft/sec} \\ 549.4285 \text{ ft/sec} \\ 162.5756 \text{ ft/sec} \end{Bmatrix}
$$

$$
\begin{Bmatrix} \dot{x}_f \\ \dot{y}_f \\ \dot{z}_f \end{Bmatrix} = \begin{Bmatrix} e_0 \\ e_x \\ e_y \\ e_z \end{Bmatrix} \otimes \begin{Bmatrix} T_0 \\ T_x \\ T_y \\ T_z \end{Bmatrix} = \begin{Bmatrix} e_0 T_x + e_x T_0 + e_y T_z - e_z T_y \\ e_0 T_y - e_x T_z + e_y T_0 + e_z T_x \\ e_0 T_z + e_x T_y - e_y T_x + e_z T_0 \end{Bmatrix} = \begin{Bmatrix} 402.43 \text{ ft/sec} \\ 882.78 \text{ ft/sec} \\ 242.37 \text{ ft/sec} \end{Bmatrix}
$$

Comparing this result with that obtained in Example 7.3.2, we see that the quaternion transformation produced exactly the same result as the Euler angle transformation in a more efficient algorithm.

11.8. Applying Rotational Constraints to the Quaternion Formulation

It is occasionally desirable to simulate aircraft motion with one or more of the rotational degrees of freedom constrained. This is normally accomplished using the Euler angle formulation. However, since the quaternion transformation is more than an order of magnitude faster than the Euler angle transformation, the ability to apply rotational constraints to the quaternion formulation may be of some interest.

When applying rotational constraints, the coordinate system in which the constraint will be applied must be specified. For example, a roll constraint is very different from a bank angle constraint. The roll can be constrained very simply, in either the Euler angle formulation or the quaternion formulation, by simply replacing the x_b-component of the angular momentum equation with the **roll constraint**

$$p = 0 \tag{11.8.1}$$

However, note that constraining the roll does not constrain the bank angle. All three Euler angles can still take any possible value with the roll completely constrained.

To constrain the bank angle, using the quaternion formulation, Eq. (11.7.9) can be employed. If the bank angle is to remain zero, Eq. (11.7.9) requires that

$$e_0 e_x + e_y e_z = 0 \tag{11.8.2}$$

and

$$e_x \dot{e}_0 + e_0 \dot{e}_x + e_z \dot{e}_y + e_y \dot{e}_z = 0 \tag{11.8.3}$$

Applying Eq. (11.5.12) to Eq. (11.8.3) produces

$$\begin{aligned} e_x(-pe_x - qe_y - re_z) + e_0(pe_0 + re_y - qe_z) \\ + e_z(qe_0 - re_x + pe_z) + e_y(re_0 + qe_x - pe_y) = 0 \end{aligned} \tag{11.8.4}$$

This can be simplified to yield the **quaternion bank angle constraint**

$$(e_0^2 + e_z^2 - e_x^2 - e_y^2)p + 2(e_0 e_y - e_x e_z)r = 0 \tag{11.8.5}$$

Thus, to constrain the bank angle, the roll and the yaw must be coordinated according to Eq. (11.8.5).

If the roll is to be constrained as well as the bank angle, then Eq. (11.8.5) requires

$$2(e_0 e_y - e_x e_z)r = 0 \tag{11.8.6}$$

From Eq. (11.7.9), this is equivalent to

$$\sin(\theta)r = 0 \tag{11.8.7}$$

Here it is observed that if both the roll and bank angles are to be constrained, then either the yaw or the elevation angle must also be constrained. With any nonzero elevation angle, any amount of yaw will produce a bank angle. This is not a function of the quaternion formulation. It is a simple kinematic fact.

The relationship between the body-fixed angular rates and the Euler angles, which was demonstrated here by using the roll and bank angles, is true in general. None of the Euler angles can be constrained by constraining only one of the body-fixed angular rates. The bank angle can be changed using only pitch and yaw, the elevation angle can be changed using only roll and yaw, and the azimuth angle can be changed using only roll and pitch. *Any motion having only one of the rotational degrees of freedom constrained, in body-fixed coordinates, allows for all three degrees of freedom in orientation.*

Similarly, the other rotational degrees of freedom can be constrained by using the **pitch constraint**

$$q = 0 \tag{11.8.8}$$

the **quaternion elevation angle constraint**

$$(e_0^2 + e_z^2 - e_x^2 - e_y^2)q - 2(e_0 e_x + e_y e_z)r = 0 \tag{11.8.9}$$

the **yaw constraint**

$$r = 0 \tag{11.8.10}$$

and the **quaternion azimuth angle constraint**

$$2(e_0 e_x - e_y e_z)q + (e_0^2 + e_y^2 - e_x^2 - e_z^2)r = 0 \tag{11.8.11}$$

Motion with only one rotational degree of freedom, in body-fixed coordinates, can also be simulated using the quaternion formulation. *Constraining any two rotational degrees of freedom in body-fixed coordinates will constrain two degrees of freedom in orientation.* For pure rolling motion,

$$q = r = 0 \tag{11.8.12}$$

Using these constraints in Eq. (11.5.12) gives

$$\begin{Bmatrix} \dot{e}_0 \\ \dot{e}_x \\ \dot{e}_y \\ \dot{e}_z \end{Bmatrix} = \frac{1}{2} \begin{bmatrix} 0 & -p & 0 & 0 \\ p & 0 & 0 & 0 \\ 0 & 0 & 0 & p \\ 0 & 0 & -p & 0 \end{bmatrix} \begin{Bmatrix} e_0 \\ e_x \\ e_y \\ e_z \end{Bmatrix} = \frac{p}{2} \begin{Bmatrix} -e_x \\ e_0 \\ e_z \\ -e_y \end{Bmatrix} \tag{11.8.13}$$

If the elevation angle and the azimuth angle are both initially zero, from Eq. (11.7.8), the initial condition is

$$\left\{\begin{array}{c} e_0 \\ e_x \\ e_y \\ e_z \end{array}\right\}_{t=0} = \left\{\begin{array}{c} \cos(\phi_0/2) \\ \sin(\phi_0/2) \\ 0 \\ 0 \end{array}\right\} \tag{11.8.14}$$

Using Eq. (11.8.14) with Eq. (11.8.13), it is noted that both e_y and e_z remain zero during this motion. Thus, **pure rolling motion** can be simulated using the constraints

$$q = r = e_y = e_z = 0 \tag{11.8.15}$$

Similarly, **pure pitching motion** can be simulated by using the quaternion constraints

$$p = r = e_x = e_z = 0 \tag{11.8.16}$$

and **pure yawing motion** results from the constraints

$$p = q = e_x = e_y = 0 \tag{11.8.17}$$

11.9. Closed-Form Quaternion Solution for Constant Rotation

It is possible to obtain a closed-form solution to the quaternion formulation for the case of constant rotation. While such a condition would frequently occur in spacecraft applications, it almost never occurs in aircraft applications. The angular rates experienced by a moving aircraft depend on the aerodynamic moments acting on the craft and are almost always changing with time. In this case, closed-form solutions to the quaternion formulation are very difficult or impossible to obtain for all but the most trivial conditions. Nevertheless, closed-form analytic solutions to the quaternion formulation for the case of constant angular rates provide an excellent mechanism for verifying the numerical algorithms used to integrate the quaternion formulation. For such verification it is useful to have a general solution that allows for rotation about any or all of the body-fixed axes. The development of such a solution, from Eq. (11.5.12), is rather straightforward.

Consider a rigid body that is rotating at a constant angular velocity. The differential equations governing the change in the components of the quaternion, $\{\mathbf{e}\}$, with time are given by Eq. (11.5.12). This system of differential equations can be written in matrix notation as

$$\{\dot{\mathbf{e}}\} - [\mathbf{M}]\{\mathbf{e}\} = 0 \tag{11.9.1}$$

$$[\mathbf{M}] = \frac{1}{2}\begin{bmatrix} 0 & -p & -q & -r \\ p & 0 & r & -q \\ q & -r & 0 & p \\ r & q & -p & 0 \end{bmatrix} \tag{11.9.2}$$

When the matrix **[M]** is constant, the solution to Eq. (11.9.1) is

$$\{e\} = [\exp([M]t)]\{e\}_{t=0} \tag{11.9.3}$$

where the matrix exponential $[\exp([M]t)]$ is computed from the matrix series definition

$$[\exp([M]t)] \equiv [i] + [M]t + \frac{[M][M]t^2}{2!} + \frac{[M][M][M]t^3}{3!} + \cdots \tag{11.9.4}$$

Here $\{e\}_{t=0}$ is the initial value of the quaternion at time $t=0$ and **[i]** is the identity matrix. If the matrix **[M]** were completely arbitrary, the matrix exponential would need to be evaluated by summing a large number of terms in the infinite series. However, in this particular case we can take advantage of a special property of the matrix that is defined in Eq. (11.9.2).

By direct multiplication it is readily shown from Eq. (11.9.2) that

$$[M][M] = -\tfrac{1}{4}\omega^2[i] \tag{11.9.5}$$

where $\omega^2 = p^2 + q^2 + r^2$. By making use of Eq. (11.9.5), the infinite series in Eq. (11.9.4) can be rearranged as

$$[\exp([M]t)] = [i]\left(1 - \frac{(\omega t/2)^2}{2!} + \frac{(\omega t/2)^4}{4!} + \cdots\right)$$

$$+ \frac{2}{\omega}[M]\left(\frac{\omega t}{2} - \frac{(\omega t/2)^3}{3!} + \frac{(\omega t/2)^5}{5!} + \cdots\right)$$

or

$$[\exp([M]t)] = [i]\cos(\omega t/2) + \frac{2}{\omega}[M]\sin(\omega t/2) \tag{11.9.6}$$

Using Eq. (11.9.6) in Eq. (11.9.3) yields the general solution for constant rotation

$$\{e\} = \left[[i]\cos(\omega t/2) + \frac{2}{\omega}[M]\sin(\omega t/2)\right]\{e\}_{t=0} \tag{11.9.7}$$

For the special initial condition where all three Euler angles are zero,

$$\begin{Bmatrix} e_0(0) \\ e_x(0) \\ e_y(0) \\ e_z(0) \end{Bmatrix} = \begin{Bmatrix} 1 \\ 0 \\ 0 \\ 0 \end{Bmatrix} \tag{11.9.8}$$

the solution given by Eq. (11.9.7) reduces to the rather obvious result

$$
\begin{Bmatrix} e_0 \\ e_x \\ e_y \\ e_z \end{Bmatrix} = \begin{Bmatrix} \cos(\omega t/2) \\ (p/\omega)\sin(\omega t/2) \\ (q/\omega)\sin(\omega t/2) \\ (r/\omega)\sin(\omega t/2) \end{Bmatrix}
\tag{11.9.9}
$$

Equations (11.9.7) and (11.9.9) provide an excellent means for testing most numerical algorithms used to integrate the quaternion formulation.

EXAMPLE 11.9.1. Consider an object moving relative to the surface of the Earth under the assumption of no atmosphere. Initially, at time $t = 0$, the primary inertial axes of the body are aligned such that the Euler angles are

$$
\begin{Bmatrix} \phi \\ \theta \\ \psi \end{Bmatrix}_{t=0} = \begin{Bmatrix} 0° \\ 90° \\ 0° \end{Bmatrix}
$$

The initial translational and rotational velocities of the body relative to this same primary body-fixed coordinate system are

$$
\begin{Bmatrix} u \\ v \\ w \end{Bmatrix}_{t=0} = \begin{Bmatrix} V_v \\ 0 \\ V_h \end{Bmatrix} \text{ and } \begin{Bmatrix} p \\ q \\ r \end{Bmatrix}_{t=0} = \begin{Bmatrix} 0 \\ \omega \\ 0 \end{Bmatrix}
$$

Using the body-fixed quaternion formulation for the case where all three primary moments of inertia are equal, obtain a solution for the position and orientation of the body as a function of time.

Solution. Because we are assuming no atmosphere, the only force is gravity and there is no moment acting about the center of gravity. Applying Newton's second law as expressed in Eq. (7.2.20) and using Eq. (11.5.7) to express the gravitational force in terms of the Euler-Rodrigues quaternion, we obtain

$$
\frac{W}{g} \begin{Bmatrix} \dot{u} \\ \dot{v} \\ \dot{w} \end{Bmatrix} = W \begin{Bmatrix} 2(e_x e_z - e_y e_0) \\ 2(e_y e_z + e_x e_0) \\ e_z^2 + e_0^2 - e_x^2 - e_y^2 \end{Bmatrix} + \frac{W}{g} \begin{Bmatrix} rv - qw \\ pw - ru \\ qu - pv \end{Bmatrix}
\tag{11.9.10}
$$

$$
\begin{Bmatrix} \dot{p} \\ \dot{q} \\ \dot{r} \end{Bmatrix} = \begin{Bmatrix} 0 \\ 0 \\ 0 \end{Bmatrix}
\tag{11.9.11}
$$

Using the kinematic transformation equations given by Eqs. (11.5.8) and (11.5.12), for the Euler-Rodrigues formulation with no wind, we have

$$
\begin{Bmatrix} \dot{x}_f \\ \dot{y}_f \\ \dot{z}_f \end{Bmatrix} = \begin{bmatrix} e_x^2 + e_0^2 - e_y^2 - e_z^2 & 2(e_x e_y - e_z e_0) & 2(e_x e_z + e_y e_0) \\ 2(e_x e_y + e_z e_0) & e_y^2 + e_0^2 - e_x^2 - e_z^2 & 2(e_y e_z - e_x e_0) \\ 2(e_x e_z - e_y e_0) & 2(e_y e_z + e_x e_0) & e_z^2 + e_0^2 - e_x^2 - e_y^2 \end{bmatrix} \begin{Bmatrix} u \\ v \\ w \end{Bmatrix} \quad (11.9.12)
$$

$$
\begin{Bmatrix} \dot{e}_0 \\ \dot{e}_x \\ \dot{e}_y \\ \dot{e}_z \end{Bmatrix} = \frac{1}{2} \begin{bmatrix} 0 & -p & -q & -r \\ p & 0 & r & -q \\ q & -r & 0 & p \\ r & q & -p & 0 \end{bmatrix} \begin{Bmatrix} e_0 \\ e_x \\ e_y \\ e_z \end{Bmatrix} \quad (11.9.13)
$$

Choosing the positive sign from Eq. (11.7.8) and using the specified initial condition in terms of Euler angles, the initial Euler-Rodrigues quaternion is

$$
\begin{Bmatrix} e_0 \\ e_x \\ e_y \\ e_z \end{Bmatrix}_{t=0} = \begin{Bmatrix} C_{\phi/2}C_{\theta/2}C_{\psi/2} + S_{\phi/2}S_{\theta/2}S_{\psi/2} \\ S_{\phi/2}C_{\theta/2}C_{\psi/2} - C_{\phi/2}S_{\theta/2}S_{\psi/2} \\ C_{\phi/2}S_{\theta/2}C_{\psi/2} + S_{\phi/2}C_{\theta/2}S_{\psi/2} \\ C_{\phi/2}C_{\theta/2}S_{\psi/2} - S_{\phi/2}S_{\theta/2}C_{\psi/2} \end{Bmatrix}_{t=0} = \begin{Bmatrix} C_{\theta/2} \\ 0 \\ S_{\theta/2} \\ 0 \end{Bmatrix}_{t=0}
$$

or

$$
\begin{Bmatrix} e_0 \\ e_x \\ e_y \\ e_z \end{Bmatrix}_{t=0} = \begin{Bmatrix} 1/\sqrt{2} \\ 0 \\ 1/\sqrt{2} \\ 0 \end{Bmatrix} \quad (11.9.14)
$$

Defining the position of the body to be measured relative to the initial position at time $t = 0$, the initial condition for position is

$$
\begin{Bmatrix} x_f \\ y_f \\ z_f \end{Bmatrix}_{t=0} = \begin{Bmatrix} 0 \\ 0 \\ 0 \end{Bmatrix} \quad (11.9.15)
$$

This completes the formulation for the 13 state variables, x_f, y_f, z_f, e_0, e_x, e_y, e_z, u, v, w, p, q, and r.

From Eq. (11.9.11), it is easily seen that the body-fixed angular rates remain constant at the initial values and from Eq. (11.9.13) with the specified initial angular rates, we have

$$\begin{Bmatrix} p \\ q \\ r \end{Bmatrix} = \begin{Bmatrix} 0 \\ \omega \\ 0 \end{Bmatrix} \tag{11.9.16}$$

and

$$\begin{Bmatrix} \dot{e}_0 \\ \dot{e}_x \\ \dot{e}_y \\ \dot{e}_z \end{Bmatrix} = \frac{1}{2} \begin{bmatrix} 0 & 0 & -\omega & 0 \\ 0 & 0 & 0 & -\omega \\ \omega & 0 & 0 & 0 \\ 0 & \omega & 0 & 0 \end{bmatrix} \begin{Bmatrix} e_0 \\ e_x \\ e_y \\ e_z \end{Bmatrix} \tag{11.9.17}$$

Since ω is constant, the solution to Eq. (11.9.17) is specified by Eq. (11.9.7). Applying Eq. (11.9.14), this gives

$$\begin{Bmatrix} e_0 \\ e_x \\ e_y \\ e_z \end{Bmatrix} = \left[\begin{bmatrix} 1 & 0 & 0 & 0 \\ 0 & 1 & 0 & 0 \\ 0 & 0 & 1 & 0 \\ 0 & 0 & 0 & 1 \end{bmatrix} \cos(\omega t/2) + \frac{2}{\omega} \frac{1}{2} \begin{bmatrix} 0 & 0 & -\omega & 0 \\ 0 & 0 & 0 & -\omega \\ \omega & 0 & 0 & 0 \\ 0 & \omega & 0 & 0 \end{bmatrix} \sin(\omega t/2) \right] \begin{Bmatrix} 1/\sqrt{2} \\ 0 \\ 1/\sqrt{2} \\ 0 \end{Bmatrix}$$

or

$$\begin{Bmatrix} e_0 \\ e_x \\ e_y \\ e_z \end{Bmatrix} = \frac{1}{\sqrt{2}} \begin{Bmatrix} \cos(\omega t/2) - \sin(\omega t/2) \\ 0 \\ \cos(\omega t/2) + \sin(\omega t/2) \\ 0 \end{Bmatrix} \tag{11.9.18}$$

Using Eqs. (11.9.16) and (11.9.18) in Eq. (11.9.10) gives

$$\begin{Bmatrix} \dot{u} \\ \dot{v} \\ \dot{w} \end{Bmatrix} = \frac{g}{2} \begin{Bmatrix} -2[\cos(\omega t/2) + \sin(\omega t/2)][\cos(\omega t/2) - \sin(\omega t/2)] \\ 0 \\ [\cos(\omega t/2) - \sin(\omega t/2)]^2 - [\cos(\omega t/2) + \sin(\omega t/2)]^2 \end{Bmatrix} + \begin{Bmatrix} -\omega w \\ 0 \\ \omega u \end{Bmatrix}$$

$$= g \begin{Bmatrix} \sin^2(\omega t/2) - \cos^2(\omega t/2) \\ 0 \\ -2\cos(\omega t/2)\sin(\omega t/2) \end{Bmatrix} + \begin{Bmatrix} -\omega w \\ 0 \\ \omega u \end{Bmatrix}$$

or

$$\begin{Bmatrix} \dot{u} \\ \dot{v} \\ \dot{w} \end{Bmatrix} + \begin{Bmatrix} \omega w \\ 0 \\ -\omega u \end{Bmatrix} = g \begin{Bmatrix} -\cos(\omega t) \\ 0 \\ -\sin(\omega t) \end{Bmatrix} \tag{11.9.19}$$

The general solution to Eq. (11.9.19) is found by adding a particular solution to the homogeneous solution. This gives

$$
\begin{Bmatrix} u \\ v \\ w \end{Bmatrix} = \begin{Bmatrix} C_1 \sin(\omega t) + C_2 \cos(\omega t) \\ C_3 \\ C_2 \sin(\omega t) - C_1 \cos(\omega t) \end{Bmatrix} + t \begin{Bmatrix} -g\cos(\omega t) \\ 0 \\ -g\sin(\omega t) \end{Bmatrix}
$$

After applying the initial condition that is specified in the problem statement, the solution for the body-fixed velocity components is

$$
\begin{Bmatrix} u \\ v \\ w \end{Bmatrix} = \begin{Bmatrix} V_v \cos(\omega t) - V_h \sin(\omega t) \\ 0 \\ V_v \sin(\omega t) + V_h \cos(\omega t) \end{Bmatrix} + t \begin{Bmatrix} -g\cos(\omega t) \\ 0 \\ -g\sin(\omega t) \end{Bmatrix} \tag{11.9.20}
$$

Using Eqs. (11.9.18) and (11.9.20) in Eq. (11.9.12) results in

$$
\begin{Bmatrix} \dot{x}_f \\ \dot{y}_f \\ \dot{z}_f \end{Bmatrix} = \begin{bmatrix} -\sin(\omega t) & 0 & \cos(\omega t) \\ 0 & 1 & 0 \\ -\cos(\omega t) & 0 & -\sin(\omega t) \end{bmatrix} \begin{Bmatrix} V_v \cos(\omega t) - V_h \sin(\omega t) - gt\cos(\omega t) \\ 0 \\ V_v \sin(\omega t) + V_h \cos(\omega t) - gt\sin(\omega t) \end{Bmatrix}
$$

or

$$
\begin{Bmatrix} \dot{x}_f \\ \dot{y}_f \\ \dot{z}_f \end{Bmatrix} = \begin{Bmatrix} V_h \\ 0 \\ -V_v + gt \end{Bmatrix} \tag{11.9.21}
$$

Equation (11.9.21) is readily integrated subject to Eq. (11.9.15) to yield

$$
\begin{Bmatrix} x_f \\ y_f \\ z_f \end{Bmatrix} = \begin{Bmatrix} V_h t \\ 0 \\ -V_v t + \frac{1}{2} gt^2 \end{Bmatrix} \tag{11.9.22}
$$

Equation (11.9.22) specifies the position of the object as a function of time. This is the well-known parabolic solution for the trajectory of a projectile in the absence of aerodynamic drag. Equation (11.9.18) gives the object's orientation as a function of time, specified in terms of the Euler-Rodrigues quaternion.

The problem in Example 11.9.1 could have been solved more easily had it been formulated using an inertial coordinate system. However, this example does demonstrate that simple problems can be solved analytically using the quaternion formulation. Such closed-form solutions are useful for verifying computer codes that have been written for numerical integration of the quaternion formulation.

11.10. Numerical Integration of the Quaternion Formulation

When integrating Eq. (11.5.12) numerically, the zeros on the diagonal can cause some problems. Integration of this system can result in very large numerical errors if a first-order method is used. For this reason, a higher-order method, such as fourth-order Runge-Kutta or fourth-order Adams-Bashforth-Moulton (see, for example, Hoffman 1992 or Gerald and Wheatley 1999) should always be used to integrate the quaternion formulation. Since most modern numerical codes use such methods, this is not at all restrictive for simulations run on modern digital computers.

Historically, aircraft flight simulations using the quaternion formulation were first commonly run using analog computers. Since these analog simulations were inherently first-order, the raw quaternion formulation produced very large integration errors. When such simulations were run, the magnitude of the quaternion would grow quite rapidly with time. Since the formulation requires the magnitude of the quaternion to remain unity in order to maintain orthogonality in the attitude transformation, these large orthogonality errors would render the simulations almost useless. To remedy this problem, techniques were developed to reduce this orthogonality error. One such method, developed by Robinson (1958a) and popularized by Mitchell and Rogers (1965), is based on the method first suggested by Corbett and Wright (1957) for maintaining orthogonality in the direction-cosine matrix. With this method, the kinematic equations in Eq. (11.5.12) were modified to include error reduction terms on the diagonal,

$$
\begin{Bmatrix} \dot{e}_0 \\ \dot{e}_x \\ \dot{e}_y \\ \dot{e}_z \end{Bmatrix} = \frac{1}{2} \begin{bmatrix} k\varepsilon & -p & -q & -r \\ p & k\varepsilon & r & -q \\ q & -r & k\varepsilon & p \\ r & q & -p & k\varepsilon \end{bmatrix} \begin{Bmatrix} e_0 \\ e_x \\ e_y \\ e_z \end{Bmatrix} \tag{11.10.1}
$$

where ε is the orthogonality error, defined as

$$
\varepsilon = 1 - (e_0^2 + e_x^2 + e_y^2 + e_z^2) \tag{11.10.2}
$$

and k is a gain coefficient, which Mitchell and Rogers (1965) said should be "set to a very high value." This error reduction scheme, called *Corbett-Wright orthogonality control*, worked quite well when used with the analog computers of that era.

While analog computers are no longer widely available, it is possible to demonstrate approximately how Corbett-Wright orthogonality control worked with an analog computer by integrating Eq. (11.10.1) using a first-order Euler integration scheme. Results of such integration are shown in Fig. 11.10.1. For this figure the total error was computed as the difference between the numerical integration result and the exact solution from Eq. (11.9.7). Notice that as the value of the Corbett-Wright gain coefficient approaches zero, which is the case of no orthogonality control, the total error becomes very large. For gain coefficients larger than about 0.5ω, the error is reduced to a more or less acceptable level. With the exception of the increase in error on the right, the results shown in Fig. 11.10.1 are very similar to the results reported by Mitchell and Rogers (1965) for a typical analog computer.

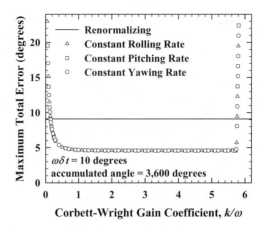

Figure 11.10.1. Effect of changing the gain coefficient on the accuracy of a first-order Euler algorithm using Corbett-Wright orthogonality control.

Unlike first-order digital integration, analog simulations were well behaved for very large values of k. The abrupt increase in error that is seen in Fig. 11.10.1 for k/ω greater than about 5.7 is a result of the well-known Euler instability that is associated with any first-order numerical integration. For Eq. (11.10.1), Fang and Zimmerman (1969) have shown that this first-order instability occurs whenever the product of the gain coefficient and the time step is greater than unity,

$$k\,\delta t > 1$$

When digital computers first became fast enough to replace analog computers for running such simulations, Corbett-Wright orthogonality control was often used in the digital solution algorithms. Furthermore, this orthogonality control is still widely used for aircraft flight simulation today. It is, however, not necessary. When a modern digital computer is used to integrate Eq. (11.5.12) with any of the prevalent fourth-order numerical methods, the orthogonality error is extremely small. Using the Corbett-Wright orthogonality control scheme with these modern numerical algorithms increases the computation time but does little to improve the accuracy of the simulation.

Even though the orthogonality error for modern numerical algorithms is very small, it can accumulate. Thus, if a simulation is to be run for long periods of time using large time steps, the quaternion should be renormalized periodically. The quaternion may be renormalized by dividing each component by the magnitude,

$$\{\mathbf{e}\}_r \;=\; \frac{\{\mathbf{e}\}}{|\{\mathbf{e}\}|} \;=\; \frac{\{\mathbf{e}\}}{\sqrt{e_0^2 + e_x^2 + e_y^2 + e_z^2}} \tag{11.10.3}$$

where the subscript r indicates renormalized. As seen in Fig. 11.10.1, renormalization is not as accurate as Corbett-Wright orthogonality control for first-order integration.

However, a different result is obtained for higher-order integration. If the time steps used with fourth-order Runge-Kutta are of a size to produce about 10 degrees of revolution per time step, it requires about 3,000,000 iterations to degrade the magnitude of the quaternion by 1 percent. Since a very long renormalization period can be used, periodic renormalization of the quaternion is much faster than using the Corbett-Wright orthogonality control scheme. However, if computation time is not a concern, Corbett-Wright orthogonality control will eliminate the growth of the orthogonality error for simulations using large time steps. If the time steps are small enough to produce about 5 degrees of rotation or less per time step, no orthogonality control whatsoever is required with fourth-order Runge-Kutta integration.

If the Corbett-Wright orthogonality control scheme is to be used for large time-step simulations, it should be remembered that this only eliminates the growth of the orthogonality error. It does not eliminate the drift error, which for modern algorithms can be much larger than the orthogonality error. In fact, when a large gain coefficient is used with fourth-order integration and Corbett-Wright orthogonality control, the drift error is increased by more than the orthogonality error is reduced, which actually increases the net error for the simulation. This is shown in Fig. 11.10.2. In contrast with the older analog simulations that used a very large gain coefficient, when Corbett-Wright orthogonality control is used with a fourth-order digital integration, a gain coefficient larger than about 2ω should not be used. As can be seen in Fig. 11.10.3, this is truly independent of the angular rate. Furthermore, as is seen in Fig. 11.10.4, decreasing the time step does not eliminate this problem. For such fourth-order numerical integration, a gain coefficient equal to the magnitude of the angular velocity will effectively eliminate the growth of the orthogonality error associated with large time steps without adversely affecting the drift error. However, for common aircraft flight simulations, the angular rates are not typically known a priori. Furthermore, the angular rate is typically changing with time and a very low angular rate requires a very low gain coefficient for accurate simulation.

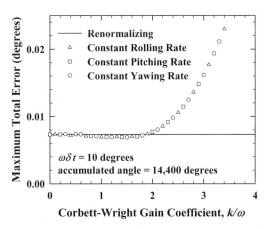

Figure 11.10.2. Effect of changing the gain coefficient on the accuracy of a fourth-order Runge-Kutta algorithm using Corbett-Wright orthogonality control.

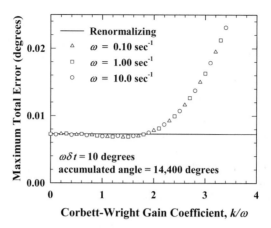

Figure 11.10.3. Effect of changing the angular rate on the accuracy of a fourth-order Runge-Kutta algorithm using Corbett-Wright orthogonality control.

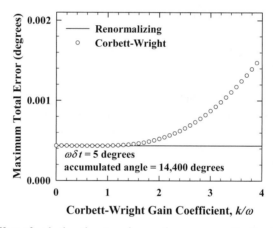

Figure 11.10.4. Effect of reducing the step size on the accuracy of a fourth-order Runge-Kutta algorithm using Corbett-Wright orthogonality control.

To avoid increasing the total error, a variable gain coefficient equal to the angular rate should be used with a fourth-order Runge-Kutta algorithm using Corbett-Wright orthogonality control,

$$k = \omega = \sqrt{p^2 + q^2 + r^2} \qquad (11.10.4)$$

The implementation of Eq. (11.10.4) adds substantially to the computational burden of Corbett-Wright orthogonality control. Thus, for greatest computational efficiency, it is much preferred to use periodic renormalization of the quaternion over any use of the Corbett-Wright orthogonality control scheme.

Fang and Zimmerman (1969) proposed a pseudo-fourth-order method for integrating Eq. (11.10.1). With conventional fourth-order Runge-Kutta the time rate of change of the quaternion is computed from Eq. (11.10.1) four times for each time step. With the Fang and Zimmerman algorithm, this same fourth-order numerical procedure is used to integrate Eq. (11.10.1), except that ε is held constant through all four computations in each time step. The orthogonality error is updated only once at the beginning of each full time step.

The Fang and Zimmerman algorithm exhibits features of both first- and fourth-order numerical integration. Although both the angular velocity and the quaternion on the right-hand side of Eq. (11.10.1) are estimated to fourth-order accuracy, ε is estimated only to first-order accuracy. Results of integrating Eq. (11.10.1) using the Fang and Zimmerman algorithm are shown in Fig. 11.10.5. Notice that the accuracy of the method for small values of the gain coefficient is comparable to that for full fourth-order integration. However, the method also exhibits an abrupt first-order instability when the gain coefficient is greater than 1 divided by the time step, $k > 1/\delta t$.

One striking observation that should be noted from Figs. 11.10.2 through 11.10.5 is that Corbett-Wright orthogonality control offers no significant reduction in simulation error for modern numerical integration algorithms. On the other hand, if the gain coefficient is not chosen judicially, Corbett-Wright orthogonality control can significantly increase the simulation error. Periodic renormalization of the quaternion does not present this problem and is more computationally efficient.

The computation time required for periodic renormalization of the quaternion can be reduced even further by recognizing that the orthogonality error is always very small for modern fourth-order integration algorithms. Equation (11.10.3) can be written as

$$\{\mathbf{e}\}_r = \frac{\{\mathbf{e}\}}{\sqrt{e_0^2 + e_x^2 + e_y^2 + e_z^2}} = \frac{\{\mathbf{e}\}}{\sqrt{1 - \varepsilon}} \qquad (11.10.5)$$

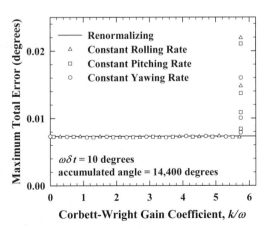

Figure 11.10.5. Effect of changing the gain coefficient on the accuracy of the numerical Fang and Zimmerman algorithm using Corbett-Wright orthogonality control.

where ε is the error in the square of the quaternion magnitude, as given by Eq. (11.10.2). Since ε is very small, Eq. (11.10.5) can be closely approximated as

$$\{e\}_r \cong \{e\}\left(1+\tfrac{1}{2}\varepsilon\right) \tag{11.10.6}$$

or after applying Eq. (11.10.2),

$$\{e\}_r \cong \{e\}\left[1.5-0.5\left(e_0^2+e_x^2+e_y^2+e_z^2\right)\right] \tag{11.10.7}$$

Periodic application of Eq. (11.10.7) is sufficient to eliminate the orthogonality error.

Renormalizing the quaternion with Eq. (11.10.7) requires nine multiplications and four additions. Using Eq. (11.10.3), on the other hand, requires eight multiplications, three additions, one square root, and one division. Typical modern hardware requires one instruction cycle for addition, one instruction cycle for multiplication, something on the order of four instruction cycles for division, and anywhere from about 4 to 30 instruction cycles for the square-root operation, depending on the processor and compiler. Therefore, renormalizing with Eq. (11.10.3) requires anywhere from 50 to 250 percent greater computation time than using Eq. (11.10.7), on typical modern hardware. Furthermore, even if future developments reduce the time required for division and square root to one instruction cycle each, Eq. (11.10.7) will be, at worst, computationally equivalent to Eq. (11.10.3).

While periodic renormalization of the quaternion using Eq. (11.10.7) provides a computationally efficient means for controlling the orthogonality error, it does not control the drift error. In fact, for higher-order algorithms, it does nothing whatsoever to reduce the total error. In Figs. 11.10.2 through 11.10.5, the points that correspond to a Corbett-Wright gain coefficient of zero were obtained using no orthogonality control at all. In each case the total error obtained with no orthogonality control was the same as that obtained when the quaternion was renormalized after each time step. From this, one may be tempted to conclude that there is no value in orthogonality control. This is not exactly true. If the orthogonality error is controlled, a small amount of drift error does not adversely affect a flight simulator, since it is constantly being corrected with virtually imperceptible pilot input. The pilot's perception of the drift error is similar to that caused by an infinitesimal change in the aerodynamics of the aircraft. The orthogonality error, on the other hand, has no counterpart in the physical world and cannot be compensated for by the pilot. When the quaternion magnitude deviates from unity, transformations obtained from Eq. (11.5.5) or Eq. (11.6.8) become nonorthogonal and vector length is not preserved through the transformation. Since rotation is an orthogonal transformation, large orthogonality errors can cause physically unrealistic distortion in the kinematic transformations. For large time-step simulations with no orthogonality control, these transformation distortions will continue to increase with time, unaffected by pilot input.

To demonstrate the effects of both orthogonality error and drift error on the visual display, Fig. 11.10.6 shows a three-dimensional image displayed after four different quaternion simulations. Each of the four simulations was carried out using fourth-order Adams-Bashforth-Moulton numerical integration. In each case all aspects of the integration, except the error control, were the same. All four images are viewed from exactly the same position and with exactly the same perspective. The differences in the images result entirely from numerical integration errors in the transformation quaternion.

The image in Fig. 11.10.6a is essentially error-free. The orthogonality error was removed using periodic renormalization, and the drift error was eliminated with pilot input. The darker image in Fig. 11.10.6b was generated using periodic renormalization without pilot input and thus contains no orthogonality error but about 17 degrees of drift error. To emphasize the error, this image has been overlaid on an error-free image in light gray. The integration used to obtain the darker image in Fig. 11.10.6c had no orthogonality control or pilot input and produced about 8 percent orthogonality error and about 17 degrees of drift error. Again, the light gray image in this figure is error-free. By comparing Fig. 11.10.6b and c, it can be seen that orthogonality error produces a scaling distortion in the visual display. Figure 11.10.6d was obtained using pilot input to eliminate the drift error, but the simulation had no orthogonality control to eliminate the scaling distortion. It can readily be shown from Eq. (11.6.8) that when orthogonality error is present, all objects in the visual display are scaled by a factor of $|\{\mathbf{e}\}|^2$.

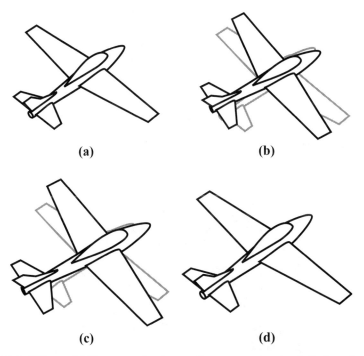

(a)   **(b)**

(c)   **(d)**

Figure 11.10.6. Effects of drift error and orthogonality error on the visual display after 2,000,000 iterations using fourth-order Adams-Bashforth-Moulton with (a) periodic renormalization, no orthogonality error, and drift error eliminated by the pilot; (b) periodic renormalization, no orthogonality error, and 17 degrees drift error; (c) no orthogonality control, 8 percent orthogonality error, and 17 degrees drift error; (d) no orthogonality control, 8 percent orthogonality error, and drift error eliminated by the pilot.

Since the orthogonality error is only a small fraction of the total error, reducing the size of the time step and/or increasing the order of the numerical integration algorithm are the only effective ways to reduce total error. The effects of time-step size on the orthogonality error and the total error for a fourth-order Runge-Kutta algorithm with no orthogonality control are shown in Figs. 11.10.7 and 11.10.8 for both single- and double-precision computations. These two figures point out a common misconception concerning the origin of the orthogonality error, which is often attributed to the *round-off error* associated with the finite word size used for numerical computation. As can be seen by comparing Fig. 11.10.7 with Fig. 11.10.8, neither the orthogonality error nor the total error is significantly affected by the computation precision for a large step size.

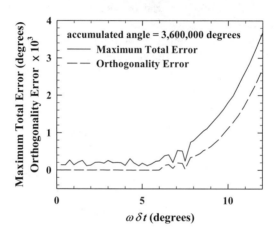

Figure 11.10.7. Effect of step size on the accuracy of a fourth-order Runge-Kutta algorithm with no orthogonality control using single-precision computations.

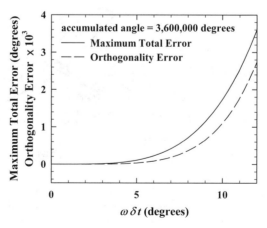

Figure 11.10.8. Effect of step size on the accuracy of a fourth-order Runge-Kutta algorithm with no orthogonality control using double-precision computations.

Furthermore, for a step size of less than about 5 degrees, the single-precision orthogonality error is exactly zero, whereas the double-precision orthogonality error is very small but finite. This is because orthogonality error does not result from the round-off error associated with word size and computation precision. It is, rather, the result of errors associated with the order of the integration algorithm. This numerical integration error is called *truncation error* because it results from truncating the Taylor series for differentiation to a finite number of terms.

In an aircraft flight simulation, computing the time derivatives of the translational and angular velocity components requires computation of the aerodynamic forces and moments. Depending on the method used, these computations can be quite time consuming. Because a digital simulation using the fourth-order Runge-Kutta algorithm requires evaluating these time derivatives four times for each time step, first- and second-order integration methods have occasionally been used in an attempt to reduce computation time. As will be demonstrated, this is not a particularly good choice.

Results of second-order Runge-Kutta integration with Corbett-Wright orthogonality control are shown in Fig. 11.10.9. Notice that as was the case for fourth-order integration, Corbett-Wright orthogonality control offers no reduction in total simulation error for this second-order integration. Furthermore, for the time step used, the accuracy of this second-order simulation is not significantly improved over that of the first-order simulation shown in Fig. 11.10.1. At first thought, this may seem counterintuitive.

From the theory of numerical methods we know that in the limit as the step size approaches zero, the global error for first-order integration should be linearly proportional to the step size, whereas the global error for the second-order integration should be proportional to the step size squared. For very small time steps, this is confirmed in Fig. 11.10.10. However, for larger time steps, the global error results primarily from the higher-order terms, which have been neglected in both the first- and second-order algorithms. Thus, for larger time steps, the second-order algorithm offers little increase in accuracy over the first-order algorithm. This can be seen in Fig. 11.10.11.

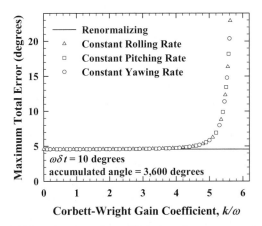

Figure 11.10.9. Effect of changing the gain coefficient on the accuracy of a second-order Runge-Kutta algorithm using Corbett-Wright orthogonality control.

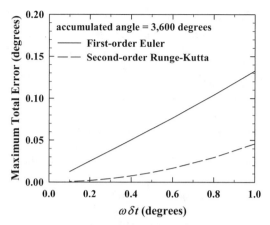

Figure 11.10.10. Comparison between first-order Euler integration and second-order Runge-Kutta integration for small time steps.

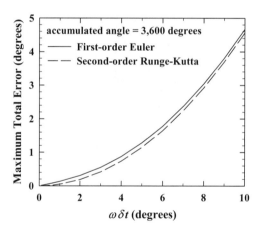

Figure 11.10.11. Comparison between first-order Euler integration and second-order Runge-Kutta integration for large time steps.

When digital computers first became practical for aircraft flight simulation, new integration algorithms were developed in an attempt to reduce computation time without sacrificing accuracy. Pope (1963) proposed one such algorithm, which was called the *exponential method*. This algorithm was first used to integrate the quaternion formulation at the NASA Langley Research Center (see Barker, Bowles, and Williams 1973). In the aircraft community, the method is commonly known as the local linearization method. Since the method has been used in aircraft flight simulation for many years but is not commonly used elsewhere, a development of this algorithm will be presented here.

The differential equations governing the change in the components of the quaternion, $\{\mathbf{e}\}$, with time are given by Eq. (11.5.12). This system of differential equations written in matrix notation is

$$\{\dot{\mathbf{e}}\} = [\mathbf{M}]\{\mathbf{e}\} \qquad (11.10.8)$$

where

$$[\mathbf{M}] = \frac{1}{2}\begin{bmatrix} 0 & -p & -q & -r \\ p & 0 & r & -q \\ q & -r & 0 & p \\ r & q & -p & 0 \end{bmatrix} \qquad (11.10.9)$$

The right side of Eq. (11.10.8) can be expanded in a Taylor series about $t = t_i$ and $\{\mathbf{e}\} = \{\mathbf{e}\}_i$,

$$\{\dot{\mathbf{e}}\} = [\mathbf{M}]_i\{\mathbf{e}\}_i + [\mathbf{M}]_i(\{\mathbf{e}\} - \{\mathbf{e}\}_i) + [\dot{\mathbf{M}}]_i\{\mathbf{e}\}_i(t - t_i)$$
$$+ \frac{1}{2!}\left(2[\dot{\mathbf{M}}]_i(\{\mathbf{e}\} - \{\mathbf{e}\}_i)(t - t_i) + [\ddot{\mathbf{M}}]_i\{\mathbf{e}\}_i(t - t_i)^2\right) \qquad (11.10.10)$$
$$+ \frac{1}{3!}\left(3[\ddot{\mathbf{M}}]_i(\{\mathbf{e}\} - \{\mathbf{e}\}_i)(t - t_i)^2 + [\dddot{\mathbf{M}}]_i\{\mathbf{e}\}_i(t - t_i)^3\right) + \cdots$$

where the subscript i indicates evaluation at time $t = t_i$.

Equation (11.10.10) forms the basis for the local linearization algorithms that have been used to integrate the quaternion formulation. These algorithms are based on the method proposed by Pope (1963), which was originally called the exponential method. Barker, Bowles, and Williams (1973) first applied this method to the quaternion formulation using a first-order approximation to Eq. (11.10.10),

$$\{\dot{\mathbf{e}}\} \cong [\mathbf{M}]_i\{\mathbf{e}\}_i + [\mathbf{M}]_i(\{\mathbf{e}\} - \{\mathbf{e}\}_i) + [\dot{\mathbf{M}}]_i\{\mathbf{e}\}_i(t - t_i) \qquad (11.10.11)$$

In an attempt to improve the local linearization algorithm, Yen and Cook (1980) used the approximation

$$\{\dot{\mathbf{e}}\} \cong [\mathbf{M}]_i\{\mathbf{e}\}_i + [\mathbf{M}]_i(\{\mathbf{e}\} - \{\mathbf{e}\}_i) + [\dot{\mathbf{M}}]_i\{\mathbf{e}\}_i(t - t_i) + [\dot{\mathbf{M}}]_i[\mathbf{M}]_i\{\mathbf{e}\}_i(t - t_i)^2 \quad (11.10.12)$$

Equation (11.10.12) includes an approximation for the second term on the right-hand side of Eq. (11.10.10). Within the second term, this approximation implies that

$$(\{\mathbf{e}\} - \{\mathbf{e}\}_i) \cong \{\dot{\mathbf{e}}\}_i(t - t_i) = [\mathbf{M}]_i\{\mathbf{e}\}_i(t - t_i)$$

and

$$[\ddot{\mathbf{M}}] \cong 0$$

Since Eq. (11.10.12) completely ignores $[\ddot{\mathbf{M}}]$ in the second-order term of Eq. (11.10.10), this equation should not be expected to give higher-order accuracy than Eq. (11.10.11). This fact was confirmed by the numerical results of Yen and Cook (1980).

From Eqs. (11.10.9) and (11.10.10), we see that a full second-order approximation for the right-hand side of Eq. (11.10.10) would require knowledge of the second derivative of the angular rate vector. While the first derivative of the angular rate vector is always available from the equations of motion, its second derivative is not directly available in a six-degree-of-freedom flight simulation. For this reason, a complete second-order approximation to the right-hand side of Eq. (11.10.10) has not been used. Since the approximation given by Eq. (11.10.12) is no more accurate than that given by Eq. (11.10.11), only the local linearization algorithm based on Eq. (11.10.11) is presented here.

The local linearization algorithm of Barker, Bowles, and Williams (1973) has been used in aircraft flight simulation for many years. This algorithm is based on the closed-form solution to Eq. (11.10.11),

$$
\begin{aligned}
\{\mathbf{e}\}_{i+1} = \ & [\exp([\mathbf{M}]_i \, \delta t)] \{\mathbf{e}\}_i \\
& + \big([\exp([\mathbf{M}]_i \, \delta t)] - [\mathbf{i}] - [\mathbf{M}]_i \, \delta t\big)[\mathbf{M}]_i^{-1}[\mathbf{M}]_i^{-1}[\dot{\mathbf{M}}]_i \{\mathbf{e}\}_i
\end{aligned}
\tag{11.10.13}
$$

where $\delta t = t_{i+1} - t_i$. Applying Eq. (11.9.6) to replace the matrix exponential on the right-hand side of Eq. (11.10.13) gives

$$
\begin{aligned}
\{\mathbf{e}\}_{i+1} = \ & \cos(\omega_i \, \delta t/2) \{\mathbf{e}\}_i + (2/\omega_i)\sin(\omega_i \, \delta t/2)[\mathbf{M}]_i \{\mathbf{e}\}_i \\
& + [\cos(\omega_i \, \delta t/2) - 1][\mathbf{M}]_i^{-1}[\mathbf{M}]_i^{-1}[\dot{\mathbf{M}}]_i \{\mathbf{e}\}_i \\
& + [(2/\omega_i)\sin(\omega_i \, \delta t/2) - \delta t][\mathbf{M}]_i [\mathbf{M}]_i^{-1}[\mathbf{M}]_i^{-1}[\dot{\mathbf{M}}]_i \{\mathbf{e}\}_i
\end{aligned}
\tag{11.10.14}
$$

From Eq. (11.10.9), it is readily shown that the inverse of $[\mathbf{M}]$ is

$$
[\mathbf{M}]^{-1} = -(4/\omega^2)[\mathbf{M}]
\tag{11.10.15}
$$

and

$$
[\mathbf{M}]^{-1}[\mathbf{M}]^{-1} = -(4/\omega^2)[\mathbf{M}][\mathbf{M}]^{-1} = -(4/\omega^2)[\mathbf{i}]
\tag{11.10.16}
$$

Thus, using Eq. (11.10.16) in Eq. (11.10.14), the local linearization algorithm can be written as

$$
\begin{aligned}
\{\mathbf{e}\}_{i+1} = \ & \cos(\omega_i \, \delta t/2) \{\mathbf{e}\}_i + (2/\omega_i)\sin(\omega_i \, \delta t/2)[\mathbf{M}]_i \{\mathbf{e}\}_i \\
& + (4/\omega_i^2)[1 - \cos(\omega_i \, \delta t/2)][\dot{\mathbf{M}}]_i \{\mathbf{e}\}_i \\
& + (4/\omega_i^2)[\delta t - (2/\omega_i)\sin(\omega_i \, \delta t/2)][\mathbf{M}]_i [\dot{\mathbf{M}}]_i \{\mathbf{e}\}_i
\end{aligned}
\tag{11.10.17}
$$

Notice that for the special case where $[\mathbf{M}]$ is constant, Eq. (11.10.17) reduces to the exact solution given by Eq. (11.9.7).

Also note that Eq. (11.10.17) is numerically indeterminate for the special case when ω_i is zero. However, using the series expansion for the sine and the cosine, it is readily shown that when ω_i is zero, Eq. (11.10.17) reduces to

$$\{\mathbf{e}\}_{i+1} \underset{\omega_i \to 0}{=} \{\mathbf{e}\}_i + \delta t\, [\mathbf{M}]_i\, \{\mathbf{e}\}_i + (\delta t^2/2)[\dot{\mathbf{M}}]_i\, \{\mathbf{e}\}_i + (\delta t^3/6)[\mathbf{M}]_i\, [\dot{\mathbf{M}}]_i\, \{\mathbf{e}\}_i \quad (11.10.18)$$

Furthermore, from Eq. (11.10.9) we see that when ω_i is zero, all 16 components of $[\mathbf{M}]$ are also zero. Thus, even though Eq. (11.10.18) is not numerically indeterminate, it can be further simplified to give

$$\{\mathbf{e}\}_{i+1} \underset{\omega_i \to 0}{=} \{\mathbf{e}\}_i + (\delta t^2/2)[\dot{\mathbf{M}}]_i\, \{\mathbf{e}\}_i \quad (11.10.19)$$

Equations (11.10.17) and (11.10.19) provide the foundation for the local linearization algorithm.

The order of the truncation error associated with the local linearization algorithm can be evaluated by using the Taylor series expansion, which is defined in Eq. (11.9.4), for the matrix exponential in Eq. (11.10.13). This gives

$$\{\mathbf{e}\}_{i+1} = \left([\mathbf{i}] + [\mathbf{M}]_i\, \delta t + \frac{[\mathbf{M}]_i [\mathbf{M}]_i\, \delta t^2}{2!} + \cdots \right) \{\mathbf{e}\}_i$$
$$+ \left(\frac{[\mathbf{M}]_i [\mathbf{M}]_i\, \delta t^2}{2!} + \cdots \right) [\mathbf{M}]_i^{-1} [\mathbf{M}]_i^{-1} [\dot{\mathbf{M}}]_i\, \{\mathbf{e}\}_i$$

or

$$\{\mathbf{e}\}_{i+1} = \{\mathbf{e}\}_i + [\mathbf{M}]_i\, \{\mathbf{e}\}_i\, \delta t + \frac{[\mathbf{M}]_i [\mathbf{M}]_i + [\dot{\mathbf{M}}]_i}{2!} \{\mathbf{e}\}_i\, \delta t^2$$
$$+ \frac{[\mathbf{M}]_i [\mathbf{M}]_i [\mathbf{M}]_i + [\mathbf{M}]_i [\dot{\mathbf{M}}]_i}{3!} \{\mathbf{e}\}_i\, \delta t^3 + \cdots \quad (11.10.20)$$

The Taylor series expansion of the exact solution is

$$\{\mathbf{e}\}_{i+1} = \{\mathbf{e}\}_i + \{\dot{\mathbf{e}}\}_i\, \delta t + \frac{\{\ddot{\mathbf{e}}\}_i}{2!} \delta t^2 + \frac{\{\dddot{\mathbf{e}}\}_i}{3!} \delta t^3 + \cdots \quad (11.10.21)$$

From Eq. (11.10.8) we have

$$\{\dot{\mathbf{e}}\}_i = [\mathbf{M}]_i\, \{\mathbf{e}\}_i \quad (11.10.22)$$

$$\{\ddot{\mathbf{e}}\}_i = [\mathbf{M}]_i\, \{\dot{\mathbf{e}}\}_i + [\dot{\mathbf{M}}]_i\, \{\mathbf{e}\}_i = \left([\mathbf{M}]_i [\mathbf{M}]_i + [\dot{\mathbf{M}}]_i \right) \{\mathbf{e}\}_i \quad (11.10.23)$$

and

$$\{\dddot{\mathbf{e}}\}_i = [\mathbf{M}]_i\, \{\ddot{\mathbf{e}}\}_i + 2[\dot{\mathbf{M}}]_i\, \{\dot{\mathbf{e}}\}_i + [\ddot{\mathbf{M}}]_i\, \{\mathbf{e}\}_i$$
$$= \left([\mathbf{M}]_i [\mathbf{M}]_i [\mathbf{M}]_i + [\mathbf{M}]_i [\dot{\mathbf{M}}]_i + 2[\dot{\mathbf{M}}]_i [\mathbf{M}]_i + [\ddot{\mathbf{M}}]_i \right) \{\mathbf{e}\}_i \quad (11.10.24)$$

Using Eqs. (11.10.22) through (11.10.24) in Eq. (11.10.21) gives

$$\{e\}_{i+1} = \{e\}_i + [M]_i\{e\}_i\,\delta t + \frac{[M]_i[M]_i + [\dot{M}]_i}{2!}\{e\}_i\,\delta t^2$$

$$+ \frac{[M]_i[M]_i[M]_i + [M]_i[\dot{M}]_i + 2[\dot{M}]_i[M]_i + [\ddot{M}]_i}{3!}\{e\}_i\,\delta t^3 + \cdots$$

(11.10.25)

Comparing the exact expansion from Eq. (11.10.25) with the expansion of the local linearization algorithm that is given by Eq. (11.10.20), the local truncation error for this algorithm is found to be

$$R_i = \frac{2[\dot{M}]_i[M]_i + [\ddot{M}]_i}{3!}\{e\}_i\,\delta t^3 + \mathcal{O}(\delta t^4) + \cdots$$

(11.10.26)

Thus, the method produces third-order truncation error with components proportional to both the first and second derivatives of the angular rate vector. Except for the special case of constant rotation, it does not provide an exact solution, as has sometimes been stated in the literature. Since the truncation error is of order δt^3, the local linearization method provides second-order accuracy in a single-point, single-step, numerical integration algorithm.

A more widely known integration method, which was also developed to reduce computation time over that required for fourth-order Runge-Kutta, is fourth-order Adams-Bashforth-Moulton (see, for example, Hoffman 1992). This commonly used numerical integration scheme is a four-point, two-step, backward-difference method. Because starting values at four different points are required, the method must be started using a single-point method such as fourth-order Runge-Kutta. However, after the first three time steps, this method requires evaluating the time derivatives only twice for each time step. This provides fourth-order accuracy similar to that provided by the fourth-order Runge-Kutta algorithm, but at a computational cost similar to that required for second-order Runge-Kutta. Since the development of the fourth-order Adams-Bashforth-Moulton algorithm is presented in most undergraduate textbooks on numerical methods, it will not be repeated here.

In general, simulation error can be reduced by either reducing the size of the time step used for the simulation or by increasing the order of the integration method. Both approaches require increased computation time per simulated second. Figure 11.10.12 shows how total error varies with step size for four different integration algorithms: second-order Runge-Kutta, second-order local linearization, fourth-order Runge-Kutta, and fourth-order Adams-Bashforth-Moulton. The results shown in Fig. 11.10.12 were obtained from numerical integration of the quaternion formulation for the special case of a sinusoidal rotation vector,

$$\begin{Bmatrix} p \\ q \\ r \end{Bmatrix} = \begin{Bmatrix} \omega_{ax}\sin(\omega_{fx}t) \\ \omega_{ay}\sin(\omega_{fy}t) \\ \omega_{az}\sin(\omega_{fz}t) \end{Bmatrix}$$

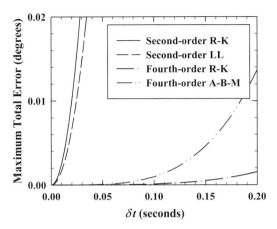

Figure 11.10.12. Comparison between second-order Runge-Kutta, second-order local linearization, fourth-order Runge-Kutta, and fourth-order Adams-Bashforth-Moulton integration.

The error was defined relative to a fourth-order Runge-Kutta solution using very small time steps. As expected, the fourth-order algorithms produce less error for a given time step than do the second-order algorithms. However, the higher-order algorithms also require a greater number of computation cycles for each time step. For this reason, it may not be directly obvious which of the four methods will result in the fastest computation for a given level of accuracy.

The second-order local linearization algorithm requires only one cycle of computation for the aerodynamic forces and moments at each time step. The second-order Runge-Kutta algorithm and the fourth-order Adams-Bashforth-Moulton algorithm each requires two computation cycles per time step, whereas the fourth-order Runge-Kutta algorithm requires four. For the same simulations that were used to produce the results in Fig. 11.10.12, the number of computation cycles per simulated second are shown as a function of step size in Fig. 11.10.13.

Because computation time depends on the method of integration as well as the step size, Fig. 11.10.12 does not present a fair comparison of the error produced by the four algorithms. Since computation time is proportional to the number of computation cycles per simulated second, a better comparison is shown in Fig. 11.10.14. In this figure, the total simulation error is plotted as a function of the number of computation cycles per simulated second. From this figure it is seen that fourth-order Adams-Bashforth-Moulton produces the smallest error for a given level of computation. Furthermore, it can be seen that a simulation using fourth-order Runge-Kutta is only slightly less efficient than one using fourth-order Adams-Bashforth-Moulton. In addition, Fig. 11.10.14 shows that the second-order algorithms are much less efficient than either of the fourth-order algorithms. These low-order algorithms should not be used, unless for some other reason very small time steps are required. The maximum total error shown in Figs. 11.10.12 and 11.10.14 was computed using

$$\omega_{ax} = \omega_{ay} = \omega_{az} = \omega_{fx} = \omega_{fy} = \omega_{fz} = 1.0 \text{ rad/sec}$$

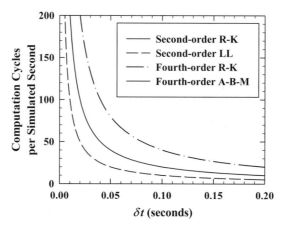

Figure 11.10.13. Computation cycles per simulated second as a function of step size for four numerical integration algorithms.

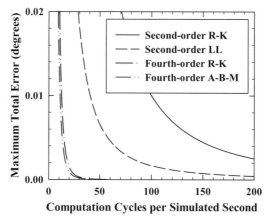

Figure 11.10.14. Simulation error as a function of computation cycles per simulated second for four numerical integration algorithms.

The absolute magnitude of the error shown in Figs. 11.10.12 and 11.10.14 depends on the angular rates. However, the relative comparison between the errors realized by the four different numerical methods is independent of these angular rates.

In some cases extremely small time steps are required for aircraft flight simulation, because of the method used to compute the aerodynamic forces and moments. When this is the case, a fourth-order integration algorithm may add unnecessarily to the computational burden. In such simulations, the second-order local linearization algorithm is by far the best alternative. Like the first-order Euler algorithm, second-order local linearization requires only one computation cycle per time step. However, it provides second-order accuracy. **In no case should a first-order Euler algorithm ever be used for aircraft flight simulation.**

11.11. Summary of the Flat-Earth Quaternion Formulation

In numerical flight simulation there is no advantage to neglecting longitudinal-lateral coupling. Thus, starting with the constant-wind Euler angle formulation presented in Eqs. (9.8.1) through (9.8.4) and expressing orientation in terms of the Euler-Rodrigues quaternion, including gyroscopic and inertial coupling, the 13 first-order differential equations that comprise this quaternion formulation for rigid-body motion are

$$
\begin{Bmatrix} \dot{u} \\ \dot{v} \\ \dot{w} \end{Bmatrix} = g \begin{Bmatrix} 2(e_x e_z - e_y e_0) \\ 2(e_y e_z + e_x e_0) \\ e_z^2 + e_0^2 - e_x^2 - e_y^2 \end{Bmatrix} + \frac{g}{W} \begin{Bmatrix} X + T_{x_b} \\ Y + T_{y_b} \\ Z + T_{z_b} \end{Bmatrix} + \begin{Bmatrix} rv - qw \\ pw - ru \\ qu - pv \end{Bmatrix} \tag{11.11.1}
$$

$$
\begin{Bmatrix} \dot{p} \\ \dot{q} \\ \dot{r} \end{Bmatrix} = \begin{bmatrix} I_{xx_b} & -I_{xy_b} & -I_{xz_b} \\ -I_{xy_b} & I_{yy_b} & -I_{yz_b} \\ -I_{xz_b} & -I_{yz_b} & I_{zz_b} \end{bmatrix}^{-1} \left(\begin{bmatrix} 0 & -h_{z_b} & h_{y_b} \\ h_{z_b} & 0 & -h_{x_b} \\ -h_{y_b} & h_{x_b} & 0 \end{bmatrix} \begin{Bmatrix} p \\ q \\ r \end{Bmatrix} \right.
$$

$$
\left. + \begin{Bmatrix} \ell + T_{z_b} y_{bp} - T_{y_b} z_{bp} + (I_{yy_b} - I_{zz_b})qr + I_{yz_b}(q^2 - r^2) + I_{xz_b} pq - I_{xy_b} pr \\ m + T_{x_b} z_{bp} - T_{z_b} x_{bp} + (I_{zz_b} - I_{xx_b})pr + I_{xz_b}(r^2 - p^2) + I_{xy_b} qr - I_{yz_b} pq \\ n + T_{y_b} x_{bp} - T_{x_b} y_{bp} + (I_{xx_b} - I_{yy_b})pq + I_{xy_b}(p^2 - q^2) + I_{yz_b} pr - I_{xz_b} qr \end{Bmatrix} \right) \tag{11.11.2}
$$

$$
\begin{Bmatrix} \dot{x}_f \\ \dot{y}_f \\ \dot{z}_f \end{Bmatrix} = \left(\begin{Bmatrix} e_0 \\ e_x \\ e_y \\ e_z \end{Bmatrix} \otimes \left(\begin{Bmatrix} 0 \\ u \\ v \\ w \end{Bmatrix} \otimes \begin{Bmatrix} e_0 \\ -e_x \\ -e_y \\ -e_z \end{Bmatrix} \right) \right) + \begin{Bmatrix} V_{wx_f} \\ V_{wy_f} \\ V_{wz_f} \end{Bmatrix} \tag{11.11.3}
$$

$$
\begin{Bmatrix} \dot{e}_0 \\ \dot{e}_x \\ \dot{e}_y \\ \dot{e}_z \end{Bmatrix} = \frac{1}{2} \begin{bmatrix} -e_x & -e_y & -e_z \\ e_0 & -e_z & e_y \\ e_z & e_0 & -e_x \\ -e_y & e_x & e_0 \end{bmatrix} \begin{Bmatrix} p \\ q \\ r \end{Bmatrix} \tag{11.11.4}
$$

Since this formulation is based on the local flat-Earth approximation, it directly applies only to short flights that do not traverse a significant fraction of the Earth's surface.

To complete the formulation, we require a model relating the aerodynamic forces, moments, and thrust to the velocity components, u, v, w, p, q, r, and their temporal derivatives. Anything from linearized aerodynamics to full CFD computations and table look-up algorithms combined with wind tunnel data can be used for this purpose.

Among some fraction of the aircraft community, the quaternion formulation has gained the somewhat undeserved reputation of being hard to understand. This reputation has more to do with the scattered nature of the literature on this topic, particularly in aircraft journals and textbooks, than it does with the raw complexity of the quaternion formulation itself.

To avoid the perceived complexity of the quaternion formulation, either the direction-cosine formulation or the Euler angle formulation has often been implemented in aircraft flight simulators. This has been done at significant computational cost. In addition to eliminating the singularity in the Euler angle formulation, the quaternion formulation is far superior to either the Euler angle formulation or the direction-cosine formulation, based on computational efficiency alone. Numerical integration of the nine component direction-cosine formulation requires more than double the computation time needed for the four component quaternion formulation. The Euler angle transformation requires about 11 times as long to evaluate as the quaternion transformation. Consequently, even in cases where the aircraft is not an all-attitude vehicle and the gimbal lock singularity is not encountered, the quaternion formulation provides important computational savings that should be seriously considered.

In order to take full advantage of the speed and accuracy of the Euler-Rodrigues quaternion formulation, it is necessary to give consideration to the numerical integration method used. Early aircraft flight simulations were run on analog computers that were inherently first-order. Numerical integration of the quaternion rate equations with a first-order method results in very large errors. Early analog programmers observed similar behavior and employed an error reduction scheme called *Corbett-Wright orthogonality control*. This orthogonality control is still found in some aircraft simulation codes that are in use today. Integration with any of today's prevalent fourth-order numerical methods produces very little orthogonality error. Using the Corbett-Wright orthogonality control scheme with modern numerical algorithms increases the computational time but does nothing to improve the accuracy of the simulation.

Even though the orthogonality error for modern numerical algorithms is very small, it can accumulate. Periodic renormalization of the quaternion eliminates this error. Corbett-Wright orthogonality control will also eliminate orthogonality error, but it does not eliminate drift error. In some cases, for certain values of the Corbett-Wright gain coefficient, the drift error is increased by more than the orthogonality error is reduced, which actually increases the total error for the simulation. Periodic renormalization of the quaternion provides a computationally efficient means for controlling orthogonality error without increasing drift error. If the orthogonality error is controlled, a small amount of drift error does not adversely affect a flight simulator, since it is constantly being corrected with virtually imperceptible pilot input. The pilot's perception of the drift error is similar to that caused by an infinitesimal change in the aerodynamics of the aircraft. Because an exact model for aircraft aerodynamics is not possible, a small amount of drift error has no significant effect on numerical flight simulation. The orthogonality error, on the other hand, has no counterpart in the physical world and cannot be compensated for by the pilot.

Reduction in size of the time step and/or increasing the order of the integration are the only effective ways to reduce total error. The fourth-order Runge-Kutta and fourth-order Adams-Bashforth-Moulton algorithms, with periodic renormalization, provide the fastest simulation for a given level of total error. If extremely small time steps are required for any reason and a single-point, single-step, algorithm is desired, the local linearization algorithm is the next best alternative. A first-order Euler algorithm should never be used for numerical integration in flight simulation.

EXAMPLE 11.11.1. Using the full 6-DOF quaternion formulation, we wish to reformulate the projectile problem in Example 7.3.3. Assume that the projectile is symmetric with respect to the pitch and yaw axes, that the products of inertia are zero, and that the rolling moment is directly proportional to the negative of the traditional dimensionless rolling rate.

Solution. From Example 7.3.3 and the symmetry of the projectile, the forces and moments are modeled as

$$
\begin{Bmatrix} X \\ Y \\ Z \end{Bmatrix} \cong \frac{1}{2}\rho u^2 S_r \begin{Bmatrix} -k_0 - k_1(\alpha^2 + \beta^2) \\ -k_2\beta \\ -k_2\alpha \end{Bmatrix} \quad \text{and} \quad \begin{Bmatrix} \ell \\ m \\ n \end{Bmatrix} \cong \frac{1}{2}\rho u^2 S_r l_r \begin{Bmatrix} -k_5 l_r \, p/u \\ -k_3\alpha - k_4 l_r \, q/u \\ +k_3\beta - k_4 l_r \, r/u \end{Bmatrix}
$$

where k_0, k_1, k_2, k_3, k_4, and k_5, are dimensionless constants, S_r is the reference area, l_r is the reference length, α is the angle of attack, and β is the sideslip angle. The aerodynamic angles will be approximated as

$$
\alpha \cong \frac{w}{u} \quad \text{and} \quad \beta \cong \frac{v}{u}
$$

Since there is no thrust and no wind, Eqs. (11.11.1) through (11.11.4) become

$$
\begin{Bmatrix} \dot{u} \\ \dot{v} \\ \dot{w} \end{Bmatrix} = \begin{Bmatrix} -K_0 u^2 - K_1(w^2 + v^2) + 2g(e_x e_z - e_y e_0) + rv - qw \\ -K_2 uv + 2g(e_y e_z + e_x e_0) + pw - ru \\ -K_2 uw + g(e_z^2 + e_0^2 - e_x^2 - e_y^2) + qu - pv \end{Bmatrix} \tag{11.11.5}
$$

$$
\begin{Bmatrix} \dot{p} \\ \dot{q} \\ \dot{r} \end{Bmatrix} = \begin{Bmatrix} -K_5 pu \\ -K_3 uw - K_4 qu + K_6 pr \\ +K_3 uv - K_4 ru - K_6 pq \end{Bmatrix} \tag{11.11.6}
$$

$$
\begin{Bmatrix} \dot{x}_f \\ \dot{y}_f \\ \dot{z}_f \end{Bmatrix} = \begin{Bmatrix} e_0 \\ e_x \\ e_y \\ e_z \end{Bmatrix} \otimes \left(\begin{Bmatrix} 0 \\ u \\ v \\ w \end{Bmatrix} \otimes \begin{Bmatrix} e_0 \\ -e_x \\ -e_y \\ -e_z \end{Bmatrix} \right) \tag{11.11.7}
$$

$$
\begin{Bmatrix} \dot{e}_0 \\ \dot{e}_x \\ \dot{e}_y \\ \dot{e}_z \end{Bmatrix} = \frac{1}{2} \begin{bmatrix} -e_x & -e_y & -e_z \\ e_0 & -e_z & e_y \\ e_z & e_0 & -e_x \\ -e_y & e_x & e_0 \end{bmatrix} \begin{Bmatrix} p \\ q \\ r \end{Bmatrix} \tag{11.11.8}
$$

where using the symmetry condition, $I_{zz_b} = I_{yy_b}$, we have

$$\begin{Bmatrix} K_0 \\ K_1 \\ K_2 \\ K_3 \\ K_4 \\ K_5 \\ K_6 \end{Bmatrix} = \begin{Bmatrix} \frac{1}{2}\rho S_r k_0\, g/W \\ \frac{1}{2}\rho S_r k_1\, g/W \\ \frac{1}{2}\rho S_r k_2\, g/W \\ \frac{1}{2}\rho S_r l_r\, k_3/I_{yy_b} \\ \frac{1}{2}\rho S_r l_r^2\, k_4/I_{yy_b} \\ \frac{1}{2}\rho S_r l_r^2\, k_5/I_{xx_b} \\ (I_{yy_b} - I_{xx_b})/I_{yy_b} \end{Bmatrix}$$

From Eq. (11.7.8) and the initial condition in Example 7.3.3, the initial condition for this formulation is

$$\begin{Bmatrix} u \\ v \\ w \\ p \\ q \\ r \\ x_f \\ y_f \\ z_f \\ e_0 \\ e_x \\ e_y \\ e_z \end{Bmatrix}_{t=0} = \begin{Bmatrix} V_0 \\ 0 \\ 0 \\ 0 \\ 0 \\ 0 \\ x_0 \\ 0 \\ z_0 \\ \cos(\theta_0/2) \\ 0 \\ \sin(\theta_0/2) \\ 0 \end{Bmatrix} \qquad (11.11.9)$$

Starting with the initial values given by Eq. (11.11.9), fourth-order Runge-Kutta can be used to integrate Eqs. (11.11.5) through (11.11.8).

EXAMPLE 11.11.2. Using the quaternion formulation from Example 11.11.1, plot the elevation of the arrow in Example 7.3.4 in inches above the starting point and its airspeed in feet per second as a function of the horizontal distance downrange in yards, from zero to 100 yards. Also plot its elevation angle and angle of attack in degrees as a function of downrange distance in yards. Compare these results with those obtained in Example 7.3.4. The aerodynamic coefficients and mass properties for this specific arrow result in $K_0 = 0.00061$ ft^{-1}, $K_1 = 0.14$ ft^{-1}, $K_2 = 0.00059$ ft^{-1}, $K_3 = 0.0016$ ft^{-2}, $K_4 = 0.0064$ ft^{-1}, $K_5 = 0.19$ ft^{-1}, and $K_6 = 0.98$. The initial values are $V_0 = 210$ ft/sec, $\alpha_0 = 0$, $\beta_0 = 0$, and $\theta_0 = 5$ degrees.

Solution. Starting with this initial condition and integrating numerically with fourth-order Runge-Kutta, using 0.01-sec time steps and applying Eq. (11.10.7) to renormalize the quaternion after each full time step, we obtain

$$
\begin{Bmatrix} t \\ u \\ v \\ w \\ p \\ q \\ r \\ x_f \\ y_f \\ z_f \\ e_0 \\ e_x \\ e_y \\ e_z \end{Bmatrix} = \begin{Bmatrix} 0.000000 \\ 210.000000 \\ 0.000000 \\ 0.000000 \\ 0.000000 \\ 0.000000 \\ 0.000000 \\ 0.000000 \\ 0.000000 \\ 0.000000 \\ 0.999048 \\ 0.000000 \\ 0.043619 \\ 0.000000 \end{Bmatrix}, \begin{Bmatrix} 0.010000 \\ 209.703259 \\ 0.000000 \\ 0.320201 \\ 0.000000 \\ -0.000535 \\ 0.000000 \\ 2.090670 \\ 0.000000 \\ -0.181301 \\ 0.999048 \\ 0.000000 \\ 0.043618 \\ 0.000000 \end{Bmatrix}, \begin{Bmatrix} 0.020000 \\ 209.406998 \\ 0.000000 \\ 0.637777 \\ 0.000000 \\ -0.002126 \\ 0.000000 \\ 4.178666 \\ 0.000000 \\ -0.359151 \\ 0.999049 \\ 0.000000 \\ 0.043612 \\ 0.000000 \end{Bmatrix}, \cdots, \begin{Bmatrix} 1.580000 \\ 176.168493 \\ 0.000000 \\ -0.141709 \\ 0.000000 \\ -0.137030 \\ 0.000000 \\ 300.310279 \\ 0.000000 \\ 11.538452 \\ 0.995957 \\ 0.000000 \\ -0.089835 \\ 0.000000 \end{Bmatrix}
$$

Using Eq. (11.7.9) to convert the components of the quaternion to Euler angles, we obtain the following results:

Time (sec)	Range (yd)	Elevation (in.)	Speed (ft/sec)	θ (deg)	α (deg)
0.000000	0.000	0.000	210.000	5.000000	0.000000
0.010000	0.697	2.176	209.704	4.999898	0.087486
0.020000	1.393	4.310	209.408	4.999185	0.174502
⋮	⋮	⋮	⋮	⋮	⋮
1.560000	98.947	−130.998	176.434	−10.148741	−0.092663
1.570000	99.525	−134.713	176.301	−10.229081	−0.070022
1.580000	100.103	−138.461	176.169	−10.308173	−0.046088

Plotting these data duplicates the results plotted in Figs. 7.3.5 and 7.3.6.

EXAMPLE 11.11.3. The vanes of the arrow in Example 11.11.2 were mounted parallel with the axis of the shaft. With the vanes mounted in this manner, no rolling moment is produced when there is no rolling rate. Often, the vanes of an arrow are mounted with some *twist*, which causes the arrow to rotate as it moves through the air. Assume that the vanes of the arrow in Example 11.11.2 are 4.0 inches long. They are mounted such that the leading edge is twisted 10 degrees around the circumference of the shaft relative to the trailing edge. The direction of this twist is that which produces a positive rolling moment with no rolling rate.

We wish to repeat the computations made in Example 11.11.2 for the case where the vanes of the arrow are mounted in this manner. Assume that the rolling moment coefficient is independent of airspeed when there is no rolling rate.

Solution. This twist in the vanes of the arrow can be modeled simply by adding a constant term to the rolling moment coefficient. Thus, the rolling moment is

$$\ell \cong \tfrac{1}{2}\rho u^2 S_r l_r (C_{\ell 0} - k_5 l_r \, p/u) \tag{11.11.10}$$

The vanes are twisted 10 degrees in 4 inches or 30 degrees per foot. With this twist, the rolling moment should be zero when the arrow is rotating 30 degrees for each foot that it moves forward through the air (i.e., $p/u = 30$ degrees per foot). Applying this to Eq. (11.11.10) gives

$$0 = C_{\ell 0} - k_5 l_r \frac{\pi}{180°} 30°/\text{ft} \quad \text{or} \quad C_{\ell 0} = k_5 \frac{\pi l_r}{6 \, \text{ft}}$$

Using this result in Eq. (11.11.10), the aerodynamic moments for this arrow are expressed as

$$\begin{Bmatrix} \ell \\ m \\ n \end{Bmatrix} \cong \tfrac{1}{2}\rho u^2 S_r l_r \begin{Bmatrix} k_5 l_r (\pi/6\,\text{ft} - p/u) \\ -k_3 \alpha - k_4 l_r \, q/u \\ +k_3 \beta - k_4 l_r \, r/u \end{Bmatrix}$$

Applying this to Eq. (11.11.2) gives

$$\begin{Bmatrix} \dot{p} \\ \dot{q} \\ \dot{r} \end{Bmatrix} = \begin{Bmatrix} K_7 u^2 - K_5 p u \\ -K_3 uw - K_4 qu + K_6 pr \\ +K_3 uv - K_4 ru - K_6 pq \end{Bmatrix} \tag{11.11.11}$$

where K_3, K_4, K_5, and K_6 are the same as in Example 11.11.2 and K_7 is given by

$$K_7 = \tfrac{1}{2}\rho S_r l_r \frac{C_{\ell 0}}{I_{xx_b}} = \frac{\pi \rho S_r l_r^2 k_5}{12\,\text{ft}\, I_{xx_b}} = \frac{\pi K_5}{6\,\text{ft}} = 0.099\,\text{ft}^{-2}$$

Replacing Eq. (11.11.6) with Eq. (11.11.11), the formulation can be integrated numerically as was done in Example 11.11.2.

Before proceeding with numerical integration, some consideration must be given to the length of the time step to be used. Because the arrow will be rotating at a substantial rate, the angular velocity must be considered in selecting the time step. From the material in Sec. 11.10, we find that a fourth-order numerical integration algorithm, with periodic renormalization of the quaternion, will provide good accuracy if the time step is selected to give about 10 steps per revolution.

For this arrow, the maximum rotation rate will be about 30 degrees per foot of travel and the maximum airspeed is 210 ft/sec. Thus, the maximum rotation rate will be somewhat less than 20 revolutions per second. With this knowledge, we shall select a time step of 0.005 second.

By using the same integration algorithm that was used in Example 11.11.2 and changing only the time step and the rolling moment equation, we obtain

$$
\begin{Bmatrix} t \\ u \\ v \\ w \\ p \\ q \\ r \\ x_f \\ y_f \\ z_f \\ e_0 \\ e_x \\ e_y \\ e_z \end{Bmatrix} =
\begin{Bmatrix} 0.000000 \\ 210.000000 \\ 0.000000 \\ 0.000000 \\ 0.000000 \\ 0.000000 \\ 0.000000 \\ 0.000000 \\ 0.000000 \\ 0.000000 \\ 0.999048 \\ 0.000000 \\ 0.043619 \\ 0.000000 \end{Bmatrix},
\begin{Bmatrix} 0.005000 \\ 209.851551 \\ 0.008189 \\ 0.160073 \\ 19.776041 \\ -0.000134 \\ 0.000007 \\ 1.045723 \\ 0.000000 \\ -0.091086 \\ 0.998722 \\ 0.025524 \\ 0.043605 \\ -0.001114 \end{Bmatrix},
\begin{Bmatrix} 0.010000 \\ 209.703258 \\ 0.061034 \\ 0.314323 \\ 35.952271 \\ -0.000526 \\ 0.000101 \\ 2.090739 \\ 0.000001 \\ -0.181306 \\ 0.994459 \\ 0.095653 \\ 0.043418 \\ -0.004176 \end{Bmatrix}, \cdots,
\begin{Bmatrix} 1.580000 \\ 175.750367 \\ -1.863796 \\ 0.988304 \\ 91.793405 \\ 0.117901 \\ -0.155337 \\ 300.030653 \\ -0.108566 \\ 11.549914 \\ -0.342722 \\ 0.935250 \\ 0.024611 \\ 0.085112 \end{Bmatrix}
$$

Time (sec)	Range (yd)	Elevation (in.)	θ (deg)	α (deg)	β (deg)
0.000000	0.000	0.000	5.000000	0.000000	0.000000
0.005000	0.349	1.093	4.999987	0.043705	0.002236
0.010000	0.697	2.176	4.999897	0.085880	0.016676
0.015000	1.045	3.248	4.999655	0.120445	0.051740
0.020000	1.393	4.310	4.999184	0.135724	0.109670
0.025000	1.741	5.361	4.998411	0.117124	0.183424
⋮	⋮	⋮	⋮	⋮	⋮
1.555000	98.567	−129.291	−9.868364	−0.725605	0.177986
1.560000	98.856	−131.135	−9.922927	−0.718113	−0.160449
1.565000	99.145	−132.988	−9.977478	−0.563182	−0.455324
1.570000	99.433	−134.850	−10.032015	−0.297676	−0.647150
1.575000	99.722	−136.720	−10.086538	0.019859	−0.699886
1.580000	100.010	−138.599	−10.141047	0.322194	−0.607610

The airspeed and elevation of the arrow do not differ substantially from those results obtained in Example 11.11.2. However, as can be seen in Fig. 11.11.1, the aerodynamic angles are significantly affected by the rotation of the arrow.

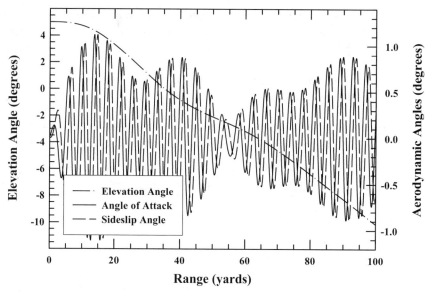

Figure 11.11.1. Effect of axial spin on the flight of an arrow.

11.12. Aircraft Position in Geographic Coordinates

The *local flat-Earth* Cartesian coordinate system (x_f, y_f, z_f) can be used to directly describe aircraft position only for relatively short flights that do not traverse a significant fraction of the Earth's surface. For longer flights, aircraft position is usually described in terms of the Earth-fixed latitude, Φ, longitude, Ψ, and geometric altitude, H. Lines of constant latitude, or parallels, are labeled in degrees north or south of the equator, with the North Pole located at 90 degrees north and the South Pole at 90 degrees south. Lines of constant longitude, or meridians, are commonly labeled in degrees east or west of the prime meridian, which runs north-south through Greenwich, England. Latitude and longitude uniquely specify any location on the Earth's surface. For example, the location of New York's John F. Kennedy International Airport (JFK) is specified in latitude and longitude as 40 degrees 38 minutes N, 73 degrees 47 minutes W.

For the purpose of numerical computation, latitude and longitude are normally expressed in radians. North latitudes are defined to be positive and south latitudes are defined as negative. By similar convention, east longitude is defined positive and west longitude is negative. Thus, the allowable range for latitude is $-\pi/2 \le \Phi \le \pi/2$ and for longitude $-\pi \le \Psi \le \pi$. With this sign convention, the location of JFK is $\Phi = 0.709185$ and $\Psi = -1.287762$.

Shape of the Earth

For the purpose of transforming the components of aircraft velocity from local flat-Earth coordinates (x_f, y_f, z_f) to geographic coordinates (Φ, Ψ, H), an *ellipsoidal-Earth model* is commonly used. This geometric model uses an ellipsoid of revolution to approximate the hypothetical surface, which is referred to as mean sea level. The actual mean-sea-level surface deviates from this reference ellipsoid primarily because mass is not distributed uniformly throughout the interior of the Earth.

The axis of revolution for the reference ellipsoid is its minor axis, which passes through the north and south poles and is referred to as the polar axis. The minor radius of the reference ellipsoid is called the polar radius and will be denoted here as R_p. Because the major radius of this reference ellipsoid lies in the equatorial plane, it is called the equatorial radius and is denoted here as R_e. A quarter section of this mean-sea-level ellipsoid, with exaggerated eccentricity, is shown in Fig. 11.12.1.

Let r_e denote the perpendicular distance from the polar axis to an arbitrary point on the surface of the reference ellipsoid and let z_e denote the perpendicular distance from the equatorial plane to this same point, as shown in Fig. 11.12.1. From the well-known equation for an ellipse, the surface of this reference ellipsoid is defined by the equation

$$\frac{r_e^2}{R_e^2} + \frac{z_e^2}{R_p^2} = 1 \qquad (11.12.1)$$

The Earth's mean-sea-level ellipsoid is often defined in terms of its equatorial radius and eccentricity rather than in terms of the equatorial and polar radii. The eccentricity of any

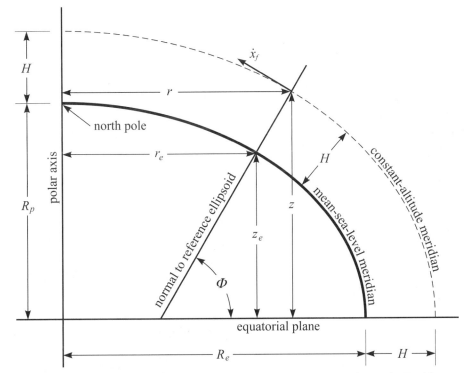

Figure 11.12.1. Quarter section of the reference ellipsoid used to approximate the Earth's mean-sea-level surface, with the eccentricity greatly exaggerated.

ellipse having a major radius a and a minor radius b is defined by $\varepsilon^2 \equiv 1 - b^2/a^2$. Thus, the polar radius can be expressed in terms of the equatorial radius and the eccentricity of the reference ellipsoid at mean sea level,

$$R_p = R_e(1 - \varepsilon^2)^{1/2} \tag{11.12.2}$$

Using Eq. (11.12.2) in Eq. (11.12.1), the surface of the reference ellipsoid is defined by the relation between r_e and z_e that must hold along this mean-sea-level surface,

$$z_e^2 = (R_e^2 - r_e^2)(1 - \varepsilon^2) \tag{11.12.3}$$

Relative to the equatorial plane, the slope of the tangent to the local meridian is found from Eq. (11.12.3) to be

$$\frac{\partial z_e}{\partial r_e} = -\frac{r_e(1 - \varepsilon^2)}{z_e} \tag{11.12.4}$$

Thus, the surface normal at an arbitrary point on the surface of the reference ellipsoid makes an angle with the equatorial plane that is given by the relation

$$\tan \Phi = \frac{\sin \Phi}{\cos \Phi} = -1 \bigg/ \frac{\partial z_e}{\partial r_e} = \frac{z_e}{r_e(1 - \varepsilon^2)} \tag{11.12.5}$$

Strictly speaking, the angle between the Earth's equatorial plane and the surface normal to the reference ellipsoid is called **geodetic latitude**. This is the latitude given on most maps and it is the latitude used in aircraft navigation and flight simulation. However, it differs from the **geocentric latitude** commonly used in astrodynamics, which is defined to be the angle between the equatorial plane and a line from the Earth's center to the point of interest. As is common practice, in this text **geodetic latitude is referred to simply as the latitude.**

Using Eq. (11.12.3) to eliminate z_e from Eq. (11.12.5) and rearranging, it is easily shown that, at any latitude, the **perpendicular distance from the Earth's polar axis to any point on the mean-sea-level ellipsoid** is

$$r_e = \frac{R_e \cos \Phi}{(1 - \varepsilon^2 \sin^2 \Phi)^{1/2}} \tag{11.12.6}$$

After using Eq. (11.12.6) to eliminate r_e from Eq. (11.12.5), it can also be shown that the **perpendicular distance from the Earth's equatorial plane to any point on the mean-sea-level ellipsoid** is

$$z_e = \frac{R_e(1 - \varepsilon^2) \sin \Phi}{(1 - \varepsilon^2 \sin^2 \Phi)^{1/2}} \tag{11.12.7}$$

The rate of change of r_e and z_e at mean sea level can be related to the rate of change of latitude by differentiating Eqs. (11.12.6) and (11.12.7). After simplifying, this yields

$$\dot{r}_e = -\frac{R_e(1-\varepsilon^2)\sin\Phi}{(1-\varepsilon^2\sin^2\Phi)^{3/2}}\dot{\Phi}$$ (11.12.8)

$$\dot{z}_e = \frac{R_e(1-\varepsilon^2)\cos\Phi}{(1-\varepsilon^2\sin^2\Phi)^{3/2}}\dot{\Phi}$$ (11.12.9)

These two velocity components can also be expressed in terms of the north component of velocity for motion along the surface of the reference ellipsoid. In terms of local flat-Earth coordinates, it is readily seen from Fig. 11.12.1 that **on the ellipsoid surface**

$$\dot{r}_e = -\dot{x}_f\sin\Phi$$ (11.12.10)

$$\dot{z}_e = \dot{x}_f\cos\Phi$$ (11.12.11)

By comparison of Eqs. (11.12.8) and (11.12.9) with Eqs. (11.12.10) and (11.12.11) it is seen that the **north component of velocity along the mean-sea-level ellipsoid is**

$$\dot{x}_f = \frac{R_e(1-\varepsilon^2)}{(1-\varepsilon^2\sin^2\Phi)^{3/2}}\dot{\Phi}$$ (11.12.12)

Because the curve defining any locus of constant latitude on the surface of the reference ellipsoid is an exact circle, the **east component of velocity along the mean-sea-level ellipsoid** is simply given by $\dot{y}_f = r_e\dot{\Psi}$, or in view of Eq. (11.12.6),

$$\dot{y}_f = \frac{R_e\cos\Phi}{(1-\varepsilon^2\sin^2\Phi)^{1/2}}\dot{\Psi}$$ (11.12.13)

Within the ellipsoidal-Earth approximation, altitude is defined to be the distance above the mean-sea-level ellipsoid measured perpendicular to this reference surface. Thus, we can write the **vertical component of velocity** as

$$\dot{z}_f = -\dot{H}$$ (11.12.14)

Combining Eqs. (11.12.12) through (11.12.14), the time derivatives of latitude, longitude, and altitude at mean sea level can be written in terms of the local flat-Earth velocity components on the surface of the reference ellipsoid,

$$\begin{Bmatrix}\dot{\Phi}\\\dot{\Psi}\\\dot{H}\end{Bmatrix} = \begin{Bmatrix}(1-\varepsilon^2\sin^2\Phi)^{3/2}\dot{x}_f/[R_e(1-\varepsilon^2)]\\(1-\varepsilon^2\sin^2\Phi)^{1/2}\dot{y}_f/(R_e\cos\Phi)\\-\dot{z}_f\end{Bmatrix}$$ (11.12.15)

Although the relations given by Eq. (11.12.15) are exact for an ellipsoidal surface, strictly speaking, they can be used only for transforming local flat-Earth velocity components at mean sea level. When the altitude is above or below the reference ellipsoid, the curvature of a constant-altitude surface differs from that of the reference ellipsoid. Furthermore, a surface that is offset everywhere normal to the mean-sea-level ellipsoid by any constant altitude is not an exact ellipsoid.

From the geometry shown in Fig. 11.12.1, it can be seen that the changes in the r- and z-coordinates associated with moving at constant latitude from mean sea level to some altitude, H, can be expressed as $\Delta r = H \cos \Phi$ and $\Delta z = H \sin \Phi$, respectively. Combining these relations with Eqs. (11.12.6) and (11.12.7), it is easily shown that the r- and z-coordinates of any point on a constant-altitude surface are given by

$$r = r_e + \Delta r = \left[\frac{R_e}{(1 - \varepsilon^2 \sin^2 \Phi)^{1/2}} + H \right] \cos \Phi \qquad (11.12.16)$$

$$z = z_e + \Delta z = \left[\frac{R_e(1 - \varepsilon^2)}{(1 - \varepsilon^2 \sin^2 \Phi)^{1/2}} + H \right] \sin \Phi \qquad (11.12.17)$$

The rate of change of r and z at constant altitude is then related to the rate of change of latitude by differentiating Eqs. (11.12.16) and (11.12.17) with respect to time at constant altitude. This gives

$$\left(\frac{\partial r}{\partial t} \right)_{H = \text{constant}} = -\left[\frac{R_e(1 - \varepsilon^2)}{(1 - \varepsilon^2 \sin^2 \Phi)^{3/2}} + H \right] \sin \Phi \; \dot{\Phi} \qquad (11.12.18)$$

$$\left(\frac{\partial z}{\partial t} \right)_{H = \text{constant}} = \left[\frac{R_e(1 - \varepsilon^2)}{(1 - \varepsilon^2 \sin^2 \Phi)^{3/2}} + H \right] \cos \Phi \; \dot{\Phi} \qquad (11.12.19)$$

Using Eqs. (11.12.16) through (11.12.19) and following a development similar to that used to obtain Eq. (11.12.15), the curvature of any constant-altitude surface is found by simply adding the altitude to the primary radii of curvature of the mean-sea-level ellipsoid for the same latitude. Thus, at any altitude, the time derivatives of latitude, longitude, and altitude can be determined from the components of aircraft velocity in local flat-Earth coordinates. With this **ellipsoidal-Earth model**, the transformation relations are

$$R_e \equiv 6{,}378.1363 \text{ km}, \quad \varepsilon^2 \equiv 0.0066943850$$

$$R_x \equiv \frac{R_e(1 - \varepsilon^2)}{(1 - \varepsilon^2 \sin^2 \Phi)^{3/2}}, \quad R_y \equiv \frac{R_e}{(1 - \varepsilon^2 \sin^2 \Phi)^{1/2}},$$

$$\begin{Bmatrix} \dot{\Phi} \\ \dot{\Psi} \\ \dot{H} \end{Bmatrix} = \begin{Bmatrix} \dot{x}_f/(R_x + H) \\ \dot{y}_f/[(R_y + H)\cos \Phi] \\ -\dot{z}_f \end{Bmatrix} \qquad (11.12.20)$$

These relations are **exact for an ellipsoidal mean-sea-level surface** and do not depend on the altitude being small compared with the primary radii of the Earth.

The quantities R_x and R_y are the *primary radii of curvature for the mean-sea-level ellipsoid* at arbitrary latitude. Within the ellipsoidal-Earth approximation, R_x and R_y are the local radii of curvature at mean sea level for north-south and east-west motion, respectively. The minimum local radius of curvature for the mean-sea-level ellipsoid occurs on the equator, where $R_x = 6{,}335.4386$ km and $R_y = R_e = 6{,}378.1363$ km. The maximum local radius of curvature occurs at the poles, where $R_x = R_y = 6{,}399.5929$ km. This is a difference of about 1 percent.

To examine how Eq. (11.12.20) could be numerically integrated over a small but finite change in aircraft position, consider a flight-path segment having initial geographic coordinates (Φ_1, Ψ_1, H_1), final geographic coordinates (Φ_2, Ψ_2, H_2), and local flat-Earth displacement components $(\Delta x_f, \Delta y_f, \Delta z_f)$. In general, the *bearing* or *ground-track angle* does not remain constant as an aircraft travels over the Earth, even along a path of zero acceleration. The bearing change that takes place along a path of zero acceleration is called the *geodesic change*, which is denoted here as $\Delta \psi_g$. Because the eccentricity of the Earth is slight, if the distance traveled over the flight segment is small compared with the local radii of curvature, we can neglect any change in curvature that takes place over the flight segment. **For small changes in latitude, longitude, and altitude**, first-order approximations for the geographic coordinates at the end of the flight segment and the geodesic change in bearing over the flight segment can be obtained from

$$\Theta_x \equiv \Delta x_f / (R_x + H_1 - \Delta z_f / 2), \quad \Theta_y \equiv \Delta y_f / (R_y + H_1 - \Delta z_f / 2)$$

$$\begin{Bmatrix} \hat{x} \\ \hat{y} \\ \hat{z} \end{Bmatrix} \equiv \begin{Bmatrix} (1 - \varepsilon^2)[\cos(\Phi_1 + \Theta_x) - \cos\Phi_1] + (1 - \varepsilon^2 \sin^2\Phi_1)\cos\Theta_y \cos\Phi_1 \\ (1 - \varepsilon^2 \sin^2\Phi_1)\sin\Theta_y \\ (1 - \varepsilon^2)[\sin(\Phi_1 + \Theta_x) - \sin\Phi_1] + (1 - \varepsilon^2 \sin^2\Phi_1)(\cos\Theta_y - \varepsilon^2)\sin\Phi_1 \end{Bmatrix}$$

$$\begin{Bmatrix} \Phi_2 \\ \Psi_2 \\ H_2 \end{Bmatrix} \cong \begin{Bmatrix} \text{atan2}[\hat{z}, (1 - \varepsilon^2)(\hat{x}^2 + \hat{y}^2)^{1/2}] \\ \Psi_1 + \text{atan2}(\hat{y}, \hat{x}) \\ H_1 \quad \Delta z_f \end{Bmatrix} \qquad (11.12.21)$$

$$\Delta \psi_g \cong (\Psi_2 - \Psi_1)\sin[(\Phi_2 + \Phi_1)/2](1 - \varepsilon^2)/(1 - \varepsilon^2 \sin^2\Phi_1)$$

Because Eq. (11.12.21) is a first-order approximation for the integral of Eq. (11.12.20), for best results it should be implemented within the fourth-order Runge-Kutta algorithm. However, because the Earth's eccentricity is extremely small and the distance traveled in one time step is much less than the local radii of curvature, implementing Eq. (11.12.21) at the end of each complete time step will usually provide satisfactory results.

The curved path described by Eq. (11.12.21) is called a *geodesic path*, which is defined to be a path of zero acceleration. Within the local flat-Earth approximation, the geodesics are straight lines. However, when the formulation is written in geographic coordinates, the geodesics are spatial curves. **Use of Eq. (11.12.21) does not assume that the aircraft follows this geodesic path in traveling from point 1 to point 2.** The transformation given by Eq. (11.12.21) simply maps the linear geodesics of the local flat-Earth approximation to the curved geodesics of the ellipsoidal-Earth approximation.

For any time step during a flight simulation the aircraft's position change in local flat-Earth coordinates can be obtained from the body-fixed velocity components and the wind vector through numerical integration of Eq. (11.11.3). The change in geographic position and the geodesic bearing change can then be determined from Eq. (11.12.21). The geodesic change in bearing $\Delta\psi_g$, which is obtained from Eq. (11.12.21), is the change in heading that takes place along a path of zero acceleration. This change in heading does not result from rotation of the aircraft. Rather, it results from a change in direction of the local meridian relative to the local geodesic. This occurs whenever the aircraft travels along any path with an east-west component, which does not exactly follow the equator. Thus, **the geodesic change in bearing that is obtained from Eq. (11.12.21) must be added to any change in heading that is predicted from integration of the aircraft equations of motion in local flat-Earth coordinates.**

At first thought, one might expect that an aircraft flying without acceleration would follow a path of constant bearing. However, this is not correct except for the special cases of north-south routes and routes that follow exactly along the equator. A path of constant geodetic bearing is called a *loxodrome* or *rhumb line*. Figure 11.12.2 shows a comparison between a geodesic path and a loxodrome on the surface of an ellipsoid, which was drawn to scale with the same eccentricity as the Earth's reference ellipsoid.

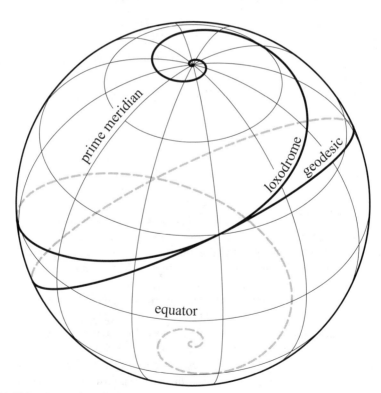

Figure 11.12.2. Comparison between a geodesic and a loxodrome on the surface of an ellipsoid with the same eccentricity as the Earth's mean-sea-level ellipsoid.

The geodesic shown in Fig. 11.12.2 was generated numerically from Eq. (11.12.21) starting at the initial geographic coordinates 30° N and 60° E with an initial heading of 70 degrees. The loxodrome shown in this figure also passes through the point 30° N and 60° E but it has a constant heading of 70 degrees. Notice that the loxodrome deviates substantially from the geodesic, especially in close proximity to the poles where the loxodrome becomes a tightly wound spiral. Solutions for both the loxodrome and the geodesic are obtained from numerical integration for the ellipsoidal-Earth model.

As mentioned previously, the Earth's actual mean-sea-level surface deviates from the reference ellipsoid primarily because mass is not distributed uniformly throughout the interior of the Earth. The vertical offset between the actual mean-sea-level surface and the reference ellipsoid is called the *undulation*, which is defined to be positive when the mean-sea-level surface is above the reference ellipsoid. The Earth's undulation varies from about −107 meters to +85 meters, which is less than 0.002 percent of the Earth's equatorial radius. Thus, the undulation is typically taken into consideration only when very precise analysis is required, and it will not be considered in this textbook. Other minor effects such as the Coriolis and centripetal accelerations associated with the Earth's rotation and gravitational forces from the sun and the moon are also ignored.

Spherical-Earth Approximation
From the scaled computer-generated geometry shown in Fig. 11.12.2, it is apparent that the eccentricity of the Earth's mean-sea-level ellipsoid is so slight that the ellipsoid cannot be visually distinguished from a sphere. The difference between the polar and equatorial radii is less than 0.34 percent and the difference between the maximum and minimum in the local radii of curvature is only about 1.0 percent. Thus, a spherical-Earth approximation is often used for aircraft flight simulation.

The historical definition of a nautical mile is the distance along a meridian on the Earth's surface that subtends 1 minute of latitude. The nautical mile is currently defined as 1,852 meters. To be consistent with the historical definition, this implies a spherical-Earth radius that is equal to $(1,852 \text{ m}) \times (10,800/\pi)$ or 6,366.707 km. This falls between the currently accepted polar radius of 6,356.7516 km and the currently accepted equatorial radius of 6,378.1363 km. Thus, a value of 6,366.707 km is commonly used for the radius of the Earth, R_E, when the spherical-Earth approximation is employed. Invoking this **spherical-Earth model**, Eq. (11.12.20) is commonly replaced with

$$R_E \equiv 6{,}366.707 \text{ km}, \quad \begin{Bmatrix} \dot{\Phi} \\ \dot{\Psi} \\ \dot{H} \end{Bmatrix} = \begin{Bmatrix} \dot{x}_f/(R_E + H) \\ \dot{y}_f/[(R_E + H)\cos\Phi] \\ -\dot{z}_f \end{Bmatrix} \qquad (11.12.22)$$

In view of the fact that available models for predicting the aerodynamic forces and moments acting on an aircraft in flight are typically less accurate than Eq. (11.12.22), the added complexity associated with using the more accurate ellipsoidal-Earth model is often difficult to justify for aircraft flight simulation. However, the ellipsoidal-Earth model is required for precisely locating positions on the Earth's surface, because terrestrial positions are nearly always given in terms of geodetic latitude, which is defined relative to the mean-sea-level ellipsoid. At the poles, the distance that subtends 1 minute of latitude at mean sea level is 1,861.566 meters. Whereas at the equator, 1 minute of latitude is subtended by only 1,842.904 meters. Although this is a difference of only 1 percent, the effect can be substantial when integrated over a long flight. Nevertheless, the spherical-Earth approximation is accurate enough for most applications of aircraft flight simulation and the algorithms used for flight simulation in geographic coordinates are easier to follow when presented in terms of the spherical-Earth approximation.

To examine how Eq. (11.12.22) can be integrated over a finite change in aircraft position, consider the three geocentric Cartesian coordinate systems that are shown in Fig. 11.12.3. The x_0-axis has its origin at the center of the Earth and points in the direction of the intersection of the equator and the local meridian at the initial position of the aircraft. The z_0-axis is aligned with the polar axis pointing in the direction of the North Pole. The y_0-axis is normal to the x_0-z_0 plane completing this right-handed Cartesian system. The x_1-y_1-z_1-coordinate system is simply a left-hand rotation of the x_0-y_0-z_0 system through an angle Φ_1 about the y_0-axis. The x_1-z_1 plane coincides with the plane of the local meridian at the initial position of the aircraft with the x_1-axis pointing outward from the center of the Earth toward the initial aircraft position. The y_1-axis is coincident with the y_0-axis. The x_2-z_2 plane is the plane of the local meridian at the final position of the aircraft with the x_2-axis pointing outward from the center of the Earth toward the final aircraft position. The y_2-axis is normal to the x_2-z_2 plane completing the right-handed orthogonal triad.

By definition, the initial aircraft position is located on the x_1-axis at an altitude H above the surface of mean-sea-level sphere. Thus, the initial aircraft position is simply described in the x_1-y_1-z_1-coordinate frame as

$$
\left\{ \begin{array}{c} x_1 \\ y_1 \\ z_1 \end{array} \right\}_{\text{initial}} = \left\{ \begin{array}{c} R_E + H \\ 0 \\ 0 \end{array} \right\} \tag{11.12.23}
$$

Consider a **local flat-Earth displacement** from this position that has a north component Δx_f, an east component Δy_f, and no change in the altitude H. The total arc length for this displacement is $(\Delta x_f^2 + \Delta y_f^2)^{1/2}$. Thus, using the **spherical-Earth approximation**, the total geocentric angle subtended by this **constant-altitude geodesic arc** is

$$
\Theta \equiv \sqrt{\Delta x_f^2 + \Delta y_f^2} \Big/ (R_E + H) \tag{11.12.24}
$$

The initial bearing or ground-track angle for this arc is the course angle at the beginning of the arc directed toward its end, as measured from north to east,

$$
\psi_{g1} \equiv \operatorname{atan2}(\Delta y_f, \Delta x_f) \tag{11.12.25}
$$

From the geometry shown in Fig. 11.12.3, we see that the x_1-coordinate of point 2 (the final aircraft position) is $(R_E + H)\cos\Theta$. We also see that the radial distance from the x_1-axis to point 2, measured parallel with the y_1-z_1 plane, is $(R_E + H)\sin\Theta$.

Because the x_1-axis is vertical at point 1 (the initial aircraft position), the x_f-y_f plane is normal to the x_1-axis and parallel with the y_1-z_1 plane. The z_1-axis points due north and the y_1-axis points due east relative to point 1. From these geometric relations, we see that the projection of the point 2 position vector on the y_1-z_1 plane has a north component of $(R_E + H)\sin\Theta\cos\psi_{g1}$ and an east component of $(R_E + H)\sin\Theta\sin\psi_{g1}$. Thus, from the geometry shown in Fig. 11.12.3, the x_1-y_1-z_1 coordinates of the final aircraft position are

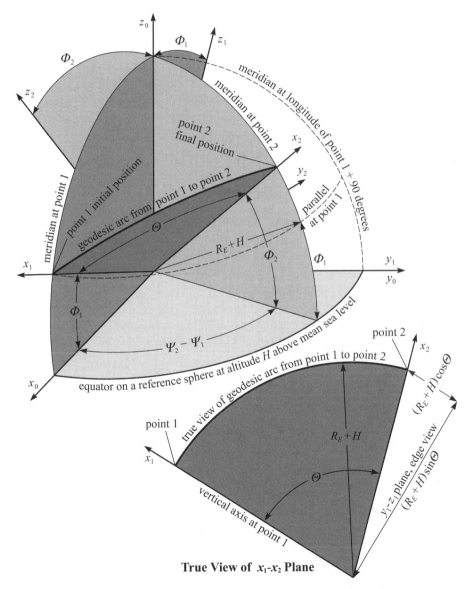

Figure 11.12.3. Isometric and true views of the geodesic arc from point 1 to point 2.

$$\left\{\begin{matrix} x_1 \\ y_1 \\ z_1 \end{matrix}\right\}_{final} = (R_E + H)\left\{\begin{matrix} \cos\Theta \\ \sin\Theta\,\sin\psi_{g1} \\ \sin\Theta\,\cos\psi_{g1} \end{matrix}\right\} \tag{11.12.26}$$

Transformation of the components of a position vector from the x_1-y_1-z_1-coordinate frame to the x_0-y_0-z_0-coordinate frame is accomplished through a single rotation,

$$\left\{\begin{matrix} x_0 \\ y_0 \\ z_0 \end{matrix}\right\} = \begin{bmatrix} \cos\Phi_1 & 0 & -\sin\Phi_1 \\ 0 & 1 & 0 \\ \sin\Phi_1 & 0 & \cos\Phi_1 \end{bmatrix}\left\{\begin{matrix} x_1 \\ y_1 \\ z_1 \end{matrix}\right\} \tag{11.12.27}$$

Defining the dimensionless coordinates

$$\left\{\begin{matrix} \hat{x} \\ \hat{y} \\ \hat{z} \end{matrix}\right\} \equiv \left\{\begin{matrix} x_0/(R_E + H) \\ y_0/(R_E + H) \\ z_0/(R_E + H) \end{matrix}\right\} \tag{11.12.28}$$

and applying Eq. (11.12.27) to the final position from Eq. (11.12.26) gives

$$\left\{\begin{matrix} \hat{x} \\ \hat{y} \\ \hat{z} \end{matrix}\right\}_{final} = \begin{bmatrix} \cos\Phi_1 & 0 & -\sin\Phi_1 \\ 0 & 1 & 0 \\ \sin\Phi_1 & 0 & \cos\Phi_1 \end{bmatrix}\left\{\begin{matrix} \cos\Theta \\ \sin\Theta\,\sin\psi_{g1} \\ \sin\Theta\,\cos\psi_{g1} \end{matrix}\right\} \tag{11.12.29}$$

The latitude and longitude for any known position are readily determined from the coordinates of the position vector in the x_0-y_0-z_0 reference frame. Using the spherical-Earth approximation and defining R to be the radial distance from the center of the Earth to the point of interest, any aircraft position can be expressed in *spherical coordinates* using the transformation

$$\left\{\begin{matrix} R \\ \Phi \\ \Psi \end{matrix}\right\} = \left\{\begin{matrix} (x_0^2 + y_0^2 + z_0^2)^{1/2} \\ \sin^{-1}(z_0/R) \\ \Psi_1 + \tan^{-1}(y_0/x_0) \end{matrix}\right\} = \left\{\begin{matrix} (x_0^2 + y_0^2 + z_0^2)^{1/2} \\ \sin^{-1}(\hat{z}) \\ \Psi_1 + \tan^{-1}(\hat{y}/\hat{x}) \end{matrix}\right\} = \left\{\begin{matrix} (x_0^2 + y_0^2 + z_0^2)^{1/2} \\ \tan^{-1}[\hat{z}/(\hat{x}^2 + \hat{y}^2)^{1/2}] \\ \Psi_1 + \tan^{-1}(\hat{y}/\hat{x}) \end{matrix}\right\} \tag{11.12.30}$$

In general, the bearing does not remain constant over the length of the arc defined by Eqs. (11.12.23) through (11.12.26). The flat-Earth bearing coincides with the geographic bearing only at the initial position, where the x_f- and y_f-axes of the local flat-Earth coordinate frame are respectively tangent to the local meridian and local parallel of the reference sphere at mean sea level. At any point along this constant-altitude geodesic arc the local bearing can be obtained from the relation

$$\tan\psi_g = \frac{\partial\Psi}{\partial\Theta}\cos\Phi \bigg/ \frac{\partial\Phi}{\partial\Theta} \tag{11.12.31}$$

The derivatives of Φ and Ψ with respect to Θ can readily be obtained by differentiating Eqs. (11.12.29) and (11.12.30),

$$\frac{\partial \Phi}{\partial \Theta} = \frac{\hat{z}'}{\cos \Phi} \tag{11.12.32}$$

and

$$\frac{\partial \Psi}{\partial \Theta} = \frac{\hat{x}\hat{y}' - \hat{y}\hat{x}'}{\hat{x}^2} \cos^2(\Psi - \Psi_1) \tag{11.12.33}$$

where

$$\begin{Bmatrix} \hat{x}' \\ \hat{y}' \\ \hat{z}' \end{Bmatrix} \equiv \frac{\partial}{\partial \Theta} \begin{Bmatrix} \hat{x} \\ \hat{y} \\ \hat{z} \end{Bmatrix} = \begin{bmatrix} \cos \Phi_1 & 0 & -\sin \Phi_1 \\ 0 & 1 & 0 \\ \sin \Phi_1 & 0 & \cos \Phi_1 \end{bmatrix} \begin{Bmatrix} -\sin \Theta \\ \cos \Theta \sin \psi_{g1} \\ \cos \Theta \cos \psi_{g1} \end{Bmatrix} \tag{11.12.34}$$

Using Eqs. (11.12.32) and (11.12.33) in Eq. (11.12.31) and applying Eqs. (11.12.29), (11.12.30), and (11.12.34) yields a relation for the bearing of a geodesic arc, like that shown in Fig. 11.12.3, as a function of the initial latitude, the initial bearing, and the geocentric angle subtended at any point along its length:

$$\tan \psi_g = \frac{\hat{x}\hat{y}' - \hat{y}\hat{x}'}{\hat{x}^2 \hat{z}'} \cos^2 \Phi \cos^2(\Psi - \Psi_1) \tag{11.12.35}$$

Thus, from Eqs. (11.12.25) and (11.12.35), the geodesic change in bearing from point 1 to point 2 on the surface of a constant-altitude sphere is given by

$$\begin{aligned} \Delta \psi_g &\equiv \psi_{g2} - \psi_{g1} \\ &= \text{atan2}[(\hat{x}\hat{y}' - \hat{y}\hat{x}')\cos^2 \Phi_2 \cos^2(\Psi_2 - \Psi_1), \hat{x}^2 \hat{z}'] \\ &\quad - \text{atan2}(\Delta y_f, \Delta x_f) \end{aligned} \tag{11.12.36}$$

For the aircraft displacement that occurs during a flight-simulation time step, we must allow for a change in altitude as well as changes in latitude and longitude. However, any change in altitude that could occur during a single flight-simulation time step is always extremely small compared with the radius of the Earth. Thus, we can use a constant-curvature approximation to compute the changes in latitude and longitude for such displacements. With this approximation, the constant-altitude relations given by Eqs. (11.12.29) and (11.12.30) are used to compute the change in latitude and longitude for each time step. However, the geocentric angle Θ subtended by the simulated flight segment is approximated from Eq. (11.12.24) by assuming a constant altitude equal to the average of the initial and final altitudes for the flight segment.

Combining the definitions from Eqs. (11.12.24) and (11.12.25) with Eqs. (11.12.29), (11.12.30), (11.12.34), and (11.12.36) provides an algorithm that can be used to compute the changes in latitude, longitude, and altitude, as well as the geodesic change in bearing, from a known displacement in local flat-Earth coordinates $(\Delta x_f, \Delta y_f, \Delta z_f)$. Starting at the initial geographic coordinates (Φ_1, Ψ_1, H_1), the algorithm is used to evaluate the final geographic coordinates (Φ_2, Ψ_2, H_2) and the geodesic bearing change $\Delta \psi_g$ from the **spherical-Earth approximation,**

$$d = \sqrt{\Delta x_f^2 + \Delta y_f^2}$$

$\text{if}\,(d < \varepsilon_t)\,\text{then}$

$\qquad \Phi_2 = \Phi_1$

$\qquad \Psi_2 = \Psi_1$

$\qquad \Delta \psi_g = 0$

else

$\qquad \Theta = d/(R_E + H_1 - \Delta z_f/2)$

$\qquad \psi_{g1} = \text{atan2}(\Delta y_f, \Delta x_f)$

$\qquad \hat{x} = \cos \Phi_1 \cos \Theta - \sin \Phi_1 \sin \Theta \cos \psi_{g1}$

$\qquad \hat{y} = \sin \Theta \sin \psi_{g1}$

$\qquad \hat{z} = \sin \Phi_1 \cos \Theta + \cos \Phi_1 \sin \Theta \cos \psi_{g1}$

$\qquad \hat{x}' = -\cos \Phi_1 \sin \Theta - \sin \Phi_1 \cos \Theta \cos \psi_{g1}$

$\qquad \hat{y}' = \quad \cos \Theta \sin \psi_{g1}$ $\qquad\qquad\qquad\qquad$ (11.12.37)

$\qquad \hat{z}' = -\sin \Phi_1 \sin \Theta + \cos \Phi_1 \cos \Theta \cos \psi_{g1}$

$\qquad \hat{r} = \sqrt{\hat{x}^2 + \hat{y}^2}$

$\qquad \Phi_2 = \text{atan2}(\hat{z}, \hat{r})$

$\qquad \Psi_2 = \Psi_1 + \text{atan2}(\hat{y}, \hat{x})$

$\qquad C = \hat{x}^2 \hat{z}'$

$\qquad S = (\hat{x} \hat{y}' - \hat{y} \hat{x}') \cos^2 \Phi_2 \cos^2 \Psi_2$

$\qquad \Delta \psi_g = \text{atan2}(S, C) - \psi_{g1}$

end if

$H_2 = H_1 - \Delta z_f$

where ε_t is a small tolerance on the order of the square root of machine precision, which is used as a test to avoid the indeterminate arctangent call for vertical flight.

The curve described by Eq. (11.12.37) is a geodesic, which was defined to be a path of zero acceleration. As discussed previously, the geodesics are straight lines within the flat-Earth approximation, but become spatial curves when the formulation is written in geographic coordinates. In the context of the spherical-Earth model, the geodesics are called *great circles*. Remember that the use of Eq. (11.12.37) does not assume that the aircraft follows a geodesic in traveling from point 1 to point 2. The transformation given by Eq. (11.12.37) simply maps the linear geodesics of the local flat-Earth approximation to the great-circle geodesics associated with the spherical-Earth approximation.

Because $\Delta\psi_g$ is the change in bearing that occurs along a path of zero acceleration, this change in heading does not result from rotation of the aircraft. It results simply from a change in direction of the local meridian relative to the local geodesic, which occurs whenever an aircraft travels any path with an east-west component that does not follow exactly along the equator. Thus, **the geodesic change in bearing that is obtained from Eq. (11.12.37) is added to any change in heading that is predicted from integration of the aircraft equations of motion in local flat-Earth coordinates.**

Because the Euler-Rodriques quaternion is used to describe aircraft orientation in Eqs. (11.11.1) through (11.11.4), we must be able to evaluate the change in the quaternion that results from a geodesic change in bearing. Using the algorithms that are presented in Sec. 11.7, this could be done by transforming the flat-Earth quaternion to Euler angles, adding the geodesic change in bearing to the azimuth angle, and then transforming the result back to a quaternion. However, this would be very inefficient. A more efficient algorithm is acquired by rearranging results obtained from Eqs. (11.6.3), (11.6.13), and (11.7.8). This yields the **quaternion transformation from flat-Earth coordinates to geographic coordinates**

$$\begin{Bmatrix} e_0 \\ e_x \\ e_y \\ e_z \end{Bmatrix}_{\text{geographic}} = \cos(\Delta\psi_g/2) \begin{Bmatrix} e_0 \\ e_x \\ e_y \\ e_z \end{Bmatrix}_{\text{flat Earth}} + \sin(\Delta\psi_g/2) \begin{Bmatrix} -e_z \\ -e_y \\ e_x \\ e_0 \end{Bmatrix}_{\text{flat Earth}} \qquad (11.12.38)$$

where $\Delta\psi_g$ is the geodesic bearing change, which could be obtained from the small-angle ellipsoidal-Earth approximation given by Eq. (11.12.21) or from the analytical solution for the spherical-Earth model that is expressed in Eq. (11.12.37). The development of Eq. (11.12.38) is left as an exercise for the student.

Within the constant-curvature spherical-Earth model, **Eqs. (11.12.37) and (11.12.38) are mathematically exact and singularity free. These relations assume only that the change in altitude over the flight segment is small compared with the radius of the Earth. They do not assume small changes in latitude and longitude.** Thus, the relations given by Eqs. (11.12.37) and (11.12.38) provide a closed-form solution for level flight. Nevertheless, numerical computations for small changes in aircraft position in geographic coordinates require evaluating small differences in large numbers. The Earth's radius is on the order of 10^8 ft. Double-precision computations provide results with 14 to 15 significant digits. Thus, double-precision computations in geographic coordinates result in a net precision on the order of 10^{-6} ft. This is approximately equivalent to the precision obtained from single-precision computations in flat-Earth coordinates. Thus, **flight simulation in geographic coordinates should always use double-precision computations and high-order numerical integration algorithms with appropriate time steps.** A good rule of thumb for fourth-order Runge-Kutta flight simulation in geographic coordinates is $0.1 \text{ ft} < V\delta t < 10 \text{ ft}$, where V is the airspeed and δt is the time step. For shorter time steps the round-off error associated with the double-precision computations is excessive and with longer time steps the truncation error for the fourth-order Runge-Kutta algorithm becomes substantial.

Navigation

Another important aspect of flight simulation is *navigation*, which also makes use of the geographic coordinate system. Navigation is the art and science of getting from one point on the Earth's surface to another. One of the simplest forms of navigation is called *rhumb-line navigation*. In reality, a rhumb line is not a line at all but a curve on the surface of the Earth that traces a path of constant bearing (i.e., a loxodrome). Rhumb-line navigation is commonly used only for traveling short distances over the Earth's surface. Even for short-distance navigation, rhumb lines are not useful for approaching locations close to either pole, because in the vicinity of a pole, rhumb lines become tightly wound spirals. This can be seen in the loxodrome example, which is shown in Fig. 11.12.2. Some rhumb-line navigation algorithms fail completely if either the start point or the destination is a pole.

The rhumb-line bearing, or ground tract, ψ_g, from any global location (Φ, Ψ, H) to any global destination (Φ_d, Ψ_d, H_d) together with the rhumb-line distance, d_r, measured over the Earth's mean-sea-level surface along this path of constant bearing, can be computed from the **spherical-Earth approximation** using the following algorithm:

$$\Delta\Psi_E = \mathrm{mod}(\Psi_d - \Psi, 2\pi); \quad \mathrm{if}(\Delta\Psi_E < 0.0)\ \Delta\Psi_E = \Delta\Psi_E + 2\pi$$

$$\Delta\Psi_W = \mathrm{mod}(\Psi - \Psi_d, 2\pi); \quad \mathrm{if}(\Delta\Psi_W < 0.0)\ \Delta\Psi_W = \Delta\Psi_W + 2\pi$$

$$\Delta\Phi = \ln\left[\tan\left(\frac{\Phi_d}{2} + \frac{\pi}{4}\right) \Big/ \tan\left(\frac{\Phi}{2} + \frac{\pi}{4}\right)\right]$$

if $\left(\left|\Phi_d - \Phi\right| < \varepsilon_t\right)$

    $r = \cos(\Phi)$

else   (11.12.39)

    $r = (\Phi_d - \Phi)/\Delta\Phi$

end if

if $(\Delta\Psi_W < \Delta\Psi_E)$

    $\psi_g = \mathrm{atan2}(-\Delta\Psi_W, \Delta\Phi); \quad \mathrm{if}(\psi_g < 0.0)\ \psi_g = \psi_g + 2\pi$

    $d_r = R_E \sqrt{(r\Delta\Psi_W)^2 + (\Phi_d - \Phi)^2}$

else

    $\psi_g = \mathrm{atan2}(\Delta\Psi_E, \Delta\Phi); \quad \mathrm{if}(\psi_g < 0.0)\ \psi_g = \psi_g + 2\pi$

    $d_r = R_E \sqrt{(r\Delta\Psi_E)^2 + (\Phi_d - \Phi)^2}$

end if

where ε_t is a small tolerance on the order of the square root of machine precision, which is used as a test to avoid the indeterminate divide by zero for east-west courses.

EXAMPLE 11.12.1. Compute the rhumb-line bearing and distance from John F. Kennedy International Airport (JFK) (40° 38′ N, 73° 47′ W) to the Los Angeles International Airport (LAX) (33° 57′ N, 118° 24′ W).

Solution. For the course from JFK to LAX the coordinates of the start point and destination are

$$\text{JFK:} \quad \Phi = (40 + 38/60)\pi/180 = 0.709185$$
$$\Psi = -(73 + 47/60)\pi/180 = -1.287762$$

$$\text{LAX:} \quad \Phi_d = (33 + 57/60)\pi/180 = 0.592539$$
$$\Psi_d = -(118 + 24/60)\pi/180 = -2.066470$$

Using these values in Eq. (11.12.39) produces

$$\left. \begin{array}{l} \Delta\Psi_E = \text{mod}(\Psi_d - \Psi, 2\pi) \\ \quad \text{if}(\Delta\Psi_E < 0.0)\ \Delta\Psi_E = \Delta\Psi_E + 2\pi \end{array} \right\} \quad \Delta\Psi_E = 5.504477$$

$$\left. \begin{array}{l} \Delta\Psi_W = \text{mod}(\Psi - \Psi_d, 2\pi) \\ \quad \text{if}(\Delta\Psi_W < 0.0)\ \Delta\Psi_W = \Delta\Psi_W + 2\pi \end{array} \right\} \quad \Delta\Psi_W = 0.778708$$

$$\Delta\Phi = \ln\left[\tan\left(\frac{\Phi_d}{2} + \frac{\pi}{4}\right) \middle/ \tan\left(\frac{\Phi}{2} + \frac{\pi}{4}\right) \right]\} \quad \Delta\Phi = -0.146801$$

$$\left. \begin{array}{l} \text{if } \left(\left|\Phi_d - \Phi\right| < \varepsilon_t\right) \\ \quad r = \cos(\Phi) \\ \text{else} \\ \quad r = (\Phi_d - \Phi)/\Delta\Phi \\ \text{end if} \end{array} \right\} \quad r = 0.794587$$

if $(\Delta\Psi_W < \Delta\Psi_E)$

$$\left. \begin{array}{l} \psi_g = \text{atan2}(-\Delta\Psi_W, \Delta\Phi) \\ \quad \text{if}(\psi_g < 0.0)\ \psi_g = \psi_g + 2\pi \end{array} \right\} \quad \psi_g = \underline{4.526057 = 259.32°}$$

$$d_r = R_E\sqrt{(r\Delta\Psi_W)^2 + (\Phi_d - \Phi)^2}\} \quad d_r = \underline{4{,}008.8 \text{ km} = 2{,}164.6 \text{ naut mi}}$$

else

$$\left. \begin{array}{l} \psi_g = \text{atan2}(\Delta\Psi_E, \Delta\Phi) \\ \quad \text{if}(\psi_g < 0.0)\ \psi_g = \psi_g + 2\pi \end{array} \right\} \quad \psi_g = 1.597459 = 91.53°$$

$$d_r = R_E\sqrt{(r\Delta\Psi_E)^2 + (\Phi_d - \Phi)^2}\} \quad d_r = 27{,}856.5 \text{ km} = 15{,}041.3 \text{ naut mi}$$

end if

Notice from Example 11.12.1 that the westerly course required to reach LAX from JFK along a path of constant bearing is not 180 degrees from the easterly course, which would accomplish the same result after traveling a much greater distance. This is because, except for the case of north-south and east-west courses, rhumb lines are not circles, even within the approximation of a spherical Earth. A rhumb line, which is not aligned east-west or north-south, is a spiral that continues to circumnavigate the globe indefinitely in coils that are ever tightening as they approach the poles (see Fig. 11.12.2). By definition, any northeasterly rhumb line will always continue in a northeasterly direction, moving farther north with each circumvolution of the Earth's axis. However, there is no way to reach the North Pole except by traveling due north. Thus, such a rhumb line on the Earth's surface is of infinite length and never intersects itself.

From this discussion, one may begin to suspect that the shortest distance between two points on the surface of the Earth is not a path of constant bearing. Of course, the shortest distance between any two points is always a straight line. However, it is not possible to travel along a straight line between two distant points on the Earth's surface without digging a substantial tunnel. If the travel must remain above ground, the shortest path connecting any two points on the Earth's surface is a segment of what is commonly called a *great circle*. Within the spherical-Earth approximation, any great circle is the intersection of the Earth's surface with a plane passing through the Earth's center. Because any three points define a unique plane, there is one and only one plane that passes through two points on the surface of the Earth as well as the Earth's center. Thus, there is one and only one great circle connecting any two points on the surface of the Earth. Within the spherical-Earth approximation, any great circle is a geodesic, similar to that shown in Fig. 11.12.2.

Since a great-circle route is the shortest path on the Earth's surface between two points, such routes are commonly used for navigation between distant locations. The initial great-circle bearing, ψ_{gc}, from any global location (Φ, Ψ, H) to any global destination (Φ_d, Ψ_d, H_d) together with the shortest great-circle distance, d_{gc}, measured over the Earth's mean-sea-level surface, can be computed from the **spherical-Earth approximation** using the following algorithm:

$$\Theta = 2\sin^{-1}\left[\sqrt{\sin^2\left(\frac{\Phi_d - \Phi}{2}\right) + \cos(\Phi)\cos(\Phi_d)\sin^2\left(\frac{\Psi_d - \Psi}{2}\right)}\right]$$

$$\text{if}\left[\sin(\Psi_d - \Psi) > 0.0\right]$$

$$\psi_{gc} = \cos^{-1}\left[\frac{\sin(\Phi_d) - \sin(\Phi)\cos(\Theta)}{\sin(\Theta)\cos(\Phi)}\right]$$

else   (11.12.40)

$$\psi_{gc} = 2\pi - \cos^{-1}\left[\frac{\sin(\Phi_d) - \sin(\Phi)\cos(\Theta)}{\sin(\Theta)\cos(\Phi)}\right]$$

end if

$$d_{gc} = R_E\Theta$$

In general, the bearing does not remain constant along such great-circle routes. The bearing varies from point to point along the route, deviating from the rhumb line toward the nearest pole. Only for exact north-south routes or a route directly along the equator does the great-circle route follow the rhumb-line route.

EXAMPLE 11.12.2. Compute the great-circle distance and the initial bearing from JFK (40° 38′ N, 73° 47′ W) to LAX (33° 57′ N, 118° 24′ W).

Solution. For the route from JFK to LAX the coordinates are

$$\text{JFK:} \quad \Phi = (40 + 38/60)\,\pi/180 = 0.709185$$
$$\Psi = -(73 + 47/60)\,\pi/180 = -1.287762$$

$$\text{LAX:} \quad \Phi_d = (33 + 57/60)\,\pi/180 = 0.592539$$
$$\Psi_d = -(118 + 24/60)\,\pi/180 = -2.066470$$

Using these values in Eq. (11.12.40) results in

$$\Theta = 2\sin^{-1}\left[\sqrt{\sin^2\left(\frac{\Phi_d - \Phi}{2}\right) + \cos(\Phi)\cos(\Phi_d)\sin^2\left(\frac{\Psi_d - \Psi}{2}\right)}\right] = 0.623585$$

$$\text{if }\left[\sin(\Psi_d - \Psi) > 0.0\right]$$
$$\psi_{gc} = \cos^{-1}\left[\frac{\sin(\Phi_d) - \sin(\Phi)\cos(\Theta)}{\sin(\Theta)\cos(\Phi)}\right]$$
$$\text{else}$$
$$\psi_{gc} = 2\pi - \cos^{-1}\left[\frac{\sin(\Phi_d) - \sin(\Phi)\cos(\Theta)}{\sin(\Theta)\cos(\Phi)}\right]$$
$$\text{end if}$$

$$\psi_{gc} = \underline{4.779727 = 273.86°}$$

$$d_{gc} = R_E\Theta = \underline{3{,}970.2 \text{ km} = 2{,}143.7 \text{ naut mi}}$$

Notice from Example 11.12.1 that the path of constant bearing from JFK to LAX is more than 10 degrees south of due west. However, from Example 11.12.2, we see that the shortest route to LAX departs JFK almost 4 degrees north of due west. For this particular example, the difference in distance between the great-circle route and the rhumb-line route is only about 21 nautical miles in more than 2,100 nautical miles. For longer routes and routes closer to the poles, the difference can be much larger. For example, the rhumb-line route from Fairbanks, Alaska (64° 50′ N, 147° 43′ W) to Moscow, Russia (55° 45′ N, 37° 35′ E) is 9,601.4 km at a bearing of 263.97 degrees. The great-circle route between these same two locations is 6,594.8 km with an initial bearing of 356.54 degrees. The rhumb-line and great-circle routes from New York to Los Angeles and from Fairbanks to Moscow are shown in Fig. 11.12.4. These course lines are drawn to scale on the surface of the mean-sea-level sphere.

Both the rhumb-line algorithm given by Eq. (11.12.39) and the great-circle algorithm given by Eq. (11.12.40) are based on the spherical-Earth approximation. Slightly more accurate results can be obtained from the ellipsoidal-Earth approximation. However, such results must be obtained from numerical integration of the differential equations governing the loxodrome and geodesic on the surface of an ellipsoid.

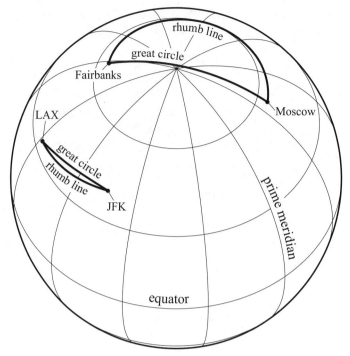

Figure 11.12.4. Comparison between rhumb-line and great-circle routes.

Since the bearing is seldom constant along a great-circle route, it is often useful to calculate the position of enroute waypoints that lie on the great circle between the start point and the destination. This is accomplished with the following algorithm:

$$\Theta = 2\sin^{-1}\left[\sqrt{\sin^2\left(\frac{\Phi_d - \Phi}{2}\right) + \cos(\Phi)\cos(\Phi_d)\sin^2\left(\frac{\Psi_d - \Psi}{2}\right)}\right]$$

$$\Theta_{wp} = d_{wp}/R_E$$

$$A = \sin(\Theta - \Theta_{wp})/\sin(\Theta)$$

$$B = \sin(\Theta_{wp})/\sin(\Theta)$$

$$x = A\cos(\Phi)\cos(\Psi) + B\cos(\Phi_d)\cos(\Psi_d)$$

$$y = A\cos(\Phi)\sin(\Psi) + B\cos(\Phi_d)\sin(\Psi_d)$$

$$z = A\sin(\Phi) + B\sin(\Phi_d)$$

$$\Phi_{wp} = \text{atan2}\left(z, \sqrt{x^2 + y^2}\right)$$

$$\Psi_{wp} = \text{atan2}(y, x)$$

(11.12.41)

where Φ_{wp} and Ψ_{wp} are, respectively, the latitude and longitude of the waypoint and d_{wp} is the distance measured along the great circle from the start point to the waypoint.

EXAMPLE 11.12.3. On the great-circle route connecting JFK and LAX, find the geographic coordinates of the waypoint that is 1,000 nautical miles from JFK.

Solution. From Example 11.12.1 or 11.12.2, the geographic coordinates are

$$\text{JFK:} \quad (0.709185, -1.287762) \qquad \text{LAX:} \quad (0.592539, -2.066470)$$

Using these values in Eq. (11.12.41) gives

$$\Theta = 2\sin^{-1}\left[\sqrt{\sin^2\left(\frac{\Phi_d - \Phi}{2}\right) + \cos(\Phi)\cos(\Phi_d)\sin^2\left(\frac{\Psi_d - \Psi}{2}\right)}\right] = 0.623585$$

$$\Theta_{wp} = d_{wp}/R_E = 0.290888$$
$$A = \sin(\Theta - \Theta_{wp})/\sin(\Theta) = 0.559283$$
$$B = \sin(\Theta_{wp})/\sin(\Theta) = 0.491144$$
$$x = A\cos(\Phi)\cos(\Psi) + B\cos(\Phi_d)\cos(\Psi_d) = -0.075245$$
$$y = A\cos(\Phi)\sin(\Psi) + B\cos(\Phi_d)\sin(\Psi_d) = -0.765932$$
$$z = A\sin(\Phi) + B\sin(\Phi_d) = 0.638503$$

$$\left.\begin{array}{l}\Phi_{wp} = \text{atan2}\left(z, \sqrt{x^2 + y^2}\right) = 0.692551 \\[2mm] \Psi_{wp} = \text{atan2}(y, x) = -1.668721\end{array}\right\} \quad \underline{(39° \ 40' \ 49'' \ \text{N}, \ 95° \ 36' \ 39'' \ \text{W})}$$

11.13. Problems

11.1. Write a computer subroutine to evaluate the quaternion product of two arbitrary quaternions. Use the general relation given by Eq. (11.6.3).

11.2. Write a computer subroutine to evaluate the components of the Euler-Rodrigues quaternion from known values of the traditional aircraft Euler angles. Use the positive sign in Eq. (11.7.8). Test your subroutine against the result that was obtained in Example 11.7.1.

11.3. Write a computer subroutine to determine the traditional aircraft Euler angles from known values of the components of the Euler-Rodrigues quaternion. Use Eq. (11.7.9) and remember that the arctangent function in most programming languages returns a value in the range $-\pi$ to π, whereas the heading angle is traditionally defined in the range 0 to 2π. Test your subroutine against the result that was obtained in Example 11.7.1.

11.4. Write a computer subroutine that uses the Euler-Rodrigues quaternion and the subroutine written for problem 11.1 to transform the components of an arbitrary vector from body-fixed coordinates to Earth-fixed coordinates. Test your subroutine against the result that was obtained in Example 11.7.1.

11.5. Modify the subroutine written for problem 11.4 to eliminate all unnecessary multiplication and assignment. Test your subroutine against the result that was obtained in Example 11.7.1.

11.6. Write a computer subroutine that uses the Euler-Rodriques quaternion and the subroutine written for problem 11.1 to transform the components of an arbitrary vector from Earth-fixed coordinates to body-fixed coordinates. Test your subroutine against the result that was obtained in Example 11.7.1.

11.7. Modify the subroutine written for problem 11.6 to eliminate all unnecessary multiplication and assignment. Test your subroutine against the result that was obtained in Example 11.7.1.

11.8. Write a computer program that uses the fourth-order Runge-Kutta numerical integration method with the Euler-Rodriques quaternion formulation to solve the problem in Example 11.9.1. Apply Eq. (11.10.7) to renormalize the quaternion after each full time step. Run this code for the case $V_v = 100$ ft/sec, $V_h = 200$ ft/sec, and $\omega = 2\pi$ sec^{-1}. Compare your numerical solution with the analytical solution obtained in Example 11.9.1.

11.9. Modify the computer program written for problem 11.8 to eliminate that part of the code, which is used to renormalize the quaternion after each full time step. Using this modified code, repeat the computations made in problem 11.8. Compare your numerical solution with the analytical solution obtained in Example 11.9.1.

11.10. Write a computer program to repeat the computations made in Example 11.11.2. Use the fourth-order Runge-Kutta numerical method to integrate the Euler-Rodriques quaternion formulation. Apply Eq. (11.10.7) to renormalize the quaternion after each full time step. Test your program against the numerical results that were presented in Example 11.11.2. To test your program even further, use an initial condition that has the arrow rotated 90 degrees about the x_b-axis. Also try an initial condition that has a heading of 225 degrees. You should be able to deduce how these solutions will differ from the one presented in Example 11.11.2.

11.11. Write a computer program to repeat the computations made in Example 11.11.3. Use the fourth-order Runge-Kutta numerical method to integrate the Euler-Rodriques quaternion formulation. Apply Eq. (11.10.7) to renormalize the quaternion after each full time step. Test your program against the numerical results that were presented in Example 11.11.3. To test your program even further, use an initial condition that has the arrow rotated 90 degrees about the x_b-axis. Also try an initial condition that has a heading of 225 degrees. You should be able to deduce how these solutions will differ from the one presented in Example 11.11.3.

11.12. Modify the computer program that was written for problem 11.11 so that it switches to the fourth-order Adams-Bashforth-Moulton numerical integration algorithm after the first three time steps. Thoroughly test your program as was done in problem 11.11.

11.13. Write a computer program to simulate the flight of an airplane with conventional controls (ailerons, elevator, and rudder). Use the fourth-order Runge-Kutta numerical integration algorithm to integrate the Euler-Rodriques quaternion formulation. Renormalize the quaternion after each full time step by applying Eq. (11.10.7). Use linearized aerodynamics to approximate the forces and moments, but be sure to design the program so that it is easy to change the mathematical model for the forces and moments without altering the rest of the program. Be sure that the inertia tensor is inverted only once, before the simulation is started.

11.14. Use the program that was written for problem 11.13 together with appropriate aerodynamic coefficients and mass properties to repeat the computations made in problem 11.11. Use the following properties for the arrow:

$$S_r = 0.15 \text{ ft}^2, \quad l_r = 2.0 \text{ ft}, \quad \rho = 2.0 \times 10^{-3} \text{ slug} \cdot \text{ft}^2, \quad W = 0.0483 \text{ lbf},$$
$$I_{xx_b} = 1.5 \times 10^{-5} \text{ slug} \cdot \text{ft}^2, \quad I_{yy_b} = I_{zz_b} = 7.5 \times 10^{-4} \text{ slug} \cdot \text{ft}^2,$$
$$k_0 = 0.0061, \quad k_1 = 1.40, \quad k_2 = 0.0059, \quad k_3 = 0.0040,$$
$$k_4 = 0.0080, \quad k_5 = 0.00475, \quad C_{\ell 0} = 0.00495$$

Thoroughly test your program as was done in problem 11.11. Be sure to try several different initial headings.

11.15. Use the program that was written for problem 11.13 together with appropriate aerodynamic coefficients and mass properties to repeat the computations made in Example 10.4.3. Be sure that the airplane is trimmed for the correct airspeed before the step change in elevator deflection is applied. Compare your numerical solution with the analytical solution presented in Example 10.4.3.

11.16. Use the program that was written for problem 11.13 together with appropriate aerodynamic coefficients and mass properties to repeat the computations made in Example 10.4.4. Be sure that the airplane is trimmed for the correct airspeed before the step change in rudder deflection is applied. Compare your numerical solution with the analytical solution presented in Example 10.4.4.

11.17. Modify the computer program written for problem 11.13 so that it switches to the fourth-order Adams-Bashforth-Moulton algorithm after the first three time steps. Also modify the program to renormalize the quaternion only after each 100 time steps. Test your modified program by using it to repeat the computations made in problems 11.14 through 11.16.

11.18. Starting with the definition of a great circle on the surface of a sphere, develop the relation for the great-circle arc length, which is expressed in Eq. (11.12.40). *Hint:* Start by developing Cartesian expressions for the vectors from the Earth's center to the start and destination points.

11.19. Starting with the quaternion rate equation given by Eq. (11.10.1), develop an expression for the time rate of change in the quaternion orthogonality error, ε, which is defined by Eq. (11.10.2).

11.20. As a first approximation for the nonlinearities associated with the forces and moments acting on an aircraft at small angles of attack, we can use the relations

$$\begin{Bmatrix} F_{x_b} \\ F_{y_b} \\ F_{z_b} \end{Bmatrix} = \begin{Bmatrix} T \\ 0 \\ 0 \end{Bmatrix} + \frac{1}{2}\rho V^2 S_w \begin{Bmatrix} -C_{D_0} - C_{D_2}C_L^2 - C_{D,\bar{q}}\bar{q} - C_{D,\delta_e}(\delta_e - \delta_{e_{\text{ref}}}) \\ C_{Y,\beta}\beta + C_{Y,\bar{p}}\bar{p} + C_{Y,\bar{r}}\bar{r} + C_{Y,\delta_a}\delta_a + C_{Y,\delta_r}\delta_r \\ -C_L \end{Bmatrix}$$

$$\begin{Bmatrix} M_{x_b} \\ M_{y_b} \\ M_{z_b} \end{Bmatrix} = \begin{Bmatrix} 0 \\ z_{T0}T \\ 0 \end{Bmatrix} + \frac{1}{2}\rho V^2 S_w \begin{Bmatrix} b_w(C_{\ell,\beta}\beta + C_{\ell,\bar{p}}\bar{p} + C_{\ell,\bar{r}}\bar{r} + C_{\ell,\delta_a}\delta_a + C_{\ell,\delta_r}\delta_r) \\ \bar{c}_w[C_{m_{\text{ref}}} + C_{m,\alpha}\alpha + C_{m,\bar{q}}\bar{q} + C_{m,\delta_e}(\delta_e - \delta_{e_{\text{ref}}})] \\ b_w(C_{n,\beta}\beta + C_{n,\bar{p}}\bar{p} + C_{n,\bar{r}}\bar{r} + C_{n,\delta_a}\delta_a + C_{n,\delta_r}\delta_r) \end{Bmatrix}$$

where

$$C_L = C_{L_{\text{ref}}} + C_{L,\alpha}\alpha + C_{L,\bar{q}}\bar{q} + C_{L,\delta_e}(\delta_e - \delta_{e_{\text{ref}}})$$

$$T = \tau(\rho/\rho_0)^a(T_0 + T'V + T''V^2)$$

and $C_{D,\bar{q}}$, C_{D,δ_e}, $C_{Y,\beta}$, $C_{Y,\bar{p}}$, $C_{Y,\bar{r}}$, C_{Y,δ_a}, C_{Y,δ_r}, $C_{L_{\text{ref}}}$, $C_{L,\alpha}$, $C_{L,\bar{q}}$, C_{L,δ_e}, $C_{\ell,\beta}$, $C_{\ell,\bar{p}}$, $C_{\ell,\bar{r}}$, C_{ℓ,δ_a}, $C_{m,\alpha}$, $C_{m,\bar{q}}$, C_{m,δ_e}, $C_{n,\beta}$, $C_{n,\bar{p}}$, $C_{n,\bar{r}}$, C_{n,δ_a}, C_{n,δ_r}, T_0, T', T'', and a are known constants, which have been determined for some level trimmed reference state where the airspeed is V_{ref}, the elevator deflection is $\delta_{e_{\text{ref}}}$, and the angle of attack, aileron, and rudder deflections are zero. The coefficients C_{D_0}, C_{D_2}, and $C_{m_{\text{ref}}}$ are constants but are not known a priori. Also known for the reference state are the drag coefficient, $C_{D_{\text{ref}}}$, and the change in drag coefficient with angle of attack, $C_{D_{\text{ref}},\alpha}$. Develop an algorithm that can be used to determine the throttle setting, τ_o, the elevator deflection, δ_{eo}, the aileron deflection, δ_{ao}, the rudder deflection, δ_{ro}, the angle of attack, α_o, and the sideslip angle, β_o, for the climbing coordinated turn described in Fig. 9.8.1. Assume that all three Euler angles are known and use the small-angle approximation for the aerodynamic angles α and β.

11.21. It is often convenient to specify a climb angle as an initial condition for flight simulation. However, if the angle of attack and/or sideslip angle are not zero, the elevation angle is not equal to the climb angle. Starting with Eq. (11.2.2), develop an expression for the elevation angle, θ, as a function of the climb angle,

γ, the bank angle, ϕ, the angle of attack, α, and the sideslip angle, β. Use the small-angle approximation for α and β. *Hint:* Equate the vertical component of Eq. (11.2.2) for the stability-axes coordinate system to that for some coordinate system where α and β are nonzero and use the relation $\cos^2\theta = 1 - \sin^2\theta$ to eliminate $\cos\theta$ from the resulting expression.

11.22. Implement the relations developed in problems 11.20 and 11.21 to obtain initial conditions for the flight simulation program written for problem 11.13. The initial conditions should start the simulation with the aircraft in steady trimmed flight at the operating condition specified by the climbing coordinated turn shown in Fig. 6.7.1. The initial airspeed, V, the initial bank angle, ϕ, and the initial climb angle, γ, should be arbitrary inputs specified by the user.

11.23. In a manner similar to that used in problem 11.21, develop an expression for the elevation angle, θ, as a function of the climb angle, γ, the bank angle, ϕ, the angle of attack, α, and the sideslip angle, β, without using the small-angle approximation for α and β. Use the definition of β that is used in Eqs. (7.1.10) through (7.1.12).

11.24. In range of linear lift, the relations given in problem 11.20 for the forces and moments acting on an aircraft at small aerodynamic angles can be extended to larger angles by using the relations

$$\begin{Bmatrix} F_{x_b} \\ F_{y_b} \\ F_{z_b} \end{Bmatrix} = \begin{Bmatrix} T \\ 0 \\ 0 \end{Bmatrix} + \frac{1}{2}\rho V^2 S_w \begin{Bmatrix} C_L\sin\alpha - C_S\sin\beta - C_D\,u/V \\ C_S\cos\beta - C_D\,v/V \\ -C_L\cos\alpha - C_D\,w/V \end{Bmatrix}$$

$$\begin{Bmatrix} M_{x_b} \\ M_{y_b} \\ M_{z_b} \end{Bmatrix} = \begin{Bmatrix} 0 \\ z_{T0}T \\ 0 \end{Bmatrix} + \frac{1}{2}\rho V^2 S_w \begin{Bmatrix} b_w C_\ell \\ \overline{c}_w C_m \\ b_w C_n \end{Bmatrix}$$

where

$$C_L = C_{L_{\text{ref}}} + C_{L,\alpha}\alpha + C_{L,\overline{q}}\overline{q} + C_{L,\delta_e}(\delta_e - \delta_{e_{\text{ref}}})$$

$$C_S = C_{Y,\beta}\beta + C_{Y,\overline{p}}\overline{p} + C_{Y,\overline{r}}\overline{r} + C_{Y,\delta_a}\delta_a + C_{Y,\delta_r}\delta_r$$

$$C_D = C_{D_0} + C_{D_2}C_L^2 + C_{D_3}C_S^2 + C_{D,\overline{q}}\overline{q} + C_{D,\delta_e}(\delta_e - \delta_{e_{\text{ref}}})$$

$$C_\ell = C_{\ell,\beta}\beta + C_{\ell,\overline{p}}\overline{p} + (C_{\ell,\overline{r}}/C_L)_{\text{ref}} C_L\overline{r} + C_{\ell,\delta_a}\delta_a + C_{\ell,\delta_r}\delta_r$$

$$C_m = C_{m_{\text{ref}}} + \left(\frac{C_{m,\alpha}}{C_{L,\alpha}}\right)_{\text{ref}}\left(C_L\frac{u}{V} - C_{L_{\text{ref}}} + C_D\frac{w}{V}\right) + C_{m,\overline{q}}\overline{q} + C_{m,\delta_e}(\delta_e - \delta_{e_{\text{ref}}})$$

$$C_n = \left(\frac{C_{n,\beta}}{C_{Y,\beta}}\right)_{\text{ref}} \left(C_S \frac{u}{V} - C_D \frac{v}{V}\right) + (C_{n,\bar{p}}/C_L)_{\text{ref}} C_L \bar{p} + C_{n,\bar{r}} \bar{r} + C_{n,\delta_a} \delta_a + C_{n,\delta_r} \delta_r$$

$$T = \tau(\rho/\rho_0)^a (T_0 + T'V + T''V^2)$$

$$V = \sqrt{u^2 + v^2 + w^2}, \quad \alpha = \tan^{-1}(w/u), \quad \beta = \tan^{-1}(v/u)$$

Modify the flight simulation program written for problem 11.22 so that it uses these relations to compute the forces and moments acting on the aircraft. Use the relations developed in problem 11.23 to specify the initial condition.

11.25. Modify the flight simulation program written for problem 11.24 so that the lift, drag, and pitching moment coefficients are computed to incorporate the effects of stall in some manner. At angles of attack below some critical value the forces and moments should be computed from the relations given in problem 11.24. At larger angles of attack, the lift and drag computations should be changed to use the algorithm that you develop. Be sure that the relations are at least first-order continuous at all angles of attack.

11.26. Starting with Eqs. (11.6.3), (11.6.13), and (11.7.8), develop the relation given in Eq. (11.12.38) for the change in the Euler-Rodriques quaternion that results from the geodesic change in bearing associated with the transformation from local flat-Earth coordinates to geographic coordinates.

11.27. Modify the flight simulation program that was written for problem 11.25 so that the aircraft position is described in terms of Earth-fixed latitude, Φ, longitude, Ψ, and geometric altitude, H. Use the spherical-Earth approximation specified by Eqs. (11.12.37) and (11.12.38).

Bibliography

Abbott, I. H., and von Doenhoff, A. E., (1949), *Theory of Wing Sections*, Dover, New York.

Abzug, M. J., (1998), *Computational Flight Dynamics*, AIAA, Reston, VA.

Abzug, M. J., and Larrabee, E. E., (1997), *Airplane Stability and Control: A History of Technologies That Made Aviation Possible*, Cambridge University Press, Cambridge.

Alley, N. R., Phillips, W. F., and Spall, R. E., (2007), "Predicting Maximum Lift Coefficient for Twisted Wings Using Computational Fluid Dynamics," *Journal of Aircraft*, **44**, 3.

Altmann, S. L., (1986), *Rotations, Quaternions, and Double Groups*, Oxford University Press, Oxford.

Altmann, S. L., (1989), "Hamilton, Rodrigues, and the Quaternion Scandal," *Mathematics Magazine*, **62**, 5.

Anderson, J. D., Jr., (2001), *Fundamentals of Aerodynamics*, 3rd ed., McGraw-Hill, New York.

Anderson, J. D., Jr., Corda, S., and Van Wie, D. M., (1980), "Numerical Lifting Line Theory Applied to Drooped Leading-Edge Wings Below and Above Stall," *Journal of Aircraft*, **17**, 12.

Anderson, R. F., (1937), *Charts for Determining the Pitching Moment of Tapered Wings with Sweepback and Twist*, NACA TR-572.

Anglin, E. L., (1965), "Analytical Study of Effects of Product of Inertia on Airplane Spin Entries, Developed Spins, and Spin Recoveries," NASA TN-D-2754.

Aramanovitch, L., (1995), "Spacecraft Orientation Based on Space Object Observations by Means of Quaternion Algebra," *Journal of Guidance, Control, and Dynamics*, **18**, 4.

Ashby, D. L., Dudley, M. R., and Iguchi, S. K., (1988), "Development and Validation of an Advanced Low-Order Panel Method," NASA TN-101024.

Ashkenas, I. L., and McRuer, D. T., (1960), "Optimization of Flight-Control, Airframe System," *Journal of Aerospace Sciences.* **27**, 3.

Babister, A. W., (1961), *Aircraft Stability and Control*, Oxford University Press, Oxford.

Babister, A. W., (1980), *Aircraft Dynamic Stability and Response*, Pergamon, Oxford.

Bairstow, L., (1939), *Applied Aerodynamics*, 2nd ed., Longmans-Green, New York.

Bar-Itzhack, I. Y., (1970), "Iterative Computation of Initial Quaternion," *Journal of Spacecraft and Rockets*, **7**, 3.

Bar-Itzhack, I. Y., (1971), "Optimum Normalization of a Computed Quaternion of Rotation," *IEEE Transactions of Aerospace and Electronic Systems*, **7**, 2.

Bar-Itzhack, I. Y., and Idan, M., (1987), "Recursive Attitude Determination from Vector Observations: Euler Angle Estimation," *Journal of Guidance, Control, and Dynamics*, **10**, 2.

Barker, L. E., Bowles, R. L., and Williams, L. H., (1973), "Development and Application of a Local Linearization Algorithm for the Integration of Quaternion Rate Equations in Real-Time Flight Simulation Problems," NASA TN D-7347.

Beatty, M. F., (1963), "Vector Representation of Rigid Body Rotation," *American Journal of Physics*, **31**, 2.

Beatty, M. F., (1967), "Kinematics of Finite, Rigid-Body Displacements," *American Journal of Physics*, **35**, 10.

Beatty, M. F., (1977), "Vector Analysis of Finite Rigid Rotations," *Journal of Applied Mechanics*, **44**, 3.

Beer, F. P., and Johnston, E. R., (1997), *Vector Mechanics for Engineers Dynamics*, 6th ed., McGraw-Hill, New York.

Bertin, J. J., (2002), *Aerodynamics for Engineers*, 4th ed., Prentice-Hall, Upper Saddle River, New Jersey.

Betz, A., (1919), "Schraubenpropeller mit geringstem Energieverlust," *Nachrichten von der Gesellschaft der Wissenschaften zu Göttingen, Mathematisch-Physikalische Klasse*, **2**.

Bloy, A. W., (1998), "Thrust Offset Effect on Longitudinal Dynamic Stability," *Journal of Aircraft*, **35**, 2.

Boyer, C. B., and Merzbach, U. C., (1989), *A History of Mathematics*, 2nd ed., Wiley, New York.

Bramwell, F. H., Fage, A., Relf, E. F., and Bryant, L. W., (1914), "Experiments on Model Propellers at the National Physical Laboratory," *Reports and Memoranda*, No. 123, British Aeronautical Research Council, London.

Broucke, R. A., (1993), "On the Use of Poincaré Surfaces of Section in Rigid Body Motion," *Journal of Astronautical Sciences*, **41**, 4.

Bryan, G. H., (1911), *Stability in Aviation*, Macmillan, London.

Bush, R. H., Power, G. D., and Towne, C. E., (1998), "WIND: The Production Flow Solver of the NPARC Alliance," AIAA-1998-0935.

Campos, L. M., (1997), "Non-linear Longitudinal Stability of a Symmetric Aircraft," *Journal of Aircraft*, **34**, 3.

Campos, L. M., Fonseca, A. A., and Azinheira, J. R., (1995), "Some Elementary Aspects of Non-linear Airplane Speed Stability in Constrained Flight," *Progress in Aerospace Science* **31**, 2.

Cayley, A., (1845), "On Certain Results Relating to Quaternions," *Philosophical Magazine*, **26**, 1.

Cheng, H., and Gupta, K. C., (1989), "An Historical Note on Finite Rotations," *ASME Journal of Applied Mathematics*, **56**, 2.

Cheng, H. K., and Meng, S. Y., (1980), "The Oblique Wing as a Lifting-Line Problem in Transonic Flow," *Journal of Fluid Mechanics*, **97**, 3.

Chou, J. C. K., (1992), "Quaternion Kinematics and Dynamic Differential Equations," *IEEE Transactions on Robotics & Automation*, **8**, 1.

Clark, T. W. K., (1913), "Effect of Side Wind on a Propeller," *Reports and Memoranda*, No. 80, British Aeronautical Research Council, London.

Cook, M. V., (1997), *Flight Dynamics Principles*, Wiley, New York.

Cooper, G. E., and Harper, R. P., Jr. (1969), "The Use of Pilot Ratings in the Evaluation of Aircraft Handling Qualities," NASA TN D-5153.

Corbett, J. P., and Wright, F. B., (1957), "Stabilization of Computer Circuits," WADC TR 57-425, E. Hochfeld, editor, Wright-Patterson Air Force Base, OH.

Crigler, J. L., and Gilman, J., (1952), "Calculation of Aerodynamic Forces on a Propeller in Pitch or Yaw," NACA TN-2685.

Croom, M., Kenney, H., Murri, D., and Lawson, K., (2000), "Research on the F/A-18E/F Using a 22%-Dynamically-Scaled Drop Model," AIAA-2000-3913.

Crowe, M. J., (1967), *A History of Vector Analysis*, Dover, New York.

Devenport, W. J., Rife, M. C., Liapis, S. I., and Follin, G. J., (1996), "The Structure and Development of the Wing-tip Vortex," *Journal of Fluid Mechanics*, **312**.

Diederich, F. W., (1951), "Charts and Tables for Use in Calculations of Downwash of Wings of Arbitrary Plan Form," NACA TN-2353.

Diehl, W. S., (1921), "The Determination of Downwash," NACA TN-42.

Duncan, W. J., (1952), *The Principles of Control and Stability of Aircraft*, Cambridge University Press, Cambridge.

Durand, W. F., and Lesley, E., (1923), "Tests on Air Propellers in Yaw," NACA TR-113.

Durand, W. F., and Lesley, E. P., (1926), "Comparison of Tests on Air Propellers in Flight with Wind Tunnel Model Tests on Similar Forms," NACA TR-220.

Dvornychenko, V. N., (1985), "The Number of Multiplications Required to Chain Coordinate Transformations," *Journal of Guidance, Control, and Dynamics*, **8**, 1.

Etkin, B., (1972), *Dynamics of Atmospheric Flight*, Wiley, New York.

Etkin, B., and Reid L. D., (1996), *Dynamics of Flight: Stability and Control*, 3rd ed., Wiley, New York.

Euler, L., (1758), "Du Mouvement de rotation des corps solides autour d'un axe variable," *Memoires de l'Académiei des Sciences de Berlin*, **14**.

Euler, L., (1770), "Problema Algebraicum ob Affectiones Prorsus Singulares Memorabile," *Novi Commentari Academiae Scientiarum Imperialis Petropolitanae*, **15**.

Euler, L., (1775a), "Formulae Generales pro Translatione Quacunque Corporum Rigidorum," *Novi Commentari Academiae Scientiarum Imperialis Petropolitanae*, **20**.

Euler, L., (1775b), "Nova Methodus Motum Corporum Rigidorum Determinandi," *Novi Commentari Academiae Scientiarum Imperialis Petropolitanae*, **20**.

Euler, L., (1862), "De Motu Corporum Circa Punctum Fixum Mobilium," *Commentatio 825 Indicis Enestroemiani, Opera Posthuma*, **2**.

Falkner, V. M., (1943), "The Calculation of Aerodynamic Loading on Surfaces of Any Shape," *Reports and Memoranda 1910*, British Aeronautical Research Council, London.

Fallon, L., (1978), "Quaternions," in *Spacecraft Attitude Determination and Control*, J. R. Wertz, editor, Kluwer Academic Publishers, Dordrecht, The Netherlands.

Fang, A. C., and Zimmerman, B. G., (1969), "Digital Simulation of Rotational Kinematics," NASA TN D-5302.

Flachsbart, O., and Kröber, G., (1930), "Experimental Investigation of Aircraft Propellers Exposed to Oblique Air Currents," NACA TM-562.

Freeman, H. B., (1932), "The Effects of Small Angles of Yaw and Pitch on the Characteristics of Airplane Propellers," NACA TR-389.

Fremaux, C. M., (1997), "Spin-Tunnel Investigation of a 1/28-Scale Model of the NASA F-18 High Alpha Research Vehicle (HARV) With and Without Tails," NASA TR-201687.

George, L., and Vernet, J. F., (1960), *La Méchanique du vol*, Béranger, Paris.

Gerald, C. F., and Wheatley, P. O., (1999), *Applied Numerical Analysis*, 6th ed., Addison Wesley Longman, Reading, MA.

Gibbs, J. W., (1901), *Vector Analysis*, E. B. Wilson, editor, Dover, New York.

Glauert, H., (1919), "The Stability Derivative of an Airscrew," *Reports and Memoranda*, No. 642, British Aeronautical Research Council, London.

Glauert, H., (1926), *The Elements of Aerofoil and Airscrew Theory*, Cambridge University Press, London.

Glauert, H., (1935), "Aeroplane Propellers, Miscellaneous Airscrew Problems," *Aerodynamic Theory*, Vol. IV, W. F. Durand, editor, Julius Springer, Berlin.

Glauert, H., (1959), *The Elements of Aerofoil and Airscrew Theory*, 2nd ed., Cambridge University Press, Cambridge.

Goett, H. J., and Delaney, N. K., (1944), "Effect of Tilt of the Propeller Axis on the Longitudinal Stability Characteristics of Single-Engine Airplanes," NACA TR-774.

Goldstein, H., (1981), *Classical Mechanics*, 2nd ed., Addison-Wesley, Reading, MA.

Goldstein, S., (1929), "On the Vortex Theory of Screw Propellers," *Proceedings of the Royal Society* **A123**, 792.

Grassman, H. G., (1862), *Die Ausdehnungslehre: vollständig und in strenger Form bearbeitet*, T. C. F. Enslin, Berlin.

Grassman, H. G., (1878), *Die lineare Ausdehnungslehre, ein neuer Zweig der Mathematik Dargestellt und durch Anwendungen auf die übrigen Zweige der Mathematik, wie auch auf die Statik, Mechanik, die Lehre vom Magnetismus und die Krystallonomie erläutert*, O. Wigand, Leipzig, Germany.

Graves, R. P., (1975), *Life of Sir William Rowan Hamilton*, Vols. 1 through 3, Arno Press, New York.

Gray, J. J., (1980), "Olinde Rodrigues' Paper of 1840 on Transformation Groups," *Archive for the History of Exact Sciences*, **21**, 4.

Green, S. I., and Acosta, A. J., (1991), "Unsteady Flow in Trailing Vortices," *Journal of Fluid Mechanics*, **227**.

Grigor'ev, V. A., and Sviatodukh, V. K., (1990), "Characteristics of the Phugoid Motion of Nonmaneuverable Aircraft," *TSAGI, Uchenye Zapiski*, **21**, 1.

Grubin, C., (1962), "Vector Representation of Rigid Body Rotation," *American Journal of Physics*, **30**, 6.

Grubin, C., (1970), "Derivation of the Quaternion Scheme via the Euler Axis and Angle," *Journal of Spacecraft and Rockets*, **7**, 10.

Grubin, C., (1979), "Quaternion Singularity Revisited," *Journal of Guidance and Control*, **2**, 3.

Hamilton, W. R., (1844), "On Quaternions: Or a New System of Imaginaries in Algebra," *Philosophical Magazine*, **25**, 3.

Hamilton, W. R., (1853), *Lectures on Quaternions*, Hodges and Smith, Dublin, Ireland.

Hamilton, W. R., (1866), *Elements of Quaternions*, Longmans-Green, London.

Hankins, T. L., (1980), *Sir William Rowan Hamilton*, Johns Hopkins University Press, Baltimore, MD.

Harris, R. G., (1918), "Forces on a Propeller Due to Sideslip," *Reports and Memoranda*, No. 427, British Aeronautical Research Council, London.

Heaviside, O., (1950), *Electromagnetic Theory*, Vol. 1, Dover, New York.

Heffley, R., and Jewell, N., (1972), "Aircraft Handling Qualities Data," NASA CR-2144.

Heller, M., Niewoehner, R. J., and Lawson, K. P., (2001), "F/A-18E/F Super Hornet High-Angle-of-Attack Control Law Development and Testing," *Journal of Aircraft*, **38**, 5.

Hibbeler, R. C., (1998), *Engineering Mechanics Dynamics*, 8th ed., Prentice Hall, Upper Saddle River, NJ.

Hoak, D. E., (1960), "USAF Stability and Control Datcom," AFWAL-TR-83-3048, Wright-Patterson AFB, OH (Revised 1978).

Hodgkinson, J., (1999), *Aircraft Handling Qualities*, AIAA, Reston, VA.

Hoffman, J. D., (1992), *Numerical Methods for Engineers and Scientists*, McGraw-Hill, New York.

Hoggard, H. P., and Hagerman, J. R., (1948), "Downwash and Wake behind Untapered Wings of Various Aspect Ratios and Angle of Sweep," NACA TN-1703.

Householder, A. S., (1964), *The Theory of Matrices in Numerical Analysis*, Blaisdell, New York.

Jacobi, D. G. J., (1969), "Bemerkungen zu einer Abhandlung Eulers über die Orthogonale Substitution," *C. G. J. Jacobis Gesammelte Werke*, Vol. III, Chelsea Publishing Company, New York.

Jiang, Y. F., and Lin, Y. P., (1991), "Error Analysis of Quaternion Transformation," *IEEE Transactions on Aerospace and Electronic Systems*, **27**, 4.

Jones, D. W., (1995), "Quaternions Quickly Transform Coordinates Without Error Buildup," *EDN*, **40**, 5.

Joukowski, N. E., (1906), "Sur les Tourbillons Adjoints," *Traraux de la Section Physique de la Societé Imperiale des Amis des Sciences Naturales*, **13**, 2.

Karamcheti, K., (1966), *Principles of Ideal-Fluid Aerodynamics*, Wiley, New York.

Katz, A., (1997), *Computational Rigid Vehicle Dynamics*, Krieger, Malabar, FL.

Katz, J., and Maskew, B., (1988), "Unsteady Low-Speed Aerodynamics Model for Complete Aircraft Configurations," *Journal of Aircraft*, **25**, 4.

Katz, J., and Plotkin, A., (2001), *Low-Speed Aerodynamics*, 2nd ed., Cambridge University Press, Cambridge.

Kline, M., (1972), *Mathematical Thoughts from Ancient to Modern Times*, Oxford University Press, Oxford.

Klumpp, A. R., (1976), "Singularity-free Extraction of a Quaternion from a Direction-Cosine Matrix," *Journal of Spacecraft and Rockets*, **13**, 12.

Klumpp, A. R., (1978), "Reply to Comment on 'Singularity Extraction of a Quaternion from a Direction-Cosine Matrix'," *Journal of Spacecraft and Rockets*, **15**, 4.

Kobayashi, O., (1992), "Static and Dynamic Flight-Path Stability of Airplanes," *Japan Society for Aeronautical and Space Sciences, Journal*, **40**.

Kokolios, A., Cook, S. P., and Niewoehner, R. J., (2005), "Use of Piloted Simulation for Evaluarion of Abrupt-Wing-Stall Characteristics," *Journal of Aircraft*, **42**, 3.

Kolk, W. R., (1961), *Modern Flight Dynamics*, Prentice-Hall, Englewood Cliffs, NJ.

Kolve, D. I., (1993), "Describing an Attitude," *Advances in the Astronautical Sciences, Proceedings of the 16th AAS Rocky Mountain Guidance and Control Conference*, Vol. 81, Univelt Inc., San Diego, CA.

Kozik, T. J., (1976), "Finite Rotations and Associated Displacement Vectors," *Journal of Applied Mechanics*, Vol. **43**, 4.

Kuethe, A. M., and Chow, C. Y., (1998), *Foundations of Aerodynamics*, 5th ed., Wiley, New York.

Kuipers, J. B., (1999), *Quaternions and Rotation Sequences: A Primer with Applications to Orbits, Aerospace and Virtual Reality*, Princeton University Press, Princeton, NJ.

Kutta, M. W., (1902), "Auftriebskräfte in Strömenden Flüssigkeiten," *Illustrierte Aeronautische Mitteilungen* **6**, 133.

Lanchester, F. W., (1908), *Aerodonetics*, Constable, London.

Lanchester, F. W., (1917), *The Flying-Machine from an Engineering Standpoint*, Constable, London.

Lecomte, P., (1962), *Méchanique du Vol*, Dunod, Paris.

Lesley, E. P., Worley, G. F., and Moy, S., (1937), "Air Propellers in Yaw," NACA TR-597.

Lotz, I., (1931), "Berechnung der Auftriebsverteilug beliebig geformter Flugel," *Zeitschrift für Flugtechnik und Motorluftschiffahrt*, **22**, 7.

Markley, F. L., (1978), "Attitude Dynamics," in *Spacecraft Attitude Determination and Control*, J. R. Wertz, editor, Kluwer Academic Publishers, Dordrecht, The Netherlands.

Markley, F. L., (1993), "New Dynamic Variables for Momentum-Bias Spacecraft," *Journal of the Astronautical Sciences*, **41**, 4.

Marks, B., Heller, M., Traven, R., and Etz, E., (2000), "Improving Single Engine Controllability with a Fly-by-Wire Flight Control System," *2000 Report to the Aerospace Profession*, Symposium Proceedings, Society of Experimental Test Pilots, Los Angeles, CA.

Mathews, J., (1976), "Coordinate-free Rotation Formalism," *American Journal of Physics*, **44**, 12.

Mattingly, J. D., (1996), *Elements of Gas Turbine Propulsion*, McGraw-Hill, New York.

Mayer, A., (1960), "Rotations and Their Algebra," *SIAM Review*, **2**, 2.

Mayo, R. A., (1979), "Relative Quaternion State Transition Relation," *Journal of Guidance and Control*, **2**, 1.

McAlister, K. W., and Takahashi, R. K., (1991), "NACA 0015 Wing Pressure and Trailing Vortex Measurements," NASA TP 3151.

McCormick, B. W., (1967, 1999), *Aerodynamics of V/STOL Flight*, Dover, Mineola, NY.

McCormick, B. W., (1995), *Aerodynamics, Aeronautics, and Flight Mechanics*, 2nd ed., Wiley, New York.

McCormick, B. W., Tangler, T. L., and Sherrieb, H. E., (1968), "Structure of Trailing Vortices," *Journal of Aircraft*, **5**, 3.

McDuffee, C. C., (1946), *The Theory of Matrices*, Chelsea Publishing Company, New York.

McFarland, R. E., (1975), "A Standard Kinematic Model for Flight Simulation at NASA-Ames," NASA CR-2497.

McLemore, H. C., and Cannon, M. D., (1954), "Aerodynamic Investigation of a Four-Blade Propeller Operating through an Angle-of-Attack Range from 0 to 180 Degrees," NACA TN-3228.

McRuer, D. T., Ashkenas, I. L., and Graham, D., (1973), *Aircraft Dynamics and Automatic Control*, Princeton University Press, Princeton, NJ.

MIL-F-8785C, (1980), *Military Specification—Flying Qualities of Piloted Airplanes*, United States Department of Defense, Philadelphia, Pennsylvania.

MIL-STD-1797A, (1995), *Military Standard—Flying Qualities of Piloted Aircraft*, United States Department of Defense, Washington DC.

Mises, R., (1945), *Theory of Flight*, Dover, New York.

Mitchell, D. G., Aponso, B., and Hoh, R. H., (1990), "Minimum Flying Qualities, Vol. I: Pilot Simulation of Multiple Axis Flying Qualities," WRDC-TR89-3125, Wright-Patterson AFB, OH.

Mitchell, E. E. L., and Rogers, A. E., (1965), "Quaternion Parameters in the Simulation of a Spinning Rigid Body," *Simulation*, **4**, 6.

Multhopp, H., (1938), "Die Berechnung der Auftriebs Verteilung von Tragflugeln," *Luftfahrtforschung*, **15**, 14.

Multhopp, H., (1950), "Method for Calculating the Lift Distribution of Wings (Subsonic Lifting Surface Theory)," *Reports and Memoranda 2884*, British Aeronautical Research Council, London.

Munk, M. M., (1924), "Analysis of W. F. Durand's and E. P. Lesley's Propellers Tests," NACA TR-175.

Munk, M. M., and Cairo, G., (1923), "Downwash of Airplane Wings," NACA TN-124.

Nelson, R. C., (1998), *Flight Stability and Automatic Control*, 2nd ed., McGraw-Hill, New York.

Niewoehner, R. J., and Kaminer, I. I., (1996a), "Integrated Aircraft-Controller Design Using Linear Matrix Inequalities," *Journal of Guidance, Control, and Dynamics*, **19**, 2.

Niewoehner, R. J., and Kaminer, I. I., (1996b), "Design of an Autoland Controller for an F-14 Aircraft Using H-Infinity Synthesis," *Journal of Guidance, Control, and Dynamics*, **19**, 3.

O'Donnell, S., (1983), *William Rowan Hamilton, Portrait of a Prodigy*, Boole Press, Dublin, Ireland.

Painlevé, P., (1910), "Étude sur le régime normal d'un avion," *Technique Aeronautique*, **1**, 1.

Pamadi, B. N., (1998), *Performance, Stability, Dynamics, and Control of Airplanes*, AIAA, Reston, VA.

Pearlman, N., (1967), "Vector Representation of Rigid-Body Rotations," *American Journal of Physics*, **35**, 12.

Pearson, H. A., and Jones, R. T., (1938), "Theoretical Stability and Control Characteristics of Wings with Various Amounts of Taper and Twist," NACA TR-635.

Pendley, R. E., (1945), "Effect of Propeller-Axis Angle of Attack on Thrust Distribution over the Propeller Disk in Relation to Wake-Survey Measurement of Thrust," NACA ARR-L5J02B.

Perkins, C. D., (1970), "Development of Airplane Stability and Control Technology," *Journal of Aircraft*, **7**, 4.

Perkins, C. D., and Hage, R. E., (1949), *Airplane Performance Stability and Control*, Wiley, New York.

Phillips, W. F., (2000a), "Phugoid Approximation for Conventional Airplanes," *Journal of Aircraft*, **37**, 1.

Phillips, W. F., (2000b), "Improved Closed-Form Approximation for Dutch Roll," *Journal of Aircraft*, **37**, 3.

Phillips, W. F., (2002a), "Propeller Momentum Theory with Slipstream Rotation," *Journal of Aircraft*, **39**, 1.

Phillips, W. F., (2002b), "An Analytical Solution for Wing Dihedral Effect," *Journal of Aircraft*, **39**, 3.

Phillips, W. F., (2002c), "Estimating the Low-Speed Sidewash Gradient on a Vertical Stabilizer," *Journal of Aircraft*, **39**, 6.

Phillips, W. F., (2004), "Lifting-Line Analysis for Twisted Wings and Washout-Optimized Wings," *Journal of Aircraft*, **41**, 1.

Phillips, W. F., and Alley, N. R., (2007), "Predicting Maximum Lift Coefficient for Twisted Wings Using Lifting-Line Theory," *Journal of Aircraft*, **44**, 3.

Phillips, W. F., Alley, N. R., and Goodrich, W. D., (2004), "Lifting-Line Analysis of Roll Control and Variable Twist," *Journal of Aircraft*, **41**, 5.

Phillips, W. F., Alley, N. R., and Niewoehner, R. J., (2008), "Effects of Nonlinearities on Subsonic Aerodynamic Center," *Journal of Aircraft*, **45**, 4.

Phillips, W. F., and Anderson, E. A., (2002), "An Analytical Approximation for the Mechanics of Airplane Spin," *Journal of Aircraft*, **39**, 6.

Phillips, W. F., Anderson, E. A., Jenkins, J. C., and Sunouchi, S., (2002), "Estimating the Low-Speed Downwash Angle on an Aft Tail," *Journal of Aircraft*, **39**, 4.

Phillips, W. F., Anderson, E. A., and Kelly, Q. J., (2003), "Predicting the Contribution of Running Propellers to Aircraft Stability Derivatives," *Journal of Aircraft*, **40**, 6.

Phillips, W. F., Fugal, S. R., and Spall, R. E., (2006), "Minimizing Induced Drag with Wing Twist, Computational-Fluid-Dynamics Validation," *Journal of Aircraft*, **43**, 2.

Phillips, W. F., Hailey, C. E., and Gebert, G. A., (2000), "A Review of Attitude Kinematics for Aircraft Flight Simulation," AIAA-2000-4302.

Phillips, W. F., Hailey, C. E., and Gebert, G. A., (2001), "A Review of Attitude Representations Used for Aircraft Kinematics," *Journal of Aircraft*, **38**, 4 (also see Errata, *Journal of Aircraft*, **40**, 1).

Phillips, W. F., Hansen, A. B., and Nelson, W. M., (2006), "Effects of Tail Dihedral on Static Stability," *Journal of Aircraft*, **43**, 6.

Phillips, W. F., Hunsaker, D. F., and Niewoehner, R. J., (2008), "Estimating the Subsonic Aerodynamic Center and Moment Components for Swept Wings," *Journal of Aircraft*, **45**, 3.

Phillips, W. F., and Niewoehner, R. J., (2006), "Effect of Propeller Torque on Minimum-Control Airspeed," *Journal of Aircraft*, **43**, 5.

Phillips, W. F., and Niewoehner, R. J., (2009), "Characteristic Length and Dynamic Time Scale Associated with Aircraft Pitching Motion," *Journal of Aircraft*, **46**, 2.

Phillips, W. F., and Santana, B. W., (2002), "Aircraft Small-Disturbance Theory with Longitudinal-Lateral Coupling," *Journal of Aircraft*, **39**, 6.

Phillips, W. F., and Snyder, D. O., (2000), "Modern Adaptation of Prandtl's Classic Lifting-Line Theory," *Journal of Aircraft*, **37**, 4.

Pistolesi, E., (1928), "Nuove Considerazioni sul Problema dell'elica in un Vento Laterale," *L'Aerotecnica*, **7**, 3.

Pope, D. A., (1963), "An Exponential Method of Numerical Integration of Ordinary Differential Equations," *Communications ACM*, **6**, 8.

Prandtl, L., (1918), "Tragflügel Theorie," *Nachricten von der Gesellschaf der Wisseschaften zu Göttingen*, Geschäeftliche Mitteilungen, Klasse.

Prandtl, L., (1921), *Applications of Modern Hydrodynamics to Aeronautics*, NACA TR-116.

Prandtl, L., and Betz, A., (1927), *Vier Abhandlungen zur Hydrodynamik und Aerodynamik*, Göttingen, Germany.

Press, W. H., Flannery, B. P., Teukolsky, S. A., and Vetterling, W. T., (1986), *Numerical Recipes—The Art of Scientific Computing*, Cambridge University Press, Cambridge.

Purser, P. E., and Campbell, J. P., (1945), "Experimental Verification of a Simplified Vee-Tail Theory and Analysis of Available Data on Complete Models with Vee Tails," NACA TR-823.

Ramaprian, B. R., and Zheng, Y., (1997), "Measurements in Rollup Region of the Tip Vortex from a Rectangular Wing," *AIAA Journal*, **35**, 12.

Rasmussen, M. L., and Smith, D. E., (1999), "Lifting-Line Theory for Arbitrary Shaped Wings," *Journal of Aircraft*, **36**. 2.

Ribner, H. S., (1943a), "Formulas for Propellers in Yaw and Charts of the Side-Force Derivative," NACA ARR-3E19.

Ribner, H. S., (1943b). "Propellers in Yaw," NACA ARR-3L09.

Ribner, H. S., (1943c), "Proposal for a Propeller Side-Force Factor," NACA RB-3L02.

Ribner, H. S., (1945a), "Propellers in Yaw," NACA TR-820.

Ribner, H. S., (1945b), "Formulas for Propellers in Yaw and Charts of the Side-Force Derivative," NACA TR-819.

Roberson, R. E., (1968), "Kinematic Equations for Bodies Whose Rotation Is Described by the Euler-Rodrigues Parameters," *AIAA Journal*, **6**, 5.

Robinson, A. C., (1958a), "On the Use of Quaternions in the Simulation of Rigid Body Motion," WADC TR 58-17, Wright-Patterson AFB, OH.

Robinson, A. C., (1958b), "On the Use of Quaternions in the Simulation of Rigid Body Motion," *Proceedings of the Simulation Council Session of the Eastern Joint Computer Conference*, Washington, DC, Macmillan, New York.

Rodrigues, O., (1840), "Des Lois géométriques qui régissent les déplacements d'un système solide dans l'espace, et de la variation des coordonnées provenant de ses déplacements consideréés indépendamment des causes qui peuvent les produire," *Journal des Mathématiques Pures et Appliquees*, **5**.

Rolfe, J. M., and Staples, K. J., (1988), *Flight Simulation*, Cambridge University Press, Cambridge.

Roskam, J., (1990), *Airplane Design Part VI: Preliminary Calculations of Aerodynamic, Thrust and Power Characteristics*, DAR Corporation, Lawrence, KS.

Roskam, J., (1998), *Airplane Flight Dynamics and Automatic Flight Control*, DAR Corporation, Lawrence, KS.

Rumph, L. B., White, R. J., and Grummann, H. R., (1942), "Propeller Forces Due to Yaw and Their Effect on Airplane Stability," *Journal of the Aeronautical Sciences*, **9**, 12.

Runckel, J. F., (1942), "The Effect of Pitch on Force and Moment Characteristics of Full-Scale Propellers of Five Solidities," NACA WR-L-446.

Saffman, P. G., (1992), *Vortex Dynamics*, Cambridge University Press, Cambridge.

Satake, I., (1975), *Linear Algebra*, Marcel Dekker, Inc., New York.

Schade, R. O., (1947), "Effects of Geometric Dihedral on the Aerodynamic Characteristics of Two Isolated Vee-Tail Surfaces," NACA TN-1369.

Schmidt, L. V., (1998), *Introduction to Aircraft Flight Dynamics*, AIAA, Reston, VA.

Shepperd, S. W., (1978), "Quaternion from Rotation Matrix," *Journal of Guidance and Control*, **1**, 3.

Shuster, M. D., (1993a), "Quaternion on the Kalman Filter," *Advances in the Astronautical Sciences*, **85**, 1.

Shuster, M. D., (1993b), "A Survey of Attitude Representations," *Journal of the Astronautical Sciences*, **14**, 4.

Shuster, M. D., and Natanson, G. A., (1993), "Quaternion Computation from a Geometric Point of View," *Journal of the Astronautical Sciences*, **41**, 4.

Sidi, M. J., (1997), *Spacecraft Dynamics and Control*, Cambridge University Press, Cambridge.

Silverstein, A., and Katzoff, S., (1939), "Design Charts for Predicting Downwash Angles and Wake Characteristics behind Plain and Flapped Wings," NACA TR-648.

Silverstein, A., Katzoff, S., and Bullivant, W., K., (1939), "Downwash and Wake behind Plain and Flapped Airfoils," NACA TR-651.

Spurrier, R. A., (1978), "Comment on 'Singularity-free Extraction of a Quaternion from a Direction-Cosine Matrix,'" *Journal of Spacecraft and Rockets*, **15**, 4.

Stevens, B. L., and Lewis, F. L., (1992), *Aircraft Control and Simulation*, Wiley, New York.

Stough, H. P., DiCarlo, D. J., and Patton, J. M., (1985), "Flight Investigation of Stall, Spin, and Recovery Characteristics of a Low-Wing, Single-Engine, T-Tail Light Airplane," NASA TR-2427.

Stuelpnagel, J., (1964), "On the Parameterization of the Three-Dimensional Rotation Group," *SIAM Review*, **6**, 4.

Tandon, G. K., (1978), "Coordinate Transformations," in *Spacecraft Attitude Determination and Control*, J. R. Wertz, editor, Kluwer Academic Publishers, Dordrecht, The Netherlands.

Van der Waerden, B. L., (1985), *A History of Algebra*, Springer-Verlag, Heidelberg, Germany.

Vathsal, S., (1991), "Derivation of the Relative Quaternion Differential Equations," *Journal of Guidance, Control, and Dynamics*, **15**, 5.

Vogeley, A. W., (1948), "Calculation of the Effect of Thrust-Axis Inclination on Propeller Disk Loading and Comparison with Flight Measurements," NACA TN-1721.

Weber, J., and Brebner, G. G., (1958), "Low-Speed Tests on 45-deg Swept-Back Wings, Part I: Pressure Measurements on Wings of Aspect Ratio 5," *Reports and Memoranda* 2882, British Aeronautical Research Council, London.

Wehage, R. A., (1984), "Quaternions and Euler Parameters—A Brief Exposition," *Computer-Aided Analysis and Optimization of Mechanical System Dynamics*,

NATO ASI Series, Vol. F9, E. J. Haug, editor, Springer-Verlag, Heidelberg, Germany.

Weil, J., and Sleeman, W. C., (1949), "Prediction of the Effects of Propeller Operation on the Static Longitudinal Stability of Single-Engine Tractor Monoplanes with Flaps Retracted," NACA TR-941.

Whittaker, E. T., (1937), *A Treatise on the Analytical Dynamics of Particles and Rigid Bodies*, 4th ed., Cambridge University Press, Cambridge.

Wilkinson, J. H., (1965), *The Algebraic Eigenvalue Problem*, Clarendon Press, Oxford.

Winograd, S., (1970), "On the Number of Multiplications Necessary to Compute Certain Functions," *Communications on Pure and Applied Mathematics*, **23**, 2.

Winograd, S., (1977), "Some Bilinear Forms Whose Multiplicative Complexity Depends on the Field of Constants," *Mathematical Systems Theory*, **10**, 2.

Xu, R., (1995), "An Exploration on the Analytical-Solution for the Longitudinal Phugoid Mode of Aircraft," *Acta Aerodynamica Sinica*, **13**.

Yen, K., and Cook, G., (1980), "Improved Local Linearization Algorithm for Solving the Quaternion Equations," *Journal of Guidance and Control*, **3**, 5.

Appendix A
Standard Atmosphere, SI Units

Geometric Altitude (m)	Geopotential Altitude (m)	Temperature (K)	Pressure (N/m^2)	Density (kg/m^3)	Speed of Sound (m/sec)
0	0	288.150	1.0133E+05	1.2250E+00	340.29
2,000	1,999	275.154	7.9501E+04	1.0066E+00	332.53
4,000	3,997	262.166	6.1660E+04	8.1935E−01	324.59
6,000	5,994	249.187	4.7218E+04	6.6011E−01	316.45
8,000	7,990	236.215	3.5652E+04	5.2579E−01	308.11
10,000	9,984	223.252	2.6500E+04	4.1351E−01	299.53
12,000	11,977	216.650	1.9399E+04	3.1194E−01	295.07
14,000	13,969	216.650	1.4170E+04	2.2786E−01	295.07
16,000	15,960	216.650	1.0353E+04	1.6647E−01	295.07
18,000	17,949	216.650	7.5652E+03	1.2165E−01	295.07
20,000	19,937	216.650	5.5293E+03	8.8910E−02	295.07
22,000	21,924	218.574	4.0475E+03	6.4510E−02	296.38
24,000	23,910	220.560	2.9717E+03	4.6938E−02	297.72
26,000	25,894	222.544	2.1884E+03	3.4256E−02	299.06
28,000	27,877	224.527	1.6162E+03	2.5076E−02	300.39
30,000	29,859	226.509	1.1970E+03	1.8410E−02	301.71
32,000	31,840	228.490	8.8906E+02	1.3555E−02	303.02
34,000	33,819	233.744	6.6341E+02	9.8874E−03	306.49
36,000	35,797	239.282	4.9852E+02	7.2579E−03	310.10
38,000	37,774	244.818	3.7714E+02	5.3665E−03	313.67
40,000	39,750	250.350	2.8714E+02	3.9957E−03	317.19
42,000	41,724	255.878	2.1997E+02	2.9948E−03	320.67
44,000	43,698	261.403	1.6950E+02	2.2588E−03	324.12
46,000	45,670	266.925	1.3134E+02	1.7141E−03	327.52
48,000	47,640	270.650	1.0229E+02	1.3167E−03	329.80
50,000	49,610	270.650	7.9779E+01	1.0269E−03	329.80
52,000	51,578	270.650	6.2228E+01	8.0097E−04	329.80
54,000	53,545	267.560	4.8491E+01	6.3137E−04	327.91
56,000	55,511	263.628	3.7657E+01	4.9761E−04	325.49
58,000	57,476	259.699	2.9137E+01	3.9085E−04	323.06
60,000	59,439	255.772	2.2460E+01	3.0592E−04	320.61
62,000	61,401	251.045	1.7246E+01	2.3931E−04	317.63
64,000	63,362	243.202	1.3150E+01	1.8837E−04	312.63
66,000	65,322	235.363	9.9405E+00	1.4713E−04	307.55
68,000	67,280	227.529	7.4447E+00	1.1399E−04	302.39
70,000	69,238	219.700	5.5204E+00	8.7534E−05	297.14

Appendix B
Standard Atmosphere, English Units

Geometric Altitude (ft)	Geopotential Altitude (ft)	Temperature (°R)	Pressure (lbf/ft²)	Density (slugs/ft³)	Speed of Sound (ft/sec)
0	0	518.670	2.1162E+03	2.3769E−03	1116.45
5,000	4,999	500.843	1.7609E+03	2.0482E−03	1097.10
10,000	9,995	483.025	1.4556E+03	1.7555E−03	1077.40
15,000	14,989	465.216	1.1948E+03	1.4962E−03	1057.36
20,000	19,981	447.415	9.7327E+02	1.2673E−03	1036.93
25,000	24,970	429.623	7.8634E+02	1.0663E−03	1016.10
30,000	29,957	411.839	6.2967E+02	8.9068E−04	994.85
35,000	34,941	394.063	4.9935E+02	7.3820E−04	973.14
40,000	39,923	389.970	3.9313E+02	5.8728E−04	968.08
45,000	44,903	389.970	3.0945E+02	4.6227E−04	968.08
50,000	49,880	389.970	2.4361E+02	3.6392E−04	968.08
55,000	54,855	389.970	1.9180E+02	2.8652E−04	968.08
60,000	59,828	389.970	1.5103E+02	2.2561E−04	968.08
65,000	64,798	389.970	1.1893E+02	1.7767E−04	968.08
70,000	69,766	392.246	9.3727E+01	1.3920E−04	970.90
75,000	74,731	394.971	7.3990E+01	1.0913E−04	974.26
80,000	79,694	397.693	5.8511E+01	8.5710E−05	977.62
85,000	84,655	400.415	4.6350E+01	6.7434E−05	980.95
90,000	89,613	403.135	3.6778E+01	5.3147E−05	984.28
95,000	94,569	405.854	2.9232E+01	4.1959E−05	987.59
100,000	99,523	408.572	2.3272E+01	3.3182E−05	990.90
105,000	104,474	411.289	1.8557E+01	2.6285E−05	994.18
110,000	109,423	418.385	1.4837E+01	2.0659E−05	1002.72
115,000	114,369	425.983	1.1912E+01	1.6290E−05	1011.79
120,000	119,313	433.578	9.6013E+00	1.2900E−05	1020.77
125,000	124,255	441.170	7.7689E+00	1.0259E−05	1029.67
130,000	129,195	448.758	6.3095E+00	8.1907E−06	1038.48
135,000	134,132	456.342	5.1427E+00	6.5650E−06	1047.22
140,000	139,066	463.923	4.2062E+00	5.2818E−06	1055.88
145,000	143,999	471.500	3.4518E+00	4.2648E−06	1064.47
150,000	148,929	479.073	2.8419E+00	3.4558E−06	1072.99
155,000	153,857	486.643	2.3471E+00	2.8097E−06	1081.43
160,000	158,782	487.170	1.9419E+00	2.3221E−06	1082.02
165,000	163,705	487.170	1.6068E+00	1.9215E−06	1082.02
170,000	168,625	487.170	1.3297E+00	1.5901E−06	1082.02
175,000	173,544	483.944	1.1000E+00	1.3242E−06	1078.43

Appendix C
Aircraft Moments of Inertia

The diagonal components of an airplane's inertia tensor are most easily estimated in the early phases of design by first estimating the radii of gyration about each of the body-fixed coordinate axes. The three moment of inertia components are then obtained from these three radii of gyration using the relations

$$I_{xx_b} = r_{xx_b}^2 W/g, \quad \text{where} \quad r_{xx_b}^2 \equiv (1/W)\iiint_W (y_b^2 + z_b^2)dW \tag{C.1}$$

$$I_{yy_b} = r_{yy_b}^2 W/g, \quad \text{where} \quad r_{yy_b}^2 \equiv (1/W)\iiint_W (x_b^2 + z_b^2)dW \tag{C.2}$$

$$I_{zz_b} = r_{zz_b}^2 W/g, \quad \text{where} \quad r_{zz_b}^2 \equiv (1/W)\iiint_W (x_b^2 + y_b^2)dW \tag{C.3}$$

For preliminary design, estimates for an airplane's three radii of gyration are commonly obtained from data for existing airplanes of similar configuration.

For example, Figs. C.1 and C.2 show nondimensional roll and pitch radii of gyration for a large number of widely different airplanes plotted as a function of gross weight. The roll radius of gyration is nondimensionalized with respect to the wingspan and the pitch radius of gyration is nondimensionalized with respect to the fuselage length. The data used to generate Figs. C.1 and C.2 were obtained from several different sources (for further details see Phillips and Niewoehner 2009). For any airplane having data available at more than one gross weight, the corresponding points in Figs. C.1 and C.2 are connected with a solid line. Note that in all such cases the pitch radius of gyration is larger for the lightest loading.

The data plotted in Figs. C.1 and C.2 can be used to estimate the roll and pitch radii of gyration for a particular type of airplane in the early phases of design. For example, the propeller-driven airplanes with fuselage-mounted engines, which are included in Figs. C.1 and C.2, have nondimensional pitch radii of gyration between 0.150 to 0.202 with an arithmetic mean of 0.176 and a standard deviation of 0.014. A lightly loaded airplane typically has a pitch radius of gyration in the upper part of this range and data for heavily loaded airplanes will usually fall in the lower part of this range.

If more is known about an airplane, its radius of gyration can often be estimated with greater precision from knowledge of the mass properties of similar airplanes. For example, the propeller-driven airplanes with fuselage-mounted engines, which are included in Figs. C.1 and C.2, are comprised of both fighters and general aviation airplanes. The nondimensional pitch radii of gyration for the fighters range from 0.150 to 0.184 with a mean of 0.171 and a standard deviation of 0.012. On the other hand, the lighter general aviation airplanes exhibit nondimensional pitch radii of gyration between 0.169 to 0.202 with a mean of 0.187 and a standard deviation of 0.013. Further details for all of the airplanes included in Figs. C.1 and C.2 are presented in Tables C.1 and C.2.

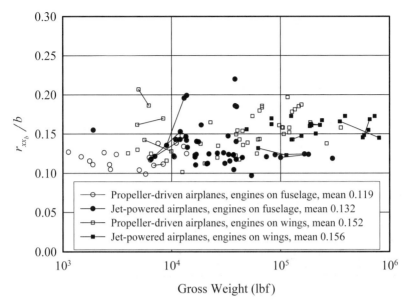

Figure C.1. Roll radius of gyration for a number of widely different airplanes nondimensionalized with respect to the wingspan. (From Phillips and Niewoehner 2009)

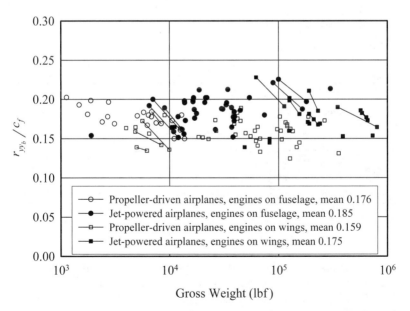

Figure C.2. Pitch radius of gyration for a number of widely different airplanes nondimensionalized with respect to the fuselage chord length. (From Phillips and Niewoehner 2009)

Airplane Type	Minimum	Maximum	Mean	Standard Deviation
Propeller-driven airplanes, engines on fuselage	**0.086**	**0.148**	**0.119**	**0.015**
general aviation airplanes	0.111	0.127	0.118	0.007
fighters	0.086	0.148	0.119	0.018
Jet-powered airplanes, engines on fuselage	**0.097**	**0.220**	**0.132**	**0.024**
fighters	0.105	0.220	0.134	0.026
business jets	0.118	0.185	0.145	0.027
transports and bombers	0.097	0.125	0.119	0.010
Propeller-driven airplanes, engines on wings	**0.102**	**0.207**	**0.152**	**0.025**
general aviation airplanes	0.102	0.207	0.143	0.032
fighters	0.138	0.165	0.151	0.014
transports and bombers	0.113	0.197	0.155	0.023
Jet-powered airplanes, engines on wings	**0.126**	**0.173**	**0.156**	**0.014**
transports and bombers, maximum weight	0.145	0.173	0.164	0.010
transports and bombers, minimum weight	0.126	0.166.	0.147	0.020

Table C.1. Nondimensional roll radius of gyration, r_{xx_b}/b.

Airplane Type	Minimum	Maximum	Mean	Standard Deviation
Propeller-driven airplanes, engines on fuselage	**0.150**	**0.202**	**0.176**	**0.014**
general aviation airplanes	0.169	0.202	0.187	0.013
fighters	0.150	0.184	0.171	0.012
Jet-powered airplanes, engines on fuselage	**0.152**	**0.226**	**0.185**	**0.020**
fighters	0.152	0.213	0.182	0.019
business jets	0.152	0.200	0.176	0.021
transports and bombers	0.180	0.226	0.200	0.018
Propeller-driven airplanes, engines on wings	**0.125**	**0.189**	**0.159**	**0.016**
general aviation airplanes	0.135	0.182	0.158	0.017
fighters	0.150	0.169	0.157	0.011
transports and bombers	0.125	0.189	0.159	0.016
Jet-powered airplanes, engines on wings	**0.153**	**0.228**	**0.175**	**0.019**
transports and bombers, maximum weight	0.153	0.177	0.164	0.008
transports and bombers, minimum weight	0.190	0.228	0.208	0.016

Table C.2. Nondimensional pitch radius of gyration, r_{yy_b}/c_f.

A method commonly used to estimate the yaw radius of gyration in the early phases of airplane design is based on correlating data nondimensionalized with the arithmetic average of the wingspan and the fuselage chord length. An alternative approach is to relate the yaw radius of gyration to the roll and pitch radii of gyration (see Phillips and Niewoehner 2009). Since most of an airplane's mass is concentrated near the x_b-y_b plane,

Eqs. (C.1) through (C.3) demonstrate that the yaw radius of gyration for an airplane is slightly less than the root square sum of the roll and pitch radii of gyration. This is demonstrated for a large number of widely different airplanes in Fig. C.3 and Table C.3.

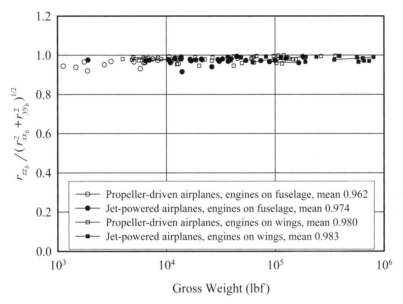

Figure C.3. Yaw radius of gyration for a number of widely different airplanes nondimensionalized with the root square sum of the roll and pitch radii of gyration. (From Phillips and Niewoehner 2009)

Airplane Type	Minimum	Maximum	Mean	Standard Deviation
Propeller-driven airplanes, engines on fuselage	**0.920**	**0.992**	**0.962**	**0.019**
general aviation airplanes	0.920	0.968	0.948	0.018
fighters	0.931	0.992	0.969	0.016
Jet-powered airplanes, engines on fuselage	**0.915**	**0.993**	**0.974**	**0.015**
fighters	0.915	0.990	0.972	0.017
business jets	0.971	0.983	0.978	0.005
transports and bombers	0.961	0.993	0.976	0.012
Propeller-driven airplanes, engines on wings	**0.946**	**0.999**	**0.980**	**0.014**
general aviation airplanes	0.954	0.988	0.979	0.010
fighters	0.974	0.989	0.980	0.008
transports and bombers	0.946	0.999	0.981	0.016
Jet-powered airplanes, engines on wings	**0.963**	**0.996**	**0.983**	**0.011**
transports and bombers, maximum weight	0.970	0.996	0.987	0.009
transports and bombers, minimum weight	0.963	0.988	0.973	0.011

Table C.3. Nondimensional yaw radius of gyration, $r_{zz_b}\big/(r_{xx_b}^2 + r_{yy_b}^2)^{1/2}$.

Nomenclature

$[\mathbf{A}]$ airfoil coefficient matrix, pg 35

$[\mathbf{A}]$ coefficient matrix in the generalized eigenproblem, pg 819

$[\mathbf{A}]_{ee}$ coefficient matrix, pg 930

$[\mathbf{A}]_{e\omega}$ coefficient matrix, pg 930

$[\mathbf{A}]_{fv}$ coefficient matrix, pg 929

$[\mathbf{A}]_{f\omega}$ coefficient matrix, pg 929

$[\mathbf{A}]_{mv}$ coefficient matrix, pg 929

$[\mathbf{A}]_{m\omega}$ coefficient matrix, pg 929

$[\mathbf{A}]_{pe}$ coefficient matrix, pg 930

$[\mathbf{A}]_{pv}$ coefficient matrix, pg 930

$[\mathbf{A}]_{w}$ coefficient matrix, pg 929

$\{\mathbf{A}\}$ arbitrary quaternion

$\mathbf{A}$ arbitrary quaternion with zero scalar component (vector)

A aftward axial force, pg 2

A complex amplitude, pg 824

$\tilde{A}$ section axial force, pgs 2, 4

A_e engine exit area

A_f amplification factor, pg 817

A_g dimensionless coefficient, pg 933

A_i engine inlet area

$A_{\ell p}$ matrix component, pg 929

$A_{\ell,\bar{p}}$ dimensionless coefficient, pg 934

$A_{\ell q}$ matrix component, pg 929

$A_{\ell r}$ matrix component, pg 929

$A_{\ell,\bar{r}}$ dimensionless coefficient, pg 934

$A_{\ell,\alpha}$ dimensionless coefficient, pg 934

$A_{\ell,\beta}$ dimensionless coefficient, pg 934

A_{mp} matrix component, pg 929

A_{mq} matrix component, pg 929

A_{mr} matrix component, pg 929

$A_{m,\bar{q}}$ dimensionless coefficient, pg 934

$A_{m,\alpha}$ dimensionless coefficient, pg 934

$A_{m,\beta}$ dimensionless coefficient, pg 934

$A_{m,\mu}$ dimensionless coefficient, pg 933

A_n coefficients in the infinite series solution for thin airfoil theory

A_n coefficients in the infinite series solution to the lifting-line equation

A_{np} matrix component, pg 930

$A_{n,\bar{p}}$ dimensionless coefficient, pg 934

A_{nq} matrix component, pg 930

A_{nr} matrix component, pg 930

$A_{n,\bar{r}}$ dimensionless coefficient, pg 934

$A_{n,\alpha}$ dimensionless coefficient, pg 934

$A_{n,\beta}$ dimensionless coefficient, pg 934

A_p area of prop circle

A_x x-component of arbitrary quaternion

$A_{x,\bar{q}}$ dimensionless coefficient, pg 934

$A_{x,\alpha}$ dimensionless coefficient, pg 934

$A_{x,\mu}$ dimensionless coefficient, pg 933

A_y y-component of arbitrary quaternion

$A_{y,\bar{p}}$ dimensionless coefficient, pg 934

$A_{y,\bar{r}}$ dimensionless coefficient, pg 934

$A_{y,\beta}$ dimensionless coefficient, pg 934

A_z z-component of arbitrary quaternion

$A_{z,\bar{q}}$ dimensionless coefficient, pg 934

$A_{z,\alpha}$ dimensionless coefficient, pg 934

$A_{z,\mu}$ dimensionless coefficient, pg 933

A_0 scalar-component of arbitrary quaternion

a speed of sound

a_N acceleration normal to the flight path

a_n planform contribution to coefficients in the infinite series solution to Prandtl's lifting-line equation

a_0 speed of sound at stagnation

a_∞ freestream speed of sound

$[\mathbf{B}]$ coefficient matrix in the generalized eigenproblem, pg 819

$[\mathbf{B}]_{iv}$ coefficient matrix, pg 929

$[\mathbf{B}]_m$ coefficient matrix, pg 929

$\{\mathbf{B}\}$ arbitrary quaternion

$\mathbf{B}$ apparent body-force vector, due to the vector sum of the acceleration of gravity and the aircraft acceleration, pg 668

$\mathbf{B}$ arbitrary quaternion with zero scalar component (vector)

B leftward side force, pg 2

B arbitrary constant for stagnation flow, pg 21

B_x x-component of arbitrary quaternion

$B_{m,\alpha'}$ dimensionless coefficient, pg 933

$B_{m,\mu'}$ dimensionless coefficient, pg 933

$B_{x,\alpha'}$ dimensionless coefficient, pg 933

$B_{x,\mu'}$ dimensionless coefficient, pg 933

B_y y-component of arbitrary quaternion

B_z z-component of arbitrary quaternion

$B_{z,\alpha'}$ dimensionless coefficient, pg 933

$B_{z,\mu'}$ dimensionless coefficient, pg 933

B_0 scalar component of arbitrary quaternion

b wingspan

b' wingtip vortex spacing

b_h span of either an aft horizontal tail or a forward canard

b_n washout contribution to coefficients in the infinite series solution to Prandtl's lifting-line equation

b_V' twice the root-to-tip semispan length for a V-tail measured parallel with the semispan dihedral

b_w span of the main wing

b_w' twice the root-to-tip semispan length measured parallel with the wing dihedral, pg 552

$[\mathbf{C}]$ coefficient matrix in the generalized linear control problem, pg 970

$[\mathbf{C}]$ direction-cosine matrix, pg 992

$[\mathbf{C}]^T$ transpose of direction-cosine matrix

$[\mathbf{C}]_f$ coefficient matrix, pg 930

$[\mathbf{C}]_m$ coefficient matrix, pg 930

C arbitrary constant

C shorthand notation for $\cos(\Theta/2)$, pg 998

C vertical components of string tension divided by the weight, pg 874

C_A axial force coefficient, pg 3

$\tilde{C}_A$ section axial force coefficient, pg 3

CAP control anticipation parameter, pg 960

$C_{A,\alpha}$ change in axial force coefficient with respect to angle of attack

$C_{A,\alpha,\alpha}$ second derivative of axial force coefficient with respect to angle of attack

C_D drag coefficient, pg 3

$\tilde{C}_D$ section drag coefficient, pg 3

C_{D_h} drag coefficient on either an aft horizontal tail or a forward canard

C_{D_i} induced drag coefficient

$C_{D_{LO}}$ drag coefficient at liftoff

$C_{D,M}$ change in C_D with respect to M

$\tilde{C}_{DM0}$ section drag coefficient at zero Mach number, pg 138

C_{Do} drag coefficient at the equilibrium reference state

$C_{D_{OC}}$ drag coefficient at the obstacle clearance altitude

C_{Dp} parasitic drag coefficient

C_{D_w} drag coefficient on the main wing

$\tilde{C}_{D_w}$ local section drag coefficient on the main wing

C_{D_0} drag coefficient at zero lift

$C_{D_{0h}}$ drag coefficient at zero lift for either an aft horizontal tail or forward canard

$C_{D_0,L}$ the linear coefficient in the parabolic relation for drag coefficient as a function of lift coefficient

C_{D_0,L_h} the linear coefficient in the parabolic relation for drag coefficient as a function of lift coefficient for either an aft horizontal tail or forward canard

C_{D_0,L_w} the linear coefficient in the parabolic relation for drag coefficient as a function of lift coefficient for the main wing

$C_{D_{0w}}$ drag coefficient at zero lift for the main wing

$C_{D,\alpha}$ change in drag coefficient with angle of attack

$\tilde{C}_{D,\alpha}$ change in section drag coefficient with angle of attack

C_{H_e} elevator hinge moment coefficient

C_{H_e0} elevator hinge moment coefficient with the local angle of attack, elevator deflection, and trim tab deflection at zero

C_{H_e,α_h} change in elevator hinge moment coefficient with local angle of attack

C_{H_e,δ_e} change in elevator hinge moment coefficient with elevator deflection

C_{H_e,δ_t} change in elevator hinge moment coefficient with trim tab deflection

C_{H_s} stabilator hinge moment coefficient

C_{H_s0} stabilator hinge moment coefficient with the local angle of attack, stabilator deflection, and elevator trim deflection at zero

C_{H_s,α_h} change in stabilator hinge moment coefficient with local angle of attack

C_{H_s,δ_s} change in stabilator hinge moment coefficient with stabilator deflection

C_{H_s,δ_e} change in stabilator hinge moment coefficient with elevator trim deflection

C_{h0} aerodynamic coefficient, pg 472

$C_{h,\alpha}$ aerodynamic coefficient, pg 472

C_{ij} components of the direction-cosine matrix

$\dot{C}_{ij}$ time rate of change of components of the direction-cosine matrix

C_L lift coefficient, pg 3

$\tilde{C}_L$ section lift coefficient, pg 3

$\tilde{C}_{L_a}$ additional section lift coefficient, pg 87

$\tilde{C}_{L_b}$ basic section lift coefficient, pg 87

C_{Ld} design lift coefficient

C_{L_h} lift coefficient on either an aft horizontal tail or a forward canard

$C_{L_h max}$ maximum lift coefficient for either an aft horizontal tail or a forward canard

C_{L_h0} lift coefficient on either an aft horizontal tail or a forward canard at zero fuselage angle of attack and zero elevator deflection

$C_{L_h,\alpha}$ lift slope for either an aft horizontal tail or a forward canard

$\tilde{C}_{L_i,\alpha}$ in situ section lift slope for the main wing, including the effects of local induced downwash, pg 551

$C_{L_{LO}}$ lift coefficient at liftoff

$C_{L,M}$ change in C_L with respect to M

$\tilde{C}_{LM0}$ section lift coefficient at zero Mach number, pg 135

$C_{L_{max}}$ maximum wing lift coefficient

$\tilde{C}_{L_{max}}$ maximum section lift coefficient

$C_{L_{OC}}$ lift coefficient at the obstacle clearance altitude

$C_{L,q}$ change in C_L with respect to q

$C_{L,\bar{q}}$ change in C_L with respect to $\bar{q}$

$C_{L,\breve{q}}$ change in C_L with respect to $\breve{q}$

C_{L_v} lift coefficient on a vertical tail

$C_{L_v,\alpha}$ lift slope for a vertical tail

C_{L_w} lift coefficient on the main wing

$C_{L_w max}$ maximum lift coefficient for the main wing

C_{L_w0} lift coefficient on the main wing at zero fuselage angle of attack and zero elevator deflection

$C_{L_w,\alpha}$ lift slope for the main wing

C_{L0} aircraft lift coefficient at zero fuselage angle of attack and zero elevator deflection

$C_{L,\alpha}$ lift slope, change in lift coefficient with angle of attack

$C'_{L,\alpha}$ stick-free change in lift coefficient with angle of attack

$\widetilde{C}_{L,\alpha}$ section lift slope, change in section lift coefficient with angle of attack

$C_{L,\alpha,\alpha}$ second derivative of lift coefficient with respect to angle of attack

C_{L,δ_e} change in aircraft lift coefficient with elevator deflection

C_ℓ aircraft rolling moment coefficient about the aircraft center of gravity, pgs 76, 497, 770

C_ℓ propeller torque coefficient, pg 182

$C_{\ell_{ac}}$ rolling moment coefficient about the aerodynamic center

$C_{\ell,M}$ change in C_ℓ with respect to M

$C_{\ell,\overline{p}}$ change in C_ℓ with respect to $\overline{p}$

$C_{\ell,\breve{p}}$ change in C_ℓ with respect to $\breve{p}$

$C_{\ell,\overline{r}}$ change in C_ℓ with respect to $\overline{r}$

$C_{\ell,\breve{r}}$ change in C_ℓ with respect to $\breve{r}$

C_{ℓ_0} rolling moment coefficient about the coordinate system origin

$C_{\ell,\beta}$ change in C_ℓ with respect to β

C_{ℓ,δ_a} change in C_ℓ with respect to δ_a

C_{ℓ,δ_r} change in C_ℓ with respect to δ_r

C_m aircraft pitching moment coefficient about the aircraft center of gravity, pgs 3, 770

$C_{\breve{m}}$ dynamic pitching moment coefficient, pg 617

$\widetilde{C}_m$ section moment coefficient, pg 3

$C_{m_{ac}}$ pitching moment coefficient about the aerodynamic center

$\widetilde{C}_{m_{ac}}$ section moment coefficient about the section aerodynamic center

$\overline{\widetilde{C}}_{m_{ac}}$ mean section moment coefficient about the section aerodynamic center, pgs 90, 127, 460

$\widetilde{C}_{m_{c/4}}$ quarter-chord section moment coefficient

C_{m_f} pitching moment coefficient for the fuselage about the aircraft center of gravity

C_{m_h} pitching moment coefficient for either an aft horizontal tail or a forward canard about its aerodynamic center

$C_{m_h 0}$ pitching moment coefficient for either an aft horizontal tail or a forward canard about its aerodynamic center at zero fuselage angle of attack and zero elevator deflection

C_{m_h,δ_e} change in the pitching moment coefficient for either an aft horizontal tail or a forward canard about its aerodynamic center with elevator deflection angle

$\widetilde{C}_{m_{le}}$ leading-edge section moment coefficient

$C_{m,M}$ change in C_m with respect to M

$\widetilde{C}_{mM0}$ section moment coefficient at zero Mach number, pg 135

$C_{m_{np}}$ aircraft pitching moment coefficient about the stick-fixed neutral point

$C_{m_{np},\alpha}$ change in aircraft pitching moment coefficient about the stick-fixed neutral point with respect to angle of attack

$C_{m,q}$ change in C_m with respect to q

$C_{m,\overline{q}}$ change in C_m with respect to $\overline{q}$

$C_{m,\breve{q}}$ change in C_m with respect to $\breve{q}$

$C_{\breve{m},\breve{q}}$ change in $C_{\breve{m}}$ with respect to $\breve{q}$

C_{m_v,δ_r} change in vertical tail moment coefficient about its aerodynamic center with rudder deflection angle

C_{m_w} wing pitching moment coefficient about its aerodynamic center

$\widetilde{C}_{m_x}$ section moment coefficient about point with axial coordinate x

$\widetilde{C}_{m_x,\alpha}$ change in section moment coefficient about point with axial coordinate x with respect to angle of attack

C_{m0} aircraft pitching moment coefficient at zero fuselage angle of attack and zero elevator deflection

C_{m_0} pitching moment coefficient about the coordinate system origin

$C_{m_0,\alpha}$ change in pitching moment coefficient about the coordinate system origin with respect to angle of attack

$C_{m_0,\alpha,\alpha}$ second derivative of pitching moment coefficient about the coordinate system origin with respect to angle of attack

$C_{m,\alpha}$ change in pitching moment coefficient with respect to angle of attack

$C'_{m,\alpha}$ stick-free change in pitching moment coefficient with respect to angle of attack

$C_{m,\hat{\alpha}}$ change in C_m with respect to $\hat{\alpha}$

$\tilde{C}_{m,\alpha}$ change in section pitching moment coefficient with angle of attack

$\tilde{C}_{m,\delta}$ change in section quarter-chord moment coefficient with flap deflection, pg 42

C_{m,δ_e} change in C_m with respect to δ_e

$C_{m,\mu}$ change in C_m with respect to μ

$C_{m,\hat{\mu}}$ change in C_m with respect to $\hat{\mu}$

C_N normal force coefficient, pg 3

$\tilde{C}_N$ section normal force coefficient, pg 3

C_{N_p} propeller normal force coefficient

$C_{N_p,\alpha}$ change in propeller normal force coefficient with propeller angle of attack

$C_{N,\alpha}$ change in normal force coefficient with respect to angle of attack

$\tilde{C}_{N,\alpha}$ change in section normal force coefficient with respect to angle of attack

$C_{N,\alpha,\alpha}$ second derivative of normal force coefficient with respect to angle of attack

C_n aircraft yawing moment coefficient about the aircraft center of gravity, pgs 78, 497, 770

C_{n_f} yawing moment coefficient for the fuselage about the aircraft center of gravity

$C_{n,\text{M}}$ change in C_n with respect to M

C_{n_p} propeller yawing moment coefficient

$C_{n_p,\alpha}$ change in propeller yawing moment coefficient with angle of attack

$C_{n,\overline{p}}$ change in C_n with respect to $\overline{p}$

$C_{n,\breve{p}}$ change in C_n with respect to $\breve{p}$

$C_{n,\overline{r}}$ change in C_n with respect to $\overline{r}$

$C_{n,\breve{r}}$ change in C_n with respect to $\breve{r}$

$C_{n,\alpha}$ change in C_n with respect to α

$C_{n,\beta}$ change in C_n with respect to β

C_{n,δ_a} change in C_n with respect to δ_a

C_{n,δ_r} change in C_n with respect to δ_r

C_P propeller power coefficient

C_p constant-pressure specific heat

C_p pressure coefficient, pg 35

C_{pl} lower surface pressure coefficient

$C_{p\text{M}0}$ pressure coefficient at zero Mach number, pg 135

C_{pu} upper surface pressure coefficient

C_T propeller thrust coefficient

C_W weight coefficient, pgs 575, 610

C_{w0} aerodynamic coefficient, pg 472

$C_{w,\alpha}$ aerodynamic coefficient, pg 472

C_X force coefficient for the x_b-component of aerodynamic force, excluding thrust

$C_{X,\text{M}}$ change in C_X with respect to M

$C_{X,\overline{q}}$ change in C_X with respect to $\overline{q}$

$C_{X,\alpha}$ change in C_X with respect to α

$C_{X,\hat{\alpha}}$ change in C_X with respect to $\hat{\alpha}$

C_{X,δ_e} change in C_X with respect to δ_e

$C_{X,\mu}$ change in C_X with respect to μ

$C_{X,\hat{\mu}}$ change in C_X with respect to $\hat{\mu}$

C_Y force coefficient for the y_b-component of aerodynamic force, pgs 497, 770

$C_{Y,\text{M}}$ change in C_Y with respect to M

C_{Y_p} propeller side force coefficient

$C_{Y,\overline{p}}$ change in C_Y with respect to $\overline{p}$

$C_{Y,\breve{p}}$ change in C_Y with respect to $\breve{p}$

$C_{Y,\overline{r}}$ change in C_Y with respect to $\overline{r}$

$C_{Y,\breve{r}}$ change in C_Y with respect to $\breve{r}$

$C_{Y_p,\beta}$ change in propeller side force coefficient with sideslip angle

$C_{Y,\beta}$ change in C_Y with respect to β

C_{Y,δ_a} change in C_Y with respect to δ_a

C_{Y,δ_r} change in C_Y with respect to δ_r

C_Z force coefficient for the z_b-component of aerodynamic force, excluding thrust

$C_{Z,\text{M}}$ change in C_Z with respect to M

$C_{Z,\overline{q}}$ change in C_Z with respect to $\overline{q}$

$C_{Z,\alpha}$ change in C_Z with respect to α

$C_{Z,\hat{\alpha}}$ change in C_Z with respect to $\hat{\alpha}$

C_{Z,δ_e} change in C_Z with respect to δ_e

$C_{Z,\mu}$ change in C_Z with respect to μ

$C_{Z,\hat{\mu}}$ change in C_Z with respect to $\hat{\mu}$

$C_{\varepsilon h}$ downwash coefficient, pg 472

$C_{\varepsilon w}$ downwash coefficient, pg 472

C_Θ shorthand notation for $\cos(\Theta)$

C_θ shorthand notation for $\cos(\theta)$

C_{θ_o} shorthand notation for $\cos(\theta_o)$

$C_{\theta/2}$ shorthand notation for $\cos(\theta/2)$

C_ϕ shorthand notation for $\cos(\phi)$

C_{ϕ_o} shorthand notation for $\cos(\phi_o)$

$C_{\phi/2}$ shorthand notation for $\cos(\phi/2)$

C_ψ shorthand notation for $\cos(\psi)$

C_{ψ_o} shorthand notation for $\cos(\psi_o)$

$C_{\psi/2}$ shorthand notation for $\cos(\psi/2)$

$C_{\Omega t}$ shorthand notation for $\cos(\Omega t)$

c section chord length, pg 27

c damping coefficient, Sec. 8.1

$\bar{c}$ geometric mean chord length, pg 83

c_b propeller blade section chord length

$\hat{c}_b$ dimensionless propeller blade section chord length ratio, pg 183

$\bar{c}_e$ mean elevator chord

c_f flap chord length, pg 40

c_f fuselage chord length, pg 473

$\bar{c}_h$ mean chord of either an aft horizontal tail or a forward canard

$\bar{c}_{mac}$ mean aerodynamic chord length, pg 84

c_n control surface contribution to the coefficients in the infinite series solution to Prandtl's lifting-line equation

c_r root chord length

c_{ref} reference chord length

c_t tip chord length

$\bar{c}_v$ mean chord of a vertical tail

c_w local chord of the main wing

$\bar{c}_w$ mean chord of the main wing

$\{\mathbf{D}\}$ control deflection vector in the generalized linear control problem, pg 970

$\{\overline{\mathbf{D}}\}$ Laplace transform of control deflection vector in the generalized linear control problem, pg 970

D drag force, pg 2

$\tilde{D}$ section drag force, pgs 2, 4

D_{add} additive drag, pg 229

D_h drag on either an aft horizontal tail or a forward canard

D_i induced drag

$\tilde{D}_i$ local section induced drag, pg 51

D_{ℓ,δ_a} dimensionless coefficient, pg 934

D_{ℓ,δ_r} dimensionless coefficient, pg 934

D_{m,δ_e} dimensionless coefficient, pg 934

D_{nac} nacelle drag, pg 229

D_{n,δ_a} dimensionless coefficient, pg 934

D_{n,δ_r} dimensionless coefficient, pg 934

D_v drag on a vertical tail

D_w drag on the main wing

D_{x,δ_e} dimensionless coefficient, pg 934

D_{y,δ_a} dimensionless coefficient, pg 934

D_{y,δ_r} dimensionless coefficient, pg 934

D_{z,δ_e} dimensionless coefficient, pg 934

E modulus of elasticity

$\mathbf{dF}$ differential aerodynamic force vector

d_f characteristic fuselage diameter, pg 473

$\mathbf{d}l$ differential vortex length vector

d_n roll contribution to the coefficients in the infinite series solution to the lifting-line equation

d_{gc} great-circle distance or arc length measured over the Earth's surface, pg 1060

d_p propeller diameter

d_r rhumb-line distance measured over the Earth's surface along a path of constant bearing

d_{wp} great-circle distance to some waypoint, pg 1062

E   Euler axis vector, pg 994

E magnitude of Euler axis vector

E endurance or length of time an aircraft can fly without refueling

E_{max} maximum endurance or maximum time an aircraft can fly without refueling

E_{ij} Euler axis variable, pg 995

E'_{ij} Euler axis variable, pg 996

E_x x_b- and x_f-component of Euler axis vector

E_y y_b- and y_f-component of Euler axis vector

E_z z_b- and z_f-component of Euler axis vector

$\{e\}$ Euler-Rodrigues quaternion

$|\{e\}|$ Euler-Rodrigues quaternion magnitude

$\{e\}^*$ Euler-Rodrigues quaternion conjugate

$\{\dot{e}\}$ time rate of change of the Euler-Rodrigues quaternion

e Oswald efficiency factor, pgs 262

e_h Oswald efficiency factor for either an aft horizontal tail or a forward canard

$\{e\}_r$ renormalized Euler-Rodrigues quaternion

e_s span efficiency factor, pgs 53, 55

e_w Oswald efficiency factor for the wing

e_x x_b- and x_f-component of the Euler-Rodrigues quaternion

$\dot{e}_x$ time rate of change of the x_b- and x_f-component of the Euler-Rodrigues quaternion

e_y y_b- and y_f-component of the Euler-Rodrigues quaternion

$\dot{e}_y$ time rate of change of the y_b- and y_f-component of the Euler-Rodrigues quaternion

e_z z_b- and z_f-component of the Euler-Rodrigues quaternion

$\dot{e}_z$ time rate of change of the z_b- and z_f-component of the Euler-Rodrigues quaternion

e_0 scalar component of the Euler-Rodrigues quaternion

$\dot{e}_0$ time rate of change of the scalar component of the Euler-Rodrigues quaternion

F   pseudo aerodynamic force vector including thrust, pg 727

F magnitude of force vector

F uninstalled thrust, pg 229

$\mathbf{F}_a$ resultant aerodynamic force vector

F_a amplitude of sinusoidal forcing function

$\overline{F}_c$ effective net force during transition, pg 357

$\mathbf{F}_j$ resultant force vector for a jet engine

$\mathbf{F}_p$ resultant pressure force vector

F_p pitch control force provided by pilot

$F_{p,n}$ longitudinal control force per g

$F_{p,V}$ control force gradient

F_r engine mount reaction force, pg 229

F_r rolling friction force

$\mathbf{F}_S$ resultant surface force vector

F_{Sx_b} x_b-component of resultant surface force

F_{Sy_b} y_b-component of resultant surface force

F_{Sz_b} z_b-component of resultant surface force

F_{x_b} x_b-component of pseudo aerodynamic force vector, including thrust

$F_{x_b o}$ equilibrium x_b-component of pseudo aerodynamic force vector, including thrust

$F_{x_b,p}$ change in x_b-component of pseudo aerodynamic force vector, including thrust, with rolling rate

$F_{x_b,q}$ change in x_b-component of pseudo aerodynamic force vector, including thrust, with pitching rate

$F_{x_b,r}$ change in x_b-component of pseudo aerodynamic force vector, including thrust, with yawing rate

$F_{x_b,u}$ change in x_b-component of pseudo aerodynamic force vector, including thrust, with x_b-component of airspeed

$F_{x_b,\dot{u}}$ change in x_b-component of pseudo aerodynamic force vector, including thrust, with $\dot{u}$

$F_{x_b,v}$ change in x_b-component of pseudo aerodynamic force vector, including thrust, with y_b-component of airspeed

$F_{x_b,w}$ change in x_b-component of pseudo aerodynamic force vector, including thrust, with z_b-component of airspeed

$F_{x_b,\dot{w}}$ change in x_b-component of pseudo aerodynamic force vector, including thrust, with $\dot{w}$

F_{y_b} y_b-component of pseudo aerodynamic force vector, including thrust

$F_{y_b 0}$ equilibrium y_b-component of pseudo aerodynamic force vector, including thrust

$F_{y_b,p}$ change in y_b-component of pseudo aerodynamic force vector, including thrust, with rolling rate

$F_{y_b,q}$ change in y_b-component of pseudo aerodynamic force vector, including thrust, with pitching rate

$F_{y_b,r}$ change in y_b-component of pseudo aerodynamic force vector, including thrust, with yawing rate

$F_{y_b,u}$ change in y_b-component of pseudo aerodynamic force vector, including thrust, with x_b-component of airspeed

$F_{y_b,v}$ change in y_b-component of pseudo aerodynamic force vector, including thrust, with y_b-component of airspeed

$F_{y_b,w}$ change in y_b-component of pseudo aerodynamic force vector, including thrust, with z_b-component of airspeed

F_{z_b} z_b-component of pseudo aerodynamic force vector, including thrust

$F_{z_b 0}$ equilibrium z_b-component of pseudo aerodynamic force vector, including thrust

$F_{z_b,p}$ change in z_b-component of pseudo aerodynamic force vector, including thrust, with rolling rate

$F_{z_b,q}$ change in z_b-component of pseudo aerodynamic force vector, including thrust, with pitching rate

$F_{z_b,r}$ change in z_b-component of pseudo aerodynamic force vector, including thrust, with yawing rate

$F_{z_b,u}$ change in z_b-component of pseudo aerodynamic force vector, including thrust, with x_b-component of airspeed

$F_{z_b,\dot{u}}$ change in z_b-component of pseudo aerodynamic force vector, including thrust, with $\dot{u}$

$F_{z_b,v}$ change in z_b-component of pseudo aerodynamic force vector, including thrust, with y_b-component of airspeed

$F_{z_b,w}$ change in z_b-component of pseudo aerodynamic force vector, including thrust, with z_b-component of airspeed

$F_{z_b,\dot{w}}$ change in z_b-component of pseudo aerodynamic force vector, including thrust, with $\dot{w}$

$\widetilde{F}_\theta$ propeller section circumferential force

f dimensionless force, pgs 341–342

f Prandtl's tip loss factor, pg 181

f' dimensionless force derivative, pg 342

$\mathbf{f}_L$ local lift force vector per unit length measured along the quarter-chord line, pg 555

f_{LO} dimensionless force, pg 343

f'_{LO} dimensionless force derivative, pg 343

f_S dimensionless force, pg 343

f'_S dimensionless force derivative, pg 343

f_T thrust fraction provided by a particular propeller, pg 507

$[\mathbf{G}]$ transfer-function matrix, pg 971

$\mathbf{G}$ dimensionless vortex strength vector

G dimensionless vortex strength, pg 100

G transfer function, pg 970

g acceleration of gravity

g_o acceleration of gravity at standard sea level

$\mathbf{H}$ angular momentum vector

H geometric altitude

$\dot{H}$ change in geometric altitude with respect to time

H_e — aerodynamic elevator hinge moment

H_s — stabilator primary hinge moment

H_{x_b} — x_b-component of angular momentum vector

H_{y_b} — y_b-component of angular momentum vector

H_{z_b} — z_b-component of angular momentum vector

h — angular momentum vector for all spinning rotors relative to the body-fixed coordinate system

h — distance above the center of gravity in a direction normal to fuselage reference line

h_h — distance above the center of gravity to aerodynamic center of either an aft horizontal tail or a forward canard

h_j — distance above the aircraft center of gravity to the jet engine thrust axis

h_{np} — distance above the center of gravity to the aircraft stick-fixed neutral point

h_{OC} — FAR takeoff obstacle clearance altitude

h_p — distance above the aircraft center of gravity to the propeller axis

h_T — distance above the center of gravity to the center of thrust

h_s — slipstream enthalpy, Chapter 2

h_v — distance above the center of gravity to aerodynamic center of a vertical tail

h_w — distance above the center of gravity to aerodynamic center of the main wing

h_w — height of the wing above the ground, used for ground effect computations

h_{x_b} — x_b-component of angular momentum vector for all spinning rotors relative to the body-fixed coordinate system

h_{y_b} — y_b-component of angular momentum vector for all spinning rotors relative to the body-fixed coordinate system

h_{z_b} — z_b-component of angular momentum vector for all spinning rotors relative to the body-fixed coordinate system

h_∞ — freestream enthalpy, Chapter 2

$[\mathbf{I}]$ — inertia tensor, pgs 725–726

I — imaginary part of complex number

I — local area moment of inertia

I_{xx_b} — moment of inertia, pgs 725–726

I_{xy_b} — product of inertia, pgs 725–726

I_{xz_b} — product of inertia, pgs 725–726

I_{yx_b} — product of inertia, pgs 725–726

I_{yy_b} — moment of inertia, pgs 725–726

I_{yz_b} — product of inertia, pgs 725–726

I_{zx_b} — product of inertia, pgs 725–726

I_{zy_b} — product of inertia, pgs 725–726

I_{zz_b} — moment of inertia, pgs 725–726

$[\mathbf{i}]$ — identity matrix

$\mathbf{i}_r$ — unit vector in the r-direction

$\mathbf{i}_x$ — unit vector in the x-direction

$\mathbf{i}_{x_b}$ — unit vector in the x_b-direction

$\mathbf{i}_{x_f}$ — unit vector in the x_f-direction

$\mathbf{i}_y$ — unit vector in the y-direction

$\mathbf{i}_{y_b}$ — unit vector in the y_b-direction

$\mathbf{i}_{y_f}$ — unit vector in the y_f-direction

$\mathbf{i}_z$ — unit vector in the z-direction

$\mathbf{i}_{z_b}$ — unit vector in the z_b-direction

$\mathbf{i}_{z_f}$ — unit vector in the z_f-direction

$\mathbf{i}_\theta$ — unit vector in the θ-direction

$\mathbf{i}_\infty$ — unit vector in the freestream direction

i — square root of -1

i — dummy index for infinite series

$[\mathbf{J}]$ — Jacobian matrix

J — propeller advance ratio, pg 183

K — propeller aerodynamic pitch-to-diameter ratio, pg 183

K_c — propeller chord-line pitch-to-diameter ratio

K_d — proportionality constant, pg 391

K_L — aeroelasticity coefficient, pg 626

K_m — aeroelasticity coefficient, pg 626

K_R — takeoff and landing coefficient, pg 342

K_T — takeoff and landing coefficient, pg 341

K_W takeoff and landing coefficient, pg 341

K_0 takeoff and landing coefficient, pg 341

K_1 takeoff and landing coefficient, pg 341

K_2 takeoff and landing coefficient, pg 341

k gain coefficient, pg 1021

k number of propeller blades

k spring constant

k_{aa} control system gain coefficient, pg 692

k_{ar} control system gain coefficient, pg 692

k_{es} stabilator-elevator linkage coefficient, pg 683

k_{ra} control system gain coefficient, pg 692

k_{rr} control system gain coefficient, pg 692

k_t torsional spring constant

L lift force, pg 2

$\widetilde{L}$ section lift force, pgs 2, 4

L_h lift on a either an aft horizontal tail or a forward canard

L_{nll} minimum allowable negative lift

L_{pll} maximum allowable positive lift

$L_{,q}$ change in L with respect to q

$L_{,\bar{q}}$ change in L with respect to $\bar{q}$

$L_{,\breve{q}}$ change in L with respect to $\breve{q}$

L_v lift on a vertical tail

L_w lift on the main wing

$L_{,\alpha}$ change in L with respect to α

l source or vortex sheet length

l distance aft of the center of gravity in a direction parallel to the fuselage reference line

ℓ propeller torque

ℓ rolling moment about the center of gravity excluding thrust, pgs 2, 498

l_f distance aft of the center of gravity to center of pressure of the fuselage

l_{Hs} stabilator hinge location, pg 685

l_h distance aft of the center of gravity to aerodynamic center of either an aft horizontal tail or a forward canard

l_{hn} distance aft of the aircraft stick-fixed neutral point to the aerodynamic center of either an aft horizontal tail or a forward canard

l_j distance aft of the aircraft center of gravity to the jet engine inlet

l_{mp} distance aft of the center of gravity to the aircraft stick-fixed maneuver point

l_{np} distance aft of the center of gravity to the aircraft stick-fixed neutral point

l'_{np} distance aft of the center of gravity to the aircraft stick-free neutral point

l_p distance aft of the aircraft center of gravity to the propeller

ℓ_p propeller torque or rolling moment

l_r global reference length

l_{ref} reference length

l_v distance aft of the center of gravity to aerodynamic center of a vertical tail

l_w distance aft of the center of gravity to aerodynamic center of the main wing

l_{wn} distance aft of the aircraft stick-fixed neutral point to the aerodynamic center of the main wing

l_{wt} distance used to compute aerodynamic derivatives with respect to translational acceleration, pgs 783, 785

M   Mach number

M_b local bending moment

M_b turbofan bypass exit Mach number

M_{cr} critical Mach number

M_e core nozzle exit Mach number

M_i inlet Mach number

M_∞ freestream Mach number

M   pseudo aerodynamic moment vector including thrust effects, pg 628

[**M**]   coefficient matrix in rate equation for Euler-Rodrigues quaternion, pg 1015

[**M**]   matrix of cofactors, pg 971

$[\mathbf{M}]^T$ transpose of matrix of cofactors, pg 971

$\mathbf{M}_a$ resultant aerodynamic moment vector

$\mathbf{M}_S$ resultant surface moment vector

M_{Sx_b} x_b-component of resultant surface moment vector

M_{Sy_b} y_b-component of resultant surface moment vector

M_{Sz_b} z_b-component of resultant surface moment vector

M_{x_b} x_b-component of pseudo aerodynamic moment vector, including all thrust effects

$M_{x_b o}$ equilibrium x_b-component of pseudo aerodynamic moment vector, including thrust effects

$M_{x_b,p}$ change in x_b-component of pseudo aerodynamic moment, including thrust effects, with rolling rate

$M_{x_b,q}$ change in x_b-component of pseudo aerodynamic moment, including thrust effects, with pitching rate

$M_{x_b,r}$ change in x_b-component of pseudo aerodynamic moment, including thrust effects, with yawing rate

$M_{x_b,u}$ change in x_b-component of pseudo aerodynamic moment, including thrust effects, with x_b-component of airspeed

$M_{x_b,v}$ change in x_b-component of pseudo aerodynamic moment, including thrust effects, with y_b-component of airspeed

$M_{x_b,w}$ change in x_b-component of pseudo aerodynamic moment, including thrust effects, with z_b-component of airspeed

M_{y_b} y_b-component of pseudo aerodynamic moment vector, including all thrust effects

$M_{y_b o}$ equilibrium y_b-component of pseudo aerodynamic moment vector, including thrust effects

$M_{y_b,p}$ change in y_b-component of pseudo aerodynamic moment, including thrust effects, with rolling rate

$M_{y_b,q}$ change in y_b-component of pseudo aerodynamic moment, including thrust effects, with pitching rate

$M_{y_b,r}$ change in y_b-component of pseudo aerodynamic moment, including thrust effects, with yawing rate

$M_{y_b,u}$ change in y_b-component of pseudo aerodynamic moment, including thrust effects, with x_b-component of airspeed

$M_{y_b,\dot{u}}$ change in y_b-component of pseudo aerodynamic moment, including thrust effects, with $\dot{u}$

$M_{y_b,v}$ change in y_b-component of pseudo aerodynamic moment, including thrust effects, with y_b-component of airspeed

$M_{y_b,w}$ change in y_b-component of pseudo aerodynamic moment, including thrust effects, with z_b-component of airspeed

$M_{y_b,\dot{w}}$ change in y_b-component of pseudo aerodynamic moment, including thrust effects, with $\dot{w}$

M_{z_b} z_b-component of pseudo aerodynamic moment vector, including thrust effects

$M_{z_b o}$ equilibrium z_b-component of pseudo aerodynamic moment vector, including thrust effects

$M_{z_b,p}$ change in z_b-component of pseudo aerodynamic moment, including thrust effects, with rolling rate

$M_{z_b,q}$ change in z_b-component of pseudo aerodynamic moment, including thrust effects, with pitching rate

$M_{z_b,r}$ change in z_b-component of pseudo aerodynamic moment, including thrust effects, with yawing rate

$M_{z_b,u}$ change in z_b-component of pseudo aerodynamic moment, including thrust effects, with x_b-component of airspeed

$M_{z_b,v}$ change in z_b-component of pseudo aerodynamic moment, including thrust effects, with y_b-component of airspeed

$M_{z_b,w}$ change in z_b-component of pseudo aerodynamic moment, including thrust effects, with z_b-component of airspeed

m mass, in Secs. 7.2 and 8.1 only

m pitching moment about the CG excluding thrust, pgs 2, 498, 729

$\widetilde{m}$ section pitching moment, pg 2

$\dot{m}$ mass flow rate

$\dot{m}$ time rate of change of aircraft mass, in Sec. 7.2 only

m_{ac} pitching moment about the aerodynamic center

$\widetilde{m}_{ac}$ section pitching moment about the aerodynamic center

$\dot{m}_b$ turbofan mass flow rate bypassing the engine core

$\dot{m}_c$ turbofan mass flow rate passing through the engine core

m_{cg_f} pitching moment for the fuselage about the aircraft center of gravity

m_{cg_j} pitching moment for the jet engine about the aircraft center of gravity

m_{cg_p} pitching moment for the propeller about the aircraft center of gravity

$\dot{m}_e$ exit mass flow rate

m_h pitching moment on either an aft horizontal tail or a forward canard about its aerodynamic center

$\dot{m}_i$ inlet mass flow rate

$\dot{m}_j$ jet engine mass flow rate

$\widetilde{m}_{le}$ leading-edge section pitching moment

$m_{,q}$ change in m with respect to q

$m_{,\overline{q}}$ change in m with respect to $\overline{q}$

$m_{,\breve{q}}$ change in m with respect to $\breve{q}$

m_v vertical tail moment about its aerodynamic center

m_w wing pitching moment about its aerodynamic center

$\widetilde{m}_x$ section pitching moment about point with axial coordinate x

m_0 pitching moment about the coordinate system origin

$m_{,\alpha}$ change in m with respect to α

$m_{,\dot{\alpha}}$ change in m with respect to $\dot{\alpha}$

N summation index for finite series

N upward normal force, pg 2

$\widetilde{N}$ section normal force, pgs 2, 4

N_j jet engine upward normal force

N_p propeller upward normal force

$[\mathbf{n}]$ null matrix, pg 930

$\mathbf{n}$ unit outward normal

n load factor, lift divided by weight

n yawing moment about the CG excluding thrust, pgs 2, 498, 729

n dummy index for infinite series

n_{cg_f} yawing moment for the fuselage about the aircraft center of gravity

n_{cg_p} yawing moment for the propeller about the aircraft center of gravity

n_{max} maximum load factor

n_{min} minimum load factor

n_{nll} negative load limit, pg 323

n_p yawing moment for the propeller about the propeller center

n_{pll} positive load limit, pg 323

$[\mathbf{P}]$ panel coefficient matrix, pg 35

P power input to a control volume

P_A available propulsive power

P_{A0} full-throttle available power at standard sea level

P_b brake power, torque multiplied by the angular velocity

P_{in} power input, Chapter 2

P_m power developed by a motor

P_R propulsive power required for steady level flight

$P_{R_{min}}$ minimum power required for steady level flight

p pressure

p rolling rate (x_b-component of angular rate vector ω_{x_b}), pgs 73, 729

$\dot{p}$ rolling acceleration (x_b-component of angular acceleration vector)

$\overline{p}$ traditional dimensionless rolling rate, pgs 75, 670

$\breve{p}$ dynamic dimensionless rolling rate, pg 670

p_b turbofan bypass exit pressure

p_d pressure just downstream of prop circle

p_e core nozzle exit pressure

p_i inlet pressure

p_l	pressure on lower surface of airfoil	$\overline{q}$	traditional dimensionless pitching rate, pgs 608, 670
p_o	equilibrium rolling rate		
p_s	ultimate slipstream pressure	$\breve{q}$	dynamic dimensionless pitching rate, pgs 616, 670
p_u	pressure just upstream of prop circle		
p_u	pressure on upper surface of airfoil	q_b	brake-power-specific fuel consumption, pg 293
p_0	stagnation pressure	$\dot{q}_c$	experimentally determined constraint, pg 617
p_{0e}	core nozzle exit stagnation pressure		
p_{0i}	inlet stagnation pressure	q_o	equilibrium pitching rate
p_{01}	compressor inlet stagnation pressure	q_P	propulsive-power-specific fuel consumption, pg 287
p_{02}	compressor outlet stagnation pressure		
p_{03}	turbine inlet stagnation pressure	q_T	thrust-specific fuel consumption, pg 304
p_{04}	turbine outlet stagnation pressure	$\mathbf{R}$	residual vector
p_{05}	afterburner outlet stagnation pressure	R	Reynolds number
p_{06}	fan outlet stagnation pressure	R	ideal gas constant
$p_{0\infty}$	freestream stagnation pressure	R	range or total distance an aircraft can fly without refueling
p_1	compressor inlet pressure		
p_2	compressor outlet pressure	R	real part of complex number
p_3	turbine inlet pressure	R	residual
p_4	turbine outlet pressure	R	turning or spin radius
p_5	afterburner outlet pressure	R_A	aspect ratio, pg 53
p_6	fan outlet pressure	R_{A_h}	aspect ratio of either an aft horizontal tail or a forward canard
p_∞	freestream pressure		
$\{\mathbf{Q}\}$	arbitrary quaternion	R_{A_w}	aspect ratio of the main wing
$\{\mathbf{Q}\}^*$	arbitrary quaternion conjugate	R_b	turbofan bypass ratio
$|\{\mathbf{Q}\}|$	arbitrary quaternion magnitude	R_{Dc}	Dutch roll coupling ratio, pg 911
$\dot{Q}$	weight of fuel consumed per unit time	R_{Dp}	Dutch roll phase-divergence ratio, pg 911
Q_x	x-component of arbitrary quaternion		
Q_y	y-component of arbitrary quaternion	R_{Ds}	Dutch roll stability ratio, pg 911
Q_z	z-component of arbitrary quaternion	R_d	phugoid pitch-damping ratio, pg 863
Q_0	scalar component of arbitrary quaternion	R_E	mean radius of the Earth
$\dot{Q}_{2-3}$	combustion chamber heat addition rate	R_e	equatorial radius of the Earth, pg 1045
$\dot{Q}_{4-5}$	afterburner heat addition rate	R_G	glide ratio, pg 308
q	pitching rate (y_b-component of angular rate vector ω_{y_b}), pgs 607, 729	R_{G0}	zero-wind glide ratio, pg 310
		R_{gx}	dimensionless coefficient, pg 791
$\dot{q}$	pitching acceleration (y_b-component of angular acceleration vector)	R_{gy}	dimensionless coefficient, pg 791
		R_L	lift residual
		$R_{L,\alpha}$	change in the lift residual with α

R_{L,δ_e} change in the lift residual with δ_e

$R_{\ell,\bar{p}}$ dimensionless coefficient, pg 791

$R_{\ell,\bar{r}}$ dimensionless coefficient, pg 791

$R_{\ell,\beta}$ dimensionless coefficient, pg 791

R_{ℓ,δ_a} dimensionless coefficient, pg 791

R_{ℓ,δ_r} dimensionless coefficient, pg 791

$R_{\max}$ maximum range or total distance an aircraft can fly without refueling

R_m moment residual

$R_{m,\bar{q}}$ dimensionless coefficient, pg 791

$R_{m,\alpha}$ change in the moment residual with α, pg 442

$R_{m,\alpha}$ dimensionless coefficient, pg 791

$R_{m,\hat{\alpha}}$ dimensionless coefficient, pg 791

R_{m,δ_e} change in the moment residual with δ_e, pg 442

R_{m,δ_e} dimensionless coefficient, pg 791

$R_{m,\mu}$ dimensionless coefficient, pg 791

$R_{m,\hat{\mu}}$ dimensionless coefficient, pg 791

$R_{n,\bar{p}}$ dimensionless coefficient, pg 791

$R_{n,\bar{r}}$ dimensionless coefficient, pg 791

$R_{n,\beta}$ dimensionless coefficient, pg 791

R_{n,δ_a} dimensionless coefficient, pg 791

R_p phugoid phase-divergence ratio, pg 863

R_p polar radius of the Earth, pg 1045

R_{n,δ_r} dimensionless coefficient, pg 791

$R_{P/W}$ dimensionless power-to-weight ratio, pg 274

R_s outer radius of ultimate slipstream

R_s phugoid stability ratio, pg 863

R_T taper ratio, pg 55

$R_{T/W}$ thrust-to-weight ratio, pg 268

R_{T_h} taper ratio of either an aft horizontal tail or a forward canard

R_{T_w} taper ratio of the main wing

$R_{V_{hw}}$ headwind velocity ratio, pg 315

R_x local lateral radius of curvature for the Earth, pg 1048

$R_{x,\bar{q}}$ dimensionless coefficient, pg 791

$R_{x,\alpha}$ dimensionless coefficient, pg 791

$R_{x,\hat{\alpha}}$ dimensionless coefficient, pg 791

R_{x,δ_e} dimensionless coefficient, pg 791

$R_{x,\mu}$ dimensionless coefficient, pg 791

$R_{x,\hat{\mu}}$ dimensionless coefficient, pg 791

R_y local longitudinal radius of curvature for the Earth, pg 1048

$R_{y,\bar{p}}$ dimensionless coefficient, pg 791

$R_{y,\bar{r}}$ dimensionless coefficient, pg 791

$R_{y,\beta}$ dimensionless coefficient, pg 791

R_{y,δ_a} dimensionless coefficient, pg 791

R_{y,δ_r} dimensionless coefficient, pg 791

$R_{z,\bar{q}}$ dimensionless coefficient, pg 791

$R_{z,\alpha}$ dimensionless coefficient, pg 791

$R_{z,\hat{\alpha}}$ dimensionless coefficient, pg 791

R_{z,δ_e} dimensionless coefficient, pg 791

$R_{z,\mu}$ dimensionless coefficient, pg 791

$R_{z,\hat{\mu}}$ dimensionless coefficient, pg 791

r   position vector

r position vector magnitude

r radial polar coordinate

r yawing rate (z_b-component of angular rate vector ω_{z_b}), pgs 670, 729

$\dot{r}$ yawing acceleration (z_b-component of angular acceleration vector)

$\bar{r}$ traditional dimensionless yawing rate, pg 670

$\bar{r}$ dimensionless variable, pg 418

$\bar{r}$ pendulum length scale, pg 875

$\tilde{r}$ dynamic dimensionless yawing rate, pg 670

r_e mean-sea-level coordinate, pg 1045

r_o equilibrium yawing rate

r_p radial position of a streamline at the propeller outlet

r_p radius of curvature for the pull-up maneuver

r_s radial position of a streamline in the ultimate slipstream

r_s specific range, pg 295

r_{xx_b} roll radius of gyration, pg 1082

r_{yy_b} pitch radius of gyration, pgs 617, 1082

r_{zz_b} yaw radius of gyration, pg 1082

S planform area

S shorthand notation for $\sin(\Theta/2)$, pg 998

$\mathcal{S}$ surface area

S_e planform area of elevator

S_f maximum cross-sectional area of the fuselage

S_h planform area of either an aft horizontal tail or a forward canard, or horizontal reference area of a V-tail

S_r global reference area

S_{ref} reference area

S_v planform area of a vertical tail, or vertical reference area of a V-tail

S_V' planform area of both semispans of a V-tail measured parallel with the semispan dihedral

S_w planform area of the main wing

S_θ shorthand notation for $\sin(\theta)$

S_{θ_o} shorthand notation for $\sin(\theta_o)$

$S_{\theta/2}$ shorthand notation for $\sin(\theta/2)$

S_ϕ shorthand notation for $\sin(\phi)$

S_{ϕ_o} shorthand notation for $\sin(\phi_o)$

$S_{\phi/2}$ shorthand notation for $\sin(\phi/2)$

S_ψ shorthand notation for $\sin(\psi)$

S_{ψ_o} shorthand notation for $\sin(\psi_o)$

$S_{\psi/2}$ shorthand notation for $\sin(\psi/2)$

$S_{\Omega t}$ shorthand notation for $\sin(\Omega t)$

s Laplace transform variable, pg 970

s spanwise coordinate

s distance or string length

$\bar{s}$ dimensionless variable, pg 418

s_a takeoff acceleration distance

s_{BF} FAR balanced field length

s_b landing braking distance

s_c distance from liftoff to obstacle clearance altitude

s_{ef} engine failure distance

s_{efg} ground roll distance with engine failure

s_f landing free-roll distance

s_{frr} failure recognition and reaction distance

s_g takeoff or landing ground roll distance

s_{OC} FAR takeoff distance from brake release to obstacle clearance altitude

s_r takeoff rotation distance

s_x sign of the x-component of the Euler-Rodrigues quaternion

s_y sign of the y-component of the Euler-Rodrigues quaternion

s_z sign of the z-component of the Euler-Rodrigues quaternion

s_0 sign of the scalar-component of the Euler-Rodrigues quaternion

$\mathbf{T}$ thrust vector

T restoring torque, pg 874

T temperature

T thrust

$\bar{T}$ average thrust

$\tilde{T}$ propeller section thrust force

T' experimental coefficient, pg 340

T'' experimental coefficient, pg 340

$\{\mathbf{T}\}$ temporary quaternion, pg 1002

T_A available thrust

T_{A0} full-throttle available thrust at standard sea level

T_b turbofan bypass exit temperature

T_e core nozzle exit temperature

T_i inlet temperature

T_{LO} thrust at liftoff

$T_{\max}$ maximum thrust

T_R thrust required for steady level flight

$T_{R_{\min}}$ minimum thrust required for steady level flight

T_S	static thrust	T_6	fan outlet temperature
T_x	x-component of temporary quaternion, pg 1002	T_∞	freestream temperature
		t	airfoil thickness, pg 27
T_{x_b}	x_b-component of thrust vector	t	time
$T_{,V}$	change in thrust with respect to airspeed	$\bar{t}$	dimensionless variable, pg 418
T_y	y-component of temporary quaternion, pg 1002	t_f	landing free-roll time
T_{y_b}	y_b-component of thrust vector	t_p	period of oscillation
T_z	z-component of temporary quaternion, pg 1002	t_r	takeoff rotation time
		$\bar{t}_0$	dimensionless variable, pg 418
T_{z_b}	z_b-component of thrust vector	t_{50}	50 percent damping time
T_θ	shorthand notation for $\tan(\theta)$	u	x_b-component of airspeed vector (V_{x_b})
T_{θ_o}	shorthand notation for $\tan(\theta_o)$	$\dot{u}$	x_b-component of aircraft acceleration vector ($\dot{V}_{x_b}$)
T_ϕ	shorthand notation for $\tan(\phi)$		
T_{ϕ_o}	shorthand notation for $\tan(\phi_o)$	$\mathbf{u}_a$	unit axial vector, pg 100
T_ψ	shorthand notation for $\tan(\psi)$	$\mathbf{u}_n$	unit normal vector, pg 100
T_{ψ_o}	shorthand notation for $\tan(\psi_o)$	u_o	equilibrium x_b-component of airspeed
T_0	experimental coefficient, pg 340	$\mathbf{u}_s$	unit normal to airfoil section, pg 100
T_0	scalar-component of temporary quaternion, pg 1002	$\mathbf{u}_\infty$	unit vector along a semi-infinite vortex
T_0	stagnation temperature	$\mathbf{V}$	velocity or airspeed vector
T_{0b}	turbofan bypass exit stagnation temperature	V	velocity magnitude or airspeed
		V_{ar}	aileron reversal airspeed
T_{0e}	core nozzle exit stagnation temperature	V_{BG}	best glide airspeed
T_{0i}	inlet stagnation temperature	V_{BG0}	zero-wind best glide airspeed
T_{01}	compressor inlet stagnation temperature	V_b	local propeller blade section relative airspeed, pg 175
T_{02}	compressor outlet stagnation temperature	$\mathbf{V}_c$	spin circumferential velocity vector
T_{03}	turbine inlet stagnation temperature	V_c	rate of climb, pg 278
T_{04}	turbine outlet stagnation temperature	$\mathcal{V}_c$	canard volume ratio, pg 426
T_{05}	afterburner outlet stagnation temperature	$\mathbf{V}_d$	spin descent velocity vector
		V_d	design airspeed
T_{06}	fan outlet stagnation temperature	V_d	spin descent velocity magnitude
$T_{0\infty}$	freestream stagnation temperature	V_d	velocity just downstream of prop circle
T_1	compressor inlet temperature	$\mathbf{V}_e$	exit velocity vector
T_2	compressor outlet temperature	V_e	exit velocity magnitude
T_3	turbine inlet temperature	V_e	propeller relative airspeed component, pg 217
T_4	turbine outlet temperature		
T_5	afterburner outlet temperature	V_{ef}	engine failure airspeed

V_{eh} — effective headwind, pg 297

V_{ex_b} — x_b-component of exit velocity vector

V_{ey_b} — y_b-component of exit velocity vector

V_{ez_b} — z_b-component of exit velocity vector

V_{frr} — failure recognition and reaction airspeed

$\mathbf{V}_g$ — ground speed vector

V_g — ground speed

V_h — horizontal velocity, pgs 309, 1017

V_h — relative wind on either an aft horizontal tail or a forward canard

V_{hw} — headwind velocity

$\overline{v}_h$ — horizontal tail volume ratio, pg 395

$\mathbf{V}_i$ — inlet velocity vector

V_i — inlet velocity magnitude

$\mathbf{V}_j$ — jet engine exit velocity vector

V_j — jet engine exit velocity magnitude

V_{LO} — liftoff airspeed for takeoff

V_M — maneuvering speed, pg 326

V_{MD} — minimum drag airspeed

V_{MDV} — minimum power airspeed

V_{MS} — minimum sink airspeed

V_{mc} — minimum-control airspeed, pg 582

V_{min} — minimum airspeed, airspeed at $C_{L\max}$

V_N — local relative normal velocity, pg 699

V_{OC} — airspeed at obstacle clearance altitude

V_o — equilibrium airspeed magnitude

$\mathbf{V}_p$ — perturbation velocity vector, pg 133

$\mathbf{V}_r$ — local relative wind vector, pg 699

V_r — r-component of velocity

V_{rs} — r-component of slipstream velocity

V_s — sink rate, pg 307

V_s — ultimate slipstream velocity, Chapter 2

V_{smgw} — level-flight stall speed at maximum gross weight

V_{stall} — stall speed, airspeed at $C_{L\max}$

V_{TD} — touchdown airspeed for landing

V_{trim} — airspeed at trim

V_v — relative wind on a vertical tail

V_v — vertical velocity, pg 1017

$\overline{v}_v$ — vertical tail volume ratio, pg 504

$\mathbf{V}_w$ — wind vector

V_w — wind speed

V_{wx_f} — x_f-component of wind vector

V_{wy_f} — y_f-component of wind vector

V_{wz_f} — z_f-component of wind vector

V_x — x-component of velocity vector

V_{x_b} — explicit notation for x_b-component of airspeed vector (u)

$\dot{V}_{x_b}$ — explicit notation for x_b-component of aircraft acceleration vector ($\dot{u}$)

V_{xd} — x-component of velocity just downstream of prop circle

V_{xi} — x-component of propeller induced velocity

V_{xs} — x-component of slipstream velocity

V_y — y-component of velocity vector

V_{y_b} — explicit notation for y_b-component of airspeed vector (v)

$\dot{V}_{y_b}$ — explicit notation for y_b-component of aircraft acceleration vector ($\dot{v}$)

V_z — z-component of velocity vector

V_{z_b} — explicit notation for z_b-component of airspeed vector (w)

$\dot{V}_{z_b}$ — explicit notation for z_b-component of aircraft acceleration vector ($\dot{w}$)

V_θ — θ-component of velocity

$V_{\theta d}$ — θ-component of velocity just downstream of prop circle

$V_{\theta i}$ — θ-component of propeller induced velocity

$V_{\theta s}$ — θ-component of slipstream velocity

V_0 — initial airspeed magnitude

$\mathbf{V}_\infty$ — freestream velocity vector

V_∞ — freestream velocity magnitude

v — y_b-component of airspeed vector (V_{y_b})

$\dot{v}$ — y_b-component of aircraft acceleration vector ($\dot{V}_{y_b}$)

$\mathbf{v}_{ji}$ dimensionless velocity vector, pg 99

$\mathbf{v}_i$ dimensionless velocity vector, pg 101

v_{ai} dimensionless axial velocity, pg 101

v_{ni} dimensionless normal velocity, pg 101

v_o equilibrium y_b-component of airspeed

v_{x_b} x_b-component of arbitrary vector

v_{x_f} x_f-component of arbitrary vector

v_{y_b} y_b-component of arbitrary vector

v_{y_f} y_f-component of arbitrary vector

v_{z_b} z_b-component of arbitrary vector

v_{z_f} z_f-component of arbitrary vector

$\mathbf{W}$ weight vector

W weight

W_e gross weight with fuel tanks empty

W_f gross weight with fuel tanks full

W_{max} maximum gross weight

W_{x_b} x_b-component of weight vector

W_{x_bo} equilibrium x_b-component of weight

W_{y_b} y_b-component of weight vector

W_{y_bo} equilibrium y_b-component of weight

W_{z_b} z_b-component of weight vector

W_{z_bo} equilibrium z_b-component of weight

$\dot{W}_{1-2}$ turbojet compressor shaft work rate

$\dot{W}_{3-4}$ turbojet turbine shaft work rate

w z_b-component of airspeed vector (V_{z_b})

$\dot{w}$ z_b-component of aircraft acceleration vector ($\dot{V}_{z_b}$)

$\mathbf{w}_i$ dimensionless velocity vector, pg 101

w_o equilibrium z_b-component of airspeed

$\{\mathbf{X}\}$ state variable vector in the generalized eigenproblem, pg 819

$\{\dot{\mathbf{X}}\}$ time derivative of state variable vector in the generalized eigenproblem, pg 819

$\{\overline{\mathbf{X}}\}$ Laplace transform of the state variable vector in the generalized linear control problem, pg 970

X x_b-component of aerodynamic force vector, excluding thrust

x aftward aerodynamic axial coordinate, pgs 2, 716

$\overline{x}$ dimensionless aftward axial coordinate, pg 412

$\overline{x}$ Laplace transform of x, pg 970

$\hat{x}$ dimensionless coordinate, pg 1049

x_a atmosphere-fixed coordinate, pg 719

x_{ac} axial coordinate of aerodynamic center

$\widetilde{x}_{ac}$ x-coordinate of local airfoil section aerodynamic center, pg 120

$\overline{x}_{ac}$ x-coordinate of wing aerodynamic center, pg 121

x_b forward body-fixed axial coordinate, pgs 8, 718

x_{bh} x_b-coordinate of the aerodynamic center of the horizontal tail or canard

x_{bv} x_b-coordinate of the aerodynamic center of the vertical tail

x_{bw} x_b-coordinate of the aerodynamic center of the main wing

x_C x-coordinate of control point

x_{CG} x-coordinate of center of gravity

$\overline{x}_c$ x-coordinate of the airfoil section aerodynamic center located at the spanwise coordinate of the wing semispan area centroid, pg 122

x_{cp} axial coordinate of center of pressure

$\overline{x}_D$ x-coordinate of the semispan aerodynamic center of drag, pg 460

x_f northward Earth-fixed coordinate, pg 718

$\dot{x}_f$ x_f-component of ground speed vector

x_h x-coordinate of the aerodynamic center of either an aft horizontal tail or a forward canard

$\overline{x}_L$ x-coordinate of the semispan aerodynamic center of lift, pg 460

x_l x-coordinate of lower surface

x_{mc} x-coordinate of maximum camber

x_{mp} x-coordinate of the aircraft stick-fixed maneuver point

x_N x-coordinate of nodal point

x_{np}	x-coordinate of the aircraft stick-fixed neutral point	y_f	eastward Earth-fixed coordinate, pg 718
x_o	dummy variable for x-integration, used in thin airfoil theory	$\dot{y}_f$	y_f-component of ground speed vector
		y_h	y-coordinate of the aerodynamic center of either an aft horizontal tail or a forward canard
x_o	equilibrium x_f-component of aircraft position vector		
x_T	x_b-coordinate of the center of thrust for the complete aircraft, pg 771	$\overline{y}_L$	y-coordinate of the semispan aerodynamic center of lift, pg 460
x_u	x-coordinate of upper surface	y_l	y-coordinate of lower surface
x_w	x-coordinate of the aerodynamic center of the main wing	y_{mc}	maximum camber
		y_N	y-coordinate of nodal point
Y	y_b-component of aerodynamic force vector, excluding thrust	y_o	equilibrium y_f-component of aircraft position vector
Y_p	y_b-component of propeller force (propeller side force)	y_t	one-half the local airfoil thickness
		y_u	y-coordinate of upper surface
y	upward aerodynamic normal coordinate, pgs 2, 716	y_w	y-coordinate of the aerodynamic center of the main wing
$\overline{y}$	dimensionless upward normal coordinate, pg 412	Z	geopotential altitude
		Z	z_b-component of aerodynamic force vector, excluding thrust
$\hat{y}$	change of variables, pg 135		
$\hat{y}$	dimensionless coordinate, pg 1049	$Z_{,\alpha}$	change in Z with respect to α
y_a	atmosphere-fixed coordinate, pg 719	z	leftward spanwise aerodynamic coordinate, pgs 2, 716
$\widetilde{y}_{ac}$	y-coordinate of local airfoil section aerodynamic center, pg 120	z'	leftward aerodynamic spanwise coordinate, pgs 519–520
$\overline{y}_{ac}$	y-coordinate of wing aerodynamic center, pg 121	$\overline{z}$	dimensionless leftward aerodynamic spanwise coordinate, pg 412
y'_b	spanwise coordinate measured parallel with the wing dihedral, pg 552	$\overline{z}'$	dimensionless leftward aerodynamic spanwise coordinate, pg 519
y_b	rightward body-fixed spanwise coordinate, pgs 8, 718	$\hat{z}$	dimensionless coordinate, pg 1049
		z_a	atmosphere-fixed coordinate, pg 719
y_{bp}	spanwise distance from the center of gravity to the propeller axis or center of thrust, positive right, pg 507	$\overline{z}_{ac}$	spanwise coordinate of wing semispan aerodynamic center, pgs 90–92
y_C	y-coordinate of control point	z_{ar}	spanwise coordinate of aileron root
y_{CG}	y-coordinate of center of gravity	z_{at}	spanwise coordinate of aileron tip
y_c	y-coordinate of camber line, pg 27	z_b	downward body-fixed normal coordinate, pgs 8, 718
$\overline{y}_c$	y-coordinate of the airfoil section aerodynamic center located at the spanwise coordinate of the wing semispan area centroid	z_{bv}	z_b-coordinate of the aerodynamic center of the vertical tail
$\overline{y}_D$	y-coordinate of the semispan aerodynamic center of drag, pg 460	$\overline{z}_c$	spanwise coordinate of wing semispan area centroid, pg 88
y_d	y-coordinate of deflected camber line	$\overline{z}_D$	z-coordinate of the semispan aerodynamic center of drag, pg 460

z_e mean-sea-level coordinate, pg 1045

z_f downward Earth-fixed coordinate, pg 718

$\dot{z}_f$ z_f-component of ground speed vector

$\bar{z}_L$ z-coordinate of the semispan aerodynamic center of lift, pg 460

z_{mac} spanwise coordinate of wing semispan mean aerodynamic chord, pg 84

z_{max} spanwise coordinate of maximum section lift coefficient, pgs 112, 117

z_o equilibrium z_f-component of aircraft position vector

z_T z_b-coordinate of the center of thrust for the complete aircraft, pg 771

z_{T0} perpendicular offset between the thrust vector and the aircraft center of gravity, pg 774

α geometric angle of attack relative to the freestream (for a complete aircraft, this is defined relative to the fuselage reference line), pgs 3–4, 719

$\dot{\alpha}$ time rate of change of angle of attack

α_b local effective propeller blade section angle of attack, including effects of downwash

α_{eff} local effective wing section angle of attack, including effects of downwash

α_f angle of attack for the minimum drag axis of the fuselage

α_h local angle of attack for either an aft horizontal tail or a forward canard

α_i induced angle of attack

α_j angle of attack for the jet engine axis

α_{L0} zero-lift angle of attack

α_{L0h} zero-lift angle of attack for either an aft horizontal tail or a forward canard

α_{L0w} zero-lift angle of attack for the main wing

α_o equilibrium angle of attack

α_p angle of attack for the propeller axis

α_T thrust angle of attack, pgs 261, 277

α_{T0} thrust angle of attack at zero fuselage angle of attack, pg 771

α_{0f} angle that the minimum drag axis of the fuselage makes with the fuselage reference line

α_{0h} mounting angle for either an aft horizontal tail or a forward canard

α_{0p} angle that the propeller axis makes with the fuselage reference line

α_{0w} mounting angle for the main wing

β general linearized definition of sideslip angle commonly used for small angle of attack, pgs 498, 501, 719–725

β propeller aerodynamic pitch angle

β_a arctangent definition of sideslip angle commonly used for analytical estimation of aerodynamic forces and moments, pgs 719–725

β_c propeller chord-line pitch angle

β_e arcsine definition of sideslip angle commonly used for experimental determination of aerodynamic forces and moments, pgs 719–725

β_p sideslip angle for the propeller axis

β_t aerodynamic pitch angle at the propeller blade tip

Γ local section circulation vector, pg 555

Γ vortex strength

Γ wing or tail dihedral angle, pgs 529–547, 550–554

Γ wing, airfoil, or propeller blade section circulation

Γ_f fuselage dihedral effect, pgs 558, 566

Γ_{wt} wingtip vortex strength

γ climb angle, pg 277

γ specific heat ratio

γ vortex sheet strength per unit spanwise length

γ_{OC} climb angle at the obstacle clearance altitude

γ_t strength of trailing vortex sheet per unit span

ΔF_{x_b} disturbance in x_b-component of pseudo aerodynamic force vector, including thrust

ΔF_{y_b} disturbance in y_b-component of pseudo aerodynamic force vector, including thrust

ΔF_{z_b} disturbance in z_b-component of pseudo aerodynamic force vector, including thrust

$\Delta \ell_p$ rolling moment increment contributed by propeller or propulsion system

ΔM_{x_b} disturbance in x_b-component of pseudo aerodynamic moment vector, including thrust effects

ΔM_{y_b} disturbance in y_b-component of pseudo aerodynamic moment vector, including thrust effects

ΔM_{z_b} disturbance in z_b-component of pseudo aerodynamic moment vector, including thrust effects

Δm_t aerodynamic twisting moment generated by aileron deflection

Δn_p yawing moment increment contributed by propeller or propulsion system

Δp disturbance in rolling rate

$\Delta \dot{p}$ disturbance in rolling acceleration

$\Delta \bar{p}$ disturbance in dimensionless rolling rate, pg 789

$\Delta \hat{\bar{p}}$ dimensionless time rate of change of disturbance in dimensionless rolling rate, pg 790

$\Delta \tilde{p}$ disturbance in congruous dimensionless rolling rate, pg 931

$\Delta \tilde{p}'$ dimensionless time rate of change of disturbance in congruous dimensionless rolling rate, pg 931

Δq disturbance in pitching rate

$\Delta \dot{q}$ disturbance in pitching acceleration

$\Delta \bar{q}$ disturbance in dimensionless pitching rate, pg 789

$\Delta \hat{\bar{q}}$ dimensionless time rate of change of disturbance in dimensionless pitching rate, pg 790

$\Delta \tilde{q}$ disturbance in congruous dimensionless pitching rate, pg 931

$\Delta \tilde{q}'$ dimensionless time rate of change of disturbance in congruous dimensionless pitching rate, pg 931

Δr disturbance in yawing rate

$\Delta \dot{r}$ disturbance in yawing acceleration

$\Delta \bar{r}$ disturbance in dimensionless yawing rate, pg 789

$\Delta \hat{\bar{r}}$ dimensionless time rate of change of disturbance in dimensionless yawing rate, pg 790

$\Delta \tilde{r}$ disturbance in congruous dimensionless yawing rate, pg 931

$\Delta \tilde{r}'$ dimensionless time rate of change of disturbance in congruous dimensionless yawing rate, pg 931

Δu disturbance in x_b-component of airspeed

$\Delta \dot{u}$ disturbance in x_b-component of aircraft acceleration vector

ΔV disturbance in velocity

$\Delta \dot{V}$ time rate of change of disturbance in velocity

Δv disturbance in y_b-component of airspeed

$\Delta \dot{v}$ disturbance in y_b-component of aircraft acceleration vector

Δw disturbance in z_b-component of airspeed

$\Delta \dot{w}$ disturbance in z_b-component of aircraft acceleration vector

ΔW_{x_b} disturbance in x_b-component of weight

ΔW_{y_b} disturbance in y_b-component of weight

ΔW_{z_b} disturbance in z_b-component of weight

Δx disturbance in position

$\Delta \dot{x}$ time rate of change of disturbance in position

Δx_c disturbance in aircraft position, pg 930

$\Delta \dot{x}_c$ time rate of change of disturbance in aircraft position, pg 930

Δx_f disturbance in x_f-component of aircraft position vector

$\Delta \dot{x}_f$ disturbance in x_f-component of ground speed vector

ΔY_p side force increment contributed by propeller or propulsion system

Δy_c disturbance in aircraft position, pg 930

$\Delta \dot{y}_c$ time rate of change of disturbance in aircraft position, pg 930

Δy_f disturbance in y_f-component of aircraft position vector

$\Delta \dot{y}_f$ disturbance in y_f-component of ground speed vector

Δz_c disturbance in aircraft position, pg 930

$\Delta \dot{z}_c$ time rate of change of disturbance in aircraft position, pg 930

Δz_f disturbance in z_f-component of aircraft position vector

$\Delta \dot{z}_f$ disturbance in z_f-component of ground speed vector

$\Delta \alpha$ disturbance in angle of attack, pg 789

$\Delta \dot{\alpha}$ time rate of change of disturbance in angle of attack

$\Delta \hat{\alpha}$ dimensionless time rate of change of disturbance in angle of attack, pg 789

$\Delta \alpha'$ congruous dimensionless time rate of change of disturbance in angle of attack, pg 931

$\Delta \beta$ disturbance in sideslip angle, pg 789

$\Delta \hat{\beta}$ dimensionless time rate of change of disturbance in sideslip angle, pg 790

$\Delta \beta'$ congruous dimensionless time rate of change of disturbance in sideslip angle, pg 931

$\Delta \delta_a$ disturbance in aileron deflection

$\Delta \delta_e$ disturbance in elevator deflection

$\Delta \delta_r$ disturbance in rudder deflection

$\Delta \varsigma_x$ congruous dimensionless disturbance in aircraft position, pg 931

$\Delta \varsigma_x'$ congruous dimensionless time rate of change of disturbance in aircraft position, pg 931

$\Delta \varsigma_y$ congruous dimensionless disturbance in aircraft position, pg 931

$\Delta \varsigma_y'$ congruous dimensionless time rate of change of disturbance in aircraft position, pg 931

$\Delta \varsigma_z$ congruous dimensionless disturbance in aircraft position, pg 931

$\Delta \varsigma_z'$ congruous dimensionless time rate of change of disturbance in aircraft position, pg 931

$\Delta \theta$ disturbance in Euler elevation angle

$\Delta \dot{\theta}$ time rate of change of disturbance in Euler elevation angle

$\Delta \hat{\theta}$ dimensionless time rate of change of disturbance in Euler elevation angle, pg 790

$\Delta \mu$ disturbance in dimensionless forward velocity, pg 789

$\Delta \hat{\mu}$ dimensionless time rate of change of disturbance in dimensionless forward velocity, pg 790

$\Delta \mu'$ congruous dimensionless time rate of change of disturbance in dimensionless forward velocity, pg 931

$\Delta \xi_x$ disturbance in dimensionless x_f-component of aircraft position, pg 789

$\Delta \hat{\xi}_x$ dimensionless time rate of change of disturbance in dimensionless x_f-component of aircraft position, pg 790

$\Delta \xi_y$ disturbance in dimensionless y_f-component of aircraft position, pg 789

$\Delta \hat{\xi}_y$ dimensionless time rate of change of disturbance in dimensionless y_f-component of aircraft position, pg 790

$\Delta \xi_z$ disturbance in dimensionless z_f-component of aircraft position, pg 789

$\Delta \hat{\xi}_z$ dimensionless time rate of change of disturbance in dimensionless z_f-component of aircraft position, pg 790

$\Delta \phi$ disturbance in Euler bank angle

$\Delta \dot{\phi}$ time rate of change of disturbance in Euler bank angle

$\Delta \hat{\phi}$ dimensionless time rate of change of disturbance in Euler bank angle, pg 790

$\Delta \psi$ disturbance in Euler azimuth angle

$\Delta \dot{\psi}$ time rate of change of disturbance in Euler azimuth angle

$\Delta \hat{\psi}$ dimensionless time rate of change of disturbance in Euler azimuth angle, pg 790

δl spatial vector, pg 102

δ flap deflection, pg 40

δ_a aileron deflection, pgs 9, 73, 378, 567

δ_{ao} equilibrium aileron deflection

$\delta_{a_{sat}}$ aileron saturation angle

δ_e elevator deflection, pgs 9, 378, 386

$\delta_{e,n}$ elevator angle per g, change in elevator deflection required to sustain each additional g of normal acceleration

δ_{eo} equilibrium elevator deflection

δ_{et} stabilator elevator-trim setting, pg 683

δ_f local section flap deflection

δ_r rudder deflection, pgs 9, 378, 501

δ_{ro} equilibrium rudder deflection

$\delta_{r_{sat}}$ rudder saturation angle

δ_s stabilator deflection, pg 682

δ_{so} equilibrium stabilator deflection

$\delta_{s,n}$ stabilator angle per g, change in stabilator deflection required to sustain each additional g of normal acceleration

δ_t trim tab deflection, pg 635

ε Earth's eccentricity, pgs 1045–1048

ε orthogonality error, pg 1021

$\widetilde{\varepsilon}_a$ local aileron section flap effectiveness, pgs 567–568

ε_b local propeller blade downwash angle, pg 174

ε_d downwash angle

ε_{dj} jet engine inlet downwash angle

ε_{dp} propeller downwash angle

ε_{d0} downwash angle at zero fuselage angle of attack

ε_{d0j} jet engine inlet downwash angle at zero fuselage angle of attack

ε_{d0p} propeller downwash angle at zero fuselage angle of attack

$\varepsilon_{d,\alpha}$ downwash gradient, the change in downwash angle with angle of attack

ε_e elevator effectiveness

ε_f flap effectiveness, pg 43

ε_{fi} ideal flap effectiveness, pg 41

ε_h downwash angle for either an aft horizontal tail or a forward canard

ε_i local propeller blade induced angle, pg 174

ε_r rudder effectiveness

ε_s sidewash angle

ε_{sp} propeller sidewash angle

ε_{s0} sidewash angle at zero sideslip angle

$\varepsilon_{s,\beta}$ sidewash gradient, the change in sidewash angle with respect to sideslip angle

ε_t small tolerance on the order of computer precision

ε_u upwash angle

ε_{u0} upwash angle at zero fuselage angle of attack

$\varepsilon_{u,\alpha}$ upwash gradient, the change in upwash angle with respect to angle of attack

ε_w downwash angle for the main wing

ε_Ω washout effectiveness, pgs 58–59

ε_∞ local propeller blade advance angle, pg 174

ζ dimensionless length vector, pg 100

ζ damping ratio, pg 815

ζ dimensionless propeller radial coordinate, pg 183

ζ dummy variable for z-integration

ζ_{sp} short-period damping ratio

η normal panel coordinate, pg 34

η_d flap deflection efficiency, pg 44

η_h flap hinge efficiency, pg 44

η_h ratio of dynamic pressure on either an aft horizontal tail or a forward canard to the freestream dynamic pressure

η_i ideal efficiency

η_O overall efficiency

η_P uninstalled propulsive efficiency

η_p propulsive efficiency

η_{p_i} ideal propulsive efficiency

η_T thermal efficiency

η_v ratio of dynamic pressure on a vertical tail to the freestream dynamic pressure

η_{xx} dimensionless coefficient, pg 933

η_{xy} dimensionless coefficient, pg 933

η_{xz} dimensionless coefficient, pg 933

η_{yx} dimensionless coefficient, pg 933

η_{yy} dimensionless coefficient, pg 933

η_{yz} dimensionless coefficient, pg 933

η_{zx} dimensionless coefficient, pg 933

η_{zy} dimensionless coefficient, pg 933

η_{zz} dimensionless coefficient, pg 933

Θ complex phase angle, pg 824

Θ Euler axis rotation angle angle, pg 994

Θ great-circle arc angle, pgs 1052, 1060

Θ_{wp} great-circle arc angle to some waypoint, pg 1062

Θ_x geographic angle, pg 1049

Θ_y geographic angle, pg 1049

θ angular polar coordinate

θ change of variables for the chordwise coordinate in thin airfoil theory, pg 29

θ change of variables for the spanwise coordinate in lifting-line theory, pg 52

θ Euler elevation angle, pgs 436, 735–737

$\dot{\theta}$ time rate of change of Euler elevation angle

θ_f change of variables for the flap chord fraction, pg 41

θ_o equilibrium Euler elevation angle

$\theta_{\max}$ spanwise coordinate of maximum section lift coefficient, pg 111

ι_{xy} dimensionless coefficient, pg 933

ι_{xz} dimensionless coefficient, pgs 791, 933

ι_{yx} dimensionless coefficient, pg 933

ι_{yz} dimensionless coefficient, pg 933

ι_{zx} dimensionless coefficient, pgs 791, 933

ι_{zy} dimensionless coefficient, pg 933

κ Goldstein's kappa factor, pg 180

κ_{ac} sweep factor in the relation for wing aerodynamic center, pgs 124–126

κ_b vortex span factor in downwash computations, pgs 412–414

κ_D planform contribution to the induced drag factor, pgs 55, 59

κ_{DL} lift-washout contribution to the induced drag factor, pg 59

κ_{Do} optimum induced drag factor, pg 62

$\kappa_{D\Omega}$ washout contribution to the induced drag factor, pg 59

κ_L lift slope factor, pgs 55, 59

κ_{Ls} stall factor, pgs 115–117

$\kappa_{L\alpha}$ sweep factor, pgs 127–128

$\kappa_{L\Lambda}$ sweep factor, pgs 115–117

$\kappa_{L\Omega}$ washout factor, pgs 111–114

κ_ℓ dihedral factor, pgs 553–554

$\kappa_{\ell\overline{p}}$ roll damping factor, pg 77

$\kappa_{M\Lambda}$ sweep factor, pgs 127–129

$\kappa_{M\Omega}$ washout factor, pgs 92–93

κ_p position factor in downwash computations, pgs 413–415

κ_s wing sweep factor in downwash computations, pgs 416–419

κ_v vortex strength factor in downwash computations, pgs 412–414

$\kappa_{Z\Lambda}$ sweep factor, pgs 117–118

$\kappa_{Z\Omega}$ washout factor, pgs 117–118

κ_β vortex sidewash factor, pg 521

κ_Γ dihedral factor, pgs 553–554

$\kappa_{\Lambda 1}$ sweep coefficient, pgs 115–117

$\kappa_{\Lambda 2}$ sweep coefficient, pgs 115–117

Λ source strength

Λ quarter-chord sweep angle

Λ_h quarter-chord sweep angle for either an aft horizontal tail or a forward canard

Λ_w quarter-chord sweep angle for the main wing

λ eigenvalue, pg 814

λ propeller aerodynamic pitch length, pg 170

λ source sheet strength per unit length

λ_c propeller chord-line pitch length, pg 170

λ_{Dr} Dutch roll eigenvalue

λ_p phugoid eigenvalue

λ_r roll mode eigenvalue

λ_s spiral mode eigenvalue

λ_{sp} short-period eigenvalue

μ Mach angle, angle that constant potential lines make with the freestream, pg 140

μ_l lower surface Mach angle

μ_u upper surface Mach angle

μ_r coefficient of rolling friction

ν kinematic viscosity

ξ surface panel coordinate, pg 34

ξ change of variables, pg 139

ρ density

ρ_0 stagnation density

ρ_∞ freestream density

σ damping rate, pg 814

σ spin axis offset angle, pg 697

σ_{Dr} Dutch roll damping rate

σ_p phugoid damping rate

σ_r roll mode damping rate

σ_s spiral mode damping rate

σ_{sp} short-period damping rate

σ_∞ Dutch roll damping rate with infinite roll damping

τ throttle setting

τ congruous dimensionless time, pg 931

τ_x dimensionless time for longitudinal motion, pg 789

τ_y dimensionless time for lateral motion, pg 789

Φ dimensionless variable, pg 25

Φ Earth-fixed latitude, pg 1044

$\dot{\Phi}$ change in latitude with respect to time

Φ_d destination latitude, pg 1058

Φ_{wp} waypoint latitude, pg 1062

ϕ Euler bank angle, pgs 319, 735–737

ϕ propeller aerodynamic angle, pg 218

ϕ velocity potential

$\dot{\phi}$ time rate of change of Euler bank angle

$\phi_{\lim}$ bank angle limit

$\phi_{\max}$ maximum bank angle

ϕ_o equilibrium Euler bank angle

ϕ_p perturbation velocity potential, pg 133

$\hat{\phi}_p$ perturbation variable, pg 133

ϕ_∞ freestream velocity potential, pg 133

φ rotation angle between two body-fixed coordinate systems, pg 798

φ_c crab angle, pg 296

φ_w wind-track angle, pg 296

$\{\chi\}$ eigenvector, pg 819

χ eigenvector component

χ normalized control surface distribution function, pg 74

$\chi_{\bar{p}}$ eigenvector rolling rate component

$\chi_{\bar{q}}$ eigenvector pitching rate component

$\chi_{\bar{r}}$ eigenvector yawing rate component

χ_α eigenvector angle of attack component

χ_β eigenvector sideslip angle component

χ_θ eigenvector elevation angle component

χ_μ eigenvector axial velocity component

χ_{ξ_x} eigenvector ξ_x-component

χ_{ξ_y} eigenvector ξ_y-component

χ_{ξ_z} eigenvector ξ_z-component

χ_ϕ eigenvector bank angle component

χ_ψ eigenvector heading component

Ψ dimensionless variable, pg 25

Ψ Earth-fixed longitude, pg 1044

Ψ phase angle, pg 818

Ψ relaxation factor, pg 102

$\dot{\Psi}$ change in longitude with respect to time

Ψ_d destination longitude, pg 1058

Ψ_{wp} waypoint longitude, pg 1062

ψ Euler azimuth angle or heading, pgs 296, 735–737

ψ propeller aerodynamic angle, pg 218

$\dot{\psi}$ time rate of change of Euler azimuth angle or heading

ψ_g ground track or rhumb-line bearing, pgs 296, 1058

ψ_{gc} great-circle bearing, pg 1060

ψ_o equilibrium Euler azimuth angle

ψ_w wind direction, pg 296

Ω vorticity vector

Ω aircraft angular velocity vector

Ω maximum total washout, geometric plus aerodynamic, pg 57

Ω turning or spin rate

$\Omega_{\max}$ maximum turning rate

Ω_{opt} optimum total washout to minimize induced drag, pg 62

ω aircraft angular rate vector

ω aircraft angular rate vector magnitude

ω control input frequency, pg 983

ω normalized washout distribution function, pgs 57, 74

ω propeller angular velocity

ω_d damped natural frequency, pg 814

$\omega_{d_{Dr}}$ Dutch roll damped natural frequency

ω_{d_p} phugoid damped natural frequency

$\omega_{d_{sp}}$ short-period damped natural frequency

ω_f frequency of sinusoidal forcing function

ω_m maximum response frequency, pg 818

ω_n undamped natural frequency, pg 815

$\omega_{n_{Dr}}$ Dutch roll undamped natural frequency

$\omega_{n_{sp}}$ short-period undamped natural frequency

ω_s ultimate slipstream angular velocity

ω_{x_b} explicit notation for x_b-component of the aircraft angular rate vector (p)

$\dot{\omega}_{x_b}$ explicit notation for x_b-component of the aircraft angular acceleration vector ($\dot{p}$)

ω_{y_b} explicit notation for y_b-component of the aircraft angular rate vector (q)

$\dot{\omega}_{y_b}$ explicit notation for y_b-component of the aircraft angular acceleration vector ($\dot{q}$)

ω_{z_b} explicit notation for z_b-component of the aircraft angular rate vector (r)

$\dot{\omega}_{z_b}$ explicit notation for z_b-component of the aircraft angular acceleration vector ($\dot{r}$)

ω_0 initial aircraft angular rate

ϖ_n dimensionless undamped natural frequency, pg 863

ϖ_∞ Dutch roll dimensionless natural frequency with infinite roll damping

Index